Aquatic Animal Health Management

Rajendran Kooloth Valappil
Iddya Karunasagar • Indrani Karunasagar
Editors

Aquatic Animal Health Management

Springer

Editors
Rajendran Kooloth Valappil
ICAR-Central Institute of Fisheries Education
Mumbai, Maharashtra, India

Indrani Karunasagar
FAO Ref Centre for Antimicrobial Resistance
Nitte University
Mangaluru, Karnataka, India

Iddya Karunasagar
FAO Ref Centre for Antimicrobial Resistance
Nitte University
Mangaluru, Karnataka, India

ISBN 978-981-96-7986-7 ISBN 978-981-96-7987-4 (eBook)
https://doi.org/10.1007/978-981-96-7987-4

This Springer imprint is published by the registered company Springer Nature Singapore Pte Ltd.
The registered company address is: 152 Beach Road, #21-01/04 Gateway East, Singapore 189721, Singapore

Preface

India is the second-largest aquaculture-producing country in the world, characterized by a wide variety of cultivation methods, including traditional pond systems, modern recirculating aquaculture systems, and integrated multi-trophic aquaculture supporting a wide range of species. Indian aquaculture encompasses a wide variety of species, including invertebrates such as crustaceans and bivalves, as well as vertebrate finfish cultivated in freshwater, brackishwater, and marine environments. However, one of the major global challenges faced by aquaculture is the economic loss caused by diseases. Approximately a decade ago, the World Bank estimated global losses due to diseases in aquaculture at approximately USD 6 billion. In India, the estimated loss is USD 2.4 billion, constituting 14.95% of the production value. Though expansion and intensification are driving the sector forward, the scale of economic losses due to diseases underscores the continued relevance of disease management in aquaculture. Therefore, it is imperative that improving aquatic animal health management using international best practices is essential for sustainable aquaculture.

India has a large network of institutions working on aquatic animal health, including institutes under the Indian Council of Agricultural Research, Colleges of Fisheries under State Agriculture Universities, and other academic and research institutions. India is one of the few countries globally to implement a systematic disease surveillance program, as the National Surveillance Programme for Aquatic Animal Diseases (NSPAAD) since 2013 and the National Network Programme on Fish Health. Research and surveillance conducted under these programs and other projects in different institutions have generated a vast repository of knowledge on aquatic animal health. However, presently, information on aquatic animal health related to different aquaculture species and systems is spread over different journal publications and reports. Therefore, this book aims to consolidate vital information on aquatic animal health into a single, accessible resource for students, academicians, industry professionals, and researchers. It is structured into five sections: (i) Pathogens and Diseases of Finfish, (ii) Pathogens and Diseases of Shellfish, (iii) Aquatic Animal Health and Environment, (iv) Health Management in Aquaculture: Concepts, Tools, Technologies and Their Application, and (v) Aquatic Animal Health Policy and Regulations. Each chapter is authored by leading experts, offering comprehensive insights into these areas.

By integrating expertise from diverse institutions and focusing on various aquaculture systems and species, this book serves as a reliable resource for the aquaculture industry, aquatic animal health professionals, and researchers. We hope it will be a valuable tool for advancing aquaculture and aquatic animal health management. We extend our gratitude to the authors for their contributions and to Saraniya Vairamuthu and her team at Springer Nature for their patience and perseverance with us.

Mumbai, Maharashtra, India — Rajendran Kooloth Valappil
Mangaluru, Karnataka, India — Iddya Karunasagar
Mangaluru, Karnataka, India — Indrani Karunasagar

Contents

Pathogens and Diseases of Finfish

Bacterial Diseases in Freshwater Fish 3
P. K. Sahoo, A. Paul, and M. Mishra

Bacterial Diseases in Brackishwater Aquaculture 31
Sujeet Kumar and S. V. Alavandi

Diseases in Coldwater Aquaculture: Pathogenesis and Control 59
Zainab Manzoor, Syed Shariq N. Qadiri, Feroz A. Shah, Shabir A. Dar, and Shahid M. Mandu

Diseases in Cage Aquaculture 85
Gayatri Tripathi and R. Bharathi Rathinam

Emerging Viral Diseases in Freshwater Fishes with Emphasis on Tilapia Lake Virus 109
T. R. Swaminathan, P. K. Pradhan, A. Sudhagar, and N. Sood

Viral Diseases in Brackishwater Fishes 139
M. Makesh, K. V. Rajendran, and S. K. Girisha

Diseases in Ornamental Fishes 175
P. Sivasankar, K. Riji John, and M. Rosalind George

Parasitic Diseases in Freshwater Fish 213
S. Mohanty and P. K. Sahoo

Pathogens and Diseases of Shellfish

Major Viral Diseases of Shrimp in India 241
M. P. Safeena, K. Sreedharan, K. U. Shyam, and K. V. Rajendran

White Spot Syndrome Virus (WSSV) 293
K. V. Rajendran, K. Sreedharan, Indrani Karunasagar, Iddya Karunasagar, and Arun K. Dhar

Microsporidians Infecting Shrimp with Special Reference to *Enterocytozoon hepatopenaei* 407
T. Sathish Kumar, K. Sreedharan, and K. V. Rajendran

Disease Syndromes in Farmed Shrimp 435
M. Poornima, K. Nalayeni, T. Sathish Kumar, and R. Vidya

Diseases of Bivalve Molluscs 451
N. K. Sanil, G. Suja, and P. K. Asokan

Aquatic Animal Health and Environment

Role of Environment in Aquatic Animal Health and Noninfectious Diseases in Finfish Aquaculture 475
S. Mohanty and P. K. Sahoo

Impact of Climate Change on Animal Health in Aquaculture Systems 507
Muralidhar M and Satheesha Avunje

Health Management in Aquaculture: Concepts, Tools, Technologies and Their Application

Health Management in Aquaculture: An Overview and Future Perspectives 531
K. Jeena, Anutosh Paria, Anirban Paul, P. K. Sahoo, and K. V. Rajendran

Health Management in Penaeid Shrimp Seed Production 543
P. Ravichandran

Disease Diagnosis in Aquaculture 567
K. Jeena, M. Makesh, Megha K. Bedekar, K. U. Shyam, Arun K. Dhar, and K. V. Rajendran

Prospects of Vaccine in Shrimp Aquaculture 625
A. S. Sahul Hameed and S. Vimal

Bacterial Vaccines for Fish 639
Megha K. Bedekar and Arya Singh

Antimicrobial Usage in Aquaculture: Potential Implications 663
Gaurav Rathore and Badireddy Madhusudana Rao

Application of Probiotics, Immunostimulants and Bacteriophages in Aquaculture 683
Subhendu Kumar Otta and Indrani Karunasagar

Bioremediators in Shrimp Aquaculture 703
K. K. Krishnani, R. Abisha, and M. Muralidhar

Finfish Cell Lines and Their Application in Viral Disease Management . . . 731
T. R. Swaminathan, R. Srinath, S. Visnuvinayagam, and A. Sudhagar

Crustacean Cell Line Development: Status, Constraints, and the Way Forward 753
C. P. Sreevidya, S. Manomi, I. S. Bright Singh, and Jayesh Puthumana

Applications of Nanotechnology in Aquatic Animal Health Management 781
Priyabrat Swain, Satya Narayan Sahoo, and Pushpa Choudhary

Application of Genetics and Genomics in Aquatic Animal Health Management 793
G. Gopikrishna

Live Feeds in Shrimp Aquaculture: An Overview 807
K. P. Sandeep, T. Sivaramakrishnan, K. Ambasankar, and K. K. Vijayan

Sustainable Intensification in Aquaculture 825
A. Panigrahi and P. S. Shyne Anand

Aquatic Animal Health Policy and Regulations

Aquatic Animal Health Regulations and Policy in India 847
J. K. Jena, N. Sood, and P. K. Pradhan

National Surveillance Programme for Aquatic Animal Diseases (NSPAAD): Enhancing Aquatic Animal Health Preparedness in India . . . 865
Kuldeep K. Lal, Neeraj Sood, T. Raja Swaminathan, Dev K. Verma, P. K. Pradhan, and J. K. Jena

Issues in International Trade of Seafood 879
Iddya Karunasagar

Building Financial Resilience in Shrimp Aquaculture: Crop Insurance to Cover Aquatic Animal Disease Losses 897
T. Ravisankar and Kuldeep K. Lal

Pathogens and Diseases of Finfish

Bacterial Diseases in Freshwater Fish

P. K. Sahoo, A. Paul, and M. Mishra

Abstract

Diseases are the major bottleneck causing low- to high-impact economic losses to farmers. Among the different aetiological agents, bacterial pathogens pose a severe threat to the aquaculture industry. With the increasing intensity of culture practices, the complexity of diseases has also increased manyfold. Although both Gram-positive and Gram-negative bacteria are responsible for causing diseases in fish, most of the bacterial infections in fish are of the Gram-negative origin. The genera *Aeromonas*, *Edwardsiella*, and *Flavobacteria* are mostly predominant in freshwater aquaculture and responsible for causing diseases. Several culture-dependent and culture-independent techniques for bacterial identification and characterization are well-established, aiding in timely diagnosis. Treatment of bacterial infections greatly depends on the use of antibiotics and other water sanitisers. In recent times, the increasing resistance commonly observed against a wide variety of antibiotics by several bacteria has rendered most treatments ineffective. Development of new-generation alternative eco-friendly means of bacterial disease management will be of immense help to the sustainable growth of the aquaculture sector.

Keywords

Bacterial diseases · Co-infection · Freshwater fish · Pathogenomic evolution · Prevention method

P. K. Sahoo (✉) · A. Paul · M. Mishra
ICAR-Central Institute of Freshwater Aquaculture, Bhubaneswar, India
e-mail: pksahoo1@hotmail.com; anirbancof@gmail.com; mitali.mishra.mm@gmail.com

R. Kooloth Valappil et al. (eds.), *Aquatic Animal Health Management*,
https://doi.org/10.1007/978-981-96-7987-4_1

1 Introduction

In fish, a disease, a complex phenomenon, occurs when susceptible fish are exposed to a virulent pathogen under unfavourable environmental conditions (Snieszko 1975). Diseases are responsible for causing loss in the range of 10–15% of the production cost. Bacterial diseases are among the most important causes of losses among fish stocks. Complete understanding of the aetiological agent, pathogenesis, biochemistry, antigenicity, epizootiology, and inter-relationship between stress-related and environmental factors is essential for the successful management and control of diseases. The reasons behind the emergence of bacterial diseases in aquaculture, although not quite clear, are the introduction of exotic or new species to an environment carrying the pathogens, the availability of more susceptible species, favourable or stressful environment, and availability of up-to-date diagnostic tools (Colorni 2004). Although recent trends in the shifting of diagnostic methods from less sensitive culture-dependent bacterial identification to more sensitive and specific culture-independent molecular methods have revolutionized newer bacterial disease incidence, detecting very low copy numbers of bacteria does not automatically imply the presence of a clinical disease. A wide range of Gram-positive and Gram-negative bacterial pathogens are associated with diseases in freshwater fish around the globe, especially in freshwater aquaculture (Table 1). Most of the bacterial pathogens are associated with generalized clinical symptoms, such as haemorrhagic septicaemia, ulcers on the body surface, fin and skin rot, and swollen and necrotized gills. Steps have been taken to develop vaccines, disease-resistant strains, and other biocontrol agents to control losses caused by bacterial diseases and reduce the use of antibiotics; however, such losses incurred due to bacterial diseases are not being minimized in aquaculture in developing and underdeveloped countries. This chapter, will serve as a standard text for students of aquaculture, veterinarians, and microbiologists, brings these aspects together and reflects the current international scenario of bacterial disease dynamics in freshwater aquaculture, especially in Indian polyculture systems.

2 Impact of Bacterial Diseases

The majority of bacterial pathogens are reported in temperate rather than in tropical aquaculture. Hence, their impact widely felt in salmonid aquaculture and economic loss has also been calculated for a few diseases, and the same is also duly addressed by the development of vaccines to prevent the disease in temperate regions. On the other hand, although high temperature serves as a favourable factor for most of the bacterial pathogens in tropical environments, the loss has been poorly addressed and so is the development of vaccinations. Recently, the emphasis on tilapia and catfish farming, besides regular carp culture with intensification and super-intensive aquaculture along with diversification through recirculating aquaculture systems (RASs), biofloc technology, and vertical farming, has increased the incidence of

Table 1 List of important bacterial pathogens causing diseases in freshwater aquaculture with their distribution, species susceptibility, clinical signs, vaccine status, and zoonotic importance

Pathogen	Distribution	Species affected	Pathology	Vaccines available	Zoonotic or not
Aeromonas hydrophila, A. sobria/veronii, and *A. caviae*	Global	Various freshwater fishes	Haemorrhages in the skin, exophthalmia ascites, ulcers	No	Yes (low)
A. salmonicida	Only temperate regions	Salmonids and other freshwater and marine fishes	Raised liquefactive muscular lesions (furuncles)	Yes	No
Edwardsiella tarda	Global	Various freshwater and marine fishes, mostly tilapia, pangas, and Indian major carps	Anorexia and pigmentation, opacity of the eyes, exophthalmia, swelling of the abdominal surface, petechial haemorrhage in the skin and fins. Internally, bloody and watery ascites inside the abdominal space	Yes	Yes (medium)
Flavobacterium columnare	Global	Various farmed freshwater fish species worldwide, like tilapia, cyprinids	Cottonmouth, skin lesions, saddleback, fin and tail rot	Yes	No
Vibrio anguillarum	Global	Various cultured freshwater/ marine fish	Red spots on the ventral and lateral areas; swollen and dark skin lesions leading to ulceration	Yes	No

(continued)

Table 1 (continued)

Pathogen	Distribution	Species affected	Pathology	Vaccines available	Zoonotic or not
Pseudomonas aeruginosa	Global	Various farmed freshwater fishes, mostly, eel, tilapia, bighead carp, silver carp, and rainbow trout	Skin ulcerations, haemorrhagic septicaemia, and fin rot An internally engorged spleen, a pale liver, and abdominal dropsy with reddish ascitic exudates	No	No
Streptococcus agalactiae/iniae	USA, Australia, Bahrain, Barbados, China, Taiwan, Thailand, Japan, Korea, Israel, Europe, Iran, India and South Africa	Various freshwater and marine fish species, tilapia, channel catfish, and rainbow trout	Exophthalmia, erratic swimming behaviour, dark colouration of the skin, corneal opacity, petechiae on the eyes and gill covers	Yes	Yes
Lactococcus garvieae	East Asia and Spain	Rainbow trout and other marine species	Anorexia, erratic swimming followed by widespread haemorrhages, a swollen abdomen, unilateral or bilateral exophthalmia, and anal prolapses	Yes	No

bacterial diseases in South-East Asia. However, their impact has never been quantified as accurately for the aquatic system as for the terrestrial system.

Estimation of the impact of fish diseases in economic terms is handicapped by the lack or inadequate information on the morbidity, mortality, growth reduction, prophylactic treatment, loss of product value, and other associated costs in fish production. Diseases can hamper reproductive performance and have a negative impact on feed conversion efficiency. However, economic losses caused by certain diseases are significant in some regions. According to the Food and Agriculture Organization (1999), diseases and environment-related problems have caused annual losses of more than USD 3000 million in the aquaculture sector in Asia. In the channel catfish (*Ictalurus punctatus*) industry in the United States, *Flavobacterium columnare* is the second most prevalent bacterium, after *Edwardsiella ictaluri*, which causes disease and mortality, with yearly losses estimated at USD 30 million (Shoemaker et al. 2011). Edwardsiellosis, caused by *Edwardsiella tarda* to olive flounder (Japanese flounder; *Paralichthys olivaceus*), caused economic loss of 489.7 billion Korean Won in South Korea in 2010 (Statistics Korea (KOSTAT): 2010). In a study conducted in China between 1990 and 1992, the annual economic loss associated with three bacterial fish pathogens (*Aeromonas hydrophila*, *Yersinia ruckeri*, and *Vibrio fluvialis*) was estimated at more than USD 120 million (Wei 2002). A recent passive surveillance-based study, conducted in a few states in India between 2014 and 2018, has revealed that bacterial diseases alone account for 12.79%, along with mixed bacterial and parasitic infections of 10.50% in freshwater aquaculture farms. Furthermore, among bacterial infections, aeromonads account for 66.66% of diseases besides a few emerging pathogens of public health significance (Sahoo et al. 2020).

2.1 Criteria for Considering Important Bacteria in Aquaculture

Bacterial diseases that affect fish can be caused by bacteria as (a) primary pathogens without the involvement of any environmental stress factors, (b) primary pathogens as a result of environmental stress factors, and (c) opportunistic pathogens as a result of environmental stress factors. The latter two conditions seem to be frequent in Indian aquaculture systems. Bacterial opportunistic pathogens are defined as microorganisms causing disease in hosts experiencing atypical environmental stressors or having stress-induced immunosuppression. Although bacteria are frequently linked to disease conditions, fulfilling Koch's postulate under experimental conditions remains challenging. Hence, in Indian aquaculture, it is essential to categorize important bacteria based on their economic importance, socio-economic impact, zoonotic potential, and degree of spread. This chapter will highlight only those bacteria that are important in Indian aquaculture, particularly in the absence of socio-economic impact data.

3 Diseases, Pathogens, Molecular Pathogenesis, and Pathology

3.1 Incidence of Bacterial Diseases

At least 13 genera of bacteria have been reported to be pathogenic to fish worldwide, including Gram-negative bacteria such as *Aeromonas*, *Edwardsiella*, *Flavobacterium*, *Francisella*, *Photobacterium*, *Piscirickettsia*, *Pseudomonas*, *Tenacibaculum*, *Vibrio*, and *Yersinia* and Gram-positive bacteria such as *Lactococcus*, *Renibacterium*, and *Streptococcus*. In a recent study conducted in Indian freshwater aquaculture, it has been noticed that the motile aeromonads, among all bacterial pathogens, were the major pathogens causing bacterial fish diseases (66.66%). Among aeromonads, *Aeromonas hydrophila* (52.94%) was found to be involved in the maximum number of diseases, followed by *Aeromonas sobria* (26.47%),

Aeromonas veronii (14.70%), *Aeromonas jandaei* (2.94%), and *Aeromonas caviae* (2.94%) (Sahoo et al. 2020). Earlier, Karunasagar et al. (1986) reported motile aeromonad infections causing mortalities in Indian major carp farms and marked it as one of the major bacterial pathogens of fishes. Other than aeromonads, many emerging bacterial pathogens have also recorded to be associated with fish mortality, such as *Klebsiella pneumoniae*, *Edwardsiella tarda*, *Proteus mirabilis*, *Plesiomonas shigelloides*, *Acinetobacter baumannii*, *Morganella morganii*, and *Serratia marcescens*. *Vibrio harveyi* and *Vibrio parahaemolyticus* have also been found to cause mortality in giant freshwater prawn (*Macrobrachium rosenbergii*) hatcheries. Furthermore, one case of ornamental fish, red eye tetra (*Moenkhausia sanctaefilomenae*), infected with acid-fast bacteria causing tumour-like growth at the head portion of the fish was also reported (Sahoo et al. 2020). A detailed description of important bacterial diseases in freshwater aquaculture systems with the aetiological agent, pathology, and their importance in the current context is provided hereunder.

4 Gram-Negative Pathogens

4.1 *Aeromonas* spp.

Aeromoniasis is the most important and common bacterial disease caused by *Aeromonas* spp. in freshwater fishes. The genus *Aeromonas* is ubiquitously found in aquatic environments and includes more than 32 known species. Based on growth characteristics and biochemical reactions, aeromonads are broadly divided into two major groupings: (i) mesophilic and (ii) psychrophilic (Janda and Duffey 1988). Mesophilic aeromonads are frequently found to cause severe disease and economic loss to farmers. Among different species identified, *A. hydrophila*, *A. sobria/veronii*, and *A. caviae* are most commonly associated with clinical conditions (Pękala-Safińska 2018; Sahoo et al. 2020). *A. hydrophila* frequently causes motile *Aeromonas* septicaemia (MAS), in a variety of fishes, which is characterized by significant internal and external haemorrhages and accumulation of body fluid and exophthalmia, often followed by heavy mortality and morbidity within several hours of disease manifestation (Camus et al. 1998; Beaz-Hidalgo et al. 2013). Gopalakrishnan (1961) described mass mortalities in Indian major carp farms due to the epizootics of *A. hydrophila* causing marked dropsy. Usual manifestations of ulcer formation and a high percentage of dropsy have been very frequently found in various fishes (Karunasagar et al. 1986; Kumar 1989). *A. sobria/veronii* and *A. caviae* are also frequently involved in carp mortalities in different culture conditions, causing similar clinical conditions like *A. hydrophila* infection (Praveen et al. 2016). Aeromonads may also become opportunistic invaders (Grizzle and Kiryu 1993), infecting fish under stressful conditions or in concert with other pathogens (Noga and Berkhoff 1990).

The virulence of aeromonads is multifactorial and involves biologically active extracellular enzymes such as lipase and aerolysin (Li et al. 2015) and endotoxins such as lipopolysaccharides (LPSs) and other cellular structures (Janda 1991; Jahid et al. 2015). A wide range of extracellular enzymes are secreted by *Aeromonas* spp., namely, amylases, lipases, chitinases, proteases, gelatinases, and nucleases. Their exact role in the pathogenicity of individual strains is yet to be determined, though they potentially play a big role in adapting to environmental changes (Tomás 2012). Among the different virulent factors, haemolysin plays a predominant role in the expression of virulence for different *Aeromonas* species. Water temperature plays an extremely important role in the expression of different virulent associated genes, and it was found that a water temperature of 28 °C is most favourable for the virulence of *A. hydrophila* (Pattanayak et al. 2020). Abreu et al. (2018) also mentioned that the virulence of *A. hydrophila* is decidedly governed by different environmental factors, such as pH and ammonia, by influencing the expression of different virulence-associated genes.

Among the psychrophilic group of aeromonads, *Aeromonas salmonicida* is the aetiological agent of a disease named 'furunculosis' in salmonids and ulcer disease in cyprinids and marine flatfish (Austin and Austin 2016). Usually, *A. salmonicida* is regarded as an obligate fish pathogen (Schubert 1974) recoverable only from clinically diseased fish. The pathogenesis of *A. salmonicida* infection is directly correlated with the stress condition of the fish and the production of extracellular and cell-associated virulence determinants (Austin and Austin 2016). The association of *A. salmonicida* in common carp infections has also been recorded in India (Adikesavalu et al. 2014).

4.2 *Edwardsiella* spp.

The genus *Edwardsiella* is basically composed of three species, *E. ictaluri*, *E. tarda*, and *Edwardsiella hoshinae* (Ewing et al. 1965; Grimont et al. 1980; Hawke et al. 1981; Mohanty and Sahoo 2007). *E. tarda* and *E. ictaluri* are mostly known to infect fishes, whereas *E. hoshinae* infection is reported in reptiles and birds (Woo and Bruno 2010). *E. tarda* can be differentiated from *E. ictaluri* based on biochemical tests, i.e. *E. tarda* tests positive for indole production, methyl red reduction, and hydrogen sulphide generation, whereas *E. ictaluri* tests negative for all these tests (Panangala et al. 2006). *E. tarda* is a Gram-negative, rod-shaped, short, facultatively anaerobic, usually motile bacterium, but the isolates from red Seabream and yellowtail of *E. tarda* are non-motile in nature (Matsuyama et al. 2005). Based on O antigen extracts, *E. tarda* is divided into four serotypes, A, B, C, and D, and, of these, serotype A has been found to be the most virulent group worldwide (Park et al. 1983).

The first case of *E. tarda* infecting fish was reported in Japanese eel, *Anguilla japonica*, where it was initially described as *Paracolobactrum anguiillimortiferum* that is associated with red disease (Hoshina 1962); thereafter *E. tarda* has been found to infect a number of fish species (Park et al. 2012). The severity and onset of *E. tarda* infections mostly depend on environmental stresses such as poor water quality, extreme temperature, high organic content, etc. (Plumb 1999; Uhland et al. 2000; Zheng et al. 2004). The pathogenesis of *E. tarda* is always multifactorial. The mechanism of its pathogenicity is poorly understood, and the attachment and penetration process is still unclear. The intestine and abraded skin seem to be the entry sites of the bacteria (Mohanty and Sahoo 2007). Infection kinetics and histopathological studies using green fluorescent protein have revealed that the gastrointestinal tract, gills, and body surface of blue gourami (*Trichogaster trichopterus*) are the sites of entry of the *E. tarda* virulent strain (Ling et al. 2001). The pathology induced by *E. tarda* is highly similar across various fish species and characterized by systemic suppurative abscesses (Miyazaki and Kaige 1985; Rashid et al. 1997; Plumb 1999; Darwish et al. 2000). Though clinical signs are diverse after the onset of the disease, *E. tarda*-infected fish show loss of appetite and pigmentation, opacity of the eyes, exophthalmia, swelling of the abdominal surface, and petechial haemorrhage in the skin and fin (Egusa 1976, Miyazaki and Kaige 1985, Sahoo et al. 1998). Bloody and watery ascitic fluid inside the abdominal space is commonly found. Congestion in internal organs, such as the kidney, liver, and spleen (Miyazaki and Egusa 1976; Miyazaki and Kaige 1985), is also common. The histopathological characteristics of edwardsiellosis in fish are suppurative hepatitis, suppurative interstitial nephritis, and purulent inflammation in the internal organs (Miyazaki and Egusa 1976; Egusa 1976; Miyazaki and Kaige 1985).

The type III secretion system (T3SS) and type VI secretion system (T6SS) play significant roles in *E. tarda* adherence, penetration, survival, and replication in host cells and phagocytes. The type VI secretion system of *E. tarda* includes 16 genes, of which 13 encoded proteins are involved in the secretion of EvpP (*E. tarda* virulence factor) (Zheng and Leung 2007). Interestingly, after invading host cells, *E. tarda* induces the expression of T3SS and inhibit flagella biosynthesis and motility by activating the qseB and qseC two-component system (Wang et al. 2011). This indicates the capability of *E. tarda* to modulate the expression of different genes for adjusting the environmental changes for adaptation to intracellular living.

4.3 *Flavobacterium* spp.

Flavobacterium spp. are yellow-pigmented, pleomorphic, motile Gram-negative chromogens. These fish pathogens are particularly associated with gill diseases like hyperplasia (swelling of the gill epithelia) and fusion of the gill lamellae. These organisms naturally occur in aquatic environments and are also components of the physiological gill microflora of healthy fishes. Three different species within the family Flavobacteriaceae, namely, *Flavobacterium psychrophilum*, *F. columnare*, and *F. branchiophilum*, are the causative agents of diseases such as bacterial cold water disease and rainbow trout fry syndrome (Nematollahi et al. 2003; Starliper 2011); columnaris disease (Shotts and Starliper 1999); and bacterial gill disease (Wakabayashi et al. 1989), respectively. There are some sporadic reports of association of other *Flavobacterium* spp. in fish diseases, such as *Flavobacterium succinicans* (Anderson and Ordal 1961), *Flavobacterium johnsoniae* (Christensen 1977), *Flavobacterium hydatis* (Strohl and Tait 1978), and other uncharacterized yellow-pigmented bacteria (Austin and Austin 2007).

Columnaris disease caused by *F. columnare* is one the most devastating diseases of freshwater fishes that can lead to mortality of up to 80% (Bernardet et al. 2002; Bernardet and Bowman 2006). The bacterium is commonly found to infect both cultured and wild fishes with marked skin lesions, fin erosion, and gill necrosis, with the gills the primary site of infection, and a high degree of morbidity and mortality (Declercq et al. 2013). The bacterium's ability to adhere to fish gills has been significantly reduced following its exposure to sodium metaperiodate or incubation with *N*-acetyl-D-glucosamine, D-glucose, D-galactose, and D-sucrose (Decostere et al. 1999). The optimum growth temperature and incubation time for *F. columnare* varies from 21 °C to 30 °C and from 24 to 48 h, respectively (Declercq et al. 2013). The abraded skin or gills are essential for bacteria to enter the bloodstream and cause systemic infections (Dalsgaard 1993). Nevertheless, *F. columnare* has also been isolated from various internal organs of fish without showing any external changes (Hawke and Thune 1992). Diagnostic level I presumptive identification of *F. columnare* can be performed by simple examination of the wet mounts of the gills where the bacterium appears as long, thin bacterial rods with a characteristic flexing or gliding motion. Environmental factors, for instance, water temperature, play a major role in disease outbreak and mortality in fishes (Holt et al. 1975). *F. columnare* has been reported to cause mortality in catla and rainbow trout in India (Verma and Rathore 2013; Singh et al. 2021).

4.4 *Vibrio* spp.

Bacteria belonging to the genus *Vibrio*, which are pathogenic species, cause 'vibriosis', one of the major bacterial diseases mainly affecting marine and brackishwater fish and occasionally freshwater fish species. These Gram-negative, facultative anaerobic, oxidase-positive rods also cause significant economic loss. The symptoms of vibriosis include lethargy, slow growth, and skin discolouration, in general, and, in severe cases, septicaemia with haemorrhage on the base of the fins and, in some cases, exophthalmia and corneal opacity. The fish-pathogenic species of *Vibrio* include *Vibrio anguillarum* ('red pest' in eels), *Vibrio ordalii* (septicaemia in salmonids), *Vibrio salmonicida* (cold water vibriosis in fishes), *Vibrio vulnificus* (warm water vibriosis in the European eel), *Vibrio viscous*, and *Vibrio wodanis* ('winter ulcer disease' in Atlantic salmon) (Gauthier 2015; Callol et al. 2015).

Among different *Vibrio* species, the most prominent pathogen infecting fishes is *V. anguillarum*, which causes huge economic loss worldwide, mostly in marine and brackishwater environments. Isolates of *V. anguillarum* are broadly classified into 23 different serogroups based on O antigens (O1–O23) (Pedersen et al. 1999); among them, serotypes O1, O2, and O3 are the most virulent strains. The remaining serotypes are non-virulent and are mostly environmental strains (Frans et al. 2011; Toranzo et al. 2017). There is also sporadic reports of *V. anguillarum* infecting freshwater fishes. The most prominent one is the report that *V. anguillarum* infection causes the deadly red pest disease in freshwater eel. This disease is characterized by red spots on the ventral and lateral areas and swollen and dark skin lesions leading to ulceration. In addition, Hacking and Budd (1971) described the outbreak of *V. anguillarum* in freshwater rainbow trout hatcheries, which caused severe fish seed mortality.

4.5 *Pseudomonas* spp.

Pseudomonas spp. are a group of psychrophilic, Gram-negative rods that cause pseudomonad haemorrhagic septicaemia in fishes. These pathogens can be isolated and identified from infected organs or affected areas using selective media like glutamate starch phenol red agar (GSP) or *Pseudomonas–Aeromonas* selective agar. Important fish-pathogenic pseudomonads include *Pseudomonas anguilliseptica*, *Pseudomonas fluorescens*, *Pseudomonas aeruginosa*, *Pseudomonas alcaligenes*, and *Pseudomonas putida*. Gross clinical signs of pseudomonad infections in fish species include extensive haemorrhagic erosions throughout the body, ulcerations on the skin, fins, and tail, septicaemia, and, in some serious cases, even exophthalmia with corneal opacity. In a few instances, it has been observed that *P. fluorescens* infections in rainbow trout result in sudden 100% mortality irrespective of the environmental conditions (Kozinska 1999). *Pseudomonas* spp. are widespread in aquatic environments and infect fish species when under environmental stress, such as poor water quality, temperature variations, overcrowding, and sub-standard nutrition (Pękala-Safińska 2018).

Among the different species of *Pseudomonas*, *P. aeruginosa* has been found to be the most pathogenic to freshwater fishes. It is a ubiquitously found fish pathogen that causes septicaemia in freshwater fish, resulting in remarkable economic loss in aquaculture sectors worldwide (Roberts 2012). The typical signs associated with *P. aeruginosa* infection include skin ulcerations, haemorrhagic septicaemia, and fin rot. Furthermore, an internally engorged spleen, a pale liver, and abdominal dropsy with reddish ascitic exudates are also prominent (Algammal et al. 2020). Algammal et al. (2020) also found that virulence-associated genes, *oprL* and *toxA*, are the most predominant virulence genes associated with *P. aeruginosa* infection and that their presence is positively correlated with the infectivity of the bacterial isolate. Among the antibiotic resistance genes, *blaCTX-M*, *blaTEM*, and *tetA* are the key genes that induce resistance against antibiotics like cefotaxime, amoxicillin, and tetracycline in *P. aeruginosa*.

4.6 Other Bacteria of Importance

Acinetobacter are Gram-negative rods that are non-motile and facultative anaerobic in nature. Though the pathogenic mechanisms of *Acinetobacter* spp. are not yet clearly understood in fish diseases, it is believed that these pathogens are responsible for enhanced antibiotic resistance in aquaculture owing to their inherent drug resistance genes. *A. baumannii* has recently emerged as a pathogen that causes severe mortality in *Channa striata* and Indian major carp in farm conditions (Rauta et al. 2011; Behera et al. 2017; Paul et al. 2020). Clinical signs of *Acinetobacter* infections include depigmentation of the skin, loss of scales, exophthalmia with congestion of the eyes, and gill petechiae in trout. Haemorrhages in the skin and gill congestion were noted in infected carp, and a post-mortem examination showed intestinal inflammation in both fish species. Disease symptoms were accompanied by mortalities ranging from 5% to 20%. Recently, with increasing diagnostic capability and surveillance in India, a few zoonotic bacteria, such as *Klebsiella pneumoniae* (Das et al. 2018), *Proteus mirabilis* (Pattanayak et al. 2018), and *Plesiomonas shigelloides* (Behera et al. 2018), have also been found to cause fish mortality events in Indian aquaculture. The above-mentioned opportunistic bacterial pathogens have been found to cause similar generalized clinical signs like any other bacterial disease affecting fish.

5 Gram-Positive Pathogens

5.1 *Streptococcus* spp.

Among the various species known to cause streptococcosis in fishes, *Streptococcus parauberis*, *Streptococcus iniae*, *Streptococcus agalactiae*, and *Streptococcus dysgalactiae* are the most prominent species regardless of geographical regions (Mishra et al. 2018). Streptococcal disease in fish was first reported in 1957, affecting cultured rainbow trout in Japan (Hoshina et al. 1958). Since then, a wide range of fish species have been reported to encounter this disease. Among the different *Streptococcus* spp., *S. iniae*, in particular, causes serious health disorders in freshwater and marine fishes including rainbow trout and other salmonids, eels, tilapia, and catfishes (Nho et al. 2009). It has been observed that environmental factors, including temperature beyond the optimal range, handling, crowding, and salinity of the aquaculture systems, play a major role in the onset of the diseases.

Virulence of different *Streptococcus* spp. varies greatly with different isolates of bacterial strains within the same species (Abdelsalam et al. 2010). The pathological signs of streptococcosis are highly similar to those of enterococcal infection that primarily include exophthalmia, erratic swimming behaviour, dark colouration of the skin, corneal opacity, and petechiae on the eyes and gill covers. Gross anatomical features include congestion at the pectoral and caudal fins, haemorrhages in the swim bladder, liver, spleen, and kidney, and stomach inflammation and meningoencephalitis in all infected fish species. In intensive aquaculture systems, streptococcal infections could be lethal in nature, with reports of around 30–50% mortality; however, in most of the cases, the disease is chronic in nature and mortalities may extend over a period of several weeks, with only a few fish dying each day (Yanong and Francis-Floyd 2002).

5.2 *Lactococcus* spp.

Lactococcosis is a systemic hyper-acute, haemorrhagic septicaemic fish disease caused by *Lactococcus garvieae*, which is an emerging zoonotic pathogen that causes great economic loss in freshwater and marine aquaculture systems (Austin and Austin 2012). *L. garvieae* is a Gram-positive, facultative anaerobic, non-motile bacterium, and its polysaccharide capsule is the major virulence factor that helps it invade the host's immune response (Eldar et al. 1996). Symptoms of the infection include anorexia and erratic swimming followed by widespread haemorrhaging. Furthermore, a swollen abdomen, unilateral or bilateral exophthalmia, and anal prolapses are also observed in severe cases (Bekker et al. 2011). At necropsy, the peritoneal cavity is mostly filled with ascitic fluid, and congestion and enlargement of internal organs such as the spleen and liver are observed (Eldar and Ghittino 1999). Kusuda and Hamaguchi (1988) reported that the extracellular toxin produced by *L. garvieae* is alone capable of reproducing clinical disease in fish. Chemotherapeutic treatment of lactococcosis in cultured fish includes macrolide antibiotics (e.g. erythromycin) along with lincomycin and oxytetracycline. However, multiple antibiotic resistance is often encountered in streptococcal infection, mostly due to plasmid-mediated transfer of antibiotic resistance genes.

6 Pathogenomic Evolution of Bacteria

Recent advances in whole genome sequencing have provided powerful insight into the genetic make-up of microbes. Genome sequencing of a number of bacterial species has increased our knowledge on the biology, host adaptation, and virulence factors of these important pathogens and uncovered several complex adaptive evolutionary strategies mediated by horizontal gene transfer, insertion sequence elements, mutations, and prophage sequences operating in fish pathogens. Furthermore, genomic information has led us to know how the genome of an environmental strain is modified into highly virulent obligatory pathogens. Comparative genomic analysis has allowed the identification of unique pathogen-specific gene clusters. There exists a wide variation in genome size and content among different genera, species, and even within strains (Sudheesh et al. 2012). Factors involved in the pathogenomic evolution of bacteria, as depicted by Sudheesh et al. (2012), are illustrated in Fig. 1.

7 Co-infections

Co-infection phenomena are highly common in nature where different opportunistic pathogens live together and, when there is any interruption of homeostasis in the host system, more than one pathogen invades and causes complex pathology. Co-infections are demarcated by infection with two or more genetically different organisms where each organism has pathogenic possessions and causes damage to

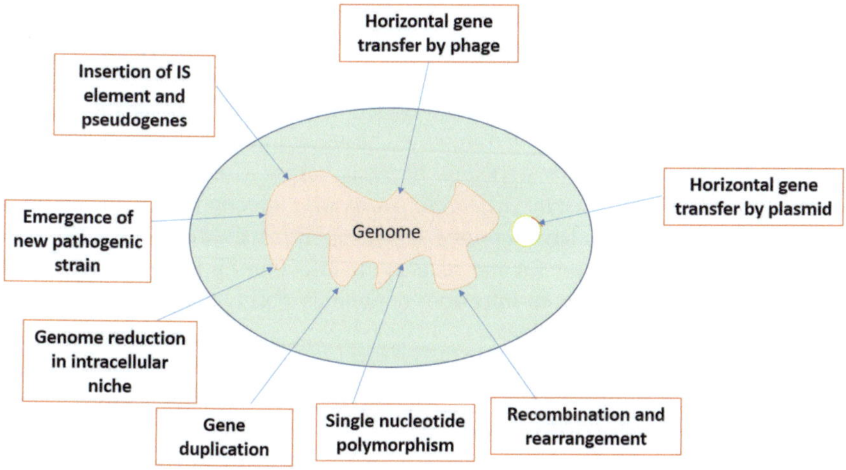

Fig. 1 Factors involved in the pathogenomic evolution of bacteria as depicted by Sudheesh et al. (2012)

the host in coincidence with other pathogens (Bakaletz 2004). During episodes of co-infection, a range of interactions among the infectious agents can be observed; the load of both pathogens may dominate: one may be increased and the other suppressed (Bakaletz 2004). During co-infection, invading pathogens compete with each other for target sites or resources important for their survival inside the same host. At times, one pathogen weakens the immune system of the host and subsequent infections by other pathogens increase the severity of disease. Pre-exposure of *Dactylogyrus intermedius* to goldfish (*Carassius auratus*) downregulates innate immune genes, namely, *TGF-β* and *C3*, which, in turn, weakens the host immune defence system, thus making them more vulnerable to *F. columnare* infection. Hence, *F. columnare* is able to invade compromised host defence easily and cause severe damage during co-infection (Zhang et al. 2015). *Dactylogyrus* spp. infection affecting cultured striped catfish (*Pangasianodon hypophthalmus*) leads to haemorrhages in both external and internal organs; furthermore, hypertrophy, hyperplasia, and fusion of the secondary lamellae have also been observed. Excessive damage to the gills and external injuries make it easy for invading secondary bacterial pathogens like *A. hydrophila*, which ultimately causes mortality (Yusoff et al. 2020). Crumlish et al. (2010) artificially co-infected Vietnamese catfish with *E. ictaluri* and *A. hydrophila* using an immersion route that led to cumulative mortalities up to 95%. On the other hand, *E. ictaluri and A. hydrophila* alone caused mortality rates of 80% and 10%, respectively. Concurrent infection of *E. ictaluri* and *F. columnare* in striped catfish (*P. hypophthalmus*) was naturally reported in Vietnam and Thailand. Experimental co-infection of *E. ictaluri* and *F. columnare* in striped catfish juveniles using immersion and injection (i.p) routes of challenge resulted in high cumulative mortality when compared to a single infection of *F. columnare* or *E. ictaluri* (Dong et al. 2015). In channel catfish, it was observed that *Ichthyophthirius multifiliis* and *E. ictaluri* have a synergetic pathogenic effect and that the susceptibility of the host and the pathogenicity of *E. ictaluri* was increased because of prior *I. multifiliis* infection (Xu et al. 2012a). Paul et al. (2020) described heavy mortality of *Channa striata* brooders because of co-infection with multiple parasitic and bacterial pathogens. Sahoo et al. (2016) reported heavy mortality of goldfish due to the co-infection of goldfish haematopoietic necrosis herpes virus (cyprinid herpesvirus-2) with multi-drug-resistant (MDR) *A. hydrophila* infection in farm conditions. Recently, Lu et al. (2021) have reported *A. hydrophila* as the key governing factor and environmental indicator for the severity of infection by tilapia lake virus in culture conditions. Different co-infections commonly observed in freshwater aquaculture are summarized in Table 2.

8 Methods of Diagnosis

Accurate and early diagnosis is the key to controlling and managing bacterial infections in aquaculture. Fish bacterial disease diagnosis has transformed drastically in the last few decades from being dependent on culture-based identification to culture-independent identification (Austin 2019) (Fig. 2). Conventional identification of bacteria that relies on culture of the pathogen from pathological samples and subsequent characterization based on the colony morphology, phenotypic characteristics, and biochemical reactions is a laborious and time-consuming process. Furthermore, the designation of bacteria up to species level based on the conventional method is also extremely difficult because of the similar reactivity potential of different bacteria. Moreover, many laboratories have adopted commercial systems for bacterial characterization, such as Analytical Profile Index (API) ZYM (API-ZYM), Biolog Gram-Negative (GN) (Biolog-GN), VITEK 2, Analytical Profile Index 20E (API 20E), API 20NE, API 50CH, API 50L, Biolog Gram-Positive (Biolog-GP), RapidID 32, and Enterotubes systems (Austin 2019). Most of these systems were originally designed for medical bacteria; hence, while using them for fish-pathogenic bacterial identification, one has to be over-cautious.

Different molecular approaches, with their high sensitivity and specificity, have become the most favoured choices in many laboratories for the diagnosis of bacterial fish diseases (Yan et al. 2018). 16S rDNA as an amplicon for the identification of fish pathogens in different molecular methods is the most favoured choice. A highly sensitive polymerase chain reaction (PCR) method for the identification of *A. salmonicida* was demonstrated by Hiney et al. (1992). This method was capable of detecting approximately two cells of bacteria. Subsequently, several other more sensitive and specific molecular techniques have been developed for the identification of bacterial pathogens, such as nested PCR, terminal restriction fragment length polymorphism (RFLP), multiplex PCR, loop-mediated isothermal amplification (LAMP), real-time PCR, real-time recombinase polymerase amplification, reverse transcription multiplex PCR, etc. (Austin 2019). For the identification of *A. hydrophila, A. caviae, A. veronii*, and *A. media* in a single reaction, Persson et al. (2015) developed a multiplex PCR method using *rpoB* and *gyrB* gene-based primers. A

Table 2 Summary of different bacterial co-infections prevalent in freshwater aquaculture

Host species	Bacterial pathogen	Co-infecting pathogen	Type of interaction during co-infection	References
Bacteria and bacterial co-infection				
P. hypophthalmus	*E. ictaluri*	*A. hydrophila*	Synergistic	Crumlish et al. (2010)
P. hypophthalmus	*E. ictaluri*	*F. columnare*	Synergistic	Dong et al. (2015)
Zebrafish	*A. hydrophila*	*A. veronii*	Synergistic	Chandrarathna et al. (2018)
Bacterial and parasitic co-infections				
Oncorhynchus mykiss	*Yersinia ruckeri*	*Myxobolus cerebralis* (Myxozoa)	Synergistic	Densmore et al. (2004)
Channa striata	*K. pneumoniae*, *A. baumannii*, *A. sobria*	*I. multifiliis*, *Epistylis* sp., *Trichodina* sp., *Argulus siamensis*, *Ergasilus* sp., and *Zschokkella auratis*	Synergistic	Paul et al. (2020)
Oreochromis niloticus	*S. iniae*	*Gyrodactylus niloticus*	Synergistic	Xu et al. (2007)
Carassius auratus	*F. columnare*	*Dactylogyrus intermedius*	Synergistic	Zhang et al. (2015)
O. mykiss	*F. columnare*	*Argulus coregoni*	Synergistic	Bandilla et al. (2006)
Ictalurus punctatus	*E. ictaluri*	*I. multifiliis*	Synergistic	Xu et al. (2012a)
O. niloticus	*S. iniae*	*I. multifiliis*	Synergistic	Xu et al. (2009)
I. punctatus	*A. hydrophila*	*I. multifiliis*	Synergistic	Xu et al. (2012b)
I. punctatus	*S. iniae* and *S. agalactiae*	*Trichdina* sp.	Synergistic	Evans et al. (2007)
P. hypophthalmus	*A. hydrophila*	*Dactylogyrus* sp.	Synergistic	Yusoff et al. (2020)
Carassius auratus	*A. hydrophila*	*Argulus* sp.	Synergistic	Shameena et al. (2021)
Bacterial and viral co-infection				
O. mykiss	*F. psychrophilum*	Infectious pancreatic necrosis virus	Synergistic	Evensen and Lorenzen (1997)
Carassius auratus	*A. hydrophila*	Cyprinid herpesvirus-2	Synergistic	Sahoo et al. (2016)
Cyprinus carpio koi	*Flavobacterium* sp.	Carp oedema virus	Synergistic	Adamek et al. (2018)
Oreochromis spp.	*A. hydrophila*	Tilapia lake virus	Synergistic	Nicholson et al. (2020)

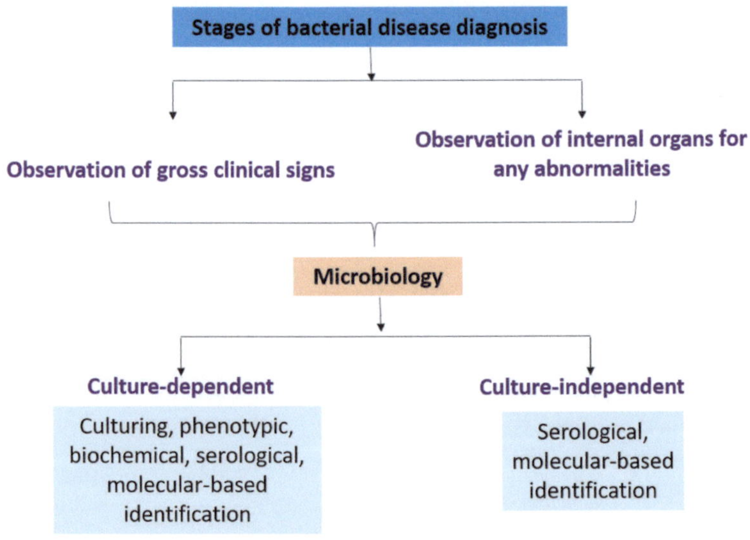

Fig. 2 Stages involved in bacterial disease diagnosis

triplex LAMP has also been developed, which can detect and discriminate *Vibrio alginolyticus*, *V. anguillarum*, and *V. harveyi* in infected fish with a high level of sensitivity in a single reaction (Yu et al. 2013).

Among the different serological methods for the identification of bacterial pathogens, enzyme-linked immunosorbent assay (ELISA) is one of the most specific and sensitive techniques commonly used. This technique involves the coating of antibody to plastic or glass surfaces, on which targeted bacterial cells or infected tissue material are added. Furthermore, an antibody-enzyme and an enzyme-substrate are added. A positive reaction in terms of a colour change would develop within 60 min (Swain and Nayak 2003). Other known important serological methods for bacterial isolation are whole-cell agglutination, latex agglutination, immuno-India ink technique, fluorescent antibody technique, immunohistochemistry, etc. (Austin 2019). However, due to inherent problems of antigenic cross-reactivity and sensitivity, serological tests are less in use for disease diagnosis, particularly in Indian aquaculture systems.

9 Prevention and Control Strategies

In aquaculture, for the control of bacterial infections, a wide range of antibiotics are used very commonly. Apart from treatment measures, recently, antibiotics have also been used as a prophylactic measure and growth promotor (Aich et al. 2018). Unregulated and unauthorized use of antibiotics has provoked the development of antibiotic-resistant fish pathogens and other aquatic bacteria. This has forced practitioners to search for alternative eco-friendly means of bacterial disease treatment.

Some of the methods of such treatment and management measures are described hereunder.

Probiotics have recently gained admiration for their ability to better manage fish health by manipulating the gut microbiota and enhancing immune responses in fish (Merrifield et al. 2010). An effective commercial probiotic should have the ability to inhibit pathogenic bacterial colonization by producing enzymes or killing harmful bacteria, and probiotic bacteria should not be pathogenic to the host (Nageswara and Babu 2006). Some probiotic bacteria cause disturbances in the quorum sensing (QS) of pathogenic bacteria in the gut microbiota by producing different quorum sensing antagonist substances (Defoirdt et al. 2004). For instance, the quorum sensing antagonist substances produced by *Lactobacillus acidophilus* were able to inhibit the QS of the *Escherichia coli* O157 gene (Medellin-Pena et al. 2007). Probiotics also interact with different immune cells, such as natural killer (NK) cells, mononuclear phagocytes (monocytes, macrophages), and polymorphonuclear leukocytes (neutrophils), and act as immunostimulants. Many probiotics are known to enhance the number of circulating macrophages, erythrocytes, granulocytes, and lymphocytes in various fishes (Irianto and Austin 2002; Kim and Austin 2006; Kumar et al. 2008). Besides their use as feed probiotics, many Gram-positive bacteria, such as *Bacillus* spp., are also used as water probiotics to improve water quality by converting organic matter into simpler compounds when directly added to water (Balcazar et al. 2006).

Phage therapy has the ability to decline the infecting bacterial population below the threshold quantity to a level where the host defence clears the remaining bacteria (Aich et al. 2018). Based on reproduction strategy, there are two types of bacteriophages: (1) lytic and (2) lysogenic. In the lytic cycle, the phage attaches to a host cell and inserts its nucleic acid into the cell, directing the host to produce offspring. In the lysogenic cycle, phage nucleic acid becomes part of the host genome and reproduces genetic material in the host cell using the host machinery. There is much experimental evidence for controlling bacterial diseases using phage-based therapeutics. Bacteriophages were found to be effective in controlling luminous bacterial diseases in shrimp hatcheries by reducing the load of *V. harveyi* in water (Karunasagar et al. 2005). Prasad et al. (2011) recorded 100% survival and reduction in the *F. columnare* bacterium in *Clarias batrachus* after 6 h of phage treatment. The main practical problem associated with field-level phage therapy development is the determination of proper dose and cost-effectiveness. Hence, isolation and characterization of those phages, which have a high infecting ability at lower doses as well as a high replication rate, will serve the purpose (Rong et al. 2014).

Vaccination is the gold standard strategy for the control and management of infectious diseases in all vertebrate animals. Before attempting any vaccination strategy in fishes, it is vital to determine when immune cells and the immune system are morphologically and functionally mature (Toranzo et al. 2009). However, the success of administering vaccines depends upon the fish species, pathogens, water temperature, and environment (Aich et al. 2018). The first commercial bacterial fish vaccine was against enteric redmouth disease (ERM) caused by *Yersinia ruckeri* for preventing vibriosis in the late 1970s (Evelyn 1996). Most of the commercial vaccines developed against fish bacterial diseases are either killed or live attenuated. Vaccination using a live attenuated antigen imitates natural infection. The antigen significantly excites the cellular immune response further, thus stimulating humoral and mucosal immunity (Clark and Cassidy-Hanley 2005). In recent times, several attempts have been made to develop subunit/recombinant vaccines against bacterial fish pathogens. The effectiveness of oral vaccine delivery depends on the stability of the antigen against exposed digestive hydrolysis so that the antigen can be absorbed well and induce a protective immune response (Vandenberg 2004). For increasing the stability of antigens, encapsulation is the most effective tool. Encapsulation has two major roles: (i) preventing antigens escaping from the feed/pellets and (ii) protecting antigens from the acidic environment in the fish stomach (Soliman et al. 2019). Bio-encapsulated feed is also a suitable method for incorporating vaccine suspensions into live feeds such as *Artemia*, copepods, and rotifers and is fed to fish fry (Lin et al. 2005). Nanoparticle-based adjuvant and efficient delivery systems have been recently proven to be effective. They enter by cellular endocytosis, which facilitates the cellular uptake of antigens and escalates its presentation ability (Vinay et al. 2018). Although commercial vaccination programmes and vaccines are available for many bacterial diseases in freshwater aquaculture in different parts of the world, currently, there are none for commercial-scale application in India. Nevertheless, many attempts have been made on a laboratory scale to develop bacterial vaccines with varied relative percentage survival (RPS) levels when challenged with homologous pathogens. For example, recombinant outer membrane protein C (ompC) injection in rohu induced a significant level of antibody production and an RPS of 66.02% against *A. hydrophila* challenge in a 140-day vaccination trial (Yadav et al. 2021) as compared to RPS rates of 44% using rompF (Yadav et al. 2018) and 82% using rompC (Dash et al. 2014) against the same bacterium in two other trials conducted in our laboratory. In a vaccination trial using formalin-killed *E. tarda*, an RPS of 92% was obtained in rohu fingerlings post-immunomodulation with a glucan-based diet when challenged with the same bacterial pathogen (Sahoo and Mukherjee 2002).

Development of a disease-resistant variety of fish through various selection approaches (classical selective breeding based on farm survival or challenge data incorporated into family, mass selection, or marker-assisted selection) renders permanent resistance for economically important diseases. A large number of selective breeding programmes in salmonids, tilapia, and carps are currently going on for various bacterial diseases, such as *A. salmonicida* infection (furunculosis) in salmonids, *Yersinia ruckeri* (ERM) in rainbow trout, *A. hydrophila* in rohu and common carp, and *V. anguillarum* in Atlantic cod (reviewed in Ødegård et al. 2011). In India, third-generation disease-resistant rohu against *A. hydrophila* infection has been produced with a selection response of 58.3% over the control population (Sahoo et al. 2013). However, undertaking a selection programme for each species and economically non-important bacterial disease is not a viable strategy.

Drug development based on antimicrobial peptides (AMPs) is a promising approach in the age of antimicrobial resistance, as they are natural compounds of the host body. Recently, different recombinant/synthetic antimicrobial peptides have been researched, and they have shown potential for application against fish-pathogenic bacteria. A recombinant antimicrobial peptide, apolipoprotein A-I of *Labeo rohita* (rApoA-I) produced in *Escherichia coli*, showed encouraging results against fish-pathogenic bacteria, *A. hydrophila*, in an in vivo challenge study. *L. rohita* administered a dose of 100 μg rApoA-I exhibited significantly higher protection (~55%) 12-h post-administration of the protein in comparison to the control group (Mohapatra et al. 2016). The recombinant AMP hepcidin from rohu (r*Lr*Hep) was found to inhibit *Edwardsiella tarda* and *A. hydrophila*, in in vitro conditions at 50 and 100 μg, respectively, at 6 h post-incubation. Furthermore, a relative percentage survival (RPS) of 73% was found when r*Lr*Hep was administered to rohu juveniles and subsequently challenged with *A. hydrophila* (Mohapatra et al. 2019).

RNA interference (RNAi) is a primeval biological machinery used to defend against different external invasions. Theoretically, RNAi sequence specifically silences any disease-related genes, thus making small interfering RNA (siRNA) a promising therapeutic tool (Hu et al. 2020). Turning off the expression of the target gene is possible at two levels: (1) transcriptional gene silencing and (2) post-transcriptional gene silencing. Therapeutic silencing of virulent genes of the invading pathogen has received maximum attention recently (Aich et al. 2018). Sarker et al. (2017) successfully demonstrated RNAi-based therapeutic proof against salmonid whirling disease by knocking down the *MyxSP-1* gene in *Tubifex tubifex*, which abrogated the *M. cerebralis* life cycle.

The amalgamation of in silico bioinformatics tools with proteomics, metabolomics, and comparative genomics is of great significance for the identification of novel drug targets against pathogens (Mondal et al. 2015). Furthermore, the in silico strategy can minimize the number of in vitro trials, making the drug discovery process less laborious and more economical (Timo et al. 2019). Recently, several encouraging results have been produced by integrating in silico and in vivo procedures for inhibiting MDR pathogens, including *Mycobacterium tuberculosis*, *Streptococcus agalactiae*, *S. aureus*, *E. coli*, and *V. cholerae* (Zang et al. 2016; Sandhaus et al. 2018; Tiwari et al. 2018; Favero et al. 2020).

10 Bacterial Diseases and Associated Human Health

There are several bacterial pathogens infecting fishes that pose a zoonotic threat to humans. However, their impact is mostly confined to gastrointestinal infections. Most zoonotic infections occur when people consume raw fish. Aeromonads are the most ubiquitous fish pathogens, and they also have the zoonotic potential. Among the different *Aeromonas* strains identified, 95.40% of those associated with human clinical cases correspond to four species, namely, *A. caviae* (37.26%), *Aeromonas dhakensis* (23.49%), *A. veronii* (21.54%), and *A. hydrophila* (13.07%) (Figueras and Beaz-Hidalgo 2015; Fernández-Bravo and Figueras 2020). Recently, Fernández-Bravo and Figueras (2020) have reported diarrhoea produced by a strain of *Aeromonas trota* in an immunocompromised individual. Another fish pathogen, *E. tarda*, is well-known for its zoonotic potential causing gas gangrene or cellulitis, gastroenteritis, and other generalized gastrointestinal infections in humans with compromised immune systems (Thune et al. 1993; Plumb 1999; Srinivasa Rao et al. 2001). The frequency of *Vibrio* spp. infections in humans increased to 141% between 1996 and 1998, 54% between 2006 and 2008, and 7% between 2011 and 2013 (CDC 2014). These infections occur in two different ways: (i) consumption of raw or undercooked fish or fish products and (ii) transmission through fish farm employees or workers who do not take precautionary measures while handling or processing fish and fish products that may be contaminated (Jahncke 2007; Iwamoto et al. 2010). The most commonly identified *Vibrio* species in the United States is *V. parahaemolyticus*, contributing to 48% of total cases (CDC 2014). *V. vulnificus* is also commonly associated with seafood-borne illness in the United States (CDC 2014). It is highly invasive and may cause acute gastroenteritis and septicaemia and can lead to death (Scallan et al. 2011). *Proteus mirabilis* is an opportunistic bacterial pathogen that causes urinary tract infections in humans (Cao et al. 2009). It has also been reported to be involved in nosocomial outbreaks in neonates in India (Jain et al. 2016). Recently, Pattanayak et al. (2018) have reported that *P. mirabilis* infection in rohu causes severe mortality in farm conditions. Furthermore, Das et al. (2018) identified and characterized the involvement of *Klebsiella pneumoniae* in the disease outbreak of cultured rohu, causing severe mortality in West Bengal, India. *K. pneumoniae* is popularly known for causing a wide range of infections in humans, including urinary tract infections, pneumonia, wound infections, and bacteraemia (Holden et al. 2016).

11 Conclusions

To feed the ever-increasing human population, intensification of aquaculture is inevitable. Other than intensification, climate change also poses several stressful environments to farmed fishes. Under different stressful conditions, fish become susceptible to several bacterial diseases. There is also pathological evolution of different bacterial pathogens, leading to the emergence of more virulent strains. Recently, co-infections with homogeneous and heterogeneous bacterial pathogens have increased manyfold. Control of bacterial fish pathogens in aquaculture has always been a challenging task from the time of inception of the sector. Conventional treatment of bacterial pathogens is restricted to the use of different drugs and chemicals. Recently, it has been observed that there is frequent development of antimicrobial resistance because of the unregulated use of different drugs and chemicals. Hence, development of new-generation, eco-friendly treatment measures against bacterial pathogens is the need of the hour. Furthermore, in silico development of different drugs may help in future control of different bacterial diseases. In the absence of any commercial bacterial vaccine, for a country like India, research needs to focus on developing effective adjuvanted economically viable vaccines, preferably oral or immersion delivery-based, for mass-scale vaccination of carp and catfish.

References

Abdelsalam M, Chen SC, Yoshida T (2010) Dissemination of streptococcal pyrogenic exotoxin G (spegg) with an IS-like element in fish isolates of *Streptococcus dysgalactiae*. FEMS Microbiol Lett 309:105–113

Abreu RE, Magalhães TC, Souza RC, Oliveira ST, Ibelli AM, Demarqui FN, Gouveia JJ, Costa MM, Gouveia GV (2018) Environmental factors on virulence of *Aeromonas hydrophila*. Aquac Int 26(2):495–507

Adamek M, Teitge F, Jung-Schroers V, Heling M, Gela D, Piackova V, Kocour M, Steinhagen D (2018) *Flavobacteria* as secondary pathogens in carp suffering from koi sleepy disease. J Fish Dis 41(11):1631–1642

Adikesavalu H, Patra A, Mondal A, Banerjee S, Abraham TJ (2014) Association of *Aeromonas salmonicida* subsp. *achromogenes* in the haemorrhagic blister of cultured carp *Cyprinus carpio* in West Bengal, India. Asian Pac J Trop Dis 4:S500–S504

Aich N, Ahmed N, Paul A (2018) Issues of antibiotic resistance in aquaculture industry and its way forward. Int J Curr Microbiol App Sci 7:26–41

Algammal AM, Mabrok M, Sivaramasamy E, Youssef FM, Atwa MH, El-Kholy AW, Hetta HF, Hozzein WN (2020) Emerging MDR-*Pseudomonas aeruginosa* in fish commonly harbor opr L and tox A virulence genes and bla TEM, bla CTX-M, and tetA antibiotic-resistance genes. Sci Rep 10(1):1–12

Anderson RL, Ordal EJ (1961) *Cytophaga succinicans* sp. n., a facultatively anaerobic, aquatic myxobacterium. J Bacteriol 81(1):130–138

Austin B (2019) Methods for the diagnosis of bacterial fish diseases. Mar Life Sci Technol 1(1):41–49

Austin B, Austin DA (2007) Bacterial fish pathogens: diseases of farmed and wild fish, 4th edn. Praxis Publishing Ltd., Chichester

Austin B, Austin AD (2012) Bacterial fish pathogens: diseases of farmed and wild fish, 5th edn. Springer/PraxisPublishing, Chichester

Austin B, Austin DA (2016) Bacterial fish pathogens: disease of farmed and wild fish, 6th edn. Springer, Dordrecht

Bakaletz LO (2004) Developing animal models for polymicrobial diseases. Nat Rev Microbiol 2(7):552–568

Balcazar JL, de Blas I, Ruiz-Zarzuela I, Cunningham D, Vendrell D, Muzquiz JL (2006) The role of probiotics in aquaculture. Vet Microbiol 114:173–186

Bandilla M, Valtonen ET, Suomalainen LR, Aphalo PJ, Hakalahti T (2006) A link between ectoparasite infection and susceptibility to bacterial disease in rainbow trout. Int J Parasitol 36(9):987–991

Beaz-Hidalgo R, Martínez-Murcia A, Figueras MJ (2013) Reclassification of *Aeromonas hydrophila* subsp. *dhakensis* Huys et al. 2002 and *Aeromonas aquariorum* Martinez-Murcia et al. 2008 as *Aeromonas dhakensis* sp. nov. comb nov. and emendation of the species *Aeromonas hydrophila*. Syst Appl Microbiol 36:171–176

Behera BK, Paria P, Das A, Bhowmick S, Sahoo AK, Das BK (2017) Molecular characterization and pathogenicity of a virulent *Acinetobacter baumannii* associated with mortality of farmed Indian major carp *Labeo rohita* (Hamilton 1822). Aquaculture 471:157–162

Behera BK, Bera AK, Paria P, Das A, Parida PK, Kumari S, Bhowmick S, Das BK (2018) Identification and pathogenicity of *Plesiomonas shigelloides* in silver carp. Aquaculture 493:314–318

Bekker A, Hugo C, Albertyn J, Boucher CE, Bragg RR (2011) Pathogenic Gram-positive cocci in South African rainbow trout, *Oncorhynchus mykiss* (Walbaum). J Fish Dis 34(6):483–487

Bernardet JF, Bowman JP (2006) The genus *Flavobacterium*. In: Dworkin M, Falkow S (eds) The prokaryotes: a handbook on the biology of bacteria: Volume 7: Proteobacteria: delta and epsilon subclasses. Deeply rooting bacteria. Springer, New York, pp 481–531

Bernardet JF, Nakagawa Y, Holmes B (2002) Proposed minimal standards for describing new taxa of the family Flavobacteriaceae and emended description of the family. Int J Syst Evol Microbiol 52(3):1049–1070

Callol A, Pajuelo D, Ebbesson L, Teles M, MacKenzie S, Amaro C (2015) Early steps in the European eel (*Anguilla Anguilla*)–*Vibrio vulnificus* interaction in the gills: Role of the RtxA13 toxin. Fish Shellfish Immunol 43(2):502–509

Camus AC, Durborow RM, Hemstreet WG, Thune RL, Hawke JP (1998) *Aeromonas* bacterial infections: motile aeromonad septicemia (No. 478). Southern Regional Aquaculture Center (Southern Regional Aquaculture Center collectively covers Alabama, Arkansas, Florida, Georgia, Kentucky, Louisiana, Mississippi, North Carolina, Oklahoma, Puerto Rico, South Carolina, Tennessee, Texas, U.S. Virgin Islands, and Virginia)

Cao B, Wang M, Liu L, Zhou Z, Wen S, Rozalski A, Wang L (2009) 16S-23S rDNA internal transcribed spacer regions in four *Proteus* species. J Microbiol Methods 77(1):109–118

CDC (2014) Foodborne Diseases Active Surveillance Network (FoodNet): food net surveillance report for 2014 (final report). U.S. Department ofHealth and Human Services, CDC, Atlanta

Chandrarathna HPSU, Nikapitiya C, Dananjaya SHS, Wijerathne CUB, Wimalasena SHMP, Kwun HJ, Heo GJ, Lee J, De Zoysa M (2018) Outcome of co-infection with opportunistic and multidrug resistant *Aeromonas hydrophila* and *A. veronii* in zebrafish: Identification, characterization, pathogenicity and immune responses. Fish Shellfish Immunol 80:573–581

Christensen PJ (1977) The history, biology, and taxonomy of the Cytophaga group. Can J Microbiol 23(12):1599–1653

Clark T, Cassidy-Hanley D (2005) Recombinant subunit vaccines: potentials and constraints. Dev Biol 121:153–163

Colorni A (2004) Diseases of Mediterranean fish species: problems, research and prospects. Bull Eur Assoc Fish Pathol 24:22–32

Crumlish M, Thanh PC, Koesling J, Tung VT, Gravningen K (2010) Experimental challenge studies in Vietnamese catfish, *Pangasianodon hypophthalmus* (Sauvage), exposed to *Edwardsiella ictaluri* and *Aeromonas hydrophila*. J Fish Dis 33(9):717–722

Dalsgaard I (1993) Virulence mechanisms in *Cytophaga psychrophila* and other Cytophaga-like bacteria pathogenic for fish. Annu Rev Fish Dis 3:127–144

Darwish A, Plumb JA, Newton JC (2000) Histopathology and pathogenesis of experimental infection with *Edwardsiella tarda* in channel catfish. J Aquat Anim Health 12(4):255–266

Das A, Acharya S, Behera BK, Paria P, Bhowmick S, Parida PK, Das BK (2018) Isolation, identification and characterization of *Klebsiella pneumoniae* from infected farmed Indian major carp *Labeo rohita* (Hamilton 1822) in West Bengal, India. Aquaculture 482:111–116

Dash P, Sahoo PK, Gupta PK, Garg LC, Dixit A (2014) Immune responses and protective efficacy of recombinant outer membrane protein R (rOmpR)-based vaccine of *Aeromonas hydrophila* with a modified adjuvant formulation in rohu (*Labeo rohita*). Fish Shellfish Immunol 39(2):512–523

Declercq AM, Haesebrouck F, Van den Broeck W, Bossier P, Decostere A (2013) Columnaris disease in fish: a review with emphasis on bacterium-host interactions. Vet Res 44(1):1–17

Decostere A, Haesebrouck F, Van Driessche E, Charlier G, Ducatelle R (1999) Characterization of the adhesion of *Flavobacterium columnare* (*Flexibacter columnaris*) to gill tissue. J Fish Dis 22(6):465–474

Defoirdt T, Boon N, Boosier P, Verstraete W (2004) Disruption of bacterial quorum sensing: an unexplored strategy to fight infections in aquaculture. Aquaculture 240:69–88

Densmore CL, Ottinger CA, Blazer VS, Iwanowicz LR, Smith DR (2004) Immunomodulation and disease resistance in post yearling rainbow trout infected with *Myxobolus cerebralis*, the causative agent of whirling disease. J Aquat Anim Health 16(2):73–82

Dong HT, Nguyen VV, Phiwsaiya K, Gangnonngiw W, Withyachumnarnkul B, Rodkhum C, Senapin S (2015) Concurrent infections of *Flavobacterium columnare* and *Edwardsiella ictaluri* in striped catfish, *Pangasianodon hypophthalmus* in Thailand. Aquaculture 448:142–150

Egusa S (1976) Some bacterial diseases of freshwater fishes in Japan. Fish Pathol 10(2):103–114

Eldar AA, Ghittino C (1999) *Lactococcus garvieae* and *Streptococcus iniae* infections in rainbow trout *Oncorhynchus mykiss*: similar, but different diseases. Dis Aquat Org 36(3):227–231

Eldar A, Ghittino C, Asanta L, Bozzetta E, Goria M, Prearo M, Bercovier H (1996) *Enterococcus seriolicida* is a junior synonym of *Lactococcus garvieae*, a causative agent of septicemia and meningoencephalitis in fish. Curr Microbiol 32(2):85–88

Evans J, Klesius P, Pasnik D, Shoemaker C (2007) Influence of natural *Trichodina* sp. parasitism on experimental *Streptococcus iniae* or *Streptococcus agalactiae* infection and survival of young channel catfish *Ictalurus punctatus* (Rafinesque). Aquac Res 38(6):664–667

Evelyn TP (1996) A historical review of fish vaccinology. Dev Biol Stand 90:3–12

Evensen Ø, Lorenzen E (1997) Simultaneous demonstration of infectious pancreatic necrosis virus (IPNV) and *Flavobacterium psychrophilum* in paraffin-embedded specimens of rainbow trout *Oncorhynchus mykiss* fry by use of paired immunohistochemistry. Dis Aquat Org 29(3):227–232

Ewing WH, McWhorter AC, Escobar MR, Lubin AH (1965) *Edwardsiella*, a new genus of Enterobacteriaceae based on a new species, *E. tarda*. Int J Syst Evol Microbiol 15:33–38

FAO (1999) The state of world fisheries and aquaculture 1999. http://www.fao.org

Favero LM, Chideroli RT, Ferrari NA, Azevedo VADC, Tiwari S, Lopera-Barrero NM, Pereira UDP (2020) In silico prediction of new drug candidates against the multidrug-resistant and potentially zoonotic fish pathogen serotype III *Streptococcus agalactiae*. Front Genet 11:1024

Fernández-Bravo A, Figueras MJ (2020) An update on the genus *Aeromonas*: taxonomy, epidemiology, and pathogenicity. Microorganisms 8(1):129

Figueras MJ, Beaz-Hidalgo R (2015) *Aeromonas* infections in humans. In: Graf J (ed) Aeromonas. Caister Academic Press, Norfolk, pp 65–108

Frans I, Michiels CW, Bossier P, Willems KA, Lievens B, Rediers H (2011) *Vibrio anguillarum* as a fish pathogen: virulence factors, diagnosis and prevention. J Fish Dis 34(9):643–661

Gauthier DT (2015) Bacterial zoonoses of fishes: a review and appraisal of evidence for linkages between fish and human infections. Vet J 203(1):27–35

Gopalakrishnan V (1961) Observation on infectious dropsy of Indian major carps and its experimental induction. J Sci Ind Res 20:357–358

Grimont PAD, Grimont F, Richard C, Sakazaki R (1980) *Edwardsiella hoshinae*, a new species of Enterobacteriaceae. Curr Microbiol 4:347–351

Grizzle JM, Kiryu Y (1993) Histopathology of gill, liver, and pancreas, and serum enzyme levels of channel catfish infected with *Aeromonas hydrophila* complex. J Aquat Anim Health 5(1):36–50

Hacking MA, Budd J (1971) Vibrio infection in tropical fish in a freshwater aquarium. J Wildl Dis 7(4):273–280

Hawke JP, Thune RL (1992) Systemic isolation and antimicrobial susceptibility of *Cytophaga columnaris* from commercially reared channel catfish. J Aquat Anim Health 4(2):109–113

Hawke JP, Mcwhorter A, Steigerwalt AG, Brenner DONJ (1981) *Edwardsiella ictaluri* sp. nov., the causative agent of enteric septicemia of catfish. Int J Syst Evol Microbiol 31:396–400

Hiney M, Dawson MT, Heery DM, Smith PR, Gannon F, Powell R (1992) DNA probe for *Aeromonas salmonicida*. Appl Environ Microbiol 58:1039–1042

Holden VI, Breen P, Houle S, Dozois CM, Bachman MA (2016) *Klebsiella pneumoniae* siderophores induce inflammation, bacterial dissemination, and HIF-1α stabilization during pneumonia. MBio 7(5):e01397–e01316

Holt RA, Sanders JE, Zinn JL, Fryer JL, Pilcher KS (1975) Relation of water temperature to *Flexibacter columnaris* infection in steelhead trout. (*Salmo gairdneri*), coho (*Oncorhynchus kisutch*) and chinook (*O. tshawytscha*) salmon. J Fish Board Can 32(9):1553–1559

Hoshina T (1962) On a new bacterium, *Paracolobactrum anguillimortiferum* n. sp. Bull Jpn Soc Sci Fish 28:162–164

Hoshina T, Sano T, Morimoto Y (1958) A *Streptococcus* pathogenic to fish. J Tokyo Univ Fish 44:57–68

Hu B, Zhong L, Weng Y, Peng L, Huang Y, Zhao Y, Liang XJ (2020) Therapeutic siRNA: state of the art. Signal Transduct Target Ther 5(1):1–25

Irianto A, Austin B (2002) Probiotics in aquaculture. J Fish Dis 25(11):633–642

Iwamoto M, Ayers T, Mahon BE, Swerdlow DL (2010) Epidemiology of seafood-associated infections in the United States. Clin Microbiol Rev 23(2):399–411
Jahid IK, Mizan MFR, Ha AJ, Ha SD (2015) Effect of salinity and incubation time of planktonic cells on biofilm formation, motility, exoprotease production, and quorum sensing of *Aeromonas hydrophila*. Food Microbiol 49:142–151
Jahncke M (2007) JIFSAN good aquacultural practices manual, Section 2 – Aquaculture Product Safety and Consumer Health, University of Maryland FDA. JIFRAN University of Maryland, College Park
Jain S, Gaind R, Kothari C, Sehgal R, Shamweel A, Thukral SS, Chellani HK (2016) VEB-1 extended-spectrum β-lactamase-producing multidrug-resistant *Proteus mirabilis* sepsis outbreak in a neonatal intensive care unit in India: clinical and diagnostic implications. JMM Case Rep 3(4):e005056
Janda JM (1991) Recent advances in the study of the taxonomy, pathogenicity, and infectious syndromes associated with the genus *Aeromonas*. Clin Microbiol Rev 4(4):397–410
Janda JM, Duffey PD (1988) Mesophilic aeromonads in human disease; current taxonomy, laboratory identification, and infectious disease spectrum. Rev Infect Dis 10:980–997
Karunasagar I, Ali PK, Jeyasekaran G, Karunasagar I (1986) Ulcerative form of *Aeromonas hydrophila* infection in *Catla Catla*. Curr Sci 55:1194–1195
Karunasagar I, Vinod MG, Kennedy B, Vijay ATNUR, Deepanjali A, Umesha KR, Karunasagar I (2005) Biocontrol of bacterial pathogens in aquaculture with emphasis on phage therapy. Diseases in Asian Aquaculture V. Fish Health Section. Asian Fisheries Society, Manila, pp 535–542
Kim DH, Austin B (2006) Cytokine expression in leucocytes and gut cells of rainbow trout, *Oncorhynchus mykiss* Walbaum, induced by probiotics. Vet Immunol Immunopathol 114(3):297–304
Kozinska A (1999) Atypical cases of disorders in cyprinid wintering caused by *Pseudomonas fluorescens* infections. Bull Eur Assoc Fish Pathol 19:216–220
Kumar D (1989) Epizootic Ulcerative Syndrome out-break in India. Summer Institute on Fish Diseases and Health Management in Freshwater Aquaculture System, 5–24 June, CIFA, Bhubaneswar
Kumar R, Mukherjee SC, Ranjan R, Nayak SK (2008) Enhanced innate immune parameters in *Labeo rohita* (Ham.) following oral administration of *Bacillus subtilis*. Fish Shellfish Immunol 24(2):168–172
Kusuda R, Hamaguchi M (1988) Extracellular and intracellular toxins of *Streptococcus* sp. isolated from yellowtail. Bull Eur Assoc Fish Pathol 8:9–10
Li F, Wang W, Zhu Z, Chen A, Du P, Wang R, Chen H, Hu Y, Li J, Kan B, Wang D (2015) Distribution, virulence-associated genes and antimicrobial resistance of *Aeromonas* isolates from diarrheal patients and water, China. J Infect 70(6):600–608
Lin JH, Yu CC, Lin CC, Yang HL (2005) An oral delivery system for recombinant subunit vaccine to fish. Dev Biol 121:175–180
Ling SH, Wang XH, Lim TM, Leung KY (2001) Green fluorescent protein-tagged *Edwardsiella tarda* reveals portal of entry in fish. FEMS Microbiol Lett 194(2):239–243
Lu TH, Chen CY, Liao CM (2021) *Aeromonas hydrophila* as an environmental indicator to detect TiLV-infected tilapia under coinfection threat. Environ Sustain Indic 11:100135
Matsuyama T, Kamaishi T, Ooseko N, Kurohara K, Iida T (2005) Pathogenicity of motile and nonmotile *Edwardsiella tarda* to some marine fish. Fish Pathol 40(3):133–135
Medellin-Pena MJ, Wang H, Johnson R, Griffiths MW (2007) Probiotic effects virulence related gene expression in *Escherichia coli* O157:H7. Appl Environ Microbiol 73:4259–4267
Merrifield DL, Dimitroglou A, Foey A, Davies SJ, Baker RTM, Bøgwald J, Castex M, Ringø E (2010) The current status and future focus of probiotic and prebiotic applications for salmonids. Aquaculture 302:1–18
Mishra A, Nam GH, Gim JA, Lee HE, Jo A, Kim HS (2018) Current challenges of *Streptococcus* infection and effective molecular, cellular, and environmental control methods in aquaculture. Mol Cells 41(6):495

Miyazaki T, Egusa S (1976) Histopathological Studies of Edwardsiellosis of the Japanese Eel (*Anguilla japonica*). Fish Pathol 11(1):33–43
Miyazaki T, Kaige N (1985) Comparative histopathology of edwardsiellosis in fishes. Fish Pathol 20(2–3):219–227
Mohanty BR, Sahoo PK (2007) Edwardsiellosis in fish: a brief review. J Biosci 32(3):1331–1344
Mohapatra A, Karan S, Kar B, Garg LC, Dixit A, Sahoo PK (2016) Apolipoprotein AI in *Labeo rohita*: cloning and functional characterisation reveal its broad spectrum antimicrobial property, and indicate significant role during ectoparasitic infection. Fish Shellfish Immunol 55:717–728
Mohapatra A, Dixit A, Garg LC, Sahoo PK (2019) Hepcidin gene of Indian major carp, *Labeo rohita*: Molecular, structural and functional characterization, and antibacterial activity of recombinant hepcidin. Aquaculture 511:734218
Mondal SI, Ferdous S, Jewel NA, Akter A, Mahmud Z, Islam MM, Afrin T, Karim N (2015) Identification of potential drug targets by subtractive genome analysis of *Escherichia coli* O157: H7: an in silico approach. Adv Appl Bioinformatics Chem: AABC 8:49–63
Nageswara PV, Babu DE (2006) Probiotics as an alternative therapy to minimize or avoid antibiotics use in aquaculture. Fish Chimes 26(1):112–114
Nematollahi A, Decostere A, Pasmans F, Haesebrouck F (2003) *Flavobacterium psychrophilum* infections in salmonid fish. J Fish Dis 26(10):563–574
Nho SW, Shin GW, Park SB, Jang HB, Cha IS, Ha MA, Kim YR, Park YK, Dalvi RS, Kang BJ, Joh SJ (2009) Phenotypic characteristics of *Streptococcus iniae* and *Streptococcus parauberis* isolated from olive flounder (*Paralichthys olivaceus*). FEMS Microbiol Lett 293(1):20–27
Nicholson P, Mon-on N, Jaemwimol P, Tattiyapong P, Surachetpong W (2020) Coinfection of tilapia lake virus and *Aeromonas hydrophila* synergistically increased mortality and worsened the disease severity in tilapia (*Oreochromis* spp.). Aquaculture 520:734746
Noga EJ, Berkhoff HA (1990) Pathological and microbiological features of *Aeromonas salmonicida* infection in the American eel (*Anguilla rostrata*). Fish Pathol 25(3):127–132
Ødegård J, Baranski M, Gjerde B, Gjedrem T (2011) Methodology for genetic evaluation of disease resistance in aquaculture species: challenges and future prospects. Aquac Res 42: 103–114
Panangala VS, Shoemaker CA, McNulty ST, Arias CR, Klesius PH (2006) Intra-and interspecific phenotypic characteristics of fish-pathogenic *Edwardsiella ictaluri* and *E. tarda*. Aquac Res 37(1):49–60
Park SI, Wakabayashi H, Watanabe Y (1983) Serotype and virulence of *Edwardsiella tarda* isolated from eel and their environment. Fish Pathol 18(2):85–89
Park SB, Aoki T, Jung TS (2012) Pathogenesis of and strategies for preventing *Edwardsiella tarda* infection in fish. Vet Res 43(1):1–11
Pattanayak S, Kumar PR, Sahoo MK, Paul A, Sahoo PK (2018) First field-based evidence of association of *Proteus mirabilis* causing large scale mortality in Indian major carp farming. Aquaculture 495:435–442
Pattanayak S, Priyadarsini S, Paul A, Kumar PR, Sahoo PK (2020) Diversity of virulence-associated genes in pathogenic *Aeromonas hydrophila* isolates and their in vivo modulation at varied water temperatures. Microb Pathog 147:104424
Paul A, Pattanayak S, Sahoo MK, Kumar PR, Kumar R, Sahoo PK (2020) Co-infection of bacterial and parasitic pathogens including the myxosporean *Zschokkella auratis* infecting brain in farmed striped murrel *Channa striata* (Bloch, 1793) causing large scale mortality. Indian J Fish 67(4):105–115
Pedersen K, Grisez L, Van Houdt R, Tiainen T, Ollevier F, Larsen JL (1999) Extended serotyping scheme for *Vibrio anguillarum* with the definition and characterization of seven provisional O-serogroups. Curr Microbiol 38(3):183–189
Pękala-Safińska A (2018) Contemporary threats of bacterial infections in freshwater fish. J Vet Res 62(3):261
Persson S, Al-Shuweli S, Yapici S, Jensen JN, Olsen KE (2015) Identification of clinical *aeromonas* species by rpoB and gyrB sequencing and development of a multiplex PCR method for detection of *Aeromonas hydrophila*, *A. caviae*, *A. veronii*, and *A. media*. J Clin Microbiol 53(2):653–656

Plumb JA (1999) *Edwardsiella* septicaemias. In: Woo PTK, Bruno DW (eds) Fish disease and disorders viral, bacterial and fungal infections, vol 3, pp 512–569

Prasad Y, Kumar D, Sharma AK (2011) Lytic bacteriophages specific to *Flavobacterium columnare* rescue catfish, *Clarias batrachus* (Linn.) from columnaris disease. J Environ Biol 32:161–168

Praveen PK, Debnath C, Shekhar S, Dalai N, Ganguly S (2016) Incidence of *Aeromonas* spp. infection in fish and chicken meat and its related public health hazards: a review. Vet World 9(1):6–11

Rashid MM, Nakai T, Muroga K, Miyazaki T (1997) Pathogenesis of experimental edwardsiellosis in Japanese flounder *Paralichthys olivaceus*. Fish Sci 63(3):384–387

Rauta PR, Kumar K, Sahoo PK (2011) Emerging new multi-drug resistant bacterial pathogen, *Acinetobacter baumannii* associated with snakehead *Channa striatus* eye infection. Curr Sci 101(4):548–553

Roberts RJ (2012) Fish pathology, 4th edn. Wiley-Blackwell, Hoboken

Rong R, Lin H, Wang J, Khan MN, Li M (2014) Reductions of *Vibrio parahaemolyticus* in oysters after bacteriophage application during depuration. Aquaculture 418:171–176

Sahoo PK, Mukherjee SC (2002) The effect of dietary immunomodulation upon *Edwardsiella tarda* vaccination in healthy and immunocompromised Indian major carp (*Labeo rohita*). Fish Shellfish Immunol 12(1):1–16

Sahoo PK, Mukherjee SC, Sahoo SK (1998) *Aeromonas hydrophila* versus *Edwardsiella tarda*: a pathoanatomical study in *Clarias batrachus*. J Aquac 6:57–66

Sahoo PK, Das S, Mahapatra KD, Saha JN, Baranski M, Ødegård J, Robinson N (2013) Characterization of the ceruloplasmin gene and its potential role as an indirect marker for selection to *Aeromonas hydrophila* resistance in rohu, *Labeo rohita*. Fish Shellfish Immunol 34(5):1325–1334

Sahoo PK, Swaminathan TR, Abraham TJ, Kumar R, Pattanayak S, Mohapatra A, Rath SS, Patra A, Adikesavalu H, Sood N, Pradhan PK (2016) Detection of goldfish haematopoietic necrosis herpes virus (Cyprinid herpesvirus-2) with multi-drug resistant *Aeromonas hydrophila* infection in goldfish: first evidence of any viral disease outbreak in ornamental freshwater aquaculture farms in India. Acta Trop 161:8–17

Sahoo PK, Paul A, Sahoo MK, Pattanayak S, Rajesh Kumar P, Das BK (2020) Incidences of infectious diseases in freshwater aquaculture farms of Eastern India: a passive surveillance based study from 2014–2018. J Aquacult Res Dev 11:579

Sandhaus S, Chapagain PP, Tse-Dinh YC (2018) Discovery of novel bacterial topoisomerase I inhibitors by use of in silico docking and in vitro assays. Sci Rep 8(1):1–9

Sarker S, Menanteau-Ledouble S, Kotob MH, El-Matbouli M (2017) A RNAi-based therapeutic proof of concept targets salmonid whirling disease in vivo. PLoS One 12(6):e0178687

Scallan E, Hoekstra RM, Angulo FJ, Tauxe RV, Widdowson MA, Roy SL, Jones JL, Griffin PM (2011) Foodborne illness acquired in the United States – major pathogens. Emerg Infect Dis 17(1):7

Schubert RH (1974) Genus II. *Aeromonas Kluyver* and van Niel 1936, 398. In: Buchanan RE, Gibbons NE (eds) Bergey's manual of determinative bacteriology, 8th edn. Williams and Wilkins, Baltimore

Shameena SS, Kumar K, Kumar S, Kumari P, Krishnan R, Karmakar S, Kumar HS, Rajendran KV, Raman RP (2021) Dose-dependent co-infection of *Argulus* sp. and *Aeromonas hydrophila* in goldfish (*Carassius auratus*) modulates innate immune response and antioxidative stress enzymes. Fish Shellfish Immunol 114:199–206

Shoemaker CA, Klesius PH, Drennan JD, Evans J (2011) Efficacy of a modified live *Flavobacterium columnare* vaccine in fish. Fish Shellfish Immunol 30:304–308

Shotts E, Starliper C (1999) Flavobacterial diseases: columnaris disease, cold-water disease and bacterial gill disease. In: Woo PTK, Bruno DW (eds) Fish diseases and disorders: viral, bacterial and fungal infections, vol 3. CABI Publishing, New York, pp 559–576

Singh S, Mallik SK, Kala K, Shahi N, Pathak R, Giri AK, Chandra S, Pant K, Patiyal RS (2021) Characterization of *Flavobacterium columnare* from farmed infected rainbow

trout, *Oncorhynchus mykiss* (Walbaum, 1792) of Central Indian Himalayan Region, India. Aquaculture:737118
Snieszko SF (1975) History and present status of fish diseases. J Wildl Dis 11:446–459
Soliman WS, Shaapan RM, Mohamed LA, Gayed SS (2019) Recent biocontrol measures for fish bacterial diseases, in particular to probiotics, bio-encapsulated vaccines, and phage therapy. Open Vet J 9(3):190–195
Srinivasa Rao PS, Lim TM, Leung KY (2001) Opsonized virulent *Edwardsiella tarda* strains are able to adhere to and survive and replicate within fish phagocytes but fail to stimulate reactive oxygen intermediates. Infect Immun 69:5689–5697
Starliper CE (2011) Bacterial coldwater disease of fishes caused by *Flavobacterium psychrophilum*. J Adv Res 2(2):97–108
Statistics Korea (KOSTAT): Republic of Korea: 2010. Survey on the status of fish culture. 2011. http://kostat.go.kr/portal/english/news/1/8/index.board?bmode=download&bSeq=&aSeq=251088&ord=1
Strohl WR, Tait LR (1978) *Cytophaga aquatilis* sp. nov., a facultative anaerobe isolated from the gills of freshwater fish. Int J Syst Evol Microbiol 28(2):293–303
Sudheesh PS, Al-Ghabshi A, Al-Mazrooei N, Al-Habsi S (2012) Comparative pathogenomics of bacteria causing infectious diseases in fish. Int J Evol Biol 2012:1–16
Swain P, Nayak SK (2003) Comparative sensitivity of different serological tests for seromonitoring and surveillance of *Edwardsiella tarda* infection of Indian major carps. Fish Shellfish Immunol 15:333–340
Thune RL, Stanley LA, Cooper RK (1993) Pathogenesis of gram-negative bacterial infections in warm water fish. Annu Rev Fish Dis 3:37–68
Timo GO, dos Reis RSSV, de Melo AF, Costa TVL, de Oliveira Magalhães P, Homem-de-Mello M (2019) Predictive power of in silico approach to evaluate chemicals against *M. tuberculosis*: a systematic review. Pharmaceuticals 12:135–142
Tiwari S, Barh D, Imchen M, Rao E, Kumavath RK, Seenivasan SP, Jaiswal AK, Jamal SB, Kumar V, Ghosh P, Azevedo V (2018) Acetate kinase (AcK) is essential for microbial growth and Betel-derived compounds potentially target AcK, PhoP and MDR proteins in *M. tuberculosis*, *V. cholerae* and Pathogenic E. coli: an in silico and in vitro Study. Curr Top Med Chem 18(31):2731–2740
Tomás JM (2012) The main *Aeromonas* pathogenic factors. International Scholarly Research Notices
Toranzo AE, Romalde JL, Magariños B, Barja JL (2009) Present and future of aquaculture vaccines against fish bacterial diseases. Options Mediterr 86:155–176
Toranzo AE, Magariños B, Avendaño-Herrera R (2017) Vibriosis: *Vibrio anguillarum*, *V. ordalii* and *Aliivibrio salmonicida*. In: Woo PTK, Cipriano RC (eds) Fish viruses and bacteria: pathobiology and protection. CABI, Wallingford, pp 314–333
Uhland FC, Helie P, Higgins R (2000) Infections of *Edwardsiella tarda* among brook trout in Quebec. J Aquat Anim Health 12(1):74–77
Vandenberg GW (2004) Oral vaccines for finfish: academic theory or commercial reality? Anim Health Res Rev 5(2):301–304
Verma DK, Rathore G (2013) Molecular characterization of *Flavobacterium columnare* isolated from a natural outbreak of columnaris disease in farmed fish, *Catla catla* from India. J Gen Appl Microbiol 59(6):417–424
Vinay TN, Bhat S, Gon Choudhury T, Paria A, Jung MH, Shivani Kallappa G, Jung SJ (2018) Recent advances in application of nanoparticles in fish vaccine delivery. Rev Fish Sci Aquac 26(1):29–41
Wakabayashi H, Huh GJ, Kimura N (1989) *Flavobacterium branchiophila* sp. nov., a causative agent of bacterial gill disease of freshwater fishes. Int J Syst Evol Microbiol 39(3):213–216
Wang X, Wang Q, Yang M, Xiao J, Liu Q, Wu H, Zhang Y (2011) QseBC controls flagellar motility, fimbrial hemagglutination and intracellular virulence in fish pathogen *Edwardsiella tarda*. Fish Shellfish Immunol 30(3):944–953
Wei Q (2002) Social and economic impacts of aquatic animal health problems in aquaculture in China. FAO Fish Tech Pap 406:55–61

Woo PTK, Bruno DW (2010) Fish diseases and disorders. Volume 3: viral, bacterial and fungal infections. In: Evans JJ, Klesius PH, Plumb JA, Shoemaker CA (eds) *Edwardsiella* septicaemias, 2nd edn. CABI International, Wallingford, pp 512–534
Xu DH, Shoemaker CA, Klesius PH (2007) Evaluation of the link between gyrodactylosis and streptococcosis of Nile tilapia, *Oreochromis niloticus* (L.). J Fish Dis 30(4):233–238
Xu DH, Shoemaker CA, Klesius PH (2009) Enhanced mortality in Nile tilapia *Oreochromis niloticus* following coinfections with ichthyophthiriasis and streptococcosis. Dis Aquat Org 85(3):187–192
Xu DH, Shoemaker CA, Klesius PH (2012a) *Ichthyophthirius multifiliis* as a potential vector of *Edwardsiella ictaluri* in channel catfish. FEMS Microbiol Lett 329(2):160–167
Xu DH, Pridgeon JW, Klesius PH, Shoemaker CA (2012b) Parasitism by protozoan *Ichthyophthirius multifiliis* enhanced invasion of *Aeromonas hydrophila* in tissues of channel catfish. Vet Parasitol 184(2–4):101–107
Yadav SK, Dash P, Sahoo PK, Garg LC, Dixit A (2018) Modulation of immune response and protective efficacy of recombinant outer-membrane protein F (rOmpF) of *Aeromonas hydrophila* in *Labeo rohita*. Fish Shellfish Immunol 80:563–572
Yadav SK, Dash P, Sahoo PK, Garg LC, Dixit A (2021) Recombinant outer membrane protein OmpC induces protective immunity against *Aeromonas hydrophila* infection in *Labeo rohita*. Microb Pathog 150:104727
Yan WL, Gu LD, Liu S, Ren W, Lyu MS, Wang SJ (2018) Identifcation of a highly specifc DNA aptamer for Vibrio vulnifcus using systematic evolution of ligands by exponential enrichment coupled with asymmetric PCR. J Fish Dis 41:1821–1829
Yanong RP, Francis-Floyd R (2002) Streptococcal infections of fish. Florida Cooperative Extension Service. IFAS, University of Florida, pp 1–5
Yu LP, Hu YH, Zhang XH, Sun BG (2013) Development of a triplex loop-mediated isothermal amplification method for rapid on-site detection of three *Vibrio* species associated with fish diseases. Aquaculture 414:267–273
Yusoff SFM, Christianus A, Matori FUAD, Talba MA, Hamid NH, Hayati R (2020) The impact of natural co-infection of *Dactylogyrus* spp. and *Aeromonas hydrophila* on behavioural, clinical, and histopathological changes of striped catfish, *Pangasianodon hypophthalmus* (Sauvage, 1878): a case study. J Sustain Sci Manag 15(7):74–82
Zang P, Gong A, Zhang P, Yu J (2016) Targeting druggable enzymome by exploiting natural medicines: an in silico–in vitro integrated approach to combating multidrug resistance in bacterial infection. Pharm Biol 54(4):604–618
Zhang C, Li DL, Chi C, Ling F, Wang GX (2015) *Dactylogyrus intermedius* parasitism enhances *Flavobacterium columnare* invasion and alters immune-related gene expression in *Carassius auratus*. Dis Aquat Org 116(1):11–21
Zheng J, Leung KY (2007) Dissection of a type VI secretion system in *Edwardsiella tarda*. Mol Microbiol 66(5):1192–1206
Zheng D, Mai K, Liu S, Cao L, Liufu Z, Xu W, Tan B, Zhang W (2004) Effect of temperature and salinity on virulence of *Edwardsiella tarda* to Japanese flounder, *Paralichthys olivaceus* (Temminck et Schlegel). Aquac Res 35(5):494–500

Bacterial Diseases in Brackishwater Aquaculture

Sujeet Kumar and S. V. Alavandi

Abstract

Bacterial diseases are a growing concern in shrimp and fish farming. During the last decade, the global pandemic of acute hepatopancreatic necrosis disease (AHPND) severely impacted the shrimp industry. *Vibrio parahaemolyticus* carrying the *Photorhabdus* insect-related (PirAB) toxin was identified as the major causative agent. However, the presence of PirAB toxin over the conjugative plasmid made it easily transmissible across several *Vibrio* and other bacterial species, making the disease control measures far more difficult. Luminescent vibriosis continues to be a major problem in shrimp hatcheries. Likewise, the increasing number of reports of photobacteriosis and black gill disease (BGD) from shrimp farms calls for the development of suitable rapid diagnostics and remedial measures. Finfish culture is taking momentous shape in the brackishwater and marine sectors. Hence, an appropriate measure to control vibriosis, photobacteriosis, and tenacibaculosis in fish farms is the need of the hour. Bacterial pathogenicity is the interplay between colonization efficiency, production of potent toxins, and the toxin secretion machinery called secretion systems. There is a need to understand these pathogenic factors to differentiate pathogenic from non-pathogenic isolates. In light of growing concerns over antimicrobial resistance (AMR), drug residues, and trade barriers, antibiotic application is highly restricted in aquaculture, especially in the shrimp sector. There is a need to find alternative therapeutic and preventive strategies for controlling bacterial diseases for the sustainable development of aquaculture.

S. Kumar (✉) · S. V. Alavandi
ICAR-Central Institute of Brackishwater Aquaculture, Chennai, India
e-mail: sujeetmicrobiol@gmail.com; shankaralavandi@hotmail.com

R. Kooloth Valappil et al. (eds.), *Aquatic Animal Health Management*,
https://doi.org/10.1007/978-981-96-7987-4_2

Keywords

Aquaculture · bacteria · disease · fish · fungi · shrimp · vibriosis · therapeutics

1 Introduction

Aquaculture is the fastest-growing food production sector. It contributes to economic and societal development, promotes nutritional security, and helps alleviate poverty (Béné et al. 2016). India is a significant player in global fisheries and aquaculture production, with a contribution of 8.0%, and stands second behind China (FAO 2024). With rising demands for aquaculture produce, culture practices have changed from age-old traditional extensive systems to super-intensive production systems. While this advancement is improving the income of the farming community, it also increases the chances for the emergence of new diseases (Chinabut and Puttinaowarat 2005). Several microbial diseases of viral, bacterial, fungal, microsporidian, and parasitic origins have emerged in the past few decades, which are threatening the growth and sustainability of the aquaculture sector.

The shrimp farming industry in several Southeast Asian countries was severely affected by the outbreaks of early mortality syndrome (EMS)/acute hepatopancreatic necrosis disease (AHPND), resulting in a significant reduction in shrimp production (Santos et al. 2020). It is caused by specific strains of *Vibrio parahaemolyticus* known to have a PirAB toxin-carrying plasmid (Prachumwat et al. 2019). Luminescent vibriosis poses a major concern to shrimp hatcheries, leading to mortality rates as high as 100% during the mysis and early post-larval stages (Kumar et al. 2021). Luminescent *Vibrio* species like *V. campbellii* and *V. harveyi* are often associated with vibriosis infection in larval shrimp (Haldar et al. 2011; Kumar et al. 2021). In finfish, vibriosis, photobacteriosis, and flexibacteriosis are the major concerns.

2 Bacterial Diseases in Shrimp

2.1 Acute Hepatopancreatic Necrosis Disease

Acute hepatopancreatic necrosis disease (AHPND), also known as early mortality syndrome (EMS), has emerged as a global threat to shrimp farming. This disease primarily affects shrimp post-larvae or juveniles within 20–30 days post-stocking, often causing mortality up to 100%. Consequently, this disease was initially designated as EMS. Since its first reported outbreak in 2009 in China (Tran et al. 2013), AHPND has been documented in several countries, including Vietnam, Malaysia, Thailand, Mexico, the Philippines, South America, Bangladesh, and the United States (Kumar et al. 2020). Between 2009 and 2016, AHPND caused severe economic losses to the shrimp farming industry, with the estimated figure reaching approximately 23 billion US dollars (Shinn et al. 2018), leading to a nearly 60% decrease in shrimp production in affected countries. Later, this figure was revised to 44 billion US dollars (Tang and Bondad-Reantaso 2019). Notably, there have been no reported cases of AHPND in India to date (Navaneeth et al. 2020; Narsale et al. 2024).

In 2013, *V. parahaemolyticus* was identified as the causative agent of AHPND (Tran et al. 2013). It was subsequently determined that this disease is exclusively associated with specific strains of *V. parahaemolyticus* carrying the pVA1 plasmid, which encodes the *Photorhabdus* insect-related (Pir) toxin, a binary toxin comprising PirA and PirB (Lai et al. 2015). Apart from the PirAB toxin, the plasmid contains five putative transposases, toxin–antitoxin system pndA, a cluster of conjugative plasmid genes, and a plasmid mobilization gene (Kumar et al. 2020). The transposase is located upstream and downstream of the PirAB operon, making the plasmid transferrable across bacterial species. The horizontal transfer of pVA1 has been documented in several *Vibrio* species, like *V. owensii* (Liu et al. 2018), *V. campbellii* (Dong et al. 2017; Soto-Rodriguez et al. 2024), *V. harveyi* (Muthukrishnan et al. 2019), and *V. punensis* (Restrepo et al. 2018), and even non-*Vibrio* spp. like *Shewanella* spp. (Wechprasit et al. 2019; Prachumwat et al. 2020).

The pathogen initially colonizes in the stomach of infected shrimp (Tran et al. 2013). Following colonization, bacteria produce the PirAB toxin, which induces the sloughing of tubular epithelial cells in the hepatopancreas (Prachumwat et al. 2019). The toxin causes damage to the resorptive (R), blister (B), fibrillar (F), and embryonic (E) cells within the hepatopancreas, resulting in physiological dysfunction and massive mortalities of the farmed shrimp (Kumar et al. 2019). Infected shrimp demonstrate symptoms such as lethargy, reduced appetite, slow growth, an empty digestive tract, and a pale-to-white appearance of the hepatopancreas. These manifestations have been corroborated through histopathological examination and polymerase chain reaction (PCR)-based techniques applying AP1, AP2, TUMSAT-Vp3, AP3, and AP4 primers noted for their enhanced sensitivity (Dangtip et al. 2015; Chandran et al. 2023).

2.2 Luminescent Vibriosis

Luminescent vibriosis is a serious challenge in shrimp hatcheries, often leading to cent per cent mortalities during mysis and early post-larval stages (Karunasagar et al. 1994; Kumar and Shekhar (2025). Among all the life stages, the zoea stage during the hatchery cycle seems to be the most susceptible (Soto-Rodríguez et al. 2006). However, there are several reports of high-level mortality from the grow-out farms of *Penaeus monodon* (De La Pena et al. 2001) and *Penaeus vannamei* (Nurhafizah et al. 2021). For long, *V. harveyi* was considered the primary causative agent of luminescent vibriosis. However, it was suggested that *V. campbellii* is primarily responsible for luminescent vibriosis in Indian shrimp hatcheries (Kumar et al. 2021). These two pathogens share similar phenotypic and genotypic characteristics. Despite their extensive similarity, they can be differentiated by their growth characteristics on a thiosulfate–citrate–bile salts–sucrose (TCBS) agar plate as *V. campbellii* forms green colonies, while *V. harveyi* mostly produces yellow colonies (Fig. 1; Kumar et al., data unpublished). The infected larvae exhibit extensive bacterial colonization on their appendages and foregut, which is subsequently followed by infection of the midgut and hepatopancreas, ultimately leading to terminal

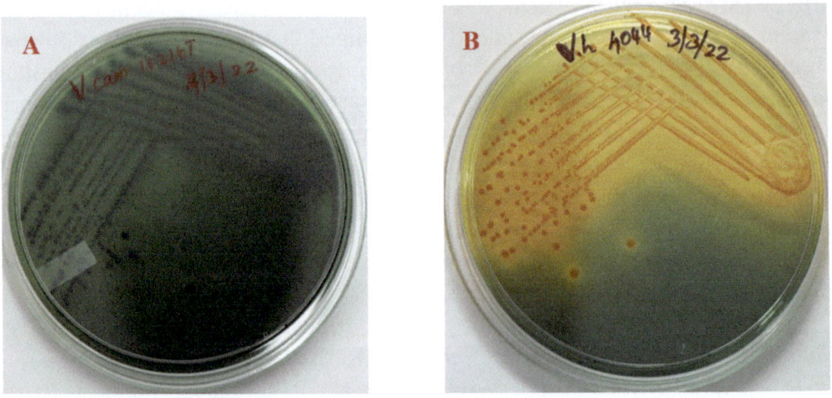

Fig. 1 Growth characteristics of type strains of *V. campbellii* LMG11216T (**a**) and *V. harveyi* LMG4044 (**b**) on TCBS agar. TCBS thiosulfate–citrate–bile salts–sucrose

septicaemia (Lavilla-Pitogo et al. 1990). The weak and moribund larvae appear with opaque, white colouration in their thoracic region, demonstrate impaired swimming behaviour, and eventually settle at the bottom of the rearing tanks (Lavilla-Pitogo et al. 1990). Notably, significant mortality rates were observed after 48 h when the population of luminescent vibrios exceeded 100 colony-forming units per millilitre (Lavilla-Pitogo et al. 1990).

2.3 Translucent Post-Larvae Disease (TPD)

Translucent post-larvae disease (TPD) has been reported as an emerging threat to *P. vannamei* farming in China (Zou et al. 2020). This disease primarily affects the post-larval stages P5–P7. Up to 90% morbidity and mortality have been reported within 3 days of the onset of symptoms. The affected larvae exhibit a pale or colourless hepatopancreas and digestive tract. Histopathological analysis of naturally and experimentally infected shrimp revealed severe necrosis and sloughing of epithelial cells of the hepatopancreas and midgut (Zou et al. 2020). TPD has been associated with specific strains of *V. parahaemolyticus* known as Vp_{TPD} (Zou et al. 2020; Liu et al. 2023). The isolated strain of *V. parahaemolyticus* could cause 100% mortality within 48 hours at 10^6 CFU doses. Two novel toxins, *Vibrio* high virulent protein, VHVP-1 and VHVP-2, have been identified as the key virulence factors associated with TPD (Liu et al. 2023). The authors suggested that the presence of VHVP-2 is essential for virulence as mutant strains lacking VHVP-2 had drastically reduced shrimp mortality. Both the toxins are encoded on a 187,791-bp plasmid, suggesting a possible threat of horizontal gene transfer (HZT) to other bacterial species. The epidemiological surveillance in Chinese farms suggests a high prevalence exceeding 50% in many areas (Jia et al. 2024). A comparative pathogenicity study revealed that *P. vannamei* and *P. japonicas* are highly susceptible to Vp_{TPD}.

2.4 Photobacteriosis

Photobacterium is a halophilic, luminescent bacterium (LB) classified under the family Vibrionaceae. It is commonly found in marine ecosystems and coastal waters. The genus has 27 species, and many of them like *P. damselae, P. phosphoreum, P. leiognathi, P. angustum, P. aquimaris, P. kishitanii, P. ganghwense*, and *P. piscicola* carry the luxCDABEGH operon and are bioluminescent (Labella et al. 2017). In recent years, *Photobacterium damselae* has received great attention as an emerging pathogen for many marine species (Gouife et al. 2022). It is categorized into two subspecies, namely, *P. damselae* subsp. *damselae* (PDD) and *P. damselae* subsp. *piscicida* (PDP) (Gauthier et al. 1995; Labella et al. 2017). PDD infection has been documented in reptiles, crustaceans, molluscs, fish, and even humans (Gouife et al. 2022). In shrimp, it has been documented in *P. monodon* (Song et al. 1993; Vaseeharan et al. 2007) and *P. vannamei* (Singaravel et al. 2020; Wang et al. 2020a). The affected shrimp show symptoms like anorexia and reduced motility. Pathologically, infected shrimp exhibit enlarged hepatopancreatic tubules along with sloughing of epithelial cells into the tubular lumen. Singaravel et al. (2020) documented the involvement of PDD in myonecrosis of affected *Litopenaeus vannamei* broodstock. The observed muscular pathology resembled lesions typically caused by infectious myonecrosis virus (IMNV). The affected shrimp exhibited opaque, whitish discolouration extensively on the distal abdominal segments and posterior appendages. However, diagnostic tests confirmed the absence of IMNV and other viral pathogens and the potential association of PDD with myonecrosis. TCBS agar is frequently employed for isolation and identification of PDD, where it characteristically forms a yellow-pigmented colony.

2.5 Necrotizing Hepatopancreatitis (NHP)

Necrotizing hepatopancreatitis (NHP) is an OIE (Office International des Epizooties)-listed bacterial disease. It severely affects shrimp farming, with mortality reaching up to 90%, but is geographically limited to Latin America and the United States (Vincent and Lotz 2007; Yu et al. 2022). This disease is caused by *Candidatus Hepatobacter penaeid*, a Gram-negative, pleomorphic, intracellular bacterium that specifically infects the hepatopancreas (Gollas-Galván et al. 2014). The pathogen is still uncultured and morphologically resembles rickettsia. It appears as coccobacilli (~0.30 μm in diameter) with a trilaminate membrane without flagella during infection. In the reproductive stage, it has a helicoidal shape (0.59–1.18 μm long × 0.36 μm wide), retains the trilaminate membrane, and develops eight periplasmic flagella (Yu et al. 2022). However, phylogenetic analysis classified NHP separately from rickettsia under the family Holosporaceae (Yu et al. 2022). The disease was first reported from a shrimp farm in Texas in the year 1985. Its occurrence is closely linked to salinity levels and water temperature, as elevated salinity and temperatures can lead to severe mortality in shrimp (Kumar et al. 2025). NHP is primarily transmitted through cannibalism of infected shrimp and the cohabitation of healthy individuals with infected ones. However, vertical transmission has also been proposed as a possible route (de Aquino Figueirôa et al. 2024). Affected shrimp show melanization and necrosis of the abdominal segments (de Aquino Figueirôa et al. 2024). Lightner et al. (2012) suggested that an NHP outbreak often happens in farms with water temperatures between 29 and 35 °C and salinities between 30 and 40‰. The mortality in the affected shrimp farms varies from 20 to 80% and can even reach up to 95% (Yu et al. 2022).

2.6 Shell Disease and Necrosis of Appendages in Crustaceans

Shell disease, also referred to as brown spot disease and black spot disease, primarily targets the exoskeleton of shrimp larvae (Lightner 1988). This disease is associated with the involvement of several bacterial genera like *Vibrio*, *Aeromonas*, and *Pseudomonas* (Jayasree et al. 2006). Symptomatically, shell disease is characterized by the presence of black spots and lesions on the exoskeleton and appendages (Fig. 2). The progressive erosion of these lesions promotes bacterial colonization and the proliferation of additional pathogens at the infected sites. In advanced stages, severe infection can result in the loss of the affected appendages, compromising the moulting and locomotion. Consequently, the affected larvae become weak and highly vulnerable, ultimately leading to mortality due to stress and energy exhaustion. In cases of tail necrosis, necrotic damage is observed in the uropods and pleopods, with diseased shrimp displaying impaired swimming capabilities and erratic gliding movements along pond margins (Jayasree et al. 2006). Currently, no effective treatment is available for shell disease. However, its management can be achieved through practices such as minimizing organic load and regularly removing dead larvae, sediment, and organic debris that may serve as reservoirs for pathogenic microbes. Additionally, avoiding high stocking densities, maintaining optimal water quality, and a proper diet are recommended for disease prevention.

2.7 Filamentous Bacterial Diseases

Filamentous bacteria such as *Leucothrix* sp., *Thiothrix* sp., *Flexibacter* sp., *Cytophaga* sp., and *Flavobacterium* sp. are known to severely affect shrimp larval

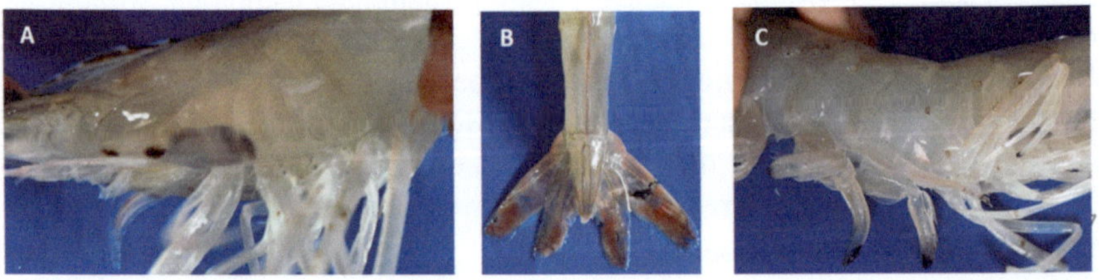

Fig. 2 Shrimp affected by the blackening of gills (**a**), tail rot (**b**), and rotting of the pleopod (**c**)

stages (Lightner 1993). They anchor themselves specifically to the gills, appendages, and body surfaces of the larvae. The long filaments of *Leucothrix* form rosettes while attached to the larval exoskeleton and respiratory surfaces. Bacterial infestation of the larval respiratory surface becomes fatal when the dissolved oxygen in the rearing water is reduced significantly. Mortality is mostly associated with high organic load and low dissolved oxygen. In wet mount larvae, the attachment of filamentous bacterial growth can be identified in the gills, appendages, and body surface. These bacterial infections can be effectively controlled by maintaining optimal water quality conditions such as optimum dissolved oxygen and low organic matter. Copper sulfate at 0.5 ppm may be used to treat filamentous bacterial infections (Baticados et al. 1990).

3 Fungal Diseases of Shrimp

3.1 Black Gill Disease (BGD)

Heavily pigmented gills, commonly referred to as black gills, have been documented in many crustaceans. The characteristic dark brown or black colouration of the gills results from melanin synthesis, a critical component of the crustacean immune system (Frischer et al. 2022). Black gill disease (BGD) was first reported in Japanese Kuruma shrimp (*P. japonicus*) in Japan, where the *Fusarium* spp. was identified to be the primary aetiological agent (Ishikawa 1968). Subsequently, the disease was also reported in black tiger shrimp (*P. monodon*), where *F. incarnatum* was implicated as the pathogen (Khoa et al. 2004). In recent years, multiple cases of BGD have been reported from *P. vannamei* farms in India associated with *Aspergillus* species like *A. awamori*, *A. Flavus*, and *A. niger* (Karthikeyan et al. 2015; Dewangan et al. 2015). Pazir et al. (2022) reported that *Aspergillus* spp. is the primary causative agent of BGD, while other microorganisms like *Vibrio* spp., are considered as secondary contributors to the disease.

The initial clinical sign of a fungal infection is a change in gill colouration. In the early phase, the gill tissues display an opaque white to yellow or brown appearance, eventually progressing to a black colouration (Frischer et al. 2022). Histopathological examination reveals the presence of fungal structures such as hyphae, conidia, and entire sporangia within the gill filaments (Yao et al. 2022). This leads to impaired respiratory function, reduced growth rate, chronic mortality, and increased susceptibility to other diseases. Environmental factors, such as poor water quality, high organic matter in water and pond bottoms, and stress from overcrowding, can contribute to the onset and severity of fungal infections. Treatment options for fungal black gill disease are limited, and prevention through good husbandry practices is often more effective. This includes improving water quality, reducing stressors, and ensuring proper sanitation in aquaculture systems. Regular monitoring of shrimp health and environmental conditions can help in mitigating the impact of fungal infections.

3.2 Larval Mycosis

Larval mycosis is a fungal infection affecting shrimp larvae, posing significant challenges in aquaculture. It is primarily caused by opportunistic fungi such as *Lagenidium*, *Fusarium*, and *Sirolpidium* species, which thrive in aquatic environments and can quickly spread through larval populations (Chong 2022). The affected larvae show lethargy, reduced feeding, abnormal swimming behaviour, and visible fungal hyphae. Infected larvae often exhibit high mortality rates, which can severely impact shrimp hatcheries and farming operations. The fungi invade and colonize the tissues of the larvae, leading to tissue necrosis and compromised physiological functions. Environmental factors play a crucial role in the occurrence and severity of larval mycosis. Poor water quality, high organic matter, and suboptimal rearing conditions create favourable environments for fungal growth and spread. Stress factors such as overcrowding and temperature fluctuations can also exacerbate the susceptibility of larvae to fungal infection.

Managing larval mycosis requires a multifaceted approach. Key preventive measures include maintaining optimal water quality through regular monitoring and filtration, ensuring proper sanitation in hatcheries, and avoiding overcrowding. Disinfection of equipment and tanks, along with the use of antifungal agents like trifuralin (Treflan) at 0.1 ppm during egg washing and dip treatment during nauplius collection stages can help control the spread of the disease (Williams et al. 1986).

4 Bacterial Disease in Finfish

4.1 Vibriosis

Vibriosis is one of the most commonly reported infectious diseases affecting marine finfish. The major species responsible for the disease are *V. harveyi, V. alginolyticus, V. anguillarum, V. salmonicida, V. ordalii, V. splendidus*, and *V. vulnificus* (Rodger 2016). In groupers (*Epinephelus* spp.), vibriosis accounts for approximately 66.7% of all reported diseases (Ina-Salwany et al. 2019). The initial clinical symptoms in infected groupers include anorexia (loss of appetite) and skin discolouration. As the disease progresses, affected fish exhibit lethargy and abnormal swimming behaviour with a tendency to swim near the surface. Skin ulcers may develop into haemorrhagic lesions with fin erosion. Additionally, exophthalmia and corneal opacity are frequently observed. Vibriosis affects all life stages of fish and can result in mortality rates of up to 50% (Ina-Salwany et al. 2019).

Cobia (*Rachycentron canadum*) culture has been frequently affected by *V. alginolyticus* (Rameshkumar et al. 2017). The affected fish exhibit abnormal, sluggish swimming behaviour near the surface and bilateral exophthalmia, leading to mortality. Histopathological analysis reveals fatty liver, ocular congestion, and acute glomerulonephritis in the kidney. Other farmed fishes susceptible to *V. alginolyticus* include Asian seabass (*Lates calcarifer*) (Krupesha Sharma et al. 2012), mullet (*Mugilidae* sp.), and brown spotted grouper.

V. anguillarum is one of the most pathogenic *Vibrio* spp. and is often considered a primary pathogen. It has worldwide distribution and affects fishes in brackishwater and marine and freshwater systems (Hickey and Lee 2018). Out of 23 reported serotypes, only O1, O2, and O3 are frequently associated with fish mortalities (Hickey and Lee 2018). This pathogen affects nearly all fish species in temperate climates including European seabass (*Dicentrarchus labrax*), gilthead seabream (*Sparus aurata*), Atlantic salmon (*Salmo salar*), Atlantic cod (*Gadus morhua*), Atlantic halibut (*Hippoglossus hippoglossus*), turbot (*Psetta maxima*), rainbow trout (*Oncorhynchus mykiss*), striped bass (*Morone saxatilis*), and ayu (*Plecoglossus altivelis*) (Mohamad et al. 2019). This disease is characterized by haemorrhagic septicaemia with clinical signs including darkening of the diseased fishes and petechial haemorrhages at the base of the fins and on the skin. The intestines become distended and filled with a clear, viscous fluid (Actis et al. 2011). The severity of infection is greater in the distal gastrointestinal tract and rectum due to a pH gradient, which is alkaline in the rectum and becomes progressively acidic towards the anterior gastrointestinal tract (Woo and Bruno 2011). Histopathological analysis of infected rainbow trout revealed that the bacterium initially colonized the spleen before spreading to the kidney, and, by the time of death, septicaemia had affected most tissues (Nelson et al. 1985).

4.2 Photobacteriosis

Photobacteriosis is a bacterial disease of marine fishes characterized by bacterial septicaemia. The genus *Photobacterium* comprises 27 species (Gouife et al. 2022). The main pathogenic species is *Photobacterium damselae*, which is classified into two subspecies: *P. damselae* subsp. *piscicida* (PDP) and *P. damselae* subsp. *damselae* (PDD) (Gomez-Gil et al. 2014). The PDP strains exclusively affect marine fishes and is the aetiological agent of pasteurellosis or photobacteriosis. The acute septicaemic form of the disease is reported in juveniles with 90–100% mortality and necrotic lesions in the liver, kidney, and spleen (Barnes et al. 2005). In older fish, the infection presents as a chronic condition characterized by whitish tubercles (0.3–0.5 mm in diameter) in multiple internal organs and so is often referred to as pseudotuberculosis (Barnes et al. 2005). PDP is an intracellular pathogen residing within macrophages and epithelial cells of different fish species (Acosta et al. 2009).

Conversely, PDD has a broader host range, affecting numerous fish species, shrimps, and even humans (Gouife et al. 2022). It affects several commercially important marine fishes, such as Asian seabass (*L. calcarifer*), ovate pompano (*Trachinotus ovatus*), eel (*Anguilla reinhardtii*), silver pomfret (*Pampus argenteus*), seabream, turbot (*Scophthalmus maximus*), rainbow trout (*O. mykiss*), and black rockfish (*Sebastes schlegelii*) (Tao et al. 2018).

4.3 Tenacibaculosis

Tenacibaculosis, formerly known as marine flexibacteriosis, is an ulcerative bacterial disease responsible for high economic losses in the culture of marine fishes worldwide (Toranzo et al. 2005). The disease is primarily attributed to *Tenacibaculum maritimum*, formerly known as *Flexibacter maritimus*. Several other species of *Tenacibaculum*, such as *T. soleae, T. dicentrarchi, T. discolor, T. gallaicum,* and *T. finnmarkense*, have also been isolated from diseased fishes and have demonstrated virulence under experimental conditions (Fernández-Álvarez and Santos 2018). They are Gram-negative, filamentous bacteria measuring 2–30 μm in length and 0.5 μm in diameter and belong to Flavobacteriaceae within the phylum Bacteroidetes (Avendaño-Herrera et al. 2006). Tenacibaculosis affects several economically important fishes like European seabass (*D. labrax*), Japanese flounder (*Paralichthys olivaceus*), Atlantic salmon (*S. salar*), red seabream (*Pagrus major*), sole, gilthead sea bream (*S. aurata*), and turbot (*S. maximus*) (Avendaño-Herrera et al. 2006). The disease severely affects juvenile fish, leading to extensive skin damage and gill erosion followed by systemic infections (Avendaño-Herrera et al. 2006). Clinical signs of infected fish include mouth erosion, necrosis, ulcerative skin lesions, fin rot, and pale internal organs (Toranzo et al. 2005). Environmental factors such as higher temperatures, increased salinity, and poor water quality serve as predisposing factors for disease outbreaks (Avendaño-Herrera et al. 2006).

4.4 Streptococcosis

Streptococcosis is a bacterial disease occurring in farmed and wild freshwater and marine fishes. This disease is zoonotic in nature, with invasive infections occurring in humans handling diseased fish. It is caused by Gram-positive cocci of *Streptococcus*, *Lactococcus*, and *Vagococcus*, especially *S. iniae, S. parauberis, S. difficilis, L. garvieae, L. piscium* and *V. salmoninarum* (Noga 2010). It has been observed in many brackishwater fishes like Asian seabass, mullet, and grouper (Toranzo et al. 2005). In seabass, this disease is characterized by massive acute mortality, with cumulative mortality commonly reaching up to 70%. Infection can occur at any stage but is often seen in large and harvestable sized fish. The common symptoms in Asian seabass include loss of equilibrium, unilateral or bilateral exophthalmia, eye opacity, haemorrhage at the base of the fins, and darkening of the skin (Amal and Zamri-Saad 2011). Additional signs of infection include septicaemia, ascites (accumulation of fluid in the body cavity), splenomegaly (enlarged spleen), and cerebral meningitis.

5 Virulence Mechanism

Pathogenic *Vibrio* species are known to possess a variety of virulence factors such as capsular polysaccharides, pili, cytotoxins, effector toxins, secretion systems, lipopolysaccharides, and flagella (Ceccarelli et al. 2013; Kumar et al. 2017, 2021; Vandeputte et al. 2024). Many of these virulence factors are chromosomally encoded, making them stable and permanent features of the bacterial genome. However, numerous virulence-associated genes in *Vibrio* spp. are frequently located on mobile genetic elements, such as plasmids, prophages, transposons, integrative and conjugative elements, and genomic islands. The acquisition and horizontal gene transfer (HZT) of these elements among pathogens contribute to the emergence of new virulent strains (Castillo et al. 2018). A notable example of such HZT is the emergence of an AHPND-causing strain of *V. parahaemolyticus*, which carries *Photorhabdus* insect-related (Pir) toxins. These toxins destroy the host cells by forming ionic pores on the cell membrane (Lee et al. 2015). The pir toxin gene is located on the pVA1 plasmid, which has an approximate size of 69 kbp. The presence of this plasmid and the *PirAB* toxin gene has also been reported from several other AHPND-causing *Vibrio* spp. such as *V. campbellii*, *V. owensii*, *V. harveyi*, and *V. punensis* (Kumar et al. 2020). This indicates that horizontal transfer to toxin-carrying plasmids plays a crucial role in the emergence of new pathogens. In *P. damselae* subsp. *damselae*, the major virulence factor is AIP56, a toxin that induces apoptosis in fish macrophages and neutrophils in European sea bass (*D. labrax*) (do Vale et al. 2005). The *aip56* gene is carried on a small plasmid pPHDP10, which has a size of 9631 bp (do Vale et al. 2005). The PDD strain produces several potent haemolysins like damselysin (Dly), and phobalysin (PhlyP), both encoded on a 153-kb plasmid known as pPHDD1. This plasmid is exclusive to the virulent strains (Gouife et al. 2022).

The first step in bacterial pathogenesis involves adherence to the host surface. Several *Vibrio* species secrete extracellular polysaccharides (EPSs) in the form of a capsule or loose slime, facilitating attachment to host cells and the mucosal layer (Chen et al. 2008). In addition to promoting colonization, the capsule plays a role in immune evasion, as encapsulated bacteria demonstrate increased resistance to phagocytosis and complement-mediated killing. The EPS production has been documented in *V. harveyi*, *V. parahaemolyticus*, *V. alginolyticus*, and *P. damselae* subsp. *piscicida* (Ina-Salwany et al. 2019).

Extracellular products (ECPs) secreted by various bacterial pathogens are important virulence factors. ECPs of *Vibrio* spp. contain haemolysins, phospholipases, proteases, cytotoxins, and siderophores (Mohamad et al. 2019). Zhang et al. (2001) examined 21 *V. harveyi* isolates and reported that the ECPs of the most virulent strain, VIB645, exhibited the highest haemolytic activity against fish erythrocytes. Later, the pathogenic strain was found to possess two closely related haemolysins (*vhhA* and *vhhB*), which share sequence homology with the thermolabile haemolysin (*tlh*) gene of *V. parahaemolyticus*. In contrast, non-pathogenic strains either possessed a single haemolysin gene or lacked it entirely (Zhang et al. 2001). Proteases represent a major class of lytic enzymes. These include metalloprotease, serine proteases, cysteine proteases, collagenases, caseinases, and gelatinases (Ruwandeepika et al. 2012). The bioluminescent bacterium *V. harveyi* produces cysteine proteases (Liu et al. 1997), serine proteases (Won and Park 2008), and metalloproteases (Teo et al. 2003) in a strain-dependent manner. The ECPs of *P. damselae* subsp. *piscicida* strains were lethal for different fish species and exhibited phospholipase, cytotoxic, and haemolytic activities as the major characteristics (Romalde 2002). Among these, phospholipases have been implicated in the pathogenesis of pasteurellosis (Romalde 2002).

Chitinases are a group of hydrolytic enzymes capable of degrading polymeric chitin into chitin oligosaccharides, diacetylchitobiose, and *N*-acetylglucosamine. Svitil et al. (1997) reported that *V. harveyi* produces 10 different types of chitinases coded by 6 different genes. Some of these chitinases are always expressed, while others are expressed in the presence of a particular kind of chitin. It is expected that chitinases help in the colonization and penetration of pathogens in the chitinous exoskeletons of crustaceans (Aguirre-Guzmán et al. 2004).

Quorum sensing is a bacterial cell-to-cell communication mechanism that involves the secretion and detection of extracellular signalling molecules called autoinducers (AIs). *V. harveyi* possesses four different quorum sensing signals, which include (a) acyl homoserine lactone-mediated autoinducer I, (b) furanosyl borate diester-mediated autoinducer 2, (c) cholera autoinducer 1, and (d) nitric oxide-sensing H-NOX protein (Henke and Bassler 2004; Plate and Marletta 2013). Quorum sensing in *V. harveyi* positively regulates bioluminescence (Henke and Bassler 2004), metalloprotease production (Mok et al. 2003), and siderophore and exopolysaccharide synthesis, while it negatively modulates the type III secretion system (T3SS) in a population density-dependent manner (Henke and Bassler 2004).

ToxR is a widely present virulence regulator in *Vibrio* species, controlling the expression of various virulence-associated genes, like *ctxAB* (cholera toxin) and *tcpA* (toxin-coregulated pilus), in *V. cholerae* and type VI secretion system 1 (T6SS1) in *V. parahaemolyticus* (Childers and Klose 2007; Zhang et al. 2018). Recently, we observed that ToxR also plays a significant role in the virulence of the luminescent pathogen *V. campbellii* (Kumar et al. 2021). Despite the presence of *toxR* gene in all the strains of *V. campbellii*, mutations at the 123rd (proline or alanine) and 150th (glutamine or proline) amino acid positions drastically affected the pathogenicity. The strains with glutamine to proline substitution at the 150th position had enhanced production of quorum sensing signal autoinducer 2, abolished protease activity, and significantly reduced pathogenicity in the challenge study (Kumar et al. 2021).

Bacteriophages, commonly referred to as phages, are viruses that infect bacteria and possess a protein coat enclosing either DNA or RNA as their genetic material. They replicate through two distinct mechanisms: the lytic cycle and the lysogenic cycle. In the lytic cycle, phage replication leads to the death of the host bacterium, whereas in the lysogenic cycle, the phage genome integrates into the bacterial chromosome, allowing the bacterium to survive and propagate the viral genetic material. *V. harveyi* is the major host of marine bacteriophages. Many lysogenic bacteriophages such as VHML (*V. harveyi* myovirus like) (Oakey and Owens 2000), VHPL (*V. harveyi* podovirus like), and siphovirus (Khemayan et al. 2006) have been reported in *V. harveyi*. Recently, a myovirus has been reported in *V. campbellii* BAA 1116 (Lorenz et al. 2016). Many such phages are known to play a role in the virulence of the pathogen.

Iron is a crucial element required for the growth and metabolism of most bacteria. Therefore, iron acquisition is widely recognized as a bacterial virulence factor (Kramer et al. 2020). A key strategy for acquiring iron is the production of siderophores, low-molecular-weight iron-chelating compounds (Kramer et al. 2020). Few PDP strains have been documented to produce the siderophore piscibactin (Souto et al. 2012), which is encoded on a 68-kb mobilizable plasmid known as pPHDP70 (Osorio et al. 2015). Curation of the pPHDP70 plasmid was found to reduce the virulence in a turbot model, indicating that the piscibactin system plays a role in the pathogenicity of these strains.

The cell wall of bacteria, especially Gram-negative bacteria, is highly complex and possesses multiple types of secretory apparatus to export metabolic products and toxins. Six different types of secretion systems, ranging from type I to type VI, have been identified in Gram-negative bacteria (Costa et al. 2015). Among these, the type III secretion system (T3SS) is a virulence factor extensively found in many Gram-negative bacterial species. The T3SS injects toxin directly into the host cell by the molecular apparatus called injectisome (Chatterjee et al. 2013). This brings changes such as the destabilization of actin molecules, preventing phagocytosis, interferes with cell signalling, and helps in colonization (Galán and Wolf-Watz 2006). *V. parahaemolyticus* possesses two sets of T3SSs: T3SS1, located on the large chromosome (chr1), and T3SS2, located on the small chromosome (chr 2) (Park et al. 2004). While T3SS1 is present in all *V. parahaemolyticus* strains, T3SS2 is exclusively found in Kanagawa phenomenon-positive strains (Park et al. 2004). As a result, T3SS2 is considered a key factor contributing to the pathogenicity of *V. parahaemolyticus* in humans. Several effector toxins have been reported in *V. parahaemolyticus*. The effectors VopQ, VopR, and VopS are secreted from T3SS1, while VopC, VopT, and VopV are from T3SS2 (Zhang and Orth 2013). The type VI secretion system (T6SS) translocates toxins to eukaryotic cells in a contact-dependent manner using a membrane-puncturing device. *V. parahaemolyticus* possesses two putative T6SS clusters (T6SS1 and T6SS2). T6SS1, predominantly present in clinical isolates, is more active at higher temperatures. Ben-Yaakov and Salomon (2019) reported that acute hepatopancreatic necrosis disease (AHPND) in shrimp caused by *V. parahaemolyticus* possesses T6SS1, while it is absent in environmental strains. They concluded that T6SSs confer a fitness advantage on AHPND-causing *V. parahaemolyticus* over competing bacteria and facilitate shrimp infection.

6 Diagnosis

The initial steps in diagnosis involve observation of clinical signs and symptoms. The cardinal feature of luminescent vibriosis in crustacean hatcheries is the greenish luminescence in the affected mysis and post-larval stages, which is quite conspicuous at night (Kumar and Shekhar 2025). AHPND-affected shrimp exhibit an empty gut and a pale atrophied hepatopancreas. The disease is usually recorded within 30 days of stocking post-larvae. The affected fishes exhibit loss of appetite (anorexia), skin discolouration, and abnormal sluggish swimming behaviour near the surface. Filamentous bacteria such as *Tenacibaculum* can be identified by fresh smears from the scrapping of the lesions. They appear as long filamentous Gram-negative bacteria under light microscopic observation.

For accurate diagnosis, bacterial species need to be isolated from target organs such as the kidney, liver, and spleen in fishes and the haemolymph and hepatopancreas in shrimp. Thiosulfate–citrate–bile salts–sucrose (TCBS) agar is the most commonly used medium for isolating *Vibrio* species. It also serves for initial differentiation based upon colour development as sucrose fermenter *Vibrio* produce yellow colonies, while non-fermenters produce green colonies. Apart from TCBS agar, several other selective mediums are also used. CHROMagar™ has been widely used for isolation and identification of the food-borne pathogen *V. parahaemolyticus*. Species like *Tenacibaculum* can be isolated using non-selective, low-nutrient media such as *Flexibacter maritimus* medium (FMM), Anacker and Ordal agar (AOA), tryptone casamino acids yeast extract agar (TCY), and tryptone yeast extract glucose agar (TYG) (Fernández-Álvarez and Santos 2018). The next step in identification involves phenotypic characterization using physiological and biochemical tests. However, *Vibrio* spp. are quite heterogeneous and pose a great obstacle in accurate identification by physiological and biochemical tests. The problem is quite common among Harveyi clade species, such as *V. harveyi* and *V. campbellii*, whose luminescence and other phenotypic characteristics are similar (Fig. 3).

Fig. 3 The luminescent bacterium *Vibrio campbellii*

The PCR-based method is a highly sensitive, rapid, and widely used technique for identifying *Vibrio* spp. As the *16S rRNA* gene exists in all the bacterial species, it serves as a reliable molecular marker for identification purposes. However, misidentification of the closely related *Vibrio* species such as *V. harveyi*, *V. campbellii*, and *V. owensii* has occurred in the past due to the limited discriminatory power of the *16S rRNA* gene (Tall et al. 2013). To enhance identification and taxonomic classification, multi-locus sequence analysis (MLSA) is employed utilizing multiple phylogenetic markers (at least five) for improved resolution. Rapid diagnosis of fish photobacteriosis is carried out by differentiating PDP from PDD strains. A multiplex PCR approach uses two primer pairs targeting the internal regions of the *16S rRNA* and *ureC* genes (Osorio et al. 2000). The PDD isolates produce two amplicons corresponding to the *16S rRNA* (267 bp) and *ureC* (448 bp) genes, whereas the PDP isolates generate only the 267-bp product. For the detection of *T. maritimum*, primers have been developed targeting the 16S rRNA gene, which differentiates it from related species (Fernández-Álvarez and Santos 2018).

Serological assays like agglutination, fluorescent antibody test (FAT), enzyme-linked immunosorbent assay (ELISA), dot blot assay, and immunohistochemistry (IHC) are commonly employed for identifying bacterial fish pathogens. The dot blot assay has been utilized for the rapid diagnosis of tenacibaculosis and typing of *Tenacibaculum* species (Fernández-Álvarez and Santos 2018). Immunohistochemistry (IHC) is particularly valuable for detecting microorganisms that are fastidious to culture or exhibit atypical morphology. Faílde et al. (2014) developed an IHC-based method for specific detection of *T. maritimum* in paraffin-embedded tissues.

7 Mode of Infection

Bacteria enter fish through the skin, gills, and gastrointestinal tract, with skin penetration regarded as the most common route of infection (Frans et al. 2011). The epidermal layer of the skin and mucus serves as the first defence against pathogen invasion. Therefore, any damage to the mucosal layer increases susceptibility to infection. Injuries caused by nets and farm equipment, along with risk factors such as overcrowding and cannibalistic behaviour of some fishes, further facilitate bacterial entry. The mucosal layer possesses immunoglobulins and various enzymes, including lysozyme, lysins, lectins, and agglutinins, which play a crucial role in preventing the penetration and adhesion of *Vibrio* species (Chen et al. 2008). Gills also function as entry points for pathogens, and their role has been well established in *V. harveyi* and *V. vulnificus* infection (Mohamad et al. 2019). Additionally, the gastrointestinal tract has also been identified as a portal of entry for *V. harveyi*, *V. alginolyticus*, and *V. anguillarum* (Kim and Lee 2017).

In crustacean hatcheries, broodstock and their rearing tanks within maturation and spawning facilities have been identified as the primary source of luminescent bacteria (LBs). Additionally, algal diets used during the zoea and mysis stages serve as another significant source for LBs. During the seed production cycle affected by LB disease, LBs were recovered from water samples from nauplii tanks. In contrast, in seed production cycles free from LB disease, LBs were detected only from the zoea stage onwards (Chrisolite et al. 2008).

In addition to the virulence of pathogenic organisms, environmental and nutritional factors play a crucial role in determining the severity of disease outbreaks (Jun and Woo 2003). The major environmental factors influencing such outcomes are water quality and levels of chemical pollutants such as ammonia-N, nitrite-N, salinity, and temperature (Huicab-Pech et al. 2016). Warmer water conditions significantly promote the survival and multiplication of *Vibrio* spp., making a substantial threat to the aquaculture industry, particularly in tropical countries (Sankar et al. 2012). Studies have indicated that an overcrowded environment, combined with water temperature exceeding 15 °C, induces stress in fish, thereby increasing their susceptibility to *Vibrio* infection (Cheng et al. 2009; Albert and Ransangan 2013). Similarly, higher salinity levels (30–35%) enhance the risk of vibriosis (Pridgeon and Klesius 2012).

8 Prevention and Control of Bacterial Diseases in Aquaculture

Adopting better management practices (BMPs), ensuring good water quality, providing optimum feed, adhering to biosecurity protocols, utilizing specific pathogen-free (SPF) stocks, and adopting an effective vaccination programme can help in controlling fish diseases (Ina-Salwany et al. 2019). Good hatchery and farm management are crucial for preventing disease outbreaks and the spread of vibriosis or any other infectious diseases. Maintaining good water quality and minimizing stress factors, such as high stocking densities, temperature and salinity fluctuations, are essential for optimal farm management (Ina-Salwany et al. 2019). One of the key measures is to use high-quality shrimp post-larvae or fish fry sourced from disease-free hatcheries. In shrimp farming, implementing biosecurity measures involves stocking seeds that have been PCR-tested for World Organisation for Animal Health (WOAH)-listed pathogens, treating water before use, optimizing larval stocking density, and maintaining optimal water quality. Additionally, continuous health monitoring during the grow-out phase and proper effluent treatment during seed production in the hatchery will ensure biosecurity.

8.1 Antibiotics

Antibiotics have been widely used in the past for treating bacterial infections in aquaculture (Gram and Ringø 2005). Several antibiotics, such as florfenicol, oxolinic acid, quinolones, and flumequine, have been employed to manage vibriosis (Shao 2001). Li et al. (2021) investigated the distribution and dynamics of 12 antibiotics and reported that oxytetracycline (OTC) and sulfamethoxazole (SMX) were most common in pond water samples. In shrimp tissue, OTC, SMX, and levofloxacin were the major residues. However, a maximum of 8 out of 12 antibiotics were detected in pond sediment, and their concentration increased five times over the culture period compared to background concentration. It is reported that approximately 75% of antibiotics used in aquaculture ponds leach into the environment through faeces and uneaten medicated feed and ultimately accumulate in the sediment (Lalumera et al. 2004). The presence of antibiotic residues in aquatic environments contributes to the emergence of antibiotic-resistant microbes, such as methicillin-resistant *Staphylococcus aureus* (Atyah et al. 2010). Additionally, antibiotics can disrupt beneficial gut microflora in fish, affecting microbial biodiversity and ecological balance. Many antimicrobials pose health risks to humans. For instance, consuming shrimp or fish containing chloramphenicol residues may lead to aplastic anaemia and severe bone marrow suppression. Nitrofuran antibiotics have been identified as mutagenic and cancer-causing agents (Citarasu 2012). Due to growing concerns, there is increasing pressure on the scientific community, policymakers, and public health advocates to restrict or ban antibiotic use other than for therapeutic purposes in humans and animals. There is a tripartite agreement between FAO, WHO, and OIE for addressing the problem of antimicrobial resistance (AMR) on a global scale (Thornber et al. 2020). Under this, the top 10 shrimp-producing countries need to develop an AMR action plan. Along with continuous monitoring of AMR, there is a need to search for alternative therapeutic strategies to support the sustainable growth of aquaculture.

8.2 Medicinal Plants and Seaweeds

Medicinal plants have been used for centuries as immunostimulants to enhance disease resistance (Van Hai 2015). Spices, herbs, herbal extracts, and seaweeds are commonly used medicinal plants (Van Hai 2015). These plants contain various bioactive compounds, such as alkaloids, phenolics, polyphenolics, quinones, lectins, terpenoids, and polypeptides (Reverter et al. 2017). Several studies suggest that they are capable of enhancing growth and survival and are, equally, safe for human consumption and the environment. For instance, Indian ginseng (*Withania somnifera*) has been reported to reduce mortality in greasy grouper (*Epinephelus tauvina*) against *V. harveyi* (Harikrishnan et al. 2012). Similarly, garlic (*Allium sativum*) has demonstrated strong antimicrobial activity against *V. harveyi* while also enhancing fish growth and survival (Talpur and Ikhwanuddin 2012).

Seaweeds serve as a rich source of bioactive compounds and exhibit a wide range of biological activities, including antimicrobial, antiviral, anti-tumour, and anti-inflammatory properties (Dash et al. 2017). Pang et al. (2006) reported that polyculture of abalone with *Gracilaria textorii* had lower *Vibrio* levels compared to monoculture systems. Additionally, ethanol extract of *G. fischeri* demonstrated enhanced antimicrobial activity in *P. monodon* juveniles infected with *V. harveyi* and stimulated the shrimps' natural defence mechanism by increasing haemocyte count, prophenol oxidase, superoxide dismutase, and superoxide anion production (Kanjana et al. 2011). Similarly, *P. monodon* post-larvae fed with red seaweed

(*Hypnea musciformis*) showed an enhanced immune response and increased survival against *V. alginolyticus* infection. Furthermore, bioactive compounds from *Sargassum wightii* exhibited potent anti-bacterial activity against *V. parahaemolyticus*, *V. anguillarum*, and *V. harveyi* (Dash et al. 2017). These findings highlight the potential of medicinal plants and seaweeds as promising sources of biologically active compounds for controlling vibriosis.

8.3 Phage Therapy

Phage therapy involves the use of bacteriophages for the treatment of bacterial infections. As natural bacterial predators, phages infect target bacteria, replicate within them, and produce numerous progeny, ultimately leading to bacterial lysis (McGrath and van Sinderen 2007). Phages are ubiquitous and can be found in environments rich in bacterial hosts, such as soil, animal intestines, and seawater (Wommack and Colwell 2000; Chrisolite et al. 2008; Thiyagarajan et al. 2011). Given their ability to target and eliminate bacteria specifically, bacteriophages represent a viable alternative to antibiotics for controlling bacterial infections (Ramos-Vivas et al. 2021; Rai et al. 2024).

Several studies have demonstrated the effectiveness of phage therapy in controlling vibriosis infection in aquaculture. For example, a lytic siphovirus was found highly effective against *V. harveyi* isolates and provided 86% survival compared to 40% survival with antibiotic treatment (Vinod et al. 2006). Lomelí-Ortega and Martínez-Díaz (2014) reported that phage therapy is effective in controlling *V. parahaemolyticus* infection in *P. vannamei*, even at a low multiplicity of infection (MOI) of 0.1. The bacteriophage CHOED conferred 100% protection against *V. anguillarum* in Atlantic salmon (*Salmo salar*) against 90% mortality in untreated fish (Silva et al. 2014). A meta-analysis of 32 studies with 67 datasets revealed that the therapeutic effect of phages is equivalent to that of antibiotics (Yang et al. 2024). The meta-analysis further suggested that MOI and treatment duration are the most significant factors influencing the therapeutic effect of bacteriophages.

Phages offer several advantages over antibiotics. One of the most significant is their ability to selectively eliminate pathogens without disturbing the beneficial microbial flora of the gut or aquatic environment (Kumar and Shekhar 2025). Additionally, they do not pose harmful effects on eukaryotic organisms. However, their application requires consideration of a restricted host range, which could be effectively tackled by developing a phage consortium having a synergistic effect (Yang et al. 2025). Furthermore, an ideal therapeutic phage should be strictly lytic, polyvalent, environment-tolerant, and free of virulence and antibiotic resistance genes (Liu et al. 2022). Moreover, parameters like phage selection, MOI, and route of administration need to be tailored to the needs of host for practical applications (Liu et al. 2022).

8.4 Probiotics

The application of probiotics has become an essential practice for enhancing growth, improving digestibility, and increasing disease resistance in farmed fish (Nayak 2010). Probiotics offer several benefits, such as the production of antimicrobial compounds, including antibiotics, bacteriocins, lysozymes, and siderophores. Probiotics also competitively exclude pathogens by competing for attachment sites, stimulating immune responses, and maintaining the balance of beneficial microbes in aquatic environments (Verschuere et al. 2000). Microbial by-products such as organic acids (e.g. formic acid, acetic acid, and lactic acid) and hydrogen peroxide have demonstrated antimicrobial properties (Nayak 2010). Probiotics also contribute to improving the feed conversion ratio and enhanced weight gain. Several probiotic species such as *Bacillus subtilus*, *B. acidophius*, *B. lichniformis*, *Pseudomonas fluorescens*, *Lactobacillus plantarum*, and *Streptomyces* have been documented for their probiotic properties and their ability to reduce vibriosis-related mortality in fish and shrimp (Buruiană et al. 2014; Gram and Ringø 2005). Recently, a study has shown that salt-tolerant *Streptomyces griseorubens* CIBA-NS1 improved the shrimp growth and provided protection against *V. campbellii* post-challenge (Sudheer et al. 2024). However, an evaluation of *Pseudomonas* sp. PM 11 and *V. fluvialis* PM 17 could not effectively boost *P. monodon* immunity (Alavandi et al. 2004).

8.5 Vaccines

Large-scale vaccination has been the critical factor behind the success of salmon culture (Sommerset et al. 2005). This boosted the interest in developing effective vaccines for other valuable marine and freshwater fishes (Gudding and Van Muiswinkel 2013). Many vaccine types, such as killed, live attenuated, recombinant protein, and DNA-based vaccines, are being investigated in fish (Ma et al. 2019; Tammas et al. 2024). Worldwide, killed whole-cell bacterin is the most popular vaccine used in aquaculture for eliciting protection in fishes (Tammas et al. 2024). Several killed vaccines are available for control of *Vibrio* species like *V. anguillarum*, *V. ordalii*, and *V. salmonicida* (Zeng et al. 2024). In photobacteriosis, the highest protection was observed with extracellular protein (ECP)-enriched bacterin applied in two booster doses and is commercially available (Ma et al. 2019). There is also a combined vaccine for vibriosis and *P. damselae* subsp. *piscicida* for application in sea bream and European sea bass culture (Ma et al. 2019). Similarly, a killed vaccine has been developed against *V. vulnificus*, which is effective in eels (Esteve-Gassent et al. 2004). However, these vaccines do not offer cross-protection against other *Vibrio* species. Recent scientific developments have led to new vaccine technologies and platforms for effective vaccine development in aquaculture. Attempts are being made to develop a live attenuated vaccine that has superior efficacy and does not need extraneous adjuvants. Applying genetic engineering, a live attenuated vaccine was developed in China for controlling *V. anguillarum* infection in turbot (Wang et al. 2020b). This applied curing of the virulence-related plasmid pEIB1 and deletion of the siderophore-related gene *aroC* resulted in 10,000-fold reduction in the virulence of wild strain *V. anguillarum* MVM425. The vaccine is commercially available in China (Wang et al. 2020b).

9 Summary

Diseases are a major constraint to aquaculture development. Bacteria belonging to the genera *Vibrio* and *Photobacterium* are halophilic and have been widely reported to cause several diseases. Over the past decade, shrimp farming in many Southast Asian countries has suffered severe losses due to AHPND. This disease is caused by the *PirAB* toxin-carrying strains of *V. parahaemolyticus*. Horizontal gene transfer facilitates the spread of *PirAB* toxin to other *Vibrio* species, thus complicating diagnosis and control measures. Recently, translucent post-larval disease has been reported from China, which causes high morbidity and mortality in shrimp post-larvae. The *V. parahaemolyticus* strain carrying a plasmid encoding the VHVP-2 toxin has been identified as the causative agent. Within shrimp hatcheries, luminescent vibriosis continues to be another major concern. With the rising intensification of shrimp farming, black gill disease, primarily caused by fungi such as *Aspergillus* and *Fusarium*, has also been increasingly reported. In Asian seabass, bacterial diseases such as vibriosis and photobacteriosis are frequently encountered. Since antibiotic use is not a viable option in aquaculture, there is an urgent need to explore alternative preventive and therapeutic measures. Probiotics, phage therapy, and sustainable efforts in vaccine development offer promising approaches for controlling bacterial infection in shrimp and finfish.

References

Acosta F, Vivas J, Padilla D, Vega J, Bravo J, Grasso V, Real F (2009) Invasion and survival of *Photobacterium damselae* subsp. *piscicida* in non-phagocytic cells of gilthead sea bream, *Sparus aurata* L. J Fish Dis 32(6):535–541

Actis LA, Tolmasky ME, Crosa JH (2011) Vibriosis. In: Fish diseases and disorders, Viral, bacterial and fungal infections, vol 3, 2nd edn. CABI, Wallingford, pp 570–605

Aguirre-Guzmán G, Mejia Ruíz H, Ascencio F (2004) A review of extracellular virulence product of *Vibrio* species important in diseases of cultivated shrimp. Aquac Res 35:1395–1404

Alavandi SV, Vijayan KK, Santiago TC, Poornima M, Jithendran KP, Ali SA, Rajan JJS (2004) Evaluation of *Pseudomonas* sp. PM 11 and *Vibrio fluvialis* PM 17 on immune indices of tiger shrimp, *Penaeus monodon*. Fish Shellfish Immunol 17:115–120

Albert V, Ransangan J (2013) Effect of water temperature on susceptibility of culture marine fish species to vibriosis. J Pure Appl Microbiol 3:48–52

Amal MNA, Zamri-Saad M (2011) Streptococcosis in tilapia (*Oreochromis niloticus*): a review. Pertanika J Trop Agric Sci 34(2):195–206

de Aquino Figueirôa LV, da Silva Teófilo T, Batista JS, Carneiro Feijó FM, Ximenes A, de Lima JT (2024) Clinical and histopathological evidence of *Hepatobacter penaei* in *Litopenaus vannamei* (Pacific white-shrimp) culture. Arch Vet Sci 29(3):1

Atyah MAS, Zamri-Saad M, Siti-Zahrah A (2010) First report of methicillin-resistant *Staphylococcus aureus* from cage-cultured tilapia (*Oreochromis niloticus*). Vet Microbiol 144:502–504

Avendaño-Herrera R, Toranzo AE, Magariños B (2006) Tenacibaculosis infection in marine fish caused by *Tenacibaculum maritimum*: a review. Dis Aquat Org 71(3):255–266

Barnes AC, Dos Santos NM, Ellis AE (2005) Update on bacterial vaccines: *Photobacterium damselae* subsp. *piscicida*. Dev Biol 121:75–84

Baticados MCL, Cruz-Lacierda ER, de la Cruz M, Duremdez-Fernandez RC, Gacutan RQ, Lavilla-Pitogo CR, Lio-po GD (1990) Diseases of penaeid shrimps in The Philippines. Aquaculture Department, Southeast Asian Fisheries Development Center, Tigbauan

Béné C, Arthur R, Norbury H, Allison EH, Beveridge M, Bush S, Campling L, Leschen W, Little D, Squires D (2016) Contribution of fisheries and aquaculture to food security and poverty reduction: assessing the current evidence. World Dev 79:177–196

Ben-Yaakov R, Salomon D (2019) The regulatory network of *Vibrio parahaemolyticus* type VI secretion system 1. Environ Microbiol 21:2248–2260

Buruiană C-T, Profir AG, Vizireanu C (2014) Effects of probiotic *Bacillus* species in aquaculture–an overview. The Annals of the University Dunarea de Jos of Galati. Fascicle VI-Food Technol 38:9–17

Castillo D, Kauffman K, Hussain F, Kalatzis P, Rørbo N, Polz MF, Middelboe M (2018) Widespread distribution of prophage-encoded virulence factors in marine *Vibrio* communities. Sci Rep 8:1–9

Ceccarelli D, Hasan NA, Huq A, Colwell RR (2013) Distribution and dynamics of epidemic and pandemic *Vibrio parahaemolyticus* virulence factors. Front Cell Infect Microbiol 3:97

Chandran A, Priya PS, Meenatchi R, Vaishnavi S, Pavithra V, Kumar TTA, Arockiaraj J (2023) Insights into molecular aspects of pathogenesis and disease management in Acute Hepatopancreatic Necrosis Disease (AHPND): an updated review. Fish Shellfish Immunol 142:109138

Chatterjee S, Chaudhury S, McShan AC, Kaur K, De Guzman RN (2013) Structure and biophysics of type III secretion in bacteria. Biochemistry 52:2508–2517

Chen Q, Yan Q, Wang K, Zhuang Z, Wang X (2008) Portal of entry for pathogenic *Vibrio alginolyticus* into large yellow croaker *Pseudosciaena crocea*, and characteristics of bacterial adhesion to mucus. Dis Aquat Org 80:181–188

Cheng A, Cheng S, Chen Y, Chen J (2009) Effects of temperature change on the innate cellular and humoral immune responses of Orange-spotted Grouper *Epinephelus coioides* and its susceptibility to *Vibrio alginolyticus*. Fish Shellfish Immunol 26:768–772

Childers BM, Klose KE (2007) Regulation of virulence in *Vibrio cholerae*: the ToxR regulon. Future Microbiol 2(3):335–344

Chinabut S, Puttinaowarat S (2005) The choice of disease control strategies to secure international market access for aquaculture products. Dev Biol (Basel) 121:255–261

Chong RSM (2022) Shrimp larval mycosis. In: Kibenge FSB, Baldisserotto B, Chong RS (eds) Aquaculture pathophysiology volume II. Crustacean and Mollusks diseases. Academic Press, London, pp 283–285

Chrisolite B, Thiyagarajan S, Alavandi SV, Abhilash EC, Kalaimani N, Vijayan KK, Santiago TC (2008) Distribution of luminescent *Vibrio harveyi* and their bacteriophages in a commercial shrimp hatchery in South India. Aquaculture 275:13–19

Citarasu T (2012) Natural antimicrobial compounds for use in aquaculture. In: Infectious disease in aquaculture. Elsevier, pp 419–456, Woodhead Publishing, Cambridge, UK

Costa TRD, Felisberto-Rodrigues C, Meir A, Prevost MS, Redzej A, Trokter M, Waksman G (2015) Secretion systems in Gram-negative bacteria: structural and mechanistic insights. Nat Rev Microbiol 13:343–359

Dangtip S, Sirikharin R, Sanguanrut P, Thitamadee S, Sritunyalucksana K, Taengchaiyaphum S, Mavichak R, Proespraiwong P, Flegel TW (2015) AP4 method for two-tube nested PCR detection of AHPND isolates of *Vibrio parahaemolyticus*. Aquac Rep 2:158–162

Dash P, Avunje S, Tandel RS, Sandeep KP, Panigrahi A (2017) Biocontrol of luminous vibriosis in shrimp aquaculture: a review of current approaches and future perspectives. Rev Fish Sci Aquac 25:245–255

De La Pena LD, Lavilla-Pitogo CR, Paner MG (2001) Luminescent vibrios associated with mortality in pond-cultured shrimp Penaeus monodon in The Philippines: species composition. Fish Pathol 36(3):133–138

Dewangan NK, Gopalakrishnan A, Kannan D, Shettu N, Singh RR (2015) Black gill disease of Pacific white leg shrimp (*Litopenaeus vannamei*) by *Aspergillus flavus*. J Coastal Life Med 3(10):761–765

Dong X, Bi D, Wang H, Zou P, Xie G, Wan X, Yang Q, Zhu Y, Chen M, Guo C, Liu Z (2017) PirAB vp-bearing *Vibrio parahaemolyticus* and *Vibrio campbellii* pathogens isolated from the same AHPND-affected pond possess highly similar pathogenic plasmids. Front Microbiol 8:1859

Esteve-Gassent MD, Fouz B, Amaro C (2004) Efficacy of a bivalent vaccine against eel diseases caused by *Vibrio vulnificus* after its administration by four different routes. Fish Shellfish Immunol 16:93–105

Faílde LD, Bermúdez R, Losada AP, Riaza A, Santos Y, Quiroga MI (2014) Immunohistochemical diagnosis of tenacibaculosis in paraffin-embedded tissues of S enegalese sole *Solea senegalensis* Kaup, 1858. J Fish Dis 37(11):959–968

FAO (2024) The state of world fisheries and aquaculture 2024. Blue Transformation in action. FAO, Rome. https://doi.org/10.4060/cd0683en

Fernández-Álvarez C, Santos Y (2018) Identification and typing of fish pathogenic species of the genus *Tenacibaculum*. Appl Microbiol Biotechnol 102(23):9973–9989

Frans I, Michiels CW, Bossier P, Willems KA, Lievens B, Rediers H (2011) *Vibrio anguillarum* as a fish pathogen: virulence factors, diagnosis and prevention. The Journal of Fish Disease 34:643–661

Frischer ME, Landers SC, Walker AN, Powell SA, Lee RF (2022) Black gill in marine decapod crustaceans: a review. Rev Fish Sci Aquac 30(4):498–519

Galán JE, Wolf-Watz H (2006) Protein delivery into eukaryotic cells by type III secretion machines. Nature 444:567–573

Gauthier G, Lafay B, Ruimy R, Breittmayer V, Nicolas JL, Gauthier M, Christen R (1995) Small-subunit rRNA sequences and whole DNA relatedness concur for the reassignment of *Pasteurella piscicida* (Snieszko et al.) Janssen and Surgalla to the genus *Photobacterium* as *Photobacterium damsela* subsp. *piscicida* comb. nov. Int J Syst Bacteriol 45(1):139–144

Gollas-Galván T, Avila-Villa LA, Martínez-Porchas M, Hernandez-Lopez J (2014) Rickettsia-like organisms from cultured aquatic organisms, with emphasis on necrotizing hepatopancreatitis bacterium affecting penaeid shrimp: an overview on an emergent concern. Rev Aquac 6(4):256–269

Gomez-Gil B, Thompson CC, Matsumura Y, Sawabe T, Iida T, Christen R et al (2014) Family Vibrionaceae (Chapter 225). In: Rosenberg E, DeLong E, Thonpson FL, Lory S, Stackebrandt E (eds) The prokaryotes, 4th edn. Springer, New York, p 88

Gouife M, Chen S, Huang K, Nawaz M, Jin S, Ma R, Wang Y, Xue L, Xie J (2022) *Photobacterium damselae* subsp. *damselae* in mariculture. Aquac Int 30(3):1453–1480

Gram L, Ringø E (2005) Prospects of fish probiotics. In: Biology of growing animals. Elsevier, Amsterdam, pp 379–417

Gudding R, Van Muiswinkel WB (2013) A history of fish vaccination: science-based disease prevention in aquaculture. Fish Shellfish Immunol 35:1683–1688

Haldar S, Chatterjee S, Sugimoto N, Das S, Chowdhury N, Hinenoya A, Asakura M, Yamasaki S (2011) Identification of *Vibrio campbellii* isolated from diseased farm-shrimps from South India and establishment of its pathogenic potential in an artemia model. Microbiology 157:179–188

Harikrishnan R, Balasundaram C, Jawahar S, Heo M-S (2012) Immunomodulatory effect of Withania somnifera supplementation diet in the giant freshwater prawn *Macrobrachium rosenbergii* (de Man) against *Aeromonas hydrophila*. Fish Shellfish Immunol 32:94–100

Henke JM, Bassler BL (2004) Three parallel quorum-sensing systems regulate gene expression in *Vibrio harveyi*. J Bacteriol 186:6902–6914

Hickey ME, Lee JL (2018) A comprehensive review of *Vibrio (Listonella) anguillarum*: ecology, pathology and prevention. Rev Fish Sci Aquac. https://doi.org/10.1111/raq.12188

Huicab-Pech ZG, Landeros-Sánchez C, Castañeda-Chávez MR, Lango-Reynoso F, López-Collado CJ, Platas Rosado DE (2016) Current state of bacteria pathogenicity and their relationship with host and environment in tilapia *Oreochromis niloticus*. J Aquac Res Dev 7:428

Ina-Salwany MY, Al-saari N, Mohamad A, Mursidi F, Mohd-Aris A, Amal MNA, Kasai H, Mino S, Sawabe T, Zamri-Saad M (2019) Vibriosis in fish: a review on disease development and prevention. J Aquat Anim Health 31:3–22

Ishikawa Y (1968) Preliminary report on black gill disease of the kuruma prawn, *Penaeus japonicus* Bate. Fish Pathol 3:34–38

Jayasree L, Janakiram P, Madhavi R (2006) Characterization of *Vibrio* spp. associated with diseased shrimp from culture ponds of Andhra Pradesh (India). J World Aquacult Soc 37:523–532

Jia T, Liu S, Yu X, Xu T, Xia J, Zhao W, Wang W, Kong J, Zhang Q (2024) Prevalence investigation of translucent post-larvae disease (TPD) in China. Aquaculture 583:740583

Jun LI, Woo NYS (2003) Pathogenicity of vibrios in fish: an overview. J Ocean Univ Qingdao 2:117–128

Kanjana K, Radtanatip T, Asuvapongpatana S, Withyachumnarnkul B, Wongprasert K (2011) Solvent extracts of the red seaweed *Gracilaria fisheri* prevent *Vibrio harveyi* infections in the black tiger shrimp *Penaeus monodon*. Fish Shellfish Immunol 30:389–396

Karthikeyan V, Selvakumar P, Gopalakrishnan A (2015) A novel report of fungal pathogen *Aspergillus awamori* causing black gill infection on *Litopenaeus vannamei* (pacific white shrimp). Aquaculture 444:36–40

Karunasagar I, Pai R, Malathi GR, Karunasagar I (1994) Mass mortality of *Penaeus monodon* larvae due to antibiotic-resistant *Vibrio harveyi* infection. Aquaculture 128:203–209

Khemayan K, Pasharawipas T, Puiprom O, Sriurairatana S, Suthienkul O, Flegel TW (2006) Unstable lysogeny and pseudolysogeny in *Vibrio harveyi* siphovirus-like phage 1. Appl Environ Microbiol 72:1355–1363

Khoa LV, Hatai K, Aoki T (2004) *Fusarium incarnatum* isolated from black tiger shrimp, *Penaeus monodon* Fabricius, with black gill disease cultured in Vietnam. J Fish Dis 27(9):507–515

Kim JY, Lee J-L (2017) Correlation of total bacterial and *Vibrio* spp. populations between fish and water in the aquaculture system. Front Marine Sci 4:147

Kramer J, Özkaya Ö, Kümmerli R (2020) Bacterial siderophores in community and host interactions. Nat Rev Microbiol 18:152–163

Krupesha Sharma SR, Rathore G, Verma DK, Sadhu N, Philipose KK (2012) Vibrio alginolyticus infection in Asian seabass (*Lates calcarifer*, B loch) reared in open sea floating cages in India. Aquac Res 44:86–92

Kumar S, Shekhar MS (2025) Status and management of bacterial diseases in shrimp farming. In: Shrimp culture technology: farming, health management and quality assurance. Singapore, Springer Nature Singapore, pp 269–282

Kumar S, Jangam AK, Akhil V, Rajendran V, Katneni VK, Rajan JJS, Grover M, Nagaleekar VK, Alavandi SV, Vijayan KK (2017) Draft genome sequence of the luminescent strain *Vibrio campbellii* LB102, isolated from a black tiger shrimp (*Penaeus monodon*) broodstock rearing system. Genome Announc 5(20):e00342–e00317

Kumar V, Bels LD, Couck L, Baruah K, Bossier P, Broeck WVD (2019) PirABVP toxin binds to epithelial cells of the digestive tract and produce pathognomonic AHPND lesions in germ-free brine shrimp. Toxins 11(12):717

Kumar R, Ng TH, Wang HC (2020) Acute hepatopancreatic necrosis disease in penaeid shrimp. Rev Aquac 12(3):1867–1880

Kumar S, Kumar CB, Rajendran V, Abishaw N, Anand PS, Kannapan S, Nagaleekar VK, Vijayan KK, Alavandi SV (2021) Delineating virulence of *Vibrio campbellii*: a predominant luminescent bacterial pathogen in Indian shrimp hatcheries. Sci Rep 11(1):1–16

Labella AM, Arahal DR, Castro D, Lemos ML, Borrego JJ (2017) Revisiting the genus *Photobacterium*: taxonomy, ecology and pathogenesis. Int Microbiol 20(1):1–10

Lai H-C, Ng TH, Ando M, Lee C-T, Chen I-T, Chuang J-C, Mavichak R, Chang S-H, Yeh M-D, Chiang Y-A (2015) Pathogenesis of acute hepatopancreatic necrosis disease (AHPND) in shrimp. Fish Shellfish Immunol 47:1006–1014

Lalumera GM, Calamari D, Galli P, Castiglioni S, Crosa G, Fanelli R (2004) Preliminary investigation on the environmental occurrence and effects of antibiotics used in aquaculture in Italy. Chemosphere 54(5):661–668

Lavilla-Pitogo CR, Baticados MCL, Cruz-Lacierda ER, Leobert D (1990) Occurrence of luminous bacterial disease of *Penaeus monodon* larvae in The Philippines. Aquaculture 91:1–13

Lee C-T, Chen I-T, Yang Y-T, Ko T-P, Huang Y-T, Huang J-Y, Huang M-F, Lin S-J, Chen C-Y, Lin S-S (2015) The opportunistic marine pathogen *Vibrio parahaemolyticus* becomes virulent by acquiring a plasmid that expresses a deadly toxin. Proc Natl Acad Sci (PNAS) 112:10798–10803

Li F, Huang J, Wang M, Chen L, Xiao Y (2021) Sources, distribution and dynamics of antibiotics in Litopenaeus vannamei farming environment. Aquaculture 545:737200

Lightner DV (1988) Diseases of cultured penaeid shrimp and prawns. Dev Aquac Fish Sci (Netherlands)

Lightner DV (1993) Diseases of cultured penaeid shrimp. In: CRC handbook of Mariculture, vol 1. CRC, Boca Raton, pp 393–486

Lightner DV, Redman RM, Pantoja CR, Tang KFJ, Noble BL, Schofield P, Mohney LL, Nunan LM, Navarro SA (2012) Historic emergence, impact and current status of shrimp pathogens in the Americas. J Invertebr Pathol 110(2):174–183

Liu P-C, Lee K-K, Tu C-C, Chen S-N (1997) Purification and characterization of a cysteine protease produced by pathogenic luminous *Vibrio harveyi*. Curr Microbiol 35:32–39

Liu L, Xiao J, Zhang M, Zhu W, Xia X, Dai X, Pan Y, Yan S, Wang Y (2018) A *Vibrio owensii* strain as the causative agent of AHPND in cultured shrimp, *Litopenaeus vannamei*. J Invertebr Pathol 153:156–164

Liu R, Han G, Li Z, Cun S, Hao B, Zhang J, Liu X (2022) Bacteriophage therapy in aquaculture: current status and future challenges. Folia Microbiol 67(4):573–590

Liu S, Wang W, Jia T, Xin L, Xu TT, Wang C, Xie G, Luo K, Li J, Kong J, Zhang Q (2023) Vibrio parahaemolyticus becomes lethal to post-larvae shrimp via acquiring novel virulence factors. Microbiol Spect 11(6):e00492–e00423

Lomelí-Ortega CO, Martínez-Díaz SF (2014) Phage therapy against *Vibrio parahaemolyticus* infection in the whiteleg shrimp (*Litopenaeus vannamei*) larvae. Aquaculture 434:208–211

Lorenz N, Reiger M, Toro-Nahuelpan M, Brachmann A, Poettinger L, Plener L, Lassak J, Jung K (2016) Identification and initial characterization of prophages in *Vibrio campbellii*. PLoS One 11:e0156010

Ma J, Bruce TJ, Jones EM, Cain KD (2019) A review of fish vaccine development strategies: conventional methods and modern biotechnological approaches. Microorganisms 7:569

McGrath S, van Sinderen D (2007) Bacteriophage: genetics and molecular biology. Caister Academic Press, Norfolk

Mohamad N, Amal MNA, Yasin ISM, Zamri Saad M, Nasruddin NS, Al-saari N, Mino S, Sawabe T (2019) Vibriosis in cultured marine fishes: a review. Aquaculture. https://doi.org/10.1016/j.aquaculture.2019.734289

Mok KC, Wingreen NS, Bassler BL (2003) *Vibrio harveyi* quorum sensing: a coincidence detector for two autoinducers controls gene expression. EMBO J 22:870–881

Muthukrishnan S, Defoirdt T, Ina-Salwany MY, Yusoff FM, Shariff M, Ismail SI, Natrah I (2019) *Vibrio parahaemolyticus* and *Vibrio harveyi* causing Acute Hepatopancreatic Necrosis Disease (AHPND) in Penaeus vannamei (Boone, 1931) isolated from Malaysian shrimp ponds. Aquaculture 511:734227

Narsale SA, Chrisolite B, Sivasankar P, Subash P, Mansoor M, Selvamagheswaran M, Debbarma S, Magesh Kumar P, Baidya S, Kadam R (2024) Isolation, characterization, virulence genes, antimicrobial resistant genes, and antibiotic susceptibility pattern of Vibrio parahaemolyticus in relation to AHPND from shrimp farms in coastal districts of Tamil Nadu. Aquac Int 32(4):3835–3851

Navaneeth KA, Bhuvaneswari T, Rajan JJS, Alavandi SV, Vijayan KK, Otta SK (2020) Characterization of *Vibrio parahaemolyticus* isolates from shrimp farms of Southeast coast of India with special reference to Acute Hepatopancreatic Necrosis Disease (AHPND) status. Aquaculture 518:734813

Nayak SK (2010) Probiotics and immunity: a fish perspective. Fish Shellfish Immunol 29:2–14

Nelson JS, Rohovec JS, Fryer JL (1985) Tissue location of vibrio bacterin delivered by intraperitoneal injection, immersion and oral routes to *Salmo gairdneri*. Fish Pathol 19:263–269

Noga EJ (2010) Diagnoses made by bacterial culture of kidney or affected organs. In: Fish disease, diagnosis and treatment, 2nd edn. Iowa State University Press, Ames, pp 199–200

Nurhafizah WWI, Lee KL, Nadirah M, Danish-Daniel M, Zainathan SC, Najiah M (2021) Virulence properties and pathogenicity of multidrug-resistant *Vibrio harveyi* associated with luminescent vibriosis in Pacific white shrimp, *Penaeus vannamei*. J Invertebr Pathol 186:107594

Oakey HJ, Owens L (2000) A new bacteriophage, VHML, isolated from a toxin-producing strain of *Vibrio harveyi* in tropical Australia. J Appl Microbiol 89:702–709

Osorio CR, Toranzo AE, Romalde JL, Barja JL (2000) Multiplex PCR assay for ureC and 16S rRNA genes clearly discriminates between both subspecies of *Photobacterium damselae*. Dis Aquat Org 40(3):177–183

Osorio CR, Rivas AJ, Balado M, Fuentes-Monteverde JC, Rodríguez J, Jiménez C, Lemos ML, Waldor MK (2015) A transmissible plasmid-borne pathogenicity Island confers piscibactin biosynthesis in the fish pathogen *Photobacterium damselae* subsp. *piscicida*. Appl Environ Microbiol 81(17):5867–5879

Pang SJ, Xiao T, Bao Y (2006) Dynamic changes of total bacteria and *Vibrio* in an integrated seaweed–abalone culture system. Aquaculture 252:289–297

Park K-S, Ono T, Rokuda M, Jang M-H, Okada K, Iida T, Honda T (2004) Functional characterization of two type III secretion systems of *Vibrio parahaemolyticus*. Infect Immun 72:6659–6665

Pazir MK, Pourmozaffar S, Mena IG, Shengjie R, Ahmadi A, Sharifpour I (2022) Black gill disease in *Litopenaeus vannamei* made by various agents. Aquac Fish 9(4):626–634

Plate L, Marletta MA (2013) Nitric oxide-sensing H-NOX proteins govern bacterial communal behavior. Trends Biochem Sci 38:566–575

Prachumwat A, Taengchaiyaphum S, Mungkongwongsiri N, Aldama-Cano DJ, Flegel TW, Sritunyalucksana K (2019) Update on early mortality syndrome/acute hepatopancreatic necrosis disease by April 2018. J World Aquacult Soc 50:5–17

Prachumwat A, Wechprasit P, Srisala J, Kriangsaksri R, Flegel TW, Thitamadee S, Sritunyalucksana K (2020) *Shewanella khirikhana* sp. nov.–a shrimp pathogen isolated from a cultivation pond exhibiting early mortality syndrome. Microb Biotechnol 13(3):781–795

Pridgeon JW, Klesius PH (2012) Major bacterial diseases in aquaculture and their vaccine development. Anim Sci Rev 7:1–16

Rai S, Kaur B, Singh P, Singh A, Benjakul S, Vijay Kumar Reddy S, Nagar V, Tyagi A (2024) Perspectives on phage therapy for health management in aquaculture. Aquac Int 32(2).1349–1393

Rameshkumar P, Nazar AKA, Pradeep MA, Kalidas C, Jayakumar R, Tamilmani G, Sakthivel M, Samal AK, Sirajudeen S, Venkatesan V, Nazeera BM (2017) Isolation and characterization of pathogenic *Vibrio alginolyticus* from sea cage cultured cobia (*Rachycentron canadum* (Linnaeus 1766)) in India. Lett Appl Microbiol 65:423–430. https://doi.org/10.1111/lam.12800

Ramos-Vivas J, Superio J, Galindo-Villegas J, Acosta F (2021) Phage therapy as a focused management strategy in aquaculture. Int J Mol Sci 22(19):10436

Restrepo L, Bayot B, Arciniegas S, Bajaña L, Betancourt I, Panchana F, Reyes Muñoz A (2018) PirVP genes causing AHPND identified in a new *Vibrio* species (*Vibrio punensis*) within the commensal Orientalis clade. Sci Rep 8(1):13080

Reverter M, Tapissier-Bontemps N, Sasal P, Saulnier D (2017) Use of medicinal plants in aquaculture. In: Diagnosis and control of diseases of fish and shellfish, vol 9. Wiley, Hoboken, pp 223–261

Rodger HD (2016) Fish disease causing economic impact in global aquaculture. In: Birkhauser advances in infectious diseases. Springer, Basel, pp 1–34. https://doi.org/10.1007/978-3-0348-0980-1_1

Romalde JL (2002) *Photobacterium damselae* subsp. *piscicida*: an integrated view of a bacterial fish pathogen. Int Microbiol 5(1):3–9

Ruwandeepika HA, Sanjeewa Prasad Jayaweera T, Paban Bhowmick P, Karunasagar I, Bossier P, Defoirdt T (2012) Pathogenesis, virulence factors and virulence regulation of vibrios belonging to the Harveyi clade. Rev Fish Sci Aquac 4:59–74

Sankar G, Saravanan J, Krishnamurthy P, Chandrakala N, Rajendran K (2012) Isolation and identification of *Vibrio* spp. in diseased *Channa punctatus* from aquaculture fish farm. Indian J Geo-Mar Sci 41:159–163

Santos HM, Tsai CY, Maquiling KRA, Tayo LL, Mariatulqabtiah AR, Lee CW, Chuang KP (2020) Diagnosis and potential treatments for acute hepatopancreatic necrosis disease (AHPND): a review. Aquac Int. https://doi.org/10.1007/s10499-019-00451-w

Shao ZJ (2001) Aquaculture pharmaceuticals and biologicals: current perspectives and future possibilities. Adv Drug Deliv Rev 50:229–243

Shinn AP, Pratoomyot J, Griffiths D, Trong TQ, Vu NT, Jiravanichpaisal P, Briggs M (2018) Asian shrimp production and the economic costs of disease. Asian Fish Sci 31S:29–58

Silva YJ, Costa L, Pereira C, Mateus C, Cunha Â, Calado R, Gomes NC, Pardo MA, Hernandez I, Almeida A (2014) Phage therapy as an approach to prevent *Vibrio anguillarum* infections in fish larvae production. PLoS One 9(12):e114197

Singaravel V, Gopalakrishnan A, Dewangan NK, Kannan D, Shettu N, Martin GG (2020) *Photobacterium damselae subsp. damselae* associated with bacterial myonecrosis and hepatopancreatic necrosis in broodstock Pacific white leg shrimp, *Litopenaeus vannamei* (Boone, 1931). Aquac Int 28:1593–1608

Sommerset I, Krossøy B, Biering E, Frost P (2005) Vaccines for fish in aquaculture. Expert Rev Vaccines 4:89–101

Song YL, Cheng W, Wang CH (1993) Isolation and characterization of Vibrio damsela infectious for cultured shrimp in Taiwan. J Invertebr Pathol 61(1):24–31

Soto-Rodríguez SA, Simoes N, Roque A, Gil BG (2006) Pathogenicity and colonization of *Litopenaeus vannamei* larvae by luminescent vibrios. Aquaculture 258(1-4):109–115

Soto-Rodriguez SA, Gomez-Gil B, Lozano-Olvera R, Aguilar-Rendón KG, González-Gómez JP (2024) Identification of new *Vibrio campbellii* strains harboring the pVA1 plasmid isolated from *Penaeus vannamei* postlarvae affected by outbreaks of acute hepatopancreatic necrosis disease (AHPND) in Mexico. Aquaculture 579:740221

Souto A, Montaos MA, Rivas AJ, Balado M, Osorio CR, Rodríguez J, Lemos ML, Jiménez C (2012) Structure and biosynthetic assembly of piscibactin, a siderophore from *Photobacterium damselae* subsp. *piscicida*, predicted from genome analysis. Eur J Org Chem 2012(29):5693–5700

Sudheer NS, Biju IF, Balasubramanian CP, Panigrahi A, Kumar TS, Kumar S, Mandal B, Das S, De D (2024) Probiotic potential of a novel endophytic *Streptomyces griseorubens* CIBA-NS1 isolated from *Salicornia* sp. against *Vibrio campbellii* infection in shrimp. Microb Pathog 191:106677

Svitil AL, Chadhain S, Moore JA, Kirchman DL (1997) Chitin degradation proteins produced by the marine bacterium *Vibrio harveyi* growing on different forms of chitin. Appl Environ Microbiol 63:408–413

Tall A, Hervio-Heath D, Teillon A, Boisset-Helbert C, Delesmont R, Bodilis J, Touron-Bodilis A (2013) Diversity of *Vibrio* spp. isolated at ambient environmental temperature in the Eastern English Channel as determined by pyrH sequencing. J Appl Microbiol 114:1713–1724

Talpur AD, Ikhwanuddin MHD (2012) Dietary effects of garlic (*Allium sativum*) on haemato-immunological parameters, survival, growth, and disease resistance against *Vibrio harveyi* infection in Asian sea bass, *Lates calcarifer* (Bloch). Aquaculture 364:6–12

Tammas I, Bitchava K, Gelasakis AI (2024) Transforming aquaculture through vaccination: a review on recent developments and milestones. Vaccine 12(7):732

Tang KFJ, Bondad-Reantaso MG (2019) Impacts of acute hepatopancreatic necrosis disease on commercial shrimp aquaculture. Rev Sci Tech 38:477–490

Tao Z, Shen C, Zhou SM, Yang N, Wang GL, Wang YJ, Xu SL (2018) An outbreak of *Photobacterium damselae* subsp. *damselae* infection in cultured silver pomfret *Pampus argenteus* in Eastern China. Aquaculture 492:201–205

Teo JW, Zhang LH, Poh CL (2003) Cloning and characterization of a novel lipase from *Vibrio harveyi* strain AP6. Gene 312:181–188

Thiyagarajan S, Chrisolite B, Alavandi SV, Poornima M, Kalaimani N, Santiago TC (2011) Characterization of four lytic transducing bacteriophages of luminescent *Vibrio harveyi* isolated from shrimp (*Penaeus monodon*) hatcheries. FEMS Microbiol Lett 325:85–91

Thornber K, Verner-Jeffreys D, Hinchliffe S, Rahman MM, Bass D, Tyler CR (2020) Evaluating antimicrobial resistance in the global shrimp industry. Rev Aquac 12(2):966–986

Toranzo AE, Magariños B, Romalde JL (2005) A review of the main bacterial fish diseases in mariculture systems. Aquaculture 246:37–61

Tran L, Nunan L, Redman RM, Mohney LL, Pantoja CR, Fitzsimmons K, Lightner DV (2013) Determination of the infectious nature of the agent of acute hepatopancreatic necrosis syndrome affecting penaeid shrimp. Dis Aquat Org. https://doi.org/10.3354/dao02621

do Vale A, Silva MT, dos Santos NM, Nascimento DS, Reis-Rodrigues P, Costa-Ramos C, Ellis AE, Azevedo JE (2005) AIP56, a novel plasmid-encoded virulence factor of *Photobacterium damselae* subsp. *piscicida* with apoptogenic activity against sea bass macrophages and neutrophils. Mol Microbiol 58(4):1025–1038

Van Hai N (2015) The use of medicinal plants as immunostimulants in aquaculture: a review. Aquaculture 446:88–96

Vandeputte M, Kashem MA, Bossier P, Vanrompay D (2024) Vibrio pathogens and their toxins in aquaculture: a comprehensive review. Rev Aquac 16(4):1858–1878

Vaseeharan B, Sundararaj S, Murugan T, Chen JC (2007) *Photobacterium damselae ssp. damselae* associated with diseased black tiger shrimp *Penaeus monodon* Fabricius in India. Lett Appl Microbiol 45(1):82–86

Verschuere L, Rombaut G, Sorgeloos P, Verstraete W (2000) Probiotic bacteria as biological control agents in aquaculture. Microbiol Mol Biol Rev 64:655–671

Vincent AG, Lotz JM (2007) Advances in research of necrotizing hepatopancreatitis bacterium (NHPB) affecting penaeid shrimp aquaculture. Rev Fish Sci 15(1-2):63–73

Vinod MG, Shivu MM, Umesha KR, Rajeeva BC, Krohne G, Karunasagar I, Karunasagar I (2006) Isolation of *Vibrio harveyi* bacteriophage with a potential for biocontrol of luminous vibriosis in hatchery environments. Aquaculture 255:117–124

Wang Z, Shi C, Wang H, Wan X, Zhang Q, Song X, Li G, Gong M, Ye S, Xie G, Huang J (2020a) A novel research on isolation and characterization of *Photobacterium damselae* subsp. *damselae* from Pacific white shrimp, *Penaeus vannamei*, displaying black gill disease cultured in China. J Fish Dis 43(5):551–559

Wang Q, Ji W, Xu Z (2020b) Current use and development of fish vaccines in China. Fish Shellfish Immunol 96:223–234

Wechprasit P, Panphloi M, Thitamadee S, Sritunyalucksana K, Prachumwat A (2019) Complete genome sequence of *Shewanella* sp. strain TH2012, isolated from shrimp in a cultivation pond exhibiting early mortality syndrome. Microbiol Resour Announc 8(13):10–1128

Williams RR, Bell TA, Lightner DV (1986) Degradation of trifluralin in seawater when used to control larval mycosis in penaeid shrimp culture. J World Aquacult Soc 17(1-4):8–12

Wommack KE, Colwell RR (2000) Virioplankton: viruses in aquatic ecosystems. Microbiol Mol Biol Rev 64:69–114

Won KM, Park SI (2008) Pathogenicity of *Vibrio harveyi* to cultured marine fishes in Korea. Aquaculture 285(1-4):8–13

Woo PT, Bruno DW (eds) (2011) Fish diseases and disorders. Volume 3: viral, bacterial and fungal infections. C.A.B. International, Wallingford

Yang L, Yang Q, Hu RG, Cong W, Li S, Kang YH (2024) The evaluation of bacteriophage therapy in aquaculture: a systematic review and meta-analysis. Aquaculture 588:740925

Yang L, Zhong W, Tang T, He M, Zhang T, Zhou B, Yin Y, Guo J, Gao Z (2025) Phage-based biocontrol strategies and application in aquatic animal disease prevention and control. Rev Aquac. https://doi.org/10.1111/raq.70024

Yao L, Wang C, Li G, Xie G, Jia Y, Wang W, Liu S, Xu T, Luo K, Zhang Q, Kong J (2022) Identification of *Fusarium solani* as a causal agent of black spot disease (BSD) of Pacific white shrimp, *Penaeus vannamei*. Aquaculture 548:737602

Yu YB, Choi JH, Kang JC, Kim HJ, Kim JH (2022) Shrimp bacterial and parasitic disease listed in the OIE: a review. Microb Pathog 166:105545

Zeng Q, Sun Y, Lai P, Chen Q, Wang H (2024) Advancements in Vibrio vaccines for aquaculture. Aquac Int 32(3):3331–3356

Zhang L, Orth K (2013) Virulence determinants for *Vibrio parahaemolyticus* infection. Curr Opin Microbiol 16:70–77

Zhang X-H, Meaden PG, Austin B (2001) Duplication of hemolysin genes in a virulent isolate of *Vibrio harveyi*. Appl Environ Microbiol 67:3161–3167

Zhang Y, Hu L, Osei-Adjei G, Zhang Y, Yang W, Yin Z, Lu R, Sheng X, Yang R, Huang X, Zhou D (2018) Autoregulation of ToxR and its regulatory actions on major virulence gene loci in *Vibrio parahaemolyticus*. Front Cell Infect Microbiol 8:291

Zou Y, Xie G, Jia T, Xu T, Wang C, Wan X, Li Y, Luo K, Bian X, Wang X, Kong J (2020) Determination of the infectious agent of translucent post-larva disease (TPD) in *Penaeus vannamei*. Pathogens 9(9):741

Diseases in Coldwater Aquaculture: Pathogenesis and Control

Zainab Manzoor, Syed Shariq N. Qadiri, Feroz A. Shah, Shabir A. Dar, and Shahid M. Mandu

Abstract

Coldwater aquaculture plays a crucial role in meeting the growing need for fish as a source of food. However, one of the major problems faced by the coldwater aquaculture industry is the frequent outbreak of infectious diseases. Poor environmental conditions, low water temperature, and bad water quality also contribute to disease outbreaks by weakening the fish immune system. The increased expansion of aquaculture operations has significantly increased the likelihood of disease outbreaks and their rapid spread. Therefore, it is crucial to understand the multiple factors that contribute to the occurrence and transmission of diseases to ensure successful management and sustainability in coldwater aquaculture. This chapter focuses on the common fish diseases prevalent in coldwater aquaculture, associated pathology, and subsequent control measures.

Keywords

Coldwater · Aquaculture · Fish Diseases · Pathogenesis · Control

1 Introduction

Aquaculture is considered to be a valuable source of animal protein and plays a critical role in global food security and nutrition by reducing the supply–demand gap for aquatic food (FAO 2022). In particular, coldwater aquaculture significantly

Z. Manzoor · S. S. N. Qadiri (✉) · F. A. Shah · S. A. Dar · S. M. Mandu
Division of Aquatic Animal Health Management, Faculty of Fisheries, Sher-e-Kashmir University of Agricultural Sciences & Technology of Kashmir, Srinagar, India
e-mail: czainabmanzoor@gmail.com; shariq@skuastkashmir.ac.in; ferozshah@skuastkashmir.ac.in; shabirsamad585@gmail.com

R. Kooloth Valappil et al. (eds.), *Aquatic Animal Health Management*,
https://doi.org/10.1007/978-981-96-7987-4_3

contributes to the global fish food market. The economic importance and the nutritional benefits linked to coldwater fish, for instance, trout and salmon, make them a much-valued commodity in the aquaculture industry (Yu et al. 2022). However, disease outbreaks are a major constraint to sustainable fish production partly due to the intensification of aquaculture. As aquaculture practices intensify, they induce stressful conditions, making fish susceptible to various pathogens and resulting in severe disease outbreaks (Elgendy et al. 2024). Coldwater fish species are prone to environmental stresses, which can negatively impact their health, thereby affecting their physiological response. The stressful conditions further compromise their immunological function and weaken their defence system. The significant mortality due to coldwater infectious diseases also poses a threat to aquaculture production and fish yield. For effective management and sustainability, it is important to comprehend the various factors that are responsible for the emergence and spread of diseases in coldwater aquaculture. One of the most critical environmental variables is water temperature, which can largely affect the susceptibility of fish to pathogens (Leung and Bates 2013). Differences in water temperature could upset the fine balance of the internal systems of the fish and make them more prone to infectious diseases. This chapter elucidates the prevalence and severity of fish diseases in coldwater aquaculture, with a major focus on bacterial, viral, fungal, and parasitic diseases. A list of commonly prevalent infectious diseases in coldwater aquaculture, along with their aetiological agent and clinical signs, are presented in Table 1.

2 Bacterial Diseases

Coldwater fishes are susceptible to a wide range of bacterial diseases that significantly affect their health, growth, and overall productivity in aquaculture. Bacterial disease outbreaks are often facilitated by various environmental stressors such as poor water quality, overcrowding, and suboptimal temperatures (Austin and Austin 2007; Qadiri et al. 2019a). The most common and prevalent coldwater bacterial diseases in aquaculture are discussed below:

2.1 Bacterial Kidney Disease (BKD)

Bacterial kidney disease (BKD) is an important fish disease found worldwide, and it mainly affects salmonid fish. This disease is also known by several other names, such as Dee disease, kidney disease, white boil disease, corynebacterial disease, and salmonid kidney disease (Plumb 2018). It was first reported in the 1930s in wild Atlantic salmon in Scotland.

BKD is caused by one of the oldest known fish pathogens, *Renibacterium salmoninarum*, which is a Gram-positive, non-motile, aerobic small diplobacillus bacterium (Sanders and Fryer 1980). This bacterium is most active at or below 15 °C and has a limited host range. It primarily affects all salmonid fishes but has also been reported in non-salmonid fish like Ayu (*Plecoglossus altivelis*) (Delghandi et al.

Table 1 Commonly prevalent infectious diseases in coldwater aquaculture

Diseases	Aetiological agent	Clinical signs	References
Bacterial diseases			
Bacterial kidney disease	*Renibacterium salmoninarum*	Petechial haemorrhages, dark pigmentation of the skin, abdominal swelling, pale gills, skin rashes, exophthalmia	Richards et al. (2017), Delghandi et al. (2020)
Columnaris disease	*Flavobacterium columnare*	Gill necrosis with a yellowish tint, fin degeneration, lateral lesions on the body	Morrison et al. (1981), Mohamed (2020)
Coldwater disease	*Flavobacterium psychrophilum*	Exophthalmia, abdominal distension, pale gills, caudal necrosis	Starliper (2011)
Bacterial gill disease	*Flavobacterium branchiophilum*	Lethargy, pipping, gasping for air due to breathing difficulty	Starliper (2012)
Rainbow trout fry syndrome	*Flavobacterium psychrophilum*	Lethargy, skin darkening, bilateral exophthalmia, tail erosion, tissue damage around the dorsal fin	Park et al. (2023)
Lactococcosis	*Lactococcus garvieae*	Anorexia, melanosis, exophthalmia, periorbital haemorrhages, anal prolapse	Bekker et al. (2011)
Motile *Aeromonas* septicaemia	*Aeromonas hydrophilia*	Haemorrhages in the skin, eyes, and fins, dropsy, exophthalmia	Austin and Austin (2007)
Enterica red mouth	*Yersinia ruckeri*	Subcutaneous haemorrhages inside and around the mouth and throat, skin darkening, anorexia, surface lethargy	Ahmed et al. (2021)
Viral diseases			
IHN	IHNV	Spiral swimming, flashing with trailing faecal casts, skin darkening, haemorrhages, exophthalmia	OIE (2021a)
VHS	VHSV	Pale gills, fin/gill/skin/eye haemorrhages, flashing and spiral swimming, skin darkening, abdominal distension.	OIE (2021b)
SVC	SVCV/*Sprivivirus cyprinus*	Pale skin, fin base/eye/gill haemorrhages, exophthalmia, swollen vent, and trailing mucoid casts	OIE (2023)
EHN	EHNV	Loss of equilibrium, flared opercula, dark body colouration	Reddacliff and Whittington (1996)
IPN	IPNV	Anorexia, pale gills, moderate exophthalmia, abdominal swelling, white/yellow casts, abnormal swimming along with ataxia (erratic corkscrew swimming)	Wood et al. (1955), Wolf (1988), Dopazo (2020)

(continued)

Table 1 (continued)

Diseases	Aetiological agent	Clinical signs	References
Fungal diseases			
Saprolegniasis	*Saprolegnia parasitica*	White or grey cotton wool-like growths on the skin and gills, red spots due to body damage, lethargy	Bruno and Wood (1999), Pachade et al. (2014)
Branchiomycosis	*Branchiomyces sanguinis*	Breathing difficulty, gill discolouration with increased mucus, feeding difficulty, gill necrosis, and inflammation	Roberts (2012), Alam et al. (2023).
Parasitic diseases			
White spot disease	*Ichthyophthirius multifiliis*	Small white spots on the fish skin and gills, lesion-like blisters on the skin/fins, anorexia, rubbing against hard surfaces	Floyd and Reed (2009)
Costiasis	*Costia necatrix*	Bluish coating on the skin with increased mucus production, respiratory distress and inflammation causing gill swelling, caudal fin erosion	Tripathi (1954), Sen and Mandal (2018), Dekari et al. (2024)
Gyrodactylosis	*Gyrodactylus salaris*	Reduced activity, scratching against walls/hard surfaces, greyish body discolouration, whitish pectoral fins	OIE (2019b)
Trichodiniasis	*Trichodina*, *Trichodinella*, *Tripartiella*	Falling of the fins, body discolouration, tissue degeneration, anorexia	Riwidiharso (2020)
Argulosis	*Argulus* spp.	Irritation, sluggish behaviour, scale loss, fin rot, pale discolouration, haemorrhage spots	Pekmezci et al. (2011), Al Mamun et al. (2021)
Whirling disease	*Myxobolus cerebralis*	Tail chasing, loss of equilibrium, erratic swimming behaviour, skeletal abnormalities	Sarker et al. (2015)
Black spot disease	*Diplostomum* sp./ heterophyids	Around 1 mm raised black nodules on the skin and fins	Bamber et al. (1983)

IHN infectious haematopoietic necrosis, *IHNV* infectious haematopoietic necrosis virus, *VHS* viral haemorrhagic septicaemia, *VHSV* viral haemorrhagic septicaemia virus, *SVC* spring viraemia of carp, *SVCV* spring viraemia of carp virus, *EHN* epizootic haematopoietic necrosis, *EHNV* epizootic haematopoietic necrosis virus, *IPN* infectious pancreatic necrosis, *IPNV* infectious pancreatic necrosis virus

2020). *R. salmoninarum* can transmit both horizontally and vertically, and it has the ability to transmit directly without the need for an intermediate host. However, it cannot survive for more than a week outside its host (Boerlage et al. 2018).

Fish with BKD often exhibit changes in their behaviour, such as loss of balance and swimming in circles. They may also develop exophthalmia (affecting one or both eyes), petechial haemorrhages around their fins and lateral lines (Richards et al. 2017), dark skin, abdominal swelling, and anaemia (pale gills). Adult infected fish often show belly rashes (shallow sores or ulcers on the skin). Histopathological changes include chronic inflammation and granulomas, along with the proliferation of melanomacrophages in the kidney, liver, and spleen tissues, glomerular degeneration, and renal corpuscle and tubule alterations (Delghandi et al. 2020).

There is currently no available effective treatment that can completely eradicate this bacterium from salmonids. To control the spread of infection, it is a common practice to cull infected fish, especially gametes from infected fish. Additionally, antibiotics such as sulfadiazine, erythromycin, azithromycin, and enrofloxacin are used to treat infected fish.

2.2 Columnaris Disease

Columnaris disease is an acute to chronic infection affecting wild and cultured fish populations of freshwater origin and leading to significant economic losses (Faisal et al. 2017). In 1917, Herbert Spencer Davis first reported the disease under the name *Bacillus columnaris*, but, later, the nomenclature was changed to *Flavobacterium columnare* (Bernardet et al. 1996). The bacterium is a Gram-negative, thin filamentous bacilli, able to cause both primary and secondary infections in fish (Wakabayashi 1991). It has the ability to cause disease year-round, but the infection is more severe and frequent in the summer months, which can lead to 100% mortality. Direct contact, indirect water transmission, and cohabitation with infected fish can easily transmit *F. columnare* (Groff and Lapatra 2000).

The infection is more severe in young fish, affecting the gills, skin, fins, and tail (Roberts 2012). Adult fish exhibit necrosis in the gills with a yellowish tint, resulting in the destruction of gill filaments (Mohamed 2020). Other signs include fin degeneration, a condition that begins as a lesion at the base of the fins, which progresses outwards and eventually extends laterally, resembling a saddleback (Fig. 1), which is why this disease is also known as saddleback disease (Morrison et al. 1981). Histopathological changes include a severe change in the structure of gill filaments, including swelling, hyperplasia, proliferation of inter lamellar cells, expansion of the epithelium, cell clubbing, and vacuolation of the primary and secondary gill lamellae (Singh et al. 2021).

The outbreak of this disease is commonly due to inadequate management practices and poor water quality. It is crucial to adhere to good husbandry practices, which include maintaining proper sanitation, ensuring a clean water supply, minimizing stocking density, and maintaining optimal dissolved oxygen (DO) levels. At early stages, when the infection is only superficial, it is advisable to allow for

Fig. 1 *Flavobacterium columnare*-infected rainbow trout with saddleback lesions and cotton wool-like patches typical of columnaris disease

effective external treatment. A recent study has shown that the bacterium shows 100% sensitivity towards ciprofloxacin and 0% sensitivity towards ampicillin, amoxicillin, and gentamycin (Elgendy et al. 2022).

2.3 Bacterial Coldwater Disease (BCWD)

Bacterial coldwater disease (BCWD) affects a wide range of fish inhabiting coldwaters, particularly farm-reared young trout and salmon, and is predominant at or below 16 °C with severity observed at or below 10 °C. The aetiological agent of this disease is *Flavobacterium psychrophilum*, which was earlier known as *Cytophaga psychrophila* and *Flexibacter psychrophilus* (Bernardet et al. 1996). This bacterium was originally recovered from the diminishing Coho Salmon population in the Pacific Northwest United States in 1948 (Borg 1960).

The initial signs of infection in fish include anorexia, eroded fin tips followed by exophthalmia, abdominal distension due to ascites, and pale gills (anaemia). Advanced infections cause necrosis in the caudal region, leading to the eventual exposure of the caudal vertebra (Fig. 2). Changes in the histopathology include focal necrosis in the spleen, liver, and kidneys with higher levels of eosinophilia and haemosiderin as well as necrosis in the muscles under skin sores (Starliper 2011). The lesion may spread to the peduncle region, giving this disease another name: 'peduncle disease' (Bernardet and Bowman 2006).

F. psychrophilum can transmit horizontally via water and fish that act as pathogen carriers, shedding bacteria and having the ability to survive outside fish hosts for longer periods of time. Vertical transmission has also been suggested, as it is recovered from ovarian fluids, milt, and egg surfaces (Cipriano 2005). Protective measures present the most effective way of controlling the adverse impact of disease and reducing the spread and include proper disinfection and sanitation in fish farms, proper carcass disposal as dead fish shed bacteria in large quantities, egg disinfection using povidone-iodine, and other good husbandry practices.

2.4 Bacterial Gill Disease (BGD)

Bacterial gill disease (BGD) was first reported by Davis in 1926. It is caused by a Gram-negative bacterium, *Flavobacterium branchiophilum* (Kimura et al. 1978), in

Fig. 2 *Flavobacterium psychrophilum*-infected fish with tail and fin rot and prominent ulcerative lesions near the caudal peduncle

young salmonids reared in hatcheries. This disease gets transmitted horizontally and is found to be highly contagious to salmonids (Starliper 2012). Many researchers consider BGD outbreaks to be associated with various predisposing factors, including springtime, low dissolved oxygen, overcrowding, overfeeding, elevated ammonia (NH_3), settlement of particulate matter in water supply pipes, etc. (Starliper 2012).

The disease, as its name suggests, affects the gills and is characterized by symptoms such as lethargy, pipping, and gasping for air on the surface of the water column due to respiratory difficulties. Surface lesions and other internal pathologies are not associated with this disease. The fish are said to appear in a soldier-like manner, swimming high in the water column. Infected fishes are unable to lay down their gills flat due to the swelling and flared opercula. Histopathology reveals gill lamellar and filament fusion, increase in size (hypertrophy and hyperplasia) of pavement cells, chloride cells, and lamellar epithelial cells along with mucus cell hyperplasia (Starliper 2012).

2.5 Rainbow Trout Fry Syndrome (RTFS)

Rainbow trout fry syndrome (RTFS) is a systemic infection mainly affecting the fry and fingerlings of hatchery-reared rainbow trout (*Oncorhynchus mykiss*). The aetiological agent of RTFS is a gliding, Gram-negative bacterium, *Flavobacterium psychrophilum*, which is also responsible for the coldwater disease of salmonids (Madetoja et al. 2003). RTFS is prominent during spring when temperatures are still below 10 °C (Faruk 2002). It causes surface lethargy, skin darkening, and bilateral exophthalmia in infected fishes, characterized by tail erosion, tissue damage around the dorsal fin, anaemia (pale gills), and systemic infection (Park et al. 2023). Mortality in fry is often associated with acute systemic infections. Internal pathology may include anorexia, splenomegaly, pale liver, and ascites, as reported by

Faruk (2002). The most prominent histopathological changes associated with RTFS include kidney, gill, spleen, and liver necrosis (pyknosis and karyolysis). In addition, vacuolation in the stomach is also observed, which may be due to insufficient food intake (Park et al. 2023).

The use of antibiotics such as oxytetracycline and amoxicillin can significantly mitigate the adversity of RTFS, but 100% control is not available (Faruk 2002). Temperature is said to play a critical role in RTFS manifestation, and this disease is generally absent above 20 °C (Park et al. 2023). Proper biosecurity measures should be established within farms, along with regular health monitoring and other good husbandry practices, to reduce the spread because of the fastidious nature of this bacterium.

2.6 Lactococcosis

Fish lactococcosis is a hyper-acute and haemorrhagic septicaemia affecting a variety of fish species globally, particularly rainbow trout (Eldar et al. 1996). This disease is caused by a Gram-negative, haemolytic, chain-forming coccus, *Lactococcus garvieae*, which was earlier known as *Streptococcus garvieae*. It is a facultative anaerobe that causes blood agar haemolysis (Vendrell et al. 2006) and is considered an emerging zoonotic pathogen in aquaculture (Meyburgh et al. 2017).

Clinical signs of the infection include loss of appetite, melanosis, erratic swimming behaviour, exophthalmia, haemorrhages in the periorbital, intraocular, and fin bases, abdominal swelling, and anal prolapse (Bekker et al. 2011). Major histological findings are associated with the spleen and kidney, including vascular changes, tubular epithelium degeneration, an increase in melanomacrophage centres, and also haematomal haemorrhages in the myocardium and bulbous arteriosa (Soltani et al. 2021). Apart from these, diffuse haemorrhages in multiple tissues (eye sclera, swim bladder serosa, skeletal muscles) and necrosis in the stomach glands with a lumen full of necrotic material are also observed in histological studies (Avci et al. 2014). Multiple drug resistance against erythromycin, lincomycin, and oxytetracycline have been frequently reported in this pathogen (Duman et al. 2020). Preventing pathogen entry into fish culture systems through sanitary procedures reduces disease outbreaks (Vendrell et al. 2006).

2.7 Motile *Aeromonas* Septicaemia (MAS)

Motile *Aeromonas* septicaemia (MAS) is one of the common diseases encountered in coldwater aquaculture. The aetiological agent is a Gram-negative, rod-shaped, motile opportunistic bacterium *Aeromonas hydrophila*, causing haemorrhagic septicaemia and ulcerative conditions in various fish species. *A. hydrophila* has epizootic potential due to its prevalence in aquaculture systems (Bhowmik et al. 2009). MAS is also known by other names such as haemorrhagic septicaemia, ulcer disease, red sore disease, and tail and fin rot. This infection is often associated with

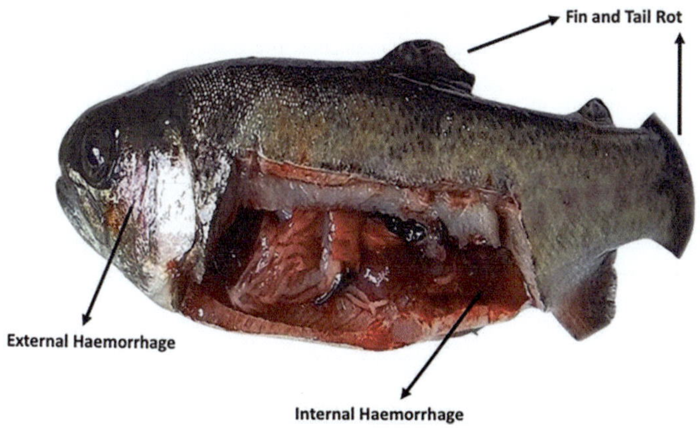

Fig. 3 *Aeromonas hydrophila*-infected fish showing extensive internal and external haemorrhages and severe fin rot

haemorrhages on the skin, eyes, fins, abdominal distension, dropsy, and exophthalmia (Fig. 3).

The disease often has an impact on the kidney, spleen, gills, liver, and skeletal muscles. The bacteria can enter the fish body through the gills, causing systemic infection (Austin and Austin 2007). It is very important to minimize stress and injury to fish so as to prevent infection in the aquaculture system. Optimizing stocking density, maintaining proper DO levels, ensuring a clean water supply, and implementing various other effective management procedures are crucial for controlling the outbreak of this infection. Various medications, like Romet-30 (sulfadimethoxine + ormetoprim), Terramycin (oxytetracycline), and Aquaflor (florfenicol), are used for the treatment of this disease (Hanson et al. 2019).

2.8 Enteric Red Mouth (ERM)

Enteric red mouth (ERM), or yersinosis, is an acute to chronic infectious systemic bacterial disease that mainly affects the farmed rainbow trout and causes cumulative mortalities that lead to huge economic losses (Kumar et al. 2015). This disease is associated with systemic haemorrhages (Rigos and Stevenson 2001) and histopathological changes in the gills, kidney, and spleen (Avci and Birincioğlu 2005). ERM was first reported in rainbow trout in the United States (Rigos and Stevenson 2001). The aetiological agent of this disease is a Gram-negative, rod-shaped, facultative anaerobe of the Enterobacteriaceae family known as *Yersinia ruckeri*. This disease can affect all age groups of susceptible fish species but causes most acute infections in fry and fingerlings (Ahmed et al. 2021). Clinical signs of the infection include anorexia, surface lethargy, exophthalmia, skin darkening, and subcutaneous haemorrhages inside and around the mouth and throat (Ahmed et al. 2021). Internally, it is characterized by splenomegaly with black discolouration; reddening of the lower intestines with an accumulation of opaque yellow fluid can be observed (Furones et al. 1993). Histopathological changes include the desquamation of epithelial cells in the secondary gill lamellae, hyperaemia, and oedema (De Lima et al. 2010), indicating that the gills are the important portal of entry for *Y. ruckeri*.

Post-infection treatment is mostly dependent on the use of antibiotics such as amoxicillin, oxolinic acid, oxytetracycline, sulphadiazine, and florfenicol but may lead to the emergence of antibiotic resistance (Defoirdt et al. 2007). ERM is the first fish disease for which an effective commercial vaccine was developed (Horne and Barnes 1999). Vaccines administered orally, by bath, or via injection can effectively elevate the immune response in rainbow trout against ERM.

3 Viral Diseases

Viruses are obligate parasites that thrive at lower water temperatures, typically between 0 °C and 15 °C, and are responsible for a variety of diseases in coldwater fishes. Viral infections can cause significant mortalities, leading to huge economic losses in the aquaculture industry. The general clinical signs in virus-infected fish include haemorrhages, anaemia, abdominal swelling, and abnormal swimming behaviours in infected fish. Some of the prevalent viral infections in coldwater fish are listed below:

3.1 Infectious Haematopoietic Necrosis

This is one of the most important OIE (Office International des Epizooties)-listed viral diseases that predominantly affect salmonid fish raised in freshwater. This disease has been reported in Europe, Asia-Pacific, Africa, and America. The aetiological agent of this disease is an enveloped, bullet-shaped virus, IHNV (infectious haematopoietic necrosis virus). The virion contains a non-segmented, single-stranded, negative-sense RNA (-ssRNA) genome of approximately 11,000 nucleotides. The viral genome encodes six proteins in the order: nucleoprotein (N), phosphoprotein (P), matrix protein (M), glycoprotein (G), a non-virion protein (NV), and polymerase (L). Due to the presence of the *NV* gene, IHNV has been placed under the genus *Novirhabdovirus*, family Rhabdoviridae, and order Mononegavirales (OIE 2021a).

The fry stage of salmonids is the most susceptible to IHNV. Adult fish are usually resistant to infection, but, even in the absence of clinical disease, they can still shed a large amount of virus in reproductive fluids and milt (Dixon et al. 2016). The target organs of IHNV are the haematopoietic tissues, and the virus is usually isolated from the kidney and spleen (OIE 2021a). IHNV causes mortality in the temperature range of 3–18 °C, but clinical disease in natural conditions occurs below 15 °C. IHNV transmission occurs horizontally via direct contact with the virus from infected fish or indirectly via water (Bootland and Leong 1999). The clinical signs associated with the IHNV infection include lethargy, spiral swimming, and flashing with trailing faecal casts. Fish may also show skin darkening, exophthalmia, abdominal distension, and haemorrhages on external surfaces. Degenerative necrosis of eosinophilic granular cells in the intestinal wall is the pathognomonic sign of this disease (Bootland and Leong 1999). The gold standard for the detection of any virus is the isolation of the same from cell lines. The recommended cell lines for IHNV isolation and detection are EPC and FHM, followed by further molecular or immunological examinations for specific identification (OIE 2021a).

There is no commercially effective treatment for IHNV while the virus can be inactivated by following various prophylactic measures, including disinfection using sodium hypochlorite, iodophor, benzalkonium chloride, and $KMnO_4$. Disinfection of eggs using iodophor can be done to avoid any egg-associated transmission. In addition, general husbandry practices should be followed, including the use of virus-free water and reducing stocking density, to mitigate the impact of IHNV infection.

3.2 Viral Haemorrhagic Septicaemia (VHS)

This is an OIE-listed viral disease affecting a wide variety of fish species, with rainbow trout being the most susceptible. It was first reported in 1938 from rainbow trout farms as infectious kidney swelling and liver degeneration disease (Schäperclaus 1938). It was also referred to as Egtved disease after Jensen confirmed its aetiology on the RTG-2 cell line in 1963 (Wolf and Quimby 1962). This disease was later renamed viral haemorrhagic septicaemia (VHS) by OIE. It has a huge negative impact on aquaculture and is currently reported in Europe, North America, and North Asia (OIE 2021b).

The aetiological agent of VHS is an enveloped, bullet-shaped virion known as the viral haemorrhagic septicaemia virus (VHSV). This virus is approximately 70 nm in diameter and 180 nm in length, belonging to the genus *Novirhabdovirus* of the family Rhabdoviridae. The virion contains a non-segmented, single-stranded, negative-sense RNA (-ssRNA) genome of approximately 11 kb or 11,000 nucleotides. The viral genome encodes six proteins in the order: nucleoprotein (N), phosphoprotein (P), matrix protein (M), glycoprotein (G), a non-virion protein (NV), and polymerase (L) (OIE 2021b). VHSV has been classified into four major genotypes (I–IV) based on the nucleotide sequence of the G-gene. Among these genotypes, I–III are reported from Europe and Japan, while genotype IV has only recently been discovered in North America, Japan, and Korea (OIE 2021b).

VHSV has the ability to cause infection and mortality in all age groups of susceptible fish, but it does not infect fish eggs. This means that VHSV transmission is only horizontal and occurs through water, either directly through the gills and skin or indirectly through urine (Smail and Snow 2011; Qadiri et al. 2020). Some vectors, like the grey heron, may carry viruses on their beaks and claws, suggesting their involvement (Olesen and Jorgensen 1982). The occurrence of infection is temperature-dependent, generally between 4 and 14 °C, with the highest mortality rate at 9–12 °C, and no natural outbreaks have been observed at above 18 °C (OIE

2021b). VHSV targets the anterior kidney, heart, and spleen, where it is found in abundance during acute infection, while during chronic infection, virus titres are high in the brain (Smail and Snow 2011). Fish infected with VHSV may undergo rapid mortality during acute infection, while during less adverse infection, fishes show different signs, including darkening of the skin, exophthalmia, pale gills indicating anaemia, and haemorrhages at the fin bases or in the gills, skin, or eyes (Fig. 4). Behavioural changes include lethargy, abnormal swimming patterns such as flashing and spiralling, and abdominal distension with the gathering of infected fish near the pond edges and outlets (OIE 2021b). VHSV is known to show a strong tropism for endothelial cells and blood cells (Qadiri et al. 2019b; Kim et al. 2019). Histopathology reveals extensive focal necrosis and degeneration in the kidney, liver, and spleen, characterized by cytoplasmic vacuoles, pyknosis, karyolysis, and lymphocytic invasion in the tissues (OIE 2021b).

The gold standard for the diagnosis of VHSV is its isolation in cell culture. Other diagnostic methods include immunohistochemistry (IHC), histopathology, reverse transcription polymerase chain reaction (RT-PCR), and in situ hybridization (Qadiri et al. 2019c). The recommended cell lines for the isolation of VHSV include BF-2, CHSE-214, EPC, FHM, and RTG-2. Recently, one-step RT-PCR has been developed and validated by Kim et al. (2018), which outperformed the old conventional RT-PCR methods with the highest detection sensitivity for all the VHSV genotypes. No commercial vaccine is available for VHSV, even though extensive research has been going on for more than 40 years. Control of a VHSV outbreak is generally dependent on various prophylactic measures, such as regular disinfection of ponds using ultraviolet (UV) light, chlorine, sodium hypochlorite, light, and iodophores, along with a reduction in stocking density, increasing temperature, appropriate feeding, use of probiotics and immunostimulants to enhance the immunity of fish, and restricting the handling procedures.

3.3 Spring Viraemia of Carp (SVC)

Spring viraemia of carp (SVC) is another important OIE-listed viral disease affecting the cyprinids, particularly common carp characterized by haemorrhages (Ahne et al. 2002). This infection usually occurs in springtime, and the outbreak temperature is generally between 11 °C and 17 °C. While European countries had long

Fig. 4 A VHSV-infected flounder with abdominal dropsy and haemorrhage on the tail fin

recorded this disease, South Africa and North America discovered it in 1998 and 2002, respectively. Asia also detected the virus for the first time in 2004. The aetiological agent of SVC is an enveloped, bullet-shaped virion commonly known as the spring viraemia of carp virus (SVCV) containing a non-segmented, single-stranded, negative-sense RNA (-ssRNA) genome. The virion was earlier called *Rhabdovirus carpio* and was included in the genus *Vesiculovirus* (Ahne et al. 2002) of the family Rhabdoviridae, but the name was eventually changed to *Sprivivirus cyprinus* and the genus to *Sprivivirus* (OIE 2023). The length of the SVCV genome is 11,019 nucleotides, encoding 5 proteins in the order: nucleoprotein (N), phosphoprotein (P), matrix protein (M), glycoprotein (G), and polymerase (L). SVCV does not contain the *NV* gene between the G- and L-genes, as in the viruses of the genus *Novirhabdovirus* (Ahne et al. 2002).

Infection due to SVCV can occur in all age groups of susceptible species but is usually more prominent with overt clinical signs in young fish up to 1 year of age with about 90% mortality (OIE 2023). The virus enters the fish via the gills and then spreads to the liver, heart, spleen, kidney, and alimentary tract. The virus titres are very high in the liver and kidney of infected fish during disease outbreak (Dixon 2008). The virus shows horizontal transmission, which may be via direct contact or indirectly via water or vectors. Vertical or egg-associated transmission has also been reported (Bekesi and Csontos 1985). Fish infected with SVCV often show lethargy and separate from shoals, gathering at the edges and inlets of ponds. Clinical signs include pale skin, fin base, eye and gill haemorrhages, skin darkening, and exophthalmia, along with swollen vents and trailing mucoid casts, with abdominal distension being the most consistent and obvious (OIE 2023). Histopathological changes can be easily observed in the entire major organs, but, due to the absence of specificity for the infection, microscopic methods are not recommended for SVCV diagnosis. SVCV can be detected by its isolation on cell culture using EPC, FHM, or GCO cell lines as recommended by OIE, followed by immunological or molecular testing (OIE 2023).

The control measures rely on various prophylactic measures such as egg disinfection using iodophor treatment, general husbandry practices such as clean water supply, proper and regular sanitation and disinfection, low stocking density during winters, and elevated temperatures above 19–20 °C. There are currently no effective vaccines for SVCV infection. A strain of common carp has been bred, which shows increased resistance to SVCV and is named 'Krasnodar' (Kirpichnikov et al. 1993).

3.4 Epizootic Haematopoietic Necrosis (EHN)

Epizootic haematopoietic necrosis (EHN) has been listed by the World Organisation for Animal Health (WOAH) (OIE) because of its international concern since its emergence in 1984. This disease has seriously affected the European perch and farmed rainbow trout with all the life stages being susceptible. The epizootic haematopoietic necrosis virus (EHNV) was first detected in 1984 in Victoria, Australia (OIE 2019a). The aetiological agent of this disease is an icosahedral virus, EHNV, belonging to the genus *Ranavirus* and family Iridoviridae. The virion is 150–180 nm in diameter, containing a linear double-stranded densovirus (DNV) (dsDNA) genome of size 150–170 Kb that can replicate both in the nucleus and cytoplasm. The genome has a high level of methylation with circular arrangements and terminal redundancy (repetitive gene sequences) (OIE 2019a).

The virus infects the kidney, spleen, and liver, with clinical signs more prominent in fingerlings and juveniles of rainbow trout and European perch. The infection is more adverse in European perch, and the prevalence of infection is extremely low in rainbow trout (OIE 2019a). The infection in rainbow trout has been observed at water temperatures between 11° and 20 °C, while the European perch outbreaks usually occur in the summer season. Transmission of the virus occurs horizontally, with no vertical transmission detected so far (OIE 2019a). There are no specific clinical signs of infection. Moribund fishes usually show loss of equilibrium, flared opercula, and dark colouration, and the liver has focal white to yellow lesions due to necrosis (Reddacliff and Whittington 1996). The virus is detected by isolating it on cell lines such as BF-2, FHM, EPC, and CHSE-214 as recommended by WOAH (OIE) at temperatures ranging from 15 to 22 °C, followed by immunological and molecular testing (OIE 2019a). There are no effective therapeutics available for EHNV infection, and control usually relies on good husbandry practices, including maintenance of a healthy environment, low stocking density, clean and high water flow, and proper sanitation.

3.5 Infectious Pancreatic Necrosis (IPN)

Infectious pancreatic necrosis (IPN) is the most prevalent viral infection, having a broad host range and global distribution, mostly affecting the salmonid family, especially the rainbow trout, and is known for its ability to survive in aquatic ecosystems for longer periods (Tapia et al. 2022). The IPN virus (IPNV) is known as the first teleost virus to be isolated and identified in cell culture. The disease was first reported in freshwater trout farms in North America in 1950 (Wood et al. 1955). The aetiological agent of this disease is a non-enveloped, icosahedral IPN virus belonging to the genus *Aquabirnavirus* and family Birnaviridae. The virion contains a bisegmented (segments A and B) double-stranded RNA (dsRNA) genome (Gibrat et al. 2013). Segment A encodes three structural proteins (VP1, VP2, and VP3) and two non-structural proteins (VP4 and VP5) (Dopazo 2020). VP2 is the major capsid protein in IPNV, contributing 60% of the total capsid proteins (Dobos 1995), and plays a critical role in the ability of the virus to infect and replicate inside a host cell. Segment B of the genome encodes a single RNA-dependent RNA polymerase (RdRp) (Skjesol et al. 2009). IPNV typically shows horizontal transmission, but there is a possibility of vertical transmission involving the transfer of the virus by seminal fluid (Ahne 1985), leading to egg-associated viral transmission from one generation to the next. Feeding on infected fish or their carcasses can lead to predatory birds and mammals becoming carriers (McAllister and Owens 1992).

Infected fish show abdominal swelling, pale gills (anaemia), skin darkening, and moderate exophthalmia, along with white/yellow casts from their anal opening (Wolf 1988). They also show behavioural changes like abnormal swimming and ataxia (Wood et al. 1955). The specific signs shown by infected fishes are erratic corkscrew swimming patterns and anorexia (Dopazo 2020). The presence of milky mucous substances in the digestive tract is the pathognomonic sign of IPNV infection (Munro and Midtlyng 2011). Histopathology reveals severe intestinal mucosal necrosis along with the death of pancreatic acinar cells and hepatic tissue damage (Dopazo 2020). The virus detection methods include isolation on cell culture, histological examination, and immunological and molecular techniques. The IPN virus can be isolated on cell culture using the CHSE-214 and BF-2 cell lines (Saint-Jean et al. 2001).

The most critical factor in controlling IPNV involves understanding its interaction with the host and identifying the target for treatment. Despite various efforts in the development of vaccines, these have not shown any significant ability to elicit an immune response against IPNV infection. Even if vaccines induce a humoral response, fish can still act as virus carriers (Munang'andu et al. 2013). Presently, control can be achieved by following various preventive measures, such as proper disinfection of eggs using iodophores and implementation of biosecurity measures.

4 Fungal Diseases

Fungi are mostly responsible for the secondary infections in fish, which are either external or internal. These diseases occur when fish are compromised or their immune system is moderated. The incidence of fungal diseases is often induced by environmental stressors such as poor water quality, low temperature, high density, and physical injuries. Some fungal infections in fish are systemic and affect the internal organs, mostly the liver, spleen, and heart. Fungal spores are known to spread rapidly in water. Efficient management techniques such as improved water quality, stress minimization, and antifungal drug treatment can limit the spread of disease to other fishes. Due to the high complexity of fungi, vaccination against these pathogens is quite challenging. Therefore, better management practices to control fungal disease outbreaks are called for within coldwater aquaculture systems. Common fungal diseases in coldwater fishes are discussed below:

4.1 Saprolegniasis

This disease was first reported in Europe in 1877 and was named salmon disease (Neish and Hughes 1980). Saprolegniasis affects both wild and farm-raised freshwater fish and has been recorded in a number of fishes, including catla, rohu, striped and spotted snakehead, walking catfish, zigzag eel, giant river catfish, and Gangetic mystus from Kolleru Lake in Andhra Pradesh (Mastan and Ahmad 2018). This disease is primarily caused by an oomycete, *Saprolegnia parasitica*, and is characterized by the growth of white or grey cotton-like mesh on the skin and gills of fish (Van West 2006). The infection starts at the head and fins of fish and then spreads to other areas of the body (Bruno and Wood 1999). A disease outbreak is usually caused by poor water quality and stress. Saprolegnia acts as a saprophyte and feeds on dead eggs, which can then transfer to healthy eggs, causing their death (Barde et al. 2020). Infected fish are often sluggish with a damaged body, identified by the presence of red spots on the body (Pachade et al. 2014). The saprolegnia fungus is responsible for the winter kill of fish due to their increased susceptibility in winter conditions (Bly and Clem 1992). Malachite green, an organic dye, is the major chemical used to treat any fungal infection. This compound has been widely used to treat infected fish and their eggs (Sudova et al. 2007). However, it has been banned for use in animals since 2002 by the US Food and Drug Administration (FDA) due to its carcinogenic and toxic properties, which can ultimately have a harmful effect on humans. Alternative chemicals that can be used include bronopol, formalin, hydrogen peroxide (H_2O_2), and iodine-free salt. Bronopol can be toxic to young fish and is thus avoided at this stage of fish life (Sudova et al. 2007).

4.2 Branchiomycosis

Branchiomycosis is a fungal infection also known as gill rot, with *Branchiomyces sanguinis* as the aetiological agent. This disease is commonly found in freshwater aquaculture systems (carp farms), affecting the gill epithelium and capillaries of the host fish (Shinn et al. 2023). The fungus enters through the gills and invades the tissues, causing severe necrosis and inflammation (Roberts 2012). The disease causes respiratory distress in infected fish, causing anaemia and eventually death. Clinical signs include gasping for air, discolouration of the gills, increased mucus production, feeding difficulty, growth retardation, and mortality (Alam et al. 2023). There is no treatment available for this infection; fish that survive the outbreak should be kept away from healthy ones as they can act as carriers of the fungus. Water quality parameters such as DO and temperature should be kept at optimum levels. Chemicals like $KMnO_4$, $CuSO_4$, NaCl and malachite green can also be used in small quantities to reduce infection severity.

5 Parasitic Diseases

Infections caused by parasites in aquaculture can have detrimental effects on the health and productivity of cultured fish, leading to significant economic losses in the industry. These parasites cause physical damage to the fish host, paving the way for other pathogen attacks. Some of the commonly prevalent coldwater fish parasitic diseases are mentioned below:

5.1 White Spot Disease

This parasitic disease is also known as ichthyotherapies or Ich disease and is caused by a protozoan ciliate parasite, *Ichthyophthirius multifiliis*, which infects the epidermis of the body surface, gills, and eyes of fish (Xu et al. 2016). Ich is a highly contagious disease, and *I. multifiliis* is an obligate parasite having three stages in its life cycle: theront, trophont, and tomont (Fig. 5). Theront is the infective stage that penetrates the host and then develops into trophont, which is the parasitic stage and feeds on fish skin and mucus (von Gersdorff Jørgensen 2016; Buchmann 2020). The mature trophont leaves the host and transforms into a reproductive tomont that forms a substratum at the bottom of the water body. The tomont then undergoes binary fission and forms a large number of tomites through reproduction via binary fission. Once the tomocyst bursts open, these tomites emerge as theronts (Matthews 2005). Theronts cannot live without a host for more than 48 hours. The life cycle of *I. multifiliis* is temperature-dependent, with warmer temperatures causing intensification and spread of infection (Floyd and Reed 2009).

The disease is characterized by the presence of small white spots (like sprinkled salt) on the skin and gills, along with lesion-like blisters on the skin or fins. Fish often show signs of irritation by rubbing against hard surfaces due to itching. Infected fish are anorexic and lethargic and show decreased activity (Floyd and Reed 2009). By microscopic examination of wet mount preparation of the skin and gills, one can easily identify the parasite. An adult parasite can be easily seen as a

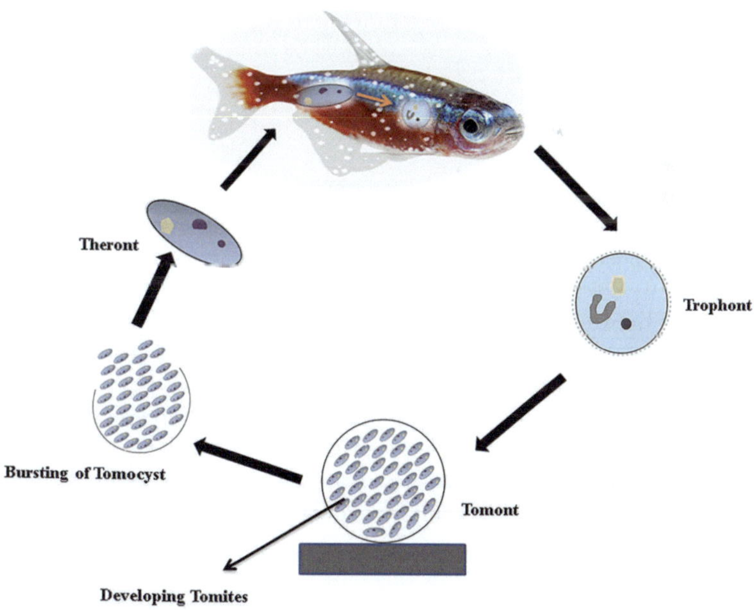

Fig. 5 Life cycle of *Ichthyophthirius multifiliis*: Theront (infective stage), trophont (mature stage), and tomont (reproductive stage)

large ciliate protozoan with a horseshoe-shaped nucleus (Floyd and Reed 2009). Ich infection can be effectively treated with malachite green, but, due to its carcinogenic nature, its use has been banned. Other alternatives that are less effective than malachite green but can control the infection include $CuSO_4$, formalin, and salt. Water management, which includes proper filtration, temperature control, and salinity, can also minimize infection. As temperature has a huge impact on the life cycle of *I. multifiliis*, its regulation can be a critical factor in preventing disease transmission (Floyd and Reed 2009).

5.2 Costiasis

Costiasis is caused by a flagellated protozoan ectoparasite, *Costia necatrix* (Frederic 1941), affecting pond fish, especially salmonid fry and fingerlings (Mitchell and Rodger 2011). Due to its extremely small size and sedentary habit, it is very difficult to identify this parasite in fish (Frederic 1941). The parasite was first described as *Bodo nectar* by Illenneguy in 1883 but was then placed within its own genus *Costia* by Le Clerq (1890) with *C. necatrix* being the only species within that genus (Frederic 1941). The disease is characterized by a bluish coating of the skin and the production of mucus in large quantities in fish, along with respiratory distress and inflammation (Tripathi 1954). Clinical signs of the infection include swollen gills, which are covered by mucus and cause breathing difficulty (Sen and Mandal 2018). Other related signs include erosion of the caudal fin, a blue-grey film, or light pink masses on the skin, leading to severe irritation and mortality in advanced cases (Dekari et al. 2024). The disease is usually treated with the use of formalin and salt.

5.3 Gyrodactylosis

This is an OIE-listed parasitic disease caused by *Gyrodactylus salaris*, which is an obligate viviparous parasite with a direct life cycle. Gyrodactylus is known to survive at water temperatures between 0 and 25 °C, causing infections in freshwater fishes and mainly affecting salmonids. All life stages are susceptible, but mortality up to 100% is prominent only in the fry and parr stages. The distribution of *G. salaris* on the fish body depends on the intensity of the infection, but it usually occurs on the fins of infected fish. The parasite can be found on other parts of the body but is less common on the gills (OIE 2019b). Infested fish are found scratching against hard substances or walls, along with reduced activity. In severe cases, fish turn greyish due to excess mucus production that may even lead to hypertrophy of the epidermis, causing the dorsal and pectoral fins to appear whitish. Erosion of the fins, particularly dorsal, tail, and pectoral fins, are observed during heavy infections due to feeding of the parasite. In addition, this infection also paves the way for secondary fungal infections by *Saprolegnia* spp. (OIE 2019b). The most commonly used chemical in the treatment of this infection is aluminium sulphate, which is highly toxic to *G. salaris* but not to fish. Other chemicals used include NaCl, formaldehyde, iodine, and chlorine compounds (OIE 2019b).

5.4 Trichodiniasis

Trichodiniasis, a major fish parasitic disease, is caused by a ciliate protozoan parasite prevalent in freshwater. This disease has significance in aquaculture because of its ability to reduce fish growth, decrease immune responses, cause chronic mortalities, changes in swimming patterns, and larval vision (Valladão et al. 2016). The aetiological agents of this disease are considered to be *Trichodina*, *Trichodinella*, and *Tripartiella* (Saha et al. 2017). The parasite has a bowl or bell shape with cilia around the body and on the ventral side and a tooth-like structure (denticle) that helps in its attachment to the host (Basson and As 2006). The pathology associated with this disease is related to gill tissues, including hyperplasia, hypertrophy, oedema, necrosis, and infiltration due to inflammation (Valladao et al. 2014). The parasite attacks the gills, fins, scales, and operculum of fish, causing discolouration of the body, loss of appetite, falling of fins, tissue degeneration in the epithelium, and mucus cell proliferation (Riwidiharso 2020). Treatment of this infestation usually involves the use of disinfectants such as formaldehyde, NaCl, formalin, and $KMnO_4$. The use of formaldehyde at 0.5 ml/l in combination with 1% NaCl can be effective in treatment. Due to the adverse effect of the parasite on the larval stages of fish, the adoption of various prophylactic measures and other management strategies is very important during larval rearing (Valladao et al. 2016).

5.5 Argulosis

Argulosis is caused by a crustacean branchiuran ectoparasite belonging to the genus *Argulus*, commonly known as fish lice (Tandel et al. 2021). *Argulus* spp. is a large obligate ectoparasite visible to the naked eye and is usually found feeding on the mucus and blood of the fish skin (Al Mamun et al. 2021). This parasite usually affects coldwater fish species, including salmon, trout, and carp (Tandel et al. 2021). It poses a major health concern to aquaculture due to its heavy infestation, which can lead to significant morbidity and mortality.

Fish infested with *Argulus* spp. show behavioural changes, including irritation, sluggish movement, and gathering near the pond edges (Al Mamun et al. 2021). Advanced cases of infestation include loss of scales, fin rot, pinpoint haemorrhages, excess mucus production, and body discolouration (pale) (Pekmezci et al. 2011). As *Argulus* is a large ectoparasite visible to the naked eye, it can be manually removed from fish using mechanical methods such as handpicking, which interrupts its egg production process and is thus an effective method to control the spread of infection. *Argulus* lays eggs on hard substrates or aquatic plants. These should be removed from ponds, reducing the spread significantly. Chemical treatments include using organophosphates, usually given two or three times at a week's interval, and chitin synthesis inhibitors such as diflubenzuron. Other chemical treatments include the use of NaCl, formaldehyde, $KMnO_4$, and formalin. Researchers have used emamectin benzoate and trichlorfon to completely eradicate the parasite (Noaman et al. 2010).

5.6 Whirling Disease

This disease is caused by a myxozoan parasite, *Myxobolus cerebralis*, in salmonids, particularly rainbow trout (Gilbert and Granath Jr 2003). It was first discovered in Europe in rainbow trout imported from North America (Hofer 1903). It got its name because of the whirling behaviour shown by fish after infection with *M. cerebralis*. The outbreak of whirling disease is dependent on many factors, particularly the presence of oligochaete hosts. *Tubifex tubifex* is considered to be the only aquatic invertebrate host that acts as a carrier for this parasite (Granath and Gilbert 2002).

The parasite penetrates and travels through the skin and gill epithelium, eventually invading the epithelial cells (El-Matbouli et al. 1999). Clinical signs include tail chasing, loss of equilibrium, erratic swimming behaviour, and skeletal abnormalities (Sarker et al. 2015). Histological changes include erosion of the cartilage and replacement by a mass of epithelioid tissue (Ksepka et al. 2020). Disease outbreaks can be controlled by following various good management practices, such as proper disinfection of equipment, hatching of eggs, and rearing of fish larvae in concrete raceways in spore-free water.

5.7 Black Sp ot Disease (BSD)

Black spot disease (BSD) or diplostomiasis is one of the commonly reported freshwater parasitic diseases caused by the encystment of metacercarial larvae of digenean trematodes (flukes) belonging to the families Diplostomidae (Niewiadomska 2002) and Heterophyidae (Duflot et al. 2021). This disease is named so because of the presence of black cysts on the external body surface of fish, which are formed due to melanin production around the encysted metacercariae (Dennis et al. 2019). The cysts appear as small, 1-mm-long black nodules on the skin and fins of fish. This disease has no adverse impact on fish, but it reduces their marketability due to its unsightly appearance on the skin. Fish act as an intermediate host for this parasite; they can get infected if cultured in water containing parasitized snails (Bamber et al. 1983). Picric acid can be used to control black spot disease in fish. Removal of aquatic weeds is also suggested as this can limit the snail population, resulting in an incomplete life cycle and thereby preventing further infestation.

6 Conclusions

The presence of fish diseases in coldwater aquaculture presents a substantial challenge to the growing aquaculture industry as it affects the well-being of fish and economic sustainability. Coldwater aquaculture is continuously affected by various infectious diseases of varying aetiologies. Implementing efficient control and prevention strategies is crucial in order to mitigate these problems. Early detection and management of disease outbreaks is important, and this can be achieved in part by using advanced diagnostic techniques and biosecurity measures. There is an intricate connection between the aetiological agent, physiological characteristics of the host, and environmental factors. Future research and diagnostic advancements are crucial for improving our comprehension and creating more efficient control measures. Effective collaboration among researchers, aquaculturists, and policymakers is essential for achieving sustainable health management practices in coldwater aquaculture.

References

Ahmed F, Soliman FM, Adly MA, Soliman HA, El-Matbouli M, Saleh M (2021) Dietary chitosan nanoparticles: potential role in modulation of rainbow trout (*Oncorhynchus mykiss*) antibacterial defense and intestinal immunity against enteric redmouth disease. Mar Drugs 19(2):72

Ahne W (1985) Studies on transmission of infectious pancreatic necrosis virus via eyed eggs and sexual products of salmonid fish. In: Ellis AE (ed) Fish and shellfish pathology. Academic Press, London, pp 262–270

Ahne W, Bjorklund HV, Essbauer S, Fijan N, Kurath G, Winton JR (2002) Spring viremia of carp (SVC). Dis Aquat Org 52(3):261–272

Al Mamun MA, Nasren S, Rathore S, Ramesh K (2021) Histopathological studies of pond reared Indian major carp, *Catla catla* infested with *Argulus japonicus* and trial for argulosis treatment. Punjab Univ J Zool 36:131–139

Alam S, Afzal G, Iqbal Z, Hussain R, Rizwan M, Afzal M, Mahmood Y, Yamin A, Raza GA, Farooq U, Iqbal S (2023) Fungal zoonotic infections in Fish an emerging threat to aquatic and terrestrial life. In: Zoonosis, vol 4. Unique Scientific Publishers, Faisalabad, pp 611–624

Austin B, Austin DA (2007) Characteristics of the diseases. In: Bacterial fish pathogens: diseases of farmed and wild fish, pp 15–46

Avci H, Birincioğlu S (2005) Pathological findings in rainbow trout (*Oncorhynchus mykiss* Walbaum, 1792) experimentally infected with *Yersinia ruckeri*. Turk J Vet Anim Sci 29(6):1321–1328

Avci H, Birincioglu SS, Tanrikul TT, Epikmen ET, Metin N, Avsever ML (2014) Experimental *Lactococcus garvieae* infection in rainbow trout, *Oncorhynchus mykiss*, W albaum 1792: a comparative histopathological and immunohistochemical study. J Fish Dis 37(5):481–495

Bamber RN, Glover R, Henderson PA, Turnpenny AWH (1983) Diplostomiasis in the sand smelt, Atherina presbyter (Cuvier), population at Fawley Power Station. J Fish Biol 23(2):201–210

Barde RD, Deshpande M, Nagthane N, Darak O, Baig MMV (2020) A review of saprolegnia infection in freshwater fishes and control of the saprolegniosis. Sustain Humanosphere 16(1):702–711

Basson L, As JV (2006) Trichodinidae and other ciliophorans (phylum Ciliophora). In: Fish diseases and disorders. Volume 1: protozoan and metazoan infections. CABI, Wallingford, pp 154–182

Bekesi L, Csontos L (1985) Isolation of spring viraemia of carp virus from asymptomatic broodstock carp, *Cyprinus carpio* L. J Fish Dis 8(5):471–472

Bekker A, Hugo C, Albertyn J, Boucher CE, Bragg RR (2011) Pathogenic gram-positive cocci in South African rainbow trout, *Oncorhynchus mykiss* (Walbaum). J Fish Dis 34(6):483–487

Bernardet JF, Bowman JP (2006) The genus flavobacterium. In: The prokaryotes, vol 7. Springer, Berlin, pp 481–531

Bernardet JF, Segers P, Vancanneyt M, Berthe F, Kersters K, Vandamme P (1996) Cutting a gordian knot: emended classification and description of the genus flavobacterium, emended description of the family Flavobacteriaceae, and proposal of *Flavobacterium hydatis* nom. nov. (basonym, *Cytophaga aquatilis* Strohl and Tait 1978). Int J Syst Evol Microbiol 46(1):128–148

Bhowmik P, Bag PK, Hajra TK, De R, Sarkar P, Ramamurthy T (2009) Pathogenic potential of *Aeromonas hydrophila* isolated from surface waters in Kolkata. India J Med Microbiol 58(12):1549–1558

Bly JE, Clem LW (1992) Temperature and teleost immune functions. Fish Shellfish Immunol 2(3):159–171

Boerlage AS, Elghafghuf A, Stryhn H, Sanchez J, Hammell KL (2018) Risk factors associated with time to first clinical case of Bacterial Kidney Disease (BKD) in farmed Atlantic Salmon (*Salmo salar* L.) in New Brunswick. Canada Prev Vet Med 149:98–106

Bootland LM, Leong JC (1999) Infectious hematopoietic necrosis virus. In: Woo PTK, Bruno DW (eds) Fish diseases and disorders, volume 3: viral, bacterial and fungal infections. CAB International, Oxon, pp 57–121

Borg AF (1960) Studies on myxobacteria associated with diseases in salmonid fishes. Wildl Dis 8:1–85

Bruno DW, Wood BP (1999) Saprolegnia and other Oomycetes. In: Fish diseases and disorders, Volume 3: Viral, bacterial and fungal infections. CAB International, UK, pp 599–643

Buchmann K (2020) Immune response to *Ichthyophthirius multifiliis* and role of IgT. Parasite Immunol 42(8):12675

Cipriano RC (2005) Intraovum infection caused by *Flavobacterium psychrophilum* among eggs from captive Atlantic salmon broodfish. J Aquat Anim Health 17(3):275–283

De Lima R, Feitosa L, Pereira ADES, De Moura MR, Aouada FA, Mattoso LHC, Fraceto LF (2010) Evaluation of the genotoxicity of chitosan nanoparticles for use in food packaging films. J Food Sci 75(6):N89–N96

Defoirdt T, Boon N, Sorgeloos P, Verstraete W, Bossier P (2007) Alternatives to antibiotics to control bacterial infections: luminescent vibriosis in aquaculture as an example. Trends Biotechnol 25(10):472–479

Dekari D, Chouhan N, Choudhary B, Saikia P, Nath RJ, Dihingia P, Chutia A, Konwar S (2024) Prevalence of blue slime disease in biofloc reared anabas testudineus fish in Golaghat district of Assam. J Exp Zool India 27(1):1423

Delghandi MR, El-Matbouli M, Menanteau-Ledouble S (2020) *Renibacterium salmoninarum*—the causative agent of bacterial kidney disease in salmonid fish. Pathogens 9(10):845

Dennis MM, Izquierdo A, Conan A, Johnson K, Giardi S, Frye P, Freeman MA (2019) Scaphanocephalus-associated dermatitis as the basis for black spot disease in Acanthuridae of St. Kitts, West Indies. Dis Aquat Org 137(1):53–63

Dixon PJR (2008) Virus diseases of cyprinids. In: Fish diseases, vol 2, pp 101–119

Dixon P, Paley R, Alegria-Moran R, Oidtmann B (2016) Epidemiological characteristics of infectious hematopoietic necrosis virus (IHNV): a review. Vet Res 47:1–26

Dobos P (1995) The molecular biology of infectious pancreatic necrosis virus (IPNV). Annu Rev Fish Dis 5:25–54

Dopazo CP (2020) The infectious pancreatic necrosis virus (IPNV) and its virulence determinants: what is known and what should be known. Pathogens 9(2):94

Duflot M, Gay M, Midelet G, Kania PW, Buchmann K (2021) Morphological and molecular identification of *Cryptocotyle lingua* metacercariae isolated from Atlantic cod (*Gadus morhua*) from Danish seas and whiting (*Merlangius merlangus*) from the English Channel. Parasitol Res 120(10):3417–3427

Duman M, Buyukekiz AG, Saticioglu IB, Cengiz M, Sahinturk PINAR, Altun SONER (2020) Epidemiology, genotypic diversity, and antimicrobial resistance of *Lactococcus garvieae* in farmed rainbow trout (*Oncorhynchus mykiss*). Iran J Fish Sci 19(1):1–18

Eldar A, Ghittino C, Asanta L, Bozzetta E, Goria M, Prearo M, Bercovier H (1996) *Enterococcus seriolicida* is a junior synonym of *Lactococcus garvieae*, a causative agent of septicemia and meningoencephalitis in fish. Curr Microbiol 32:85–88

Elgendy MY, Abdelsalam M, Mohamed SA, Ali SE (2022) Molecular characterization, virulence profiling, antibiotic susceptibility, and scanning electron microscopy of *Flavobacterium columnare* isolates retrieved from Nile tilapia (*Oreochromis niloticus*). Aquac Int 30(2):845–862

Elgendy MY, Ali SE, Dayem AA, Khalil RH, Moustafa MM, Abdelsalam M (2024) Alternative therapies recently applied in controlling farmed fish diseases: mechanisms, challenges, and prospects. Aquac Int 32:9017–9078

El-Matbouli M, Hoffmann RW, Schoel H, McDowell TS, Hedrick RP (1999) Whirling disease: host specificity and interaction between the actinosporean stage of *Myxobolus cerebralis* and rainbow trout *Oncorhynchus mykiss*. Dis Aquat Org 35(1):1–12

Faisal M, Diamanka A, Loch TP, LaFrentz BR, Winters AD, García JC, Toguebaye BS (2017) Isolation and characterization of *Flavobacterium columnare* strains infecting fishes inhabiting the Laurentian Great Lakes basin. J Fish Dis 40(5):637–648

FAO (2022) The state of world fisheries and aquaculture. Towards blue transformation. Food and Agriculture Organization

Faruk MAR (2002) A review on rainbow trout fry syndrome (RTFS). Pak J Biol Sci 5(2):230–233

Fish FF (1941) Notes on *Costia necatrix*. Trans Am Fish Soc 70(1):441–445

Floyd RF, Reed P (2009) *Ichthyophthirius multifiliis* (white spot) infections in fish. Fisheries and Aquatic Sciences Department, Florida Cooperative Extension Service, Institute of Food and Agricultural Sciences, University of Florida, CIR, p 920

Furones MD, Rodgers CJ, Munn CB (1993) *Yersinia ruckeri*, the causal agent of enteric redmouth disease (ERM) in fish. Annu Rev Fish Dis 3:105–125

von Gersdorff Jørgensen L (2016) The dynamics of neutrophils in zebrafish (Danio rerio) during infection with the parasite *Ichthyophthirius multifilis*. Fish Shellfish Immunol 55:159–164

Gibrat JF, Mariadassou M, Boudinot P, Delmas B (2013) Analyses of the radiation of birnaviruses from diverse host phyla and of their evolutionary affinities with other double-stranded RNA and positive strand RNA viruses using robust structure-based multiple sequence alignments and advanced phylogenetic methods. BMC Evol Biol 13:1–17

Gilbert MA, Granath WO Jr (2003) Whirling disease of salmonid fish: life cycle, biology, and disease. J Parasitol 89(4):658–667

Granath WO, Gilbert MA (2002) The role of *Tubifex tubifex* (Annelida: Oligochaeta: Tubificidae) in the transmission of *Myxobolus cerebralis* (Myxozoa: Myxosporea: Myxobolidae). In American Fisheries Society Symposium, p 79–86

Groff JM, Lapatra SE (2000) Infectious diseases impacting the commercial culture of salmonids. J Appl Aquac 10(4):17–90

Hanson LA, Hemstreet WG, Hawke JP (2019) Motile Aeromonas septicemia (MAS) in fish. SRAC Publication No. 0478

Hofer B (1903) Uber die drehkrankheit der regenbogenforelle. Allgemeine Fischerei-Zeitung 28(1):7–8

Horne MT, Barnes AC (1999) Enteric redmouth disease (*Yersinia ruckeri*). In: Woo PTK, Bruno DW (eds) Fish diseases and disorders, vol. 3: viral, bacterial and fungal infections. CAB Int'l, Wallingford, pp 455–477

Kim HJ, Cuenca A, Olesen NJ (2018) Validation of a novel one-step reverse transcription polymerase chain reaction method for detecting viral haemorrhagic septicaemia virus. Aquaculture 492:170–183

Kim SJ, Qadiri SSN, Kim JO, Oh MJ (2019) Kinetics of infectious virus and viral RNA copy number in the blood of olive flounder infected with viral hemorrhagic septicemia virus (VHSV). Virus Res 267:16–20

Kimura N, Wakabayashi H, Kudo S (1978) Studies on bacterial gill disease in salmonids—I selection of bacterium transmitting gill disease. Fish Pathol 12(4):233–242

Kirpichnikov VS, Ilyasov I, Shart LA, Vikhman AA, Ganchenko MV, Ostashevsky AL, Simonov VM, Tikhonov GF, Tjurin VV (1993) Selection of Krasnodar common carp (*Cyprinus carpio* L.) for resistance to dropsy: principal results and prospects. In: Genetics in aquaculture. Elsevier, pp 7–20

Ksepka SP, Rash JM, Simcox BL, Besler DA, Dutton HR, Warren MB, Bullard SA (2020) An updated geographic distribution of *Myxobolus cerebralis* (Hofer, 1903) (Bivalvulida: Myxobolidae) and the first diagnosed case of whirling disease in wild-caught trout in the South-Eastern United States. J Fish Dis 43(7):813–820

Kumar G, Menanteau-Ledouble S, Saleh M, El-Matbouli M (2015) *Yersinia ruckeri*, the causative agent of enteric redmouth disease in fish. Vet Res 46:1–10

Leung TL, Bates AE (2013) More rapid and severe disease outbreaks for aquaculture at the tropics: implications for food security. J Appl Ecol 50:215–222

Madetoja J, Nystedt S, Wiklund T (2003) Survival and virulence of *Flavobacterium psychrophilum* in water microcosms. FEMS Microbiol Ecol 43(2):217–223

Mastan SA, Ahmad MO (2018) Observations on Saprolegnia infection in freshwater fishes of Lake Kolleru. Int J Fish Aquat Stud 6(1):102–105

Matthews RA (2005) *Ichthyophthirius multifiliis* Fouquet and ichthyophthiriosis in freshwater teleosts. Adv Parasitol 59:159–241

McAllister PE, Owens WJ (1992) Recovery of infectious pancreatic necrosis virus from the faeces of wild piscivorous birds. Aquaculture 106(3–4):227–232

Meyburgh CM, Bragg RR, Boucher CE (2017) *Lactococcus garvieae*: an emerging bacterial pathogen of fish. Dis Aquat Org 123(1):67–79

Mitchell SO, Rodger HD (2011) A review of infectious gill disease in marine salmonid fish. J Fish Dis 34(6):411–432

Mohamed SA (2020) Bacteriological and molecular studies on *Flexibacter columnaris* isolated from fishes. Ph.D. Thesis, Benha University

Morrison C, Cornick J, Shum G, Zwicker B (1981) Microbiology and histopathology of 'saddleback'disease of underyearling Atlantic salmon, *Salmo salar* L. J Fish Dis 4(3):243–258

Munang'andu HM, Sandtrø A, Mutoloki S, Brudeseth BE, Santi N, Evensen Ø (2013) Immunogenicity and cross protective ability of the central VP2 amino acids of infectious pancreatic necrosis virus in Atlantic salmon (*Salmo salar* L.). PLoS One 8(1):54263

Munro ES, Midtlyng PJ (2011) Infectious pancreatic necrosis and associated aquatic birnaviruses. In: Fish diseases and disorders. Volume 3: viral, bacterial and fungal infections. CABI, Wallingford, pp 1–65

Neish GA, Hughes GC (1980) Fungal diseases of fishes. THF Publication, Neptune, p 346

Niewiadomska K (2002) Family Diplostomidae Poirier, 1886. In: Keys to the Trematoda, vol 1. Cabi Publishing, Wallingford, pp 167–196

Noaman V, Chelongar Y, Shahmoradi AH (2010) The first record of *Argulus foliacesus* (Crustacea: Branchiura) infestation on lionhead goldfish (*Carassius auratus*) in Iran. Iran J Parasitol 5(2):71

Olesen NJ, Jorgensen PV (1982) Can and do herons serve as vector for Egtved virus? Bull Eur Assoc Fish Pathol 3:48

Pachade GR, Bhatkar NV, Hande DV (2014) Incidences of mycotic infections in *Channa punctatus* of Wadali lake, Amravati, MS, India. Int Res J Biol Sci 3(11):47–50

Park J, Roh H, Lee Y, Lee JY, Kang HY, Seong MJ, Kang Y, Heo YU, Kim BS, Park CI, Kim DH (2023) Characterization and pathogenicity of *Flavobacterium psychrophilum* isolated from rainbow trout (*Oncorhynchus mykiss*) in Korea. Microorganisms 11(10):2546

Pekmezci GZ, Yardimci B, Bolukbas CS, Beyhan YE, Umur S (2011) Mortality due to heavy infestation of *Argulus foliaceus* (Linnaeus, 1758) (Branchiura) in pond-reared carp, *Cyprinus carpio* L., 1758 (Pisces). Crustaceana 84(5–6):553–557

Plumb JA (2018) Health maintenance of cultured fishes: principal microbial diseases. Wiley-Blackwell, Hoboken

Qadiri SSN, Makesh M, Rajendran KV, Rathore G, Purushothaman CS (2019a) Specific immune response in mucosal and systemic compartments of *Cirrhinus mrigala* vaccinated against *Edwardsiella tarda*: in vivo kinetics using different antigen delivery routes. J World Aquacult Soc 50(4):856–865

Qadiri SSN, Kim SJ, Krishnan R, Kim JO, Kole S, Kim WS, Oh MJ (2019b) Localization and tissue tropism of viral haemorrhagic septicemia virus (VHSV) in experimentally infected juvenile olive flounder, *Paralichthys olivaceus*: an in situ hybridization and immunohistochemical study. Aquaculture 505:242–252

Qadiri SSN, Kim SJ, Krishnan R, Kim JO, Kim WS, Oh MJ (2019c) Development of an in-situ hybridization assay using riboprobes for detection of viral haemorrhagic septicemia virus (VHSV) mRNAs in a cell culture model. J Virol Methods 264:1–10

Qadiri SSN, Kim SJ, Krishnan R, Kole S, Oh MJ (2020) Early detection and localization of viral haemorrhagic septicemia virus (VHSV) genomic mRNAs in external tissues of juvenile olive flounder (*Paralichthys olivaceus*) following an immersion challenge. Aquaculture 518:734859

Reddacliff LA, Whittington RJ (1996) Pathology of epizootic haematopoietic necrosis virus (EHNV) infection in rainbow trout (*Oncorhynchus mykiss* Walbaum) and redfin perch (*Perca fluviatilis* L). J Comp Pathol 115(2):103–115

Richards CA, Murphy CA, Brenden TO, Loch TP, Faisal M (2017) Detection accuracy of *Renibacterium salmoninarum* in Chinook salmon, *Oncorhynchus tshawytscha* (Walbaum) from non-lethally collected samples: effects of exposure route and disease severity. Prev Vet Med 145:110–120

Rigos G, Stevenson R (2001) The effect of antibiotic treatment on the establishment of persistent infection with *Yersinia ruckeri* Serovar II in rainbow trout *Oncorhynchus mykiss* (Walbaum). Aquac Int 9(3):247–253

Riwidiharso E (2020, November) Diversity of Trichodina spp. on the larvae of freshwater fish in Banyumas and surrounding areas of Central Java. IOP Conf Series Earth Environ Sci 593(1):012039

Roberts RJ (2012) Fish pathology, 4th edn. W.B. Saunders, Philadelphia, pp 391–392

Saha M, Bandyopadhyay PK, Göçmen BAYRAM (2017) First record of ectoparasitic ciliates, of genus Trichodina (Ciliophora: Trichodinidae) parasiting cultured Oranda Gold Fish (*Carassius auratus auratus* L.) in India. Zootaxa 4319(1):128–142

Saint-Jean SR, Borrego JJ, Perez-Prieto SI (2001) Comparative evaluation of five serological methods and RT-PCR assay for the detection of IPNV in fish. J Virol Methods 97(1–2):23–31

Sanders JE, Fryer JL (1980) *Renibacterium salmoninarum* gen. nov., sp. nov., the causative agent of bacterial kidney disease in salmonid fishes. Int J Syst Evol Microbiol 30(2):496–502

Sarker S, Kallert DM, Hedrick RP, El-Matbouli M (2015) Whirling disease revisited: pathogenesis, parasite biology and disease intervention. Dis Aquat Org 114(2):155–175

Schäperclaus W (1938) Die schädigungen der deutschen fischerei durch fischparasiten und fischkrankheiten. Fischerei-Zeitung 41:267–270

Sen K, Mandal R (2018) Fresh-water fish diseases in West Bengal, India. Int J Fish Aquat Stud 6(5):356–362

Shinn AP, Avenant-Oldewage A, Bondad-Reantaso MG, Cruz-Laufer AJ, García-Vásquez A, Hernández-Orts JS, Kuchta R, Longshaw M, Metselaar M, Pariselle A, Pérez-Ponce de León G (2023) A global review of problematic and pathogenic parasites of farmed tilapia. Rev Aquac 15:92–153

Singh S, Mallik SK, Kala K, Shahi N, Pathak R, Giri AK, Chandra S, Pant K, Patiyal RS (2021) Characterization of *Flavobacterium columnare* from farmed infected rainbow trout, *Oncorhynchus mykiss* (Walbaum, 1792) of Central Indian Himalayan region. India Aquaculture 544:737118

Skjesol A, Aamo T, Hegseth MN, Robertsen B, Jørgensen JB (2009) The interplay between infectious pancreatic necrosis virus (IPNV) and the IFN system: IFN signaling is inhibited by IPNV infection. Virus Res 143(1):53–60

Smail DA, Snow M (2011) Viral haemorrhagic septicaemia. In: Fish diseases and disorders. Volume 3: viral, bacterial and fungal infections. CABI, Wallingford, pp 110–142

Soltani M, Baldisserotto B, Hosseini Shekarabi SP, Shafiei S, Bashiri M (2021) Lactococcosis a re-emerging disease in aquaculture: disease significant and phytotherapy. Vet Sci 8(9):181

Starliper CE (2011) Bacterial Coldwater disease of fishes caused by *Flavobacterium psychrophilum*. J Adv Res 2(2):97–108

Starliper CE (2012) 1.2. 1 Bacterial gill disease. In: American Fisheries Society (AFS) fish health section blue book

Sudova E, Machova J, Svobodova Z, Vesely T (2007) Negative effects of malachite green and possibilities of its replacement in the treatment of fish eggs and fish: a review. Vet Med 52(12):527

Tandel RS, Chanu KV, Hussain Bhat RA, Dash P, Shah TK, Thakuria D (2021) Morphometric and molecular identification of *Argulus japonicus* (Thiele 1900) in vulnerable Himalayan snow trout, *Schizothorax richardsonii* (Gray 1832). Aquac Res 52(12):6770–6778

Tapia D, Eissler Y, Reyes-Lopez FE, Kuznar J, Yáñez JM (2022) Infectious pancreatic necrosis virus in salmonids: molecular epidemiology and host response to infection. Rev Aquac 14(2):751–769

Tripathi YR (1954) Studies of parasites of Indian fishes: protozoa (Mastigophora and Ciliophora). Rec Indian Museum *52(2/4)*:221–230

Valladao GMR, Gallani SU, De Padua SB, Laterça M, Pilarski F (2014) *Trichodina heterodentata* (Ciliophora) infestation on Prochilodus lineatus larvae: a host–parasite relationship study. Parasitology 141(5):662–669

Valladão GMR, Alves LDO, Pilarski F (2016) Trichodiniasis in Nile tilapia hatcheries: diagnosis, parasite: host-stage relationship and treatment. Aquaculture 451:444–450

Van West P (2006) *Saprolegnia parasitica*, an oomycete pathogen with a fishy appetite: new challenges for an old problem. Mycologist 20(3):99–104

Vendrell D, Balcázar JL, Ruiz-Zarzuela I, De Blas I, Gironés O, Múzquiz JL (2006) *Lactococcus garvieae* in fish: a review. Comp Immunol Microbiol Infect Dis 29(4):177–198

Wakabayashi H (1991) Effect of environmental conditions on the infectivity of *Flexibacter columnaris* to fish. J Fish Dis 14(3):279–290

WOAH (World Health Organisation for Animal Health) (2019a) Aquatic manual for EHNV. OIE, Paris

WOAH (World Health Organisation for Animal Health) (2019b) Aquatic manual for Gyrodactylus. OIE, Paris

WOAH (World Health Organisation for Animal Health) (2021a) Aquatic manual for IHNV. OIE, Paris

WOAH (World Health Organisation for Animal Health) (2021b) Aquatic manual for VHSV. OIE, Paris

WOAH (World Health Organisation for Animal Health) (2023) Aquatic manual for SVCV. OIE, Paris

Wolf K (1988) Infectious hematopoietic necrosis. In: Fish viruses and fish viral diseases. Cornell University Press, Ithaca, pp 83–114

Wolf K, Quimby MC (1962) Established eurythermic line of fish cells in vitro. Science 135(3508):1065–1066

Wood EM, Snieszko SF, Yasutake WT (1955) Infectious pancreatic necrosis in brook trout. Arch Pathol 60:26–28

Xu DH, Zhang QZ, Zhang D (2016) Two in vitro methods for screening potential parasiticides against *Ichthyophthirius multifiliis* using *Tetrahymena thermophila*. J Fish Dis 39(3):285–294

Yu SE, Dong SL, Zhang ZX, Zhang YY, Sarà G, Wang J, Dong YW (2022) Mapping the potential for offshore aquaculture of salmonids in the Yellow Sea. Mar Life Sci Tech 4(3):329–342

Diseases in Cage Aquaculture

Gayatri Tripathi and R. Bharathi Rathinam

Abstract

Cage aquaculture in inland waters is being regarded as an opportunity to utilize prevailing reservoirs with great production potential to augment fish production. Indian reservoirs with a water spread area of 3.51 million ha offer an enormous opportunity to enhance productivity through enclosure culture. This untapped potential can be utilized in a well-planned manner to increase aquaculture production. Intensification of aquaculture production results in an increased level of stress in fishes, which ultimately increases the disease susceptibility of fishes. Diseases are the major constraint to the aquaculture industry, as they lead to huge losses to farmers. Two broad categories of diseases affect fish: infectious and non-infectious diseases. Bacterial, fungal, viral, and parasitic diseases are of an infectious origin and are widespread among freshwater fish populations; if not controlled early on, they may significantly reduce fish production. Aquatic animal diseases also depend on abiotic stress factors such as adverse environmental conditions and management practices. Cage environments are rich in nutrients that are a good substrate for heterotrophic bacterial and fungal growth, especially when hygiene practices are not carried out adequately. High stocking density, limited fish movement, and poor water quality parameters are the major stressors that fish encounter in the cage culture system. A comprehensive summary of the infectious and non-infectious diseases commonly encountered in cage-farmed freshwater fishes is provided here.

G. Tripathi (✉) · R. Bharathi Rathinam
ICAR-Central Institute of Fisheries Education, Mumbai, India
e-mail: gayatrit@cife.edu.in; rbharathi@cife.edu.in

R. Kooloth Valappil et al. (eds.), *Aquatic Animal Health Management*,
https://doi.org/10.1007/978-981-96-7987-4_4

Keywords

Cage culture · Infectious diseases · Freshwater aquaculture · Pathogens · Disease incidences

1 Introduction

Originally, cages were used for holding and transporting fish in small enclosures for a short duration. The earliest evidence of cage fish farming was recorded in the inland water bodies of Kampuchea in the late 1800s (Pantulu 1979). However, of late, it has emerged as the key to achieving India's Blue Revolution (NFDB 2016). Marine cage fish farming commenced with the successful commercial culture of yellow tail in Japan in the 1960s. Subsequently, Thailand, Malaysia, and Korea established cage culture as a successful farming industry in the 1970s, 1980s, and 1990s, respectively (Gopakumar 2009). It is estimated that 80–90% of marine fish culture production in Asia possibly comes from cage farming. However, it is difficult to determine the exact production data from cage farming. Freshwater cage fish farming seems to have originated independently in at least two countries, Cambodia and Indonesia. The traditional method of cage culture practised in Cambodia was widely adopted by South-East Asian countries such as Vietnam, Thailand, and other Indo-Chinese countries. On the other hand, in Indonesia, floating bamboo cages have been used to culture cyprinid fry collected from the wild since 1922 in Jambi (Reksalegora 1979). Besides, Vass and Sachlan (1957) reported a distinct method of cage culture in Java, Indonesia, where they have been using submerged bamboo cages as enclosures for carp culture since the early 1940s. At present, it is the most predominant culture system of freshwater aquaculture in Asia (De Silva and Phillips 2007). Freshwater cage fish farming is also reported as an efficient culture system compared to pond, tank, and pen farming systems in Malaysia (Iliyasu et al. 2016).

Cage culture is considered a high-return, low-impact farming practice (Rao et al. 2013). Indian reservoirs with a water spread area of 3.51 million ha have enormous scope for enhancing fish production through enclosure culture (Hassan et al. 2017). The predominance of reservoirs in the peninsular states, namely, Tamil Nadu, Karnataka, Andhra Pradesh, Kerala, Orissa, and Maharashtra, accounts for more than 56% of the total reservoir area in the country. This untapped potential can be utilized in a well-planned manner to increase aquaculture production.

In India, cage culture was introduced in 2007 by ICAR-CMFRI, and within a decade, it was widely accepted by the fishers and farmers of Kerala in open seas and brackish waters. ICAR-Central Inland Fisheries Research Institute (ICAR-CIFRI) has also developed cage farming techniques and installed cages in reservoirs, rivers, and lake systems in the northern states of India. Moreover, the National Fisheries Development Board (NFDB) has also implemented several projects on cage culture and provides financial assistance to encourage this practice in various states of

India. Success stories of commercial cage farming of pangasius and tilapia in the Tilaiya and Chandil reservoirs in India have attracted farmers towards inland cage fish farming. Due to its ability to attain higher production levels and the scope for wider prospects, cage culture is considered the "future of fisheries" to increase overall fish production in India (NFDB 2017–2018).

2 Diseases in Cage Culture Systems

Disease outbreaks in aquaculture are of great concern as they can cause substantial economic losses to farmers. The intensification and diversification of aquaculture practices introduce various health problems during the culture period, which may develop into a significant constraint. In general, the relative frequency of disease outbreaks was found to be higher in cage-cultured fishes than in pond-raised ones (Masser et al. 1991). Cage environments are rich in nutrients that can be a good substrate for heterotrophic bacterial and fungal growth, especially when hygiene practices are not carried out adequately. High stocking density, limited fish movement, and poor water quality are the major stressors that fish encounter in the cage culture system. Moreover, such an environment is suitable for the emergence, establishment, and transmission of new pathogens (Murray and Peeler 2005).

Transmission of infections from cultured to wild populations has been observed to be more widespread in cage farms, as the contact between fishes is greater in cage sites (Morton et al. 2005). Two types of diseases can occur in cage-cultured fishes: first, diseases brought into the system via candidate culture species, and, second, already existing diseases that get exacerbated by the environmental conditions within the system or mechanical defects of the cages (Okaeme et al. 1990). Mishra et al. (2017) reported that bacterial and fungal diseases are more prevalent in cage culture systems compared to parasitic and viral diseases. In this chapter, bacterial, parasitic, viral, fungal, and non-infectious diseases reported in freshwater cage farming systems are discussed in detail.

3 Bacterial Diseases

Due to their rich nutrient profile, cage environments support the growth of numerous heterotrophic bacteria. At the same time, most of the heterotrophic bacteria associated with the aquaculture system are opportunistic pathogens that act as disease-causing agents under stress conditions. It is reported that the major disease outbreaks in cage culture systems are due to bacterial and fungal infections (Chitmanat et al. 2016; Mishra et al. 2017). In general, infections caused by Gram-negative bacterial pathogens are more common in cage-cultured fishes than those caused by Gram-positive pathogens (Table 1).

Table 1 Bacterial diseases reported in freshwater cage fish farming

Disease	Causative agent	Host species	Clinical symptoms
Gram-negative bacterial infections			
Bacillary necrosis of pangasius (BNP)/ enteric septicaemia of catfish (ESC)	*Edwardsiella ictaluri*	Channel catfish, pangasius	Exophthalmia, pale gills, abdominal distension, multifocal white lesions on several organs, haemorrhages, and necrosis of the internal organs
Motile *Aeromonas* septicaemia	Group of motile, *Aeromonas* spp. (*A. hydrophila*, *A. sobria*, and *A. caviae*)	Pangasius, tilapia, channel catfish, sand goby	Haemorrhagic septicaemia, lethargy and loss of appetite, haemorrhage on the mouth, fins and caudal peduncle, and around the eyes, exophthalmia, and distended liver
Columnaris (mouth fungus)	*Flavobacterium columnare*	Tilapia, common carp, channel catfish	Gill rot and yellow erosions, paleness of the gills and internal organs, gasping on the surface
Carp erythrodermatitis	*Aeromonas hydrophila*	Common carp	Skin ulcers, dropsy, loss of scales, exophthalmia, haemorrhages on the skin, gill, and fin base
Furunculosis	*Aeromonas salmonicida*	Salmonids	Boils like furuncles on the dorsum and haemorrhagic areas
Enteric red mouth	*Yersinia ruckeri*	Salmonids, rainbow trout	Dark colouration, blindness, lethargy, ulceration and haemorrhage around the oral cavity, congregation of fishes near the edges of ponds
Bacterial coldwater disease/peduncle disease	*Flavobacterium psychrophilum*	Rainbow trout, salmonids	Spiral swimming behaviour, pale gills, haemorrhage and dark pigmentation on the caudal peduncle
Edwardsiellosis	*Edwardsiella tarda*	Channel catfish, Nile tilapia, cyprinids	Haemorrhagic ulcerations on the body, swollen abdomen and lesions on the caudal peduncle
Pseudomonas septicaemia	*Pseudomonas fluorescens* and *Plesiomonas shigelloides*	Channel catfish, Nile tilapia, cyprinids	Ascites, exophthalmia, and ulcers. Similar clinical symptoms during *Aeromonas* septicaemia infection

(continued)

Table 1 (continued)

Disease	Causative agent	Host species	Clinical symptoms
Gram-positive bacterial infections			
Streptococcal septicaemia/ meningoencephalitis	*Streptococcus agalactiae*, *Streptococcus Constellatus*, *Streptococcus iniae*, *Streptococcus ictaluri*, *Streptococcus difficile*, *Streptococcus shiloi*	Red tilapia, Mozambique tilapia	Petechial haemorrhages on the skin, abdominal swelling with ascites, darkening and discolouration of unilateral and bilateral exophthalmia without corneal opacity
Streptococcosis	*Streptococcus agalactiae*, *S. iniae*, *Streptococcus dysgalactiae*	Tilapia	Haemorrhagic septicaemia, lethargy, haemorrhagic eyes, corneal opacity in the eyes, exophthalmia, swollen kidney and erratic spiral swimming motion
Bacterial kidney disease	*Renibacterium salmoninarum*	Salmonids	Exophthalmia, lethargy, ascetic fluid accumulation, dark discolouration and haemorrhages at the fin bases
Clostridium sp.	*Clostridium* sp.	Pangasius catfish	Haemorrhages along the ventral fin base, swollen kidney and spleen, and many necrotic nodules in the liver, kidney, and spleen
Acid-fast bacterial infections			
Francisellosis	Francisella-like organism	Tilapia, catfish, cyprinids	Clinical signs are not reported properly; white foci at the internal organs are observed
Salmonid rickettsial syndrome/ piscirickettsiosis	*Piscirickettsia salmonis*	Salmonids	Skin ulcers and haemorrhage
Nocardiosis	*Nocardia* spp.	Freshwater fishes	Skin ulceration, nodules over the skin, opercular erosion and white nodules at the gill filaments

3.1 Motile *Aeromonas* Septicaemia (MAS)

Motile *Aeromonas* septicaemia (MAS), caused by *Aeromonas hydrophila*, belongs to the family Aeromonadaceae, which is regarded as the most common and troublesome disease in the intensive culture system. Scale protrusion disease, red spot

Fig. 1 Fish exhibiting clinical signs of aeromoniasis

disease (RSD) or red pest, and haemorrhagic disease are the other names of motile *Aeromonas* septicaemia (Fig. 1). Aeromonads are opportunistic pathogens, in that they cause infection in fish that are already suffering from other diseases. They also form a part of the intestinal flora of healthy fish.

A. hydrophila is motile, has a polar flagellum, and is ubiquitous in freshwater ecosystems; however, it is abundant in water with a high organic load. Wounds are considered a major predisposing factor (Zhang et al. 2016) along with stress factors such as overcrowding, low oxygen, poor water quality, heavy parasitic burden, pH disturbances, high temperature, rough handling and transport, etc., which can lead to a disease outbreak. Fry and fingerlings are more susceptible to *A. hydrophila* infection; however, it can happen in all age groups. Likewise, warm water and temperate species are more likely to be affected than cold water fishes. Exophthalmia, abdominal distension, pale gills, and oedema in scale pockets are the clinical manifestations associated with *A. hydrophila* infection. This infection can cause 100% mortality under certain circumstances. Pathological conditions such as dermal ulceration, tail or fin rot, ocular ulcerations, and erythrodermatitis have also been noticed in infected fishes. Carp, catfish, eel, goldfish, tilapia, etc. are the commercially important fishes that are mostly affected by this bacterial infection. Chemical treatment with potassium permanganate ($KMnO_4$) is occasionally effective against the infection, and medicated feed such as Terramycin and Romet (antibiotic-based) are also commercially available. However, identifying and eliminating the stress factors associated with the disease will help overcome it (Camus et al. 1998).

3.2 Edwardsiellosis

Edwardsiella tarda is the aetiological agent of edwardsiellosis. It belongs to the family Enterobacteriaceae. *Edwardsiella piscicida* is a newly described species that differentiates fish-pathogenic strains from those isolated from humans. *Edwardsiella ictaluri* is another major pathogen in this family that is reported to cause severe infections in fish. It is on the list of the important pathogens associated with the freshwater cage culture system (Wamala et al. 2018). *E. ictaluri* is the causative agent of bacillary necrosis of pangasius, which is considered a significant pathogen that causes economic loss to the catfish industry (Ferguson et al. 2001). It is also responsible for enteric septicaemia of catfish (ESC) or hole-in-the-head disease,

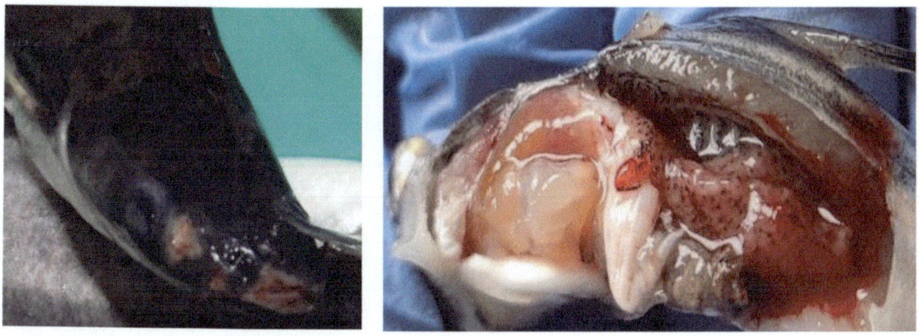

Fig. 2 Fish infected with edwardsiellosis

which is a highly fatal, seasonal, and systemic disease. The bacterium is facultatively anaerobic, Gram-negative, straight rod, motile with flagella, and can also be intracellular during infection. It is mainly associated with high water temperatures and organic pollution. Although fish of all ages are susceptible to *E. ictaluri* infection, high mortalities have been reported among fingerlings and juveniles.

Fish affected by edwardsiellosis exhibit ulcerative abscesses in the internal organs, haemorrhagic ulcers on the skin, fin base, and body surface, rectal protrusion, and anaemia. The external signs of the infection are lesions on the skin and mouth, pale gills, excessive mucus secretion, scale erosion, ulcers, and swelling and bleeding of the anus (Fig. 2). The bacterium causing edwardsiellosis can be found in pond water, mud, and fish intestines. Freshwater fishes such as *Pangasius hypophthalmus*, tilapia (*Oreochromis niloticus*), and farm-raised catfish (*Ictalurus punctatus*) are highly infection-prone to edwardsiellosis (Mohanty and Sahoo 2007; Park et al. 2012).

3.3 *Pseudomonas* Septicaemia

Pseudomonas fluorescens, *Pseudomonas aeruginosa*, and *Pseudomonas putida* are the causative agents of haemorrhagic septicaemia of freshwater fishes. Pseudomonads are opportunistic bacterial pathogens associated with the cage culture system. *Pseudomonas* septicaemia is an acute septicaemic bacterial disease characterized by high mortality in acute conditions and ascites in chronic conditions. All fish species are susceptible to this infection irrespective of age. *Pseudomonas aeruginosa*, a known human pathogenic organism, is responsible for considerable economic losses in the commercial cultivation of tilapia (Thomas et al. 2014). It was also found to be responsible for causing ulcerative lesions in catfish (Khairnar et al. 2013). Some of the predisposing factors of this bacterial infection are overcrowding, reduced dissolved oxygen (DO) levels, excess pH, and the presence of physical stress. In acute cases, fish lose equilibrium and muscle control and float in an upward position. As the disease progresses, large light-coloured patches on the body surface develop into large ulcerous lesions that penetrate the musculature (Fig. 3). Common

Fig. 3 Fish infected with *Pseudomonas* septicaemia

clinical signs include fin erosion at the bases, haemorrhages and lesions on the body surface, loosening of scales, mucous secretion, and shallow-to-deep ulcers. Preventive treatment, along with good management practices, will help control *Pseudomonas* septicaemia (Austin and Austin 2007).

3.4 Streptococcosis

Streptococcus iniae, capable of causing invasive disease outbreaks in aquaculture farms, is the causative agent of streptococcosis. In some cases, *Streptococcus agalactiae* and *Lactococcus garvieae* have also been identified as the causative agents. The Streptococcaceae family comprises a diverse group of bacterial pathogens that can infect a wide range of hosts. *Streptococcus acidominimus*, *Streptococcus agalactiae*, *Streptococcus constellatus*, *Streptococcus dysgalactiae*, *Streptococcus suis*, and *Streptococcus uberis* have also been isolated from cage-cultured tilapia (Ismail et al. 2016). The bacterium is a Gram-positive coccus in chains and has zoonotic importance as well. Tilapia, yellow tail, pangasius, cyprinids, cichlids, characids, and monodactylids are the common hosts of this pathogen.

The most pronounced external sign of this disease is exophthalmia, due to which it is also known as "pop eye" (Fig. 4). Ascites in the body cavity, fish gathering in the corners of the cage, erratic swimming, pale liver, discoloured gills, an enlarged spleen, external haemorrhages, and eye abnormalities, such as blindness, corneal opaqueness, and sunken eyes, are also reported as clinical symptoms. Streptococcosis has been reported in several fish species worldwide, contributing to an annual loss of about USD 150 million (Amal and Zamri-Saad 2011). It has also been reported in cage-cultured freshwater and marine fishes in different countries. Streptococcosis caused by several *Streptococcus* species, viz., *S. agalactiae* (70%), *Leuconostoc* spp., and *Streptococcus constellatus* (30%), has also been found to be involved in the infection of cage-cultured red tilapia in Malaysia (Baker 2008).

Fig. 4 Fish infected with streptococcosis showing exophthalmia

Fig. 5 Fish infected with columnaris disease

3.5 Columnaris Disease

Columnaris is the major disease caused by the pathogenic bacterium *Flavobacterium columnare*. Saddleback disease, tail rot, cotton mouth, and mouth fungus are the common names of this disease. It is considered the second important reason for mass mortality in catfish cultured in ponds in the United States. Durborow et al. (1998) reported that the frequency of columnaris disease is higher in fishes farmed in intensive culture systems such as cage culture and closed recirculation systems. The pathogen, *F. columnare*, causes acute-to-chronic infections and typically affects the gills, skin, and fins (Fig. 5). It is widely distributed in freshwater environments and can infect wild and farmed freshwater fish species such as carp, channel catfish, goldfish, eel, perch, salmonids, and tilapia. It usually occurs when the water temperature exceeds 15 °C. The key clinical signs are lesions on the body surface causing skin erosions and gill necrosis. Characteristic "saddleback" lesions can be found in advanced cases, especially in catfish. A preliminary, presumptive diagnosis can be based on the observation of long, slender, flexing rods forming "haystacks" or columns in wet mount preparations made from lesions (Declercq et al. 2013).

4 Viral Diseases

Viral diseases are known to cause heavy mortality in fish farms, especially in fry and fingerlings. Plumb (1999) listed the factors that assist the development of viral diseases, such as high stocking density, water temperature, poor water quality, handling stress, age of fish, and deprived nutrition. However, the influence of water

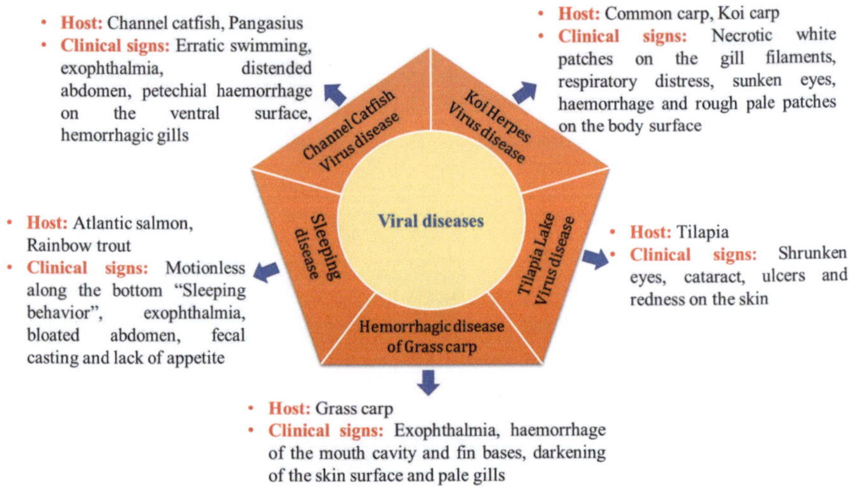

Fig. 6 Viral diseases reported in freshwater cage fish farming

quality, temperature, and crowded conditions also facilitate viral epizootics in open water culture systems. Tilapia lake virus (TiLV) is the recently reported viral disease that has been causing heavy mortality in tilapia cage farming. Disease outbreaks of the koi herpesvirus (KHV) have been reported to cause massive mortality in cage-cultured carp. The important viral diseases encountered in freshwater cage aquaculture are given in Fig. 6.

4.1 Tilapia Lake Virus (TiLV)

Tilapia lake virus (TiLV) is an emerging virus causing disease in farmed tilapia species (*Oreochromis niloticus*). It is a negative-sense RNA virus with 10 segments (Orthomyxo-like) belonging to the family Orthomyxoviridae and has been reported in Israel, Ecuador, Colombia, and Egypt (FAO 2018). Syncytial hepatitis of tilapia (SHT) and tilapia 1-month mortality syndrome are the other names of the disease caused by TiLV. It is considered a significant threat to the tilapia aquaculture industry worldwide, causing almost 90% mortality (Surachetpong et al. 2017). It is reported that TiLV disease can occur in water temperatures ranging from 15 to 30 °C; however, the favourable water temperature associated with disease spread is 25 °C.

All age groups are susceptible to this infection. The clinical manifestations noted are loss of appetite, lethargy and reduced schooling behaviour, darkening of the skin, ulcers on the skin surface, loss of scales or protruding scales, pale gills, opacity in the eyes, and fluid accumulation in the body (Fig. 7). Occasionally, infected fish may exhibit "pop eye" and a reddened or ulcerated skin surface, often leading to secondary bacterial infections. Typical histopathological alterations observed in the brain and liver of TiLV-infected fishes are haemorrhages, congestion of blood

Fig. 7 Fishes infected with the tilapia lake virus

vessels, and infiltration of inflammatory cells within the affected tissues. Apart from this, the presence of syncytial cells (multinucleated giant cells) is also seen in the liver tissue (Tattiyapong et al. 2017). Currently, there are no effective treatments, vaccination protocols, or medications to control the impact of a TiLV outbreak. Good management practices, genetic selection, and proper quarantine measures may help reduce the spread of this disease.

4.2 Koi Herpesvirus (KHV)

The koi herpesvirus or carp interstitial nephritis and gill necrosis virus (CNGV) is an enveloped dsDNA virus belonging to the family Alloherpesviridae (Dishon et al. 2005). A disease outbreak due to KHV was originally reported in Israel in 1998. Mass mortality of cage-farmed common carp due to KHV disease was observed in Indonesia and later in Japan (Sunarto et al. 2005). Water temperature is regarded as the major external factor involved in the pathogenesis of this virus. The optimum temperature range in which KHV is reported to cause mortality is 18–25 °C. Clinical symptoms noticed are necrotized white patches on the gills, sunken eyes, excess mucus production, and haemorrhage on the body surface. Major histopathological alterations are observed in the gill tissue, including inclusion bodies, hyperplasia, lamellar fusion, focal necrosis, and epithelial cell degeneration. Generally, 80–90% of mortality is observed in naturally infected susceptible fishes. Standardized vaccination protocols have proven effective against KHV infection in fishes. Live attenuated licensed vaccine (KoVax Ltd., Israel) (Dhar et al. 2014) and formalin-inactivated vaccine formulations are commercially available (Yasumoto et al. 2006).

5 Parasitic Diseases

The rapid development of fish culture in cages is also associated with the emergence of parasitic diseases. There is a paucity of information on the parasitic diseases encountered in cage culture systems. Disease outbreaks and mass mortality due to infections caused by trematodes, viz., *Gyrodactylus* spp. and *Dactylogyrus* spp., are commonly reported in cage culture (Mishra et al. 2017). Apart from this, pathogenic protistans, which are commonly reported in freshwater fishes reared in cages, include myxosporeans, trichodinids, and dinoflagellates. A list of the parasitic diseases observed in freshwater cage-cultured fishes is provided in Table 2.

5.1 Dactylogyrosis

As a monogenean trematode parasite, *Dactylogyrus* sp. Is among the most common parasites associated with fish culture (Fig. 8). It can cause mass mortality in earlier life stages of freshwater fishes, most frequently in cyprinids. Severe gill damage occurs due to the attachment of monogenean, which paves the way for secondary bacterial infections, leading to massive mortality. It is an oviparous, spindle-shaped parasite with or without 4 eyespots, 2 anchors, and 14 marginal hooks in its haptor (Fig. 8). These morphological features are used to identify the organism. Rubbing the body against a cage net, devoid of scales at the forehead (above the eyes), pale to greyish gills, anorexia, lethargy, swimming on the water surface and gasping for air, loss of appetite, and excess mucus production are the non-specific clinical signs associated with dactylogyrosis infection in fishes. Dip or bath treatment in salt, formalin, and potassium permanganate are used as a chemotherapeutic method to treat (Willomitzer 1980) this infection. The dose and concentration depend upon the host, the physico-chemical parameters of water, and the parasitic species. However, the effectiveness of such treatments in large water bodies is less compared to that in pond culture. Limiting stocking density, procuring healthy hatchlings from hatcheries, providing good quality feed, and disinfecting nets will help farmers reduce the infection level (Woo and Gregory 2014).

Table 2 List of parasitic diseases reported in freshwater cage fish farming

Disease	Causative agent	Host species	Clinical symptoms
Trichodinosis	*Trichodina* infection	Tilapia, channel catfish, pangasius	Flashing, mottling skin, rapid breathing, weakness, epidermal injuries, and high mortality during intensive stocking
Myxozoan infection	*Hennegoides berlandi, Hennegoides malayensis, Hennegoides pangasii, Henneguya shariffi, Myxobolus baskai*, and *Myxobolus pangasii*	Pangasius, Asian red tail catfish, channel catfish, mad barb	Plasmodia in the multi-layered gill filaments, haemorrhagic gills, increased mucus production, and respiratory distress
Whirling disease	*Myxobolus cerebralis*	Salmonids	Skeletal deformity and blackening of the tail
Ceratomyxosis	*Ceratomyxa shasta*	Salmonids	Anorexia, exophthalmia, darkening, and lethargy
Epistylis infection	*Epistylis* sp.	Pangasius catfish	Skin and gill wounds
Ciliate protozoan infection	*Balantidium, Ichthyonyctus*	Pangasius, grass carp	Emaciation and poor growth
Parasitic isopod	*Alitropus typus*	Cyprinids, catfish, tilapia	Poor growth and increased fish mortality
Coccidiosis	*Eimeria coccidia*	Cyprinids, catfish, gouramies	Poor growth and emaciation
Dactylogyrulosis	*Dactylogyrus* sp.	Common carp, catfish	Haemorrhagic gills, respiratory dysfunction, pale to greyish gills, anorexia, and gasping on the surface
Gyrodactylosis	*Gyrodactylus salaris*	Salmonids, common carp, channel catfish	Flashing, increased mucus production, increased epidermis thickening, and tissue necrosis
Bothiro cephalosis	*Bothriocephalus acheilognathi*	Common carp	Distended abdomen, ascetic fluid accumulation, and retarded growth
Fish borne trematodes	*Haplorchis pumilio, Centrocestus armatus*	Pangasius catfish	Excessive mucus secretion, gasping on the surface
Costiasis	*Lchthyobodo necatrix*	Channel catfish	Gasping on the water surface, excess mucus secretions, and pale gills

(continued)

Table 2 (continued)

Disease	Causative agent	Host species	Clinical symptoms
Plerocercoid stage of tapeworms	*Triaenophorus nodulosus*	Common carp, catfish	Not reported
Cestodes	*Diphyllobothrium dendriticum*, *Eubothrium crassum*	Salmonids, grass carp, pangasius	Retarded growth, erratic swimming behaviour, dropsy, ascites accumulation, and distended abdomen
Ichthyophthiriasis	*Ichthyophthirius multifiliis*	Tilapia, channel catfish	Anorexia, lethargy, white spots on the skin and gills
Velvet disease	*Piscinoodinium pillulare*	Cyprinids	Yellow-to-rust-coloured skin, dense mucus covering, anorexia, and skin ulcers
Oodinium infection	*Amyloodinium*	Catfish, tilapia	Gasping on the water surface, excess mucus secretions, and pale gills
Proliferative kidney disease	*Tetracapsuloides bryosalmonae*	Salmonids	Distended abdomen, exophthalmia, pale gills, respiratory distress, and dark colouration
Ergasilosis	*Ergasilus* sp.	Catfish	Gill dysfunction, poor growth, and white or green threads hanging out of the fish
Lernaeosis	*Lernaea cyprinacea*	Cyprinids, snakeheads, catfish	Haemorrhages on the skin surface, muscle and dermal necrosis at places, and blindness
Argulosis	*Argulus japonicus*, *Argulus foliaceus*, *Argulus indicus*, *Argulus siamensis*	Cyprinids, tilapia, anabantids	Lethargy, cease to feed, rubbing on the substrate, and haemorrhagic inflamed wounds

5.2 Trichodiniasis

Trichodiniasis is caused by urecolariid ciliates, which are protozoans that mainly attach to the gills and skin of fish, where they feed on epithelial cells, bacteria, and erythrocytes. *Trichodina*, *Trichodinella*, *Chilodinella*, *Glossatella*, and *Tripartiella* (Basu and Haldar 1998) are the different genera of parasitic fauna reported to be associated with trichodiniasis. Indian major carps, along with other carps, are

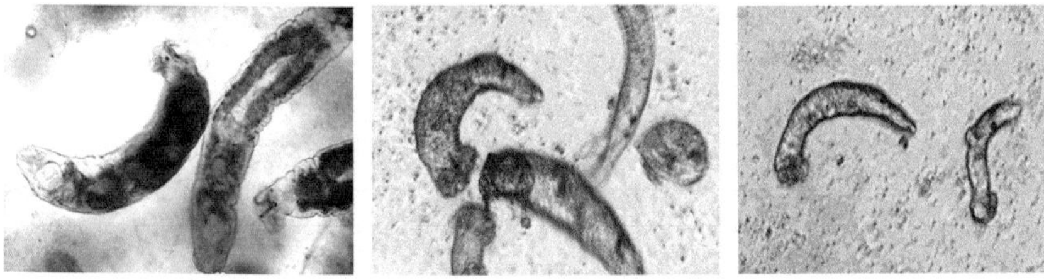

Fig. 8 *Dactylogyrus* sp. isolated from infected fish

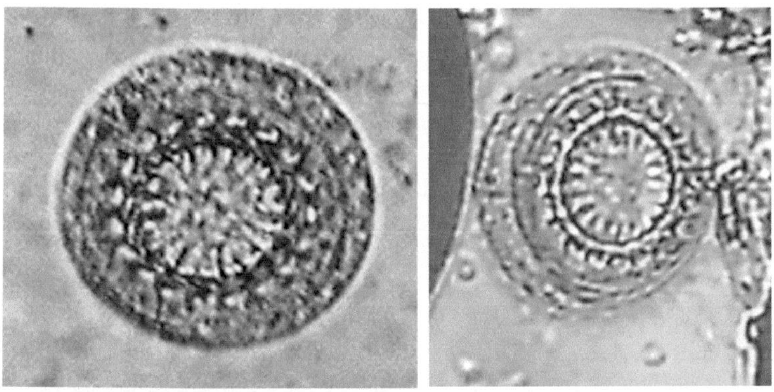

Fig. 9 *Trichodina* sp. isolated from infected fish

majorly affected by this infection. These parasites are prevalent in open water bodies. High stocking density in cages increases the host's availability to the parasite. Mcardle (1984) recorded trichodina as a cause of mortality in cage-cultured salmon and rainbow trout. It is a disc-shaped parasite, ubiquitous in most fishes in both freshwater and saltwater (Fig. 9). It is transmitted via contact and contaminated water. Clinical signs noted in infected fishes are lethargy, anorexia, dark colouration, necrosis and ulcers on different parts of the body, detached scales, frayed fins, erratic swimming, scratching against the wall, and excessive mucus accumulation (Tantry et al. 2016). Moreover, the epidermal injuries caused by *Trichodina* spp. increase the possibility of secondary (opportunistic) bacterial infections. Heavy infestation leads to extremely high mortality in young fish.

5.3 Ichthyophthiriasis

"Ich" or "white spot" disease is caused by the ciliate protozoan parasite *Ichthyophthirius multifiliis*. It produces numerous pinhead-sized white spots on the skin, fins, and gills; hence, it is commonly called white spot disease. It results in large-scale mortalities in freshwater fishes and causes significant economic damage to fish farmers (Jørgensen and Buchmann 2020). *I. multifiliis* is a holotrichous ciliate protozoan that burrows into the skin via its rotating cilia. It stays within the epidermis of the host until it leaves the fish. The trophont, or mature stage of the parasite, has a large "horseshoe"-shaped nucleus, and the entire surface of the parasite is covered with cilia. This structure can be seen in the microscopic examination of biopsy material from the skin or gills, which is the confirmatory diagnosis of this infection.

Mortalities due to ichthyophthiriasis are highly common in ornamental fishes than in food fishes. However, it has a worldwide distribution and can infect a broad range of freshwater fishes. *I. multifiliis* is considered a significant pathogen in commercial aquaculture of cyprinids, catfish, and trout. A predominant clinical sign of the infection is the presence of white spots on the body, fins, and gills. Apart from this, excessive slime on the body, difficulty in breathing, clamped fins, and loss of appetite are the major symptoms associated with this disease. Dip treatment using a mixture of 0.1 ppm of malachite green and 50 ppm of formalin has been observed to be most effective in controlling this disease. Quinine hydrochloride at 30 mg per litre can also be used. A prolonged bath in a 5% solution of methylene blue (1–2 drops) is also recommended in chronic conditions (Sharma et al. 2012).

5.4 Argulosis

Argulosis is caused by *Argulus* sp., commonly called fish lice, which belongs to the class Crustacea. They are mobile macroscopic ectoparasites and are easily observable on the skin and fins of fish (Fig. 10). They feed on the flesh, mucous layer, and by imbibing fish blood, which leads to localized tissue damage. Strong attachment to the skin surface and prolonged feeding of the parasite with its sucker results in the destruction of the epithelium, leading to the formation of lesions. Argulosis is included in the list of the major health problems in cage culture in Latin America, the Caribbean, and Africa (Blow and Leonard 2007). *Argulus japonicus*, *Argulus foliaceus*, *Argulus Indicus*, and *Argulus siamensis* are the four economically

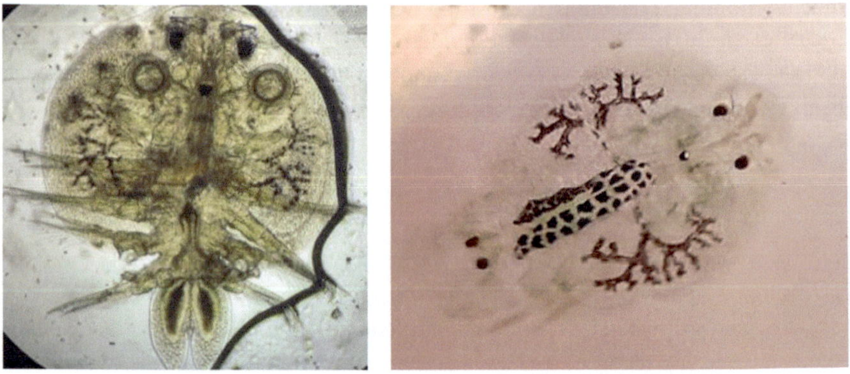

Fig. 10 *Argulus* sp. isolated from cage-cultured fish

significant parasitic species observed in freshwater aquaculture. Argulosis can infect a wide range of fish hosts, such as cyprinids, anabantids, channids, tilapia, siluriformes, etc., and young fishes are more susceptible to it. Recently, Sriwongpuk (2020) has reported *A. indicus* infestation in cage-cultured red tilapia in Thailand.

Heavily infected fishes exhibit restless behaviour, rubbing of the body on the substrate, lethargy, reduced feeding, and lesions on the external body surface. They may also allow for secondary bacterial and fungal infections of the skin and muscles, often leading to large-scale mortality in broodstock ponds (Sahoo et al. 2013). Anaemia and epidermal hyperplasia are also observed in infected fishes. There are more than 150 species of *Argulus* reported from all over the world. Dip treatment for a minute in 500-ppm potassium permanganate helps in healing the lesions caused by this parasite, while pond treatment with malathion at 0.25 ppm (three consecutive doses at weekly intervals) produces satisfactory results in controlling the disease. For the removal of *Argulus* sticky eggs, the common practice in our country is to place small bamboo sticks in the water area and then remove them after a week to kill the eggs.

6 Pseudo-fungal Diseases/Oomycete Infections

Oomycetic or pseudo-fungal infections are commonly called water mould infections or cotton tuft disease. The pathogens that cause these infections can tolerate a wide temperature range, and the intensity of disease outbreaks is reported to be higher during the winter season. However, during favourable environmental conditions, all the fish species are susceptible to fungal infections. Organic matter accumulation due to the decay of fouling organisms, unused feed, excreta released in the system, and decomposed feed enhances the rapid growth of fungal pathogens, leading to heavy mortality in cage systems. Saprolegniasis is mainly prevalent in cage culture systems during winter, when the cages are blocked with fouling agents (Mishra et al. 2017). Epizootic ulcerative syndrome (EUS) has been causing significant economic losses in freshwater aquaculture in South-East Asian countries. The fungal/oomycete diseases reported in freshwater cage farming are depicted in Fig. 11.

6.1 Epizootic Ulcerative Syndrome (EUS)

Epizootic ulcerative syndrome (EUS) was first reported in Ayu fish in 1971 in Japan. It is also known as red spot disease (RSD) and mycotic granulomatosis (MG). It has a complex infectious aetiology. It is primarily caused by the oomycete *Aphanomyces invadens* (= *A. piscicida*) but is also associated with bacteria like *Aeromonas hydrophila* and viruses like Rhabdovirus (Woo and Gregory 2014). In general, it is characterized by the presence of an invasive oomycete agent called *Aphanomyces invadens*, necrotized ulcerative lesions, and typical granulomatous lesions. It is regarded as a disease of high economic importance in fish culture. When the disease

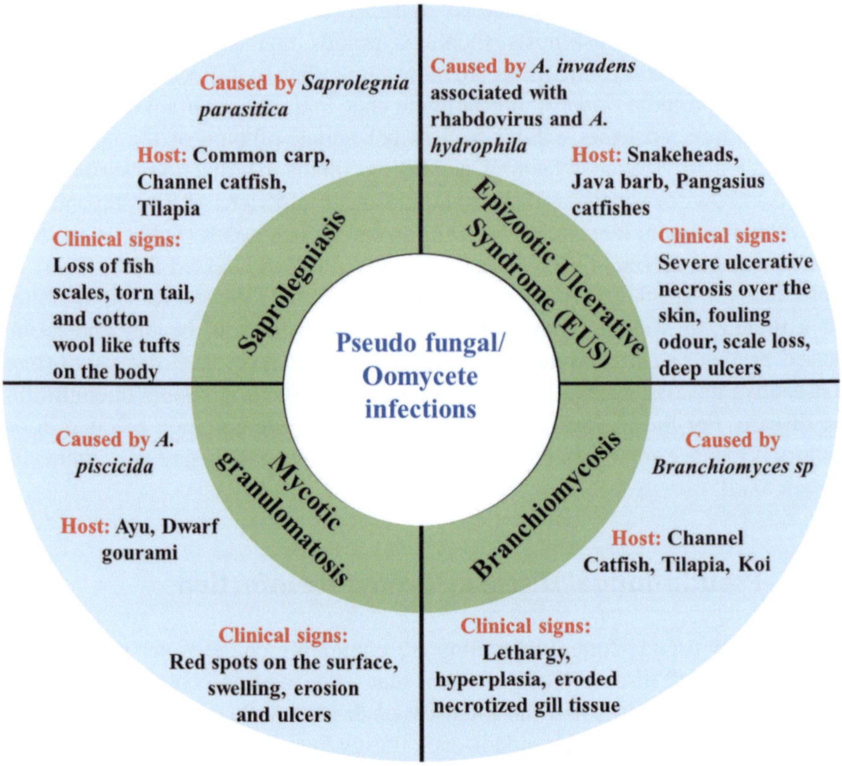

Fig. 11 Pseudo-fungal/oomycete diseases reported in cage farms

outbreak occurred in freshwater aquaculture, it caused huge losses in South-East Asia, particularly India. Due to its severity, mixed infectious aetiology, massive mortality, and economic impact on farmed freshwater fish culture, it is listed as a notifiable disease by WOAH (World Organization for Animal Health). EUS was noticed in all kinds of freshwater systems, including streams, rivers, ponds, lakes, reservoirs, rice paddies, and irrigation canals. Snakeheads cultured in cages are very much susceptible to this disease in the Philippines (Lio-Po et al. 1992), and it has also been reported in cage-cultured silver barb and mad barb (Christensen 1989).

Haemorrhages and ulcers on the body surface are found to be the major clinical signs of this disease (Fig. 12). It has a broad host fish range, including *Catla*, *Labeo rohita*, *Cirrhinus mrigala*, *Cyprinus carpio*, *Channa* sp., *Puntius* sp., and *Gudusia chapra*. At first, reddened lesions, often with haemorrhages, are seen on the skin surface, which progressively develop into deeper forms of ulcers. Infected fishes become weak and drift on the water surface. Even though a diverse variety of organisms have been isolated from the wounds and internal organs of the fish, *A. invadens* is believed to be the primary causative organism (Bagum et al. 2013). ICAR-Central Institute of Freshwater Aquaculture (ICAR-CIFA) developed a chemical formulation, CIFAX, which controls EUS and other bacterial infections (Anon 2017).

Fig. 12 Fish exhibiting clinical signs of epizootic ulcerative syndrome

6.2 Saprolegniasis

Saprolegniasis is caused by *Saprolegnia parasitica*, *Saprolegnia declina*, and *Saprolegnia saprolytica*, commonly called water moulds belonging to the family Saprolegniaceae. They can cause a disease outbreak when there is stress or mechanical injury to fish. It is also reported that they cause an important disease condition in channel catfish during the winter season, called "winter saprolegniasis" (Durborow et al. 2003). *S. parasitica* infects brooder fishes and incubates eggs and is considered a serious problem in hatcheries and seed production units. It is assessed that 10% of all salmon hatchlings are infected by saprolegniasis (epizootics), leading to major financial loss in the salmon industry (Phillips et al. 2008). Karunasagar et al. (2003) reported the incidence of saprolegniasis in tropical aquaculture systems in Asia. Infection by *S. parasitica* results in high mortality in cyprinid fishes cultured in ponds and cages (Jha et al. 1984).

Clinical signs frequently associated with saprolegniasis include cotton wool-like tufts in eggs, on the skin surface, and in gills. Initially, circular, grey, or whitish skin lesions are noted, which develop rapidly and cause damage to the epidermis layer (Fig. 13). Lethargy, impaired osmoregulation, and loss of equilibrium are also noticed. Destruction of the epidermis may lead to secondary bacterial infections as well. Controlling and treating saprolegniasis in intensive open water culture systems is difficult. It can be transmitted through direct contact with contaminated fish, eggs, water supply, farm equipment, cage netting, etc. Overcrowding, pollution, inappropriate handling, poor water quality, parasitism, and increased organic load are the predisposing factors of this disease (Bruno et al. 2011). Traditionally, malachite green has been used to control saprolegniasis; however, it has been banned due to its carcinogenic effects. Currently, several molecular and genomic approaches are available to develop an effective fish vaccine against this pathogen.

Fig. 13 Fish infected with saprolegniasis

7 Non-infectious Diseases

Non-infectious diseases are non-contagious, caused by nutritional deficiencies, environmental problems, and genetic anomalies.

7.1 Water Quality Parameters

Water quality management is extremely crucial for any aquaculture system. Deterioration of water quality causes stress to cultured fishes, which makes them susceptible to various diseases (Devi et al. 2017). Effendie et al. (2005) reported that water quality fluctuations in Lake Cirata (Indonesia) resulted in mass mortality of fish cultured in cages. Water temperature is reported to influence the growth of fish in open water culture systems. Significant changes from the optimal range can even result in heavy mortality. In Malaysia, high mortality of cage-cultured tilapia was observed when the water temperature was >30 °C (Siti-zahrah et al. 2008). Dissolved oxygen is an important attribute in aquaculture production. Low DO concentration and high ammonia levels are commonly observed in cage culture systems due to high stocking density (Karnatak and Kumar 2014). Yambot (2000) reported massive mortality of cage-cultured fishes in Lake Taal, Philippines, due to low dissolved oxygen.

7.2 Nutritional Diseases

Diseases that are caused by excess or deficiency of specific nutrients are called nutritional diseases or disorders. These generally result in reduced growth, poor fecundity, loss of appetite, increased susceptibility to diseases, and high mortality. High stocking intensity increases the probability of nutritional diseases in cage-cultured common carp (Hasan et al. 2007; Hoole et al. 2011). Lipidosis is reported in cage-cultured grouper in the Philippines and Thailand (Amar and Lavilla-Pitogo 2004). However, very little information is available on the nutrition-related diseases reported in freshwater cage-cultured fishes.

8 Conclusions

Freshwater cage farming is a great opportunity to utilize the untapped production potential of existing open water resources. It is practised at both semi-intensive and intensive levels in South-East Asian countries, including India. In general, diseases that occur in the cage culture system as well as in ponds are similar. However, the impact of disease outbreaks on fish health is more significant in the former. Generating a database on notable fish diseases in the cage culture system will provide the essential information to implement effective health management strategies. There are reports on bacterial and parasitic pathogens having zoonotic potential associated with the cage system, posing a human health risk. Hence, basic research on identifying various biotic and abiotic stress factors in the cage culture system, understanding their effects on cultured animals, and investigating their zoonotic importance will help achieve a sustainable cage farming industry in India.

References

Amal MNA, Zamri-Saad M (2011) Streptococcosis in tilapia (*Oreochromis niloticus*): a review. Pertanika J Trop Agric Sci 34:195–206

Amar EC, Lavilla-Pitogo CR (2004) Nutritional diseases. In: Diseases of cultured groupers. Aquaculture Department, Southeast Asian Fisheries Development Center, pp 59–66

Anon (2017) CIFAX fish medicine. Agrawal Trading Company, Raipur

Austin B, Austin DA (2007) Characteristics of the diseases. In: Bacterial fish pathogens: diseases of farmed and wild fish, pp 15–46

Bagum N, Monir MS, Khan MH (2013) Present status of fish diseases and economic losses due to incidence of disease in rural freshwater aquaculture of Bangladesh. J Innov Dev Strategy 7:48–53

Baker (2008) Multiple streptococcal species infection in cage-cultured red tilapia but showing similar clinical signs. Fisheries Research Institute, Batu Maung, Malaysia, Azila Abdullah, Rimatulhana Rimly

Basu S, Haldar DP (1998) Comparative study on prevalence of protozoan parasites in pure and hybrid carps. Environ Ecol 16:584–587

Blow P, Leonard S (2007) A review of cage aquaculture: sub-Saharan Africa

Bruno DW, West PV, Beakes GW (2011) Saprolegnia and other oomycetes. In: Fish diseases and disorders. Volume 3: viral, bacterial and fungal infections, 2nd edn, pp 669–720

Camus AC, Durborow RM, Hemstreet WG, Thune RL, Hawke JP (1998) Aeromonas bacterial infections-motile aeromonad septicemia, vol 478. Southern Regional Aquaculture Center

Chitmanat C, Lebel P, Whangchai N, Promya J, Lebel L (2016) Tilapia diseases and management in river-based cage aquaculture in Northern Thailand. J Appl Aquac 28:9–16

Christensen MS (1989) Techniques and economics of intensive cultivation of jelawat and lempam carp in floating cages. A handbook for extension workers and farmers

De Silva S, Phillips M (2007) A review of cage aquaculture: Asia (excluding China). FAO Fish Tech Pap 498:18–48

Declercq AM, Haesebrouck F, Van den Broeck W, Bossier P, Decostere A (2013) Columnaris disease in fish: a review with emphasis on bacterium-host interactions. Vet Res 44:1–17

Devi PA, Padmavathy P, Aanand S, Aruljothi K (2017) Review on water quality parameters in freshwater cage fish culture. Int J Appl Res 3:114–120

Dhar AK, Manna SK, Allnutt FT (2014) Viral vaccines for farmed finfish. Virus Dis 25:1–17

Durborow RM, Thune RL, Hawke JP, Camus JC (1998) Columnaris disease: a bacterial infection caused by *Flavobacterium columnare*. Southern Regional Aquaculture Center, Publication No. 479. US Department of Agriculture, Stoneville

Durborow RM, Wise DJ, Terhune JS (2003) Saprolegniasis (winter fungus) and branchiomycosis of commercially cultured channel catfish. Southern Regional Aquaculture Center

Effendie I, Nirmala K, Hasan Saputra U, Sudrajat AO, Zairin M Jr, Kurokura H (2005) Water quality fluctuations under floating net cages for fish culture in Lake Cirata and its impact on fish survival. Fish Sci 71:972–977

Food and Agriculture Organization of the United Nations (FAO) (2018) Tilapia lake virus expert knowledge elicitation risk assessment. FAO animal health risk analysis – assessment, Issue no. 7. FAO, Rome

Ferguson HW, Turnbull JF, Shinn AP, Thompson K, Dung TT, Crumlish M (2001) Bacillary necrosis in farmed *Pangasius hypophthalmus* (Sauvage) from the Mekong Delta, Vietnam. J Fish Dis 24:509–513

Gopakumar G (2009) History of cage culture, cage culture operations, advantages and disadvantages of cages and current global status of cage farming

Hasan MR, Hecht T, De Silva SS, Tacon AG (2007) Study and analysis of feeds and fertilizers for sustainable aquaculture development

Hassan MA, Mishal P, Karnatak G, Sharma AP (2017) Towards the blue revolution in India: prospects for inland open waters. World Aquacult 48:25–28

Hoole D, Bucke D, Burgess P, Wellby I (2011) Diseases of carp and other cyprinid fishes. England. Fishing News Books. 264 pp

Iliyasu A, Mohamed ZA, Terano R (2016) Comparative analysis of technical efficiency for different production culture systems and species of freshwater aquaculture in Peninsular Malaysia. Aquacult Rep 3:51–57

Ismail NIA, Amal MNA, Shohaimi S, Saad MZ, Abdullah SZ (2016) Associations of water quality and bacteria presence in cage cultured red hybrid tilapia, *Oreochromis niloticus*× *O. mossambicus*. Aquacult Rep 4:57–65

Jha BC, Pisolkar MD, Rao YR (1984) Occurrence of fungal infestation in silver carp, *Hypophthalmichthys molitrix* (Val.) in cages. J Inland Fish Soc India 13:90

Jørgensen LVG, Buchmann K (2020) 18 Ichthyophthiriosis (*Ichthyophthirius* sp). Climate Change and Infectious Fish Diseases 349

Karnatak G, Kumar V (2014) Potential of cage aquaculture in Indian reservoirs. Int J Fish Aquat Stud 1:108–112

Karunasagar I, Karunasagar I, Otta SK (2003) Disease problems affecting fish in tropical environments. J Appl Aquac 13:231–249

Khairnar K, Raut MP, Chandekar RH, Sanmukh SG, Paunikar WN (2013) Novel bacteriophage therapy for controlling metallo-beta-lactamase producing *Pseudomonas aeruginosa* infection in Catfish. BMC Vet Res 9:264

Lio-Po GD, Albright LJ, Alapide-Tendencia EV (1992) *Aeromonas hydrophila* in the epizootic ulcerative syndrome (EUS) of snakehead, *Ophicephalus striatus*, and catfish, *Clarias batrachus*: quantitative estimation in natural infection and experimental induction of dermo-muscular necrotic lesion. In: Diseases in Asian Aquaculture, I: Proceedings of the first Symposium on Diseases in Asian Aquaculture: 26–29 November 1990, Bali, Indonesia, pp 461–474

Masser M, Duarte S, Frutal T (1991) Catfish cage farming survey: timely information. Alabama Cooperative Extension Service, Auburn University, Auburn

McArdle JF (1984) *Trichodina* as a cause of mortalities in cage reared rainbow trout (*Salmo gairdneri*) and salmon (*Salmo salar*)

Mishra SS, Das R, Choudhary P, Debbarma J, Sahoo SN (2017) Present status of fisheries and impact of emerging diseases of fish and shellfish in Indian aquaculture. J Aquacult Res Mar Sci:5–26

Mohanty BR, Sahoo PK (2007) Edwardsiellosis in fish: a brief review. J Biosci 32:1331–1344

Morton A, Routledge RD, Williams R (2005) Temporal patterns of sea louse infestation on wild Pacific salmon in relation to the fallowing of Atlantic salmon farms. N Am J Fish Manag 25:811–821

Murray AG, Peeler EJ (2005) A framework for understanding the potential for emerging diseases in aquaculture. Prev Vet Med 67:223–235

NFDB (2016) Guidelines for cage culture in inland open water bodies of India, Hyderabad

NFDB, Hyderabad (2018) NFDB annual report 2017–18

Pantulu VR (1979) Floating cage culture of fish in the Lower Mekong Basin. In: Pillay TVR, Dill WMA (eds) Advances in aquaculture. Fishing News Books Ltd, Farnham, pp 423–427

Okaeme AN, Olufemi B, Amubode F (1990) A review of diseases associated with cage culture systems: diagnosis and control in small-scale fish farming. In: 13th annual conference of the Fisheries Society of Nigeria (FISON), pp 92–96

Park SB, Aoki T, Jung TS (2012) Pathogenesis of and strategies for preventing *Edwardsiella tarda* infection in fish. Vet Res 43:67

Phillips AJ, Anderson VL, Robertson EJ, Secombes CJ, Van West P (2008) New insights into animal pathogenic oomycetes. Trends Microbiol 16:13–19

Plumb JA (1999) Health maintenance and principal microbial diseases of cultured fishes. Lowa State University Press, Ames, p 328

Rao GS, Joseph I, Philipose KK, Mojjada S (2013) Cage aquaculture in India. Central Marine Fisheries Research Institute, Kochi, p 240

Reksalegora O (1979) Fish cage culture in the town of Jambi, Indonesia. In: Proceedings of the international workshop on Pen Cage Culture of Fish, Tigbauan, Iloilo, Philippines. Aquaculture Department, Southeast Asian Fisheries Development Center, pp 51–53

Sahoo PK, Mohanty J, Garnayak SK, Mohanty BR, Kar B, Prasanth H, Jena JK (2013) Estimation of loss due to argulosis in carp culture ponds in India. Indian J Fish 60:99–102

Sharma M, Shrivastav AB, Sahni YP, Pandey G (2012) Overview of the treatment and control of common fish diseases

Siti-zahrah A, Padilah B, Azila R, Rimatulhana R, Shahidan H (2008) Multiple streptococcal species infection in cage-cultured Red Tilapia but showing similar clinical signs, 313–320

Sriwongpuk S (2020) A new report of Argulus indicus (crustacea: branchiura) infestation in red tilapia (*Oreochromis niloticus x Oreochromis mossambicus*) in Thailand. Int J Geomat 18:182–187

Sunarto A, Rukyani A, Itami T (2005) Indonesian experience on the outbreak of koi herpesvirus in koi and carp (*Cyprinus carpio*). Bull Fish Res Agency 2:15–22

Surachetpong W, Janetanakit T, Nonthabenjawan N, Tattiyapong P, Sirikanchana K, Amonsin A (2017) Outbreaks of tilapia lake virus infection, Thailand, 2015–2016. Emerg Infect Dis 23:1031–1033

Tantry TA, Nazir R, Chishti MZ, Ahmad F, Dar GH, Dar JS (2016) A report on the incidence of *Trichodina heterodentata* from fishes of Jammu, J&K India. J Parasit Dis 40:524–527

Tattiyapong P, Dachavichitlead W, Surachetpong W (2017) Experimental infection of Tilapia Lake Virus (TiLV) in Nile tilapia (*Oreochromis niloticus*) and red tilapia (*Oreochromis* spp.). Vet Microbiol 207:170–177

Thomas J, Thanigaivel S, Vijayakumar S, Acharya K, Shinge D, Seelan TSJ, Mukherjee A, Chandrasekaran N (2014) Pathogenicity of *Pseudomonas aeruginosa* in *Oreochromis mossambicus* and treatment using lime oil nanoemulsion. Colloids Surf B: Biointerfaces 116:372–377

Vass KR, Sachlan M (1957) Cultivation of common carp in running water in West Java. In: Proceedings of Indo-Pacific Fisheries Council, 6th session (1956), sections II and III, pp 187–196

Wamala SP, Mugimba KK, Mutoloki S, Evensen Ø, Mdegela R, Byarugaba DK, Sørum H (2018) Occurrence and antibiotic susceptibility of fish bacteria isolated from *Oreochromis niloticus* (Nile tilapia) and *Clarias gariepinus* (African catfish) in Uganda. Fish Aquat Sci 21:6

Willomitzer J (1980) Therapy of major ectoparasitoses in grass carp (*Ctenopharyngodon idella*) fry and fingerlings. Acta Vet Brno 49:279–282

Woo PT, Gregory DWB (eds) (2014) Diseases and disorders of finfish in cage culture. CABI

Yambot AV (2000) Problems and issues of Nile tilapia cage farming in Taal Lake, Philippines. In: Liao CI, Lin KC (eds) Cage aquaculture in Asia: proceedings of the first international symposium on cage aquaculture in Asia. Asian Fisheries Society, Quezon City, pp 241–252

Yasumoto S, Kuzuya Y, Yasuda M, Yoshimura T, Miyazaki T (2006) Oral immunization of common carp with a liposome vaccine fusing Koi Herpesvirus Antigen. Fish Pathol 41:141–145

Zhang D, Xu DH, Shoemaker C (2016) Experimental induction of motile Aeromonas septicemia in channel catfish (*Ictalurus punctatus*) by waterborne challenge with virulent *Aeromonas hydrophila*. Aquacult Rep 3:18–23

Emerging Viral Diseases in Freshwater Fishes with Emphasis on Tilapia Lake Virus

T. R. Swaminathan, P. K. Pradhan, A. Sudhagar, and N. Sood

Abstract

The increasing demand for fish and the stagnation in capture fisheries drive the aquaculture industry to grow around the world to meet the global demand. However, over the decades, the growth of the aquaculture industry has been accompanied by the rise of new aquatic animal diseases, many of which have spread to various countries through the transboundary movements of pathogens. Furthermore, various environmental stressors, farming inputs, and global trade have led to the emergence of new pathogens. Stress due to intensive culture practices at the farm level makes animals susceptible to various new pathogens by compromising their immune system and facilitating rapid transmission of diseases. Particularly, viral pathogens, which have either been established for decades or are newly emerging, are difficult to control due to limitations in the availability of efficacious treatments, including effective viral vaccines. Here, we review a few significant viral pathogens reported from various finfishes, including cyprinid herpesvirus-2 (CyHV-2), carp edema virus (CEV), viral nervous necrosis virus (VNNV), infectious spleen and kidney necrosis virus (ISKNV), and tilapia lake virus (TiLV), which had caused massive disease outbreaks and severe economic losses in the aquaculture and ornamental fish industries. Furthermore, these diseases/pathogens have been recorded during national-level surveillance undertaken by the National Surveillance Programme on Aquatic

T. R. Swaminathan (✉) · A. Sudhagar
Centre for Peninsular Aquatic Genetic Resources, ICAR National Bureau of Fish Genetic Resources, Kochi, India
e-mail: rajathanga@yahoo.co.in; arunsaqua@gmail.com

P. K. Pradhan · N. Sood
Exotics and Aquatic Animal Health Division, ICAR National Bureau of Fish Genetic Resources, Lucknow, India
e-mail: pradhanpk1@gmail.com; nsoodnbfgr@gmail.com

R. Kooloth Valappil et al. (eds.), *Aquatic Animal Health Management*,
https://doi.org/10.1007/978-981-96-7987-4_5

Animal Diseases (NSPAAD) in India since 2013. Tilapia is one of the most important aquaculture species, second to carp, in the world and a primary source of protein in many developing countries. Tilapia lake virus (TiLV) is an emerging virus that is rapidly spreading across the world among all tilapia fishes. Over the past 6 years, TiLV outbreaks have been reported in at least 16 countries, spanning 3 continents, including Asia, Africa, and America, and the presence of TiLV is anticipated in many countries. This chapter provides an overview of emerging viral finfish diseases in India; however, considering the growing significance of its impact and spread, here, the emphasis is given to TiLV.

Keywords

Aquaculture · Virus · Pathogens · Surveillance

1 Introduction

Aquaculture is an important food production sector that addresses global food and nutritional security, providing livelihoods to the growing human population, and, thereby, plays an important role in achieving the United Nation's Sustainable Development Goals by 2030 (Belton et al. 2018; FAO 2020). About 85% of the world's aquaculture produce comes from Asia, and it can meet the protein demand of 15–20% of the world's human population. There has been consistent growth in the total world fish production during the past few years, and it reached 178.5 million tonnes in 2018. In spite of the stable global capture fish production in the last three decades, global aquaculture production has increased by more than 527% since 1990 (FAO 2020). This increase in aquaculture fish production was achieved by shifting the culture practice from traditional extensive methods to semi-intensive and intensive production systems. However, this shift in fish culture practice has opened up the risk of infectious disease outbreaks caused by bacteria, oomycetes, parasites, and viruses in cultured fish (Lafferty and Hofmann 2016). Furthermore, these pathogenic agents can spill over to wild fish stocks, which can significantly affect their population size.

Apart from the intensification of aquaculture practices, international trade of live fish and fishery products has increased the spread and outbreaks of new emerging viral diseases like koi herpesvirus (KHV), cyprinid herpesvirus-2 (CyHV-2), spring viraemia of carp virus (SVCV), viral nervous necrosis (VNN), Red Sea bream iridovirus disease (RSIVD), carp edema virus (CEV), and tilapia lake virus (TiLV). Moreover, climate change can further exacerbate the severity and spread of these emerging diseases (CIAT 2014). Viral disease outbreaks have already led to huge economic losses in aquaculture operations in many countries (Walker and Winton 2010). The majority of viruses that can infect and cause diseases in teleost fish are herpesviruses, iridoviruses, reoviruses, birnaviruses, and rhabdoviruses. Research studies into the development of new fish cell lines for virus isolation and extensive fish disease surveillance have led to the discovery of several new pathogenic viruses that were not known before. In 1988, only 59 infectious fish viruses were known

(Wolf 2019), and, presently, there are more than 125 different viruses that can cause serious disease outbreaks in fish (Noga 2010). Furthermore, with the advent of new, improved diagnostic techniques, this number is expected to increase in the coming years.

The occurrence of viral diseases remains a significant limiting factor for aquaculture production and for the sustainability of biodiversity in the natural environment. Moreover, viral diseases are often challenging to control in aquaculture operations due to the lack of effective therapeutics and vaccines. Viruses are transmitted either horizontally from one fish to another or vertically from parent to offspring. Stringent biosecurity measures and appropriate husbandry practices like avoidance of excessive stocking and feeding can reduce viral disease outbreaks. Moreover, long-term programmes such as the development of specific pathogen-free (SPF) and specific pathogen-resistant (SPR) fish stocks can significantly reduce the impact of viral diseases in aquaculture operations. In addition, it is important to have strict regulatory measures in place when screening for diseases during international live fish trade.

In the past few decades, finfish viral diseases have been overlooked in India due to the lack of active surveillance and the non-availability of cell lines for virus isolation. After the establishment of the National Repository of Fish Cell lines at the Indian Council of Agricultural Research (ICAR)-National Bureau of Fish Genetic Resources (NBFGR), Lucknow, and the implementation of the National Surveillance Programme for Aquatic Animal Diseases (NSPAAD), many transboundary and emerging finfish viruses have been reported in India. In this chapter, we discuss the emerging finfish viruses in India with a special emphasis on TiLV.

2 Cyprinid Herpesvirus-2 (CyHV-2)

Cyprinid herpesvirus-2 (CyHV-2) is known to cause herpesviral haematopoietic necrosis disease (HVHND) in goldfish (*Carassius auratus*) and other cyprinids (Panicz et al. 2019). The mortality rate is quite high (50–100%) among affected fish. This disease was first reported in Japan in 1992, leading to mass mortality among farmed goldfish (Fig. 1). At present, CyHV-2 has been reported in 16 countries across the globe, including India (Sahoo et al. 2016). CyHV-2 belongs to the family Alloherpesviridae and the genus *Cyprinivirus*. It has an icosahedral capsid with linear double-stranded DNA, and its genome size is about 290.3 kb (Thangaraj et al. 2021). Furthermore, two genotypes of CyHV-2 have been proposed, namely, C and J genotypes representing China and Japan, respectively (Li et al. 2015). They are enveloped viruses (170–220 nm) located within the cellular vesicles of the cytoplasm and the extracellular spaces of the infected cell. Moreover, typical spherical or hexagonal intranuclear inclusion bodies are formed by the virus with the nucleocapsid having an edge-to-edge diameter of 115–117 nm (Jung and Miyazaki 1995) (Fig. 2). CyHV-2 replicates and assembles in the gills and haematopoietic organs such as the spleen and kidney of affected fish (Hedrick et al. 2000). Within the infected host cell, the virus matures in the Golgi apparatus (Wu et al. 2013).

Fig. 1 Heavy mortality observed in goldfish affected by cyprinid herpesvirus-2 (CyHV-2)

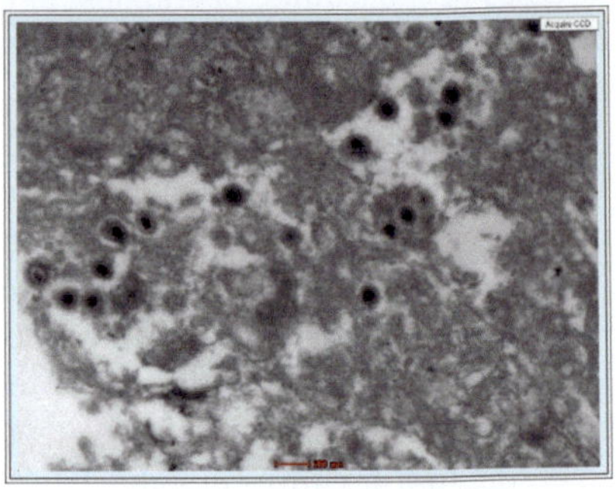

Fig. 2 Transmission electron microscopy (TEM) of a goldfish gill infected by CyHV-2. Viral capsids in the cytoplasm with electron-dense cores and capsid surrounded by electron-dense material (arrows). Scale bar = 200 nm. CyHV-2, cyprinid herpesvirus-2

This virus has a high specificity towards its host and is highly pathogenic to goldfish, crucian carp (*Carassius carassius*), and goldfish x carp hybrids (Hedrick et al. 2006; Davison et al. 2013). All stages of goldfish, including egg, fry, fingerling, and adult, are susceptible to CyHV-2 infection (Groff et al. 1998). Horizontal transmission is the most common way that CyHV-2 spreads; however, vertical transmission from parent to offspring is also possible (Goodwin et al. 2009). This virus can remain as a latent infection within its host and can become an overt infection due to water temperature fluctuation or stress (Goodwin et al. 2009).

Clinical signs of HVHND include lethargy, erratic swimming, anorexia, pale skin, bilateral exophthalmia, haemorrhages on the skin, pustules in the fin, decaying white gill filaments with bleached appearance, hyperaemia on the submaxilla and abdomen, inflammation of vent, and anaemia (Groff et al. 1998; Goodwin et al. 2006; Jeffery et al. 2007; Daněk et al. 2012; Wang et al. 2012; Fichi et al. 2013; Wu et al. 2013; Xu et al. 2013; Sahoo et al. 2016; Adamek et al. 2018). In addition, pale discolouration of the liver, ascites, splenomegaly, renal hypertrophy, empty intestine, and petechial haemorrhaging of the swim bladder and other internal organs have also been noticed (Jung and Miyazaki 1995; Jeffery et al. 2007; Wang et al. 2012; Fichi et al. 2013). Moreover, extensive necrosis could be noticed in the gills, spleen, and kidney of infected fish (Jung and Miyazaki 1995; Jeffery et al. 2007).

Even though many experimental vaccine trials have been conducted against CyHV-2, until now, there is no commercial vaccine or effective treatment available for HVHND (Thangaraj et al. 2021). The virus remains a serious threat to the Indian ornamental fish trade and food fish aquaculture. Effective biosecurity measures and regulated fish trade are necessary to control the spread of this virus. In addition, intensive research investments will be required for the development of rapid and cheap diagnostics, therapeutics, and prophylaxis to prevent CyHV-2 infection.

3 Carp Edema Virus (CEV)

Carp edema virus (CEV) causes carp edema virus disease (CEVD) or koi sleepy disease (KSD) in common carp (*Cyprinus carpio*) and its variant koi carp. This disease was first reported in Japan in the 1970s; however, it has spread to other countries due to the recent boom of the global live ornamental fish trade (Rehman et al. 2020). Apart from Japan, CEV has been reported in the United States, the Czech Republic, Austria, India, Germany, the Netherlands, Italy, China, Brazil, Poland, Hungary, and South Korea (Rehman et al. 2020). Mortality rates can even be up to 100%, and hence, this virus is considered a major threat to the common carp and koi aquaculture industry. In India, CEV was reported for the first time in the year 2016 in cultured ornamental koi carp (Swaminathan et al. 2016).

KSD-affected fish exhibit typical sleepy behaviour and lethargy, congregate near the water inlet or surface of the pond, and later tend to lie at the bottom of the pond and die of anoxia (Miyazaki et al. 2005). Furthermore, clinical signs of KSD are congestion and necrosis of the gills, enophthalmos, oedema, ulcers on the skin, and excessive mucus production (Grimmett et al. 2006; Lewisch et al. 2015) (Fig. 3). CEV infection may lead to the host's immunosuppression, leading to secondary bacterial and ectoparasitic infections (Adamek et al. 2019). Histopathological observations suggest thickening and adhesion of the secondary gill lamellae, oedema and abscission in the epithelial cells of the gill filaments, and infiltration of mononuclear cells, eosinophilic granular cells, and few erythrocytes (Miyazaki et al. 2005). Furthermore, degeneration of renal tubules, cytoplasmic vacuolization, infiltration of inflammatory cells, and multifocal necrosis have been observed in the kidney. In addition, oedema in the dermal connective tissues, atrophy of the splenic

Fig. 3 Carp edema virus (CEV)-affected koi carp with a swollen body and the gills showing a pale appearance, swelling, and abundant mucus

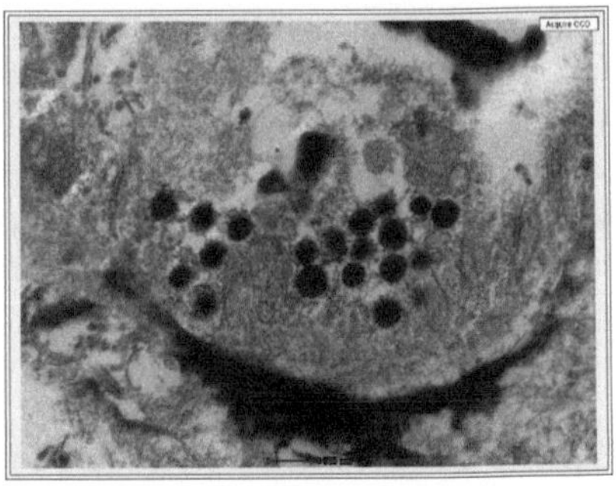

Fig. 4 A transmission electron photomicrograph of a koi gill epithelial cell containing large intracytoplasmic spheroid CEV virions (arrows) (bar = 500 nm). CEV, carp edema virus

pulp, enlarged hepatocytes, and muscle fibres have also been observed in infected fish (Rehman et al. 2020). The virus transmits horizontally from one fish to another; however, vertical transmission is not known. In addition, the virus can persist in its host without showing any clinical infection. Until now, there are no treatments or prophylactic measures available for KSD. However, immersion of infected fish in 0.5% salt has been found to be effective to a certain extent.

CEV is an unclassified double-stranded DNA virus and morphologically resembles the family Poxviridae. The size of the immature and mature virions is about 416–450 nm and 300–400 × 250–400 nm diameters, respectively (Adamek et al. 2017) (Fig. 4). Two lineages of CEV are characterized i.e. lineage 1 from koi carp (resembles the original Japanese CEV) and lineage 2 from common carp (Way and Stone 2013). Furthermore, studies suggest the existence of five genogroups of CEV, namely, genogroup I, IIa, IIb, IIIa and IIIb, based on the phylogenetic analysis of the partial core protein p4a (Soliman et al. 2019). The disease outbreak due to CEV is temperature-dependent and commonly occurs between 15 and 25 °C. However, when the water temperature is increased up to 28 °C, a remarkable reduction in the mortality and morbidity of the fish could be observed.

Common carp and koi carp are widely cultured in India for food and recreational purposes, respectively. With reports of CEV in India, extreme precautionary measures and regulations are necessary during live fish trade in order to prevent further spread of the virus (Swaminathan et al. 2016). Furthermore, a detailed investigation is necessary to study the survival of the virus outside the host and the susceptibility of other fish species to CEV infection apart from common carp and koi carp.

4 Viral Encephalopathy and Retinopathy (VER)

Viral encephalopathy and retinopathy (VER) or viral nervous necrosis (VNN) is considered an emerging disease by the World Organization for Animal Health (OIE), and it can infect many marine, brackishwater, and freshwater fish species. The virus causing this disease is known to infect more than 120 different fish species, and mortality may reach up to 100% in infected larvae, juveniles, and, in some cases, adults (Chi et al. 2003; Costa and Thompson 2016). This virus mainly infects larvae and juveniles, leading to large-scale mortalities of up to 100%, and the infected adult fish serve as carriers. The disease was first reported in Asian seabass in Australia in 1990 (Glazebrook et al. 1990) and subsequently reported in many countries across the globe, including India in 2005 (Azad et al. 2005).

The viral aetiological agent of VER is classified under the family Nodaviridae and the genus *Betanodavirus*. It is a non-enveloped, spherical-shaped virus with a diameter of about 25–30 nm. The genome of *Betanodavirus* consists of two or three molecules of single-stranded (ss) positive-sense RNA, where RNA1 is about 3.1 kb that codes for replicase (RNA-dependent RNA polymerase), RNA2 with a size of about 1.4 kb codes for coat protein (42 kDa), and RNA3 of size 0.4 kb encodes for the protein B2 (Furusawa et al. 2006; OIE 2019b). RNA3 is a sub-genomic transcript of RNA1 and is found only in the host cell nucleus infected by the virus but not in the virion (Iwamoto et al. 2001; Kuo et al. 2011). Scientists have identified five genotypes and three serotypes of *Betanodavirus*, namely, striped jack nervous necrosis virus (SJNNV), tiger puffer nervous necrosis virus (TPNNV), barfin flounder nervous necrosis virus (BFNNV), red-spotted grouper nervous necrosis virus (RGNNV), and turbot *Betanodavirus* strain (TNV). SJNNV and TPNNV belong to serotype A and serotype B, respectively, whereas BFNNV and RGNNV belong to serotype C (Zorriehzahra 2020).

Infected fish exhibit erratic swimming behaviour like whirling, spiralling and circular movements, and reduced or no feed intake (Nopadon et al. 2009). Typical external signs are not seen in this disease, and, in a few cases, progressive change in pigmentation could be observed, and in a few cases,hyperinflation of the swim bladder could be observed in the larval stage (Mori et al. 1992). Necrosis and vacuolation are prominent in the spinal cord, brain, and retina of infected larvae and juveniles. However, microscopic changes are rarely seen in symptomatic adult infected fish. The virus replicates in the brain, spinal cord, and retina of infected fish. In many cases, it is found in the gonad after stress (Mushiake et al. 1994). However, intracytoplasmic inclusions could be observed in the affected organs. Viral particles can be visualized through electron microscopy in the cytoplasm or the endoplasmic reticulum membrane of the infected cell.

Horizontal transmission is the common mode of infection spread for *Betanodavirus*, and, moreover, the vertical mode of transmission is also possible. Water temperature plays a vital role in the manifestation of the disease, particularly in seabass, and the outbreaks are common during the summer season (Iwamoto et al. 2000). The virus can survive in aquatic environments for a long time, particularly when the water temperature is low (Frerichs et al. 2000). However, when dried, >99% of the virus gets inactivated in 7 days at 21 °C (Maltese and Bovo 2007). In addition, the virus can survive freezing and hence can be potentially transmitted through raw fish used for feeding (Mori et al. 2005). The virus could persist in recovered fish for a long period of time and can serve as a carrier (Johansen et al. 2002). Moreover, it could spread between farms through contaminated personnel and farm equipment, and, hence, adequate biosecurity measures should be ensured to prevent its spread.

Iodine, sodium hypochlorite, benzalkonium chloride, and hydrogen peroxide can inactivate betanodaviruses (Frerichs et al. 2000). Moreover, ozone treatment and ultraviolet (UV) exposure can reduce the viral load on the fish egg surface (Grotmol and Totland 2000) and virus-contaminated water (Frerichs et al. 2000), respectively. Recently, ICAR-Central Institute of Brackishwater Aquaculture (ICAR-CIBA) has developed a recombinant vaccine for VNN under the commercial name Nodavac-R. With the augmentation of marine and brackishwater finfish aquaculture in India, VER has become a potential threat to the industry. Hence, more research prioritization should be focused on this disease in India.

5 Red Sea Bream Iridovirus (RSIV)

Red Sea bream iridoviral disease (RSIVD) is caused by a large double-stranded DNA virus named Red Sea bream iridovirus (RSIV). This disease is listed as a notifiable disease by the World Organization for Animal Health (OIE), Paris. RSIV can infect more than 30 different fish species belonging to the orders Perciformes and Pleuronectiformes (Matsuoka et al. 1996; Kawakami and Nakajima 2002). It is a major threat and can cause huge mortality among farmed marine fish, especially Red Sea bream (*Pagrus major*). This disease was first identified in the year 1990 among farmed Red Sea breams of Japan (Inouye et al., 1992) and is quite prevalent in South-East Asian countries such as China, Taiwan, Hong Kong, South Korea, Malaysia, Singapore, the Philippines, and Thailand. Recently, RSIV has been reported for the first time in India in Asian sea bass (*Lates calcarifer*) cultures in open estuarine cages (Girisha et al. 2020). Mortality can range between 0% and 100% depending on the affected fish species, the size and age of the fish, the temperature of the water, and husbandry practices. RSIVD outbreaks generally occur when the water temperature rises above 25 °C in the summer season.

RSIV belongs to the family Iridoviridae and the genus *Megalocytivirus*. Furthermore, scientists have identified three genotypes of *Megalocytivirus*, including RSIV. The other two genotypes are infectious spleen and kidney necrosis virus (ISKNV) and turbot reddish body iridovirus (TRBIV) (Subramaniam et al. 2012). RSIV is an enveloped virus with icosahedral symmetry and a diameter of 200–240 nm. It is a double-stranded DNA virus with a genome size of 111.15 KB (Shiu et al. 2018; Puneeth et al. 2021).

The virus can infect all life stages of fish from juvenile to adult; however, higher susceptibility is observed in juveniles than in adults. Furthermore, the *Oplegnathus* sp. has a higher susceptibility to infection than other species. The horizontal mode of transmission is the principal transmission route in the case of RSIV. However, vertical transmission from parent to offspring is not known. Infected fish exhibit lethargic swimming behaviour and conspicuous respiratory patterns. Clinical signs include pale discolouration and petechiae in the gills, anaemia, splenomegaly, and low haematocrit value (Inouye et al., 1992; Jung et al. 1997; Nakajima and Maeno 1998). The virus is observed to infect the gills, kidney, spleen, heart, and intestines of fish. Moreover, in histological observations, abnormally enlarged basophilic cells could be noticed in the tissues of the heart, spleen, kidney, intestines, and gills of the affected fish (Inouye et al., 1992). Enlarged cells can also be observed in Giemsa-stained spleen smears.

Commercial formalin-killed vaccines are available for RSIVD and are protective in many commercially farmed marine fish species. However, these vaccines do not offer much protection in the fish species belonging to the genus *Oplegnathus* sp. The recent report of RSIV in India has suggested that the virus might have already existed in this region before and had been overlooked (Girisha et al. 2020). With the active promotion of commercial mariculture in India, there is a potential threat from this virus that can hamper the industry. Hence, active research must be carried out on RSIV for its prevention and spread in India.

6 Infectious Spleen and Kidney Necrosis Virus (ISKNV)

Infectious spleen and kidney necrosis virus (ISKNV) is a double-stranded DNA virus closely related to RSIV, which has a wide host range and global distribution. ISKNV belongs to the family Iridoviridae and the genus *Megalocytivirus*. This virus can infect and cause mass mortality in at least 50 different fish species from marine, brackish, and freshwater habitats (Whittington et al. 2010; Sahoo and Goodwin 2012; Subramaniam et al. 2012). ISKNV was first reported in 1994 in China in mandarin fish (*Siniperca chuatsi*) (He et al. 2002). Since then, it has been detected in various fish species in many countries such as Japan, Taiwan, Indonesia, Korea, Singapore, Malaysia, the United States, Australia, Germany, Brazil, Ghana, and recently in India (Lancaster et al. 2003; Go and Whittington 2006; Tanaka et al. 2014; Jung-Schroers et al. 2016; Yanong and Waltzek 2010; Girisha et al. 2020, Pereira Figueiredo et al. 2022; Ramírez-Paredes et al. 2021; Sukenda et al. 2020). Initially, ISKNV was endemic to South-East Asia, and due to the transboundary trade movement of live ornamental fish, the virus spread to other locations of the world (Whittington and Chong 2007; Jeong et al. 2008).

ISKNV-infected fish exhibit erratic swimming behaviour, discolouration of the skin, petechial haemorrhages, and reduced feed intake (Fig. 5). Moreover, gross clinical signs of an enlarged spleen and kidney are also observed. Histological observation of the kidney and spleen of ISKNV-affected fish showed basophilic cytoplasmic inclusion bodies with hypertrophic cells (He et al. 2002). The mortality

Fig. 5 Clinical signs of an infected giant gourami, *Osphronemus goramy* showing darkened skin, excess mucous, skin haemorrhages, and a distended abdomen

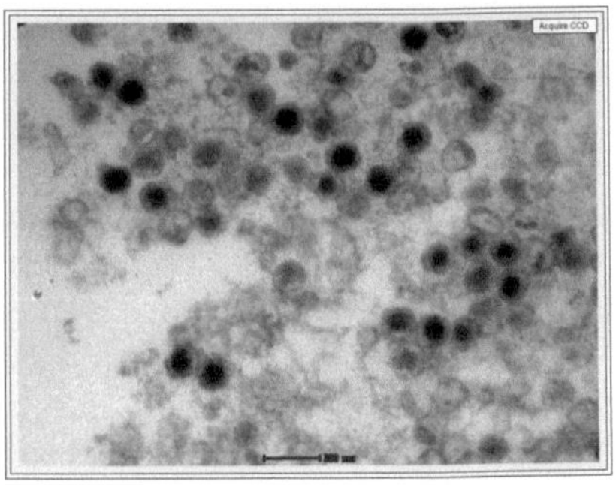

Fig. 6 A transmission electron microscopic image of the kidney showing large numbers of cytoplasmic polygonal mature ISKNV virus particles in the cytoplasm. ISKNV, infectious spleen and kidney necrosis virus

percentage can range up to 100%, and at certain times, infected fishes become asymptomatic carriers that aid in the spread of the virus. The principal mode of transmission is by horizontal means, and the virus can infect all stages of fish, from juveniles to adults. The genome size of ISKNV is 111 kb, and it has about 124 open reading frames (ORFs) (He et al. 2001). The virus is icosahedral in shape, encapsulated with a nucleocapsid layer, and ranges in size from 110 to 150 nm (Shi et al. 2004; Jung-Schroers et al. 2016) (Fig. 6). Standard immunological and polymerase chain reaction (PCR)-based molecular techniques are well-established for the diagnosis of ISKNV. Interestingly, the monoclonal antibody developed for the detection of RSIV can also detect ISKNV (OIE 2019a).

Recently, two separate outbreaks of ISKNV have occurred in the state of Kerala, India, leading to mass mortality (100%) of giant gourami (*Osphronemus goramy*) and Oscar fish (*Astronotus ocellatus*) (Swaminathan et al. 2021a; Raj et al. 2024). In addition, ISKNV was also reported in 10 different ornamental fish species belonging to the families Poeciliidae, Osphronemidae, Cichlidae, and Cyprinidae, from various retail outlets located in three districts of southern Karnataka. In addition, the domestic ornamental fish trade involves an inter-state supply chain across India. Hence, this global and domestic transboundary movement of ornamental fish may potentially favour the spread of ISKNV. Sometimes, the infection could be asymptomatic with no apparent clinical signs (Girisha et al. 2021). These asymptomatic fishes can act as carriers for further spread of the virus. Undoubtedly, the transboundary movement of fish, high infectivity rate, and wide host range of ISKNV make it a potential threat to the endemic fishes of India.

7 Tilapia Lake Virus Disease (TiLVD)

Tilapia lake virus disease, caused by *Tilapia tilapinevirus*, is associated with lethargy, haemorrhages, abdominal swelling, exophthalmia, and severe mortalities in farmed and wild populations of tilapia and characterized by syncytia formation in the liver of affected fish. This disease is also known as syncytial hepatitis of tilapia (Ferguson et al. 2014), tilapia one-month mortality syndrome (Tattiyapong et al. 2017), or summer mortality syndrome (Fathi et al. 2017).

7.1 Aetiology

Tilapia lake virus disease (TiLVD) is caused by an enveloped, negative-sense, single-stranded (ss) RNA virus with 10 segments. Each of the 10 segments contains an open reading frame flanked by non-coding regions. Segment 1 shares a weak sequence homology with the RNA-dependent RNA polymerase subunit of PB1 of the influenza C virus, whereas other segments show no homology to any other known viruses. All 10 segments have 13 inverted partially complementary sequences at their 5′ and 3′ noncoding termini, a genome organization consistent with all known orthomyxoviruses. Based on a segmented, negative-sense, ssRNA genome with organization similar to that of an orthomyxovirus, TiLV was initially touted as an 'orthomyxo-like virus'. However, some of the characteristics atypical of orthomyxovirus include the formation of syncytia and the presence of virions within the perinuclear cisternae and free virions in the cytoplasm (Del-Pozo et al. 2017). Subsequently, this virus is now classified as a new species, *Tilapia tilapinevirus*, under the genus *Tilapinevirus* and the family Amnoonviridae, on the basis of sequence analysis. This virus has a diameter of 55–110 nm, with a genome size of 10,323 nucleotides. A total of 20 complete TiLV genome sequences from six different countries, including 10 from Thailand, three from Bangladesh, two each from the United States and Israel, and one each from Peru, Ecuador, and India, are available (Verma et al. 2022). The replication and transcription of the virus take place in the nucleus of fish cells.

7.2 Geographic Distribution

A new RNA virus was found to be responsible for the decline in the catch of tilapia in Lake Kinneret, also known as the Sea of Galilee, Israel, in the year 2014 (Eyngor et al. 2014). The virus was named tilapia lake virus based on the host species (tilapia) and the site from which it was detected for the first time (a lake). Around the same time, Ferguson et al. (2014) described a case of mass mortality of farmed tilapia (*Oreochromis niloticus*) in Ecuador with histopathological lesions typical of syncytial hepatitis. Later, these mortalities were confirmed to be caused by the same virus as the one responsible in Israel.

To date, 17 countries across 4 continents (Asia, Africa, and South and North America) have reported TiLV, comprising Colombia (Kembou Tsofack et al. 2017), Ecuador (Bacharach et al. 2016; Ferguson et al. 2014), Egypt (Fathi et al. 2017; Nicholson et al. 2017), Israel (Eyngor et al. 2014), India (Behera et al. 2018), Indonesia (Koesharyani et al. 2018), Malaysia (Amal et al. 2018), Peru (OIE 2018c; Pulido et al. 2019), the Philippines (OIE 2017a), Thailand (Dong et al. 2017a; Surachetpong et al. 2017), Tanzania and Uganda (Mugimba et al. 2018), Chinese Taipei (OIE 2017b), Mexico, the United States (Ahasan et al. 2020; OIE 2018a, 2018b), Bangladesh (Chaput et al. 2020), and Vietnam (Thao et al. 2020) (Fig. 1). Importantly, data from the screening of archived samples (fertilized eggs, yolk-sac larvae, fry, and fingerlings) collected from 2012 to 2017 from Thailand hatcheries experiencing unexplained mortalities of tilapia fry revealed that these were positive for TiLV. During this period, more than 40 countries worldwide imported tilapia fry and fingerlings from affected hatcheries and, therefore, were potentially at a high risk of TiLV spread. In addition, three countries were at a lower risk of TiLV spread as they had imported fish from these hatcheries before 2012 (Dong et al. 2017a). Of the countries at a higher risk of TiLV spread, 10, including Bangladesh, India, Indonesia, Malaysia, Vietnam, Mexico, the Philippines, Tanzania, Uganda, and the United States, have already confirmed the presence of TiLV.

7.3 Hosts

The tilapia lake virus can cause disease in several species, both naturally and via experimental exposure to infection. Natural infections of TiLV have been reported in *Sarotherodon galilaeus*, *Tilapia zillii*, *Oreochromis aureus*, *Tristramella simonis intermedia*, Nile tilapia (*O. niloticus*), grey tilapia (*O. niloticus* × *O. aureus*), red tilapia (*Oreochromis* sp.), and giant gourami (*Osphronemus goramy*) (Eyngor et al. 2014; Fathi et al. 2017; Ferguson et al. 2014; Mugimba et al. 2018; Surachetpong et al. 2017; Chiamkunakorn et al. 2019). Importantly, strain variation in the susceptibility of Nile tilapia to infection with TiLV has been observed, whereby the Chitralada strain is reported to be more susceptible to TiLV infection in comparison to the genetically all-male tilapia (GMT) strain (Eyngor et al. 2014). Although TiLV genomic RNA has been detected in wild river barb (*Barbonymus schwanenfeldii*) and farmed barramundi (*Lates calcarifer*) (Abdullah et al. 2018; Piamsomboon and Wongtavatchai 2021), the mere detection of TiLV nucleic acid does not confirm the susceptibility of these species, therefore warranting experimental infection studies.

Tilapia lake virus disease (TiLVD) has been reproduced experimentally in *O. niloticus*, Mozambique tilapia (*Oreochromis mossambicus*), Malawi cichlid, *Aulonocara* spp., and giant gourami (Behera et al. 2018; Eyngor et al. 2014; Jaemwimol et al. 2018; Tattiyapong et al. 2017; Waiyamitra et al. 2021; Yamkasem et al. 2021). On the other hand, most warm water fish including snakeskin gourami (*Trichogaster pectoralis*), iridescent shark (*Pangasianodon hypophthalmus*), walking catfish (*Clarias macrocephalus*), striped snakehead fish (*Channa striata*), climbing perch (*Anabas testudineus*), common carp (*Cyprinus carpio*), silver barb (*Barbodes gonionotus*), Asian sea bass (*Lates calcarifer*), rohu (*Labeo rohita*), catla (*Catla catla*), mrigal (*Cirrhinus mrigala*), grass carp (*Hypophthalmichthys molitrix*), pearl spot (*Etroplus suratensis*), orange chromide (*Pseudetroplus maculatus*), and Canara pearl spot (*Etroplus canarensis*) are resistant to the virus as determined in experimental infection studies (Jaemwimol et al. 2018; Pradhan et al. 2020, 2021; Thangaraj et al. 2022).

7.4 Transmission

TiLV can be transmitted both horizontally and vertically. Horizontal transmission is considered to be the major route for the spread of TiLV, with the virus being transmitted to healthy fish through direct contact with the skin and mucus of infected fish or through cannibalism of moribund or dead fish (Liamnimitr et al. 2018; Tang et al. 2021). It has been suggested that TiLV-infected fish exhibiting no clinical signs could transmit the virus to susceptible fish (Senapin et al. 2018). Vertical transmission of TiLV has also been observed in natural cases of infection with TiLV and following experimental infection (Dong et al. 2020; Yamkasem et al. 2019). Under experimental conditions, the disease has been successfully reproduced by cohabitation, intraperitoneal challenges, and the intragastric route (Eyngor et al. 2014; Liamnimitr et al. 2018). On the other hand, frozen tilapia fillets have been suggested to offer minimal risk of transmission of TiLV (Thammatorn et al. 2019). The role of vectors such as birds, molluscs, water insects, and invertebrates in transmission of the virus needs further investigation.

7.5 Tissue Tropism

The target tissues of TiLV include the liver, kidney, spleen, gills, brain, eye, and connective tissues of muscles, ovary, and testis. Dong et al. (2017b) reported TiLV hybridization signals in multiple organs of naturally diseased fish, including the liver, kidney, brain, gills, spleen, and connective tissue in the muscle. The predilection of TiLV for the liver and gastrointestinal tract has been reported in samples originating from Ecuador (Del-Pozo et al. 2017). TiLV has been detected in the mucus and gonads of naturally infected tilapia, suggesting horizontal and vertical transmissions of TiLV, and, importantly, the virus levels have been reported to correspond to those observed in the liver (Liamnimitr et al. 2018; Yamkasem et al. 2019). On the other hand, TiLV has not been detected in the mucus of apparently healthy TiLV-infected fish, suggesting that these fish cannot spread the virus (Chiamkunakorn et al. 2019). Furthermore, TiLV has not been detected or isolated from faeces of naturally infected and experimentally challenged tilapia (Liamnimitr et al. 2018). Mugimba et al. (2018) reported a high prevalence of TiLV nucleic acids in the head kidney and spleen, a low presence in the liver, and absence in the brain of sub-clinically infected tilapia from the Tanzanian and Ugandan parts of Lake Victoria, suggesting that lymphoid organs could be ideal for screening the presence of TiLV nucleic acids during surveillance. In addition, blood and liver biopsy samples have been reported to be appropriate for screening sub-clinically infected populations of cichlids (Chiamkunakorn et al. 2019).

7.6 Pathogenesis

The pathogenesis of TiLV disease is still not clear. It has been hypothesized that TiLV enters naïve tilapia through direct contact with the skin or mucus of infected fish or via the oral route, which is corroborated by the successful reproduction of the disease by cohabitation and intragastric challenges (Eyngor et al. 2014; Liamnimitr et al. 2018; Pierezan et al. 2019). The host cell receptors during TiLV entry are not known. Recently, in a study, it has been found that TiLV does not agglutinate turkey, Atlantic salmon, or tilapia red blood cells (RBCs), and therefore, it was concluded that TiLV uses mechanisms different from those used orthomyxoviruses for entry into host cells (Chengula et al. 2019). Once the virus crosses the epithelial barrier, it enters the bloodstream, as evidenced by its detection in the blood (Chiamkunakorn et al. 2019). Subsequently, the virus is distributed to internal organs like the kidney, spleen, liver, brain, gills, and muscles and starts its replication cycle (Del-Pozo et al. 2017; Mugimba et al. 2018). Inside the cells, replication and transcription of the virus have been reported to take place in the nucleus (Bacharach et al. 2016). Moreover, the replication of TiLV is not affected by ammonium chloride (NH_4Cl), a lysosomotropic agent, as has been reported for orthomyxoviruses. Ferguson et al. (2014) found that severe liver damage and gastrointestinal tract disruption would have led to protein-losing enteropathy, thereby resulting in ascites and exophthalmia in affected fish. Moreover, co-infections of *Oreochromis* spp. with TiLV and *Aeromonas hydrophila* have been reported to exacerbate the disease condition, leading to increased severity and higher mortality in tilapia (Nicholson et al. 2020).

7.7 Predisposing Factors

There is paucity of scientific knowledge regarding the epidemiological aspects of TiLVD. Importantly, TiLV has been detected in fish without clinical signs (Mugimba et al. 2018; Senapin et al. 2018), and it has been associated with variable mortality levels (9–90%) (Jansen et al. 2019; Surachetpong et al. 2020). Therefore, the appearance of clinical signs and wide variation in observed mortalities due to TiLVD could be due to the fact that diseases in aquaculture are the result of multifaceted and dynamic interactions between the environment, the condition of the host, husbandry practices, biosecurity measures, and the presence of multiple pathogens (Kibenge 2016; Subasinghe 2005). High mortalities due to TiLVD occur during the hot season, especially when the water temperature is above 25 °C, suggesting that increased water temperature is a risk factor (Eyngor et al. 2014; Ferguson et al. 2014; Nicholson et al. 2017). Besides, a high vulnerability of tilapia juveniles to TiLVD has been reported (Amal et al. 2018; Behera et al. 2018; Del-Pozo et al. 2017; Dong et al. 2017a, 2017b; Ferguson et al. 2014; Surachetpong et al. 2017). Kabuusu et al. (2018) studied production-level risk factors for syncytial hepatitis in farmed tilapia and found that the risk of mortality was two times higher in the all-male Chitralada population, in comparison to the genetically improved farm tilapia (GIFT) and genetically male tilapia (GMT) strains of Nile tilapia. They concluded that controlling the number of pond production cycles per year, improving the weight of fish at the time of transfer, and lowering stocking density at the time of seeding the pond may be more practical approaches for reducing mortalities due to TiLVD.

7.8 Co-infection

TiLV infection in tilapia has been found to be associated with parasitic infestations, namely, monogenean and ciliated protozoa (Surachetpong et al. 2017; Jansen et al. 2019). Similarly, co-infection of TiLV with bacteria has also been reported. The bacterial species isolated from tilapia concurrently infected with TiLV include *Aeromonas hydrophila*, *Aeromonas veronii*, *Aeromonas enteropelogenes*, *Aeromonas ichthiosma*, *Flavobacterium* sp., *Streptococcus agalactiae*, and *Lactococcus garvieae* (Nicholson et al. 2017; Dong et al. 2017b; Amal et al. 2018; Rao et al. 2021; Basri et al. 2020; Surachetpong et al. 2017; Swaminathan et al. 2021b). Recently, tilapia parvovirus co-infection along with TiLV infection in red hybrid tilapia has been reported in Thailand (Yamkasem et al. 2021). In cases of co-infection, determining the primary causative pathogen can be challenging. However, the co-infecting parasites or microbes are likely to have a significant impact on the outcome of the disease, particularly with respect to disease severity and progression (Abdel-Latif and Khafaga 2020).

7.9 Diagnosis

7.9.1 Clinical Signs and Gross Lesions

The clinical signs of TiLVD are non-specific and vary depending on the geographic region and age of the fish. However, in most cases, these include lethargy, anorexia, abnormal swimming behaviour, skin erosions, protrusion or loss of scales, discolouration, gill pallor, and abdominal distension (Ferguson et al. 2014; Behera et al. 2018; Tattiyapong et al. 2017). In addition, ulcers or open wounds have been reported in a few cases of TiLV infection (Fathi et al. 2017; OIE 2018c). During post-mortem, fluid in the abdominal cavity, a pale liver, and an enlarged gall bladder and spleen along with fluid-filled intestines can be observed (Tang et al. 2021) (Fig. 7a, b).

7.9.2 TiLV Detection in Cell Lines

A number of fish cell lines have been used for the isolation and propagation of TiLV. These include E-11 (Eyngor et al. 2014); CFF (Behera et al. 2018; Thangaraj et al. 2018); OmB and TmB (Kembou Tsofack et al. 2017), TiB (Wang et al. 2018); OnlB, OnlL, AFF, and AOF (Thangaraj et al. 2018); TFC#10 (Mugimba et al. 2018); TG, TH, and TE (Nanthini et al. 2019); OnH (Yadav et al. 2021), and RTE, RTS, and RTK (Suryakodi et al. 2021). The cytopathic effects are usually observed after 4–12 days of inoculation of cell lines, and these include plaque formation, aggregation, rounding, and cell detachment (Fig. 8a, b). On the other hand, no cytopathic

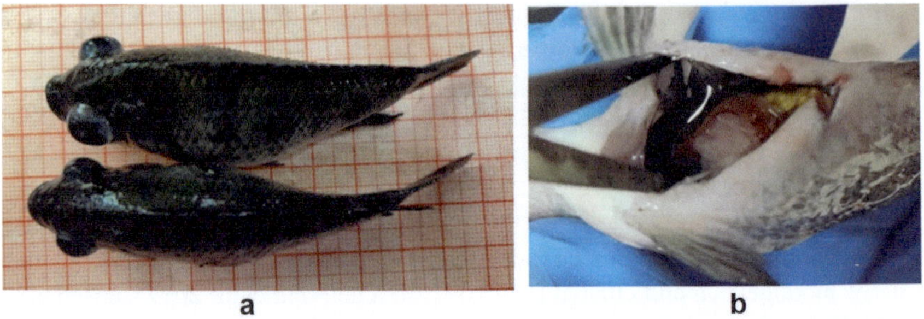

Fig. 7 Tilapia juveniles showing darkening of the body, bilateral exophthalmia and abdominal distension (**a**), and accumulation of fluid in abdominal cavity (**b**) following experimental infection with tilapia lake virus

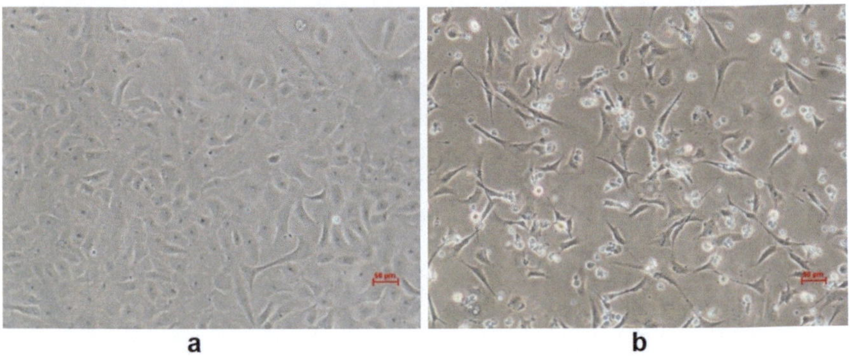

Fig. 8 A control CFF cell line (**a**) and the cytopathic effects on CFF cell line following inoculation of a filtered tissue homogenate from TiLV-infected tilapia (**b**)

effects have been reported in a number of cell lines, namely, CHSE-214, BF-2, BB, EPC, FHM, and RTG-2 (Eyngor et al. 2014); PHF (Soni et al. 2018); PSF, RTF, CCKF, HBF, and FtGF (Thangaraj et al. 2018); and TO-2 (Kembou Tsofack et al. 2017) following inoculation of TiLV, which suggests that these cell lines are not suitable for isolation of TiLV.

7.9.3 Histopathological Alterations

The major histopathological lesions of TiLVD have been reported in internal organs, namely, the liver, gastrointestinal tract, brain, eyes, spleen, and kidney (Ferguson et al. 2014; Eyngor et al. 2014). One of the most characteristic histopathological alterations in the liver of affected tilapia has been reported to be syncytia formation besides multiple foci of hepatocyte necrosis (Ferguson et al. 2014; Dong et al. 2017a; Kembou Tsofack et al. 2017; Amal et al. 2018; Behera et al. 2018). The cytoplasm of hepatocytes is foamy and may contain brightly eosinophilic or brown lipoproteinaceous material (Ferguson et al. 2014). Intracytoplasmic inclusions have also been reported in the hepatocytes of experimentally infected tilapia (Tattiyapong

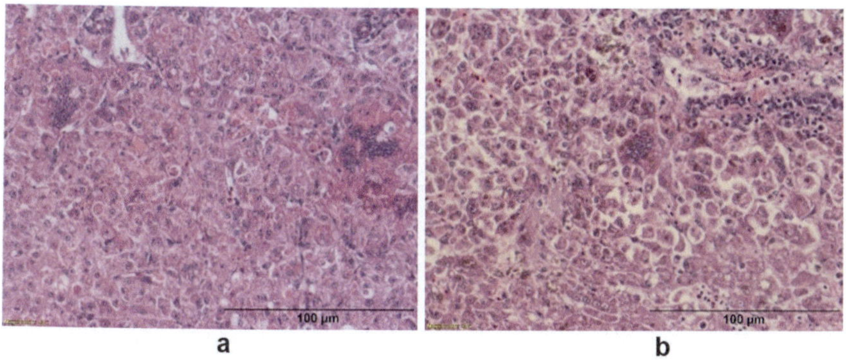

Fig. 9 Syncytial cells in a section of the liver (**a**, **b**) along with individualization and necrosis of hepatocytes (**b**) in TiLV-infected tilapia. TiLV, tilapia lake virus

et al. 2017) (Fig. 9a, b). Extensive necrosis of gastric glands with the presence of eosinophilic material has been reported in affected tilapia from Ecuador (Ferguson et al. 2014). Histopathological lesions in the brain include oedema, focal haemorrhages in leptomeninges, congestion in white and grey matter, foci of gliosis, and perivascular cuffing along with neuronal degeneration (Eyngor et al. 2014; Surachetpong et al. 2017). Ocular lesions include an undulating, thin, and ruptured lenticular capsule surrounded by circular fibroplasia and fibrosis with multiple synechiae to the iris and ciliary body. There are cataractous changes in the lens characterized by eosinophilic homogeneous spherical structures, enlarged lens epithelial cells (Eyngor et al. 2014). In the spleen, proliferating lymphocytes surround the ellipsoids along with an increase in the number and size of melanomacrophage centres (Eyngor et al. 2014). In the kidneys, moderate congestion and interstitial haemorrhages have been reported (Eyngor et al. 2014; Fathi et al. 2017).

7.9.4 Transmission Electron Microscopy

Under a transmission electron microscope, in the affected cells, nuclei often have dispersed chromatin and mitochondria appear to be enlarged and, in some cases, have lost cristae. Most virions are round, measure 60–100 nm in diameter, and can be observed in the cytoplasm of hepatocytes and the space of Disse (Ferguson et al. 2014). In most cases, virions are present within membranous intracytoplasmic structures morphologically consistent with endosomes located below the plasma membrane and at the central region of the cytoplasm (Ferguson et al. 2014; Yadav et al. 2021). On the other hand, in perinuclear areas, virions are contained within prominent membrane-bound anastomosing cisternae. Enveloped virions are 70–110-nm wide, have a moderately electron-dense core, a trilaminar capsid-like structure, and a 10- to 15-nm-thick moderately electron-dense external band (envelope) containing regularly spaced electron-dense structures (Del-Pozo et al. 2017) (Fig. 10a, b).

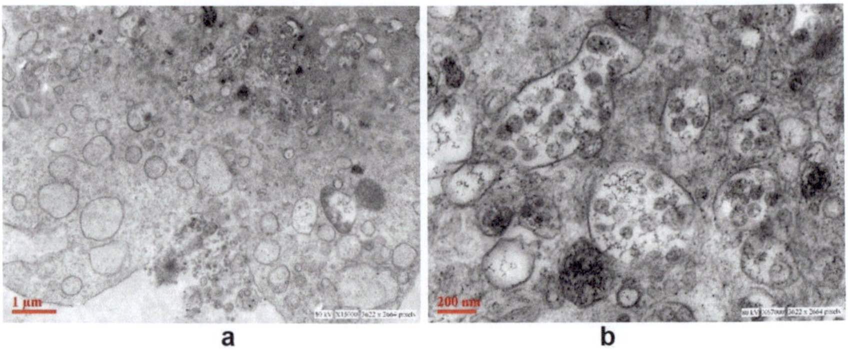

Fig. 10 TiLV virions in cytoplasmic vesicles and outside OnH cells following inoculation of TiLV. Some of the mitochondria show swelling and loss of cristae (**a**). Higher magnification of the cell showing virions in cytoplasmic vacuoles more clearly (**b**)

7.9.5 Molecular Assays

A number of reverse transcription PCR (RT-PCR) and RT quantitative PCR (RT-qPCR) methods are being used for the detection of TiLV (Table 1). Currently, most of the assays being used are based on the detection of segment 3 of TiLV. One-step RT-PCR targeting TiLV segment 3 developed by Eyngor et al. (2014) was subsequently modified as nested PCR (Kembou Tsofack et al. 2017), and the resulting sensitivity improved from 70,000 virus copies in one-step RT-PCR to 7 virus copies by nested RT-PCR. The internal primers used in nested PCR were also used for SYBR Green-based RT-qPCR with a sensitivity of 70 TiLV copies (Kembou Tsofack et al. 2017). Due to the non-specific amplification observed in external primers, Dong et al. (2017b) developed a semi-nested RT-PCR with a detection sensitivity of 7.5 virus copies, and this assay was later used for the detection of TiLV in clinically healthy adult and fingerling tilapia with no overt signs of disease (Senapin et al. 2018). Subsequently, Mugimba et al. (2018) used primers targeting segment 2 of TiLV and used the developed assay for the detection of TiLV in apparently healthy tilapia from Lake Victoria. A SYBR Green-based RT-qPCR assay for detecting a 112-bp fragment of segment 3 was developed by Tattiyapong et al. (2018). This assay can detect up to two virus copies per microlitre, which has been used to detect TiLV nucleic acids in experimentally injected carp (Pradhan et al. 2020, 2021). In addition, a TaqMan RT-qPCR assay based on primers and probe targeting segment 3 has been developed for TiLV detection (Waiyamitra et al. 2018). Two RT loop-mediated isothermal amplification (RT-LAMP) assays based on colorimetric changes have been developed to target segments 3 and 1 of TiLV (Phusantisampan et al. 2019; Yin et al. 2019), and these offer quick, cost-effective, and easy-to-interpret results.

Table 1 RT-PCR and RT-qPCR methods currently being used for detection of tilapia lake virus

S. no.	Primer	Primer sequence (5′-3′)	Product (bp)	Cycling conditions	Segment and sensitivity	References
1	NE1	TATGCAGTACTTTCCCTGCC	491	35 cycles at 95 °C for 30 s, 56 °C for 60 s, and 72 °C for 60 s; final amplification at 72 °C for 7 min	3 (70,000 virus copies)	Eyngor et al. (2014); Kembou Tsofack et al. (2017)
	NE2	TTGCTCTGAGCAAGAGTACC				
	ME1	GTTGGGCACAAGGCATCCTA	250	35 cycles at 95 °C for 30 s, 56 °C for 60 s, and 72 °C for 60 s; final amplification at 72 °C for 7 min	3 (7 virus copies)	
	ME2	TATCACGTGCGTACTCGTTCAGT				
2	NE1	TATGCAGTACTTTCCCTGCC	415	25 cycles at 94 °C for 30 s, 60 °C for 30 s, 72 °C for 30 s, followed by 1 cycle of 72 °C for 7 min	3	Dong et al. (2017b)
	ME1	GTTGGGCACAAGGCATCCTA				
	ME1	GTTGGGCACAAGGCATCCTA	250	25 cycles at 94 °C for 30 s, 60 °C for 30 s, 72 °Cfor 30 s, followed by 1 cycle of 72 °C for 7 min	3 (7.5 virus copies)	
	ME2	TATCACGTGCGTACTCGTTCAGT				
3	TiLV-112F	CTGAGCTAAAGAGGCAATATGGATT	112	Denaturation at 95 °C for 3 min followed by 40 cycles of 95 °C for 10 s and 60 °C for 30 s	3 (2 virus copies μL^{-1})	Tattiyapong et al. (2017)
	TiLV-112R	CGTGCGTACTCGTTCAGTATAAGTTCT				

(continued)

Table 1 (continued)

S. no.	Primer	Primer sequence (5′-3′)	Product (bp)	Cycling conditions	Segment and sensitivity	References
4	TiLV-93F	AGCCTGCCACACAGAAG	93	Initial denaturation step at 95 °C for 2 min, followed by 40 cycles of 95 °C for 5 s and 56 °Cfor 30 s	3 (can detect TiLV in 100 pg of RNA from infected fish tissues)	Waiyamitra et al. (2018)
	TiLV-93R	CTGCTTGAGTTGTGCTTCT				
	TiLV-93 probe	FAM-CTCTACCAGCTAGTGCCCCA-Iowa Black-3′				
5	TiLV-Seg2-F	GTCCAGGGCGGTATGTATTG	834		2[a]	Mugimba et al. (2018)
	TiLV-Seg2-R	CTTACGGCTGACAAGTCTCTAAG				

RT-PCR reverse transcription polymerase chain reaction, *RT-qPCR*, reverse transcription quantitative polymerase chain reaction
[a]Sensitivity not mentioned

7.10 Prevention and Control

TiLVD is an emerging disease and is considered to be a threat to the global tilapia industry. This disease is under consideration for inclusion in the list of finfish diseases reportable to the World Organisation for Animal Health (OIE). Since it is a viral disease, there is no effective treatment for it. The only effective measures for control of the disease include a combination of biosecurity measures, breeding of fish with improved resistance, and vaccination, as reported for other finfish viral diseases (Hjeltnes et al. 2017). The movement of live-infected tilapia poses the greatest risk for the spread of the virus. Therefore, restrictions on the movement of tilapia from TiLV-affected compartment, zone, and country to unaffected regions should be imposed. If required, such movements should always be accompanied by a health certificate stating that the consignment of the tilapia has been screened and found to be free of the virus. In a TiLV-endemic area, stocking only with TiLV-free seed and/or fry/fingerlings should be practised. As TiLV has been shown to be transmitted vertically, screening of the broodstock for TiLV employing non-lethal methods must be undertaken in tilapia hatcheries. Besides, there is wide variation in the mortality rate due to TiLVD. Therefore, efforts should be taken to minimize the stress on the fish besides maintaining the water quality parameters in the optimum range for tilapia (Tang et al. 2021).

Fish surviving TiLV outbreaks have been found to resist subsequent outbreaks, which suggests a mounting host immune response against TiLV infection (Eyngor et al. 2014). This indicates that vaccinations could be an appropriate strategy for disease control. A patent (US 2016/0354458 A1) for an attenuated TiLV vaccine has been filed in the United States. Recently, in a study, β-propiolactone-inactivated TiLV has been found to offer higher protection against TiLV challenge in comparison to formaldehyde inactivation. In addition, a β-propiolactone-inactivated TiLV vaccine coupled with the adjuvant Montanide IMS 1312 VG and booster immunization at 3 weeks provided a high level of protection from TiLV challenge (Zeng et al. 2021b). Similarly, a VP20 (encoded by segment 8)-based vaccine has been reported to elicit both humoral and cellular immune responses and significantly protect the fish from TiLV challenge. The DNA prime-protein boost immunization strategy, comprising priming with pVoptiVP20 followed by a booster with rVP2 protein in adjuvant at 3 weeks showed the potential advantage over a single vaccination (Zeng et al. 2021a). A DNA vaccine developed against TiLV (pcDNA3.1–ORF10) significantly induced protective immunity and reduced the viral load in tilapia (Yu et al. 2022).

Furthermore, efforts are underway to develop a nanoparticle tilapia lake virus vaccine.

8 Conclusions

Although TiLV is considered a threat to the global tilapia industry at present, there is lack of information regarding TiLV infection, particularly with respect to the host receptors, virus multiplication, pathogenesis, epidemiology, and control. There is a need to identify the host receptors involved in the uptake of the virus in addition to investigating the kinetics of the virus in fish organs, and this information is vital for developing control strategies. The structural and non-structural genes encoded by different segments and the role of viral genes that regulate host immune responses require further investigation. It is still unknown if there are different genotypes of TiLV in circulation in different countries and whether these are associated with varying mortality levels. The host genes that confer protection on some strains of tilapia need to be identified as these can lead to undertaking a marker-assisted breeding programme for TiLV-resistant strains of tilapia. Besides, information regarding the susceptibility and possible carrier status of non-tilapiines, especially ornamental cichlids, is crucial for preventing the spread and developing control strategies for TiLVD. Moreover, as large variations in mortalities due to TiLV have been reported in tilapia populations from different regions, there is a need to undertake epidemiological studies to identify the risk factors associated with higher mortalities. Such risk factors, including the effect of environmental parameters with particular emphasis on water temperature and co-infecting organisms on the disease outcome, need to be elucidated. Although a few diagnostic methods have been developed and are used for the detection of TiLV, there is a need for validation of these methods and finding out the method that is sensitive, specific, and universally applicable. Furthermore, the development of point-of-care diagnostics can help in the quick detection of infection with TiLV at the pond site and thus would be extremely useful for minimizing losses due to TiLV infection. Although several attempts are being made to develop vaccines against TiLV, no vaccine is available commercially. Therefore, there is a need to develop effective, cheap, and easy-to-administer vaccines. The above points emphasize the need for undertaking focused efforts to address the knowledge gaps and for developing control measures to minimize the losses due to TiLVD.

References

Abdel-Latif HM, Khafaga AF (2020) Natural co-infection of cultured Nile tilapia *Oreochromis niloticus* with *Aeromonas hydrophila* and *Gyrodactylus cichlidarum* experiencing high mortality during summer. Aquac Res 51:1880–1892

Abdullah A, Ramly R, Ridzwan MM, Sudirwan F, Abas A, Ahmad K, Murni M, Kua BC (2018) First detection of tilapia lake virus (TiLV) in wild river carp (*Barbonymus schwanenfeldii*) at Timah Tasoh Lake, Malaysia. J Fish Dis 41:1459–1462

Adamek M, Oschilewski A, Wohlsein P, Jung-Schroers V, Teitge F, Dawson A, Gela D, Piackova V, Kocour M, Adamek J, Bergmann SM (2017) Experimental infections of different carp strains with the carp edema virus (CEV) give insights into the infection biology of the virus and indicate possible solutions to problems caused by koi sleepy disease (KSD) in carp aquaculture. Vet Res 48:12

Adamek M, Hellmann J, Jung-Schroers V, Teitge F, Steinhagen D (2018) CyHV-2 transmission in traded goldfish stocks in Germany—a case study. J Fish Dis 41:401–404

Adamek M, Teitge F, Steinhagen D (2019) Quantitative diagnostics of gill diseases in common carp: not as simple as it seems. Dis Aquat Org 134:197–207

Ahasan MS, Keleher W, Giray C, Perry B, Surachetpong W, Nicholson P, Al-Hussinee L, Subramaniam K, Waltzek TB (2020) Genomic characterization of tilapia lake virus isolates recovered from moribund Nile tilapia (*Oreochromis niloticus*) on a farm in the United States. Microbiol Resour Announc 9:e01368–e01319

Amal MNA, Koh CB, Nurliyana M, Suhaiba M, Nor-Amalina Z, Santha S, Diyana-Nadhirah KP, Yusof MT, Ina-Salwany MY, Zamri-Saad M (2018) A case of natural co-infection of tilapia Lake virus and *Aeromonas veronii* in a Malaysian red hybrid tilapia (*Oreochromis niloticus* x *O. mossambicus*) farm experiencing high mortality. Aquaculture 485:12–16

Azad IS, Shekhar MS, Thirunavukkarasu AR, Poornima M, Kailasam M, Rajan JJS, Ali SA, Abraham M, Ravichandran P (2005) Nodavirus infection causes mortalities in hatchery produced larvae of *Lates calcarifer*: first report from India. Dis Aquat Org 63:113–118

Bacharach E, Mishra N, Briese T, Zody MC, Kembou Tsofack JE, Zamostiano R, Berkowitz A, Ng J, Nitido A, Corvelo A, Toussaint NC (2016) Characterization of a novel orthomyxo-like virus causing mass die-offs of tilapia. MBio 7:1110–1128

Basri L, Nor RM, Salleh A, Md. Yasin IS, Saad MZ, Abd. Rahaman NY, Barkham T, Amal MNA (2020) Co-infections of tilapia lake virus, *Aeromonas hydrophila* and *Streptococcus agalactiae* in farmed red hybrid tilapia. Animals 10:2141

Behera BK, Pradhan PK, Swaminathan TR, Sood N, Paria P, Das A, Verma DK, Kumar R, Yadav MK, Dev AK, Parida PK (2018) Emergence of tilapia lake virus associated with mortalities of farmed Nile tilapia *Oreochromis niloticus* (Linnaeus 1758) in India. Aquaculture 484:168–174

Belton B, Bush SR, Little DC (2018) Not just for the wealthy: rethinking farmed fish consumption in the Global South. Glob Food Sec 16:85–92

Chaput DL, Bass D, Alam MM, Al Hasan N, Stentiford GD, van Aerle R, Moore K, Bignell JP, Haque MM, Tyler CR (2020) The segment matters: probable reassortment of tilapia lake virus (TiLV) complicates phylogenetic analysis and inference of geographical origin of new isolate from Bangladesh. Viruses 12:258

Chengula AA, Mutoloki S, Evensen Ø, Munang'andu HM (2019) Tilapia lake virus does not hemagglutinate avian and piscine erythrocytes and NH4Cl does not inhibit viral replication in vitro. Viruses 11:1152

Chi SC, Shieh JR, Lin SJ (2003) Genetic and antigenic analysis of betanodaviruses isolated from aquatic organisms in Taiwan. Dis Aquat Org 55:221–228

Chiamkunakorn C, Machimbirike VI, Senapin S, Khunrae P, Dong HT, Rattanarojpong T (2019) Blood and liver biopsy for the non-destructive screening of tilapia lake virus. J Fish Dis 42:1629–1636

CIAT (2014) Climate change and aquatic animal disease. Retrieved June 6, 2021, from https://unfccc.int/files/documentation/submissions_from_non-party_stakeholders/application/pdf/517.pdf

Costa JZ, Thompson KD (2016) Understanding the interaction between Betanodavirus and its host for the development of prophylactic measures for viral encephalopathy and retinopathy. Fish Shellfish Immunol 53:35–49

Daněk T, Kalous L, Veselý T, Krásová E, Reschová S, Rylková K, Kulich P, Petrtýl M, Pokorová D, Knytl M (2012) Massive mortality of Prussian carp *Carassius gibelio* in the upper Elbe basin associated with herpesviral hematopoietic necrosis (CyHV-2). Dis Aquat Org 102:87–95

Davison AJ, Kurobe T, Gatherer D, Cunningham C, Korf I, Fukuda H, Hedrick RP, Waltzek TB (2013) Comparative genomics of carp herpesviruses. J Virol 87:2908–2922

Del-Pozo J, Mishra N, Kabuusu R, Cheetham S, Eldar A, Bacharach E, Lipkin WI, Ferguson HW (2017) Syncytial hepatitis of tilapia (*Oreochromis niloticus* L.) is associated with orthomyxovirus-like virions in hepatocytes. Vet Pathol 54:164–170

Dong HT, Ataguba GA, Khunrae P, Rattanarojpong T, Senapin S (2017a) Evidence of TiLV infection in tilapia hatcheries from 2012 to 2017 reveals probable global spread of the disease. Aquaculture 479:579–583

Dong HT, Siriroob S, Meemetta W, Santimanawong W, Gangnonngiw W, Pirarat N, Khunrae P, Rattanarojpong T, Vanichviriyakit R, Senapin S (2017b) Emergence of tilapia lake virus in Thailand and an alternative semi-nested RT-PCR for detection. Aquaculture 476:111–118

Dong HT, Senapin S, Gangnonngiw W, Nguyen VV, Rodkhum C, Debnath PP, Delamare-Deboutteville J, Mohan CV (2020) Experimental infection reveals transmission of tilapia lake virus (TiLV) from tilapia broodstock to their reproductive organs and fertilized eggs. Aquaculture 515:734541

Eyngor M, Zamostiano R, Kembou Tsofack JE, Berkowitz A, Bercovier H, Tinman S, Lev M, Hurvitz A, Galeotti M, Bacharach E, Eldar A (2014) Identification of a novel RNA virus lethal to tilapia. J Clin Microbiol 52:4137–4146

FAO (2020) The state of world fisheries and aquaculture 2020. FAO https://doi.org/10.4060/ca9229en

Fathi M, Dickson C, Dickson M, Leschen W, Baily J, Muir F, Ulrich K, Weidmann M (2017) Identification of tilapia Lake virus in Egypt in Nile tilapia affected by 'summer mortality'syndrome. Aquaculture 473:430–432

Ferguson HW, Kabuusu R, Beltran S, Reyes E, Lince JA, del Pozo J (2014) Syncytial hepatitis of farmed tilapia, *Oreochromis niloticus* (L.): a case report. J Fish Dis 37:583–589

Fichi G, Cardeti G, Cocumelli C, Vendramin N, Toffan A, Eleni C, Siemoni N, Fischetti R, Susini F (2013) Detection of cyprinid herpesvirus 2 in association with an *Aeromonas sobria* infection of *Carassius carassius* (L.), in Italy. J Fish Dis 36:823–830

Figueiredo HCP, Tavares GC, Dorella FA, Rosa JCC, Marcelino SAC, Pierezan F, Pereira FL (2022) First report of infectious spleen and kidney necrosis virus in Nile tilapia in Brazil. Transbound Emerg Dis 69:3008–3015

Frerichs GN, Tweedie A, Starkey WG, Richards RH (2000) Temperature, pH and electrolyte sensitivity, and heat, UV and disinfectant inactivation of sea bass (*Dicentrarchus labrax*) neuropathy nodavirus. Aquaculture 185:13–24

Furusawa R, Okinaka Y, Nakai T (2006) Betanodavirus infection in the freshwater model fish medaka (*Oryzias latipes*). J Gen Virol 87:2333–2339

Girisha SK, Puneeth TG, Nithin MS, Kumar BN, Ajay SK, Vinay TN, Suresh T, Venugopal MN, Ramesh KS (2020) Red Sea bream iridovirus disease (RSIVD) outbreak in Asian seabass (*Lates calcarifer*) cultured in open estuarine cages along the west coast of India: first report. Aquaculture 520:734712

Girisha SK, Kushala KB, Nithin MS, Puneeth TG, Naveen Kumar BT, Vinay TN, Suresh T, Ajay SK, Venugopal MN, Ramesh KS (2021) First report of the infectious spleen and kidney necrosis virus (ISKNV) infection in ornamental fishes in India. Transbound Emerg Dis 68:964–972

Glazebrook JS, Heasman MP, De Beer SW (1990) Picorna-like viral particles associated with mass mortalities in larval barramundi, *Lates calcarifer* Bloch. J Fish Dis 13:245–249

Go J, Whittington R (2006) Experimental transmission and virulence of a megalocytivirus (family Iridoviridae) of dwarf gourami (*Colisa lalia*) from Asia in Murray cod (*Macculochella peelii peelii*) in Australia. Aquaculture 258:140–149

Goodwin AE, Khoo L, LaPatra SE, Bonar C, Key DW, Garner M, Lee MV, Hanson L (2006) Goldfish hematopoietic necrosis herpesvirus (cyprinid herpesvirus 2) in the USA: molecular confirmation of isolates from diseased fish. J Aquat Anim Health 18:11–18

Goodwin AE, Sadler J, Merry GE, Marecaux EN (2009) Herpesviral haematopoietic necrosis virus (CyHV-2) infection: case studies from commercial goldfish farms. J Fish Dis 32:271–278

Grimmett SG, Warg JV, Getchell RG, Johnson DJ, Bowser PR (2006) An unusual koi herpesvirus associated with a mortality event of common carp *Cyprinus carpio* in New York State, USA. J Wildl Dis 42:658–662

Groff JM, LaPatra SE, Munn RJ, Zinkl JG (1998) A viral epizootic in cultured populations of juvenile goldfish due to a putative herpesvirus etiology. J Vet Diagn Invest 10:375–378

Grotmol S, Totland GK (2000) Surface disinfection of Atlantic halibut *Hippoglossus hippoglossus* eggs with ozonated sea-water inactivates nodavirus and increases survival of the larvae. Dis Aquat Org 39:89–96

He JG, Deng M, Weng SP, Li Z, Zhou SY, Long QX, Wang XZ, Chan SM (2001) Complete genome analysis of the mandarin fish infectious spleen and kidney necrosis iridovirus. Virology 291:126–139

He JG, Zeng K, Weng SP, Chan SM (2002) Experimental transmission, pathogenicity and physical–chemical properties of infectious spleen and kidney necrosis virus (ISKNV). Aquaculture 204:11–24

Hedrick RP, Gilad O, Yun S, Spangenberg JV, Marty GD, Nordhausen RW, Kebus MJ, Bercovier H, Eldar A (2000) A herpesvirus associated with mass mortality of juvenile and adult koi, a strain of common carp. J Aquat Anim Health 12:44–57

Hedrick RP, Waltzek TB, McDowell TS (2006) Susceptibility of koi carp, common carp, goldfish, and goldfish× common carp hybrids to cyprinid herpesvirus-2 and herpesvirus-3. J Aquat Anim Health 18:26–34

Hjeltnes B, Bang-Jensen B, Bornø G, Haukaas A, Walde C (2017) The health situation in Norwegian aquaculture 2016. Norwegian Veterinary Institute, 127

Inouye K, Yamano K, Maeno Y, Nakajima K, Matsuoka M, Wada Y, Sorimachi M (1992) Iridovirus infection of cultured red sea bream, *Pagrus major*. Fish Pathol 27:19–27

Iwamoto T, Nakai T, Mori KI, Arimoto M, Furusawa I (2000) Cloning of the fish cell line SSN-1 for piscine nodaviruses. Dis Aquat Org 43:81–89

Iwamoto, T., Mise, K., Mori, K.I., Arimoto, M., Nakai, T. And Okuno, T., 2001. Establishment of an infectious RNA transcription system for Striped jack nervous necrosis virus, the type species of the betanodaviruses. J Gen Virol 82, 2653–2662

Jaemwimol P, Rawiwan P, Tattiyapong P, Saengnual P, Kamlangdee A, Surachetpong W (2018) Susceptibility of important warm water fish species to tilapia lake virus (TiLV) infection. Aquaculture 497:462–468

Jansen MD, Dong HT, Mohan CV (2019) Tilapia lake virus: a threat to the global tilapia industry. Rev Aquac 11:725–739

Jeffery KR, Bateman K, Bayley A, Feist SW, Hulland J, Longshaw C, Stone D, Woolford G, Way K (2007) Isolation of a cyprinid herpesvirus 2 from goldfish, *Carassius auratus* (L.), in the UK. J Fish Dis 30:649–656

Jeong JB, Kim HY, Jun LJ, Lyu JH, Park NG, Kim JK, Do Jeong H (2008) Outbreaks and risks of infectious spleen and kidney necrosis virus disease in freshwater ornamental fishes. Dis Aquat Org 78:209–215

Johansen R, Ranheim T, Hansen MK, Taksdal T, Totland GK (2002) Pathological changes in juvenile Atlantic halibut *Hippoglossus hippoglossus* persistently infected with nodavirus. Dis Aquat Org 50:161–169

Jung SJ, Miyazaki T (1995) Herpesviral haematopoietic necrosis of goldfish, *Carassius auratus* (L.). J Fish Dis 18:211–220

Jung S, Miyazaki T, Miyata M, Danayadol Y, Tanaka S (1997) Pathogenicity of iridovirus from Japan and Thailand for the red sea bream *Pagrus major* in Japan, and histopathology of experimentally infected fish. Fish Sci 63:735–740

Jung-Schroers V, Adamek M, Wohlsein P, Wolter J, Wedekind H, Steinhagen D (2016) First outbreak of an infection with infectious spleen and kidney necrosis virus (ISKNV) in ornamental fish in Germany. Dis Aquat Org 119:239–244

Kabuusu RM, Aire AT, Stroup DF, Macpherson CNL, Ferguson HW (2018) Production-level risk factors for syncytial hepatitis in farmed tilapia (*Oreochromis niloticus* L). J Fish Dis 41:61–66

Kawakami H, Nakajima K (2002) Cultured fish species affected by Red Sea bream Iridoviral disease from 1996 to 2000. Fish Pathol 37:45–47

Kembou Tsofack JE, Zamostiano R, Watted S, Berkowitz A, Rosenbluth E, Mishra N, Briese T, Lipkin WI, Kabuusu RM, Ferguson H, Del Pozo J (2017) Detection of tilapia lake virus in clinical samples by culturing and nested reverse transcription-PCR. J Clin Microbiol 55:759–767

Kibenge FSB (2016) Determinants of emergence of viral diseases in aquaculture. In: Aquaculture virology. Academic Press, Massachusetts, pp 95–116

Koesharyani I, Gardenia L, Widowati Z, Khumaira K, Rustianti D (2018) Studi kasus infeksi tilapia lake virus (TiLV) pada ikan nila (*Oreochromis niloticus*). Jurnal Riset Akuakultur 13:85–92

Kuo HC, Wang TY, Chen PP, Chen YM, Chuang HC, Chen TY (2011) Real-time quantitative PCR assay for monitoring of nervous necrosis virus infection in grouper aquaculture. J Clin Microbiol 49:1090–1096

Lafferty KD, Hofmann EE (2016) Marine disease impacts, diagnosis, forecasting, management and policy. Philos Trans R Soc B Biol Sci 371:20150200

Lancaster MJ, Williamson MM, Schroen CJ (2003) Iridovirus-associated mortality in farmed Murray cod (*Maccullochella peelii peelii*). Aust Vet J 81:633–634

Lewisch E, Gorgoglione B, Way K, El-Matbouli M (2015) Carp edema virus/koi sleepy disease: an emerging disease in Central-East Europe. Transbound Emerg Dis 62:6–12

Li L, Luo Y, Gao Z, Huang J, Zheng X, Nie H, Zhang J, Lin L, Yuan J (2015) Molecular characterisation and prevalence of a new genotype of cyprinid herpesvirus 2 in mainland China. Can J Microbiol 61:381–387

Liamnimitr P, Thammatorn W, Sonicha U, Tattiyapong P, Surachetpong W (2018) Non-lethal sampling for tilapia Lake virus detection by RT-qPCR and cell culture. Aquaculture 486:75–80

Maltese C, Bovo G (2007) Viral encephalopathy and retinopathy. Ittiopatologia 4:93–146

Matsuoka S, Inouye K, Nakajima K (1996) Cultured fish species affected by red sea bream iridoviral disease from 1991 to 1995. Fish Pathol 31:233–234

Miyazaki, T., Isshiki, T. And Katsuyuki, H., 2005. Histopathological and electron microscopy studies on sleepy disease of koi *Cyprinus carpio koi* in Japan. Dis Aquat Org 65, 197–207

Mori KI, Nakai T, Muroga K, Arimoto M, Mushiake K, Furusawa I (1992) Properties of a new virus belonging to nodaviridae found in larval striped jack (*Pseudocaranx dentex*) with nervous necrosis. Virology 187:368–371

Mori K, Sugaya T, Nishiocat G, Fujinamy DK, Oka Y, Arimoto M, Nakai T (2005) Detection of Betanodaviruses from feed fish used in marine aquaculture. Abstract O-142. In: 12th international conference diseases of fish and shellfish, European Association of Fish Pathologists, pp 11–16

Mugimba KK, Chengula AA, Wamala S, Mwega ED, Kasanga CJ, Byarugaba DK, Mdegela RH, Tal S, Bornstein B, Dishon A, Mutoloki S (2018) Detection of tilapia lake virus (TiLV) infection by PCR in farmed and wild Nile tilapia (*Oreochromis niloticus*) from Lake Victoria. J Fish Dis 41:1181–1189

Mushiake K, Nishizawa T, Nakai T, Furusawa I, Muroga K (1994) Control of VNN in Striped Jack: selection of Spawners based on the detection of SJNNV gene by polymerase chain reaction (PCR). Fish Pathol 29:177–182

Nakajima K, Maeno Y (1998) Pathogenicity of Red Sea bream Iridovirus and other fish Iridoviruses to Red Sea bream. Fish Pathol 33:143–144

Nanthini R, Abdul Majeed S, Vimal S, Taju G, Sivakumar S, Santhosh Kumar S, Pillai D, Sneha KG, Rakesh CG, Sahul Hameed AS (2019) In vitro propagation of tilapia lake virus in cell lines developed from *Oreochromis mossambicus*. J Fish Dis 42:1543–1552

Nicholson P, Fathi MA, Fischer A, Mohan C, Schieck E, Mishra N, Heinimann A, Frey J, Wieland B, Jores J (2017) Detection of Tilapia Lake virus in Egyptian fish farms experiencing high mortalities in 2015. J Fish Dis 40:1925–1928

Nicholson P, Mon-On N, Jaemwimol P, Tattiyapong P, Surachetpong W (2020) Coinfection of tilapia lake virus and *Aeromonas hydrophila* synergistically increased mortality and worsened the disease severity in tilapia (*Oreochromis* spp.). Aquaculture 520:734746

Noga EJ (2010) Fish disease: diagnosis and treatment. Wiley, p 519

Nopadon P, Aranya P, Tipaporn T, Toshihiro N, Takayuki K, Masashi M, Makoto E (2009) Nodavirus associated with pathological changes in adult spotted coral groupers (*Plectropomus maculatus*) in Thailand with viral nervous necrosis. Res Vet Sci 87:97–101

OIE (2017a) Immediate notifications and follow-ups. Tilapia lake virus, the Philippines. Retrieved from https://www.oie.int/wahis_2/public/wahid.php/Reviewreport/Review?page_refer=MapFullEventReport&reportid=25278

OIE (2017b) Immediate notifications and follow-ups. Tilapia lake virus, Chinese Taipei. Retrieved from https://www.oie.int/wahis_2/public/wahid.php/Reviewreport/Review?reportid=24033

OIE (2018a) Immediate notification. Tilapia lake virus, USA. Retrieved from https://www.oie.int/wahis_2/public/wahid.php/Reviewreport/Review?page_refer=MapFu llEventReport&reportid=30412

OIE (2018b) Immediate notification. Tilapia lake virus, Mexico. Retrieved from https://www.oie.int/wahis_2/public/wahid.php/Reviewreport/Review?page_refer=MapFu llEventReport&reportid=27650

OIE (2018c) Immediate notification. Tilapia lake virus, Peru. Retrieved from https://www.oie.int/wahis_2/public/wahid.php/Reviewreport/Review?reportid=26818

OIE (2019a) Red Sea bream iridoviral disease, Vol. 2.3.8 In: Manual of diagnostic tests for aquatic animals. OIE, Paris

OIE (2019b) Viral encephalopathy and retinopathy, p. 2.3.12. In: Manual of diagnostic tests for aquatic animals. OIE, Paris

Panicz R, Sadowski J, Eljasik P (2019) Detection of Cyprinid herpesvirus 2 (CyHV-2) in symptomatic ornamental types of goldfish (*Carassius auratus*) and asymptomatic common carp (*Cyprinus carpio*) reared in warm-water cage culture. Aquaculture 504:131–138

Phusantisampan T, Tattiyapong P, Mutrakulcharoen P, Sriariyanun M, Surachetpong W (2019) Rapid detection of tilapia lake virus using a one-step reverse transcription loop-mediated isothermal amplification assay. Aquaculture 507:35–39

Piamsomboon P, Wongtavatchai J (2021) Detection of tilapia Lake virus (TiLV) in healthy fish from the pre-existing disease environment using different RT-PCR methods. Turk J Fish Aquat Sci 21:205–209

Pierezan F, Yun S, Surachetpong W, Soto E (2019) Intragastric and intracoelomic injection challenge models of tilapia lake virus infection in Nile tilapia (*Oreochromis niloticus* L.) fingerlings. J Fish Dis 42:1301–1307

Pradhan PK, Paria A, Yadav MK, Verma DK, Gupta S, Swaminathan TR, Rathore G, Sood N, Lal KK (2020) Susceptibility of Indian major carp *Labeo rohita* to tilapia lake virus. Aquaculture 515:734567

Pradhan PK, Verma DK, Yadav SC, Dev AK, Swaminathan TR, Paria A, Valappil RK, Sood N (2021) Carps, *Catla catla, Cirrhinus mrigala* and *Hypophthalmichthys molitrix* are resistant to experimental infection with tilapia lake virus (TiLV). Aust Fish 6:56

Pulido LLH, Mora CM, Hung AL, Dong HT, Senapin S (2019) Tilapia lake virus (TiLV) from Peru is genetically close to the Israeli isolates. Aquaculture 510:61–65

Puneeth TG, Baliga P, Girisha SK, Shekar M, Nithin MS, Suresh T, Kumar BN (2021) Complete genome analysis of a red seabream iridovirus (RSIV) isolated from Asian seabass (*Lates calcarifer*) in India. Virus Res 291:198199

Raj NS, Sudhagar A, Swaminathan TR, Mohandas SP, Sood N, Pradhan PK, Basheer VS, Sarkar UK (2024) Unveiling the ISKNV menace: disease outbreak investigations in the Oscar fish (*Astronotus ocellatus*) farms of Kerala, India. J Aquac 33:79–93

Ramírez-Paredes JG, Paley RK, Hunt W, Feist SW, Stone DM, Field TR, Haydon DJ, Ziddah PA, Nkansa M, Guilder J, Gray J (2021) First detection of infectious spleen and kidney necrosis virus (ISKNV) associated with massive mortalities in farmed tilapia in Africa. Transbound Emerg Dis 68:1550–1563

Rao M, Kumar SH, Kumar S, Bedekar MK, Tripathi G, Kooloth Valappil R (2021) Microbiological investigation of tilapia lake virus–associated mortalities in cage-farmed *Oreochromis niloticus* in India. Aquac Int 29:511–526

Rehman T, Yin L, Latif MB, Zhou Y, Wang K, Geng Y, Huang X, Chen D, Fang J, Chen Z, Guo H (2020) Current findings on carp edema virus, control challenges, and future outlook. Aquac Int 28:2015–2026

Sahoo, P.K. and Goodwin, A.E., 2012. Viruses of freshwater finfish in the Asian–Pacific region. Indian J Virol 23, 99–105

Sahoo PK, Swaminathan TR, Abraham TJ, Kumar R, Pattanayak S, Mohapatra A, Rath SS, Patra A, Adikesavalu H, Sood N, Pradhan PK (2016) Detection of goldfish haematopoietic necrosis herpes virus (cyprinid herpesvirus-2) with multi-drug resistant *Aeromonas hydrophila* infection in goldfish: first evidence of any viral disease outbreak in ornamental freshwater aquaculture farms in India. Acta Trop 161:8–17

Senapin S, Shyam KU, Meemetta W, Rattanarojpong T, Dong HT (2018) Inapparent infection cases of tilapia lake virus (TiLV) in farmed tilapia. Aquaculture 487:1–55

Shi CY, Wang YG, Yang SL, Huang J, Wang QY (2004) The first report of an iridovirus-like agent infection in farmed turbot, *Scophthalmus maximus*, in China. Aquaculture 236:11–25

Shiu JY, Hong JR, Ku CC, Wen CM (2018) Complete genome sequence and phylogenetic analysis of megalocytivirus RSIV-Ku: a natural recombination infectious spleen and kidney necrosis virus. Arch Virol 163:1037–1042

Soliman H, Lewisch E, El-Matbouli M (2019) Identification of new genogroups in Austrian carp edema virus isolates. Dis Aquat Org 136:193–197

Soni P, Pradhan PK, Swaminathan TR, Sood N (2018) Development, characterization and application of a new epithelial cell line from caudal fin of *Pangasianodon hypophthalmus* (Sauvage 1878). Acta Trop 182:215–222

Subasinghe RP (2005) Epidemiological approach to aquatic animal health management: opportunities and challenges for developing countries to increase aquatic production through aquaculture. Prev Vet Med 67:117–124

Subramaniam K, Shariff M, Omar AR, Hair-Bejo M (2012) Megalocytivirus infection in fish. Rev Aquac 4:221–233

Sukenda S, Gardenia L, Zairin M Jr, Lusiastuti A, Alimudin A (2020) Identification of giant gourami iridovirus (GGIV): a new infectious spleen and kidney necrosis virus (ISKNV) from natural outbreak in cultured *Osphronemus goramy*. Aquac Int 28:1069–1082

Surachetpong W, Janetanakit T, Nonthabenjawan N, Tattiyapong P, Sirikanchana K, Amonsin A (2017) Outbreaks of tilapia lake virus infection, Thailand, 2015–2016. Emerg Infect Dis 23:1031

Surachetpong W, Roy SRK, Nicholson P (2020) Tilapia lake virus: the story so far. J Fish Dis 43:1115–1132

Suryakodi S, Majeed SA, Taju G, Vimal S, Sivakumar S, Ahmed AN, Shah FA, Bhat SA, Sarma D, Begum A, Hameed AS (2021) Development and characterization of novel cell lines from kidney and eye of rainbow trout, *Oncorhynchus mykiss* for virological studies. Aquaculture 532:736027

Swaminathan TR, Kumar R, Dharmaratnam A, Basheer VS, Sood N, Pradhan PK, Sanil NK, Vijayagopal P, Jena JK (2016) Emergence of carp edema virus in cultured ornamental koi carp, *Cyprinus carpio* koi, in India. J Gen Virol 97:3392–3399

Swaminathan TR, Raj NS, Preena PG, Pradhan PK, Sood N, Kumar RG, Sudhagar A, Sood NK (2021a) Infectious spleen and kidney necrosis virus-associated large-scale mortality in farmed giant gourami, *Osphronemus goramy*, in India. J Fish Dis 44:2043–2053

Swaminathan TR, Nithyanantham SR, Narendrakumar L, Dharmaratnam A, Sood N, Pradhan PK, Ramachandra KSS, Lal KK (2021b) Co-infection of *Lactococcus garvieae* and Tilapia lake virus (TiLV) in Nile tilapia *Oreochromis niloticus* cultured in India. Dis Aquat Org 147:127–140

Tanaka N, Izawa T, Kuwamura M, Higashiguchi N, Kezuka C, Kurata O, Wada S, Yamate J (2014) The first case of infectious spleen and kidney necrosis virus (ISKNV) infection in aquarium-maintained mandarin fish, *Siniperca chuatsi* (Basilewsky), in Japan. J Fish Dis 37:401–405

Tang KF, Bondad-Reantaso MG, Surachetpong W, Dong HT, Fejzic N, Wang Q, Wajsbrot N, Hao B (2021) Tilapia lake virus disease strategy manual. FAO Fisheries and Aquaculture Circular No. 1220. FAO, Rome, p 48

Tattiyapong P, Dachavichitlead W, Surachetpong W (2017) Experimental infection of tilapia Lake virus (TiLV) in Nile tilapia (*Oreochromis niloticus*) and red tilapia (*Oreochromis* spp.). Vet Microbiol 207:170–177

Tattiyapong P, Sirikanchana K, Surachetpong W (2018) Development and validation of a reverse transcription quantitative polymerase chain reaction for tilapia lake virus detection in clinical samples and experimentally challenged fish. J Fish Dis 41:255–261

Thammatorn W, Rawiwan P, Surachetpong W (2019) Minimal risk of tilapia lake virus transmission via frozen tilapia fillets. J Fish Dis 42:3–9

Thangaraj RS, Ravi C, Kumar R, Dharmaratnam A, Saidmuhammed BV, Pradhan PK, Sood N (2018) Derivation of two tilapia (*Oreochromis niloticus*) cell lines for efficient propagation of Tilapia Lake virus (TiLV). Aquaculture 492:206–214

Thangaraj RS, Nithianantham SR, Dharmaratnam A, Kumar R, Pradhan PK, Thangalazhy Gopakumar S, Sood N (2021) Cyprinid herpesvirus-2 (CyHV-2): a comprehensive review. Rev Aquac 13:796–821

Thangaraj RS, Nithianantham SR, Narendrakumar L, Johny TK, Sood N, Pradhan PK, Lal KK (2022) Cichlids endemic to India are not susceptible to Tilapia Lake virus infection. Aquaculture 548:737589

Thao HPN, Triet HT, Vy HTN, Thao TTL, Hieu NCB, Huong TTT, Mai HCN, Huy HTL, Linh NT (2020) Modified semi-nested RT-PCR detecting tilapia lake virus (TiLV) and initial isolation of TiLV virus from Vietnamese tilapia. J Mekong Fish 18:45–60

Verma DK, Sood N, Paria A, Swaminathan TR, Mohan CV, Rajendran KV, Pradhan PK (2022) Reassortment and evolutionary dynamics of tilapia lake virus genomic segments. Virus Res 308:198625

Waiyamitra P, Tattiyapong P, Sirikanchana K, Mongkolsuk S, Nicholson P, Surachetpong W (2018) A TaqMan RT-qPCR assay for tilapia lake virus (TiLV) detection in tilapia. Aquaculture 497:184–188

Waiyamitra P, Piewbang C, Techangamsuwan S, Liew WC, Surachetpong W (2021) Infection of tilapia tilapinevirus in Mozambique tilapia (*Oreochromis mossambicus*), a globally vulnerable fish species. Viruses 13:1104

Walker PJ, Winton JR (2010) Emerging viral diseases of fish and shrimp. Vet Res 41:51

Wang L, He J, Liang L, Zheng X, Jia P, Shi X, Lan W, Xie J, Liu H, Xu P (2012) Mass mortality caused by Cyprinid Herpesvirus 2 (CyHV-2) in Prussian carp (*Carassius gibelio*) in China. Bull Eur Assoc Fish Pathol 32:164–173

Wang Y, Wang Q, Zeng W, Yin J, Li Y, Ren Y, Shi C, Bergmann SM, Zhu X (2018) Establishment and characterization of a cell line from tilapia brain for detection of tilapia lake virus. J Fish Dis 41:1803–1809

Way K, Stone D (2013) Emergence of carp edema virus-like (CEV-like) disease in the UK. CEFAS Finfish News 15:32–34

Whittington RJ, Chong R (2007) Global trade in ornamental fish from an Australian perspective: the case for revised import risk analysis and management strategies. Prev Vet Med 81:92–116

Whittington RJ, Becker JA, Dennis MM (2010) Iridovirus infections in finfish–critical review with emphasis on ranaviruses. J Fish Dis 33:95–122

Wolf K (2019) Fish viruses and fish viral diseases. Cornell University Press, p 480

Wu T, Ding Z, Ren M, An L, Xiao Z, Liu P, Gu W, Meng Q, Wang W (2013) The histo-and ultra-pathological studies on a fatal disease of Prussian carp (*Carassius gibelio*) in mainland China associated with cyprinid herpesvirus 2 (CyHV-2). Aquaculture 412:8–13

Xu J, Zeng L, Zhang H, Zhou Y, Ma J, Fan Y (2013) Cyprinid herpesvirus 2 infection emerged in cultured gibel carp, *Carassius auratus gibelio* in China. Vet Microbiol 166:138–144

Yadav MK, Rastogi A, Joaquin MPC, Verma DK, Rathore G, Swaminathan TR, Paria A, Pradhan PK, Sood N (2021) Establishment and characterization of a continuous cell line from heart of Nile tilapia *Oreochromis niloticus* and its susceptibility to tilapia lake virus. J Virol Methods 287:113989

Yamkasem J, Tattiyapong P, Kamlangdee A, Surachetpong W (2019) Evidence of potential vertical transmission of tilapia lake virus. J Fish Dis 42:1293–1300

Yamkasem J, Piewbang C, Techangamsuwan S, Pierezan F, Soto E, Surachetpong W (2021) Susceptibility of ornamental African cichlids *Aulonocara* spp. to experimental infection with tilapia lake virus. Aquaculture 542:736920

Yanong RP, Waltzek TB (2010) Megalocytivirus infections in fish, with emphasis on ornamental species. University of Florida EDIS. FA182

Yin J, Wang Q, Wang Y, Li Y, Zeng W, Wu J, Ren Y, Tang Y, Gao C, Hu H, Bergmann SM (2019) Development of a simple and rapid reverse transcription–loopmediated isothermal amplification (RT-LAMP) assay for sensitive detection of tilapia lake virus. J Fish Dis 42:817–824

Yu NT, Zeng WW, Xiong Z, Liu ZX (2022) A high efficacy DNA vaccine against Tilapia lake virus in Nile tilapia (*Oreochromis niloticus*). Aquacult Rep 24:101166

Zeng W, Wang Y, Chen X, Wang Q, Bergmann SM, Yang Y, Wang Y, Li BO, Lv Y, Li H, Lan W (2021a) Potency and efficacy of VP20-based vaccine against tilapia lake virus using different prime-boost vaccination regimens in tilapia. Aquaculture 539:736654

Zeng W, Wang Y, Hu H, Wang Q, Bergmann SM, Wang Y, Li BO, Lv Y, Li H, Yin J, Li Y (2021b) Cell culture-derived tilapia lake virus-inactivated vaccine containing montanide adjuvant provides high protection against viral challenge for tilapia. Vaccine 9:86

Zorriehzahra MJ (2020) Viral nervous necrosis disease. In: Emerging and reemerging viral pathogens. Academic Press, Massachusetts, pp 673–703

Viral Diseases in Brackishwater Fishes

M. Makesh, K. V. Rajendran, and S. K. Girisha

Abstract

Brackishwater fish production has increased in recent years and is poised to narrow the gap between production and demand since there are limitations in the expansion of freshwater aquaculture. Increased culture intensity and species diversification have resulted in increased disease outbreaks in brackishwater aquaculture, especially those caused by viruses. The important viral diseases affecting brackishwater fishes include infectious haematopoietic necrosis, infectious salmon anaemia, pancreas disease or sleeping disease, viral haemorrhagic septicaemia, infectious pancreatic necrosis, viral nervous necrosis and red sea bream iridoviral disease. Most of these diseases cause serious economic losses and are OIE (World Organisation for Animal Health) notifiable diseases. Of these, viral nervous necrosis and red sea bream iridoviral disease have been recorded from brackishwater fishes in India. The nature of the disease, its causative agent, transmission mechanism, pathogenesis, diagnosis and control measures, including the recent advances in vaccine development, are reviewed in this chapter. As no therapeutic drugs are available for these diseases, their control depends on adopting good management practices and strict biosecurity protocols to prevent pathogen introduction into the facility, including quarantine and screening of stocks before introduction into a hatchery or culture facility. Further,

M. Makesh (✉)
ICAR-Central Institute of Brackishwater Aquaculture, Chennai, India
e-mail: mmakesh@yahoo.com

K. V. Rajendran
ICAR-Central Institute of Fisheries Education, Mumbai, India
e-mail: rajendrankv@hotmail.com

S. K. Girisha
KVAFSU-College of Fisheries, Mangalore, India
e-mail: skgirisha@gmail.com

R. Kooloth Valappil et al. (eds.), *Aquatic Animal Health Management*,
https://doi.org/10.1007/978-981-96-7987-4_6

notifying disease-free zones, preventing the introduction of stocks from endemic areas, and vaccinating susceptible populations with a potent vaccine are also crucial. As vaccination by injection is not feasible in smaller fish, the development of mucosal vaccines, which can be administered orally, by immersion or through the nasal route, stimulating mucosal immunity, thereby preventing pathogen entry through these entry portals, is essential. The development and application of immunostimulants is also vital. As a long-term strategy, selective breeding for disease resistance is of key significance in preventing viral diseases in brackishwater aquaculture.

Keywords

Viral diseases · Brackishwater fishes · Diagnosis · Control

1 Introduction

Global aquaculture production in 2022 reached a record 130.9 million tonnes, of which aquatic animals accounted for 94.4 million tonnes (FAO 2024). Of the total aquaculture production, 35.3 million tonnes were produced from marine resources, including brackishwater. The increase in aquaculture production mostly stems from the intensification of culture systems and the diversification of farmed species. The increasing competition for freshwater for agricultural, industrial and human use will limit the expansion of freshwater aquaculture. Further, marine capture fish production has plateaued over the last couple of decades. Hence, the vast unused brackishwater and coastal marine waters are looked upon as potential sources for increasing aquaculture production. The use of open brackishwater resources such as estuaries, creeks, backwaters and lagoons for aquaculture will reduce the pressure on freshwater aquaculture systems to a considerable extent (Kumaran et al. 2021). Many aquaculture research institutes across the globe are trying to breed marine fish under captive conditions, especially in brackishwater, to meet the global need for increased food production for the growing population. As a result, the number of fish species farmed in brackishwater has increased steadily.

With the expansion of farming areas, intensification of farming practices and species diversification, the incidence of diseases, especially viral diseases in brackishwater aquaculture, is increasing. The causative agents of these diseases are temperature-dependent and cause disease in geographical regions with a particular range of water temperatures. However, environmental changes like global warming increase the risk of the spread of the virus to new geographical areas and host species previously not known to be infected. A thorough understanding of the disease and the viruses affecting brackishwater fishes is essential to control the incidence and the consequent economic losses. This chapter briefly discusses the important viral diseases of brackishwater fishes, the causative agent, modes of transmission, its pathogenesis, diagnosis and control. Some of the viruses that affect both marine and freshwater fish species are also included. A summary of important characteristics of the causative agents of viral diseases of brackishwater fishes is given in Table 1.

2 Viral Diseases of Brackishwater Fishes

2.1 Infectious Haematopoietic Necrosis

Infectious haematopoietic necrosis (IHN) is an economically important OIE (World Organisation for Animal Health) notifiable disease of salmonids. Although outbreaks of the disease occurred in the 1940s and 1950s, the viral aetiology of the outbreaks was first recorded in 1951 in sockeye salmon (*Oncorhynchus nerka*) in the USA (Rucker et al. 1953), and the virus isolation using Sockeye salmon cells was first reported in 1969 (Wingfield et al. 1969). The disease affects a wide variety of salmonids, especially Atlantic salmon (*Samo salar*) and rainbow trout (*Oncorhynchus mykiss*). The disease was initially reported in western North America, but it is spread subsequently to many European and Asian countries in the northern hemisphere (OIE 2019). The disease is prevalent in both farmed and wild fish populations

2.1.1 Aetiology

The disease is caused by infectious haematopoietic necrosis virus (IHNV) classified under the family *Rhabdoviridae,* subfamily *Gammarhabdovirinae*, genus *Novirabhdovirus* and species Salmonid novirhabdovirus. IHNV is an enveloped bullet-shaped virus measuring 50–80 nm in diameter and 110–200 nm in length. The genome consists of a negative sense, single-stranded, non-segmented linear RNA of 11.1–11.5 kb (Walker et al. 2018). The genome codes for five major structural proteins viz., L (150–225 kDa), G (63–80 kDa), N (38–47 kDa), P (22–26 kDa) and M (17–22 kDa). In addition to the structural proteins, the novirhabdoviruses code for a non-structural, non-virion, NV (12–14 kDa) protein, which is unique to the Novirhabdoviruses (Kurath and Leong 1985) and, hence, the name Novirhabdovirus (Walker et al. 2018). The genes coding for these proteins are arranged in the order 3′-N-P-M-G-NV-L-5′ (Kurath and Leong 1985). The NV protein aminoacid sequences are less conserved between species compared to other proteins (Kurath et al. 1997). The NV protein helps in the pathogenicity of the virus by inhibiting apoptosis and interfering with the host immune response (Ammayappan et al. 2011).

2.1.2 Transmission

The virus is usually transmitted through contaminated water and feed. Transmission is usually horizontal, as infected juveniles shed the virus in high titres (OIE 2019). IHNV is also transmitted by birds and invertebrates. Further, experimental transmission of the virus through the vector salmon louse, *Lepeophtheirus salmonis,* has been reported (Jakob et al. 2011). Higher stocking density increases the chances of horizontal transmission of the virus, as the susceptible fish has increased chances of

Table 1 Causative agents of viral diseases of brackishwater fishes and their characteristics

Disease	Host species	Pathogen/classification	Viral morphology	Viral genome	References
Diseases caused by RNA viruses					
Infectious haematopoietic necrosis	Salmons and trout	Infectious haematopoietic necrosis virus Genus: *Novirhabdovirus* Family: *Rhabdoviridae*	Enveloped, bullet-shaped 150–190 × 65–75 nm	Non-segmented, negative-sense, ssRNA, 11 kb	Dixon et al. (2016); Yong et al. (2019)
Infectious salmon anaemia	Atlantic salmon	Infectious salmon anaemia virus Genus: *Isavirus* family: *Orthomyxoviridae*	Enveloped, 100–120 nm in diameter	Eight single-stranded, negative-sense RNA segments	Dannevig et al. (1995a)
Pancreas disease (PD) and sleeping disease (SD)	Atlantic salmon, common dab, rainbow trout and Arctic charr	Salmonid alphavirus Genus: *Alphavirus* family: *Togaviridae*	Enveloped, spherical, 60–70 nm in diameter	Single-stranded, positive-sense RNA virus, ~12 kb	Mcloughlin and Graham (2007)
Viral haemorrhagic septicaemia	Rainbow trout, turbot, Japanese flounder as well as a broad range of wild freshwater and marine species	Viral haemorrhagic septicaemia virus Genus: *Novirhabdovirus* Family: *Rhabdoviridae*	Bullet-shaped measuring about 180 nm long and 60 nm in diameter	12 kb negative-sense ssRNA	OIE (2019)
Infectious pancreatic necrosis	Rainbow trout, brook trout, brown trout, Atlantic salmon	Infectious pancreatic necrosis virus Genus: *Aquabirnavirus* Family: *Birnaviridae*	Non-enveloped, icosahedral measuring about 70 nm in diameter	Bi-segmented, dsRNA, segment A: ~3.0 kbp; segment B: 2.4 kbp	Munro and Midtlyng (2011); Lago et al. (2016); Delmas et al. (2019)

Viral nervous necrosis (viral encephalopathy and retinopathy)	Asian seabass, European seabass, turbot, halibut, Japanese parrotfish, red-spotted grouper, and striped jack	Nervous necrosis virus Genus: *Betanodavirus* family: *Nodaviridae*	Icosahedral, non-enveloped, 25–30 nm in diameter	Two positive-sense ssRNA molecules-RNA1 (3.1 kb) and RNA2 (1.4 kb)	Mori et al. (1992)
Diseases caused by DNA viruses					
Red seabream iridoviral disease	Red seabream and more than 30 other marine cultured species	Red seabream iridovirus Genus: *Megalocytivirus* Family: *Iridoviridae*	Non-enveloped, icosahedral, 200–240 nm in diameter	dsDNA, 112 kb	Jancovich et al. (2012)

exposure to the virus shed from an infected fish in addition to increased susceptibility caused by the deterioration of water quality at high stocking densities (Ogut and Reno 2004). Egg-associated transmission of IHNV to offspring has been reported (Mulcahy and Pascho 1985). However, it is not clear whether the transmission is vertical or surface-associated. Although egg disinfection with 100 ppm iodophor for 60 min at 10 °C does not inactivate the virus completely (Goldes and Mead 1995), the disinfection procedure decreases the probability of transmission of IHNV to the progeny (Oidtmann et al. 2014).

2.1.3 Pathogenesis

Fish up to 2 months of age are more susceptible to the infection, while adult fish do not exhibit clinical signs or mortality (LaPatra 1998). IHNV infects fish at a water temperature of 3–18 °C (Bootland and Leong 1999). The major portals of entry of the virus are the gills and the gastrointestinal tract. The virus exhibits specific tropism for connective tissue. Following entry through the gill and gastrointestinal tract, the virus enters the circulation, resulting in systemic viraemia. The virus initially replicates in the epithelial cells lining the oral cavity, pharynx, oesophagus, stomach, pyloric caeca and intestine and subsequently infects the adjacent connective tissue of each organ (Drolet et al. 1994). The virus reaches the haematopoietic tissue of the kidney through blood, where the virus replication is intense. Virus replication is also prominent in the lamina propria lining the stomach, especially when the virus entry is through the GI tract. Virus replication is also found in the retinal cells of the eye, resulting in inflammation and exophthalmia, a prominent sign of IHN (Drolet et al. 1994). Degenerative necrosis is prominent in haematopoietic tissues such as the kidney and spleen and in the liver, pancreas and digestive tract. Eosinophilic granular cell necrosis in the intestinal wall is pathognomonic of IHN (Bootland and Leong 1999).

2.1.4 Diagnosis

Clinical signs of IHN include lethargy, loss of appetite, abnormal swimming, skin darkening, distended abdomen, exophthalmia and petechiae in internal organs and skin. Whole fry or viscera of fingerlings or kidney and spleen of bigger fish are ideal for diagnostic purposes. The gold standard for diagnosis of IHN is the isolation of the virus in cell culture followed by identification of the virus using molecular or immunological tools (OIE 2019). Cell lines such as EPC and FHM may be infected with the virus inoculum and incubated at 15 °C for 7–10 days (Lorenzen et al. 1999). Upon observation of CPE, the virus may be identified using virus neutralization test or indirect fluorescent antibody test or ELISA (OIE 2019). CPE observed with IHNV are pyknotic cells, rounding and cell detachment from the substrate. Serum neutralization tests can be used to quantify specific antibodies in serum (Louboutin et al. 2021). RT-PCR using primers targeting the mid-region of the G gene can be used for the diagnosis of IHN in clinical samples (Emmenegger et al. 2000). Other diagnostics tests, such as loop-mediated isothermal amplification (LAMP) (Gunimaladevi et al. 2005) and quantitative PCR (Overturf et al. 2001), have also been reported for IHNV detection.

2.1.5 Control

Disinfection of eggs and adoption of strict biosecurity protocols to prevent virus entry are essential for the prevention of IHN. An autogenous killed vaccine and a DNA vaccine have been licensed for commercial use in Atlantic salmon in North America (OIE 2019). The DNA vaccine for IHN (apex-IHN) is the first DNA vaccine approved for use in aquaculture. The vaccine encoding the glycoprotein (G) against IHNV is commercially available in Canada and the USA, and no outbreak of IHNV has been reported in the vaccinated fish (Salonius et al. 2007). Many other patents have been obtained for DNA vaccines for IHN. Vaccination through injection results in stress to the fish, and hence, other routes of administration have been explored. To avoid the stress due to injection vaccination, Ballesteros et al. (2015a) reported an orally administered DNA vaccine expressing the glycoprotein (G) of IHNV, encapsulated into alginate microspheres, which offered immune response and protection in a dose-dependent manner. Another oral vaccine for IHN using improved yeast surface display technology has been reported (Zhao et al. 2017). The advantage of oral vaccines is that they can be administered to all life stages of fish, including larvae. Vaccination through other mucosal routes, such as the nasal route, was also demonstrated where a live-attenuated IHNV vaccine administered through the nares was found to be safe (LaPatra et al. 2015; Salinas et al. 2015; Larragoite et al. 2016). A bivalent vaccine encoding the G protein of IHNV and VP2–VP3 of IPNV offered significant protection against both IHNV and IPNV (Xu et al. 2017; Zhao et al. 2019).

2.2 Infectious Salmon Anaemia

Infectious salmon anaemia (ISA) is an OIE notifiable highly infectious disease of farmed Atlantic salmon, *Salmo salar,* with a high cumulative mortality rate (Plumb and Hanson 2011). The disease is characterized by severe anaemia, as the name indicates, and is associated with haemorrhages of varying degrees and necrosis in many visceral organs (OIE 2019). The disease was first reported in Norway (Thorud and Djupvik 1988) and was subsequently reported in farmed Atlantic Salmon in Canada, USA, the United Kingdom and Chile. The species susceptible to ISA are brown trout (*Salmo trutta*), Atlantic salmon (*Salmo salar*) and rainbow trout (*Oncorhynchus mykiss*) (OIE 2019).

2.2.1 Aetiology

The disease is caused by Salmon Isavirus, also called infectious salmon anaemia virus (ISAV), belonging to the family *Orthomyxoviridae* and genus *Isavirus*. The virus is enveloped, measuring 100–120 nm in diameter (Dannevig et al. 1995a). The genome consists of eight segments of negative sense, single-stranded RNA of a total size of approximately 14.5 kb (Mjaaland et al. 1997). Segments 1, 2 and 4 encode for PB2, PB1 and PA, respectively, while segment 3 encodes the 68 kDa nucleoprotein; segment 5 encodes a 50 kDa surface glycoprotein (G); segment 6 encodes a receptor binding and receptor destroying enzyme haemagglutinin-esterase (HE) of

42 kDa; segment 7 codes for a nuclear export protein and a protein with type-I interferon antagonistic properties; and segment 8 codes for the matrix protein (22 kDa) and an RNA-binding structural protein with type I interferon antagonistic properties. Two phenotypically distinct variants exist: the virulent ISAV devoid of the highly polymorphic region (HPR) of haemagglutinin esterase, ISAV-HPRΔ and the non-pathogenic ISAV with non-deleted HPR, ISAV-HPR0 (Rimstad and Markussen 2019). The full-length ISAV variant, HPR0, is suggested to be the precursor of the virulent HPR-deleted variant, HPRΔ. HPR0 variant is reported in both wild and farmed asymptomatic Atlantic salmon. It is endemic in wild Atlantic salmon as a seasonal transient infection of the gill epithelial cells (Christiansen et al. 2011) and is not associated with clinical disease. The HPR0 variant has not been isolated in cell culture, and thus, the in vitro and in vivo studies using the isolated virus and the possibility of the HPR0 variant giving rise to the HPRΔ variant have not been established.

2.2.2 Transmission

The disease is transmitted horizontally through water, infected live fish and vectors. Transmission from infected neighbouring farms accounts for about 50% of the infections (Aldrin et al. 2021). The major route of infection is through gills and, to a lesser extent, through the skin and intestine. The virus is excreted through urine, faeces and mucous. Atlantic salmon smolts, after transferring from freshwater to seawater, get infected through wild fish. Asymptomatic wild brown trout serve as a source of infection for cage-cultured salmons (Nylund and Jakobsen 1995). Vertical transmission of the virus through eggs and embryos cannot be ruled out, especially when biosecurity measures are not adequate. The presence of infectious virus in eggs and ovarian fluid from asymptomatic adult broodstock has been demonstrated (Marshall et al. 2014). HPR0 variant may give rise to an HPR-deleted variant, although the chances are low (Lyngstad et al. 2012).

2.2.3 Pathogenesis

The virus infects all stages of fish, from yolk-sac fry to adults. The virus enters the host through the gills. The virus replicates in the endothelial cells lining the blood vessels, resulting in the onset of viraemia. Endothelial cells in other visceral organs, viz., kidney, heart, liver and spleen, also get infected (Aamelfot et al. 2012). The virus is excreted through natural excretions and secretions (OIE 2019). The virus can also be transmitted by sea lice, *Caligus rogercresseyi* mechanically, and the presence of sea lice in the culture system increases the risk of the spread of the virus (Oelckers et al. 2014).

2.2.4 Diagnosis

Clinical signs suggestive of ISA are pale gills, exophthalmia, haemorrhages in the skin and a distended abdomen. Although there are no gross or microscopic lesions pathognomonic to ISA, splenomegaly, pinpoint haemorrhages, accumulation of erythrocytes and necrosis in the affected organs are consistent in diseased fish (OIE 2019). Blood, heart and mid-kidney are ideal for virus isolation and detection. ISA can be diagnosed by isolating the virus in ASK or SHK-1 cells, resulting in cytopathic effects such as vacuolation, cell rounding and detachment from the substrate (Dannevig et al. 1995b). The optimum temperature for the replication of the virus in cell culture is 10–15 °C (Falk et al. 1997). The cell culture supernatant may be tested by nucleic acid or antibody-based tests for confirmation. Samples from areas endemic for IPNV should be incubated with IPNV antiserum for one hour before infecting the cells to neutralize any IPNV present. An RT-PCR (Emmenegger et al. 2000) targeting the G (glycoprotein) gene and real-time RT-PCR using SYBR Green chemistry (Munir and Kibenge 2004) and TaqMan probe (Mjaaland et al. 2005) both targeting segment 8 of ISAV have been reported.

2.2.5 Control

Selective breeding and vaccination can reduce the probability of infection in the susceptible population (Chase-Topping et al. 2021). Screening and culling of stocks when the virus is detected will reduce the incidence of the disease (Aldrin et al. 2021). Another control measure adopted is to cull infected eggs and fish to prevent further spread of the virus. Disinfection of eggs reduces the incidence of the disease in newly hatched larvae. The haemaglutinin-esterase (HE) protein is one of the most variable proteins and is responsible for receptor-binding activity and generates neutralizing antibodies. Hence, the HE protein is commonly used for developing subunit or DNA vaccines (Mikalsen et al. 2005; Gomez-Casado et al. 2011; Robertsen et al. 2016). Many vaccines against ISAV have been licensed for use in Norway, Chile, Ireland, Finland and Canada (Gomez-Casado et al. 2011). Many of them are inactivated, injectable vaccines, either monovalent or multivalent, along with inactivated bacterial pathogens. Inactivated vaccines, when injected along with adjuvants, elicit higher immune responses and offer protection in a dose-dependent manner (Lauscher et al. 2011). An oral vaccine for ISA consisting of recombinant HE and fusion protein induced a high level of specific IgM antibodies and protection in Atlantic salmon (Caruffo et al. 2016). Several oral immunizations are required in the field to maintain high antibody levels and protection (Tobar et al. 2015). A salmonid alphavirus-based replicon vaccine expressing HE of ISAV was found to be efficacious and was able to protect Atlantic salmon against the ISAV challenge (Wolf et al. 2013, 2014).

2.3 Pancreas Disease and Sleeping Disease

Pancreas disease (PD) and sleeping disease (SD) are infectious diseases of Atlantic salmon caused by salmonid alphavirus (SAV) subtypes and rainbow trout in freshwater caused by salmon alphavirus subtype 2 (SAV 2), respectively. PD was first observed in Atlantic salmon in the 1970s in Scotland (Deperasińska et al. 2018) and, subsequently, in Ireland and Norway. The disease is confined exclusively to Europe but for a lone report of PD in pen-reared Atlantic salmon in North America (Kent and Elston 1987). SD was first reported in France (Boucher et al. 1994), and the causative agent was identified by Castric et al. (1997) by isolating the virus. The disease was subsequently reported in England, Scotland (Graham et al. 2003), Germany, Italy, Spain, Poland, Croatia, France, Norway and Switzerland (OIE 2019). The causative agents of both PD and SD are closely related members of the species salmonid alphavirus.

2.3.1 Aetiology

Salmonid alphavirus belongs to the genus *Alphavirus* and the family *Togaviridae.* It is an enveloped virus with icosahedral symmetry measuring approximately 60–70 nm in diameter. The genome is positive-sense, single-stranded RNA of about 12 kb. The genome codes for four structural (E1, E2, E3 and 6 K) and four non-structural proteins (nsP1, nsP2, nsP3 and nsP4). Although the virus causing PD and SD exhibit minor differences in their antigenicity, the virus neutralization test indicates that they belong to the same serotype within the salmonid alphavirus species (Weston et al. 2002). Glycoprotein E1 is more conserved, whereas glycoprotein E2 has more neutralizing epitopes (Nelson et al. 1995). Glycoprotein E2 binds to the host cell receptor, aids in the virus entry and has a major role in envelope formation and release of the virus by budding (Fringuelli et al. 2008). Six genotypes (SAV1–6) of SAV have been identified based on the E2 and nsP3 gene sequence (Fringuelli et al. 2008). However, good cross-neutralization was observed between the genotypes, suggesting that they are closely related members of the species salmonid alphavirus (Graham et al. 2014).

2.3.2 Transmission

SAV is transmitted horizontally through water from infected fish to susceptible fish. The risk of vertical transmission is negligible but cannot be ruled out (Kongtorp et al. 2010). Fish surviving PD become asymptomatic life-long carriers of the virus and pose a risk for other susceptible populations. The spread of the virus to previously unaffected areas is probably through water currents and the transport of live infected fish (Kristoffersen et al. 2009).

2.3.3 Pathogenesis

PD is more common in Atlantic salmon smolts during their first year in sea cages. PD is characterized by anorexia, lethargy, increased yellowish faecal casts and mortality. Histologically, pancreatic acinar cell necrosis is observed in the early stages, and complete loss of acinar cells is observed 2–5 weeks post-infection. Multifocal cardiomyocytic necrosis in heart musculature and red and white muscle fibre degeneration in skeletal muscle are consistent in PD (McLoughlin et al. 2002). SD is characterized by affected fish lying on their side at the bottom of the tank due to extensive skeletal muscle necrosis and, hence, the name sleeping disease. Histological lesions are similar to PD (Boucher and Baudin-Laurencin 1996).

2.3.4 Diagnosis

Diagnosis of PD and SD is based on clinical signs, histopathological lesions, virus isolation, nucleic acid and antibody-based techniques. The ideal organs for PCR detection are pseudobranch and the heart ventricle (Andersen et al. 2007). Fish with acute necrosis of exocrine pancreatic tissue can be demonstrated by immunohistochemistry (Taksdal et al. 2007). SAV can be isolated in CHSE-214 or other cell lines such as EPC, BF-2, SHK-1, FHM and CHH-1 (OIE 2019). Samples from areas endemic for IPNV should be incubated with indigenous IPNV serotype antisera before inoculation to cell culture to prevent cytopathic effects due to IPNV. The samples should be passaged at least once before concluding the sample as negative for SAV. Cell culture supernatants should also be examined by immunofluorescence, as the cytopathic effects may not always be evident (OIE 2019). RT-PCR targeting the E-2 gene (Fringuelli et al. 2008) and real-time RT-PCR targeting nsP1 (Hodneland and Endresen 2006) have been recommended for diagnosis of SAV infection by OIE. A serum neutralization test to titrate the presence of neutralizing antibodies in infected fish can also be used to diagnose PD and SD. An indirect double-antibody sandwich has been reported to detect multiple subtypes of SAV simultaneously (Gao et al. 2021).

2.3.5 Control

PD and SD can be controlled by following general biosecurity and hygiene practices, including stocking good quality disease-free seeds, regular cleaning of tanks, removal of dead fish and controlling ectoparasites. Two inactivated vaccines, NORVAX® Compact PD and ALPHA JECT micro® 1 PD and a DNA vaccine, Clynav, for use in Atlantic Salmon in the European Union for PD, are available commercially. Inactivated SAV-1 and SAV-3 genotypes provided protection to Atlantic salmon against challenge with wild-type virus (Lopez-Doriga et al. 2001; Karlsen et al. 2012). A comparison of the efficacy of inactivated whole-virus vaccine, subunit vaccine and DNA vaccine expressing E1 and E2 spike proteins of salmonid alphavirus subtype 3 (SAV-3) revealed that the immunogenicity of inactivated whole virus was superior to subunit or DNA vaccines (Xu et al. 2012). On the contrary, Thorarinsson et al. (2021) reported that the DNA vaccine produced significantly higher neutralizing antibodies, lower viraemia, reduced transmission and higher weight gain compared to inactivated vaccine.

2.4 Viral Haemorrhagic Septicaemia

Viral haemorrhagic septicaemia (VHS) is an OIE-listed disease affecting farmed rainbow trout, turbot, Japanese flounder and several species of wild freshwater and marine species. The disease was first reported in rainbow trout in a village, Egtved, in Denmark in the early 1950s, and hence, the causative agent is also called the Egtved virus. The viral aetiology was first proposed by Schaperclaus by experimentally transmitting the disease using bacteria-free filtrate. However, the isolation of the virus was first reported in 1963 using trout cell culture (Jensen 1965). The disease is prevalent in approximately 80 species of freshwater and marine wild fish in the temperate regions of the Northern Hemisphere (Skall et al. 2005), spanning North America, Europe and Asia. VHS is a serious disease in farmed rainbow trout in Europe, causing up to 100% mortality, especially in rainbow trout fry.

2.4.1 Aetiology

The causative agent of the disease is viral haemorrhagic septicaemia virus (VHSV), a *Piscine rhabdovirus* of the genus *Novirhabdovirus*, under the family *Rhabdoviridae*. The virus is enveloped and is bullet-shaped, measuring 80 nm × 180 to 200 nm (Kimura et al. 1986). The genome consists of a single molecule of negative sense, ssRNA, of about 11.1–11.5 kb (Hoffmann et al. 2005). Novirhabdovirus codes for five major structural proteins viz., L (150–225 kDa), G (63–80 kDa), N (38–47 kDa), P (22–26 kDa), and M (17–22 kDa). The virus also codes for a non-virion protein designated as NV, having a molecular weight of 12–14 kDa. The NV protein is expressed in virus-infected cells but not present in purified virus. NV protein is found to localize in infected cell nuclei, inhibit cell apoptosis and interfere with host innate immunity (Ammayappan and Vakharia 2011). The VHSV isolates have been grouped into four genotypes based on the gene sequence of N, G and NV genes. The genotype I is further classified into several sub-lineages. Genotype IV has two lineages, IVa and IVb (OIE 2019).

2.4.2 Transmission

Virus transmission is usually horizontal through infected fish and water. Infected fish shed the virus in urine (Wolf 1988) and reproductive fluids (OIE 2019). Virus transmission occurs readily at a temperature range of 1–15 °C, and infection usually occurs between 4 and 14 °C. At low temperatures, between 1 and 5 °C, the onset of the disease is slow, with a prolonged course and high cumulative mortality. The disease occurs in all seasons, but outbreaks are high during spring when the water temperature fluctuates (OIE 2019). There is no report on vertical transmission of the virus.

2.4.3 Pathogenesis

The virus enters the host through the gills and replicates in the kidney and spleen, which are the target organs (Brudeseth et al. 2002). The virus replicates optimally at 15–20 °C (Kimura et al. 1986) and can remain infective for one year at 4 °C (Hawley and Garver 2008). VHSV infects and causes mortality in fish of all life stages, although the disease is common in young populations previously not exposed to the virus. Surviving fish become long-term carriers of the virus (OIE 2019). Although the virus is abundant in the kidney, heart and spleen, in the viraemic stage, the virus is present in all the organs. In chronic cases, the virus load is high in the brain (Wolf 1988). Ovarian fluid and milt can also be used for diagnosis, where non-lethal sampling is required.

2.4.4 Diagnosis

Clinical signs include lethargy, skin darkening, exophthalmia, anaemia and haemorrhages at the base of fins, gills and eyes. Petechial haemorrhage in the skin and muscles is characteristic of VHS. Abnormal swimming with spiralling and flashing movement and distended abdomen are also common. Post-mortem findings include a swollen spleen, dark red kidney, petechial haemorrhage in the dorsal musculature, and mottled and pale liver. The virus can be isolated using cell lines BF-2 and

RTG-2, which have higher sensitivity compared to other cell lines (Lorenzen et al. 1999). VHSV isolated using cell lines can be confirmed by immunoassays or nucleic acid assays (OIE 2019). Antibody-based detection of viral antigens using techniques like immunofluorescence test or ELISA can also be used for diagnosis. The target organs for antigen detection are the heart, kidney and spleen. RT-PCR (Snow et al. 2004) and quantitative RT-PCR (Garver et al. 2011; Jonstrup et al. 2013) can also be used for the diagnosis of VHS. Sequencing of the PCR products will give additional information on the genotype involved. Antibodies appear 3–4 weeks after infection and persist for more than six months (Lorenzen and Lapatra 1999; Fregeneda-Grandes and Olesen 2007) and, hence, antibody detection assays can be used for surveillance.

2.4.5 Control

As no commercial vaccine is available, control of the disease is achieved through surveillance followed by measures such as stamping-out and fallowing. This method has helped the UK, Norway and Denmark to eradicate acute disease outbreaks (OIE 2019). Several experimental vaccines have been reported for VHS. A thermoresistant, live-attenuated VHS vaccine induced high levels of protection against the virulent virus (DeKinkelin and Bearzotti 1981). Another live-attenuated VHS vaccine for oral administration to rainbow trout provided significant protection against VHSV (Adelmann et al. 2008). Recombinant/subunit vaccines expressing the G protein in *Escherichia coli* (Lorenzen et al. 1993) or *Aeromonas salmonicida* (Noonan et al. 1995) did not offer protection due to the lack of post-translational modifications in the recombinant proteins produced in the prokaryotes. Although many DNA vaccines against VHSV have been reported to protect fish against homologous strains (Heppell et al. 1998; Fernandez-Alonso et al. 2001; Pereiro et al. 2012; Lazarte et al. 2017; Sepúlveda et al. 2019), their efficacy is limited against heterologous strains. Further, DNA vaccines are not approved in Europe, as the DNA-vaccinated fish may be considered as genetically modified organisms (Gomez-Casado et al. 2011). Inactivated PLGA-conjugated and chitosan nanoparticles-conjugated VHSV vaccine administered through mucosal routes (immersion followed by oral route) provided 73.3% and 66.6% RPS, respectively (Kole et al. 2019a, 2019b).

2.5 Infectious Pancreatic Necrosis

Infectious pancreatic necrosis (IPN) is a disease of young salmonids and is highly contagious. The disease is characterized by abnormal corkscrew swimming, anorexia, darkening of the skin, exophthalmia and pinpoint haemorrhages on the ventral side. The disease was earlier called acute catarrhal enteritis but was subsequently named IPN based on the histopathological findings of massive necrosis of the pancreas in brook trout (Wood et al. 1955). The mortality can reach up to 100% depending on the host, viral load, infection route and age of the fish. Young fry are the most susceptible, with a high mortality rate.

2.5.1 Aetiology

The causative agent of the disease is infectious pancreatic necrosis virus (IPNV), a non-enveloped icosahedral virus belonging to the genus *Aquabirnavirus* and family *Birnaviridae* (Delmas et al. 2019) measuring 57–74 nm (Dobos and Roberts 1983). IPNV is the first fish virus to be isolated in vitro using cell culture (Wolf et al. 1960). The genome consists of two segments of double-stranded RNA. Segment A is about 2962–3097 bp and codes for two proteins, a non-structural protein VP5 and a polyprotein of 106 kDa, which on internal proteolytic cleavage results in two structural proteins VP2 and VP3 and a non-structural protein VP4 (Dopazo 2020). Segment B is about 2777 bp and codes for RNA-dependent RNA polymerase, VP1 (94 kDa) (Shivappa et al. 2004). VP2 has the host cell attachment sites and virus-neutralizing epitopes and is responsible for the antigenicity of the virus (Caswell-Reno et al. 1986; Dopazo 2020). Two hypervariable regions exist on VP2 based on which IPNV is classified into seven genogroups (I–VII) (Nishizawa et al. 2005).

2.5.2 Transmission

Fish recovering an epizootic often become life-long asymptomatic carriers of the virus. The virus can withstand a pH as low as 3 and a temperature of 60 °C and, hence, can survive in the environment for a prolonged period. Virus transmission is horizontal, predominantly through water. Although the virus is transmitted vertically through reproductive fluids, the prevalence, mortality and virus titre are low in the progeny (Bootland et al. 1991).

2.5.3 Pathogenesis

The virus enters through the gills, intestinal tract and skin from infected water. Subsequently, the virus is detected in blood leucocytes during the viraemic phase, resulting in the virus being distributed to several organs. The virus is detected in the kidney, pancreas, spleen, skin, liver, heart, intestine, brain and reproductive organs. Viral load is very high in the head kidney, pancreas and intestine. The virus causes necrosis of pancreatic acinar cells, intestinal mucosa and hepatocytes (Dopazo 2020).

2.5.4 Diagnosis

A presumptive diagnosis can be made based on clinical signs, although they are not apparent in many cases. Histological lesions such as necrosis of the pancreas, heart and intestines may be suggestive of IPN. IPN can be diagnosed by isolating the virus in cell lines such as BF-2 and CHSE-214 (Lorenzen et al. 1999), followed by confirmation of the virus by ELISA or indirect fluorescent antibody test (IFAT) or RT-PCR. The presence of the virus inside blood leukocytes in an asymptomatic carrier state hampers virus isolation. The problem can be overcome by using a mitogen to initiate cell division (Knott and Munro 1986) or by lysing blood leukocytes (Munro et al. 2004), which increases the sensitivity of the assay. IFAT and immunohistochemistry of sections of affected organs using IPNV-specific antibodies are useful methods for diagnosis. This technique can also be used to diagnose IPN in archived fixed tissue samples. RT-PCR and nested RT-PCR have been reported for diagnosing IPN using primers spanning the VP2 gene or the junction of the VP3 and

VP4 coding regions (Lopez-Lastra et al. 1994; Ahmadi et al. 2013). A real-time RT-PCR, said to be 100 times more sensitive than RT-PCR, has been reported by Orpetveit et al. (2010).

2.5.5 Control

IPNV can be controlled by early diagnosis, restricting trade movements, screening wild fish and vaccination (Dopazo 2020). Owing to the economic impact of the disease, several vaccines, either monovalent or polyvalent, along with other bacterial and viral antigens, are available commercially. However, most of these vaccines fail to protect fish at early stages against IPNV because the fish are not immunocompetent when they are most susceptible to IPNV (Gomez-Casado et al. 2011). The efficacy of inactivated whole-virus vaccines is reported to be better than subunit or DNA vaccines (Munang'andu et al. 2012; Tamer et al. 2021). Recombinant vaccines containing capsid protein (VP2) of IPNV either alone or as a fusion protein with VP3 administered to rainbow trout by injection elicited higher levels of specific IgM and reduced viral load in challenged fish (Dadar et al. 2015; Guo et al. 2018). When administered intraperitoneally, virus-like particles (VLPs) of IPNV offered excellent protection to fish against IPNV (Martinez-Alonso et al. 2012). The administration of vaccines by injection is stressful and is not feasible for small fry. Hence, many oral vaccines have been developed to protect salmonid fries from IPNV (Min et al. 2012; Chen et al. 2020; Li et al. 2020a). Many DNA vaccines containing plasmid constructs expressing VP2 of IPNV for administration through injection or orally through feed after conjugating with alginate microspheres or chitosan and tripolyphosphate nanoparticles have been reported (de Las Heras et al. 2009, 2010; Cuesta et al. 2010; Ballesteros et al. 2014, 2015b; Ahmadivand et al. 2017, 2018). The efficacy of the DNA vaccines was dose-dependent (Ahmadivand et al. 2017), stimulated specific IgM response resulting in low viral load on challenge and upregulated several immune-related genes. A recombinant vaccine with IHNV as the backbone vector expressing VP2 of IPNV provided better protection and survival when challenged with IPNV and IHNV, demonstrating the potential of recombinant viruses to protect fish against two or more viruses (Zhao et al. 2019).

2.6 Viral Nervous Necrosis

Viral nervous necrosis (VNN), also called viral encephalopathy and retinopathy (VER), is a serious viral disease affecting several species of marine, brackishwater and freshwater fishes. The disease was first reported in hatchery-reared larvae and juveniles of Japanese parrotfish, *Oplegnathus fasciatus*, in Japan (Yoshikhosi and Inoue 1990) and in the larvae of barramundi, *Lates calcarifer*, in Australia (Glazebrook et al. 1990). The disease has a vast host range, and the list keeps expanding with new species being reported to be susceptible. Currently, 177 species are reported to be susceptible, and 62 of them had epizootic outbreaks, including Atlantic halibut, barfin flounder, Atlantic and Pacific cod, Asian sea bass, Japanese flounder, European sea bass, red spotted grouper, sevenband grouper, gilthead

Fig. 1 Grey mullet swimming with belly-up due hyperinflation of swim bladder

seabream, turbot and tiger puffer (Bandín and Souto 2020). The disease is characterized by erratic swimming, such as whirling movements, darting, swimming with belly-up due to hyperinflation of the swim bladder, skin darkening and loss of appetite (Fig. 1) (Bandín and Souto 2020).

The disease was first reported in India in 2005, and the possibility of vertical transmission of the virus was proposed due to the presence of the virus in the ovarian tissue (Azad et al. 2005, 2006). The virus was isolated in SSN-1 cells, and sequencing of the coat protein gene revealed that the Indian isolates belonged to the RGNNV genotype (Binesh et al. 2013; Riji John et al. 2014). Outbreaks of VNN were reported in marine clownfish *Amphiprion sebae* Bleeker (Binesh et al. 2013). The virus was shown to infect freshwater mosquitofish *Gambusia affinis* experimentally, and a cell line derived from the skin of *G. affinis* could be used to isolate the virus (Praveenraj et al. 2018; Jyotsna et al. 2020).

2.6.1 Aetiology

The disease is caused by nervous necrosis virus (NNV), a betanodavirus belonging to the family *Nodaviridae*. NNV is non-enveloped and measures about 25–30 nm in diameter with T = 3 icosahedral symmetry (Mori et al. 1992). The genome consists of two single-stranded positive-sense RNA molecules, RNA1 and RNA2, of size 3.1 kb and 1.4 kb, respectively, both lacking polyA tail while their 3′ ends are capped (Mori et al. 1992). The RNA1 codes for protein A, an RNA-dependent RNA polymerase (RdRp) of size 110 kDa. The RNA1 also contains a subgenomic RNA3 of size ranging from 371 to 378 bp, coding for two non-structural proteins, B1 and B2. The protein B1 has anti-necrotic properties (Sommerset and Nerland 2004), while the protein B2 has cellular RNA interference suppression activity (Fenner et al. 2006). The RNA2 codes for the capsid protein of molecular weight of about

40–42 kDa (Mori et al. 1992; Nishizawa et al. 1997; Nallala et al. 2021). Four genotypes of the virus based on the T4 variable region of RNA2 have been reported: barfin flounder NNV (BFNNV), striped jack NNV (SJNNV), tiger puffer NNV (TPNNV) and red-spotted grouper NNV (RGNNV) (Nishizawa et al. 1997). The genome of the virus being bi-segmented, natural reassortants such as RGNNV/SJNNV and SJNNV/RGNNV have also been reported (Toffan et al. 2017). Betanodavirus is classified into three serotypes based on virus neutralization tests; SJNNV and TPNNV are classified as serotype A and B, respectively, while BFNNV and RGNNV belong to the serotype C (Morit et al. 2003).

2.6.2 Geographical Distribution

The host range and geographical distribution of the genotypes depend on their thermotolerance. RGNNV is the most widely distributed genotype affecting the highest number of species in the temperature range of 25–30 °C. RGNNV affects farmed and wild marine, brackishwater and some freshwater species (Bandín and Souto 2020). RGNNV is reported in Asia, Australia and Europe, while there is a lone report of RGNNV in the USA (Curtis et al. 2001). BFNNV is found in Europe, North America and Japan, affecting cold water fishes in the temperature range of 15–20 °C. SJNNV and its reassortants are found in Europe and Japan, affecting fishes in the temperature range of 20–25 °C, while TFNNV is limited to Japan, infecting tiger puffer at 20 °C (Bandín and Souto 2020). RGNNV is the only genotype reported in India.

2.6.3 Transmission

The spread of the virus is mostly by the transport of infected larvae and juveniles for aquaculture. The prevalence of the virus in wild fish also serves as a source of the virus in farmed fish (Nallala et al. 2021). Viral particles excreted from infected fish, cannibalism and dead fish are sources for horizontal transmission of the virus within the farm. The virus is also persistent in frozen fish for a long period and serves as a source of infection when infected trash fish is used for feeding. Water is another important abiotic vector for the transmission of the virus. Invertebrates such as molluscs, gastropods, shrimps and crabs are said to be carriers of the virus and can be a potential source of infection to susceptible populations (Bandín and Souto 2020). Live-feed organisms such as artemia and rotifer can also harbour the virus and aid in its transmission (Skliris and Richards 1998). NNV is relatively stable and can withstand 37 °C for four days and 25 °C for three months. The virus can remain infective for over six months at 15 °C and can be transmitted horizontally to other fishes during this period (Frerichs et al. 2000). The virus is also shed through eggs and milt resulting in vertical transmission to the offspring. Infected adult fish remain as subclinical carriers of the virus, transmitting the virus to susceptible populations (Johansen et al. 2002).

2.6.4 Pathogenesis

The virus enters the host through the epithelial covering of gills, skin, nasal and oral cavity (Mladineo 2003; Johansen et al. 2004; Tanaka et al. 2004), resulting in initial viraemia during which the virus spreads through the circulatory system followed by neuronal spread (Korsnes et al. 2009). The virus is neurotrophic, and after entering through the epithelial layers of the mucosal surfaces, it reaches the CNS and retina through the peripheral nerves (Costa and Thompson 2016). The virus replicates in the brain, retina and spinal cord, causing degeneration of the neurons, resulting in extensive intracytoplasmic vacuoles (Tanaka et al. 2004). The spread of the virus greatly varies with genotype, host species and the route of infection. Typically, the clinical signs appear between 3 and 10 days post-infection in experimentally infected fish. In early life stages, the virus causes hyperplasia and epithelial degeneration in the gill operculum, oral cavity and skin (Nguyen et al. 1996). No macroscopic lesions are found in dead fish except for the inflation of the swim bladder. Histology reveals extensive neural degeneration of the CNS and extensive vacuolation of the brain and retina (Nguyen et al. 1996).

2.6.5 Diagnosis

The ideal samples for diagnosis of VNN are whole larvae or the brain, spinal cord and retina of fingerlings and adult fish, which are the organs of predilection. The ovarian fluid of brooders can also be used for the non-lethal diagnosis of broodstock (Satyanarayana et al. 2021). NNV can be isolated in cell lines such as SSN-1 (Fig. 2a, b) (Frerichs et al. 1996), GF-1 (Chi et al. 1999), E-11, a clone of SSN-1 (Iwamoto et al. 2000), SISK, SISS and SIGE (Sarath Babu et al. 2013). VNN can also be diagnosed by histology, electron microscopy and fluorescent antibody technique (Fig. 3a–c) (Nguyen et al. 1996), ELISA (Huang et al. 2001), in situ hybridization (Comps et al. 1996), antigen capture ELISA (Shieh and Chi 2005) and immunohistochemistry (Thiery et al. 1997). A nanoparticle-based lateral flow biosensor for the visual detection of fish NNV has also been reported (Toubanaki et al. 2015). Several RT-PCR and nested RT-PCR assays have been reported to detect either RNA-1 or RNA-2 (Nishizawa et al. 1994, 1995; Thiery et al. 1999; Dalla Valle et al. 2000; Rajan et al. 2016). A SYBR green dye-based (Kim et al. 2016) and probe-based (Panzarin et al. 2010; Hodneland et al. 2011; Baud et al. 2015) real-time PCR have been developed to detect NNV. Screening for nodavirus by indirect

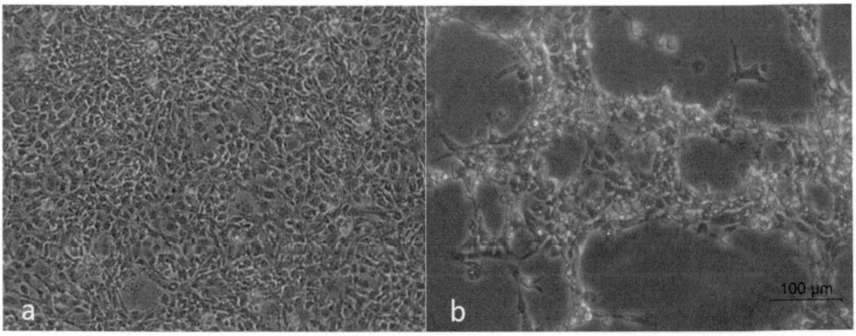

Fig. 2 SSN-1 cells: uninfected (**a**), infected with RGNNV showing vacuolation, rounding and detachment (**b**)

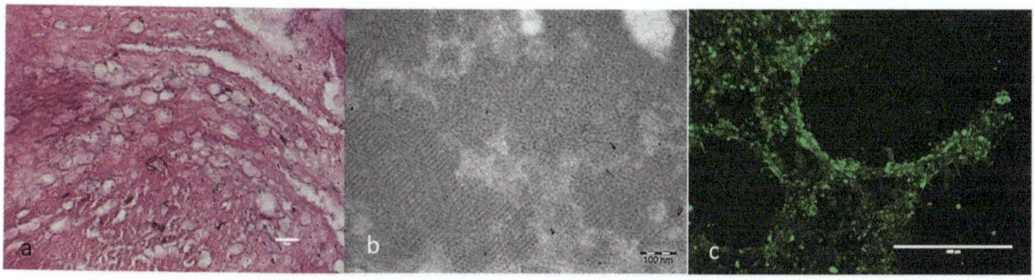

Fig. 3 Histological section of Asian seabass brain showing vacuolations (**a**); Transmission Electron micrograph of RGNNV infected SSN-1 cell showing virus particles; and Fluorescent antibody test for RGNNV infected SSN-1 cells showing specific fluorescence (**c**)

ELISA has been described in European seabass broodstock (Breuil and Romestand 1999), barfin flounder (Watanabe et al. 2000), barramundi, *Lates calcarifer* (Huang et al. 2001; Jaramillo et al. 2016a; Satyanarayana et al. 2021), Seven band Grouper, *Epinephelus septemfasciatus* (Choi et al. 2014) and by competitive ELISA in Australian bass, *Macquaria novemaculeata* (Jaramillo et al. 2016b).

2.6.6 Control

It is very difficult to control the disease once it is established in a farm or hatchery. The only practical way of preventing disease outbreaks in hatcheries and nursery-rearing facilities is to screen and use virus-free brooders and seeds and to vaccinate the brooders and juveniles. Vaccination of Asian seabass (*Lates calcarifer*) (Pakingking et al. 2018) and grouper (*Epinephelus tukula*) (Kai et al. 2010) broodstocks has been reported to prevent or reduce the risk of vertical transmission of the virus. Two inactivated vaccines are commercially available in the European market, viz., ALPHA JECT micro® 1 Noda and Icthiovac® VNN. Vaccines using subunit and recombinant proteins expressed in *Escherichia coli* (Vimal et al. 2014a; Gonzalez-Silvera et al. 2019), yeast (Cho et al. 2017) and plants (Cho et al. 2018) are also reported to protect fish against VNN. Immersion vaccine consisting of a single-walled carbon nanotube as the carrier of the antigen produced RPS of up to 70.76% (Liu et al. 2021). Recombinant protein vaccines offer high protection in vaccinated fish, with RPS of 72% in grouper (Lin et al. 2018) and more than 80% in Atlantic seabass (Thiery et al. 2006). DNA vaccines have been used to vaccinate Asian seabass, European seabass, turbot and grouper (Vimal et al. 2014b, 2016; Chen et al. 2015). However, the protection offered by DNA vaccines against VNN varies widely between species (Bandín and Souto 2020). DNA vaccines also pose ethical issues due to the possibility of classifying vaccinated fish as genetically modified organisms due to the risk of integration of the plasmids into the host genome.

2.7 Red Seabream Iridoviral Disease

Red sea bream iridoviral disease (RSIVD) is an OIE notifiable disease of several marine and brackishwater fish species in East and Southeast Asian countries. The

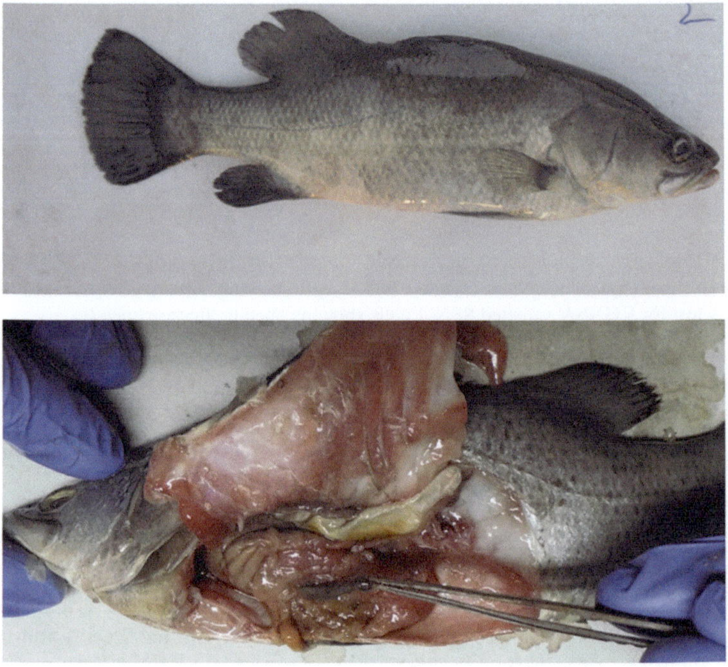

Fig. 4 Asian seabass infected with RSIV (**a**), showing enlarged spleen (**b**)

disease was first reported in Japan in 1990 in farmed red sea bream (*Pagrus major*) (Inouye et al. 1992). More than 40 species are reported to be affected by this disease. The disease mainly affects species belonging to the order Perciformes, but a few species belonging to the order Tetraodontiformes and Pleuronectiformes are also affected (Matsuoka et al. 1996; Kawakami and Nakajima 2002). The disease is characterized by lethargy, severe anaemia, petechiae of gills and splenomegaly (Fig. 4a, b). Histopathology reveals enlarged basophilic cells with inclusion bodies in the spleen, kidney, heart, liver and gills (Fig. 5a, b).

In India, a disease outbreak due to RSIV was first reported in Asian seabass (*Lates calcarifer*) cultured in open estuarine cages in 2018 (Girisha et al. 2020). The experimentally infected Asian seabass fishes showed typical clinical signs of RSIV and 100% mortality by 6 days post-infection. The isolated virus was propagated in SISS cells (Fig. 6a, b), and complete genome sequencing revealed that the Indian isolate belonged to RSIV Genotype II (Puneeth et al. 2021).

2.7.1 Etiology

The causative agent of the disease is red sea bream iridovirus (RSIV) (Inouye et al. 1992). Another virus, infectious spleen and kidney necrosis virus (ISKNV) (He et al. 2001; Oseko et al. 2004), which is related to but distinct from RSIV, also causes RSIVD. Both viruses belong to the genus *Megalocytivirus* and are under the family *Iridoviridae.* Turbot reddish body iridovirus (TRBIV) and rockbream

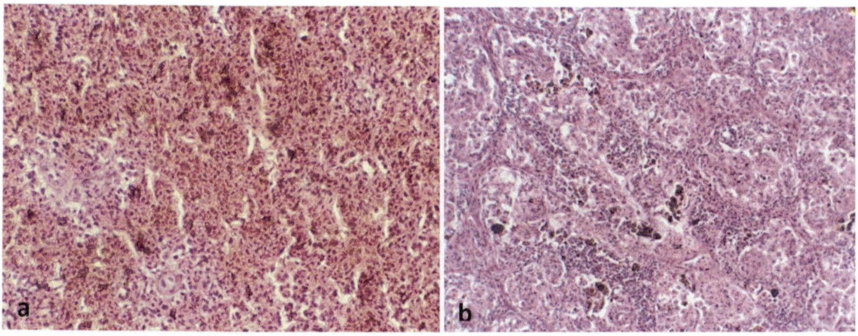

Fig. 5 Histological section of Asian seabass spleen: Uninfected spleen (**a**), infected spleen showing inclusion bodies (**b**)

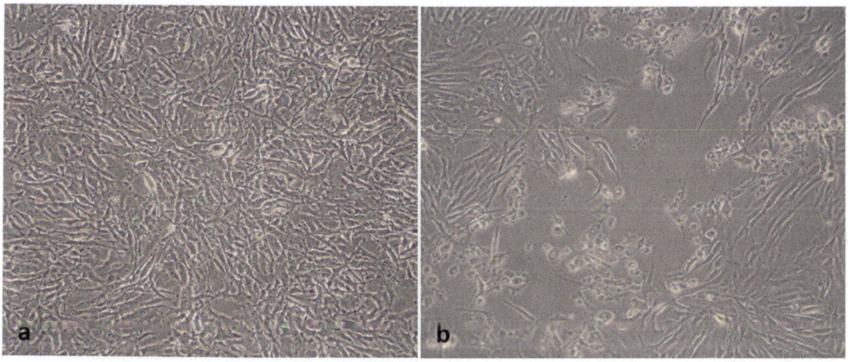

Fig. 6 SISS cells: uninfected (**a**), infected with RSIV showing cytopathic effect: enlargement, rounding of cells and detachment (**b**)

iridovirus (RBIV) are closely related viruses that have a high sequence similarity with RSIV. Further investigation needs to be carried out to decide whether these viruses should be included under RSIV. The virions are large with icosahedral symmetry measuring 120–200 nm. The genome is a linear, double-stranded DNA of 103 to 220 kbp long, which is highly methylated, circularly permutated and terminally redundant with 97–221 open reading frames (ORFs) (Chou et al. 1998; Eaton et al. 2010; Kurita and Nakajima 2012).

2.7.2 Transmission

The principal mode of transmission is horizontal through water (He et al. 2002). The movement of infected carrier fish and ornamental fish trade is responsible for the spread of the virus to previously uninfected areas (Go and Whittington 2006; Go et al. 2006). Detection of ISKNV in the follicular cells surrounding the oocytes and a higher level of infection in the fry stages suggest that vertical transmission can also happen (Suebsing et al. 2016).

2.7.3 Pathogenesis

The disease occurs in summer when the water temperature is above 25 °C. The virus binds to the cell surface receptors on the skin and enters the host cell by clathrin-mediated endocytosis or macropinocytosis in a pH-dependent manner (Wang et al. 2014) and targets the dermal fibroblasts. The envelope protein plays an important role in viral attachment and fusion with the host receptors. Following the host cell entry, immediate early (IE) and delayed early (DE) gene transcripts of the viral genome are produced (D'Costa et al. 2001, 2004; Williams et al. 2005), followed by viral genomic DNA synthesis as large concatameric structures in the nucleus. The viral DNA is released into the host cytoplasm by breaking the nuclear membrane (Chao et al. 2004). Viral-encoded cytosine DNA methyltransferase aids viral DNA methylation in the cytoplasm of the host cell (Thompson et al. 1987). After the initiation of viral DNA replication, late (L) genes coding for viral structural proteins are expressed and are involved in the formation of virion (Williams et al. 2005). The newly synthesized viral particles mature within the virus assembly sites, and freshly produced virions are released either by budding from the plasma membrane (Williams et al. 2005) or arranged as paracrystalline arrays in the host cytoplasm and released by lysis of the cell (Liu et al. 2016).

2.7.4 Diagnosis

A presumptive diagnosis can be made based on clinical signs, post-mortem findings and histopathology. Affected fish are lethargic; gills are pale with petechial haemorrhage and enlarged spleen. The spleen, heart, kidney, intestine and gills have abnormally enlarged cells. The virus can be isolated using GF cell line grown at 25 °C from spleen and kidney tissues. The cytopathic effects include enlarged and rounded cells. However, continuous passage of the virus in these cell lines decreases the infectivity of the virus (Kurita and Nakajima 2012). Following observation of CPE, the virus should be identified by IFAT or nucleic acid-based tests. PCR to detect both RSIV and ISKNV, described by Kurita et al. (1998), has been recommended by OIE for the diagnosis of RSIVD. A monoclonal antibody (MAb) developed against RSIV (Nakajima and Sorimachi 1995) and an indirect fluorescent antibody test (IFAT) developed by Nakajima et al. (1995) can detect both RSIV and ISKNV. Various other techniques, such as nested PCR (Oshima et al. 1996, 1998; Kurita et al. 1998; Jeong et al. 2004, 2006; Lao et al. 2009), enzyme-linked immunosorbent assay (ELISA) (Huang et al. 2012), real-time PCR (Lin et al. 2017) and LAMP assay (Ding et al. 2010; Suebsing et al. 2016) have been reported for the diagnosis of RSIVD. LAMP combined with a lateral flow technique (Ding et al. 2010) has been used for the rapid and visual detection of RSIV (Caipang et al. 2004). A recombinase polymerase amplification (RPA) and RPA combined with lateral flow dipsticks (RPA-LFD), which is rapid and highly sensitive for the detection of ISKNV, can be used for on-site testing (Li et al. 2020b).

2.7.5 Control

Selective breeding using marker-assisted selection combined with DNA-based family selection for resistance to RSIV can be used to develop RSIV-resistant population (Sawayama et al. 2019). A formalin-inactivated vaccine is commercially available in Japan for RSIVD for red seabream (*Pagrus major)*, striped jack (*Pseudocaranx dentex*), Malabar grouper (*Epinephelus malabaricus*) and orange-spotted grouper (*Epinephelus coioides*) (Nakajima et al. 1997, 1999), and this is the first commercially available viral vaccine for marine fish (Kurita and Nakajima 2012). The vaccine, when administered intraperitoneally to juvenile red seabream, resulted in higher survival upon challenge (Nakajima et al. 1997). This vaccine did not offer protection to the genus *Oplegnathus*, as it is highly susceptible to RSIV (Kurita and Nakajima 2012). Intraperitoneal injection of a formalin-killed recombinant *Escherichia coli* expressing the 351R capsid protein of RSIV in fusion with GAPDH of *Edwardsiella tarda* resulted in significantly higher survival on challenge (Shimmoto et al. 2010). A DNA vaccine consisting of plasmids encoding the major capsid protein (MCP) and a transmembrane domain against RSIV could provide better protection to red seabream (Caipang et al. 2006). The efficacy of small interfering RNAs (siRNAs) targeting the MCP gene of RSIV was proved in vitro, indicating the potential of siRNA in protecting farmed fish against RSIVD (Dang et al. 2008). A commercial oil-adjuvanted vaccine, AQUAVAC® IridoV is available for tilapia and Asian seabass for intraperitoneal injection to reduce mortality due to iridovirus.

3 Conclusion

Viral diseases cause huge economic losses to the aquaculture industry. Besides, some of the viruses remain in the hosts without any clinical manifestations, thereby persisting in the susceptible population. This makes it more difficult to control and prevent disease outbreaks. No potent therapeutic measure is available for any of the viral diseases affecting brackishwater fishes. Hence, control of the diseases relies on adopting good management practices, following strict biosecurity measures, quarantining and screening fish before stocking, culling of affected stocks and vaccinating the susceptible population. Most of the vaccines available are injectable vaccines that can be used for fish weighing above 20 g, while most viral diseases cause high mortality in larval and early juvenile stages. Further, mucosal organs such as skin, gills, intestine and olfactory organs possess lymphoid cells such as B and T lymphocytes and dendritic-like cells, which, upon stimulation by extracellular antigens, elicit cell-mediated and humoral immunity (Somamoto and Nakanishi 2020). Hence, it is important to develop vaccines for mucosal delivery that can be used to administer the vaccine to all life stages of the fish. As the mucosal sites such as gills, skin and intestinal tract are the portals of entry for most of the pathogens, mucosal vaccines produce immunity at these sites, thereby preventing pathogen entry. Although a few vaccines that can be administered through these mucosal routes are reported, more work on effective delivery mechanisms, including the use of potent adjuvants, especially for mucosal delivery of vaccines to enhance the immune response, is required. The development of immunostimulants to boost the innate and adaptive immunity of fish will also help the fish ward off the disease. Selective breeding for disease resistance is another area requiring attention to produce disease-resistant stocks. Surveillance for viral diseases and declaring geographical areas free of viruses will also help to prevent the introduction of stocks from endemic areas.

References

Aamelfot M, Dale OB, Weli S, Koppang EO, Falk K (2012) Expression of 4-O-acetylated sialic acids on Atlantic salmon endothelial cells correlates with cell tropism of infectious salmon anemia virus. J Virololgy 86:10571–10578

Adelmann M, Kollner B, Bergmann SM, Fischer U, Lange B, Weitschies W et al (2008) Development of an oral vaccine for immunisation of rainbow trout (*Oncorhynchus mykiss*) against viral haemorrhagic septicaemia. Vaccine 26(6):837–844

Ahmadi N, Oryan A, Akhlaghi M, Hosseini A (2013) Tissue distribution of infectious pancreatic necrosis virus serotype Sp in naturally infected cultured rainbow trout, *Oncorhynchus mykiss* (Walbaum): an immunohistochemical and nested-PCR study. J Fish Dis 36:629–637

Ahmadivand S, Soltani M, Behdani M, Evensen O, Alirahimi E, Hassanzadeh R, Soltani E (2017) Oral DNA vaccines based on CS-TPP nanoparticles and alginate microparticles confer high protection against infectious pancreatic necrosis virus (IPNV) infection in trout. Dev Comp Immunol 74:178–189

Ahmadivand S, Soltani M, Behdani M, Evensen O, Alirahimi E, Soltani E, Hassanzadeh R, Ashrafi-Helan J (2018) VP2 (PTA motif) encoding DNA vaccine confers protection against lethal challenge with infectious pancreatic necrosis virus (IPNV) in trout. Mol Immunol 94:61–67

Aldrin M, Huseby RB, Bang Jensen B, Jansen MD (2021) Evaluating effects of different control strategies for infectious Salmon Anaemia (ISA) in marine salmonid farming by scenario simulation using a disease transmission model. Prev Vet Med 191:105360

Ammayappan A, Vakharia VN (2011) Nonvirion protein of novirhabdovirus suppresses apoptosis at the early stage of virus infection. J Virol 85:8393–8402

Ammayappan A, Kurath G, Thompson TM, Vakharia VN (2011) A reverse genetics system for the Great Lakes strain of viral hemorrhagic septicemia virus: the NV gene is required for pathogenicity. Marine Biotechnol (New York, N.Y.) 13:672–683

Andersen L, Bratland A, Hodneland K, Nylund A (2007) Tissue tropism of salmonid alphaviruses (subtypes SAV1 and SAV3) in experimentally challenged Atlantic salmon (*Salmo salar* L.). Arch Virol 152:1871–1883

Azad IS, Shekhar MS, Thirunavukkarasu AR, Poornima M, Kailasam M, Rajan JJS, Ali SA, Mathew Abraham SA, Ravichandran P (2005) Nodavirus infection causes mass mortalities in hatchery produced larvae of Asian sea- bass, *Lates calcarifer*: first report from India. Dis Aquat Org 63:113–118

Azad IS, Jithendran KP, Shekhar MS, Thirunavukkarasu AR, de la Pena LD (2006) Immunolocalisation of nervous necrosis virus indicates vertical transmission in hatchery produced Asian sea bass (*Lates calcarifer* Bloch)—a case study. Aquaculture 255:39–47

Ballesteros NA, Rodriguez SS, Perez-Prieto SI (2014) Food pellets as an effective delivery method for a DNA vaccine against infectious pancreatic necrosis virus in rainbow trout (*Oncorhynchus mykiss*, Walbaum). Fish Shellfish Immunol 37:220–228

Ballesteros NA, Alonso M, Saint-Jean SR, Perez-Prieto SI (2015a) An oral DNA vaccine against infectious haematopoietic necrosis virus (IHNV) encapsulated in alginate microspheres induces dose-dependent immune responses and significant protection in rainbow trout (*Oncorrhynchus mykiss*). Fish Shellfish Immunol 45:877–888

Ballesteros NA, Saint-Jean SR, Perez-Prieto SI (2015b) Immune responses to oral pcDNA-VP2 vaccine in relation to infectious pancreatic necrosis virus carrier state in rainbow trout *Oncorhynchus mykiss*. Vet Immunol Immunopathol 165:127–137

Bandín I, Souto S (2020) Betanodavirus and VER disease: a 30-year research. Pathogens 9:106. https://doi.org/10.3390/pathogens9020106

Baud M, Cabon J, Salomoni A, Toffan A, Panzarin V, Bigarré L (2015) First generic one step real-time Taqman RT-PCR targeting the RNA1 of betanodaviruses. J Virol Methods 211:1–7

Binesh CP, Renuka K, Malaichami N, Greeshma C (2013) First report of viral nervous necrosis-induced mass mortality in hatchery-reared larvae of clownfish, *Amphiprion sebae* Bleeker. J Fish Dis 36:1017–1020

Bootland LM, Leong JC (1999) Infectious hematopoietic necrosis virus. In: Woo PTK, Bruno DW (eds) Fish diseases and disorders, volume 3: viral, bacterial and fungal infections. CAB International, Oxon, pp 57–121

Bootland LM, Dobos P, Stevenson RMW (1991) The IPNV carrier state and demonstration of vertical transmission in experimentally infected brook trout. Dis Aquat Org 10:13–21

Boucher P, Baudin-Laurencin F (1996) Sleeping disease and pancreas disease: comparative histopathology and acquired cross protection. J Fish Dis 19:303–310

Boucher P, Castric J, Baudin Laurencin F (1994) Observations of virus-like particles in rainbow trout *Oncorhynchus mykiss* infected with sleeping disease virulent material. Bull Eur Assoc Fish Pathol 14:215–216

Breuil G, Romestand B (1999) A rapid ELISA method for detecting specific antibody level against nodavirus in the serum of the sea bass, (*Dicentrarchus labrax* L.): application to the screening of spawners in a sea bass hatchery. J Fish Dis 22:45–52

Brudeseth BE, Castric J, Evensen Ø (2002) Studies on pathogenesis following single and double infection with viral hemorrhagic septicemia virus and infectious hematopoietic necrosis virus in rainbow trout (*Oncorhynchus mykiss*). Vet Pathol 39:180–189

Caipang CMA, Haraguchi I, Ohira T, Hirono I, Aoki T (2004) Rapid detection of a fish iridovirus using loop-mediated isothermal amplification (LAMP). J Virol Methods 121:155–161

Caipang CMA, Takano T, Hirono I, Aoki T (2006) Genetic vaccines protect red seabream, *Pagrus major*, upon challenge with red seabream iridovirus (RSIV). Fish Shellfish Immunol 21:130–138

Caruffo M, Maturana C, Kambalapally S, Larenas J, Tobar JA (2016) Protective oral vaccination against infectious salmon anaemia virus in Salmo salar. Fish Shellfish Immunol 54:54–59

Castric J, Baudin Laurencin F, Bremont M, Jeffroy J, Le Ven A, Bearzotti M (1997) Isolation of the virus responsible for sleeping disease in experimentally infected rainbow trout (*Oncorhynchus mykiss*). Bull Eur Assoc Fish Pathol 17:27–30

Caswell-Reno P, Reno PW, Nicholson B (1986) Monoclonal antibodies to infectious pancreatic necrosis virus: analysis of viral epitopes and comparison of different isolates. J Gen Virol 67:2193–2205

Chao CB, Chen CY, Lai YY, Lin CS, Huang HT (2004) Histological, ultrastructural, and in situ hybridization study on enlarged cells in grouper Epinephelus hybrids infected by grouper iridovirus in Taiwan (TGIV). Dis Aquat Org 58:127–142

Chase-Topping ME, Pooley C, Moghadam HK, Hillestad B, Lillehammer M, Sveen L, Doeschl-Wilson A (2021) Impact of vaccination and selective breeding on the transmission of infectious salmon anemia virus. Aquaculture 535:736365

Chen NC, Yoshimura M, Guan HH, Wang TY, Misumi Y, Lin CC, Chuankhayan P, Nakagawa A, Chan SI, Tsukihara T et al (2015) Crystal structures of a piscine betanodavirus: mechanisms of capsid assembly and viral infection. PLoS Pathog 11:1–25

Chen Y, Hua X, Ren X, Duan K, Gao S, Sun J, Feng Y, Zhou Y, Guan X, Li D, Wang N, Li J, Yang J, Xia D, Shi W, Liu M (2020) Oral immunization with recombinant *lactobacillus casei* displayed AHA1- CK6 and VP2 induces protection against infectious pancreatic necrosis in rainbow trout (*Oncorhynchus mykiss*). Fish Shellfish Immunol 100:18–26

Chi SC, Hu WW, Lo BJ (1999) Establishment and characterization of a continuous cell line (GF-1) derived from grouper, *Epinephelus coioides* (Hamilton): a cell line susceptible to grouper nervous necrosis virus (GNNV). J Fish Dis 22:173–182

Cho SY, Kim HJ, Lan NT, Han H-J, Lee D-C, Hwang JY, Kwon M-G, Kang BK, Han SY, Moon H et al (2017) Oral vaccination through voluntary consumption of the convict grouper

Epinephelus septemfasciatus with yeast producing the capsid protein of red-spotted grouper nervous necrosis virus. Vet Microbiol 204:159–164

Cho HS, Seo JY, Park SI, Kim TG, Kim TJ (2018) Oral immunization with recombinant protein antigen expressed in tobacco against fish nervous necrosis virus. J Vet Med Sci 80:272–279

Choi B, Gye HJH, Oh MM-J, Nishizawa T (2014) Cell culture medium inhibits antigen binding used in an ELISA for detection of antibodies against nervous necrosis virus. J Aquat Anim Health 26:168–172

Chou HY, Hsu CC, Peng TY (1998) Isolation and characterization of a pathogenic iridovirus from cultured grouper (Epinephelus sp.) in Taiwan. Fish Pathol 33:201–206

Christiansen DH, Ostergaard PS, Snow M, Dale OB, Falk K (2011) A low-pathogenic variant of infectious salmon anemia virus (ISAV-HPR0) is highly prevalent and causes a non-clinical transient infection in farmed Atlantic salmon (*Salmo salar* L.) in The Faroe Islands. J Gen Virol 92:909–918

Comps M, Trindade M, Delsert C (1996) Investigation of fish encephalitis viruses (FEV) expression in marine fishes using DIG-labelled probes. Aquaculture 143:113–121

Costa JZ, Thompson KD (2016) Understanding the interaction between Betanodavirus and its host for the development of prophylactic measures for viral encephalopathy and retinopathy. Fish Shellfish Immunol 53:35–49

Cuesta A, Chaves-Pozo E, de las Heras AI, Saint-Jean SR, Pérez-Prieto S, Tafalla C (2010) An active DNA vaccine against infectious pancreatic necrosis virus (IPNV) with a different mode of action than fish rhabdovirus DNA vaccines. Vaccine 28:3291–3300

Curtis PA, Drawbridge M, Iwamoto T, Nakai T, Hedrick RP, Gendron AP (2001) Nodavirus infection of juvenile white seabass, Atractoscion nobilis, cultured in southern California: first record of viral nervous necrosis (VNN) in North America. J Fish Dis 24:263–271

D'Costa SM, Yao H, Bilimoria SL (2001) Transcription and temporal cascade in Chilo iridescent virus infected cells. Arch Virol 146:2165–2178

D'Costa SM, Yao HJ, Bilimoria SL (2004) Transcriptional mapping in Chilo iridescent virus infections. Arch Virol 149:723–742

Dadar M, Memari HR, Vakharia VN, Peyghan R, Shapouri MS, Mohammadian T, Hasanzadeh R, Ghasemi M (2015) Protective and immunogenic effects of *Escherichia coli*-expressed infectious pancreatic necrosis virus (IPNV) VP2-VP3 fusion protein in rainbow trout. Fish Shellfish Immunol 47:390–396

Dalla Valle L, Zanella L, Patarnello P, Paolucci L, Belvedere P, Colombo L (2000) Development of a sensitive diagnostic assay for fish nervous necrosis virus based on RT-PCR plus nested PCR. J Fish Dis 23:321–327

Dang LT, Kondo H, Hirono I, Aoki T (2008) Inhibition of red seabream iridovirus (RSIV) replication by small interfering RNA (siRNA) in a cell culture system. Antivir Res 77:142–149

Dannevig BH, Falk K, Namork E (1995a) Isolation of the causal virus of infectious salmon anemia (ISA) in a long-term cell line from Atlantic salmon head kidney. J Gen Virol 76:1353–1359

Dannevig BH, Falk K, Press CM (1995b) Propagation of infectious salmon anemia (ISA) virus in cell culture. Vet Res 26:438–442

de Las Heras A, Perez PS, Rodriguez SS (2009) In vitro and in vivo immune responses induced by a DNA vaccine encoding the VP2 gene of the infectious pancreatic necrosis virus. Fish Shellfish Immunol 27:120–129

de Las Heras AI, Saint-Jean SR, Perez-prieto SI (2010) Immunogenic and protective effects of an oral DNA vaccine against infectious pancreatic necrosis virus in fish. Fish Shellfish Immunol 28:562–570

DeKinkelin P, Bearzotti M (1981) Immunization of rainbow trout against viral hemorrhagic septicaemia (VHS) with a thermoresistant variant of the virus. Dev Biol Stand 49:431–439

Delmas D, Attoui H, Ghosh S, Malik YS, Mundt E, Vakharia VN (2019) ICTV report consortium, ICTV virus taxonomy profile: Birnaviridae. J Gen Virol 100:5–6

Deperasińska I, Schulz S, Siwicki AK (2018) Salmonid alphavirus (SAV). J Veterin Res 62:1–6

Ding WC, Chen J, Shi YH, Lu XJ, Li MY (2010) Rapid and sensitive detection of infectious spleen and kidney necrosis virus by loop-mediated isothermal amplification combined with a lateral flow dipstick. Arch Virol 155:385–389
Dixon P, Paley R, Alegria-Moran R, Oidtmann B (2016) Epidemiological characteristics of infectious hematopoietic necrosis virus (IHNV): a review. Vet Res 47:63
Dobos P, Roberts TE (1983) The molecular biology of infectious pancreatic necrosis virus: a review. Can J Microbiol 29:377–384
Dopazo CP (2020) The infectious pancreatic necrosis virus (IPNV) and its virulence determinants: what is known and what should be known. Pathogens 9:94
Drolet BS, Rohovec JS, Leong JC (1994) The route of entry and progression of infectious haematopoietic necrosis virus in *Oncorhynchus mykiss* (Walbaum): a sequential immunohistochemical study. J Fish Dis 17:337–347
Eaton HE, Ring BA, Brunetti CR (2010) The genomic diversity and phylogenetic relationship in the family iridoviridae. Viruses 2:1458–1475
Emmenegger EJ, Meyers TR, Burton TO, Kurath G (2000) Genetic diversity and epidemiology of infectious hematopoietic necrosis virus in Alaska. Dis Aquat Org 40:163–176
Falk K, Namork E, Rimstad E, Mjaaland S, Dannevig BH (1997) Characterization of infectious salmon anemia virus, an orthomyxo-like virus isolated from Atlantic salmon (*Salmo salar* L.). J Virol 71:9016–9023
FAO (2024) The state of world fisheries and aquaculture 2024—blue transformation in action, Rome
Fenner BJ, Thiagarajan R, Chua HK, Kwang J (2006) Betanodavirus B2 is an RNA interference antagonist that facilitates intracellular viral RNA accumulation. J Virol 80:85–94
Fernandez-Alonso M, Rocha A, Coll JM (2001) DNA vaccination by immersion and ultrasound to trout viral haemorrhagic septicaemia virus. Vaccine 19:3067–3075
Fregeneda-Grandes JM, Olesen NJ (2007) Detection of rainbow trout antibodies against viral haemorrhagic septicaemia virus (VHSV) by neutralisation test is highly dependent on the virus isolate used. Dis Aquat Org 74:151–158
Frerichs GN, Rodger HD, Peric Z (1996) Cell culture isolation of piscine neuropathy nodavirus from juvenile sea bass, *Dicentrarchus labrax*. J Gen Virol 77:2067–2071
Frerichs GN, Tweedie A, Starkey WG, Richards RH (2000) Temperature, pH, and electrolyte sensitivity, and heat, UV and disinfectant inactivation of sea bass (*Dicentrarchus labrax*) neuropathy nodavirus. Aquaculture 185:13–24
Fringuelli E, Rowley HM, Wilson JC, Hunter R, Rodger HD, Graham DA (2008) Phylogenetic analyses and molecular epidemiology of European salmonid alphaviruses (SAV) based on partial E2 and nsP3 gene nucleotide sequences. J Fish Dis 31:811–823
Gao S, Wang N, Yang J, Sun J, Wang Y, Xia D, Tian J, Zhao Y, Feng Y, Zhou Y, Guan X, Shi W, Liu M (2021) Development and evaluation of indirect double-antibody sandwich ELISA for rapid detection of salmonid alphavirus using Baculoviridae expressed E1 protein. Aquaculture 544:737081
Garver KA, Hawley LM, Mcclure CA, Schroeder T, Aldous S, Doig F, Snow M, Edes S, Baynes C, Richard J (2011) Development and validation of a reverse transcription quantitative PCR for universal detection of viral hemorrhagic septicemia virus. Dis Aquat Org 95:97–112
Girisha SK, Puneeth TG, Nithin MS, Kumar BTN, Ajay SK, Vinay TN, Suresh T, Venugopal MN, Ramesh KS (2020) Red Sea bream iridovirus disease (RSIVD) outbreak in Asian seabass (*Lates calcarifer*) cultured in open estuarine cages along the west coast of India: first report. Aquaculture 520:734712
Glazebrook JS, Heasman MP, de Beer SW (1990) Picorna-like viral particles associated with mass mortalities in larval barramundi, *Lates calcarifer* Bloch. J Fish Dis 13:245–249
Go J, Whittington R (2006) Experimental transmission and virulence of a megalocytivirus (family Iridoviridae) of dwarf gourami (*Colisa lalia*) from Asia in Murray cod (*Maccullochella peelii peelii*) in Australia. Aquaculture 258:140–149
Go J, Lancaster M, Deece K, Dhungyel O, Whittington R (2006) The molecular epidemiology of iridovirus in Murray cod (*Maccullochella peelii peelii*) and dwarf gourami (*Colisa lalia*) from

distant biogeographical regions suggests a link between trade in ornamental fish and emerging iridoviral diseases. Mol Cell Probes 20:212–222

Goldes SA, Mead SL (1995) Efficacy of iodophor disinfection against egg surface-associated infectious hematopoietic necrosis virus. Prog Fish-Cult 57:26–29

Gomez-Casado E, Estepa A, Coll JM (2011) A comparative review on European-farmed finfish RNA viruses and their vaccines. Vaccine 29:2657–2671

Gonzalez-Silvera D, Guardiola FA, Espinosa C, Chaves-Pozo E, Esteban MÁ, Cuesta A (2019) Recombinant nodavirus vaccine produced in bacteria and administered without purification elicits humoral immunity and protects European sea bass against infection. Fish Shellfish Immunol 88:458–463

Graham, D.A., Rowley, H.M., Walker, I.W., Weston, J.H., Branson, E.J., Todd,D., 2003. First isolation of sleeping disease virus from rainbow trout, *Oncorhynchus mykiss* (Walbaum), in the United Kingdom. J Fish Dis 26, 691–694

Graham DA, Rowley HR, Frost P (2014) Cross-neutralization studies with salmonid alphavirus subtype 1-6 strains: results with sera from experimental studies and natural infections. J Fish Dis 37:683–691

Gunimaladevi I, Kono T, Lapatra SE, Sakai M (2005) A loop mediated isothermal amplification (LAMP) method for detection of infectious hematopoietic necrosis virus (IHNV) in rainbow trout (*Oncorhynchus mykiss*). Arch Virol 150:899–909

Guo M, Shi W, Wang Y, Wang Y, Chen Y, Li D, Ren X, Hua X, Tang L, Li Y, Liu M (2018) Recombinant infectious hematopoietic necrosis virus expressing infectious pancreatic necrosis virus VP2 protein induces immunity against both pathogens. Fish Shellfish Immunol 78:187–194

Hawley LM, Garver KA (2008) Stability of viral hemorrhagic septicemia virus (VHSV) in freshwater and seawater at various temperatures. Dis Aquat Org 82:171–178

He JG, Deng MS, Weng P, Li Z, Zhou SY, Long QX, Wang XZ, Chang SM (2001) Complete genome analysis of the mandarin fish infectious spleen and kidney necrosis iridovirus. Virology 291:126–139

He JG, Zeng K, Weng SP, Chan SM (2002) Experimental transmission, pathogenicity and physical–chemical properties of infectious spleen and kidney necrosis virus (ISKNV). Aquaculture 204:11–24

Heppell J, Niels Lorenzen N, Armstrong NK, Wu T, Lorenzen E, Einer-Jensen K, Schorr J, Davis HL (1998) Development of DNA vaccines for fish: vector design, intramuscular injection and antigen expression using viral haemorrhagic septicaemia virus genes as model. Fish Shellfish Immunol 8:271–286

Hodneland K, Endresen C (2006) Sensitive and specific detection of salmonid alphavirus using real-time PCR (TaqMan). J Virolog Meth 131:184–192

Hodneland K, Garcıa R, Balbuena JA, Zarza C, Fouz B (2011) Real-time RT-PCR detection of betanodavirus in naturally and experimentally infected fish from Spain. J Fish Dis 34:189–202

Hoffmann B, Beer M, Schutze H, Mettenleiter TC (2005) Fish rhabdoviruses: molecular epidemiology and evolution. Curr Top Microbiol Immunol 292:81–117

Huang B, Tan C, Chang SF, Munday B, Mathew JA, Ngoh GH, Kwang J (2001) Detection of nodavirus in barramundi, Lates calcarifer (Bloch), using recombinant coat protein-based ELISA and RT-PCR. J Fish Dis 24:135–141

Huang Z, Tang J, Li M, Fu Y, Dong C, Zhong JF, He J (2012) Immunological evaluation of *Vibrio alginolyticus*, *Vibrio harveyi*, *Vibrio vulnificus* and infectious spleen and kidney necrosis virus (ISKNV) combined-vaccine efficacy in *Epinephelus coioides*. Vet Immunol Immunopathol 150:61–68

Inouye K, Yamano K, Maeno Y, Nakajima K, Matsuoka M, Wada Y, Sorimachi M (1992) Iridovirus infection of cultured red sea bream, *Pagrus major*. Fish Pathol 27:19–27

Iwamoto T, Nakai T, Mori K, Arimoto M, Furusawa I (2000) Cloning of the fish cell line SSN-1 for piscine nodaviruses. Dis Aquat Org 43:81–89

Jakob E, Barker DE, Garver KA (2011) Vector potential of the salmon louse Lepeophtheirus salmonis in the transmission of infectious haematopoietic necrosis virus (IHNV). Dis Aquat Org 97:155–165

Jancovich JK, Chinchar VG, Hyatt A, Miyazaki T, William T, Zang QY (2012) Family Iridoviridae. In: King AMQ, Adams MJ, Carstens EB, Lefkowitz EJ (eds) Virus taxonomy: ninth report of the international committee on taxonomy of viruses. Elsevier Academic Press, San Diego, pp 193–210

Jaramillo D, Dürr S, Hick P, Whittington R (2016a) Bayesian estimation of diagnostic sensitivity and specificity of a nervous necrosis virus antibody ELISA. Prev Vet Med 123:138–142

Jaramillo D, Hick P, Deece K, Tweedie A, Kirkland P, Arzey E, Whittington RJ (2016b) Comparison of ELISA formats for detection of antibodies specific for nervous necrosis virus (Betanodavirus) in the serum of immunized barramundi *Lates calcarifer* and Australian bass *Macquaria novemaculeata*. Aquaculture 451:33–38

Jensen MH (1965) Research on the virus of Egtved disease. Ann N Y Acad Sci 126:422–426

Jeong JB, Park KH, Kim HY, Hong SH, Kim KH, Chung JK, Komisar JL, Do Jeong H (2004) Multiplex PCR for the diagnosis of red sea bream iridoviruses isolated in Korea. Aquaculture 235:139–152

Jeong JB, Jun LJ, Park KH, Kim KH, Chung JK, Komisar JL, Do Jeong H (2006) Asymptomatic iridovirus infection in various marine fishes detected by a 2-step PCR method. Aquaculture 255:30–38

Johansen R, Ranheim T, Hansen MK, Taksdal T, Totland GK (2002) Pathological changes in juvenile Atlantic halibut *Hippoglossus hipoglossus* persistently infected with nodavirus. Dis Aquat Org 50:161–169

Johansen R, Rove S, Svendsen NAK, Modahl I, Dannevig B (2004) A sequential study of pathological findigs in Atlantic halibut, *Hippoglossus hippoglossus* (L), throughout one year after an acute outbreak of viral encephalopathy and retinopathy. J Fish Dis 27:327–341

Jonstrup SP, Kahns S, Skall HF, Boutrup TS, Olesen NJ (2013) Development and validation of a novel Taqman-based real-time RT-PCR assay suitable for demonstrating freedom from viral haemorrhagic septicaemia virus. J Fish Dis 36(1):9–23

Jyotsna, Vijayakumar P, Ravi M, Sudhakaran R, Mekata T, Rajaswaminathan T (2020) Development and characterization of a skin cell line (SGA) from the mosquitofish *Gambusia affinis* and its susceptibility to fish Betanodavirus. Aquaculture 520:734778

Kai YH, Su HM, Tai KT, Chi SC (2010) Vaccination of grouper broodfish (*Epinephelus tukula*) reduces the risk of vertical transmission by nervous necrosis virus. Vaccine 28:996–1001

Karlsen M, Tingbø T, Solbakk I-T, Evensen O, Furevik A, Aas-Eng A (2012) Efficacy and safety of an inactivated vaccine against salmonid alphavirus (family Togaviridae). Vaccine 30:5688–5694

Kawakami H, Nakajima K (2002) Cultured fish species affected by red sea bream iridoviral disease from 1996 to 2000. Fish Pathol 37:45–47

Kent ML, Elston RA (1987) Pancreas disease in pen-reared Atlantic salmon in North America. Bull Eur Assoc Fish Pathol 7:29–31

Kim J-O, Kim J-O, Kim W-S, Oh M-J (2016) Development and application of quantitative detection method for nervous necrosis virus (NNV) isolated from sevenband grouper *Hyporthodus septemfasciatus*. Asian Pac J Trop Med 9:742–748

Kimura T, Yoshimizu M, Gorie S (1986) A new rhabdovirus isolated in Japan from cultured hirame (Japanese flounder) *Paralichthys olivaceus* and ayu *Plecoglossus altivelis*. Dis Aquat Org 1:209–217

Knott RM, Munro AL (1986) The persistence of infectious pancreatic necrosis virus in Atlantic salmon. Vet Immunol Immunopathol 12:359–364

Kole S, Qadiri SSN, Shin S-M, Kim W-S, Lee J, Jung S-J (2019a) PLGA encapsulated inactivated-viral vaccine: formulation and evaluation of its protective efficacy against viral haemorrhagic septicaemia virus (VHSV) infection in olive flounder (*Paralichthys olivaceus*) vaccinated by mucosal delivery routes. Vaccine 37:973–983

Kole S, Qadiri SSN, Shin S-M, Kim W-S, Lee J, Jung S-J (2019b) Nanoencapsulation of inactivated-viral vaccine using chitosan nanoparticles: evaluation of its protective efficacy and immune modulatory effects in olive flounder (*Paralichthys olivaceus*) against viral haemorrhagic septicaemia virus (VHSV) infection. Fish Shellfish Immunol 91:136–147

Kongtorp RT, Stene A, Andreassen PA, Aspehaug V, Graham DA, Lyngstad TM, Olsen AB, Olsen RS, Sandberg M, Santi N, Wallace C, Breck O (2010) Lack of evidence for vertical transmission of SAV 3 using gametes of Atlantic salmon, *Salmo salar* L., exposed by natural and experimental routes. J Fish Dis 33:879–888

Korsnes K, Karlsbakk E, Devold M, Nerland AH, Nylund A (2009) Tissue tropism of nervous necrosis virus (NNV) in Atlantic cod, *Gadus morhua* L., after intraperitoneal challenge with a virus isolate from diseased Atlantic halibut, *Hippoglossus hippoglossus* (L.). J Fish Dis 32:655–665

Kristoffersen AB, Viljugrein H, Kongtorp RT, Brun E, Jansen PA (2009) Risk factors for pancreas disease (PD) outbreaks in farmed Atlantic salmon and rainbow trout in Norway during 2003–2007. Prev Vet Med 90:127–136

Kumaran M, Kumaraguru Vasagam KP, Kailasam M, Subburaj R, Anand PR, Ravisankar T, Sendhilkumar R, Santhanakumar J, Vijayan KK (2021) Three-tier cage aquaculture of Asian Seabass (*Lates calcarifer*) fish in the coastal brackishwaters – a techno-economic appraisal. Aquaculture 543:737025

Kurath G, Leong JC (1985) Characterization of infectious hematopoietic necrosis virus mRNA species reveals a nonvirion rhabdovirus protein. J Virol 53:462–468

Kurath G, Higman KH, Bjorklund HV (1997) Distribution and variation of NV genes in fish rhabdoviruses. J Gen Virol 78:113–117

Kurita J, Nakajima K (2012) Megalocytiviruses. Viruses 4:521–538

Kurita J, Nakajima K, Hirono I, Aoki T (1998) Polymerase chain reaction (PCR) amplification of DNA of red sea bream iridovirus (RSIV). Fish Pathol 33:17–23

Lago L, Rodríguez JF, Bandín I, Dopazo CP (2016) Aquabirnavirus polyploidy: a new strategy to modulate virulence? J Gen Virol 97:1168–1177

Lao H, Ye X, Zou W, Bai J, Tan A (2009) Detection of infectious spleen and kidney necrosis virus (ISKNV) of mandarin fish (*Siniperca chuatsi*) by nested PCR. South China Fish Sci 5(4):69–72

LaPatra SE (1998) Factors affecting pathogenicity of infectious hematopoietic necrosis virus (IHNV) for salmonid fish. J Aquat Anim Health 10:121–131

LaPatra SE, Kao S, Erhardt EB, Salinas I (2015) Evaluation of dual nasal delivery of infectious hematopoietic necrosis virus and enteric red mouth vaccines in rainbow trout (*Oncorhynchus mykiss*). Vaccine 33:771–776

Larragoite ET, Tacchi L, LaPatra SE, Salinas I (2016) An attenuated virus vaccine appears safe to the central nervous system of rainbow trout (*Oncorhynchus mykiss*) after intranasal delivery. Fish Shellfish Immunol 49:351–354

Lauscher A, Krossøy B, Frost P, Grove S, König M, Bohlin J, Falk K, Austbø L, Rimstad E (2011) Immune responses in Atlantic salmon (*Salmo salar*) following protective vaccination against infectious salmon anemia (ISA) and subsequent ISA virus infection. Vaccine 29:6392–6401

Lazarte JMS, Kim YR, Lee JS, Im SP, Kim SW, Jung JW, Kim J, Lee WJ, Jung S (2017) Enhancement of glycoprotein-based DNA vaccine for viral hemorrhagic septicemia virus (VHSV) via addition of the molecular adjuvant, DDX41. Fish Shellfish Immunol 62:356–365

Li S, Hu Y, Li X, Han S, Zhang B, Yan Z, Xue R, Qiang Gao Q, Wu J, Zhao X, Liu J (2020a) Development of a live vector vaccine against infectious pancreatic necrosis virus in rainbow trout. Aquaculture 524:735275

Li H, Yuan G, Luo Y, Yu Y, Ai T, Su J (2020b) Rapid and sensitive detection of infectious spleen and kidney necrosis virus by recombinase polymerase amplification combined with lateral flow dipsticks. Aquaculture 519:734926

Lin Q, Fu X, Liu L, Liang H, Guo H, Yin S, Kumaresan V, Huang Z, Li N (2017) Application and development of a TaqMan real-time PCR for detecting infectious spleen and kidney necrosis virus in *Siniperca chuatsi*. Microb Pathog 107:98–105

Lin C-FF, Jiang H-KK, Chen N-CC, Wang T-YY, Chen T-YY (2018) Novel subunit vaccine with linear array epitope protect giant grouper against nervous necrosis virus infection. Fish Shellfish Immunol 74:551–558

Liu Y, Tran BN, Wang F, Ounjai P, Wu J, Hew CL (2016) Visualization of assembly intermediates and budding vacuoles of Singapore grouper iridovirus in grouper embryonic cells. Sci Rep 6:18696

Liu G-Y, Li J, Zhang Z-Y, Liu J-Y, Zhang C, Zhu B, Wang G-X (2021) An immersion subunit vaccine loaded by single-walled carbon nanotube protects pearl gentian grouper from viral nervous necrosis virus. Aquaculture 541:736813

Lopez-Doriga MV, Smail DA, Smith RJ, Domenech A, Castric J, Smith PD, Ellis AE (2001) Isolation of salmon pancreas disease virus (SPDV) in cell culture and its ability to protect against infection by the wild type agent. Fish Shellfish Immunol 11:505–522

Lopez-Lastra M, Gonzalez M, Jashes M, Sandino AM (1994) A detection method for infectious pancreatic necrosis virus (IPNV) based on reverse transcription (RT)-polymerase chain reaction (PCR). J Fish Dis 17:269–282

Lorenzen N, Lapatra SE (1999) Immunity to rhabdoviruses in rainbow trout: the antibody response. Fish Shellfish Immunol 9:345–360

Lorenzen N, Olesen NJ, Jorgensen PEV, Etzerodt M, Holtet TL, Thogersen HC (1993) Molecular-cloning and expression in *Escherichia coli* of the glycoprotein gene of VHS virus, and immunization of rainbow-trout with the recombinant protein. J Gen Virol 74:623–630

Lorenzen E, Carstensen B, Olesen NJ (1999) Inter-laboratory comparison of cell lines for susceptibility to three viruses: VHSV, IHNV and IPNV. Dis Aquat Organ 37:81–88

Louboutin L, Cabon J, Vigouroux E, Morin T, Danion M (2021) Comparative analysis of the course of infection and the immune response in rainbow trout (*Oncorhynchus mykiss*) infected with the 5 genotypes of infectious hematopoietic necrosis virus. Virology 552:20–31

Lyngstad TM, Kristoffersen AB, Hjortaas MJ, Devold M, Aspehaug V, Larssen RB, Jansen PA (2012) Low virulent infectious salmon anaemia virus (ISAV-HPR0) is prevalent and geographically structured in Norwegian salmon farming. Dis Aquat Org 101:197–206

Marshall SH, Ramírez R, Labra A, Carmona M, Muñoz C (2014) Bona fide evidence for natural vertical transmission of infectious salmon anemia virus in freshwater brood stocks of farmed Atlantic salmon (*Salmo salar*) in Southern Chile. J Virol 88:6012–6018

Martinez-Alonso S, Vakharia VN, Saint-Jean SR, Perez-Prieto S, Tafalla C (2012) Immune responses elicited in rainbow trout through the administration of infectious pancreatic necrosis virus-like particles. Dev Comp Immunol 36:378–384

Matsuoka S, Inouye K, Nakajima K (1996) Cultured fish species affected by red sea bream iridoviral disease from 1991 to 1995. Fish Pathol 31:233–234

Mcloughlin MF, Graham DA (2007) Alphavirus infections in salmonids—a review. J Fish Dis 30:511–531

McLoughlin MF, Nelson RN, McCormick JI, Rowley HM, Bryson DB (2002) Clinical and histopathological features of naturally occurring pancreas disease in farmed Atlantic salmon, *Salmo salar* L. J Fish Dis 25:33–43

Mikalsen AB, Sindre H, Torgersen J, Rimstad E (2005) Protective effects of a DNA vaccine expressing the infectious salmon anemia virus hemagglutinin-esterase in Atlantic salmon. Vaccine 23:4895–4905

Min L, Li-Li Z, Jun-Wei G, Xin-Yuan Q, Yi-Jing L, Di-Qiu L (2012) Immunogenicity of Lactobacillus-expressing VP2 and VP3 of the infectious pancreatic necrosis virus (IPNV) in rainbow trout. Fish Shellfish Immunol 32:196–203

Mjaaland S, Rimstad E, Falk K, Dannevig BH (1997) Genomic characterisation of the virus causing infectious salmon anemia in Atlantic salmon (*Salmo salar* L.): an orthomyxo-like virus in a teleost. J Virol 71:7681–7686

Mjaaland S, Markussen T, Sindre H, Kjøglum S, Dannevig BH, Larsen S, Grimholt U (2005) Susceptibility and immune responses following experimental infection of MHC compatible Atlantic salmon (*Salmo salar* L.) with different infectious salmon anaemia virus isolates. Arch Virol 150:2195–2216

Mladineo I (2003) The immunohistochemical study of nodavirus changes in larval, juvenile and adult sea bass tissue. J Appl Ichthyol 19:366–370

Mori K, Nakai T, Muroga K, Arimoto M, Mushiake K, Furusawa I (1992) Properties of a new virus belonging to nodaviridae found in larval striped jack (*Pseudocaranx dentex*) with nervous necrosis. Virology 187:368–371

Morit K, Mangyoku T, Iwamoto T, Arimoto M, Tanaka S, Nakai T (2003) Serological relationships among genotypic variants of betanodavirus. Dis Aquat Organ 57:19–26

Mulcahy D, Pascho RJ (1985) Vertical transmission of infectious hematopoietic necrosis virus in sockeye salmon, *Oncorhynchus nerka* (Walbaum)—isolation of virus from dead eggs and fry. J Fish Dis 8:393–396

Munang'andu HM, Fredriksen BN, Mutoloki S, Brudeseth B, Kuo T-Y, Marjara IS, Dalmo RA, Evensen O (2012) Comparison of vaccine efficacy for different antigen delivery systems for infectious pancreatic necrosis virus vaccines in Atlantic salmon (*Salmo salar* L.) in a cohabitation challenge model. Vaccine 30:4007–4016

Munir K, Kibenge FSB (2004) Detection of infectious salmon anaemia virus byreal-time RT-PCR. J Virol Methods 117:37–47

Munro ES, Midtlyng PJ (2011) Infectious pancreatic necrosis virus and associated aquatic birnavirus. In: Woo PTK, Bruno DW (eds) Fish diseases and disorders Vol. 3: viral, bacterial and fungal infections, vol 2011, 2nd edn. CAB International, Cambridge, MA, pp 1–65

Munro ES, Gahlawat SK, Ellis AE (2004) A sensitive non-destructive method for detecting IPNV carrier Atlantic salmon, *Salmo salar* L., by culture of virus from plastic adherent blood leucocytes. J Fish Dis 27:129–134

Nakajima K, Sorimachi M (1995) Production of monoclonal antibodies against red sea bream iridovirus. Fish Pathol 30:47–52

Nakajima K, Maeno Y, Fukudome M, Fukuda Y, Tanaka S, Matsuoka S, Sorimachi M (1995) Immunofluorescence test for the rapid diagnosis of red sea bream iridovirus infection using monoclonal antibody. Fish Pathol 30:115–119

Nakajima K, Maeno Y, Kurita J, Inui Y (1997) Vaccination against red sea bream iridoviral disease in red sea bream. Fish Pathol 32:205–209

Nakajima K, Maeno Y, Honda A, Yokoyama K, Tooriyama T, Manabe S (1999) Effectiveness of a vaccine against red sea bream iridoviral disease in a field trial test. Dis Aquat Org 36:73–75

Nallala VS, Makesh M, Radhika K, Sathish Kumar T, Raja P, Subburaj R, Kailasam M, Vijayan KK (2021) Characterization of red-spotted grouper nervous necrosis virus isolated from ovarian fluids of asymptomatic wild Asian seabass, *Lates calcarifer*. Aquaculture 542:736846

Nelson RT, McLoughlin MF, Rowley HM, Platten MA, McCormick JI (1995) Isolation of a toga-like virus from farmed Atlantic salmon *Salmo salar* with pancreas disease. Dis Aquat Org 22:25–32

Nguyen HD, Nakai T, Muroga K (1996) Progression of striped jack nervous necrosis virus (SJNNV) infection in naturally and experimentally infected striped jack *Pseudocaranx dentex* larvae. Dis Aquat Org 24:99–105

Nishizawa T, Mori K, Nakai T, Furusawa I, Muroga K (1994) Polymerase chain reaction (PCR) amplification of RNA of striped jack nervous necrosis virus (SJNNV). Dis Aquat Org 18:103–107

Nishizawa T, Mori K, Furuhashi M, Nakai T, Furusawa I, Muroga K (1995) Comparison of the coat protein genes of five fish nodaviruses, the causative agents of viral nervous necrosis in marine fish. J Gen Virol 76:1563–1569

Nishizawa T, Furuhashi M, Nagai T, Nakai T, Muroga K (1997) Genomic classification of fish nodaviruses by molecular phylogenetic analysis of the coat protein gene. Appl Environ Microbiol 63:1633–1636

Nishizawa T, Kinoshita S, Yoshimizu M (2005) An approach for genogrouping of Japanese isolates of aquabirnaviruses in a new genogroup, VII, based on the VP2/NS junction region. J Gen Virol 86:1973–1978

Noonan B, Enzmann PJ, Trust TJ (1995) Recombinant infectious necrosis virus and viral hemorrhagic septicemia virus glycoprotein epitopes expressed in *Aeromonas salmonicida* induce protective immunity in rainbow trout (*Onchorynchus mykiss*). Appl Environ Microbiol 61:3586–3591

Nylund A, Jakobsen P (1995) Sea trout as a carrier of infectious salmon anaemia virus (ISAV). J Fish Biol 47:174–176

Oelckers K, Vike S, Duesund H, Gonzalez J, Wadsworth S, Nylund A (2014) *Caligus rogercresseyi* as a potential vector for transmission of infectious Salmon Anaemia (ISA) virus in Chile. Aquaculture 420–421:126–132

Ogut H, Reno PW (2004) Effects of fish density on spread of infectious hematopoietic necrosis virus (IHNV) in rainbow trout, *Oncorhynchus mykiss*. Isr J Aquacult Bamidgeh 56:218–225

Oidtmann BC, Peeler EJ, Thrush MA, Cameron AR, Reese RA, Pearce FM, Dunn P, Lyngstad TM, Tavornpanich S, Brun E, Stärk KDC (2014) Expert consultation on risk factors for introduction of infectious pathogens into fish farms. Prev Vet Med 115:238–254

OIE (World Organization for Animal Health) (2019) Manual of diagnostic tests for aquatic animals 2019. Organisation Mondiale de la Santé Animale, France

Orpetveit I, Mikalsen AB, Sindre H, Evensen O, Dannevig BH, Midtlyng PJ (2010) Detection of infectious pancreatic necrosis virus in subclinically infected Atlantic salmon by virus isolation in cell culture or real-time reverse transcription polymerase chain reaction: influence of sample preservation and storage. J Vet Diagn Invest 22:886–895

Oseko N, Chuah TT, Palamisamy V, Maeno Y, Kurita J (2004) Iridovirus isolated from diseased sea bass *Lates calcarifer* and red drum *Sciaenops ocellatus* causing mass mortality in Malaysia. In: 7th Asian fisheries forum 04 abstracts, Penang, Malaysia, 127

Oshima S, Hata JI, Segawa C, Hirasawa N, Yamashita S (1996) A method for direct DNA amplification of uncharacterized DNA viruses and for development of a viral polymerase chain reaction assay: application to the red sea bream iridovirus. Anal Biochem 242:15–19

Oshima S, Hata JI, Hirasawa N, Ohtaka T, Hirono I, Aoki T, Yamashita S (1998) Rapid diagnosis of red sea bream iridovirus infection using the polymerase chain reaction. Dis Aquat Org 32:87–90

Overturf K, Lapatra SE, Powell M (2001) Real-time PCR for the detection and quantitative analysis of IHNV in salmonids. J Fish Dis 24:325–333

Pakingking R, de Jesus-Ayson EG, Reyes O, Brian Bautista N (2018) Immunization regimen in Asian sea bass (*Lates calcarifer*) broodfish: a practical strategy to control vertical transmission of nervous necrosis virusduring seed production. Vaccine 36:5002–5009

Panzarin V, Patarnello P, Mori A, Rampazzo E, Cappellozza E, Bovo G, Cattoli G (2010) Development and validation of a real-time TaqMan PCR assay for the detection of betanodavirus in clinical specimens. Arch Virol 155:1193–1203

Pereiro P, Martinez-Lopez A, Falco A, Dios S, Figueras A, Coll JM, Novoa B, Estepa A (2012) Protection and antibody response induced by intramuscular DNA vaccine encoding for viral haemorrhagic septicaemia virus (VHSV) G glycoprotein in turbot (*Scophthalmus maximus*). Fish Shellfish Immunol 32:1088–1094

Plumb JA, Hanson LA (2011) Health maintenance and principal microbial diseases of cultured fishes, 3rd edn. Blackwell Publishing Ltd, Iowa, pp 106–170

Praveenraj J, Ezhil Praveena P, Bhuvaneswari T, Navaneeth Krishnan A, Jithendran KP (2018) Experimental infection of Betanodavirus in freshwater fish *Gambusia affinis* (Baird and Girard, 1853)—a potential infection model for viral encephalopathy and retinopathy. Aquac Int 26:617–627

Puneeth TG, Baliga P, Girisha SK, Shekar M, Nithin MS, Suresh T, Naveen Kumar BT (2021) Complete genome analysis of a red seabream iridovirus (RSIV) isolated from Asian seabass (*Lates calcarifer*) in India. Virus Res 291:198199

Rajan JJS, Ezhil Praveena P, Bhuvaneswari T, Jithendran KP (2016) Design and evaluation of reverse transcription nested PCR primers for the detection of betanodavirus in finfish. Virus Dis 27:123–129

Riji John K, Rosalind George M, Jeyatha B, Saravanakumar R, Sundar P, Jithendran KP, Koppang EO (2014) Isolation and characterization of Indian betanodavirus strain from infected farm-reared Asian seabass *Lates calcarifer* (Bloch, 1790) juveniles. Aquac Res 45:1481–1488

Rimstad E, Markussen T (2019) Infectious salmon anaemia virus—molecular biology and pathogenesis of the infection. J Appl Microbiol 129:85–97

Robertsen B, Chang C-J, Bratland L (2016) IFN-adjuvanted DNA vaccine against infectious salmon anemia virus: antibody kinetics and longevity of IFN expression. Fish Shellfish Immunol 54:328–332

Rucker RR, Whipple WJ, Parvin JR, Evans CA (1953) Contagious disease of salmon possibly of virus origin. Fish Bull Fish Wildlife Serv 54:35–46

Salinas I, LaPatra SE, Erhardt EB (2015) Nasal vaccination of young rainbow trout (*Oncorhynchus mykiss*) against infectious hematopoietic necrosis and enteric red mouth disease. Dev Comp Immunol 53:105–111

Salonius K, Simard N, Harland R, Ulmer JB (2007) The road to licensure of a DNA vaccine. Curr Opin Investig Drugs 8:635–641

Sarath Babu V, Abdul Majeed S, Nambi KSN, Taju G, Madan N, Sundar Raj N, Sahul Hameed AS (2013) Comparison of betanodavirus replication efficiency in ten Indian fish cell lines. Arch Virol 158:1367–1375

Satyanarayana NV, Makesh M, Sain A, Jayaprakash NS, Kailasam M, Vijayan KK (2021) Non-lethal screening of Asian seabass (*Lates calcarifer*) by monoclonal antibody based indirect enzyme linked immunosorbent assay for viral nervous necrosis. Fish Shellfish Immunol Rep 2:100011

Sawayama E, Kitamura S-I, Nakayama K, Ohta K, Okamoto H, Ozaki A, Takagi M (2019) Development of a novel RSIVD-resistant strain of red sea bream (*Pagrus major*) by marker-assisted selection combined with DNA-based family selection. Aquaculture 506:188–192

Sepúlveda D, Lorenzen E, Rasmussen JS, Einer-Jensen K, Collet B, Secombes CJ, Lorenzen N (2019) Time-course study of the protection induced by an interferon-inducible DNA vaccine against viral haemorrhagic septicaemia in rainbow trout. Fish Shellfish Immunol 85:99–105

Shieh J, Chi S (2005) Production of monoclonal antibodies against grouper nervous necrosis virus (GNNV) and development of an antigen capture ELISA. Dis Aquat Org 63:53–60

Shimmoto H, Kawai K, Ikawa T, Oshima SI (2010) Protection of red sea bream *Pagrus major* against red sea bream iridovirus infection by vaccination with a recombinant viral protein. Microbiol Immunol 54:135–142

Shivappa R, Song H, Yao K, Aas-Eng A, Evensen Ø, Vakharia VN (2004) Molecular characterization of Sp serotype strains of infectious pancreatic necrosis virus exhibiting differences in virulence. Dis Aquat Org 61:23–32

Skall HF, Olesen NJ, Mellergaard S (2005) Viral haemorrhagic septicaemia virus in marine fish and its implications for fish farming – a review. J Fish Dis 28:509–529

Skliris GP, Richards RH (1998) Assessment of the susceptibility of the brine shrimp *Artemia salina* and rotifer *Brachionus plicatilis* to experimental nodavirus infections. Aquaculture 169:133–141

Snow M, Bain N, Black J, Taupin V, Cunningham CO, King JA, Skall HF, Raynard RS (2004) Genetic population structure of marine viral haemorrhagic septicaemia virus (VHSV). Dis Aquat Org 61:11–21

Somamotoa T, Nakanishi T (2020) Mucosal delivery of fish vaccines: local and systemic immunity following mucosal immunisations. Fish Shellfish Immunol 99:199–207

Sommerset I, Nerland A (2004) Complete sequence of RNA1 and subgenomic RNA3 of Atlantic halibut nodavirus (AHNV). Dis Aquat Org 58:117–125

Suebsing R, Pradeep PJ, Jitrakorn S, Sirithammajak S, Kampeera J, Turner WA, Saksmerprome V, Withyachumnarnkul B, Kiatpathomchai W (2016) Detection of natural infection of infectious spleen and kidney necrosis virus in farmed tilapia by hydroxynapthol blue-loop-mediated isothermal amplification assay. J Appl Microbiol 121:55–67

Taksdal T, Olsen AB, Bjerkaas I, Hjortaas MJ, Dannevig BH, Graham DA, Mcloughlin MF (2007) Pancreas disease in farmed Atlantic salmon, *Salmo salar* L., and rainbow trout, *Oncorhynchus mykiss* (Walbaum), in Norway. J Fish Dis 30:545–558

Tamer C, Cavunt A, Durmaz Y, Ozan E, Kadi H, Kalayci G, Ozkan B, Isidan H, Albayrak H (2021) Inactivated infectious pancreatic necrosis virus (IPNV) vaccine and *E. coli*-expressed recombinant IPNV-VP2 subunit vaccine afford protection against IPNV challenge in rainbow trout. Fish Shellfish Immunol 115:205–211

Tanaka S, Takagi M, Miyazaki T (2004) Histopathological studies on viral nervous necrosis of sevenband grouper, *Epinephelus septemfasciatus* Thunberg, at the grow-out stage. J Fish Dis 27:385–399

Thiery R, Peducasse S, Castric J, Leven A, Jeffroy J, Laurenci FB (1997) Experimental transmission of viral encephalopathy and retinopathy to juvenile sea bass (*Dicentrarchus labrax*). Fish Pathol 17:118–122

Thiery R, Raymond J-C, Castric J (1999) Natural outbreak of viral encephalopathy and retinopathy in juvenile sea bass, *Dicentrarchus labrax*: study by nested reverse transcriptase–polymerase chain reaction. Virus Res 63:11–17

Thiery R, Cozien J, Cabon J, Lamour F, Baud M, Schneemann A (2006) Induction of a protective immune response against viral nervous necrosis in the European sea bass *Dicentrarchus labrax* by using betanodavirus virus-like particles. J Virol 80:10201–10207

Thompson JP, Granoff A, Willis DB (1987) Infection with frog virus 3 allows transcription of DNA methylated at cytosine but not adenine residues. Virology 160:275–277

Thorarinsson R, Wolf JC, Inami M, Phillips L, Jones G, Macdonald AM, Rodriguez JF, Sindre H, Skjerve E, Rimstad E, Evensen O (2021) Effect of a novel DNA vaccine against pancreas disease caused by salmonid alphavirus subtype 3 in Atlantic salmon (*Salmo salar*). Fish Shellfish Immunol 108:116–126

Thorud KE, Djupvik HO (1988) Infectious anaemia in Atlantic salmon (*Salmo salar* L.). Bull Eur Assoc Fish Pathol 8:109–111

Tobar I, Arancibia S, Torres C, Vera V, Soto P, Carrasco C, Alvarado M, Neira E, Arcos S, Tobar JA (2015) Successive oral immunizations against *Piscirickettsia salmonis* and infectious salmon anemia virus are required to maintain a long-term protection in farmed salmonids. Front Immunol 6:244

Toffan A, Pascoli F, Pretto T, Panzarin V, Abbadi M, Buratin A, Quartesan R, Gijon D, Padros F (2017) Viral nervous necrosis in gilthead sea bream (*Sparus aurata*) caused by reassortant betanodavirus RGNNV/SJNNV: an emerging threat for Mediterranean aquaculture. Sci Rep 7:1–12

Toubanaki DK, Margaroni M, Karagouni E (2015) Nanoparticle-based lateral flow biosensor for visual detection of fish nervous necrosis virus amplification products. Mol Cell Probes 29:158–166

Vimal S, Madan N, Farook MA, Nambi KSN, Abdul Majeed S, Rajkumar T, Venu S, Thirunavukkarasu AR, Sahul Hameed AS (2014a) Production of recombinant vaccine using capsid gene of nodavirus to protect Asian sea bass, *Lates calcarifer* (Bloch, 1790). Aquaculture 418–419:148–154

Vimal S, Abdul Majeed S, Nambi KSN, Madan N, Farook MA, Venkatesan C, Taju G, Venu S, Subburaj R, Thirunavukkarasu AR et al (2014b) Delivery of DNA vaccine using chitosan–tripolyphosphate (CS/TPP) nanoparticles in Asian sea bass, *Lates calcarifer* (Bloch, 1790) for protection against nodavirus infection. Aquaculture 420–421:240–246

Vimal S, Farook MA, Madan N, Abdul Majeed S, Nambi KSN, Taju G, Sundar Raj N, Venu S, Subburaj R, Thirunavukkarasu AR et al (2016) Development, distribution and expression of a DNA vaccine against nodavirus in Asian seabass, *Lates calcarifier* (Bloch, 1790). Aquac Res 47:1209–1220

Walker PJ, Blasdell KR, Calisher CH, Dietzgen RG, Kondo H, Kurath G, Longdon B, Stone DM, Tesh RB, Tordo N, Vasilakis N, Whitfield AE, ICTV Report Consortium (2018) ICTV virus taxonomy profile: Rhabdoviridae. J Gen Virol 99:447–448

Wang S, Huang X, Huang Y, Hao X, Xu H, Cai M, Wang H, Qin Q (2014) A novel marine DNA virus (Singapore grouper iridovirus, SGIV) entry into host cells occurs via clathrin-mediated endocytosis and macropinocytosis in a pH-dependent. J Virol 88:13047–13063

Watanabe K, Nishizawa T, Yoshimiru M (2000) Selection of brood stock candidates of barfin flounder using an ELISA system with recombinant protein of barfin flounder nervous necrosis virus. Dis Aquat Org 41:219–223

Weston JH, Villoing S, Brémont M, Castric J, Pfeffer M, Jewhurst VA, McLoughlin MF, Rødseth O, Christie KE, Koumans J, Todd D (2002) Comparison of two aquatic alphaviruses, salmon pancreas disease virus, and sleeping disease virus, by using genome sequence analysis, monoclonal reactivity, and cross-infection. J Virol 76:6155–6163

Williams T, Barbosa-Solomieu V, Chinchar VG (2005) A decade of advances in iridovirus research. Adv Virus Res 65:173–248

Wingfield WH, Fryer JL, Pilcher KS (1969) Properties of the sockeye salmon virus (Oregon strain). Proc Soc Exp Biol Med 130(4):1055–1059

Wolf K (1988) Viral hemorrhagic septicemia. In: Fish viruses and fish viral diseases. Cornell University Press, Ithaca, pp 217–249

Wolf K, Dunbar CE, Snieszko SF (1960) Infectious pancreatic necrosis of trout: I. A tissue-culture study. Progress Fish-Cult 22:64–68

Wolf A, Hodneland K, Frost P, Braaen S, Rimstad E (2013) A hemagglutinin-esterase-expressing salmonid alphavirus replicon protects Atlantic salmon (*Salmo salar*) against infectious salmon anemia (ISA). Vaccine 31(4):661–669

Wolf A, Hodneland K, Frost P, Hoeijmakers M, Rimstad E (2014) Salmonid alphavirus-based replicon vaccine against infectious salmon anemia (ISA): impact of immunization route and interactions of the replicon vector. Fish Shellfish Immunol 36:383–392

Wood EM, Snieszko SF, Yasutake WT (1955) Infectious pancreatic necrosis in brook trout. AMA Arch Pathol 60:26–28

Xu C, Mutoloki S, Evensen O (2012) Superior protection conferred by inactivated whole virus vaccine over subunit and DNA vaccines against salmonid alphavirus infection in Atlantic salmon (*Salmo salar* L.). Vaccine 30:3918–3928

Xu L, Zhao J, Liu M, Ren G, Jian F, Yin J, Feng J, Liu H, Lu T (2017) Bivalent DNA vaccine induces significant immune responses against infectious hematopoietic necrosis virus and infectious pancreatic necrosis virus in rainbow trout. Sci Rep 7(1):5700

Yong CY, Ong HK, Tang HC, Yeap SK, Omar AR, Ho KL, Tan WS (2019) Infectious hematopoietic necrosis virus: advances in diagnosis and vaccine development. PeerJ 7:e7151

Yoshikhosi K, Inoue K (1990) Viral nervous necrosis in hatchery-reared larvae and juveniles of Japanese parrotfish, *Oplegnathus fasciatus* (Temminck & Schlegel). J Fish Dis 13:69–77

Zhao J-Z, Xu L-M, Liu M, Cao Y-S, LaPatra SE, Yin J-S, Liu H-B, Lu T-Y (2017) Preliminary study of an oral vaccine against infectious hematopoietic necrosis virus using improved yeast surface display technology. Bull Eur Assoc Fish Pathol 85:196–204

Zhao J-Z, Liu M, Xu L-M, Zhang Z-Y, Cao Y-S, Shao Y-Z, Yin J-S, Liu H-B, Lu T-Y (2019) A chimeric recombinant infectious hematopoietic necrosis virus induces protective immune responses against infectious hematopoietic necrosis and infectious pancreatic necrosis in rainbow trout. Mol Immunol 116:180–190

Diseases in Ornamental Fishes

P. Sivasankar, K. Riji John, and M. Rosalind George

Abstract

Aquarium fishkeeping is considered to be one of the most well-liked pastimes worldwide. The ornamental fish industry currently boasts a trade of about 5.9 billion U.S. dollars. However, this sector is vulnerable to various disease problems due to pathogenic microorganisms and parasites. The outbreaks of infectious diseases of ornamental fishes are the major challenges for the aquarists as they are associated with high economic losses through reduced growth performance and increased mortality. In the case of ornamental fishes, infectious diseases could impair the performance of aquarium fish through colour change, scale loss, different pathological signs, morbidity and mortality. Bacterial, viral, fungal and parasites are common infectious pathogens of ornamental fishes. Tropical ornamental fishes are more susceptible to parasitic infections caused by various fish parasites, including protozoans, trematodes, cestodes, nematodes, acanthocephalans, crustaceans, etc. Many of the fungi are also considered opportunistic pathogens that affect several ornamental fishes. It mainly attacks the fish when stressed due to the unfavourable environmental conditions. These infectious diseases can be prevented and controlled by environmental management to minimize stress, maintaining water quality parameters, quarantine of new fish and live foods before introducing to the ornamental unit, regular monitoring of the behaviour of fish, etc.

P. Sivasankar
Fisheries College and Research Institute, Tamil Nadu Dr. J. Jayalalithaa Fisheries University, Thoothukudi, Tamil Nadu, India
e-mail: sanknag@gmail.com

K. Riji John (✉) · M. Rosalind George
Kerala University of Fisheries and Ocean Studies, Kochi, India
e-mail: rijijohn@gmail.com; rosalindmg@yahoo.com

R. Kooloth Valappil et al. (eds.), *Aquatic Animal Health Management*,
https://doi.org/10.1007/978-981-96-7987-4_7

Keywords

Aquarium · Infectious diseases · Ornamental fish · Prevalence

1 Introduction

The ornamental fish setting was initiated as a hobby by showcasing goldfish since the eighteenth century. In the twentieth century, the setting up of aquariums and rearing ornamental fish gained international prominence. India has vast resources of ornamental fishes, both freshwater and marine, found in the Western Ghats, North-eastern states and Gulf of Mannar. Approximately, 600 species have been identified as potentially valuable ornamental fishes. The two major groups of ornamental fishes are oviparous (egg layers) and live-bearers (viviparous, e.g. mollies, platies, guppies and swordtail). The egg layers are further categorized as guarders (egg layers with care), non-guarders (egg layers with no care), mouth-incubators, nest builders, egg buriers, egg-carriers, egg-scatterers with non-adhesive eggs, egg-scatterers with adhesive eggs, egg-depositors etc. (Moyle and Cech 2003; Helfman et al. 2009; Pandey 2013).

According to the MPEDA (Marine Products Export Development Authority), there are about one million aquarium hobbyists in India. The Indian market can be categorized into two segments: the domestic market, which is contributed by exotic fishes, and the export market, in which indigenous ornamental fishes dominate. The size of the ornamental fish market globally is estimated to be USD 5881.2 million in 2022, which is expected to double to USD 11305.7 million by 2030. The corresponding figures for India stand at 181.7 and 388.5, respectively (Horizon Grand View Research 2024). Being one of the most widely traded commodities, it is essential to control the transboundary movement of pathogens during the export and import of ornamental fishes through adequate surveillance and quarantine facilities. Additionally, aquatic fishes are sensitive, delicate and prone to a variety of diseases under stressful circumstances much faster than cultivable fishes that are maintained in big tanks and ponds with large quantities of water. Numerous disease conditions of ornamental fish arise due to temperature fluctuation, suboptimal water quality, improper handling, low-quality diet, etc. The ornamental fishes are more susceptible to diseases such as fin and tail rot, white spot disease and columnaris disease. The diseased fish may exhibit general clinical signs such as loss of pigment, sluggishness, irregular swimming activity, loss of appetite, etc. Dropsy, popeye disease, slimy skin disease, velvet disease, ectoparasite infection caused by fluke attack, anchor worm and fish lice are also found in many circumstances (Helfman et al. 2009; Madhuri 2013; Santos et al. 2017). Viral pathogens such as viral nervous necrosis, koi herpesvirus and lymphocystis disease virus are some of the major threats to ornamental fish production. If left untreated, the disease may become systemic and kill the infected fish. This section discusses the major infectious diseases that affect ornamental fishes.

2 Viral Diseases of Ornamental Fish

2.1 Viral Nervous Necrosis (VNN)

A highly devastating disease that affects over 40 species of marine fish worldwide is caused by fish nodaviruses, which belong to the genus *Betanodavirus* of *Nodaviridae* family (Munday et al. 2002; Maltese and Bovo 2007). Nodaviruses are small (25–30 nm), single-stranded, icosahedral, non-enveloped viruses with positive-sense RNA genome that is capped but not polyadenylated. The 100 kDa RNA-dependent RNA polymerase (protein A) is encoded by the 3.1 kb RNA1, while the 1.4 kb RNA2 encodes 42 kDa coat protein (α) (Mori et al. 1992; Nagai and Nishizawa 1999). It has also been discovered that infected cells contain a subgenomic RNA3, which is produced during RNA replication of RNA1 (Sommerset and Nerland 2004). In addition to being a transactivator during RNA2 replication (Eckerle and Ball 2002), it encodes a B2 protein with an unclear function that acts as an RNA interference antagonist that promotes intracellular viral RNA accumulation (Fenner et al. 2006). Viral encephalopathy and retinopathy (VER) or viral nervous necrosis (VNN) is caused by the virus infecting the central nervous system (CNS) of fish. Although adults may also be affected, larvae and juveniles are the primary victims of VNN infection, which can cause great mortality rates of up to 100% (Munday et al. 2002; Pirarat et al. 2009).

In India, neurological system-related signs have been found in some species of ornamental fish, viz. goldfish (*Carassius auratus*), rainbow shark (*Epalzeorhynchos frenatum*) and albino rainbow shark (Jithendran et al. 2011). The primary clinical manifestations of VNN are dropsy, descaling, anorexia and aberrant swimming behaviour. The most distinctive lesions of VNN include necrosis and vacuolation of the retina and nerve tissues of the central nervous system (brain and spinal cord) (Gomez et al. 2006). An experimentally infected guppy (*Poecilia reticulata*) exhibited clinical signs with substantial mortality beginning at 15 dpi, which rose to 100% at 30 dpi (Nazari et al. 2014).

2.2 Lymphocystis Disease Virus (LCDV)

Lymphocystis disease virus, or Lymphocystivirus, is an iridovirus belonging to the family *Iridoviridae* that causes Lymphocystis. It is a chronic disease that affects both marine and freshwater fish species (Wolf 1988). The LCDV has a double-layered, icosahedral capsid about 200 ± 50 nm in diameter with an exterior envelope and a fringe of external protrusions that resemble fibrils (Walker 1962; Zwillenberg and Wolf 1968; Wolf 1988). The LCDV genome is a linear dsDNA molecule with terminal redundancy and circular permutation (Darai et al. 1983, 1985; Schnitzler and Darai 1993). The two sequenced viruses, such as LCDV-1 and LCDV-C, have differed widely in their genome sizes, which were 102.6 kbp and 186 kbp, respectively.

The LCDV can infect a variety of fish species including gouramies (*Osphronemidae*), cichlids (*Cichlidae*), killifishes (*Cyprinodontidae*), sunfishes (*Centrarchidae*), gobies (*Gobiidae*), damselfish/clownfish (*Pomacentridae*), butterflyfishes (*Chaetodontidae*), snooks (*Centropomidae*) and dottybacks (*Pseudochromidae*). (Yanong 2010). The virus infection appears irregular, small to moderately large, nodular, wart-like growths on the body surface viz. skin, fins or gills. The lymphocystis nodules can be cream or grey, but they can also be varied if they show up under pigmented areas (Yanong 2010). If the infection progresses and increases in size, it may also cause popeye (exophthalmia) in fish when tissues behind the eye are infected and the eyes start protruding. Lymphocystis causes cells in the skin, fins or internal tissues of fish to undergo prominent enlargement, and the infected cells, known as fibroblasts, are part of the connective tissues. Once the lymphocystis cells reside on their host for 4 weeks or more, they burst or fall off, dispersing the infected cells throughout the water. After spreading, the cells either drop to the bottom of the tank and lie dormant or enter another host through a broken skin, fins or gills. The lymphocystis infection can occur without any visible symptoms because fish may have fewer diseased cells or may have limited internal organs infected. In such circumstances, a wet mount of suspected tissues from the infected fishes would have to be subjected to microscopic examination for identification of lymphocystis.

2.3 Infectious Spleen and Kidney Necrosis Virus (ISKNV)

ISKNV is the type species of the genus *Megalocytivirus* belonging to the family *Iridoviridae* (Guo et al. 2012), and it was first discovered in 1994 in Chinese mandarin fish (He et al. 2002). The virion size is 140–200 nm in diameter with icosahedral symmetry and a genome size of 111 kb. The virus has a wide host range and can infect nearly 50 distinct species of freshwater, brackishwater and marine fishes (Jung-Schroers et al. 2016). The virus has been found to be infecting various fish species, including *Carassius auratus, Astronotus ocellatus, Poecilia sphenops, Trichopodus trichopterus, Maylandia lombardoi, Pterophyllum scalare, Mikrogeophagus ramirezi, Cyprinus carpio koi, Poecilia reticulata and Puntius titteya* (Girisha et al. 2021).

Infected fish exhibit pale colouration, petechial haemorrhages, anorexia and irregular swimming (He et al. 2002). Internally, affected fish show gross changes, including swelling of the spleen and kidney. Histopathology of the kidney, spleen, gill tissue, cranial connective tissue and endocardium has shown hypertrophic cells with large basophilic cytoplasmic inclusions (He et al. 2002). When the silver gourami, *Trichopodus microlepis* and pearl gourami, *T. leeri* were exposed to experimental infection with a tissue homogenate of ISKNV, it resulted in 20% and 70% cumulative mortalities, respectively (Jeong et al. 2008). In 2018–19, there were notable deaths of ornamental fish species due to ISKNV, accompanied by a variety of clinical symptoms such as slow and erratic swimming behaviour, mucus discharge, fin rot, haemorrhage and body discolouration, along with unceasing mortality in India (Girisha et al. 2021).

2.4 Koi Herpes Virus (KHV)

Koi herpes virus was first documented in common carp in Israel and USA in 1998 (Hedrick et al. 2000; Perelberg et al. 2003) and in Germany in 1997–1998 (Bretzinger et al. 1999). KHV was also reported in preserved tissue samples in England during 1996 (Haenen et al. 2004). Since then, KHV outbreaks have been reported in Israel, Indonesia, Malaysia, Thailand, Japan, Taiwan and China. KHV possess a large, linear dsDNA genome with an icosahedral core measuring 100–110 nm (Hedrick et al. 2000). The virus is also known as cyprinid herpesvirus-3 (CyHV-3) by keeping with the nomenclature of other cyprinid herpesviruses, such as CyHV-1 (carp pox virus, fish papillomavirus) and CyHV-2 (goldfish haematopoietic necrosis virus) (Waltzek et al. 2005). Varieties of common carp such as koi, crucian carp, grass carp leather, mirror and ghost koi are susceptible to KHV infection (Haenen et al. 2004; Hedrick et al. 2000; Kempter and Bergmann 2007). Russian sturgeon and Atlantic sturgeon, hybrids of goldfish, carp and koi, have also been found to be susceptible to KHV (Hedrick et al. 2006; El-Matbouli et al., 2007; Kempter et al. 2009). Currently, KHV is classified under the family *Alloherpesviridae* and genus *Cyprinivirus* (Davison et al. 2009). Infection of CyHV-2 in gold fish has been reported causing extensive mortality in goldfish (Sahoo et al. 2016). The goldfish exhibited large scale haemorrhages on the body, fins and gills, lepidorthosis, necrosed gills, protruded anus and shrunken eyes. Internally the fish had white nodular necrotic foci in spleen and kidneys, along with necrosis and fusion of gill lamellae.

KHV can affect fish of all ages and cause 80–100% mortality in vulnerable populations when the water temperatures range from 16° to 25 °C (Haenen et al. 2004). Affected fish turn abnormal and swim erratically with rapid breathing, enlarged gills and partial skin lesions (Gray et al. 2002; Pokorova et al. 2005). Outward manifestations of KHV infection may include sunken eyes, pale skin patches and bleeding, swollen, pale or patchy gills (Goodwin 2012; Oh et al. 2001). Histological observations of infected fish show widely proliferated gill epithelium, and infected cells with intranuclear inclusions coupled with degenerative and necrotic alterations. Microscopic investigations of the kidney, liver, spleen and gastrointestinal tract reveal a large number of macrophages carrying ingested cellular debris and necrosis of parenchyma cells (Hedrick et al. 2000).

2.5 Koi Ranavirus (KIRV)

Ranaviruses are a genetically diverse group of viruses that cause systemic necrotizing infections in lower vertebrates. They come under the family *Iridoviridae* and are enveloped and icosahedral, having a size of around 150 nm with a genome size of about 105 kb. Ranaviruses are detected in the ornamental fishes exported from

Southeast Asia to the USA (Hedrick and McDowell 1995). Koi ranavirus was primarily identified and isolated in India from moribund koi that had persistent mortality, swimming irregularities, sporadic surfacing and skin darkening (George et al. 2015). Infected cell cultures were found to contain icosahedral virus particles that were 100–120 nm in size and budded from the cell membrane. Major capsid protein gene sequence analysis revealed a 99.9% identical match with the largemouth bass virus (LMBV) obtained from North America. The KIRV can be transmitted horizontally and exhibit typical clinical signs like uncoordinated swimming, turning over and vertical hanging prior to death (George et al. 2015).

2.6 Similar Damselfish Virus (SRDV)

Similar damselfish virus was documented in marine ornamental fish, Similar Damselfish (*Pomacentrus similis* Allen 1991) in India (Sivasankar et al. 2017). The virus has been described to belong to the family *Iridoviridae*, specifically to the genus *Ranavirus*. The virus can thrive well in freshwater, and marine fish cell lines established from snakehead and seabass. The virus has a size of 120–130 nm, with a double-stranded DNA genome and icosahedral symmetry. Similar damselfish fingerlings infected experimentally with SRDV showed cumulative mortality rates as high as 93.33% and exhibited clinical signs including skin discolouration, unbalanced swimming, lack of appetite demonstrated by no feed intake and lying at the bottom of the tank, followed by death. The experimentally infected fish also showed skin ulceration in the dorsal region behind the head (Sivasankar et al. 2017).

2.7 Carp Edema Virus (CEV)

Carp edema viral disease primarily affects three continents: Asia, North America and Europe. It causes widespread mortality in both wild and farmed species of common carp (*Cyprinus carpio*), including koi (Hesami et al. 2015). Since the initial discovery of CEV in Japanese koi in 1974, its infection has spread throughout Japan (Ono et al. 1986; Amita et al. 2002). The viral infection has also been reported in India (Swaminathan et al. 2016). CEV is a large, dsDNA virus belonging to *Poxviridae* family. Based on the phylogenetic analysis of 478 nucleotide segments (partial) of the CEV 4a gene, it has been divided into three different genogroups (Matras et al. 2017). Carp pox, which is caused by a herpesvirus (Cyprinid herpesvirus 1) and causes wart-like growths on the skin of common carp types, is different from CEV infection (Hartman et al. 2013). Since the CEV could not be cultivated in cell cultures, visualization is by electron microscopy imaging, which shows the virus as poxvirus-like particles in infected gill epithelial cells (Ono et al. 1986; Oyamatsu et al. 1997; Miyazaki et al. 2005).

The disease was first known as viral oedema of carp due to haemorrhagic lesions or skin erosions with oedema of the subcutaneous tissues (Oyamatsu et al. 1997). Since the affected fish exhibit behavioural abnormalities such as lethargy, unresponsiveness and weak refluxes, CEVD is also called as koi sleepy disease (KSD). The infected fish typically lie motionless on the sides of the tank, only moving briefly when agitated or disturbed (Miyazaki et al. 2005). Comparatively, juveniles are more vulnerable to this viral infection, and weakened juveniles swim slowly along the tank's edges or may hang just below the water's surface and die (Hesami et al. 2015). Skin lesions around the mouth and base of the fins, exophthalmia, congested and oedematous gills with hyperplasia, lamellar fusion, clubbing and necrosis are the clinical indications of CEV infection (Ono et al. 1986; Miyazaki et al. 2005). The impact of temperature on the onset of CEV infections has been the subject of contradictory reports. Nevertheless, research suggests that temperature stress does not contribute to the development of CEV infections in fish (Oyamatsu et al. 1997; Miyazaki et al. 2005; Sahoo et al. 2020).

3 Bacterial Diseases of Ornamental Fish

The aquatic environment is abound with bacteria that act as opportunistic pathogens, and these cause widespread infections in ornamental fish (Lewbart 2001). Bacterial pathogens can be either obligate or opportunistic in nature. Four main categories of bacterial pathogens that affect fish are systemic Gram-negative bacteria (*Vibrio*, *Aeromonas*, *Pseudomonas*, *Edwardsiella* and *Flavobacterium*), external Gram-negative bacteria (*Flexibacter maritimus, Flavobacterium columnarae*, yellow-pigmented bacteria, and *Cytophaga* spp.), systemic Gram-positive bacteria (*Streptococcus* spp. and allied species) and slow-growing acid-fast bacteria (*Mycobacterium* spp) and one more discovered group of pathogens (intracellular rickettsial pathogen) (Roberts et al. 2009). Gram-negative bacteria like *Aeromonas, Citrobacter, Edwardsiella, Flavobacterium, Pseudomonas* and *Vibrio* are the major groups responsible for the numerous bacterial infections of fish. Depending on the specific pathogen and other host-related variables, clinical manifestations of the bacterial disease can be acute, chronic or peracute (Roberts et al. 2009).

3.1 Furunculosis

The disease, furunculosis, is caused by *Aeromonas salmonicida*, a Gram-negative, non-motile, rod-shaped bacterium and is a common obligate bacterial pathogen of freshwater and marine fish. Furunculosis is named for the sub-acute or chronic phase of the disease, which is identified by the presence of furuncles, or boil-like lesions, in the muscles. It is considered one of the most significant fish diseases due to its extensive distribution and vast host range, especially among the valuable salmonids and koi (*Cyprinus carpio*). Since the disease condition is very prevalent in koi, it is also called 'koi ulcerative disease' or simply 'ulcer disease'. *A. salmonicida* was also described and characterized in goldfish *C. auratus* in 1980 (Shotts and Talkington 1980; Elliott and Shotts Jr 1980). *A. salmonicida* can cause a cutaneous ulcerative disease in goldfish, and such a condition is called as ulcer disease, which is a stress-related disorder linked to the atypical form of the pathogen (Dror et al. 2006). *A. salmonicida* caused about 50% of mortalities among fish reared in an aquarium (Wada et al. 2010). The infected fish had abrasions on the lower jaw but no other significant body surface abnormalities. However, the kidney was enlarged and had white surface nodules.

Lethargy, mild exophthalmia, blood-shot fins, bloody discharge from the nares and vent and many haemorrhages in the muscle and other tissues are the hallmarks of furunculosis disease (Snieszko 1958; McCarthy and Roberts 1980). Internally, there may be kidney necrosis, splenic enlargement and liver haemorrhage. The disease may be associated with a low mortality rate, although fish that survive this type of sickness typically have scar tissue near their furuncles (McCarthy 1975). In the ulcerative version of the disease, the skin often bleeds and necrotises, exposing the underlying muscle and bone. As a result of these ulcers, the fish may be vulnerable to invasion by secondary bacterial, fungal, metazoan and protozoan diseases (Lewbart 2001).

3.2 Pseudomoniasis

Pseudomonas spp. is the common bacterial flora of freshwater, which often becomes an opportunistic pathogen for fish in stressful situations. Bacterial disease outbreaks are triggered by a number of factors, but the most prevalent ones in aquariums are stress, overcrowding and poor water quality (Sasmal et al. 2004). Sekiten byo (red spot) disease is caused by fish pathogenic pseudomonads that include *P. anguilliseptica, P. fluorescens, P. chlororaphis, P. baetica, P. plecoglossicida, P. luteola, P. putida, P. pseudoalcaligenes* (Wakabayashi and Egusa 1972). Among these pseudomonad fish pathogens, *P. anguilliseptica* is regarded as the most important pathogen, which was identified as the aetiological agent of red-spot disease of Japanese eels (*Anguilla japonica*) (Wakabayashi and Egusa 1972). *Pseudomonas* was also found to be infecting different fish species such as *C. auratus, D. rerio, X. helleri, L. caeruleus, X. maculatus, T. trichopterus, G. ternetzi* and *P. innesi* (Sasmal et al. 2004; Walczak et al. 2017). *P. fluorescens*, is responsible for pseudomonas septicaemia, a condition that is almost the same as MAS and was formerly thought to be one of the causes of 'haemorrhagic septicaemia' (Plumb and Hanson, 2011).

Clinical indicators of infections caused by Pseudomonas species include haemorrhaging on the body surface and bloated abdomen, fin/tail rot with or without external haemorrhage, haemorrhagic septicaemia, extensive skin lesions, bacterial haemorrhagic ascites, exophthalmia with external ulceration etc. (Austin and Austin 2012). Internal clinical signs included a pale liver, a haemorrhaged kidney and an intestine filled with yellowish exudate. Black spot sea bream in Spain were found to have low-level mortalities, even though there were no outward symptoms of infection (López-Romalde et al. 2003).

3.3 Columnaris Disease

Flexibacter columnare, formerly known as *Cytophaga columnaris*, is the causative agent of columnaris disease (Bernardet 1989). It is a strictly aerobic bacterium, a Gram-negative rod that is 4–8 and 0.5–0.7 mm in size and glides. This infection, known as "cotton wool disease" in the aquarium business, affects platies (*Xiphophorus maculatus*) and black mollies (*Poecilia sphenops*) (Decostere et al. 1998), goldfish (Mitiku 2018) and other ornamental fishes. *F. columnare* is an opportunistic pathogen that is associated with risk factors such as crowding, poor water quality, transportation, poor nutrition and parasitic or other bacterial diseases.

The skin and gill epithelial tissues are often the only areas affected by columnaris, though severe infections can spread throughout the body (Noga 2000). As the infection primarily affects the gill, the blood vessels supplying the gills become congested or blocked, causing the surface epithelium of the lamellae to detach from the capillary bed (Pacha and Ordal 1967). Lesions may develop on the skin, gills or muscles of adult fish. Neon tetras (*Paracheirodon innesi*) infected with columnaris disease have exhibited muscular lesions and skin discolouration (Michel et al. 2002). Typically, gill lesions are composed of yellow-orange patches of necrosis that begin at the outer edge of the gills and spread out to the gill arch base. Large-scale erosion may eventually destroy the gill filament (Pacha and Ordal 1970). The presence of a cotton-like mass or plaque on the head, fins or tail is another way to identify columnaris disease in fish. Areas of haemorrhage often surround these lesions. Deep ulcers may develop as a result of ongoing infections (Lewbart 2001).

3.4 Motile Aeromonad Disease

Freshwater fishes are most frequently affected by motile aeromonad disease (MAD) or motile aeromonad septicaemia (MAS), which is caused by *Aeromonas hydrophila*. Other species, such as *A. sobria, A. caviae* and *A. veronii* have also been connected with MAD (Noga 2000). These organisms are ubiquitous and opportunistic pathogens that exploit the fish under environmental stress conditions, viz. extreme temperature, poor nutrition, crowding and transport (Lewbart 2001). In addition to *A. hydrophila* and *A. veronii*, other species, such as *A. media*, have also been reported from several ornamental fish species (Walczak et al. 2017). The afflicted fish may exhibit sloughing of the skin, fins, and scales, stomach distention, exophthalmia, engorged gill lamellae, petechiation and haemorrhage. Following systemic infection, haemorrhagic septicaemia causes inflammation and necrosis of the gastrointestinal tract, kidney, muscle and spleen (Lewbart 2001).

3.5 Edwardsiellosis

The aetiological agents of Edwardsiellosis are *Edwardsiella tarda* and *E. ictaluri*, which are Gram-negative rods, motile, facultative anaerobic bacteria. Both *E.*

ictaluri and *E. tarda* have been reported from fighting fish (*Betta splendens*) (Vandepitte et al. 1983). The zebrafish (*D. rerio*), pink barbs (*Puntius conchonius*) and green knife fish (*Eigenmannia virescens*) were also found to be afflicted by *E. ictaluri* (Kent and Lyons 1982; Blazer et al. 1985). *E. tarda* has been isolated from angel fish (*Pterophyllum scalare*) that were dying without obvious symptoms other than appetite loss (Turgay 2020). It has been reported to inhabit in aquatic invertebrates, including sea urchins (Sasaki and Aita 1975) and freshwater aquarium snails (Bartlett and Trust 1976).

Clinical manifestations, including lethargy, ulcers and respiratory distress, were found during the infection with both species (Lewbart 2001). The mild infection of *E. tarda* in fish appears clinically as small cutaneous lesions (size from 3 to 5 mm in diameter) in the postero-lateral region of the body. As a result of the progression, abscesses in the muscles of the body and tail would be developed. These abscesses may expand and become gas-filled hollow areas (Meyer and Bullock 1973). Several clinical symptoms, such as irregular swimming, spiral movement, floating close to the water's surface, exophthalmia, and pigmentation loss, can be seen in fish infected with Edwardsiellosis by *E. tarda* (Woo and Bruno 2010; Park et al. 2012). Lesions in internal organs are the other clinical signs found in a parrot cichlid (Walczak et al. 2017). Several virulence elements are involved in the pathogenesis of edwardsiellosis (Park et al. 2012). It has also been established that *E. tarda* is capable of intracellular replication and systemic infection using various factors, including type III secretion systems (Wang et al. 2011). This would negate the effect of the application of antibiotics, if need be, in case of *E. tarda* infections.

3.6 Vibriosis

Vibriosis is caused by members of the genus *Vibrio,* which represents normal environmental bacterial flora and are considered to be opportunistic pathogens. Water quality variations and crowding are among the factors that cause the incidence of vibriosis in cultured fish due to increased stress (Lewbart 2001; Kumar et al. 2015). Many *Vibrio* species are associated with infectious diseases of fish, which include *Vibrio anguillarum, V. harveyi, V. alginolyticus, V. parahaemolyticus, V. cholerae, V. vulnificus, V. ordalii, V. damsela (now classified as Photobacterium damselae subsp. damselae), V. salmonicida* and *V. mimicus* (Colwell and Grimes 1984; Egidius et al. 1986; Saeed 1995; Mohamad et al. 2019). Among these species, *V. anguillarum* and *V. ordalii* are the most commonly identified bacterium from marine ornamental fish (Barker 2001). Reports indicate the occurrence of *V. anguillarum in* guppy (*Lebistes reticulatus)* (Hacking and Budd 1971), *V. carachariae* in brown shark *(Carchminus plumbeus)* (Grimes et al. 1984) and *V. cholerae* in goldfish (*C. auratus)* (Reddacliff et al. 1993).

Fish with vibriosis show a variety of macroscopic clinical abnormalities such as exophthalmia, dermal ulceration, mouth-head ulceration, myonecrosis, myositis, corneal damage, panophthalmitis, splenomegaly, congestion and necrosis, heart haemorrhage, gill with anaemia, fin-tail necrosis and fin-skin erythema (Austin and

Austin 1999). Vibriosis-affected fish exhibit clinical symptoms and lesions that resemble those of *A. salmonicida* (Lewbart 2001). Vibriosis affected fish would have dark, listless, anorexic appearances, and in severe cases, ulcers may be found on the skin. In chronic cases, granulomas may form in different muscle regions of the body. Numerous virulence factors, such as haemolysin, type III and type VI secretion systems, adhesion factors, iron uptake systems, lipopolysaccharide, protease and outer membrane proteins, determine the pathogenicity of vibrios (Li et al. 2019).

3.7 Streptococcosis

Farmed and wild populations, freshwater and marine fish, including ornamental fishes, are susceptible to streptococcosis, which is a septicaemic disease. *Streptococcus* spp., the causative agents of streptococcosis were found to be infecting various fish species including kuhli loach (*Acanthophthalmus kuhli*), pearl danios *(Brachydanio albolineatus),* tiger and green tiger barbs *(Barbus tetrazona),* rosy barbs *(B. conchonius),* bloodfin tetras *(Aphyocharax anisitsi),* blue tetras *(Boehlkea fredcochui),* red-eyed tetras *(Moenhhausia sanctaefilomenae),* high-fin leopard danios *(Brachydanio redo),* rainbow sharks *(Epalzeorhynchus frenatus)* and red-tailed black sharks *(E. bicolor)* (Lewbart 2001). Generally, streptococcosis is caused by *Streptococcus* species, including Lancefield group B or D *Streptococcus, S. faecalis* and *S. faecium* belonging to the *Enterococcus* group (Inglis et al. 1993), *S. iniae,* (Soltani et al. 2005) and *S. agalactiae* (Bowater et al. 2012).

Clinical signs of streptococcus infection include exophthalmia, aberrant swimming, ocular haemorrhage and darkened body colour, loss of buoyancy control, lethargy, ascites, ulcerations, haemorrhages in or around the base of the fins, gill plate, eye, vent/anus or elsewhere on the body (Yanong and Francis-Floyd 2002). Necropsy findings include liver haemorrhage, pale gills, spleen congestion, renal haemorrhages and liquefactive necrosis. Streptococcosis incidence typically happens in the summer when the water temperature is 28–29 °C (Chong 2015).

3.8 Mycobacteriosis

Since the causative agent of mycobacteriosis is taxonomically similar to *Mycobacterium tuberculosis,* this disease condition was originally known as 'fish tuberculosis' or 'piscine tuberculosis'. Bacterial pathogens responsible for mycobacteriosis are classified into different species including *M. aurum, M. anabanti, M. chelonei* sub sp. *piscarium, M. chelonei, M. marinum, M. fortuitum, M. piscium, M.platypoecilus, M. parafortuitum, M. poriferae, M. salmoniphilum, M. ranae, M. simiae, M. simiae M. triplex* and *M. scrofulaceum.* In vertebrates, the majority of pathogenic mycobacteria are intracellular parasites of phagocytes (Gauthier and Rhodes 2009). The species *M. marinum, M. abscessus, M. chelonae* and *M. fortuitum* are the most frequently detected in fish (Austin and Austin 2012). Mycobacterial infections are among the significant bacterial diseases that affect aquarium or pet fish, especially freshwater ornamental fish (Chong 2015). The infection is also known to infect a wide range of tropical fish species (Smith 1997). Many aquarium fishes are often diagnosed with chronic and systemic subclinical bacterial infections brought about by different species of Mycobacterium.

Mycobacterium spp. infection is known to occur in goldfish (*C. auratus*), discus (*Symphysodon discus*), neon tetra (*P. innesi*), black acaras *(Cichlasoma bimaculatum),* oscars *(Astronotus ocellatus),* guppies *(L. reticularis),* firemouth cichlids *(Cichlasoma meeki)* and napoleon fish *(Cheilinus undulatus)* (Chong 2015; Wada et al. 1993; Conroy and Conroy 1999). Mycobacterial infections were found to manifest as granulomas in ornamental fishes, including pearl gourami *(T. leeri),* angelfish *(P. scalare),* cacatoo dwarf cichlid (*Apistogramma cacatuoides),* three spot gourami *(Trichogaster trichopterus)* and cardinal tetra *(Paracheirodon axelrodi)* (Lescenko et al. 2003).

Affected fish show exophthalmia, ulcers, subcutaneous masses, visceral nodules and weight loss (Lewbart 2001). The bacterial infection may appear in all tissues, including gills, eyes, fins, visceral organs and musculature. Infected ornamental fish exhibited pigmentary changes, emaciation, abdominal distention, petechia around the opercula, hyperaemic gills, scale loss, skin erosions and ulcerations, widening of the spleen, liver and kidney, congested liver with inflammation, multifocal hepatic necrosis, spinal defects, etc. (Chinabut 1999; Astrofsky et al. 2000; Gauthier and Rhodes 2009). Histopathologically, mycobacteriosis in fish is characterized by granulomatous inflammation and typical white or grey nodules in internal organs (Chinabut 1999).

3.9 Nocardiosis

Nocardiosis caused by *Nocardia* sp. is a lethal granulomatous disease, affecting a variety of freshwater and saltwater teleosts and shellfish. The first incidence of nocardiosis caused by *N. asteroides* was documented in neon tetra (*Hyphessobrycon innesi*) (Valdez and Conroy 1963; Conroy 1964). Four *Nocardia* species have been reported from diseased fish and shellfish, viz. *Nocardia seriolae, N. asteroides, N. crassostreae* and *N. salmonicida* (Maekawa et al. 2018). The infection with *Nocardia* sp. was recorded from various geographical regions, including the United States, Germany, Argentina, China, Spain, Japan and Taiwan (Chinabut 1999). It is a systemic disease that affects fish with nodular lesions characteristic of granulomata, which are found in the integument and other internal organs. The infection is typically chronic and intermittent, and young fish appear to be more vulnerable than older fish (Lewbart 2001). Although the Nocardia infection is typically less acute, the clinical signs are similar to those of mycobacteriosis (Austin and Austin 2007). Fish affected may exhibit scale loss and exophthalmia with opaque eyes. In addition, yellowish-white nodules appear dispersed throughout the different tissues, including gills, spleen, liver, heart, muscle, ovary, and mesenteries (Roberts 2012a). Blood pools are also observed beneath the epithelium of the oral and operculum cavities. Infected fish may first exhibit a swollen abdomen, exophthalmia, emaciation and slow or flashing behaviour. Later on, nodular and diffuse granulomatous lesions may develop in gills, skeletal muscle and visceral organs (Lee et al. 2007). Nocardiosis infections with signs of lethargy, clamped fin and slow motion have also been reported from Oscar (*Astronotus ocellatus*), goldfish (*C. auratus*) and shark (*Myxocyprinus asiaticus*) (Sheikhzadeh et al. 2020).

4 Parasitic Diseases of Ornamental Fish

4.1 Protozoan Parasites

The parasitic protozoa are classified as the members of the phyla Ciliophora, Myxozoa, Microspora, Sarcomastigophora and Apicomplexa (Smith and Noga 1993). Protozoan parasites infecting ornamental fish include *Trichodina* sp., *Ichthyophthirius multifiliis*, *Chilodonella* sp., *Tetrahymena pyriformis, Ichthyobodo necator*, etc. (Towers 2013; Iqbal and Haroon 2014). These parasites elicit various clinical signs, such as white patches or papules on the fish's skin, epithelial hyperplasia and excessive mucus secretion on the body and gills.

4.1.1 Ich Disease or White Spot Disease

Ich disease, or white spot disease, is caused by the ciliates of the phylum Ciliophora, which include *Ichthyophthirius multifiliis* and *Cryptocaryon irritans,* which are the most common parasites in freshwater and marine ornamental fishes, respectively. There are significant variations between the two parasites in terms of life cycle duration and resistance to salinity. Both parasites are obligate pathogens infecting the body surface, gills and fins and cause rapid mortality. *I. multifiliis* is roughly 1 mm in size, with a large 'horseshoe'-shaped nucleus, which can be seen through biopsy examination under the microscope as a pathognomonic diagnostic pointer for the species (Wootten 2012). *C. irritans,* unlike *Ichthyophthirius,* does not have the characteristic large C-shaped nucleus (Towers 2013).

The parasite causes hyperplasia of the epidermal cells by feeding on tissue fluids and epithelial cells. The infection appears as several pinhead-shaped white spots on the entire body surface, viz. skin, fins and gills. Additional signs include flashing activity, changes in skin colour, excessive mucus production and increased death rates (Meyers et al. 2019). The parasite has a direct life cycle comprised of three stages: (i) infective theront; (ii) obligate, fish-associated trophont and (iii) waterborne reproductive tomont. The theront is around 40-um long and oblong (pyriform) in shape. When the infectious pelagic theront comes into contact with a susceptible host, it penetrates through the skin or gill surface epithelia and transforms into a feeding trophont and grows rapidly.

The trophonts emerge through the epidermis, inflicting major damage to the epithelium and ultimately killing the fish, most likely as a result of osmoregulatory failure (Hines and Spira 1974). Once mature, the parasite leaves the host and transforms into a motile tomont. It swims freely for about an hour and attaches to some vegetation where the parasite undergoes continuous division, forming a number of cells or tomites, which later differentiate into infective theronts. Generally, 500–1000 parasites are produced following nine to ten divisions (Matthews 2005) and the life cycle is completed in approximately 16-18 h at 22-25°C (Dickerson, 2006) .

4.1.2 Chilodonellosis

Some species of the genus *Chilodonella* coming under the phylum Ciliophora are obligate parasites of fish and are oval and dorsoventrally flattened with a slightly convex dorsal side. The parasite has a cosmopolitan distribution, and it infects the skin and gill of the host. The parasite directly feeds on the epithelium by piercing the host cells, exploiting its cyclostome and consuming the contents (Wiles et al. 1985). The infection can later lead to ulcers that are aggravated by secondary bacterial infections and considerable mortality. Clinical signs of the infected fish include increased mucus production, with dark, slimy, patchy or mottled grey appearance that later on may lead to detachment of epithelial cell debris from the skin surface (Basson and Van-As 2006). Moribund fish show hypoxia and uncoordinated swimming and are found to be emaciated. Infection on the gill causes hyperplasia of the gill epithelium, leading to lamellar fusion and reduced respiratory efficiency. *Brooklynella* is the marine counterpart of *Chilodonella*, which is capable of causing severe gill damage and significant mortalities in marine aquarium fish.

4.1.3 Trichodinosis

Trichodinids are ectoparasites of skin and gills. These parasites cause relatively insignificant infections, which produce chronic mortality. *Trichodina* is characterized by an aboral attachment disc and cup to a cylindrical body with denticles. The parasite adheres to the skin surface and damages the epithelium by constant circular movement and dislodging the epithelial layers of cells that form the food of the parasite (Noga 2010). Heavy trichodinid infection causes excessive mucus secretion and peeled epithelia, and fish may reveal a greyish-blue colour. While the prolific epithelial growth is supposed to protect the fish, the same is fed by the parasite. The infected fish become listless, anorexic, swimming just beneath the water surface or close to the water edge and express low-scale mortality (Basson and Van-As 2006). Secondary bacterial infections often lead to escalation of debilitation and mortality.

4.1.4 Velvet Disease

Amyloodinium and *Piscinoodinium* are the two dinoflagellate epithelial parasites causing 'velvet' or 'rust' or 'gold dust disease' in marine and freshwater fishes, respectively (Lom and Dykova 1992; Gratzek 1993; Cheung 1993; Noga 2010). *A. ocellatum* is thought to be the most dangerous parasite that causes severe epizootics in marine aquariums cultured in warm water (Noga and Levy 2006). Because the organism may reproduce rapidly in fish held in crowded conditions, particularly in closed environments, it can result in devastating disease and mortality. One of the most prevalent infections affecting marine aquarium fish is velvet disease. There have been severe outbreaks of the disease in marine clownfish (*Amphiprion* sp.) (Francis-Floyd and Floyd 2011). Clinically, the fish would cease feeding, flash and flush water back across its gills. Heavily infected fish's skin appears dusty, dull gold or brown sheen. A closer look at the skin may show uneven mucus accumulation and scale loss (Reed and Francis-Floyd 1994). The infection of *Piscinodinium* sp. has been found in the gill of ornamental fish, both guppy and goldfish (Dewi and Fadhilla 2018).

The parasitic infection causes complete gill epithelial hyperplasia and secondary lamellar fusion (Paperna 1980). Heavy infestations can cause gill hyperplasia, haemorrhage and necrosis. Hypoxia due to gill damage and osmoregulatory failure due to severe epithelial damage can result in mortality of the fish. The parasite's life cycle starts with a free-swimming infectious stage called a dinospore that adheres to the skin or gills and converts into a feeding stage called a trophont that uses its rhizoids to pierce epithelial cells (Noga 2010). Trophont, after feeding, becomes a tomont and divides to produce infective dinospores. Since *Piscinoodinium* has well-developed chloroplasts and a different attachment mechanism than *Amyloodinium*, it likely gets at least some of its nourishment from photosynthesis (Noga and Levy 2006).

4.1.5 Costiasis

Costiosis, or Ichthyobodosis, is produced by one of the smallest Zoomastigophoran fish ectoparasites known as *Ichthyobodo necator* (*Costia necatrix*). *I. necator* is a flagellated protozoan ectoparasite that damages the skin and gills of both freshwater and marine fishes by penetrating epidermal epithelial cells with its flagella (Gratzek 1988). The infected fish may show gill filaments with lamellar epithelial hyperplasia and fusion of the lamellae. Poor water quality and crowding are responsible for the rapid reproduction of parasites and overpowering of the host defences.

In the free-swimming stage, this parasite is ovoid to kidney-shaped and divides by binary fission (Noga 2010). Of the two pairs of flagella, the parasite swims with a jerky movement with the help of the longer pair of flagella. When attached to the host, the parasite elongates to a cuneiform shape, and it uses a kind of holdfast organ to grasp a host cell by its pointed anterior end. Following penetration, the parasite feeds on the fish, which causes the malpighian cells to proliferate and the goblet cells at the location to get exhausted, resulting in spongiosis and epidermal sloughing (Robertson et al. 1981). The parasitic and free-swimming stages are thought to encyst in unfavourable environmental circumstances (Woo 2006). The cysts in the water increase the risk of infection and help the organism spread to new regions.

4.1.6 Hexamitosis or Hole in the Head Disease

Hexamita and *Spironucleus* are the major diplomonad pathogens in freshwater angelfish and discus production (Gratzek 1988; Francis-Floyd et al. 1993). *Hexamita* and *Spironucleus* are small pyriform organisms that have one posterior pair and three anterior pairs of flagella. Since flagellated protozoa are normally found in fish digestive tracts, they can cause serious gastrointestinal disorders if present in high numbers. Hole in the Head Disease or Head and Lateral Line Erosion caused by these parasites is commonly found in oscars and other cichlids (Gratzek 1988). Diplomonad flagellate infection has been recorded in some ornamental fishes in the family *Cichlidae,* i.e. Oscar (*A. ocellatus*), angelfish (*P. scalare*) and blue mbuna (*Labeotropheus fuelleborni*) (Mankhakhet et al. 2012). The infestation of *Hexamita* has been found in the intestine of *C. auratus* and *P. scalare* (Kayis et al. 2013). When parasites are in high numbers, fish may display signs of abdominal distention, emaciation, anorexia, weakness, listlessness, faecal casts, pale gills, haemorrhagic vent, exophthalmia and/or body darkening (Meyers et al. 2019). Histopathological changes in infected fish were inflammation of the intestine, granulomatous liver and excess numbers of melanomacrophage in the spleen (Mankhakhet et al. 2012). In case of heavy infection, histopathological changes like gastritis can be observed in the intestine and perforations in the serosal granuloma in the stomach (Ferguson 1989).

Hexamitid trophozoites are typically found extracellularly in the intestinal lumen, internal organs or surface lesions (Woo 2006). Infection usually occurs through the ingestion of cysts or trophozoites. Infection can later lead to parasitaemia in immunocompromised hosts with trophozoites in the blood and many internal organs, even in the cranium and skeletal tissues (Ferguson 1989). Fish with hole-in-the-head disease would have tiny, hole-like lesions on their bodies that would exude cheesy, yellow mucus threads. These can be confined to the head region or extended posteriorly to the lateral line (Paull and Matthews 2001). Bilateral ulcers can also develop in the cranial cartilage due to large erosions (Ferguson 1989).

4.1.7 Neon Tetra Disease

Microsporidia are unicellular organisms that live as internal parasites and are widely distributed in freshwater, estuarine and marine environments. They have a direct life cycle in a range of invertebrates and are extensively found in teleosts. Microsporidians were previously thought of as protozoa, but new research has shown that they belong in the kingdom Fungi as a distinct phylum called Zygomycota (Dykov 2006). Pleistophora is a significant microsporidian parasite of fish that causes ‘Neon tetra disease’ in zebra danios, cichlids and cyprinids (Towers 2013). Microsporidians infect by penetrating the host gastrointestinal epithelium with a tiny, thick-walled spore that has a tubular polar filament (Harms 1996). Pathology of this parasitic infection is related to the location and quantity of xenomas developed by hypertrophied infected host cells. The timely removal of afflicted or dead fish is crucial because mature spores can be discharged from lesions on the body surface or during the decomposition of a dead host (Harms 1996).

Single-walled spores containing sporoplasm are the hallmark of microsporidia. The extrudable polar tube found within the spore is used to inject the sporoplasm into the host cell once it is inside the host. The sporoplasm produces a large number of meronts after reaching the target organs by binary or multiple fission (Noga 2010). Meronts later transform into sporonts, which produce a large number of mature spores through sporogony that are released up to the death of the host.

4.1.8 Whirling Disease

Two of the major myxopsporean pathogens coming under phylum Myxozoa affecting fish are *Myxobolus cerebralis* (earlier *Myxosoma cerebralis*), which causes whirling disease, and *Ceratomyxa shasta* that causes fatal intestinal pathology

(Hallett and Bartholomew 2012). Whirling disease is a chronic and highly debilitating disease that takes months to show clinical signs, which include blacktail, whirling behaviour, stunted growth, skeletal deformities and death. Myxospores spread through the faeces of piscivorous birds. These spores have to be ingested by an oligochaete host (*Tubifex tubifex*), where the sporoplasm of the parasite develops into an actinospore. There are two infectious phenotypes of *M. cerebralis* spores: (i) a triactinomyxon-type actinospore and (ii) a myxobolus-type myxospore (Hallett and Bartholomew 2012).

Invertebrate host tubificids release actinospores, which adhere to the fish's epithelial surfaces and pierce the epidermis. They proliferate and migrate to the central nervous system through the peripheral nerves (El-Matbouli et al. 1999). Behavioural alterations exhibited as repeated episodes of circular swimming, and anterior body contractions occur due to the compression of the brain stem caused by the granulomatous inflammation that compresses the spinal cord when the parasite develops in the tissue (Rose et al. 2000). Skeletal deformities are noticed in the infected fish that arise due to cartilage damage and associated inflammation resulting from disrupted osteogenesis (MacConnell and Vincent 2002). These could include a crooked spine, misaligned jaws, a decreased nose, an indented skull and a shorter operculum. The trophozoites of *M. cerebralis* are present in the gill arches, auditory capsules and the growing cranial cartilages around the brain. They lead to the gradual destruction of cartilage, which is then replaced by granulation tissue (Wootten 2012). Mortality may occur directly by physical harm or indirectly by the incapacity to feed or avoid predators (Hedrick et al. 1998; Steinbach et al. 2009).

4.1.9 Trypanosomiasis (Blood Parasites)

Globally, fish trypanosomes are found in a variety of freshwater and marine fish species (Lom and Dykova 1992; Fujimoto et al. 2013). Haemoflagellates of the genera *Trypanosoma* and haematozoic *Trypanoplasma* are known to spread to fish through the invertebrate, most likely a blood-sucking aquatic leech (Woo 2006). In some fish like *Plecostomus,* the clinical significance of low-level trypanosome infection is not yet determined, as both healthy and moribund have been found to harbour the parasites. It has been observed that goldfish (*C. auratus*) (Nazrul Islam and Woo 1991a) and catfish (*Heteropneustes fossilis*) (Bhaskar et al. 1984) exhibit significant hemoglobinemia and anaemia due to trypanosomiasis. Anorexia was produced in goldfish that were experimentally infected with *Trypanosma danilewskyi* (Nazrul Islam and Woo 1991b).

Trypanosomiasis infection is associated with various haematological alterations (Fujimoto et al. 2013). In the blood, the parasite multiplies quickly through binary fission, and it develops in the leeches as well. The parasite grows in the leech's gut after being consumed with blood meal. Trypomastigotes split into epimastigotes, which are then duplicated by binary fission, with the ultimate appearance of slender metatrypanosomes (Woo 2006). These migrate and accumulate in the proboscis of leeches and are transmitted to the fish upon feeding, which is then infective for the fish.

4.2 Metazoan Parasites

4.2.1 Monogenean Infections (Skin/Gill Flukes)

Trematode parasites of fish are classified as monogeneans and digeneans, which come under the phylum Platyhelminthes. Most monogeneans are fish ectoparasites that live on the skin, fins, gills, mouth and nostrils (Buchmann and Bresciani 2006). Most of the species are small worms less than 3-cm long, having a direct life cycle without any intermediate hosts (Wootten 2012). They possess a posterior haptor as an attachment organ armed with sclerotinized structures such as hooks (hamuli), clamps or suckers. (Buchmann and Bresciani 2006). They also have an anterior part, prohaptor with attaching capacities (adhesive pads and cephalic openings). Based on the existence of eyespots, the reproduction method and the shape of the posterior attachment organ (opisthohaptor), there are two groups of monogeneans (Noga 2010). Monopisthocotylea, which includes dactylogyrids, gyrodactylids and capsalids, are characterized by a single opisthohaptor with marginal hooklets and massive sclerotised hooks positioned in the centre. On the other hand, in Polyopisthocotylea, the opisthohaptor is made up of a collection of tiny, muscular, sticky clamps or suckers that have cuticular sclerites.

The outermost layers of the skin and gills are the primary food source for monopisthocotyleans. The feeding activity of monopisthocotyleans irritates the fish, which also results in focal reddening from haemorrhage, epithelial hyperplasia or excessive mucus production (Buchmann and Bresciani 2006). On the other hand, polyopisthocotyleans mostly consume blood, which can result in severe anaemia. The monogeneans have a direct life cycle with a single host, whereas the digeneans have an indirect life cycle with an intermediate host (Smith and Noga 1993; Gratzek 1988). Monogeneans are hermaphroditic flatworms that have an ovary and testes.

Gyrodactylidae and *Dactylogyridae* are the two most important monogenean families. Dactylogyrids are oviparous, have four eye spots and are typically found on the gills, whereas Gyrodactylids are viviparous, have no eye spots and mostly live on the skin. The eggs laid by adults hatch and release free-swimming larvae called oncomiracidia, which infect suitable host fish in a matter of hours (Wootten 2012). When the oncomiracidia is attached to an appropriate host, it matures into adults. In contrast, gyrodactylids are viviparous, meaning they do not have an intermediary stage and produce offspring that are identical to their parents. In most cases, gyrodactylids spread from one fish host to another by direct physical contact. They destroy the host's epithelium by feeding activities with an amazing bunch of hooks on the opisthaptor, the posterior attachment organ (Harms 1996). Monogeneans also cause localized haemorrhages and rupture the attachment site. They simultaneously consume ruptured blood and tissue cells (Abidi et al. 2011).

There have been reports of *Gyrodactylus* sp. and *Dactylogyrus* sp. infecting the skin and gills of ornamental fish, including koi carp, goldfish and guppy (Dewi and Fadhilla 2018; Petchimuthu et al. 2018). Seven *Dactylogyrus* spp. including *D. vastator* and three *Gyrodactylus* spp. from goldfish; *Dactylogyrus ostraviensis* from rosy barb, while two *Trianchoratus* spp. from three-spot gourami and pearl gourami have been recorded (Trujillo-González et al. 2018). Poor sanitation and deteriorating water quality caused by overcrowding, excessive ammonia or nitrite, organic pollution or low oxygen are responsible for the increase in monogenean infections. Under these circumstances, they reproduce quickly and double their number in 24 h.

4.2.2 Digenean Infections

Fish digeneans are endoparasites that have at least one intermediate host during their life cycle. These are some of the major taxa that belong to the phylum Platyhelminthes. The majority of digeneans have two suckers located at the anterior end and also on the ventral side in the anterior half (Wootten 2012). Adult digeneans are parasitic to fish and other vertebrates, such as birds and mammals. They are hermaphrodites and produce eggs in the final host and are released through the gut. The eggs hatch into miracidium larvae, which infect a mollus, where it develops into cercaria (Noga 2010). Infected mollusc releases the cercaria, which enters the fish host and differentiates into a metacercaria, which becomes encysted. When the final host consumes the fish, the parasite matures into an adult. Therefore, fish act as intermediate and definitive hosts for digenean trematodes (Smith and Noga 1993; Gratzek 1988). Encysted digenean metacercaria are frequently found in the epidermis and underlying tissues of tropical fish and are occasionally seen in the coelomic cavity (Lewbart 2001). Freshwater fish are reported to have encysted or free metacercariae of numerous species, such as *Diplostomum, Tylodelphys* and *Apatemon* (Wootten 2012). Metacercariae are noticed within the vitreous humour, retina and lens of the eye. *D. spathaceum*, also known as the 'eye fluke', is one of the most significant species and has been found in over 125 different fish species (Höglund 1991).

The gastrointestinal tract is the primary site of adult digeneans, and it can infrequently infect the circulatory system, ovaries, peritoneal cavity, swim bladder or urinary bladder (Noga 2010). Cercarial migration can harm the host, resulting in inflammation, necrosis and bleeding along the migration route (Sommerville 1981). In contrast to *D. spathaceum* metacercariae, which can induce eye pathology and related morbidity, metacercariae are typically harmless, and fish carry high worm burdens without any apparent negative effects (Noga 2010). However, they frequently cause disfigurement and can cause ornamental fishes to lose their colouration, which can make them look unappealing in aquariums. Digeneans that live in the gastrointestinal tract as adults rarely harm the body, whereas blood flukes can seriously harm the gill vasculature (Harms 1996). Other intestinal parasites, such as sanguinicolids and aporocotyles (blood flukes), have the potential to be harmful, seriously harming the gills and affecting respiration. (Paperna and Dzikowski 2006). Entry of cercaria and massive encystment of metacercaria in different tissues such as gills, heart, kidney, viscera and eyes by various species of digenean cause morbidities and chronic mortalities in infected host fish.

4.2.3 Cestode Infection

Cestodes are endoparasitic, which belong to the phylum Platyhelminthes and have at least one intermediate host during their life cycle. The body of the adult worms is generally segmented with internal and external segmentation (proglottids), white in colour, elongated and provided with a scolex (Wootten 2012). They are endoparasites in the intestine of the fish, and the larval forms are usually encysted in the viscera and musculature. Cestodes are oviparous and have one intermediate host, a copepod or fish itself. The adults release eggs, which either enter directly into the intermediate copepod host and hatch inside or develop as free-swimming larvae named coracidium. In the intermediate host, the larvae developed from the egg or the coracidium larvae, penetrate the gut wall and develop further into a procercoid. When the procercoid enters the fish, the fish acts as the second intermediate host, and the larvae develop into a plerocercoid, which later develops into an adult cestode when taken by a final host, either fish, bird or mammal (Noga 2010). If the fish ingests the plerocercoid from the copepod, the fish acts as the final host, and the adult cestode is formed.

Most disease-producing cestodes belong to one of the three orders: *Caryophyllidea* (*Caryophyllaeus* and *Khawia*), *Proteocephalidea* (*Proteocephalus* spp.) and *Pseudophyllidea* (*Ligula*, *Diphyllobothrium*, *Schistocephalus* and *Triaenophorus*). The Asian tapeworm, *Bothriocephalus acheilognathi*, is one of the most dangerous adult cestodes that can infect fish (Dick et al. 2006). Severe damage and adhesions of viscera or muscle in fish are caused by the pressure necrosis of the migrating plerocercoids (Noga 2010). However, unless they are in great quantities, mature cestodes are largely asymptomatic and infect the intestine or pyloric ceca. Heavy infections in the gut can cause mechanical obstruction and distension of the intestine, and the fish can become emaciated and anaemic. Pathologically, cestode infection causes abdominal distension, growth retardation, physiological alteration and dislocation and morphological changes of internal organs in host fish (Arme et al. 1983). The physiological alterations associated with infectious pathology include decreased liver glycogen, muscle proteins, carbohydrates, blood amino acids, lipids and inhibition of gametogenesis.

4.2.4 Nematodes

Nematodes are roundworms with cylindrical, elongated bodies that taper at both ends. They are sexually dimorphic, bilaterally symmetrical and possess a gut with oesophagus and intestine (Wootten 2012). The Nematoda families include *Spiruroidea, Trichuroidea, Dioctophymoidea* and *Ascaridoidea.* Nematodes complete their life history through an invertebrate intermediate host, usually an arthropod. Adults release eggs or larvae that are ingested by the intermediate host, where the development of the next larval stage takes place. The parasite completes its life cycle when the fish consumes the intermediate host. If the fish acts as a paratenic host, the larva roams into the viscera and tissues where it encysts, and it develops into an adult only in a final suitable host, either predator fish, bird or mammal when they consume the paratenic host.

Most of the parasitic nematodes are found to infect the intestine of the fish. Internally migrating larvae would certainly cause pathology, while enteric adults would be responsible for mechanical injuries, obstruction of the alimentary canal, damage of blood vessels and different ducts, tissue atrophy, toxicity arising from their metabolic products, gastrointestinal inflammation depriving host nutrition and emaciation (Molnar et al. 2006). Nematodes can be found in fish as larvae of *Contracaecum* spp., *Eustrongyloides* spp., and as adult forms such as *Capillaria* spp. and the familiar zoonotic *Anisakis* spp. (Harms 1996). Nematodes of the genus *Camallanus* are widely distributed in ornamental fish all over the world. They are mostly found in the intestines of poecilids (Kim et al. 2002). Massive infections of *Capillaria* and *Capillostrongyloides* were found attached and feeding on the intestinal mucosa of aquarium-reared neotropical cichlids and siluroids that were attributed to emaciation and mortalities in these fishes (Moravec and Gut 1982; Moravec et al. 1987). The nematodes could infect anywhere in the fish body, but it is discernible when they hang out of the anus, causing emaciation, stunted growth, lethargy and death (Magada and Mercy 2019).

4.2.5 Crustacean Parasites

Crustaceans that act as parasites on fish, belonging to the phylum Arthropoda, encompass branchiurans, copepods and isopods. They exhibit bilateral symmetry and have segmented bodies along with jointed appendages. The adult stages are usually noticeable on the host (Harms 1996). Several crustacean parasites invade the skin and gills of tropical ornamental fishes (Towers 2013). Both freshwater and marine crustacean parasites can infest fish. The majority are ectoparasites that target the skin, gills and oral cavity.

4.2.6 Lernaea

Parasitic copepods can exhibit a body form that resembles free-living copepods, such as *Ergasilus* spp., or may be highly adapted for a parasitic lifestyle, like *Lernaea spp*, known as 'anchor worms'. *Lernaea cyprinacea* is a significant crustacean parasite that affects freshwater fish in tropical regions. Only the adult female lernaeids and ergasilids lead a parasitic lifestyle, while males and juvenile forms of both genders live independently (Harms 1996). Female adults measure between 12 and 16 mm in length, excluding their egg sacs. *Lernaea* is the primary genus of lernaeid copepods, although other genera like *Opistolernaea* and *Lernaeagiraffa* also play significant roles in tropical ecosystems (Paperna 1991).

Lernaea possesses a direct life cycle that includes three nauplius stages and five copepodid stages, with the final stage producing male and female cyclopoids (Avenant-Oldewage 2012). They have non-feeding and free-swimming naupliar stages (Shields and Tidd 1974). After the third naupliar stage, the organism goes through five additional copepodid stages, during which it consumes epidermal and dermal tissues of fish gill, before finally maturing into adult males and females (Goodwin 1999). The males perish following mating, while the females embed themselves permanently by tunnelling into the host's skin. The head region of the parasite subsequently transforms, resulting in the development of lateral structures that assist in anchoring and embedding into tissues, via the skin into the muscles or occasionally into the gills or oral cavity (Avenant-Oldewage 2012). The parasite consumes the host's tissues, and after 4 days of attachment, the egg sac forms externally on the host, dispersing the eggs into the surrounding environment. Under ideal conditions (above 25 °C), a solitary female can generate hundreds of larvae every fortnight for a maximum of 16 weeks (Noga 2010). Small fish may die when the parasite invades their internal organs. *Lernaea* infestations produce skin ulcers and haemorrhages in the host, which can facilitate secondary infections by bacteria, viruses and fungi (Lester and Hayward 2006).

4.2.7 Argulus

Argulus is a holoparasite that targets the surface of freshwater fish, belongs to the Branchiura group, is characterized by a flattened dorsal-ventral structure and can reach lengths of up to 1 cm (Piasecki and Avenant-Oldewage 2008). Often referred to as fish lice, these organisms consume the bodily fluids of fish by piercing a retractable pre-oral spine (stylet) into the host using their proboscis-like mouth. *Argulus* is commonly found attached to the skin or fins of the host, using modified ventral appendages that resemble hooks and suckers. They navigate the host's surface effortlessly and swim effectively using their swimming legs, even if they become separated from the host (Lester and Hayward 2006). They can live for multiple days without being attached to a host. The parasite's life cycle is brief, lasting approximately 30 days or longer, and is influenced by the surrounding temperature. Adult parasites deposit their eggs on plants or various surfaces, and the larvae emerge within 10–50 days (Paperna 1991). Juveniles that are 1–3 mm in size enter the host within 2–3 days and mature into adults. At a temperature of 24 °C, the eggs of *A. foliaceus* typically hatched after an average incubation time of 27 days, with a range of 19–39 days, while at 14 °C, the eggs took 67 days to hatch, ranging from 60–75 days (Hunt and Cable 2020). There was no significant difference in hatching success between eggs incubated at 24 °C and those incubated at 14 °C, with rates ranging from 57.7 to 63.7%.

Infested fish exhibit irregular swimming patterns and unusual behaviours due to irritation from the parasite, resulting from the mechanical damage caused by hooks and suckers. *Argulus* induces a significant inflammatory response at the location where its stylet penetrates and has been associated with the spread of various viral, bacterial and hemoprotozoal infections, as well as localized bleeding and increased pigmentation (Shimura et al. 1983; Pfeil-Putzien 1978; Harms 1996; Lester and Hayward 2006). *Argulus foliaceus* typically infests cyprinids, and a low rate of infection has been noted in *C. auratus* (Iqbal and Haroon 2014). The pathological lesions induced by the parasite become necrotic and ulcerated, leading to secondary infections, and animals often become lethargic and anaemic (Wootten 2012). Mortalities due to *Argulus* infection usually occur when the fish carry a large number of parasite loads (Shimura et al. 1983). Such mortalities occur due to the loss of ionic and osmotic homeostasis consequent to the damage to the epithelial layer and invasion of secondary bacterial and fungal pathogens (Lester and Hayward 2006).

4.2.8 Leeches

Leeches may sometimes invade ornamental and pond fish species. Normally, they do not cause notable health issues. However, when a fish is heavily infested with leeches, it can lead to serious anaemia and possibly death. They can also serve as carriers for blood parasites like *Hemogregarina*, *Trypanoplasma, Trypanosoma* and others. Skin lesions may lead to secondary infections of bacteria and fungi (Harms 1996; Towers 2013).

Leeches, known scientifically as Hirudinea, infest both freshwater and saltwater fishes (Smith and Noga 1993; Gratzek 1988). Leeches are often noticed on the surfaces of the body, fins, mouth, gill chamber and eyes. They may also parasitize ornamental fish and those found in ponds. In an aquarium, leeches can be observed moving on the gravel and will stretch out in response to the movements of a fish (Gratzek 1988). Aquatic leeches primarily serve as carriers of disease-causing organisms. Freshwater and marine leeches can also serve as primary vectors for blood parasites, including *Hemogregarina, Trypanoplasma, Trypanosoma* and others (Burreson 2006). Aquatic leeches act as secondary intermediate hosts for digeneans, containing metacercariae from various species. Generally, leeches do not cause significant pathology. Leeches typically draw blood from their host and then detach, falling to the bottom of the aquarium where they reproduce. The pathology is generally localised; significant infestations may lead to serious skin erosion and could even cause death due to excessive blood loss or complications arising from numerous feeding wounds (Burreson 2006). The skin lesions can become infected later with bacteria and fungi (Harms 1996; Towers 2013).

5 Fungal Diseases of Ornamental Fishes

Most fish fungal pathogens are part of the Ascomycota phylum, which are characterized by thick-walled, non-motile spores (Hibbett et al. 2007). In addition, fish diseases have also been linked to species classified under the Zygomycota.

Mesomycetozoea is a group of fungi that encompasses several fish pathogenic fungi, falling into the orders Dermocystida and Ichthyophonida. Notable examples within this group include *Dermocystidium* sp., *Ichthyophonus hoferi* and *Sphaerothecum destruens* (Gozlan et al. 2014). The Oomycetes, belonging to the phylum Oomycota, constitute the fourth group of fish pathogenic fungi, with a saprolegnialean or peronosporalean lineage. These water moulds, which are part of the oomycetes but taxonomically distinct, are parasitic in nature, affecting a wide range of fish species (Beakes et al. 2012). The order Saprolegniales consists of genera that are responsible for causing infections in fish, including *Achlya, Aphanomyces and Saprolegnia.* The significant genus under the peronsporalean lineage is *Pythium.*

5.1 Saprolegniasis (Water Mould Infection or Oomycete Infection)

The infection is mainly caused by the species *Saprolegnia parasitica,* although several other oomycetes are also implicated in water mould infection. These opportunistic pathogens are found to cause infection in immunosuppressed animals (Noga 2010). The two most generally isolated species from fish among Oomycetes are *S. parasitica* and *S. diclina*. They are commonly known as the *S. diclina–S. parasitica* complex (Noga 1993).

Clinically, saprolegniasis is characterized by grey to white patches on the skin surface with cotton wool-like hyphae extending into the water (Roberts 2012). The mould also infects the gills, and internal infections are also occasionally reported. Pathogenesis of the disease occurs due to breaches in the epidermis. The lesions usually begin as small focal infections and spread rapidly over a wider area in a short time. The new lesions would generally be white in colour, which over time turn to brown or green depending on the sediment, debris, algae, etc., that get trapped on to the mycelial filaments. While water mould infections are superficial on the skin and rarely penetrate the underlying muscles, the infection can turn fatal if the infection is on the gills or covers a large area of the skin. Morbidity and mortality will result from the loss of serum electrolytes and proteins if the affected skin area is large enough to induce osmotic imbalance (Richards and Pickering 1979). Oomycetes are generally secondary pathogens invading the wounds created by other pathogens. Dead fish and other submerged organic matter are the important sites on which water mould settles and colonises. Lower than optimum temperature for specific fish species is one of the most important predisposing factors that cause *Saprolegnia* infection (Roberts 2012).

Water moulds are usually transmitted by motile zoospores that are sporulated from the vegetative hyphae through asexual reproduction, though reproduction by gemma can also occur (Noga 1993). Sexual reproduction is by means of oospore. Saprolegnia is also found to invade incubating fish eggs, with the infection starting from dead eggs and later spreading to live ones. Poor water conditions, such as stagnant water, insufficient oxygen, high ammonia concentrations and excessive organic waste, are often associated with these infections.

5.2 Aphanomyces Infection

Aphanomyces invadans, the sole species within the genus *Aphanomyces*, is a pathogenic, clonal species involved in one of the most destructive diseases, Epizootic ulcerative syndrome (EUS), that has affected fish populations worldwide (Willoughby et al. 1995; Lilley et al. 1997; Blazer et al. 2002). EUS, a disease marked by distinctive fungal granulomas, originated in Queensland, Australia, and Japan in 1971 and has subsequently spread to fish populations across Asia, Europe, Africa and North America (Roberts 2012). This disease can infect a diverse range of estuarine and freshwater fish species, leading to severe skin ulcers and high mortality rates. Snakehead fish in the wild are often affected by this infection. Clinically, it is identified by the presence of fungal hyphae that invade the muscle tissue and internal organs of infected fish (Kamilya and John 2020). While the proteolytic fungi can inflict significant ulcerative damage that can kill many host species, it is often observed that other microorganisms, such as protozoa, bacteria and additional fungi, can contribute to the death of the animals (Chinabut and Roberts 1999). *A. invadans* differs from other pathogenic oomycete fungi by lacking a sexual stage and reproducing solely through asexual means, producing biflagellate zoospores (Iberahim et al. 2018).

The infection is believed to be initiated by certain environmental factors that make fish more susceptible to EUS (Baldock et al. 2005). Undamaged skin is not typically susceptible to *A. invadans* infection, suggesting that skin damage or a weakened immune system may be necessary to trigger an outbreak (Noga 2010; Iberahim et al. 2018). Infection by *A. invadans* typically happens when the temperature is low (around 18–22 ° C in tropical areas) after heavy rainfall, a shift in pH and other factors that support *A. invadans* sporulation. The low temperature can suppress the immune system, hindering the inflammatory response to oomycete infections (Catap and Munday 1998).

5.3 Branchiomycosis

Branchiomyces sanguinis and *B. demigrans* are the two fungi that infect fish under ecological stress, cause Branchiomycosis, a disease also known as 'Gill Rot'. A variety of fish from various families are known to be impacted by the fungi. Infarctive gill necrosis is the defining feature of the disease, caused by the growth of *Branchiomyces spp* within blood vessels (Roberts 2012). The respiratory system pathology is characterized by hyperplasia, gill lamellar fusion, the obstruction of blood vessels by fungal hyphae, telangiectasis and vascular necrosis. As the name indicates, *B. sanguinis* has high tropism to blood vessels, while hyphae of *B. demigrans* penetrate the adjacent necrotic tissues (Wolke 1975). Due to thrombosis and ischemia of the gills, it appears mottled with intermittent dark and light areas (Noga 2010). Infection is highly fatal, with high early morbidity and massive mortality. The disease usually occurs due to bad hygiene, and water quality maintenance is essential to prevent the infection.

5.4 Ichthyophonosis

The pathogen responsible for ichthyophonus disease is *Ichthyophonus hoferi,* classified within a new group of organisms known as Mesomycetozoea, whose evolutionary lineage is closely related to that of animals (Mendoza et al. 2002). Ichthyophonosis is a systemic disease characterized by the formation of granulomas and affects both freshwater and saltwater fish (Kocan et al. 2010; Hamazaki et al. 2013). Upon entering the host, *Ichthyophonus* spreads systemically by disseminating its infective stages via the blood or lymphatic systems throughout the organism. Once it becomes trapped in the capillaries, the fungus starts to multiply by discharging the infective stages into the bloodstream. *Ichthyophonus* infections have been observed in two varieties of ornamental fish that include tiger barb (*Puntius tetrazona*) and black tetra (*Gymnocorymbus ternetzi*). The affected fish exhibited symptoms such as unusual abdominal swelling, erratic swimming behaviour, fatigue and a low mortality rate (Zadeh et al. 2014). Although affected fish sometimes fail to exhibit external signs of disease, well-vascularized internal tissues get affected mostly, and granulomas encapsulating the pathogen with intense fibrotic reaction are observed in the kidney, liver, spleen and heart (McVicar and McLay 1985; Sitja and Alvarez-Pellitero 1990).

The development and life cycle of *Ichthyophonus* is said to fluctuate based on salinity, pH levels, carbon dioxide concentrations and glucose availability (Okamoto et al. 1985; Spanggaard et al. 1995; Spanggaard and Huss 1996). They generate both single and multinucleated endospores (Spanggaard et al. 1995; Franco-Sierra and Alvarez-Pellitero 1999) and amoeboid forms that are emitted at particular pH levels (Okamoto et al. 1985). Although *Ichthyophonus* is an obligate pathogen, its resting spores can remain viable in seawater for up to 2 years (McVicar 1999).

Different stages are observed during the life cycle of the pathogen, with spherical or oval-shaped resting stage or spore stage having multiple nuclei being most common. These spores that are present in different organs germinate into non-septate hyphae, and new spores or hyphal bodies are formed from the hyphae (McVicar 1999). Transmission of the fungus is not very clear but is most likely through the intestine by feeding the infected tissues that carry amoeboid stages, which could be infective (McVicar 1982; Kocan et al. 2013). These invade the mucosal layer and, upon introducing into the bloodstream, are transported to various organs like muscles, where new cysts develop. These cysts then continue to infect nearby tissues by producing plasmodia or endospores (Roberts 2012). In certain instances, the cysts develop in the gastrointestinal tract and trigger a localized infection that could propagate to other areas of the body.

The clinical signs may encompass alterations in behaviour and indications related to organ dysfunction, including fatigue, weight loss, discolouration, fluid retention, neurological issues and a rise in death rates (McVicar 1999). Liver or kidney infections can lead to bulging eyes and abdominal swelling. When the swim bladder is damaged due to infection, the fish typically remains near the bottom of the pond and eventually dies. An infection in the brain can cause major alterations in fish behaviour (McVicar 1999). Affected fish exhibit white or creamy nodular lesions ranging from 1 to 5 mm in size across various tissues (McVicar and McLay 1985). Nodules discovered in the kidneys, spleen, heart, liver and brain contain cellular debris and fungus. Fish suffering from severe infections may exhibit rough or granulomatous skin and have white to grey-white lesions in various internal organs and body parts. Infections can lead to a textured skin appearance known as the 'sandpaper effect', which results from the loss of epithelial tissue over developing dermal fungal granulomas that are roughly 1 mm in size and elevated above the skin (Roberts 2012). A notable characteristic of parasite germination that takes place following the death of the host has been proposed as a specific diagnostic marker for identifying *Ichthyophonus* (McVicar 1982).

6 Conclusion

As the list of pathogens infecting the ornamental fish is huge and the infections are sometimes fatal, appropriate management measures for the fish in the aquarium and ponds/tanks are essential. Without special diagnostic tests, infectious diseases are very hard to distinguish from each other only because of the clinical signs. The earliest sign of an internal infection would be anorexia, which should give the first warning that the fish is under stress due to either biotic or abiotic factors. Viral pathogens are considered major threats that affect ornamental fish production. Mortality due to bacterial and fungal pathogens is slow to progress, and clinical signs start emerging as the days pass by. Needless to say, water quality degradation and unjustifiable crowding are the main reasons for opportunistic pathogens to invade the host and manifest the disease. Controlling the infections would be difficult if the diseases pass the threshold of fish immune status.

At least in controlled facilities like aquariums or small holding tanks, the fish can be treated with various therapeutic measures, but it may not always help to recover the fish beyond a specific stage of infection. Particularly, viral infections are to be kept at bay by all possible means as there are no pragmatic ways of treatment. Internal infections of bacterial origin can be treated with medicated feeds containing antibiotics, while external infection requires water medication in the early stage of onset of infection. Parasitic infestations can be easily distinguished and, hence, can be controlled to a certain extent by adequate intervention. Most of the parasites that infest the gills and skin of ornamental fish cause irritation, which can be detected early. Standard chemicals like formalin, potassium permanganate and copper sulphate can be used to manage parasite infections. Fungal pathogens are another group of infectious agents, and their spores are common in the aquatic environment, but they do not usually cause disease in healthy fishes. The fungal outbreak is mostly found in fish when they are stressed or infected with external pathogens like bacteria and parasites. As a result of fungal diseases, the fish may exhibit brown, matted patches or cottony growth. External fungal infections can be controlled with chemicals like formalin, bronopol and salt.

Instead of resorting to therapeutic measures using chemicals and antibiotics for treatment, following correct biosecurity measures and prophylactic methods is recommended. Providing a nutritionally balanced diet, monitoring and maintaining the water quality regularly and following a correct schedule for cleaning, water exchange, filter replacements, etc., would help keep the fish healthy. Checking for behavioural changes is a useful diagnostic approach. Fish should be regularly and closely monitored during feeding for any change in behaviour or appearance. Changes in colour pattern, breathing, change in swimming patterns, frequent surfacing, flashing, etc., are signs of distress to the fish and the start of infections. Many marine and freshwater fish species obtained from the wild and farmed would be expected to have some external pathogens. As a biosecurity measure, newly obtained fish should never be introduced or stocked directly into an aquarium. Such fishes should be quarantined with close observation for a specific time to avert the introduction or spreading of infectious diseases in an aquarium.

References

Abidi R, Khan GE, Chauhan UK (2011) Monogenean infestations among freshwater ornamental fishes: an overview. J Ecophysiol Occup Health 11(3/4):199

Allen GR 1991 Damselfishes of the World. Mergus Publishers, Melle, Germany. 1–271

Amita K, Oe M, Matoyama H, Yamaguchi N, Fukuda H (2002) A survey of koi herpesvirus and carp edema virus in colorcarp cultured in Niigata Prefecture, Japan. Fish Pathol 37(4):197–198

Arme C, Bridges JF, Hoole D (1983) Pathology of cestode infections in the vertebrate host. In: Arme C, Pappas PW (eds) Biology of the Eucestoda, vol 2. Academic, London, pp 499–538

Astrofsky KM, Schrenzel MD, Bullis RA, Smolowitz RM, Fox JG (2000) Diagnosis and management of atypical Mycobacterium spp. infections in established laboratory zebrafish (*Brachydanio rerio*) facilities. Comp Med 50(6):666–672

Austin B, Austin DA (1999) Bacterial fish pathogens, diseases of farmed and wild fish. Springer-Praxis Publishing

Austin B, Austin DA (2007) Bacterial fish pathogens: diseases in farmed and wild fish. Ellis Horwood, Chichester, p 552

Austin B, Austin DA (2012) Bacterial fish pathogens. Springer, Dordrecht, p 482

Avenant-Oldewage A (2012) 21 *Lernaea cyprinacea* and related species. In: Woo PTK, Buchmann K (eds) Fish parasites: pathobiology and protection. CABI, Oxfordshire, pp 337–349

Baldock FC, Blazer V, Callinan R, Hatai K, Mohan CV, Bondad-Reantaso GM (2005) Outcomes of a short expert consultation on epizootic ulcerative syndrome (EUS): re – examination of causal factors, case definition and nomenclature. In: Walker P, Laster R, Bondad-Reantaso MG (eds) Diseases in Asian aquaculture V, fish health section. Asian Fisheries Society, Manila, pp 555–585

Barker G (2001) Bacterial diseases. In: Wildgoose WH (ed) BSAVA manual of ornamental fish, 2nd edn. British Small Animal Veterinary Association, Gloucester, pp 185–193

Bartlett KH, Trust TJ (1976) Isolation of salmonellae and other potential pathogens from the freshwater aquarium snail, *Ampullaria*. Appl Environ Microbiol 31:635–639

Basson L, Van-As JG (2006) Trichodinidae and other ciliophorans (phylum Ciliophora). In: Woo PTK (ed) Fish diseases and disorders: protozoan and metazoan infections, vol 1, 2nd edn. CABI Publishing, Oxfordshire, pp 154–182

Beakes GW, Glockling SL, Sekimoto S (2012) The evolutionary phylogeny of the oomycete "fungi". Protoplasma 249(1):3–19

Bernardet JF (1989) *Flexibacter columnaris*: first description in France and comparison with bacterial strains from other origins. Dis Aquat Org 6:37–44

Bhaskar RT, Bhaskar Rao S, Isreal K (1984) Haemoglobin level of the fish *Heteropneustes fossilis* due to *Trypanosoma* infection. Curr Sci 53(6):340

Blazer VS, Shotts EB, Waltman WD (1985) Pathology associated with *Edwardsiella ictaluri* in catfish, *Ictalurus punctatus* Rafinesque, and *Danio devario* (Hamilton-Buchanan, 1822). J Fish Biol 27(2):167–175

Blazer VS, Lilley JH, Schill WB, Kiryu Y, Densmore CL, Panyawachira V, Chinabut S (2002) *Aphanomyces invadans* in Atlantic menhaden along the east coast of the United States. J Aquat Anim Health 14(1):1–10

Bowater RO, Forbes-Faulkner J, Anderson IG, Condon K, Robinson B, Kong F, Gilbert GL, Reynolds A, Hyland S, McPherson G, O'Brien J, Blyde D (2012) Natural outbreak of *Streptococcus agalactiae* (GBS) infection in wild giant Queensland grouper, *Epinephelus lanceolatus* (Bloch) and other wild fish in northern Queensland, Australia. J Fish Dis 35(3):173–186

Bretzinger ACHIM, Fischer-Scherl T, Oumouna M, Hoffmann R, Truyen U (1999) Mass mortalities in Koi carp, *Cyprinus carpio*, associated with gill and skin disease. Bull Eur Assoc Fish Pathol 19(5):182–185

Buchmann K, Bresciani J (2006) Monogenea (phylum Platyhelminthes). In: Woo PTK (ed) Fish diseases and disorders: protozoan and metazoan infections, vol 1, 2nd edn. CABI Publishing, Oxfordshire, pp 297–344

Burreson EM (2006) Phylum Annelida: Hirudinea as vectors and disease agents. In: Woo PTK (ed) Fish diseases and disorders: protozoan and metazoan infections, vol 1, 2nd edn. CABI Publishing, Oxfordshire, pp 566–591

Catap ES, Munday BL (1998) Effects of variations of water temperature and dietary lipids on the expression of experimental epizootic ulcerative syndrome (EUS) in sand whiting, *Sillago ciliata*. Fish Pathol 33:327–335

Cheung P (1993) Parasitic diseases of marine tropical fishes. In: Stoskopf MK (ed) Fish medicine. Saunders, Philadelphia, pp 646–658

Chinabut S (1999) Mycobacteriosis and nocardiosis. In: Woo PTK, Bruno DW (eds) Fish diseases and disorders, vol 3. CABI Publishing, New York, pp 319–340

Chinabut S, Roberts RJ (1999) Pathology and histopathology of epizootic ulcerative syndrome (EUS). Aquatic animal Health Research Institute, Department of Fisheries. Royal Thai Government, Bangkok, p 33. ISBN 974-7604-55-8

Chong RSM (2015) Pathology of cultured fish in Hong Kong. Agriculture, Fisheries and Conservation Department, pp 1–229

Colwell RR, Grimes DJ (1984) *Vibrio* diseases of marine fish populations. Helgoländer Meeresun 37:265–287

Conroy DA (1964) Nocardiosis as a disease of tropical fish. Vet Rec 76:676

Conroy G, Conroy DA (1999) Acid-fast bacterial infection and its control in guppies (*Lebistes reticulatus*) reared on an ornamental fish farm in Venezuela. Vet Rec 144(7):177–178

Darai G, Anders K, Koch HG, Delius H, Gelderblom H, Samalecos C, Flügel RM (1983) Analysis of the genome of fish lymphocystis disease virus isolated directly from epidermal tumours of pleuronectes. Virology 126(2):466–479

Darai G, Delius H, Clarke J, Apfel H, Schnitzler P, Flügel RM (1985) Molecular cloning and physical mapping of the genome of fish lymphocystis disease virus. Virology 146(2):292–301

Davison AJ, Eberle R, Ehlers B, Hayward GS, McGeoch DJ, Minson AC, Pellett PE, Roizman B, Studdert MJ, Thiry E (2009) The order Herpesvirales. Arch Virol 154:171–177

Decostere A, Haesebrouck F, Devriese LA (1998) Characterization of four *Flavobacterium columnare* (*Flexibacter columnaris*) strains isolated from tropical fish. Vet Microbiol 62(1):35–45

Dewi, R. R. and Fadhilla, A., 2018. The prevalence of parasites in ornamental fish from fish market in Medan. In: IOP conference series: earth and environmental science, IOP Publishing, 122(1), 012110)

Dick TA, Chambers C, Isinguzo I (2006) Cestoidea (Phylum Platyhelminthes). In: Woo PTK (ed) Fish diseases and disorders: protozoan and metazoan infections, vol 1, 2nd edn. CABI Publishing, Oxfordshire, pp 391–416

Dickerson HW (2006) *Ichthyophthirius multifiliis* and *Cryptocaryon irritans* (phylum Ciliophora). In: Woo PTK (ed) Fish diseases and disorders: protozoan and metazoan infections, vol 1, 2nd edn. CABI Publishing, Oxfordshire, pp 116–153

Dror M, Sinyakov MS, Okun E, Dym M, Sredni B, Avtalion RR (2006) Experimental handling stress as infection-facilitating factor for the gold fish ulcerative disease. Vet Immunol Immunopathol 109(3–4):279–287

Dykov AI (2006) Phylum Microspora. In: Woo PTK (ed) Fish diseases and disorders: protozoan and metazoan infections, vol 1, 2nd edn. CABI Publishing, Oxfordshire, pp 205–229

Eckerle LD, Ball LA (2002) Replication of the RNA segments of a bipartite viral genome is coordinated by a transactivating subgenomic RNA. Virology 296(1):165–176

Egidius E, Wiik R, Andersen K, Hoff KA, Hjeltnes B (1986) *Vibrio salmonicida* sp. nov., a new fish pathogen. Int J Syst Bacteriol 36:518–520

Elliott DG, Shotts EB Jr (1980) Aetiology of an ulcerative disease in goldfish *Carassius auratus* (L.): microbiological examination of diseased fish from seven locations. J Fish Dis 3(2):133–143

El-Matbouli M, Hoffmann RW, Schoel H, McDowell TS, Hedrick RP (1999) Whirling disease: host specificity and interaction between the actinosporean stage of *Myxobolus cerebralis* and rainbow trout *Oncorhynchus mykiss*. Dis Aquat Org 35(1):1–12

El-Matbouli M, Rucker U, Soliman H (2007) Detection of cyprinid herpes virus-3 DNA in infected fish tissues by nested polymerase chain reaction. Dis Aquat Org 78:23–28

Fenner BJ, Thiagarajan R, Chua HK, Kwang J (2006) Betanodavirus B2 is an RNA interference antagonist that facilitates intracellular viral RNA accumulation. J Virol 80(1):85–94

Ferguson HW (1989) Systemic pathology of fish. A text and atlas of comparative tissue responses in diseases of teleosts. Iowa State University Press

Francis-Floyd R, Floyd MR (2011) Amyloodinium ocellatum, an important parasite of cultured marine fish. Southern Regional Aquaculture Center, publication, No. 4705

Francis-Floyd R, Bolon B, Reed P, Poynton S (1993) Pathology and clinical management of diplomonad flagellate infections of angel fish. In: Proceedings of the international association of aquatic animal medicine, Chicago, pp 66–67. https://www.vin.com/apputil/content/defaultadv1.aspx?pId=11257andcatId=32245andid=3864039

Franco-Sierra A, Alvarez-Pellitero P (1999) The morphology of *Ichthyophonus* sp. In their mugilid hosts (Pisces: Teleostei) and following cultivation in vitro. A light and electron microscopy study. Parasitol Res 85:562–575

Fujimoto RY, Neves MS, Santos RF, Souza NC, Do Couto MV, Lopes JN, Diniz DG, Eiras JC (2013) Morphological and hematological studies of *Trypanosoma* spp. infecting ornamental armored catfish from Guamá River-PA, Brazil. Anais da Academia Brasileira de Ciȩncias 85(3):1149–1156

Gauthier DT, Rhodes MW (2009) Mycobacteriosis in fishes: a review. Vet J 180:33–47. https://doi.org/10.1016/j.tvjl.2008.05.01

George MR, John KR, Mansoor MM, Saravanakumar R, Sundar P, Pradeep V (2015) Isolation and characterization of a ranavirus from koi, *Cyprinus carpio* L., experiencing mass mortalities in India. J Fish Dis 38(4):389–403

Girisha SK, Kushala KB, Nithin MS, Puneeth TG, Naveen Kumar BT, Vinay TN, Suresh T, Ajay SK, Venugopal MN, Ramesh KS (2021) First report of the infectious spleen and kidney necrosis virus (ISKNV) infection in ornamental fishes in India. Transbound Emerg Dis 68(2):964–972

Gomez DK, Lim DJ, Baeck GW, Youn HJ, Shin NS, Youn HW, Hwang JH, Park SC (2006) Detection of betanodaviruses in apparently healthy aquarium fishes and invertebrates. J Vet Sci 7(4):369–374

Goodwin AE (1999) Massive *Lernaea cyprinacea* infestations damaging the gills of channel catfish polycultured with bighead carp. J Aquat Anim Health 11(4):406–408

Goodwin A (2012) Herpesviruses in fish. Southern Regional Aquaculture Center (SRAC) Publication No 4710

Gozlan RE, Marshall W, Lilje O, Jessop C, Gleason FH, Andreou D (2014) Current ecological understanding of fungal-like pathogens of fish: what lies beneath? Front Microbiol 5:62

Gratzek JB (1988) Parasites associated with ornamental fish. Vet Clin N Am Small Anim Pract 18(2):375–399

Gratzek JB (1993) Parasites associated with freshwater tropical fish. In: Stoskopf MK (ed) Fish medicine. Saunders, Philadelphia, pp 573–590

Gray WL, Mullis L, LaPatra SE, Groff JM, Goodwin A (2002) Detection of koi herpesvirus DNA in tissues of infected fish. J Fish Dis 25(3):171–178

Grimes DJ, Stemmler J, Hada H, May EB, Maneval D, Hetrick FM, Jones RT, Stoskopf M, Colwell RR (1984) Vibrio species associated with mortality of sharks held in captivity. Microb Ecol 10(3):271–282

Guo CJ, Wu YY, Yang LS, Yang XB, He J, Mi S, Jia KT, Weng SP, Yu XQ, He JG (2012) Infectious spleen and kidney necrosis virus (a fish iridovirus) enters Mandarin fish fry cells via caveola-dependent endocytosis. J Virol 86(5):2621–2631

Hacking MA, Budd J (1971) Vibrio infection in tropical fish in a freshwater aquarium. J Wildl Dis 7(4):273–280

Haenen OLM, Way K, Bergmann SM, Ariel E (2004) The emergence of koi herpesvirus and its significance to European aquaculture. Bull Eur Assoc Fish Pathol 24(6):293–307

Hallett SL, Bartholomew JL (2012) *Myxobolus cerebralis* and *Ceratomyxa shasta*. In: Woo PTK, Buchmann K (eds) Fish parasites: pathobiology and protection. CABI, Oxfordshire, pp 131–162

Hamazaki T, Kahler E, Borba BM, Burton T (2013) PCR testing can be as accurate as culture for diagnosis of *Ichthyophonus hoferi* in Yukon River Chinook salmon *Oncorhynchus tshawytscha*. Dis Aquat Org 105:21–25

Harms CA (1996) Treatments for parasitic diseases of aquarium and ornamental fish. Sem Av Ex Pet Med 2(5):54–63

Hartman KH, Yanong RP, Pouder DB, Petty BD, Francis-Floyd R, Riggs AC, Waltzek TB (2013) Koi herpesvirus (KHV, CyHV3) disease. University of Florida. EDIS publication VM113

He JG, Zeng K, Weng SP, Chan SM (2002) Experimental transmission, pathogenicity and physical–chemical properties of infectious spleen and kidney necrosis virus (ISKNV). Aquaculture 204(1–2):11–24

Hedrick RP, McDowell TS (1995) Properties of iridoviruses from ornamental fish. Vet Res 26(5–6):423–427

Hedrick RP, El-Matbouli M, Adkison MA, MacConnell E (1998) Whirling disease: re-emergence among wild trout. Immunol Rev 166:365–376

Hedrick RP, Gilad O, Yun S, Spangenberg JV, Marty GD, Nordhausen RW, Kebus MJ, Bercovier H, Eldar A (2000) A herpesvirus associated with mass mortality of juvenile and adult koi, a strain of common carp. J Aquat Anim Health 12(1):44–57

Hedrick RP, Waltzek TB, McDowell TS (2006) Susceptibility of koi carp, common carp, goldfish, and goldfish× common carp hybrids to cyprinid herpesvirus-2 and herpesvirus-3. J Aquat Anim Health 18(1):26–34

Helfman G, Collette FD, Bowen BW (2009) The diversity of fishes: biology, evolution, and ecology. Wiley-Blackwell, Chichester

Hesami S, Viadanna P, Steckler N, Spears S, Thompson P, Kelley K, Yanong R, Francis-Floyd R, Shelley J, Groff J, Goodwin A, Haenen O, Waltzek T (2015) Carp edema virus disease (CEVD)/koi sleepy disease (KSD). University of Florida. EDIS Publication FA189 [Internet]. https://www.vin.com/apputil/content/defaultadv1.aspx?id=3843980&pid=11131&print=1

Hibbett DS, Binder M, Bischoff JF, Blackwell M, Cannon PF, Eriksson OE, Huhndorf S, James T, Kirk PM, Lücking R, Lumbsch HT (2007) A higher-level phylogenetic classification of the fungi. Mycol Res 111:509–547

Hines RS, Spira DT (1974) Ichthyophthiriasis in the mirror carp *Cyprinus carpio* (L.) V. Acquired immunity. J Fish Biol 6(4):373–378

Höglund J (1991) Ultrastructural observations and radiometric assay on cercarial penetration and migration of the digenean *Diplostomum spathaceum* in the rainbow trout *Oncorhynchus mykiss*. Parasitol Res 77(4):283–289

Horizon Grand View Research (2024). https://www.grandviewresearch.com/horizon/outlook/ornamental-fish-market-size/global

Hunt R, Cable J (2020) Life in the fast lane: temperature, density and host species impact survival and growth of the fish ectoparasite *Argulus foliaceus*. J Therm Biol 92:102687

Iberahim NA, Trusch F, Van West P (2018) *Aphanomyces invadans*, the causal agent of epizootic ulcerative syndrome, is a global threat to wild and farmed fish. Fungal Biol Rev 32(3):118–130

Inglis I, Roberts RJ, Bromage NR (1993) Bacterial diseases of fish. Blackwell Science, pp 196–210

Iqbal Z, Haroon F (2014) Parasitic infections of some freshwater ornamental fishes imported in Pakistan. Pak J Zool 46(3):651–656

Jeong JB, Kim HY, Jun LJ, Lyu JH, Park NG, Kim JK, Do Jeong H (2008) Outbreaks and risks of infectious spleen and kidney necrosis virus disease in freshwater ornamental fishes. Dis Aquat Org 78(3):209–215

Jithendran KP, Shekhar MS, Kannappan S, Azad IS (2011) Nodavirus infection in freshwater ornamental fishes in India-diagnostic histopathology and nested RT-PCR. Asian Fish Sci 24:12–19

Jung-Schroers V, Adamek M, Wohlsein P, Wolter J, Wedekind H, Steinhagen D (2016) First outbreak of an infection with infectious spleen and kidney necrosis virus (ISKNV) in ornamental fish in Germany. Dis Aquat Org 119(3):239–244

Kamilya D, John KR (2020) Epizootic ulcerative syndrome (*Aphanomyces invadans*). In: Woo PTK, Leong JA, Buchmann K (eds) Climate change and infectious fish diseases. CABI, Wallingford, pp 291–302

Kayis S, Balta F, Serezli R, Er A (2013) Parasites on different ornamental fish species in Turkey. J Fish Sci 7(2):114

Kempter J, Bergmann SM (2007) Detection of koi herpesvirus (KHV) genome in wild and farmed fish from northern Poland. Aquaculture 272:S275

Kempter J, Sadowski J, Schutze H, Fischer U, Dauber M, Fichtner D, Panicz R, Bergmann SM (2009) Koi herpesvirus: do Acipenserid restitution programs pose a threat to carp farms in the disease free zones? Acta Ichthyol Piscat 39(9):119–126

Kent ML, Lyons JM (1982) *Edwardsiella ictaluri* in the green knife fish, *Eigenmania virescens*. Fish Health News 11(1–2):ii

Kim JH, Hayward CJ, Heo GJ (2002) Nematode worm infections (*Camallanus cotti, Camallanidae*) in guppies (*Poecilia reticulata*) imported to Korea. Aquaculture 205(3–4):231–235

Kocan RM, Gregg JL, Hershberger PK (2010) Release of infectious cells from epidermal ulcers in *Ichthyophonus* sp.-infected pacific herring (*Clupea pallasii*): evidence for multiple mechanisms of transmission. J Parasitol 96:348–352

Kocan R, LaPatra S, Hershberger P (2013) Evidence for an amoeba-like infectious stage of *Ichthyophonus* sp. and description of a circulating blood stage: a probable mechanism for dispersal within the fish host. J Parasitol *99*(2):235–240

Kumar P, Thirunavukkarasu AR, Subburaj R, Thiagarajan G (2015) Concept of stress and its mitigation in aquaculture. In: Perumal SART, Pachiappan P (eds) Advances in marine and Brackishwater Aquaculture. Springer, New Delhi

Lee DC, Cho MY, Park SI (2007) A review of nocardial infection in fishes. J Fish Pathol 20(1):1–23

Lescenko P, Matlova L, Dvorska L, Bartos M, Vavra O, Navratil S, Novotny L, Pavlik I (2003) Mycobacterial infection in aquarium fish. Vet Med 48(3):71–78

Lester RJG, Hayward CJ (2006) Phylum Arthropoda. In: Woo PTK (ed) Fish diseases and disorders: protozoan and metazoan infections, vol 1, 2nd edn. CABI Publishing, Oxfordshire, pp 466–565

Lewbart GA (2001) Bacteria and ornamental fish. Semin Avian Ex Pet Med 10(1):48–56

Li L, Meng H, Gu D, Li Y, Ji M (2019) Molecular mechanisms of *Vibrio parahaemolyticus* pathogenesis. Microbiol Res 222:43–51

Lilley JH, Hart D, Richards RH, Roberts RJ, Cerenius L, Soderhall K (1997) Pan-Asian spread of single fungal clone results in large scale fish kills. Vet Rec 140:653–654

Lom J, Dykova I (1992) Protozoan parasites of fish. Developments in aquaculture and fisheries science, vol 26. Elsevier, New York, p 315

López-Romalde S, Magariños B, Núñez S, Toranzo AE, Romalde JL (2003) Phenotypic and genetic characterization of *pseudomonas anguilliseptica* strains isolated from fish. J Aquat Anim Health 15(1):39–47

MacConnell E, Vincent ER (2002) The effects of Myxobolus cerebralis on the salmonid host. In: Bartholomew JL, Wilson JC (eds) Whirling disease: reviews and current topics. Symposium 29. American Fisheries Society, Bethesda, pp 95–107

Madhuri S (2013) A manual on the aquaculture development of freshwater fish and shellfish. Narendra Publishing House, Delhi, p 117p

Maekawa S, Yoshida T, Wang PC, Chen SC (2018) Current knowledge of nocardiosis in teleost fish. J Fish Dis 41(3):413–419

Magada S, Mercy TVA (2019) Health management in ornamental fish farming. In: BMP guidelines for ornamental fish production. National Fisheries Development Board, Hyderabad

Maltese C, Bovo G (2007) Viral encephalopathy and retinopathy. Ittiopatologia 4:93–146

Mankhakhet S, Suanyuk N, Tantikitti C, Phromkunthong W, Kiriratnikom S, Lerssutthichawal T, Viriyapongsutee B (2012) Diplomonad flagellates of some ornamental fish cultured in Thailand. *Songklanakarin J Sci Technol* 34(5):487–494

Matras M, Borzym E, Stone D, Way K, Stachnik M, Maj-Paluch J, Palusińska M, Reichert M (2017) Carp edema virus in polish aquaculture–evidence of significant sequence divergence and a new lineage in common carp *Cyprinus carpio* (L.). J Fish Dis 40(3):319

Matthews RA (2005) *Ichthyophthirius multifiliis* Fouquet and ichthyophthiriosis in freshwater teleosts. Adv Parasitol 59:159–241

McCarthy DH (1975) Detection of *Aeromonas salmonicida* antigen in diseased fi sh tissue. J Gen Microbiol 88:185–187

McCarthy DH, Roberts RJ (1980) Furunculosis of fish – the present state of our knowledge. In: Droop MA, Jannasch HW (eds) Advances in aquatic microbiology. Academic Press, London, pp 293–341

McVicar AH (1982) *Ichthyophonus* infections in fishes. In: Roberts RJ (ed) Microbial disease of fish. Academic, London, pp 243–270

McVicar AH (1999) *Ichthyophonus* and related organisms. In: Woo PTK, Bruno D (eds) Fish diseases and disorders, viral, bacterial and fungal infections, vol 3. CABI, Oxford, pp 661–668

McVicar AH, McLay HA (1985) Tissue response of plaice, haddock, and rainbow trout to the systemic fungus *Ichthyophonus*. In: Ellis AE (ed) Fish and shellfish pathology. Academic Press, London, pp 329–346

Mendoza L, Taylor JW, Ajello L (2002) The class Mesomycetozoea: a group of microorganisms at the animal-fungal boundary. Ann Rev Microbiol 56:315–344

Meyer FP, Bullock GL (1973) *Edwardsiella tarda,* a new pathogen of channel cat fish *(Ictalurus punctatus)*. Appl Microbiol 25(1):155–156

Meyers TR, Bentz C, Ferguson J, Stewart D, Starkey N, Burton T (2019) Diseases of wild and cultured fishes in Alaska. Alaska Department of Fish and Game, Fish Pathology Laboratories, Juneau, p 128

Michel C, Bernardet JF, Daniel P, Chilmonczyk S, Urdaci M, De Kinkelin P (2002) Clinical and aetiological aspects of a summer enteritic syndrome associated with the sporulating segmented filamentous bacterium, '*Candidatus* Arthromitus' in farmed rainbow trout, *Oncorhynchus mykiss* (Walbaum). J Fish Dis 25(9):533–543

Mitiku MA (2018) A review on Columnaris disease in freshwater farmed fish. Int J Adv Res Biol Sci 5:79–90

Miyazaki T, Isshiki T, Katsuyuki H (2005) Histopathological and electron microscopy studies on sleepy disease of koi *Cyprinus carpio* koi in Japan. Dis Aquat Org 65(3):197–207

Mohamad N, Amal MNA, Yasin ISM, Zamri Saad M, Nasruddin NS, Al-saari N, Mino S, Sawabe T (2019) Vibriosis in cultured marine fishes: a review. Aquaculture 512:734289. https://doi.org/10.1016/j.aquaculture.2019.734289

Molnar K, Buchmann K, Csaba S (2006) Phylum nematoda. In: Woo PTK (ed) Fish diseases and disorders: protozoan and metazoan infections, vol 1, 2nd edn. CABI Publishing, Oxfordshire, pp 417–443

Moravec F, Gut J (1982) Morphology of the nematode *Capillaria pterophylli* Heinze, 1933, a pathogenic parasite of some aquarium fishes. Folia Parasitol 29(3):227–231

Moravec F, Gelnar M, Rehulka J (1987) Parasite of aquarium fishes in Europe. Folia Parasitol 34:157–161

Mori KI, Nakai T, Muroga K, Arimoto M, Mushiake K, Furusawa I (1992) Properties of a new virus belonging to *Nodaviridae* found in larval striped jack (*Pseudocaranx dentex*) with nervous necrosis. Virology 187(1):368–371

Moyle PB, Cech JJ (2003) Fishes, an introduction to ichthyology, 5th edn. Benjamin Cummings, San Francisco

Munday BL, Kwang J, Moody N (2002) Betanodavirus infections of teleost fish: a review. J Fish Dis 25(3):127–142

Nagai T, Nishizawa T (1999) Sequence of the non-structural protein gene encoded by RNA1 of striped jack nervous necrosis virus. The DDBJ accession number of the sequence reported in this paper is AB025018. J Gen Virol 80(11):3019–3022

Nazari A, Hassan MD, Bovo G, Zorriehzahra MJ, Azmi TI, Arshad SS (2014) Pathogenicity of viral nervous necrosis virus for Guppy fish, *Poecilia reticulata*. Iran J Fish Sci 13(1):168–177

Nazrul Islam AKM, Woo PTK (1991a) Anorexia in goldfish *Carassius auratus* infected with *Trypanosoma danilewskyi*. Dis Aquat Org 11:45–48
Nazrul Islam AKM, Woo PTK (1991b) Anemia and its mechanism in goldfish *Carassius auratus* infected with *Trypanosoma danilewskyi*. Dis Aquat Org 11:3743
Noga EJ (1993) Fungal diseases of marine and estuarine fish. In: Couch JA, Fournie JW (eds) Pathobiology of marine and estuarine organisms. CRC Press, Boca Raton, pp 85–110
Noga EJ (2000) Fish disease: diagnosis and treatment. Iowa State University Press. 367 p
Noga EJ (2010) Fish disease: diagnosis and treatment, 2nd edn. Wiley-Blackwell, Ames, p 519
Noga EJ, Levy MG (2006) Phylum Dinoflagellata. In: Woo PTK (ed) Fish diseases and disorders: protozoan and metazoan infections, vol 1, 2nd edn. CABI Publishing, Oxfordshire, pp 16–45
Oh MJ, Jung SJ, Choi TJ, Kim HR, Rajendran KV, Kim YJ, Park MA, Chun SK (2001) A viral disease occurring in cultured carp *Cyprinus carpio* in Korea. Fish Pathol 36(3):147–151
Okamoto N, Nakase K, Suzuki H, Nakai Y, Fujii K, Sano T (1985) Life history and morphology of *Ichthyophonus hoferi in vitro*. Fish Pathol 20:273–285
Ono SI, Nagai A, Sugai N (1986) A histopathological study on juvenile colorcarp, *Cyprinus carpio*, showing edema. Fish Pathol 21(3):167–175
Oyamatsu T, Hata N, Yamada K, Sano T, Fukuda H (1997) An etiological study on mass mortality of cultured colorcarp juveniles showing edema. Fish Pathol 32(2):81–88
Pacha RE, Ordal EJ (1967) Histopathology of experimental columnaris disease in young salmon. J Comp Pathol 77:419–423
Pacha RE, Ordal EJ (1970) Myxobacterial diseases of salmonids. In: Snieszko SF (ed) A symposium on diseases of fishes and Shell fishes. American Fisheries Society, Special Publication No. 5, Washington D.C., pp 243–257
Pandey G (2013) Treatment and control of ornamental fish infection. Univers J Pharm 02(02):52–56
Paperna I (1980) *Amyloodinium ocellatum* (Brown, 1931) (Dinoflagellida) infestations in cultured marine fish at Eilat, Red Sea: epizootiology and pathology. J Fish Dis 3(5):363–372
Paperna I (1991) Diseases caused by parasites in the aquaculture of warm water fish. Annu Rev Fish Dis 1:155–194
Paperna I, Dzikowski R (2006) Digenea (Phylum Platyhelminthes). In: Woo PTK (ed) Fish diseases and disorders: protozoan and metazoan infections, vol 1, 2nd edn. CABI Publishing, Oxfordshire, pp 345–390
Park SB, Aoki T, Jung TS (2012) Pathogenesis of and strategies for preventing *Edwardsiella tarda* infection in fish. Vet Res 43(1):1–11
Paull GC, Matthews RA (2001) *Spironucleus vortens*, a possible cause of hole-in-the-head disease in cichlids. Dis Aquat Org 45(3):197–202
Perelberg A, Smirnov M, Hutoran M, Diamant A, Bejerano Y, Kotler M (2003) Epidemilogical description of a new viral disease afflicting cultured *Cyprinus carpio* in Israel. Isr J Aquacult—Bamid 55:5–12
Petchimuthu M, George MR, John KR, Kaviarasu D, Dinesh R (2018) Occurrence of parasites in aquaculture systems of South Tamil Nadu. J Entomol Zool Stud 6(2):283–286
Pfeil-Putzien C (1978) Experimental transmission of spring viraemia of carp through carp lice (*Angulus foliaceus*). J Vet Med Ser B 25(4):319–323
Piasecki W, Avenant-Oldewage A (2008) Diseases caused by crustacea. In: Eiras JC, Segner H, Wahli T, Kapoor BG (eds) Fish diseases. Science Publishers, New Hampshire, pp 1115–1200
Pirarat N, Ponpornpisit A, Traithong T, Nakai T, Katagiri T, Maita M, Endo M (2009) Nodavirus associated with pathological changes in adult spotted coral groupers (*Plectropomus maculatus*) in Thailand with viral nervous necrosis. Res Vet Sci 87:97–101
Plumb JA, Hanson LA (2011) Health maintenance and principal microbial diseases of cultured fishes, 3rd edn. Blackwell Publishing Ltd., p 492
Pokorova D, Vesely T, Piackova V, Reschova S, Hulova J (2005) Current knowledge on koi herpesvirus (KHV): a review. Vet Med 50(4):139–147
Reddacliff GL, Hornitzky M, Carson J, Petersen R, Zelski R (1993) Mortalities of goldfish, *Carassius auratus* (L.), associated with vibrio cholerae (non-01) infection. J Fish Dis 16(5):517–520

Reed P, Francis-Floyd R (1994) Amyloodinium infections of marine fish. University of Florida/ Institute of Food and Agricultural Sciences. Extension 1–3

Richards RH, Pickering AD (1979) Changes in serum parameters of Saprolegnia-infected brown trout, *Salmo trutta* L. J Fish Dis 2(3):197–206

Roberts RJ (2012) The mycology of teleosts. In: Roberts R (ed) Fish pathology, 4th edn. Wiley-Blackwell, London, pp 383–401

Roberts RJ (2012a) The bacteriology of teleosts. In: Roberts R (ed) Fish pathology, 4th edn. Wiley-Blackwell, London, pp 339–382

Roberts HE, Palmeiro B, Weber ES III (2009) Bacterial and parasitic diseases of pet fish. Vet Clin North Am: Exot Anim Pract 12(3):609–638

Robertson DA, Roberts RJ, Bullock AM (1981) Pathogenesis and autoradiographic studies of the epidermis of salmonids infested with *Ichtyobodo necator* (Henneguy, 1883). J Fish Dis 4(2):113–125

Rose JD, Marrs GS, Lewis C, Schisler G (2000) Whirling disease behavior and its relation to pathology of brain stem and spinal cord in rainbow trout. J Aquat Anim Health 12(2):107–118

Saeed MO (1995) Association of *Vibrio harveyi* with mortalities in cultured marine fish in Kuwait. Aquaculture 136:21–29

Sahoo PK, Swaminathan TR, Abraham TJ, Kumar R, Pattanayak S, Mohapatra A, Rath SS, Patra A, Adikesavalu H, Sood N, Pradhan PK, Das BK, Jayasankar P, Jena JK (2016) Detection of goldfish haematopoietic necrosis herpes virus (Cyprinid herpesvirus-2) with multi-drug resistant Aeromonas hydrophila infection in goldfish: First evidence of any viral disease outbreak in ornamental freshwater aquaculture farms in India. Acta Trop 161:8–17

Sahoo PK, Pattanayak S, Paul A, Sahoo MK, Kumar PR (2020) Carp edema virus in ornamental fish farming in India: a potential threat to koi carps but not to co-cultured Indian major carp or goldfish. Indian J Exp Biol 58:254–262

Santos MA, Jerônimo GT, Cardoso L, Tancredo KR, Medeiros PB, Ferrarezi JV, Gonçalves ELT, Assis GC, Martins ML (2017) Parasitic fauna and histopathology of farmed freshwater ornamental fish in Brazil. Aquaculture *470*:103–109

Sasaki T, Aita A (1975) A study on fi sh diseases. Part 1. *Edwardsiella* isolated from the seaurchin. Jpn J Microbiol *30*:368

Sasmal D, Banerjee T, Bandyopadhyay S, Abraham TJ (2004) Antibiotic sensitivity of bacterial flora associated with ornamental fish. Indian J Fish 51:245–249

Schnitzler P, Darai G (1993) Identification of the gene encoding the major capsid protein of fish lymphocystis disease virus. J Gen Virol 74(10):2143–2150

Sheikhzadeh N, Fakheri T, Nofouzi KN (2020) Nocardiosis in some ornamental fishes in Iran. J Zoonotic Dis 4(1):28–34

Shields RJ, Tidd WM (1974) Site selection on hosts by copepods of *Lernaea cyprinacea* L. (Copepoda). Crustaceana 27:225–230

Shimura S, Inoue K, Kudo M, Egusa S (1983) Studies on effects of parasitism of *Argulus coregoni* (Crustacea: *Branchiura*) on furunculosis of *Oncorhynchus masou* (*Salmonidae*). Fish Pathol 18(1):37–40

Shotts EB, Talkington FD (1980) Aetiology of an ulcerative disease in goldfish, *Carassius auratus* (L.): characterization of the causative agent. J Fish Dis 3:181–186

Sitja A, Alvarez-Pellitero P (1990) First report of Ichthyophonus disease in wild and cultured sea bass *Dicentrarchus labrax* from the Spanish Mediterranean area. Dis Aquat Org 8:145–150

Sivasankar P, John KR, George MR, Mageshkumar P, Manzoor MM, Jeyaseelan MJP (2017) Characterization of a virulent ranavirus isolated from marine ornamental fish in India. Virus Dis 28(4):373–382

Smith SA (1997) Mycobacterial infections in pet fish. Semin Avian Exot Pet Med 6(1):40–45

Smith SA, Noga E (1993) General parasitology. In: Stoskopf MK (ed) Fish medicine. PA, Saunders, Philadelphia, pp 132–148

Snieszko SF (1958) Fish furunculosis. In: *Fishery leaflet* 467. United States fish and wildlife service, Washington, D.C., p 4

Soltani M, Jamshidi S, Sharifpour I (2005) Streptococcosis caused by *Streptococcus iniae* in farmed rainbow trout (*Oncorhynchys mykiss*) in Iran: biophysical characteristics and pathogenesis. Bull Eur Assoc Fish Pathol 25(3):95–106

Sommerset I, Nerland AH (2004) Complete sequence of RNA1 and subgenomic RNA3 of Atlantic halibut nodavirus (AHNV). Dis Aquat Org 58(2–3):117–125

Sommerville C (1981) A comparative study of the tissue response to invasion and encystment by *Stephanochasmus baccatus* (Nicoll, 1907) (*Digenea*: *Acanthocolpidae*) in four species of flat fish. J Fish Dis 4(1):53–68

Spanggaard B, Huss HH (1996) Growth of the fish parasite *Ichthyophonus hoferi* under food relevant conditions. Int J Food Sci Technol 31(5):427–432

Spanggaard B, Huss HH, Bresciani J (1995) Morphology of *Ichthyophonus hoferi* assessed by light and scanning electron microscopy. J Fish Dis 18:567–577

Steinbach ELC, Stromberg KE, Ryce EKN, Bartholomew JL (2009) Whirling disease in the United States. A summary of Progress in Research and management 2009. Trout Unlimited Whirling Disease Foundation, Bozeman, p 61

Swaminathan TR, Kumar R, Dharmaratnam A, Basheer VS, Sood N, Pradhan PK, Sanil NK, Vijayagopal P, Jena JK (2016) Emergence of carp edema virus in cultured ornamental koi carp, *Cyprinus carpio* koi, in India. J Gen Virol 97(12):3392–3399

Towers L (2013) Parasitic diseases of tropical and ornamental fish. https://thefishsite.com/articles/parasitic-diseases-of-tropical-and-ornamental-fish

Trujillo-González A, Becker JA, Vaughan DB, Hutson KS (2018) Monogenean parasites infect ornamental fish imported to Australia. Parasitol Res 117(4):995–1011

Turgay E (2020) Edwardsiellosis in freshwater angelfish (*Pterophyllum scalare*). Vol 72, 2020 October 08, https://doi.org/10.46989/001c.19140

Valdez IE, Conroy DA (1963) The study of a tuberculosis-like condition in neon tetras (*Hyphessobrycon innesi*) II. Characteristics of the bacterium isolated. Microbiol Esp 16:249–253

Vandepitte J, Lemmens P, De Swert L (1983) Human edwardsiellosis traced to ornamental fish. J Clin Microbiol 17(1):165–167

Wada S, Hatai K, Tanaka E, Kitahara T (1993) Mixed infection of an acid-fast bacterium and an imperfect fungus in a Napoleon fish (*Cheilinus undulatus*). J Wildl Dis 29(4):591–595

Wada S, Ozeki K, Kurata, O, Hatai K, Ishida A and Ui K 2010 Atypical Aeromonas salmonicida infection in sailfin sandfish Arctoscopus japonicus. Fish Pathology, 45(2):92–95

Wakabayashi H, Egusa S (1972) Characteristics of a *Pseudomonas* sp. from an epizootic of pond-cultured eels (*Anguilla japonica*). Bull Jpn Soc Sci Fish 38(6):577–587

Walczak N, Puk K, Guz L (2017) Bacterial flora associated with diseased freshwater ornamental fish. J Vet Res 61(4):445–449

Walker R (1962) Fine structure of lymphocystis virus of fish. Virology 18:503–505

Waltzek TB, Kelley GO, Stone DM, Way K, Hanson L, Fukuda H, Hirono I, Aoki T, Davison AJ, Hedrick RP (2005) Koi herpesvirus represents a third cyprinid herpesvirus (CyHV-3) in the family Herpesviridae. J Gen Virol 86(6):1659–1667

Wang X, Wang Q, Yang M, Xiao J, Liu Q, Wu H, Zhang Y (2011) QseBC controls flagellar motility, fimbrial hemagglutination and intracellular virulence in fish pathogen *Edwardsiella tarda*. Fish Shellfish Immunol 30:944–953

Wiles M, Cone DK, Odense PH (1985) Studies on *Chilodonella cyprini* and *C. hexasticha* (Protozoa, Ciliata) by scanning electron microscopy. Can J Zool 63(10):2483–2487

Willoughby LG, Roberts RJ, Chinabut S (1995) *Aphanomyces invaderis* sp. nov., the fungal pathogen of freshwater tropical fish affected by epizootic ulcerative syndrome. J Fish Dis 18(3):273–275

Wolf K (1988) Lymphocystis disease. In: Wolf K (ed) Fish viruses and fish viral diseases. Cornell University Press, Ithaca, pp 268–291

Wolke RE (1975) Pathology of bacterial and fungal disease affecting fish. In: Ribelin WE, Migaki G (eds) Fish pathology. University Wisconsin Press, Madison, pp 33–116

Woo PTK (2006) Diplomonadida (Phylum Parabasalia) and Kinetoplastea (Phylum Euglenozoa). In: Woo PTK (ed) Fish diseases and disorders: protozoan and metazoan infections, vol 1, 2nd edn. CABI Publishing, Oxfordshire, pp 46–115

Woo PTK, Bruno DW (2010) Fish diseases and disorders. Volume 3: viral, bacterial and fungal infections. In: Evans JJ, Klesius PH, Plumb JA, Shoemaker CA (eds) Edwardsiella septicaemias. CABI international, Wallingford, pp 512–534

Wootten R (2012) The parasitology of teleosts. In: Roberts R (ed) Fish pathology, 4th edn. Wiley-Balckwell, London, pp 292–338

Yanong RPE (2010) Lymphocystis disease in fish. Tropical aquaculture laboratory, school of forest resources and conservation. In: Program in fisheries and aquatic sciences, vol 33570. UF/IFAS extension, Ruskin

Yanong RP, Francis-Floyd R (2002) Streptococcal infections of fish. Florida Cooperative Extension Service. IFAS, University of Florida, pp 1–5

Zadeh MJ, Peyghan R, Manavi SE (2014) The detection of Ichthyophonus hoferi in naturally infected fresh water ornamental fishes. J Aquac Res Dev 5(7):1–5

Zwillenberg LD, Wolf K (1968) Ultrastructure of lymphocystis virus. J Virol 2:393–399

Parasitic Diseases in Freshwater Fish

S. Mohanty and P. K. Sahoo

Abstract

The freshwater ecosystem and its parasitic diversity are poorly addressed. Besides their pathogenicity to farmed fish, parasites are excellent ecological players and indicators. Failing good husbandry practices, including optimum stocking, nutrition, and animal welfare, fishes are invariably encountered by parasites, which opens windows for disease incidences and also results in concomitant opportunistic pathogens to establish. Furthermore, the transmissions of these parasites are strongly influenced by critical environmental parameters such as water temperature, alkalinity, and overcrowding, which necessitate and underline the importance of management and prophylaxis. Parasites cause substantial pathological debilities in the host at the cost of its energy and growth, leading to huge mortality and economic losses. The emergence of parasites is a real challenge due to significant changes induced by global warming, pollution, and shifts in aquaculture practices. Appropriate diagnosis and treatments are essential to minimize the impact of parasitic infections. In this background, this chapter provides information on parasite diversity, prevalence, and life cycle, as well as a few research thrust areas of commonly encountered parasites in freshwater aquaculture.

Keywords

Parasitic disease · Freshwater environment · Pathogenesis · Management

S. Mohanty · P. K. Sahoo (✉)
ICAR-Central Institute of Freshwater Aquaculture, Bhubaneswar, Odisha, India
e-mail: msnatashree@yahoo.com; pksahoo1@hotmail.com

R. Kooloth Valappil et al. (eds.), *Aquatic Animal Health Management*,
https://doi.org/10.1007/978-981-96-7987-4_8

1 Introduction

Global climatic change and intensification in the aquaculture system are two major factors predicted to be responsible for increased incidences of diseases. Increased anthropogenic activities coupled with urbanization, industrialization, and intensification of aquaculture have encouraged a huge quantum of nutrients into the culture system that facilitates the process of eutrophication. The generalist parasites' virulence and intensity increase in eutrophied conditions mostly (Johnson et al. 2010). These effects are mostly determined by the presence of a large number of intermediate hosts, efficiently changing the parasite life cycle (McKenzie and Townsend 2007). Each fish species probably hosts a parasite during its life cycle (Paladini et al. 2017). Parasites frequently exist in a dynamic equilibrium with their hosts, and any modulation in the environment can alter the parasite/host equilibrium, leading to disease incidence. The parasites can bring mechanical changes at the tissue level that are reflected in terms of proliferation and fusion of gill lamellae, physiological changes including cell hyperplasia, immunomodulation, and changes in the fish behaviour and reproductive capacity (Buchmann and Lindenstrom 2002). Parasitic diseases cause serious socio-economic, ecological, and welfare consequences and direct financial loss in terms of mortality.

Diseases cause huge economic losses to aquaculture around the globe (Mustafa et al. 2001; Johnson et al. 2004; Sahoo et al. 2013; Monir et al. 2015). The economic loss due to diseases is estimated to be US$ 1.05 to US$ 9.58 billion/year worldwide (Shinn et al. 2015). In a study conducted during 2014–2018 in India with a total of 219 cases of fish mortalities investigated under a passive surveillance programme, the parasites were marked as the major disease-causing agents (Sahoo et al. 2020). Out of 164 cases of parasitic infections, mixed parasitic infections (37.80%) were found to be primary aetiological agents leading to mortality, followed by single infection with *Argulus* sp. (19.51%), *Dactylogyrus* sp. (18.90%), myxosporeans (12.80%), trichodinosis (3.04%), and other parasites (4.88%) that include *Lernaea, Epistylis, Glossiphonia complanata, Metanophrys sinensis,* and *Zoothamnium* spp. Many disease processes are complicated by the complex interaction of several factors, making it difficult to calculate the actual costs attributable to each pathogen. Hence, active and passive surveillance can help in estimating economic loss due to parasites in Indian aquaculture.

Parasites with direct or simple life cycles with a water-borne infectious stage have a tendency to govern in aquaculture. This is evident by infections with protistans, monogeneans, parasitic crustaceans, myxozoans, and digeneans. Parasites also encourage co-infections simultaneously, and multiple infections increase morbidity and mortality. The present chapter is an attempt to offer an account of different classes of parasites affecting freshwater aquaculture.

2 Major Parasites, Their Life Cycle, and Diagnosis

2.1 Protista

Protista comprises major groups of eukaryotic parasites with definite morphological characteristics and infection patterns. Having wide adaptability to the environment, protists significantly affect the health of aquatic animals. Contact and cohabitation are the major transmission routes for Protistans with simple life cycles and reproduction by binary fission. One interesting fact about the *Trypanosoma* life cycle is that it utilizes leeches as vector and/or intermediate hosts. Diagnosis for this parasite is based on microscopy that seeks expertise in fixing, staining, and observation under bright-field and differential interference contrast microscopy, phase contrast, and electron microscopy. Nevertheless, molecular tools can also add confirmation of this parasite.

Here are a few groups of Protista with examples as described by Paladini et al. (2017).

Flagellates including *Amyloodinium*, *Cryptobia*, *Gymnodinium*, *Hexamita*, *Ichthyobodo*, *Piscinoodinium*, *Spironucleus*, *Trypanosoma*, and *Trypanoplasma* are characterized by the presence of flagella for movement.

Amoebae are protists that use pseudopodia for movement. For example, *Vermamoeba* (formerly *Hartmannella*), *Paramoeba*, and *Thecamoeba.*

Alveolates include *Perkinsus* that have small cavities (alveoli) beneath their cell membranes providing flexibility to the pellicle and also taking part in osmoregulation.

Apicomplexans include *Aggregata*, *Eimeria*, *Goussia*, *Haemogregarina*, and *Margolisiella* (formerly *Pseudoklossia*) are typically parasitic protists with an apical complex that facilitated them to invade host cells.

Cercozoans are a diverse group of protists that includes *Haplosporidium*, *Bonamia*, *Marteilia*, and *Paramarteilia*.

Ciliates like *Anophryoides*, *Apiosoma*, *Capriniana*, *Chilodonella*, *Cryptocaryon*, *Epistylis*, *Ichthyophthirius*, *Tetrahymena*, and *Trichodina* have hair like structures called cilia, which are used for their movement.

2.2 Myxozoa

Myxozoans belong to the phylum Cnidaria, which comprises of two important classes, i.e., Myxosporea and Malacosporea, which are obligate endoparasites that cause significant mortalities and economic loss. Approximately 2500 species under 64 genera have been described in hosts from different regions (Atkinson et al. 2015). The major species found in Indian aquaculture include *Myxoblous* (Fig. 1a–c), *Thelohanellus* and *Henneguya* spp.

The life cycle of a few myxozoans is interesting, typically involving a vertebrate and an invertebrate host, and many are still unknown. However, a few species can

Fig. 1 (**a** and **b**) Myxoboliasis in carp. (**c**) Spores of *Myxobolous* spp.

transmit without the involvement of a second host (Estensoro et al. 2010). Their typical development within a host is through the production of spores, which actually act as the source of infection following cell-within-cell division called pluricellularity. Sexual reproduction occurs in invertebrate hosts, including oligochaetes, polychaetes, and bryozoans. Vertebrate hosts typically consist of fish and amphibians (Canning and Okamura 2003).

Diagnosis of Myxozoans requires microscopic analysis of infected fish with gross signs of discoloration and lesions. The presence of spores (responsible for infection) is the diagnostic marker for myxozoan infection. Phase contrast or differential interference microscopy can be employed to examine fixed or fresh spores. Blood and tissue smears or imprints stained with May–Grunwald–Giemsa, Giemsa, or silver nitrate are effective for detecting both spores and extrasporogonic stages (Parker and Warner 1970; Lom et al. 1983; Reed et al. 2002; Longshaw et al. 2003). Scanning electron microscopy (SEM) can also be helpful for absolute observation of surface morphology, for example, species like *Chloromyxum*. Additionally, molecular diagnostic tools like ribosomal or specific gene PCR, phylogeny analysis, RFLP (Restriction Fragment Length Polymorphism), or LAMP-PCR are potential tools for diagnosing myxozoan infections.

2.3 Monogenea

Monogeneans belong to phylum Platyhelminthes and are important parasites infecting fish with approximately 750 genera and 5500 species. These parasitic flatworms cause substantial economic loss in fish, commonly infecting skin and gills. These parasites have two subclasses i.e., Monopisthocotylea and Polyopisthocotylea. Monogeneans exhibit identical reproductive traits and distinct adaptations for secured attachment to the host. They can give birth to young ones (viviparous) or produce eggs (oviparous). Individual parasite possesses both male and female reproductive organs. However, many monogeneans are protandrous, in which the individual first functions as male followed by female. Specialized attachment features make them unique from other parasites. The muscular posterior attachment organ is called the opisthaptor, consisting of sclerotized hooks, connecting bars, and clamps for continuous nourishment from the host. Two genera, *Dactylogyrus* (gill fluke) and *Gyrodactylus* (skin fluke) are major monogeneans that are common in freshwater aquaculture systems in India, causing infection in fish. Monogeneans are mostly host-specific, and their sizes vary from 200 μm to 4 cm in length. They mostly graze on the epidermis; however, *Dactylogyrus* (gill fluke) is a blood feeder.

These parasites are mostly oviparous with an exception, Gyrodactylidae, which are viviparous and have a single definitive host in their life cycles. The eggs may be operculated with a lid or non-operculated. They may bear one or two polar processes or filaments which help in attachment and movement. Eggs may possess hooks terminally, bear spines, or sticky droplets for adherence. Eggs display different buoyancy patterns for transmission and can be laid singly, in chains, or deposited as egg masses. They may be sinking or anchored to the substrate. The eggs hatch into oncomiracidium, which may be eyed and ciliated; it swims fast to find a suitable host on which it matures. They are not zoonotic and, hence, do not pose any risk to human health. The shapes and configuration of the opisthaptor and copulatory apparatus are important keys in the identification of species.

Monogeneans can be diagnosed from freshly collected gills or skin samples under a microscope with or without the aid of stains like haematoxylin or carmine-based stains. However, molecular tools, as described earlier, seem to be more promising now-a-days for species-level differentiation.

2.4 Digenea

2.4.1 Digeneans

Digeneans belong to the class trematodes of phylum Platyhelminthes, commonly called flukes/flatworms. These are the largest group of metazoan endoparasites, mainly infecting the digestive system or associated organs. More than 10,000 digeneans are reported worldwide. A few important digenetic trematodes are described below:

(i) *Diplostomum spathaceum* was initially described as *Distoma spathaceum* from the intestine of gulls, subsequently termed eye fluke. It causes visual impairments and blindness in fish affecting overall survival and susceptibility of fish.
(ii) *Posthodiplostomum cuticola* causes black spot (Fig. 2) disease in muscle and tissue.
(iii) *Centrocestus formosanus*: This fluke targets the gills of cultured tropical fish, potentially causing respiratory stress and secondary infections.
(iv) *Clinostomum complanatum* leads to the development of yellow cysts in the body cavity and muscles (Aohagl et al. 1992).

These parasites are major threats to aquaculture and reduce the growth and marketability of infected fish. They have an indirect life cycle of 2–4 hosts. The developmental stages include eggs, miracidia, sporocysts, rediae, cercariae, and

Fig. 2 Black grub disease (Diplostomiasis) in fish

metacercariae. Adult digeneans reside in the gastrointestinal tract of the final vertebrate host, mostly fish, and release eggs into the environment. These eggs hatch into ciliated larvae called miracidia, which search for the first intermediate host, a mollusc, typically a snail. Miracidia penetrate snails and develop into asexual stages, either sporocysts or rediae, which multiply and produce more larvae. The rediae/sporocysts are further developed into cercariae, which are highly motile and search for a second intermediate host that may be another fish or a crustacean. In the second intermediate host, cercariae migrate and encyst to form metacercariae, which ultimately infect the final vertebrate host. However, fish to serve as intermediate or definitive host, it is determined by the parasite species and appropriate host availability to successfully complete the life cycle. When fish are intermediate hosts, the metacercarial stages infect multiple organs like eyes, gills, skin, musculature, heart, and kidney.

Preservation, fixation, and staining are important for appropriate diagnosis. Ideally, it is advised to examine these parasites alive following the collection of samples. Ethanol (70–95%) can be used as a fixative for subsequent morphological and molecular studies. Several staining methods can be used to visualize their internal anatomy, including iron acetocarmine (Georgiev et al. 1986; Kostadinova et al. 2003), Semichon's acetocarmine or Ehrlich's haematoxylin (Choudhury and Nelson 2000), and van Cleave's haematoxylin (Overstreet 1976). After staining, specimens are dehydrated using a graded series of alcohol to remove water. The specimen is then subjected to a clearing medium like clove oil, which makes the specimen more transparent and easier to observe. Mounting was finally done with flattened specimens and permanently mounted in Canada balsam or DPX, often used as the best mountant to preserve the specimen for long-term study. These methods, with suitable precautions, can help parasitologists accurately identify parasites and also set a package of management strategies for their control.

2.5 Cestoda

Cestodes or tapeworms are widely recognized and distributed due to their zoonotic potential to infect humans following consumption of undercooked food. These are endoparasites. They belong to the family Platyhelminthes and class cestode that includes 13 orders and approximately 5000 species, among which the Bothriocephalidea, Tetraphyllidea, Diphyllobothriidea, Lecanicephalidea, Proteocephalidea, Caryophyllidea, Pseudophyllidea, and Trypanorhyncha are of major concern. They have a unique and distinct anatomy with a flattened body, anterior scolex, a neck, and the strobila. The head-like structure is called a scolex that attaches to the host's intestine with suckers or hooks. The neck is a short segment behind the scolex, where new proglottids are produced. Strobila is the long, segmented body, made up of many proglottids having both male and female reproductive organs. Cestodes feed through tegument consisting of microtriches and filtering tissue.

The most important and harmful parasite group of zoonotic importance is Diphyllobothriidea (*D. latum* and *D. dendriticum*). Dephyllobothriasis is a fish-borne and parasitic disease in humans. The cestode life cycles are invariably heteroxenous. This means it requires the involvement of multiple hosts, requiring a definitive host and at least one or two intermediate hosts. Fish can act either as an intermediate host, where fish can produce different parasite larval stages (coracidium, procercoid, and plerocercoid), or as a definitive host, where the adult cestode secures itself and lays eggs. Several tapeworm species can have different tissue tropisms. The first intermediate hosts are invertebrate organisms, mainly aquatic crustaceans. The adult parasites prefer to live in the fish intestine and negatively affect metabolism, causing damage to the intestinal mucosa and immune suppression.

The cestodes can be fixed in 70–99% ethanol. The parasite can be identified by recording scolex, number and shapes of suckers, hooks, and bothria. However, molecular tools are more advantageous for species identification.

2.6 Arthropoda

Arthropoda is a diverse phylum that represents mainly crustaceans like copepods and isopods, and causes significant infections and subsequent mortality. The subphylum Crustacea includes 971 families, 9598 genera, and around 64,663 extant species, of which an estimated 8000+ species have parasitic stages. The parasite most studied under the phylum is the genus *Argulus,* which is responsible for the huge loss in freshwater aquaculture. There are approximately 129 species of *Argulus* found in freshwater and brackishwater environments worldwide. *A. siamensis,* and *A. japonicus* are the two important species observed to infest almost all major farmed carps in India. Argulosis has proven costly to carp producers in India and is capable of causing significant economic loss unless controlled by timely intervention. The estimated loss due to *Argulus* infestation amounts to approximately Rs.

29,524/– ha/year in the Indian subcontinent. It is evident that rohu (*Labeo rohita*) and catla (*Catla catla*) are the most susceptible species, whereas bata (*Labeo bata*), silver carp (*Hypophthalmichthys molitrix*), and grass carps are moderately resistant species to *Argulus* infestation in polyculture system. Mrigal and other medium carps are also found to be susceptible. Although this infestation is mostly noticed in yearlings in grow-out ponds, the parasite invariably infects all size ranges starting from fry onwards. The attached parasites pierce the skin of the host by their modified mouth parts and feed on blood, mucus, and dead cells while simultaneously releasing toxic anticoagulants into the fish. Subsequent secondary infections with bacteria or fungus at the feeding site make the fish more vulnerable to stress, leading to growth suppression and heavy mortality.

2.6.1 Life Cycle

(i) **Parasitic copepods**

The parasite does not require an intermediate host to complete its life cycle; hence, it has a direct life cycle. However, their successful transmission and spread of disease depends on the mobile parasitic stage and the availability of the host. The parasitic copepods are capable of moving over the surface of their host. This mobility allows the parasite to search for suitable places on the host, either to attach or embed itself. The parasites are attached to the host by specialized attachment structures (Fig. 3).

(ii) **Isopods**

Isopods are small crustaceans, and some can be parasitic especially the aquatic isopods. Among 95 known families of Isopoda, a few, including Bopyridae, Cryptoniscidae, Cymothoidae, Dajidae, Entoniscidae, Gnathiidae, and Tridentellidae, are parasitic. They get attached to fins, gills, skins, and also the mouth, causing lesions and open ulcers. Adult isopods are dorsoventrally flattened crustaceans that lack a carapace. They have a head fused to the first thoracic somite. Isopods have sessile eyes and a thorax with seven somites. Each somite is equipped

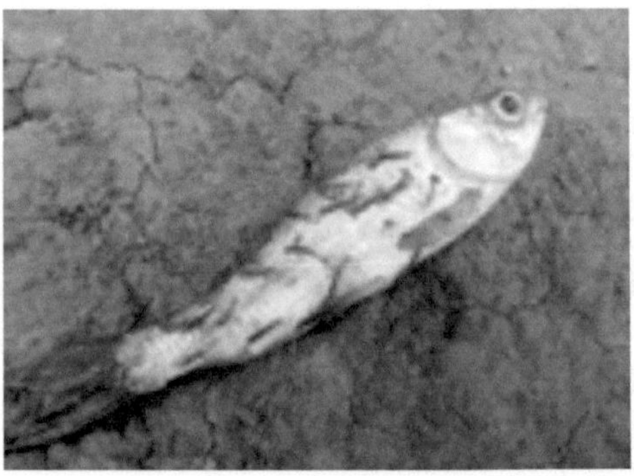

Fig. 3 Infestation of *Lernaea* spp. (anchor worm)

Fig. 4 *Argulus* spp. on the body surface of *Labeo rohita*

with a pair of uniramous swimming legs. Isopods are usually dioecious, with the exception of cymothoid isopods (primarily marine) that are protandrous hermaphrodites. The first male parasite attached to fish changes into a female, which may then prevent the development of other males into females. Gravid females release eggs into a brood pouch called a marsupium, within which the eggs embryonate, hatch, and undergo two moults to juvenile stages.

(iii) **Branchiurans**

The branchiurans such as *Argulus* (Fig. 4) are highly mobile and temporary parasites of dioecious nature (separate sexes), moving on, off, and between their hosts to feed and lay eggs. Argulids are ovoid with a dorsoventrally flattened carapace and a pair of first maxillae, which are modified suctorial organs. They possess compound eyes, preoral stylet used for piercing, four pairs of swimming legs, and a tubular mouth with denticulated mandibles. Branchiurans feed on mucus, tissue fluids, and the blood of their hosts. Branchiuran lays eggs in rows on hard substrate available in the water bodies, including dyke grass, dead fish, or snails. Egg development is temperature-dependent and cohorts of eggs may overwinter. The first larval stage of Argulus is the metanauplius stage, which parasitizes a host within 2–3 days of hatching or else die. The parasite then passes through 11–12 moults until it reaches sexual maturity.

2.7 Nematoda

Nematodes (roundworms), although widespread in nature, are less infectious in aquaculture systems. Invertebrates including molluscs and crustaceans, serve as intermediate hosts bearing the larval stages of parasites. Adult nematodes are mostly found in the intestine of fish, but some species of *Philometra* are found in gonads; *Hysterothylacium fabri* in the gall bladder and pyloric caeca of migratory fish, and *Anguillicoloides crassus* in the swim bladder of eels. A diverse range of organ tropism has been marked in the larval stages of roundworms. The roundworms have cylindrical bodies with an anterior cephalic extremity consisting of structures such as lips, teeth, and/or papillae, a posterior tail, an elongated body that can be covered by spines and/or ridges, and sensory structures called phasmids. The syncytial epidermis secreting cuticle that covers the body. The pseudocoelomic cavity contains a digestive and an excretory system. Nematodes show sexual dimorphism and unique morphological features in both sexes that help in their identification (Paladini et al. 2017). *Gnathostoma* and *Capillaria* spp. are of zoonotic importance that spread the infection through eating of raw or undercooked fish.

Nematodes have a complex life cycle that involves a fish as either an intermediate or definitive host. When the definitive host is a fish, typically, there is only a single intermediate host. If the final host is not a fish, then there are typically two intermediate hosts, one of which is a fish. A few fish species can also act as a paratenic host, in which the parasite does not develop, but the mechanism contributes to the spread of the parasite in the aquatic environment. A few nematode species, such as *Capillaria* spp., can transmit directly through fish-to-fish contact.

2.8 Acanthocephala

Acanthocephalans are commonly called thorny-headed/spiny-headed worms. They have unique biological features. Anatomically, they have proboscis with multiple hooks, which ensures a strong attachment to the host intestine. They lack a digestive system and take nutrients directly through the cuticular layer. Acanthocephalans have distinct male and female morphological characteristics and, hence, are sexually dimorphic. Adult parasites reproduce sexually within the intestine of their definitive host. Fish also play a role as definitive hosts to species such as *Pomphorhynchus laevis* in trout and *A. anguillae* in European eels, *Anguilla anguilla* (Schmidt and Nickol 1985). Fertilized eggs develop inside the female to the acanthor stage, then released within its host's faecal waste into the aquatic environment. Embryonated eggs are ingested by a suitable intermediate host, which hatches into an acanthella stage and subsequently encysts in its host's tissues to develop larval cystacanth. The fish can act as a paratenic host/non-essential host for transmission of infections like *Acanthocephalus anguillae*.

3 Co-infection: A Current Trend in Increasing Mortality in Fish

Co-infection, synonymously called multiple infections, dual infections, secondary infections, polymicrobial diseases, complicated infections, concurrent infections, mixed infections (Fig. 5), and superinfections due to contact with two or different pathogens either by simultaneous or after succeeding disease episodes of the primary pathogen, i.e., as secondary infections is a natural phenomenon (Bakaletz 2004). Co-infections can dictate the disease course and the severity of the disease through synergistic and, more rarely, antagonistic interactions (Hany et al. 2020). Xu et al. (2007) reported the presence of *Gyrodactylus* and *Streptococcus iniae* in tilapia co-infected by both parasite and bacteria, leading to 42.2% mortality, which

Fig. 5 Mixed parasitic infection in carp

was much higher than 6.7% loss in the non-parasitized fish exposed to *S. iniae* challenge. Many reports have proved that the presence of a primary pathogen reduces the resistance to infection to a secondary pathogen. For example, in rainbow trout, primary infection with *Myxobolus cerebralis* leads to immunosuppression, leading to secondary infection by *Yersinia ruckeri* (Densmore et al. 2004). In another study, when channel catfish were co-infected by *I. multifiliis* and fluorescent *E. ictaluri*, all tomonts were found to carry the fluorescent bacteria. *E. ictaluri* could able to survive and multiply inside the tomonts, resulting in higher mortalities in infected fish due to their synergistic action (Xu et al. 2012). In a survey conducted by Sahoo et al. (2020) in freshwater aquaculture systems of two states of India, mixed parasitic infections (37.80%) were found to be the major cause of disease, followed by other pathogens. *I. multifiliis* is a ciliated ectoparasitic fish protozoan and causes significant losses in freshwater fish worldwide. It increases the chances of secondary bacterial invasion and fish mortality by damaging the epithelium of the gills and skin. Shoemaker et al. (2012) explored the effect of *I. multifiliis* parasitism on survival, haematology and bacterial burden of channel catfish exposed one day before to *Edwardsiella ictaluri,* the causative agent of enteric septicemia of catfish. Higher bacterial loads in different organs with higher mortalities were detected in the co-infected group (71.1%) as compared to single-infected groups.

Sahoo et al. (2020), while investigating the root cause of large-scale mortality of koi carps (*Cyprinus carpio* var. *koi*) in ornamental fish farms, recorded *Trichodina* infestation in the gills and bacterium *Aeromonas hydrophila* in kidney tissue of moribund fish along with carp edema virus (CEV). Further, the authors concluded that CEV is the primary etiological agent that causes Koi sleepy disease, and secondary infection by bacterium and parasites aggravates the condition. While looking into a mass mortality event in adults of *Channa striata* heavily infected with *Ichthyophthirius multifiliis* on the body surface, *Zschokkella auratis* (a myxosporean) in the brain tissue, *Acinetobacter baumannii* in eye fluid/tissue and *Aeromonas sobria* and *Klebsiella pneumoniae* in blood samples were detected (Paul et al. 2020). Interestingly, samples from infected fish also revealed multiple parasitic pathogens, viz., *Epistylis* sp., *Trichodina* sp., *Argulus siamensis,* and *Ergasilus* sp. Similarly, in a disease outbreak in adult koi carp recorded by Swaminathan et al. (2016), CEV-infected fish gills also revealed *Dactylogyrus* sp. and bacteria, viz.

Aeromonas sobria, Klebsiella pneumoniae, Proteus penneri, Shewanella decolorationis and *Escherichia coli*. Sahoo et al. (2016) similarly noticed cyprinid herpes virus-2 (CyHV-2) in association with multidrug-resistant *Aeromonas hydrophila* infection in goldfish *Carassius auratus*, from commercial farms. The presence of these opportunistic bacteria or parasites can never be ignored as their association may enhance the pathogenesis when host immunity is compromised. Hence, co-infections by multiple parasites have a great impact on the host-parasite ecology and can alter the duration and the severity of the disease (Klemme et al. 2016), and it may even reduce the efficacy of vaccination. A list of different types of co-infections (homologous and heterologous pathogens) reported in freshwater aquaculture is summarised in Tables 1, 2 and 3.

Table 1 Co-infections of homologous parasitic pathogens in different freshwater fishes

Host species	First pathogen	Second pathogen	Type of interaction	References
Farmed brown trout	Five myxozoan species (*Tetracapsuloides bryosalmonae, Sphaerospora truttae, Chloromyxum schurovi, Chloromyxum truttae*)	Three myxozoan species: *T. bryosalmonae, S. truttae* and *C. schurovi*	Antagonistic	Holzer et al. (2006)
Farmed brown trout, *Salmo trutta*	*Tetracapsuloides bryosalmonae* (Myxozoa)	*Chloromyxum schurovi* (Myxozoa)	Antagonistic	Peeler et al. (2008)
Wild brown trout	*Tetracapsuloides bryosalmonae*	*Raphidascaris acus*	Synergistic	Schmidt-Posthaus et al. (2012)
Carassius carassius	*Trypanosoma carassii*	*Trypanoplasma*	Synergistic	Anastasiia et al. (2014)
Carassius carassius	*Dactylogyrus intermedius*	*Flavobacterium columnare*	Synergistic	Zhang et al. (2015)
Abramis brama	*Trypanosoma abramidis*	*Trypanoplasma*	Antagonistic	Anastasiia et al. (2014)
Scardinius erythrophthalmus	*Trypanosoma scardinii*	*Trypanoplasma*	Antagonistic	Anastasiia et al. (2014)
Cobitis taenia	*Trypanosoma cobitis*	*Trypanoplasma*	Synergistic	Anastasiia et al. (2014)
Silurus glanis	*Trypanosoma markewitschi*	*Trypanoplasma*	Antagonistic	Anastasiia et al. (2014)
Sander lucioperca	*Trypanosoma lucioperca*	*Trypanoplasma*	Antagonistic	Anastasiia et al. (2014)
Perca fluviatilis	*Trypanosoma percae*	*Trypanoplasma*	Synergistic	Anastasiia et al. (2014)
Channa striata	*Epistylis* sp., *Trichodina* sp.,	*Argulus siamensis, Ergasilus* sp.	Synergistic	Paul et al. (2020)

Table 2 Co-infections of heterologous pathogens with parasites in different freshwater fishes

Host species	First pathogen	Second pathogen	Type of interaction	References
Rainbow trout, *Oncorhynchus mykiss*	*Myxobolus cerebralis* (Myxozoa)	*Yersinia ruckeri*	Synergistic	Densmore et al. (2004)
Nile tilapia, *Oreochromis niloticus*	*Gyrodactylus niloticus (Helminth)*	*Streptococcus iniae*	Synergistic	Xu et al. (2007)
Goldfish, *Carassius auratus*	*Dactylogyrus intermedius* (Helminth)	*F. columnare*	Synergistic	Zhang et al. (2015)
Rainbow trout, *Oncorhynchus mykiss*	*Argulus coregoni* (Fish louse)	*F. columnare*	Synergistic	Bandilla et al. (2006)
Channel catfish, *Ictalurus punctatus*	*Ichthyophthirius multifiliis* (protozoa)	*E. ictaluri*	Synergistic	Xu et al. (2012)
Nile tilapia, *Oreochromis niloticus*	*I. multifiliis*	*S. iniae*	Synergistic	Xu et al. (2009)
Channel catfish, *Ictalurus punctatus*	*I. multifiliis*	*A. hydrophila*	Synergistic	Xu et al. (2012)
Channel catfish, *Ictalurus punctatus*	*E. ictaluri*	*I. multifiliis*	Synergistic	Shoemaker et al. (2012)
Channel catfish, *Ictalurus punctatus*	*S. iniae or S. agalactiae*	*Trichodina* sp.	Synergistic	Evans et al. (2007)
Channel catfish	Trematode *Bolbophorus* spp.	*Edwardsiella ictaluri*	Synergistic	Labrie et al. (2004)
Channa striata	*Ichthyophthirius multifiliis, Zschokkella auratis,*	*Acinetobacter baumannii*	Synergistic	Paul et al. (2020)
Tilapia guineensis	*Clinostomum tilapiae, C. complanatum* and *Euclinostomum heterostomum*	–	Synergistic	Echi et al. (2009)
Tilapia guineensis	*Clinostomum tilapiae, C. complanatum* and *Euclinostomum heterostomum*	–	Synergistic	Echi et al. (2009)

(continued)

Table 2 (continued)

Host species	First pathogen	Second pathogen	Type of interaction	References
Tilapia zilli	*C. complanatum, C. tilapiae* and *Euclinostomum hetereostomum*	–	Synergistic	Echi et al. (2012)
Nile tilapia	Four specimens were infected with *Centrocestus formosanus* and *Austrodiplostomum compactum*. One specimen with *A. compactum* and *Drepanocephalus* sp. One specimen with *A. compactum* and *Ribeiroia* sp.	–	Synergistic	Pinto et al. (2014)
	G. cichlidarum and *C. sclerosus*	–	Synergistic	Zhi et al. (2018)
	G. niloticus and *S. iniae*	–	Synergistic	Xu et al. (2007)
	I. multifiliis and *S. iniae*	–	Synergistic	Xu et al. (2009)
	F. columnare and *Myxobolus tilapiae*	–	Synergistic	Eissa et al. (2010)
	Ectoparasites such as *Trichodina heterodentata, G. cichlidarum* and *I. multifilis* Bacteria such as *S. iniae*	–	Synergistic	Martins et al. (2011)
Hybrid tilapia (*O. niloticus* × *O. aureus*)	*F. columnare* after parasitism by *I. multifiliis*	–	Synergistic	Xu et al. (2014)
Hybrid *tilapia (O. niloticus* × *O. aureus)*	*F. columnare* and *Trichodina* sp.	–	Synergistic	Xu et al. (2015)
Nile tilapia	Bacteria such as *V. alginolyticus, V. harveyii, E. faecalis* and *A. hydrophila* Ectoparasites such as *I. multifiliis, Trichodina* sp. and *Gyrodactylus* sp	–	Synergistic	Nofal and Abdel-Latif (2017)
Nile tilapia	*A. hydrophila* and *G. cichlidarum*	–	Synergistic	Hany et al. (2020)

Table 3 Parasitic and viral co-infections

Salmonids	*Tetracapsuloides bryosalmonae*	Viral haemorrhagic septicaemia (VHS)	Antagonist	Gorgoglione et al. (2019)
Koi carps *(Cyprinus carpio koi)*	Carp edema virus (CEV)	*Trichodina* sp., *Aeromonas hydrophila*	Synergistic	Sahoo et al. (2020)
Cyprinus carpio koi	Carp edema virus (CEV)	*Dactylogyrus* sp. and bacteria viz. *Aeromonas sobria, Klebsiella pneumoniae, Proteus penneri, Shewanella decolorationis, and Escherichia coli* isolated from the gills	Synergistic	Swaminathan et al. (2016)

3.1 Parasitic Co-infections

3.2 Parasitic and Bacterial Co-infections

4 Climate Change and Other Stress Factors for Parasite Dispersal

According to the Intergovernmental Panel on Climate Change (IPCC), mean surface temperatures are expected to rise significantly over the coming decades due to increased greenhouse gas emissions. The projected temperature increase of 1.4° to 5.8 °C by 2100 is going to be the greatest challenge for mitigation efforts (Houghton et al. 2001). As fishes are poikilothermic animals, their physiology is directly affected by the ambient temperature; hence, water temperature changes can cause stress. Changes in the climate or water temperature can lead to changes in the pathogenicity or virulence of aquatic pathogens. Further, climatic changes can lead to changes in the host range of the pathogens or changes in the geographic distribution of the host and the pathogens. The multiplication/replication rate of parasites and other pathogens is greatly affected by temperature. It is predicted that climate change has an important effect on parasitic diseases in fresh- and marine water (Marcogliese 2008). Rising temperature increases parasite virulence and transmission rates (Kutz et al. 2005; Larsen and Mouritsen 2014). Hence, climate change coupled with multiple stressors during culture practices might affect the life cycle of parasites and host fitness (Paaijmans et al. 2009). Parasite infection and transmission depend on host health status, water quality, temperature, and the presence of intermediate hosts (Marcogliese 2001). It is expected that parasites having a direct life cycle will become more threatening as they will complete their life cycle quickly. Temperature considerably influences various stages of parasitic life cycles, including egg release, development, and infectivity in both intermediate and definitive hosts. Higher temperature fastens the life cycle, hence it is one of the approaches to control parasites to close the cycle and further prevent the spread of the parasite population. Similarly, temperature has a role in controlling host behaviour and physiology, including feeding patterns, resistance to infection, and the ecological dynamics between the parasite and its host as per their requirement (Marcogliese 2001). However, if the conditions are not suitable, parasitic infection is prolonged, leading to more generations of parasites (Callaway et al. 2012). As temperature directly affects the developmental rates of *Argulus* spp., higher water temperatures lead to more rapid population cycles. In addition, depending on the ambient environmental temperature, female *A. coregoni* produce two types of eggs: eggs that are laid during the early reproductive season and hatch immediately and diapausing eggs, which are produced later in the season and show delayed hatching (Hakalahti et al. 2006). Likewise, temperature also controls the transmission and development of *Diplostomum spathaceum* at all stages of the life cycle, except during sexual reproduction in the bird (Hakalahti et al. 2006). Another factor, like water depth, also decides parasite richness in a water body. Low water depth displays a low flow rate, and this allows more free-swimming infective stages. Thus, thermal stress, in combination with low water levels, increases the parasite incidences and immunosuppression of the host (Marcogliese 2001). Climate change has potential effects on parasites' life cycle, transmission, fitness, and hatching directly and indirectly due to stressed fish hosts. A summary of the effect of temperature on the parasite life cycle, host, and transmission process is provided in Table 4.

5 Parasite Management

5.1 Chemical Parasiticides

Chemical parasiticides provide an immediate effect on controlling parasites through different modes and targets of action. Most of the parasiticides target the nerve or muscle of the parasite, causing paralysis, growth inhibition, moult retardation, fecundity modulation, or direct killing by unknown mechanisms. The commonly used parasiticides are classified as acetylcholinesterase (AChE)-inhibitors that include carbamates and organophosphates (e.g., azamethiphos, triclorfon), sodium channel modulators, such as pyrethroids and pyrethrins (e.g., cypermethrin, deltamethrin), glutamate-gated chloride channel (GluCl) allosteric modulators (e.g., emamectin benzoate), nicotinic acetylcholine receptor (NAChR) competitive modulators such as neonicotinoids, NAChR allosteric modulators (e.g., active ingredient spinosad) and NAChR channel blockers, GABA-gated chloride channel blockers, which mostly act on nerve and muscle of parasites. The second group of compounds that affect growth and development, for example, are chitin synthesis inhibitors, juvenile hormone mimics, inhibitors of mite growth factors, disruptors of moulting process, ecdysone receptor agonists, inhibitors of acetyl CoA carboxylase, etc. The albendazole group of drugs is commonly used for endoparasites (Martin et al. 1997).

Table 4 Effect of temperature on parasite life cycle, host, and transmission process

Effect on parasite	Effect on host	Effect on transmission process	References
Rapid embryonic development and hatching	Altered feeding and ecology	Earlier reproduction	Marcogliese (2008)
Fast development and maturation	Altered behaviour	More generation	Marcogliese (2008)
Decreased longevity of larvae and adult	Reduced host resistance	Potential transmission	Marcogliese (2008)
Increased mortality after tolerance limit	–	–	Marcogliese (2008)
Parasite metabolism	–	Higher number of transmission stages, which lead to higher parasite fitness and quick transmission in a single outbreak.	Lohmus and Bjorklund (2015)
Rapid spread of parasites	–	–	Lohmus and Bjorklund (2015)
Less occurrence of disease when temperature in not optimum	–	–	Lohmus and Bjorklund (2015)
Warming increases prevalence of parasites	–	–	Harvell et al. (2002); Marcogliese (2008)
Increase in water temperature at higher latitudes could result in an increase in parasite-associated mortality	'Involuntary' over-exploitation of the host.	–	Lohmus and Bjorklund (2015)
–	Increased water temperature increases major histocompatibility complex (MHC) class IIB, involved in heat tolerance	–	Lohmus and Bjorklund (2015)

5.2 Alternative Treatments Against Fish Parasites

Though chemotherapy is an option to control parasites immediately in aquaculture, indiscriminate use of chemicals leads to the selection of resistant organisms and damage the environment. The transfer of chemical residues to the human chain through food is a big challenge. Thus, consumption of farmed fish that have been treated with chemicals and drugs may pose human health risks (Hashimoto et al.

2016; Soares et al. 2016). Therefore, the use of chemical pesticides to control fish diseases is often restricted in many countries (Ling et al. 2015). Hence, alternative approaches using Essential Oils (EOs) are used as potential alternative sources of therapeutics. The use of EOs as bioactive products is a promising alternative source to control fish parasites. More than 3000 EOs are known, and among those, very less approx. 0.4% have been tested against fish parasites. Thus, more studies are required to explore using varieties of EOs for controlling parasites (Tavares-Dias 2018). Therapeutic baths in *Lippia sidoides* and *Mentha piperita* by Hashimoto et al. (2016) against monogenean gill parasites *Cichlidogyrus tilapiae, Cichlidogyrus thurstonae, Cichlidogyrus halli, and Scutogyrus longicornis* in Nile tilapia has been successfully studied. The use of medicinal plants and their bioactive compounds has been a sustainable approach to curing fish ectoparasites. Recently, Puspha et al. (2019) evaluated the efficacy of aqueous and organic solvent extracts of *Azadirachta indica* leaf against *Argulus japonicus* in *Carassius auratus.*

5.2.1 Non-chemical Approaches

Non-chemical approaches are basically part of management strategies and husbandry practices, which can minimise the occurrence and establishment of parasites and also its spread. Integrated pest management aims to identify the source of infestation and management during routine cultural practices.

Cleaning and brushing of tank systems and proper and frequent disinfection practices can prevent the growth of biofilms and organic loading. This approach can minimize the proliferation of parasites or ensure the dislodge of different stages of parasites like *Ichthyophthirius multifiliis*. Biosecurity and bird fencing can prevent the entry of disease carriers and as ingress of pathogens. Proper feed management can prevent stress and provide immunity to fight against parasite establishment and further propagation. Dietary supplements, herbal extracts, and natural immunostimulants can also prevent disease on set due to improved immunity and resistance (Harikrishnan et al. 2011). Handling stress and changes in water parameters coupled with the breakdown of immunity can encourage various opportunistic bacterial and protistan pathogens to grow. Biological control, like grass carps, ducks are used to control vegetation and snail populations, respectively, in ponds. This helps in cleaning vegetation and also parasite carriers, which are responsible for the spread of argulus. Other physical control measures include the periodical removal of bamboo or wooden frames onto which parasites, such as *Argulus*, can lay their eggs. A wide variety of cleaner fish, like wrasse species, are used to control sea lice infections on farmed marine salmon. Likewise, the maintenance of dykes and banks around pond sites to withstand flood events helps to minimize the likelihood of stock loss, mixing, or the introduction of undesirable species. The use of mechanical filters can maintain optimal water quality. The use of certified quality seeds is a prerequisite for maintaining a healthy environment, and so is the quarantine in the process of disease prevention.

5.3 Vaccination

Following the onset of diseases, the cost incurred in diagnostics, labour, chemicals, and production loss due to mortality may be significant. In many cases, when fish are sick, they become off-feed, and treatment becomes difficult. Hence, the prevention of diseases through optimum husbandry and immune prophylaxis by vaccination are promising approaches. A vaccine is any biologically based preparation intended to provide protective immunity to a particular disease or group of diseases. In the advent of adverse effects associated with anti-parasitic drugs, vaccination is being widely adopted to prevent diseases. However, no commercial vaccine is available as of now against any of the fish parasites. Several trial studies on vaccine development have been made for different fish parasites such as *Ichthyophthirius multifiliis, Cryptocaryon irritans, Cryptobia salmositica, Lepeophtheirus salmonis,* and *Argulus siamensis* through different approaches (Das et al. 2018). This approach can certainly reduce reliance on chemicals and can also be ecologically amicable. There are three generations of parasite vaccines where: the first generation includes whole-organism vaccines, live-attenuated vaccines, killed, and subunit vaccines; the second generation comprises protein antigens, and recombinant protein components; and the third generation includes DNA-based vaccines. A few important attempts in parasitic vaccine development are discussed here.

5.3.1 Vaccination Against *Ichthyophthirius multifiliis*

I. multifiliis, is a highly contagious ciliated protozoan parasite that causes white spot disease in freshwater fishes. The parasite is single-celled, popularly known as ich, that causes white cysts in fins and skin. It has been associated with mortality and reduced growth. Several attempts to treat the parasite using vaccines have been made; however, to date, there is no commercially available vaccine against the parasite. Live theronts, when injected intraperitoneally, confer immunity with significantly higher anti-Ich antibodies compared to those of control fish (Jorgensen et al. 2008), although there are concerns about the cultivation of the parasite for large-scale production. Various studies encompassing vaccine development against ich are under process. Killed and inactivated vaccine (theronts and trophonts), DNA vaccines or recombinant vaccines have proven to give better protection (Xu et al. 2004). Experimental trials with formalin-inactivated theronts and trophonts induced both serum and cutaneous antibody responses in channel catfish and rainbow trouts (Dalgaard et al. 2002). Vaccine preparations for the genus *Tetrahymena*, a closely related ciliate to Ich, used in experimental vaccination trials conferred protection against other species of the protozoan. The most efficient candidate is immobilization antigen; i-antigen-derived vaccines, which include either purified i-antigens or passive Ab transfer raised against i-antigens. A total of 17 i-antigen genes have been identified in the *I. multifiliis* genome (Coyne et al. 2011) and five i-antigen serotypes, A-E, have been characterized. Vaccines containing purified i-Ag together with FCA induce protection but only against the same serotype of the parasite bearing homologous i-Ag on their surface (Xu et al. 2009). i-Ag vaccine adjuvant by parasitic HSP 70C in fish against cryptocaryonosis caused by *Cryptocaryon irritans* in marine fishes has been attempted that enhanced the levels of mucosal IgM, which efficiently immobilized live theronts (Josepriyaa et al. 2015). A major limitation of this method is specific serotype protection with purified i-antigen. Currently, DNA vaccine or recombinant vaccine approaches against parasitic *I. multifiliis* have been found to be very effective in inducing protective immunity.

5.3.2 Vaccination Against *Argulus*

Argulosis caused by *Argulus* sp. is currently the major concern in the freshwater aquaculture industry. The ribosomal protein P0 involved in protein synthesis has been explored as a vaccine candidate. The keyhole limpet hemocyanin (KLH)-conjugated peptide has been used in the immunization of rohu at a dose of 1.5 μg/g body weight. No significant difference in parasite load was found in vaccinated fish; however, a delayed mortality of 59% in the immunized group was noticed as compared to 75% mortality in the control group (15 days post-infection). The partial protection noticed in the study indicated the need for further dose optimization of this molecule to develop into a vaccine candidate (Kar et al. 2017). The whole antigens of *A. siamensis* (50 μg protein/fish) along with adjuvant (FCA) and three doses of intraperitoneal boosters (at 14 days' intervals) followed by challenge with 250 numbers of metanauplii of *Argulus* showed 35.42% of low-grade infection and 22.92% high-grade infection in immunized fish as compared to 14.58% and 41.67%, respectively, in control fish. Further, the immunized fish showed reduced haemorrhages on the body surface along with higher antibody response as compared to controls. The results revealed the possibility of control of this parasite through effective vaccines for long-term protection (Das et al. 2018).

The parasite's structural complexity and variable life stages are of concern for developing effective vaccines. The large-quantity production for vaccine trials through in vitro multiplication remains a big challenge. Further, parasites transform with several phases or stages, even on or in the body of the host during the infection process, in which their habitat, mode of nutrition, reproduction, and antigenic profile may change. In such infection models, designing effective anti-parasitic vaccines becomes a great challenge due to the immune-evasion strategies. Additionally, the multiplication of parasites in laboratory conditions, which often requires the natural hosts obligatorily, is also a limiting factor. However, the modulation of the immune system in fish during parasite interaction has generated a better understanding of responses involved in parasitic infections (Sitja-Bobadilla 2008). Hence, the identification and production of protective antigens is probably the most feasible strategy in parasite vaccine development.

5.4 Selective Breeding and Disease Resistance

The lack of any vaccine as a permanent prevention strategy, along with the evolvement of drug resistance among the parasites, has led to looking into alternative approaches. Among the several management approaches, selective breeding for resistant traits is one of the most reliable methods and provides long-term control in solving disease issues (Das and Sahoo 2014). Moderate heritability has been seen for traits associated with parasitic resistance, indicating the potential for undertaking genetic improvement programmes through selective breeding. Lin et al. (2020) tested the possibility of genomic selection (GS) on resistance to heterobothriosis, caused by the monogenean parasite *Heterobothrium okamotoi*, an economically important parasite of the tiger pufferfish *Takifugu rubripes.* For certain ectoparasites (e.g., salmon lice), simply counting the number of parasites attached to the fish represents the primary disease-resistance phenotype used for selection (Tsai et al. 2016). Though this can be a promising approach to avoid susceptibility and commencement of parasitic infection, very few reports on selective breeding programmes have been documented.

6 Conclusion

Fishes, being aquatic animals, continually get exposed to a diverse range of pathogens which can lead to disease development. This gets intensified when fish are under any kind of stress that leads to an immuno-compromised state. Biotic and abiotic stressors, availability of vectors, host–parasite interaction, parasite life stages and transmission ecology decide the degree of parasitic infestation. The harmful effects of parasites not only restrict or exhaust the cellular fluid and nutrients of the host but also invite opportunistic pathogens, causing open ulcers. Besides the stress imposed by present-day farming practices and water quality deterioration, the emergence of disease caused by parasites may get aggravated due to climate change. Therefore, there is a need for concerted research efforts on understanding host–parasite interactions, the epidemiology of diseases, and developing new prophylaxis and other control measures for parasites. Nevertheless, the pathogenesis of parasites may be exaggerated due to pollution, climate change, and shifts in the selection of different hosts due to expansion in aquaculture practices. Preparedness, wide awareness, good husbandry, and suitable quarantine of new stock can minimise parasite prevalence, risk and further transmission.

References

Anastasiia G, Alexander L, Alexei YK, Julius L, Vyacheslav Y (2014) High prevalence of trypanosome co-infections in freshwater fishes. Folia Parasitol 61:495–504

Aohagl Y, Shibahara T, Machlda N, Vamaga V, Kagota K (1992) *Clinostomum complanatum* (Trematoda: Clinostomatidae) in five new fish hosts in Japan. J Wildl Dis 28:467–469

Atkinson SD, Bartosova-Sojkova P, Whipps CM, Bartholomew JL (2015) Approaches for characterising myxozoan species. In: Okamura B, Gruhl A, Bartholomew JL (eds) Myxozoan evolution, ecology and development 02. Springer, Cham, Switzerland, pp 111–123

Bakaletz LO (2004) Developing animal models for polymicrobial diseases. Nat Rev Microbiol 2:552–568

Bandilla M, Valtonen ET, Suomalainen L-R, Aphalo PJ, Hakalahti T (2006) A link between ectoparasite infection and susceptibility to bacterial disease in rainbow trout. Int J Parasitol 36:987–991

Buchmann K, Lindenstrom T (2002) Interactions between monogenean parasites and their fish hosts. Int J Parasitol 32:309–319

Callaway R, Shinn AP, Grenfell SE et al (2012) Review of climate change impacts on marine aquaculture in the UK and Ireland. Aquat Conserv Mar Freshwat Ecosyst 22:389–421

Canning EU, Okamura B (2003) Biodiversity and evolution of the Myxozoa. Adv Parasitol 56:43–131

Choudhury A, Nelson PA (2000) Redescription of *Crepidostomum opeongoensis* Caira, 1985 (Trematoda: Allocreadiidae) from fish hosts *Hiodon alosoides* and *Hiodon tergisus* (Osteichthyes: Hiodontidae). J Parasitol 86:1305–1312

Coyne RS, Hannick L, Shanmugam D, Hostetler JB, Brami D, Joardar VS, Johnson J, Radune D, Singh I, Kumar U et al (2011) Comparative genomics of the pathogenic ciliate *Ichthyophthirius multifiliis*, its free-living relatives and a host species provide insights into adoption of a parasitic lifestyle and prospects for disease control. Genome Biol 12:R100. http://genomebiology.com/2011/12/10/R100

Dalgaard M, Buchmann K, Li A (2002) Immunization of rainbow trout fry with *Ichthyophthirius multifiliis* sonicate: protection of host and immunological changes. Bull Eur Assoc Fish Pathol 22:288–297

Das P, Mohanty J, Badhe MR, Sahoo PK, Sardar KK, Parija SC (2018) Assessment of protective response induced by whole antigens of fish ectoparasite, *Argulus siamensis* in *Labeo rohita*. J Entomol Zool Stud 6(3):1751–1755

Das S, Sahoo PK (2014) Markers for selection of disease resistance in fish: a review. Aquac Int 22:1793–1812

Densmore CL, Ottinger CA, Blazer VS, Iwanowicz LR (2004) Immunomodulation and disease resistance in post yearling rainbow trout infected with *Myxobolus cerebralis*, the causative agent of whirling disease. J Aquat Anim Health 16:73–82

Echi PC, Eyo JE, Okafor FC, Onyishi GC, Ivoke N (2012) First record of co–infection of three Clinostomatid parasites in Cichlids (Osteichthyes: Cichlidae) in a tropical freshwater lake. Iran J Public Health 41:86–90

Echi PC, Okafor FC, Eyo JE (2009) Co-infection and morphometrics of three Clinostomatids (Digenea: Clinostomatidae) in *Tilapia guinensis* Bleeker, 1862 from Opi lake, Nigeria. Bio-Res 7(1):432–436

Eissa AE, Zaki M, Baiomy AAA (2010) *Flavobacterium columnare*/*Myxobolus tilapiae* concurrent infection in the earthen pond reared Nile tilapia (*Oreochromis niloticus*) during the early summer. Interdiscipl Bio Central 2(2):1–10

Estensoro I, Redondo MJ, Alvarez-Pellitero P, Sitjà-Bobadilla A (2010) Novel horizontal transmission route for *Enteromyxum leei* (Myxozoa) by anal intubation of gilthead sea bream *Sparus aurata*. Dis Aquat Org 92:51–58

Evans JJ, Klesius PH, Pasnik DJ, Shoemaker CA (2007) Influence of natural *Trichodina* sp. parasitism on experimental *Streptococcus iniae* or *Streptococcus agalactiae* infection and survival of young channel catfish *Ictalurus punctatus* (Rafinesque). Aquac Res 38:664–667

Georgiev B, Biserkov V, Genov T (1986) In toto staining method for cestodes with iron acetocarmine. Helminthologia 23:279–281

Gorgoglione B, Taylor NGH, Holland JW, Feist SW, Secombes CJ (2019) Immune response modulation upon sequential heterogeneous co-infection with *Tetracapsuloides bryosalmonae* and VHSV in brown trout (*Salmo trutta*). Fish Shellfish Immunol 88:375–390

Hakalahti T, Karvonen A, Valtonen ET (2006) Climate warming and disease risks in temperate regions—*Argulus coregoni* and *Diplostomum spathaceum* as case studies. J Helminthol 80:93–98

Hany MR, Abdel-Latif, Mahmoud AO, Dawood, Simon M, Mansour E (2020) The nature and consequences of co-infections in tilapia: a review. J Fish Dis 43:651–664

Harikrishnan R, Balasundaram C, Heo MS (2011) Impact of plant products on innate and adaptive immune system of cultured finfish and shellfish. Aquaculture 317:1–15

Harvell CD, Mitchell CE, Ward JR, Altizer S, Dobson AP, Ostfeld RS, Samuel MD (2002) Climate warming and disease risks for terrestrial and marine biota. Science 296:2158–2162

Hashimoto GSO, Marinho NF, Ruiz ML, Acchile M, Chagas EC, Chaves FCM, Martins ML (2016) Essential oils of *Lippia sidoides* and *Mentha piperita* against monogenean parasites and their influence on the hematology of Nile tilapia. Aquaculture 450:182–186

Holzer AS, Sommerville C, Wootten R (2006) Molecular studies on the seasonal occurrence and development of five myxozoans in farmed *Salmo trutta* L. Parasitology 132:193–205

Houghton JT, Ding Y, Griggs DJ, Noguer M, van der Linden PJ, Dai X, Maskell K, Johnson, CA (2001) (Eds.). Climate Change 2001: The Scientific Basis. Contribution of Working Group I to the Third Assessment Report of the Intergovernmental Panel on Climate Change. Cambridge University Press, pp 1–893

Johnson PTJ, Townsend AR, Cleveland CC, Glibert PM, Howarth RW, McKenzie VJ, Rejmankova E, Ward MH (2010) Linking environmental nutrient enrichment and disease emergence in humans and wildlife. Ecol Appl 20:16–29

Johnson SC, Treasurer JW, Bravo S, Nagasawa K, Kabata Z (2004) A review of the impact of parasitic copepods on marine aquaculture. Zool Stud 43:229–243

Jorgensen LG, Nemli E, Heinecke RD, Raida MK, Buchmann K (2008) Immune-relevant genes expressed in rainbow trout following immunisation with a live vaccine against *Ichthyophthirius multifiliis*. Dis Aquat Org 80(3):189–197

Josepriyaa TA, Chiena K, Lina H, Huang H, Wub C, Songa Y (2015) Immobilization antigen vaccine adjuvanted by parasitic heat shock protein 70C confers high protection in fish against cryptocaryonosis. Fish Shellfish Immunol 45:517–527

Kar B, Mohapatra A, Mohanty J, Sahoo PK (2017) Evaluation of ribosomal P0 peptide as a vaccine candidate against *Argulus siamensis* in *Labeo rohita*. Open Life Sci 12:99–108

Klemme I, Louhi KR, Karvonen A (2016) Host infection history modifies co-infection success of multiple parasite genotypes. J Anim Ecol 85:591–597

Kostadinova A, Power AM, Fernández M, Balbuena JA, Raga JA, Gibson DI (2003) Three species of *Magnibursatus naidenova*, 1969 (Digenea: Derogenidae) from Atlantic and Black Sea marine teleosts. Folia Parasitol 50:202–210

Kutz SJ, Hoberg EP, Polley L, Jenkins EJ (2005) Global warming is changing the dynamics of Arctic host–parasite systems. Proc R Soc B 272:2571–2576

Labrie L, Komar C, Terhune J, Camus A, Wise D (2004) Effect of sublethal exposure to the trematode *Bolbophorus* spp. on the severity of enteric septicemia of catfish in channel catfish fingerlings. J Aquat Anim Health 16:231–237. https://doi.org/10.1577/H04-011.1

Larsen MH, Mouritsen KN (2014) Temperature–parasitism synergy alters intertidal soft-bottom community structure. J Exp Mar Biol Ecol 460:109–119

Lin Z, Hosoya S, Sato M, Mizuno N, Kobayashi Y, Itou T, Kikuchi K (2020) Genomic selection for heterobothriosis resistance concurrent with body size in the tiger pufferfish, *Takifugu rubripes*. Sci Rep 10(1):1–13

Ling F, Jiang C, Liu G, Li M, Wang G (2015) Anthelmintic efficacy of cinnamaldehyde and cinnamic acid from cortex cinnamon essential oil against *Dactylogyrus intermedius*. Parasitology 142:1744–1750

Lohmus M, Bjorklund M (2015) Climate change: what will it do to fish–parasite interactions? Biol J Linn Soc 116:397–411

Lom J, Dyková I, Pavlaskova M (1983) Unidentified' mobile protozoans from the blood of carp and some unsolved problems of myxosporean life cycles. J Protozool 30:497–508

Longshaw M, Frear P, Feist SW (2003) *Myxobolus buckei* sp. n. (Myxozoa), a new pathogenic parasite from the spinal column of three cyprinid fishes from the United Kingdom. Folia Parasitol 50:251–262

Marcogliese DJ (2001) Implications of climate change for parasitism of animals in the aquatic environment. Can J Zool 79(8):1331–1352

Marcogliese DJ (2008) The impact of climate change on the parasites and infectious diseases of aquatic animals. Rev Sci Tech 27(2):467–484

Martin RJ, Robertson AP, Bjorn H (1997) Target sites of anthelmintics. Parasitology 114(7):111–124

Martins ML, Shoemaker CA, Xu D, Klesius PH (2011) Effect of parasitism on vaccine efficacy against *Streptococcus iniae* in Nile tilapia. Aquaculture 314:18–23

McKenzie VJ, Townsend AR (2007) Parasitic and infectious disease responses to changing global nutrient cycles. EcoHealth 4:384–396

Monir S, Bagum N, Rahman S, Ashaf-Ud-Doulah M, Bhadra A, Borty SC (2015) Parasitic diseases and estimation of loss due to infestation of parasites in Indian major carp culture ponds in Bangladesh. Int J Fish Aquat Stud 2:118–122

Mustafa A, Rankaduwa W, Campbell P (2001) Estimating the cost of sea lice to salmon aquaculture in eastern Canada. Can Vet J 41:54–56

Nofal MI, Abdel-Latif HM (2017) Ectoparasites and bacterial co-infections causing summer mortalities among cultured fishes at Al-Manzala with special reference to water quality parameters. Life Sci J 14(6):72–83

Overstreet RM (1976) A redescription of Crassicutis archosargi, a digenean exhibiting an unusual tegumental attachment. J Parasitol 702–708

Paaijmans KP, Read AF, Thomas MB (2009) Understanding the link between malaria risk and climate. Proc Natl Acad Sci USA 106:13844–13849

Paladini G, Longshaw M, Gustinelli A, Andrew P, Shinn A (2017) Parasitic diseases in aquaculture: their biology, diagnosis and control. Chapter. April 2017. https://doi.org/10.1002/9781119152125.ch4

Parker JD, Warner MC (1970) Effects of fixation, dehydration and staining on dimensions of myxosporidan and microsporidan spores. J Wildl Dis 6:448–456

Paul A, Pattanayak S, Sahoo MK, Kumar RP, Kumar R, Sahoo PK (2020) Co-infection of bacterial and parasitic pathogens including the myxosporean *Zschokkella auratis* infecting brain in farmed striped murrel *Channa striata* (Bloch, 1793) causing large scale mortality. Indian J Fish 67(4):105–115

Peeler EJ, Feist SW, Longshaw M, Thrush MA, St-Hilaire S (2008) An assessment of the variation in the prevalence of renal myxosporidiosis and hepatitis in wild brown trout, *Salmo trutta* L., within and between rivers in south–West England. J Fish Dis 31:719–728

Pinto HA, Mati VLT, Melo AL (2014) Metacercarial infection of wild Nile tilapia (*Oreochromis niloticus*) from Brazil. Sci World J:7807492. https://doi.org/10.1155/2014/807492

Puspha K, Saurav K, Ramesh M, Shameena S, Ashutosh DD, Rajendran KV, Raman RP (2019) Antiparasitic effect of aqueous and organic solvent extracts of *Azadirachta indica* leaf against *Argulus japonicus* in *Carassius auratus*. Aquaculture 511:634175

Reed CC, Basson L, Van As LL (2002) *Myxobolus* species (Myxozoa), parasites of fishes in the Okavango River and Delta, Botswana, including descriptions of two new species. Folia Parasitol 49:81–88

Sahoo PK, Mohanty J, Garnayak SK, Mohanty BR, Kar B, Prasanth H, Jena JK (2013) Estimation of loss due to argulosis in carp culture ponds in India. Indian J Fish 60:99–102

Sahoo PK, Paul A, Sahoo MK, Pattanayak S, Rajesh Kumar P, Das BK (2020) Incidences of infectious diseases in freshwater aquaculture farms of eastern India: a passive surveillance based study from 2014–2018. J Aquac Res Dev 11:1–5. https://doi.org/10.35248/2155-9546.20.11.579

Sahoo PK, Swaminathan TR, Jawahar T, Abraham KR, Pattanayak S, Mohapatra A, Rath SS, Patra A, Adikesavalu H, Sood N, Pradhan PK, Das BK, Jayasankar P, Jena JK (2016) Detection of goldfish haematopoietic necrosis herpes virus (cyprinid herpesvirus-2) with multi-drug resistant *Aeromonas hydrophila* infection in goldfish: first evidence of any viral disease outbreak in ornamental freshwater aquaculture farms in India. Acta Trop 161:8–17

Schmidt GD, Nickol BB (1985) Development and life cycles. In: Crompton DWT, Nickol BB (eds) Biology of the Acanthocephala. Cambridge University Press, Cambridge, pp 273–305

Schmidt-Posthaus H, Polkinghorne A, Nufer L, Schifferli A, Zimmermann DR, Segner H, Steiner P, Vaughan L (2012) A natural freshwater origin for two chlamydial species, *Candidatus Piscichlamydia salmonis* and *Candidatus Clavochlamydia salmonicola*, causing mixed infections in wild brown trout (*Salmo trutta*). Environ Microbiol 14:2048–2057

Shinn A, Pratoomyot J, Bron J, Paladini G, Brooker E, Brooker A (2015) Economic impacts of aquatic parasites on global finfish production. GAA:58–61

Shoemaker CA, Martins ML, Xu DH, Klesius PH (2012) Effect of *Ichthyophthirius multifiliis* parasitism on the survival, hematology and bacterial load in channel catfish previously exposed to *Edwardsiella ictaluri*. Parasitol Res 111:2223–2228

Sitja-Bobadilla A (2008) Living off a fish: a trade-off between parasites and the immune system. Fish Shellfish Immunol 25:358–372

Soares BV, Neves LR, Oliveira MSB, Chaves FCM, Dias MKR, Chagas EC, Tavares-Dias M (2016) Antiparasitic activity of the essential oil of *Lippia alba* on ectoparasites of *Colossoma macropomum* (tambaqui) and its physiological and histopathological effects. Aquaculture 452:107–114

Swaminathan TR, Kumar R, Dharmaratnam A, Basheer VS, Sood N, Pradhan PK, Sanil NK, Vijayagopal P, Jena JK (2016) Emergence of carp edema virus in cultured ornamental koi carp, *Cyprinus carpio* koi, in India. J Gen Virol 97(12):3392–3399

Tavares-Dias M (2018) Current knowledge on use of essential oils as alternative treatment against fish parasites. Aquat Living Resour 31:13

Tsai HY, Hamilton A, Tinch AE, Guy DR, Bron JE, Taggart JB, Gharbi K, Stear M, Matika O, Pong-Wong R, Bishop SC, Houston RD (2016) Genomic prediction of host resistance to sea lice in farmed Atlantic salmon populations. Genet Sel Evol 48:1–11

Xu DH, Klesius PH, Shelby RA (2004) Immune responses and host protection of channel catfish, *Ictalurus punctatus* (Rafinesque), against *Ichthyophthirius multifiliis* after immunization with live theronts and sonicated trophonts. J Fish Dis 27:135–141

Xu DH, Klesius PH, Shoemaker CA (2009) Effect of immunization of channel catfish with inactivated trophonts on serum and cutaneous antibody titers and survival against *Ichthyophthirius multifiliis*. Fish Shellfish Immunol 26:614–618

Xu DH, Shoemaker CA, Klesius PH (2012) *Ichthyophthirius multifiliis* as a potential vector of *Edwardsiella ictaluri* in channel catfish. FEMS Microbiol Lett 329:160–167

Xu D-H, Shoemaker CA, LaFrentz BR (2014) Enhanced susceptibility of hybrid tilapia to *Flavobacterium columnare* after parasitism by *Ichthyophthirius multifiliis*. Aquaculture 430:44–49

Xu D-H, Shoemaker CA, Zhang D (2015) Treatment of *Trichodina* sp. reduced load of *Flavobacterium columnare* and improved survival of hybrid tilapia. Aquac Rep 2:126–131

Xu D-H, Shoemaker CA, Klesius PH (2007) Evaluation of the link between gyrodactylosis and streptococcosis of Nile tilapia, *Oreochromis niloticus* (L.). J Fish Dis 30:233–238

Zhang C, Li DL, Chi C, Ling F, Wang GX (2015) *Dactylogyrus intermedius* parasitism enhances *Flavobacterium columnare* invasion and alters immune-related gene expression in *Carassius auratus*. Dis Aquat Org 116:11–21

Zhi T, Xu X, Chen J, Zheng Y, Zhang S, Peng J, Yang T (2018) Expression of immune-related genes of Nile tilapia *Oreochromis niloticus* after *Gyrodactylus cichlidarum* and *Cichlidogyrus sclerosus* infections demonstrating immunosuppression in coinfection. Fish Shellfish Immunol 80:397–404

Pathogens and Diseases of Shellfish

Major Viral Diseases of Shrimp in India

M. P. Safeena, K. Sreedharan, K. U. Shyam,
and K. V. Rajendran

Abstract

Viral diseases have become an important limiting factor in the growth of shrimp aquaculture worldwide due to their ability to cause high mortality and lack of curative measures. More than 20 viral pathogens are known to infect penaeid shrimps, and five of them are listed by the World Organization for Animal Health (WOAH). Other than the most deleterious white spot diseases (WSD) caused by white spot syndrome virus, viruses such as Decapod Penstyldensovirus 1 (PstDV1)/Infectious hypodermal and hematopoietic necrosis virus (IHHNV), Decapod hepanhamaparvovirus 1 (DHPV-1)/Hepatopancreatic parvovirus (HPV), Penaeus monodon-type baculovirus/Monodon baculovirus (MBV) and Infectious myonecrosis virus (IMNV) are reported from Indian shrimp aquaculture. This chapter provides an overview of these viruses.

Keywords

Infectious hypodermal and hematopoietic necrosis virus (IHHNV) · Hepatopancreatic parvovirus (HPV) · Monodon baculovirus (MBV) · Infectious myonecrosis virus (IMNV)

M. P. Safeena
Kerala University of Fisheries and Ocean Studies, Kochi, India
e-mail: safeena.mp@kufos.ac.in

K. Sreedharan · K. V. Rajendran (✉)
ICAR-Central Institute of Fisheries Education, Mumbai, India
e-mail: sreedharan@cife.edu.in; rajendrankv@hotmail.com

K. U. Shyam
DTU Aqua – National Institute of Aquatic Resources, Silkeborg, Denmark
e-mail: sku23207@gmail.com

R. Kooloth Valappil et al. (eds.), *Aquatic Animal Health Management*,
https://doi.org/10.1007/978-981-96-7987-4_9

1 Introduction

Shrimp farming holds a prominent position in the aquaculture sector and has experienced substantial expansion in recent times. Currently, penaeid shrimps account for approximately 7.9 million tonnes (62.2%) of total crustacean production, making them the most economically significant farmed aquatic species. However, the rapid expansion of shrimp farming to keep pace with the increasing demand has been accompanied by the incidence of diseases. Of all the diseases, virus epizootics have a major impact on shrimp farming as they are the most varied, highly infectious and widespread pathogens. To date, there are almost 20 viruses known to infect penaeid shrimps, many of which cause mass mortalities in aquaculture. In India, in addition to WSSV, viruses such as Decapod Penstyldensovirus 1, Decapod hepanhamaparvovirus 1, *Penaeus monodon*-type baculovirus and Infectious myonecrosis virus (IMNV) are of considerable concern due to their ability to cause economic losses to shrimp culture. This chapter provides a brief overview of each virus.

2 Decapod Penstyldensovirus 1 (PstDV1)/Infectious Hypodermal and Hematopoietic Necrosis Virus (IHHNV)

2.1 Pathogen and the Disease

Decapod Penstyldensovirus 1, initially known as IHHNV/*Penaeus stylirostris* densovirus (*Pst*DNV), was first identified from the post-larvae of the blue shrimp, *P. stylirostris,* and juveniles that were brought into Hawaii from Costa Rica and Ecuador in 1981 and led to significant mortality of up to 90% (Lightner et al. 1983a, b). Later, the virus has been recorded from various locations worldwide, including Taiwan in 1986 (Bell and Lightner 1987), Australia in 2008 (Molthathong et al. 2013) and India in 2009 (Rai et al. 2009). While it was reported to cause acute mortality in *P. stylirostris*, it did not result in large-scale mortality in *P. vannamei* and *P. monodon;* instead, it led to runt deformity syndrome (RDS). RDS causes stunted growth and accounts for approximately 50% of losses to the shrimp industry (Lee et al. 2022). PstDV1 was also found to infect *M. rosenbergii* juveniles but did not lead to substantial mortality (Hsieh et al. 2006). Considering the morbidity and mortality, the likelihood of international spread and other criteria, WOAH has listed the virus since 1995 (Chai et al. 2016).

PstDV1 is a non-enveloped, icosahedral symmetrical, single-stranded, linear DNA virus (3.9 kb long) with a diameter of 22–23 nm and is considered the tiniest penaeid shrimp virus (Bonami et al. 1990) (Fig. 1). Initially, this virus was classified as *Penaeus stylirostris* densovirus (PstDNV), the Penstyldensovirus genus of the subfamily Densovirinae under the Parvoviridae family. The virus was reclassified in July 2019 by the ICTV into the family of Parvoviridae, Hamaparvovirinae

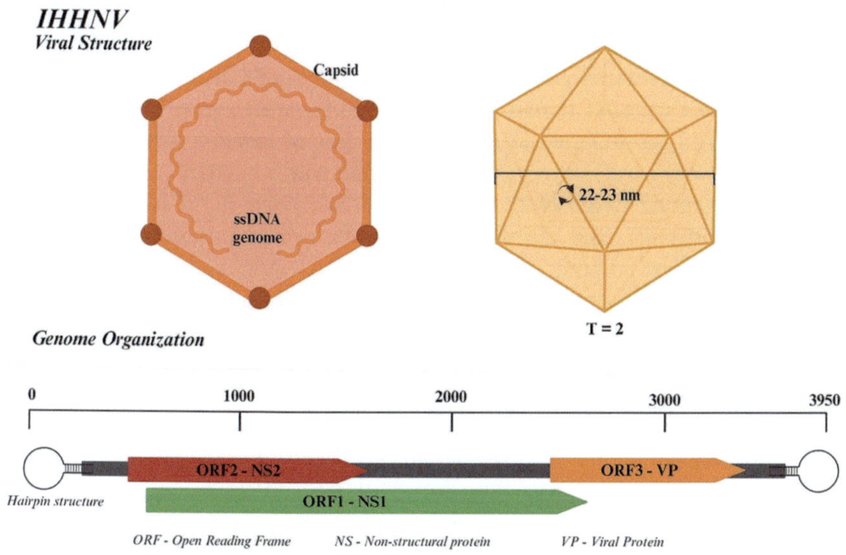

Fig. 1 IHHNV virus structure and genome organization

subfamily, and included in the Penstylhamaparvovirus genus. Presently, the virus is named decapod penstyldensovirus 1(PstDV1), according to the ICTV's recommendation (Lee et al. 2022; Kibenge et al. 2024).

Despite being DNA viruses, parvoviruses have high mutation rates, leading to the emergence of diverse genotypes in various shrimp culture provinces. There are presently five documented genotypes, of which three are infectious-genotypes (I, II and III) and two are non-infectious genotypes A and B. Both these non-infectious types of IHHNV-related sequences have been found integrated into the gene of *P. monodon* recorded from Africa, Mozambique and Tanzania (Tang and Lightner 2006; Tang et al. 2007). The Type I genotype is prevalent in Australia, and genotype II is mainly found in Southeast Asia and America. In contrast, the type III genotype of IHHNV is predominant in East Asia. The phylogenetic tree constructed for the IHHNV whole genome sequences (Shen et al. 2015) indicated that the non-infectious IHHNV is different from the infectious types and occurs as endogenous viral elements (EVEs) integrated into the *P. monodon* chromosomal DNA (Lee et al. 2024). The nucleotide sequence study of the IHHNV genome discloses three major coding regions: (1) a left ORF (NS1), (2) a mid ORF (NS2) and (3) a right ORF (VP), whereas the purified virions comprise four different molecular weight polypeptides of 74, 47, 39 and 37.5 kDa (Afanasiev et al. 1991; Boublik et al. 1994; Shike et al. 2000; Rai et al. 2011).

2.2 Clinical Signs

In acute epidemics, IHHNV infection results in substantial deaths in juveniles and also in sub-adult stages of *P. stylirostris* (Bell and Lightner 1984). Though IHHNV infections have been reported in all the life stages of penaeid shrimps, the post-larvae and juveniles of shrimp are more vulnerable to IHHNV. The vertically infected early larval stages of *P. stylirostris* do not show any clinical symptoms. However, the juveniles and late stages may exhibit evident disease signs, resulting in mass mortalities. Although the gross signs of IHHNV are non-specific, acute IHHNV infection may lead to reduced feed intake, followed by changes in appearance, as detailed by Lightner and Redman (1992) and Owens et al. (1992). Shrimp infected with acute IHHNV become motionless, exhibit rolling-over behaviour and sink slowly to the bottom of the pond, where healthy shrimps cannibalize them. In the acute infection stage, white or buff-coloured spots are also reported to develop in the cuticular epidermis of *P. stylirostris,* such as at the abdomen tergal plates junction, resulting in a spotted appearance (Lightner 1996).

IHHNV infection does not show major clinical signs, but the diseased juveniles and post-larvae of *P. monodon* and *P. vannamei* showed runt-deformity syndrome (RDS) (Brock et al. 1983; Lightner et al. 1983b; Bell and Lightner 1984; Tang and Lightner 2002; Rai et al. 2012; Zhao et al. 2024). The major clinical signs of RDS are deformation of the rostrum, appearance of wrinkles in the antenna, roughness in the cuticle and distortion of the sixth abdominal segment and tail fan of infected shrimp. IHHNV infection in *P. vannamei* was reported to cause growth retardation between 30% and 90%, which led to 10% to 50% economic loss (Wyban et al. 1992). In infected *P. monodon*, abnormalities were observed in the rostrum, antennae and uropods, and a change in shell colour and hardness was also noticed throughout the body (Bell and Lightner 1984; Primavera and Quinitio 2000). Further, Rai et al. (2009) reported slow growth in farmed *P. monodon* due to IHHNV infection.

2.3 Host Range

Early in the 1980s, *P. stylirostris* and *P. vannamei* were found to have IHHNV in America. Later, based on the IHHNV infection reported in *P. monodon* from the Asian region, it was inferred that the virus was introduced through the transportation of live black tiger shrimp (Lightner et al. 1983a; Lightner and Redman 1992). IHHNV has been found in most penaeid shrimps from different shrimp-producing nations since it was first discovered in farmed shrimp in Hawaii in 1981 (Lightner 1996). Worldwide, penaeid shrimps such as *P. vannamei*, *P. stylirostris*, *P. monodon, P. occidentalis*, *P. semisulcatus*, *P. californiensis*, *P. schmitti* and *P. japonicus* have all been reported to be naturally infected (Felix and Devaraj 1993; Lightner and Redman 1991; Morales-Covarrubias et al. 1999; Pantoja et al. 1999). In southern Taiwan (Hsieh et al. 2006) and Malaysia (Hazreen et al. 2012), reports of IHHNV natural infection in post-larvae and sub-adult *M. rosenbergii* have been recorded.

Additionally, *P. setiferus*, *P. aztecus* and *P. duorarum* have been reported to develop experimental infections (Lightner 1996). Yang et al. (2007) discovered this virus in 2003 while collecting Harry-clawed shore crabs (*Hemiarapsus penicillatus*) from northern China's culture zones. Native wild *F. subtilis* specimens from the Pacoti River estuary on the northeastern Brazilian coast were the subject of an experimental investigation that revealed their vulnerability to IHHNV (Coelho et al. 2009). In Argentina, *Artemesia longinaris* was identified as a novel wild host for IHHNV (Martorelli et al. 2010). The natural occurrence of IHHNV in *Neohelice granulata* crab has been reported by Cavalli et al. (2013). Macías-Rodríguez et al. (2014) identified *H. penicillatus* and *Callinectes arcuatus* (*C. arcuatus*) as carriers of IHHNV, despite the absence of clinical symptoms. Additionally, bivalves have been found to carry IHHNV, with positive samples noticed in the gills, muscle tissues and gonads of *M. edulis* through PCR analysis (Wei et al. 2017). Furthermore, IHHNV-positive samples were found in six other bivalves, although no apparent pathogenic effects were detected in infection experiments (Pang et al. 2018). As significant carriers, bivalves have major risks for the spread of IHHNV. In addition, the virus has been reported from freshwater Red Claw Crayfish (*Cherax quadricarinatus*) imported into South Korea and also from *Procambarus clarkii* of the Middle and Lower Reaches of the Yangtze River in China (Lee et al. 2021; Xu et al. 2023). Polychaete worms (*Perinereis helleri*) collected from IHHNV-infected shrimp ponds were also found to be infected with the virus (Liu et al. 2020).

2.4 Mortality Pattern

The mortality pattern of IHHNV infection may vary based on the genotype, as they may exhibit different virulence, susceptibility of the host species and the growing stage of the diseased shrimp. During acute infection, diseased post-larvae and juveniles of *P. stylirostris* sink to the tank bottom without swimming. Within a short duration, it can cause up to 90% mortality (Rai et al. 2012). Chronic infection does not cause mass mortality but is reported to cause RDS in *P. vannamei* and *P. monodon* and sub-adults of *M. rosenbergii* juveniles.

2.5 Target Organs and Histopathology

IHHNV mainly targets ectodermal tissues like cuticular epidermis, fore and hind gut hypodermal epithelium, nerve cord and nerve ganglia, and mesodermal tissues such as haematopoietic tissues, antennal gland, tubule epithelium, gonads, lymphoid organ, connective tissue and striated muscles. In contrast, it does not infect tissues that are of endodermal origin. IHHNV infection is diagnosed histologically by identifying significant eosinophilic Cowdry Type A inclusion bodies surrounded by marginated chromatin in hypertrophied nuclei of cells in ectodermal and mesodermal tissues (Rai et al. 2012). Similar histological manifestations were reported in IHHNV-infected *P. monodon* (Rai et al. 2009) and *Procambarus clarkii* (Chen et al. 2017). A recent study on tissue tropism of IHHNV in *P. monodon* using real-time PCR detected high levels of viral copies in gill tissue, followed sequentially in the hindgut, pleopod, hepatopancreas, lymphoid organ, ventral nerve cord and abdominal muscle (Arbon et al. 2024).

2.6 Geographical Distribution

IHHNV has been documented in several parts of the world, including the Caribbean, North and South America, Central America and Indo-Pacific (Fig. 2). *P. stylirostris*, imported from South and Central America, was the source of the first detection of IHHNV in 1981 in Hawaii (Lightner et al. 1983a) and from Costa Rica (Parajeles-Mora et al. 2021). Post-larvae of *P. vannamei* imported to Mexico were found to be infected with the virus (Lightner 1992). It was discovered in 1986 in imported *P. vannamei* confined stocks in Taiwanese shrimp-growing facilities (Lightner et al. 1987). According to Owens et al. (1992), IHHNV naturally occurs in the Indo-West Pacific area of Australia, and subsequently, Saksmerprome et al. (2010) confirmed the presence of an infectious variant of IHHNV in farmed *P. monodonin* in the country. According to Tang and Lightner (2002), *P. stylirostris* imported for aquaculture to Guam and French Polynesia was infected with IHHNV. East and Southeast Asian nations, including Singapore, Malaysia, Thailand, Indonesia, Taiwan, Philippines and India, have reported cases of the virus from wild and cultivated *P. monodon* (Rai et al. 2012). An epidemiological investigation in shrimp and crab farming areas in

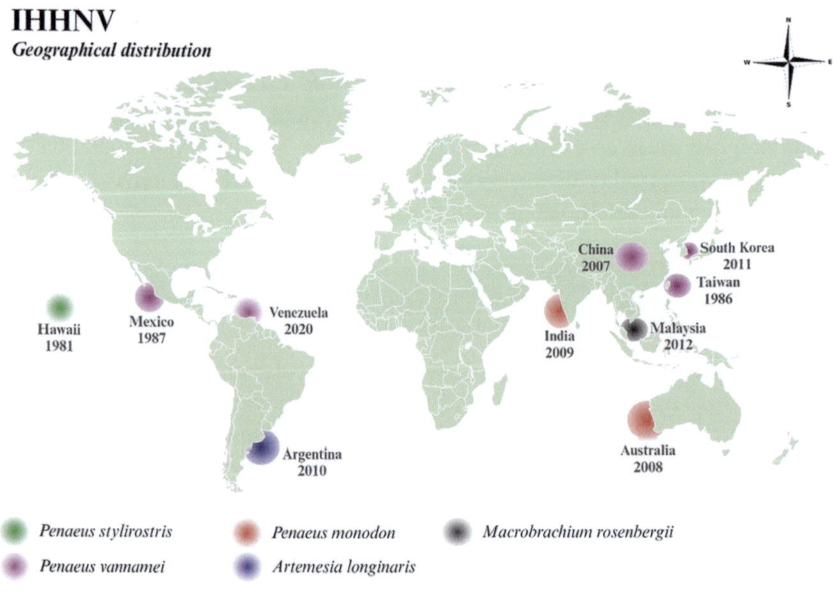

Fig. 2 Geographical distribution of IHHNV (first reports of the virus) across the world

China showed a high prevalence of IHHNV (Yang et al. 2007). Northeastern Brazilian shrimp farms also reported a high prevalence of IHHNV (dos Santos Braz et al. 2009). Notable prevalence has been recorded in the wild *P. monodon* in Brunei Darussalam (Claydon et al. 2010). A relatively high prevalence (30%) of the virus has been detected in wild *Artemesia longinaris* collected from Argentina (Martorelli et al. 2010). However, a low prevalence of the virus has been recorded in farmed *P. vannamei* in Venezuela (Boada et al. 2008). Kim et al. (2011) have reported the occurrence of IHHNV in *P. vannamei* in South Korea. Further, the Korean isolate of IHHNV showed a high nucleotide similarity to the Ecuador strain. IHHNV infection was also observed in wild berried broodstock of freshwater prawn of *M. rosenbergii* in Malaysia (Hazreen Nita et al. 2012). Aly et al. (2021) reported the presence of the virus in farmed *P. semisulcatus* collected from Egyptian provinces along the Mediterranean coast. Co-infection of IHHNV and IMNV in cultivated *P. vannamei* in northeastern Brazil was documented by Teixeira-Lopes et al. (2011). Similarly, in India, co-infection of IHHNV and WSSV has been reported from wild-caught crustaceans from the Andaman and Nicobar Archipelago (Saravanan et al. 2021). Wild-caught *P. monodon* in the Gulf of Mexico and the Northwestern Atlantic Ocean also showed co-infection of WSSV and IHHNV (Krol et al. 2024).

2.7 Transmission

In aquaculture environments, horizontal transmission is the major route of virus infection, which occurs when shrimp consume infected carcasses or come in contact with virus-containing water. The infection can also be transmitted vertically by the release of viral particles by the broodstock during spawning time and subsequent ingestion by the larvae during their early feeding. Further, the transmission can occur directly from eggs to larvae (Lotz 1997). The vertical transmission of IHHNV through female *P. vannamei* has been confirmed by PCR testing of the embryos and larvae derived from IHHNV-infected female shrimp mated with uninfected males (Motte et al. 2003). Further, the ovarian tissues and fertilized eggs of *P. chinensis* reported the presence of IHHNV (Zhang and Sun 1997). Besides, the infection can be transmitted by potential carriers. Viable IHHNV particles have been detected in the excrement of laughing gulls (*Larus atricilla*) and domestic chickens (*Gallus domesticus*), which consumed the virus-infected shrimp tissues (Vanpatten et al. 2004). It is also reported that *P. sytlirostris,* which survived IHHNV infection, can become lifelong carriers of the virus, and it can cause vertical and horizontal transmission.

2.8 Diagnosis

Polymerase chain reaction (PCR) is a reliable and sensitive method used to detect IHHNV infection in shrimp. Various PCR techniques, such as single-step, nested, duplex, multiplex and real-time PCR, etc., have been used worldwide to detect

IHHNV in shrimp samples (Lee et al. 2022, 2024; Gao et al. 2023). A notable early method was developed in 2000 by Nunan et al. which can amplify a 1681 bp IHHNV coat protein fragment using primers of the Hawaii-based IHHNV genome. This method successfully detected IHHNV in *P. stylirostris*, *P. monodon* and *P. vannamei* across multiple geographical locations, such as Panama, the Philippines, Mexico, Ecuador, Hawaii and Texas. Tang et al. (2007) further refined the detection by designing primers (IHHNV392F/R) that amplified a non-structural protein 392 bp segment of the coding region. Other innovations include the development of multiplex PCR techniques, such as those developed by Khawsak et al. (2008) for detecting six shrimp viruses and by Yang et al. (2006) for simultaneous detection of IHHNV and WSSV. Chen et al. (2012) later improved the sensitivity of multiplex PCR for detecting IHHNV and WSSV. Real-time PCR techniques have also been widely applied to detect IHHNV. Tang and Lightner (2001) used a TaqMan probe to detect IHHNV, amplifying an 81 bp portion from a CDS region of ORF1. A duplex real-time PCR assay was developed by Dhar et al. (2001) using SYBR Green chemistry for the detection of both WSSV and IHHNV, while Xie et al. (2008) advanced a real-time multiplex PCR for WSSV, IHHNV and TSV, demonstrating significantly enhanced sensitivity compared to conventional PCR. According to Sritunyalucksana et al. (2023), though many PCR-based methods have been developed for the confirmation of IHHNV, no single PCR test is 100% reliable. Therefore, besides PCR testing, confirmation by conventional techniques like histology is required for the parvoviruses (Sritunyalucksana et al. 2023). However, Dhar et al. (2024) have developed a Taqman assay for more accurate detection of IHHNV.

Immunodetection methods have also been explored for the detection of IHHNV. Poulos et al. (1994b) used murine monoclonal antibodies to develop an ELISA, and more recent techniques have incorporated monoclonal antibodies against recombinant IHHNV capsid proteins (Sithigorngul et al. 2009). Gene probe techniques, such as dot blot and in situ hybridization (ISH), have also been employed for specific and sensitive detection of IHHNV. Mari et al. (1993a) developed a first-generation gene probe, BS4.5, which showed high specificity for IHHNV. Yang et al. (2007) later introduced a DIG-labelled probe for non-lethal screening of pathogen-free shrimp. A probe-based real-time enzymatic recombinase amplification (ERA) was developed for rapid detection of IHHNV (Arbon et al. 2022). Loop-mediated isothermal amplification (LAMP) techniques have emerged as powerful diagnostic tools. Sun et al. (2006) introduced a LAMP assay for IHHNV detection that exhibited a 100-fold sensitivity increase compared to traditional PCR. Arunrut et al. (2011) further found that rapid visual identification enhanced LAMP diagnostics with a chromatographic lateral flow dipstick (LFD). Multiplex LAMP (mLAMP) assays have been developed to detect IHHNV and WSSV with superior sensitivity compared to nested and traditional PCR (He and Xu 2011). Arunrut et al. (2024) have recently developed a real-time triplex LAMP assay that could detect IHHNV with high specificity and sensitivity.

Emerging methods such as recombinase polymerase amplification (RPA) combined with lateral flow dipstick (Jaroenram and Owens 2014), real-time enzymatic recombinase amplification assay (Zhang et al. 2023) and long-range PCR

(Imsonpang et al. 2024) offer promising improvements in sensitivity and specificity. The integration of RPA with CRISPR technology, as in the Cas9-RPA-Cas12a (CRPAC) system, further enhances detection accuracy, particularly for distinguishing IHHNV from endogenous viral elements (Aiamsa-at et al. 2025). A summary of the diagnostic techniques used in detecting IHHNV is given in Table 1.

2.9 Host–Pathogen Interaction

Limited studies have been done on IHHNV infection and pathogenic processes in shrimp, and many aspects of the disease are still unknown. Although a number of proteins and immune-related enzymes have been recognized as important participants in the infection process, their precise functions are still unknown.

Table 1 Different diagnostic techniques used in the detection of IHHNV

Technique name	Species names	Viruses	
Polymerase chain reaction (PCR)	*P. stylirostris*, *P. monodon*, *P. vannamei*	IHHNV	Nunan et al. (2000)
Nested PCR	*P. stylirostris*, *P. monodon*, *P. vannamei*	IHHNV	Lee et al. (2022, 2024)
Duplex PCR	*P. vannamei*	IHHNV, WSSV	Yang et al. (2006)
	P. vannamei	IHHNV, EHP	Gao et al. (2023)
Real-time PCR	*P. stylirostris*, *P. monodon*, *P. vannamei*	IHHNV	Dhar et al. (2001, 2024)
	Procambarus clarkii	IHHNV	Chen et al. (2018)
Multiplex reverse transcription-PCR (mRT-PCR)	*P. vannamei*	IHHNV, six distinct viruses	Khawsak et al. (2008)
Loop-mediated isothermal amplification (LAMP)	*P. vannamei*	IHHNV	Sun et al. (2006)
Multiplex LAMP	*P. vannamei*	IHHNV, WSSV	He and Xu (2011)
Real-time triplex LAMP	*P. vannamei*	IHHNV	Arunrut et al. (2024)
Recombinase polymerase amplification (RPA) with LFD	*P. vannamei*	IHHNV	Jaroenram and Leigh Owens, (2014)
Enzyme-linked immunosorbent assay (ELISA)	*P. vannamei*	IHHNV	Sithigorngul et al. (2009)
Long-range PCR	*P. vannamei*	IHHNV	Imsonpang et al. (2024)
Real-time enzymatic recombinase amplification (ERA)	*P. vannamei*	IHHNV	Zhang et al. (2023)

Prophenoloxidase-2, HSP90, STAT and Na+/K+-ATPase, for example, are four proteins on *P. vannamei* gill membranes that bind to the IHHNV capsid protein and may facilitate viral entry, as reported by Yin and Fei (2015). Similar interactions between IHHNV capsid proteins and actin from *P. japonicus* and troponin I from *P. vannamei* have been discovered, underscoring the significance of these host factors in promoting infection. The investigation of IHHNV invasion routes has been made possible in part by Hou and Lin's (2015) discovery of fluorescent virus-like particles, or fVLPs.

LvRab5B and BCCIP are the two host proteins involved in immunological signalling and viral replication and interact with the virus to modify the shrimp's immune response to IHHNV (Liu et al. 2018; Zeng et al. 2019). Yin et al. (2022) reported that the putative host cell receptor, NN23, present in the gill membrane of shrimp, identifies the capsid protein of IHHNV. The infection caused by IHHNV has an impact on the activity of immune-related enzymes, like superoxide dismutase (SOD) and acid phosphatase (ACP), which are found to be enhanced in infected shrimp (Hu et al. 2015). Nevertheless, IHHNV also inhibits several immunological pathways, such as PPAF (prophenoloxidase-activating factor), which may contribute to the survival and dissemination of the virus in the host (Zhao et al. 2014). These findings highlight the intricacy of the host-virus relationships and the necessity of more research on the immune responses brought on by IHHNV infection.

2.10 Management

Shrimp aquaculture is severely threatened by the Infectious hypodermal and hematopoietic necrosis virus (IHHNV), leading to the need for efficient management and preventative techniques. In order to reduce the spread of viruses, biosecurity measures, including good husbandry practices and limiting the introduction of new animals, are crucial. It is essential to regularly monitor shrimp populations for the presence of IHHNV to provide early identification and prompt response to outbreaks. Moreover, research has been done to determine whether immunization and antiviral drugs might strengthen shrimp resistance to IHHNV. IHHNV-specific double-stranded RNA (dsRNA) and recombinant proteins were developed and used to reduce the spread of the virus. It shows the potential of recombinant proteins in antiviral defence by demonstrating their ability to efficiently reduce viral load in shrimp challenged with IHHNV (Praveen 2013). Chen et al. (2024) used RNAi technology to interfere *Litopenaeus vannami* Caspace-3 gene and found that the susceptibility of the shrimp to IHHNV was increased. This finding also paved the way for developing a new antiviral strategy by inducing this gene and further research is exploring the utilization of RNA interference methodologies to impede viral replication (Aranguren Caro et al. 2022).

Herbal extracts are reported to have antiviral activities against viral infections in shrimp. Ameer et al. (2024) recently reported the antiviral effects of leaf extract of *Rhizophora mucronata* against *Penaeus stylirostris* penstyldensovirus 1 (PstDVI)/IHHNV. The impact of IHHNV on shrimp health and productivity can be further reduced by implementing sustainable measures, such as the use of pathogen-resistant shrimp strains and enhanced farming methods. Improvements in the fight against IHHNV include breeding initiatives to produce IHHNV-resistant shrimp lines, like those produced by Super Shrimp, Inc., which have demonstrated over 80% survival in IHHNV-positive cases (Lightner 1992; Nunan et al. 2001). However, no effective vaccines or chemotherapeutic measures have been confirmed to date. A novel technique, such as the use of polychaete-assisted sand filters and tangential flow ultrafiltration methods, has shown promise in decreasing the transmission of IHHNV from water (Palmer et al. 2021; Wang et al. 2022).

3 Decapod Hepanhamaparvovirus 1 (DHPV-1)/ Hepatopancreatic Parvovirus (HPV)

3.1 Pathogen and the Disease

Hepatopancreatic parvovirus (HPV), also known as Decapod hepanhamaparvovirus 1 (DHPV-1), *Penaeus merguiensis* densovirus (PmeDNV) and *P. chinensis* densovirus (PchDNV), causes significant economic losses in the shrimp farming industry (Flegel 2006; Jeeva et al. 2012; Pollak et al. 2022). *P. chinensis* from South Korea, *P. merguiensis* from Singapore, *P. semisulcatus* from Kuwait and *P. monodon* from the Philippines were the shrimp species from which the virus was recorded (Chong and Loh 1984; Lightner and Redman 1985). This virus is reported to cause mortality in larval and postlarval stages of shrimp and stunted growth in juvenile stages in grow-out ponds (Lightner et al. 1993; Spann et al. 1997; Flegel et al. 1999; Limsuwan 1999). However, as it was observed that DPHV-1 was no longer linked to any appreciable detrimental economic effects on the aquaculture sector, it was delisted from the World Organization for Animal Health (WOAH). Nevertheless, this virus continues to be detected in both wild and cultivated penaeid shrimps from many countries such as Australia, Brazil, China, El Salvador, India, Indonesia, Israel, Kenya, Madagascar, Malaysia, Mexico, New Caledonia, Philippines, Singapore, South Korea, Taiwan, Tanzania, Thailand and the United States of America (Kibenge et al. 2024). Like all other parvoviruses, DHPV-1 infection also appears to be limited in its pathogenicity and economic relevance by numerous infections with other helper viruses (Chantanachookin et al. 1993; Cotmore et al. 2019; Singaravel et al. 2021).

DHPV-1 is classified under the genus Hepanhamaparvovirus and subfamily Hamaparvovirinae of the family Parvoviridae, which may infect both vertebrates and invertebrates, according to the properties of the virion (Cotmore et al. 2019). DHPV-1 is an icosahedral non-enveloped virus having a diameter of 22–24 nm. Its single-stranded linear DNA is around 6 kb in size. Like all other parvoviruses, the DHPV1 genome possesses hairpin structures on both ends of the linear DNA. Three primary open reading frames (ORFs), ORF1, ORF2 (NS1) and ORF3 (VP gene), make up the DHPV1 genome, which codes for the NS2 gene, NS1 gene and VP gene from the 3′ to the 5′ end and a short nucleotide sequence region overlaps

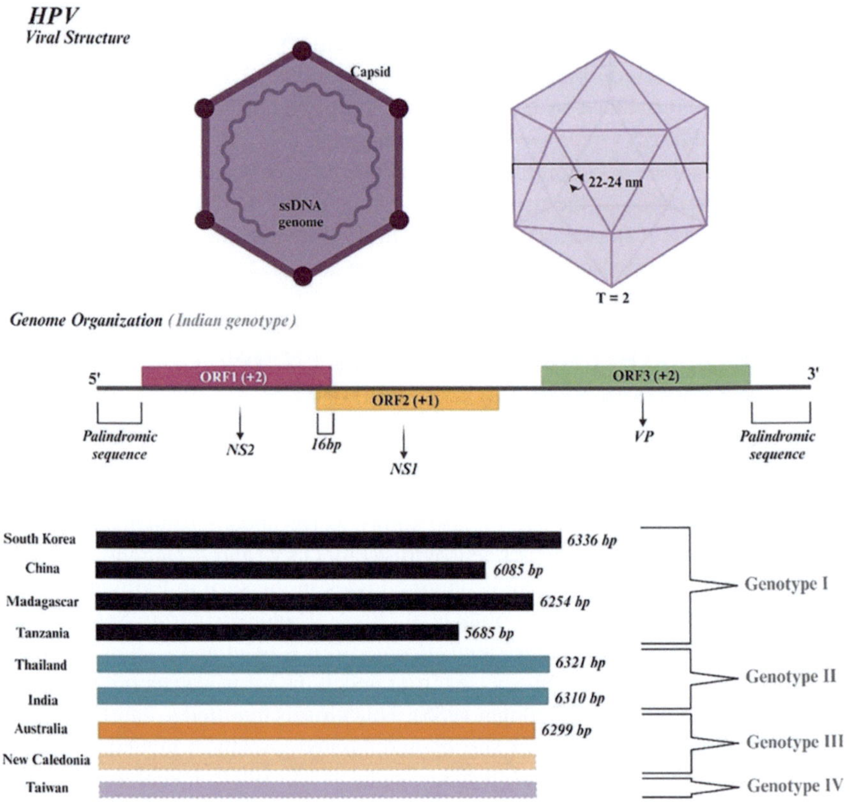

Fig. 3 HPV virus structure and genome organization. The illustration also depicts available genotypes compared to their genome sizes

between ORF1 and ORF2 (Safeena et al. 2012) (Fig. 3). So far four genotypes of DHPV1 have been recognized based on the structural protein gene sequence. Genotype I comprised of strains from South Korea, China, Madagascar and Tanzania; Genotype II includes strains from countries like India, Indonesia and Thailand; Genotype III comprised of strains from Australia and New Caledonia, whereas Genotype IV comprised of the novel strain from Taiwan (Lee et al. 2022) (Fig. 3).

3.2 Clinical Signs

Owing to the lack of distinct clinical symptoms, HPV/DHPV-1's significance as a causative agent of the disease is underestimated, and the disease's effects on world shrimp aquaculture are not quantified. Early larval and postlarval phases have been linked to high levels of HPV infections (Flegel 1995; Lightner et al. 1993; Spann et al. 1997). This virus causes stunted growth in juvenile stages and has been linked to 100% fatalities during outbreaks (Flegel et al. 1992, 1999; Limsuwan 1999). Poor development rate, hepatopancreatic atrophy, anorexia, reduced preening activity, higher surface and gill fouling with epicommensal organisms and opaqueness of the tail muscles are a few non-specific clinical symptoms of HPV-infected animals (Chen 1992; Lightner 1992; Sukhumsirichart et al. 1999). In addition to their lower size, HPV-infected shrimp frequently exhibit usual colouring and antennal length, and they have also been found in wild shrimps and in hatchery-raised larvae that appear healthy (Manjanaik et al. 2005; Umesha et al. 2003). However, severely infected shrimp may also show symptoms like whitish hepatopancreas, poor growth, anorexia, gill fouling and reduced preening activity (Pantoja and Lightner 2001; Manivannan et al. 2002). DHPV-1 has been blamed for significant crop losses in farms, as the afflicted tiny shrimp will not have any market value. Stress and circumstances related to overcrowding have been linked to higher death rates in shrimp ponds. As DHPV-1 is typically seen in association with other diseases and is rarely seen alone in epizootics, it is challenging to record the fatalities caused by these infections.

3.3 Host Range

HPV/DHPV-1 is a widespread virus that causes slow growth in shrimp and is a serious concern when wild-caught shrimp seeds are introduced into culture ponds. This virus affects an increasing number of both cultivated and wild shrimps, including *P. merguiensis*, *P. semisulcatus*, *P. chinensis*, *P. esculentus*, *P. monodon*, *P. indicus*, *P. penicillatus*, *P. japonicus*, *P. stylirostris* and *P. vannamei* (Lightner and Redmann 1985; Roubal et al. 1989; Lightner 1996; Paynter et al. 1985; Spann et al. 1997; La Fauce et al. 2007a). A similar virus has been reported in post-larvae of *Macrobrachium rosenbergii* (Anderson et al. 1990; Amala et al. 2025) and crayfish (La Fauce and Owens 2007). Additionally, DHPV from Australia has been detected in the haemolymph of the mud crab *Scylla serrata* (Owens et al. 2010). In India, DHPV-1 has been identified in *P. monodon*, *P. indicus* and *P. semisulcatus* and wild shrimps such as *Parapenaeopsis stylifera*, *P. japonicus*, *Metapenaeus monoceros*, *M. affinis*, *M. elegans*, *M. dobsoni*, *M. ensis* and *Solenocera choprai* using nested PCR (Manjanaik et al. 2005).

3.4 Mortality Pattern

The infection with HPV/DHPV-1 is often seen with slow growth in juvenile shrimp; hence, the mortality pattern of DHPV-1 is not well understood (Phuthaworn et al. 2016). However, it was reported that DHPV-1 infection has been associated with chronic mortality during the early larval and postlarval phases of shrimp (Pantoja and Lightner 2001; Manivannan et al. 2002). DHPV-1-associated cumulative mortality of 50–100% was reported in juvenile *P. merguiensis* after 4–8 weeks of infection (Lightner and Redman 1985). However, the exact mortality rates vary depending on shrimp species, age and environmental conditions. In outdoor ponds, DHPV-1 can cause low, continuous mortality in affected shrimp throughout the culture period, with cumulative mortality potentially reaching 80%. However, in indoor farming conditions, the mortality rate may be lower, suggesting that the farming environment can influence the lethal capacity of DHPV-1 (Liu et al. 2022).

DHPV-1 infection often occurs as a co-infection with other hepatopancreatic pathogens, and the cumulative impact of multiple viruses may contribute to increased mortality (Manivannan et al. 2002; Safeena et al. 2012). Mass mortality of post-larvae due to co-infection by MBV, HPV and WSSV was also observed (Manivannan et al. 2002). Environmental factors, co-infections and the severity of DHPV-1 infection, etc., may play a role in determining the impact on shrimp survival.

3.5 Target Organs and Histopathology

The development of large solitary basophilic intranuclear inclusion bodies in the hypertrophied nuclei of hepatopancreatic tubules and neighbouring midgut cells is characteristic of DHPV 1 infection (Lightner 1998). When DHPV 1 inclusion bodies are fully developed, the host cell shows nuclear hypertrophy, compression and chromatin margination. The cells that are actively proliferating (E- and F-cells) near the distal ends of the hepatopancreatic tubules contain the majority of DHPV 1 inclusion bodies typical of autonomous parvoviruses (Fig. 4).

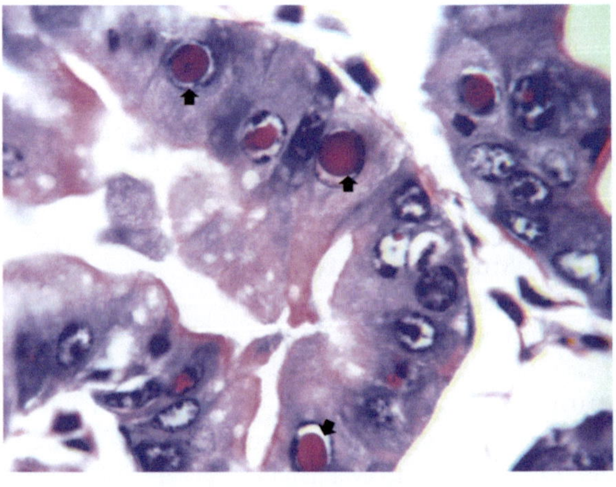

Fig. 4 Histological section of hepatopancreas of freshwater prawn, *M. rosenbergii* infected with HPV

3.6 Geographical Distribution

After DHPV 1 was initially identified in Asia, all reports of HPV/DHPV 1-positive shrimps were from Asia or the Indo-Pacific area, leading experts to believe that the virus originated there. With the live shrimp transportation from Asia for aquaculture, the virus eventually spread to wild shrimps in the Americas (Lightner 1996). The virus has been recorded in countries including Australia, Asia, Africa, China, Korea, Taiwan, the Philippines, Indonesia, Malaysia, Singapore, Kenya, Israel, Kuwait, North and South America and India. With the transport of infected *P. vannamei*, HPV was introduced into North and South America, the Pacific coast of western Mexico and the coastal regions of El Salvador (Lightner 1996). The virus entered Israel and Hawaii via the importation of diseased *P. penicilliatus* from Kenya and *P. chinensis* from Korea, respectively. The first reports of HPV were recorded in Australia in *P. esculentus* from the Gulf of Carpentaria and Moreton Bay (Paynter et al. 1985). Further, *P. merguiensis*, *P. monodon* and *P. japonicus* from Australia were later shown to have the infection (Lightner 1996; Roubal et al. 1989; Spann et al. 1997). The earliest reports of HPV infection in *P. monodon* were from Thailand in 1992 (Flegel et al. 1992; Flegel and Sriurairatana 1993, 1994). Umesha et al. (2003) reported a 34% prevalence of HPV in *P. monodon* postlarval samples from India. A new type of DHPV 1 has been reported from the Madagascar coast, which did not show any histopathological signs but gave positive results by PCR analysis (Cruz-Flores et al. 2024). These reports of HPV infection from different geographical areas imply that HPV is extensively present in wild, cultivated and hatchery-raised shrimp (Kibenge et al. 2024). The geographical distribution of this virus is depicted in Fig. 5.

3.7 Transmission

In 2003, Catap et al. reported the effective horizontal spread of HPV in *P. monodon* post-larvae by oral challenge. According to Lightner and Redman (1992), HPV may spread both ways vertically and horizontally. An experiment on the horizontal infection of DHPV 1 from diseased Artemia to healthy *P. monodon* post-larvae through feeding trials clearly proved the horizontal transmission of this virus through a vector (Sivakumar et al. 2009), and one of the significant concerns about the spread of HPV in shrimp farms is due to cannibalism. La Fauce and Owens (2007) observed the susceptibility of crayfish (*C. quadricarinatus*) to hepatopancreatic parvovirus (PmergDNV) infection and mortality after the first day of infection. The vertical transmission of HPV contamination from parental adults to F1 offspring of *P. chinensis* has been demonstrated by Lightner (1996) and further confirmed by HPV infections in hatchery-reared post-larvae (Manivannan et al. 2002; Umesha et al. 2003).

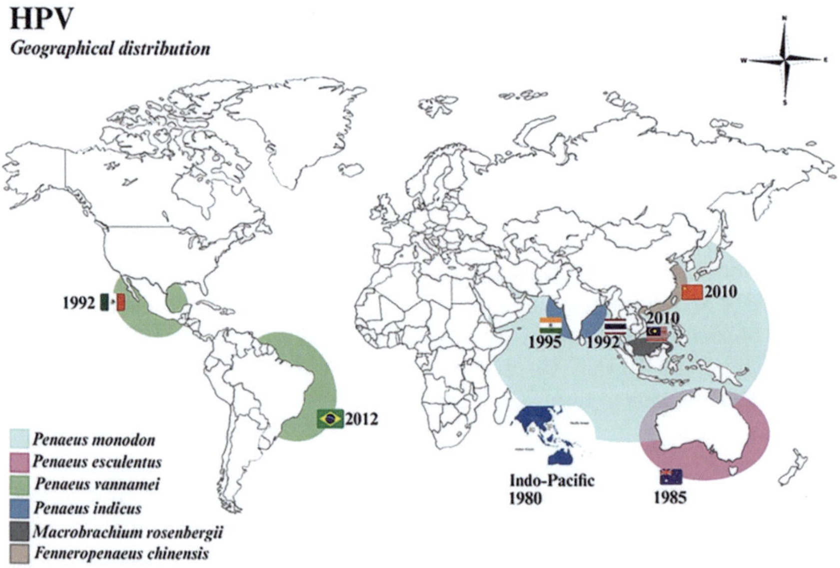

Fig. 5 Geographical distribution of HPV detected with their host species

3.8 Diagnosis

Initially, DHPV 1 diagnosis relied on conventional methods, such as squash preparation and histopathology of hepatopancreas, as there were no distinct gross signs of infection (Lightner 1996). The hallmark of HPV infection is the presence of single, large basophilic intranuclear inclusion bodies in the hypertrophied nuclei of hepatopancreatic epithelial cells and adjacent midgut cells (Lightner 1996). A simple diagnostic method involves preparing smears of hepatopancreatic tissue in 10% formalin with 2.8% NaCl, followed by H&E staining to visualize HPV inclusion bodies. Transmission electron microscopy (TEM) reveals these inclusion bodies as ovoid to spherical, with diameters ranging from 5 to 11 μm, containing electron-dense granular material and virion particles. This process also leads to chromatin margination with virions and lateral displacement of nucleoli.

Molecular diagnostic techniques have improved HPV detection, offering quick, sensitive and specific alternatives. These techniques facilitate regular surveillance of shrimp larvae, broodstock and potential HPV carriers, enabling early diagnosis in aquaculture systems. Among the molecular methods, polymerase chain reaction (PCR) has become the most commonly used technique due to its speed and sensitivity. However, PCR can yield false negatives due to nucleotide variations, necessitating multiple primer sets to detect different virus strains (Lightner et al. 1994; Phromjai et al. 2001). Srisala et al. (2021b) have developed a semi-nested universal PCR method to screen DHPV 1 in both *M. rosenbergii* and *P. monodon.* TaqMan-based real-time PCR, targeting the capsid protein gene, has been developed for the Australian HPV isolate (La Fauce et al. 2007b). A generic TaqMan test designed by

Yan et al. (2010) could detect HPV strains from nine geographic areas with a sensitivity of detecting a single DNA copy of HPV. A duplex real-time PCR assay has also been developed for concurrent detection of HPV and MBV (Tang and Lightner 2011). Furthermore, a highly sensitive multiplex real-time PCR assay coupled with high-resolution melting (HRM) analysis has been designed for the detection of selected viruses in shrimp (Panichareon et al. 2011). A type-specific TaqMan real-time PCR assay was developed to quantify HPV Type III infections, revealing the widespread distribution of HPV Types I and III in Korea (Jang et al. 2011). Further, Kim et al. (2025) developed a qPCR method for the rapid detection of the newly emerged novel genotype of Decapod hepandensovirus 1 (nDHPV1) in *P. vannamei*. By the use of Laser microdissection and next-generation sequencing, Cruz-Flores and Dhar (2024) could identify a new DHPV 1 strain, which PCR could not detect. This advanced technique can be used for studying the evolution of viruses as well as for their prompt diagnosis at the early stages of infection.

Though immunological methods are less sensitive than PCR, they are practical, precise and time-efficient, offering cost-effective pathogen screening for large sample sizes and were developed for the detection of HPV. A monoclonal antibody-based assay targeting HPV capsid protein epitopes (CP-N and CP-C) has been developed to enhance the sensitivity of HPV detection (Srisuk et al. 2011). For detecting low-grade HPV infections, PCR-ELISA has been developed, utilizing non-radioactively labelled genomic probes (Sukhumsirichart et al. 2002). Initially, in situ hybridization and electron microscopy techniques were employed to elucidate the infection cycle and development of HPV (Mari et al. 1995; Pantoja and Lightner 2001). The Loop-mediated isothermal amplification (LAMP) method, coupled with Lateral flow dipstick (LFD), provides a sensitive diagnostic option (Nimitphak et al. 2008) with sensitivity comparable to nested-PCR (Manjanaik et al. 2005). Additionally, the recombinase polymerase amplification (RPA) lateral flow technique has emerged as a rapid and accurate HPV diagnostic tool with high sensitivity and specificity (Pollak et al. 2022). This technique further enhances the efficiency of HPV detection in shrimp aquaculture. A summary of the diagnostic techniques used in detecting IHHNV is given in Table 2.

3.9 Host–Pathogen Interaction

Like other Parvoviruses, HPV/DHPV 1 also multiplies in the nucleus and produces large intranuclear inclusion bodies. DHPV 1 infection begins with attachment to the microvilli, and then it enters the host cell by pinocytosis (Mari et al. 1995). HPV matures inside the nucleoplasm and accumulates in the major inclusion bodies (Pantoja and Lightner 2001). The virus replicates inside the nucleus of the host by a unidirectional single-strand displacement mechanism called 'rolling hairpin replication', using the viral self-priming hairpin telomeres, cellular polymerases, ligases and other replication factors (Söderlund-Venermo 2019). Many parvoviruses contain phospholipase A2 (PLA2) enzymatic domain, which is inevitable for virus cell entry, present in the extended N-terminus of the VP1 protein (Kibenge et al. 2024).

Table 2 Different diagnostic techniques used in the detection of HPV

Technique name	Species names	Viruses	References
Non-nested PCR	*Penaeus monodon, P. indicus, P. semisulcatus*	HPV	Manjanaik et al. (2005)
Nested-PCR	*Parapenaeopsis stylifera, Penaeus japonicus, Metapenaeus monoceros, M. affinis, M. elegans, M. dobsoni, M. ensis and Solenocera choprai*	HPV	Manjanaik et al. (2005)
TaqMan-based real-time PCR	*P. merguiensis*	HPV	La Fauce et al. (2007b)
	Fenneropenaeus chinensis	HPV	Yan et al. (2010)
	P. chinensis	HPV	Jang et al. (2011)
Duplex real-time PCR	*P. monodon*	HPV, MBV	Tang and Lightner (2011)
Multiplex real-time PCR	NA	PmDNV, WSSV, YHV	Panichareon et al. (2011)
qPCR	*P. vannamei*		Kim et al. (2025)
LAMP-LFD	*P. monodon*	HPV	Nimitphak et al. (2008)
RPA	*Fenneropenaeus merguiensis*	HPV	Pollak et al. (2022)
Monoclonal antibody	*P. monodon*	HPV	Srisuk et al. (2011)
PCR-ELISA	*P. monodon*	HPV	Sukhumsirichart et al. (2002)

As the size of the DHPV1 genome is so small, HPV also lacks a viral DNA polymerase and requires actively dividing host cells for replication. Mature parvovirus particles build up during the last step of the replication cycle, which disrupts the nuclear membrane and releases viruses into the cytoplasm (Kawase 1990). The spreading of this virus to new host cells starts with the release of intranuclear inclusion bodies into the lumens of the hepatopancreatic tubules (Pantoja and Lightner 2001). To date, no HPV-susceptible shrimp cell lines have been developed to study the pathology, modes of transmission and virulence mechanism of the virus. However, an insect model, *Acheta domesticus,* has been used successfully for conducting a bioassay of PmergDNV (La Fauce and Owens 2008).

3.10 Management

As there are no antiviral treatments, following strict biosecurity measures and selecting disease-free or resistant strains remain the only basis for disease management. Virus transmission can be controlled by selecting shrimp broodstock from ponds without showing any HPV infections or by selecting lineages or families known to have lower levels of HPV (Knibb et al. 2015). Further, the gene silencing approach using RNA interference has been described for HPV (Attasart et al. 2011). Similarly, studies on the development of protein vaccines and HPV gene-specific dsRNA have shown that HPV replication in experimentally challenged shrimp is inhibited (Safeena 2014). However, oral administration of these antiviral proteins and molecules remains a great challenge, and the usefulness of these vaccination techniques is yet to be realized in shrimp farming.

4 *Penaeus Monodon*-Type Baculovirus/Monodon Baculovirus (MBV)

4.1 Pathogen and the Disease

MBV is one of the comprehensively studied viral pathogens of shrimp. The first case of MBV was reported in Taiwan in 1977 (Liao 1977; Lightner et al. 1983c). Subsequently, MBV has been reported to occur in various commercially cultivated penaeid shrimp worldwide (Anderson et al. 1987; Nash et al. 1988a; Lightner and Redman 1992). Although it was reported that the Taiwanese shrimp industry collapsed in 1988 due to an MBV-associated epizootic (Lin 1989), further research showed that managerial and environmental factors also played a role (Liao et al. 1992). Although high levels of infection may induce delayed growth and thus declined production at harvest, MBV often does not cause mortality in shrimp (Fegan et al. 1991; Flegel et al. 2004). Around 85–100% of shrimp mortalities in hatcheries have been attributed to this virus (Lightner et al. 1983c; de la Peña et al. 2005; Natividad et al. 2006). The co-infection of MBV with other shrimp pathogens, including WSSV, IHHNV and HPV, has been documented (Anshary et al. 2017; Abraham et al. 2020).

MBV is a double-stranded DNA baculovirus that possesses a rod-like structure and a single envelope. The virus replicates inside the nucleus and produces intranuclear occlusion bodies. In a review, Brock and Lightner (1990) provided extensive documentation of the ultrastructure of the virions and occlusion bodies (OBs). The development of the cellular membrane structure of penaeid shrimp cells infected with MBV has been studied by Lu et al. (1996). Nevertheless, the evidence relied solely on morphological information obtained from examinations of the infected cells using transmission electron microscopy. Mari et al. (1993b) used sucrose gradient ultracentrifugation to isolate MBV from infected hepatopancreas of *P. monodon*. By collecting various fractions, they purified virions, envelopes, nucleocapsids and polyhedron occlusion body subunits. The authors claim that undamaged virion particles have dimensions of 68–77 nm in diameter and 265–282 nm in length. Each enveloped particle possesses a single capsid. The surface of the envelope consists of a tiny, uniformly sized granular structure interspersed with tiny spikes that are visible at the intact particle's tip. The application of PTA staining revealed the appendages at the extremities of the envelope, which are believed to be the extensions of the envelope. Nucleocapsids have helical rows of surface features that are separated by roughly 6 nm. These features give the nucleocapsid surface a 'cross-hatched' appearance because they are oriented perpendicular to the nucleocapsid axis. The non-enveloped nucleocapsids are 250–269 nm long and 62–68 nm in diameter (Fig. 6).

Ramasamy et al. (2000) reported two varieties of MBV OB in *P. monodon* from India. While type 1 OB had occluded and non-occluded rod-shaped virions with an envelope consisting of two layers, type 2 OB featured an envelope with only a single layer. Although most OB subunits are spherical, some have six sides. The occlusion body subunits (OBS) assume a tri-dimensional crystalline structure, explained by the apparent cubic symmetry and the six-sided shape of MBV occlusion body

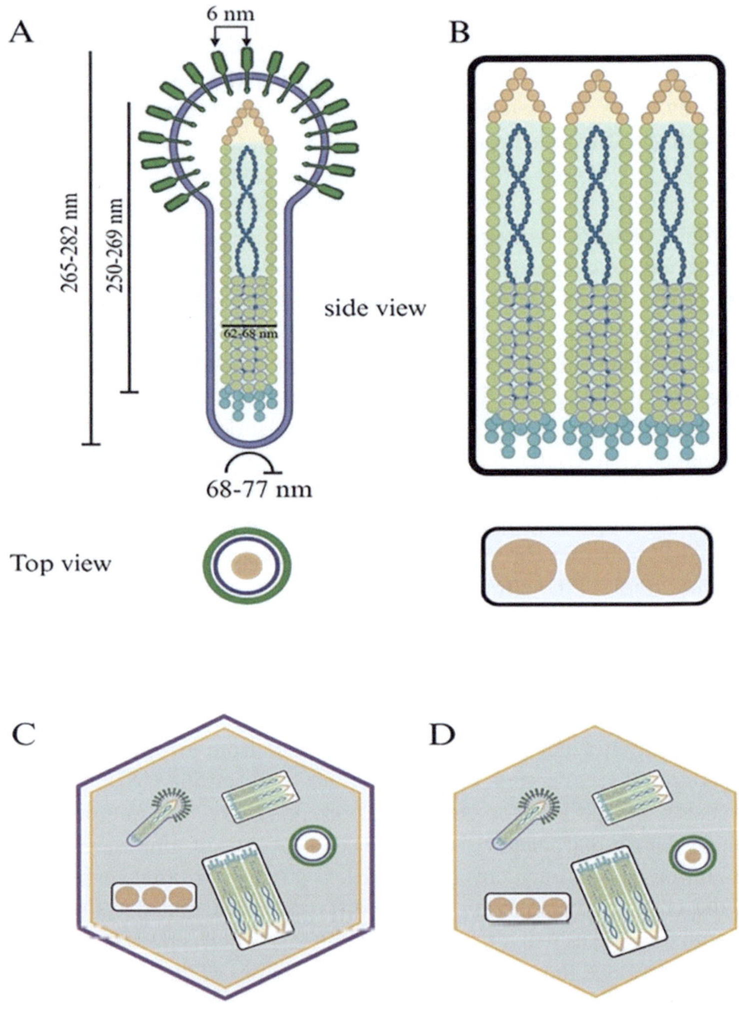

Fig. 6 MBV virus structure. (**a**) Detailed view of a single MBV particle, where both cross-sectional side views and top views are illustrated, (**b**) MBV occluded viral particles, (**c**) Type 1 occlusion body and (**d**) Type II occluded body

subunits (OBS) (Mari et al. 1993b). Among the three viral families, Bacilloviridae, Reoviridae and Poxviridae, that contain polyhedrins, MBV possesses one of the largest OBSs (Mari et al. 1993a, b). MBV has been shown to affect all the life stages of *Penaeus monodon*, with the most susceptible being late larval, post-larval and young juvenile shrimp. However, MBV infection has not been reported from the early larval stages, and OBs were not detectable in larvae prior to the PL2 stage (Chen et al. 1989a).

According to a report, the larger the animal, the less severe the infection, and vice versa (Lightner et al. 1983c). Three stages in MBV infection have been proposed by Lightner et al. (1983c). Stage 1 is noticeable by mildly hypertrophied nuclei with a small number of mature virions without OBs; stage 2 is marked by hypertrophied nuclei with complete virions and developing occlusions; and stage 3 is characterized by mature occlusions and completely formed virions. The final phase consists of cytolysis and cell necrosis.

MBV status has only been established by morphological traits and phylogenetic study of incomplete genomic sequences, and because of this, the virus's nomenclature and classification have undergone numerous revisions. As MBV possesses occlusion bodies with many virions and a single nucleocapsid, it has been categorized within the subgenus SNPV and given the name PmSNPV (singly enveloped nuclear polyhedrosis virus from *P. monodon*) (Mari et al. 1993a, b). Later, in its eighth report, which was included in the 9th ICTV report (2012), the International Committee on Taxonomy of Viruses (ICTV) suggested MBV as a 'tentative species' called PemoNPV in the genus Nucleopolyherdovirus (Theilmann et al. 2005). According to the phylogenetic study, the virus was renamed *P. monodon* nudivirus, PmNV (Wang and Jehle 2009) and reallocated to the genus Nudivirus.

Furthermore, MBV was shown to differ significantly from viruses grouped in the Baculoviridae based on the comparison of the chemical characteristics of the polyhedrin generated by the virus and the phylogenetic relationship of the entire genome sequence. According to Yang et al. (2014), it has been suggested to rename it as *Penaeus monodon* nudivirus (PmNV) under the family Nudiviridae under a unique new genus called *Gammanudivirus*. However, the scientific literature, notably the Manual of Diagnostic Tests for Aquatic Animals (OIE 2019), frequently refers to the virus as Penaeus monodon-type baculovirus (MBV).

4.2 Clinical Signs

Lethargy, decreased feeding and preening behaviour, and reduced development rate are all gross clinical signs of MBV infection. Additionally, it has been elucidated that MBV infection has been linked to a high prevalence of bacterial infection and gill and surface-fouling organisms (Lightner 1988; Vijayan et al. 1995). Localized 'shell disease-type' lesions of the gills or cuticle or bacterial septicemia have resulted from this. Although mild or severe MBV infection resulted in the development of smaller and darker-coloured *P. monodon* post-larvae than the normal ones,

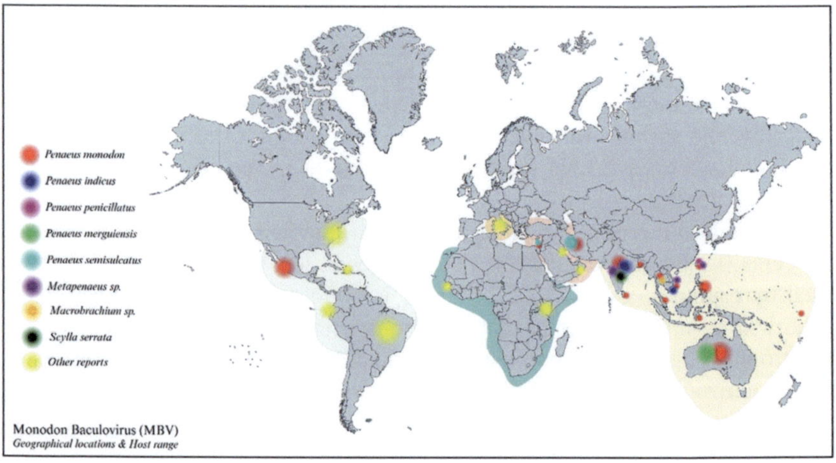

Fig. 7 Geographical distribution of MBV. The illustration highlights the host species of the MBV detection, and the continental areas are marked with coloured selections

the study by Flegel et al. (1999) found no statistical relationship between MBV infection and stunted growth.

4.3 Host Range

The incidence of MBV has been reported from a wide range of farmed, captured or imported penaeid shrimp species globally (Fig. 7). Even though MBV primarily infects wild *P. monodon*, it has also been documented from *P. penicillatus, P. indicus, P. semisulcatus, P. merguiensis, P. japonicus, P. kerathurus, P. esculentus, Litopenaeus vannamei, Metapenaeus ensis* and *M. lysianassa* (Sano et al. 1981; Doubrovsky et al. 1988; Chen et al. 1989b; Colorni 1989; Thikiew 1990; Natividad 1991; Vickers et al. 1993; Vijayan et al. 1995; Hao 1999; Manivannan et al. 2004; Tokhmafshan et al. 2004). Furthermore, freshwater prawn, *Macrobrachium rosenbergii*, has been known to harbour MBV (Gangnonngiw et al. 2010). By using PCR detection, histological analysis and immunohistochemistry, it has been demonstrated that *P. vannamei* is resistant to MBV infection (Gangnonngiw and Kanthong 2019). Recently, Rajendran et al. (2022) reported MBV from the mud crab *Scylla serrata*.

4.4 Geographical Distribution

MBV displays a wide geographical distribution. After a detailed investigation, Lightner et al. (1987) and Chen et al. (1989b) showed that MBV was present and widely distributed throughout the Taiwanese shrimp cultivation area, even though it was initially reported in *P. monodon* imported from Taiwan to Mexico (Lightner and Redman 1981; Lightner et al. 1983c). Subsequently, other research works found a strong correlation between MBV and disease-related losses in numerous shrimp culture facilities in Taiwan and other Indo-Pacific culture regions, including Japan (Baticados et al. 1990, 1991; Colorni et al. 1987; Lester et al. 1987; Doubrovsky et al. 1988; Nash et al. 1988a, b; Chen et al. 1989a, b; Fegan et al. 1991; Natividad 1991; Vogt 1992; Turnbull et al. 1994). Since then, assessments of shrimp farming facilities and wild shrimp have shown widespread dissemination of MBV from various global locations (Natividad and Lightner 1992; Felix and Devaraj 1993; Ramasamy et al. 1995; Alavandi et al. 1995).

According to reports, MBV is enzootic in wild penaeids throughout the Indo-Pacific region of Asia, where it has demonstrated a broad geographic range (Lightner 1996). Many reports have documented the distribution of the virus globally, including Americas (Mexico, Ecuador, Brazil, Puerto Rico and several South-eastern US states), the Middle East, (Kuwait, Oman, Iran and Israel), the Mediterranean (Italy), Africa (Kenya, Gambia), the Pacific (Tahiti, Hawaii) and the Indo-Pacific (Peoples Republic of China, Vietnam, Taiwan, Philippines, Malaysia, Singapore, Thailand, Sri Lanka, India, Indonesia, Bengladesh and Australia) (Alavandi et al. 1995; Ramasamy et al. 1995, 2000; Vijayan et al. 1995; Lightner 1996; Sundararaj et al. 1996; Karunasagar et al. 1998; Hao 1999; Vickers et al. 2000; Manivannan et al. 2002, 2004; Ambipillai et al. 2003; Otta et al. 2003; Mishra et al. 2005; Oanh and Phuong 2005; Uma et al. 2007; Umesha et al. 2003, 2006; Vaseeharan and Ramasamy 2004; Gunalan et al. 2010; Suganthi et al. 2012; Joseph et al. 2015; Manickavel et al. 2018; Chakroborty et al. 2020; Abraham et al. 2020; Haq et al. 2022; Khatun et al. 2023) (Fig. 7).

4.5 Mortality Pattern

It has been reported that during the early stages of MBV infection in the post-larvae of *M. rosenbergii,* the lesions produced by the virus in the hepatopancreas are comparable to those of *P. monodon.* If other farming conditions are favourable, shrimp can survive MBV infection effectively (Lightner et al. 1987; Fegan et al. 1991). Flegel et al. (1999) have also reported that *P. monodon* is tolerant to MBV. In cultivated shrimp, severe infection with high incidence has been observed without causing significant harm to the shrimp's health (Vijayan et al. 1995). However, in *P. monodon* cultured in raceways, facultative infections and crowding may make the infection more severe and common (Lightner et al. 1983c). On the other hand, shrimp infected with MBV would be more susceptible to subsequent infections (Natividad and Lightner 1992). The lack of significant effects in animals is because the disease does not manifest until an infected animal experiences a stressor that triggers the disease development.

Significant mortality from transfer stress was seen in *P. monodon* broodstock and *P. monodon* and *P. penicillatus* larvae infected with MBV (Chen et al. 1989b). On the other hand, the reports indicated that the hepatopancreas of the animals infected with MBV had a significant bacterial infection. Lin (1989), Tangtongpiroj (1989) and Thikiew (1990) have all documented high mortality rates linked to MBV infection. Although MBV is thought to reduce productivity, it has been found that it does not result in a high mortality rate (Flegel et al. 1992). Since then, it has been observed that *P. monodon*, which exhibits slow-growth disease, has a significant prevalence of co-infection with both HPV and MBV (Chayaburakul et al. 2004). According to Flegel and Pasharawipas (1998), MBV is one of the non-lethal viral infections that shrimp seem to withstand during disease epidemics linked to other viruses, such as YHV.

In addition to increasing the mortality rate of infected shrimp by predisposing them to other pathogens, MBV infection may result in decreased growth, harvest value and general animal health. Conversely, it is also hypothesized that inadequate rearing conditions could cause the virus proliferation, leading to a secondary infection and eventual death (Flegel 2006). It was reported that MBV-infected seeds can be successfully cultivated at optimum stocking densities (6 PL/m2) with appropriate feed and water management (Gunalan et al. 2010). Although MBV is not a serious disease, Flegel (2006) thought it should be eliminated from the farming system because it is unlikely that shrimp would have a high infection level without experiencing any adverse effects.

4.6 Target Organs and Histopathology

The anterior midgut epithelium of very young post-larvae and the hepatopancreatic tubule and duct epithelium of post-larvae, juveniles and adults are among the organs and tissues that MBV targets (Lightner et al. 1983c). Shrimp juveniles and adults' duct epithelium and hepatopancreatic tubular epithelial cells' nucleus, as well as the anterior midgut epithelium of young post-larvae, are the sites of virus reproduction (Couch 1981; Anderson et al. 1987; Johnson and Lightner 1988; Brock and Lightner 1990; Lightner 1996). The appearance of many spherical OBs in the hepatopancreas and anterior midgut epithelial cells is the most obvious clinical indication of the infection. Major histopathological changes include necrosis, lysis and sloughing-off of hepatopancreatic tubular cells into the lumen (Lightner and Redman 1981; Baticados et al. 1991; Vijayan et al. 1995). In addition to chromatin shrinkage and margination, the MBV-infected hepatopancreatic cells had hypertrophied nuclei with one or more eosinophilic occlusion bodies (Ramasamy et al. 2000; Gangnonngiw et al. 2010).

4.7 Transmission

This virus spreads mainly through horizontal methods. Contaminated post-larvae (PL) from wild broodstock that is grossly normal and infected with PmNV are the primary source of PmNV infection in rearing ponds. Virus particles are embedded in occlusion bodies (OBs), which are composed of polyhedrin and protein matrix and are responsible for virus transmission between animals. Both infected broodstock and larvae have been found to release OBs through their faeces, which can be consumed by other larvae MBV has been shown to transmit horizontally through oral contact with contaminated tissues, OBs, free viruses or fomites. Although vertical routes can transmit the virus from an MBV-infected mother to offspring, transovarian transfer of infection has not been previously established.

However, according to a study by Kanjanasopa et al. (2015), MBV can spread from the broodstock to its progeny through the ovarian route instead of just through faecal contamination. It was discovered that the shrimp's oocytes, eggs and nauplii contained polyhedrin, the MBV's protein. Brock and Lightner (1990) have suggested that the failure of experimental horizontal transmission of MBV might be because virions are inactivated by freezing and slow thawing. However, Chen et al. (1992) and Paynter et al. (1992) demonstrated that MBV can spread horizontally through ingestion of both the free and occluded virions. It was also reported that viruses can spread directly through the water column in a horizontal direction (Paynter et al. 1992).

4.8 Diagnosis

The simplest way to detect the infection caused by MBV is by direct microscopic examination. The presence of OBs in the freshly sampled hepatopancreas, visible through larvae and post-larvae carapace, is a typical hallmark of MBV infection. When hepatopancreas tissues are prepared as squash mounts and stained with 0.05% malachite green, clustered OBs can be seen in the hypertrophied nucleus (Lightner 1985). The hepatopancreatic squash preparation by haematoxylin-eosin staining can also be used to identify the OBs in which the eosinophilic OBs can be observed. The midgut epithelial cells and non-lethal sampling of fresh faecal strings can also be used to detect OBs (Lightner 1996). The OBs can also be visualized by the microscopic examination of the stained histological sections of the hepatopancreas and anterior midgut.

The sensitivity of detection of MBV can be enhanced by either acridine orange staining or epifluorescence microscopy (Diamant and Colorni 1987; Colorni 1989), fluorescent microscopic examination of aqueous phloxine-stained hepatopancreas and faecal samples (squash mount preparation), and tissue sections (Thurman et al. 1990; Fegan et al. 1991). In contrast to the pale green background, the OB is seen as brilliant yellow-green (Lightner 1996).

MBV can also be detected by dot blot hybridization (Chang et al. 1993; Mishra et al. 2005) and in situ hybridization in fixed tissue slices (Poulos et al. 1994a, b; Lu et al. 1995; Lightner 1996). MBV can be confirmed by the presence of OBs under an electron microscope (Fegan et al. 1991). The electron microscopic characteristics of MBV have been investigated by many researchers (Lightner and Redman 1981; Johnson and Lightner 1988; Lightner et al. 1983c). There have also been reports of polyclonal or monoclonal antibodies produced against MBV polyhedrin

(Satidkanitkul et al. 2005; Boonsanongchokying et al. 2006; Sridulyakul et al. 2011), as well as other techniques, including ELISA (Hsu et al. 2000).

A monoclonal antibody-based immunochromatographic strip test has been developed to detect the polyhedrin protein of MBV (Wangman et al. 2012). A common method for diagnosing MBV is the polymerase chain reaction (PCR). There have been reports of PCR using primers that target the DNA polymerase gene (Hsu et al. 2000) and polyhedrin gene of insect baculovirus (Lu et al. 1993) of MBV. Even though the first primer could amplify other baculoviruses, the latter primer could amplify only MBV. A nested PCR methodology was designed by Belcher and Young (1998). An improved PCR method for the detection of MBV was made by Surachetpong et al. (2005). A duplex PCR for sensitively detecting MBV and WSSV, as well as MBV and HPV, has been reported (Natividad et al. 2006; Tang and Lightner 2011). A multiplex RT-PCR was developed to detect six main shrimp viruses, including MBV (Khawsak et al. 2008). Subsequently, a multiplex PCR for concurrently detecting WSSV, MBV and IHHNV has been developed by Sibonga et al. (2013). The real-time PCR employing MBV-specific TaqMan probe and SYBR Green has been developed for diagnosis and quantification (Yan et al. 2009; Kumar et al. 2014).

Furthermore, Chaivisuthangkura et al. (2009) and Caipang et al. (2011) reported Loop-mediated isothermal amplification (LAMP). A LAMP method coupled with a flow dipstick has also been developed (Nimitphak et al. 2010). Different diagnostic methods employed to detect MBV are summarized in Table 3.

4.9 Management

MBV epizootics are hatchery and/or pond management issues that are not specific to any location or time of year. Although stress exacerbates the disease, *P. monodon*

Table 3 Different diagnostic methods used in the detection of MBV

Diagnostic method	Detection limit	References
PCR	100 copies of plasmid DNA	Surachetpong et al. (2005)
Multiplex PCR	1.5 pg	Khawsak et al. (2008)
LAMP	50 viral copies per ng	Chaivisuthangkura et al. (2009)
LAMP-LFD	0.1 pg	Nimitphak et al. (2010)
LAMP	10 pg of DNA mL^{-1}	Caipang et al. (2011)
Real-time PCR (TaqMan probe)	1 copy of plasmid DNA	Yan et al. (2009)
Real-time PCR (SYBR Green)	12 copies	Kumar et al. (2014)
Duplex PCR	0.1 fg of plasmid DNA	Natividad et al. (2006)
	1 copy	Tang and Lightner (2011)
MAb to polyhedrin	0.2 μg/mL^{-1}	Boonsanongchokying et al. (2006)

tolerates it effectively under ideal conditions. The best way to control the MBV infection is to eradicate the contaminated broodstock and adhere to stringent disinfection protocols, as the virus is spread horizontally from infected individual shrimp. Formalin, chlorine and iodophore are among the chemical agents that have been suggested for use in sanitizing shrimp hatcheries due to their purported effectiveness in inactivating MBV. In order to eradicate MBV infection in *P. monodon* larval stages, Chen et al. (1992) explored a number of washing techniques. They found that a combination of formalin and iodophore washing could totally eradicate the virus from nauplii or fertilized eggs.

Due to its demonstrated resistance to these chemical treatments compared to other shrimp viruses, MBV requires higher concentrations of the chemicals to be rendered inactive (Spann et al. 1993). Given this, the iodine dose previously recommended will not be sufficient for the complete inactivation of MBV (Manisseri et al. 1999), and the authors recommended that hatchery disinfection solutions contain at least 10,000 parts per million of iodine. Given these facts, a viable strategy for eliminating MBV should include a variety of techniques, including the elimination of contaminated stocks and stringent hatchery disinfection protocols (Brock 1983; Wyban and Sweeny 1991; Chen et al. 1992) that employ various chemicals in accurate dosage rates.

Chotipuntu et al. (2015) reported using an upwelling flow-through system to decontaminate MBV in shrimp eggs. The authors revealed that the system they designed could delay the occurrence of MBV in shrimp larvae. As a management option against MBV, silencing the MBV structural gene of p74 was undertaken, and dsRNA was delivered into *P. monodon* post-larvae using a suitable nanoparticle carrier system. The findings showed that *P. monodon* had an effective gene-silencing process and a noteworthy survival rate of 86.63% (Kumar et al. 2016).

5 Infectious Myonecrosis Virus (IMNV)/Penaeid Shrimp Myonecrosis Virus (PsIMNV)

5.1 Pathogen and the Disease

The infectious myonecrosis virus (IMNV), also referred to as the Penaeid shrimp myonecrosis virus (PsIMNV), is the cause of infectious myonecrosis (IMN), a serious disease that affects farmed shrimp. The virus was previously recognized as idiopathic myonecrosis when the condition was first reported in Brazil (Lightner et al. 2004; Coelho et al. 2009). Due to its transboundary spread and significant effects on a number of shrimp species, both the World Organization for Animal Health (WOAH) and the Food and Agriculture Organization (FAO)/Network of Aquaculture Centers in Asia-Pacific (NACA) added IMNV to the list of reportable diseases in the year 2005 and 2006, respectively (Prasad et al. 2017). According to Andrade et al. (2007), IMNV had a substantial economic impact in Brazil, with estimates as high as USD 440 million at the end of 2005. Brazil and Indonesia had lost around USD1 billion by the end of 2011 (Lightner et al. 2012).

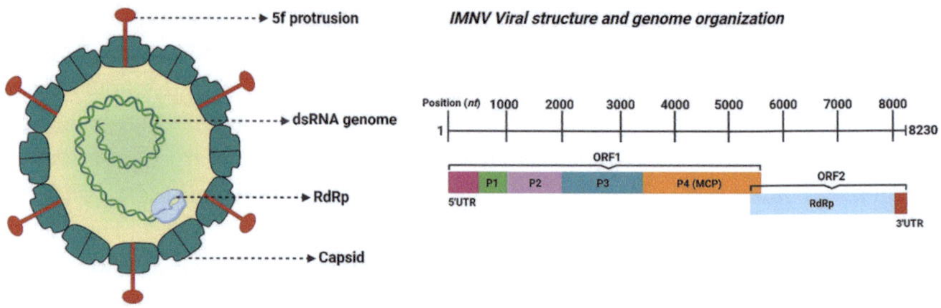

Fig. 8 Virus structure and genome organization of Infectious myonecrosis virus: 5′ and 3′ UTRs (untranslated regions), ORFs (open reading frames), major capsid protein (MCP) and RNA-dependent RNA polymerase (RdRp). (Reproduced from Tang et al. 2019)

Despite having a lower death rate than WSSV and YHV, IMNV is reported to be a stress-dependent virus that troubles *P. vannamei* when water quality parameters, including temperature, pH, dissolved oxygen and plankton, are suddenly altered (Jha et al. 2021). As IMNV infection progresses slowly, it has significant financial implications because of the high feed conversion ratio caused by a massive increase in feed intake by the infected animals (Lightner et al. 2004).

All life stages of shrimp are infected by the virus, including post-larvae, juveniles and adults; nevertheless, mortality was noted in both juveniles and adults (Nunes et al. 2004). It has been reported that IMNV was found to be co-infected with other pathogens such as IHHNV (Teixeira-Lopes et al. 2011; Aly et al. 2021), EHP (Jithendran et al. 2021), *Mr*NV (Senapin et al. 2013), DIV1 (Srisala et al. 2021a, b), WSSV (Feijo et al. 2013) and *Vibrio harveyi* (Oktaviana et al. 2014). Suryakodi et al. (2022) published the first report on the co-occurrence of WSSV, IMNV and EHP in farmed *P. vannamei*.

IMNV possess double-stranded RNA with a genome size of 7561–8230 bp and contains two open reading frames (ORF1 and ORF2) that partially overlap. This 40 nm-diameter, non-enveloped icosahedral virus has surface protrusions that resemble fibres. ORF1 encodes major capsid protein (MCP), whereas ORF2 encodes RNA-dependent RNA polymerase (RdRp) (Fig. 8). Based on the phylogenetic study of RdRp, IMNV was placed under the family Totiviridae.

5.2 Clinical Signs

Shrimp infected with IMNV exhibited general behavioural signs, such as sluggishness during or shortly after stressful situations like feeding and netting, prompt changes in the water temperature and salinity drops. Under these stressful circumstances, the mortality rate rises (Lightner et al. 2004; Nunes et al. 2004; Poulos et al. 2006; Lightner 2011). The production loss was usually reported from the farms after 40 days of culture or more (Taukhid and Nuraini 2008; Flegel 2009). The infected shrimps showed decreased feed consumption, which elevated the feed conversion ratio (FCR) and caused the survival rate to fall to 21% (Tang et al. 2005; Coelho et al. 2009). The acute stage of infection is when the clinical symptoms are most noticeable. The signs of IMNV-infected *P. vannamei* included focal to widespread white necrotic patches in the striated (skeletal) muscle, especially in the tail fan and segments at the distal part of the body. Redness may develop in some infected shrimp, giving the impression that they are cooked (Nunes et al. 2004). Translucence is lost in the infected shrimp body, and the disease begins at the second or third segment and progresses to the telson (Coelho et al. 2009).

5.3 Host Range

The penaeid shrimps are the primary host of IMNV. Pacific white leg shrimp, *P. vannamei* and Southern brown shrimp, *Farfantepenaeus subtilis*, have been reported to be naturally infected by IMNV, while *P. monodon* and *P. stylirostris* have been demonstrated to be experimentally infected (Lightner et al. 2004; Tang et al. 2005; Coelho et al. 2009). Although *P. monodon* has not been known to die from IMNV infection, it may be a possible viral carrier. The presence of IMNV was reported in farmed *P. semisulcatus* (Aly et al. 2021) and wild-captured broodstock of *P. monodon* (Srisala et al. 2021a, b). According to reports, under the experimental conditions, *Artemia franciscana* acts as a vector for the transmission of IMNV. However, because infected *Artemia* may be the source of IMNV in grow-out ponds, mass mortalities were not seen in *P. vannamei* fed with it (de Silva et al. 2015). Without confirming that they are actual vectors or carriers, bivalves and polychaete worms from the infected pond have also been reported to be positive for IMNV infection. The cause of the virus's presence in these organisms might be attributed to the intake of infected tissues or water from the contaminated sites (Andrade and Lightner 2009).

5.4 Mortality Pattern

While IMNV infection progresses slowly and has a low mortality rate during the grow-out period, cumulative mortality in ponds during harvest may reach 70% (Senapin et al. 2011). However, varying mortality rates (20–50%) have been recorded in IMNV-infected shrimp (Nunes et al. 2004; Hameed et al. 2017). Further, a cumulative mortality of up to 80% has been documented in Brazil due to IMNV disease outbreaks (Andrade et al. 2022). Experimental challenge with IMNV resulted in 100% mortality of healthy shrimp (Suryakodi et al. 2022; Wan et al. 2023). Although the virus caused mortality in intramuscularly challenged shrimp, the oral method failed to cause mortality.

5.5 Target Organs and Histopathology

The skeletal muscle is the primary target organ; however, the presence of IMNV could be detected in haemolymph, gill, hepatopancreas and lymphoid organ. Histopathological manifestations are characterized by coagulative necrosis of skeletal muscles, with haemocytic infiltration in striated muscles, fibrosis, fluid accumulation in muscle fibre (oedema) and hypertrophy of lymphoid organs induced by the build-up of lymphoid organ spheroids (LOS) (Lightner et al. 2004; Poulos et al. 2006; Puspitasari et al. 2023).

5.6 Geographical Distribution

After the first report of IMNV in farmed *P. vannamei* in north-eastern Brazil, the disease was subsequently reported in Indonesia (Poulos et al. 2006; Senapin et al. 2007), Brazil (Teixeira-Lopes et al. 2011), India (Hameed et al. 2017), Malaysia (2018) and subsequently from China (Wan et al. 2023). The presence of IMNV was reported in farmed *P. semisulcatus* from Egyptian provinces (Aly et al. 2021). The presence of IMNV had also been identified from the wild captured broodstock of *P. monodon* from the north-eastern Indian Ocean (from Thailand) (Srisala et al. 2021a, b) (Fig. 9). The transboundary movement of the infected or carrier broodstock and post-larvae has been implicated in the disease transmission from Brazil to other areas.

In India, shrimp farms in Purba Medinipur District, West Bengal, were the first to report the presence of IMNV (Hameed et al. 2017). The disease was characterized by a gradual increase in mortality, and the cumulative mortality ranged from

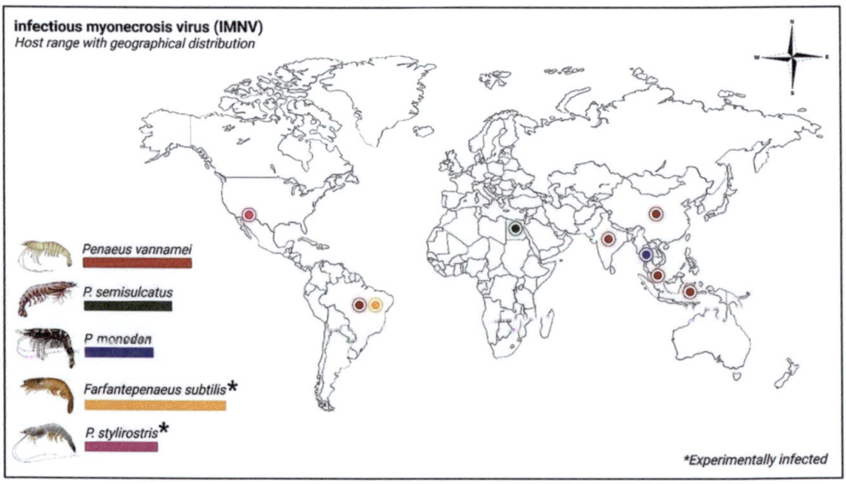

Fig. 9 Geographical distribution and host species of IMNV

20% to 50%. Subsequently, the disease was recorded in Andhra Pradesh and Tamil Nadu (Jithendran et al. 2021). In another study (Suryakodi et al. 2022), IMNV was found to be co-infected with EHP and WSSV in freshwater-reared *P. vannamei* in Tamil Nadu.

5.7 Transmission

Although the precise mode of IMNV transmission is uncertain, infected shrimp may spread horizontally by cannibalism or through water. It has been reported that the ovaries of IMNV-infected female shrimp displayed 100% positive results, and naturally infected males showed a declined survival rate of sperm cells, which indicated that IMNV is vertically transmitted. The ability of IMNV to sustain infectivity in the gastrointestinal tract and faecal matter of seabirds that consume dead or dying shrimp infected with IMNV has been presumed (Vanpatten et al. 2004).

5.8 Diagnosis

The primary methods for diagnosing IMNV infection are immunological, molecular and histopathological, apart from clinical signs. The major diagnostic techniques include nested reverse transcriptase polymerase chain reaction assay (RT-PCR), quantitative real-time RT-PCR (Taqman and SYBR green qPCR), and in situ hybridization (ISH) (Tang et al. 2005; Poulos and Lightner 2006; Andrade et al. 2007). Poulos and Lightner (2006) designed a nested reverse-transcriptase polymerase chain reaction (RT-PCR) to rapidly and accurately detect IMNV. Furthermore, a sensitive real-time PCR (Taqman qPCR) was designed by Andrade (2007) using primers and probes targeting ORF-1 of IMNV.

A more sensitive RT-PCR technique has been developed by Senapin et al. (2007) by employing primers targeting the RdRp region. The capsid protein of IMNV was used in the RT-PCR method developed by Coelho-Mello et al. (2014). Puthawibool et al. (2009) established an RT-LAMP and LFD combination for detecting IMNV. The sensitivity of the RT-LAMP NALF method developed by Andrade and Lightner (2009) has been found to be higher than one-step RT-PCR and RT-LAMP (two primer pairs), respectively, while demonstrating a similar sensitivity to RT-LAMP (using three primer pairs). Arunrut et al. (2013) developed a gold nanoparticle probe and RT-LAMP combination for IMNV detection. A new SYBR green-based qPCR was developed by Kokkattunivarthil et al. (2018). Two new enzymatic recombinase amplification (ERA)-based tests for IMNV were developed by Zhang et al. (2024): real-time RT-ERA and a combination of RT-ERA and lateral flow dipsticks assay (RT-ERA-LFD). Further, a novel technology employing shrimp MultiPath™ (SMP) for the detection of IMNV and other 12 potential pathogens in *Penaeus vannamei* was carried out by Genz et al. (2024).

A polyclonal antibody that can detect IMNV using western blotting has been produced (Mello et al. 2011). Using a variety of immunological techniques,

Kunanopparat et al. (2011) developed two monoclonal antibodies (mAbs) against the capsid protein of IMNV that could identify spontaneous IMNV infections in *P. vannamei*. It was observed that the MAbs' diagnostic sensitivities ranged from 6 to 8 fmol/spot. These MAbs are specific to IMNV and have no cross-reactivity with non-related viruses or tissues, but their combined sensitivity was approximately ten times less than that of a single-step RT-PCR using the same material. The IMNV100 kDa major capsid protein is the target of four MAbs that were produced by Seibert et al. (2010). For the quick diagnosis of IMNV, immuno-chromatographic test strips were also produced (Chaivisuthangkura et al. 2013; Wangman et al. 2016). A rapid lateral flow immunochromatographic assay (LFA) was developed to diagnose IMNV from tissue homogenates (in 20 minutes) with good specificity and sensitivity (Kumar et al. 2022) (Table 4).

5.9 Host–Pathogen Interactions

Significant host immune responses induced by IMNV, especially at a late stage of infection, include a significant decrease in haemocytes, an increase in apoptosis, an increased release of reactive oxygen species (ROS), and a change in the activity of immune effector molecules, have been reported (Costa et al. 2009). Interestingly, the *P. vannamei* laminin receptor (Lamr) is reported to participate in viral infection by binding exclusively to the IMNV capsid protein. Thus, it may be possible to decrease viral infection by preventing the capsid protein and receptor from interacting (Busayarat et al. 2011).

Table 4 Summary of different diagnostic techniques used in the detection of IMNV

Sl. no.	Diagnostic method	Sensitivity of detection	References
1	RT-PCR	10 viral genome copies/μL RNA	Poulos and Lightner (2006)
2	Real-time PCR (Taqman qPCR)	10 IMNV copies/μL RNA	Andrade et al. (2007)
3	RT-LAMP and LFD combination	Similar to nested RT-PCR detection	Puthawibool et al. (2009)
4	RT-LAMP NALF	100 and 10-fold sensitive than single-step RT-PCR and RT-LAMP	Andrade and Lightner (2009); Andrade (2007)
5	RT-LAMP and gold nanoparticle probe	10 copies of in vitro transcribed IMNV RNA	Arunrut et al. (2013)
6	Real-time PCR (SYBR Green)	10 viral copies	Kokkattunivarthil et al. (2018)
7	LFA	103 virus particles	Kumar et al. (2022)
8	Real-time RT-ERA	101 copies/μL	Zhang et al. (2024)
9	RT-ERA-LFD	101 copies/μL	Zhang et al. (2024)
10	Shrimp MultiPath™ (SMP)	LOD 6.6 copies/μL of total nucleic acid sample	Genz et al. (2024)

Baladrat et al. (2022) studied the effect of stocking and IMNV infection on the immune response of *P. vannamei* and found that the total haemocyte count declined drastically at higher stocking densities (400 shrimp m^{-2}). Another study demonstrated that different stocking densities and IMNV infection resulted in the enhanced expression of lectin and translationally controlled tumor protein (TCTP) and downregulation of Toll receptors (Nurhudah et al. 2024). After the IMNV challenge, Viana et al. (2024) observed an increase in total haemocyte count (THC) 12 hours post-infection (hpi) and upregulated gene profiles of lectins, particularly in the hepatopancreas and gills. A transcriptomic study in *P. vannamei* identified IMNV-induced gene expression changes in shrimp (Rong et al. 2024). Early-stage infection study showed that IMNV colonized and spread by blocking hemagglutination and genes linked to apoptosis and by activating the JAK-STAT pathway. Amino acid and glucose metabolism pathways were enriched during the mid-stage of infection. Further, the advanced stage of infection resulted in elevated genes for the cytoskeleton and extracellular matrix and downregulated pathways for autophagy and apoptosis.

5.10 Management

As there is no effective way to treat or prevent the spread of IMNV, the best strategies are management, prevention and rapid diagnosis. *P. vannamei* was shown to have a 20% mortality rate from experimental IMNV infections, but *P. stylirostris* and *P. monodon* showed no mortality. Therefore, a practical way to reduce productivity losses may be stocking *P. monodon* and *P. stylirostris* resistant to IMNV in ponds. Other management measures include disinfection of eggs and larvae to avoid vertical transmission, adopting suitable quarantine measures, stocking SPF (specific pathogen-free) shrimp, reducing stocking densities and stress, administering immunostimulants, etc.

Loy et al. (2012) demonstrated that administration of IMNV-specific dsRNA in *P. vannamei* significantly improves the survival rate of shrimp and was found to be resistant to subsequent viral challenges. Further, administration of IMNV-dsRNA5 (0.5 μg/shrimp) at 48 h following the IMNV challenge led to 50% survival in comparison to the other groups (Loy et al. 2013). Feijo et al. (2015) reported that the administration of dsRNA (0.5–5 μg per shrimp) two days before the IMNV challenge could considerably lower the IMNV-associated mortalities in shrimp. The nano-delivery of encapsulated IMNV-dsRNA (100 μg) reported to protect *P. vannamei* (70% survival) (Phanse et al. 2015).

Several herbal extracts, immunostimulants and probiotic administrations have been found to be effective against IMNV. The inclusion of *Phyllanthus niruri* extract (20 mg kg^{-1}) in the diet significantly enhanced the survival rate (86.7%) of IMNV-infected shrimp (Nuryati and Sari 2011). After being challenged with IMNV, *P. vannamei*'s survival and immunological response were markedly improved by the oral administration of a symbiotic consisting of *Vibrio alginolitycus* SKT-bR (probiotic) and sweet potato (*Ipomoea batatas* L.) oligosaccharide (prebiotics) (Oktaviana et al. 2014). Widanami et al. (2014) found that oral administration of *Bacillus* NP5 rendered shrimp improved resistance against IMNV. After oral challenge with IMNV, *P. vannamei*'s survival was enhanced when 1000 mg kg^{-1} of β-1,3/1,6-glucan was administered orally (Neto and Nunes 2015). Vieira-Girão et al. (2017) investigated the in vitro anti-IMNV activity of Ctn (15–34), an eicosapeptide derived from an antimicrobial peptide belonging to the cathelicidin family. They discovered that Ctn (15–34) at a concentration of less than 12.5 μM was adequate to counteract the harmful effects of IMNV. The oral administration of a blend of herbal oil from different plants was tested for anti-IMNV activity, and reduced cumulative mortality of shrimp was seen in treatment groups (Babikian et al. 2017). The oral administration of aqueous extract from *Cynodon dactylon* (20% inclusion in the diet) showed a protective role against IMNV infection in *P. vannamei* (Jha et al. 2020). The immunological response and survival rate of *P. vannamei* infected with IMNV were markedly enhanced by the injection of k-carrageenan (20 g kg^{-1}) (Azhar et al. 2023). Yang et al. (2024) showed that combined disinfection (10 ppm) using peroxymonosulfate (PMS) and povidone-iodine with an exposure time of minimum 2 days could successfully decline the IMNV load in seawater samples.

5.11 Conclusion

Viral diseases cause considerable production loss in shrimp aquaculture worldwide. Intensification of aquaculture, species diversification and international trade of live and frozen aquatic products are some of the major drivers of the emergence of diseases in aquaculture. Over the years, there has been an increase in the emergence of viral diseases, either an existing virus emerging in a new geographical location or in a new host species or the virus exhibiting increased virulence. Further, new viruses continue to emerge and put the production system at increased risk, and the economic losses due to direct mortality and morbidity are huge, besides other consequences. Besides, the most virulent virus, WSSV, viruses such as Infectious hypodermal and haematopoietic necrosis virus (IHHNV), Hepatopancreatic parvovirus (HPV), Monodon baculovirus (MBV) and Infectious myonecrosis virus (IMNV) have been reported in Indian shrimp aquaculture. Though these viruses have not been attributed to massive mortality in shrimp, the majority of them can cause production loss due to their impact on growth. However, unlike other viruses discussed in the chapter, IMNV, an RNA virus, causes considerable mortality. Further, it is to be noted that most of these viruses often appear as co-infection and can lead to increased morbidity. As the complete cure or elimination of these viruses from aquatic systems is difficult, strict adherence to pathogen-specific and overall health management measures is the best way to minimize the risk of viral diseases.

References

Abraham TJ, Mondal A, Patra A, Adikesavalu H (2020) Prevalence of white spot virus and monodon baculovirus in shrimp culture systems of West Bengal. Indian J Fish 8(2):861–864

Afanasiev BN, Galyov EE, Buchatsky LP, Kozlov YV (1991) Nucleotide sequence and genomic organization of *Aedes densonucleosis* virus. Virology 185(1):323–336

Aiamsa-at P, Sunantawanit S, Chumroenvidhayakul R, Nakarin F, Sanguanrat P, Sritunyalucksana K, Chaijarasphong T (2025) A novel dual CRISPR-Cas assay for detection of infectious hypodermal and hematopoietic necrosis virus (IHHNV) in penaeid shrimp without false positives from its endogenous viral elements (EVEs). Aquaculture 595:741452

Alavandi SV, Vijayan KK, Rajendran KV (1995) Shrimp diseases, their prevention and control. Central Institute of Brackishwater Aquaculture (CIBA). Bulletin 3:1–17

Aly SM, Mansour SM, Thabet RY, Mabrok M (2021) Studies on infectious myonecrosis virus (IMNV) and infectious hypodermal and hematopoietic necrosis virus (IHHNV) in cultured penaeid shrimp in Egypt. Dis Aquat Org 143:57–67

Amala J, Sanil NK, Poojary N, Kezhedath J, Sreedharan K, Bedekar MK, Rajendran KV (2025) First report of dual infection of hepatopancreatic parvovirus and Enterocytozoon hepatopenaei in wild and farmed Macrobrachium rosenbergii. J Fish Dis, 0:e14165. https://doi.org/10.1111/jfd.14165

Ambipillai L, Sobhana KS, George KC, Sanil NK (2003) Histopathological survey of cultured shrimps in Cochin, Kerala. J Marine Biol Assoc India 45(2):178–185

Ameer A, V DDD, Gopakumar L, Murray F, Saranya M, V SP, Singh IB (2024) Detection of antiviral moieties in the leaves of Rhizophora mucronata functional on the shrimp pathogen *Penaeus stylirostris* penstyldensovirus 1 (Pst DVI)/IHHNV as the prospective drugs of choice in aquaculture. Aquac Int 32:1–18

Anderson IG, Shariff M, Nash G, Nash M (1987) Mortalities of juvenile shrimp, *Penaeus monodon*, associated with *Penaeus monodon* baculovirus, cytoplasmic reo-like virus, and rickettsial and bacterial infections, from Malaysian brackishwater ponds. Asian Fish Sci (Philippines) 1:47–64

Anderson IG, Law AT, Shariff M, Nash G (1990) A parvo-like virus in the giant freshwater prawn, *Macrobrachium rosenbergii*. J Invertebr Pathol 55(3):447–449

Andrade T (2007) New hybrid assay can detect IMNV in resource-poor settings. Global Aquaculture Advocate:68–69

Andrade TPD, Lightner DV (2009) Development of a method for the detection of infectious myonecrosis virus by reverse transcription loop-mediated isothermal amplification and nucleic acid lateral flow hybrid assay. J Fish Dis 32:911–924

Andrade TPD, Srisuvan T, Tang K, Lightner DV (2007) Real-time reverse transcription polymerase chain reaction assay using TaqMan probe for detection and quantification of infectious myonecrosis virus (IMNV). Aquaculture 264:9–15

Andrade TP, Cruz-Flores R, Mai HN, Dhar AK (2022) Novel infectious myonecrosis virus (IMNV) variant is associated with recent disease outbreaks in *Penaeus vannamei* shrimp in Brazil. Aquaculture 554:738159

Anshary H, Sriwulan, Sukenda, Baxa DV (2017) Multiple viral pathogens occurrence in tiger shrimp (*Penaeus monodon*) broodstock from Sulawesi coastal waters. Aquacult Aquar Conserv Legisl Bioflux 10(4):936–950

Aranguren Caro LF, Gomez-Sanchez MM, Piedrahita Y, Mai HN, Cruz-Flores R, Alenton RRR, Dhar AK (2022) Current status of infection with infectious hypodermal and hematopoietic necrosis virus (IHHNV) in the Peruvian and Ecuadorian shrimp industry. PLoS One 17(8):e0272456

Arbon PM, Condon K, Martinez MA, Jerry DR (2022) Molecular detection of six viral pathogens from Australian wild sourced giant black tiger shrimp (*Penaeus monodon*) broodstock. Aquaculture 548:737651

Arbon PM, Martinez MA, Garrett M, Jerry DR, Condon K (2024) Determining patterns of tissue tropism for IHHNV, GAV and YHV-7 infection in giant black tiger shrimp (*Penaeus monodon*) using real-time RT-qPCR. Aquaculture 584:740680

Arunrut N, Prombun P, Saksmerprome V, Flegel TW, Kiatpathomchai W (2011) Rapid and sensitive detection of infectious hypodermal and hematopoietic necrosis virus by loop-mediated isothermal amplification combined with a lateral flow dipstick. J Virol Methods 171(1):21–25

Arunrut N, Kampeera J, Suebsing R, Kiatpathomchai W (2013) Rapid and sensitive detection of shrimp infectious myonecrosis virus using a reverse transcription loop-mediated isothermal amplification and visual colorogenic nanogold hybridization probe assay. J Virol Methods 193(2):542–547

Arunrut N, Jitrakorn S, Tondee B, Saksmerprome V, Kiatpathomchai W (2024) Real-time triplex loop-mediated isothermal amplification (LAMP) using a turbidimeter for detection of shrimp infectious hypodermal and hematopoietic necrosis virus (IHHNV). J Aquat Anim Health 36(3):205–219

Attasart P, Kaewkhaw R, Chimwai C, Kongphom U, Panyim S (2011) Clearance of *Penaeus monodon* densovirus in naturally pre-infected shrimp by combined ns1 and vp dsRNAs. Virus Res 159(1):79–82

Azhar F, Mukhlis A, Lestari DP, Marzuki M (2023) Application of kappa-carrageenan as immunostimulant agent in non-specific defense system of vannamei shrimp. AACL Bioflux 16(1):616

Babikian HY, Jha RK, Babikian YH, Wisoyo D, Asih Y, Srisombat S, Jiaravanon B (2017) Effectiveness of natural herbal oil formulation against infectious myonecrosis virus in whiteleg shrimp *Penaeus vannamei*. Scientia Ricerca 1(2):50–56

Baladrat NK, Nurhudah M, Utari HB (2022) Immune response of White Shrimp (*Litopenaeus vannamei*) to different density and IMNV challenge. Sci J Fish Marine/Jurnal Ilmiah Perikanan dan Kelautan 14(1):83–92

Baticados MCL, Cruz-Lacierda ER, de la Cruz MC, Duremdez-Fernandez RC, Gacutan RQ, Lavilla-Pitogo CR, Lio-Po GD (1990) Diseases of penaeid shrimps in The Philippines. Aquaculture extension manual no. 16. Aquaculture Department, Southeast Asian Fisheries Development Center (SEAFDEC), Tigbauan, Iloilo, Philippines, p 46

Baticados MCL, Lavilla-Pitogo CR, Paner MG, de la Pena LD, Tendencia EA (1991) Occurrence and pathology of *Penaeus monodon* baculovirus infection in hatcheries and ponds in Philippines. Bamidegh 43:35–41

Belcher CR, Young PR (1998) Colourimetric PCR-based detection of monodon baculovirus in whole *Penaeus monodon* postlarvae. J Virol Methods 74:21–29

Bell TA, Lightner DV (1984) IHHN virus: infectivity and pathogenicity studies in *Penaeus stylirostris* and *Penaeus* vannamei. Aquaculture 38(3):185–194

Bell TA, Lightner DV (1987) IHHN disease of *Penaeus stylirostris*: effects of shrimp size on disease expression. J Fish Dis 10(3):165–170

Boada M, De Donato M, Rodulfo H (2008) Detection of the Infectious Hypodermal and Hematopoietic Necrosis Virus (IHHNV) in asymtomatic cultured White Shrimp, *Litopenaeus* vannamei (Boone), in Venezuela. Revista Científica 18(1):07–11

Bonami JR, Trumper B, Mari J, Brehelin M, Lightner DV (1990) Purification and characterization of the infectious hypodermal and haematopoietic necrosis virus of penaeid shrimps. J Gen Virol 71(11):2657–2664

Boonsanongchokying C, Sang-oum W, Sithigorngul P, Sriurairatana S, Flegel TW (2006) Production of monoclonal antibodies to polyhedrin of monodon baculovirus (MBV) from shrimp. Sci Asia 32:371–376

Boublik Y, Jousset FX, Bergoin M (1994) Complete nucleotide sequence and genomic organization of the *Aedes albopictus* parvovirus (AaPV) pathogenic for Aedes aegypti larvae. Virology 200(2):752–763

Brock JA (1983) Pathogen considerations for peaneid shrimp imported into Hawai. Aquaculture Development Program, Hawai

Brock JA, Lightner DV (1990) Diseases of crustacea. Diseases caused by microorganisms. Dis Marine Anim 3:245–349

Brock JA, Lightner DV, Bell TA (1983) A review of four virus (BP, MBV, BMN and IHHNV) diseases of penaeid shrimp with particular reference to clinical significance, diagnosis and control in shrimp aquaculture. In: Proceedings of the 71st international. Council for the Exploration of the Sea, CM, pp 1–18

Busayarat N, Senapin S, Tonganunt M, Phiwsaiya K, Meemetta W, Unajak S, Jitrapakdee S, Lo CF, Phongdara A (2011) Shrimp laminin receptor binds with capsid proteins of two additional shrimp RNA viruses YHV and IMNV. Fish Shellfish Immunol 31(1):66–72

Caipang CMA, Sibonga G, JSMFJ, Amar MJA, Calpe AT (2011) Effective health management in shrimp hatcheries: use of loop-mediated isothermal amplification (LAMP) for the rapid detection of Philippine isolates of *Penaeus monodon*-type baculovirus (MBV). ABAH Bioflux 3(1):78–85

Catap ES, Lavilla-Pitogo CR, Maeno Y, Traviña RD (2003) Occurrence, histopathology and experimental transmission of hepatopancreatic parvovirus infection in *Penaeus monodon* postlarvae. Dis Aquat Org 57(1–2):11–17

Cavalli LS, Batista CR, Nornberg BF, Mayer FQ, Seixas FK, Romano LA, Marins LF, Abreu PC (2013) Natural occurrence of White spot syndrome virus and infectious hypodermal and hematopoietic necrosis virus in *Neohelice granulata* crab. J Invertebr Pathol 114(1):86–88

Chai C, Liu Y, Xia X, Wang H, Pan Y, Yan S, Wang Y (2016) Prevalence and genomic analysis of infectious hypodermal and hematopoietic necrosis virus (IHHNV) in *Litopenaeus vannamei* shrimp farmed in Shanghai, China. Arch Virol 161:3189–3201

Chaivisuthangkura P, Srisuk C, Rukpratanporn S, Longyant S, Sridulyakul P, Sithigorngul P (2009) Rapid and sensitive detection of *Penaeus monodon* nucleopolyhedrovirus by loop-mediated isothermal amplification. J Virol Methods 162(1–2):188–193

Chaivisuthangkura P, Senapin S, Wangman P, Longyant S, Sithigorngul P (2013) Simple and rapid detection of infectious myonecrosis virus using an immunochromatographic strip test. Ach Virol 158:1925–1930

Chakrobortty D, Ali MR, Dey BK, Gupta N, Islam SS, Sui L (2020) Viral contamination of tiger shrimp *Penaeus monodon* broodstock in Bangladesh. Aquac Int 28:2161–2172

Chang PS, Lo CF, Kou GH, Lu CC, Chen SN (1993) Purification and amplification of DNA from *Penaeus monodon* type baculovirus (MBV). J Invertebr Pathol 62:116–120

Chantanachookin C, Boonyaratpalin S, Kasornchandra J, Direkbusarakom S, Ekpanithanpong U, Supamataya K, Sriurairatana S, Flegel TW (1993) Histology and ultrastructure reveal a new granulosis-like virus in *Penaeus monodon* affected by yellow-head disease. Dis Aquat Org 17:145–145

Chayaburakul K, Nash G, Pratanpipat P, Sriurairatana S, Withyachumnarnkul B (2004) Multiple pathogens found in growth-retarded black tiger shrimp *Penaeus monodon* cultivated in Thailand. Dis Aquat Org 60(2):89–96

Chen D (1992) An overview of the disease situation, diagnostic techniques, treatments and preventatives used on shrimp farms in China. In: Diseases of cultured Penaeid Shrimp in Asia and the Unites States. The Oceanic Institute, Hawaii, pp 47–55

Chen SN, Chang PS, Kou GH, Lightner DV (1989a) Studies on virogenesis and cytopathology of *Penaeus monodon* baculovirus (MBV) in the giant tiger prawn (*Penaeus monodon*) and the red tail prawn (*Penaeus penecillatus*). Fish Pathol 24:89–100

Chen SN, Lo CF, Liu SM, Kou GH (1989b) The first identification of *Penaeus monodon* baculovirus (MBV) in the cultured sand shrimp *Metapenaeus ensis*. Bull Eur Assoc Fish Pathol 9:62–64

Chen SN, Chang PS, Kou GH (1992) Infection route and eradication of *Penaeus monodon* baculovirus (MBV) in larval giant tiger prawn *Penaeus monodon*. In: Fulks W, Main KL (eds) Diseases of cultured shrimp in Asia and United States. Honolulu, The Oceanic Institute, pp 177–184

Chen Y-B, Zhou J-F, Wan X-H, Song G (2012) Establishment of a multiplex PCR and an investigation of co-infection rate of WSSV and IHHNV in *Penaeus vannamei* in northern of Jiangsu. J Anim Vet Adv 11(2):181–185

Chen BK, Dong Z, Liu DP, Yan YB, Pang NY, Nian YY, Yan DC (2017) Infectious hypodermal and haematopoietic necrosis virus (IHHNV) infection in freshwater crayfish *Procambarus clarkii*. Aquaculture 477:76–79

Chen BK, Dong Z, Pang NY, Nian YY, Yan DC (2018) A novel real-time PCR approach for detection of infectious hypodermal and haematopoietic necrosis virus (IHHNV) in the freshwater crayfish *Procambarus clarkii*. J Invertebr Pathol 157:100–103

Chen CY, Yang N, He C, Gao Y, Chang LR, Li T, Si LJ, Yan DC (2024) The immune response of Caspase-3 gene in infectious hypodermal and hematopoietic necrosis virus (IHHNV) infection of *Litopenaeus vannamei*. Aquaculture 581:740472

Chong YC, Loh H (1984) Hepatopancreas chlamydial and parvoviral infections of farmed marine prawns in Singapore. Singap Vet J 8(9):51–56

Chotipuntu P, Wuthisuthimethavee S, Direkbusrakom S (2015) Decontamination of monodon baculovirus in marine shrimp eggs using upwelling flow-through system. Walailak J Sci Technol 12(6):527–532

Claydon K, Tahir RAH, Said HM, Lakim MH, Tamat W (2010) Prevalence of shrimp viruses in wild *Penaeus monodon* from Brunei Darussalam. Aquaculture 308(3–4):71–74

Coelho MG, Silva AC, Nova CMV, Neto JM, Lima AC, Feijó RG, Apolinário DF, Maggioni R, Gesteira TC (2009) Susceptibility of the wild southern brown shrimp (*Farfantepenaeus subtilis*) to infectious hypodermal and hematopoietic necrosis (IHHN) and infectious myonecrosis (IMN). Aquaculture 294(1–2):1–4

Coelho-Mello MV, Guedes MIF, Rodriguez-Málaga S, Almeida LMD, Moreira MDF, Oliveira TRD (2014) Molecular characterization of infectious myonecrosis virus (IMNV) isolated from the shrimp *Litopenaeus vannamei* farmed in Ceará State, Brazil. Lat Am J Aquat Res 42(3):649–652

Colorni A (1989) Penaeid pathology in Israel: problems and research. In: Advances in tropical aquaculture, Tahiti. Actes de Colloque 9 AQUACOP IFREMER, pp 89–96

Colorni A, Samocha T, Colorni B (1987) Pathogenic viruses introduced into mariculture systems by imported penaeid shrimp. Bamidgeh 39:21–28

Costa AM, Buglione CC, Bezerra FL, Martins PC, Barracco MA (2009) Immune assessment of farm-reared *Penaeus vannamei* shrimp naturally infected by IMNV in NE Brazil. Aquaculture 291(3–4):141–146

Cotmore SF, Agbandje-McKenna M, Canuti M, Chiorini JA, Eis-Hubinger AM, Hughes J, Mietzsch M, Modha S, Ogliastro M, Pénzes JJ, Pintel DJ (2019) ICTV virus taxonomy profile: Parvoviridae. J Gen Virol 100(3):367–368

Couch JA (1981) Viral diseases of invertebrates other than insects. In: Davidson EW (ed) Pathogenesis of invertebrate microbial diseases. Allanheld, Osmun Publisher, Totowa, pp 127–160

Cruz-Flores R, Dhar AK (2024) Tracking the emergence of a novel genotype of Decapod hepanhamaparvovirus in shrimp using laser microdissection and next generation sequencing. PLoS One 19(10):e0311592

Cruz-Flores R, Siewiora HM, Kanrar S, Le Groumellec M, Dhar AK (2024) Identification and characterization of a novel strain of Decapod hepanhamaparvovirus in black tiger shrimp (*Penaeus monodon*) from Madagascar that does not cause histological lesions. Aquaculture 588:740904

da Silva SMBC, Lavander HD, Luna MMS, da Silva AOME, Gálvez AO, Coimbra MRM (2015) *Artemia franciscana* as a vector for infectious myonecrosis virus (IMNV) to *Litopenaeus vannamei* juvenile. J Invertebr Pathol 126:1–5

de la Peña LD, Lavilla-Pitogo CR, Villar CBR, Paner MG (2005) Prevalence of economically-important viral diseases in wild shrimp (*Penaeus monodon*) in The Philippines. In: Nagasawa K (ed) Recent advances in diagnosis and control of fish and shrimp diseases in Southeast Asia. SEAFDEC Aquaculture Department, Iloilo

Dhar AK, Roux MM, Klimpel KR (2001) Detection and quantification of infectious hypodermal and hematopoietic necrosis virus and white spot virus in shrimp using real-time quantitative PCR and SYBR Green chemistry. J Clin Microbiol 39(8):2835–2845

Dhar AK, Cruz-Flores R, Mai HN, Warg J (2024) Comparison of Polymerase Chain Reaction (PCR) assay performance in detecting Decapod penstylhamaparvovirus 1 in penaeid shrimp. J Virol Methods 323:114840

Diamant A, Colorni A (1987) Viral infections of penaeid shrimp in Israel and use of acridine orange fluorescence as a rapid diagnostic method. In: Third international EAFP conference, Bergen, Norway, 31 Aug–3 Sept, 1987

dos Santos Braz RDF, Reis LG, Martins PCC, de Sales MP, Meissner RV (2009) Prevalence of infectious hypodermal and hematopoietic necrosis virus (IHHNV) in *Penaeus vannamei* cultured in North-eastern Brazil. Aquaculture 288(1–2):143–146

Doubrovsky A, Paynter JL, Sambhi SK, Atherton JG, Lester RJG (1988) Observations on the ultrastructure of baculovirus in Australian *Penaeus monodon* and *Penaeus merguiensis*. Aust J Mar Freshwat Res 39:743–749

Fegan DF, Flegal TW, Sriurairatana S, Waiakrutra M (1991) The occurrence, development and histopathology of monodon baculovirus in *Penaeus monodon* in Southern Thailand. Aquaculture 96:205–217

Feijo RG, Kamimura MT, Oliveira-Neto JM, Vila-Nova CMVM, Gomes ACS, Coelho MDGL, Vasconcelos RF, Gesteira TCV, Marins LF, Maggioni R (2013) Infectious myonecrosis virus and white spot syndrome virus co-infection in Pacific white shrimp (*Litopenaeus vannamei*) farmed in Brazil. Aquaculture 380–383:1–5

Feijo RG, Maggioni R, Martins PCC, de Abreu KL, Oliverira-Neto JM, Guertler C, Justino EB, Perazzolo LM, Marins LF (2015) RNAi-based inhibition of infectious myonecrosis virus replication in Pacific white shrimp *Litopenaeus vannamei*. Dis Aquat Org 114:89–98

Felix S, Devaraj M (1993) Incidence of destructive MBV and IHHNV in a commercial shrimp hatchery. Seaf Export J 25:13–18

Flegel TW (1995) Environmental control of infectious diseases in Thailand. In: Diseases in Asian aquaculture II: proceedings of the second symposium on diseases in Asian aquaculture. Fish Health Section, Asian Fisheries Society

Flegel TW (2006) Detection of major penaeid shrimp viruses in Asia, a historical perspective with emphasis on Thailand. Aquaculture 258(1–4):1–33

Flegel TW (2009) Current status of viral diseases in Asian shrimp aquaculture. Israel J Aquacult Bamidgeh 61(3):229–239

Flegel TW, Pasharawipas T (1998) Active viral accommodation: a new concept for crustacean response to viral pathogens. In: Flegel TW (ed) Advances in shrimp biotechnology. National Center for Genetic Engineering and Biotechnology, Bangkok, pp 245–250

Flegel TW, Sriurairatana S (1993) Black tiger prawn diseases in Thailand. In: Akiyama DM (ed) Technical bulletin of the American Soybean Association. American Soybean Association, Singapore, pp 1–31

Flegel TW, Sriurairatana S (1994) Shrimp health management: an environmental approach. Malaysian Fisheries Society, c/o Faculty of Fisheries and Marine Science, Universiti Pertanian Malaysia

Flegel TW, Fegan DF, Kongsom S, Vuthikornudomkit S, Sriurairatana S, Boonyaratpalin S, Chantanachookhin C, Vickers J, Macdonald OD (1992) Occurrence, diagnosis and treatment of shrimp diseases in Thailand. In: Fulks W, Main KL (eds) Diseases of cultured penaeid shrimp in Asia and the United States. Honolulu, Oceanic Institute, pp 57–112

Flegel TW, Thamavit V, Pasharawipas T, Alday-Sanz V (1999) Statistical correlation between severity of hepatopancreatic parvovirus infection and stunting of farmed black tiger shrimp (*Penaeus monodon*). Aquaculture 174(3–4):197–206

Flegel TW, Nielsen L, Thamavit V, Kongtim S, Pasharawipas T (2004) Presence of multiple viruses in non-diseased, cultivated shrimp at harvest. Aquaculture 240(1–4):55–68

Gangnonngiw W, Kanthong N (2019) Evidence that the whiteleg shrimp *Penaeus (Litopenaeus) vannamei* is refractory to infection by *Penaeus monodon* nudivirus (PmNV), also known as MBV. Aquaculture 499:290–294

Gangnonngiw W, Laisutisan K, Sriurairatana S, Senapin S, Chuchird N, Limsuwan C, Chaivisuthangkura P, Flegel TW (2010) Monodon baculovirus (MBV) infects the freshwater prawn *Macrobrachium rosenbergii* cultivated in Thailand. Virus Res 148:24–30

Gao Y, Chen CY, Cao Z, Yuan RQ, Chang LR, Li T, Si LJ, Yan DC, Li F (2023) Development of a duplex PCR for the simultaneous detection of EHP and IHHNV and analysis of the correlation between these two pathogens. J Invertebr Pathol 201:108013

Genz B, Gerszon J, Franz LM, Salgadu A, Firestone SM, Sellars MJ, Moser RJ (2024) Detection of infectious myonecrosis virus in *Penaeus vannamei* using the multiplexed PCR platform shrimp MultiPath™. Aquaculture 584:740613

Gunalan B, Soundarapandian P, Dinakaran GK (2010) Effect of different stocking densities on the MBV infected seeds of black tiger shrimp, *Penaeus monodon* (Fabricius). Asian J Agric Sci 2:5–8

Hameed ASS, Majeed AS, Vimal S, Madan N, Rajkumar T, Santhoshkumar S, Sivakumar S (2017) Studies on the occurrence of infectious myonecrosis virus in pond-reared *Litopenaeus vannamei* (Boone, 1931) in India. J Fish Dis 40(12):1823–1830

Hao NV (1999) Study on diseases in *Penaeus monodon* cultured in TraVinh Province, Vietnam. Report of UNDP Project (VIE/96/025)

Haq MAB, Meetei KB, Ravichandran S (2022) Surveillance of WSSV and MBV in *Penaeus monodon* and SPF *Litopenaeus vannamei* broodstock from hatcheries in Tamil Nadu and Andhra Pradesh. In: Trends in aquaculture. Weser Books, pp 44–56

Hazreen Nita MK, Kua BC, Bhassu S, Othman RY (2012) Detection and genetic profiling of infectious hypodermal and haematopoietic necrosis virus (IHHNV) infections in wild berried freshwater prawn, *Macrobrachium rosenbergii* collected for hatchery production. Mol Biol Rep 39:3785–3790

He L, Xu HS (2011) Development of a multiplex loop-mediated isothermal amplification (mLAMP) method for the simultaneous detection of white spot syndrome virus and infectious hypodermal and hematopoietic necrosis virus in penaeid shrimp. Aquaculture 311(1–4):94–99

Hou LH, Lin FY (2015) Construction of fluorescent virus-like particles of infectious hypodermal and hematopoietic necrosis virus. J Appl Oceanograp 34(3):411–418

Hsieh CY, Chuang PC, Chen LC, Tu C, Chien MS, Huang KC, Kao HF, Tung MC, Tsai SS (2006) Infectious hypodermal and haematopoietic necrosis virus (IHHNV) infections in giant freshwater prawn, *Macrobrachium rosenbergii*. Aquaculture 258(1–4):73–79

Hsu YL, Wang KH, Yang YH, Tung MC, Hu CH, Lo CF, Wang CH, Hsu T (2000) Diagnosis of *Penaeus monodon*-type baculovirus by PCR and by ELISA of occlusion bodies. Dis Aquat Org 40:93–99

Hu WenJuan HW, Fang WenHong FW, Jiang Min JM, Zhang JianLong ZJ, Wang Yuan WY, Gu DePing GD, Hu WeiGuo HW, Yu ZhongLi YZ, Zhou JunFang ZJ (2015) Investigation and analyses of IHHNV-WSSV co-infection in *Penaeus vannamei* cultured in freshwater and its effects on immune-related enzymes. J Shanghai Ocean Univ 24(5):685–693

Imsonpang S, Pudgerd A, Chotwiwatthanakun C, Srisala J, Sanguanrut P, Kasamechotchung C, Sritunyalucksana K, Taengchaiyaphum S, Vanichviriyakit R (2024) Confirmatory test of active IHHNV infection in shrimp by immunohistochemistry and IHHNV-LongAmp PCR. J Fish Dis 47(3):13905

Jang IK, Suriakala K, Kim JS, Meng XH, Choi TJ (2011) A TaqMan real-time PCR assay for quantifying type III hepatopancreatic parvovirus infections in wild broodstocks and hatchery-reared postlarvae of *Fenneropenaeus chinensis* in Korea. J Microbiol Biotechnol 21(11):1109–1115

Jaroenram W, Owens L (2014) Recombinase polymerase amplification combined with a lateral flow dipstick for discriminating between infectious *Penaeus stylirostris* densovirus and virus-related sequences in shrimp genome. J Virol Methods 208:144–151

Jeeva S, Kang SW, Lee YS, Jang IK, Seo HC, Choi TJ (2012) Complete nucleotide sequence analysis of a Korean strain of hepatopancreatic parvovirus (HPV) from *Fenneropenaeus chinensis*. Virus Genes 44:89–97

Jha RK, Wisoyo SD, Kristina, Srisombat S (2020) Study on the efficacy of *Cynodon dactylon* against Infectious Myonecrosis Virus (IMNV) in *Penaeus vannamei*. bioRxiv, 2020-11

Jha RK, Babikian H, Kristina SS, Srisombat S (2021) Managing infectious myonecrosis virus (IMNV) in Vannamei shrimp culture: learning by doing. Int J Fish Aquat Stud 9(1):385–391

Jithendran KP, Krishnan AN, Jagadeesan V, Anandaraja R, Praveena PE, Anushya S, Amarnath CB, Bhuvaneswari T (2021) Co-infection of infectious myonecrosis virus and *Enterocytozoon hepatopenaei* in *Penaeus vannamei* farms in the east coast of India. Aquac Res 52(10):4701–4710

Johnson PT, Lightner DV (1988) The rod-shaped nuclear viruses of crustaceans: gut-infecting species. Dis Aquat Org 4:123–141

Joseph TC, James R, Rajan LA, Surendran PK, Lalitha KV (2015) Occurrence of viral pathogens in *Penaeus monodon* postlarvae from aquaculture hatcheries. Data Brief 4:170–176

Kanjanasopa D, Pongtippatee P, Vanichviriyakit R, Wongtripop S, Pradeep PJ, Withyachumnarnkul B (2015) An evidence on trans-ovarian transmission of monodon baculovirus (MBV) infection in *Penaeus monodon*. Aquaculture 443:5–10

Karunasagar I, Otta SK, Karunasagar I (1998) Monodon baculovirus and bacterial speticaemia associated with mass mortality of cultivated shrimp (*Penaeus monodon*) from the east coast of India. Indian J Virol 14:27–30

Kawase S (1990) Insect parvovirus disease. In: Handbook of parvoviruses, vol 2, pp 213–228

Khatun K, Khatun M, Ghosh AK, Hasan MA, Ray S, Sarower MG (2023) Molecular diagnosis of white spot syndrome virus (WSSV), infectious hypodermal and hematopoietic necrosis virus (IHHNV), monodon baculovirus (MBV), and acute hepatopancreatic necrosis disease (AHPND) causative agent in black tiger shrimp hatcheries. ELBA Bioflux 15(1):12–23

Khawsak P, Deesukon W, Chaivisuthangkura P, Sukhumsirichart W (2008) Multiplex RT-PCR assay for simultaneous detection of six viruses of penaeid shrimp. Mol Cell Probes 22(3):177–183

Kibenge F, Kibenge M, Montes de Oca M, Godoy M (2024) Parvoviruses of aquatic animals. Pathogens 13(8):625

Kim JH, Choresca CH Jr, Shin SP, Han JE, Jun JW, Han SY, Park SC (2011) Detection of infectious hypodermal and hematopoietic necrosis virus (IHHNV) in *Litopenaeus vannamei* shrimp cultured in South Korea. Aquaculture 313(1–4):161–164

Kim B, Lee C, Jeon HJ, Piamsomboon P, Kim JH, Han JE (2025) Development and applications of the real-time polymerase chain reaction method for rapid quantitative detection of the newly emerged novel genotype of Decapod hepandensovirus 1 in *Penaeus vannamei*. Aquaculture 595:741568

Knibb W, Quinn J, Kuballa A, Powell D, Remilton C, Nguyen NH (2015) Yearly, pond, lineage and family variation of hepatopancreatic parvo-like virus (HPV) copy number in banana shrimp *Fenneropenaeus merguiensis*. J Invertebr Pathol 128:73–79

Kokkattunivarthil S, Krishnan R, Kezhedath J, Prasad KP (2018) New set of PCR primers for SYBR green-based qPCR detection of IMNV in India. Aquaculture 495:726–730

Krol JD, Hill JM, Smith PRK, Kendrick MR, Gooding EL, Fuchs C, Whelan NV, Bullard SA (2024) First detection of white spot syndrome virus (WSSV) and infectious hypodermal and hematopoietic necrosis virus (IHHNV) from wild caught giant tiger prawn, *Penaeus monodon* (Fabricius), 1798 (Penaeoidea: Penaeidae) from the Gulf of Mexico and Northwestern Atlantic Ocean. BioInvas Rec 13(1):121

Kumar DR, Sanjuktha M, Rajan JJS, Bharathi AR, Santiago TC, Alavandi SV, Poornima M (2014) Development of SYBR Green based real time PCR assay for detection of monodon baculovirus in *Penaeus monodon*. J Virol Methods 205:81–86

Kumar DR, Elumalai R, Raichur AM, Sanjuktha M, Rajan JJ, Alavandi SV, Vijayan KK, Poornima M, Santiago TC (2016) Development of antiviral gene therapy for monodon baculovirus using dsRNA loaded chitosan-dextran sulfate nanocapsule delivery system in *Penaeus monodon* postlarvae. Antivir Res 131:124–130

Kumar SS, Jamalpure S, Ahmed AN, Taju G, Vimal S, Majeed SA, Suryakodi S, Rahamathulla S, Paknikar KM, Rajwade JM, Hameed AS (2022) An indigenous, field-deployable, lateral flow immunochromatographic assay rapidly detects infectious Myonecrosis in Shrimp, *Litopenaeus vannamei*. Mar Biotechnol 24(6):1110–1124

Kunanopparat A, Chaivisuthangkura P, Senapin S, Longyant S, Rukpratanporn S, Flegel TW, Sithigorngul P (2011) Detection of infectious myonecrosis virus using monoclonal antibody specific to N and C fragments of the capsid protein expressed heterologously. J Virol Methods 171:141–148

La Fauce K, Owens L (2007) Investigation into the pathogenicity of *Penaeus merguiensis* densovirus (PmergDNV) to juvenile *Cherax quadricarinatus*. Aquaculture 271(1–4):31–38

La Fauce KA, Owens L (2008) The use of insects as a bioassay for *Penaeus merguiensis* densovirus (Pmerg*DNV*). J Invertebr Pathol 98(1):1–6

La Fauce KA, Elliman J, Owens L (2007a) Molecular characterisation of hepatopancreatic parvovirus (PmergDNV) from Australian *Penaeus merguiensis*. Virology 362(2):397–403

La Fauce KA, Layton R, Owens L (2007b) TaqMan real-time PCR for detection of hepatopancreatic parvovirus from Australia. J Virol Methods 140(1–2):10–16

Lee C, Choi SK, Jeon HJ, Lee SH, Kim YK, Park S, Park JK, Han SH, Bae S, Kim JH, Han JE (2021) Detection of infectious hypodermal and hematopoietic necrosis virus (IHHNV, Decapod Penstylhamaparvovirus 1) in commodity red claw crayfish (*Cherax quadricarinatus*) imported into South Korea. J Marine Sci Eng 9(8):856

Lee D, Yu YB, Choi JH, Jo AH, Hong SM, Kang JC, Kim JH (2022) Viral shrimp diseases listed by the OIE: a review. Viruses 14(3):585

Lee Y, Vijayan J, Roh H, Park J, Lee JY, Nguyen TL, Kim HJ, Kim WS, Dhar AK, Park CI, Kim DH (2024) Nucleic acid amplification-based methods for diagnosis of shrimp viral diseases. Rev Aquac 16(2):892–922

Lester RJG, Doubrovsky A, Paynter JL, Sambhi SK, Atherton JG (1987) Light and electron microscope evidence of baculovirus infection in the prawn *Penaeus plebejus*. Dis Aquat Org 3:217–219

Liao IC (1977) A culture study on grass prawn, *Penaeus monodon*, in Taiwan-the patterns, the problems and the prospects. J Fish Soc Taiwan 5(2):11–29

Liao IC, Su MS, Chang CF (1992) Diseases of *Penaeus monodon* in Taiwan: a review from 1977 to 1991. In: Fulks W, Main KL (eds) Diseases of cultured penaeid shrimp in Asia and the United States. Oceanic Institute, Honolulu, pp 113–137

Lightner DV (1985) A review of the diseases of cultured penaeid shrimps and prawns with emphasis on recent discoveries and developments. In: Proceedings of the first international conference on the culture of penaeid prawns/shrimps, Iloilo City, Philippines, pp 79–103

Lightner DV (1988) Diseases of cultured penaeid shrimp and prawns. In: Sindermann CJ, Lightner DV (eds) Disease diagnosis and control in North American Marine aquaculture, pp 8–127

Lightner DV (1992) New developments in penaeid virology: application of biotechnology in research and disease diagnosis for shrimp viruses of concern in the Americas. Disease of Cultured Penaeid Shrimp in Asia and the United States. pp 233–253

Lightner DV (1996) A handbook of shrimp pathology and diagnostic procedures for diseases of cultured penaeid shrimp. World Aquaculture Society, Baton Rouge, p 256

Lightner DV (1998) Strategies for the control of viral diseases of shrimp in the Americas. Fish Pathol 33(4):165–180

Lightner DV (2011) Virus diseases of farmed shrimp in the Western Hemisphere (the Americas): a review. J Invertebr Pathol 106(1):110–130

Lightner DV, Redman RM (1981) A baculovirus-caused disease of the penaeid shrimp, *Penaeus monodon*. J Invertebr Pathol 38(2):299–302

Lightner DV, Redman RM (1985) A parvo-like virus disease of penaeid shrimp. J Invertebr Pathol 45(1):47–53

Lightner DV, Redman RM (1991) Hosts, geographic range and diagnostic procedures for the penaeid virus diseases of concern to shrimp culturists in the Americas. Dev Aquac Fish Sci 22:173–196

Lightner DV, Redman R (1992) Penaeid virus diseases of the shrimp culture industry of the Americas. In: Marine shrimp culture: principles and practices, pp 569–588

Lightner DV, Redman RM, Bell TA (1983a) Infectious hypodermal and hematopoietic necrosis, a newly recognized virus disease of penaeid shrimp. J Invertebr Pathol 42(1):62–70

Lightner DV, Redman RM, Bell TA, Brock JA (1983b) Detection of IHHN virus in *Penaeus stylirostris* and *P. vannamei* imported into Hawaii. J World Maricult Soc 14(1–4):212–225

Lightner DV, Redman RM, Bell TA (1983c) Observations on the geographic distribution, pathogenesis and morphology of the baculovirus from *Penaeus monodon* Fabricius. Aquaculture 32(3–4):209–233

Lightner DV, Hedrick RP, Fryer JL, Chen SN, Liao IC, Kou GH (1987) A survey of cultured penaeid shrimp in Taiwan for viral and other important diseases. Fish Pathol 22(3):127–140

Lightner DV, Redman RM, Moore DW, Park MA (1993) Development and application of a simple and rapid diagnostic method to studies on hepatopancreatic parvovirus of penaeid shrimp. Aquaculture 116(1):15–23

Lightner DV, Redman RM, Poulos BT, Mari JL, Bonami JR, Shariff M (1994) Distinction of HPV-type viruses in *Penaeus chinensis* and *Macrobrachium rosenbergii* using a DNA probe. Asian Fish Sci 7(4):267–272

Lightner DV, Pantoja CR, Poulos BT, Tang KFJ, Redman RM, Pasos-de-Andrade T, Bonami JR (2004) Infectious Myonecrosis–new disease in Paciic White Shrimp. Glob Aquacult Advoc 7(5):85

Lightner DV, Redman RM, Pantoja CR, Tang KFJ, Noble BL, Schofield P, Mohney LL, Nunan LM, Navarro SA (2012) Historic emergence, impact and current status of shrimp pathogens in the Americas. J Invertebr Pathol 110:174–183

Limsuwan C (1999) Shrimp culture in Thailand toward year 2000. AAHRI Newslett 8(1):5–6

Lin CK (1989) Prawn culture in Taiwan-what went wrong? World Aquacult 20:19–20

Liu QingYun LQ, Li QiangYong LQ, Wang ShaoShao WS, Zhu WeiLin ZW, Peng Min PM, Zeng DiGang ZD, Chen XiuLi CX, Yang ChunLing YC, Zhao YongZhen ZY, Peng JinXia PJ, He PingPing HP (2018) Interaction between *Litopenaeus vannamei* Rab5b protein (LvRab5B) and IHHNV proteins. J Fish China 42(11):1829–1839

Liu S, Rao M, Cowley JA, Morgan JA, Barnes AC, Palmer PJ (2020) Polychaetes (*Perinereis helleri*) reared in sand beds filtering nutrients from shrimp (*Penaeus monodon*) culture ponds can transiently carry IHHNV. Aquaculture 528:735560

Liu S, Xia J, Tian Y, Yao L, Xu T, Li X, Li X, Wang W, Kong J, Zhang Q (2022) Investigation of pathogenic mechanism of covert mortality Nodavirus infection in *Penaeus vannamei*. Front Microbiol 13:904358

Lotz JM (1997) Viruses, biosecurity and specific pathogen-free stocks in shrimp aquaculture. World J Microbiol Biotechnol 13:405–413

Loy JD, Mogler MA, Loy DS, Janke B, Kamrud K, Scura ED, Harris DL, Bartholomay LC (2012) dsRNA provides sequence-dependent protection against infectious myonecrosis virus in *Litopenaeus vannamei*. J Gen Virol 93:880–888

Loy JD, Loy DS, Mogler MA, Janke B, Kamrud K, Harris DL, Bartholomay LC (2013) Sequence-optimized and targeted double-stranded RNA as a therapeutic antiviral treatment against infectious myonecrosis virus in *Litopenaeus vannamei*. Dis Aquat Org 105:57–64

Lu CC, Tang FJK, Kou GH, Chen SN (1993) Development of a *Penaeus monodon*-type baculovirus (MBV) DNA probe by polymerase chain reaction and sequence analysis. J Fish Dis 16:551–559

Lu CC, Tang KFJ, Kou GH, Chen SN (1995) Detection of *Penaeus monodon*-type baculovirus (MBV) infection in *Penaeus monodon* Fabricius by *in situ* hybridization. J Fish Dis 18:337–345

Lu CC, Tang KFJ, Chen SN (1996) Morphogenesis of the membranous labyrinth in penaeid shrimp cells infected with *Penaeus monodon* baculovirus (MBV). J Fish Dis 19:357–364

Macías-Rodríguez NA, Mañón-Ríos N, Romero-Romero JL, Camacho-Beltrán E, Magallanes-Tapia MA, Leyva-López NE, Hernández-López J, Magallón-Barajas FJ, Perez-Enriquez R, Sánchez-González S, Méndez-Lozano J (2014) Prevalence of viral pathogens WSSV and IHHNV in wild organisms at the Pacific Coast of Mexico. J Invertebr Pathol 116:8–12

Manickavel N, Haq MAB, Baharlooeian M, Manie EAM, Vaitheeswari S, Bakyaraj R (2018) Surveillance of WSSV and MBV in *Penaeus monodon* and SPF *Litopenaeus vannamei* brood stock from hatcheries in Tamil Nadu and Andhra Pradesh, Southeast coast of India. Int J New Innov Eng Technol 9(1):94–111

Manisseri MK, Spann KM, Lester RJG (1999) Iodine as a disinfectant against monodon baculovirus (MBV). Asian Fish Sci 12:105–111

Manivannan S, Otta SK, Karunasagar I, Karunasagar I (2002) Multiple viral infection in *Penaeus monodon* shrimp postlarvae in an Indian hatchery. Dis Aquat Org 48(3):233–236

Manivannan S, Kennedy B, Karunasagar I, Karunasagar I (2004) Prevalence of monodon baculovirus in wild *Metapenaeus* species along the southwest coast of India. Aquaculture 232:63–67

Manjanaik B, Umesha KR, Karunasagar I, Karunasagar I (2005) Detection of hepatopancreatic parvovirus (HPV) in wild shrimp from India by nested polymerase chain reaction (PCR). Dis Aquat Org 63(2–3):255–259

Mari J, Bonami JR, Lightner D (1993a) Partial cloning of the genome of infectious hypodermal and haematopoietic necrosis virus, an unusual parvovirus pathogenic for penaeid shrimps; diagnosis of the disease using a specific probe. J Gen Virol 74(12):2637–2643

Mari J, Bonami JR, Poulos B, Lightner DV (1993b) Preliminary characterization and partial cloning of the genome of a baculovirus from *Penaeus monodon* (PmSNPV = MBV). Dis Aquat Org 16:207–215

Mari J, Lightner DV, Poulos BT, Bonami JR (1995) Partial cloning of the genome of an unusual shrimp parvovirus (HPV): use of gene probes in disease diagnosis. Dis Aquat Org 22:129–134

Martorelli SR, Overstreet RM, Jovonovich JA (2010) First report of viral pathogens WSSV and IHHNV in Argentine crustaceans. Bull Mar Sci 86(1):117–131

Mello MVC, de Aragão MEF, Torres-Franklin MLP, de Oliviera Neto JM, Guedes MIF (2011) Purification of infectious myonecrosis virus (IMNV) in species of marine shrimp *Litopenaeus vannamei* in the State of Ceará. J Virol Methods 177(1):10–14

Mishra SS, Shekhar MS, Azad IS (2005) Concurrent infection with WSSV and MBV in Tiger prawn, *Penaeus monodon* (Fabricius) in West Bengal and their detection using PCR and DNA Dot-blot hybridization technique. Indian J Biotechnol 4(4):506–515

Molthathong S, Jitrakorn S, Joyjinda Y, Boonchird C, Witchayachamnarnkul B, Pongtippatee P, Flegel T, Saksmerprome V (2013) Persistence of *Penaeus stylirostris* densovirus delays mortality caused by white spot syndrome virus infection in black tiger shrimp (*Penaeus monodon*). BMC Vet Res 9:1–7

Morales-Covarrubias MS, Chavez-Sanchez C (1999) Histopathological studies on wild broodstock of white shrimp *Penaeus vannamei* in the Platanitos area, adjacent to San Blas, Nayarit, Mexico. J World Aquacult Soc 30(2):192–200

Motte E, Yugcha E, Luzardo J, Castro F, Leclercq G, Rodríguez J, Miranda P, Borja O, Serrano J, Terreros M, Montalvo K (2003) Prevention of IHHNV vertical transmission in the white shrimp *Litopenaeus vannamei*. Aquaculture 219(1–4):57–70

Nash G, Poemomo A, Nash MB (1988a) Baculovirus infection in brackishwater pond cultured *Penaeus monodon* (Fabricius) in Indonesia. Aquaculture 73:1–6

Nash M, Nash G, Anderson IG, Shariff M (1988b) A reo-like virus observed in the tiger prawn, *Penaeus monodon* (Fabricius), from Malaysia. J Fish Dis 11:531–535

Natividad JM (1991) The *Penaeus monodon* baculovirus (MBV): its epizootiology, prevention and control in penaeid shrimp hatcheries and grow-out ponds in the Philippines. Ph.D. Dissertation, School of Renewable Natural Resources, University of Arizona. 1991

Natividad JM, Lightner DV (1992) Susceptibility of the different larval and postlarval stages of the black tiger prawn, *Penaeus monodon* Fabricius, to monodon baculovirus (MBV). In: Diseases in Asian aquaculture. Fish Health Section, Asian Fisheries Society, Manila, pp 111–125

Natividad KDT, Migo MVP, Albaladejo JD, Magbanua JPV, Nomura N, Matsumura M (2006) Simultaneous PCR detection of two shrimp viruses (WSSV and MBV) in postlarvae of *Penaeus monodon* in The Philippines. Aquaculture 257:142–149

Neto HS, Nunes AJP (2015) Performance and immunological resistance of *Litopenaeus vannamei* fed a β-1.3/1.6-glucan-supplemented diet after per os challenge with the infectious myonecrosis virus (IMNV). Rev Bras Zootec 44:165–173

Nimitphak T, Kiatpathomchai W, Flegel TW (2008) Shrimp hepatopancreatic parvovirus detection by combining loop-mediated isothermal amplification with a lateral flow dipstick. J Virol Methods 154(1–2):56–60

Nimitphak T, Meemetta W, Arunrut N, Senapin S, Kiatpathomchai W (2010) Rapid and sensitive detection of *Penaeus monodon* nucleopolyhedrovirus (PemoNPV) by loop-mediated isothermal amplification combined with a lateral-flow dipstick. Mol Cell Probes 24(1):1–5

Nunan LM, Poulos BT, Lightner DV (2000) Use of polymerase chain reaction for the detection of infectious hypodermal and hematopoietic necrosis virus in penaeid shrimp. Mar Biotechnol 2:319–328

Nunan LM, Arce SM, Staha RJ, Lightner DV (2001) Prevalence of infectious hypodermal and hematopoietic necrosis virus (IHHNV) and white spot syndrome virus (WSSV) in *Litopenaeus vannamei* in the Pacific Ocean off the coast of Panama. J World Aquacult Soc 32(3):330–334

Nunes AJP, Cunha-Martins P, Vasconselos-Gesteira TC (2004) Carcinicultura ameacada. Rev Panoram Aquic, (in Portuguese) 83:37–51

Nurhudah M, Baladrat NK, Rahardjo S, Utari HB, Wiradana PA (2024) Effect of stocking density on stress-related gene expression of Pacific white shrimp (*Litopenaeus vannamei*) infected with infectious myonecrosis virus (IMNV). J Med Veterin 7(2):276–291

Nuryati S, Sari IR (2011) Administration of *Phyllanthus niruri* to control IMNV (myonecrosis infectious virus) infection white shrimp *Litopenaeus vannamei*. J Akuakultur Indones 10(2):192–202

Oanh DTH, Phuong NT (2005) Prevalence of white spot syndrome virus (WSSV) and monodon baculovirus (MBV) infection in *Penaeus monodon* postlarvae in Vietnam. In: Walker P, Lester R, Bondad-Reantaso MG (eds) Diseases in Asian aquaculture V. Fish Health Section, Asian Fisheries Society, Manila, pp 395–404

OIE (2019) Manual of diagnostic tests for aquatic animals. World Organisation for Animal Health, Paris

Oktaviana A, Widanarni, Yuhana M (2014) The use of synbiotics to prevent IMNV and *Vibrio harveyi* co-infection in *Litopenaeus vannamei*. Hayati J Biosci 21:127–134

Otta SK, Karunasagar I, Karunasagar I (2003) Detection of monodon baculovirus and white spot syndrome virus in apparently healthy *Penaeus monodon* postlarvae from India by polymerase chain reaction. Aquaculture 220:59–67

Owens L, Anderson IG, Kenway M, Trott L, Benzie JAH (1992) Infectious hypodermal and haematopoietic necrosis virus (IHHNV) in a hybrid penaeid prawn from tropical Australia. Dis Aquat Org 14(3):219–228

Owens L, Liessmann L, La Fauce K, Nyguyen T, Zeng C (2010) Intranuclear bacilliform virus and hepatopancreatic parvovirus (*Pmerg*DNV) in the mud crab *Scylla serrata* (Forskal) of Australia. Aquaculture 310(1–2):47–51

Palmer PJ, Rao M, Cowley JA (2021) Reduced transmission of IHHNV to *Penaeus monodon* from shrimp pond wastewater filtered through a polychaete-assisted sand filter (PASF) system. Aquaculture 535:736359

Pang XM, Wei YW, Fan DD, Li CH, Miao L, Chen J (2018) Application of the infectious hypodermal and hematopoietic necrosis virus (IHHNV) to bivalve shellfish. Oceanologia et Limnologia Sinica 49:883–888

Panichareon B, Khawsak P, Deesukon W, Sukhumsirichart W (2011) Multiplex real-time PCR and high-resolution melting analysis for detection of white spot syndrome virus, yellow-head virus, and *Penaeus monodon* densovirus in penaeid shrimp. J Virol Methods 178(1–2):16–21

Pantoja CR, Lightner DV (2001) Detection of hepatopancreatic parvovirus (HPV) of penaeid shrimp by *in situ* hybridization at the electron microscope level. Dis Aquat Org 44(2):87–96

Pantoja CR, Lightner DV, Holtschmit KH (1999) Prevalence and geographic distribution of infectious hypodermal and hematopoietic necrosis virus (IHHNV) in wild blue shrimp *Penaeus stylirostris* from the Gulf of California, Mexico. J Aquat Anim Health 11(1):23–34

Parajeles-Mora JF, Peña-Navarro N, Solorzano-Morales A, Dolz G (2021) Detection of IHHNV in *Litopenaeus vannamei* farms in Costa Rica. Agronomía Mesoamericana 32(2):587–598

Paynter JL (1985) Prawn virus from juvenile *Penaeus esculentus*. In: Second Australian prawn seminar. CSIRO

Paynter JL, Vickers JE, Lester RJG (1992) Experimental transmission of *Penaeus monodon* type baculovirus (MBV). In: Shariff M, Subasinghe RP, Arthur JR (eds) Diseases in Asian aquaculture I. Fish Health Section, Asian Fisheries Society, Manila, pp 97–110

Phanse Y, Puttamreddy S, Loy D, Ross K, Ramirez JV, Narasimhan B, Bartholomay L (2015) Efficacious polyanhydride nanovaccines against viral infections in shrimp, vol 37. Society for Biomaterials Annual Meeting and Exposition. Driving Biomaterial Innovation and the Race to Translation, Charlotte, North Carolina, USA, p 44

Phromjai J, Sukhumsirichart W, Pantoja C, Lightner DV, Flegel TW (2001) Different reactions obtained using the same DNA detection reagents for Thai and Korean hepatopancreatic parvovirus of penaeid shrimp. Dis Aquat Org 46(2):153–158

Phuthaworn C, Nguyen NH, Quinn J, Knibb W (2016) Moderate heritability of hepatopancreatic parvovirus titre suggests a new option for selection against viral diseases in banana shrimp (*Fenneropenaeus merguiensis*) and other aquaculture species. Genet Sel Evol 48:1–10

Pollak NM, Fais O, Kristoffersen J, Phuthaworn C, Knibb W, Macdonald J (2022) Rapid sample preparation and low-resource molecular detection of hepatopancreatic parvoviruses (HPV) by recombinase polymerase amplification lateral flow detection assay in shrimps (*Fenneropenaeus merguiensis*). PLoS One 17(11):0276164

Poulos BT, Lightner DV (2006) Detection of infectious myonecrosis virus (IMNV) of penaeid shrimp by reverse-transcriptase polymerase chain reaction (RT-PCR). Dis Aquat Org 73:69–72

Poulos BT, Mari J, Bonami JR, Redman R, Lightner DV (1994a) Use of non-radioactively labelled DNA probes for the detection of a baculovirus from *Penaeus monodon* by *in situ* hybridization on fixed tissue. J Virol Methods 49:187–194

Poulos BT, Lightner DV, Trumper B, Bonami JR (1994b) Monoclonal antibodies to a penaeid shrimp parvovirus, infectious hypodermal and hematopoietic necrosis virus (IHHNV). J Aquat Anim Health 6(2):149–154

Poulos BT, Tang KFJ, Pantoja CP, Bonami JR, Lightner DV (2006) Purification and characterization of infectious myonecrosis virus of penaeid shrimp. J Gen Virol 87:987–996

Prasad KP, Shyam KU, Banu H, Jeena K, Krishnan R (2017) Infectious Myonecrosis Virus (IMNV)–an alarming viral pathogen to Penaeid shrimps. Aquaculture 477:99–105

Praveen R (2013) Recombinant protein and double stranded RNA (dsRNA) as immune molecule against infectious hypodermal and hematopoietic necrosis virus (IHHNV) infection of shrimp (Doctoral dissertation, Karnataka Veterinary, Animal and Fisheries Sciences University, Bidar), pp 5–130

Primavera JH, Quinitio ET (2000) Runt-deformity syndrome in cultured giant tiger prawn *Penaeus monodon*. J Crustac Biol 20(4):796–802

Puspitasari I, Suseno DN, Yudana IG, Anwar MK (2023, December) Histopathological condition of *Litopenaeus vannamei* infected with Infectious Myonecrosis Virus (IMNV) on different days of culture collected from Aceh, Sidoarjo, Probolinggo and Lampung Indonesia. IOP Conf Ser Earth Environ Sci 1273(1):012083

Puthawibool T, Senapin S, Kiatpathomchai W, Flegel TW (2009) Detection of shrimp infectious myonecrosis virus by reverse transcription loop-mediated isothermal amplification combined with a lateral flow dipstick. J Virol Methods 156:27–31

Rai P, Pradeep B, Karunasagar I, Karunasagar I (2009) Detection of viruses in *Penaeus monodon* from India showing signs of slow growth syndrome. Aquaculture 289(3–4):231–235

Rai P, Safeena MP, Karunasagar I, Karunasagar I (2011) Complete nucleic acid sequence of *Penaeus stylirostris* densovirus (PstDNV) from India. Virus Res 158(1–2):37–45

Rai P, Safeena MP, Krabsetsve K, La Fauce K, Owens L, Karunasagar I (2012) Genomics, molecular epidemiology and diagnostics of infectious hypodermal and hematopoietic necrosis virus. Indian J Virol 23:203–214

Rajendran KV, Pagare S, Raut S, Prasad KP, Pathan MA (2022) Monodon baculovirus (MBV) infects wild mud crab, *Scylla serrata*. J Invertebr Pathol 187:107701

Ramasamy P, Brennan GP, Jayakumar R (1995) A record and prevalence of monodon baculovirus from postlarval *Penaeus monodon* in Madras. India Aquacult 130:129–135

Ramasamy P, Rajan PR, Purushothaman V, Brennan GP (2000) Ultrastructure and pathogenesis of monodon baculovirus (PmSNPV) in cultured larvae and natural brooders of *Peneaus monodon*. Aquaculture 184:45–66

Rong Y, Qiao Y, Cao X, Jiang G, Cheng J, Jiang H, Shen H (2024) Transcriptomic analysis reveals dynamic host-pathogen interactions and potential antiviral targets during infectious myonecrosis virus infection in *Litopenaeus vannamei*. Aquacult Rep 39:102369

Roubal FR, Paynter JL, Lester RJG (1989) Electron microscopic observation of hepatopancreatic parvo-like virus (HPV) in the penaeid prawn, *Penaeus merguiensis* de Man, from Australia. J Fish Dis 12(2):199

Safeena MP (2014) Efficacy of Recombinant Protein and Double Stranded RNA (DsRNA) for the Protection of Shrimp against Hepatopancreatic Parvovirus (HPV) (Doctoral dissertation, Karnataka Veterinary, Animal and Fisheries Sciences University, Bidar), 5–162

Safeena MP, Rai P, Karunasagar I (2012) Molecular biology and epidemiology of hepatopancreatic parvovirus of penaeid shrimp. Indian J Virol 23:191–202

Saksmerprome V, Puiprom O, Noonin C, Flegel TW (2010) Detection of infectious hypodermal and haematopoietic necrosis virus (IHHNV) in farmed Australian *Penaeus monodon* by PCR analysis and DNA sequencing. Aquaculture 298(3–4):190–193

Sano T, Nishimura T, Oguma K, Momoyama K, Takeno N (1981) Baculovirus infection of cultured Kuruma shrimp, *Penaeus japonicus*, in Japan. Fish Pathol 15:185–191

Saravanan K, Praveenraj J, Kiruba-Sankar R, Devi V, Biswas U, Kumar TS, Sudhagar A, El-Matbouli M, Kumar G (2021) Co-infection of infectious hypodermal and hematopoietic necrosis virus (IHHNV) and white spot syndrome virus (WSSV) in the wild crustaceans of Andaman and Nicobar Archipelago, India. Viruses 13(7):1378

Satidkanitkul A, Sithigorngul P, Sang-oum W, Rukpratanporn S, Sriurairatana S, Withayachumnarnkul B, Flegel TW (2005) Synthetic peptide used to develop antibodies for detection of polyhedrin from monodon baculovirus (MBV). Dis Aquat Org 65:79–84

Seibert HC, Borsa M, Rosa RD, Cargnin-Ferreira E, Pereira AML, Grisard EC, Zanetti CR, Pinto AR (2010) Detection of major capsid protein of infectious myonecrosis virus in shrimps using monoclonal antibodies. J Virol Methods 169:169–175

Senapin S, Phewsalya K, Briggs M, Flegel TW (2007) Outbreaks of infectious myonecrosis virus (IMNV) in Indonesia confirmed by genome sequencing and use of an alternative RT-PCR detection method. Aquaculture 266:32–38

Senapin S, Phiwsaiya K, Gangnonngiw W, Flegel TW (2011) False rumours of disease outbreaks caused by infectious myonecrosis virus (IMNV) in the whiteleg shrimp in Asia. J Negat Results Biomed 10:1–5

Senapin S, Phewsalya K, Gangnonngiw W, Briggs M, Sithigorngul P, Flegel TW (2013) Dual infections of IMNV and *Mr*NV in cultivated *Penaeus vannamei* from Indonesia. Aquaculture 372-375:70–73

Shen H, Zhang W, Shao S (2015) Phylogenetic and recombination analysis of genomic sequences of IHHNV. J Basic Microbiol 55(8):1048–1052

Shike H, Dhar AK, Burns JC, Shimizu C, Jousset FX, Klimpel KR, Bergoin M (2000) Infectious hypodermal and hematopoietic necrosis virus of shrimp is related to mosquito brevidensoviruses. Virology 277(1):167–177

Sibonga MF, Geduspan JS, Caipang CM (2013) Optimized multiplex PCR assays for simultaneous detection of viruses infecting hatchery-reared shrimp, *Penaeus monodon* in The Philippines. Human Veterin Med 5(3):142–145

Singaravel V, Gopalakrishnan A, Martin GG (2021) Multiple infections of *Entrerocytozoon hepatopenaei* and Hepatopancreatic parvovirus in pond-reared *Penaeus vannamei* in India. Aquaculture 545:737232

Sithigorngul P, Hajimasalaeh W, Longyant S, Sridulyakul P, Rukpratanporn S, Chaivisuthangkura P (2009) Simple immunoblot and immunohistochemical detection of *Penaeus stylirostris* densovirus using monoclonal antibodies to viral capsid protein expressed heterologously. J Virol Methods 162(1–2):126–132

Sivakumar VK, Sarathi M, Venkatesan C, Sivaraj A, Hameed AS (2009) Experimental exposure of *Artemia* to hepatopancreatic parvo-like virus and subsequent transmission to post-larvae of *Penaeus monodon*. J Invertebr Pathol 102(3):191–195

Söderlund-Venermo M (2019) Emerging human parvoviruses: the rocky road to fame. Ann Rev Virol 6:71–91

Spann KM, Lester RJG, Paynter JL (1993) Efficiency of chlorine as a disinfectant against monodon baculovirus (MBV). Asian Fish Sci 6:295–301

Spann TP, Moir RD, Goldman AE, Stick R, Goldman RD (1997) Disruption of nuclear lamin organization alters the distribution of replication factors and inhibits DNA synthesis. J Cell Biol 136(6):1201–1212

Sridulyakul P, Suwannaka T, Chaivisuthangkura P, Longyant S, Rukpratanporn S, Sithigorngul P (2011) *Penaeus monodon* nucleopolyhedrovirus detection using monoclonal antibodies specific to recombinant polyhedrin protein. Aquaculture 321:216–222

Srisala J, Sanguanrut P, Thaiue D, Laiphrom S, Siriwattano J, Khudet J, Powtongsook S, Flegel TW, Sritunyalucksana K (2021a) Infectious myonecrosis virus (IMNV) and Decapod iridescent virus 1 (DIV1) detected in captured, wild *Penaeus monodon*. Aquaculture 545:737262

Srisala J, Thaiue D, Sanguanrut P, Aldama-Cano DJ, Flegel TW, Sritunyalucksana K (2021b) Potential universal PCR method to detect decapod hepanhamaparvovirus (DHPV) in crustaceans. Aquaculture 541:736782

Srisuk C, Chaivisuthangkura P, Sukhumsirichart W, Sridulyakul P, Longyant S, Rukpratanporn S, Sithigorngul P (2011) Improved immunodetection of *Penaeus monodon* densovirus with monoclonal antibodies raised against recombinant capsid protein. Aquaculture 311(1–4):19–24

Sritunyalucksana K, Sanguanrut P, Srisala J, Thawonsuwan J, Saleetid N, Vanichviriyakit R, Chotwiwatthanakun C, Flegel TW, Taengchaiyaphum S (2023) No single PCR test is sufficient to determine parvovirus IHHNV presence in or impact on farmed shrimp production. bioRxiv, 2023-09

Suganthi G, Uma A, Rebecca G, Saravanabava K (2012) Strain variations in monodon baculovirus (MBV) infecting *Penaeus monodon* (tiger shrimp) in India. CIBTech J Biotechnol 1(1):22–26

Sukhumsirichart W, Wongteerasupaya C, Boonsaeng V, Panyim S, Sriurairatana S, Withyachumnarnkul B, Flegel TW (1999) Characterization and PCR detection of hepatopancreatic parvovirus (HPV) from *Penaeus monodon* in Thailand. Dis Aquat Org 38(1):1–10

Sukhumsirichart W, Kiatpathomchai W, Wongteerasupaya C, Withyachumnarnkul B, Flegel TW, Boonseang V, Panyim S (2002) Detection of hepatopancreatic parvovirus (HPV) infection in *Penaeus monodon* using PCR-ELISA. Mol Cell Probes 16(6):409–413

Sun ZF, Hu CQ, Ren CH, Shen Q (2006) Sensitive and rapid detection of infectious hypodermal and hematopoietic necrosis virus (IHHNV) in shrimps by loop-mediated isothermal amplification. J Virol Methods 131(1):41–46

Sundararaj A, Manohar MBB, Sheela RRP, Selvaraj D, Ravishankar B (1996) Occurrence of monodon baculovirus infection in shrimp in Tamil Nadu. Indian J Fish 43:103–105

Surachetpong W, Poulos BT, Tang KFJ, Lightner DV (2005) Improvement of PCR method for the detection of monodon baculovirus (MBV) in penaeid shrimp. Aquaculture 249:69–75

Suryakodi S, Nafeez Ahmed A, Badhusha A, Santhosh Kumar S, Sivakumar S, Abdul Majeed S, Taju G, Rahamathulla S, Sahul Hameed AS (2022) First report on the occurrence of white spot syndrome virus, infectious myonecrosis virus and *Enterocytozoon hepatopenaei* in *Penaeus vannamei* reared in freshwater systems. J Fish Dis 45(5):699–706

Tang KF, Lightner DV (2001) Detection and quantification of infectious hypodermal and hematopoietic necrosis virus in penaeid shrimp by real-time PCR. Dis Aquat Org 44(2):79–85

Tang KF, Lightner DV (2002) Low sequence variation among isolates of infectious hypodermal and hematopoietic necrosis virus (IHHNV) originating from Hawaii and the Americas. Dis Aquat Org 49(2):93–97

Tang KFJ, Lightner DV (2006) Infectious hypodermal and hematopoietic necrosis virus (IHHNV)-related sequence in the genome of the black tiger prawn *Penaeus monodon* from Africa and Australia. Virus Res 118:185–191

Tang KF, Lightner DV (2011) Duplex real-time PCR for detection and quantification of monodon baculovirus (MBV) and hepatopancreatic parvovirus (HPV) in *Penaeus monodon*. Dis Aquat Org 93(3):191–198

Tang KFJ, Pantoja CP, Poulos BT, Redman RM, Lightner DV (2005) In situ hybridization demonstrates that *Litopenaeus vannamei*, *L. stylirostris* and *Penaeus monodon* are susceptible to experimental infection with infectious myonecrosis virus (IMNV). Dis Aquat Org 63:261–265

Tang KFJ, Navarro SA, Lightner DV (2007) PCR assay for discriminating between infectious hypodermal and hematopoietic necrosis virus (IHHNV) and virus-related sequences in the genome of *Penaeus monodon*. Dis Aquat Org 74:165–170

Tang KF, Bondad-Reantaso MG, Arthur JR (2019) Shrimp infectious myonecrosis strategy manual. FAO Fisheries and Aquaculture Circular C1187:1–43

Tangtongpiroj JA (1989) Virus disease of black tiger prawn. Fish News 13:25–29

Taukhid, Nuraini YL (2008) Infectious myonecrosis virus (IMNV) in Pacific white shrimp (*Litopenaeus vannamei*) in Indonesia. Indones Agric J 3(2):139–146

Teixeira-Lopes MA, Vieira-Girão PRN, da Cruz Freire JE, Rocha ÍRCB, Costa FHF, Rádis-Baptista G (2011) Natural co-infection with infectious hypodermal and hematopoietic necrosis virus (IHHNV) and infectious myonecrosis virus (IMNV) in *Litopenaeus vannamei* in Brazil. Aquaculture 312(1–4):212–216

Theilmann DA, Blissard GW, Boning B, Jehle JA, O'Reilly DR, Rohrmann GF, Thiem S, Vlak JM (2005) Baculoviridae. In: Fauquet CM, Mayo MA, Maniloff J, Desselberger U, Ball LA (eds) Virus taxonomy, VIIIth report of the International Committee on taxonomy of viruses, pp 177–185

Thikiew N (1990) A serious MBV outbreak in the black tiger prawn. Livestock Prod Magaz 7:13–16

Thurman RB, Lightner DV, Bell TA, Hazanow S (1990) Unique physiochemical properties of the occluded penaeid shrimp baculoviruses and their use in diagnosis of infections. J Aquat Anim Health 2:128–131

Tokhmafshan M, Mohammad SH, Hassan D, Wang YG (2004) Identification of *Penaeus monodon* baculovirus (MBV) in cultured *Penaeus semisulcatus* in Islamic Republic of Iran. Iran J Fish Sci 4(1):25–44

Turnbull JF, Larkins PE, Matondang R, Mc Padden C (1994) Histopathological disease survey of cultured shrimps in North-East Sumatra, Indonesia. J Fish Dis 17:57–65

Uma A, Ramanathan N, Saravanabava K (2007) A study by PCR on the prevalence of white spot syndrome virus, monodon baculovirus and hepatopancreatic parvovirus infection in *Penaeus monodon* postlarvae in Tamil Nadu, southeast coast of India. Asian Fish Sci 20:227–239

Umesha RK, Uma A, Otta SK, Karunasagar I, Karunasagar I (2003) Detection by PCR of hepatopancreatic parvovirus (HPV) and other viruses in hatchery-reared *Penaeus monodon* postlarvae. Dis Aquat Org 57:141–146

Umesha KR, Kennedy B, Dass M, Naik BM, Venugopal MN, Karunasagar I, Karunasagar I (2006) High prevalence of dual and triple viral infections in black tiger shrimp ponds in India. Aquaculture 258(1–4):91–96

Vanpatten KA, Nunan LM, Lightner DV (2004) Seabirds as potential vectors of penaeid shrimp viruses and the development of a surrogate laboratory model utilizing domestic chickens. Aquaculture 241(1–4):31–46

Vaseeharan B, Ramasamy P (2004) Abundance of potentially pathogenic micro-organisms in *Penaeus monodon* larvae rearing systems in India. Microbiol Res 158:299–308

Viana JT, dos Santos Rocha R, Maggioni R (2024) Immunological lectins in shrimp *Penaeus vannamei* challenged with infectious myonecrosis virus (IMNV) under low-salinity conditions. Fish Shellfish Immunol 148:109471

Vickers JE, Paynter JL, Spradbrow PB, Lester RJG (1993) An impression smear method for rapid detection of *Penaeus monodon*-type baculovirus (MBV) in Australian prawns. J Fish Dis 16:507–511

Vickers JE, Webb R, Young PR (2000) Monodon baculovirus from Australia: ultrastructural observations. Dis Aquat Org 39:169–176

Vieira-Girão PRN, Falcão CB, Rocha IRCB, Lucena HMR, Costa FHF, Rádis-Baptista G (2017) Antiviral activity of Ctn (15-34), a cathelicidin-derived eicosapeptide, against infectious myonecrosis virus in *Litopenaeus vannamei* primary hemocyte cultures. Food Environ Virol 9(3):277–286

Vijayan KK, Alavandi SV, Rajendran KV, Alagarswami K (1995) Prevalence and histopathology of monodon baculovirus (MBV) infection in *Penaeus monodon* and *P. indicus* in shrimp farms in the South-east coast of India. Asian Fish Sci 8:267–272

Vogt G (1992) Transformation of anterior midgut and hepatopancreas cells by monodon baculovirus (MBV) in *Penaeus monodon* postlarvae. Aquaculture 107:239–248

Wan X, Xie G, Wang C, Xu T, Zhang Q (2023) A confirmed case of infectious myonecrosis virus (IMNV) infection in cultured *Penaeus vannamei* in China. Aquaculture 577:739953

Wang Y, Jehle JA (2009) Nudiviruses and other large, double-stranded circular DNA viruses of invertebrates: new insights on an old topic. J Invertebr Pathol 101:187–193

Wang N, Yin W, Zhang N, Yang J, Jia P, Fang B, Yue Z, Xie Y, Li J, Zhao J, Liu H (2022) Establishment and clinical application of a detection method for the infectious hypodermal and hematopoietic necrosis virus in water. Aquaculture 546:737228

Wangman P, Longyant S, Chaivisuthangkura P, Sridulyakul P, Rukpratanporn S, Sithigorngul P (2012) *Penaeus monodon* nucleopolyhedrovirus detection using an immunochromatographic strip test. J Virol Methods 183:210–214

Wangman P, Longyant S, Utari HB, Senapin S, Pengsuk C, Sithigorngul P, Chaivisuthangkura P (2016) Sensitivity improvement of immunochromatographic strip test for infectious myonecrosis virus detection. Aquaculture 453:163–168

Wei YW, Fan DD, Chen J (2017) The mussel *Mytilus edulis* L. as an important reservoir of infectious hypodermal and hematopoietic necrosis virus (IHHNV). Aquac Res 48(3):1346–1350

Widanarni, Yuhana M, Muhammad DA (2014) *Bacillus* NP5 improves growth performance and resistance against infectious Myonecrosis virus in White Shrimp (*Litopenaeus vannamei*). ILMU KELAUTAN 19(4):211–218

Wyban JA, Sweeny JN (1991) Shrimp health maintenance. In: Intensive shrimp production technology. The Oceanic Institute, Hawaii, pp 103–112

Wyban JA, Swingle J, Sweeney JN, Pruder GD (1992) Development and commercial performance of high health shrimp from SPF broodstock Penaeus vannamei. In: Wyban J (ed) Proceedings of the special session on shrimp farming. World Aquaculture Society, Baton Rouge, LA, USA, pp 254–260

Xie Z, Xie L, Pang Y, Lu Z, Xie Z, Sun J, Deng X, Liu J, Tang X, Khan M (2008) Development of a real-time multiplex PCR assay for detection of viral pathogens of penaeid shrimp. Arch Virol 153:2245–2251

Xu F, Wei Y, Lu J, Chen J (2023) Prevalence of infectious hypodermal and hematopoietic necrosis virus (IHHNV) in farmed *Procambarus clarkii* of the middle and lower reaches of the Yangtze river in China. Pathogens 12(8):1038

Yan D, Tang KFJ, Lightner DV (2009) Development of a real-time PCR assay for detection of monodon baculovirus (MBV) in penaeid shrimp. J Invertebr Pathol 102(2):97–100

Yan DC, Tang KFJ, Lightner DV (2010) A real-time PCR for the detection of hepatopancreatic parvovirus (HPV) of penaeid shrimp. J Fish Dis 33(6):507–511

Yang B, Song XL, Huang J, Shi CY, Liu QH, Liu L (2006) A single-step multiplex PCR for simultaneous detection of white spot syndrome virus and infectious hypodermal and haematopoietic necrosis virus in penaeid shrimp. J Fish Dis 29(5):301–305

Yang B, Song XL, Huang J, Shi CY, Liu L (2007) Evidence of existence of infectious hypodermal and hematopoietic necrosis virus in penaeid shrimp cultured in China. Vet Microbiol 120(1–2):63–70

Yang YT, Lee DY, Wang Y, Hu JM, Li WH, Leu JH, Chang GD, Ke HM, Kang ST, Lin SS, Kou GH, Lo CF (2014) The genome and occlusion bodies of marine *Penaeus monodon* nudivirus (PmNV, also known as MBV and PemoNPV) suggest that it should be assigned to a new nudivirus genus that is distinct from the terrestrial nudiviruses. BMC Genomics 15:628

Yang M, Wang B, Xu R, Li P, Xia J, Jia T, Li W, Zhang Y, Zhang Q (2024) Evaluation of combined disinfection methods of infectious myonecrosis virus (IMNV) in seawater using Pacific white shrimp (*Penaeus vannamei*). Aquaculture 590:741063

Yin XT, Fei RM (2015) Preliminary screening for infectious hypodermal and hematopoietic necrosis virus capsid protein receptors from *Litopenaeus vannamei* gill membrane proteins. J Fish Sci China 22(6):1160–1166

Yin X, Wang X, Sun H, Fei R (2022) NM23 is a CP-binding protein involved in infectious hypodermal and hematopoietic necrosis virus infection in shrimp. Animals 12(5):621

Zeng D, Peng M, Yang Q, Yang C, Liao Z, Li Q, Liu Q, Zhu W, Wang H, Li M, Chen X (2019) The *Penaeus stylirostris* densovirus capsid protein interacts with the *Litopenaeus vannamei* BCCIP protein. Fish Shellfish Immunol 88:198–206

Zhang JX, Sun XQ (1997) A preliminary study on the virus in the eggs of *Penaeus chinensis*. J Oceanograp Huanghai Bohai Seas 15:48–51

Zhang L, Liu K, Liu M, Hu J, Bao Z, Wang M (2023) Development of a real-time enzymatic recombinase amplification assay for rapid detection of infectious hypodermal and hematopoietic necrosis virus (IHHNV) in shrimp *Penaeus vannamei*. J Invertebr Pathol 201:108024

Zhang L, Zhou Q, Liu J, Liu M, Hu J, Bao Z, Wang M (2024) Development of recombinase amplification assays for the rapid detection of infectious myonecrosis virus. J Invertebr Pathol 108143:108143

Zhao YongZhen ZY, Chen XiuLi CX, Zeng DiGang ZD, Yang ChunLing YC, Peng Min PM, He PingPing HP, Li YongMei LY, Peng JinXia PJ, Wei PinYuan WP, Chen XiaoHan CX (2014) Cloning and analysis of full-length cDNA of prophenoloxidase activating factor in *Litopenaeus vannamei*. J Fish China 38(8):1068–1078

Zhao C, Dewangan NK, Gopalakrishnan A, Sravani S, John AS, Singh RR, Xu W, Guo G, Xia S, Tiwari S, Li X (2024) Deformities in *Litopenaeus vannamei* caused by infectious hypodermal and hematopoietic necrosis viral (IHHNV) infection in intensive grow-out ponds of South east coast of India. Israel J Aquacult Bamidgeh 76(3):217–228

White Spot Syndrome Virus (WSSV)

K. V. Rajendran, K. Sreedharan, Indrani Karunasagar, Iddya Karunasagar, and Arun K. Dhar

Abstract

White spot syndrome virus (WSSV) is a highly infectious virus causing 100% mortality in commercially farmed crustaceans within a few days. The viral epizootics result in substantial financial loss to the farmers, threatening the sustainability of the shrimp industry worldwide. WSSV is a bacilliform, non-occluded, enveloped virus with a circular dsDNA genome and belongs to the family *Nimaviridae*. It is one of the largest known animal viruses. The white spot disease (WSD), caused by WSSV, is manifested in the form of white spots/patches underneath the exoskeleton of shrimp, along with other clinical signs. A range of highly sensitive and specific diagnostic techniques have been developed for the detection and quantification of the virus. The virus has been reported in several Southeast Asian countries, North American and Latin American countries, Africa and Australia. Various geographical isolates with genotypic variability have been identified in the virus. The virus possesses a wide host range and affects all life stages of farmed penaeid shrimps. Additionally, it affects several species of prawns, crabs, crayfish and other aquatic invertebrates. The transmission of the virus has been reported to be through horizontal and vertical means. Several methods, including gross clinical signs, histopathological manifestations, in situ hybridization, and immunological and DNA-based molecular methods, were

K. V. Rajendran (✉) · K. Sreedharan
ICAR-Central Institute of Fisheries Education, Mumbai, India
e-mail: rajendrankv@hotmail.com; sreedharan@cife.edu.in

I. Karunasagar · I. Karunasagar
FAO Ref Centre for Antimicrobial Resistance, NITTE University, Mangaluru, India
e-mail: indrani.karunasagar@nitte.edu.in; iddya.karunasagar@nitte.edu.in

A. K. Dhar
The University of Arizona, Tucson, AZ, USA
e-mail: adhar@arizona.edu

R. Kooloth Valappil et al. (eds.), *Aquatic Animal Health Management*,
https://doi.org/10.1007/978-981-96-7987-4_10

established for the laboratory and field-level diagnosis of WSSV. Prophylactic measures against WSSV have been explored, including diverse vaccination methods and RNAi technology targeting different viral genes. Stocking specific pathogen-free (SPF) post-larvae and implementing stringent biosecurity measures at farming facilities are crucial to prevent the introduction and spread of WSSV.

Keywords

White spot syndrome virus (WSSV) · White spot disease (WSD) · Shrimp disease

1 Introduction

A variety of infectious agents cause disease conditions in farmed shrimp, and among them, viral diseases are the major limiting factor in aquaculture, causing approximately 60% of the production losses (Flegel 2001, 2012). Accordingly, the World Organisation for Animal Health (WOAH) listed five viral diseases infecting crustaceans as notifiable diseases (Table 1). White spot syndrome virus (WSSV), which causes white spot disease (WSD), is one of the most lethal shrimp viruses, infecting a wide range of crustaceans. The carapace of the infected shrimp displays white spots, and the virus causes heavy mortality in several commercially important crustacean species. The virus caused widespread disease outbreaks, resulting in large-scale mortalities and enormous financial losses in most shrimp-growing countries since the disease was first reported in 1992. Additionally, WSD outbreaks have been attributed to the change in the farming practices of different penaeid species. For example, until 2000, the black tiger shrimp (*Penaeus monodon*) was the leading farmed shrimp species in Asia, with an estimated global production of 631,000 tonnes. The total production of the species remained at approximately 717,000 tonnes in 2020. On the other hand, the production of *Penaeus vannamei* during the same period had increased from 155,000 tonnes to 5.8 million tonnes. The decline or the stagnation in production of *P. monodon* was primarily due to WSD outbreaks in farmed *P. monodon* that eventually led to the introduction of domesticated specific pathogen (WSSV)-free (SPF) *P.vannamei* in Asia.

2 White Spot Disease (WSD) and White Spot Syndrome Virus (WSSV)

2.1 History and Geographical Distribution

White spot disease caused by WSSV was first described in farmed shrimp in northern Taiwan in 1992 (Chou et al. 1995). Subsequently, it was reported in Japan (Inouye et al. 1994; Nakano et al. 1994). The disease quickly disseminated to other Asian countries, including Korea, Japan, Vietnam, Thailand, Malaysia, Indonesia,

Table 1 Viral diseases infecting penaeid shrimp listed by the World Organization for Animal Health (WOAH 2023)

Sl. no.	Disease/syndrome	Pathogen	Distribution	Status in India
1	Infection with infectious hypodermal and hematopoietic necrosis virus	Infectious hypodermal and hematopoietic necrosis virus (IHHNV) (Linear single-stranded (ss) DNA virus)	Worldwide distribution	Reported
2	Infection with infectious myonecrosis virus	Infectious myonecrosis virus (Double-stranded (ds) RNA virus)	Countries in the Americas, Asia, and Africa (Brazil, Indonesia, India, and Egypt)	Reported
3	Infection with Taura syndrome	Taura syndrome virus (TSV) (Linear, positive-sense single-stranded (ss+) RNA virus)	Americas, Southeast Asia (People's Republic of China, Thailand, Malaysia, and Indonesia), and the Middle East (Saudi Arabia)	Not reported
4	Infection with white spot syndrome virus	White spot syndrome virus (WSSV) (Double-stranded DNA virus)	People's Republic of China, Japan, Republic of Korea, Southeast Asia, South Asia, the Indian Continent, the Mediterranean, the Middle East, Australia, and the Americas	Reported
5	Infection with yellow head virus genotype 1	Yellow head virus (YHV) genotype 1 (Positive-sense single-stranded (ss+) RNA genome)	Chinese Taipei, Indonesia, Malaysia, The Philippines, Sri Lanka, Thailand, and Mexico	Not reported

and India (Inouye et al. 1994; Nakano et al. 1994; Mohan et al. 1998; Park et al. 1998). Incidence of WSSV infection has also been reported in other countries such as Madagascar, Mozambique, Ecuador, Brazil, Sri Lanka, Iran, Saudi Arabia, Iraq, Bangladesh, and several other parts of the world (Galavíz-Silva et al. 2004; Peña et al. 2007; Tang et al. 2012, 2013; Afsharnasab et al. 2014; Hossain et al. 2014; Amarakoon et al. 2016). WSSV was not recorded in Australia until recently; however, the virus was first detected in Southeast Queensland in 2016, wherein shrimp farms reported the disease from late 2016 to early 2017. Further, the country also recorded WSSV in shrimps and crabs from the wild collection in early 2017 (https://www.outbreak.gov.au/current-responses-to-outbreaks/white-spot-disease).

In India, though the disease was encountered in 1993, the first report was published in 1995, and the pathogen was identified as systemic ectodermal and mesodermal baculovirus (SEMBV) (Annon 1995). Subsequently, several studies on the virus were reported, including infections in shrimps and crabs (Manohar et al. 1996; Karunasagar et al. 1997; Rajendran et al. 1998; Sen et al. 1999; Madhavi et al. 2002;

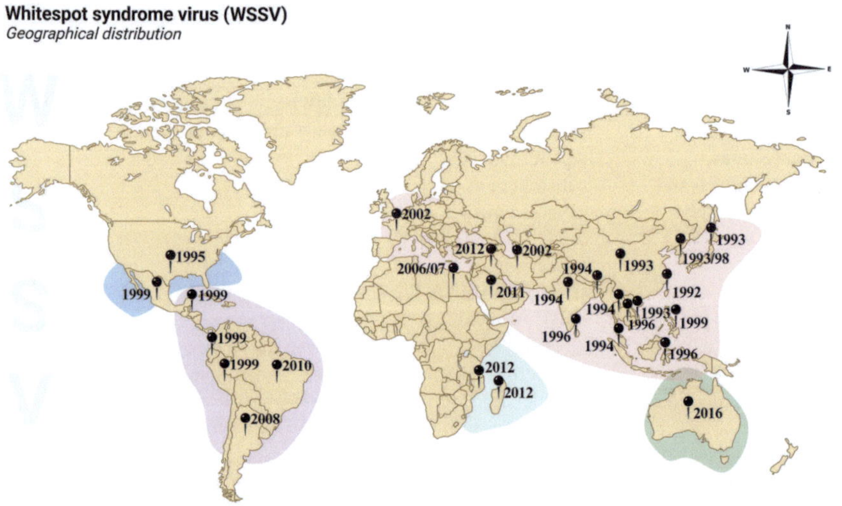

Fig. 1 Geographical distribution of white spot syndrome virus based on the published reports

Manivannan et al. 2002; Selvam et al. 2012; Joseph et al. 2015a, b; Ahmad et al. 2017; Dewangan et al. 2017; Raja et al. 2017; Saravanan et al. 2017a, b; Chandrakala and Mekala 2018; Gunasekaran et al. 2018; Rathipriya et al. 2019; Abraham et al. 2020; Atnur et al. 2021; Babu et al. 2021). The current geographical distribution of WSSV is depicted in Fig. 1.

2.2 Economic Impact

Since its emergence in 1993, WSSV has been causing widespread disease outbreaks and significant economic losses in farmed shrimp. The virus continues to be one of the most significant threats and limiting factors to shrimp production as of now, despite the introduction and widespread farming of domesticated specific pathogen-free (SPF) Pacific whiteleg shrimp, *P. vannamei* (Shinn et al. 2018; Rajendran et al. 2021). Reports describing the estimated disease-associated economic loss in different countries/regions are available. According to Patil et al. (2021), the economic cost of diseases could be either direct or indirect. The loss of stock due to disease and the expenditure involved in controlling/managing the infection will be the direct cost, and the indirect cost includes ancillary losses involving supporting sectors of the industry, jobs, markets, etc. (Briggs et al. 2004; Patil et al. 2021).

Various assessments on the yearly loss of production in shrimp aquaculture due to infectious diseases have been reported from different countries such as Thailand, China, and Brazil (Flegel et al. 2008; FAO 2016; Shinn et al. 2018). For example, in Asia, direct economic losses due to the yellow head virus (YHV) since 1992 and

WSSV since 1994 together were estimated at USD 1 billion per year (Briggs et al. 2004). Similarly, in Latin America, the economic loss, initially due to Taura syndrome virus (TSV) since 1993 and later, WSSV from 1999, was approximately USD 0.5 billion per year (Briggs et al. 2004). Regional estimates on WSSV outbreaks showed a revenue loss of USD 4–6 billion in Asia and USD 1 billion in America over 10 years since its emergence (Lightner 2003). In a later report, the estimated economic loss over a decade due to shrimp diseases in Asian countries was USD 20 billion (Davies 2016). Lightner et al. (2012) were the first to collate data and summarize the economic loss due to some of the OIE-listed viruses infecting shrimp since their emergence/discovery.

Lightner et al. (2012) noted that the reported losses due to WSSV (~USD 46 billion in Asia and ~USD 1–2 billion in the Americas) might be much less than the actual impact of the disease, and the total economic losses due to WSSV could be ~USD 15 billion. This is nearly one-tenth of the shrimp production globally (Stentiford et al. 2012), with an estimated annual loss of ~USD 1 billion (Flegel et al. 2008; Stentiford et al. 2012). In 2015, WSSV caused a loss of 80–100% shrimp stock in 5370 hectares of shrimp farms that included both *P. monodon* and *P. vannamei* in Vietnam and resulted in a cumulative loss of ~USD 11.02 million (Shinn et al. 2018). However, this estimate was noted to be different from the previous report on the WSSV-associated losses (USD 55.58 million) reported by Hien et al. (2016). In Indonesia, the annual loss due to WSSV was estimated to be USD 191 million (Hastuti and Haryadi 2016). It has been reported that shrimp diseases in India resulted in an economic loss of up to USD 100 million (Mohan and Basavarajappa 2001). However, in a study undertaken later during 2006–2008, Kalaimani et al. (2013) estimated that various shrimp diseases in *P. monodon* in India resulted in a gross national loss of 48,717 metric tonnes of shrimp amounted to INR 1022.1 crore, and 2.15 million man-days per year. The authors reported that WSSV alone, with the highest probability of occurrence (0.196 on a 0–1 scale), resulted in the loss of 23,318.51 metric tonnes of shrimp along with a loss of farm income to the tune of INR 489.25 crore and employment of 1,032,214 man-days per year (Kalaimani et al. 2013). In a study investigating the economic consequences of WSSV and *Enterocytozoon hepatopenaei* (EHP), Patil et al. (2021) estimated that WSSV caused higher production loss (ton/ha/crop) compared to EHP and was estimated to be 2.58 ± 0.32. WSSV was also found to have a higher probability of occurrence (25%) than EHP (17%). The loss of production caused by WSSV was estimated to be 0.33 M tons, with a resultant economic loss of INR 1670 crore (USD 238.33 million). The study also reported that with a high probability of occurrence of infectious diseases (49%), the annual loss of shrimp production in India was estimated to be 0.21 million tonnes, valued at USD 1.02 billion. Further, this led to a huge employment loss of 1.65 million man-days valued at USD 7.07 million (Patil et al. 2021).

3 The Disease

3.1 Clinical Signs, Target Tissues, and Pathological Responses

Animals infected with WSSV develop white spots ranging from 0.5 to 3.0 mm in diameter on the cuticle (Lo et al. 1996; Kasornchandra et al. 1998; Wu et al. 2001), which is reported to be due to the dysfunctioning of the integument by which calcium salts get accumulated within the cuticle, thus, appearing as white spots (Wang et al. 2000a) (Fig. 2). However, observation of white spots alone cannot be considered a confirmatory sign for preliminary diagnosis of this disease (Sánchez-Paz 2010). Other signs of WSSV infection comprise reduced preening ability (Wongteerasupaya et al. 1995; Durand et al. 1997), sudden drop in feed intake (Chou et al. 1995; Durand et al. 1996; Flegel 1997), reddening of appendages and body surface (Lightner and Redman 1998; Nadala Jr et al. 1998), and loose cuticle (Lo et al. 1996; Durand et al. 1997), bulging of branchiostegites owing to the build-up of fluid (Otta et al. 1999), hypertrophy and yellowing of the hepatopancreas (Hameed et al. 1998), and thinning and enhanced clotting time of haemolymph (Wang et al. 2000a; Kiatpathomchai et al. 2001). In the farm, WSSV-infected shrimp congregate near the corner of the pond and exhibit disease signs within 1–2 days prior to the occurrence of the first mortalities (Kou et al. 1998), and the mortality rate may reach 100% in 10 days after the initiation of infection (Durand et al. 1996; Karunasagar et al. 1997; Lotz and Soto 2002).

The virus targets mainly the cells of organs of ectodermal and mesodermal origin, including those of the epidermis, gills, foregut, hindgut (Wongteerasupaya et al. 1995; Chang et al. 1996), antennal gland, lymphoid organ (Durand et al. 1996; Chang et al. 1998a), muscle, eye-stalk, heart (Kou et al. 1998), gonads (Lo et al. 1997), haematopoietic cells and cells associated with the nervous system (Rajendran et al. 1999; Wang et al. 2000c; Wongprasert et al. 2003). Epithelial cells of endodermal origin, such as the hepatopancreas, mid-gut caeca (anterior and posterior) and mid-gut trunk, are not infected (Hameed et al. 1998). However,

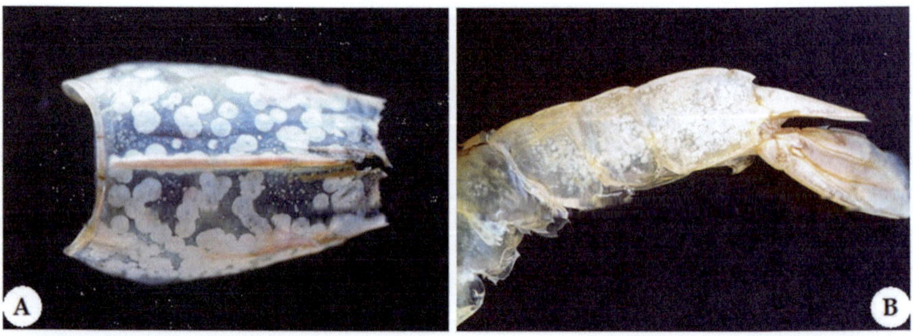

Fig. 2 (**a**) White spots/patches observed on the carapace of WSSV-infected shrimp; (**b**) A heavily infected shrimp showing white spots in the tail region

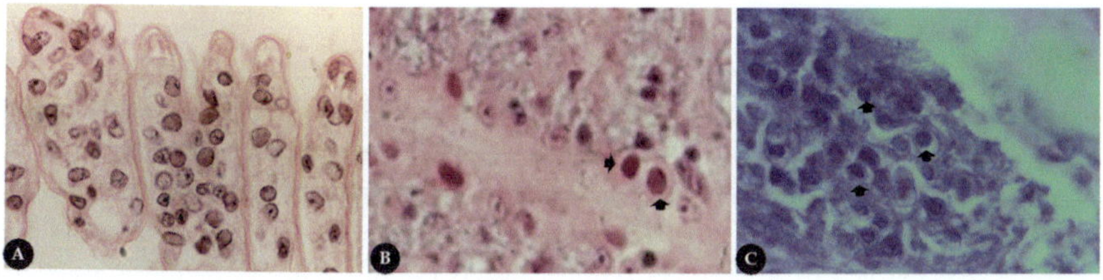

Fig. 3 (**a**) Histological section of gills of WSSV-infected shrimp showing hypertrophied nuclei and chromatin margination; (**b**) Connective tissue underlying the cuticular epidermis showing Cowdry-A-type inclusions (arrows); (**c**) Subcuticular ectodermal layer showing large basophilic inclusions (arrows) indicating the late stage of infection. Note the severe necrotic changes in the tissue

in the late stages of infection, the virus causes severe damage in the epithelia of the stomach, gills and integument (Chang et al. 1996; Wang et al. 2000a) and may cause dysfunction of multiple organs leading to the death of the animals. The infected cell nuclei appear hypertrophied with marginated chromatin (Durand et al. 1997; Wang et al. 2000a) and contain eosinophilic to basophilic inclusion bodies (Flegel 2006). The early stage of infection results in the formation of Cowdry A-type inclusions (Flegel 2006) (Fig. 3). A study on the ultrastructure of haemolymph tissue of *Cherax quadricarinatus* following WSSV infection revealed pronounced pathological changes, including damaged haematopoietic tissue, distorted golgiosomes, indistinct damage in the mitochondrial structures and deformation and breakage of the nuclear membrane. Moreover, plenty of heterochromatin originating in the nucleus and oedematous conditions in the cytoplasm were observed (Ning et al. 2016).

Many authors noticed alterations in different haematological and physiological parameters, activity of antioxidant and membrane-bound enzymes and expression of genes associated with diverse functions in WSSV-infected shrimp (Yoganandhan et al. 2003b; Mohankumar and Ramasamy 2006a, b; Rameshthangam and Ramasamy 2006; Pais et al. 2007; Mathew et al. 2009; Sivakamavalli and Vaseeharan 2014; Pemula and Krishnan 2015; Peruzza et al. 2019). Infection with WSSV also results in pronounced metabolic changes in different tissues of shrimp (Chen et al. 2016; Wu et al. 2017; Alfaro et al. 2021). The early stage of WSSV infection results in the enhanced expression of the female reproductive genes such as vitellogenin and thrombospondin; however, it is downregulated at a later stage (Tomy et al. 2015). There is a large body of information on the impact of WSSV infection on the various immune genes of crustaceans (Antony et al. 2011a, b, c; Kiruthiga et al. 2012; Paria et al. 2012; Deepika et al. 2014; Kulkarni et al. 2014a; Vidya et al. 2014, 2016; Sreedharan et al. 2017, 2018; Boonrawd et al. 2017; Li et al. 2018; Peruzza et al. 2019; Shekhar et al. 2019; Dechamma et al. 2020; Rajeish et al. 2021). WSSV also affects the composition of the intestinal microbiota in *Litopenaeus vannamei* (Wang et al. 2019; Valente et al. 2020).

3.2 Transmission

The transmission of WSSV can be either horizontal, through the ingestion of infected tissue, direct exposure to virion-containing water, presumably through the gills (Chou et al. 1998), by cannibalism of infected animals (Chang et al. 1996) and horizontal (Corsin et al. 2001) or vertical from parent to offspring (Durand and Lightner 2002). Wu et al. (2001) demonstrated cannibalism as a major route of transmission in experimental conditions. The digestive tract and gills were usually believed to be the portal of entry of WSSV into shrimp (Chou et al. 1998). However, it has been demonstrated that the antennal gland can be a portal of access to WSSV to elicit disease in shrimp (Liu et al. 2021a, b, c, d). Further, the study showed that salinity-induced stress could stimulate the proliferation of WSSV in the antennal gland and aggravate the disease process (Liu F et al. 2021a). Vertical transmission has been recognized as the major route of disease transmission in hatcheries (Mushiake et al. 1999). There are many sources through which WSD can get into the aquaculture system, including the introduction of live shrimp or other host species from infected to non-infected sites. Other possible means of WSSV transmission encompass anthropogenic activities, migration of seabirds or other animals, contaminated frozen food products, pond sediments and aquaculture gears, and untreated infected shrimp by-products from processing plants (Durand et al. 1997; Supamattaya et al. 1998). Asymptomatic vectors or carriers (like crabs) also take part in disease spread in farms (Chou et al. 1995; Rajendran et al. 1999), while the major risk is associated with arthropod vectors as they are common in shrimp culture areas (Supamattaya et al. 1998; Rajendran et al. 1999). Horizontal transmission dynamics of WSSV in shrimp through cohabitation were studied by Tuyen et al. (2014), and it was found that the mortality rate in *P. monodon* was comparatively high than that in *P. vannamei*. The evidence of trans-ovarian transmission of WSSV in *P. indicus* was reported by Vijayan et al. (2024), wherein the authors detected the virus in both ovaries and spermatophores.

3.3 Host Range

One of the most remarkable features of WSSV is its wide host range. WSSV infects all the penaeid shrimps and a wide range of other decapods. At least 18 farmed and wild penaeid shrimps have been reported to be susceptible (Durand et al. 1997; Lo et al. 1997; Lightner and Redman 1998; Park et al. 1998; Rajendran et al. 1999). Moreover, WSSV displayed susceptibility to different freshwater prawn species (Rajendran et al. 1999; Hameed et al. 2000, 2001; Chakraborty et al. 2002). Apart from shrimps, WSSV affects other crustaceans, including caridean (Hameed et al. 2000; Shi et al. 2000; Kiran et al. 2002), lobster (Chang et al. 1998a, b; Rajendran et al. 1999), crayfish (Wang et al. 1998; Corbel et al. 2001; Jiravanichpaisal et al. 2001; Edgerton 2004a, b; Jiravanichpaisal et al. 2004) and crabs (Rajendran et al. 1999; Das et al. 2016; Bharathi et al. 2017; Norizan et al. 2019; Pratapa et al. 2023).

A few non-decapod crustaceans were also found to carry WSSV (Supamattaya et al. 1998; Zhang et al. 2008; Priyangha et al. 2020).

Various non-crustaceans serve as vectors or reservoirs of WSSV and play a vital role in the ecological dispersal and retention of the virus (Desrina et al. 2022). These include chaetognaths and rotifers (Ramırez-Douriet et al. 2005), polychaetes (Supak et al. 2005; Vijayan et al. 2005; Desrina et al. 2013, 2015), and larvae of some aquatic insects (Lo et al. 1996), piscivorous birds (Vanpatten et al. 2004), molluscs (Chang et al. 2011) and oysters (Vazquez-Boucard et al. 2010). In recent years, the application of metagenome sequencing identified the WSSV genome in the faecal virome of domestic chickens (*Gallus gallus domesticus*) (Chauhan et al. 2023). Among the various hosts of WSSV, piscivorous birds are significant from the perspective of ecological dispersal of the virus, considering the wide-ranging migratory and foraging routes of the birds. The sensitivity of these hosts to WSSV infection differs considerably and, in some hosts, results in non-destructive or latent infection, making them potential carriers and a key point of origin of infection in a culture system (Rajendran et al. 1999; Pradeep et al. 2012). Several host species of WSSV were reported (Porchas-Cornejo et al. 2018; Bandeira et al. 2019; Zhu et al. 2019; Priyangha et al. 2020; Xu et al. 2021), and the list of hosts (both natural and experimental) is provided in Table 2.

4 The Pathogen

4.1 Nomenclature

The virus causing white spot syndrome had different nomenclatures until it was finally designated as WSSV by the International Committee on Taxonomy of Viruses (ICTV). The names from the early days of its discovery were *Pm*NOBII, as the second non-occluded baculovirus reported from *P. monodon* (Wang et al. 1995; Karunasagar et al. 1997), hypodermal and haematopoietic necrosis baculovirus (HHNBV) (Durand et al. 1997), rod-shaped nuclear virus of *P. japonicus* (RV-PJ) (Inouye et al. 1994), penaeid rod-shaped DNA virus (PRDV) (Venegas et al. 2000), SEMBV (Wongteerasupaya et al. 1995; Hameed et al. 1998) and white spot baculovirus (Chou et al. 1995; Lightner 1996). However, the viral agent involved in all the reported incidences of the disease was found to be the same. Eventually, the name WSSV was given (reviewed by Escobedo-Bonilla et al. 2008). Later, the International Committee on Taxonomy of Viruses (ICTV) assigned WSSV as the only member of the family *Nimaviridae* under the genus *Whispovirus* based on its molecular characteristics (Lo et al. 2012). Currently, there is at least one virus, Chionoecetes opilio bacilliform virus, infecting the snow crab, *Chionoecetes opilioare*, that could be classified as a member of the family *Nimaviridae* (Kawato et al. 2023), and there are five other endogenous sequences that have similarities to nimaviruses (Rampaul et al. 2024).

Table 2 List of hosts which are reported to be susceptible (natural or experimental) to WSSV

Sl. no.	Name of host	Nature of infection	References	Detection method	Status[a]
Shrimp/prawn					
1	*Alpheus brevicristatus*	N	Takahashi et al. (2003)	Nested PCR	PCR detection
2	*Alpheus lobidens*	N	Takahashi et al. (2003)	Nested PCR	PCR detection
3	*Alpheus distinguendus*	N	Xu et al. (2021)	LAMP, PCR	ND
4	*Alpheus digitalis*		Xu et al. (2021)	LAMP, PCR	PCR detection
5	*Alpheus japonicus*	N	Xu et al. (2021)	LAMP, PCR	PCR detection
6	*Aristeus* sp.	N	Chakraborty et al. (2002)	Nested PCR	ND
7	*Artemesia longinaris*	N	Martorelli et al. (2010)	qPCR	ND
8	*Austinogebia edulis*	N	Zhu et al. (2019)	PCR, EM	PCR detection
9	*Callianassa* sp.	N	Lo et al. (1996)	PCR	ND
10	*Crangon affinis*	N	Xu et al. (2021)	LAMP, PCR	PCR detection
11	*Farfantepenaeus aztecus*	N, E	Lightner (1996); Chapman et al. (2004)	PCR, HIST	Susceptible host
12	*Farfantepenaeus duorarum*	E	Lightner (1996)	HIST	Susceptible host
13	*Fenneropenaeus penicillatus*	N	Chou et al. (1995); Lo et al. (1996); Wang et al. (1998)	PCR	Susceptible host
14	*Heterocarpus* sp.	N	Chakraborty et al. (2002)	Nested PCR	ND
15	*Latreutes planirostris*	N	Xu et al. (2021)	LAMP, PCR	PCR detection
16	*Latreutes anoplonyx*	N	Xu et al. (2021)	LAMP, PCR	ND
17	*Leptochela gracilis*	N	Xu et al. (2021)	LAMP, PCR,	PCR detection
18	*Litopenaeus schmitii*	E	Bustamante et al. (2004)	HIST, ISH	Susceptible host
19	*Litopenaeus vannamei*	N, E	Lightner (1996); Haq et al. (2015)	TEM, PCR, HIST	Susceptible host
20	*Litopenaeus setiferus*	N, E	Lightner (1996); Chapman et al. (2004)	PCR, Histo	Susceptible host
21	*Marsupenaeus japonicus*	N, E	Takahashi et al. (1994); Lo et al. (1996); Wang et al. (1998); Zhan et al. (1998)	TEM, HIST	Susceptible host
22	*Metapenaeus ensis*	N, E	Lo et al. (1996); Chang et al. (1998a, b); Wang et al. (1998)	PCR, DNA probe	Susceptible host

(continued)

Table 2 (continued)

Sl. no.	Name of host	Nature of infection	References	Detection method	Status[a]
23	*Metapenaeus dobsoni*	N, E	Rajendran et al. (1999); Hossain et al. (2001b); Chakraborty et al. (2002)	HIST, PCR	Susceptible host
24	*Metapenaeus lysianassa*	N	Hao et al. (1999)	PCR, HIST	Susceptible host
25	*Metapenaeus monoceros*	N, E	Rajendran et al. (1999); Hossain et al. (2001a)	PCR, Probe, HIST	Susceptible host
26	*Metapenaeus elegans*	N	Chakraborty et al. (2002)	Nested PCR	Susceptible host
27	*Metapenaeus brevicornis*	N	Hossain et al. (2001a); Mohan (2008)	PCR	Susceptible host
28	*Metapenaeus affinis*	N	Mohan (2008)	Nested PCR	Susceptible host
29	*Parapenaeopsis stylifera*	N	Hossain et al. (2001b); Chakraborty et al. (2002)	PCR, probe, Nested PCR	Susceptible host
30	*Penaeus monodon*	N, E	Chou et al. (1995); Lo et al. (1996); Hossain et al. (2001a); Chakraborty et al. (2002); Wongteerasupaya et al. (2003)	Nested PCR, TEM, HIST	Susceptible host
31	*Penaeus indicus*	N, E	Flegel (1997); Rajendran et al. (1999); Rajan et al. (2000); Hossain et al. (2001a)	TEM, HIST, PCR	Susceptible host
32	*Penaeus paulensis*	N	Cavalli et al. (2011)	PCR, HIST	Susceptible host
33	*Penaeus merguiensis*	N	Flegel (1997)	TEM, HIST	Susceptible host
34	*Penaeus semiculcatus*	N, E	Lo et al. (1996); Wang et al. (1998); Rajendran et al. (1999); Hossain et al. (2001b)	PCR, HIST	Susceptible host
35	*Penaeus canaliculatus*	N	Mohan (2008)	Nested PCR	Susceptible host
36	*L. schmitti*	E	Bustamante et al. (2004)	Histo, ISH	Susceptible host
37	*Penaeus chinensis*	N, E	Cai et al. (1995); Flegel (1997); Zhan et al. (1998); Xu et al. (2021)	EM, HIST	Susceptible host
38	*Penaeus stylirostris*	E, N	Lightner (1996); Galavíz-Silva et al. (2004)	TEM, PCR, HIST	Susceptible host
39	*Solenocera crassicornis*	N	Hossain et al. (2001a)	PCR	PCR detection
40	*Solenocera indica*	N	Hossain et al. (2001b)	Nested PCR, DNA probe	NA

(continued)

Table 2 (continued)

Sl. no.	Name of host	Nature of infection	References	Detection method	Status[a]
41	*Seranopsis longipes*	N	Mohan (2008)	Nested PCR	ND
42	*Solenocera choprai*	N	Mohan (2008)	Nested PCR	ND
43	*Squilla* sp.	N, E	Hameed et al. (2000); Hossain et al. (2001b); Chakraborty et al. (2002)	Nested PCR	ND
44	*Trachypenaeus curvirostris*	N, E	Chang et al. (1998a, b); Wang et al. (1998); Xu et al. (2021)	LAMP, PCR, EM	ND
45	*Macrobrachium idella*	E	Rajendran et al. (1999); Hameed et al. (2000)	Western blot, Histo	Incomplete evidence
46	*Macrobrachium lamarrei*	E	Hameed et al. (2000)	Western blot, HIST	Incomplete evidence
47	*M. nipponense*	E	Yun et al. (2014); Zhao et al. (2017)	qPCR, HIST	Susceptible host
48	*Macrobrachium rosenbergii*	N, E	Lo et al. (1996); Flegel (1997); Wang et al. (1998); Chang et al. (1998a, b); Rajendran et al. (1999); Hameed et al. (2000); Hossain et al. (2001a, b); Chakraborty et al. (2002)	RT-PCR, Southern blot, HIST	Incomplete evidence
49	*Exopalaemon carinicauda*	N	Zhiwen et al. (2003)	Dot blot, ISH	Susceptible host
50	*Exopalaemon orientalis*	N, E	Chang et al. (1998a, b); Flegel (1997); Wang et al. (1998)	Nested PCR, DNA probe	ND
51	*Exopalaemon orientis*	E	Chang et al. (1998a, b)		Susceptible host
52	*Palaemon adspersus*	E	Corbel et al. (2001)	TEM, Dot blot, ISH, PCR	Incomplete evidence
53	*Palaemon gravieri*	N	Xu et al. (2021)	LAMP, PCR	PCR detection
54	*Palaemon macrodactylus*	N	Martorelli et al. (2010)	qPCR	PCR detection
55	*Palaemonetes pugio*	N, E	Muhammad and Lotz (2015); Muhammad et al. (2020)	qPCR,	PCR detection
56	*Palaemon ritteri*	N	Sánchez-Paz et al. (2015)	RT-PCR	Susceptible host
57	*Palaemon pandaliformis*	N	Bandeira et al. (2019)	LAMP	PCR detection
58	*Palaemon styliferus*	N	Flegel (1997); Lo et al. (1996); Hossain et al. (2001a)	PCR	ND

(continued)

Table 2 (continued)

Sl. no.	Name of host	Nature of infection	References	Detection method	Status[a]
Crab					
1	*Atergatis integerrimus*	E	Hameed et al. (2003)	PCR, HIST	PCR detection
2	*Armases cinereum*	N	Muhammad et al. (2020)	qPCR	PCR detection
3	*Ashtoret miersii*	E	Hameed et al. (2003)	PCR	PCR detection
4	*Cancer pagurus*	E	Corbel et al. (2001); Bateman et al. (2012b)	TEM, ISH, Histo, PCR	Susceptible host
5	*Calappa lophos*	N, E	Flegel (1997); Chakraborty et al. (2002)	PCR	PCR detection
6	*Calappa philargius*	E	Hameed et al. (2003)	PCR, HIST	PCR detection
7	*Callinectes arcuatus*	N	Galavíz-Silva et al. (2004)	PCR	ND
8	*Callinectes sapidus*	N, E	Chang et al. (2001); Powell et al. (2015); Blaylock et al. (2019)	TaqMan PCR, ISH, PCR, Immuno-strip	ND
9	*Carcinus maenas*	E	Corbel et al. (2001); Bateman et al. (2012b)	TEM, Dot blot, ISH, PCR	Incomplete evidence
10	*Charybdis annulata*	N, E	Hossain et al. (2001b); Hameed et al. (2003)	PCR, HIST	ND
11	*Charybdis cruciata*	N	Hossain et al. (2001b); Chakraborty et al. (2002)	Nested PCR	ND
12	*Charybdis feriata*	N	Lo et al. (1996)	Nested PCR	ND
13	*Charybdis granulata*	E	Chang et al. (1998a, b)	DNA probe	ND
14	*Charybdis hoplites*	N	Chakraborty et al. (2002); Mohan (2008)	Nested PCR	ND
15	*Charybdis lucifera*	N, E	Chakraborty et al. (2002); Hameed et al. (2003)	PCR, HIST	ND
16	*Charybdis natator*	N	Flegel (1997)	PCR, HIST	ND
17	*Charybdis riversandersoni*	N	Mohan (2008)	Nested PCR	ND
18	*Chhapgarus intermedius*	N	Hossain et al. (2001a, b)	PCR	PCR detection
19	*Chasmagnathus granulata*	N	Marques et al. (2011)	Nested PCR	ND
20	*Circulium rotundatum*	N	Otta et al. (1999)	PCR	PCR detection
21	*Cyrtograpsus angulatus*	N	Martorelli et al. (2010)	qPCR	PCR detection

(continued)

Table 2 (continued)

Sl. no.	Name of host	Nature of infection	References	Detection method	Status[a]
22	*Demania splendida*	E	Hameed et al. (2003)	PCR, HIST	PCR detection
23	*Diogenes nitidimanus*	E	Chang et al. (2012)	PCR	PCR detection
24	*Doclea hybrida*	E	Hameed et al. (2003)	PCR, HIST	ND
25	*Doclea muricata*	E	Hameed et al. (2003)	PCR	PCR detection
26	*Eriocheir sinensis*	E, N	Bateman et al. (2012b); Ding et al. (2015)	Nested PCR, TEM, HIST	Susceptible host
27	*Gelasimus marionis nitidus*	N	Hossain et al. (2001b)	Nested PCR	ND
28	*Gelasimus vocans*	N	Hossain et al. (2001a)	PCR	PCR detection
29	*Grapsus albolineatus*	E	Hameed et al. (2003)	PCR, HIST	PCR detection
30	*Halimede ochtodes*	E	Hameed et al. (2003)	PCR, HIST	PCR detection
31	*Helice tridens*	N	Lo et al. (1996)	Nested PCR	PCR detection
32	*Helice tientsinensis*	N	Xu et al. (2007)	Dot Blot, ISH	Incomplete evidence
33	*Leptuca panacea*	N, E	Muhammad et al. (2020)	TaqMan RT-PCR	PCR detection
34	*Leptuca spinicarpa*	N	Muhammad et al. (2020)	TaqMan RT-PCR	PCR detection
35	*Leptuca pugilator*	E	Kanchanaphum et al. (1998)	PCR, ISH	Incomplete evidence
36	*Liagore rubromaculata*	E	Hameed et al. (2003)	PCR, HIST	PCR detection
37	*Liocarcinus depurator*	E	Corbel et al. (2001)	TEM, Dot blot, ISH, PCR	Susceptible host
38	*Liocarcinus puber*	E	Corbel et al. (2001)	TEM, Dot blot, ISH, PCR	ND
39	*Lithodes maja*	E	Hameed et al. (2003)	PCR, HIST	PCR detection
40	*Macrophthalmus sulcatus*	N	Hossain et al. (2001b)	Nested PCR	PCR detection
41	*Macrophthalmus (Mareotis) japonicus*	N	Xu et al. (2007)	Dot blot, ISH	Incomplete evidence
42	*Mantura* sp.	N	Flegel (1997)	PCR	ND
43	*Matuta miersi*	E	Hameed et al. (2003)	PCR, HIST	ND

(continued)

Table 2 (continued)

Sl. no.	Name of host	Nature of infection	References	Detection method	Status[a]
44	*Matuta planipes*	N	Otta et al. (1999)	Nested PCR	PCR detection
45	*Metopograpsus messor*	N, E	Rajendran et al. (1999); Hossain et al. (2001b)	Nested PCR	Incomplete evidence
46	*Menippe rumphii*	E	Hameed et al. (2003)	PCR, HIST	PCR detection
47	*Minuca longisignalis*	N	Muhammad et al. (2020)	TaqMan RT-PCR	PCR detection
48	*Minuca minax*	N	Muhammad et al. (2020)	TaqMan RT-PCR	PCR detection
49	*Minuca rapax*	N	Muhammad et al. (2020)	TaqMan RT-PCR	PCR detection
50	*Neohelice granulata*	N	Cavalli et al. (2013); Moser and Marques (2023)	PCR and sequencing	PCR detection
51	*Pagurus benedicti*	N, E	Chang et al. (2012)	PCR	Susceptible host
52	*Pagurus angustus*	E	Chang et al. (2012)	PCR	PCR detection
53	*Paradorippe granulata*	E	Hameed et al. (2003)	PCR, HIST	PCR detection
54	*Parthenope prensor*	E	Hameed et al. (2003)	PCR, HIST	PCR detection
55	*Parathelphusa hydrodromous*	E	Hameed et al. (2001); Ahmad et al. (2017)	PCR, HIST, RT-PCR, ELISA	Incomplete evidence
56	*Parathelphusa (Vela) pulvinata*	E	Hameed et al. (2001)	PCR, Histo	Incomplete evidence
57	*Philyra syndactyla*	E	Hameed et al. (2003)	PCR, HIST	PCR detection
58	*Podophthalmus vigil*	E	Hameed et al. (2003)	PCR, HIST	ND
59	*Portunus pelagicus*	N, E	Lo et al. (1996); Supamattaya et al. (1998); Chakraborty et al. (2002); Mohan (2008)	DNA probe, HIST, EM, Nested PCR	Susceptible host
60	*Portunus sanguinolentus*	N, E	Lo et al. (1996); Chakraborty et al. (2002); Hameed et al. (2003); Mohan (2008)	PCR, HIST, Nested PCR	Susceptible host
61	*P. trituberculatus*	N, E	Meng et al. (2009); Muhammad and Lotz (2015)	HIST, RT-PCR, TaqMan realtime PCR	Susceptible host
62	*Pseudograpsus intermedius*	N	Hossain et al. (2001a); Chakraborty et al. (2002)	Nested PCR	ND

(continued)

Table 2 (continued)

Sl. no.	Name of host	Nature of infection	References	Detection method	Status[a]
63	*Scylla serrata*	N, E	Lo et al. (1996); Supamattaya et al. (1998); Kanchanaphum et al. (1998); Rajendran et al. (1999); Chen et al. (2000); Hossain et al. (2001a); Chakraborty et al. (2002); Hameed et al. (2003); Bharathi et al. (2017); Norizan et al. (2019); Pratapa et al. (2023)	DNA probe, PCR, HIST, ISH	ND
64	*Scylla tranquebarica*	N, E	Rajendran et al. (1999); Gopalakrishnan et al. (2011)	PCR, SEM, HIST	ND
65	*Scylla olivacea*	E	Somboonna et al. (2010); Pratapa et al. (2023)	qPCR, HIST, IHC	ND
66	*Scylla paramamosain*	E	Somboonna et al. (2010)	qPCR, HIST, IHC	ND
67	*Sesarma* sp	N, E	Flegel (1997); Kanchanaphum et al. (1998); Rajendran et al. (1999); Das et al. (2016)	PCR, ISH	Incomplete evidence
68	*Sesarma reticulatum*	N	Muhammad et al. (2020)	PCR	Incomplete evidence
69	*Somanniathelpusa* sp.	E	Flegel (1997)	ND	ND
70	*Thalamita danae*	N, E	Flegel (1997); Hameed et al. (2003)	PCR, HIST	ND
71	*Uca spinicarpa*	N, E	Muhammad et al. (2020)	qPCR	ND
Lobster					
1	*Homarus gammarus*	E	Bateman et al. (2012a, b)	Nested PCR, TEM, HIST	Susceptible host
2	*H. americanus*	E	Clark et al. (2013)	HIST, TEM, qPCR	ND
3	*Nephrops norvegicus*	E	Bateman et al. (2012b)	Nested PCR, TEM, HIST	Susceptible host
4	*Panulirus homarus*	E	Rajendran et al. (1999); Musthaq et al. (2006a)	PCR, HIST, Western blot	Susceptible host
5	*Panulirus longipes*	E	Flegel (1997)	Nested PCR	Susceptible host
6	*Panulirus ornatus*	E	Flegel (1997); Rajendran et al. (1999); Musthaq et al. (2006a)	PCR, HIST, Western blot	Susceptible host

(continued)

Table 2 (continued)

Sl. no.	Name of host	Nature of infection	References	Detection method	Status[a]
7	*Panulirus penicillatus*	E	Chang et al. (1998a, b); Wang et al. (1998)	Nested PCR, DNA probe	Susceptible host
8	*Panulirus polyphagus*	E	Rajendran et al. (1999)	HIST	Susceptible host
9	*Panulirus versicolor*	E	Chang et al. (1998a, b); Wang et al. (1998)	Nested PCR, DNA probe	Susceptible host
10	*Panulirus argus*	E	Ross et al. (2019)	qPCR	Susceptible host
11	*Scyllarus arctus*	E	Corbel et al. (2001)	PCR, TEM, Dot blot, ISH	Incomplete evidence
Crayfish					
1	*Astacus leptodactylus*	E	Corbel et al. (2001)	TEM, Dot blot, ISH, PCR	Susceptible host
2	*A. astacus*	E	Jiravanichpaisal et al. (2006)	PCR	PCR detection
3	*Austropotamobius pallipes*	E	Edgerton (2004b); Bateman et al. (2012b)	PCR, HIST	Susceptible host
4	*Cherax destructor albidus*	E	Edgerton (2004a)	DNA probe, HIST	NA
5	*C. quadricarinatus*	E, N	Lee et al. (2021)	TEM, ISH, IHC, Nested PCR	Susceptible host
6	*Orconectes (Faxonius) limosus*	E	Corbel et al. (2001)	TEM, Dot blot, ISH, PCR	Susceptible host
7	*O. punctimanus*	N	Lo et al. (1999); Wang et al. (2000b)	PCR, DNA probe	PCR detection
8	*Pacifastacus leniusculus*	E	Jiravanichpaisal et al. (2001); Bateman et al. (2012b)	Nested PCR, TEM, HIST	Susceptible host
9	*Procambarus clarkii*	E, N	Chang et al. (1998a, b); Baumgartner et al. (2009); Xiaoli et al. (2011); Lee et al. (2021)	HIST, Nested PCR, qPCR, TEM	Susceptible host
10	*P. zonangulus*	N	Baumgartner et al. (2009)	Nested PCR, qPCR	Susceptible host
Copepod					
1	*Apocyclops royi*	E	Chang et al. (2011)	Nested PCR, qPCR	PCR detection
2	*Calanus pacificus californicus*	E	Mendoza-Cano and Sánchez-Paz (2013)	qPCR	Susceptible host

(continued)

Table 2 (continued)

Sl. no.	Name of host	Nature of infection	References	Detection method	Status[a]
3	*Nitocra* sp.	E	Zhang et al. (2008)	Nested PCR	PCR detection
4	*Schmackeria dubia*	N	Lo et al. (1996)	Nested PCR	Reservoir host
Other crustaceans					
1	*Octolasmis neptuni*	N	Priyangha et al. (2020)	Nested PCR	NA
2	*Euphausia pacifica*	N	Xu et al. (2021)	LAMP, PCR	PCR detection
3	*Acetes* sp. (Krill)	E	Supamattaya et al. (1998)	EM, ISH, HIST	Incomplete evidence
4	*A. chinensis*	N	Xu et al. (2021)	LAMP, PCR	Incomplete evidence
5	*A. franciscana*	E	Li et al. (2003); Haq et al. (2011)	Nested PCR	PCR detection
6	*Artemia* sp.	E	Hameed et al. (2002)	Nested PCR	PCR detection
7	*Ergasilus manicatus*	E	Overstreet et al. (2009)	qPCR	Incomplete evidence
8	*Calanus pacificus californicus*	E	Mendoza-Cano et al. (2014)	qPCR	ND
Polychaete					
1	*Dendronereis* spp	N	Desrina et al. (2015)	Nested PCR, IHC, RT-PCR	Susceptible host
2	*Marphysa gravelyi*	N	Vijayan et al. (2005)	Nested PCR	PCR detection
3	*Perenereis cultrifera*	N, E	Banu et al. (2015); Shalini et al. (2016, 2020)	PCR	ND
4	*P. nunita*	E	Supak et al. (2005)	Nested PCR	ND
Rotifer					
1	*Brachionus urceus*	E, N	Yan et al. (2004); Zhang et al. (2006)	Nested PCR, Dot blot	PCR detection
2	*B. plicatilis*	E	Corre Jr et al. (2012)	Nested PCR	ND
3	*B. urceolaris*	N	Yan et al. (2004)	PCR	PCR detection
Mulluscs					
1	*Melanoides tuberculatus*	N	Bandeira et al. (2019)	LAMP	PCR detection
2	*Pomacea lineata*	N	Bandeira et al. (2019)	LAMP	ND
3	*Meretrix lusoria*	E	Chang et al. (2011)	Nested PCR, qPCR, RT-PCR	Incomplete evidence

(continued)

Table 2 (continued)

Sl. no.	Name of host	Nature of infection	References	Detection method	Status[a]
4	*Crassostrea (Magallana) gigas*	N	Vazquez-Boucard et al. (2010)	Nested PCR	PCR detection
5	*Haliotis diversicolor*	N	Wei et al. (2018)	MET	ND
6.	*Meretrix casta*	N	Mohan (2008)	Nested PCR	ND
7.	*Perna viridis*	N	Mohan (2008)	Nested PCR	ND
Echinoderms					
1	*Holothuria scabra*	E	Dela Pena et al. (2024)	Nested PCR	ND
Insects					
1	*Ephydridae* sp.	N	Lo et al. (1996)	Nested PCR	ND
2	*Gerries*	N	Mohan (2008)	Nested PCR	ND
3	*Notonecta*	N	Mohan (2008)	Nested PCR	ND
4	*Nepa*	N	Mohan (2008)	Nested PCR	ND
5	*Ranatra*	N	Mohan (2008)	Nested PCR	ND
6	Dragon nymphfly	N	Mohan (2008)	Nested PCR	ND
Birds					
1	*Gallus gallus domesticus*	E, N	Vanpatten et al. (2004), Chauhan et al. (2023)	HIST, PCR, MET	ND
2	*Larus atricilla*	E	Vanpatten et al. (2004)	HIST, PCR	ND
Plankton					
1	*Chlorella* sp.	E	Liu B et al. (2007a)	Nested PCR	ND
2	*Alexandrium tamarense*	E	Jiang (2012)	DNA probe	ND
3	*A. minutum*	E	Jiang (2012)	DNA probe	ND

N natural, *E* experimental, *ND* not determined, *MET* metagenomic, *TEM* transmission electron microscopy, *SEM* scanning electron microscopy, *PCR* polymerase chain reaction, *RT* reverse transcriptase, *ISH* in situ hybridization, *IHC* immunohistochemistry, *LAMP* loop mediated isothermal amplification, *EM* electron microscopy, *HIST* histology, *ELISA* enzyme-linked immunosorbent assay

[a]The status is on the basis of the listing of different host species by WOAH (World Organization for Animal Health), as susceptible host species, species with incomplete evidence for susceptibility and species showing positive PCR reaction

4.2 Virus Structure

WSSV is a bacilliform, non-occluded, enveloped virus (Chou et al. 1995; Wang et al. 1995; Wongteerasupaya et al. 1995). The virus possesses a long envelope extension at one extremity, whose function and structure have not been determined (Durand et al. 1997). The size of the virion ranges between 210–420 nm in length and 70–167 nm in diameter (Chang et al. 1996; Guoxing et al. 1997; Flegel and Alday-Sanz 1998; Park et al. 1998; Rajendran et al. 1999). The viral envelope is a lipidic bilayer membrane and is 6–7 nm thick (Wongteerasupaya et al. 1995; Durand et al. 1997; Nadala Jr et al. 1998). The nucleocapsid size ranges between 180 and

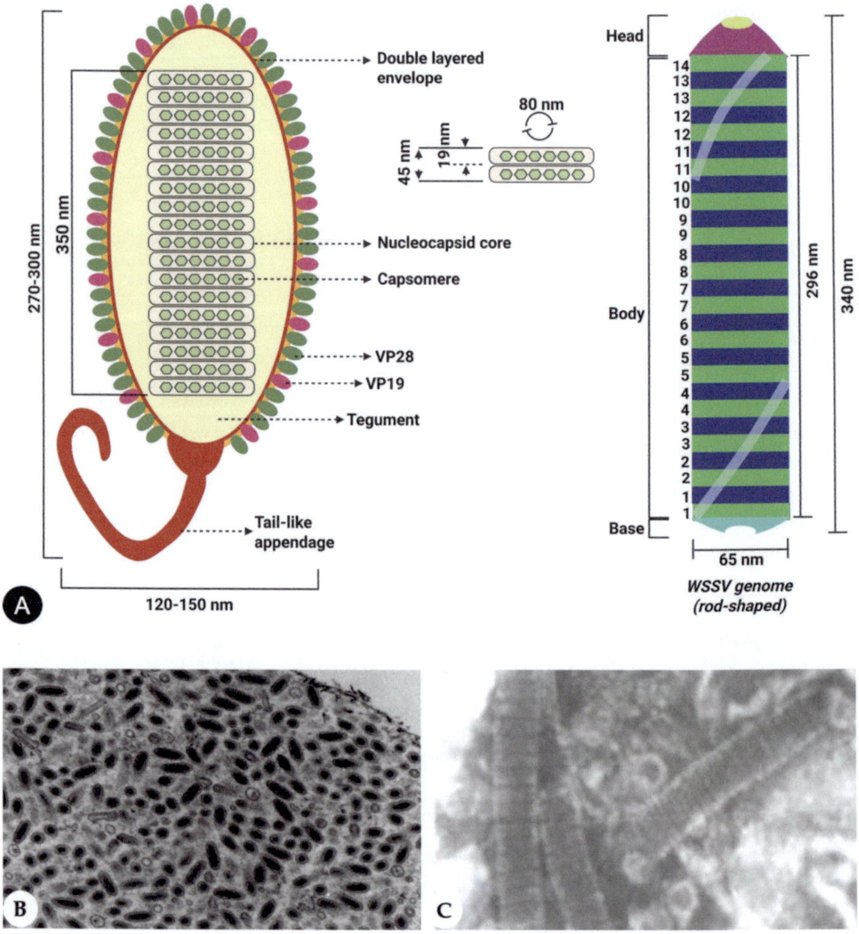

Fig. 4 (**a**) Diagrammatic representation of WSSV structure. (**b**, **c**) Transmission electron micrographs. (**b**) Stomach epithelium of WSSV-infected shrimp showing rod-shaped virions; (**c**) Virus particles purified from WSSV-infected shrimp haemolymph showing rod-shaped nucleocapsids. Note the cross-hatched or striated appearance

420 nm in length and is packed inside the virion (Durand et al. 1997; Hameed et al. 1998; Kasornchandra et al. 1998; Rajendran et al. 1999). Also, the nucleocapsid surface consists of 15 vertical helices, and each helix is comprised of 13–15 capsomers (Durand et al. 1997; Nadala Jr et al. 1998) with a space of 7 nm in-between. The cross-hatched or striated material that is structured as stacked rings of subunits is a characteristic feature of WSSV (Huang et al. 2001) (Fig. 4).

4.3 Genome

The virus possesses a circular dsDNA (approximately 300 kbp) genome (Leu et al. 2009). It is one of the largest known animal viruses with a genome size of 280–314 kb (Krishnan et al. 2023) with 180 open reading frames (ORFs) (van Hulten et al. 2001a; Raoult et al. 2004). A large number of geographical isolates of the virus with genotypic variability have been identified. Full genome sequences of WSSV have been reported in China (Yang et al. 2001), Thailand (van Hulten et al. 2000), Taiwan (Tsai et al. 2000), Korea (Chai et al. 2013), Mexico (Rodriguez-Anaya et al. 2016), Australia (Oakey et al. 2019), India (Sivakumar et al. 2018), Brazil (Dantas et al. 2018) and Ecuador (Restrepo et al. 2018). Five isolates of the virus were reported from farmed shrimp (*P. vannamei* and *P. monodon*) in India (WSSV-IN-CIBA2 (MH883318), WSSV-IN-CIBA3 (MH883319), WSSV-IN-LS, WSSV-IN-NS and WSSV-IN-AP4RU). Additionally, three complete genome sequences have been reported, of which two are from shrimp farmed on the West coast of India and the third one from the East coast (Rajendran et al. 2021).

Genome sequences of WSSV isolates from different countries showed minor differences in the genome size, as shown here: Taiwan (307 kb), China (305 kb), Korea (295 kb), Mexico (293 kb) and Thailand (292 kb). Overall, the sequence identity among different geographic isolates of WSSV ranged between 97% and 99% (Marks et al. 2004). Nine homologous regions (hr 1–9) with different patterns of repeats were identified in the WSSV genome (van Hulten et al. 2001a) that may play an important role in DNA replication (Marks et al. 2004). The difference in the genome size of different geographical isolates of WSSV was attributed to deletions/insertion or the presence of variable number of tandem repeats (VNTRs) within the genome (Simrouni et al. 2014). Variable regions in the coding regions of WSSV genome have been reported, especially the genomic deletions occurring between ORFs wsv461/wsv464 (14/15) and ORFs wsv77/wsv502 (23/24) and VNTRs within wsv129 (ORF75), wsv178 (ORF94) and wsv249 (ORF125) (Hoa et al. 2005). These variable regions have been employed as molecular markers to identify diversity among WSSV geographic isolates (Zwart et al. 2010; Muller et al. 2010; Gao et al. 2014; Gudkovs et al. 2014; Li et al. 2017). Many researchers have reported genotypic variations within the genome sequence of different WSSV isolates and concluded that these variations could be used as genetic markers for determining diversity among WSSV isolates (Dieu et al. 2004; Musthaq et al. 2006b; Pradeep et al. 2008; Muller et al. 2010; John et al. 2010; Ramos-Paredes et al. 2012; Shekar et al. 2012; Simrouni et al. 2014; Rodriguez-Anaya et al. 2016; Patil et al. 2017; Parrilla-Taylor et al. 2018; Piamsomboon et al. 2018; Siddique et al. 2018).

Recently, a novel strain of WSSV affecting *P. vannamei* in Indian shrimp farms (WSSV-NS) with a genome size of 279,797 bp containing a deletion of 13,170 bp was reported (Sivakumar et al. 2018). Another strain (WSSV-LS) with a genome length of 292,967 bp but without deletions has also been reported by the same authors (Sivakumar et al. 2018). Similarly, the full genome sequence of Australian WSSV (WSSV-AU) has been published with a genome length of 285,973 bp, which showed 91–97% homology with other reported genome sequences (Oakey and

Smith 2018). Substantial deletions were observed in WSSV-AU, similar to WSSV-NS. According to Oakey and Smith (2018), significant deletions were detected in the genome regions that were earlier designated as genomic markers. Besides, the region that codes for the envelope protein also showed a deletion. The two WSSV genome sequences reported recently from the west coast of India, WSSV-IN-CWG3 and WSSV-IN-PG1, have a genome length (nucleotide) of 281,060 bp and 285,601 bp, respectively. In contrast, the genome sequence of a WSSV isolate from the east coast of India (WSSV-IN-DBA1182) consists of 281,813 bp (Rajendran et al. 2021). Despite a difference in genome length, phylogenetic analysis showed a close similarity of WSSV-IN-DBA1182 and WSSV-IN-CWG3 with three other genotypes from India. However, two other WSSV isolates (WSSV-IN-PG1 and WSSV-IN-LS) shared the same clade with a Mexican strain of the virus (KU216744) (Rajendran et al. 2021). This highlights the genomic variations that exist among the circulating genotypes of WSSV in India, although the implication of this genomic variation remains obscure.

Zeng (2021) carried out a global epidemiological study of WSSV based on ORF75, ORF14/15 and ORF23/24 regions in the viral genome. Based on the ORF14/15 variable region, WSSV isolates were categorized into three groups: (i) isolate from Thailand (WSSV-TH-96-II; representative isolates of type I-clade); (ii) isolate of Indian origin (WSSV-TH-96-II; type II-clade); and (iii) another Indian isolate (WSSV-IN-06-I; type III-clade) (Zeng 2021). Based on the ORF75 region, WSSV isolates were divided into subtypes: (i) isolates identified from Thailand (WSSV-TH and TH-96-II; representative isolates of subtype I) and (ii) isolates reported from China and Taiwan (WSSV-CN and WSSV-TW; representative isolates of subtype IIa and subtype IIb). The author concluded that most strains of WSSV spread across the world from an origin in the Indian Ocean, though the origin of Australian WSSV remains unknown (Zeng 2021; Oakey et al. 2019).

4.4 Genes and Proteins

WSSV genes have been categorized into three groups on the basis of their temporal expression profile: immediate-early (IE), early and late genes (Marks et al. 2005). In all, 21 IE genes have been reported from WSSV (Li F et al. 2009a; Lin F et al. 2011a; Liu et al. 2005), and of them IE1 is the most extensively investigated gene (Liu et al. 2008). Some of the genes are found to be associated with latency and anti-apoptotic activities (Table 3).

The core DNA of WSSV virion is encircled by three structural layers: an outer envelope, a tegument and a nucleocapsid (Tsai et al. 2006; Li SZ et al. 2007b). More than 58 WSSV structural proteins have been identified, and a minimum of 35 envelope proteins have been reported (Li et al. 2007b; Sánchez-Paz 2010). The envelope proteins expressed on the surface play a vital role in binding the virus to receptors and fusion, besides interacting with the host immune system (Chang et al. 2010), whereas the envelope proteins expressed within the virion help in the transmission of nucleocapsid to the host's cell nucleus (Wan et al. 2008). VP19, VP24,

Table 3 A list of WSSV genes, their temporal expression and putative functions

Type of gene/ protein	Examples	Reference
Immediate early genes	Ie1 (wsv069), ie2, ie3, wsv051, wsv056, wsv078, wsv079, wsv080, wsv083, wsv091, wsv094, wsv098, wsv099, wsv100, wsv101, wsv103, wsv108, wsv178, wsv249, wsv403 and wsv465	Li et al. (2009a, b)
Early genes	wsv477, VP11 (WSSV394)	Han et al. (2007), Liu et al. (2014)
Late genes	WSSV-DNA polymerase, VP15 and VP28	Nupan et al. (2011)
Latency-related genes	Orf427, Orf151 and Orf366, orf89	Khadijah et al. (2003); Hossain et al. (2004a)
Anti-apoptotic genes	WSSV134 and WSSV322, orf390	Wang et al. (2004); Yan et al. (2010); Lertwimol et al. (2014)
Structural proteins	VP95, VP507, VP53B, VP357, VP337, VP32, VP320, VP216, VP387, VP53C, VP13A, VP14, VP184, VP75, VP11, ORF390, VP136B, VP362, VP448	Leu et al. (2009)
Structural proteins and envelope	VP1684, Collagen-like, VP53A, VP150, VP110, VP187, VP124, VP41A, VP51A, VP52A, VP300, VP41B, VP281 (VP36B, VP33), VP384 (VP51B, V52B), VP38A (VP38), VP13B (VP16), VP60A (VP56), VP90, VP39B (VP39), VP31, VP68 (VP12B), VP19, VP28	Tsai et al. (2006); Leu et al. (2009)
Structural proteins and tegument	VP24, VP36A, VP39A, VP26, VP95	Tsai et al. (2006); Leu et al. (2009)
Structural proteins and capsid	VP160B, VP15, VP76 (VP73), VP136 (VP136A), VP160A (VP190), VP466 (VP51C), VP51, VP664	Tsai et al. (2006); Leu et al. (2009)

VP26 and VP28 are major envelope proteins (Momtaz et al. 2018), and among them, VP28 and VP26 are the major proteins, accounting for approximately 60% of the envelope (Tang et al. 2007). VP28 plays a vital role in the initial steps of systemic infection (van Hulten et al. 2001b) and serves as an attachment protein that helps in the binding of the virus to the host cell receptor(s) and facilitates viral entry into the cell (Yi et al. 2004). VP24 was identified as a chitin-binding protein (Li et al. 2016) and mediates infection by binding to chitin structures in the gastrointestinal tract. VP26 can bind actin or actin-associated proteins and aid WSSV in reaching the site of transcription and replication (Xie et al. 2005). VP19 is involved in the envelope coating of WSSV (Li et al. 2019). Another protein, VP22, acts as a linker protein and is located between the envelope and nucleocapsid. VP15 is an important nucleocapsid protein that helps the genome arrangement of WSSV (reviewed by Sánchez-Paz 2010).

Some structural proteins of WSSV interact with other WSSV structural proteins and are involved in the infection process. It has been reported that VP28 interacts with VP24 and VP26 (Xie et al. 2006; Xie and Yang 2006); VP24 interacts with

WSV010 (Chen J et al. 2007a); VP37 interacts with VP28 and VP26 (Liu et al. 2009); VP51A interacts with VP26 (Chang et al. 2008) and VP90 interacts with VP26 and VP28 (Li et al. 2012). The interaction of the envelope proteins VP11 and VP51B with several other proteins has been reported (Liu et al. 2014; Lee and Chen 2017). WSSV envelope proteins can form a protein complex known as infectosome that interacts with the host chitin-binding protein (CBP) and promotes the infection process (Huang PY et al. 2014b). Many other WSSV proteins interact with host receptors and innate immune molecules and result in various vital activities that will benefit the virus or the host (Table 4). A VP24-truncated isolate of WSSV was reported, but it failed to elicit infection (Han et al. 2007).

4.5 Virus Replication

Initially, a model of WSSV life cycle and morphogenesis correlating with the cellular lesions comprising six stages was proposed by Escobedo-Bonilla et al. (2008). Subsequently, other researchers have reported more details on the virus replication process (Sánchez-Paz 2010; Verbruggen et al. 2016). However, a complete model of the WSSV replication cycle using cell cultures from the lymphoid organ of *P. vannamei* was hypothesized by Li et al. (2015). According to these researchers, WSSV replication undergoes the following stages: the virus particle enters the cell through endocytosis after the attachment with the cell membrane surface; upon entry, disassembly of the virus particle occurs; nucleocapsid is released after the virus envelope fuses with the endosome membrane. According to experimental evidence, after attachment, virus particles enter the cells within 60 min post-infection (pi), and uncoating occurs within the same period. The nucleocapsid subsequently moves closer to the nucleus, and the viral genome is introduced into the nucleus through the nuclear pore. Within the nucleus, mRNAs of immediate-early genes are transcribed. The mRNAs are subsequently transported to the cytoplasm, where they are translated into proteins, especially major capsid proteins, by free ribosomes. This way, within 1-h post-infection (hpi), major capsid protein VP664 gets expressed. Early and late proteins are activated by the immediate-early proteins. From 3 hpi, transcription of the envelope protein genes such as VP28 occurs. According to the hypothesis, the VP28 protein is expressed in the rough endoplasmic reticulum (rER), transported to the inner nuclear membrane and proliferates within the nucleus. The next step is the multiplication of the viral genome inside the nucleus and the assembly of capsids around the viral genome to form the nucleocapsid. This process occurs from 6 hpi. This is followed by the budding of nucleocapsid at the inner nuclear membrane and free virus particles in the nucleus. The final stage in the virus replication is the release of virus particles by cell lysis (Li et al. 2015) (Fig. 5).

Table 4 Summary of the WSSV-encoded proteins and their interactions with host proteins

Sl. no.	Viral protein	Host protein	Species	Effect	References
1	ORF427	Protein phosphatase	*Penaeus monodon*	Regulation of WSSV life cycle	Lu and Kwang (2004)
2	WSSV249 (E3 ligase, RING-H2 protein)	PvUbc	*Litopenaeus vannamei*	Involved in ubiquitination pathway	Wang et al. (2005)
3	VP26	Actin	*Procambarus clarkii*	Mediates early stage of infection	Xie and Yang (2005)
4	WSSV222 (anti-apoptosis protein)	Tumour suppressor-like protein	*L. vannamei*	Ubiquitin-mediated proteolysis of the protein	He et al. (2006)
5	VP28	Rab7	*P. monodon*	Mediates WSSV infection	Sritunyalucksana et al. (2006)
6	wsv069	STAT	*P. monodon*	Promotes viral gene expression	Liu et al. (2007a, b)
7	VP187	Integrin	*Marsupenaeus japonicus*	Promotes infection	Li et al. (2007a, b)
8	ICP11	Histone-binding	*P. monodon*	Interferes with the nucleosome assembly	Wang et al. (2008)
9	VP9	Pm-RACK1	*P. monodon*	Mediates intracellular function	Tonganunt et al. (2009)
10	VP24, VP53A, VP110, VP53B, VP337, VP32, VP124, VP41A, VP51B, VP60A and VP39B	Chitin-binding protein (PmCBP)	*P. monodon*	Recognize and facilitate WSSV infection	Chen et al. (2007a, b, 2009)
11	VP15, VP26, VP28	PlgC1qR	*Pacifastacus leniusculus*	Prevent WSSV replication	Watthanasurorot et al. (2010)
12	VP95, VP28, VP26, VP24, VP19, VP14	LvCTL1	*L. vannamei*	Antiviral immune response	Zhao et al. (2009)
13	VP28	FcLec3	*Fenneropenaeus chinensis*	Antiviral immune response	Wang et al. (2009)
14	VP28, VP26, VP281	MjLecA, MjLecB, MjLecC	*M. japonicus*	Antiviral immune response	Song et al. (2010)
15	WSSV ie1	LvRelish	*L. vannamei*	Enhances the expression of viral gene	Huang et al. (2010)

(continued)

Table 4 (continued)

Sl. no.	Viral protein	Host protein	Species	Effect	References
16	WSSV449	Caspase	*P. monodon*	Prevents apoptosis	Leu et al. (2010)
17	WSSV069 (ie1), WSSV449, WSSV303, WSSV371	TLR-mediated NF-kB pathway activation	*L. vannamei*	TLR-mediated NF-kB pathway activation	Wang et al. (2011)
18	WSV083 (immediate-early gene)	Focal adhesion kinase	*M. japonicus*	Promotes integrin-mediated signal transduction	Lu et al. (2011)
19	DNA polymerase gene	Fortilin	*P. monodon*	Inhibition of the virus gene expression	Nupan et al. (2011)
20	VP15	Fortilin	*P. monodon*	Inhibition of expression	Nupan et al. (2011)
21	VP28	Fortilin	*P. monodon*	Inhibition of expression	Nupan et al. (2011)
22	VP38	Caspase	*M. japonicus*	Transcriptional repression of the host gene	Zuo et al. (2011)
23	VP41B	Caspase	*M. japonicus*	Transcriptional activation of the host gene	Zuo et al. (2011)
24	VP15	Immunophilin like protein (PmFKBP46)	*P. monodon*	Involved in genome packaging during virion assembly	Sangsuriya et al. (2011)
25	WSSV ie1	TATA box-binding protein	*P. monodon*	Promotes the transcriptional activity of ie1	Liu et al. (2011)
26	VP466	Rab6, tropomyosin	*M. japonicus*	Antiviral phagocytosis	Ye et al. (2012a, b)
27	*wsv477*	miR-7	*M. japonicus*	Inhibit virus infection	Huang and Zhang et al. (2012)
28	VP24, VP28, VP31, VP32, VP39B, VP51B and VP53A	Glucose transporter 1 (Glut1)	*P. monodon* *L. vannamei*	Inhibition of WSSV infection	Huang et al. (2012)
29	VP28, VP26, and VP24	PcPHB1 (Prohibitin)	*P. clarkii*	Prevent infection	Lan et al. (2013)
30	WSV477	Kazal serine proteinase inhibitor SPI*Pm*2	*P. monodon*	Reduces WSSV multiplication	Ponprateep et al. (2013)
31	VP292	F0ATP synthase b-chain	*L. vannamei*	Promotes infection	Li et al. (2013)

32	VP37	F(1)ATPase β-subunit	*L. vannamei*	Prevents infection	Zhan et al. (2013)
33	VP28	LdlrLec1, LdlrLec2	*M. japonicus*	Antiviral immune response	Xu et al. (2014)
34	VP28	MjsvCL	*M. japonicus*	Antiviral immune response	Wang et al. (2014)
35	Protein Kinase 1	Ferritin	*L. vannamei*	Inhibits host defense mechanism	Lin et al. (2015)
36	WSSV-miR-N24	Caspase 8	*M. japonicus*	Represses the apoptosis of shrimp hemocytes	Huang et al. (2014a, b, c)
37	VP31	Triosephosphate isomerase (TPI)	*F. chinensis*	Promotes infection	Zhao et al. (2014)
38	IE1 and WSSV304	LvKLF	*L. vannamei*	Transcriptional regulation	Huang et al. (2014a, b, c)
39	VP26, VP31, VP37, VP90, VP136	Integrin	*L. vannmei*	Mediates WSSV infection	Zhang et al. (2014)
40	WSSV108	PmKLF	*P. monodon*	Replication	Liu WJ et al. (2015)
41	VP51	LvRPL7	*L. vannamei*	Promotes infection	Liu et al. (2015a, b, c)
42	WSSV-miR-22	STAT	*M. japonicus*	Mediates infection	Ren et al. (2015)
43	vp26, vp28, and wssv102	miR-10a	*L. vannamei*	Enhance the translation of viral proteins	Huang JY et al. (2017)
44	VP31	Arginine kinase	*F. chinensis*	Aids in infection	Ma et al. (2016)
45	VP26 and VP37	LvCCAP17 clathrin coat AP17	*L. vannamei*	Promote infection	Wang et al. (2016)
46	VP37, VP32, VP38A, VP39B, and VP41A	Peritrophin-Like Protein (LvPT)	*L. vannamei*	Promote infection	Xie et al. (2015)
47	VP19	MjSRC	*M. japonicus*	Restricts viral proliferation	Yang et al. (2016)
48	VP31	PmLamr	*P. monodon*	Enhances attachment to host surface	Liu et al. (2016)
49	WSSV186, WSSV189, WSSV395, WSSV458, and WSSV471	ALFPm3	*P. monodon*	Antiviral immune response	Suraprasit et al. (2014); Methatham et al. (2017)
50	WSSV453	PmproPPAE2	*P. monodon*	Anti-melanization mechanism	Sutthangkul et al. (2017)
51	WSSV-miR-N13 and WSSV-miR-N23	Dorsal	*M. japonicus*	Promotes infection	Ren et al. (2017)

(continued)

Table 4 (continued)

Sl. no.	Viral protein	Host protein	Species	Effect	References
52	IE1	LvYY1 Yin Yang 1 (YY1)	*L. vannamei*	Promotes lytic activation and viral replication.	Huang et al. (2017a, b)
53	wsv023	LvGCP2	*L. vannamei*	Promotes infection	Chen et al. (2017)
54	VP15, VP39A and VP28	FmLC, FmLC6	*Fenneropenaeus merguiensis*	Antiviral immune response	Runsaeng et al. (2018)
55	VP28, tegument VP39A and also capsid VP15	FmLdlr Lectin	*F. merguiensis*	Enhance phagocytosis	Kwankaew et al. (2018)
56	WSSV134	Hemocyte homeostasis-associated protein (PmHHAP)	*P. monodon*	Suppresses caspase-mediated apoptotic cell death during infection	Apitanyasai et al. (2018)
57	WSSV164	PmproPO1 and PmproPO2	*P. monodon*	Anti-melanization mechanism	Sangsuriya et al. (2018)
58	WSSV-miR-315	*PPAE3*	*P. monodon*	Attenuates Prophenoloxidase activation	Jaree et al. (2018)
59	WSSV-miR-N46	DCP1 gene	*M. japonicus*	Inhibits replication	Sun and Zhang (2019)
60	*wsv191* and *wsv407*	Shrimp miR-1000	*M. japonicus*	Inhibit virus replication	Gong et al. (2018)
61	VP24	Polymeric immunoglobulin receptor (pIgR)	*M. japonicus*	Involves in cellular entry of virus	Niu GJ et al. (2019a)
62	WSSV164 and WSSV189	*Pm*GILT	*P. monodon*	Anti-WSSV activity	Thipwong et al. (2019)
63	*wsv001*	mja-miR-6489-3p	*L. vannamei*	Mediates infection	Shekhar et al. (2019)
64	VP19, VP24, and VP28	MjRPS27	*M. japonicus*	Inhibit virus proliferation by preventing virion assembly	Diao et al. (2019)
65	VP41B	Glycosylase protein, PmENGase	*P. monodon*	Promotes infection	Huang JY et al. (2020)
66	VP24	PEN2	*P. monodon*	Prevents viral entry	Xiao et al. (2020)

67	VP28	BigPEN	*P. monodon*	Prevents viral entry	Xiao et al. (2020)
68	IE1, wsv056, wsv249, wsv403	MKK4, JNK/c-Jun	*L. vannamei*	Promote viral replication	Wang et al. (2020)
69	VP26	MjPellino	*M. japonicus*	Increases the immune response to virus	Zhang et al. (2020)
70	VP51B and VP150	F1 ATP synthase β subunit	*P. monodon*	Promote infection	Liu et al. (2021a, b, c, d)
71	wsv006	Constitutive photomorphogenic-like protein subunit 5 (CSN5)	*L. vannamei*	Involved in virus infection	Zou et al. (2021)
72	VP26	Hemocyte-specific small protein (LvHSSP)	*P. vannamei*	Mediates virus adhesion and entry into host cells	Sun et al. (2023)
73	WSV056	Sodium-coupled monocarboxylate transporter 1 (SMCT1)	*P. vannamei*	Involved in viral replication	Wang W et al. (2023)
74	wsv108	Heat shock protein 70 (HSP70)	*L. vannamei*	Promotes apoptosis	Yang et al. (2024)
75	WSSV322	Heat shock protein 90 (*LvHSP90*)	*L. vannamei*	Modulates haemocyte apoptosis during viral infection	Yingsunthonwattana et al. (2024)
76	WSV403	Stat92E	*P. vannamei*	Suppresses the JAK/STAT pathway and its downstream genes	Wang et al. (2024)
77	IE1	Integrin-α5	*P. vannamei*	Inhibits integrin-FAK signaling pathway	Lu et al. (2024)

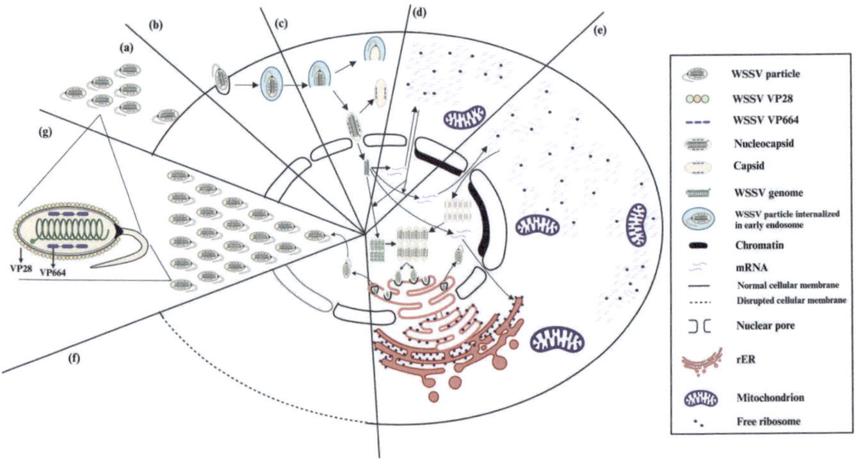

Fig. 5 An illustrative model of the replication cycle of WSSV proposed by Li et al. (2015) (detailed descriptions of different stages are provided in the text above)

4.6 Diagnosis

Many levels of diagnosis have been employed for specific and sensitive diagnosis and virus detection, such as gross clinical signs, histological analysis, electron microscopy, immunological methods, in situ hybridization techniques and polymerase chain reaction (PCR) techniques. As bacterial infections have also been found to be associated with the development of white spots on the cuticle, it would be prudent not to rely only on gross clinical signs in the diagnosis of WSD (Pradeep et al. 2012). In cases when WSD is suspected based on clinical signs, histological methods can be employed as confirmatory tests (Takahashi et al. 1994; Wang et al. 2000c). A rapid, non-lethal histopathological method using paraffin and cryosections of tissues like eye stalk and pleopod for WSSV detection was developed by Rajendran et al. (2005). However, histopathological methods are often unsuitable for detecting WSSV infection at a subclinical stage. PCR is, therefore, considered the gold standard for WSD diagnosis. Due to its high sensitivity, specificity, speed and robustness, PCR-based assays are now widely used for WSSV detection (Crockford 2008; Haq et al. 2012a, b). Many modifications of PCR have been employed for detecting WSSV either singly (Lo et al. 1996; Takahashi et al. 1996; Otta et al. 1999, 2003, 2014; Tang and Lightner 2000; Thakur et al. 2002; Yoganandhan et al. 2003a, b; Vaseeharan et al. 2003; Hossain et al. 2004a, b, 2014; Mishra and Shekhar 2005; Ayub et al. 2008; Dante et al. 2008; Park et al. 2013; Guha et al. 2015; Haq et al. 2015; Jayabarath et al. 2017; Park et al. 2020; Venkadesh et al. 2021) or coupled with other pathogens as in multiplex PCR (Tsai et al. 2002; Yang et al. 2006; Khawsak et al. 2008; Liang et al. 2010; Jeeva et al. 2014;

Thamizhvanan et al. 2019). A modified PCR was developed based on nested PCR due to the problems associated with the earlier protocol (Nunan and Lightner 2011). In situ hybridization with DNA probe offers high specificity, and the technique has also been employed to detect WSSV (Durand et al. 1997; Nunan and Lightner 1997; Chang et al. 1998a; Wen-Bin et al. 2000; Tang et al. 2013). Real-time PCR technique for WSSV detection has been reported by many authors either singly or as multiplex PCR with other viruses (Dhar et al. 2001; Durand and Lightner 2002; Xie et al. 2008; Jang et al. 2009; Panichareon et al. 2011; Haq et al. 2012a, b; Leal et al. 2014). The loop-mediated isothermal amplification (LAMP) method has been developed by several researchers for the sensitive and rapid diagnosis of WSSV (Kono et al. 2004; Mekata et al. 2009; Jaroenram et al. 2009; Chou et al. 2011; Nicolasora et al. 2014). Further, a CRISPR-based diagnostic assay capable of rapid, single-copy detection of WSSV, which can lend itself to field-level diagnosis, has been reported (Sullivan et al. 2019).

The development of immunological detection methods in shrimp has not progressed compared to PCR-based assays because of the difficulties in getting purified antigens for immunization. Many authors have reported monoclonal antibodies specific to different viral proteins, and their diagnostic sensitivity was found to be comparable to single-step PCR (Anil et al. 2002; Liu et al. 2002; Chaivisuthangkura et al. 2004, 2006, 2010; Makesh et al. 2006a; Siriwattanarat et al. 2013; Vaniksampanna et al. 2017; Inchara et al. 2018). Detection of WSSV employing polyclonal antibodies specific to the envelope protein has also been reported (You et al. 2002; Yoganandhan et al. 2004). Few researchers have developed lateral flow assays using gold nanoparticles for the rapid detection of WSSV (Sithigorngul et al. 2006; Cheng et al. 2007; Jaroenram et al. 2009; Kulabhusan et al. 2017a, b). Other published reports of immunological assays include latex agglutination assay (Okumura et al. 2004, 2005; Makesh et al. 2006b), electrochemical method (Natarajan et al. 2017, 2019; Takemura et al. 2020), affinity immunosensor (Loyprasert-Thananimit et al. 2012; Waiyapoka et al. 2015), label-free piezoelectric microcantilever sensors (Capobianco et al. 2010), colorimetric assay (Quéré et al. 2002; Loyprasert-Thananimit et al. 2014), immunomagnetic reduction assay (Liu et al. 2015a, b, c), ELISA (Zhan et al. 2004; Tang et al. 2018; Murugan and Sankaran 2018), western blot (Nadala Jr et al. 1998), protein marker-based (Sathish et al. 2004b) and flow-through immunoassay (Patil et al. 2013). Dot blot and immunochromatographic tests have also been used for rapid detection of the virus (Nadala and Loh 2000; Zhiwen et al. 2003; Powell et al. 2006; Wang and Zhan 2006; Sithigorngul et al. 2006, 2011; Patil et al. 2008; Shekhar and Gopikrishna 2010; Thiruppathiraja et al. 2011; Wangman et al. 2017; Bunsanong et al. 2018; Sheng et al. 2018). A capacitive biosensor with high sensitivity that can detect as low as one copy of WSSV in shrimp pond water has been developed by Samanman et al. (2011). Tangential flow ultrafiltration has also been used to enhance the detection rate of WSSV (Alavandi et al. 2015). Various immunological and molecular techniques reported for the diagnosis of WSD are summarized in Table 5.

Table 5 Summary of different methods for diagnosing WSSV

Name of the assay	WSSV target gene	Sensitivity	Reference
Immunological techniques			
Immunodot			
	VP28	1.0 × 10^5 viral DNA copies	Shekhar and Gopikrishna (2010)
	VP26	10^5 copies μL^{-1} of WSSV	Bunsanong et al. (2018)
Monoclonal antibodies	VP19	1.2 fmol/spot	Chaivisuthangkura et al. (2010)
	VP26	7–14 fmol/spot of the rVP26F109	Vaniksampanna et al. (2017)
	VP28	400 pg of purified WSSV, 20 pg of r-28	Liu et al. (2002)
	VP28	$10(^4)$ copies/mg tissue	Hou et al. (2011)
Polyclonal antiserum	VP28	5 ng of total haemolymph protein	Yoganandhan et al. (2004)
Flow-through immunoassay (FTA)	VP28 Vp18	0.312 $\mu g\ mL^{-1}$ WSSV protein 8–10 min	Patil et al. (2013)
Immuno-based colorimetric assay	VP26	1.6 × $10(^4)$ and 1.6 × $10(^7)$ copies μL^{-1} of WSSV (sensitivity)	Loyprasert-Thananimit et al. (2014)
Polyclonal antibody-based biosensor	VP26	160–160,000 copies of WSSV sensitivity	Loyprasert-Thananimit et al. (2012)
Label-free piezoelectric microcantilever sensors (lead magnesium niobate-lead titanate)	VP28 VP664	10 virions^{-1} 100 μL	Capobianco et al. (2010)
Immunomagnetic reduction (IMR) assay	ICP11	2 × 10^{-3} ng mL^{-1}, detection limit	Liu et al. (2015a, b, c)
Label-free immunosensor	VP26	1.6 × $10(^1)$–1.6 × $10(^6)$ copies μL^{-1}	Waiyapoka et al. (2015)
Lateral flow immunoassay (gold nanoparticles conjugated to a polyclonal antibody)	VP28	10^3 copies of WSSV	Kulabhusan et al. (2017a)
Lateral flow immunoassay using phage-displayed peptide as bio-recognition probe	VP28	12.5 $\mu g\ mL^{-1}$ of the virus, 20 min	Kulabhusan et al. (2017a, b)

(continued)

Table 5 (continued)

Name of the assay	WSSV target gene	Sensitivity	Reference
Electrochemical detection conductive silicon rubber (CSR) composed of graphene quantum dots and gold nanoparticle-embedded polyaniline nanowires	VP28	Low as 48.4 DNA copies mL^{-1}	Takemura et al. (2020)
CRISPR-based	VP108	~1.06 copies mL^{-1}	Sullivan et al. (2019)
Molecular techniques			
Cross-priming amplification (CPA)	Conserved sequence of WSSV	10 copies μL^{-1}	Yang et al. (2014)
Recombinase polymerase amplification (RPA)	VP28	10 molecules 6.41 ± 0.17 min at 39 °C	Xia et al. (2014)
RPA-CRISPR	VP28	10^1 copies/reaction	Wang et al. (2023a, b, c)
Coupled recombinase polymerase amplification (RPA) and *Pyrococcus furiosus* Argonaute (*Pf*Ago)	VP28	Single copy per reaction	Wang et al. (2023a, b, c)
Multiplex loop-mediated isothermal amplification (mLAMP)	VP28	10^2	He and Xu (2011)
FRET LAMP	ORF191	10^2 copies of the target DNA to less than 70 min	Chou et al. (2011)
LAMP-AuNPs	ORF191	2×10^2	Seetang-Nun et al. (2013)
LAMP-based microfluidic chip	VP28	10^2 copies/reaction 40 min at 63 °C	Hu et al. (2023)
LAMP	VP28	Six copies of WSSV plasmid DNA 55 min at 65 °C	Kumar et al. (2025)
Nested PCR	VP28	1 fg	Hameed et al. (2005)
Multiplex RT PCR	VP28	100 pg	Xie et al. (2007)
		5.00 pg	Khawsak et al. (2008)
		1×10^3 copy number	Chen et al. (2012)
		21.4 copies	Jeeva et al. (2014)
Real-time multiplex polymerase chain reaction (rtm-PCR)	VP28	2000 copies	Xie et al. (2008)
		500 fg	Panichareon et al. (2011)

(continued)

Table 5 (continued)

Name of the assay	WSSV target gene	Sensitivity	Reference
Duplex real-time PCR	VP28	Two copies of WSSV control plasmid	Leal et al. (2014)
Real-time PCR	VP28	12 copies per PCR reaction	Mendoza-Cano and Sánchez-Paz (2013)
Real-time PCR	VP664 and VP28	1.15 copies of WSSV plasmid	Kim et al. (2023)
Real-time enzymatic recombinase amplification (RT-ERA)	VP28	2.3 copies μL^{-1} 11.91 ± 0.12 min at 42 °C	Zhang et al. (2023)
Capacitive biosensor	Glutathione S-transferase-white spot binding protein (GST-WBP)	1 copy μL^{-1} of WSSV in shrimp pond water	Samanman et al. (2011)
Electrochemical sensor with gold nanoparticles	VP28	2.38 ng/mL	Thuy et al. (2023)

NA not available

5 Disease Management

WSD, caused by WSSV, results in enormous economic losses to the shrimp aquaculture industry globally and remains the most significant limiting factor in the sustainable production of farmed shrimp. Numerous efforts have been undertaken to develop effective prevention and control strategies for WSD since the disease was first reported in the early 1990s. An overview of these management strategies is presented in the following section.

5.1 Environmental and Pond-Level Control Measures

Considering that the treatment and control of the virus infection in the rearing facilities (hatcheries and grow-out ponds) are extremely difficult or next to impossible, preventing the introduction of the virus from shrimp-rearing facilities is a recommended alternative (Menasveta 2002; Pan et al. 2017). Lo and Kou et al. (1998) recommended that prior to stocking animals, the pond should be sterilized by drying, as the virus is not viable for long periods in dried pond sediments (Kumar et al. 2013, 2019). The authors also suggested that feeding shrimp with dried commercial feed instead of fresh arthropods would minimize the risk of WSSV introduction in a culture facility. Other practices such as restricting water exchange, raising water temperature and maintaining an optimum level of water quality parameters like pH, DO, and salinity could help in containing the virus to some extent (Cock et al. 2009;

Gunalan et al. 2010; Raj et al. 2012; Selvam et al. 2012; Vaseeharan et al. 2013; Kakoolaki et al. 2015). A recent review reported the effect of different physicochemical parameters on the incidence of WSSV outbreaks (Millard et al. 2021). The virulence of WSSV was found to be higher at temperatures between 25 and 28 °C, and exposure to elevated temperatures (above 30 °C) has been reported to protect shrimp (Raj et al. 2012; Jiang et al. 2019). Moreover, shrimp reared at suboptimal salinities were found to aggravate WSSV infectivity (Lin et al. 2012; Ramos-Carreño et al. 2014). The information on the influence of dissolved oxygen concentration on the WSSV outbreak is lacking, and an earlier study indicated that enough aeration in the culture ponds during the initial culture period reduced the WSSV-induced shrimp mortalities (Ruiz-Velazco et al. 2010). Studies reporting contradictory results on the influence of pH on WSSV infection are also published (Gunalan et al. 2010; Gao et al. 2011; Kathyayani et al. 2019). High pH, in combination with lower temperature, was found to be associated with WSSV outbreaks (Gunalan et al. 2010). Prolonged exposure of shrimp to nitrogenous wastes in water has been shown to increase the susceptibility of the animals to WSSV (Kathyayani et al. 2019).

With the farming of specific pathogen-free (SPF) *P. vannamei*, the introduction of WSSV through shrimp larvae to the grow-out system is significantly reduced. However, the introduction of the virus into the grow-out ponds can occur through various routes and potential vectors/carriers such as crabs, polychaete worms and other benthic organisms are a major concern. Therefore, the exclusion of these organisms from the culture system before stocking is recommended. This can be achieved by ploughing the pond bottom, removing sludge and drying the pond before stocking or by chemically treating the pond bottom using lime. However, the vectors/carriers and their eggs can get into the system through incoming water, and this can be prevented by filtering the water using a 300 μm mesh screen. Further, crab and bird fencing, foot dips and restricted entry to culture facilities can eliminate vectors of WSSV in most shrimp farms (World Bank 2014). Rigorous sanitation practices (Sánchez-Paz 2010) and use of SPF or PCR-negative seed stocks, development and use of virus-resistant species, avoiding stocking during winter, lowering the stocking density of the shrimp, reducing environmental stress, use of biosecure water and culture systems or combination of these practices may serve to effectively minimize the threat of WSD epidemics (OIE 2019). Minimal water exchange and implementation of technologies such as greenhouse and polyculture have been suggested to control WSSV (Dey et al. 2019). The greenhouse technology involves rearing the animals in an indoor facility where a higher temperature is maintained compared to the surrounding environment (Dey et al. 2019). This is based on the evidence that lower temperature favours viral replication besides compromising immunity. Vidal et al. (2001) reported the absence of WSSV infection in *P. vannamei* reared at 35 °C, while the disease develops when the animals are reared at 26 °C because at higher temperatures (32–33 °C), the virus replication is reduced (Rahman et al. 2006b). Rearing larvae in the indoor nursery is beneficial as it provides a biosecure environment and the advantage of keeping the animals at a higher temperature. In a recent study, it has been reported that WSSV remains viable and infective in pond sediments for a considerable period, even after sun-drying.

Therefore, the adoption of better management practices has been suggested for the successful completion of the crop cycle (Kumar et al. 2019).

A cost-effective biosecurity programme can be implemented in shrimp aquaculture production systems through a variety of management strategies. Biosecurity involves a set of practices to minimize the probability of pathogen introductions and their subsequent spread. For an effective biosecurity programme, various risks need to be identified, and appropriate biosecurity procedures need to be adopted to diminish those risks. The key elements for minimizing the risk of disease introduction and spread within a facility involve stocking SPF stock, pre-treatment of source water and sanitizing various farm machinery using disinfectants, frequent monitoring of animals' health and periodic screening of the stock using sensitive diagnostic tools. Best management practices should be followed to create a sustainable, non-stressful environment for shrimp culture.

5.2 Physical and Chemical Methods for Inactivation of Virus

Inactivation of WSSV employing different physical methods (heat, UV light, drying, pH, ozone, gamma irradiation and desiccation) and chemical agents (sodium hypochlorite, povidone-iodine, benzalkonium chloride, formalin, ethanol and sodium chloride) have been reported. However, due to the limitations of physical inactivation methods, chemical inactivation is preferred, and among them, chemical disinfectants, such as sodium hypochlorite, a halogen, have been reported to be the most effective in inactivating the virus even at a low dosage (Balasubramanian et al. 2006). Different virus-inactivating agents used against WSSV, the dosage of application and the experimental species in which the efficacy of these chemical agents are provided in Table 6.

5.3 Use of Immunostimulants

Immunostimulants of various categories have been used widely to enhance the innate (non-specific immune system) defense response and thereby increasing resistance to a particular pathogen (Sakai 1999). Dietary supplementation of immunostimulant β-glucan from various sources was found to protect shrimp from WSSV (Chang et al. 1999; Sajeevan et al. 2009a, b; Sajeevan and Rosamma 2009; Sukumaran et al. 2010; Pacheco et al. 2011; Ghaednia et al. 2012; Thitamadee et al. 2014; Bai et al. 2014). Different types of microbial immunostimulants have been reported to reduce WSSV infection in shrimp (Rahman et al. 2006a, b; Rohini et al. 2012; Subashini et al. 2012; Peraza-Gómez et al. 2014; Genio et al. 2015; Fierro-Coronado et al. 2019a, b; Pilotto et al. 2019). Alternative therapeutic measures using various herbs and medicinal plants have been found to augment disease resistance in shrimp against WSSV (Achuthankutty and Desai 2004; Citarasu et al. 2006, 2015; Balasubramanian et al. 2007, 2008a, b; Rameshthangam and Ramasamy 2007; Pholdaeng and Pongsamart 2010; Yogeeswaran et al. 2012; Velmurugan et al.

Table 6 Different methods reported for the inactivation of WSSV

Sl. no.	Agent	Condition/dose	Animal model	Reference
1	Heat	55 °C for 90 min, 70 °C for 5 min	*Penaeus monodon*	Chang et al. (1998b)
		60 min at 50 °C, 1 min at 60 °C and 0.2 min at 70 °C	*P. japonicus*	Nakano et al. (1998)
		55 °C and higher for 20 min	*P. monodon* *P. japonicus*	Maeda et al. (1998)
		60 °C for 20 min or 70 °C for 10 min	*P. indicus* *Paratelphusa hydrodromous*	Balasubramanian et al. (2006)
2	Desiccation	1, 1.5 or 2 h at 30 °C with a water content of 3.4% in the filter paper	*P. monodon* *P. japonicus*	Maeda et al. (1998)
		3 h at 26 °C	*P. japonicus*	Nakano et al. (1998)
3	pH	pH 1 for 10 min, pH 3 for 60 min pH 12 for 10 min	*P. monodon*	Chang et al. (1998b)
		pH 12 for 10 min	*P. indicus* *Paratelphusa hydrodromous*	Balasubramanian et al. (2006)
4	Ultraviolet light	9.30×10^5 μW s/cm^2 for 60 min	*P. monodon*	Chang et al. (1998b)
		1×10^4 μW s/cm^2	*P. japonicus*	Nakano et al. (1998)
		3×10^4 μW s/cm^2	*P. monodon*	Oseko et al. (2006)
		3.1×10^5 μW s/cm^2 4.6×10^5 μW s/cm^2	*P. indicus* *Paratelphusa hydrodromous*	Balasubramanian et al. (2006)
5	Ozone	Total residual oxidants concentration of 0.5 μg mL^{-1} for 10 min	*P. monodon*	Chang et al. (1998b)
6	Formalin	0.25 ppm	*P. monodon*	Oseko et al. (2006)
		5 ppm for 10 min	*P. japonicus*	Nakano et al. (1998)
7	Diethyl ether	20%	*P. monodon* *P. japonicus*	Maeda et al. (1998)

(continued)

Table 6 (continued)

Sl. no.	Agent	Condition/dose	Animal model	Reference
8	Sodium hypochlorite	Total free chlorine concentration of 100 ppm for 10 min	*P. monodon*	Chang et al. (1998b)
		1 ppm for 10 min	*P. japonicus*	Nakano et al. (1998)
		0.5% effective chloride	*P. monodon*	Oseko et al. (2006)
		10, 25 and 50 ppm for either 10 or 30 min	*P. monodon* *P. japonicus*	Maeda et al. (1998)
		200 ppm for 10 min	*P. indicus* *Paratelphusa hydrodromous*	Balasubramanian et al. (2006)
		100 ppm for 10 min and 10 ppm for 30 min	NA	Department of agriculture (2013)
9	Benzalkonium chloride	100 ppm for 10 min	*P. indicus* *Paratelphusa hydrodromous*	Balasubramanian et al. (2006)
		75 ppm for 10 min	*P. monodon*	Chang et al. (1998b); Department of agriculture (2013)
10	Iodine containing compound	Povidone-iodine 100 ppm for 10 min	*P. monodon*	Chang et al. (1998b)
		10, 50 and 100 ppm for 30 min povidone-iodine	*P. monodon* *P. japonicus*	Maeda et al. (1998)
		2.5 ppm effective povidone-iodine (Isodine®)	*P. monodon*	Oseko et al. (2006)
		Povidone-iodine 100 ppm for 10 min and 10 ppm for 30 min	NA	Department of agriculture (2013)
11	Ethanol	More than 30%	*P. monodon*	Oseko et al. (2006)
		30% for 1 min	*P. japonicus*	Nakano et al. (1998)

(continued)

Table 6 (continued)

Sl. no.	Agent	Condition/dose	Animal model	Reference
12	Sodium chloride	15.0% or higher in 24 h at 25 °C	*P. monodon* *P. japonicus*	Maeda et al. (1998)
		0–10% (not effective)	*P. monodon*	Chang et al. (1998b)
		25% within 24 h	*P. japonicus*	Nakano et al. (1998)
13	Trimethylammoniummethylene chloride	25 ppm for 10 min	*P. japonicus*	Nakano et al. (1998)
14	Sodium carbonate peroxyhydrate	5000 ppm for 60 min	*P. japonicus*	Nakano et al. (1998)
15	Chloroform	Within 15 min	*P. japonicus*	Nakano et al. (1998)
16	Monopersulfate (MPS)	4.8 ppm	*Marsupenaeus japonicus*	Gunasekara et al. (2020)

NA not available

2012, 2013, 2015a; Medina-Beltrán et al. 2012; Peraza-Gómez et al. 2009, 2014; Chithanbaran and David 2014; Ghosh et al. 2014; Kumaran et al. 2014; Bindhu et al. 2014; Wu et al. 2015; Sundaram et al. 2016; Trejo-Flores et al. 2016; Rajkumar et al. 2017; Palanikumar et al. 2018; Fierro-Coronado et al. 2019a, b; Vibin et al. 2020; Shiny et al. 2020; Sujithra et al. 2020; Ghosh et al. 2024; Zhang et al. 2023a, b, 2024). Many studies have reported the immunostimulatory activity of extracts and derivatives from different seaweeds and microalgae (*Dunaliella salina*, *Gracilaria* sp., *Sargassum* sp., *Porphyra* sp., *Undaria* sp., *Ulva* sp., *Hypnae* sp., *Turbinaria* sp., etc.) (Chotigeat et al. 2004; Manilal et al. 2009; Immanuel et al. 2010, 2012a, b; Huynh et al. 2011; Lin et al. 2011a, b; Madhumathi and Rengasamy 2011; Sirirustananun et al. 2011; Sivagnanavelmurugan et al. 2012, 2018; Declarador et al. 2014; Dinesh et al. 2014; Rudtanatip et al. 2014; Wongprasert et al. 2014; Arizo et al. 2015; Sinurat et al. 2016; Zahra et al. 2017; Schleder et al. 2018; Meenakshi et al. 2019; Yudiati et al. 2019; Niu et al. 2019a, b; Klongklaew et al. 2020; Jin et al. 2021; Mariot et al. 2021; Elshopakey et al. 2024), mangroves (*Ceriops tagal*, *Sonneratia alba* and *Bruguiera gymnorrhiza*) and seagrass (*Halophila ovalis*) against WSSV in shrimp (Sudheer et al. 2012; Yuvaraj and Arul 2017; Muliani and Susianingsih 2018; Muliani and Kadriah 2020). Even though various immunostimulant preparations from diverse sources have been experimentally tested, only a few have rendered satisfactory results (Table 7). These varying responses could be attributed to different factors such as dose and frequency of immunostimulant used, size of the experimental animal, mode of administration and pathogen challenge. Promising results were shown using combinations of microbial immunostimulant (MI) and powdered plant (PP) (Peraza-Gómez et al. 2014), extract from the plant *Cyanodon dactylon* (Balasubramanian et al. 2008a, b; Howlader et al. 2020) and mangrove, *Ceriops tagal* (Sudheer et al. 2012).

Table 7 Details of immunostimulants used for controlling white spot disease in shrimp

Type	Host/size	Mode of administration/ duration/challenge method	Doses	SR/MR/RPS/CM (%)	Parameters studied	Reference
Microbial						
Beta-1,3-glucan (*Schizophyllum commune*)	*Penaeus monodon* (PL15)	Oral, 15 days Challenge: immersion, 2 h	2 g kg^{-1}	12.2% (SR), 6 dpi	NA	Chang et al. (1999)
	(5.5 ± 0.5 g)	Oral, 10 days Challenge: IM injection		0% (SR), 6 dpi		
		Oral, 20 days Challenge: IM injection		20% (SR), 6 dpi		
Beta-1,3-glucan (*S. commune*)	*P. monodon* (6.5 ± 0.4 g)	Oral, 20 days Challenge: IM injection	1 g kg^{-1}	0% (SR), 12 dpi	THC, PA, SOA, PO, SOD	Chang et al. (2003)
			2 g kg^{-1}	30% (SR), 12 dpi		
			10 g kg^{-1}	Around 42.2% (SR), 12 dpi		
			20 g kg^{-1}	24.4% (SR), 12 dpi		
Candida sake S165 (CSY)	*Fenneropenaeus indicus* (20 ± 2 g)				THC, SOA, PO, PD	Sajeevan et al. (2009a)
Whole biomass		Oral, 40 days Challenge: oral	10% (w/w)	64% ± 5.3% (SR), 10 dpi		
Glucan		Oral, 40 days, once in 7 days Challenge: oral	0.2%	36% ± 4% (SR), 10 dpi		

Candida sake S165 glucan	*F. indicus* (PL)	Dip, 3 h			NA	Sajeevan and Rosamma (2009)
		Challenge: 10th day of post-dip treatment.	0.1 mg mL^{-1}	16% ± 5% (SR), 7 dpi		
			0.5 mg mL^{-1}	36% ± 5% (SR), 7 dpi		
			1 mg mL^{-1}	36% ± 5% (SR), 7 dpi		
			2 mg mL^{-1}	8% ± 2% (SR), 7 dpi		
		Challenge: 15th day of post-dip treatment.	0.1 mg mL^{-1}	18% ± 2% (SR), 7 dpi		
			0.5 mg mL^{-1}	29% ± 2% (SR), 7 dpi		
			1 mg mL^{-1}	31% ± 3% (SR), 7 dpi		
			2 mg mL^{-1}	13% ± 1% (SR), 7 dpi		
		Challenge: 20th day of post-dip treatment.	0.1 mg mL^{-1}	14% ± 2% (SR), 7 dpi		
			0.5 mg mL^{-1}	11% ± 2% (SR), 7 dpi		
			1 mg mL^{-1}	14% ± 1% (SR), 7 dpi		
			2 mg mL^{-1}	15% ± 1% (SR), 7 dpi		

(continued)

Table 7 (continued)

Type	Host/size	Mode of administration/duration/challenge method	Doses	SR/MR/RPS/CM (%)	Parameters studied	Reference
Vibrio alginolyticus (formalin-killed)	*Marsupenaeus japonicus* (4 ± 2 g)	Oral, 45 days Challenge: oral	17 mL/kg (1 × 10^8 cells kg^{-1})	66.66%, 7 dpi (SR)	PO	Subashini et al. (2012)
Beta-1,3-glucan (*Saccharomyces cerevisiae*)	*F. indicus* (11.32 ± 1.20 g)	Oral, 20 days Challenge: IM injection	2 g kg^{-1} feed	Below 40% (SR), 21 dpi	THC, TPP, BCA, PA	Ghaednia et al. (2012)
			10 g kg^{-1} feed	53.32% ± 5.77% (SR), 21 dpi		
			20 g kg^{-1} feed	48.32% ± 5.77% (SR), 21 dpi		
Candida aquaetextoris S527	*P. monodon* (30 ± 3 g)	Oral, 30 days Challenge: oral	10%		TP, THC, PO, SOA, ALP, ACP Genes: ALF, CRU-1, CRU-2, CR-3, PEN-3 and PEN-5, α2M, AST, PX, PO, TG	Babu et al. (2013)
		Daily				
		Once in 3 days				
		Once in 7 days		65% ± 6.68% (SR), 9 dpi		
		Once in 10 days				
Microbial immunostimulant (MI) and powdered plant (PP) mixture	*Litopenaeus vannamei* (3.6 ± 0.58 g)	Oral, 21 days Challenge: oral at 7th and 10th days	PP (1.0 g kg^{-1}) + MI (2.1 g kg^{-1}) PP (2.0 g kg^{-1}) + MI (4.2 g kg^{-1}) PP (4.0 g kg^{-1}) + MI (8.4 g kg^{-1}).	100% SR in all the treatments	PO, THC, SOA WSSV Prevalence: decreased to 8.3% in treatment I, and to 0% in II and III.	Peraza-Gómez et al. (2014)
Beta-1,3-glucan	*P. monodon* (10–15 g)	IM injection (single dose)	5 μg g^{-1}	43% (MR), 3 dpi	PO, H_2O_2	Thitamadee et al. (2014)
		IM injection (double dose)		60% (MR), 3 dpi		

Marine yeast glucans	*P. monodon* (PL)	Oral, 45 days, once in 7 days Challenge: oral	0.2%		ALF, CRU-1, CRU-2, CRU-3, PEN-3, PEN-5, α2M, AST, CAS, CAT, GPx, GST, HCN, PX, CATC, PO, Rab-7, SOD, TG	Wilson et al. (2015)
Candida parapsilosis R20				66.66% (SR), 7 dpi		
Hortaea werneckii R23				70.27% (SR), 7 dpi		
Candida spencermartinsiae R28				60.97% (SR), 7 dpi		
Candida haemulonii R63				43.64% (SR), 7 dpi		
Candida oceani R89				58.53% (SR), 7 dpi		
Debaryomyces fabryi R100				54.05% (SR), 7 dpi		
Debaryomyces nepalensis R305				45.7% (SR), 7 dpi		
Meyerozyma guilliermondii R340				43.24% (SR), 7 dpi		
Cyanobacteria *Arthrospira platensis*	*L. vannamei* (12 ± 3 g)	Oral, 40 days Challenge: oral	0.6%	76% (SR), 20 dpi	THC, TPC, PO, SAT Gene: LvProPO2, LvToll2, LvDorsal, LvSTAT, LvPen 3, LvALF-A, LvALF-C, LvFrep, LvPPAE2, LvIMD, LvRelish, Lvα2M-2	Pilotto et al. (2019)

(continued)

Table 7 (continued)

Type	Host/size	Mode of administration/duration/challenge method	Doses	SR/MR/RPS/CM (%)	Parameters studied	Reference
Seaweed						
Fucoidan from *Sargassum polycystum* (crude)	*P. monodon* (5–8 g)	Oral, 4 days Challenge: immersion	100 g kg^{-1}	4.4% (SR), 10 dpi	PA	Chotigeat et al. (2004)
			200 g kg^{-1}	14% (SR), 10 dpi		
			400 g kg^{-1}	44% (SR), 10 dpi		
	(12–15 g)		100 g kg^{-1}	42% (SR), 10 dpi		
			200 g kg^{-1}	93% (SR), 10 dpi		
Polysaccharide (*Acrosiphonia orientalis*)	*P. monodon* (2–5 g)	Oral, 14 days Challenge: oral	2 g kg^{-1}	Around 80% (SR), 14 dpi	THC, PO	Manilal et al. (2009)
			4 g kg^{-1}	88% (SR), 14 dpi		
			6 g kg^{-1}	65% (SR), 14 dpi		
			8 g kg^{-1}	Around 70% (SR), 14 dpi		
Gracilaria tenuistipitata	*L. vannamei*	Immersion, 3 h Challenge: IM injection	400 mg L^{-1}	60% (SR), 5 dpi	THC, PO, RBA, SOD, LA	Lin et al. (2011a, b)
			600 mg L^{-1}	80% (SR), 5 dpi		
Gracilaria tenuistipitata	*L. vannamei* (11.2 ± 2.1 g)	Oral, 14 days Challenge: IM injection	0.5 g kg^{-1}	Around 30% (SR), 6 dpi	THC, PO, RBA, SOD, GPx, LA	Sirirustananun et al. (2011)
			1.0 g kg^{-1}	Around 40% (SR), 6 dpi		
			2.0 g kg^{-1}	Around 60% (SR), 6 dpi		
Fucoidan from *Sargassum wightii*	*P. monodon* (PL 20)	Oral, 45 days Challenge: IM injection	0.1%	72% (CM), 21 dpi	THC, PO, RBA, SOD, PA	Immanuel et al. (2012a)
			0.2%	62% (CM), 21 dpi		
			0.3%	51% (CM), 21 dpi		

Sodium alginate (*Sargassum wightii*) powder	*P. monodon* PL (15)	Oral (*Artemia* enriched), 20 days Challenge: immersion	100 mg L^{-1}	0% (SR), 21 dpi	NA	Immanuel et al. (2012b)
			200 mg L^{-1}	9% (SR), 21 dpi		
			300 mg L^{-1}	16% (SR), 21 dpi		
			400 mg L^{-1}	25% (SR), 21 dpi		
Sodium alginate (*Sargassum wightii*) beads			100 mg L^{-1}	2% (SR), 21 dpi		
			200 mg L^{-1}	14% (SR), 21 dpi		
			300 mg L^{-1}	23% (SR), 21 dpi		
			400 mg L^{-1}	32% (SR), 21 dpi		
Fucoidan (*Sargassum wightii*)	*P. monodon* (PL)	Oral (enriched with *Artemia* nauplii), 20 days Challenge: immersion	100 mg L^{-1}	94% (MR), 21 dpi	NA	Sivagnanavelmurugan et al. (2012)
			200 mg L^{-1}	82% (MR), 21 dpi		
			300 mg L^{-1}	71% (MR), 21 dpi		
			400 mg L^{-1}	60% (MR), 21 dpi		
Ulvan	*P. monodon* (2.39 ± 0.38 g)	Oral, 14 days Challenge: oral	500 mg kg^{-1}	Below 10% (SR), 5 dpi	THC, RBA, PO	Declarador et al. (2014)
			1000 mg kg^{-1}	Around 20% (SR), 5 dpi		
			1500 mg kg^{-1}	0% (SR), 5 dpi		
Sulfated galactans (SG) (*Gracilaria fisheri*)	*P. monodon* (5–8 g)	Oral through *Artemia salina* enriched with SG 7 days Challenge: IM injection	100 µg mL^{-1}	63.5% (CM), 14 dpi	THC, PO, SOD, SOA	Wongprasert et al. (2014)
			200 µg mL^{-1}	39.5% (CM), 14 dpi		

(continued)

Table 7 (continued)

Type	Host/size	Mode of administration/ duration/challenge method	Doses	SR/MR/RPS/CM (%)	Parameters studied	Reference
Fucoidan (*Sargassum binderi Sonder*)	*L. vannamei* (6–8 g)	Oral, 15 days 300 mg kg^{-1} weight shrimp Challenge: oral	LMW	40% (MR), 3 dpi	THC Gene expression: PO, SOD, TG more in treatments	Sinurat et al. (2016)
			HMW	10% (MR), 3 dpi		
			CF	45% (MR), 3 dpi		
			PF	55% (MR), 3 dpi		
Fucoidan	*L. vannamei* (10.5 ± 0.5 g)	Injection; Challenge: IM injection at 48 h & 144 h	2 μg g^{-1}	Below 40% (SR), 6 dpi	THC, PO, RBA, SOD Gene expression: LGBP, PX, POI, PO II, AST, HHAP	Chen et al. (2016)
			6 μg g^{-1}	Above 40% (SR), 6 dpi		
			10 μg g^{-1}	Above 40% (SR), 6 dpi		
Gracilaria verrucosa	*L. vannamei* (6.07 ± 0.10 g)	Oral, 14 days Challenge: IM injection	2 g kg^{-1}	22.62% ± 7.43% (RPS), 10 dpi	THC, PA, RBA, PO	Zahra et al. (2017)
			3 g kg^{-1}	27.38% ± 2.06% (RPS), 10 dpi		
			4 g kg^{-1}	41.07% ± 3.09% (RPS), 10 dpi		
			5 g kg^{-1}	13.69% ± 1.03% (RPS), 10 dpi		
Dietary macro algae	*L. vannamei* (0.65 ± 0.01 g)	Oral, 8 weeks Challenge: IM injection	2%	CM lower in all treatments (6 dpi)	PO, SOD, TAS Genes: PO, IkBα, TGF-β, TNF-α, IL1β, IL6, and IL8.	Niu et al. (2019a, b)
Porphyra haitanensis						
Undaria pinnatifida						
Saccharina japonica						
Gracilaria lemaneiformis						

Sodium alginate from *Sargassum siliquosum*	*L. vannamei* (3.1–4.0 g)	Oral 30 days Challenge: IM injection	4.0 g kg^{-1}	Above 30% (SR), 5 dpi	THC, PA, SOD, PO increased; No change in TPP Toll and lectin: varying response	Yudiati et al. (2019)
			6.0 g kg^{-1}.	Around 20% (SR), 5 dpi		
Fucoidan (*Turbinaria decurrens*)	*L. vannamei* (5–8 g)	Oral, 15 days Challenge: immersion at 5th day	100 mg kg^{-1}	5.8% (SR), 10 dpi	NA	Meenakshi et al. (2019)
			200 mg kg^{-1}	19.3% (SR), 10 dpi		
			400 mg kg^{-1}	61% (SR), 10 dpi		
	L. vannamei (12–15 g)		100 mg kg^{-1}	54% (SR), 10 dpi		
			200 mg kg^{-1}	97% (SR), 10 dpi		
HWCEs	*L. vannamei* (22.57 ± 2.41 g)	Injection with extract–virus mixtures	*U. intestinalis* 1 mg mL^{-1}	5.0% ± 7.1% (MR), 10 dpi	NA	Klongklaew et al. (2020)
			5 mg mL^{-1}	6.1% ± 10.5% (MR), 10 dpi		
			10 mg mL^{-1}	10.0% ± 10.0% (MR), 10 dpi		
			U. rigida 1 mg mL^{-1}	100.0% ± 0.0% (MR), 10 dpi		
			5 mg mL^{-1}	96.7% ± 5.8% (MR),10 dpi		
			10 mg mL^{-1}	100.0% ± 0.0% (MR), 10 dpi		
			C. lentillifera 1 mg mL^{-1}	0.0% ± 0.0% (MR), 10 dpi		
			5 mg mL^{-1}	6.7% ± 11.5% (MR), 10 dpi		
			10 mg mL^{-1}	0.0% ± 0.0% (MR), 10 dpi		

(continued)

Table 7 (continued)

Type	Host/size	Mode of administration/duration/challenge method	Doses	SR/MR/RPS/CM (%)	Parameters studied	Reference
Cystoseira trinodis fucoidan (FCT)	*L. vannamei*	Oral, 60 days Challenge: immersion	0.1%, 0.2%, and 0.4% FCT	100% mortality in control within 10 days. FCT-treated shrimps: Mortality increase from 43.33% to 75.00% during 20 days of infection.	NA	Salehpour et al. (2021)
Gracilaria verrucosa	*P. vannamei* (15.26 ± 1.29 g)	Oral: 21 days Challenge: IM injection	0 (control), 2, 4, and 8 g kg diet^{-1}	More than 60% survival in 4 g and 8 g/kg	THC, PA, LA, ROS, PO, CAT, ACP, ALP, SOD, GPx, MDA Genes: ALF1, CRU1, PEN4, CTL, TRAF6, STAT, TLR1, NOS	Elshopakey et al. (2024)
Medicinal plants/herbs						
Medicinal plants	*P. monodon* (8.0 ± 0.5 g)	Oral, 25 days Challenge: IM injection	100 mg kg^{-1}	0% (SR), 13 dpi	CT, THC, OxH, PO, SOA	Citarasu et al. (2006)
			200 mg kg^{-1}	Around 40% (SR), 20 dpi		
			400 mg kg^{-1}	Around 60% (SR), 20 dpi		
			800 mg kg^{-1}	74% (SR), 20 dpi		

Medicinal plants	*P. monodon* (10–15 g)	IM injection with different concentrations of plant extract and virus	*Cynodon dactylon* (AE, 100 mg kg^{-1}) and *Momordica charantia* (ME, 150 mg kg^{-1})	No MR	NA	Balasubramanian et al. (2007)
			Lantana camara (AE), *Phyllanthus amarus* (AE) and *Aegle marmelos* (ME, 150 mg kg^{-1})	40% MR	NA	
Cyanodon dactylon	*P. monodon* (10–15 g)	Oral: 5 days Challenge: oral on 6th to day	2%	100% (SR), 30 dpi	PO, SOA, SOD, NOA, THC, CT	Balasubramanian et al. (2008a, b)
Asparagus racemosus (Used as immune-adjuvant)	*P. monodon* (10.0 ± 1.0 g)	Oral: 30 days Challenge: IM injection	IgY (control)	100% (CM)	THC, LA, PO, SOA	Kumaran et al. (2010)
			Anti-WSSV IgY raised without adjuvant	Reduction in CM up to 50%		
			Anti-WSSV IgY raised with adjuvant	Reduction in CM up to 80%		

(continued)

Table 7 (continued)

Type	Host/size	Mode of administration/duration/challenge method	Doses	SR/MR/RPS/CM (%)	Parameters studied	Reference
Powder of plants (PP) (*Echinacea purpurea* and *Uncaria tomentosa*)	*L. vannamei* (10.9 ± 1.09 g)	Oral, 30 days Challenge: immersion and oral at 16th and 17th to day	0.5 g kg^{-1}	86.6% (SR), 13 dpi WSSV prevalence: 83.3%	THC, PO	Medina-Beltrán et al. (2012)
			1.0 g kg^{-1}	90% (SR), 13 dpi WSSV prevalence: 50.0%		
			2.0 g kg^{-1}	93.3% (SR), 13 dpi WSSV prevalence: 16.6%		
			4.0 g kg^{-1}	93.3% (SR), 13 dpi WSSV prevalence: 8.3%		
Momordica charantia	*L. vannamei* (6–8 g)	Oral, 7 days Challenge: oral	1500 mg kg^{-1}	Above 80% (SR), 15 dpi	NA	Ghosh et al. (2014)
			1250 mg kg^{-1}	Above 80% (SR), 15 dpi		
			1000 mg kg^{-1}	Above 80% (SR), 15 dpi		
			750 mg kg^{-1}	86% (SR), 15 dpi		
			500 mg kg^{-1}	63% (SR), 15 dpi		
			250 mg kg^{-1}	18% (SR), 15 dpi		

Rhizophora mucronata	*L. vannamei* (6–8 g)	Oral, 7 days Challenge: oral	1500 mg kg^{-1}	Above 80% (SR), 7 dpi	NA	Chakraborty et al. (2014)
			1250 mg kg^{-1}	Above 80% (SR), 7 dpi		
			1000 mg kg^{-1}	85% (SR), 7 dpi		
			750 mg kg^{-1}	Above 60% (SR), 7 dpi		
			500 mg kg^{-1}	Around 60% (SR), 7 dpi		
			250 mg kg^{-1}	Below 40% (SR), 7 dpi		
Agathi grandiflora	*F. indicus* (9.0 ± 0.5 g)	Oral, 30 days Challenge: IM injection	100 mg kg^{-1}	100% (CM), 10 dpi	THC, PO, SOD, LA	Bindhu et al. (2014)
			200 mg kg^{-1}	Below 80% (CM), 10 dpi		
			300 mg kg^{-1}	Around 40% (CM), 10 dpi		
			400 mg kg^{-1}	Around 20% (CM), 10 dpi		
Herbal extract mixture *Agati grandiflora*, *Justicia tranquebariensis*, *Eclipta erecta*	*P. monodon* (8.0 ± 0.5 g)	Oral, 60 days Challenge: IM injection	100 mg kg^{-1}	100% (MR), 10 dpi	THC, OxH, CT, PO	Kamaran et al. (2014)
			200 mg kg^{-1}	80% (MR), 10 dpi		
			400 mg kg^{-1}	Above 30% (MR), 10 dpi		
			800 mg kg^{-1}	Above 40% (MR), 10 dpi		

(continued)

Table 7 (continued)

Type	Host/size	Mode of administration/duration/challenge method	Doses	SR/MR/RPS/CM (%)	Parameters studied	Reference
Java grass *Cyperus rotundus*	*F. indicus* (9.0 ± 0.5 g)	Oral, 30 days Challenge: IM injection	100 mg kg^{-1}	Around 70% (CM), 10 dpi	THC, CA, PO, SOA,	Citarasu et al. (2015)
			200 mg kg^{-1}	Around 60% (CM), 10 dpi		
			400 mg kg^{-1}	21% (CM), 10 dpi		
			800 mg kg^{-1}	25% (CM), 10 dpi		
Gynura bicolor	*L. vannamei* (10.1 ± 1.5 g)	Oral, 28 days Challenge: IM injection	0.5 g kg^{-1}	23.3%, 6 dpi (SR)	THC, PO, RBA, SOD, LA Gene: PX, LGBP	Wu et al. (2015)
			1.0 g kg^{-1}	36.7%, 6 dpi (SR)		
			2.0 g kg^{-1}	53.3%, 6 dpi (SR)		
Leaf of Aloe vera	*L. vannamei* (80 ± 5 mg)	Oral, 26 days, every 2 days Challenge: oral at 5th day	1 g kg^{-1}	90% (SR), 21 dpi		Trejo-Flores et al. (2016)
			2 g kg^{-1}	83% (SR), 21 dpi		
			4 g kg^{-1}	80% (SR), 21 dpi		
Cynodon dactylon (EE)	*L. vannamei* (13.8 ± 0.86 g)	Oral, Challenge by IM injection	1%	100% (CM), 240 h	THC, AA, PO	Junior et al. (2017)
			2%	38% (CM), 240 h		
			4%	100% (CM), 240 h		
Argemone mexicana	*L. vannamei* (9.0 ± 0.5 g)	Oral, 30 days Challenge: IM injection	100 mg kg^{-1}	30% (SR), 10 dpi	THC, OxH, PO, LA	Palanikumar et al. (2018)
			200 mg kg^{-1}	45% (SR), 10 dpi		
			300 mg kg^{-1}	75% (SR), 10 dpi		
			400 mg kg^{-1}	79% (SR), 10 dpi		

Powdered plants (PP)	*P. vannamei* (528 ± 31.7 mg)	Oral, 10 days Oral challenge at 7th day of feeding	1 g kg^{-1}	Around 70% (SR), 3 dpi	NA	Fierro-Coronado et al. (2019a, b)
			2 g kg^{-1}	Around 85% (SR), 3 dpi		
			4 g kg^{-1}	Around 95% (SR), 3 dpi		
Cynodon dactylon (EE)	*P. vannamei* (1.0 g)	Oral 2 weeks Challenge: oral	2%	Around 80% (CM), 11 dpi	NA	Jha et al. (2020)
			3%	0% (CM), 11 dpi		
Cynodon dactylon (CE)		Oral, 2 weeks Challenge: oral	10%	51% (CM), 11 dpi		
			15%	51% (CM), 11 dpi		
			20%	0% (CM), 11 dpi		
Herbal extracts	*F. indicus* (9 ± 1 g)	Oral, 30 days Challenge: IM injection	D1	100% (CM), 5 dpi	THC, CA, OxH, PO	Shiny et al. (2020)
			D2	Around 40% (CM), 5 dpi		
			D3	50% (CM), 5 dpi		
Olive (*Olea europaea*)	*P. vannamei* (11.1 ± 0.2 g)	Oral, 14 days Challenge: oral	50 mg kg^{-1}	40% (SR) , 22 dpi	TP, GLU, OxH, SOD, PO, CT, THC	Gholamhosseini et al. (2020)
			100 mg kg^{-1}	40% (SR), 22 dpi		
			200 mg kg^{-1}	55% (SR), 22 dpi		
Cynodon dactylon	*P. monodon* (13.541 ± 2.927 g)	IM injection Challenge: IM injection	75 mg kg^{-1}	85% (SR), 15 dpi	NA	Howlader et al. (2020)
			100 mg kg^{-1}	100% (SR), 15 dpi		
			150 mg kg^{-1}	100% (SR), 15 dpi		
Annona muricata	*F. indicus* (8.5 ± 3 g)	Oral, 40 days	1 g 100 g^{-1}	70% (CM), 20 dpi	CT, THC, OxH, PO	Sujithra et al. (2020)
Phyllanthus niruri (ME)	*P. monodon* (PL20)	Oral, 90 days Challenge: IM injection	1%	84% (MR), 21 dpi	Gene: PO	Vibin et al. (2020)
			2%	73% (MR), 21 dpi		
			3%	62% (MR), 21 dpi		

(continued)

Table 7 (continued)

Type	Host/size	Mode of administration/duration/challenge method	Doses	SR/MR/RPS/CM (%)	Parameters studied	Reference
Zingiber officinale and *Aegle marmelos*	*P. monodon* (6 g)	Oral: 4 weeks	0% (control), 0.05%, 0.1%	Lowest CM (20% in T1 of both the plants)	THC, CT, PO, SOD Gene: LA, PEN, PO	Ghosh et al. (2024)
Mangrove						
Ceriops tagal	*P. monodon* (6–8 g)	Oral, 7 days Challenge: oral	500 mg kg body weight^{-1} day^{-1}	100% (SR), 7 dpi	PO, AST, PX, α2M, TG, HCN, CRU, PEN-3	Sudheer et al. (2012)
			250 mg kg body weight^{-1} day^{-1}	33.3% (SR), 7 dpi		
			125 mg kg body weight^{-1} day^{-1}	44.4% (SR), 7 dpi		
Sonneratia alba and *Bruguiera gymnorrhiza*	*P. monodon* (6 ± 1 g)	IM injection with extract and virus	ME: 0.1 g/0.01 L NTE buffer		THC, DHC	Muliani and Susianingsih (2018)
			WE, BE, DEE: 0.05 g/0.01 L NTE buffer			
S. alba			WE	40% (MR), 10 dpi		
			BE	6.67% (MR), 10 dpi		
			DEE	0% (MR), 10 dpi		
			ME	0% (MR), 10 dpi		
B. gymnorrhiza			WE	60% (MR), 10 dpi		
			BE	0% (MR), 10 dpi		
			DEE	6.67% (MR), 10 dpi		
			ME	100% (MR), 10 dpi		

Different extracts of *S. alba*, *S. caseolaris*, *S. lanceolata*, *B. gymnorrhiza*	*P. monodon* (5–7 g)	IM injection with extract and virus	500 mg/100 mL of NTE buffer		THC, DHC, PO	Muliani and Kadriah (2020)
S. alba (BE)				57% (RPS), 10 dpi		
S. alba (DEE)				86% (RPS), 10 dpi		
S. caseolaris (BE)				53% (RPS), 10 dpi		
S. lanceolata (BE)				20% (RPS), 10 dpi		
B. gymnorrhiza (DEE)				40% (RPS), 10 dpi		
B. gymnorrhiza (BE)				57% (RPS), 10 dpi		
Sea grass						
Halophila ovalis	*P. monodon* (15 ± 2 g)	Oral, 25 days Challenge: IM injection	0.25 g kg^{-1}	0% (SR), 10 dpi	THC, HTP, PO, RBA	Yuvaraj and Arul (2017)
			0.5 g kg^{-1}	13% (SR), 21 dpi		
			1.0 g kg^{-1}	41% (SR), 21 dpi		

Abbreviations: *THC* total haemocyte count, *PA* phagocytic activity, *RBA* respiratory burst activity, *LYS* lysosyme activity, *SOA* superoxide anion, *SOD* superoxide dismutase, *PO* phenoloxidase activity, *PD* peroxidase activity, *PPO* prophenoloxidase, *BA* bactericidal activity, *THP* total haemocyte protein, *SAA* serum antibacterial activity, *BCA* bacterial clearance activity, *RPS* relative percentage of survival, *SR* survival rate, *MR* mortality rate, *CM* cumulative mortality, *BCE* bacterial clearance efficiency, *TG* transglutaminase, *HCN* haemocyanin, *NOS* nitric oxide synthase activity, *NO* nitric oxide, *SHA* serum haemolytic activity, *AT* agglutination titre, *LGBP* lipopolysaccharide and β-1,3-glucan binding protein, *PEN* penaeidin, *SP* serine protease, *cMnSOD* cytoplasmic manganese SOD, *AST* astakine, *GPx* glutathione peroxidase, *GST* glutathione S-transferase, *PX* peroxinectin, *CATC* cathepsin C, *CPE* cytopathic effect, *CRU* crustin, *SAT* serum agglutination titer, *NOA* nitric oxide activity, *TGA* transglutaminase activity, *LA* lysozyme activity, *TAS* total antioxidant status, *CR* clotting rate, *POD* peroxidase activity, *SMLD* serum malonaldehyde, *DHC* differential haemocyte count, *HHAP* haemocyte homeostasis-associated protein, *BE* butanol extract, *DEE* diethyl extract, *WE* water extract, *ME* methanol extract, *BPE* banana peel extract, *EE* ethanol extract, *CE* crude extract, *HWCEs* hot water crude extracts, *CT* coagulation time, *ACP* acid phosphatase, *ALP* alkaline phosphatase, *CAS* caspase, *CATH* cathepsin, *CTL* C-type lectin, *CAT* catalase, *MDA* maleic dialdehyde, *GR* glutathione reducase, *A2M* alpha(2)-macroglobulin, *TP* total protein, *IM* intramuscular, *PL* post-larvae, *TPP* total plasma protein, *TR* thioredoxin, *ALF* anti-lipopolysaccharide factor, *MPM* mucine-like peritrophin, *STAT* signal transducer and activator of transcription, *TSP* total serum protein, *dpi* day post-infection, *LMW* low molecular weight, *HMW* high molecular weight, *CF* crude fucoidan, *PF* purified fucoidan, *NA* not available

5.4 Use of Probiotics

The protective effect of different probiotic preparations in enhancing the immune response and protection against WSD has been investigated by many researchers. The dietary administration of *Vibrio alginolyticus*, marine actinomycetes and different species of *Bacillus* resulted in enhanced protection in shrimp and crayfish following WSSV infection (Kumar et al. 2006; Rodríguez et al. 2007; Li et al. 2009a, b; Tseng et al. 2009; Chai et al. 2016; Sanchez-Ortiz et al. 2016; Sandeepa and Ammani 2018; Sekar et al. 2019; Lai et al. 2020). The use of a probiotic mixture in protecting shrimp against WSSV was reported by Peraza-Gómez et al. (2009). Management of WSSV infection using *Pediococcus pentosaceus* or *Staphylococcus hemolyticus* has been reported in *L. vannamei.* (Leyva-Madrigal et al. 2011). The use of different synbiotic preparations has also been found to be effective against WSSV infection (Partida-Arangure et al. 2013; Febrianti et al. 2016; Yunarty et al. 2016; Widanarni et al. 2020). The details of various probiotics and their efficacy against WSSV are summarized in Table 8.

5.5 Use of Vaccines

Several vaccine formulations have been developed to manage WSSV infection in shrimp aquaculture over the years. Major categories of vaccines include inactivated vaccines, recombinant subunit vaccines, DNA vaccines and RNA-based vaccines that employ RNAi technology. Various methods such as heat (Namikoshi et al. 2004; Zhu et al. 2009), binary ethylenimine (BEI) (Zhu et al. 2009), formalin (Namikoshi et al. 2004; Singh et al. 2005; Amar and Faisan 2011; Yogeeswaran et al. 2012; Kulkarni et al. 2014b) or gamma irradiation (Heidarieh et al. 2014; Sedeh et al. 2015; Boreyri et al. 2017) have been employed for producing inactivated virus for using as candidate vaccines. When orally administered with formalin-inactivated virus preparation (IVP) for 7 days, *Fenneropenaeus indicus* was completely protected against WSSV exposure (Singh et al. 2005). Further, an evaluation of the inactivated WSSV vaccine preparation using formalin (a combination of oral and immersion methods) in grow-out ponds was undertaken and found effective in improving the survival rates (Amar et al. 2021).

Different viral protein subunits such as VP28 (Namikoshi et al. 2004; Jha et al. 2006; Caipang et al. 2008; Satoh et al. 2008; Kono et al. 2009; Syed et al. 2009; Fu et al. 2010; Syed and Kwang 2011; Kulkarni et al. 2013, 2014b; Akhila et al. 2014; Feng et al. 2014; Nguyen et al. 2014; Valdez et al. 2014; Pham et al. 2017; Solís-Lucero et al. 2016; Thomas et al. 2016; Taengchaiyaphum et al. 2017; Zhai et al. 2019; Kiataramgul et al. 2020; Boonyakida et al. 2020; Zhu et al. 2020), VP24 (Thomas et al. 2014; Boonyakida et al. 2020), VP26 (Namikoshi et al. 2004; Satoh et al. 2008; Valdez et al. 2014; Boonyakida et al. 2020), VP19 (Jha et al. 2006; Witteveldt et al. 2004a; Ha et al. 2008; Boonyakida et al. 2020; Zhu et al. 2020), VP15 (Boonyakida et al. 2020) and other proteins have been used as antigen for the preparation of the recombinant vaccine. Apart from *Escherichia coli*, other prokaryotic and

Table 8 Details of various probiotic used in the management of white spot disease

Name of the product/bacterial species	Host name/ size	Mode of administration	Dose	Survival	Immune parameters studied	Reference
Synbiotic (Inulin + *Bacillus* sp.)	*Litopenaeus vannamei* (12.8 ± 1.8 g)	Oral, 21 days Oral challenge on 8th day	LAB (1 × 10^5 CFU g^{-1} feed)	88.9% (SR)	THC, LA	Partida-Arangure et al. (2013)
			LAB + Inulin (8.0 g kg^{-1} feed)	100% (SR)		
Bacillus stratosphericus + *Arthrobacter defluvii* sp + *Lactobacillus* sp	*L. vannamei* (12.8 ± 1.8 g)	Oral, 21 days Challenge: oral at 8th day	5 × 10^6 CFU mL^{-1}	46% (SR)	NA	Karthik et al. (2015)
Actinomycetes	*P. monodon* (1 ± 0.3 g)	Oral, 2 weeks Challenge: oral	5 mL of concentrated culture into 10 g feed	35% (SR), 22 dpi	NA	Velmurugan et al. (2015a, b)
Synbiotic microcapsule (*Bacillus* NP5 RfR and mannan oligosaccharide combination)	*L. vannamei* (4.411 ± 0.395 g)	Oral, 30 days Challenge: IM injection	0.5% (w/w)	All treatments showed higher SR than control	THC, PPO, RBA	Febrianti et al. (2016)
			1%			
			2%			
Synbiotic microcapsule (*Bacillus* NP5 and mannan oligosaccharide)	*L. vannamei* (5.44 ± 0.44 g)	Oral, 30 days Challenge: IM injection	0.5% (w/w)	All treatments showed higher SR than control	THC, PO, RBA	Yunarty et al. (2016)
			1%			
			2%			
Bacillus PC465	*L. vannamei* (1.1 ± 0.2 g)	Oral, 30 days	10^7 CFU g^{-1}	29.9% ± 8.7% (RPS), 18 dpi	PEN-3a, PX, Lec-3, TR, HCN, PPO, CRU, Rab-6	Chai et al. (2016)
		Injection challenge	10^9 CFU g^{-1}	26.2% ± 3.3% (RPS), 18 dpi		
		Natural infection challenge	10^7 CFU g^{-1}	36.2% ± 2.3% (RPS), 22 dpi		
			10^9 CFU g^{-1}	40.1% ± 1.7% (RPS), 22 dpi		

(continued)

Table 8 (continued)

Name of the product/bacterial species	Host name/ size	Mode of administration	Dose	Survival	Immune parameters studied	Reference
Mixed *Bacillus* sp.	*L. vannamei* (1 ± 0.1 g)	Oral Challenge: oral at 12th and 24th day	1 × 10^6 CFU g^{-1} feed	98.9% (SR), 32nd culture WSSV Prevalence: 25%	SOD, PO, LvToll, TG, Hsp70	Sanchez-Ortiz et al. (2016)
			2 × 10^6 CFU g^{-1} feed	100% (SR), 32nd culture WSSV Prevalence: 67%		
			4 × 10^6 CFU g^{-1} feed	98.9% (SR), 32nd culture WSSV Prevalence: 67%		
			6 × 10^6 CFU g^{-1} feed	98.9% (SR), 32nd culture WSSV Prevalence: 42%		
LAB	*L. vannamei* (6 g)	Oral, 28 days Challenge, oral	5%		THC, PO, SOD, CAT	Sandeepa and Ammani (2018)
			10%			
			15%			
B. subtilis KA1	*L. vannamei* (6 g)	Immersion: 2 days Challenge: immersion	2 × 10^8 CFU mL^{-1}	84% (SR), 26th day	NA	Sekar et al. (2019)
B. subtilis KA3				28% (SR), 26th day		

Bacillus sp. NP5	*L. vannamei* (1.8 ± 0.06 g)	Oral, 8 weeks Challenge: IM injection			THC, PO, PA, RBA	Widanarni et al. (2020)
		Prebiotic (honey)	0.5% (v/w)	68.89% ± 3.84% 5 dpi		
		Probiotic	1% (v/w)	62.2% ± 3.14%		
		Synbiotic	NA	68.89% ± 3.84%		
Lactobacillus johnsonii KD1	*L. vannamei* (0.03 g)	Immersion	4.5×10^8 CFU/2 L seawater	Around 80% (CM) 5th day	PX, SP, PO, ALF-1	Dekham et al. (2023)
			4.5×10^{10} CFU/2 L seawater	Aroiund 60% (CM) 5th day		
Pediococcus pentosaceus MR001	*L. vannamei* (4–5 g)	Oral Challenge: IM 50 μL suspension of WSSV (2.2×10^4 copies/μL).	1×10^7, CFU g^{-1} feed	0%	PO, LvToll, SOD, HCN, STAT	Wanna et al. (2023)
			1×10^8, CFU g^{-1} feed	0%		
			1×10^9, CFU g^{-1} feed	Above 20% (SR) 15th day		

Abbreviations: as given in Table 7

eukaryotic expression systems (spores of Gram-positive bacterium, *Bacillus subtilis*, yeast *Pichia pastoris*, phage baculovirus vector, algae *Chlamydomonas*) have also been reported by many authors to improve the efficacy of vaccination (Jha et al. 2006; Valdez et al. 2014; Solís-Lucero et al. 2016; Kiataramgul et al. 2020). Out of all these vaccination strategies, complete protection was obtained by dietary administration of VP26 and VP28, wherein protection was sustained for 10 days following oral challenge with WSSV (Satoh et al. 2008). Likewise, recombinant spores of *B. subtilis* displaying CotC::Vp26 fusion protein on its surface provided 100% protection to *L. vannamei* (Valdez et al. 2014). It was reported that when *Saccharomyces cerevisiae* displaying fusion protein VP28–VP24 was administered, it completely protected shrimp with a proportionate decline in the viral load (Lei et al. 2021).

Different viral genes have been used for developing DNA vaccines (Rout et al. 2007; Kumar et al. 2008; Kono et al. 2009; Ning et al. 2009; Li et al. 2010; Mu et al. 2012; Ravi et al. 2020). Of the various approaches employed, those using VP28 brought about heightened protection (Kumar et al. 2008).

Many studies have employed RNAi targeted towards different WSSV genes and recorded different levels of protection against WSD in shrimp (Robalino et al. 2004; Das et al. 2010; Ahanger et al. 2014; Akhila et al. 2015; Thammasorn et al. 2015; Puneeth et al. 2017; Binayanti et al. 2019; Amarnath et al. 2019; Ramos-Carreño et al. 2021; Riet et al. 2021; Weerachatyanukul et al. 2021; Yang et al. 2024; Hong and Kim 2024). Recently, a peptide vaccine was formulated using different truncated proteins developed based on the structural protein VP15. This protein expressed in *E. coli* expression system showed improved survival against WSD in *Marsupenaeus japonicus* after 20 days of infection. Interestingly, four truncated versions of the protein provided higher protection than the whole VP15 protein. The study revealed that one of the truncated proteins, SR11, could significantly improve the survival rate in shrimp (Boonyakida et al. 2021).

Oral administration of vaccines would be the most feasible method in shrimp aquaculture; however, the vaccine's instability in the animal gut is a limitation. The degradation of vaccine in the extreme gastrointestinal environment of the host could be reduced by encapsulating vaccine in formulation using chitosan (Kumar et al. 2008; Sarathi et al. 2008; Pathan et al. 2013; Vimal et al. 2013; Ufaz et al. 2018), glucan (Zhu and Zhang 2012), liposomes (Mavichak et al. 2010) and chitosan/tripolyphosphate nanoparticles (Ravi et al. 2020). More recently, VLPs derived from plants and IHHNV have been used for the efficient delivery of dsRNA (Ramos-Carreño et al. 2021; Weerachatyanukul et al. 2021). Further, the protection against WSSV in shrimp has been reported to be enhanced by supplementing the vaccine with different adjuvants, immunostimulants and probiotics (Amar and Faisan 2011; Yogeeswaran et al. 2012; Kono et al. 2015; Sedeh et al. 2015). Various anti-WSSV vaccination approaches used in shrimp are summarized in Tables 9 and 10.

Table 9 Different vaccination methods reported for WSSV

Shrimp species	Vaccine	Type/mode of application	Response		Reference
			Duration	RPS (%)[a]	
Penaeus japonicus	Live virus	Sub-lethal injection of live virus	1–3 month	6–67	Wu et al. (2002)
P. japonicus	Inactivated virus	Injection of heat-inactivated virus	10–30 days	15–30	Namikoshi et al. (2004)
		Injection of formalin-inactivated virus	10–30 days	5–55	
	Recombinant protein	Injection of VP26	10 days	57	
		Injection and booster of VP26	10 days	60	
		Injection of VP28	10 days	17	
		Injection and booster of VP28	10 days	95	
Penaeus monodon	Recombinant protein (*Escherichia coli*)	Injection of VP19	2–25 days	33–57	Witteveldt et al. (2004a)
		Injection of VP19 + VP28	2–25 days	33–41	
		Injection of VP28	2–25 days	21–44	
	Recombinant protein (*Escherichia coli*)	Oral vaccination with VP19	2–3 days	44–64	Witteveldt et al. (2004b)
		Oral vaccination with VP28	7–21 days	29–77	

(continued)

Table 9 (continued)

Shrimp species	Vaccine	Type/mode of application	Response		Reference
			Duration	RPS (%)[a]	
Penaeus indicus	Inactivated	Oral vaccination with formalin-inactivated virus	1, 5, 10, 15 days	0, 100, 100, 0	Singh et al. (2005)
Procambarus clarkii	Recombinant protein (*Pichia pastoris*)	Injection and booster of VP28 (challenged at 3rd day post-vaccination)	25 days	91	Jha et al. (2007)
		Injection and booster of VP19 + VP28 (challenged at 3rd day post-vaccination)		84	
		Injection and booster of VP19 (challenged at 3rd day post-vaccination)		45	
		Injection and booster of VP28 (challenged at 21st day post-vaccination)		78	
		Injection and booster of VP19 + VP28 (challenged at 21st day post-vaccination)		76	
		Injection and booster of VP19 (challenged at 21st day post-vaccination)		17	
	Recombinant protein (*E. coli*)	Oral vaccination of VP19 (challenged at 3rd and 21st day post-vaccination)	25 days	76–82	Jha et al. (2006)
		Oral vaccination of VP28 (challenged at 3rd and 21st day post-vaccination)		43.2–49.0	
		Injection vaccination of VP19 (challenged at 3rd and 21st day post-vaccination)		52.91–56.46	
		Injection vaccination of VP28 (challenged at 3rd and 21st day post-vaccination)		39.6–39.83	
		Immersion vaccination of VP19 (challenged at 3rd and 21st day post-vaccination)		83–86.3	
		Immersion vaccination of VP28 (challenged at 3rd and 21st day post-vaccination)		46.3–46.5	
		Oral vaccination of VP28 + VP19 (challenged at 3rd and 21st day post-vaccination)		51.0–53.2	
		Injection vaccination of VP28 + VP19 (challenged at 3rd and 21st day post-vaccination)		36.5–36.5	
		Immersion vaccination of VP28 + VP19 (challenged at 3rd and 21st day post-vaccination)		53.0–56.5	

P. monodon	Recombinant protein (*E. coli*)	Injection of VP292	30 days	41	Vaseeharan et al. (2006)
		Injection and booster of VP292	30 days	52	
P. monodon	DNA vaccine	Injection of VP15	25 days	0	Rout et al. (2007)
			50 days	0	
		Injection of VP28	25 days	51	
			50 days	30	
		Injection of VP35	25 days	0	
			50 days	0	
		Injection of VP281	25 days	46	
			50 days	34	
Penaeus chinensis	Recombinant protein (Baculovirus)	Injection of VP19	22 days	50	Ha et al. (2008)
		Oral vaccination of VP19	25 days	50	
		Oral vaccination of VP466	25 days	48	
P. japonicus	Recombinant protein expressed in *Brevibacillus brevis*	Oral vaccination of VP28	3 days	81	Caipang et al. (2008)
			2 weeks	24	
P. japonicus	Recombinant protein (*E. coli*)	Oral vaccination of VP26 (oral, injection and immersion challenge)	10 days	100, 61, 71	Satoh et al. (2008)
		Oral vaccination VP28 (oral, injection and immersion challenge)		100, 34, 70	
P. monodon	DNA vaccine	Injection of VP28	7 days	90	Kumar et al. (2008)
			14 days	77	
			21 days	67	
			30 days	57	
P. clarkii	Recombinant protein (*Bacillus subtilis*)	Injection of VP28 (transgenic spores)	27 days	78.3	Fu et al. (2008)
		Injection of VP28 (vegetative cells)		51.7	

(continued)

Table 9 (continued)

Shrimp species	Vaccine	Type/mode of application	Response		Reference
			Duration	RPS (%)[a]	
P. clarkii	DNA vaccine (*Salmonella typhimurium*)	Oral vaccination of VP28 (challenged at 7th, 15th and 25th days post-vaccination)	15 days	83.3, 66.7, 56.7	Ning et al. (2009)
P. monodon	Recombinant baculovirus with IE1 promoter	Injection and booster of VP28	3 days	86	Syed et al. (2009)
			15 days	74	
P. japonicus	DNA vaccine	Injection of VP28	7 days	70	Kono et al. (2009)
P. clarkii	Inactivated vaccine	Injection with BEI-inactivated WSSV (2 mM) (challenge on 7th and 21st day post-vaccination)		77, 66	Zhu et al. (2009)
		Injection with BEI-inactivated WSSV (3 mM) (challenge on 7th and 21st day post-vaccination)		63, 30	
		Injection with heat inactivated WSSV (challenge on 7th and 21st day post-vaccination)		0, 0	
P. chinensis	Recombinant protein (*B. subtilis*)	Oral vaccination of VP28	14 days	83	Fu et al. (2010)
P. japonicus	Recombinant protein	Oral vaccination with liposome adjuvant of VP28	3 days	68	Mavichak et al. (2010)
		Oral vaccination with VP28	3 days	50	
Litopenaeus vannamei	DNA vaccine	Injection of VP28	7 days	53	Li et al. (2010)
			14 days	20	
P. monodon	Inactivated WSSV vaccine + nutritional additives	Injection of FIV (intramuscular challenge on 3rd 15th and 30th days post-vaccination)	15 days	83, 67, 33	Amar and Faisan (2011)
		Immersion of FIV (immersion challenge on 3rd and 45th days post-vaccination)		886, 66	
		Oral vaccination of FIV (immersion challenge on 45th and 60th days post-vaccination)		63.7, 8	
		FIV + wheat grass (oral)		72.7 (CS)	
		FIV + MSM (oral)		73.3 (CS)	

P. clarkii	Recombinant protein (*B. subtilis*)	Oral administration of CotB-VP28 (immediate challenge, challenge on 7th and 21st days post-vaccination)	16 days	37.9, 46.4, 30	Ning et al. (2011)
		Oral administration of CotBC-VP28 (immediate challenge, challenge on 7th and 21st days post-vaccination)		44.8, 50, 33.3	
P. clarkii	Recombinant protein (silk worm pupae)	Oral vaccination of VP28 (10 g kg^{-1}, 50 g kg^{-1}, 100 g kg^{-1}, 200 g kg^{-1})	20 days	82.2 , 94.4, 48.2, 56.6 (CM)	Wei and Yang (2011)
L. vannamei	Recombinant protein	Oral vaccination of VP19	21 days	66.7 (CM)	Choi et al. (2011)
		Oral vaccination of VP28		41.7 (CM)	
P. monodon	Recombinant protein from baculovirus expression system (Bac)	Feeding Bac-VP28 coated feed: Immersion challenge	3 days	81	Syed and Kwang (2011)
			15 days	76	
		Immersion in Bac-VP28 containing sea water (3 times): Immersion challenge	3 days	75	
			15 days	68	
L. vannamei	Recombinant protein (*E. coli*)	Injection of VP28 (1 μg) (challenge with 10^3 or 10^4 dilutions of WSSV)	12 days	100 (CM)	Yang et al. (2012)
		Injection of VP28 (5 μg) (challenge with 10^4 dilutions of WSSV)		0 (CM)	
		Injection of VP28 (5 μg) (challenge with 10^3 dilutions of WSSV)		100 (CM)	
		Injection of 36B (5 μg) (challenge with 10^4 dilutions of WSSV)		100 (CM)	
P. clarkii	Recombinant protein (*E. coli*)	Oral administration of rTAT-VP28 (for 7 and 14 days)		63.3–67.8	Zhang et al. (2012)
		Oral administration of rVP28 (for 7 and 14 days)		44.4–53.6	
L. vannamei	Recombinant protein	Oral vaccination of VP28C (challenge on 7th and 14th days post feeding)	14 days	23.3–47.5	Qiu et al. (2012)
		Oral vaccination of VP28N (challenge on 7th and 14th days post feeding)		12.5–16.7	

(continued)

Table 9 (continued)

Shrimp species	Vaccine	Type/mode of application	Response		Reference
			Duration	RPS (%)[a]	
P. monodon	DNA vaccine	Intramuscular injection of PmAV chitosan nanoparticles	15 days	83.3	Pathan et al. (2013)
L. vannamei	Recombinant protein	Intramuscular and oral administration of rVP39	20 days	60	Akhila et al. (2014)
L. vannamei	Recombinant protein (*B. subtilis*)	Oral administration of CotB-VP28	14 days	35 (CM)	Nguyen et al. (2014)
P. monodon	Recombinant vaccine (*E. coli*)	Oral administration of VP24	15 days	36 (CM)	Thomas et al. (2014)
P. japonicus	Recombinant protein (prepared by cell-free technology)	Injection of VP28	25 days	94	Kono et al. (2014)
		Oral vaccination of VP28	20 days	100	
L. vannamei	*B. subtilis* recombinant spores	Feeding the pelleted feed coated with bacterial spores expressing cotC:VP 26 fusion proteins	16 days	100	Valdez et al. (2014)
		Expressing cotC:VP 28 fusion proteins		90	
Crayfish	Transgenic *Dunaliella salina*	Lysed cells of transgenic green algae mixed with feed and fed to crayfish	20 days	59	Feng et al. (2014)
P. monodon	Inactivated vaccine (formalin)	Oral administration (challenge on 1st day post-vaccination)	8 days	100	Sudheer et al. (2015)
		Oral administration (challenge on 5th day post- vaccination)		50	
L. vannamei	Inactivated vaccine	Gamma irradiated WSSV vaccine (oral)	10 days	73.33	Sedeh et al. (2015)
		GI-WSSV vaccine + GI-V.P (oral)		86.66	
L. vannamei	Recombinant vaccine (M13 phage)	Injection of VP28	5 days	49.99	Solís-Lucero et al. (2016)
L. vannamei	Recombinant vaccine (*Anabaena* sp.)	Injection of VP28		68	Jia et al. (2016)

P. monodon	*B. subtilis* recombinant spores	Feeding the pelleted feed coated with bacterial spores expressing cotC::VP 28 fusion proteins	14 days	75	Pham et al. (2017)
P. monodon	Recombinant vaccine	Injection of 1XrVP28	10 days	67	Taengchaiyaphum et al. (2017)
		Injection of 4XrVP28		81	
Macrobrachium rosenbergii	Recombinant vaccine (Baculovirus vector)	Oral administration of recombinant bivalent vaccine fused with flagellin 2			Cho et al. (2017)
		Ac-ie1VP28, Ac-ie1VP19, Ac-VP19-ie1VP28, Ac-VP28- ie1VP19, Ac-VP19-ie1VP28FL2, Ac-VP28-ie1VP19FL2	10 days	28.9, 34.2, 21.2, 23.7, 15.8, 10.5 (CM)	
L. vannamei	Transgenic *Synechocystis* sp. PCC6803	Oral administration of VP28	7 days	88.42	Zhai et al. (2019)
			10 days	62.5	
			40 days of vaccination	80	
L. vannamei	Transgenic *Synechococcus sp.* PCC 7942	Oral administration of VP19, VP28, VP19 + 28	10 days	57.8, 62.2, 71.1	Zhu et al. (2020)
L. vannamei	Transgenic *Chlamydomonas reinhardtii*	Oral vaccination with VP28	14 days	87	Kiataramgul et al. (2020)
P. japonicus	Recombinant vaccine (in *E. coli*)	Injection and booster of VP15, VP19, VP24, VP26, VP28	20 days	78, 27, 56, 44, 50	Boonyakida et al. (2020)
	Recombinant vaccine (in silkworm larvae)	Injection of VP15, VP19	10 days	45, 60	
Marsupenaeus japonicus	Recombinant vaccine (in silkworm pupae)	Oral administration of VP15		More than 70	Boonyakida et al. (2022)

Abbreviations: *RPS* relative percentage of survival, *CS* cumulative survival, *CM* cumulative mortality

[a] The results depicts the Relative percentage of survival in most instances. The values in the paranthesis in few contexts indicate either Cumulative mortlaity or cumulative survival

Table 10 A summary of the major RNAi trials reported in shrimp and prawns for the management of white spot disease

Shrimp species/size	Target gene	Delivery method/type of RNAi	Protection	Duration	Reference
Litopenaeus vannamei (3–4 g)	RR2	Intramuscular injection of dsRNA	78%	10 days	Robalino et al. (2005)
	DNA Polymerase		44%		
	VP28		85%		
P. monodon (2.5 ± 0.6 g)	VP28	Intramuscular injection of siRNA	50%	13 days	Westenberg et al. (2005)
	VP15		42%		
	GFP		33%		
	GFP		50%		
Penaeus chinensis (5–6 g)	VP28	Intramuscular injection with long dsRNAs (challenge with LD_{50} of WSSV)	0% (CM)	7 days	Kim et al. (2007)
	VP281		20% (CM)		
	protein kinase		6.7% (CM)		
	GFP		13.3% (CM)		
	VP28	Intramuscular injection with long dsRNAs (challenge with 2 X LD_{50} of WSSV)	0% (CM)		
	VP281		46.7% (CM)		
	protein kinase		6.7% (CM)		
	GFP		33.3% (CM)		
P. monodon (8–10 g)	VP28	Intramuscular injection with dsRNA	100%	30 days	Sarathi et al. (2007)
P. monodon (8–10 g)	VP28	Feed pellets coated with *E. coli* expressing dsRNA	68%	30 days	Sarathi et al. (2008)
		Feed pellet coated with chitosan-dsRNA complex	37%		
P. monodon (10–12 g)	VP28 (10 μg)	Intramuscular injection of long-hairpin RNA (lhRNA)	50%	20 days	Krishnan et al. (2009)
	VP19 (10 μg)		29%		
	VP28 (25 μg)		58%		
	VP28 (35 μg)		75%		
	VP28 (50 μg)		79%		

P. monodon (20 g)	VP19, VP28, ICP11	Intramuscular injection of long-hairpin RNA		15 days	Das et al. (2010)
		T1: ICP11-LH; 35 μg	67%		
		T2: VP28 35 μg	75%		
		T3: VP28 and VP19 17.5 μg each	75%		
		T4: VP28 and VP19 25 μg:10 μg	58%		
		T5: ICP11, VP28 and VP19 11.5 μg each	58%		
Marsupenaeus japonicus (10.0 g)	VP28	Intramuscular injection with β-1,3-D-glucan encapsulated vp28-siRNA particles	Shrimp mortality was significantly delayed	7 days	Zhu and Zhang (2012)
P. monodon (1.5 ± 0.5 g)	vp28	Intramuscular injection of dsRNA (6 μg/g body weight)	90%	14 days	Sanjuktha et al. (2012)
	vp281		66.7%		
	rr1		93.3%		
	dnapol		76.6%		
L. vannamei (8–10 g)	VP28	IM injection with *E. coli* expressing antisense RNA (10 μg)	50%	9 days	Akhila et al. (2015)
	VP39		60%		
	WSV477		90%		
	VP28 + VP39		50%	10 days	
	VP28 + wsv477		50%	12 days	
	VP39 + wsv477		50%	14 days	
P. monodon (25 ± 2 g)	VP24, VP28, thymidylate synthase (TS), ribonucleotide reductase-2 (RR2)	IM injection with *E. coli* expressing antisense RNA under the control of shrimp histone-3 (H3) or penaedin (Pn) promoter. VP24 (pH3–VP24, pPn–VP24), VP28 (pH3–VP28, pPn–VP28), TS (pH3–TS, pPn–TS), and RR2 (pH3–RR2) (25 μg)	pH3–VP24: 80% pH3–P28: 90% pH3–TS: 80% pH3–RR2: 70%	16 days	Ahanger et al. (2014)

(continued)

Table 10 (continued)

Shrimp species/ size	Target gene	Delivery method/type of RNAi	Protection	Duration	Reference
P. monodon	VP28	IM injection with dsRNA expressed in virus like particles	44.5%	7 days	Jariyapong et al. (2014)
L. vannamei (23.3 ± 4.2 g)	VP 28	IM injection with *E. coli* expressing dsRNA (4 μg)	10% (CM)	10 days	Escobedo-Bonilla et al. (2015)
	VP26		21% (CM)		
	ORF89		10% (CM)		
	WSSV191		81% (CM)		
L. vannamei (PL45)	VP28, VP19, RR2 and TK-TMK	IM injection with *E. coli* expressing dsRNA	95%	28 days	Kumar et al. (2015)
L. vannamei (3–5 g)	VP28, WSSV051	Feed pellets coated with *E. coli* expressing dsRNA Oral administration with four doses (1.2 and 2.4 μg dsRNA for single and double dose per meal) 6 mg of WSSV051-dsRNA 6 mg of VP28-dsRNA 6 mg of multi-WSSV dsRNA 12 mg of multi-WSSV dsRNA	dsWSSV051: 60% (CM)	7 days	Thammasorn et al. (2015)
Macrobrachium rosenbergii	VP9 (ICP11)	Injection of dsRNA	80%	25 days	Alenton et al. (2016)
L. vannamei (10.0 g)	Bac-VP28-dsrr2	Intramuscular injection of dsRNA	33% (CM)	14 days	Rattanarojpong et al. (2016)
	Bac-VP28		64% (CM)		

P. monodon (7–10 g)	Injection of VP24	Oral and injection with dsRNA (2.5 μg/g of shrimp)	50%	15 days	Puneeth et al. (2017)
	Injection of VP26		60%		
	Injection of RR1		80%		
	Injection of WSSV477		90%		
	Oral feeding of VP24		50% (CM)		
	Oral feeding of VP26		50% (CM)		
	Oral feeding of RR1		30% (CM)		
	Oral feeding of WSSV477		20% (CM)		
L. vannamei (2.618 ± 0.527 g)	VP28	IM injection with *E. coli* expressing dsRNA (≥12 μg dsRNA/g)	97.5%	10 days	Nilsen et al. (2017)
L. vannamei (5 ± 0.75 g)	ORF89	IM injection with dsRNA expressed in yeast *Yarrowia lipolytica* (12 μg)	25%	30 h	Alvarez-Sanchez et al. (2018)
L. vannamei (10.5 ± 3.5 g)	VP19	IM injection with *E. coli* expressing dsRNA Dose 1: 0.02 μg; Dose 2: 0.2 μg; Dose 3: 2 μg	Dose 3 showed significant effect on granular hemocytes	8 days	Binayanti et al. (2019)
L. vannamei (7–10 g)	VP28	Oral feeding with *Bacillus subtilis* expressing dsRNA	83%	14 days	Saelim et al. (2020)

5.6 Other Management Strategies Used to Control White Spot Disease in Crustaceans

Ochoa-Meza et al. (2019) reported that the administration of a single dose of silver nanoparticles protected WSSV-infected *L. vannamei.* Romo-Quiñonez et al. (2020) studied the efficacy of Argovit-4, a silver nanoparticle-based feed formulation, in reducing the mortality of WSSV-challenged shrimp. It was found that the efficacy of different formulations against WSSV increased when encapsulated with chitosan nanoparticles (Taju et al. 2018). The protective abilities of natural herbal oil formulation (Jha et al. 2016; Babikian et al. 2020), biosurfactant (Monica et al. 2017), synthetic drug (Reddy et al. 2016), glycerol monolaurate (Qian and Zhu 2021), melanin, Phytocee™, sodium butyrate and kefir containing feedstuffs (Thang et al. 2019; Selvam et al. 2020; Choi et al. 2020; Xiao et al. 2021) against WSD have been reported. The efficacy of anti-WSSV compounds from different *Streptomyces* sp. and *Nocardiopsis* sp. has been elucidated (Jenifer et al. 2018, 2019; Manimaran et al. 2018; Rajkumar et al. 2018). Instead of the standard antiviral methods of blocking the viral receptors using recombinant viral proteins or virus antiserum, Huang et al. (2020a, b) used polyamine carbon quantum dots (polyamine CQDs) to inhibit the virus replication. The study demonstrated the ability of CQDs to bind to the viral envelope and hinder replication and the consequent reduction in mortality of WSSV-challenged animals (Huang et al. 2020a, b). The efficacy of different derivatives of coumarin and novel amino-nitrophenyl and amino-fluorophenyl medicines against WSSV has also been evaluated and found effective in reducing the mortality in WSSV-infected *P. vanammei* (Liu et al. 2021a, b; Shan et al. 2021). The oral administration of a low dose of metabolite from endophytic fungi, *Nodulisporium* sp. KT29 was found to improve the innate immunity and survival of *P. vannamei* post-larvae following co-infection with WSSV and *Vibrio harveyi* (Wahjuningrum et al. 2022). The inhibitory activity of cuminaldehyde against WSSV has been reported in which 40 μmol/L could significantly inhibit WSSV replication (Zhang et al. 2023a, b). The anti-WSSV effect of 2-amino-4-(4-chlorophenyl)-5-oxo-4H,5H-pyrano[3,2-*c*]chromene-3-carbonitrile (C7) has also been studied and found that a dose of 50 mg/kg significantly inhibited WSSV proliferation in adult shrimp (Shan et al. 2024). The in vitro anti-WSSV activity of nanoparticles synthesized by seaweed (*Sargassum swartzii*) and bacteria (*Brevibacterium casei*) has been studied recently (Namitha et al. 2024a, b).

6 Disease Resistance Against WSSV

Selective breeding has been considered a long-term disease control strategy (Bedier et al. 1998; Roch 1999; Bachere 2000) and has been applied in shrimp aquaculture (especially in *P. vannamei* and *P. stylirostris*) as a tool in genetic improvement for economically important traits such as growth and disease resistance with positive results. Over the years, researchers have developed genetic lines of *P. vannamei* with enhanced WSSV resistance (Gitterle et al. 2005; Gitterle and Diener 2014;

Cuéllar-Anjel et al. 2012). A study aimed to produce specific pathogen-resistant (SPR) *P. chinensis* lines to WSSV recorded improved survival in the third and fourth generations (Kong et al. 2003). The association between microsatellite DNA markers and WSSV resistance in shrimp has also been investigated (Dong et al. 2008; Chakrabarty et al. 2014). WSSV resistance in *L. vannamei*, based on the analysis of SNPs associated with selected immune genes, has been studied, and SNPs located on *TRAF6*, *Cu/Zn SOD* and *nLvALF2* displayed noteworthy influence on WSSV resistance in the animal (Yu et al. 2017). Further, Trang et al. (2019) conducted a study on the genetic variation in disease resistance against WSSV in *L. vannamei*. Some of the studies indicated that resistance to WSSV is heritable, as observed in some experimental populations of *P. vannamei*, and through selective breeding, genetic improvement in WSSV resistance can be attained. Lillehammer et al. (2020) reported enhanced survival in shrimp following the WSSV challenge using genomic selection techniques.

7 Conclusion

White spot syndrome virus (WSSV), which causes white spot disease (WSD), is a highly virulent shrimp pathogen, causing significant economic losses to shrimp farmers globally. The disease affects all major farmed shrimp species and causes rapid and high mortality within a few days. The virus is transmitted vertically and horizontally, infecting a wide array of crustaceans and other aquatic organisms. Many of the hosts have the potential to be either an active carrier or a passive reservoir and pose serious challenges to both farmed and wild crustaceans. This poses a serious challenge in controlling the disease in a shrimp farm. Over the past three decades, tremendous scientific efforts have been made in studying this this virus and the disease, and today, we have a much better understanding of the biology and pathology of the disease. Though several prophylactic measures, including immunostimulants, probiotics and vaccines for managing the disease, have been reported, none of these approaches provides assured protection against WSD. As a result, disease prevention through implementing biosecurity measures and best management practices (BMPs) continues to remain as the best possible option for minimizing the risk of losses due to WSD. Further, the farming of specific pathogen-free (WSSV-free) *P. vannamei* has significantly reduced the risk of the virus being introduced in the culture system through post-larvae and has helped in increasing shrimp production throughout Asia. Despite these efforts, the disease remains a serious challenge to sustainable shrimp production worldwide. Currently, domesticated *P. vannamei* lines with enhanced resistance to WSD, as well as acute hepatopancreatic necrosis disease (AHPND), are available in some countries in Latin America. However, these genetic lines are not widely available. A large-scale commercial availability of WSSV-resistant line of shrimp and an effective vaccine and/or therapeutics that can be administered orally at the field level will be of great significance in overcoming the challenges posed by WSD.

References

Abraham TJ, Mondal A, Patra A, Adikesavalu H (2020) Prevalence of white spot virus and monodon baculovirus in shrimp culture systems of West Bengal, India. J Fish 8(2):861–864

Achuthankutty CT, Desai UM (2004) Treatment of white spot syndrome virus (WSSV) in penaeid shrimp aquaculture using plant extract. In: Proceedings of the national seminar on new frontiers in marine biosciences research, January 22–23

Afsharnasab M, Kakoolaki S, Afazli F (2014) The status of white spot syndrome virus (WSSV) in Islamic Republic of Iran. Iran J Fish Sci 13(4):1021–1055

Ahamed AS, Haq MAB (2016) WSSV transmission in *Paratelphusa hydrodomous* fresh water rice field crab concerning with WSSV-419 like protein gene expression studies. Int J Res Appl Sci Eng Technol 4(12):408–413

Ahanger S, Sandaka S, Ananad D, Mani MK, Kondadhasula R, Reddy CS, Marappan M, Valappil RK, Majumdar KC, Mishra RK (2014) Protection of shrimp *Penaeus monodon* from WSSV infection using antisense constructs. Mar Biotechnol 16:63–73

Ahmad T, Sanyal KB, Mukherjee D, Abraham TJ, Dash G (2017) Detection of white spot virus (WSV) in *Litopenaeus vannamei* from shrimp aquaculture farms in East Midnapore district, West Bengal (India). Int J Fish Aquat Stud 5(2):200–204

Akhila DS, Rai P, Mani MK, Ballamoole KK, Karunasagar I, Karunasagar I (2014) Protection of *Litopenaeus vannamei* against white spot syndrome virus using bacterially expressed recombinant envelope proteins VP39 and VP28. Isr J Aquac 64:1116. https://doi.org/10.46989/001c.20729

Akhila DS, Madhu MK, Rai P, Condon K, Owens L, Karunasagar I (2015) Antisense RNA mediated protection from white spot syndrome virus (WSSV) infection in Pacific white shrimp *Litopenaeus vannamei*. Aquaculture 435:306–309

Alavandi SV, Ananda Bharathi R, Kumar SS, Dineshkumar N, Saravanakumar C, Rajan JJS (2015) Tangential flow ultrafiltration for detection of white spot syndrome virus (WSSV) in shrimp pond water. J Virol Methods 218:7–13

Alenton RRR, Kondo H, Hirono I, Maningas MBB (2016) Gene silencing of VP9 gene impairs WSSV infectivity on *Macrobrachium rosenbergii*. Virus Res 214:65–70

Alfaro AC, Nguyen TV, Bayot B, Leon JAR, Domínguez-Borbor C, Sonnenholzner S (2021) Metabolic responses of whiteleg shrimp to white spot syndrome virus (WSSV). J Invertebr Pathol 180:107545. https://doi.org/10.1016/j.jip.2021.107545

Alvarez-Sanchez AR, Romo-Quinones C, Rosas-Quijano R, Reyes AG, Barraza A, Magallon-Barajas F, Angulo C, Mejia-Ruiz CH (2018) Production of specific dsRNA against white spot syndrome virus in the yeast *Yarrowia lipolytica*. Aquac Res 49(1):480–491

Amar EC, Faisan JP (2011) Efficacy of an inactivated vaccine and nutritional additives against white spot syndrome virus (WSSV) in shrimp (*Penaeus monodon*). Isr J Aquac 63:9. https://doi.org/10.46989/001c.20593

Amar EC, Faisan JP, Gapasin RSJ (2021) Field efficacy evaluation of a formalin-inactivated white spot syndrome virus (WSSV) vaccine for the preventive management of WSSV infection in shrimp grow-out ponds. Aquaculture 531:735907. https://doi.org/10.1016/j.aquaculture.2020.735907

Amarakoon AADGU, Wijegoonawardane PKM, Wicraamasinghe WAL (2016) Detection and prevalence of white spot syndrome disease (WSSV) in shrimp farms. World Sci News 56:239–251

Amarnath BC, Saravanan P, Otta SK (2019) Determining the dosage and time of injection for WSSV VP28 double stranded RNA to *Penaeus indicus* in providing effective protection against WSSV. In: Jithendran, K.P.; Saraswathy, R.; Balasubramanian, C.P.; Kumaraguru Vasagam, K.P.; Jayasankar, V.; Raghavan, R.; Alavandi, S.V., and Vijayan, K.K. (eds.), *BRAQCON 2019: World Brackishwater Aquaculture Conference*. J Coast Res 86(SI):102–106. Coconut Creek (Florida), ISSN 0749-0208

Anil TM, Shankar KM, Mohan CV (2002) Monoclonal antibodies developed for sensitive detection and comparison of white spot syndrome virus isolates in India. Dis Aquat Org 51(1):67–75

Annon (1995) SEMBV – an emerging viral threat to cultured shrimp in Asia. Asian Shrimp News 3:2–3

Antony SP, Philip R, Joseph V, Singh ISB (2011a) Anti-lipopolysaccharide factor and crustin-III, the anti-white spot virus peptides in *Penaeus monodon*: control of viral infection by up-regulation. Aquaculture 319(1–2):11–17

Antony SP, Singh ISB, Jose RM, Kumar PRA, Philip R (2011b) Antimicrobial peptide gene expression in tiger shrimp, *Penaeus monodon* in response to gram-positive bacterial probionts and white spot virus challenge. Aquaculture 316(1–4):6–12

Antony SP, Singh ISB, Sudheer NS, Vrinda S, Priyaja P, Philip R (2011c) Molecular characterization of a crustin-like antimicrobial peptide in the giant tiger shrimp, *Penaeus monodon*, and its expression profile in response to various immunostimulants and challenge with WSSV. Immunobiology 216(1–2):184–194

Apitanyasai K, Amparyup P, Charoensapsri W, Sangsuriya P, Tassanakajon A (2018) Shrimp hemocyte homeostasis-associated protein (PmHHAP) interacts with WSSV134 to control apoptosis in white spot syndrome virus infection. Fish Shellfish Immunol 76:174–182

Arizo MAM, Simeon EC, Layosa MJT, Mortel RMM, Pineda CMB, Lim JJE, Maningas MBB (2015) Crude fucoidan from *Sargassum polycystum* stimulates growth and immune response of *Macrobrachium rosenbergii* against white spot syndrome virus (WSSV). Aquac Aquar Conserv Legis 8(4):532–543

Atnur VS, Vijayakumar S, Khavi M, Karunasagar I (2021) Assessment of white spot syndrome virus (WSSV) prevalence along the west coast of India. J Entomol Zool Stud 9(1):524–527

Ayub F, Sarker MY, Alam MS (2008) Prevalence of white spot syndrome virus infection detected by one-step and nested PCR in selected tiger shrimp (*Penaeus monodon*) hatcheries. Aquac Int 16:405–415

Babikian H, Babikyan Y, Jha RK, Wisoyo SD, Asih Y, Srisombat S (2020) Performance of natural oil blend formulation (NOBF) against white spot syndrome virus (WSSV) agent in *Penaeus vannamei* Boone, 1931. Int J Fish Aquat Stud 8(6):200–208

Babu DT, Antony SP, Joseph SP, Singh ISB, Philip R (2013) Marine yeast *Candida aquaetextoris* S527 as a potential immunostimulant in black tiger shrimp *Penaeus monodon*. J Invertebr Pathol 112(3):243–252

Babu B, Sathiyaraj G, Mandal A, Kandan S, Biju N, Palanisamy S, You SG, Nisha RG, Prabhu NM (2021) Surveillance of disease incidence in shrimp farms located in the east coastal region of India and *in vitro* antibacterial efficacy of probiotics against *Vibrio parahaemolyticus*. J Invertebr Pathol 179:107536. https://doi.org/10.1016/j.jip.2021.107536

Bachere E (2000) Shrimp immunity and disease control. Aquaculture 191:3–11

Bai N, Gu M, Zhang W, Xu W, Mai K (2014) Effects of β-glucan derivatives on the immunity of white shrimp *Litopenaeus vannamei* and its resistance against white spot syndrome virus infection. Aquaculture 426–427:66–73

Balasubramanian G, Sudhakaran R, Musthaq SS, Sarathi M, Hameed ASS (2006) Studies on the inactivation of white spot syndrome virus of shrimp by physical and chemical treatments, and seaweed extracts tested in marine and freshwater animal models. J Fish Dis 29(9):569–572

Balasubramanian G, Sarathi M, Kumar RS, Hameed ASS (2007) Screening the antiviral activity of Indian medicinal plants against white spot syndrome virus in shrimp. Aquaculture 263(1–4):15–19

Balasubramanian G, Sarathi M, Venkatesan C, Thomas J, Hameed ASS (2008a) Studies on the immunomodulatory effect of extract of *Cyanodon dactylon* in shrimp, *Penaeus monodon*, and its efficacy to protect the shrimp from white spot syndrome virus (WSSV). Fish Shellfish Immunol 25(6):820–828

Balasubramanian G, Sarathi M, Venkatesan C, Thomas J, Hameed ASS (2008b) Oral administration of antiviral plant extract of *Cynodon dactylon* on a large scale production against White spot syndrome virus (WSSV) in *Penaeus monodon*. Aquaculture 279(1–4):2–5

Bandeira J, Morais R, Silva R, Mendes E, Silva S, Santos F (2019) First report of white spot syndrome virus in wild crustaceans and mollusks in the Paraíba River, Brazil. Aquac Res 50(2):680–684

Banu MN, Haq MA, Rao MV (2015) Experimental challenge of white spot syndrome virus (WSSV) from polychaete *Pereneris cultifera* to peppermint shrimp *Lysmata wurdemanni* in captivity. Int J Res Eng Appl Sci 5(9):194–205

Bao W, Tang KFJ, Alcivar-Warren A (2023) The complete genome of an endogenous nimavirus (Nimav-1_LVa) from the Pacific whiteleg shrimp *Penaeus* (*Litopenaeus*) *vannamei*. Genes 11:94. https://doi.org/10.3390/genes11010094

Bateman KS, Munro J, Uglow B, Small HJ, Stentiford GD (2012a) Susceptibility of juvenile European lobster *Homarus gammarus* to shrimp products infected with high and low doses of white spot syndrome virus. Dis Aquat Org 100(2):169–184

Bateman KS, Tew I, French C, Hicks RJ, Martin P, Munro J, Stentiford GD (2012b) Susceptibility to infection and pathogenicity of white spot disease (WSD) in non-model crustacean host taxa from temperate regions. J Invertebr Pathol 110(3):340–351

Baumgartner WA, Hawke JP, Bowles KP, Varner W, Hasson KW (2009) Primary diagnosis and surveillance of white spot syndrome virus in wild and farmed crawfish (*Procambarus clarkii*, *P. zonangulus*) in Louisiana, USA. Dis Aquat Org 85:15–22

Bedier E, Cochard JC, Le Moullac G, Patrois J (1998) Selective breeding and pathology in penaeid shrimp culture: the genetic approach to pathogen resistance. World Aquacult 29:46–51

Bharathi VT, Chakravarty MS, Ganesh PRC (2017) Mud crab *Scylla serrate* (Forskal) as carrier of white spot syndrome virus (WSSV) in Coringa Mangroves, Andhra Pradesh. Int J Adv Res 5:2699–2708

Binayanti, Anshary H, Parenrengi A, Malina AC (2019) Use of dsRNA WSSV VP19 with different dose for infection control white spot syndrome virus (WSSV) in vaname shrimp *Litopenaeus vannamei*. Int J Sci Res 9(8):197–203

Bindhu F, Velmurugan S, Donio MBS, Michaelbabu M, Citarasu T (2014) Influence of *Agathi grandiflora* active principles inhibit viral multiplication and stimulate immune system in Indian white shrimp *Fenneropenaeus indicus* against white spot syndrome virus infection. Fish Shellfish Immunol 41(2):482–492

Blaylock RB, Curran SS, Lotz JM (2019) White spot syndrome virus (WSSV) in cultured juvenile blue crabs *Callinectes sapidus*: oral versus injection exposure, and feeding frequency effects. Dis Aquat Org 133(2):147–156

Boonrawd S, Mani R, Ponprateep S, Supungul P, Masrinoul P, Tassanakajon A, Rimphanitchayakit V (2017) Characterization of PmSpätzle 1 from the black tiger shrimp *Peneaus monodon*. Fish Shellfish Immunol 65:88–95

Boonyakida J, Xu J, Satoh J, Nakanishi T, Mekata T, Kato T, Park EY (2020) Antigenic properties of VP15 from white spot syndrome virus in kuruma shrimp *Marsupenaeus japonicus*. Fish Shellfish Immunol 101:152–158

Boonyakida J, Xu J, Satoh J, Nakanishi T, Mekata T, Kato T, Park EY (2021) Identification of antigenic domains and peptides from VP15 of white spot syndrome virus and their antiviral effects in *Marsupenaeus japonicus*. Sci Rep 11:12766. https://doi.org/10.1038/s41598-021-92002-8

Boonyakida J, Nakanishi T, Satoh J, Shimahara Y, Mekata T, Park EY (2022) Immunostimulation of shrimp through oral administration of silkworm pupae expressing VP15 against WSSV. Fish Shellfish Immunol 128:157–167

Boreyri D, Afsharnasab M, Motalebi AA, Haghigh A (2017) Immunological findings in shrimp *Litopenaeus vannamei* exposed to attenuated WSSV vaccine produced by gamma irradiation. Iran J Fish Sci 16(1):382–396

Briggs M, Funge-Smith S, Subasinghe R, Phillips M (2004) Introductions and movement of *Penaeus vannamei* and *Penaeus stylirostris* in Asia and the Pacific, RAP publication 10. FAO, Bangkok

Bunsanong N, Chotigeat W, Deachamag P, Thananimit S (2018) Semiquantitative dot-blot immunogold assay for specific detection of white spot syndrome virus. Biotechnol Appl Biochem 65(4):586–593

Bustamante MLU, Cofficny RS, Prieto AA, Guzmán GA, Juárez RV (2004) Susceptibilidad de *Litopenaeus schmitti* y *Cherax quadricarinatus* al virus del síndrome de la mancha blanca (WSSV). Cienc Mar 30(4):537–545

Cai S, Huang J, Wang C, Song X, Sun X, Yu J, Yan Z, Coughai Y (1995) Epidemiological studies on explosive epidemic disease of prawn 1993–1994. J Fish China 19:112–117

Caipang CMA, Verjan N, Ooi EL, Kondo H, Hirono I, Aoki T, Kiyono H, Yuki Y (2008) Enhanced survival of shrimp, *Penaeus* (*Marsupenaeus*) *japonicus* from white spot syndrome disease after oral administration of recombinant VP28 expressed in *Brevibacillus brevis*. Fish Shellfish Immunol 25:315–320

Capobianco JA, Shih WH, Leu JH, Lo GCF, Shih WY (2010) Label free detection of white spot syndrome virus using lead magnesium niobate-lead titanate piezoelectric microcantilever sensors. Biosens Bioelectron 26(3):964–969

Cavalli LS, Romano LA, Marins LF, Abreu PC (2011) First report of White spot syndrome virus in farmed and wild penaeid shrimp from Lagoa dos Patos estuary, southern Brazil. Braz J Microbiol 42:1176–1179

Cavalli LS, Batista CR, Nornberg BF, Mayer FQ, Seixas FK, Romano LA, Marins LF, Abreu PC (2013) Natural occurrence of white spot syndrome virus and infectious hypodermal and hematopoietic necrosis virus in Neohelice granulata crab. J Invertebr Pathol 114(1):86–88

Chai CY, Yoon J, Lee YS, Kim YB, Choi TJ (2013) Analysis of the complete nucleotide sequence of a white spot syndrome virus isolated from Pacific white shrimp. J Microbiol 51:695–699

Chai PC, Song XL, Chen GF, Xu H, Huang J (2016) Dietary supplementation of probiotic *Bacillus* PC465 isolated from the gut of *Fenneropenaeus chinensis* improves the health status and resistance of *Litopenaeus vannamei* against white spot syndrome virus. Fish Shellfish Immunol 54:602–611

Chaivisuthangkura P, Tangkhabuanbutra J, Longyant S, Sithigorngul W, Rukpratanporn S, Menasveta P, Sithigorngul P (2004) Monoclonal antibodies against a truncated viral envelope protein (VP28) can detect white spot syndrome virus (WSSV) infections in shrimp. Sci Asia 30:359–363

Chaivisuthangkura P, Phattanapaijitkul P, Thammapalerd N, Rukpratanporn S, Longyant S, Sithigorngul W, Sithigorngul P (2006) Development of a polyclonal antibody specific to VP19 envelope protein of white spot syndrome virus (WSSV) using a recombinant protein preparation. J Virol Methods 133(2):180–184

Chaivisuthangkura P, Longyant S, Rukpratanporn S, Srisuk C, Sridulyakul P, Sithigorngul P (2010) Enhanced white spot syndrome virus (WSSV) detection sensitivity using monoclonal antibody specific to heterologously expressed VP19 envelope protein. Aquaculture 299(1/4):15–20

Chakrabarty U, Dutta S, Mallik A, Mandal N (2014) White spot syndrome virus (WSSV) and prevalence of disease resistance in a commercially cultured population of *Penaeus monodon* Fabricius, 1798 (Decapoda, Dendrobranchiata). Crustaceana 87(14):1593–1605

Chakraborty A, Otta SK, Joseph B, Kumar S, Hossain MS, Karunasagar I, Venugopal MN, Karunasagar I (2002) Prevalence of white spot syndrome virus in wild crustaceans along the coast of India. Curr Sci 82(11):1392–1397

Chakraborty S, Ghosh U, Balasubramanian T, Das P (2014) Screening, isolation and optimization of anti–white spot syndrome virus drug derived from marine plants. Asian Pac J Trop Biomed 4:S107–S117

Chandrakala N, Mekala R (2018) Comparative study on the indicating factors and prevalence of WSSV outbreak in culture and wild *Penaeus monodon* (Fab) along the South East Coast of Tamil Nadu. Int J Appl Res 4(1):69–72

Chang PS, Lo CF, Wang YC, Kou GH (1996) Identification of white spot syndrome associated baculovirus (WSBV) target organs in the shrimp *Penaeus monodon* by in situ hybridization. Dis Aquat Org 27(2):131–139

Chang PS, Chen HC, Wand YC (1998a) Detection of white spot syndrome associated baculovirus WSBV in experimentally infected wild shrimps, crabs and lobsters by in situ hybridization. Aquaculture 164:23–242

Chang PS, Chen LJ, Wang YC (1998b) The effect of ultraviolet irradiation, heat, pH, ozone, salinity and chemical disinfectants on the infectivity of white spot syndrome baculovirus. Aquaculture 166(1–2):1–17

Chang CF, Su MS, Chen HY, Lo CF, Kou GH, Liao IC (1999) Effect of dietary β-1,3-glucan on resistance to white spot syndrome virus (WSSV) in postlarval and juvenile *Penaeus monodon*. Dis Aquat Org 36:163–168

Chang YS, Peng SE, Wang HC, Hsu HC, Ho CH, Wang CH, Wang SY, Lo CF, Kou GH (2001) Sequencing and amplified restriction fragment length polymorphism analysis of ribonucleotide reductase large subunit gene of the white spot syndrome virus in blue crab (*Callinectes sapidus*) from American coastal waters. Mar Biotechnol 3:163–171

Chang CF, Su MS, Chen HY, Liao C (2003) Dietary beta-1,3-glucan effectively improves immunity and survival of *Penaeus monodon* challenged with white spot syndrome virus. Fish Shellfish Immunol 15(4):297–310

Chang YS, Liu WJ, Chou TL, Lee YT, Lee TL, Huang WT, Kou GH, Lo CF (2008) Characterization of white spot syndrome virus envelope protein VP51A and its interaction with viral tegument protein VP26. J Virol 82(24):12555–12564

Chang YS, Liu WJ, Lee CC, Chou TL, Lee YT, Wu TS, Huang JY, Huang WT, Lee TL, Kou GH, Wang AHJ, Lo CF (2010) A 3D model of the membrane protein complex formed by the white spot syndrome virus structural proteins. PLoS One 5(5):e10718. https://doi.org/10.1371/journal.pone.0010718

Chang YS, Chen TC, Liu WJ, Hwang JS, Kou GH, Lo CF (2011) Assessment of the roles of copepod *Apocyclops royi* and bivalve mollusk *Meretrix lusoria* in white spot syndrome virus transmission. Mar Biotechnol 13:909–917

Chang YS, Liu WJ, Chen TC, Chan TY, Liu KF, Chuang JC, Kou GH, Lo CF, Wang HC (2012) Feeding hermit crabs to shrimp broodstock increases their risk of WSSV infection. Dis Aquat Org 98(3):193–199

Chapman RW, Browdy CL, Savin S, Prior S, Wenner E (2004) Sampling and evaluation of white spot syndrome virus in commercially important Atlantic penaeid shrimp stocks. Dis Aquat Org 59(3):179–185

Chauhan RP, Fogel R, Limson J (2023) Nanopore MinION sequencing generates a white spot syndrome virus genome from a pooled cloacal swab sample of domestic chickens in South Africa. Microorganisms 11(11):2802

Chen LL, Lo CF, Chiu YL, Chang CF, Kou GH (2000) Natural and experimental infection of white spot syndrome virus (WSSV) in benthic larvae of mud crab *Scylla serrata*. Dis Aquat Org 40:157–161

Chen J, Li Z, Hew CL (2007a) Characterization of a novel envelope protein WSV010 of shrimp white spot syndrome virus and its interaction with a major viral structural protein VP24. Virology 364:208–213

Chen LL, Lu LC, Wu WJ, Lo CF, Huang WP (2007b) White spot syndrome virus envelope protein VP53A interacts with *Penaeus monodon* chitin-binding protein (PmCBP). Dis Aquat Org 74(3):171–178

Chen KY, Hsu TC, Huang PY, Kang ST, Lo CF, Huang WP, Chen LL (2009) *Penaeus monodon* chitin-binding protein (PmCBP) is involved in white spot syndrome virus (WSSV) infection. Fish Shellfish Immunol 27(3):460–465

Chen YB, Zhou JF, Wan XH, Gao S (2012) Establishment of a multiplex PCR and an investigation of co-infection rate of WSSV and IHHNV in *Penaeus vannamei* in northern of Jiangsu. J Anim Vet Adv 11(2):181–185

Chen YY, Kitikiew S, Yeh ST, Chen JC (2016) White shrimp *Litopenaeus vannamei* that have received fucoidan exhibit a defense against *Vibrio alginolyticus* and WSSV despite their recovery of immune parameters to background levels. Fish Shellfish Immunol 59:414–426

Chen Y, Bi H, Li X, Zhang Z, Yue H, Weng S, He J (2017) Wsv023 interacted with *Litopenaeus vannamei* γ-tubulin complex associated proteins 2, and decreased the formation of microtubules. R Soc Open Sci 4:160379. https://doi.org/10.1098/rsos.160379

Cheng QY, Meng XL, Xu JP, Lu W, Wang J (2007) Development of lateral-flow immunoassay for WSSV with polyclonal antibodies raised against recombinant VP (19+28) fusion protein. Virol Sin 22:61–67

Chithanbaran S, David S (2014) Antiviral property and growth promoting potential of Punarnava, *Boerhaavia diffusa* in tiger prawn culture. Indian J Geo Mar Sci 43(12):2236–2243

Cho H, Park NH, Jang Y, Gwon YD, Cho Y, Heo YK, Park KH, Lee HJ, Choi TJ, Kim YB (2017) Fusion of flagellin 2 with bivalent white spot syndrome virus vaccine increases survival in freshwater shrimp. J Invertebr Pathol 144:97–105

Choi, Ran M, Kim YJ, Jang JS, Kim SK (2011) Transcriptional analysis for oral vaccination of recombinant viral proteins against white spot syndrome virus (WSSV) in *Litopenaeus vannamei*. J Microbiol Biotechnol 21(2):170–175

Choi W, Choi CW, Son DB, Jeong BC, Kim HC, Lee H, Suh JW (2020) Effects of fermented Kefir as a functional feed additive in *Litopenaeus vannamei* farming. Fermentation 6:0118. https://doi.org/10.3390/fermentation6040118

Chotigeat W, Tongsupa S, Supamataya K, Phongdara A (2004) Effect of fucoidan on disease resistance of black tiger shrimp. Aquaculture 233(1–4):23–30

Chou H, Huang C, Wang C, Chiang H, Lo CF (1995) Pathogenicity of a baculovirus infection causing white spot syndrome in cultured penaeid shrimp in Taiwan. Dis Aquat Org 23(3):165–173

Chou HY, Huang CY, Lo CF, Kou GH (1998) Studies on transmission of white spot syndrome associated baculovirus (WSBV) in *Penaeus monodon* and *P. japonicus* via waterborne contact and oral ingestion. Aquaculture 164:263–276

Chou PH, Lin YC, Teng PH, Chen CL, Lee PY (2011) Real-time target-specific detection of loop-mediated isothermal amplification for white spot syndrome virus using fluorescence energy transfer-based probes. J Virol Methods 173(1):67–74

Citarasu T, Sivaram V, Immanuel G, Rout N, Murugan V (2006) Influence of selected Indian immunostimulant herbs against white spot syndrome virus (WSSV) infection in black tiger shrimp. Fish Shellfish Immunol 21(4):372–384

Citarasu T, Michaelbabu M, Vakharia VN (2015) Influence of *Cyperus rotundus* active principles inhibit viral multiplication and stimulate immune system in Indian white shrimp *Fenneropenaeus indicus* against white spot syndrome virus infection. World Acad Sci Eng Technol Int J Anim Vet Sci 9(4):360–364

Clark KF, Greenwood SJ, Acorn AR, Byrne PJ (2013) Molecular immune response of the American lobster (*Homarus americanus*) to the white spot syndrome virus. J Invertebr Pathol 114(3):298–308

Cock J, Gitterle T, Salazar M, Rye M (2009) Breeding for disease resistance of penaeid shrimps. Aquaculture 286:1–11

Corbel V, Zuprizal, Shi Z, Huang Z, Arcier JM, Bonami JR (2001) Experimental infection of European crustaceans with white spot syndrome virus (WSSV). J Fish Dis 24:377–382

Corre V Jr, Faisan J Jr, Carton-Kawagoshi RJ, Elle BJ, Traifalgar RF, Caipang CM (2012) Evidence of WSSV transmission from the rotifer (*Brachionus plicatilis*) to the black tiger shrimp (*Penaeus monodon*) postlarvae and means to control rotifer resting eggs using industrial disinfectants. Aquac Aquar Conserv Legis 5(2):64–68

Corsin F, Turnbull JF, Hao NV, Mohan CV, Phi TT, Phuoc LH, Tinh NTN, Morgan, KL (2001) Risk factors associated with white spot syndrome virus infection in a Vietnamese riceshrimp farming system. Dis Aquat Organ 47:1–12

Crockford M (2008) White spot disease. Australia and New Zealand standard diagnostic procedures. Australian Government. https://www.agriculture.gov.au/sites/default/files/sitecollectiondocuments/animal/ahl/ANZSDP-White-spot-syndrome-virus.pdf. Accessed 25 Sept 2018

Cuéllar-Anjel J, White-Noble B, Schofield P, Chamorro R, Lightner DV (2012) Report of significant WSSV-resistance in the Pacific white shrimp, *Litopenaeus vannamei*, from a Panamanian breeding program. Aquaculture 368:36–39

Dantas MDA, Teixeira DG, Silva-Portela RCB, Soares PET, Lima JPMS, Agnez-Lima LF, Lanza DCF (2018) Direct sequencing of the white spot syndrome virus from Brazil: genome assembly and new insights on phylogeny. Virus Res 245:52–61

Dante K, Natividad T, Nomura N, Matsumura M (2008) Detection of white spot syndrome virus DNA in pond soil using a 2-step nested PCR. J Virol Methods 149(1):28–34

Das R, Karthireddy S, Gireesh-Babu P, Reddy AK, Krishna G, Chaudhari A (2010) Protection of *Penaeus monodon* from infection of white spot syndrome virus by DNA construct expressing long hairpin-RNA against ICP11 gene. Indian J Virol 21(2):95–102

Das S, Biswas G, Ghoshal TK (2016) Occurrence of pathogenic shrimp viruses in selected wild crab species of Sunderban, India. Indian J Fish 63(2):143–146

Davies R (2016) Disease has cost Asia shrimp sector over $20b. Under Current News. https://www.undercurrentnews.com/2016/09/09/disease-has-cost-asia-shrimp-sector-over20bn/

Dechamma MM, Mani MK, Rajeish M, Santhosh KS, Akhila DS, Maiti B, Bossier P, Karunasagar I, Karunasagar I (2020) Differential expression of *akirin* gene in black tiger shrimp *Penaeus monodon* in response to immunostimulant administration and infections with *Vibrio harveyi* and white spot syndrome virus. J World Aquacult Soc 51(4):1054–1065

Declarador RS, Serrano AEJ, Corre VLJ (2014) Ulvan extract acts as immunostimulant against white spot syndrome virus (WSSV) in juvenile black tiger shrimp *Penaeus monodon*. Aquac Aquar Conserv Legis 7(3):153–161

Deepika A, Sreedharan K, Paria A, Makesh M, Rajendran KV (2014) Toll-pathway in tiger shrimp (*Penaeus monodon*) responds to white spot syndrome virus infection: evidence through molecular characterisation and expression profiles of MyD88, TRAF6 and TLR genes. Fish Shellfish Immunol 41(2):441–454

Dekham K, Jones SM, Jitrakorn S, Charoonnart P, Thadtapong N, Intuy R, Dubbs P, Siripattanapipong S, Saksmerprome V, Chaturongakul S (2023) Functional and genomic characterization of a novel probiotic *Lactobacillus johnsonii* KD1 against shrimp WSSV infection. Sci Rep 13(1):21610

Dela Pena LD, Caber DB, Villacastin AJ, Cabillon NA, Arboleda J, Castellano JLA, Watanabe S (2024) Sandfish (*Holothuria scabra*) as potential reservoir of white spot syndrome virus (WSSV) when co-cultured with black tiger prawn (*Penaeus monodon*). Bull Eur Assoc Fish Pathol 44:1–14

Department of Agriculture (2013) Disease strategy: white spot disease (Version 2.0). In: Australian aquatic veterinary emergency plan (AQUAVETPLAN). Australian Government Department of Agriculture, Canberra

Desrina H, Verreth JAJ, Prayitno SB, Rombout JHWM, Vlak JM, Verdegem MCJ (2013) Replication of white spot syndrome virus (WSSV) in the polychaete *Dendronereis* spp. J Invertebr Pathol 114(1):7–10

Desrina H, Verreth JAJ, Verdegem MCJ, Vlak JM (2015) Transmission of white spot syndrome virus (WSSV) from *Dendronereis* spp. (Peters) (Nereididae) to penaeid shrimp. J Fish Dis 38(5):419–428

Desrina, Prayitno SB, Verdegem MC, Verreth JA, Vlak JM (2022) White spot syndrome virus host range and impact on transmission. Rev Aquac 14(4):1843–1860

Dewangan NK, Ayyaru G, Kuzhanthaivel R, Thirugnanasambandan S, Martin GG, Daniel K, Ramakrishna RK (2017) Incidence of simultaneous infection of infectious hypodermal and haematopoietic necrosis virus (IHHNV) and white spot syndrome virus (WSSV) in *Litopenaeus vannamei*. Aquaculture 471:1–7

Dey B, Dugassa G, Hinzano S, Bossier P (2019) Causative agent, diagnosis and management of white spot disease in shrimp: a review. Rev Aquac 12:822–865

Dhar AK, Roux MM, Klimpel KR (2001) Detection and quantification of infectious hypodermal and hematopoietic necrosis virus and white spot virus in shrimp using real-time quantitative PCR and SYBR Green chemistry. J Clin Microbiol 39(8):2835–2845

Diao MQ, Li C, Xu JD, Zhao XF, Wang JX (2019) RPS27, a sORF-encoded polypeptide, functions antivirally by activating the NF-κB pathway and interacting with viral envelope proteins in shrimp. Front Immunol 10:2763. https://doi.org/10.3389/fimmu.2019.02763

Dieu BTM, Marks H, Siebenga JJ, Goldbach RW, Zuidema D, Duong TP, Vlak JM (2004) Molecular epidemiology of white spot syndrome virus within Vietnam. J Gen Virol 85:3607–3618

Dinesh S, Karmarkar M, Vinodhini S, Vidhya G, Hemalatha K, Sudhakaranm R (2014) Confirmation of anti-WSSV activity from red algae *Hypnae spinella* in freshwater crab *Paratelphusa hydrodomous*. Int J ChemTech Res 6(8):4022–4026

Ding Z, Yao Y, Zhang F, Wan J, Sun M, Liu H, Zhou G, Tang J, Pan J, Xue H, Zhao Z (2015) The first detection of white spot syndrome virus in naturally infected cultured Chinese mitten crabs, *Eriocheir sinensis* in China. J Virol Methods 220:49–54

Dong S, Kong J, Meng X, Zhang Q, Zhang T, Wang R (2008) Microsatellite DNA markers associated with resistance to WSSV in *Penaeus* (*Fenneropenaeus*) *chinensis*. Aquaculture 282:138–141

Durand S, Lightner D (2002) Quantitative real time PCR for the measurement of white spot syndrome virus in shrimp. J Fish Dis 25:381–389

Durand S, Lightner DV, Nunan LM, Redman RM, Mari J, Bonami JR (1996) Application of gene probes as diagnostic tools for the White Spot Baculovirus (WSBV) of penaeid shrimps. Dis Aquat Org 27:59–56

Durand S, Lightner DV, Redman RM, Bonami JR (1997) Ultrastructure and morphogenesis of white spot syndrome baculovirus (WSSV). Dis Aquat Org 29:205–211

Edgerton BF (2004a) Susceptibility of the Australian freshwater crayfish *Cherax destructor albidus* to white spot syndrome virus (WSSV). Dis Aquat Org 59:187–193

Edgerton B (2004b) Studies on the susceptibility of the European white-clawed freshwater crayfish, *Austropotamobius pallipes* (Lereboullet), to white spot virus for analysis of the likelihood of introduction and impact on European freshwater crayfish populations. Freshw Crayfish 14:228–235

Elshopakey GE, Abdelwarith AA, Younis EM, Davies SJ, Elbahnaswy S (2024) Alleviating effects of *Gracilaria verrucosa* supplement on non-specific immunity, antioxidant capacity and immune-related genes of pacific white shrimp (*Litopenaeus vannamei*) provoked with white spot syndrome virus. BMC Vet Res 20(1):487

Escobedo-Bonilla CM, Alday-Sanz V, Wille M, Sorgeloos P, Pensaert MB, Nauwynck HJ (2008) A review on the morphology, molecular characterization, morphogenesis and pathogenesis of white spot syndrome virus. J Fish Dis 31(1):1–18

Escobedo-Bonilla CM, Vega-Pena S, Mejia-Ruiz CH (2015) Efficacy of double-stranded RNA against white spot syndrome virus (WSSV) non-structural (orf89, wsv191) and structural (vp28, vp26) genes in the Pacific white shrimp *Litopenaeus vannamei*. J King Saud Univ 27:182–188

FAO (2016) National aquaculture sector overview. Fisheries and Aquaculture Department, Food and Agriculture Organization of the United. Nations for a world without hunger, India. http://www.fao.org/fishery/countrysector/naso_india/en

Febrianti D, Yuhana M, Widanarni. (2016) Dietary synbiotic microcapsule influences the immune responses, growth performance and microbial populations to white spot syndrome virus in Pacific white shrimp (*Litopenaeus vannamei*). J Fish Aquat Sci 11(1):28–42

Feng S, Feng W, Zhao L, Gu H, Li Q, Shi K, Guo S, Zhang N (2014) Preparation of transgenic *Dunaliella salina* for immunization against white spot syndrome virus in crayfish. Arch Virol 159:519–525

Fierro-Coronado JA, Luna-González A, Cáceres-Martínez CJ, Álvarez-Ruiz P, Escamilla-Montes R, González-Ocampo HA, Peraza-Gómez V (2019a) Effect of microbial immunostimulants on WSSV infection percentage and the expression of immune-related genes in white shrimp (*Litopenaeus vannamei*). Rev Colomb Cienc Pecu 32(3):221–231

Fierro-Coronado JA, Luna-González A, Cáceres-Martínez CJ, Ruiz-Verdugo CA, Escamilla-Montes R, Diarte-Plata G, Flores-Miranda MDC, Álvarez-Ruiz P, Peraza-Gómez V (2019b) Effect of medicinal plants on the survival of white shrimp (*Penaeus vannamei*) challenged with WSSV and *Vibrio parahaemolyticus*. Lat Am J Aquat Res 47(2):377–381

Flegel TW (1997) Major viral diseases of the black tiger prawn (*Penaeus monodon*) in Thailand. World J Microbiol Biotechnol 13:433–442

Flegel TW (2001) The shrimp response to viral pathogens. In: Browdy CL, Jory DE (eds) The new wave. Proceedings of the special session on sustainable shrimp aquaculture. World Aquaculture Society, Boca Raton, pp 190–214

Flegel TW (2006) Detection of major penaeid shrimp viruses in Asia, a historical perspective with emphasis on Thailand. Aquaculture 258(1–4):1–33

Flegel TW (2012) Historic emergence, impact and current status of shrimp pathogens in Asia. J Invertebr Pathol 110(2):166–173
Flegel TW, Alday-Sanz V (1998) The crisis in Asian shrimp aquaculture: current status and future needs. J Appl Ichthyol 14:269–273
Flegel TW, Lightner DV, Lo CF, Owens L (2008) Shrimp disease control: past, present and future. In: Bondad-Reantaso MG, Mohan CV, Crumlish M, Subasinghe RP (eds) Diseases in Asian aquaculture VI. Fish Health Section, Asian Fisheries Society, Manila, pp 355–378, 505p
Fu LL, Li WF, Du HH, Dai W, Xu ZR (2008) Oral vaccination with envelope protein VP28 against white spot syndrome virus in *Procambarus clarkii* using *Bacillus subtilis* as delivery vehicles. Lett Appl Microbiol 46(5):581–586
Fu LL, Shuai JB, Xu ZR, Li JR, Li WF (2010) Immune responses of *Fenneropenaeus chinensis* against white spot syndrome virus after oral delivery of VP28 using *Bacillus subtilis* as vehicles. Fish Shellfish Immunol 28:49–55
Galavíz-Silva L, Molina-Garza Z, Alcocer-González JM, Rosales-Encinas JL, Ibarra-Gámez C (2004) White spot syndrome virus genetic variants detected in Mexico by a new multiplex PCR method. Aquaculture 242(1–4):53–68
Gao H, Kong J, Li Z, Xiao G, Meng X (2011) Quantitative analysis of temperature, salinity and pH on WSSV proliferation in Chinese shrimp *Fenneropenaeus chinensis* by real-time PCR. Aquaculture 312:26–31
Gao M, Li F, Xu L, Zhu X (2014) White spot syndrome virus strains of different virulence induce distinct immune response in *Cherax quadricarinatus*. Fish Shellfish Immunol 39(1):17–23
Genio JE, Traifalgar RF, Corre VJ (2015) Application frequency of dietary *Vibrio harveyi* lipopolysaccharide (LPS) on growth and white spot syndrome virus resistance of post larvae *Penaeus monodon*. Isr J Aquac 67:1–14
Ghaednia B, Mirbakhsh M, Sharifpour I, Mehrabi MR, Yeghaneh V, Shamsiyan S (2012) Dietary administration of yeast β-1,3 1,6 glucan on immunity and survival rate of white Indian shrimp, *Fennerpenaeus indicus* challenged with white spot syndrome disease. J Adv Vet Res 2:24–31
Gholamhosseini A, Kheirandish MR, Shiry N, Akhlaghi M, Soltanian S, Roshanpour H, Banaee M (2020) Use of a methanolic olive leaf extract (*Olea europaea*) against white spot virus syndrome in *Penaeus vannamei*: comparing the biochemical, hematological and immunological changes. Aquaculture 528:735556. https://doi.org/10.1016/j.aquaculture.2020.735556
Ghosh U, Chakraborty S, Balasubramanian T, Das P (2014) Screening, isolation and optimization of anti-white spot syndrome virus drug derived from terrestrial plants. Asian Pac J Trop Biomed 4(Suppl 1):S118–S128
Ghosh AK, Ahmmed SS, Islam HR, Hasan MA, Banu GR, Panda SK, Schoofs L, Luyten W (2024) Oral administration of *Zingiber officinale* and *Aegle marmelos* extracts enhances growth and immune functions of the shrimp *Penaeus monodon* against the white spot syndrome virus (WSSV). Aquac Int 32(1):613–632
Gitterle T, Diener J (2014) Genetic improvement programs for shrimp vary between Asia and America. Global Aquaculture Advocate, pp 44–46
Gitterle T, Salte R, Gjerde B, Cock J, Johansen H, Salazar M, Lozano C, Rye M (2005) Genetic (co)variation in resistance to white spot syndrome virus (WSSV) and harvest weight in *Penaeus* (*Litopenaeus*) *vannamei*. Aquaculture 246:139–149
Gong Y, Ju C, Zhang X (2018) Shrimp miR-1000 functions in antiviral immunity by simultaneously triggering the degradation of two viral mRNAs. Front Immunol 9:2999. https://doi.org/10.3389/fimmu.2018.02999
Gopalakrishnan A, Rajkumar M, Sun J, Wang M, Kumar KS (2011) Mud crab, *Scylla tranquebarica* (Decapoda: Portunidae), a new host for the white spot syndrome virus. Aquac Res 42(2):308–312
Gudkovs N, Murwantoko M, Walker PJ (2014) Stability of the WSSV ORF94 VNTR genotype marker during passage in marine shrimp, freshwater crayfish and freshwater prawns. Dis Aquat Org 111(3):249–257
Guha B, Hasan BMA, Datta S (2015) Application of PCR screened fry for increasing production of *Penaeid* shrimp in growout ponds of West Bengal, India. Int J Fish Aquat Stud 2(6):113–118

Gunalan B, Soundarapandian P, Dinakaran GK (2010) The effect of temperature and pH on WSSV infection in cultured marine shrimp *Penaeus monodon* (Fabricius). Middle-East J Sci Res 5(1):28–33

Gunasekara CWR, Kim SR, Rajapaksha LGTG, Wimalasena SHMP, Pathirana HNKS, Shin GW (2020) Disinfectant effect of monopersulfate (MPS) compound to white spot syndrome virus (WSSV) of shrimp. J Fish Pathol 33(2):139–143

Gunasekaran T, Gopalakrishnan A, Deivasigamani B, Seralathan MV, Kathirkaman P (2018) Spontaneous white spot syndrome virus (WSSV) infection in mud crab (*Scylla serrata* Forskal 1775) fattening pens farm of south east coast of India. Comp Clin Pathol 27:413–419

Guoxing Z, Yalin S, Kai Z, Zhi C (1997) Biacilliform virus infection in cultured Chinese shrimp, Penaeus orientalis, in China. J Mar Biotechnol 5(2):113–118

Ha YM, Soo-Jung G, Thi-Hoai N, Ra CH, Kim KH, Nam YK, Kim SK (2008) Vaccination of shrimp (*Penaeus chinensis*) against white spot syndrome virus (WSSV). J Microbiol Biotechnol 18:964–967

Hameed ASS, Anilkumar M, Stephen ML, Jayaraman R (1998) Studies on the pathogenicity of systemic ectodermal and mesodermal baculovirus and its detection in shrimp by immunological methods. Aquaculture 160(1–2):31–45

Hameed ASS, Charles XM, Anilkumar M (2000) Tolerance of *Macrobrachium rosenbergii* to white spot syndrome virus. Aquaculture 183:207–213

Hameed ASS, Yoganandhan K, Sathish S, Murugan V, Rasheed M, Jayaraman K (2001) Experimental pathogenicity of white spot syndrome virus (WSSV) in two freshwater crabs (*Partelphusa hydrodomous* and *P. pulvinata*). Aquaculture 201:179–186

Hameed ASS, Murthi BLM, Rasheed M, Sathish S, Yoganandhan K, Murugan V, Jayaraman K (2002) An investigation of *Artemia* as a possible vector for white spot syndrome virus (WSSV) transmission to *Penaeus indicus*. Aquaculture 204:1–10

Hameed ASS, Balasubramanian G, Musthaq SS, Yoganandhan K (2003) Experimental infection of twenty species of Indian marine crabs with white spot syndrome virus (WSSV). Dis Aquat Org 57:157–161

Hameed ASS, Parameswaran V, Musthaq SS, Sudhakaran R, Balasubramanian G, Yoganandhan K (2005) A simple PCR procedure to detect white spot syndrome virus (WSSV) of shrimp, *Penaeus monodon* (Fabricious). Aquac Int 13:441–450

Han F, Xu J, Zhang X (2007) Characterization of an early gene (wsv477) from shrimp white spot syndrome virus (WSSV). Virus Genes 34:193–198

Hao NV, Thuy DT, Loan LT, Phi TT, Phuoc LH, Duong HH, Corsin F, Chanratchakool P (1999) Presence of the two viral pathogens WSSV and MBV in three wild shrimp species (*Penaeus indicus*, *Metapenaeus ensis* and *Metepenaeus lysianassa*) cultured in the mangrove forest of Ca Mau Province. Asian Fish Sci 12(4):309–325

Haq MAB, Ali HA, Nazar AR, Ghosh U, Shalini R, Chakraborty S (2011) Assessment of *Artemia franciscana* as a probable vector for WSSV transmission to *Macrobrachium idella idella* (Hilgendorf, 1898). Int J Chem Anal Sci 2(9):1159–1170

Haq MAB, Kavitha N, Vignesh R, Shalini R (2012a) Identification and sequence based detection of WSSV infecting SPF *Litopeneaus vannamei* (Boone, 1931) in captivity. Int J Pharm Bio Sci 3(1):547–559

Haq MAB, Priya KK, Rajaram R, Vignesh R, Srinivasan M (2012b) Real time PCR quantification of WSSV infection in specific pathogen free (SPF) *Litopenaeus vannamei* (Boone, 1931) exposed to antiviral nucleotide. Asian Pac J Trop Biomed 2(2):S1120–S1129

Haq MAB, Durgadevi K, Banu MN, Ahamed S, Tiwary C, Srinivasan M (2015) Detection of white spot syndrome virus (WSSV) in the pacific white shrimp *Litopenaeus vannamei* in southern India using PCR, SEM and histological techniques. Indian J Biotechnol 14:369–375

Hastuti MS, Haryadi D (2016) Current status of acute hepatopancreatic necrosis disease (AHPND) and othertransboundary diseases of farmed shrimps in Indonesia. In: Pakingking RV Jr, de Jesus-Ayson EGT, Acosta BO (eds) Addressing acute hepatopancreatic necrosisdisease (AHPND) and other transboundary diseases for improved aquatic animal health in Southeast Asia, Proceedings of the ASEAN Regional Technical Consultation on EMS/AHPND and other transboundary diseases for improved aquatic animal health in Southeast Asia, 22–24 February 2016, Makati City, Philippines. Southeast Asian Fisheries Development Center-Aquaculture Department, Tigbauan, Iloilo, pp 37–43

He L, Xu HS (2011) Development of a multiplex loop-mediated isothermal amplification (mLAMP) method for the simultaneous detection of white spot syndrome virus and infectious hypodermal and hematopoietic necrosis virus in penaeid shrimp. Aquaculture 311:94–99

He F, Fenner BJ, Godwin AK, Kwang J (2006) White spot syndrome virus open reading frame 222 encodes a viral E3 ligase and mediates degradation of a host tumor suppressor via ubiquitination. J Virol 80(8):3884–3892

Heidarieh M, Sedeh FM, Soltani M, Rajabifar S, Afsharnasab M, Dashtiannasab A (2014) White spot syndrome virus inactivation study by using gamma irradiation. Chin J Oceanol Limnol 32(5):1024–1028

Hien NT, Huong NT, Chuong VD, Nga NTV, Quang PH, Hang BTV, Long NV (2016) Status of acute hepatopancreatic necrosis disease (AHPND) and other emerging diseases of penaeid shrimps in Viet Nam. In: Pakingking RV Jr, de Jesus-Ayson EGT, Acosta BO (eds) Addressing acute hepatopancreatic necrosis disease (AHPND) and other transboundary diseases for improved aquatic animal health in Southeast Asia, Proceedings of the ASEAN Regional Technical Consultation on EMS/AHPND and other transboundary diseases for improved aquatic animal health in Southeast Asia, 22–24 February 2016, Makati City, Philippines. Southeast Asian Fisheries Development Center-Aquaculture Department, Tigbauan, Iloilo, pp 88–95

Hoa TT, Hodgson RA, Oanh DT, Phuong NT, Preston NJ, Walker PJ (2005) Genotypic variations in tandem repeat DNA segments between ribonucleotide reductase subunit genes of white spot syndrome virus (WSSV) isolates from Vietnam. In: Diseases in Asian aquaculture V. Fish Health Section, Asian Fisheries Society, Manila, pp 339–351

Hong SJ, Kim KH (2024) RNA interference targeting WSSV ribonucleotide reductase 2 provides long-term protection against infection in *Litopenaeus vannamei*. Dis Aquat Org 159:71–78

Hossain MS, Otta SK, Karunasagar I, Karunasagar I (2001a) Detection of white spot syndrome virus (WSSV) in wild captured shrimp and in non-cultured crustaceans from shrimp ponds in Bangladesh by polymerase chain reaction. Fish Pathol 36(2):93–95

Hossain MS, Chakraborty A, Joseph B, Otta SK, Karunasagar I, Karunasagar I (2001b) Detection of new hosts for white spot syndrome virus of shrimp using nested polymerase chain reaction. Aquaculture 198(1–2):1–11

Hossain MS, Khadijah S, Kwang J (2004a) Characterization of ORF89—a latency-related gene of white spot syndrome virus. Virology 325(1):106–115

Hossain M, Otta SK, Chakraborty A, Kumar S, Karunasagar I, Karunasagar I (2004b) Detection of WSSV in cultured shrimps, captured brooders, shrimp postlarvae and water samples in Bangladesh by PCR using different primers. Aquaculture 237:59–71

Hossain A, Sanyal SK, Siddique M, Biswas R, Sultana M, Hossain M (2014) Envelope protein gene VP466-a target for PCR detection of white spot syndrome virus in shrimp. Bangladesh J Microbiol 31:65–68

Hou CL, Cao Y, Xie RH, Wang YZ, Du HH (2011) Characterization and diagnostic use of a monoclonal antibody for VP28 envelope protein of white spot syndrome virus. Virol Sin 26(4):260–266

Howlader P, Ghosh AK, Islam SS, Bir J, Banu GR (2020) Antiviral activity of *Cynodon dactylon* on white spot syndrome virus (WSSV)-infected shrimp: an attempt to mitigate risk in shrimp farming. Aquac Int 28:1725–1738

Hu K, Li Y, Wang F, Liu J, Li Y, Zhao Q, Zheng X, Zhu N, Yu X, Fang S, Huang J (2023) A loop-mediated isothermal amplification-based microfluidic chip for triplex detection of shrimp pathogens. J Fish Dis 46(2):137–146

Huang T, Zhang X (2012) Functional analysis of a crustacean microRNA in host-virus interactions. J Virol 86(23):12997–13004

Huang CH, Zhang LR, Zhang JH, Xiao LC, Wu QJ, Chen DH, Li JKK (2001) Purification and characterization of white spot syndrome virus (WSSV) produced in an alternate host: crayfish, *Cambarus clarkii*. Virus Res 76(2):115–125

Huang XD, Zhao L, Zhang HQ, Xu XP, Jia XT, Chen YH, Wang PH, Weng SP, Yu XQ, Yin ZX, He JG (2010) Shrimp NF-κB binds to the immediate-early gene *ie1* promoter of white spot syndrome virus and upregulates its activity. Virology 406(2):176–180

Huang HT, Leu JH, Huang PY, Chen LL (2012) A putative cell surface receptor for white spot syndrome virus is a member of a transporter superfamily. PLoS One 7:e33216. https://doi.org/10.1371/journal.pone.0033216

Huang PH, Lu SC, Yang SH, Cai PS, Lo CF, Chang LK (2014a) Regulation of the immediate-early genes of white spot syndrome virus by *Litopenaeus vannamei* kruppel-like factor (LvKLF). Dev Comp Immunol 46(2):364–372

Huang PY, Leu JH, Chen LL (2014b) A newly identified protein complex that mediates white spot syndrome virus infection via chitin-binding protein. J Gen Virol 95:1799–1808

Huang T, Cui Y, Zhang X (2014c) Involvement of viral microRNA in the regulation of antiviral apoptosis in shrimp. J Virol 88(5):2544–2554

Huang JY, Kang ST, Chen IT, Chang LK, Lin SS, Kou GH, Chu CY, Lo CF (2017a) Shrimp miR-10a is co-opted by white spot syndrome virus to increase viral gene expression and viral replication. Front Immunol 8:1084. https://doi.org/10.3389/fimmu.2017.01084

Huang PH, Huang TY, Cai PS, Chang LK (2017b) Role of *Litopenaeus vannamei* Yin Yang 1 in the regulation of the white spot syndrome virus immediate early gene *ie1*. J Virol 91(6):e02314–e02316

Huang HT, Lin HJ, Huang HJ, Huang CC, Lin JHY, Chen LL (2020a) Synthesis and evaluation of polyamine carbon quantum dots (CQDs) in *Litopenaeus vannamei* as a therapeutic agent against WSSV. Sci Rep 10:7343. https://doi.org/10.1038/s41598-020-64325-5

Huang JY, Wang HC, Chen YC, Wang PS, Lin SJ, Chang YS, Liu KF, Lo CF (2020b) A shrimp glycosylase protein, PmENGase, interacts with WSSV envelope protein VP41B and is involved in WSSV pathogenesis. Dev Comp Immunol 108:103667. https://doi.org/10.1016/j.dci.2020.103667

Huynh TG, Yeh ST, Lin YC, Shyu JF, Chen LL, Chen JC (2011) White shrimp *Litopenaeus vannamei* immersed in seawater containing *Sargassum hemiphyllum* var. Chinense powder and its extract showed increased immunity and resistance against *Vibrio alginolyticus* and white spot syndrome virus. Fish Shellfish Immunol 31:286–296

Immanuel G, Sivagnanavelmurugan M, Balasubramanian V, Palavesam A (2010) Effect of hot water extracts of brown seaweeds Sargassum spp. on growth and resistance to white spot syndrome virus in shrimp Penaeus monodon postlarvae. Aquaculture research 41(10):e545–e553

Immanuel G, Sivagnanavelmurugan M, Marudhupandi T, Radhakrishnan S, Palavesam A (2012a) The effect of fucoidan from brown seaweed *Sargassum wightii* on WSSV resistance and immune activity in shrimp *Penaeus monodon* (Fab). Fish Shellfish Immunol 32(4):551–564

Immanuel G, Sivagnanavelmurugan M, Balasubramanian V, Palavesam A (2012b) Sodium alginate from *Sargassum wightii* retards mortalities in *Penaeus monodon* postlarvae challenged with white spot syndrome virus. Dis Aquat Org 99(3):187–196

Inchara UB, Sathish RP, Shankar KM, Abhiman PB, Prakash P (2018) Evaluation of the sensitivity of the flow through assay for detection of white spot syndrome virus (WSSV) using a cocktail of monoclonal antibodies. J Immunol Methods 456:54–60

Inouye K, Miwa S, Oseko N, Nakano H, Kimura T, Momoyama K, Hiraoka M (1994) Mass mortalities of cultured kuruma shrimp *Penaeus japonicus* in Japan in 1993: electron microscopic evidence of the causative virus. Fish Pathol 29(2):149–158

Jang IK, Meng XH, Seo HC, Cho YR, Kim BR, Ayyaru G, Kim JS (2009) A TaqMan real-time PCR assay for quantifying white spot syndrome virus (WSSV) infections in wild broodstock and hatchery-reared postlarvae of fleshy shrimp, *Fenneropenaeus chinensis*. Aquaculture 287(1–2):40–45

Jaree P, Wongdontri C, Somboonwiwat K (2018) White spot syndrome virus -induced shrimp miR-315 attenuates prophenoloxidase activation via *PPAE3* gene suppression. Front Immunol 9:2184. https://doi.org/10.3389/fimmu.2018.02184

Jariyapong P, Chotwiwatthanakun C, Somrit M, Jitrapakdee S, Xing L, Cheng HR, Weerachatyanukul W (2014) Encapsulation and delivery of plasmid DNA by virus-like nanoparticles engineered from *Macrobrachium rosenbergii* nodavirus. Virus Res 179:140–146

Jaroenram W, Kiatpathomchai W, Flegel TW (2009) Rapid and sensitive detection of white spot syndrome virus by loop-mediated isothermal amplification combined with a lateral flow dip-stick. Mol Cell Probes 23(2):65–70

Jayabarath J, Uma MR, Maanusha J, Bhavanisowndharya B (2017) Detection of WSS virus on Banana prawn, *Peneaus mergueinsis* by PCR technique. J Chem Pharm Sci 10(4):1636–1638

Jeeva S, Kim NI, Jang IK, Choi TJ (2014) Development of a multiplex PCR system for the simultaneous detection of white spot syndrome virus and hepatopancreatic parvovirus infection. Aquac Res 45(6):1073–1083

Jenifer A, Remya R, Velmurugan S, Michaelbabu M, Citarasu T (2018) *Streptomyces castaneoglobisporus* AJ9, A haloalkaliphilic actinomycetes isolated from solar salt works in southern India and its pharmacological properties. Indian J Geo Mar Sci 47(2):475–488

Jenifer JSCA, Michaelbabu M, Lelin C, Thirumalaikumar E, Nisha SRJ, Uma G, Citarasu T (2019) Antimicrobial potential of haloalkaliphilic *Nocardiopsis* sp. AJ1 isolated from solar salterns in India. J Basic Microbiol 59(3):288–301

Jha RK, Xu ZR, Shen J, Bai SJ, Sun JY, Li WF (2006) The efficacy of recombinant vaccines against white spot syndrome virus in *Procambarus clarkii*. Immunol Lett 105:68–76

Jha RK, Xu ZR, Bai SJ, Sun JY, Li WF, Shen J (2007) Protection of *Procambarus clarkii* against white spot syndrome virus using recombinant oral vaccine expressed in *Pichia pastoris*. Fish Shellfish Immunol 22(4):295–307

Jha RK, Babikian YH, Babikian HY, Wisoyo SD, Asih Y, Srisombat S, Jiaravanon B (2016) Effectiveness of natural herbal oil formulation against white spot syndrome virus in *Penaeus vannamei*. J Pharmacogn Nat Prod 2:4. https://doi.org/10.4172/2472-0992.1000123

Jha RK, Wisoyo SD, Kristina, Srisombat S (2020) *Cynodon dactylon* as potential anti-white spot syndrome virus agent in cultured shrimp *Penaeus vannamei*. Int J Fish Aquat Stud 8(6):192–199

Jia XH, Zhang CL, Shi DJ, Zhuang MM, Wang X, Jia R, Zhang ZY, Huang J, Sun YH, Qian WY, Peng GH, He PM (2016) Oral administration of *Anabaena*-expressed VP28 for both drug and food against white spot syndrome virus in shrimp. J Appl Phycol 28:1001–1009

Jiang G (2012) Can white spot syndrome virus be transmitted through the phytoplankton → rotifer → artemia → shrimp pathway? Afr J Biotechnol 11(5):1277–1282

Jiang N, Pan X, Gu Z, Liu W, Si K, Zhou Y, Zhou Y, Zhai L, Fan Y, Zeng L (2019) Proliferation dynamics of WSSV in crayfish, *Procambarus clarkii*, and the host responses at different temperatures. J Fish Dis 42:497–510

Jin W, Lai Y, Zhu F (2021) Effect of dietary fucoidan on innate immune response of *Procambarus clarkii* and disease resistance against white spot syndrome virus. Aquaculture 534:736233. https://doi.org/10.1016/j.aquaculture.2020.736233

Jiravanichpaisal P, Bangyeekhun E, Söderhäll K, Söderhäll I (2001) Experimental infection of white spot syndrome virus in freshwater crayfish *Pacifastacus leniusculus*. Dis Aquat Org 47(2):151–157

Jiravanichpaisal P, Söderhäll K, Söderhäll I (2004) Effect of water temperature on the immune response and infectivity pattern of white spot syndrome virus (WSSV) in freshwater crayfish. Fish Shellfish Immunol 17(3):265–275

Jiravanichpaisal P, Soderhall K, Soderhall I (2006) Characterization of white spot syndrome virus replication in in vitro-cultured haematopoietic stem cells of freshwater crayfish, *Pacifastacus leniusculus*. J Gen Virol 87:847–854

John KR, George MR, Iyappan T, Thangarani AJ, Jeyaseelan MJP (2010) Indian isolates of white spot syndrome virus exhibit variations in their pathogenicity and genomic tandem repeats. J Fish Dis 33(9):749–758

Joseph TC, James R, Rajan LA, Surendran PK, Lalitha KV (2015a) Occurrence of viral pathogens in *Penaeus monodon* post-larvae from aquaculture hatcheries. Data Brief 4:170–176

Joseph TC, James R, Rajan LA, Surendran PK, Lalitha KV (2015b) White spot syndrome virus infection: threat to crustacean biodiversity in Vembanad Lake, India. Biotechnol Rep 7:51–54

Junior OT, Kuhn F, Padilha PJM, Vicente LRM, Costa SW, Silva BC, Schleder DD, Boligon AA, Scapinello J, Nesi CN, Magro DJ, Castellví SL (2017) Effect of *Cynodon dactylon* extract on white spot virus-infected *Litopenaeus vannamei*. Aquac Int 25:1107–1122

Kakoolaki S, Afsharnasab M, Sharifpour I (2015) The relation between temperature and salinity with WSSV occurrence in shrimp farms in Iran: an article review. J Surv Fish Sci 2:31–41

Kalaimani N, Ravisankar T, Chakravarthy N, Raja S, Santiago TC, Ponniah AG (2013) Economic losses due to disease incidences in shrimp farms of India. Fish Technol 50:80–86

Kanchanaphum P, Wongteerasupaya C, Sitidilokratana N, Boonsaeng V, Panyim S, Tassanakajon A, Withyachumnarnkul B, Flegel TW (1998) Experimental transmission of white spot syndrome virus (WSSV) from crabs to shrimp *Penaeus monodon*. Dis Aquat Org 34(1):1–7

Karthik R, Pushpam AC, Vanitha MCN (2015) Development of marine derived probiotic bacterial consortium for the sustainable management of *Litopenaeus vannamei* culture. Int J Adv Res Eng Technol 6(10):62–75

Karunasagar I, Otta SK, Karunasagar I (1997) Histopathological and bacteriological study of white spot syndrome of *Penaeus monodon* along the west coast of India. Aquaculture 153(1–2):9–13

Kasornchandra J, Boonyaratpalin S, Itami T (1998) Detection of white-spot syndrome in cultured penaeid shrimp in Asia: microscopic observation and polymerase chain reaction. Aquaculture 164(1–4):243–251

Kathyayani SA, Poornima M, Sukumaran S, Nagavel A, Muralidhar M (2019) Effect of ammonia stress on immune variables of Pacific white shrimp *Penaeus vannamei* under varying levels of pH and susceptibility to white spot syndrome virus. Ecotoxicol Environ Saf 184:109626. https://doi.org/10.1016/j.ecoenv.2019.109626

Kawato S, Fujishima S, Nozaki R, Kondo H, Isshiki T, Hirono I (2023) Genome sequence of Chionoecetes opilio bacilliform virus, a nimavirus infecting the snow crab *Chionoecetes opilio*. Arch Virol 168:101. https://doi.org/10.1007/s00705-023-05731-8

Khadijah S, Neo S, Hossain M, Miller L, Mathavan S, Kwang J (2003) Identification of white spot syndrome virus latency-related genes in specific-pathogen-free shrimps by use of a microarray. J Virol 77:10162–10167

Khawsak P, Deesukon W, Chaivisuthangkura P, Sukhumsirichart W (2008) Multiplex RT-PCR assay for simultaneous detection of six viruses of penaeid shrimp. Mol Cell Probes 22(3):177–183

Kiataramgul A, Maneenin S, Purton S, Areechon N, Hirono I, Brocklehurst TW, Unajak S (2020) An oral delivery system for controlling white spot syndrome virus infection in shrimp using transgenic microalgae. Aquaculture 521:735022. https://doi.org/10.1016/j.aquaculture.2020.735022

Kiatpathomchai W, Boonsaeng V, Tassanakajon A, Wongteerasupaya C, Jitrapakdee S, Panyim S (2001) A nonstop, single tube, semi-nested PCR technique for grading the severity of white spot syndrome virus infections in *Penaeus monodon*. Dis Aquat Org 47:235–239

Kim CS, Kosuke Z, Nam YK, Kim SK, Kim KH (2007) Protection of shrimp (*Penaeus chinensis*) against white spot syndrome virus (WSSV) challenge by double-stranded RNA. Fish Shellfish Immunol 23:242–246

Kim HJ, Shyam KU, Oh MJ, Lee JH, Rajendran KV, Kim DH, Kim HJ, Kim WS (2023) A multiplex real-time polymerase chain reaction (qPCR) kit targeting VP664 and VP28 genes of white spot syndrome virus (WSSV). Aquaculture 577:739968

Kiran RBP, Rajendran KV, Jung SJ, Oh MJ (2002) Experimental susceptibility of different life-stages of the giant freshwater prawn, *Macrobrachium rosenbergii* (de Man), to white spot syndrome virus (WSSV). J Fish Dis 25(4):201–207

Kiruthiga C, Rajesh S, Rashika V, Priya R, Narayanan RB (2012) Molecular cloning, expression analysis and characterization of peroxiredoxin during WSSV infection in shrimp *Fenneropenaeus indicus*. J Invertebr Pathol 109(1):52–58

Klongklaew N, Praiboon J, Tamtin M, Srisapoome P (2020) Antibacterial and antiviral activities of local Thai green macroalgae crude extracts in Pacific white shrimp (*Litopenaeus vannamei*). Mar Drugs 18:140. https://doi.org/10.3390/md18030140

Kong J, Zhang Q, Liu P, Yue Z, Wang Q (2003) Selective breeding raises WSSV resistance in fleshy prawns. Global Aquaculture Advocate, pp 25–26

Kono T, Savan R, Sakai M, Itami T (2004) Detection of white spot syndrome virus in shrimp by loop-mediated isothermal amplification. J Virol Methods 115(1):59–65

Kono T, Sonoda K, Kitao Y, Mekata T, Itami T, Sakai M (2009) The expression analysis of innate immune-related genes in kuruma shrimp *Penaeus japonicus* after DNA vaccination against penaeid rod-shaped DNA virus. Fish Pathol 44:94–97

Kono T, Fall J, Korenaga H, Sudhakaran R, Biswas G, Mekata T, Itami T, Sakai M (2014) Recombinant VP28 produced by cell-free technique confers protection in kuruma shrimp (*Marsupenaeus japonicus*) against white spot syndrome virus. Turk J Fish Aquat Sci 14:547–555

Kono T, Biswas G, Fall J, Mekata T, Hikima J, Itami T, Sakai M (2015) Adjuvant effects of poly I:C and imiquimod on the immunization of kuruma shrimp (*Marsupenaeus japonicus*) with a recombinant protein, VP28 against white spot syndrome virus. Aquaculture 446:236–241

Kou GH, Peng SE, Chiu YL, Lo CF (1998) Tissue distribution of white spot syndrome virus (WSSV) in shrimp and crabs. In: Flegel TW (ed) Advances in shrimp biotechnology. National Center for Genetic Engineering and Biotechnology, Bangkok, pp 267–271

Krishnan P, Babu PG, Saravanan S, Rajendran KV, Chaudhari A (2009) DNA constructs expressing long-hairpin RNA (lhRNA) protect *Penaeus monodon* against white spot syndrome virus. Vaccine 27:3849–3855

Krishnan K, Katneni VK, Prabhudas SK, Kaikkolante N, Jangam AK, Katneni UK, Hauton C, Peruzza L, Mudagandur SS, Koyadan VK, Poochirian JK (2023) MRF: a tool to overcome the barrier of inconsistent genome annotations and perform comparative genomics studies for the largest animal DNA virus. Virol J 20(1):72

Kulabhusan PK, Rajwade JM, Sugumar V, Taju G, Hameed ASS, Paknikar KM (2017a) Field-usable lateral flow immunoassay for the rapid detection of white spot syndrome virus (WSSV). PLoS One 12(1):e0169012. https://doi.org/10.1371/journal.pone.0169012

Kulabhusan PK, Rajwade JM, Hameed ASS, Paknikar KM (2017b) Lateral flow assay for rapid detection of white spot syndrome virus (WSSV) using a phage-displayed peptide as bio-recognition probe. Appl Microbiol Biotechnol 101(11):4459–4469

Kulkarni AD, Rombout JHWM, Singh ISB, Sudheer NS, Vlak JM, Caipang CMA, Brinchmann MF, Kiron V (2013) Truncated VP28 as oral vaccine candidate against WSSV infection in shrimp: an uptake and processing study in the midgut of *Penaeus monodon*. Fish Shellfish Immunol 34(1):159–166

Kulkarni AD, Caipang CMA, Kiron V, Rombout JHWM, Fernandes JMO, Brinchmann MF (2014a) Evaluation of immune and apoptosis related gene responses using an RNAi approach in vaccinated *Penaeus monodon* during oral WSSV infection. Mar Genomics 18 Pt A:55–65

Kulkarni AD, Kiron V, Rombout JHWM, Brinchmann MF, Fernandes JMO, Sudheer NS, Singh ISB (2014b) Protein profiling in the gut of *Penaeus monodon* gavaged with oral WSSV-vaccines and live white spot syndrome virus. Proteomics 14(13–14):1660–1673

Kumar SS, Philip R, Achuthankutty CT (2006) Antiviral property of marine actinomycetes against white spot syndrome virus in penaeid shrimps. Curr Sci 91(6):807–811

Kumar RS, Ahamed I, Sarathi A, Basha NA, Hameed ASS (2008) Immunological responses of *Penaeus monodon* to DNA vaccine and its efficacy to protect shrimp against white spot syndrome virus (WSSV). Fish Shellfish Immunol 24:467–478

Kumar SS, Bharathi RA, Rajan JJS, Alavandi SV, Poornima M, Balasubramanian CP, Ponniah AG (2013) Viability of white spot syndrome virus (WSSV) in sediment during sun-drying (drainable pond) and under non-drainable pond conditions indicated by infectivity to shrimp. Aquaculture 402–403:119–126

Kumar A, Laramore S, Alexander P, Allnutt FCT, Sayre RT (2015) Double stranded RNA simultaneously targeting four white spot syndrome virus (WSSV) genes provides protection against WSSV in *Litopenaeus vannamei*. Int J Mar Sci Technol 22(2):5–10

Kumar KV, Shekhar MS, Otta SK, Karthic K, Kumar A, Gopikrishna G, Vijayan KK (2018) First report of a complete genome sequence of white spot syndrome virus from India. Genome Announc 6(8):e00055-18. https://doi.org/10.1128/genomeA.00055-18

Kumar SS, Bharathi RA, Rajan JJS, Chitra V, Muralidhar M, Alavandi SV (2019) Viability of white spot syndrome virus (WSSV) in shrimp pond sediments with reference to physicochemical properties. Aquac Int 27:1369–1382

Kumar TS, Suvetha S, Rajeshwar BN, Makesh M, Shekhar MS, Lal KK (2025) Portable DNA extraction and loop-mediated isothermal amplification (LAMP) for the on-site detection of white spot syndrome virus (WSSV). Aquaculture 596:741760

Kumaran T, Babu MM, Selvaraj T, Dhas SA, Citarasu T (2010) Production of anti WSSV IgY edible antibody using herbal immunoadjuvant *Asparagus racemosus* and its immunological influence against WSSV infection in *Penaeus monodon*. J Aquac Feed Sci Nutr 2(1):1–5

Kumaran T, Viji VT, Velmurugan S, Citarasu T (2014) Influence of selected antiviral herbal active principles against shrimp white spot syndrome virus (WSSV). Int J Mol Biol Biochem 2(1):41–49

Kwankaew P, Praparatana R, Runsaeng P, Utarabhand P (2018) An alternative function of C-type lectin comprising low-density lipoprotein receptor domain from *Fenneropenaeus merguiensis* to act as a binding receptor for viral protein and vitellogenin. Fish Shellfish Immunol 74:295–308

Lai Y, Luo M, Zhu F (2020) Dietary *Bacillus amyloliquefaciens* enhance survival of white spot syndrome virus infected crayfish. Fish Shellfish Immunol 102:161–168

Lan JF, Li XC, Sun JJ, Gong J, Wang XW, Shi XZ, Shi LJ, Weng YD, Zhao XF, Wang JX (2013) Prohibitin interacts with envelope proteins of white spot syndrome virus and prevents infection in the red swamp crayfish, *Procambarus clarkia*. J Virol 87(23):12756–12765

Leal CAG, Carvalho AF, Leite RC, Figueiredo HCP (2014) Development of duplex real-time PCR for the detection of WSSV and PstDV1 in cultivated shrimp. BMC Vet Res 10:150. https://doi.org/10.1186/1746-6148-10-150

Lee YJ, Chen LL (2017) WSSV envelope protein VP51B links structural protein complexes and may mediate virus infection. J Fish Dis 40(4):571–581

Lee C, Kim JH, Choi SK, Jeon HJ, Lee SH, Kim BK, Kim YK, Lee KJ, Han JE (2021) Detection of infectious white spot syndrome virus in red claw crayfish (*Cherax quadricarinatus*) and red swamp crayfish (*Procambarus clarkii*) imported into Korea. Aquaculture 544:737117

Lei H, Li S, Lu X, Ren Y (2021) Oral administration of *Saccharomyces cerevisiae* displaying VP28-VP24 confers protection against white spot syndrome virus in shrimp. Virus Res 302:198467. https://doi.org/10.1016/j.virusres.2021.198467

Leutwimol T, Sangsuriya P, Phiwsaiya K, Senapin S, Phongdara A, Boonchird C, Flegel TW (2014) Two new anti-apoptotic proteins of white spot syndrome virus that bind to an effector caspase (PmCasp) of the giant tiger shrimp *Penaeus* (*Penaeus*) *monodon*. Fish Shellfish Immunol 38(1):1–6

Leu JH, Yang F, Zhang X, Xu X, Kou GH, Lo CF (2009) Whispovirus. In: Van Etten JL (ed) Lesser known large dsDNA viruses. Current topics in microbiology and immunology, vol 328. Springer, Berlin, Heidelberg. https://doi.org/10.1007/978-3-540-68618-7_6

Leu JH, Chen LL, Lin YR, Kou GH, Lo CF (2010) Molecular mechanism of the interactions between white spot syndrome virus anti-apoptosis protein AAP-1 (WSSV449) and shrimp effector caspase. Dev Comp Immunol 34(10):1068–1074

Leyva-Madrigal KY, Luna-González A, Escobedo-Bonilla CM, Fierro-Coronado JA, Maldonado-Mendoza IE (2011) Screening for potential probiotic bacteria to reduce prevalence of WSSV and IHHNV in whiteleg shrimp (*Litopenaeus vannamei*) under experimental conditions. Aquaculture 322–323:16–22

Li Q, Zhang J, Chen Y, Yang F (2003) White spot syndrome virus (WSSV) infectivity for *Artemia* at different developmental stages. Dis Aquat Org 57:261–264

Li D, Zhang MC, Yang H, Zhu YB, Xu X (2007a) Beta-integrin mediates WSSV infection. Virology 368:122–132

Li SZ, Lin Q, Chen J, Wu JL, Lim TK, Loh SS, Tang X, Hew CL (2007b) Shotgun identification of the structural proteome of shrimp white spot syndrome virus and iTRAQ differentiation of envelope and nucleocapsid subproteomes. Mol Cell Proteomics 6(9):1609–1620

Li F, Li M, Ke W, Ji Y, Bian X, Yan X (2009a) Identification of the immediate-early genes of white spot syndrome virus. Virology 385(1):267–274

Li JQ, Tan BP, Mai KS (2009b) Dietary probiotic *Bacillus* OJ and isomaltooligosaccharides influence the intestine microbial populations, immune responses and resistance to white spot syndrome virus in shrimp (*Litopenaeus vannamei*). Aquaculture 291:35–40

Li X, Liu QH, Hou L, Huang J (2010) Effect of VP28 DNA vaccine on white spot syndrome virus in *Litopenaeus vannamei*. Aquac Int 18:1035–1044

Li Q, Liu QH, Huang J (2012) VP90 of white spot syndrome virus interacts with VP26 and VP28. Acta Virol 56(1):57–62

Li Q, Liu QH, Huang J (2013) F0ATP synthase b-chain of *Litopenaeus vannamei* involved in white spot syndrome virus infection. Virus Genes 47(1):42–48

Li W, Desmarets LM, De Gryse GM, Theuns S, Van Tuan V, Van Thuong K, Bossier P, Nauwynck HJ (2015) Virus replication cycle of white spot syndrome virus in secondary cell cultures from the lymphoid organ of *Litopenaeus vannamei*. J Gen Virol 96:2844–2854

Li Z, Li F, Han Y, Xu L, Yang F (2016) VP24 is a chitin-binding protein involved in white spot syndrome virus infection. J Virol 90(2):842–850

Li F, Gao M, Xu L, Yang F (2017) Comparative genomic analysis of three white spot syndrome virus isolates of different virulence. Virus Genes 53(2):249–258

Li H, Yin B, Wang S, Fu Q, Xiao B, Lǚ K, He J, Li C (2018) RNAi screening identifies a new Toll from shrimp *Litopenaeus vannamei* that restricts WSSV infection through activating Dorsal to induce antimicrobial peptides. PLoS Pathog 14(9):e1007109. https://doi.org/10.1371/journal.ppat.1007109

Li L, Hong Y, Qiu H, Yang F, Li F (2019) VP19 is important for the envelope coating of white spot syndrome virus. Virus Res 270:197666. https://doi.org/10.1016/j.virusres.2019.197666

Liang T, Ji H, Lian L, Wu T, Gu W, Wang W (2010) A rapid assay for simultaneous detection of *Spiroplasma eriocheiris* and white spot syndrome virus in *Procambarus clarkii* by multiplex PCR. Lett Appl Microbiol 51:532–538

Lightner DV (1996) A handbook of pathology and diagnostic procedures for diseases of penaeid shrimp. Special publication of the World Aquaculture Society, Baton Rouge

Lightner DV (2003) The penaeid shrimp viral pandemics due to IHHNV, WSSV, TSV and YHV: history in the Americas and current status. In: Proceedings of the 32nd joint UJNR aquaculture panel symposium, Davis and Santa Barbara, CA, USA, pp 17–20

Lightner DV, Redman RM (1998) Shrimp diseases and current diagnostic methods. Aquaculture 164(1–4):201–220

Lightner DV, Redman RM, Pantoja CR, Tang KFJ, Noble BL, Schofield P, Mohney LL, Nunan LM, Navarro SA (2012) Historic emergence, impact and current status of shrimp pathogens in the Americas. J Invertebr Pathol 110:174–183

Lillehammer M, Bangera R, Salazar M, Vela S, Erazo EC, Suarez A, Cock J, Rye M, Robinson NA (2020) Genomic selection for white spot syndrome virus resistance in whiteleg shrimp boosts survival under an experimental challenge test. Sci Rep 10:20571. https://doi.org/10.1038/s41598-020-77580-3

Lin F, Huang H, Xu L, Li F, Yang F (2011a) Identification of three immediate-early genes of white spot syndrome virus. Arch Virol 156:1611–1614

Lin YC, Yeh ST, Li CC, Chen LL, Cheng A, Chen J (2011b) An immersion of *Gracilaria tenuistipitata* extract improves the immunity and survival of white shrimp *Litopenaeus vannamei* challenged with white spot syndrome virus. Fish Shellfish Immunol 31:1239–1246

Lin YC, Chen JC, Li CC, Morni WZ, Suhaili ASN, Kuo YH, Chang YH, Chen LL, Tsui WC, Chen YY, Huang CL (2012) Modulation of the innate immune system in white shrimp *Litopenaeus vannamei* following long-term low salinity exposure. Fish Shellfish Immunol 33:324–331

Lin SJ, Lee DY, Wang HC, Kang ST, Hwang PP, Kou GH, Huang MF, Chang GD, Lo CF (2015) White spot syndrome virus protein kinase 1 defeats the host cell's iron-withholding defense mechanism by interacting with host ferritin. J Virol 89(2):1083–1093

Liu W, Wang YT, Tian DS, Yin ZC, Kwang J (2002) Detection of white spot syndrome virus (WSSV) of shrimp by means of monoclonal antibodies (MAbs) specific to an envelope protein (28 kDa). Dis Aquat Org 49(1):11–18

Liu WJ, Chang YS, Wang CH, Kou GH, Lo CF (2005) Microarray and RT-PCR screening for white spot syndrome virus immediate-early genes in cycloheximide-treated shrimp. Virology 334:327–341

Liu B, Yu Z, Song X, Guan Y (2007a) Studies on the transmission of WSSV (white spot syndrome virus) in juvenile *Marsupenaeus japonicus* via marine microalgae. J Invertebr Pathol 95(2):87–92

Liu WJ, Chang YS, Wang AH, Kou GH, Lo CF (2007b) White spot syndrome virus annexes a shrimp STAT to enhance expression of the immediate-early gene ie1. J Virol 81(3):1461–1471

Liu WJ, Chang YS, Wang HC, Leu JH, Kou GH, Lo CF (2008) Transactivation, dimerization, and DNA-binding activity of white spot syndrome virus immediate-early protein IE1. J Virol 82:11362–11373

Liu QH, Ma CY, Chen WB, Zhang XL, Liang Y, Dong SL, Huang J (2009) White spot syndrome virus VP37 interacts with VP28 and VP26. Dis Aquat Org 85:23–30

Liu WJ, Chang YS, Huang WT, Chen IT, Wang KCHC, Kou GH, Lo CF (2011) *Penaeus monodon* TATA box-binding protein interacts with the white spot syndrome virus transactivator IE1 and promotes its transcriptional activity. J Virol 85(13):6535–6547

Liu WJ, Shiung HJ, Lo CF, Leu JH, Lai YJ, Lee TL, Huang WT, Kou GH, Chang YS (2014) Characterization and interactome study of white spot syndrome virus envelope protein VP11. PLoS One 9(1):e85779. https://doi.org/10.1371/journal.pone.0085779

Liu BH, Lin YC, Ho CS, Yang CC, Chang YT, Chang JF, Li CY, Cheng CS, Huang JY, Lee YF, Hsu MH, Lin FC, Wang HC, Lo CF, Yang SY, Wang HC (2015a) A novel detection platform for shrimp white spot syndrome virus using an ICP11-dependent immunomagnetic reduction (IMR) assay. PLoS One 10(9):e0138207. https://doi.org/10.1371/journal.pone.0138207

Liu QH, Ma FF, Guan GK, Wang XF, Li C, Huang J (2015b) White spot syndrome virus VP51 interact with ribosomal protein L7 of *Litopenaeus vannamei*. Fish Shellfish Immunol 44(1):382–388

Liu WJ, Lo CF, Kou GH, Leu JH, Lai YJ, Chang LK, Chang YS (2015c) The promoter of the white spot syndrome virus immediate-early gene WSSV108 is activated by the cellular KLF transcription factor. Dev Comp Immunol 49:7–18

Liu WJ, Li YC, Kou GH, Lo CF (2016) Laminin receptor in shrimp is a cellular attachment receptor for white spot syndrome virus. PLoS One 11(6):e0156375. https://doi.org/10.1371/journal.pone.0156375

Liu F, Li S, Yu Y, Zhang C, Li F (2021a) Antennal gland of shrimp as an entry for WSSV infection. Aquaculture 530:735932

Liu WJ, Chang YS, Chen PY, Wu SP (2021b) F1 ATP synthase β subunit is a putative receptor involved in white spot syndrome virus infection in shrimp by binding with viral envelope proteins VP51B and VP150. Dev Comp Immunol 114:103810. https://doi.org/10.1016/j.dci.2020.103810

Liu L, Shan LP, Yan MC, Liu GL, Chen J (2021c) Infection of WSSV shows potential promise of a novel antiviral amino-nitrophenyl medicine for application in culture of shrimp seedling. Aquaculture 534:736283. https://doi.org/10.1016/j.aquaculture.2020.736283

Liu L, Shan LP, Zhou Y, Chen J (2021d) Small molecule inhibitors of white spot syndrome virus: promise in shrimp seedling. Int J Mol Sci 22:3450. https://doi.org/10.3390/ijms22073450

Lo CF, Kou GH (1998) Virus-associated white spot syndrome of shrimp in Taiwan: a review. Fish Pathol 33:365–371

Lo CF, Ho CH, Peng SE, Chen CH, Hsu HC, Chiu YL, Chang CF, Liu KF, Wang CH, Kou GH (1996) White spot syndrome baculovirus (WSBV) detected in cultured and captured shrimp, crabs and other arthropods. Dis Aquat Org 27:215–225

Lo CF, Ho CH, Chen CH, Liu KF, Chiu YL, Yeh PY, Peng SE, Hsu HC, Liu HC, Chang CF, Su MS, Wang CH, Kou GH (1997) Detection and tissue tropism of white spot syndrome baculovirus (WSBV) in captured brood ers of *Penaeus monodon* with a special emphasis on reproductive organs. Dis Aquat Org 30:53–72

Lo CF, Hsu HC, Tsai MF, Ho CH, Peng SE, Kou GH, Lightner DV (1999) Specific genomic DNA fragment analysis of different geographical clinical samples of shrimp white spot syndrome virus. Dis Aquat Org 35(3):175–185

Lo CF, Aoki T, Bonami JR, Flegel TW, Leu JH, Lightner DV, Stentiford G, Soderhall K, Walker PW, Wang HC, Xun X, Yang F, Vlak JM (2012) Nimaviridae. In: King AMQ, Adams MJ, Carstens EB, Lefkowitz EJ (eds) Virus taxonomy: classification and nomenclature of viruses: ninth report of the international committee on taxonomy of viruses. Elsevier Academic Press, San Diego, pp 229–234

Lotz JM, Soto MA (2002) Model of white spot syndrome virus (WSSV) epidemics in *Litopenaeus vannamei*. Dis Aquat Org 50(3):199–209

Loyprasert-Thananimit S, Saleedang A, Kanatharana P, Thavarungkul P, Chotigeat W (2012) Production of a polyclonal antibody to the VP26 nucleocapsid protein of white spot syndrome virus (WSSV) and its use as a biosensor. Front Chem Sci Eng 6(2):216–223

Loyprasert-Thananimit S, Saleedang A, Deachamag P, Waiyapoka T, Neulplub M, Chotigeat W (2014) Development of an immuno-based colorimetric assay for white spot syndrome virus. Biotechnol Appl Biochem 61(3):357–362

Lu L, Kwang J (2004) Identification of a novel shrimp protein phosphatase and its association with latency-related ORF427 of white spot syndrome virus. FEBS Lett 577(1–2):141–146

Lu H, Ruan L, Xu X (2011) An immediate-early protein of white spot syndrome virus modulates the phosphorylation of focal adhesion kinase of shrimp. Virology 419(2):84–89

Lu K, Chen X, Zhang H, Zhu J, Zhao Y, Chen X, Zhang Y, Yao D (2024) White spot syndrome virus IE1 protein blocks the integrin-FAK signaling to enhance viral infection in shrimp. Fish Shellfish Immunol 157:110073

Ma C, Gao Q, Liang Y, Li C, Liu C, Huang J (2016) Shrimp arginine kinase being a binding protein of WSSV envelope protein VP31. Chin J Oceanol Limnol 34(6):1287–1296

Madhavi R, Janakiram P, Jayasree L, Murthy PSN (2002) Occurrence of concurrent infections with multiple viruses in *Penaeus monodon* from culture ponds of north coastal Andhra Pradesh. Curr Sci 82(11):1397–1400

Madhumathi M, Rengasamy R (2011) Antioxidant status of *Penaeus monodon* fed with *Dunaliella salina* supplemented diet and resistance against WSSV. Int J Eng Sci Technol 3(10):7249–7259

Maeda M, Kasornchandra J, Itami T, Suzuki N, Hennig O, Kondo M, Albaladejo JD, Takahashi Y (1998) Effect of various treatments on white spot syndrome virus (WSSV) from *Penaeus japonicus* (Japan) and *P. monodon* (Thailand). Fish Pathol 33(4):381–387

Makesh M, Koteeswaran A, Chandran DJ, Manohar BM, Ramasamy V (2006a) Development of monoclonal antibodies against VP28 of WSSV and its application to detect WSSV using immunocomb. Aquaculture 261(1):64–71

Makesh M, Koteeswaran A, Chandran DJ, Manohar BM, Ramasamy V, Chandramohan A, Raman RP, Mukherjee SC (2006b) Rapid detection of white spot syndrome virus (WSSV) of *Penaeus monodon* by latex agglutination test using monoclonal antibodies. J Indian Fish Assoc 33:169–174

Manilal A, Sujith S, Selvin J, Kiran GS, Shakir C (2009) *In vivo* antiviral activity of polysaccharide from the Indian Green Alga, *Acrosiphonia orientalis* (J. Agardh): potential implication in shrimp disease management. World J Fish Mar Sci 1(4):278–282

Manimaran M, Rajkumar T, Vimal S, Taju G, Majeed SA, Hameed ASS, Kannabiran K (2018) Antiviral activity of 9(10H)-Acridanone extracted from marine *Streptomyces fradiae* strain VITMK2 in *Litopenaeus vannamei* infected with white spot syndrome virus. Aquaculture 488:66–73

Manivannan S, Otta SK, Karunasagar I, Karunasagar I (2002) Multiple viral infection in *Penaeus monodon* shrimp postlarvae in an Indian hatchery. Dis Aquat Org 48(3):233–236

Manohar BM, Soundararaj A, Selvaraj D, Sheela DRR, Chidambaram P, Mohan AC, Ravishankar B (1996) An outbreak of SEMBV and MBV infections in cultured Penaeus monodon in Tamil Nadu. Indian J Fish 43 (4): 403–406

Mariot LV, Bolívar N, Coelho JDR, Goncalves P, Colombo SM, Nascimento FV, Schleder DD, Hayashi L (2021) Diets supplemented with carrageenan increase the resistance of the Pacific white shrimp to WSSV without changing its growth performance parameters. Aquaculture 545:737172. https://doi.org/10.1016/j.aquaculture.2021.737172

Marks H, Goldbach RW, Vlak JM, van Hulten MC (2004) Genetic variation among isolates of white spot syndrome virus. Arch Virol 149:673–697

Marks H, Vorst O, van Hollwelingen A, van Hulten MCW, Vlak JM (2005) Gene-expression profiling of white spot syndrome virus in vivo. J Gen Virol 86:2081–2100

Marques JS, Müller IC, Moser JR, Sincero TC, Marques MRF (2011) Wild captured crab, *Chasmagnathus granulata* (Dana, 1851), a new host for white spot syndrome virus (WSSV). Aquaculture 318(1–2):20–24

Martorelli SR, Overstreet RM, Jovonovich JA (2010) First report of viral pathogens WSSV and IHHNV in Argentine crustaceans. Bull Mar Sci 86(1):117–131

Mathew S, Kumar KA, Anandan R, Nair PGV, Devadasan K (2009) Haematological alterations in *Penaeus monodon* artificially infected with white spot syndrome virus (WSSV). Asian Fish Sci 22:1005–1014

Mavichak R, Takano T, Kondo H, Hirono I, Wada S, Hatai K, Inagawa H, Takahashi Y, Yoshimura T, Kiyono H, Yuki Y, Aoki T (2010) The effect of liposome-coated recombinant protein VP28 against white spot syndrome virus in kuruma shrimp, *Marsupenaeus japonicus*. J Fish Dis 33:69–74

Medina-Beltrán V, Luna-González A, Fierro-Coronado JA, Campa-Córdova AI, Peraza-Gómez V, Flores-Miranda MC, Gutiérrez-Rivera JN (2012) *Echinacea purpurea* and *Uncaria tomentosa* reduce the prevalence of WSSV in witheleg shrimp (*Litopenaeus vannamei*) cultured under laboratory conditions. Aquaculture 358:164–169

Meenakshi S, Saravanan R, Balasubramanian T, Palavesam A (2019) *In vivo* administration of fucoidan from *Turbinaria decurrens* protects shrimps from white spot syndrome virus. Indian J Geo Mar Sci 48(2):212–216

Mekata T, Sudhakaran R, Kono T, Supamattaya K, Linh H, Sakai M, Itami T (2009) Real-time quantitative loop-mediated isothermal amplification as a simple method for detecting white spot syndrome virus. Lett Appl Microbiol 48:25–32

Menasveta P (2002) Improved shrimp growout systems for disease prevention and environmental sustainability in Asia. Rev Fish Sci 10:391–402

Mendoza Cano F, Sánchez-Paz A (2013) Development and validation of a quantitative real-time polymerase chain assay for universal detection of the white spot syndrome virus in marine crustaceans. Virol J 10:186. https://doi.org/10.1186/1743-422X-10-186

Mendoza-Cano F, Sánchez-Paz A, Terán-Díaz B, Galván-Alvarez D, Trinidad EG, Enríquez-Espinoza T, López JH (2014) The endemic copepod *Calanus pacificus californicus* as a potential vector of white spot syndrome virus. J Aquat Anim Health 26:113–117

Meng XH, Jang IK, Seo HC, Cho YR (2009) White spot syndrome virus quantification in blue crab *Portunus trituberculatus* hatchery-produced larvae and wild populations by TaqMan real-time PCR, with an emphasis on the relationship between viral infection and crab health. Aquaculture 291(1–2):18–22

Methatham T, Boonchuen P, Jaree P, Tassanakajon A, Somboonwiwat K (2017) Antiviral action of the antimicrobial peptide ALFPm3 from *Penaeus monodon* against white spot syndrome virus. Dev Comp Immunol 69:23–32

Millard RS, Ellis RP, Bateman KS, Bickley LK, Tyler CR, Aerle RV, Santos EM (2021) How do abiotic environmental conditions influence shrimp susceptibility to disease? A critical analysis focussed on white spot disease. J Invertebr Pathol 186:107369. https://doi.org/10.1016/j.jip.2020.107369

Mishra SS, Shekhar MS (2005) White spot syndrome virus isolates of tiger shrimp *Penaeus monodon* (Fabricious) in India are similar to exotic isolates as revealed by polymerase chain reaction and electron microscopy. Indian J Exp Biol 43(7):654–661

Mohan SA (2008) Study on natural reservoirs of shrimp pathogenic viruses. MSc thesis, Department of Fishery Microbiology, Karnataka Veterinary, Animal and Fisheries Sciences University, College of Fisheries, Mangalore

Mohan CV, Basavarajappa HN (2001) India. In: Subasinghe RP, Arthur JR, Phillips MJ, Reantaso MB (eds) Thematic review on management strategies for majordiseases in shrimp aquaculture, A component of the WB/NACA/WWF/FAO Program on Shrimp Farming and the Environment. Report of the workshop held in Cebu, Philippines from 28–30 November, 1999. WB/NACA/WWF/FAO Consortium Programme on Shrimp Farming and the Environment, pp 51–58

Mohan CV, Shankar KM, Kulkarni S, Sudha PM (1998) Histopathology of cultured shrimp showing gross signs of yellow head syndrome and white spot syndrome during 1994 Indian epizootics. Dis Aquat Org 34:9–12

Mohan CV, Corsin F, Thakur PC, Padiyar PA, Madhusudan M, Turnbull JF, Hao NV, Morgan KL (2002) Usefulness of dead shrimp specimens in studying the epidemiology of white spot syndrome virus (WSSV) and chronic bacterial infection. Dis Aquat Org 50(1):1–8

Mohankumar K, Ramasamy P (2006a) White spot syndrome virus infection decreases the activity of antioxidant enzymes in *Fenneropenaeus indicus*. Virus Res 115(1):69–75

Mohankumar K, Ramasamy P (2006b) Activities of membrane bound phosphatases, transaminases and mitochondrial enzymes in white spot syndrome virus infected tissues of *Fenneropenaeus indicus*. Virus Res 118(1–2):130–135

Momtaz F, Foysal J, Rahman MM, Fotedar R (2018) Design of epitope based vaccine against shrimp white spot syndrome virus (WSSV) by targeting the envelope proteins: an immunoinformatic approach. Turk J Fish Aquat Sci 19:149–159

Monica M, Priyanka T, Akshaya M, Rajeswari V, Sivakumar L, Somasundaram ST, Shenbhagarathai R (2017) The efficacy of poly-β-hydroxy butyrate (PHB)/biosurfactant derived from *Staphylococcus hominis* against white spot syndrome virus (WSSV) in *Penaeus monodon*. Fish Shellfish Immunol 71:399–410

Moser JR, Marques MRF (2023) Susceptibility of *Neohelice granulata* (Decapoda, Varunidae) to white spot syndrome virus (WSSV). Aquac Int 31(2):975–996

Mu Y, Lan JF, Zhang XW, Wang XW, Zhao XF, Wang JX (2012) A vector that expresses VP28 of WSSV can protect red swamp crayfish from white spot disease. Dev Comp Immunol 36(2):442–449

Muhammad M, Lotz JM (2015) Prevalence and infectivity of white spot syndrome virus in the daggerblade grass shrimp *Palaemonetes pugio*. The Korean Society of Fisheries and Aquatic Science, p 467

Muhammad M, Lotz JM, Blaylock RB, Curran SS (2020) White spot syndrome virus in decapods from Mississippi Sound, USA, and susceptibility of *Palaemonetes pugio* and *Uca panacea* to a Chinese isolate. Dis Aquat Org 138:121–131

Muliani N, Kadriah IAK (2020) The effect of several types of mangrove exctracs on tiger shrimp *Penaeus monodon* survival rate challenged with white spot syndrome virus (WSSV). IOP Conf Ser Earth Environ Sci 564:012054. https://doi.org/10.1088/1755-1315/564/1/012054

Muliani, Susianingsih E (2018) The effectiveness of methanol extract and fractionations from mangrove leaves *Sonneratia alba* and *Bruguiera gymnorrhiza* to prevent white spot syndrome virus (WSSV) infection in black tiger shrimp *Penaeus monodon*. J Akuakult Indones 17(1):43–52

Muller IC, Andrade TPD, Tang-Nelson KFJ, Marques MRF, Lightner DV (2010) Genotyping of white spot syndrome virus (WSSV) geographical isolates from Brazil and comparison to other isolates from the Americas. Dis Aquat Org 88:91–98

Murugan V, Sankaran K (2018) Bacterial lipid modification of ICP11 and a new ELISA system applicable for WSSV infection detection. Mar Biotechnol 20(3):375–384

Mushiake K, Shimizu K, Satoh J, Mori K, Arimoto M, Ohsumi S, Imaizumi K (1999) Control of penaeid acute viremia (PAV) in Penaeus japonicus: selection of eggs based on the PCR detection of the causative virus (PRDV) from receptaculum seminis of spawned broodstock. Fish Pathol 34:203–207

Musthaq SS, Sudhakaran R, Balasubramanian G, Hameed ASS (2006a) Experimental transmission and tissue tropism of white spot syndrome virus (WSSV) in two species of lobsters, *Panulirus homarus* and *Panulirus ornatus*. J Invertebr Pathol 93(2):75–80

Musthaq SS, Sudhakaran R, Ahmed VPI, Balasubramanian G, Hameed ASS (2006b) Variability in the tandem repetitive DNA sequences of white spot syndrome virus (WSSV) genome and suitability of VP28 gene to detect different isolates of WSSV from India. Aquaculture 256(1–4):34–41

Nadala ECB Jr, Tapay LM, Loh PC (1998) Characterization of a non-occluded baculovirus-like agent pathogenic to penaeid shrimp. Dis Aquat Organ 33:221–229

Nadala JEC, Loh PC (2000) Dot-blot nitrocellulose enzyme immunoassays for the detection of white-spot virus and yellow-head virus of penaeid shrimp. J Virol Methods 84(2):175–179

Nakano H, Koube H, Umezawa S, Momoyama K, Hiraoka M, Inouye K, Oseko S (1994) Mass mortalities of cultured kuruma shrimp, Penaeus japonicus, in Japan in 1993: epizootiological survey and infection trials. Fish Pathol 29:135–139

Nakano H, Hiraoka M, Sameshima M, Kimura T, Momoyama K (1998) Inactivation of penaeid rod-shaped DNA virus (PRDV), the causative agent of penaid acute viremia (PAV), by some chemical and physical treatments. Fish Pathol 33(2):65–71

Namikoshi A, Wu JL, Yamashita T, Nishizawa T, Nishioka T, Arimoto M, Muroga K (2004) Vaccination trials with *Penaeus japonicus* to induce resistance to white spot syndrome virus. Aquaculture 229:25–35

Namitha R, Anoop BS, Dominic DD, Ameer A, Santhiya P, Singh IB, Dharani G, Govindaraju K (2024a) Seaweed assisted preparation of Fe2O3 nanoparticles and their antiviral studies against white spot syndrome virus (WSSV): in vitro and in silico approach. Aquaculture 593:741365

Namitha R, Abirami B, Anoop BS, Dominic DD, Ameer A, Manigundan K, Radhakrishnan M, Santhiya P, Bhaskar PV, Govindaraju K, Singh IB (2024b) Synthesis and characterization of gold nanoparticles using *Brevibacterium casei* (SOSIST-06) isolated from Southern Ocean water samples and their in vitro and in silico anti-WSSV activity. Aquaculture 579:740205

Natarajan A, Devi KSS, Raja S, Kumar AS (2017) An elegant analysis of white spot syndrome virus using a graphene oxide/methylene blue based electrochemical immunosensor platform. Sci Rep 7:46169. https://doi.org/10.1038/srep46169

Natarajan A, Kumar AS, Sudhakaran R (2019) Validation of white spot syndrome virus infection in penaeid shrimp of different geographical regions using electrochemical immunosensor. Int J Innov Technol Explor Eng 8(7):2475–2481

Nguyen AT, Pham CK, Pham HT, Pham HL, Nguyen AH, Dang LT, Huynh HA, Cutting SM, Phan TN (2014) *Bacillus subtilis* spores expressing the VP28 antigen: a potential oral treatment to protect *Litopenaeus vannamei* against white spot syndrome. FEMS Microbiol Lett 358(2):202–208

Nicolasora ADD, Maralit BA, Caipang CMA, Santos MD, Calpe A, Maningas BB (2014) Utilization of loop-mediated isothermal amplification (LAMP) technology for detecting white spot syndrome virus (WSSV) and Vibrio spp. in *Litopenaeus vannamei* in selected sites in the Philippines. Philipp Sci Lett 7(2):309–316

Nilsen P, Karlsen M, Sritunyalucksana K, Thitamadee S (2017) White spot syndrome virus VP28 specific double-stranded RNA provides protection through a highly focused siRNA population. Sci Rep 7(1):1028. https://doi.org/10.1038/s41598-017-01181-w

Ning JF, Zhu W, Xu JP, Zheng CY, Meng XL (2009) Oral delivery of DNA vaccine encoding VP28 against white spot syndrome virus in crayfish by attenuated *Salmonella typhimurium*. Vaccine 27(7):1127–1115

Ning D, Leng X, Li Q, Xu W (2011) Surface-displayed VP28 on *Bacillus subtilis* spores induces protection against white spot syndrome virus in crayfish by oral administration. J Appl Microbiol 111(6):1327–1336

Ning J, Zhang MM, Tong QQ, Wang DL, Zhao YL (2016) Effect of white spot syndrome virus (WSSV) infection on immune enzyme activity and ultrastructure in the haemolymph tissue of *Cherax quadricarinatus* (Decapoda, Astacidea). Crustaceana 89(6–7):669–684

Niu GJ, Wang S, Xu JD, Yang MC, Sun JJ, He ZH, Zhao XF, Wang JX (2019a) The polymeric immunoglobulin receptor-like protein from *Marsupenaeus japonicus* is a receptor for white spot syndrome virus infection. PLoS Pathog 15(2):e1007558. https://doi.org/10.1371/journal.ppat.1007558

Niu J, Xie JJ, Guo TY, Fang HH, Zhang YM, Liao SY, Xie SW, Liu YJ, Tian LX (2019b) Comparison and evaluation of four species of macro-algae as dietary ingredients in *Litopenaeus vannamei* under normal rearing and WSSV challenge conditions: effect on growth, immune response, and intestinal microbiota. Front Physiol 9:1880. https://doi.org/10.3389/fphys.2018.01880

Norizan N, Shaharom-Harrison F, Hassan M, Musa N, Wahid MEA, Zainathan SC (2019) First detection of white spot syndrome virus (WSSV) in wild mud crab *Scylla* spp. (de Haan, 1883) from Setiu Wetlands, Malaysia. Songklanakarin J Sci Technol 41(1):45–52

Nunan LM, Lightner DV (1997) Development of a non-radioactive gene probe by PCR for detection of white spot syndrome virus (WSSV). J Virol Methods 63(1–2):193–201

Nunan LM, Lightner DV (2011) Optimized PCR assay for detection of white spot syndrome virus (WSSV). J Virol Methods 171(1):318–321

Nupan B, Phongdara A, Saengsakda M, Leu JH, Lo CF (2011) Shrimp Pm-fortilin inhibits the expression of early and late genes of white spot syndrome virus (WSSV) in an insect cell model. Dev Comp Immunol 35(4):469–475

Oakey HJ, Smith CS (2018) Complete genome sequence of a white spot syndrome virus associated with a disease incursion in Australia. Aquaculture 484:152–159

Oakey J, Smith C, Underwood D, Afsharnasab M, Alday-Sanz V, Dhar A, Sivakumar S, Hameed ASS, Beattie K, Crook A (2019) Global distribution of white spot syndrome virus genotypes determined using a novel genotyping assay. Arch Virol 164(8):2061–2082

Ochoa-Meza AR, Álvarez-Sánchez AR, Romo-Quiñonez CR, Barraza A, Magallón-Barajas FJ, Chávez-Sánchez A, García-Ramos JC, Toledano-Magaña Y, Bogdanchikova N, Pestryakov A, Mejía-Ruiz CH (2019) Silver nanoparticles enhance survival of white spot syndrome virus infected *Penaeus vannamei* shrimps by activation of its immunological system. Fish Shellfish Immunol 84:1083–1089

OIE (2019) Chapter 2.2.8: Infection with white spot syndrome virus. In: Mannual of diagnostic tests for aquatic animals. OIE, Paris

Okumura T, Nagai F, Yamamoto S, Yamano K, Oseko N, Lnouye K, Oomura H, Sawada H (2004) Detection of white spot syndrome virus from stomach tissue homogenate of the kuruma shrimp (*Penaeus japonicus*) by reverse passive latex agglutination. J Virol Methods 119(1):11–16

Okumura T, Nagai F, Yamamoto S, Yamano K, Oomura H, Lnouye K, Ito M, Sawada H (2005) Detection of white spot syndrome virus (WSSV) from hemolymph of penaeid shrimps *Penaeus japonicus* by reverse passive latex agglutination assay using high-density latex particles. J Virol Methods 124(1–2):143–148

Oseko N, Chuah TT, Maeno Y, Kua BC, Palanasamy V (2006) Examination for viral inactivation of WSSV (white spot syndrome virus) isolated in Malaysia using black tiger prawn (*Penaeus monodon*). Jpn Agric Res Q 40(1):93–97

Otta SK, Shubha G, Joseph B, Chakraborty A, Karunasagar I, Karunasagar I (1999) Polymerase chain reaction (PCR) detection of white spot syndrome virus (WSSV) in cultured and wild crustaceans in India. Dis Aquat Org 38:67–70

Otta SK, Karunasagar I, Karunasagar I (2003) Detection of monodon baculovirus and whitespot syndrome virus in apparently healthy *Penaeus monodon* postlarvae from India by polymerase chain reaction. Aquaculture 220:59–67

Otta SK, Arulraj R, Praveena PE, Manivel R, Panigrahi A, Bhuvaneswari T, Ravichandran P, Jithendran KP, Ponniah AG (2014) Association of dual viral infection with mortality of Pacific white shrimp (*Litopenaeus vannamei*) in culture ponds in India. Virus Dis 25:63–68

Overstreet RM, Jovonovich J, Ma H (2009) Parasitic crustaceans as vectors of viruses, with an emphasis on three penaeid viruses. Integr Comp Biol 49(2):127–141

Pacheco R, Ascencio F, Zarain M, Gómez G, Campa A (2011) Enhancement of superoxide dismutase and catalase activity in juvenile brown shrimp, *Farfantepenaeus californiensis* (Holmes, 1900), fed β-1.3 glucan vitamin E, and β-carotene and infected with white spot syndrome virus. Lat Am J Aquat Res 39(3):534–543

Pais R, Shekar M, Karunasagar I, Karunasagar I (2007) Hemagglutinating activity and electrophoretic pattern of hemolymph serum proteins of *Penaeus monodon* and *Macrobrachium rosenbergii* to white spot syndrome virus injections. Aquaculture 270(1–4):529–534

Palanikumar P, Benitta DJD, Lelin C, Thirumalaikumar E, Michaelbabu M, Citarasu T (2018) Effect of *Argemone mexicana* active principles on inhibiting viral multiplication and stimulating immune system in Pacific white leg shrimp *Litopenaeus vannamei* against white spot syndrome virus. Fish Shellfish Immunol 75:243–252

Pan X, Zhang Y, Sha X, Wang J, Li J, Dong P, Liang X (2017) Highly sensitive detection of low-abundance white spot syndrome virus by a pre-amplification PCR method. J Microbiol Biotechnol 27(3):471–479

Panichareon B, Khawsak P, Deesukon W, Sukhumsirichart W (2011) Multiplex real-time PCR and high-resolution melting analysis for detection of white spot syndrome virus, yellow-head virus, and *Penaeus monodon* densovirus in penaeid shrimp. J Virol Methods 178(1–2):16–21

Paria A, Greeshma SS, Chaudhari A, Makesh M, Purushothaman CS, Rajendran KV (2012) Nonspecific effect of double-stranded (DS) RNA on prophenoloxidase (proPO) expression in *Penaeus monodon*. Appl Biochem Biotechnol 169(1):281–289

Park JH, Lee YS, Lee S, Lee Y (1998) An infectious viral disease of penaeid shrimp newly found in Korea. Dis Aquat Org 34:71–75

Park JY, Kim KI, Joh SJ, Kang JY, Kwon JH, Lee HS, Kwon YK (2013) Development of a highly sensitive single-tube nested PCR protocol directed toward the sequence of virion envelope proteins for detection of white spot syndrome virus infection: improvement of PCR methods for detection of WSSV. Aquaculture 410–411:225–229

Park SC, Choi SK, Han SH, Park S, Jeon HJ, Lee SC, Kim KY, Lee YS, Kim JH, Han JE (2020) Detection of infectious hypodermal and hematopoietic necrosis virus and white spot syndrome virus in whiteleg shrimp (*Penaeus vannamei*) imported from Vietnam to South Korea. J Vet Sci 21(2):e31. https://doi.org/10.4142/jvs.2020.21.e31

Parrilla-Taylor DP, Vibanco-Perez N, Duran-Avelar MDJ, Gomez-Gil B, Llera-Herrera R, Vazquez-Juarez R (2018) Molecular variability and genetic structure of white spot syndrome virus strains from northwest Mexico based on the analysis of genomes. FEMS Microbiol Lett 365(20):216. https://doi.org/10.1093/femsle/fny216

Partida-Arangure BO, Luna-González A, Fierro-Coronado JA, Flores-Miranda MDC, González-Ocampo HA (2013) Effect of inulin and probiotic bacteria on growth, survival, immune response, and prevalence of white spot syndrome virus (WSSV) in *Litopenaeus vannamei* cultured under laboratory conditions. Afr J Biotechnol 12(21):3366–3375

Pathan M, Gireesh-Babu P, Pavan-Kumar A, Jeena K, Sharma R, Makesh M, Prasad KP, Krishna G (2013) In vivo therapeutic efficacy of recombinant *Penaeus monodon* antiviral protein (rPmAV) administered in three different forms to WSSV infected *Penaeus monodon*. Aquaculture 376–379:64–67

Patil R, Palaksha KJ, Anil TM, Guruchannabasavanna, Patil P, Shankar KM, Mohan CV, Sripada A (2008) Evaluation of an immunodot test to manage white spot syndrome virus (WSSV) during cultivation of the giant tiger shrimp *Penaeus monodon*. Dis Aquat Org 79(2):157–161

Patil R, Shankar KM, Kumar BTN, Kulkarni A, Patil P, Moger N (2013) Development of a monoclonal antibody-based flow-through immunoassay (FTA) for detection of white spot syndrome virus (WSSV) in black tiger shrimp *Penaeus monodon*. J Fish Dis 36(9):753–762

Patil PK, Solanki HG, Otta SK, Patel K, Saravanan P, Chavali G, Pillai SM (2017) Genetic variations among white spot syndrome virus (WSSV) isolates from shrimp farms in Gujarat, northwest coast of India. Indian J Fish 64(4):125–128

Patil PK, Geetha R, Ravisankar T, Avunje S, Solanki HG, Abraham TJ, Vinoth SP, Jithendran KP, Alavandi SV, Vijayan KK (2021) Economic loss due to diseases in Indian shrimp farming with special reference to *Enterocytozoon hepatopenaei* (EHP) and white spot syndrome virus (WSSV). Aquaculture 533:736231. https://doi.org/10.1016/j.aquaculture.2020.736231

Pemula AK, Krishnan S (2015) Temporal analysis of molecular changes in shrimp (*Penaeus vannamei*) tissues with respect to white spot disease. J Food Sci Technol 52:7236–7244

Peña LDDL, Lavilla-Pitogo CR, Villar CBR, Paner MG, Sombito CD, Capulos GC (2007) Prevalence of white spot syndrome virus (WSSV) in wild shrimp *Penaeus monodon* in the Philippines. Dis Aquat Org 77(3):175–179

Peraza-Gómez V, Luna-González A, Campa-Córdova AI, López-Meyer M, Fierro-Coronado JA, Álvarez-Ruiz P (2009) Probiotic microorganisms and antiviral plants reduce mortality and prevalence of WSSV in shrimp (*Litopenaeus vannamei*) cultured under laboratory conditions. Aquac Res 40(13):1481–1489

Peraza-Gómez V, Luna-González A, González-Prieto JM, Fierro-Coronado JA, González-Ocampo HA (2014) Protective effect of microbial immunostimulants and antiviral plants against WSSV in *Litopenaeus vannamei* cultured under laboratory conditions. Aquaculture 420–421:160–164

Peruzza L, Shekhar MS, Kumar KV, Swathi A, Karthic K, Hauton C, Vijayan KK (2019) Temporal changes in transcriptome profile provide insights of white spot syndrome virus infection in *Litopenaeus vannamei*. Sci Rep 9:13509. https://doi.org/10.1038/s41598-019-49836-0

Pham KC, Tran HTT, Doan CV, Le PH, Nguyen ATV, Nguyen HA, Hong HA, Cutting SM, Phan TN (2017) Protection of *Penaeus monodon* against white spot syndrome by continuous oral administration of a low concentration of *Bacillus subtilis* spores expressing the VP28 antigen. Lett Appl Microbiol 64(3):184–191

Pholdaeng K, Pongsamart S (2010) Studies on the immunomodulatory effect of polysaccharide gel extracted from *Durio zibethinus* in *Penaeus monodon* shrimp against *Vibrio harveyi* and WSSV. Fish Shellfish Immunol 28(4):555–561

Piamsomboon P, Chaweepack T, Han JF, Tang KF, Wongtavatchai J (2018) Genetic variation of white spot syndrome virus (WSSV) in Pacific white shrimp *Litopenaeus vannamei* (Boone 1931) culture of Thailand. Aquac Res 49(1):265–272

Pilotto MR, Milanez S, Moreira RT, Rosa RD, Perazzolo LM (2019) Potential immunomodulatory and protective effects of the Arthrospira-based dietary supplement on shrimp intestinal immune defenses. Fish Shellfish Immunol 88:47–52

Ponprateep S, Phiwsaiya K, Tassanakajon A, Rimphanitchayakit V (2013) Interaction between Kazal serine proteinase inhibitor SPIPm2 and viral protein WSV477 reduces the replication of white spot syndrome virus. Fish Shellfish Immunol 35(3):957–964

Porchas-Cornejo M, Alvarez-Ruiz P, Alvarez Tello FJ, Martinez-Porchas M, Martinez-Cordova LR, Martinez J, García-Morales R (2018) Detection of the white spot syndrome virus in zooplankton samples collected off the coast of Sonora, Mexico. Aquac Res 49:48–56

Powell J, Burge E, Browdy C, Shepard E (2006) Efficiency and sensitivity determination of Shrimple®, an immunochromatographic assay for white spot syndrome virus (WSSV), using quantitative real-time PCR. Aquaculture 257:167–172

Powell JW, Browdy CL, Burge EJ (2015) Blue crabs *Callinectes sapidus* as potential biological reservoirs for white spot syndrome virus (WSSV). Dis Aquat Org 113(2):163–167

Pradeep B, Shekar M, Gudkovs N, Karunasagar I, Karunasagar I (2008) Genotyping of white spot syndrome virus prevalent in shrimp farms of India. Dis Aquat Org 78(3):189–198

Pradeep B, Rai P, Mohan SA, Shekhar MS, Karunasagar I (2012) Biology, host range, pathogenesis and diagnosis of white spot syndrome virus. Indian J Virol 23(2):161–174

Pratapa MG, Kumar S, Bedekar MK, Kumar HS, Rajendran KV (2023) Experimental infection reveals mud crab, *Scylla serrata* is less susceptible than *Scylla olivacea* and shrimp, *Penaeus vannamei* to white spot syndrome virus (WSSV). Aquaculture 577:739877

Priyangha SJ, Gopalakrishnan A, Muhil Vannan S, Gunasekaran T, Somasundaram ST (2020) First report of pedunculate barnacle (*Octolasmis neptuni*), as potential asymptomatic carrier of white spot syndrome virus (WSSV). Comp Clin Pathol 29:631–638

Puneeth TG, Akhila DS, Dechamma MM, Shreeharsha JM, Shivakumar SK, Venugopal MN (2017) Comparative efficacy of dsRNA VP24, VP26, RR1 and WSV477 gene against WSSV infection in *Penaeus monodon*. Int J Curr Microbiol App Sci 6(2):665–674

Qian X, Zhu F (2021) Use of glycerol monolaurate as a treatment against white spot syndrome virus in crayfish (*Procambarus clarkii*). Aquaculture 541:736853. https://doi.org/10.1016/j.aquaculture.2021.736853

Qiu Z, Liu Q, Huang J (2012) Efficiency of two fragments of VP28 against white spot syndrome virus in *Litopenaeus vannamei*. Aquaculture 338:2–12

Quéré R, Commes T, Marti J, Bonami JR, Piquemal D (2002) White spot syndrome virus and infectious hypodermal and hematopoietic necrosis virus simultaneous diagnosis by miniarray system with colorimetry detection. J Virol Methods 105(2):189–196

Rahman MM, Escobedo-Bonilla CM, Wille M, Sanz VA, Audoorn L, Neyts L, Pensaert MB, Sorgeloos P, Nauwynck HJ (2006a) Clinical effect of cidofovir and a diet supplemented with *Spirulina platensis* in white spot syndrome virus (WSSV) infected specific pathogen-free *Litopenaeus vannamei* juveniles. Aquaculture 255(1–4):600–605

Rahman MM, Escobedo-Bonilla CM, Corteel M, Dantas-Lima JJ, Wille M, Sanz VA, Pensaert MB, Sorgeloosa P, Nauwynck HJ (2006b) Effect of high water temperature (33°C) on the clinical and virological outcome of experimental infections with white spot syndrome virus (WSSV) in specific pathogen-free (SPF) *Litopenaeus vannamei*. Aquaculture 261(3):842–849

Raj S, Vijayan KK, Alavandi SV, Balasubramanian CP, Santiago TC (2012) Effect of temperature and salinity on the infectivity pattern of white spot syndrome virus (WSSV) in giant tiger shrimp *Penaeus monodon* (Fabricius, 1837). Indian J Fish 59(3):109–115

Raja RA, Panigrahi A, De D, Kumar S (2017) Investigations on white spot disease outbreak in *Penaeus monodon* (Fabricius, 1798) in association with *Vibrio mimicus* infection in the Sunderbans, West Bengal, India. Indian J Fish 64(1):56–60

Rajan PR, Ramasamy P, Purushothaman V, Brennan GP (2000) White spot baculovirus syndrome in the Indian shrimp *Penaeus monodon* and *P. indicus*. Aquaculture 184(1–2):31–44

Rajeish M, Mani MK, Dechamma MM, Bossier P, Rai P, Karunasagar I, Maiti B (2021) Expression profile of heat shock protein 70 in lymphoid organs of *Penaeus monodon* in response to white spot syndrome virus infection. Aquac Res 52(3):1316–1320

Rajendran KV, Vijayan KK, Alavandi SV, Rangaswamy SC, Srinivasagam S, Rajababu D (1998) White spot disease outbreak in confined rain-fed shrimp culture ponds of Chilka Lake: a case study. J Aquac 6:43–47

Rajendran KV, Vijayan KK, Santiago TC, Krol RM (1999) Experimental host range and histopathology of white spot syndrome virus (WSSV) infection in shrimp, prawns, crabs and lobsters from India. J Fish Dis 22(3):183–191

Rajendran KV, Vijayan KK, Santiago TC, Rajan JJS (2005) White spot syndrome virus (WSSV) infection in tiger shrimp *Penaeus monodon*: a non-lethal histopathological rapid diagnostic method using paraffin and frozen sections. Aquac Int 13:341–349

Rajendran KV, Deepika A, Kulkarni A, Mahapatra M, Kumar SH, Bedekar MK, Kollanoor RJ, Mulloorpeedikayil RG, Mohideenpitchai MM, Muthumariappan S, Panchavarnam S, Devaraj K, Bass D, van Aerle R (2021) Three draft genome sequences of white spot syndrome virus from India. Microbiol Resour Announc 10:e00579-21. https://doi.org/10.1128/MRA.00579-21

Rajkumar T, Taju G, Majeed SA, Sajid MS, Kumar SS, Sivakumar S, Thamizhvanan S, Vimal S, Hameed ASS (2017) Ontogenetic changes in the expression of immune related genes in response to immunostimulants and resistance against white spot syndrome virus in *Litopenaeus vannamei*. Dev Comp Immunol 76:132–142

Rajkumar T, Manimaran M, Taju G, Vimal S, Majeed SA, Kannabiran K, Sivakumar S, Kumar KM, Madhan S, Hameed ASS (2018) Antiviral viral compound from *Streptomyces ghanaensis* like strain against white spot syndrome virus (WSSV) of shrimp. bioRxiv:1–59. https://doi.org/10.1101/340265

Rameshthangam P, Ramasamy P (2006) Antioxidant and membrane bound enzymes activity in WSSV-infected *Penaeus monodon* Fabricius. Aquaculture 254(1–4):32–39

Rameshthangam P, Ramasamy P (2007) Antiviral activity of bis (2-methylheptyl) phthalate isolated from *Pongamia pinnata* leaves against white spot syndrome virus of *Penaeus monodon* Fabricius. Virus Res 16(1–2):38–44

Ramırez-Douriet C, Silva-Dávila RD, Méndez-Lozano J, Urias DE, Leyva-Arana I, Lopez-Meyer M (2005) White spot syndrome virus detection in zooplankton of coastal lagoons and shrimp commercial ponds in Sinaloa, Mexico. 135th annual meeting of the American Fisheries Society

Ramos-Carreño S, Valencia-Yáñez R, Correa-Sandoval F, Ruíz-García N, Díaz-Herrera F, Giffard-Mena I (2014) White spot syndrome virus (WSSV) infection in shrimp (*Litopenaeus vannamei*) exposed to low and high salinity. Arch Virol 159:2213–2222

Ramos-Carreño S, Giffard-Mena I, Zamudio-Ocadiz JN, Nuñez-Rivera A, Valencia-Yañez R, Ruiz-Garcia J, Viana MT, Cadena-Nava RD (2021) Antiviral therapy in shrimp through plant virus VLP containing VP28 dsRNA against WSSV. Beilstein J Org Chem 17:1360–1373

Ramos-Paredes J, Grijalva-Chon JM, Rosa-Vélez JDL, Enríquez-Paredes LM (2012) New genetic recombination in hypervariable regions of the white spot syndrome virus isolated from *Litopenaeus vannamei* (Boone) in northwest Mexico. Aquac Res 43:339–348

Rampaul M, Argenta N, Bojko J, Fazelan Z, Clark KF (2024) Nimaviruses of crustaceans. In Aquaculture Virology 547–575. Academic Press

Raoult D, Audic S, Robert C, Abergel C, Renesto P, Ogata H, La Scola B, Suzan M, Claverie JM (2004) The 1.2-megabase genome sequence of Mimivirus. Science 306(5700):1344–1350

Rathipriya A, Anand T, Uma A, Marx KK (2019) Prevalence of diseases caused by white spot virus and *Enterocytozoon hepatopenaei* in *Penaeus vannamei* shrimp farms in Nagapattinam district, Tamil Nadu, India. Int J Curr Microbiol App Sci 8(6):2616–2622

Rattanarojpong T, Khankaew S, Khunrae P, Vanichviriyakit R, Poomputsa K (2016) Recombinant baculovirus mediates dsRNA specific to rr2 delivery and its protective efficacy against WSSV infection. J Biotechnol 229:44–52

Ravi M, Sudhakar T, Haran SH, Sudhakaran R, Dhas TS (2020) Nanoparticles based DNA vaccine in marine water crabs (*Scylla serrata*) for protection against white spot syndrome virus (WSSV). Biocatal Agric Biotechnol 28:101764. https://doi.org/10.1016/j.bcab.2020.101764

Reddy CBR, Dinesh S, Anusha N, Itami T, Reddy SR, Sudhakaran R (2016) Antiviral activity of 3-(1-chloropiperidin-4-yl)-6-fluoro benzisoxazole 2 against white spot syndrome virus in freshwater crab, *Paratelphusa hydrodomous*. Aquac Res 47(8):2677–2681

Ren Q, Huang Y, He Y, Wang W, Zhang X (2015) A white spot syndrome virus microRNA promotes the virus infection by targeting the host STAT. Sci Rep 5:18384. https://doi.org/10.1038/srep18384

Ren Q, Huang X, Cui Y, Sun J, Wang W, Zhang X (2017) Two white spot syndrome virus microRNAs target the dorsal gene to promote virus infection in *Marsupenaeus japonicus* shrimp. J Virol 91(8):e02261-16

Restrepo L, Reyes A, Bajaña L, Betancourt I, Bayot B (2018) Draft genome sequence of a white spot syndrome virus isolate obtained in Ecuador. Genome Announc 6(26):e00605-18. https://doi.org/10.1128/genomeA.00605-18

Riet J, Costa-Filho J, Agno LD, Medeiros L, Azevedo R, Nogueira LF, Maggioni R, Pedrosa VF, Romano LA, Altenbuchner J, Wasielesky J, Marins LF (2021) *Bacillus subtilis* expressing double-strand RNAs (dsRNAs) induces RNA interference mechanism (RNAi) and increases survival of WSSV-challenged *Litopenaeus vannamei*. Aquaculture 541:736834. https://doi.org/10.1016/j.aquaculture.2021.736834

Robalino J, Browdy CL, Prior S, Metz A, Parnell P, Gross P, Warr G (2004) Induction of antiviral immunity by double-stranded RNA in a marine invertebrate. J Virol 78:10442–10448

Robalino J, Bartlett T, Shepard E, Prior S, Jaramillo G, Scura E, Chapman RW, Gross PS, Browdy CL, Warr GW (2005) Double-stranded RNA induces sequence specific antiviral silencing in addition to nonspecific immunity in a marine shrimp: convergence of RNA interference and innate immunity in the invertebrate antiviral response? J Virol 79:13561–13571

Roch P (1999) Defense mechanisms and disease prevention in farmed marine invertebrates. Aquaculture 172:125–145

Rodríguez J, Espinosa Y, Echeverría F, Cárdenas G, Román R, Stern S (2007) Exposure to probiotics and β-1,3/1,6-glucans in larviculture modifies the immune response of Penaeus vannamei juveniles and both the survival to white spot syndrome virus challenge and pond culture. Aquaculture 273:405–415

Rodriguez-Anaya LZ, Gonzalez-Galaviz JR, Casillas-Hernandez R, Lares-Villa F, Estrada K, Ibarra-Gamez JC, Sanchez-Flores A (2016) Draft genome sequence of white spot syndrome virus isolated from cultured *Litopenaeus vannamei* in Mexico. Genome Announc 4:e01674-15. https://doi.org/10.1128/genomeA.01674-15

Rohini B, Fredrick WS, Sandhya N, Subhashini D (2012) Immunostimulatory effect of various frequencies of the marine *Vibrio alginolyticus* on juvenile *Penaeus monodon* against WSSV and its biochemical properties. World J Fish Mar Sci 4(4):360–368

Romo-Quiñonez CR, Álvarez-Sánchez AR, Álvarez-Ruiz P, Chávez-Sánchez MC, Bogdanchikova N, Pestryakov A, Mejia-Ruiz CH (2020) Evaluation of a new Argovit as an antiviral agent included in feed to protect the shrimp *Litopenaeus vannamei* against white spot syndrome virus infection. PeerJ 8:e8446. https://doi.org/10.7717/peerj.8446

Ross EP, Behringer DC, Bojko J (2019) White spot syndrome virus and the Caribbean spiny lobster, *Panulirus argus*: susceptibility and behavioral immunity. J Invertebr Pathol 162:1–9

Rout N, Kumar S, Jaganmohan S, Murugan V (2007) DNA vaccines encoding viral envelope proteins confer protective immunity against WSSV in black tiger shrimp. Vaccine 25:2778–2786

Rudtanatip T, Asuvapongpatana S, Withyachumnarnkul B, Wongprasert K (2014) Sulfated galactans isolated from the red seaweed *Gracilaria fisheri* target the envelope proteins of white spot syndrome virus and protect against viral infection in shrimp haemocytes. J Gen Virol 95:1126–1134

Ruiz-Velazco JMJ, Hernández-Llamas A, Gomez-Muñoz VM, Magallon FJ (2010) Dynamics of intensive production of shrimp *Litopenaeus vannamei* affected by white spot disease. Aquaculture 300:113–119

Runsaeng P, Kwankaew P, Utarabhand P (2018) FmLC6: an ultimate dual-CRD C-type lectin from *Fenneropenaeus merguiensis* mediated its roles in shrimp defense immunity towards bacteria and virus. Fish Shellfish Immunol 80:200–213

Saelim H, Loprasert S, Phongdara A (2020) *Bacillus subtilis* expressing dsVP28 improved shrimp survival from WSSV challenge. Sci Asia 46S:19–26

Sajeevan TP, Rosamma P (2009) Effect of treatment with β-1,3 glucan from the yeast *Candida sake* S165 on protection to *Fenneropenaeus indicus* postlarvae against WSSV infection. J Mar Biol Assoc India 51(2):227–230

Sajeevan TP, Lowman DW, Williams DL, Selven S, Anas A, Rosamma P (2009a) Marine yeast diet confers better protection than its cell wall component (1-3)-β-D-glucan as an immunostimulant in *Fenneropenaeus indicus*. Aquac Res 40(15):1723–1730

Sajeevan TP, Philip R, Singh ISB (2009b) Dose/frequency: a critical factor in the administration of glucan as immunostimulant to Indian white shrimp *Fenneropenaeus indicus*. Aquaculture 287(3–4):248–252

Sakai M (1999) Current status of fish immunostimulants. Aquaculture 172:63–92

Salehpour R, Biuki NA, Mohammadi M, Dashtiannasab A, Ebrahimnejad P (2021) The dietary effect of fucoidan extracted from brown seaweed, *Cystoseira trinodis* (*C. Agardh*) on growth and disease resistance to WSSV in shrimp *Litopenaeus vannamei*. Fish Shellfish Immunol 119:84–95

Samanman S, Kanatharana P, Chotigeat W, Deachamag P, Thavarungkul P (2011) Highly sensitive capacitive biosensor for detecting white spot syndrome virus in shrimp pond water. J Virol Methods 173(1):75–84

Sanchez-Ortiz AC, Angulo C, Luna-Gonzalez A, Alvarez-Ruiz P, Mazon-Suastegui JM, Campa-Cordova AI (2016) Effect of mixed-*Bacillus* spp isolated from pustulose ark *Anadara tuberculosa* on growth, survival, viral prevalence and immune-related gene expression in shrimp *Litopenaeus vannamei*. Fish Shellfish Immunol 59:95–102

Sánchez-Paz A (2010) White spot syndrome virus: an overview on an emergent concern. Vet Res 41(6):43. https://doi.org/10.1051/vetres/2010015

Sánchez-Paz A, Terán-Díaz B, Enríquez-Espinoza T, Encinas-Garcia T, Vázquez-Sánchez I, Mendoza-Cano F (2015) The tidepool shrimp, *Palaemon ritteri* Holmes, constitutes a novel host to the white spot syndrome virus. J Fish Dis 38(7):613–620

Sandeepa GM, Ammani K (2018) Evaluation of probiotic characteristics of bacteria isolated from aquaponds of Nellore district and the assessment of its efficacy in white leg shrimp infected with virus and water quality parameters in experimental conditions. Int J Microbiol Res 10(9):1341–1347

Sangsuriya P, Senapin S, Huang WP, Lo CF, Flegel TW (2011) Co-interactive DNA-binding between a novel, immunophilin-like shrimp protein and VP15 nucleocapsid protein of white spot syndrome virus. PLoS One 6(9):e25420. https://doi.org/10.1371/journal.pone.0025420

Sangsuriya P, Charoensapsri P, Sutthangkul J, Senapin S, Hirono I, Tassanakajon A, Amparyup P (2018) A novel white spot syndrome virus protein WSSV164 controls prophenoloxidases, PmproPOs in shrimp melanization cascade. Dev Comp Immunol 86:109–117

Sanjuktha M, Stalin RV, Aravindan K, Alavandi SV, Poornima M, Santiago TC (2012) Comparative efficacy of double-stranded RNAs targeting WSSV structural and nonstructural genes in controlling viral multiplication in *Penaeus monodon*. Arch Virol 157:993–998

Sarathi M, Ahmed VPI, Venkatesan C, Balasubramanian G, Prabavathy J, Hameed ASS (2007) Comparative study on immune response of *Fenneropenaeus indicus* to *Vibrio alginolyticus* and white spot syndrome virus. Aquaculture 271(1–4):8–20

Sarathi M, Basha AN, Ravi M, Venkatesan C, Kumar BS, Hameed ASS (2008) Clearance of white spot syndrome virus (WSSV) and immunological changes in experimentally WSSV-injected *Macrobrachium rosenbergii*. Fish Shellfish Immunol 25(3):222–230

Saravanan K, Kumar PP, Praveenraj J, Baruah A, Sivaramakrishnan T, Kumar TS, Kumar SP, Sankar RK, Roy SD (2017a) Investigation and confirmation of white spot syndrome virus (WSSV) infection in wild caught penaeid shrimps of Andaman and Nicobar Islands, India. Virus Dis 28(4):368–372

Saravanan K, Lini RM, Otta SK (2017b) Screening of pathogens from a biosecured pacific white shrimp (*Penaeus vannamei*) farm in Kattur, Tamil Nadu. Indian J Fish 64(Special Issue):229–234

Sathish S, Musthaq SS, Hameed ASS, Narayanan RB (2004a) Production of recombinant structural proteins from the Indian WSSV isolate. Aquaculture 242(1–4):69–80

Sathish S, Selvakumar C, Hameed ASS, Narayanan RB (2004b) 18-kDa protein as a marker to detect WSSV infection in shrimps. Aquaculture 238(1–4):39–50

Satoh J, Nishizawa T, Yoshimizu M (2008) Protection against white spot syndrome virus (WSSV) infection in kuruma shrimp orally vaccinated with WSSV rVP26 and rVP28. Dis Aquat Org 82:89–96

Schleder DD, Peruch LGB, Poli MA, Ferreira TH, Silva CP, Andreatta ER, Hayashi L, Vieira FDN (2018) Effect of brown seaweeds on Pacific white shrimp growth performance, gut morphology, digestive enzymes activity and resistance to white spot virus. Aquaculture 495:359–365

Sedeh FM, Afsharnasab M, Heidarieh M (2015) Immunization of *Litopenaeus vannamei* shrimp against white spot syndrome virus (WSSV) by gamma-irradiated WSSV plus *Vibrio parahaeomolyticus*. Vaccine Res 2(5):107–112

Seetang-Nun Y, Jaroenram W, Sriurairatana S, Suebsing R, Kiatpathomchai W (2013) Visual detection of white spot syndrome virus using DNA-functionalized gold nanoparticles as probes combined with loop-mediated isothermal amplification. Mol Cell Probes 27(2):71–79

Sekar A, Kim M, Jeon H, Kim K (2019) Screening and selection of bacteria inhibiting white spot syndrome virus infection to *Litopenaeus vannamei*. Biochem Biophys Rep 19:100663. https://doi.org/10.1016/j.bbrep.2019.100663

Selvam DG, Rahiman KMM, Hatha AAM (2012) An investigation into occasional white spot syndrome virus outbreak in traditional paddy cum prawn fields in India. Sci World J 2012:340830. https://doi.org/10.1100/2012/340830

Selvam R, Marimuthu S, D'Souza P (2020) Impact of phytocee™, a phytogenic feed additive on survivability of *Litopenaeus vannamei* shrimps under WSSV and salinity stress challenges. Int J Fish Aquat Stud 8(1):261–265

Sen A, Singh ISB, Rengarajan K, Philip R, Kumar GS (1999) Evidence of a Bacilliform virus causing outbreaks of whitespot disease in Penaeus monodon H. Milne Edwards in India. Asian Fish Sci 12:41–47

Shalini R, Haq MAB, Ahamed AS, Kumar P, Sedhuraman V, Banu MN (2016) WSSV transmission studies on polychaete *Pereneris cultifera* to Pacific white shrimp SPF *Litopenaeus vannamei* in captivity. Int J Pure Appl Biosci 4(6):59–75

Shalini R, Haq MB, Vaitheeswari S, Sureandiran B, Vinod B (2020) WSSV transmission studies on polychaete *Pereneris cultifera* to Pacific white shrimp SPF *Litopenaeus vannamei* in captivity. Int J Adv Sci Technol 29(2):2076–2091

Shan LP, Hu LH, Zhao Q, Zhang LN, Liu L, Chen J, Chen JP (2021) Difference in medication pattern potentially enhances antiviral efficiency of a novel amino-fluorophenyl compound on WSS risk in shrimp seedling culture. Aquaculture 539:736670. https://doi.org/10.1016/j.aquaculture.2021.736670

Shan LP, Zhang X, Hu Y, Liu L, Chen J (2024) Potential application of a small molecule compound in aquaculture against WSSV infection by maintaining the homeostasis of intestinal microbiota. Aquaculture 582:740562

Shekar M, Pradeep B, Karunasagar I (2012) White spot syndrome virus: genotypes, epidemiology and evolutionary studies. Indian J Virol 23:175–183

Shekhar MS, Gopikrishna G (2010) Development of immunodot blot assay for the detection of white spot syndrome virus infection in shrimps (*Penaeus monodon*). Aquac Res 41(11):1683–1690

Shekhar MS, Karthic K, Kumar KV, Kumar JA, Swathi A, Hauton C (2019) Comparative analysis of shrimp (*Penaeus vannamei*) miRNAs expression profiles during WSSV infection under experimental conditions and in pond culture. Fish Shellfish Immunol 93:288–295

Sheng X, Tang Q, Zhang L, Tang X, Xing J, Zhan W (2018) Development and application of a rapid semiquantitative immunochromatographic test strip to detect white spot syndrome virus. Aquaculture 495:773–779

Shi Z, Huang C, Zhang J, Chen D, Bonami JR (2000) White spot syndrome virus (WSSV) experimental infection of the freshwater crayfish, *Cherax quadricarinatus*. J Fish Dis 23(4):285–288

Shinn AP, Pratoomyot J, Griffiths D, Trong TQ, Vu NT, Jiravanichpaisal P, Briggs M (2018) Asian shrimp production and the economic costs of disease. Asian Fish Sci 31:29–58

Shiny DB, Jasmin JV, Sujithra S, Tamizharasi MJ, Kumaran T (2020) Phytochemical screening and immunostimulant efficiency of selected medicinal plant extracts against white spot syndrome virus. Plant Arch 20(2):7021–7026

Siddique MA, Haque MIM, Sanyal SK, Hossain A, Nandi SP, Alam ASMRU, Sultana M, Hasan M, Hossain MA (2018) Circulatory white spot syndrome virus in south-west region of Bangladesh from 2014 to 2017: molecular characterization and genetic variation. AMB Express 8:25. https://doi.org/10.1186/s13568-018-0553-z

Simrouni MM, Mesbah M, Memari RH, Shapoori SAMR, Boroujeni PMM (2014) Sequence analysis of ORF94 in different white spot syndrome virus (WSSV) isolates of Iran. Iran J Fish Sci 13(2):492–502

Singh ISB, Manjusha M, Pai SS, Philip R (2005) *Fenneropenaeus indicus* is protected from white spot disease by oral administration of inactivated white spot syndrome virus. Dis Aquat Org 66:265–270

Sinurat E, Saepudin E, Peranginangin R, Hudiyono S (2016) Immunostimulatory activity of brown seaweed-derived fucoidans at different molecular weights and purity levels towards white spot syndrome virus (WSSV) in shrimp *Litopenaeus vannamei*. J Appl Pharm Sci 6(10):82–91

Sirirustananun N, Chen C, Lin YC, Yeh ST, Liou CH, Chen LL, Sim SS, Chiew SL (2011) Dietary administration of a *Gracilaria tenuistipitata* extract enhances the immune response and resistance against *Vibrio alginolyticus* and white spot syndrome virus in the white shrimp *Litopenaeus vannamei*. Fish Shellfish Immunol 31(6):848–855

Siriwattanarat R, Longyant S, Chaivisuthangkura P, Wangman P, Vaniksampanna A, Sithigorngul P (2013) Improvement of immunodetection of white spot syndrome virus using a monoclonal antibody specific for heterologously expressed icp11. Arch Virol 158(5):967–979

Sithigorngul W, Rukpratanporn S, Pecharaburanin N, Longyant S, Chaivisuthangkura P, Sithigorngul P (2006) A simple and rapid immunochromatographic test strip for detection of white spot syndrome virus (WSSV) of shrimp. Dis Aquat Org 72(2):101–106

Sithigorngul P, Rukpratanporn S, Chaivisuthangkura P, Sridulyakul P, Longyant S (2011) Simultaneous and rapid detection of white spot syndrome virus and yellow head virus infection in shrimp with a dual immunochromatographic strip test. J Virol Methods 173(1):85–91

Sivagnanavelmurugan M, Marudhupandi T, Arunachalam P, Immanuel G (2012) Antiviral effect of fucoidan extracted from the brown seaweed, *Sargassum wightii*, on shrimp *Penaeus monodon* postlarvae against white spot syndrome virus. J World Aquacult Soc 43(5):697–706

Sivagnanavelmurugan M, Radhakrishnan S, Palavesam A, Arul V, Immanuel G (2018) Characterization of alginic acid extracted from Sargassum wightii and determination of its antiviral activity on shrimp *Penaeus monodon* postlarvae against white spot syndrome virus. Int J Curr Res Life Sci 7(4):1863–1872

Sivakamavalli J, Vaseeharan B (2014) Variations in biochemical and histological characteristics of WSSV infected green tiger shrimp *Penaeus semisulcatus*. J Recept Signal Transduct Res 34(5):386–395

Sivakumar S, Vimal S, Majeed A, Kumar S, Taju G, Madan N, Rajkumar T, Thamizhvanan S, Shamsudheen K, Scaria V, Sivasubbu S, Hameed ASS (2018) A new strain of white spot syndrome virus affecting *Litopenaeus vannamei* in Indian shrimp farms. J Fish Dis 41:1129–1146

Solís-Lucero G, Manoutcharian K, Hernández-López J, Ascencio F (2016) Injected phage-displayed-VP28 vaccine reduces shrimp *Litopenaeus vannamei* mortality by white spot syndrome virus infection. Fish Shellfish Immunol 55:401–406

Somboonna N, Mangkalanan S, Udompetcharaporn A, Krittanai C, Sritunyalucksana K, Flegel TW (2010) Mud crab susceptibility to disease from white spot syndrome virus is species-dependent. BMC Res Notes 3:315. http://www.biomedcentral.com/1756-0500/3/315

Song KK, Li DF, Zhang MC, Yang HJ, Ruan LW, Xu X (2010) Cloning and characterization of three novel WSSV recognizing lectins from shrimp *Marsupenaeus japonicus*. Fish Shellfish Immunol 28:596–603

Sreedharan K, Deepika A, Paria A, Babu PPS, Makesh M, Rajendran KV (2017) Ontogeny and expression analysis of tube (interleukin-1 receptor-associated kinase-4 homolog) from *Penaeus monodon* in response to white spot syndrome virus infection and on exposure to ligands. Agri Gene 3:21–31

Sreedharan K, Deepika A, Paria A, Bedekar MK, Makesh M, Rajendran KV (2018) Ontogenetic and expression of different genes involved in the Toll pathway of black tiger shrimp (*Penaeus monodon*) following immersion challenge with *Vibrio harveyi* and white spot syndrome virus (WSSV). Agri Gene 8:63–71

Sritunyalucksana K, Wannapapho W, Lo CF, Flegel TW (2006) PmRab7 is a VP28-binding protein involved in white spot syndrome virus infection in shrimp. J Virol 80(21):10734–10742

Stentiford GD, Neil DM, Peeler EJ, Shields JD, Small HJ, Vlak FTW (2012) Disease will limit future food supply from the global crustacean fishery and aquaculture sectors. J Invertebr Pathol 110(2):141–157

Subashini A, Sethi SN, Revathi K (2012) Effect of immunostimulant on enhancement of the immune response of kuruma shrimp *Marsupenaeus japonicus* and its resistance against white spot syndrome virus (WSSV). E-Planet 9(1):23–30

Sudheer NS, Philip R, Singh ISB (2012) Anti-white spot syndrome virus activity of *Ceriops tagal* aqueous extract in giant tiger shrimp *Penaeus monodon*. Arch Virol 157:1665–1675

Sudheer NS, Poulose G, Thomas A, Viswanath K, Kulkarni A, Narayanan RB, Philip R, Singh ISB (2015) Expression profile of bio-defense genes in *Penaeus monodon* gills in response to formalin inactivated white spot syndrome virus vaccine. Antivir Res 117:60–68

Sujithra S, Rajila R, Kumari SG, Shiny DB, Kumaran T (2020) Effects of immunostimulant Annona muricata extract against white spot syndrome virus infection. Eur J Mol Clin Med 7(11):2399–2412

Sukumaran V, Lowman DW, Sajeevan TP, Philip R (2010) Marine yeast glucans confer better protection than that of baker's yeast in *Penaeus monodon* against white spot syndrome virus infection. Aquac Res 41(12):1799–1805

Sullivan TJ, Dhar AK, Cruz-Flores R, Bodnar AG (2019) Rapid, CRISPR-based, field-deployable detection of white spot syndrome virus in shrimp. Sci Rep 9(1):19702. https://doi.org/10.1038/s41598-019-56170-y

Sun Y, Zhang X (2019) Role of DCP1-DCP2 complex regulated by viral and host microRNAs in DNA virus infection. Fish Shellfish Immunol 92:21–30

Sun M, Li S, Yu Y, Zhang X, Li F (2023) A novel hemocyte-specific small protein participates in white spot syndrome virus infection via binding to viral envelope protein. Viruses 15(1):227

Sundaram D, Kesavan K, Kumaravel H, Mohammed RF, Tohru M, Toshiaki I, Raja S (2016) Protective efficacy of active compounds from *Phyllanthus amarus* against white spot syndrome virus in freshwater crab (*Paratelphusa hydrodomous*). Aquac Res 47(7):2061–2067

Supak LSL, Boonnat A, Poltana P, Kanchanaphum P, Gangnonngiw W, Nash G, Withyachumnarnkul AB (2005) Infectivity of white spot syndrome virus (WSSV) to the polychaete Pereneis nuntia and a possibility of WSSV transmission from the polychaete to the black tiger shrimp *Penaeus monodon*. In: Walker PJ, Lester RG, Bondad-Reantaso MG (eds) Diseases in Asian aquaculture V. Fish Health Section, Asian Fisheries Society, Manila, pp 353–361

Supamattaya K, Hoffmann RW, Boonyaratpalin S, Kanchanaphum P (1998) Experimental transmission of white spot syndrome virus (WSSV) from black tiger shrimp *Penaeus monodon* to the sand crab *Portunus pelagicus*, mud crab *Scylla serrata* and krill *Acetes* sp. Dis Aquat Org 32:79–85

Suraprasit S, Methatham T, Jaree P, Phiwsaiya K, Senapin S, Hirono I, Lo CF, Tassanakajon A, Somboonwiwat K (2014) Anti-lipopolysaccharide factor isoform 3 from *Penaeus monodon* (ALFPm3) exhibits antiviral activity by interacting with WSSV structural proteins. Antivir Res 110:142–150

Sutthangkul J, Amparyup P, Eum JH, Strand MR, Tassanakajon A (2017) Anti-melanization mechanism of the white spot syndrome viral protein, WSSV453, via interaction with shrimp proPO-activating enzyme, PmproPPAE2. J Gen Virol 98(4):769–778

Syed MS, Kwang J (2011) Oral vaccination of baculovirus-expressed VP28 displays enhanced protection against white spot syndrome virus in *Penaeus monodon*. PLoS One 6:e26428. https://doi.org/10.1371/journal.pone.0026428

Syed MS, Madhan S, Hameed ASS, Kwang J (2009) Localization of VP28 on the baculovirus envelope and its immunogenicity against white spot syndrome virus in *Penaeus monodon*. Virol J 391:315–324

Taengchaiyaphum S, Nakayama H, Srisala J, Khiev R, Aldama-Cano DJ, Thitamadee S, Sritunyalucksana K (2017) Vaccination with multimeric recombinant VP28 induces high protection against white spot syndrome virus in shrimp. Dev Comp Immunol 76:56–64

Taju G, Kumar DV, Majeed SA, Vimal S, Tamizhvanan S, Kumar SS, Sivakumar S, Basha AN, Haribabu P, Kannabira K, Hameed ASS (2018) Delivery of viral recombinant VP28 protein using chitosan tripolyphosphate nanoparticles to protect the whiteleg shrimp, Litopenaeus vannamei from white spot syndrome virus infection. Int J Biol Macromol 107(Pt A):1131–1141

Takahashi Y, Itami T, Kondom M, Maeda M, Fujii R, Tomonaga S, Supamattaya K, Boonyaratpalin S (1994) Electron microscopic evidence of bacilliform virus infection in kuruma shrimp (*Penaeus japonicus*). Fish Pathol 29:121–125

Takahashi Y, Itami T, Maeda M, Suzuki N, Kasornchandra J, Supamattaya K, Khongpradit R, Boonyaratpalin S, Kondo M, Kawai K, Kusuda R, Hirono I, Aoki T (1996) Polymerase chaln reaction (PCR) amplification of bacilllform virus (RV PJ) DNA in *Penaeus japonicus* Bate and systemic ectodermal and mesodermal baculovirus (SEMBV) DNA in *Penaeus monodon* Fabricus. J Fish Dis 19:399–403

Takahashi Y, Fukoda K, Kondo M, Chongthaleong A, Nishi K, Nishimura M, Ogata K, Shinya I, Takise K, Fujishima K, Matsumura M (2003) Detection and prevention of WSSV infection in cultured shrimp. Asian Aquaculture Magazine, pp 25–27

Takemura K, Satoh J, Boonyakida J, Park S, Chowdhury AD, Park EY (2020) Electrochemical detection of white spot syndrome virus with a silicone rubber disposable electrode composed of graphene quantum dots and gold nanoparticle-embedded polyaniline nanowires. J Nanobiotechnol 18(1):152. https://doi.org/10.1186/s12951-020-00712-4

Tang KFJ, Lightner DV (2000) Quantification of white spot syndrome virus DNA through a competitive polymerase chain reaction. Aquaculture 189:11–21
Tang X, Wu J, Sivaraman J, Hew CL (2007) Crystal structures of major envelope proteins VP26 and VP28 from white spot syndrome virus shed light on their evolutionary relationship. J Virol 81(12):6709–6717
Tang KFJ, Navarro SA, Pantoja CR, Aranguren FL, Lightner DV (2012) New genotypes of white spot syndrome virus (WSSV) and Taura syndrome virus (TSV) from the Kingdom of Saudi Arabia. Dis Aquat Org 99(3):179–185
Tang KFJ, Pantoja CR, Redman RM, Lightner DV (2013) A histological variant of white spot syndrome virus (WSSV) from the Kingdom of Saudi Arabia. J Invertebr Pathol 113(1):82–85
Tang X, Liang Q, Liu L, Sheng X, Xing J, Zhan W (2018) An optimized double-antibody sandwich ELISA for quantitative detection of WSSV in artificially infected crayfish. J Virol Methods 251:133–138
Thakur PC, Corsin F, Turnbull JF, Shankar KM, Hao NV, Padiyar PA, Madhusudhan M, Morgan KL, Mohan CV (2002) Estimation of prevalence of white spot syndrome virus (WSSV) by polymerase chain reaction in *Penaeus monodon* postlarvae at time of stocking in shrimp farms of Karnataka, India: a population-based study. Dis Aquat Org 49(3):235–243
Thamizhvanan S, Sivakumar S, Kumar SS, Kumar DV, Suryakodi S, Balaj K, Rajkumar T, Vimal S, Majeed SA, Taju G, Hameed ASS (2019) Multiple infections caused by white spot syndrome virus and *Enterocytozoon hepatopenaei* in pond-reared *Penaeus vannamei* in India and multiplex PCR for their simultaneous detection. J Fish Dis 42(3):447–454
Thammasorn T, Sangsuriya P, Meemetta W, Senapin S, Jitrakorn S, Rattanarojpong T, Saksmerprome V (2015) Large-scale production and antiviral efficacy of multi-target double-stranded RNA for the prevention of white spot syndrome virus (WSSV) in shrimp. BMC Biotechnol 15:110. https://doi.org/10.1186/s12896-015-0226-9
Thang ND, Tu LD, Na NTL, Trang NT, Nghia PT (2019) Melanin-containing feedstuffs protect *Litopenaeus vannamei* from white spot syndrome virus. Int Aquat Res 11:303–310
Thipwong J, Saelim H, Panrat T, Phongdara A (2019) *Penaeus monodon* GILT enzyme restricts WSSV infectivity by reducing disulfide bonds in WSSV proteins. Dis Aquat Org 135(1):59–70
Thiruppathiraja C, Kumar S, Murugan V, Adaikkappan P, Sankaran K, Alagar M (2011) An enhanced immuno-dot blot assay for the detection of white spot syndrome virus in shrimp using antibody conjugated gold nanoparticles probe. Aquaculture 318(3–4):262–267
Thitamadee S, Srisala J, Taengchaiyaphum S, Sritunyalucksana K (2014) Double-dose β-glucan treatment in WSSV-challenged shrimp reduces viral replication but causes mortality possibly due to excessive ROS production. Fish Shellfish Immunol 40(2):478–484
Thomas A, Sudheer NS, Viswanathan K, Kiron V, Singh ISB, Narayanan RB (2014) Immunogenicity and protective efficacy of a major white spot syndrome virus (WSSV) envelope protein VP24 expressed in *Escherichia coli* against WSSV. J Invertebr Pathol 123:17–24
Thomas A, Sudheer NS, Viswanathan K, Kiron V, Singh ISB, Narayanan RB (2016) Expression profile of key immune-related genes in *Penaeus monodon* juveniles after oral administration of recombinant envelope protein VP28 of white spot syndrome virus. Microb Pathog 96:72–79
Thuy LHT, Dang PN, Cao H, Nguyen HA, Chiao JC, Jen CP, Do Quang L, Duc TC, Bui TT (2023) A highly sensitive immunosensor for white spot syndrome virus (WSSV) envelope protein VP28 detection based on electrochemical impedance spectroscopy technique. In: 2023 IEEE sensors. IEEE, Vienna, pp 1–4. https://doi.org/10.1109/SENSORS56945.2023.10325273
Tomy S, Otta SK, Kumar KV, Ponniah AG (2015) Possible effect of white spot syndrome virus (WSSV) infection on reproduction associated genes in the Indian white shrimp, *Fenneropenaeus indicus* (H. Milne-Edwards, 1837). Indian J Geo Mar Sci 44(12):1880–1883
Tonganunt M, Saelee N, Chotigeat W, Phongdara A (2009) Identification of a receptor for activated protein kinase C1 (Pm-RACK1), a cellular gene product from black tiger shrimp (*Penaeus monodon*) interacts with a protein, VP9 from the white spot syndrome virus. Fish Shellfish Immunol 26(3):509–514

Trang TT, Hung NH, Ninh NH, Knibb W, Nguyen NH (2019) Genetic variation in disease resistance against white spot syndrome virus (WSSV) in *Liptopenaeus vannamei*. Front Genet 10:264

Trejo-Flores JV, Luna-González A, Álvarez-Ruíz P, Escamilla-Montes R, Peraza-Gómez V, Diarte-Plata G, Esparza-Leal HM, Campa-Córdova AI, Gámez-Jiménez C, Rubio-Castro A (2016) Protective effect of *Aloe vera* in *Litopenaeus vannamei* challenged with *Vibrio parahaemolyticus* and white spot syndrome virus. Aquaculture 465:60–65

Tsai MF, Yu HT, Tzeng HF, Leu JH, Chou CM, Huang CJ, Wang CH, Lin JY, Kou GH, Lo CF (2000) Identification and characterization of a shrimp white spot syndrome virus (WSSV) gene that encodes a novel chimeric polypeptide of cellular-type thymidine kinase and thymidylate kinase. Virology 277(1):100–110

Tsai JM, Shiau LJ, Lee HH, Chan PW, Lin CY (2002) Simultaneous detection of white spot syndrome virus (WSSV) and Taura syndrome virus (TSV) by multiplex reverse transcription-polymerase chain reaction (RT-PCR) in pacific white shrimp *Penaeus vannamei*. Dis Aquat Org 50(1):9–12

Tsai JM, Wang HC, Leu JH, Wang AHJ, Zhuang Y, Walker PJ, Kou GH, Lo CH (2006) Identification of the nucleocapsid, tegument, and envelope proteins of the shrimp white spot syndrome virus virion. J Virol 80(6):3021–3029

Tseng DY, Ho PL, Huang SY, Cheng SC, Shiu YL, Chiu CS, Liu CH (2009) Enhancement of immunity and disease resistance in the white shrimp, *Litopenaeus vannamei*, by the probiotic, *Bacillus subtilis* E20. Fish Shellfish Immunol 26:339–344

Tuyen NX, Verreth J, Vlak JM, Jong DMC (2014) Horizontal transmission dynamics of white spot syndrome virus by cohabitation trials in juvenile *Penaeus monodon* and *P. vannamei*. Prev Vet Med 117:286–294

Ufaz S, Balter A, Tzror C, Einbender S, Koshet O, Shainsky-Roitman J, Yaari Z, Schroeder A (2018) Anti-viral RNAi nanoparticles protect shrimp against white spot disease. Mol Syst Des Eng 3:38–48

Valdez A, Yepiz-Plascencia G, Ricca E, Olmos J (2014) First *Litopenaeus vannamei* WSSV 100% oral vaccination protection using CotC::Vp26 fusion protein displayed on *Bacillus subtilis* spores surface. J Appl Microbiol 117:347–357

Valente CDS, Rodiles A, Marques MRF, Merrifield DL (2020) White spot syndrome virus (WSSV) disturbs the intestinal microbiota of shrimp (*Penaeus vannamei*) reared in biofloc and clear seawater. Appl Microbiol Biotechnol 104(18):8007–8023

Velmurugan S, Raman K, Viji VT, Donio MBS, Jenifer JA, Babu MM, Citarasu T (2013) Screening and characterization of antimicrobial secondary metabolites from Halomonas salifodinae MPM-TC and its in vivo antiviral influence onIndian white shrimp Fenneropenaeus indicus against WSSV challenge. J King Saud Univ – Sci 25 (3): 181–190

van Hulten MC, Tsai MF, Schipper CA, Lo CF, Kou GH, Vlak JM (2000) Analysis of a genomic segment of white spot syndrome virus of shrimp containing ribonucleotide reductase genes and repeat regions. J Gen Virol 81:307–316

van Hulten M, Witteveldt J, Peters S, Kloosterboer N, Tarchini R, Fiers M, Sandbrink H, Lankhorst R, Vlak J (2001a) The white spot syndrome virus DNA genome sequence. Virology 286:7–22

van Hulten MC, Witteveldt J, Snippe M, Vlak JM (2001b) White spot syndrome virus envelope protein VP28 is involved in the systemic infection of shrimp. Virology 285(2):228–233

Vaniksampanna A, Longyant S, Wangman P, Sithigorngul P, Chaivisuthangkura P (2017) Enhancement and confirmation of white spot syndrome virus detection using monoclonal antibody specific to VP26. Aquac Res 4(4):1699–1710

Vanpatten KA, Nunan LM, Lightner DV (2004) Seabirds as potential vectors of penaeid shrimp viruses and the development of a surrogate laboratory model utilizing domestic chickens. Aquaculture 241(1–4):31–46

Vaseeharan B, Jayakumar R, Ramasamy P (2003) PCR-based detection of white spot syndrome virus in cultured and captured crustaceans in India. Lett Appl Microbiol 37(6):443–447

Vaseeharan B, Anand PT, Murugan T, Chen JC (2006) Shrimp vaccination trials with the VP292 protein of white spot syndrome virus. Lett Appl Microbiol 43:137–142

Vaseeharan B, Ramasamy P, Wesley SG, Chen JC (2013) Influence of acute salinity changes on biochemical, hematological and immune characteristics of Fenneropenaeus indicus during white spot syndrome virus challenge. Microbiol Immunol 57:463–469

Vazquez-Boucard C, Alvarez-Ruiz P, Escobedo-Fregoso C, Anguiano-Vega G, de Jesus Duran-Avelar M, Pinto VS, Escobedo-Bonilla CM (2010) Detection of white spot syndrome virus (WSSV) in the Pacific oyster *Crassostrea gigas*. J Invertebr Pathol 104(3):245–247

Velmurugan S, Viji VT, Babu MM, Punitha MJ, Citarasu T (2012) Antimicrobial effect of *Calotropis procera* active principles against aquatic microbial pathogens isolated from shrimp and fishes. Asian Pac J Trop Biomed 2(2):S812–S817

Velmurugan S, Jerin N, Babu MM, Bindhu F, Dhas SA, Citarasu T (2015a) Screening and characterization of antiviral compounds from *Enteromorpha flexuosa* against white spot syndrome virus (WSSV) and its in vivo influence on Indian white shrimp *Fenneropenaeus indicus*. Aquac Int 23:65–80

Velmurugan S, John ST, Nagaraj DS, Ashine TA, Kumaran S, Pugazhvendan SR (2015b) Isolation of Actinomycetes from shrimp culture pond and antagonistic to pathogenic *Vibrio* spp. and WSSV. Int J Curr Microbiol Appl Sci 4(7):82–92

Venegas CA, Nonaka L, Mushiake K, Nishizawa T, Muroga K (2000) Quasi-immune response of Penaeus japonicus to penaeid rod-shaped DNA virus (PRDV). Dis Aquat Org 42(2):83–89

Venkadesh B, Bharathkumar A, Anuanandhi K, Samundeeshwari B, Srilakshmi L, Manikandan M, Kumar DK, Parimalam M, Venkadesh B (2021) A study on the occurrence of white spot syndrome virus (WSSV) in *Scylla seratta* (Forsskal, 1755) around Kasimedu, Vanagaram, Amjikarai. Int J Entomol Res 6(2):84–87

Verbruggen B, Bickley LK, van Aerle R, Bateman KS, Stentiford GD, Santos EM, Tyler CR (2016) Molecular mechanisms of white spot syndrome virus infection and perspectives on treatments. Viruses 8(1):23. https://doi.org/10.3390/v8010023

Vibin A, Sivagnanavelmurugan M, Ramasubburayan R, Palavesam A, Immanuel G (2020) Immunostimulatory effect and ProPo gene expression status of methanolic extract of *Phyllanthus niruri* against WSSV in shrimp *Penaeus monodon* (Fab). J Aquat Biol Fish 8:26–36

Vidal OM, Granja CB, Aranguren F, Brock JA, Salazar M (2001) A profound effect of hyperthermia on survival of *Litopenaeus vannamei* juveniles infected with white spot syndrome virus. J World Aquacult Soc 32:364–372

Vidya R, Paria A, Deepika A, Sreedharan K, Makesh M, Purushothaman CS, Chaudhari A, Babu PG, Rajendran KV (2014) Toll-like receptor of mud crab, *Scylla serrata*: molecular characterisation, ontogeny and functional expression analysis following ligand exposure, and bacterial and viral infections. Mol Biol Rep 41:6865–6877

Vidya R, Makesh M, Purushothaman CS, Chaudhari A, Babu PG, Rajendran KV (2016) Report of leucine-rich repeats (LRRs) from *Scylla serrata*: ontogeny, molecular cloning, characterization and expression analysis following ligand stimulation, and upon bacterial and viral infections. Gene 590(1):159–168

Vijayan KK, Raj VS, Balasubramanian CP, Alavandi SV, Sekhar VT, Santiago TC (2005) Polychaete worms—a vector for white spot syndrome virus (WSSV). Dis Aquat Org 63:107–111

Vijayan KK, Anand PS, Balasubramanian CP, Rajan JS, Praveena PE, Aravind R, Sudheer NS, Francis B, Panigrahi A, Otta SK (2024) Vertical transmission and prevalence of white spot syndrome virus (WSSV) in the wild spawning population of the Indian white shrimp, *Penaeus indicus*. J Invertebr Pathol 203:108058

Vimal S, Majeed AS, Taju G, Nambi KSN, Raj NS, Madan N, Farook MA, Rajkumar T, Gopinath D, Hameed ASS (2013) Chitosan tripolyphosphate (CS/TPP) nanoparticles: preparation, characterization and application for gene delivery in shrimp. Acta Trop 128(3):486–493

Wahjuningrum D, Hariati S, Yuhana M, Effendi I, Citarasu T, Utami DAS, Tarman K (2022) Low dose of *Nodulisporium* sp. KT29 metabolite promotes production performance and innate immunity of Pacific white leg shrimp (*Litopenaeus vannamei*) against co-infection of white spot syndrome virus and Vibrio harveyi. Aquac Int 30(5):2611–2628

Waiyapoka T, Deachamag P, Chotigeat W, Bunsanong N, Kanatharana P, Thavarungkul P, Loyprasert-Thananimit S (2015) Application of a label-free immunosensor for white spot syndrome virus (WSSV) in shrimp cultivation water. Appl Biochem Biotechnol 177(4):821–830

Wan Q, Xu L, Yang F (2008) VP26 of white spot syndrome virus functions as a linker protein between the envelope and nucleocapsid of virions by binding with VP51. J Virol 82(24):12598–12601

Wang X, Zhan W (2006) Development of an immunochromatographic test to detect white spot syndrome virus of shrimp. Aquaculture 255(1):196–200

Wang CH, Lo CF, Leu JH, Chou CM, Yeh PY, Chou HY, Tung MC, Chang CF, Su MS, Kou GH (1995) Purification and genomic analysis of baculovirus associated with white spot syndrome (WSBV) of *Penaeus monodon*. Dis Aquat Org 23(3):239–242

Wang YC, Lo CF, Chang PS, Kou GH (1998) Experimental infection of white spot baculovirus in some cultured and wild decapods in Taiwan. Aquaculture 164:221–231

Wang Q, Nunan LM, Lightner DV (2000a) Identification of genomic variations among geographic isolates of white spot syndrome virus using restriction analysis and southern blot hybridization. Dis Aquat Org 43:175–181

Wang Q, Poulos BT, Lightner DV (2000b) Protein analysis of geographic isolates of shrimp white spot syndrome virus. Arch Virol 145:263–274

Wang YG, Mohd Daud H, Shariff M, Zamri SM, Chen X (2000c) Histopathology and cytopathology of white spot syndrome virus (WSSV) in cultured *Penaeus monodon* from peninsular Malaysia with emphasis on pathogenesis and the mechanism of white spot formation. Dis Aquat Org 39:1–11

Wang Z, Hu L, Yi G, Xu H, Qi Y, Yao L (2004) ORF390 of white spot syndrome virus genome is identified as a novel anti-apoptosis gene. Biochem Biophys Res Commun 325(3):899–907

Wang Z, Chua HK, Gusti AARA, He F, Fenner B, Manopo I, Wang H, Kwang J (2005) RING-H2 protein WSSV249 from white spot syndrome virus sequesters a shrimp ubiquitin-conjugating enzyme, PvUbc, for viral pathogenesis. J Virol 79(14):8764–8772

Wang HC, Wang HC, Ko TP, Lee YM, Leu JH, Ho CH, Huang WP, Lo CF, Wang AHJ (2008) White spot syndrome virus protein ICP11: a histone-binding DNA mimic that disrupts nucleosome assembly. Proc Natl Acad Sci USA 105(52):20758–20763

Wang XW, Xu WT, Zhang XW, Zhao XF, Yu XQ, Wang JX (2009) A C-type lectin is involved in the innate immune response of Chinese white shrimp. Fish Shellfish Immunol 27(4):556–562

Wang PH, Gu ZH, Wan DH, Zhang MY, Weng SP (2011) The shrimp NF-kB pathway is activated by white spot syndrome virus (WSSV) 449 to facilitate the expression of WSSV069 (ie1), WSSV303 and WSSV371. PLoS One 6(9):e24773. https://doi.org/10.1371/journal.pone.0024773

Wang XW, Xu YH, Xu JD, Zhao XF, Wang JX (2014) Collaboration between a soluble C-type lectin and calreticulin facilitates white spot syndrome virus infection in shrimp. J Immunol 193:2106–2117

Wang XF, Liu QH, Wu Y, Huang J (2016) *Litopenaeus vannamei* clathrin coat AP17 involved in white spot syndrome virus infection. Fish Shellfish Immunol 52:309–316

Wang J, Huang Y, Xu K, Zhang X, Sun H, Fan L, Yan M (2019) White spot syndrome virus (WSSV) infection impacts intestinal microbiota composition and function in *Litopenaeus vannamei*. Fish Shellfish Immunol 84:130–137

Wang S, Li H, Weng S, Li C, He J (2020) White spot syndrome virus establishes a novel IE1/JNK/c-Jun positive feedback loop to drive replication. iScience 23(1):100752. https://doi.org/10.1016/j.isci.2019.100752

Wang P, Yang L, Guo B, Shang Z, Gao S, Zhang X (2023a) Highly sensitive detection of white spot syndrome virus with an RPA-CRISPR combined one-pot method. Aquaculture 567:739296

Wang W, Chen J, Yu C, Pan C, Lin Z, Chen J (2023b) WSV056-mediated downregulation of PvSMCT1 contributes to immune evasion of white spot syndrome virus in the infection of *Penaeus vannamei*. Aquaculture 563:739010

Wang Y, Chen Y, Tang Y, Wang Y, Gao S, Yang L, Wang P (2023c) A recombinase polymerase amplification and *Pyrococcus furiosus* Argonaute combined method for ultra-sensitive detection of white spot syndrome virus in shrimp. J Fish Dis 46(12):1357–1365

Wang W, Pan C, Lai Q, Zhang Q, Chen J (2024) WSV403 suppresses JAK/STAT signaling to impair antiviral resistance in Pacific white shrimp *Penaeus vannamei*. Aquaculture 579:740139

Wangman P, Siriwattanarat R, Longyant S, Pengsuk C, Sithigorngul P, Chaivisuthangkura P (2017) High sensitivity immunochromatographic strip test (ICP11 strip test) for white spot syndrome virus detection using monoclonal antibodies specific to ICP11 non-structural protein. Aquaculture 470:25–31

Wanna W, Aucharean C, Kaitimonchai P, Jaengkhao W (2023) Effect of dietary *Pediococcus pentosaceus* MR001 on intestinal bacterial diversity and white spot syndrome virus protection in Pacific white shrimp. Aquacult Rep 30:101570

Watthanasurorot A, Jiravanichpaisal P, Soderhall I, Soderhall K (2010) A gC1qR prevents white spot syndrome virus replication in the freshwater crayfish *Pacifastacus leniusculus*. J Virol 84:10844–10851

Weerachatyanukul W, Chotwiwatthanakun C, Jariyapong P (2021) Dual VP28 and VP37 dsRNA encapsulation in IHHNV virus-like particles enhances shrimp protection against white spot syndrome virus. Fish Shellfish Immunol 113:89–95

Wei KQ, Yang JX (2011) Histological alterations and immune response in the crayfish *Procambarus clarkii* given rVP28-incorporated diets. Fish Shellfish Immunol 31(6):1122–1128

Wei HY, Huang S, Yao T, Gao F, Jiang JZ, Wang JY (2018) Detection of viruses in abalone tissue using metagenomics technology. Aquac Res 49(8):2704–2713

Wen-Bin Z, Yuan-Hong W, Zhi-Dong Z (2000) Detection of white spot syndrome virus (WSSV) of *Penaeus chinensis* by *in situ* hybridization. Chin J Oceanol Limnol 18(3):241–246

Westenberg M, Heinhuis B, Zuidema D, Vlak JM (2005) siRNA injection induces sequence-independent protection in *Penaeus monodon* against white spot syndrome virus. Virus Res 114:133–139

Widanarni, Rahmi D, Gustilatov M, Sukenda, Utami DAS (2020) Immune responses and resistance of white shrimp *Litopenaeus vannamei* administered *Bacillus* sp. NP5 and honey against white spot syndrome virus infection. J Akuakult Indones 19(2):118–130

Wilson W, Lowman D, Antony SP, Puthumana J, Singh ISB, Philip R (2015) Immune gene expression profile of *Penaeus monodon* in response to marine yeast glucan application and white spot syndrome virus challenge. Fish Shellfish Immunol 43(2):346–356

Witteveldt J, Vlak JM, van Hulten MCW (2004a) Protection of *Penaeus monodon* against white spot syndrome virus using a WSSV subunit vaccine. Fish Shellfish Immunol 16:571–579

Witteveldt J, Cifuentes CC, Vlak JM, van Hulten MCW (2004b) Protection of *Penaeus monodon* against white spot syndrome virus by oral vaccination. J Virol 78:2057–2061

Wongprasert K, Khanobdee K, Glunukarn S, Meeratana P, Withyachumnarnkul B (2003) Time-course and levels of apoptosis in various tissues of black tiger shrimp *Penaeus monodon* infected with white-spot syndrome virus. Dis Aquat Org 55:3–10

World Organisation for Animal Health (WOAH) (2023) Manual of diagnostic tests and vaccines for terrestrial animals. Available at: (https://www.woah.org/en/what-wedo/standards/codes-and-manuals/#ui-id-2)

Wongprasert K, Rudtanatip T, Praiboon J (2014) Immunostimulatory activity of sulfated galactans isolated from the red seaweed *Gracilaria fisheri* and development of resistance against white spot syndrome virus (WSSV) in shrimp. Fish Shellfish Immunol 36(1):52–60

Wongteerasupaya C, Vickers JE, Sriurairatana S, Nash GL, Akarajamorn A, Boonsaeng V, Panyim S, Tassanakajon A, Withyachumnarnkul B, Flegel TW (1995) A non-occluded, systemic baculovirus that occurs in cells of ectodermal and mesodermal origin and causes high mortality in the black tiger prawn *Penaeus monodon*. Dis Aquat Org 21(1):69–77

Wongteerasupaya C, Pungchai P, Withyachumnarnkul B, Boonsaeng V, Panyim S, Flegel TW, Walker PJ (2003) High variation in repetitive DNA fragment length for white spot syndrome virus (WSSV) isolates in Thailand. Dis Aquat Org 54:253–257

World Bank (2014) Reducing diseases in aquaculture. Agriculture and environmental services discussion paper no. 9. World Bank, Washington, DC

Wu JL, Namikoshi A, Nishizawa T, Mushiake K, Teruya K, Muroga K (2001) Effects of shrimp density on transmission of penaeid acute viremia in *Penaeus japonicus* by cannibalism and the waterborne route. Dis Aquat Org 47:129–135

Wu JL, Nishioka T, Mori K, Nishizawa T, Muroga K (2002) A time-course study on the resistance of *Penaeus japonicus* induced by artificial infection with white spot syndrome virus. Fish Shellfish Immunol 13:391–403

Wu CC, Chang YP, Wang JJ, Liu CH, Wong SL, Jiang CM, Hsieh SL (2015) Dietary administration of *Gynura bicolor* (Roxb. Willd.) DC water extract enhances immune response and survival rate against *Vibrio alginolyticus* and white spot syndrome virus in white shrimp *Litopeneaus vannamei*. Fish Shellfish Immunol 42(1):25–33

Wu H, Zhang J, He Y, Zhou J, Yan J, Jiang M (2017) A metabolic study in hepatopancreas of *Litopenaeus vannamei* response to white spot syndrome virus. Int Aquat Res 9:195–201

Xia X, Yu Y, Weidmann M, Pan Y, Yan S, Wang Y (2014) Rapid detection of shrimp white spot syndrome virus by real time, isothermal recombinase polymerase amplification assay. PLoS One 9(8):e104667. https://doi.org/10.1371/journal.pone.0104667

Xiao B, Fu Q, Niu S, Zhu P, He J, Li C (2020) Penaeidins restrict white spot syndrome virus infection by antagonizing the envelope proteins to block viral entry. Emerg Microbes Infect 9(1):390–412

Xiao C, Zhang Y, Zhu F (2021) Effect of dietary sodium butyrate on the innate immune response of *Procambarus clarkii* and disease resistance against white spot syndrome virus. Aquaculture 541:736784. https://doi.org/10.1016/j.aquaculture.2021.736784

Xiaoli X, Xiuzhen S, Wenbin Z (2011) Development and application of antibody microarray for white spot syndrome virus detection in shrimp. Chin J Oceanol Limnol 29:930–941

Xie X, Yang F (2005) Interaction of white spot syndrome virus VP26 protein with actin. Virology 336:93–99

Xie X, Yang F (2006) White spot syndrome virus VP24 interacts with VP28 and is involved in virus infection. J Gen Virol 87(Pt 7):1903–1908

Xie ZH, Pang YS, Liu JB, Deng XW, Tang XF (2005) Studies on development of two-temperature multiplex PCR for WSSV and IHHNV. Mar Sci 29:9–12

Xie X, Xu L, Yang F (2006) Proteomic analysis of the major envelope and nucleocapsid proteins of white spot syndrome virus. J Virol 80(21):10615–10623

Xie Z, Pang Y, Deng X, Tang X, Liu J, Lu Z, Khan MI (2007) A multiplex RT-PCR for simultaneous differentiation of three viral pathogens of penaeid shrimp. Dis Aquat Org 76:77–80

Xie Z, Xie L, Pang Y, Lu Z, Xie Z, Sun J, Deng X, Liu J, Tang X, Khan M (2008) Development of a real-time multiplex PCR assay for detection of viral pathogens of penaeid shrimp. Arch Virol 153:2245–2251

Xie S, Zhang X, Zhang J, Li F, Xiang J (2015) Envelope proteins of white spot syndrome virus (WSSV) interact with *Litopenaeus vannamei* peritrophin-like protein (LvPT). PLoS One 10(12):e0144922. https://doi.org/10.1371/journal.pone.0144922

Xu WJ, Sheng XZ, Shi H, Wang ZF, Hu ZH (2007) Artificial infection for *Portunus trituberculatus* with WSSV and histopathological observation. J Shanghai Fish Univ 16(1):33–39

Xu YH, Bi WJ, Wang XW, Zhao YR, Zhao XF, Wang JX (2014) Two novel C-type lectins with a low-density lipoprotein receptor class A domain have antiviral function in the shrimp *Marsupenaeus japonicus*. Dev Comp Immunol 42:323–332

Xu T, Shan X, Li Y, Yang T, Teng G, Wu Q, Wang C, Tang KF, Zhang Q, Jin X (2021) White spot syndrome virus (WSSV) prevalence in wild crustaceans in the Bohai Sea. Aquaculture 542:736810

Yan DC, Dong SL, Huang J, Yu XM, Feng MY (2004) White spot syndrome virus (WSSV) detected by PCR in rotifers and rotifer resting eggs from shrimp pond sediments. Dis Aquat Org 59:69–73

Yan F, Xia D, Lv S, Qi Y, Xu H (2010) Functional analysis of the orf390 gene of the white spot syndrome virus. Virus Res 151(1):39–44

Yang F, He J, Lin X, Li Q, Pan D, Zhang X, Xu X (2001) Complete genome sequence of the shrimp white spot bacilliform virus. J Virol 75:11811–11820

Yang B, Song XL, Huang J, Shi CY, Liu QH, Liu L (2006) A single-step multiplex PCR for simultaneous detection of white spot syndrome virus and infectious hypodermal and haematopoietic necrosis virus in penaeid shrimp. J Fish Dis 29(5):301–305

Yang JY, Chang CI, Liu KF, Hseu JR, Chen LH, Tsai JM (2012) Viral resistance and immune responses of the shrimp *Litopenaeus vannamei* vaccinated by two WSSV structural proteins. Immunol Lett 148(1):41–48

Yang HL, Huang J, Yang B, Liu F, Zhang QL (2014) The establishment and application of isothermal cross-priming amplification techniques in detecting penaeid shrimp white spot syndrome virus. Lett Appl Microbiol 59(2):200–206

Yang MC, Shi XZ, Yang HT, Sun JJ, Xu L, Wang XW, Zhao XF, Wang JX (2016) Scavenger receptor C mediates phagocytosis of white spot syndrome virus and restricts virus proliferation in shrimp. PLoS Pathog 12(12):e1006127. https://doi.org/10.1371/journal.ppat.1006127

Yang L, Huang S, Wang ZA, Han D, Gan Y, Geng R, Zuo H, Guo Z, Weng S, He J, Xu X (2024) Oral delivery of bacteria expressing wsv108 gene-specific dsRNA protects shrimp from white spot syndrome virus (WSSV) infection. Int J Biol Macromol 261:129840

Ye T, Zong R, Zhang X (2012a) Involvement of interaction between viral VP466 and host tropomyosin proteins in virus infection in shrimp. Gene 505(2):254–258

Ye T, Zong R, Zhang X (2012b) The role of white spot syndrome virus (WSSV) VP466 protein in shrimp antiviral phagocytosis. Fish Shellfish Immunol 33(2):350–358

Yi G, Wang Z, Qi Y, Yao L, Qian J, Hu L (2004) Vp28 of shrimp white spot syndrome virus is involved in the attachment and penetration into shrimp cells. BMB Rep 37(6):726–734

Yingsunthonwattana W, Sangsuriya P, Supungul P, Tassanakajon A (2024) *Litopenaeus vannamei* heat shock protein 90 (LvHSP90) interacts with white spot syndrome virus protein, WSSV322, to modulate hemocyte apoptosis during viral infection. Fish Shellfish Immunol 151:109695

Yoganandhan K, Narayanan RB, Hameed ASS (2003a) Larvae and early post-larvae of *Penaeus monodon* (Fabricius) experimentally infected with white spot syndrome virus (WSSV) show no significant mortality. J Fish Dis 26(7):385–391

Yoganandhan K, Sathish S, Murugan V, Narayanan RB, Hameed ASS (2003b) Screening the organs for early detection of white spot syndrome virus in penaeid shrimp by histopathology and PCR techniques. Aquaculture 215:21–29

Yoganandhan K, Musthaq SS, Narayanan RB, Hameed ASS (2004) Production of polyclonal antiserum against recombinant VP28 protein and its application for the detection of white spot syndrome virus in crustaceans. J Fish Dis 27(9):517–522

Yogeeswaran A, Velmurugan S, Punitha SMJ, Babu MM, Selvaraj T, Kumaran T, Citarasu T (2012) Protection of *Penaeus monodon* against white spot syndrome virus by inactivated vaccine with herbal immunostimulants. Fish Shellfish Immunol 32(6):1058–1067

You Z, Nadala ECBJ, Yang J, Hulten MCWV, Loh PC (2002) Production of polyclonal antiserum specific to the 27.5 kDa envelope protein of white spot syndrome virus. Dis Aquat Org 51(1):77–80

Yu Y, Liu J, Li F, Zhang X, Zhang C, Xiang J (2017) Gene set based association analyses for the WSSV resistance of Pacific white shrimp *Litopenaeus vannamei*. Sci Rep 7:40549. https://doi.org/10.1038/srep40549

Yudiati E, Isnansetyo A, Murwantoko, Triyanto, Handayani CR (2019) Alginate from *Sargassum siliquosum* simultaneously stimulates innate immunity, upregulates immune genes, and enhances resistance of Pacific white shrimp (*Litopenaeus vannamei*) against white spot syndrome virus (WSSV). Mar Biotechnol 21:503–514

Yun JM, Kim BS, Hwang SM, Kim YB, Choi WB, Choi TJ (2014) Artificial infection of the native Korean freshwater prawn *Macrobrachium nipponense* (De Haan, 1849) (Decapoda, Palaemonidae) with white spot syndrome virus (WSSV). Crustaceana 87(7):866–880

Yunarty, Yuhana M, Widanarni (2016) Effects of microencapsulated synbiotic administration at different dosages against heavy co-infection of white spot disease (WSD) and *Vibrio harveyi* in Pacific white shrimp (*Litopenaeus vannamei*). Indones J Mar Sci 21(4):169–176

Yuvaraj N, Arul V (2017) Immunomodulatory effects of seagrass Halophila ovalis polysaccharide mixed feed in adult black tiger shrimp *Penaeus monodon* and its protective efficacy against white spot syndrome virus infection. Iran J Fish Sci 16(3):993–1007

Zahra A, Sukenda S, Wahjuningrum D (2017) Extract of seaweed Gracilaria verrucosa as immunostimulant to controlling white spot disease in Pacific white shrimp *Litopenaeus vannamei*. J Akuakult Indones 16(2):174–183

Zeng Y (2021) Molecular epidemiology of white spot syndrome virus in the world. Aquaculture 537:736509. https://doi.org/10.1016/j.aquaculture.2021.736509

Zhai YF, Shi DJ, He PM, Cai CE, Yin R, Jia R (2019) Effect of trans-*vp28* gene *Synechocystis* sp. PCC6803 on growth and immunity of *Litopenaeus vannamei* and defense against white spot syndrome virus (WSSV). Aquaculture 512:734306. https://doi.org/10.1016/j.aquaculture.2019.734306

Zhan WB, Wang YH, Fryer JL, Yu KK, Fukuda H, Meng QX (1998) White spot syndrome virus infection of cultured shrimp in China. J Aquat Anim Health 10:405–410

Zhan W, Wang X, Chen J, Xing J, Fukuda H (2004) Elimination of shrimp endogenous alkaline phosphatase background and development of enzyme immunoassays for the detection of white spot syndrome virus (WSSV). Aquaculture 239(1–4):15–21

Zhan W, Wang X, Chi Y, Tang X (2013) The VP37-binding protein F1ATP synthase β subunit involved in WSSV infection in shrimp *Litopenaeus vannamei*. Fish Shellfish Immunol 34(1):228–235

Zhang JS, Dong SL, Tian XL, Dong YW, Liu XY, Yan DC (2006) Studies on the rotifer (*Brachionus urceus* Linnaeus, 1758) as a vector in white spot syndrome virus (WSSV) transmission. Aquaculture 261:1181–1185

Zhang JS, Dong SL, Dong YW, Tian XL, Hou CQ (2008) Bioassay evidence for the transmission of WSSV by the harpacticoid copepod *Nitocra* sp. J Invertebr Pathol 97:33–39

Zhang Y, Ning JF, Qu XQ, Meng XL, Xu JP (2012) TAT-mediated oral subunit vaccine against white spot syndrome virus in crayfish. J Virol Methods 181(1):59–67

Zhang JY, Liu QH, Huang J (2014) Multiple proteins of white spot syndrome virus involved in recognition of beta-integrin. J Biosci 39(3):381–388

Zhang H, Cheng W, Zheng J, Wang P, Liu Q, Li Z, Shi T, Zhou Y, Mao Y, Yu X (2020) Identification and molecular characterization of a pellino protein in kuruma prawn (*Marsupenaeus japonicus*) in response to white spot syndrome virus and *Vibrio parahaemolyticus* infection. Int J Mol Sci 21:1243. https://doi.org/10.3390/ijms21041243

Zhang L, Wang Y, Hu J, Bao Z, Wang M (2023) Rapid detection of white spot syndrome virus in *Penaeus vannamei* based on real-time enzymatic recombinase amplification. Aquaculture 566:739196

Zhang X, Hu L, Liu HW, Song DW, Shen YF, Liu L, Hu Y, Chen J (2023a) Cinnamaldehyde, a major component of *Cinnamomum cassia* Presl ethanol extract, has the potential to unlock the outbreak of WSSV. Aquaculture 575:739761

Zhang X, Hu LH, Song DW, Hu Y, Chen J (2023b) Evaluation on prevention and treatment of cum inaldehyde in culture of shrimp against white spot syndrome virus. Aquaculture 562:738760

Zhang X, Hu L, Hu Y, Liu HW, Liu L, Chen J (2024) Quercetin as a major component of *Punica granatum* L. effectively controls WSSV infection and transmission in the larval shrimp culture stage. Aquaculture 578:740134

Zhao ZY, Yin ZX, Xu XP, Weng SP, Rao XY, Dai ZX, Luo YW, Yang G, Li ZS, Guan HJ, Li SD, Chan SM, Yu XQ, He JG (2009) A novel C-type lectin from the shrimp *Litopenaeus vannamei* possesses anti-white spot syndrome virus activity. J Virol 83(1):347–356

Zhao JM, Tang XQ, Zhan WB (2014) Interaction between WSSV envelope protein VP31 and triosephosphate isomerase in haemocytes of the Chinese shrimp, *Fenneropenaeus chinensis* (Osbeck, 1765) (Decapoda, Penaeoidea). Crustaceana 87(4):401–413

Zhao C, Fu H, Sun S, Qiao H, Zhang W, Jin S, Jiang S, Xiong Y, Gong Y (2017) Experimental inoculation of oriental river prawn *Macrobrachium nipponense* with white spot syndrome virus (WSSV). Dis Aquat Org 126(2):125–134

Zhiwen L, Jie H, Yan L, Li L, Kaikang U (2003) Detecting white spot syndrome virus (WSSV) in *Exopalaemon carinicauda* by dot blot hybridization and in situ hybridization. J Anhui Agric Coll 30(1):82–85

Zhu F, Zhang X (2012) Protection of shrimp against white spot syndrome virus (WSSV) with β-1, 3-D-glucan-encapsulated vp28-siRNA particles. Mar Biotechnol 14:63–68

Zhu F, Du H, Miao ZG, Quan HZ, Xu ZR (2009) Protection of *Procambarus clarkii* against white spot syndrome virus using inactivated WSSV. Fish Shellfish Immunol 26(5):685–690

Zhu F, Twan WH, Tseng LC, Peng SH, Hwang JS (2019) First detection of white spot syndrome virus (WSSV) in the mud shrimp *Austinogebia edulis* in Taiwan. Sci Rep 9:18572. https://doi.org/10.1038/s41598-019-54837-0

Zhu C, Shi D, Liao S, He P, Jia R (2020) Effects of *Synechococcus* sp. PCC 7942 harboring *vp19, vp28, and vp (19 + 28)* on the survival and immune response of *Litopenaeus vannamei* infected WSSV. Fish Shellfish Immunol 99:1–8

Zou RF, Cai M, Liu QH (2021) LvCSN5 is involved in WSSV infection via interaction with wsv006. Dev Comp Immunol 114:103870. https://doi.org/10.1016/j.dci.2020.103870.1

Zuo H, Chen C, Gao Y, Lin J, Jin C, Wang W (2011) Regulation of shrimp PjCaspase promoter activity by WSSV VP38 and VP41B. Fish Shellfish Immunol 30(4–5):1188–1191

Zwart MP, Dieu BTM, Hemerik L, Vlak JM (2010) Evolutionary trajectory of white spot syndrome virus (WSSV) genome shrinkage during spread in Asia. PLoS One 5(10):e13400. https://doi.org/10.1371/journal.pone.0013400

Microsporidians Infecting Shrimp with Special Reference to *Enterocytozoon hepatopenaei*

T. Sathish Kumar, K. Sreedharan, and K. V. Rajendran

Abstract

Microsporidians are unicellular eukaryotic obligate intracellular parasites that cause infections in a wide range of metazoan hosts from terrestrial and aquatic environments. A few genera of microsporidia have been reported to infect shellfish, including shrimp. Further, several aquatic microsporidians have been associated with severe economic impacts in the aquaculture industry, most commonly the members of the Enterocytozoon Group Microsporidia. The disease caused by the microsporidian parasite *Enterocytozoon hepatopenaei* (EHP), termed hepatopancreatic microsporidiosis (HPM), has become a major problem in the shrimp farming sector in many Southeast Asian countries. Recently, the shrimp microsporidian *Enterocytozoon hepatopenaei* has been redescribed as *Ecytonuceospora hepatopenaei* (EHP) based on morphological and molecular analysis. However, to avoid confusion, the microsporidian is referred to in this chapter as *Enterocytozoon hepatopenaei*. EHP mainly targets the hepatopancreas resulting in growth retardation and increased size variation in whiteleg shrimp *P. vannamei*. Many aquatic macrofauna, including artemia, molluscs, mussels, polychaetes, crabs, and freshwater prawns, have tested positive for EHP and act as carriers of the disease. The major route of transmission is reported to be horizontal through cannibalism and cohabitation. Several diagnostic approaches, including advanced molecular methods, have evolved for the rapid detection and quantification of the parasite. Several researchers have elucidated the host–

T. Sathish Kumar
ICAR-Central Institute of Brackishwater Aquaculture, Chennai, India
e-mail: sathishkumar.t@icar.org.in

K. Sreedharan · K. V. Rajendran (✉)
ICAR-Central Institute of Fisheries Education, Mumbai, India
e-mail: sreedharan@cife.edu.in; rajendrankv@hotmail.com

R. Kooloth Valappil et al. (eds.), *Aquatic Animal Health Management*,
https://doi.org/10.1007/978-981-96-7987-4_11

pathogen interaction through metabolomic, proteomic, metagenomic, and transcriptomic studies. The adoption of better management practices, biosecurity at farms and hatcheries, and standardized therapeutics have been proposed as effective preventive measures for managing EHP.

Keywords

Microsporidian · *Enterocytozoon hepatopenaei* (EHP) · *Ecytonuceospora hepatopenaei* · Shrimp aquaculture · Hepatopancreatic microsporidiosis (HPM)

1 Introduction

Of the crustaceans farmed around the world, penaeid shrimps constitute the major aquaculture commodity in tropical and sub-tropical countries, including China, India, Bangladesh, Southeast Asia, Tropical America, and Middle Eastern and Northern Africa (Valderrama and Anderson 2011). The major cultivated penaeid shrimps include *Penaeus vannamei, P. monodon, P. merguiensis, P. japonicus, P. chinensis, P. stylirostris, and P. indicus* (Rosenberry 2004). Along with the expansion and intensification of shrimp culture practices, infectious diseases have become persistent, causing significant economic losses. One example is the breakdown of *P. monodon* cultivation in India and other countries because of white spot syndrome virus (WSSV) infection. However, Pacific whiteleg shrimp (*P. vannamei*), with the availability of specific pathogen-free stock and culture-amenable qualities such as fast growth rate, the possibility of high stocking density, and resistance to low salinity and low temperature, has replaced *P. monodon* as the major farmed species. At the species level, with a production of 6.8 million tonnes, *P. vannamei* was the top species produced in 2022 (FAO 2024).

In spite of having various advantages, farmed *P. vannamei* faces disease challenges such as white gut/faeces syndrome, early mortality syndrome, and slow growth (Schofield et al. 2021; Rajendran et al. 2021; Sathish Kumar et al. 2022b). *Enterocytozoon hepatopenaei* (EHP) is one of the serious pathogens of penaeid shrimp, which has been prevalent in shrimp cultivation systems for an extended period, even eclipsing other destructive diseases like white spot disease and acute hepatopancreatic necrosis disease (AHPND)/early mortality syndrome (EMS). Despite causing low mortalities (Sathish Kumar et al. 2022a), this microsporidian parasite has a serious impact, resulting in slow growth and non-gainful farming. EHP, causing hepatopancreatic microsporidiosis in penaeid shrimps, has come to the spotlight after its rapid spread in the majority of shrimp farming countries. Apart from directly affecting the hepatopancreas, the organ engaged with absorption and metabolic activities in shrimp, EHP is implicated as a contributing agent of diseases such as EMS and white gut/faeces syndrome (Aranguren et al. 2017; Sathish Kumar et al. 2022b).

2 Microsporidians

Microsporidians are spore-forming unicellular eukaryotic obligate intracellular parasites with a cosmopolitan distribution across a wide range of hosts. Several aquatic microsporidians have been associated with severe economic impacts in aquaculture industry, most commonly the members of the Enterocytozoon Group Microsporidia (abbreviated as EGM) (Stentiford et al. 2019), such as *Nucleospora salmonis* (syn. *Enterocytozoon salmonis*) (Chilmonczyk et al. 1991), *Enterospora nucleophila* (Palenzuela et al. 2014), and *Enterocytozoon hepatopenaei* (EHP) (Rajendran et al. 2016).

2.1 Microsporidians of Crustaceans and Shrimp

Some microsporidian genera infect shellfish, including *Agmasoma, Ameson, Nosema, Pleistophora, Tuzetia, Thelohania, Flabelliforma, Glugoides, Vavraia, Ordospora, Nadelspora, Perizia,* and *Enterospora* (Tourtip et al. 2009; Canning et al. 2002). The first microsporidian recognised to infect shrimp was *Thelohania duorara*, described by Iversen and Manning (1959) as musculature infections in the pink shrimp in *P. duorarum*. Kelly (1979) reported the pathology and tissue specificity of three microsporidian species, *Thelohania duorara, Agmasoma penaei,* and *Pleistophora* sp., in *P. duorarum*. Subsequently, *Agmasoma penaei,* which was discovered as an infection in the muscles and connective tissues, was reported from *P. merguiensis* and *P. monodon* (Pasharawipas et al. 1994). Later, another microsporidian named *Tuzetia weidneri* was reported to infect the musculature of *P. setiferus* and *P. aztecus* (Canning et al. 2002). Further, a novel microsporidian *Perizia* sp. was detected in cotton shrimp disease-affected *P. monodon* and *P. indicus* (Han et al. 2016). However, even before *Agmasoma penaei*, a microsporidian infecting the hepatopancreatic tubules and midgut epithelium of *P. monodon* was reported by Anderson et al. (1988), though not fully characterized at that point in time. Though not identified taxonomically, a similar hepatopancreatic microsporidian was reported from *P. japonicus* in Australia (Hudson et al. 2001). Later, Chayaburakul et al. (2004) discovered a microsporidian exclusively infecting the hepatopancreas of *P. monodon* farmed in Thailand affected by Monodon slow growth syndrome (MSGS). Subsequently, Tourtip et al. (2009) named the new microsporidian *Enterocytozoon hepatopenaei* (EHP) based on ultrastructural studies.

2.2 *Enterocytozoon hepatopenaei* (EHP)

Enterocytozoon hepatopenaei, a generalist microsporidian known to switch hosts (Stentiford et al. 2016), has been reported from different penaeid species such as *P. monodon, P. vannamei*, and *P. Indicus*. EHP was also detected in *P. stylirostris* collected from Brunei (Tang et al. 2015). The experimental transmission of EHP has been demonstrated in the Indian white shrimp *P. indicus* (Sathish Kumar et al.

2019). EHP can infect all life stages of *P. vannamei,* including post-larvae and broodstocks. EHP mainly affects the hepatopancreas (HP) (Tourtip et al. 2009), an organ that performs vital functions such as digestion, absorption, and detoxification of xenobiotics. Damage to the hepatopancreas affects digestion, physiology, carbohydrate and lipid metabolism, and immunity (Sathish Kumar et al. 2022a). Apart from the hepatopancreas, EHP infects the midgut epithelial cells (Tang et al. 2016). EHP hinders the growth of shrimp, causing substantial stunting, such that the stunted shrimps are about half the size of uninfected shrimp at a specific age (Santhoshkumar et al. 2017). The growth retardation is indicated by poor pond performance with respect to reduced average daily growth (ADG) and average body weight (ABW) together with high size variation, i.e., high % coefficient of variance (%CV) (Chaijarasphong et al. 2021). The reason for the rapid spread of EHP as a pathogen of concern in shrimp cultivation is not known. One of the possibilities is that evolutionary adaptations might have permitted new parasite strains to thrive well in shrimp ponds, thereby increasing the rate of transmission to the new host population (Tang et al. 2016). EHP, like most other parasites, is an opportunistic pathogen that infects animals under biotic and abiotic stress (Tourtip et al. 2009). EHP infection disturbs the cells of the hepatopancreatic tubules and permits different pathogens like *Vibrio spp* to colonize the necrotized tubular epithelium.

Recently, the *Enterocytozoon hepatopenaei* from *Penaeus vannamei* (EHP_{Pv}) was redescribed based on several features, including morphology of the spore, life cycle, pathology, and molecular characteristics. New features have been discovered in EHP_{Pv} in contrast to the *E. hepatopenaei* isolated from *P. monodon* (EHP_{Pm}), described by Tourtip et al. (2009). Based on morphological and molecular analyses, a new genus, *Ecytonucleospora,* has been proposed, transferring *Enterocytozoon hepatopenaei* to *Ecytonucleospora hepatopenaei* n. gen. et comb (Wang et al. 2023).

2.2.1 Spore Morphology and Life Cycle

The spores of EHP are characterized by an ovoid shape with a smooth surface, dent-like curvature and ridge-like elevations on the spore wall (Rajendran et al. 2016). The spore size of EHP was determined as 1.65 ± 0.15 μm × 0.92 ± 0.05 μm by Wang et al. (2023) (Figs. 1 and 2). Earlier, Rajendran et al. (2016) also reported a similar spore size of 1.7 × 1.0 μm using scanning electron microscopy. Spores of various aquatic microsporidians possess mucous layers, sporal hairs, tubules, or tails formed as an outgrowth of the exospore layer (Canning and Vavra 2000; Vávra et al. 2005), which help in holding the spores in the water column and aiding their availability to take-up by the host (Vávra and Lukeš 2013). However, no such appendages are recorded in EHP spores.

Generally, the spores of microsporidians have a two-layered chitinaceous wall embedded with proteins that are reported to be involved in host cell attack and tissue recognition. The outer spore coat of microsporidian spores is an electron-thick layer, while the inward coat is electron-lucent (Huger 1960) (Fig. 3). The spore wall in EHP is made of a cell layer inside an electron-lucent endospore (40–60 nm) and an electron-dense exospore (10–15 nm). Internally, the spore is equipped with a remarkable expulsion apparatus finishing in an anchoring disc at the anterior end of

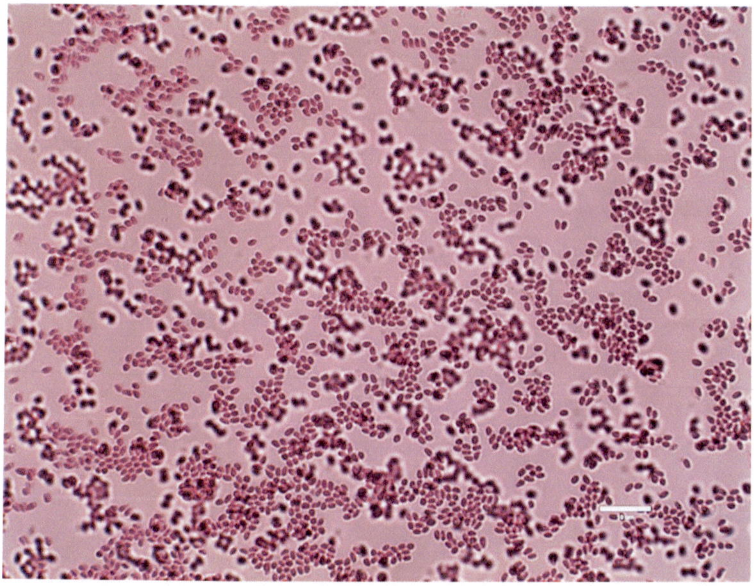

Fig. 1 Microscopic view of purified spores of *Enterocytozoon hepatopenaei* (EHP) stained with Phloxine B

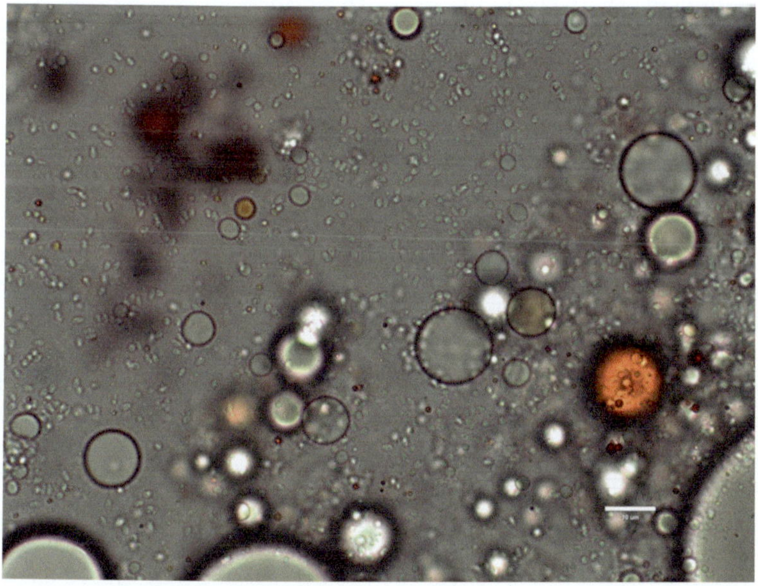

Fig. 2 Wet mount preparation of hepatopancreatic tissue smear showing the presence of *Enterocytozoon hepatopenaei* (EHP) spores

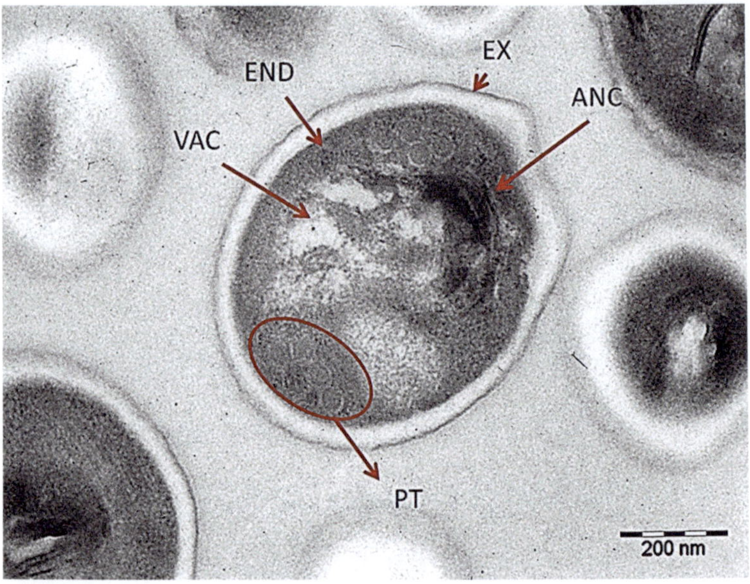

Fig. 3 Transmission electron micrograph of an *Enterocytozoon hepatopenaei* (EHP) spore revealing key ultrastructural features

the spore. Under suitable conditions inside a susceptible host, the polar tube is discharged from the apical end of the spore, which enters the host cell and is used to inject the infective sporoplasm into the host cell. The polar filament of EHP spores makes 4–5 coils (Wang et al. 2023), and that can penetrate the host cells once extruded from the spores. Several studies have shown that microsporidians can invade the host cell either by dynamic intrusion using the polar tube to enter the host cell or by exocytosis of the spore into the host cell, followed by expulsion of the polar tube into the host cell (Franzen 2008). However, the strategy for invasion adopted by EHP is yet to be elucidated.

Although the life cycle of EHP is not fully understood, it is accepted that it is completed in a single host. This has been demonstrated by a cohabitation bioassay by Salachan et al. (2017), which showed that EHP spores could infect another uninfected host directly without any secondary/alternative hosts. Like other microsporidians, EHP does not have any developmental stage outside the host cells. Once inside a host cell, the sporoplasm gets released from the polar cylinder and starts the proliferative advancement stage called merogony (Franzen 2004). In the hepatopancreatic tubular epithelial cells, the sporoplasm is believed to become meronts. The early uninucleate meronts of EHP develop into early and late-stage merogonial plasmodia, which are multinucleated as confirmed by transmission electron microscopy (TEM) (Wang et al. 2023). The merogonial plasmodia developed directly in the cytoplasm of host hepatopancreatic epithelial cells.

The merogony stage is followed by sporogony, where meronts transform into sporogonal plasmodia (sporonts) (Franzen 2004). The sporogonic stage is characterised by the formation of electron-dense discs in the parasite cytoplasm. The multinucleated sporogonial stages are strongly nucleophilic towards the host cell nuclei (Wang et al. 2023). The plasmodia undergo several cytoplasmic divisions through plasma membrane invagination resulting in numerous uninucleated sporoblasts. After separation, each early sporoblast contains a set of assembled organelles. As development continues, the late sporoblast develops into immature spores. The sporoblasts undergo morphogenesis and develop into mature spores. Various organelles such as anchoring disk, polar tube, posterior vacuole, and spore wall are gradually mature during this process (Wang et al. 2023).

Mature spores are oval-shaped, containing a single nucleus (monokaryotic spore), with a polar filament coiled 4–5 times at the posterior end and ending in an anchoring disc at the apical end (Wang et al. 2023). Thus, EHP can transmit the infection horizontally by direct ingestion of spores via cohabitation or cannibalism. Several studies indicate that EHP infection spreads quickly on farms because of the cannibalistic nature of shrimp by feeding on infected shrimp (Santhoshkumar et al. 2017) and through the release of spores with the faeces by infected shrimp (Tangprasittipap et al. 2013).

2.2.2 Genome and Pathogenesis

The genetic material of EHP is 3260 kbp long DNA comprising 2540 genes, yet lacks functional genes for the glycolytic pathway, pentose phosphate pathway and most other genes associated with fatty acid and trehalose metabolism (Boakye et al. 2017). Evidently, the microsporidian possesses no intrinsic pathways for ATP production and is dependent on host cell nutrients to meet its ATP necessities (Boakye et al. 2017).

The presence of EHP in the hepatopancreas harms the hepatopancreatic epithelial cells by denying the ATP required for cell functioning. Host mitochondria often surround the EHP spores to facilitate ATP intake (Fig. 4) (Sathish Kumar et al. 2022a; Boakye et al. 2017). The genome study by Boakye et al. (2017) uncovered the mechanism by which EHP uses the host cell ATP. There is also proof to suggest

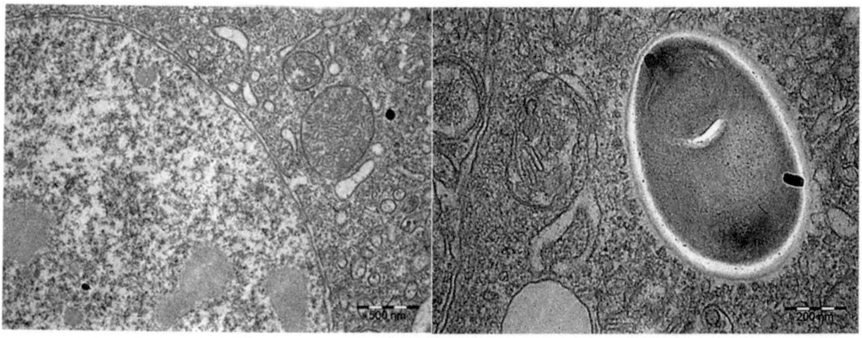

Fig. 4 Transmission electron micrograph of EHP-infected hepatopancreas section revealing the early plasmodial stage and mature spore alongside the host mitochondria

that the spores of the EHP could attack individual host cells by endocytosis, as the hepatocytes in shrimp are engaged with phagocytosis. This would bring the microsporidian in contact with the host ATP supply. The system by which the EHP uses the host ATP can be through an increased ADP/ATP ratio inside the EHP cell (brought about by truncated glycolytic pathway), which in turn coordinates the microsporidian ATP translocases to transport ATP from the surrounding host cell cytoplasm into the microsporidian cell (Boakye et al. 2017). This causes the EHP-infected HP cells to exhaust the cell supplements, leading to death and rupture of the tubules (Aranguren et al. 2017), and will also weaken the shrimp's innate immune system due to insufficient nutrients (Pascual et al. 2006). The genomic studies have additionally uncovered that glucose-6-phosphate dehydrogenase and hexokinase pathways are retained in *Enterocytozoonidae,* including EHP, and those pathways are proposed to be engaged in protecting the microsporidian from host reactive oxygen species damage (Boakye et al. 2017). Thus, EHP has the genetic capacity to harm the host and protect itself from the host's immune system, making it a successful parasite.

Besides denying the host of essential supplements, EHP is additionally suspected to aggravate the hormonal imbalance and lead to slow growth. It was recorded from the proteomics and metabolomics study that the EHP infection brings about the upregulation of ecdysteroid-regulated protein, and this could be a result of the downregulated ecdysteroids to prevent the shrimp from moulting (Ning et al. 2019). These physiological changes driven by the microsporidian infection will inevitably decrease the growth rate, which is the general clinical implication of EHP infection. In this way, the multiplying parasite will subvert a high proportion of energy from the host, thereby negatively affecting the shrimp's growth rate. EHP infection also affects the digestive enzymes and alters nutrient metabolism such as protein, carbohydrate and lipid metabolism (Duan et al. 2021; Sathish Kumar et al. 2022a). Thus, decreased levels of nutrient intake, absorption, and immunity will affect the shrimp's growth. In laboratory conditions, after 90 days, the mean growth rate of EHP-challenged shrimp (12.17 ± 0.80 g) was significantly lower than the control shrimp (19.27 ± 0.5 g) (Sathish Kumar et al. 2022a). EHP infection resulted in a significant reduction in total protein, glycogen, ATP and glucose. The infection also altered the key genes in glucose metabolism and induced the formation of oxidative damage products such as lipid peroxide (LPO) and malondialdehyde (MDA) (Cao et al. 2023). Zhang et al. (2023) studied the interplay of metabolic and transcriptional responses in shrimp during the early and late stages of EHP infection and elucidated the crucial role of polyamine in EHP development.

In addition to hindering growth, EHP can also put the animal under stress by affecting physiology and immune functions. EHP affects the nonspecific immune parameters such as alkaline phosphatase, catalase, total antioxidant capacity, phenoloxidase, superoxide anion and total haemocyte count (Sathish Kumar et al. 2022a). Consequently, alkaline phosphatase and monoester hydrolase, which play a significant role in the immunity of shrimp, get downregulated (Duan et al. 2021). In this manner, the shrimp under stress due to EHP infection ends up defenceless to infection by other opportunistic pathogens and, therefore, EHP infections are frequently reported to be associated with co-infection with bacteria (acute hepatopancreatic necrosis disease (AHPND), septic hepatopancreatic necrosis (SHPN)), viruses (Tobacco Streak Virus (TSV)), and white faeces syndrome (WFS) (Tang et al. 2016; Aranguren et al. 2017; Tang et al. 2017). Co-infection of EHP with HPV (hepatopancreatic parvovirus), infectious myonecrosis virus (IMNV), and white spot syndrome virus (WSSV) has also been reported (Babu et al. 2021; Singaravel et al. 2021; Jithendran et al. 2021; Lee et al. 2022). Apart from weakened immune response, such co-infections might be due to disturbance of the EHP-infected HP tubular epithelial cells, which in turn become the substrate for colonization by the inhabitant pathogenic *Vibrio* spp. Although not directly causing mortality (Tangprasittipap et al. 2013), co-infection of EHP and different pathogens can cause mild to severe mortality of shrimp stocks (Chayaburakul et al. 2004; Tang et al. 2015).

Severe EHP infection can also result in rapid damage of hepatopancreatic tubule epithelial cells with subsequent release of the sloughed midgut and tubular epithelial cells and spores into the faeces. Sriurairatana et al. (2014) stated that aggregated transformed microvilli (ATM) are responsible for the cause of WFS, though they could not identify any associated pathogen. However, a study conducted by Sathish Kumar et al. (2022b) demonstrated that experimental feeding of WFS-affected hepatopancreatic tissue to shrimp results in the development of ATM and WFS along with EHP spores. The study concluded that WFS is a clinical condition caused by EHP in combination with an unknown agent. Aranguren et al. (2021) reproduced the WFS under experimental conditions by challenging shrimp with EHP and *Vibrio parahaemolyticus*. Munkongwongsiri et al. (2022), through microbiome analysis, identified *Propionigenium* and *Vibrio* genera along with EHP as potential component causes of WFS. Similarly, WFS was reproduced by challenging shrimp with EHP and *Vibrio* spp. (*V. harveyi* and *V. alginolyticus*) (Subash et al. 2023a). Thus, there is good evidence that EHP has a contributory role in WFS. The clinical manifestation of WFS is more complicated and similar to a syndromic clinical condition, such as diarrhoea, and might be caused by EHP in combination with *Vibrio* spp/other pathogens.

2.2.3 Geographical Distribution and Prevalence

Hepatopancreas-infecting shrimp microsporidians have been reported in Thailand, Vietnam, Indonesia, China, Malaysia, and India (Chayaburakul et al. 2004; Tang et al. 2015, 2016; Thitamadee et al. 2016; Rajendran et al. 2016) and have been recorded in eight species of penaeid shrimp around the world (Salachan et al. 2017). Though there is no genetic evidence, EHP is believed to be endemic to Australasia (Salachan et al. 2017) because of its morphological similarity with the earlier reported unidentified microsporidians from hepatopancreatic tissue of *P. monodon* in Malaysia (Anderson et al. 1988) and *P. japonicus* in Australia (Hudson et al. 2001). The first incidence of EHP was recorded in Thailand as a microsporidian affecting the hepatopancreas of farmed *P. monodon* in 2004 (Chayaburakul et al. 2004). Following its first characterization by Tourtip et al. (2009), EHP was reported in Vietnam in 2010 as infections in *P. monodon* affected by white faeces syndrome. In 2013–2014, a high prevalence of EHP was recorded in *P. vannamei* farming from

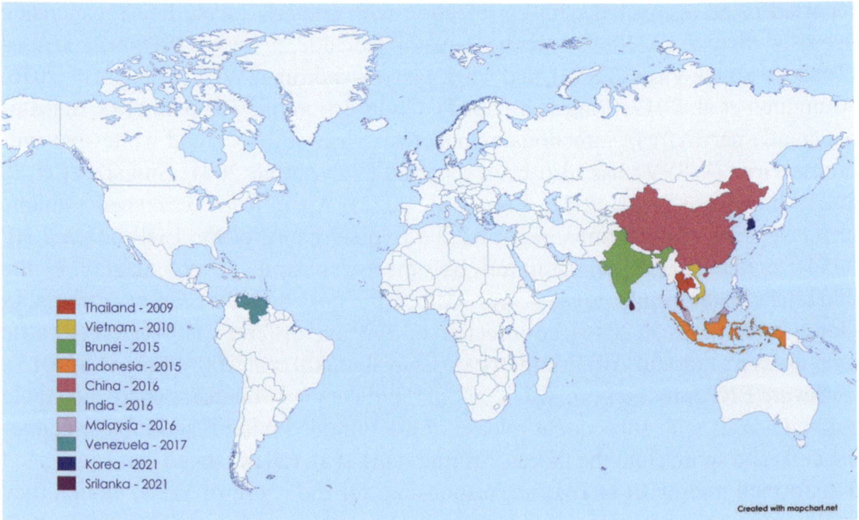

Fig. 5 Geographical distribution of EHP

Thailand (Sanguanrut et al. 2018). Following the widespread development of *P. vannamei* farming in Asia, EHP infection spread quickly and thus was reported in shrimp-growing countries such as China, Indonesia, Malaysia, and India (Tang et al. 2015, 2016; Thitamadee et al. 2016; Rajendran et al. 2016). Infection with EHP has also been reported in farmed *P. vannamei* from Korea (Kim et al. 2022) and Sri Lanka (Ananda et al. 2022) and, co-infection of HPV and EHP has been reported in *P. vannamei* from Taiwan (Lee et al. 2022). The currently known geographical distribution of EHP is illustrated in Fig. 5.

In India, among the infectious diseases, EHP was reported to be more prevalent (27%) than WSSV (6%) and IMNV (3%) (CIBA 2019). The disease has been recorded in both the east and west coasts of India, and in shrimp farms located in northern inland saline areas. It is reported that shrimp seed originating from hatcheries located on the East Coast could be a major source of transmission of EHP in the West Coast and inland saline areas (CIBA 2019). EHP-infected shrimp farms recorded size variation/growth retardation, white faeces syndrome and loose shell (Rajendran et al. 2016; Biju et al. 2016). The production loss due to EHP in Indian shrimp farms was estimated at 1.80 ± 0.24 tonnes/hectare/crop with a total production loss of 0.77 M tonnes and a revenue loss of US$ 567.62 M during 2018–19 (Patil et al. 2021). Of late, the prevalence of EHP has been widely observed in *P. monodon* culture on both the east and west coasts of India (Rajeshwar et al. 2024).

2.2.4 Host Range and Transmission

Apart from penaeid shrimp, EHP has been recorded in giant freshwater prawns (*Macrobrachium rosenbergii*) and ridgetail white prawn (*Palaemon carinicauda*) (Jang et al. 2022; Hu et al. 2023a). Many researchers reported that several cohabiting aquatic fauna, such as artemia, polychaetes, clams, mussels, and crabs, served as potential carriers of EHP in transmitting infection mainly through horizontal routes (Kummari et al. 2018; Piamsomboon et al. 2019; Desrina et al. 2020; Karthikeyan and Sudhakaran 2020; Krishnan et al. 2021; Munkongwongsiri et al. 2021; Mondal et al. 2023). EHP has also been reported in several aquatic macrofauna belonging to Arthropoda, Mollusca, and Chordata (Sajiri et al. 2023) (Table 1).

Table 1 Different host species and their status in the transmission of EHP

Sl. no	Species	Host	Reference
1	*P. monodon*	Natural host	Biju et al. (2016), Rajeshwar et al. (2024)
2	*P.vannamei*	Natural host	Rajendran et al. (2016)
3	*P.stylirostris*	Natural host	Tang et al. (2015)
4	*P.japonicus*	Natural host	Hudson et al. (2001)
5	*P.indicus*	Natural host	Mondal et al. (2023)
6	*Palaemon carinicauda*	Experimental host	Hu et al. (2023a)
7	*Macrobrachium rosenbergii*	Natural host	Jang et al. (2022)
8	Dragonfly (*Anax parthenope*, *Pantala flavescens*, and *Ischnura senegalensis*)	Natural host	Dewangan et al. (2023)
9	Mysid shrimp	Carrier/vector	CIBA (2022)
10	Polychaetes	Carrier/vector	Krishnan et al. (2021)
11	*Artemia*	Possible Reservoir	Guo et al. (2024)
12	*Mytilopsis leucophaeata*	Carrier/vector	Munkongwongsiri et al. (2022)
13	*Scylla serrata*	Carrier/vector	Mani et al. (2022), Mondal et al. (2023)
14	*S. olivacea*	Carrier/vector	Mani et al. (2022)
15	*Portunus pelagicus*	Carrier/vector	Mani et al. (2022)
16	*Portunus sanquinolentus*	Carrier/vector	Mani et al. (2022), Mondal et al. (2023)
17	*Ocypode quadrata*	Carrier/vector	Mani et al. (2022)
18	*Paracleistostoma depressum*	Carrier/vector	Sajiri et al. (2023)
19	*Parasesarma eumolpe*	Carrier/vector	Sajiri et al. (2023)
20	*Metaplax elegans*	Carrier/vector	Sajiri et al. (2023)
21	*Varuna yui*	Carrier/vector	Sajiri et al. (2023)
22	*Episesarma* sp.	Carrier/vector	Sajiri et al. (2023)
23	*Mytella strigata*	Carrier/vector	Sajiri et al. (2023)
24	*Pirenella cingulata*	Carrier/vector	Sajiri et al. (2023)
25	*Ranatra* sp.	Carrier/vector	Sajiri et al. (2023)
26	*Telescopium telescopium*	Carrier/vector	Sajiri et al. (2023)
27	*Mugilogobius chulae*	Carrier/vector	Sajiri et al. (2023)
28	*Gerrid* sp.—aquatic insect	Carrier/vector	CIBA (2022)
29	*Belostoma* sp.—aquatic insect	Carrier/vector	CIBA (2022)
30	*Dysticus* sp.—aquatic insect	Carrier/vector	Mondal et al. (2023)

Salachan et al. (2017) described the transmission of EHP by cohabitation by placing EHP-infected shrimp in perforated plastic containers along with healthy shrimp. After 14 days, all the normal shrimp were found positive for EHP by PCR. Incidence of EHP in aquatic macrofauna such as crabs, polychaetes, non-penaeids, and shrimp from aquaculture ponds has been reported (Kummari et al. 2018). The possibility of vertical transmission of EHP in shrimp was demonstrated by Khac et al. (2018), where SPF broodstocks were infected by cohabitation with infected shrimp and by feeding with the hepatopancreas of infected shrimp. However, vertical transmission via trans-gonadal tissue was reported to be unlikely (Chaijarasphong et al. 2021; CIBA 2021).

Mai et al. (2020) studied the experimental infection of EHP by different challenge methods such as intramuscular injection, oral administration, cohabitation, intra-hepatopancreatic injection and reverse gavage. They found that injection in the hepatopancreas and reverse gavage were more effective in producing infection. The experimental transmission of EHP in the larvae of *P. vannamei* via oral and immersion challenge carried out by Karthikeyan and Sudhakaran (2019) revealed that the samples collected post-challenge were positive for EHP. Further, the potential of *Artemia salina* in transmitting EHP in shrimp was explored. *Artemia* immersed in EHP-infected tissue homogenate tested EHP-positive by PCR, and subsequent feeding of the infected *Artemia* to shrimp resulted in EHP infection (Karthikeyan and Sudhakaran 2020). However, Guo et al. (2024) found that *Artemia* was not susceptible to EHP and asserted that *Artemia* was a possible EHP reservoir. Similarly, polychaete worms fed with the faecal matter of EHP-infected shrimp showed EHP infection, and feeding of these polychaetes to healthy shrimp resulted in EHP infection in the shrimp (Krishnan et al. 2021). Munkongwongsiri et al. (2021) and Dewangan et al. (2023) noticed EHP-positive samples of false mussels (*Mytilopsis leucophaeata*) from EHP-infected shrimp culture ponds, and subsequent cohabitation bioassays confirmed that they served only as passive carriers of EHP without getting infected, as shown by different diagnostic methods. The role of dragonflies (*Anax parthenope*, *Pantala flavescens*, and *Ischnura senegalensis*) as natural hosts of EHP through cohabitation and oral administration challenge experiments has been explored and reported the possibility of horizontal transmission of EHP to insects (Dewangan et al. 2023). The experimental infection of five different marine crabs with EHP-positive indicated no mortality; however, the crabs were found EHP-positive in many tissues, indicating the potential role of marine crabs as vectors of EHP (Mani et al. 2022).

2.2.5 Diagnosis

There are no specific clinical signs of EHP infection, and currently, it can be diagnosed through histological examination and molecular methods (Tangprasittipap et al. 2013; Tang et al. 2015). EHP can remain as an undetectable chronic infection in shrimp (Rajendran et al. 2016). In more advanced stages, EHP-infected shrimp typically display soft shells, lethargy, reduced feed intake, and an empty midgut (Aranguren et al. 2017). It has been reported that the EHP-infected post-larvae, when stocked, showed a reduced growth rate after the initial culture period of

25 days and reduced feed intake (50–70%) along with discolouration of the hepatopancreas (Tang et al. 2015). Under severe infections, animals exhibit a white gut with discoloured hepatopancreas, and white faeces can be seen floating in the water column in the ponds (Rajendran et al. 2016). Yet, these clinical signs cannot be used as a confirmatory test for detecting EHP, as the causal association with EHP has not been established.

Inconclusive clinical signs of EHP have led to reliance upon microscopic diagnostic techniques such as light microscopic observation of hepatopancreatic tissue smears (Figs. 1 and 2), histology and in situ hybridization (ISH) to detect the EHP infection. For microscopic observation, hepatopancreatic tissue smears can be stained with hematoxylin and eosin or Giemsa's to visualise the microsporidian spores (Sritunyalucksana et al. 2014b; Rajendran et al. 2016). Plasmodia without any observable spores cannot be used as a confirmatory test for EHP infection (Sritunyalucksana et al. 2014b). Light infection of EHP can go undetected if only light microscopy is used for the detection, as only heavily infected hepatopancreas showed the presence of spores in the smear preparation under the microscope (Rajendran et al. 2016). Modified trichrome stain was also used for the microscopic detection of EHP (Praveena et al. 2018). The calcofluor white (CWF) stain has been used for the easy identification of EHP spores in infected tissues and faeces (Zhao et al. 2020) (Fig. 6). Double staining with calcofluor white and acridine orange has been used to distinguish different life stages of EHP (Gao et al. 2020). Non-lethal diagnosis of EHP spores in faeces and hepatopancreas has been demonstrated for EHP detection (Cruz-Flores et al. 2019).

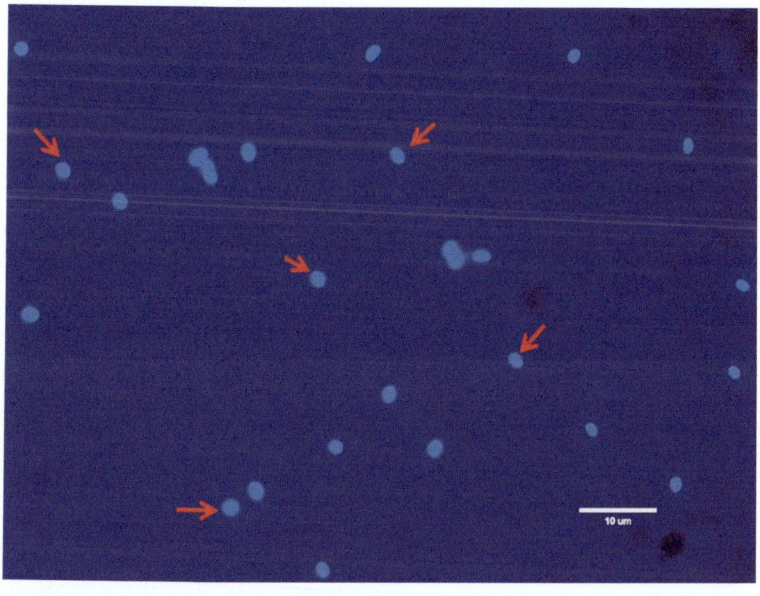

Fig. 6 EHP-infected hepatopancreatic tissue smear stained with calcofluor white (CFW) stain showing EHP spores

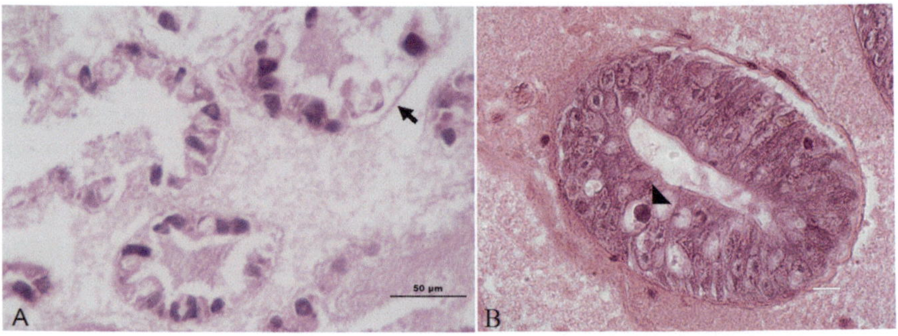

Fig. 7 Histological section of EHP-infected hepatopancreas (**a**) HP section showing severe necrosis in the tubules and epithelial cells, detachment of epithelium (arrow) and dilation of tubules. (**b**) Deeply basophilic inclusion-like body (arrowhead) of the presumptive developmental stage of EHP

Histology is another reliable diagnostic tool to detect microsporidian infections in the hepatopancreas (Fig. 7). Prominent histopathological signs of EHP infection include the presence of presumptive developmental stages of EHP plasmodia as eosinophilic to deeply basophilic inclusion-like bodies either on the vacuole or in the cytoplasm, nuclear hypertrophy and necrosis of epithelial cells of the hepatopancreatic tubules, and presence of basophilic granulation in the tubular epithelial cells with the absence of any haemocyte infiltration. As infections progress, the tubular epithelia can detach from the underlying connective tissue, leading to severe necrosis of hepatopancreatic tubules. In advanced stages, no remnants of tubular epithelium may be visible, and the lumen may contain aggregations of EHP spores surrounded by flattened haemocytic cells. Histopathological changes are more prominent in the medial, proximal and distal regions of hepatopancreas (Rajendran et al. 2016; Aranguren et al. 2017). Some of the histological changes observed in the EHP-infected shrimp show similarity with *Enterospora canceri* reported from crabs (Stentiford et al. 2007; Rajendran et al. 2016). Other disadvantages of using microscopic techniques for EHP detection are that they are sacrificial in nature, and it is difficult to resolve EHP spores through this method.

In situ hybridization can provide a more reliable and accurate assessment of the severity of EHP infection than histology (Salachan et al. 2017). PCR and in situ hybridization methods for EHP were initially described in 2009 (Tourtip et al. 2009). Molecular techniques such as in situ hybridization (ISH), Polymerase chain reaction (PCR), real-time PCR, Loop-mediated isothermal amplification (LAMP), Recombinase polymerase amplification (RPA), and CRISPR (clustered regularly interspaced short palindromic repeats)-cas techniques have been developed for the diagnosis of EHP. The PCR detection method based on the SSU rRNA gene was developed for diagnosis (Tangprasittipap et al. 2013). However, primers based on SSU rRNA showed cross-reactivity with several closely related aquatic microsporidians, resulting in false-positive results. Subsequently, more specific PCR detection targeting the spore wall protein gene (SWP) was developed, which could differentiate EHP from other closely related microsporidians (Jaroenlak et al. 2016)

(Fig. 7). An alternative ISH and PCR detection using EHP-510 F/R primers targeting the 18S rRNA segment has also been developed (Tang et al. 2015). Further, a PCR method employing β-tubulin was also developed, which could be used to monitor the low-level EHP infection in shrimp (Han et al. 2018). Koiwai et al. (2018) designed a multiplex PCR-DNA chromatography method for the simultaneous detection of several pathogens, including EHP. Hou et al. (2021) reported a new one-step PCR assay that could detect 20 copies of EHP plasmid DNA. A combination of direct PCR technology and capillary electrophoresis has been designed for the detection of EHP, and the detection limit was 25 copies/PCR reaction (Lin et al. 2022). Further, the nested PCR using the SWP gene was modified to screen a unique strain of EHP from *Macrobrachium rosenbergii,* and the redesigned internal primer was able to detect EHP from both *P. vannamei* and *M. rosenbergii* (Wang et al. 2022).

Real-time PCR assays have been reported for the quantification of EHP. These include a fluorescence SYBR Green assay targeting SSU rRNA developed by Liu et al. (2018) and EvaGreen real-time PCR assay developed by Dong et al. (2024). A real-time LAMP method using the DNA-binding dye of SYTO-16 targeting the SSU rRNA gene was established by Ma et al. (2019). Wang et al. (2020) established an integrated SYBR Green 1 fluorescence qPCR and Fluorescent Brightener 28 staining method, which showed a direct correlation between the number of spores observed under microscopy and the number of copies present in the template. A number of quantitative and visual LAMP protocols have also been described (Karthikeyan et al. 2017; Cai et al. 2018; Sathish Kumar et al. 2018; Ma et al. 2019). Furthermore, colorimetric visual LAMP protocols using the SWP gene based on the nanogold method (Arunrut et al. 2021) and simple indicator dyes (Sathish Kumar et al. 2021) have been developed. Chen et al. (2022) developed a LAMP-based onsite rapid testing device with a highly sensitive detector for EHP detection.

A real-time RPA-based detection method has been developed for the rapid detection of EHP (35–42 °C in 15 min) with a detection limit of 10 copies/μL EHP (Li et al. 2021). Kanitchinda et al. (2020) developed the RPA-Cas12a rapid detection method, in which the result could be visualized in 1 h and with a detection limit of 50 copies of DNA per reaction. A duplex RPA-LFD assay for the simultaneous visual detection of AHPND and EHP infections in shrimp was developed by Wang et al. (2021). A sensitive LFD-RPA has been developed with a detection limit of 10 copies/μL of plasmid within 20 min at 40 °C (Pang et al. 2022). Yang et al. (2023) have developed a rapid and sensitive farm-level detection of EHP employing a combination of the nuclease *Pyrococcus furiosus* Argonaute and RPA method. Furthermore, a real-time RPA has been developed targeting the SWP gene of EHP by Pang et al. (2022).

A duplex droplet digital PCR for the simultaneous detection of EHP and *Vibrio parahaemolyticus* has been developed by Zhang et al. (2022), which has a tenfold higher sensitivity than the qPCR assay. Further, a LAMP-based microfluidic chip has been developed for the simultaneous detection of DIV1, WSSV, and EHP with a limit of detection (LOD) of 10 copies/reaction for EHP (Hu et al. 2023b). Subsequently, a combination of digital microfluidic (DMF) platforms with colorimetric LAMP was also developed by Xie et al. (2023). A summary of available diagnostic methods for EHP and their attributes is provided in Table 2.

Table 2 Summary of reported molecular diagnostic techniques for EHP detection

Method	Duration	Sensitivity	Target gene	Reference
PCR	1 h 30 s	NA	SSU rRNA	Tang et al. (2015)
Nested PCR	1 h 38 min	10 copies	SWP	Jaroenlak et al. (2016)
PCR	1 h 20 min	20 copies	SSU rRNA	Hou et al. (2021)
Nested PCR	1 h 42 min	1 ng/ul	β-tubulin	Han et al. (2018)
PCR-DNA chromatography detection	29 min	10 copies	SSU rRNA	Koiwai et al. (2018)
Nested PCR	36 min	1000 Copies	SWP	Wang et al. (2022)
Direct PCR	50 min	25 copies	SWP	Lin et al. (2022)
Duplex droplet digital PCR	1 h 20 min	2.3 copies	SWP	Zhang et al. (2022)
RT-PCR (Taqman)	1 h 2 min	40 copies	SSU rRNA	Liu et al. (2018)
RT-PCR (EvaGreen)	34 min	100 copies	SWP	Dong et al. (2024)
RT-PCR (SYBR Green I)	32 min	10 copies	PTP	Wang et al. (2020)
Real-time LAMP	38 min	NA	SSU rRNA	Karthikeyan et al. (2017)
Real-time LAMP	45 min	NA	SSU rRNA	Cai et al. (2018)
LAMP—Colorimetric	1 h	10 copies	SSU rRNA	Sathish Kumar et al. (2018)
Real-time LAMP	40 min	10 copies	SSU rRNA	Ma et al. (2019)
LAMP—Colorimetric	50 min	10 copies	SWP	Arunrut et al. (2021)
LAMP—Colorimetric	40 min	10 copies	SWP	Sathish Kumar et al. (2021)
LAMP-based microfluidic chip	40 min	10 copies	SSU rRNA	Hu et al. (2023a, b)
LAMP—Digital microfluidic platform	45 min	10 copies	PTP	Xie et al. (2023)
Real-time RPA	15 min	10 copies	SSU rRNA	Li et al. (2021)
Real-time RPA	3–7 min	13 copies	SWP	Ma et al. (2021)
RPA-CRISPR Cas	1 h	50 copies	PTP	Kanitchinda et al. (2020)
RPA—lateral flow	20 min	10 copies	SWP	Pang et al. (2022)
RPA-PfAgo*	1.5 h	1 copies	SWP	Yang et al. (2023)
Duplex RPA	35 min	10 copies	SWP	Wang et al. (2021)

NA Not Available, *includes sample processing time

2.2.6 Prevention and Control

Microsporidians lack mitochondria and oxidative phosphorylation and depend on other metabolic pathways for energy. The enzymes involved in those pathways could be key therapeutic targets against microsporidian. Also, the microtubules of microsporidians are formed by polymerization of tubulin protein. Benzimadazole drugs, such as albendazole, cambendazole, benomyl, etc., can inhibit tubulin polymerization (Han and Weiss 2018). Methionine aminopeptidase 2 (MetAP2) is very important for tissue repair, protein degradation, and angiogenesis in microsporidians. Drugs such as ovalicin and fumagillin are anti-angiogenics, which are reported to act against microsporidian. However, fumagillin was found to be ineffective in reducing or eliminating EHP (Tang et al. 2016). Chitin is essential for spore infectivity and extracellular survival. Nikomycina and polyoxin can inhibit the activity of chitin synthesis. Spore germination is reported to be calcium-dependent; hence, calcium antagonists such as verapamil and lanthanum and calcium channel blockers such as nifedipine can inhibit germination (Han and Weiss 2018). Since the first incidence of slow growth syndrome in commercial penaeid aquaculture, farmers have resorted to various treatment measures, including the use of certain feed additives with bioactive substances of medicinal value, such as garlic powder or paste, which contains allicin, to treat EHP infection (Tang et al. 2016). Unfortunately, no treatment measures were successful. Chemotherapeutants successful in treating microsporidian infection in other animals, such as fumagillin used to treat *Nosema* in silkworms, were not effective in treating EHP in shrimp (Tang et al. 2016). As phagocytosis is believed to be a possible mode of invasion by EHP, it may require pathogen-ligand binding to phagocytic receptors to gain access into the phagocytic hepatopancreatic cells. Thus, identifying and blocking the appropriate ligands may inhibit pathogen entry into the host cell and provide a means of controlling infection (Boakye et al. 2017). However, such molecular approaches are still in their infancy.

Spore germination is an important event through which the microsporidian can infect the host (Fig. 8). Induced germination of EHP spores under laboratory conditions has been demonstrated successfully (Aldamo-Cano et al. 2018; Sathish Kumar et al. 2022c). EHP spore germination can be completely inhibited by different disinfectants such as potassium permanganate at 15 ppm for 15 min, active chlorine

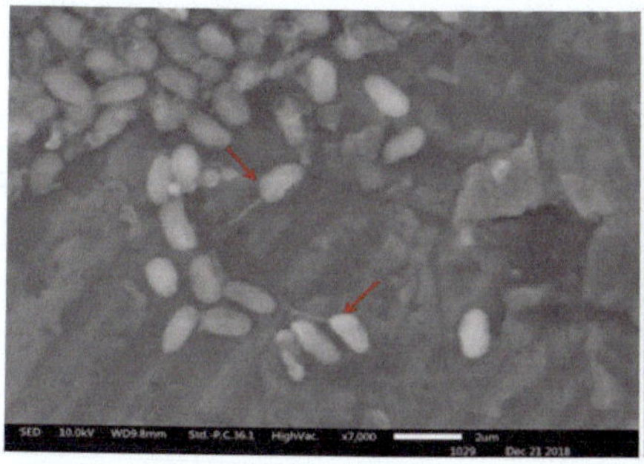

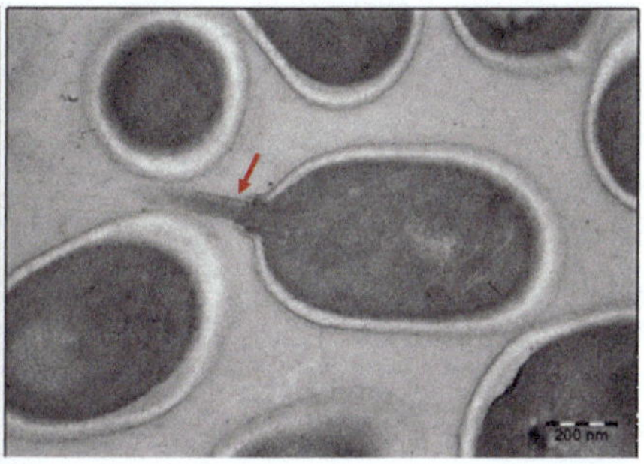

Fig. 8 Electron micrographs of EHP spores showing germination with ejecting polar tubule (arrow)

(65%) at 40 ppm for 15 min and 20% ethanol for 15 min (Aldama-Cano et al. 2018). Also, 100 ppm formalin, 3% sodium hydroxide, and 3% hydrogen peroxide completely inhibited the polar tubule extrusion (CIBA 2019). EHP spore germination was completely inhibited at −20 °C for 2 h (Aldama-Cano et al. 2018). Munkongwongsiri et al. (2021) demonstrated that EHP loses its viability if heated at 75 °C for at least 1 min, during which there was no polar tube extrusion.

Ning et al. (2021) investigated the role of linoleic acid (LNA) in improving the growth performance and immune status of EHP-infected shrimp and found that incorporation of LNA in the feed increased the length and weight of the shrimp, apart from upregulating genes associated with immunity and metabolism. Pattarayingsakul et al. (2022) found that treating EHP with 20 ppm active chlorine promptly eliminates infectivity. The efficacy of neem aqueous extract in managing EHP infection in shrimp aquaculture was investigated, and an increase in the innate immune response in animals and a reduction in the hepatopancreatic spore development were noticed (Madesh et al. 2024). The supplementation of 5-aminolaevulinic acid (5-ALA) significantly reduced mortality in *P. vannamei* and increased biomass (Kongplong et al. 2023). It has also been reported that 75 mg/L albendazole (an inhibitor of β-tubulin polymerization) in 24-h intervals prevented EHP infection in shrimp (Subash et al. 2023b). RNAi approach was also tested against EHP, in which dsRNA targeting polar tube protein 2 of EHP was found to reduce the EHP copy numbers in infected shrimp significantly (Yuanlae et al. 2024). 'CIBA EHP cura I', a phytochemical-based therapeutic developed by ICAR-CIBA, significantly reduced the EHP load and enhanced the feed intake, immunity, growth, and survival of shrimp (CIBA 2023). Various prophylactic/therapeutic measures reported against EHP are given in Table 3.

The control of EHP requires preventing the spread of infection from infected stocks to uninfected stocks, which is made difficult by the ability of the microsporidian to transmit horizontally through cannibalism and through spores released in faeces. Such a transmission strategy of EHP makes early harvesting of the crop the only way to curb the spread of infection from infected ponds to uninfected culture facilities. Following Best management practices (BMPs) and biosecurity measures is the viable option to prevent EHP from entering commercial farming

Certain live feeds like Artemia and polychaetes used for feeding the broodstocks in the hatchery facilities are found susceptible to EHP infection, and feeding such infected live feeds can act as a source of infection to the broodstocks. Hence, these live feeds can be used only after freezing (expected to inactivate EHP), pasteurisation, or gamma irradiation of the frozen feeds (Sritunyalucksana et al. 2014a). Another recommended biosecurity measure is the PCR screening of faeces from broodstock and the post-larvae for EHP infection, and eradicating the infected stocks from the system. It should also be noted that the *P. vannamei* stocks are not given the opportunity to develop resistance against the infection, as the broodstocks used in the commercial hatcheries are mostly the SPF brooders rather than the survivors of prior EHP exposure (Tang et al. 2016).

Another source of EHP infection is the carrier organisms, which may be present within the culture systems and, thus, carry over the infection to the new stocks in the

Table 3 Summary of various prophylactic/therapeutic measures and their mode of action reported for EHP

Sl no	Treatment measures	Role	Reference
	Disinfectants		
1	Potassium permanganate at 15 ppm for 15 min	Spore germination inhibition	Aldama-Cano et al. (2018)
2	Active chlorine (65%) at 40 ppm for 15 min	Spore germination inhibition	Aldama-Cano et al. (2018)
3	20% ethanol for 15 min	Spore germination inhibition	Aldama-Cano et al. (2018)
4	100 ppm formalin	Spore germination inhibition	CIBA (2019)
5	3% sodium hydroxide	Spore germination inhibition	CIBA (2019)
6	3% hydrogen peroxide	Spore germination inhibition	CIBA (2019)
Physical			
7	−20 °C for 2 h	Spore inactivation	Aldama-Cano et al. (2018)
8	75 °C for at least 1 min	Spore inactivation	Munkongwongsiri et al. (2021)
Nutraceuticals/therapeutics			
9	Linoleic acid (LNA)	Growth↑, Immunity↑	Ning et al. (2021)
10	Neem aqueous extract	Immunity↑, spore count↓	Madesh et al. (2024)
11	5-Aminolaevulinic acid (5-ALA)	Mortality↓	Kongplong et al. (2023)
12	75 mg/L albendazole	Prevention	Subash et al. (2023b)
13	dsRNA targeting polar tube protein 2	Spore count↓	Yuanlae et al. (2024)
14	CIBA EHP cura I	Immunity↑, spore count, Growth↑, Surivival↑	CIBA (2023)

subsequent culture. Thus, a thorough biosecurity measure has to be followed to prevent the entry of any susceptible carriers of EHP, such as Artemia and polychaetes, into the system (Newman 2015). The spores can also spread through white faeces, which are usually found in some of the EHP-infected ponds. They contain infective spores, and they can sink to the pond bottom to act as a reservoir of infection. Shrimp, being the benthic feeders, become exposed to reinfection through such spores (Tang et al. 2016). Therefore, the removal of any visible floating white feces is also recommended.

Other environmental sources of infection are residual spores that may be present in previously infected ponds (Jaroenlak et al. 2016), and the spread of the disease can be controlled by following proper disinfection measures. Sritunyalucksana et al. (2014) recommended using CaO (quick lime) at the rate of 6 tonnes/ha for disinfection of EHP-infected earthen ponds. They recommended the application of lime by ploughing in the dry sediments and filling the water one week after the application to allow the conversion of applied CaO to $CaCO_3$ (calcium carbonate) and the resultant fall in the pH to the normal range, which would have risen after the application of quick lime. For disinfection of brooder tanks, along with all equipment, filters, reservoirs, and pipes, cleaning with 2.5% sodium hydroxide solution (25 g NaOH/L freshwater), giving 3 h contact time and immediate washing, is recommended (Newman 2015). The tanks should be dried for 7 days after the NaOH washing and rinsed with 200 ppm acidified chlorine to disinfect the brooder tanks (Sritunyalucksana et al. 2014) . In an in vivo challenge study conducted, potassium permanganate at 15 ppm, calcium hypochlorite at 30 ppm, and hydrogen peroxide at 3% were found to be effective against EHP spores in soil (CIBA 2019). Application of >15 ppm potassium permanganate and >40 ppm calcium hypochlorite is also recommended for inactivation of spores in pond soil (NACA 2019).

3 Conclusion

Among the microsporidians infecting shrimp, *Enterocytozoon hepatopenaei* (EHP) has seen a rapid increase in prevalence among tropical shrimp farming countries and is being viewed as a serious pathogen causing significant economic losses. Several research attempts have been made to understand this discrete fungal-related protist during the past decade after its first discovery, and the significant findings of these studies are reviewed in this article. However, although being intensively studied, research gaps still exist in understanding the life cycle and the mode of invasion adopted by EHP. Understanding the prevalence, histopathology, natural and experimental host range, mode of transmission through experimental infection, and development of sensitive diagnostic tests have been the foci of studies with limited attempts to find therapeutic measures. However, more attention should be given to understanding the mode of invasion in the host (either by active invasion or endocytosis), which will aid in preventing and treating hepatopancreatic microsporidiosis, for which no treatment is currently available. It has been shown that proper biosecurity measures, starting with PCR screening of seeds and brooders, avoidance of susceptible live feed, and proper drying cycles and disinfection measures, can prevent EHP from infecting new crops. In light of new insights into the pathogenesis and host-parasite relationships, future research needs to be oriented to understand more about the microbiome changes brought about by EHP infection in the shrimp and environment, as EHP, in most cases, occurs as co-infections and are most often linked to the diseases of obscure aetiology like white gut/white faeces syndrome.

References

Aldama-Cano DJ, Sanguanrut P, Munkongwongsiri N, Ibarra-Gámez JC, Itsathitphaisarn O, Vanichviriyakit R, Flegel TW, Sritunyalucksana K, Thitamadee S (2018) Bioassay for spore polar tube extrusion of shrimp *Enterocytozoon hepatopenaei* (EHP). Aquaculture 490:156–161

Ananda KLN, De Silva DSG, Dhanapala DMPM, Chandrarathna WPR, Lakmal JAS, Gunawardena TA, Kalupahana AW, Ranathunga RMNPK, Asoka JM, Wijesundera RRMKK, Anupama NMT (2022) First report on *Enterocytozoon hepatopenaei* (EHP) infection in Pacific whiteleg shrimp

(*Litopenaeus vannamei*) cultured in Sri Lanka. National Aquatic Resources Research and Development Agency (NARA), Scientific Sessions 2022

Anderson IG, Shamsudin MN, Shariff M, Nash G (1988) Bacterial septicemia in juvenile tiger shrimp, *Penaeus monodon*, cultured in Malaysian brackishwater ponds. Asian Fish Sci 2:93–108

Aranguren Caro LF, Mai HN, Cruz-Florez R, Agcalao Marcos FL, Alenton RRR, Dhar AK (2021) Experimental reproduction of White Feces Syndrome in whiteleg shrimp, *Penaeus vannamei*. PLoS One 16(12):e0261289. https://doi.org/10.1371/journal.pone.0261289

Aranguren LF, Han JE, Tang KF (2017) *Enterocytozoon hepatopenaei* (EHP) is a risk factor for acute hepatopancreatic necrosis disease (AHPND) and septic hepatopancreatic necrosis (SHPN) in the Pacific white shrimp *Penaeus vannamei*. Aquaculture 471:37–42

Arunrut N, Tondee B, Khumwan P, Kampeera J, Kiatpathomchai W (2021) Rapid and sensitive colorimetric detection of microsporidian *Enterocytozoon hepatopenaei* (EHP) based on spore wall protein (SWP) gene using loop-mediated isothermal amplification combined with DNA functionalized gold nanoparticles as probes. Aquaculture 533:736–206

Babu B, Sathiyaraj G, Mandal A, Kandan S, Biju N, Palanisamy S, You S, Nisha RG, Prabhu NM (2021) Surveillance of disease incidence in shrimp farms located in the east coastal region of India and in vitro antibacterial efficacy of probiotics against *Vibrio parahaemolyticus*. J Invertebr Pathol 179:107536

Biju N, Sathiyaraj G, Raj M, Shanmugam V, Baskaran B, Govindan U, Kumaresan G, Kasthuriraju KK, Chellamma TSRY (2016) High prevalence of *Enterocytozoon hepatopenaei* in shrimps *Penaeus monodon* and *Litopenaeus vannamei* sampled from slow growth ponds in India. Dis Aquat Org 120:225–230

Boakye DW, Jaroenlak P, Prachumwat A, Williams TA, Bateman KS, Itsathitphaisarn O, Sritunyalucksana K, Paszkiewicz KH, Moore KA, Stentiford GD, Williams BA (2017) Decay of the glycolytic pathway and adaptation to intranuclear parasitism within Enterocytozoonidae microsporidia. Environ Microbiol 19:2077–2089

Cai SX, Kong FD, Xu SF, Yao CL (2018) Real-time loop mediated isothermal amplification for rapid detection of *Enterocytozoon hepatopenaei*. PeerJ 6:e5993

Canning EU, Vavra J (2000) Phylum Microsporidia Balbiani 1882. In: Lee JJ, Leedale GF, Bradbury P (eds) An Illustrated Guide to the Protozoa, 2nd edition. Society of Protozoologists, pp 39–126

Canning EU, Curry A, Overstreet RM (2002) Ultrastructure of *Tuzetia weidneri sp. n.* (Microsporidia: Tuzetiidae) in skeletal muscle of *Litopenaeus setiferus* and *Farfantepenaeus aztecus* (Crustacea: Decapoda) and new data on *Perezia nelsoni* (Microsporidia: Pereziidae) in *L. setiferus*. J Invertebr Pathol 80:184–194

Cao Z, Chen C, Wang C, Li T, Chang L, Si L, Yan D (2023) *Enterocytozoon hepatopenaei* (EHP) infection alters the metabolic processes and induces oxidative stress in *Penaeus vannamei*. Animals 13(23):3661

Chaijarasphong T, Munkongwongsiri N, Stentiford GD, Aldama-Cano DJ, Thansa K, Flegel TW, Sritunyalucksana K, Itsathitphaisarn O (2021) The shrimp microsporidian *Enterocytozoon hepatopenaei* (EHP): Biology, pathology, diagnostics and control. J Invertebr Pathol 186:107458

Chayaburakul K, Nash G, Pratanpipat P, Sriurairatana S, Withyachumnarnkul B (2004) Multiple pathogens found in growth-retarded black tiger shrimp *Penaeus monodon* cultivated in Thailand. Dis Aquat Org 60:89–96

Chen Z, Zhao K, He Z, Luo X, Qin Z, Tan Y, Zheng X, Wu Z, Deng Y, Chen H, Guo Y (2022) Development and evaluation of a thermostatic nucleic acid testing device based on magnesium pyrophosphate precipitation for detecting *Enterocytozoon hepatopenaei*. Chin Chem Lett 33(8):4053–4056

Chilmonczyk S, Cox WT, Hedrick RP (1991) *Enterocytozoon salmonis n. sp.*: an intranuclear microsporidium from salmonid fish. J Protozool 38(3):264–269

CIBA (2019) Annual report 2019. Central Institute of Brackishwater Aquaculture, Chennai, p 244p

CIBA (2021) Annual report 2021. Central Institute of Brackishwater Aquaculture, Chennai, pp 96–97

CIBA (2022) Annual report 2022. Central Institute of Brackishwater Aquaculture, Chennai, pp 124–125

CIBA (2023) Annual report 2023. Central Institute of Brackishwater Aquaculture, Chennai, p 135p

Cruz-Flores R, Mai HN, Noble BL, Schofield PJ, Dhar AK (2019) Detection of *Enterocytozoon hepatopenaei* using an invasive but non-lethal sampling method in shrimp (*Penaeus vannamei*). J Microbiol Methods 162:38–41

Desrina D, Prayitno SB, Haditomo AHC, Latritiani R, Sarjito S (2020) Detection of *Enterocytozoon hepatopenaei* (EHP) DNA in the polychaetes from shrimp ponds suffering white feces syndrome outbreaks. Biodivers J Biol Diversity 21(1)

Dewangan NK, Pang J, Zhao C, Cao C, Yin B, Weng S, He J (2023) Host and transmission route of *Enterocytozoon hepatopenaei* (EHP) between dragonfly and shrimp. Aquaculture 574:739642

Dong X, Chen Y, Lou H, Wang G, Zhou C, Wang L, Li X, Luo J, Huang J (2024) Development of a melting curve-based triple Eva Green Real-Time PCR Assay for simultaneous detection of three shrimp pathogens. Animals 14(4):592

Duan Y, Chen H, Wang J, Zeng S, Wang Y, Mo Z, Dan X, Li Y (2021) Response signatures of *Litopenaeus vannamei* to natural *Enterocytozoon hepatopenaei* infection revealed by the integration of the microbiome and transcriptome. Aquaculture 542:736885. https://doi.org/10.1016/J.AQUACULTURE.2021.736885

FAO (2024) The Food and Agriculture Organization of the United Nations. The State of World Fisheries and Aquaculture (SOFIA), Blue transformation in action

Franzen C (2004) Microsporidia: how can they invade other cells? Trends Parasitol 20:275–279

Franzen C (2008) Microsporidia: a review of 150 years of research. Open Parasitol J 2(1):1

Gao Y, Yang J, Chen L, Wu J, Zhou Z (2020) Double staining with calcofluor white and acridine orange to distinguish different life stages of *Enterocytozoon hepatopenaei*. Fish Shellfish Immunol 98:47–54

Guo XM, Gao W, Chen X, Wang HL, Zhao RH, Xie GS, Huang J (2024) Is Artemia susceptible to the microsporidium *Ecytonucleospora hepatopenaei* (EHP) infection? Aquaculture 590:741014

Han B, Weiss LM (2018) Therapeutic targets for the treatment of microsporidiosis in humans. Expert Opin Ther Targets 22:903. https://doi.org/10.1080/14728222.2018.1538360

Han JE, Tang KF, Pantoja CR, Lightner DV, Redman RM, Le Groumellec M (2016) Detection of a new microsporidium Perezia sp. in shrimps *Penaeus monodon* and *P. indicus* by histopathology, in situ hybridization and PCR. Dis Aquat Org 120(2):165–171

Han JE, Tang KF, Kim JH (2018) The use of beta-tubulin gene for phylogenetic analysis of the microsporidian parasite *Enterocytozoon hepatopenaei* (EHP) and in the development of a nested PCR as its diagnostic tool. Aquaculture 495:899–902

Hou ZH, Yu JY, Wang JJ, Li T, Chang LR, Fang Y, Yan DC (2021) Development of a PCR assay for the effective detection of *Enterocytozoon hepatopenaei* (EHP) and investigation of EHP prevalence in Shandong Province, China. J Invertebr Pathol 184:107653

Hu G, Wang W, Xu K, Wang C, Liu D, Xu J, Yan B, Nanjing J, Huan G (2023a) Transcriptomic and metabolomic analyses of Palaemon carinicauda hepatopancreas in response to *Enterocytozoon hepatopenaei* (EHP) infection. Aust Fish 8:92. https://doi.org/10.3390/fishes8020092

Hu K, Li Y, Wang F, Liu J, Li Y, Zhao Q, Zheng X, Zhu N, Yu X, Fang S, Huang J (2023b) A loop-mediated isothermal amplification-based microfluidic chip for triplex detection of shrimp pathogens. J Fish Dis 46(2):137–146

Hudson DA, Hudson NB, Pyecroft SB (2001) Mortalities of *Penaeus japonicus* prawns associated with microsporidean infection. Aust Vet J 79:504–505

Huger A (1960) Electron microscope study on the cytology of a microsporidian spore by means of ultrathin sectioning. J Insect Pathol 2(81):105

Iversen ES, Manning RB (1959) A new microsporidian parasite from the pink shrimp (*Penaeus duorarum*). Trans Am Fish Soc 88:130–132

Jang GI, Kim SM, Oh YK, Lee SJ, Hong SY, Lee HE, Kwon MG, Kim BS (2022) First report of *Enterocytozoon hepatopenaei* infection in giant freshwater prawn (*Macrobrachium rosenbergii* de Man) cultured in the Republic of Korea. Animals 12(22):3149

Jaroenlak P, Sanguanrut P, Williams BAP, Stentiford GD, Flegel TW, Sritunyalucksana K, Itsathitphaisarn O (2016) A nested PCR assay to avoid false positive detection of the microsporidian *Enterocytozoon hepatopenaei* (EHP) in environmental samples in shrimp farms. PLoS One 11:e0166320. https://doi.org/10.1371/journal.pone.0166320

Jithendran KP, Krishnan AN, Jagadeesan V, Anandaraja R, Praveena PE, Anushya S, Amarnath CB (2021) Co-infection of infectious myonecrosis virus and *Enterocytozoon hepatopenaei* in *Penaeus vannamei* farms in the east coast of India. Aquac Res 52(10):4701–4710. https://doi.org/10.1111/ARE.15304

Kanitchinda S, Srisala J, Suebsing R, Prachumwat A, Chaijarasphong T (2020) CRISPR-Cas fluorescent cleavage assay coupled with recombinase polymerase amplification for sensitive and specific detection of *Enterocytozoon hepatopenaei*. Biotechnol Rep 27:e00485. https://doi.org/10.1016/j.btre.2020.e00485

Karthikeyan K, Sudhakaran R (2019) Experimental horizontal transmission of *Enterocytozoon hepatopenaei* in post-larvae of whiteleg shrimp, *Litopenaeus vannamei*. J Fish Dis 42(3):397–404

Karthikeyan K, Sudhakaran R (2020) Exploring the potentiality of Artemia salina to act as a reservoir for microsporidian *Enterocytozoon hepatopenaei* of penaeid shrimp. Biocatal Agric Biotechnol 25:101607

Karthikeyan K, Sharma A, Mekata T, Itami T, Sudhakaran R (2017) Rapid and sensitive real-time loop-mediated isothermal amplification for the detection of *Enterocytozoon hepatopenaei* of shrimp. Aquaculture 481:119–123. https://doi.org/10.1016/j.aquaculture.2017.08.036

Kelly JF (1979) Tissue specificities of *Thelohania duorara*, *Agmasoma penaei*, and Pleistophora sp., microsporidian parasites of pink shrimp, *Penaeus duorarum*. J Invertebr Pathol 33(3):331–339

Khac HV, Thanh TNT, Thu GNT, Le CH, Nguyen VD (2018) Vertical transmission and early diagnosis of the microsporidian *Enterocytozoon hepatonaei* in whiteleg shrimp *Penaeus vannamei*. J Pure Appl Microbiol 12(3):1125

Kim JH, Lee C, Jeon HJ, Kim BK, Lee NK, Choi SK, Han JE (2022) First report on *Enterocytozoon hepatopenaei* (EHP) infection in Pacific white shrimp (*Penaeus vannamei*) cultured in Korea. Aquaculture 547:737525

Koiwai K, Kodera T, Thawonsuwan J, Kawase M, Kondo H, Hirono I (2018) A rapid method for simultaneously diagnosing four shrimp diseases using PCR-DNA chromatography method. J Fish Dis 41(2):395–399

Kongplong S, Kanjanasopa D, Pongtippatee P, Vanichviriyakit R, Withyachumnarnkul B (2023) 5-Aminolaevulinic acid reduced the mortality of the Pacific white shrimp *Litopenaeus vannamei* infected with *Enterocytozoon hepatopenaei*. Aquaculture 568:739322

Krishnan AN, Kannappan S, Aneesh PT, Praveena PE, Jithendran KP (2021) Polychaete worm: a passive carrier for *Enterocytozoon hepatopenaei* in shrimp. Aquaculture 545:737187

Kummari S, Haridas DV, Handique S, Peter S, Rakesh CG, Sneha KG, Manojkumar B, Pillai D (2018) Incidence of hepatopancreatic microsporidiasis by *Enterocytozoon hepatopenaei* (EHP) in *Penaeus vannamei* culture in Nellore District, Andhra Pradesh, India and the role of management in its prevention and transmission. J Curr Microbiol Appl Sci 7:2125–2134

Lee CF, Chang YC, Chiou HY, Chang HW (2022) Concurrent infection of a novel genotype of hepatopancreatic parvovirus and *Enterocytozoon hepatopenaei* in *Penaeus vannamei* in Taiwan. J Fish Dis 45(8):1201–1210

Li G, Cong F, Cai W, Li J, Wu M, Xiao L, Hu X, Zeng W, He D (2021) Development of a recombinase polymerase amplification (RPA) fluorescence assay for the detection of *Enterocytozoon hepatopenaei* (EHP). Aquac Rep 19:100584

Lin HY, Yen SC, Tsai SK, Shen F, Lin JHY, Lin HJ (2022) Combining direct PCR technology and capillary electrophoresis for an easy-to-operate and highly sensitive infectious disease detection system for shrimp. Life 12(2):276

Liu YM, Qiu L, Sheng AZ, Wan XY, Cheng DY, Huang J (2018) Quantitative detection method of *Enterocytozoon hepatopenaei* using TaqMan probe real-time PCR. J Invertebr Pathol 151:191–196. https://doi.org/10.1016/j.jip.2017.12.010

Ma B, Yu H, Fang J, Sun C, Zhang M (2019) Employing DNA binding dye to improve detection of *Enterocytozoon hepatopenaei* in real-time LAMP. Sci Rep 9:1–7. https://doi.org/10.1038/s41598-019-52459-0

Ma C, Fan S, Wang Y, Yang H, Qiao Y, Jiang G, Lyu M, Dong J, Shen H, Gao S (2021) Rapid detection of *Enterocytozoon hepatopenaei* infection in shrimp with a real-time isothermal recombinase polymerase amplification assay. Front Cell Infect Microbiol 11:1. https://doi.org/10.3389/fcimb.2021.631960

Madesh S, Sudhakaran G, Sreekutty AR, Kesavan D, Almutairi BO, Arokiyaraj S, Dhanaraj M, Seetharaman S, Arockiaraj J (2024) Exploring neem aqueous extracts as an eco-friendly strategy to enhance shrimp health and combat EHP in aquaculture. Aquac Int 32(3):3357–3377

Mai HN, Cruz-Flores R, Aranguren Caro LF, White BN, Dhar AK (2020) A comparative study of *Enterocytozoon hepatopenaei* (EHP) challenge methods in *Penaeus vannamei*. J Invertebr Pathol 171:107336. https://doi.org/10.1016/j.jip.2020.107336

Mani R, Raja S, Kesavan K, Vijay P, Babu VS, Dhas DS, Velu K (2022) Experimental infection of *Enterocytozoon hepatopenaei* parasite (EHP) of penaeid shrimp in Indian marine crabs. Arch Microbiol 204(7):416

Mondal S, Deepika A, Hundare S, Poojary N, Abraham TJ, Sreedharan K, Rajendran KV (2023) A study on the natural host range of *Enterocytozoon hepatopenaei* in different species of shrimp and co-habiting aquatic fauna. Indian J Agric Res 70(3):98–106

Munkongwongsiri N, Aldama-Cano DJ, Suebsing R, Thaiue D, Prasartset T, Itsathitphaisarn O, Sritunyalucksana K (2021) Microsporidian *Enterocytozoon hepatopenaei* (EHP) spores are inactivated in 1 min at 75 °C. Aquaculture 533:736178. https://doi.org/10.1016/J.AQUACULTURE.2020.736178

Munkongwongsiri N, Prachumwat A, Eamsaard W, Lertsiri K, Flegel TW, Stentiford GD, Sritunyalucksana K (2022) *Propionigenium* and *Vibrio* species identified as possible component causes of shrimp white feces syndrome (WFS) associated with the microsporidian *Enterocytozoon hepatopenaei*. J Invertebr Pathol 192:107784. https://doi.org/10.1016/j.jip.2022.107784

Network of Aquaculture Centres in Asia-Pacific (NACA) (2019) Fact sheet on *Enterocytozoon hepatopenaei*, a microsporidian parasite of shrimp. Retrieved from https://enaca.org/?id=1064

Newman SG (2015) Microsporidian impacts shrimp production – industry efforts address control, not eradication. Global Aquaculture Advocate, 16–17 (March/April)

Ning J, Huang J, Li C, Zhang X, Wang R, Wu Q (2019) Proteomic and metabolomic responses of Litopenaeus vannamei to *Enterocytozoon hepatopenaei* infection. Fish Shellfish Immunol 92:177–187. https://doi.org/10.1016/j.fsi.2019.06.026

Ning M, Bi J, Sun W, Xie X, Huang Y, Gu W, Wang W, Qiao Y, Jiang G, Shen H, Meng Q (2021) Linolenic acid improves growth performance and immune status of *Penaeus vannamei* infected by *Enterocytozoon hepatopenaei*. Aquaculture 535:736397

Palenzuela O, Redondo MJ, Cali A, Takvorian PM, Alonso-Naveiro M, Alvarez-Pellitero P, Sitjà-Bobadilla A (2014) A new intranuclear microsporidium, *Enterospora nucleophila* n. sp., causing an emaciative syndrome in a piscine host (*Sparus aurata*), prompts the redescription of the family Enterocytozoonidae. Int J Parasitol 44(3–4):189–203

Pang J, Zhao C, Su H, Liu Z, Lu Q, He X, Weng S, He J (2022) Rapid visual detection of *Enterocytozoon hepatopenaei* by recombinase polymerase amplification combined with lateral flow dipstick. Front Mar Sci 9:985314

Pascual M, Ahumada JA, Chaves LF, Rodó X, Bouma M (2006) Malaria resurgence in the East African highlands: temperature trends revisited. Proc Natl Acad Sci USA 103(15):5829–5834

Pasharawipas T, Flegel TW, Chaiyaroj S, Mongkolsuk S, Sirisinha S (1994) Comparison of amplified RNA gene sequences from microsporidian parasites (*Agmasoma* or *Thelohania*) in *Penaeus merguiensis* and *P. monodon*. Asian Fish Sci 7:169–178

Patil PK, Geetha R, Ravisankar T, Avunje S, Solanki HG, Abraham TJ, Vinoth SP, Jithendran KP, Alavandi SV, Vijayan KK (2021) Economic loss due to diseases in Indian shrimp farming with special reference to *Enterocytozoon hepatopenaei* (EHP) and white spot syndrome virus (WSSV). Aquaculture 533:736231

Pattarayingsakul W, Munkongwongsiri N, Thitamadee S, Sritunyalucksana K, Aldama-Cano DJ (2022) Shrimp microsporidian EHP spores in culture water lose activity in 10 days or can be inactivated quickly with chlorine. Aquaculture 548:737665

Piamsomboon P, Choi SK, Hanggono B, Nuraini YL, Wati F, Tang KF, Park S, Kwak D, Rhee MH, Han JE, Kim JH (2019) Quantification of *Enterocytozoon hepatopenaei* (EHP) in penaeid shrimps from Southeast Asia and Latin America using taqman probe-based quantitative PCR. Pathogens 8(4):233

Praveena PE, Bhuvaneswari T, Krishnan AN, Jagadeesan V, Rajan JJS, Jithendran KP (2018) An improved microscopic method for the rapid diagnosis of the emerging microsporidian parasite *Enterocytozoon hepatopenaei* in shrimp. Curr Sci 115(5):758–763

Rajendran KV, Shivam S, Ezhil Praveena P, Babu SV, Sathish S, Avunje S, Jithendran KP, Sanjukta R, Sood N, Vijayan KK (2016) Emergence of *Enterocytozoon hepatopenaei* (EHP) in farmed *Penaeus (Litopenaeus) vannamei* in India. Aquaculture 454:272–280

Rajendran KV, Stentiford GD, Bass D (2021) The rise of the syndrome – sub-optimal growth disorders in farmed shrimp. Rev Aquac 13(4):1888–1906. https://doi.org/10.1111/raq.12550

Rajeshwar BN, Kumar TS, Jithendran KP, Bedekar MK, Rajendran KV (2024) Prevalence of *Enterocytozoon hepatopenaei* (EHP) in *Penaeus monodon* farms in India. Aquac Int 32:1–22

Rosenberry B (2004) Major cultivated penaeid shrimp species: a review. Aquac Res 35(12):1140–1154

Sajiri S, Limsuwan C, Panchompoo J, Somsiri T (2023) Detection of *Enterocytozoon hepatopenaei* in various aquatic macrofauna species from different phyla: Arthropoda, Mollusca, and Chordata. Aquac Res 54(2):250–260. https://doi.org/10.1111/are.16034

Salachan PV, Jaroenlak P, Thitamadee S, Itsathitphaisarn O, Sritunyalucksana K (2017) Laboratory cohabitation challenge model for shrimp hepatopancreatic microsporidiosis (HPM) caused by *Enterocytozoon hepatopenaei* (EHP). BMC Vet Res 13:9

Sanguanrut P, Munkongwongsiri N, Kongkumnerd J, Thawonsuwan J, Thitamadee S, Boonyawiwat V, Tanasomwang V, Flegel TW, Sritunyalucksana K (2018) A cohort study of 196 Thai shrimp ponds reveals a complex etiology for early mortality syndrome (EMS). Aquaculture 493:26–36. https://doi.org/10.1016/j.aquaculture.2018.04.033

Santhoshkumar S, Sivakumar S, Vimal S, Abdul Majeed S, Taju G, Haribabu P, Uma A, Sahul Hameed AS (2017) Biochemical changes and tissue distribution of *Enterocytozoon hepatopenaei* (EHP) in naturally and experimentally EHP-infected whiteleg shrimp, *Litopenaeus vannamei* (Boone, 1931), in India. J Fish Dis 40:529–539

Sathish Kumar T, Navaneeth Krishnan A, Joseph Sahaya Rajan J, Makesh M, Jithendran KP, Alavandi SV, Vijayan KK (2018) Visual loop-mediated isothermal amplification (LAMP) for the rapid diagnosis of *Enterocytozoon hepatopenaei* (EHP) infection. Parasitol Res 117:1485. https://doi.org/10.1007/s00436-018-5828-4

Sathish Kumar T, Radhika K, Patil PK, Makesh M, Alavandi SV, Vijayan KK (2019) Experimental infection and horizontal transmission of *Enterocytozoon hepatopenaei* (EHP) in penaeid shrimps. Abstract presented at Maricon 2019, 16–20 Dec 2019, Cochin University of Science and Technology (CUSAT), Kerala, India. Abstract volume ISBN: 978-81922264

Sathish Kumar T, Radhika K, Joseph Sahaya Rajan J, Makesh M, Alavandi SV, Vijayan KK (2021) Closed-tube field-deployable loop-mediated isothermal amplification (LAMP) assay based on spore wall protein (SWP) for the visual detection of *Enterocytozoon hepatopenaei* (EHP). J Invertebr Pathol 183:107624. https://doi.org/10.1016/J.JIP.2021.107624

Sathish Kumar T, Praveena PE, Sivaramakrishnan T, Rajan JJS, Makesh M, Jithendran K (2022a) Effect of *Enterocytozoon hepatopenaei* (EHP) infection on physiology, metabolism, immunity, and growth of *Penaeus vannamei*. Aquaculture 553:738105

Sathish Kumar T, Makesh M, Alavandi S, Vijayan K (2022b) Clinical manifestations of White feces syndrome (WFS), and its association with *Enterocytozoon hepatopenaei* in *Penaeus vannamei* grow-out farms: a pathobiological investigation. Aquaculture 547:737463

Sathish Kumar T, Praveena PE, Rajendran KV, Jithendran KP (2022c) Artificial germination of *Enterocytozoon hepatopenaei* (EHP) spores induced by ions under the scanning electron microscope. J Invertebr Pathol 188:107699

Schofield PJ, Noble BL, Aranguren Caro LF, Mai HN, Padilla TJ, Millabas J, Dhar AK (2021) Pathogenicity of Acute Hepatopancreatic Necrosis Disease (AHPND) on the freshwater prawn, Macrobrachium rosenbergii, and Pacific White Shrimp, Penaeus vannamei, at various salinities. Aquac Res 52(4):1480–1489. https://doi.org/10.1111/are.15001

Singaravel V, Gopalakrishnan A, Martin GG (2021) Multiple infections of *Enterocytozoon hepatopenaei* and Hepatopancreatic parvovirus in pond-reared *Penaeus vannamei* in India. Aquaculture 545:737232. https://doi.org/10.1016/J.AQUACULTURE.2021.737232

Sritunyalucksana K, Sanguanrut P, Salachan PV, Thitamadee S, Flegel TW (2014a) Urgent appeal to control the spread of the shrimp microsporidian parasite Enterocytozoon hepatopenaei (EHP). Network of Aquaculture Centres in Asia-Pacific (NACA), May 9, pp 1–5

Sriurairatana S, Boonyawiwat V, Gangnonngiw W, Laosutthipong C, Hiranchan J, Flegel TW (2014b) White feces syndrome of shrimp arises from transformation, sloughing and aggregation of hepatopancreatic microvilli into vermiform bodies superficially resembling gregarines. PLoS One 9:e99170. https://doi.org/10.1371/journal.pone.0099170

Stentiford GD, Bateman KS, Longshaw M, Feist SW (2007) *Enterospora canceri* n. gen., n. sp., intranuclear within the hepatopancreatocytes of the European edible crab *Cancer pagurus*. Dis Aquat Org 75:61–72

Stentiford GD, Becnel JJ, Weiss LM, Keeling PJ, Didier ES, Bjornson S, Freeman MA, Brown MJF, Roesel K, Sokolova Y, Snowden KF (2016) Microsporidia–emergent pathogens in the global food chain. Trends Parasitol 32:336–348

Stentiford GD, Bass D, Williams BA (2019) Ultimate opportunists—the emergent Enterocytozoon group microsporidia. PLoS Pathog 15(5):e1007668

Subash P, Chrisolite B, Sivasankar P, Rosalind George M, Vijay Amirtharaj K, Padmavathy P, Rani V, Sankar Sri Balaje R, Gowtham S, Mageshkumar P (2023a) White feces syndrome in *Penaeus vannamei* is potentially an *Enterocytozoon hepatopenaei* (EHP) associated pathobiome origin of Vibrio spp. J Invertebr Pathol 198:107932. https://doi.org/10.1016/j.jip.2023.107932

Subash P, Uma A, Ahilan B, Kannan SS (2023b) In vivo and in silico investigations on the efficacy of albendazole against *Enterocytozoon hepatopenaei* (EHP) infecting Penaeus vannamei. Aquaculture 575:739801

Tang KFJ, Pantoja CR, Redman RM, Han JE, Tran LH, Lightner DV (2015) Development of in situ hybridization and PCR assays for the detection of *Enterocytozoon hepatopenaei* (EHP), a microsporidian parasite infecting penaeid shrimp. J Invertebr Pathol 130:37–41. https://doi.org/10.1016/j.jip.2015.06.009

Tang KF, Han JE, Aranguren LF, White-Noble B, Schmidt MM, Piamsomboon P, Risdiana E, Hanggono B (2016) Dense populations of the microsporidian *Enterocytozoon hepatopenaei* (EHP) in feces of *Penaeus vannamei* exhibiting white feces syndrome and pathways of their transmission to healthy shrimp. J Invertebr Pathol 140:1–7

Tang KF, Aranguren LF, Piamsomboon P, Han JE, Maskaykina IY, Schmidt MM (2017) Detection of the microsporidian *Enterocytozoon hepatopenaei* (EHP) and Taura syndrome virus in *Penaeus vannamei* cultured in Venezuela. Aquaculture 480:17–21

Tangprasittipap A, Srisala J, Chouwdee S, Somboon M, Chuchird N, Limsuwan C, Srisuvan T, Flegel TW, Sritunyalucksana K (2013) The microsporidian Enterocytozoon hepatopenaei is not the cause of white feces syndrome in whiteleg shrimp *Penaeus (Litopenaeus) vannamei*. BMC Vet Res 9:139

Thitamadee S, Prachumwat A, Srisala J, Jaroenlak P, Salachan PV, Sritunyalucksana K, Flegel TW, Itsathitphaisarn O (2016) Review of current disease threats for cultivated penaeid shrimp in Asia. Aquaculture 452:69–87

Tourtip S, Wongtripop S, Stentiford GD, Bateman KS, Sriurairatana S, Chavadej J, Sritunyalucksana K, Withyachumnarnkul B (2009) *Enterocytozoon hepatopenaei* sp. nov. (Microsporida: Enterocytozoonidae), a parasite of the black tiger shrimp *Penaeus monodon* (Decapoda: Penaeidae): fine structure and phylogenetic relationships. J Invertebr Pathol 102:21–29

Valderrama D, Anderson JL (2011) Shrimp production review. In: Conference presentation at Global Outlook for Aquaculture Leadership (GOAL), Santiago, Chile, pp 6–9

Vávra J, Lukeš J (2013) Microsporidia and 'The Art of Living Together'. Adv Parasitol 82:253–319. https://doi.org/10.1016/B978-0-12-407706-5.00004-6

Vávra J, Hylis M, Oborník M, Vossbrinck CR (2005) Microsporidia in aquatic microcrustacea: the copepod microsporidium Marssoniella elegans Lemmermann, 1900 revisited. Folia Parasitol (Praha) 52:163–168

Wang L, Lv Q, He Y, Gu R, Zhou B, Chen J, Fan X, Pan G, Long M, Zhou Z (2020) Integrated qPCR and staining methods for detection and quantification of *Enterocytozoon hepatopenaei* in shrimp *Litopenaeus vannamei*. Microorganisms 8(9):1366

Wang P, Ma C, Liao L, Yu J, Yi L, Qiao Y, Liu X, Gao S, Shen H, Lu Q (2021) Simultaneous visual diagnosis of acute hepatopancreatic necrosis disease and *Enterocytozoon hepatopenaei* infection in shrimp with duplex recombinase polymerase amplification. J Fish Dis 44(11):1753–1763

Wang Y, Zhou J, Yin M, Ying N, Xiang Y, Liu W, Ye J, Li X, Fang W, Tan H (2022) A modification of nested PCR method for detection of *Enterocytozoon hepatopenaei* (EHP) in giant freshwater prawn *Macrobrachium rosenbergii*. Front Cell Infect Microbiol 12:1013016

Wang Y, Chen J, Na Y, Li XC, Zhou JF, Fang WH, Tan HX (2023) *Ecytonucleospora hepatopenaei* n. gen. et comb. (Microsporidia: Enterocytozoonidae): a redescription of the *Enterocytozoon hepatopenaei* (Tourtip et al., 2009), a microsporidian infecting the widely cultivated shrimp *Penaeus vannamei*. J Invertebr Pathol 201:107988

Xie M, Chen T, Cai Z, Lei B, Dong C (2023) A digital microfluidic platform coupled with colorimetric loop-mediated isothermal amplification for on-site visual diagnosis of multiple diseases. Lab Chip 23(12):2778–2788

Yang L, Guo B, Wang Y, Zhao C, Zhang X, Wang Y, Tang W, Shen H, Wang P, Gao S (2023) *Pyrococcus furiosus* argonaute combined with recombinase polymerase amplification for rapid and sensitive detection of *Enterocytozoon hepatopenaei*. J Agric Food Chem 11:944–951. https://doi.org/10.1021/acs.jafc.2c06582

Yuanlae S, Prasartset T, Reamtong O, Munkongwongsiri N, Panphloi M, Preechakul T, Suebsing R, Thitamadee S, Prachumwat A, Itsathitphaisarn O, Taengchaiyaphum S (2024) Shrimp injection with dsRNA targeting the microsporidian EHP polar tube protein reduces internal and external parasite amplification. Sci Rep 14(1):4830

Zhang H, Gong HY, Cao WW, Que MY, Ye L, Shi L (2022) Duplex droplet digital PCR method for the detection of *Enterocytozoon hepatopenaei* and *Vibrio parahaemolyticus* acute hepatopancreatic necrosis disease. J Fish Dis 45(6):761–769

Zhang L, Zhang S, Qiao Y, Cao X, Cheng J, Meng Q, Shen H (2023) Dynamic interplay of metabolic and transcriptional responses in shrimp during early and late infection stages of *Enterocytozoon hepatopenaei* (EHP). Int J Mol Sci 24(23):16738

Zhao RH, Gao W, Qiu L, Chen X, Dong X, Li C, Huang J (2020) A staining method for detection of *Enterocytozoon hepatopenaei* (EHP) spores with calcofluor white. J Invertebr Pathol 172:107347

Disease Syndromes in Farmed Shrimp

M. Poornima, K. Nalayeni, T. Sathish Kumar, and R. Vidya

Abstract

Globally, shrimp aquaculture is a rapidly growing, vibrant enterprise. In India, since the introduction of the exotic Pacific white shrimp, *Penaeus vannamei,* in 2009, substantial progress has been realized by shrimp aquaculture during the past decade and has made a significant contribution to the national economy. The growing demand from the farming sector has led to a trend in the intensification of Indian shrimp culture. However, the escalation of culture practice has aggravated various disease problems, and since 2011, *Penaeus vannamei* farms in India have been facing continuous mortalities and morbidities due to various diseases. The increasing incidences of diseases are significant challenges to shrimp rearing systems, and the success of shrimp farming is threatened due to continuous epidemics in the shrimp farming sector. In addition to infectious diseases, shrimp aquaculture in India has been confronting certain syndromes with unknown etiology that are constantly causing mortality or reduced growth, thereby incurring severe economic losses to shrimp farmers. Syndromes such as white feces syndrome (WFS), Running mortality syndrome (RMS), Black gill syndrome, and White muscle syndrome (WMS) affecting productivity are being reported from shrimp culture ponds. As the literature on disease syndromes is scant, more efforts are needed on researchable issues related to these syndromes to develop preventive health management strategies to curtail production losses in shrimp aquaculture settings. Hence, in this chapter, details on disease syndromes reported from shrimp farming are presented.

M. Poornima (✉) · K. Nalayeni · T. Sathish Kumar · R. Vidya
ICAR-Central Institute of Brackishwater Aquaculture, Chennai, India
e-mail: modempoornima@gmail.com; knalayeni@gmail.com; manofpm@gmail.com; vidyaumadevi@gmail.com

R. Kooloth Valappil et al. (eds.), *Aquatic Animal Health Management*,
https://doi.org/10.1007/978-981-96-7987-4_12

Keywords

LSS · RMS · Shrimp · Syndrome · WFS

1 Introduction

Aquaculture is the fastest food-generating segment, and it has accomplished an all-time high production of 114.5 million tonnes globally (FAO 2020). Shrimp farming is a vibrant aquaculture sector that produces high-quality food and provides livelihood and sustenance to millions of people. The global aquaculture of shrimp is dominated by two species, *Penaeus monodon* (black tiger shrimp) and *P. vannamei* (Pacific white shrimp). In India, shrimp culture practices were started traditionally in natural water bodies such as "bheries" in West Bengal and "Pokkali" fields in Kerala. In the mid-90 s, Indian shrimp culture was wholly dominated by black tiger shrimp, *P. monodon*. During this period, the intensive farming practices and viable phase of shrimp culture prospered in India. As a result, India's black tiger shrimp culture attained the maximum production of 1.44 lakh tones in 2006–2007 (www.mpeda.com). However, the intensification of aquaculture practices led to the occurrence of disease epizootics, and the *P. monodon* culture became unproductive. Subsequently, the Pacific white shrimp *P. vannamei,* an exotic species, was introduced into the Indian shrimp production system. The *P. vannamei,* with SPF status, extravagant growth rate, and tolerance to a range of salinity, was immediately adopted by Indian shrimp farmers, and thus, *P. vannamei* became the dominant species of shrimp culture. However, with more intensification, the incidence of diseases emerged as the major limitation to the progress and sustainability of the shrimp industry. Severe economic losses and mass mortality due to different infectious diseases have been reported. While white spot syndrome virus (WSSV) continues to be the single most devastating virus due to severe economic loss, in recent times, many new emerging diseases and syndromes with unknown etiology are constantly causing mortality or reduced growth, thereby resulting in severe economic losses to shrimp farmers. This chapter focuses on information related to disease syndromes with unknown etiology associated with shrimp culture.

2 Diseases of Syndromic Etiology

2.1 White Feces Syndrome

Recently, in key shrimp farming countries, white feces syndrome (WFS) has been causing severe distress to shrimp culture. The syndrome is termed WFS, as the affected shrimp defecate white fecal strings. It has been reported that the production loss due to WFS was assessed to be about 10–15% in 2010 in Thailand (Sriurairatana et al. 2014). Since 2015, in India, 17% of shrimp farms on the East Coast were reported to be affected by WFS. This disease has been described in black tiger shrimp and Pacific white shrimp. Due to reduced productivity, WFS can cause

Fig. 1 Whitish fecal strings floating on the water surface of WFS-affected shrimp culture pond

moderate to severe economic loss (Tang et al. 2007). WFS-affected shrimp excrete white fecal strings and show white/golden brown intestine. Thus, in WFS-affected shrimp farms, whitish fecal strings floating on the pond surface are apparent for a period of 10 days to 45 days or more (Fig. 1). WFS usually becomes evident after 30–40 days of culture. Feed consumption in the WFS-affected ponds is significantly reduced, and the affected shrimp display clinical signs of growth reduction, size variation, loose shells, elevated FCR, and daily mortalities. Loose shell-affected shrimps become inactive and can be seen sluggishly swimming at the surface of the pond water. The occurrence of WFS has been reported to be concomitant with gregarine parasite, ATM structures, vibriosis, *Enterocytozoon hepatopenaei*, bacteria such as *Candidatus, Bacilloplasma, and Phascolarcto*, blue-green algae, and fungi (Tangprasittipap et al. 2013; Sriurairatana et al. 2014; Chaweepack et al. 2015; Hou et al. 2018). It was later found that gregarines had no role in WFS (Sriurairatana et al. 2014).

Many pathogens and pathological conditions have been recorded in shrimp affected with WFS. It was reported that WFS was due to aggregated transformed microvilli (ATM) in the HP tubule (Sriurairatana et al. 2014). The hepatopancreatic microsporidian *Enterocytozoon hepatopenaei* (EHP) is significantly found to be associated with WFS-affected shrimp, according to Tang et al. (2016) and Caro et al. (2020). However, based on the metagenomic studies conducted at ICAR-CIBA (unpublished), no *Vibrio* spp. was found to be dominant among the WFS-affected shrimp. The disease is diagnosed by squash preparation and histopathological examination of the hepatopancreas. Squash preparation of WFS-affected hepatopancreas discloses ATM structures or non-motile vermiform bodies in the lumen of the hepatopancreas (HP) tubules (Sriurairatana et al. 2014). ATM structures inside the tubule lumen of HP are clearly seen in the HP smears stained with hematoxylin-eosin. Histological sections of the hepatopancreas of HP reveal ATM structures in the tubular lumen and dilated HP tubules with sloughed HP epithelial cells, accompanied by severe necrosis. The occurrence of WFS can be reduced by following good farm management practices. The feed should be reduced, and white fecal threads with dense EHP spores should be removed on a regular basis from the WFS-affected shrimp ponds. As the shrimp hepatopancreas has regeneration capability, medications/additives (e.g., taurine) may be used to improve epithelial cell proliferation. Further, as EHP is prevalent in WFS-affected shrimp, control/preventive methods suggested against EHP may be followed. In this, a higher dose of lime is recommended as EHP spores will get destroyed only by raising the pH of the soil to 12 or more. Treatment of pond sediments by application of CaO (quick lime) @ 6 tons per ha has been recommended. It is advised to till the CaO into the dry pond sediment (10–12 cm) and then drench the sediment to activate the lime. NACA suggested >15 ppm $KMnO_4$ or >40 ppm chlorine to inactivate spores in the soil. Best management practices (BMPs), such as drying and disinfection of ponds after each harvest to ensure that the EHP spores, along with the carriers, are destroyed, should be carried out.

2.2 White Muscle Syndrome (WMS)

White muscle syndrome (WMS) associated with muscle necrosis and low mortalities in the *P. vannamei* grow-out cultures was reported from Ecuador. The WMS-affected shrimp show focal to extensive necrotic regions in striated muscle tissues, exhibiting a white, opaque appearance (Fig. 2). Similar lesions have been described in shrimp infected with Infectious myonecrosis (IMN) (Poulos et al. 2006), penaeid white tail disease (PWTD) (Tang et al. 2007). WMS in shrimp can also be triggered by the progressive infection of microsporidians belonging to the genera *Ameson* and *Agmasoma*, dietary deficiency of selenium and sudden variations in water quality parameters like salinity, temperature, and dissolved oxygen. Histologically, WMS-affected shrimp samples with macroscopic lesions revealed destruction of sarcomere structure, coagulative muscle necrosis and hemocytic infiltration (Melena et al. 2012). Though histological lesions seen in the suspected sample are indistinguishable from those reported for IMNV and PWTD in *P. vannamei,* the absence of cytoplasmic inclusions in the skeletal muscle is a characteristic feature of WMS. As WMS-affected samples were devoid of Infectious myonecrosis virus (IMNV) and *Penaeus vannamei* nodavirus (PvNV) infections, it was suggested that the etiological agent of WMS could be either a novel infectious agent or a variant strain of IMNV (Melena et al. 2012).

Fig. 2 White, opaque, necrotic, patchy areas in abdominal muscle seen in WMS-affected farmed shrimps

2.3 Running Mortality Syndrome (RMS)

Running mortality syndrome (RMS) is an idiopathic syndrome that causes constant low-level mortalities every day throughout the culture period. The term Running mortality syndrome (RMS) was coined by the shrimp farmers in India. Since 2011, the disease has been extensively reported in the *P. vannamei* farming areas in Andhra Pradesh (AP) and Tamil Nadu (TN), India. The only apparent clinical sign of RMS is the presence of whitish patches in the second to fourth abdominal segments in affected shrimp (Alavandi et al. 2019). Animals affected with RMS were reported to be free from WSSV, IHHNV, IMNV, TSV, YHV, MBV, HPV, and PvNV. However, *Vibrio* spp., viz., *V. parahaemolyticus* and *V. azureus* are predominantly observed in hemolymph of RMS-affected shrimp. In RMS-affected farms, usually daily mortalities are reported to happen 30–40 days of culture (DOC). Though the mortality due to RMS was slow (<1%/day), the cumulative damage was reported to be high (70%) over time and, therefore, farmers opt for an early or premature harvest, which results in severe economic loss. The muscle tissue had severe necrosis accompanied by hemocytic infiltration, while hepatopancreas from most of the affected animals appeared normal on histopathological examination (Alavandi et al. 2019). However, some of the hepatopancreas samples of RMS-affected shrimp showed increased inter-hepatopancreatic tubular space with hemocytic infiltration and enlargement of the nucleus. Experiments on oral feeding and cohabitation with the infected animals failed to reproduce the RMS condition (Alavandi et al. 2019). Curiously, experimentally RMS-affected shrimp exhibited recovery and appeared healthy and active after 6–7 days with optimal water parameters. Though the exact cause of RMS is unidentified, the condition is reported to be linked to high stocking densities, high nitrite-N, and high turbidity. Accordingly, it is stated that RMS is a pond management-associated syndrome and can be prevented by best management practices (Alavandi et al. 2019).

2.4 Black Gill Disease (BGD)

The darkened gill tissue condition observed in shrimp is commonly known as black gill disease/burned gills/blackspot disease in crustaceans. Black gill disease due to *Fusarium* infection in *Penaeus japonicus* was first reported in Japan by Ishikawa (1967). Black gill severely impacted the shrimp culture sector in Georgia since the first report in 1996 (Page 2012; Frischer et al. 2018). The primary species affected are *Litopenaeus setiferus* and *Farfantepenaeus aztecus,* with the condition described in both the wild and cultured shrimp (Bower et al. 1994; Stentiford et al. 2012). There has been an epidemic of shrimp black gill in wild penaeid shrimp populations off the coasts of Georgia and South Carolina. This may have contributed to the decline of the shrimp fishery in Georgia (Gambill et al. 2015). The geographic distribution of black gill disease is ubiquitous, and all penaeid shrimp are known to be affected. The gills of affected shrimps appear reddish, brownish to black in color and with the advancement of the disease, gill filaments become black (Fig. 3). In the affected shrimp, the disease causes destruction and dysfunction of gills, secondary infections, and slow growth associated with persistent and increased mortalities due to predation (Frischer et al. 2018).

Histological observations reported were considerable hemocyte infiltration, tissue necrosis and melanin deposition in the affected areas of gills. Secondary

Fig. 3 Brownish to black discolouration of the gills of shrimp with black gill disease

bacteria, fungi, and protozoa infections were reported to be associated with BGD (Frischer et al. 2017). The contributing agents of black gill disease are reported to be numerous, including various pathogens, parasites, contaminants, and nutrient deficiencies (Cerenius et al. 2010; Frischer et al. 2017, 2018). It is most probable that stress factors, such as increased water temperature and/or high salinities, combined with high densities of animals in ponds, aggravate the disease condition (Frischer et al. 2018). Nutrient discrepancies like ascorbic acid deficiency, high stocking density, insufficient aeration, more plankton in the water, toxic pollutants like cadmium, copper, potassium permanganate, ozone, ammonia, and nitrite in nurturing water, heavy siltation, and high organic load of mud in the pond bottom are the predisposing factors (Fontaine and Lightner 1975; Magarelli et al. 1979; Rao and Doughtie 1984; Cerenius et al. 2010; Frischer et al. 2017). The prevalence of shrimp affected by black gill is correlated with water temperature and salinity (Fowler et al. 2018). Since 1996, in coastal Georgia and South Carolina, black gill outbreaks with an average prevalence of 6.5–45.7% occurred, with the highest prevalence in the late summer and fall (August to December) (Gambill et al. 2015; Frischer et al. 2017). In Georgia and South Carolina, black gill outbreaks display a consistent seasonal pattern with signs coinciding with extreme water temperatures and the least dissolved oxygen concentrations (Frischer et al. 2017).

Black gill condition in shrimp is the consequence of a general innate immune response triggered by a wide variety of pathogens activating the biochemical process called "melanization" which gradually develops into gill discolouration or "blackened gill" condition, eventually leading to the death of affected species (Khoa et al. 2004; Cerenius et al. 2010). Once a parasite gains entry into the shrimp, recognition molecules in the hemolymph begin to synthesize and activate immune factors, activate the prophenoloxidase (proPo) system, including the proPO cascade of serine proteinases and phenoloxidase (PO), the enzyme involved in melanin formation (Soderhall and Cerenius 1998; Aguirre-Guzmán et al. 2009). Melanin, a long-lived pigment with antibacterial properties, forms a nodule and encapsulates the extraneous material or microbes (Holmblad and Söderhäll 1999; Cerenius et al. 2008). The thick capsule of melanin formed around the microbe produces visible black spots that are caused by the color of the melanin in the gill lamellae (Aguirre-Guzmán et al. 2009), hence the term "black gill." Once the nodule is formed, short-lived cytotoxic reactive intermediates of oxygen and nitrogen involved in PO-mediated melanization break down the foreign material or organism (Charoensapsri et al. 2014). When moulting occurs, the old exoskeleton containing the nodules is shed, where the reactive intermediates break down the microbes, thereby removing the foreign invader. However, when host tissue is exposed for extended periods or at excess levels, these cytotoxic constituents can cause host tissue damage (Charoensapsri et al. 2014). It is reasonably conceivable that melanization in gill tissue interferes with critical gill functions, including respiration and ion regulation, leading to a lessened flow of hemolymph through the gill vasculature and affecting the metabolic activities of shrimp (Burnett and Burnett 2015). The prevention or control of the disease may be possible if the cause can be identified.

Also, avoiding heavy metal discharges into the rearing facilities and exclusion of black soil after each harvest by scraping is recommended as preventive measures.

2.5 Softshell Syndrome

The occurrence of Soft-shell syndrome, Soft-shelling, or chronic soft-shell syndrome was described in Indonesia, Malaysia, and Thailand. The disease was reported in juvenile and adult stages of *P. monodon*. The gross signs are the persistence of a soft shell with dark, rough, and wrinkled surfaces. Soft-shelled shrimp become weak, susceptible to cannibalism and surface fouling, and fetch only a low price upon harvest (Baticados et al. 1986). It is reported to be caused by pesticide contamination, nutritional insufficiency, poor pond water, and soil conditions. The incidence of soft-shell was reported to be associated with conditions such as infrequent water exchange, high soil pH (>6), low water phosphate (<1 ppm), and low (<7%) organic matter content (Baticados et al. 1990a). Also, a considerably high incidence of soft shells correlated with improper feeding practices like rancid or low-quality feeds, inadequate nourishment, lack of supplementary diets, and relatively higher stocking densities (Baticados et al. 1990a). The substantial dissimilarities in tissue calcium and phosphorus levels of soft-shelled shrimp compared to normal ones are perhaps due to impaired mechanisms in the mobilization of calcium and phosphorus from the hepatopancreas to the exoskeleton (Baticados et al. 1987). Reversal of soft shell condition with improvement in growth and shell quality upon supplementary feed, like mussel meat at 8–14% of the body weight daily for 2–4 weeks, was observed by Baticados et al. (1986). The disease preventive methods include adequate feeding of shrimp with good-quality feeds, thoroughly flushing ponds when using chemical pesticides, providing a supplementary diet containing a 1:1 ratio of calcium to phosphorus, and maintaining good quality pond water and soil with regular water exchange.

2.6 Red Disease Syndrome

The Red disease syndrome was first detected in black tiger shrimp in Taiwan (Liao et al. 1977). The disease has been documented in both cultured and wild-caught *P. monodon* in the Philippines and in farmed *P. stylirostris* in Hawaii (Lightner 1985). The red disease occurs mainly due to β-carotene and other carotenoids (hepatopancreatic carotenoids) pigments released into hemolymph due to atrophy and necrosis of hepatopancreatic tubules (Goodwin 1960). Affected shrimp appear red due to the deposition and distribution of β-carotene in tissues through hemolymph. Affected hepatopancreas shows severe necrosis and sloughed epithelial linings in tubules with hemocyte encapsulation and hemocyte infiltration in inter-tubular spaces (Lightner 1985). The disease progression was reported to have four stages. In the beginning, the affected shrimp show yellowish-green discolouration and remain active. Within 2 to 4 days, the animals turn reddish with reddish pleopods.

Subsequently, in the next 5 to 7 days, the affected shrimp become reddish. In the final stages, affected shrimp become lethargic, anorexic, and exhibit excessive surface fouling (Liao et al. 1977). Expressively, the presence of Gram-positive bacteria was observed in most of the red disease cases of *P. monodon* (Lightner 1985). On the other hand, the Gram-positive bacteria were not present in red disease syndrome-affected *P. stylirostris* in Hawaii. It was reported that the red disease might be a generalized syndrome caused by more than one infective agent. But the disease was observed only in *P. monodon* polycultured with *P. penicillatus* and *P. semisulcatus*, and studies showed that the red disease failed to transmit to unaffected *P. monodon,* and the disease was refractory to antibiotic therapy. Hence, it was concluded that red disease is not due to infectious agents (Liao et al. 1977). Further, red disease was reported to be associated with feeding rancid fish or exclusive artificial diets and thick anaerobic detritus deposits in manure-fertilized ponds (Lightner 1985). The histopathology of the hepatopancreas of *P. stylirostris* with experimental acute and subacute aflatoxicosis was found strikingly akin to red disease (Lightner et al. 1982). These observations supported the idea that mycotoxins or microbial toxins possibly present in rotten fish and pond detritus may be the reason for red disease incidences (Lightner 1985).

2.7 Loose Shell Syndrome

Loose shell syndrome (LSS) is a chronic disease condition first reported during 1998 in farmed *P. monodon* in India (Alavandi et al. 2008). LSS, also reported in juveniles and adults of *P. vannamei* grow-out ponds, is prevalent in Tamil Nadu, Andhra Pradesh, Kerala, Orissa, Gujarat, and West Bengal (Raja et al. 2015). The affected shrimps have flaccid, spongy abdomens due to muscular dystrophy with loose exoskeletons. LSS condition causes low survival, chronic mortalities, and poor-quality meat, leading to production and economic losses in shrimp farming. Though the presence of WSSV and MBV in LSS-affected ponds was reported, it was not conclusively related to the incidence of LSS in shrimp ponds (AftabUddin et al. 2018). As the exact etiological agent of LSS is not known, manifestation of LSS is attributed to poor management practices, soil and water quality, and nutritional imbalances (AftabUddin et al. 2018). The clinical symptoms are mainly observed in juvenile shrimp weighing 20–35 g, which include weak and lethargic movement, off-feeding, soft muscles, exoskeleton with colloidal mat due to impaired moulting, weak response to stimuli, and erratic swimming near the dyke (Gopalakrishnan and Parida 2005; Loka et al. 2012). Gross lesions observed are the gaps between the abdominal muscle and exoskeleton, melanised and shrunken hepatopancreas, and empty guts. Microscopic lesions like degraded hepatopancreatic tubules with enlarged hemal spaces, loss of muscle fibres, expansion of sarcomere space, and damage to gill filaments were also reported (Gopalakrishnan and Parida 2005; Alavandi et al. 2008). Microbial studies conducted to rule out the causative agent of LSS demonstrated *Vibrio*, *Flavobacterium*, *Pseudomonas*, *Acinetobacter*, and *Moraxella* in the ventral sinuses of LSS-affected shrimp. Among the prevalent bacterial pathogens, *Vibrio alginolyticus* was the dominant species, followed by *V. harveyi* and *V. vulnificus* (Loka et al. 2012).

2.8 Black Death Disease

The Black death disease or shrimp scurvy in penaeid shrimp is due to the deficiency of ascorbic acid in the diet (Lightner 1985). The disease is primarily seen in shrimps reared in a system that lacks algae and ascorbic acid in the diet. The ailment had been observed in postlarvae and juveniles of penaeid shrimp (Lightner et al. 1979). Stress factors like handling and mechanical injury are reported to intensify disease conditions. In severe cases, the animals stop feeding, and mortality occurs within 24–72 hours. In general, 1–5% mortality per day was recorded due to this disease (Lightner et al. 1979). Affected shrimp show melanized hemocytic lesions in the epithelial and connective tissues on the cuticular surface (Lightner 1993). The blackened lesions were formed due to the deposition of melanin in the affected areas, such as subcuticular tissues of body parts, antennae, pereopods, stomach, hindgut, gill cover, and tail-end and more lesions in collagen-rich tissues (Lightner et al. 1977, 1979; Hunter et al. 1979; Spaargaren 1996). Histopathologically, areas of severe necrosis and hemocytic infiltration are observed. The inclusion of L-ascorbic acid in the shrimp diet and rearing the shrimp in an algae-rich environment are reported to prevent the black death disease (Lightner 1985).

2.9 Blue Shell Syndrome

The Blue shell syndrome, alternatively called the Blue disease or Sky blue shrimp disease, was first seen in 1978 in broodstock with heavy mortalities from Tahiti (Aquacop 1983; Latscha 1989). The disease has been recorded in Indonesia, Malaysia, Taiwan, the Philippines, and Thailand (Baticados et al. 1990b). The gross signs in affected shrimp include lethargy and sky blue coloration of the body with a soft, thin, rough cuticular surface. The disease is generally observed in intensive grow-out ponds toward the end of the culture period. Juvenile and adult stages were reported to be affected (Lightner 1993; Aquacops 1984). Though the cause of the disease is implicit, it is assumed to be due to a nutritional deficiency, most possibly a lack of carotenoids like astaxanthin or its precursors in the diet. Research on blue shrimps has consistently revealed substantially low levels of total carotenoid content (4 to 7 ppm in the exoskeleton) and an unusual blue color, which is an indicator of nutritional carotenoid deficiency (Latscha 1989; Howell and Matthews 1991). Low levels of the carotenoid astaxanthin and soil, water quality glitches such as low dissolved oxygen levels, high organic wastes, and acid-sulphate soil are reported to flare up the disease incidences (Lightner 1985; Aquacop 1984; Baticados et al. 1990b; Lavilla-Pitogo et al. 2000a). Histopathological changes observed are disruption of the hepatopancreatic tubules (Baticados et al. 1990b). Preventive approaches include avoiding high stocking density, supplementation of diets with Vitamin A or carotenoid sources 45 days post-stocking, and frequent or daily exchange of 10 to 15% pond water to reduce the hydrogen sulfide burden in bottom layers (Baticados et al. 1990b; Howell and Matthews 1991; Lavilla-Pitogo et al. 2000a).

2.10 Gas Bubble Disease (GBD)

The supersaturated levels of total dissolved gas (TDG) in water cause Gas bubble disease (GBD) condition. GBD was first described in *Penaeus aztecus* and, later, has been recorded in other penaeid species such as *P. brasiliensis* and *P. paulensis* (Lightner et al. 1974; Brisson 1985). Gas-bubble disease in penaeids is known to occur when the oxygen levels in seawater reach or surpass 250% of normal saturation (Supplee and Lightner 1976). The effects of GBD can vary from mild to fatal, depending on the level of supersaturation of total dissolved gases (TDG), the life stage, species affected, animal health status, distribution depth, and water temperature (Smiley et al. 2011). In the affected shrimp, gas bubbles appear in the gills, under the cuticle, or in the gut. The gills appear whitish, the shrimp exhibit erratic swimming, and dead shrimp float on the pond surface (Lightner 1993). Once the disease is set in, immediate mechanical aeration should be given to reduce dissolved oxygen levels (Supplee and Lightner 1976). Precautions to avoid GBD include a visual inspection of water samples for the presence of small air bubbles (<0.5 mm) generally during low temperatures, prevention of air leaks in the water circulating system, regular measurement of the dissolved oxygen levels at different points in the system and frequent cleaning of bio-fouling inside the pipes (Colt 1986).

2.11 Muscle Cramp Syndrome (MCS)

Muscle cramp syndrome (MCS) is also known as Cramped–muscle syndrome (CMS) and Cramped muscle and tail syndrome (CTS) (Lightner 1993). The MCS has been reported in Taiwan, the Philippines, China, Malaysia, Thailand, and Indonesia (Liao et al. 1977; Liao 1984; Flegel 2009). Juvenile and adult stages of *P. vannamei* and *P. monodon* are reported to be affected. The affected, cramped shrimp body becomes rigid and curved with a partial or complete bent tail (Fig. 4) and with variable mortalities (Apud 1985; Baticados et al. 1990c). An acceptable level of CMS in shrimp ponds would be 5 per cent or less (Clifford and Cook 2002). The reason for the cramping is not known. However, the disease has been proposed to be prompted by environmental stressors or nutritional factors (Erazo-Pagador 2001). Besides, CMS seems to be caused by mineral imbalance and ecological stresses such as reduced dietary and environmental potassium, low DO, high temperature, vibriosis, presence of toxins in the water, and high stocking density. Body cramp problems are usually encountered during handling, transfer, or harvesting during hot days (Baticados et al. 1990c; Lavilla-Pitogo et al. 2000b). While entirely cramped shrimp lie on their sides in the pond, the partially cramped animals swim with a humped abdomen (Erazo-Pagador 2001). The condition predisposes cramped

Fig. 4 Muscle-cramped shrimp displaying rigid body and humped abdomen

shrimp to cannibalism. MCS causes the whitening of abdominal muscle tissue that typically begins in the middle of the muscle. Histopathologically, the characteristic lesions of MCS are muscle necrosis without accumulation of hemocytes. This usually occurs spontaneously in stressed *P. vannamei* and, if the cramps are not resolved quickly, leads to mortality (Lightner 1996). Abstention of handling or transferring shrimp during hot weather conditions helps to preclude the disease (Apud 1985; Brock and Main 1994).

2.12 Hemocytic Enteritis

The Hemocytic enteritis condition, caused by blooms of filamentous blue-green algae belonging to the family Oscillatoriceae, was reported in both freshwater and marine penaeid shrimp (Lightner 1985). The disease mainly infects the juveniles and sub-adult life stages of penaeid shrimps. Experimental infection with blue-green algae (*Schizothrix calcicola*) is reported to cause the disease (Mckee 1981). *S. calcicole,* present in both freshwater and marine culture practices, is reported to have a potent endotoxin (Keleti et al. 1979). Consequently, the ingested algae release the endotoxin into the shrimp gut, which in turn causes severe necrosis, hemocytic inflammation in the mucosal epithelium of the midgut, degeneration, and necrosis of hepatopancreatic epithelial cells (Lightner 1978; Lightner and Redman 1985). Mortality in shrimp is mainly caused by secondary bacterial infections, malabsorption of nutrients, and osmotic imbalance. Also, *V. alginolyticus* is reported to be found predominantly in the affected animals, accompanied by septic hemolymph of shrimp (Lightner 1985). Mortality rates are generally <20%, but in a raceway-reared *P. stylirostris,* mortality as high as 85% was reported (Lightner 1978). The disease can be prevented by replacing the growth of toxic benthic blue-green algae with the beneficial planktonic bloom at optimal density and water depth (Baticados 1988).

3 Conclusion

Diseases are major limitations in the expansion and sustainability of shrimp aquaculture. Disease syndromes with unknown etiologies are emerging as a concern in shrimp aquaculture production. More research efforts are needed for the identification of causative agents of disease syndromes. Unlike the major diseases, these disease syndromes do not cause mass mortality but result in reduced growth, increased FCR, and, in turn, lead to severe economic losses. Disease syndromes such as nutritional deficiency and environmental stress can be prevented and treated by maintaining optimal water quality parameters and balanced nutrient diets. The implementation of better management practices and optimal culture parameters may aid in circumventing the shrimp disease syndromes at the farm level.

References

AftabUddin S, Roman WU, Hasan CK, Ahmed M, Rahman H, Siddique MAM (2018) First incidence of loose-shell syndrome disease in the giant tiger shrimp *Penaeus monodon* from the brackish water ponds in Bangladesh. J Appl Anim Res 46(1):210–217

Aguirre-Guzmán G, Sánchez-Martinez JG, Campa-Córdova AI, Luna-González A, Ascencio F (2009) Penaeid shrimp immune system. Thai J Vet Med 39:205–215

Alavandi SV, Babu TD, Abhilash KS, Vijayan KK, Kalaimani N, Santiago TC (2008) Loose shell syndrome of farmed *Penaeus monodon* in India is caused by a filterable agent. Dis Aquat Org 81:163–171

Alavandi SV, Muralidhar M, Dayal JS, Rajan JS, Praveena PE, Bhuvaneswari T, Otta SK (2019) Investigation on the infectious nature of Running Mortality Syndrome (RMS) of farmed Pacific white leg shrimp, *Penaeus vannamei* in shrimp farms of India. Aquaculture 500:278–289

Apud FD (1985) Extensive and semi-intensive culture of prawn and shrimp in The Philippines. In: Taki Y, Primavera JH, Llobrera JA (eds) Proceedings of the first international conference on the culture of Penaeid prawns/shrimps, pp 105–113

Aquacop (1983) Constitution of broodstock, maturation, spawning and hatching systems for Penaeid shrimps in the center Oceanlogique du Paci¢que. In: Mcvey JP (ed) Handbook of mariculture 1 crustacean aquaculture. CRC Press, Boca Raton, pp 105–121

Aquacop (1984) Review of ten years of experimental Penaeid shrimp culture in Tahiti and New Caledonia (South Paci¢c). J World Maric Soc 15:73–91

Baticados MCL (1988) Diseases of *Penaeus monodon*, Biology and Culture of *Penaeus monodon*, pp 139–178

Baticados MCL, Coloso RM, Duremdez RC (1986) Studies on the chronic soft-shell syndrome in the tiger prawn, *Penaeus monodon* Fabricius, from brackishwater ponds. Aquaculture 56:271–285

Baticados MCL, Coloso RM, Duremdez RC (1987) Histopathology of the chronic soft-shell syndrome in the tiger prawn *Penaeus monodon*. Dis Aquat Org 3(1):13–28

Baticados MCL, Cruz-Lacierda ER, de la Cruz MC, Duremdez Fernandez RC, Gacutan RQ, Lavilla-Pitogo CR, Lio-Po GD (1990a) Diseases of Penaeid shrimps in The Philippines. Aquaculture Department, SEAFDEC, Tigbauan, Iloilo, Philippines, p 24

Baticados MCL, Cruz-Lacierda ER, de la Cruz MC, Duremdez Fernandez RC, Gacutan RQ, Lavilla-Pitogo CR, Lio-Po GD (1990b) Diseases of Penaeid shrimps in The Philippines. Aquaculture Department, SEAFDEC, Tigbauan, Iloilo, Philippines, pp 27–28

Baticados MCL, Cruz-Lacierda ER, de la Cruz MC, DuremdezFernandez RC, Gacutan RQ, Lavilla-Pitogo CR, Lio-Po GD (1990c) Diseases of Penaeid shrimps in The Philippines. Aquaculture Department, SEAFDEC, Tigbauan, Iloilo, Philippines, p 32

Bower SM, McGladdery SE, Price IM (1994) Synopsis of infectious diseases and parasites of commercially exploited shellfish. Annu Rev Fish Dis 4:1–199

Brisson S (1985) Gas-bubble disease observed in pink shrimps, *Penaeus brasiliensis* and *Penaeus paulensis*. Aquaculture 47(1):97–99

Brock JA, Main KL (1994) A guide to the common problems and diseases of cultured *Penaeus vannamei*. World Aquaculture Society, Baton Rouge, p 242

Burnett KG, Burnett LE (2015) Respiratory and metabolic impacts of crustacean immunity: are there implications for the insects? Integr Comp Biol 55:856–868

Caro LFA, Mai HN, Pichardo O, Cruz-Flores R, Hanggono B, Dhar AK (2020) Evidences supporting *Enterocytozoon hepatopenaei* association with white feces syndrome in farmed *Penaeus vannamei* in Venezuela and Indonesia. Dis Aquat Org 141:71–78. https://doi.org/10.3354/dao03522

Cerenius L, Lee BL, Söderhäll K (2008) The proPO-system: pros and cons for its role in invertebrate immunity. Trends Immunol 29(6):263–271

Cerenius L, Kawabata S-I, Lee BL, Nonaka M, Söderhäll K (2010) Proteolytic cascades and their involvement in invertebrate immunity. Trends Biochem Sci 35(10):575–583

Charoensapsri W et al (2014) Melanization reaction products of shrimp display antimicrobial properties against their major bacterial and fungal pathogens. Dev Comp Immunol 47:150–159. https://doi.org/10.1016/j.dci.2014.07.010

Chaweepack T, Muenthaisong B, Chaweepack S, Kamei K (2015) The potential of galangal (*Alpinia galangal* Linn.) extract against the pathogens that cause white feces syndrome and acute hepatopancreatic necrosis disease (AHPND) in Pacific white shrimp (*Litopenaeus vannamei*). Int J Biol 7:8–17

Clifford HC, Cook HL (2002) Disease management in shrimp culture ponds-part 3. Aquac Mag-Akansas 28(4):20–29

Colt J (1986) Gas supersaturation on the design and operation of aquatic systems. Aquac Eng 5:49–85

Erazo-Pagador GE (2001) Environmental and other non infectious disease. In: Lavilla CR, Cruz-Lacierda ER (eds) Health management in aquaculture in Lio-Po, G.D. Southeast Asian Fisheries Development Center, Tigbauan, Iloilo, Philippines, pp 75–81

FAO (2020) The state of world fisheries and aquaculture 2020. Sustain Action. https://doi.org/10.4060/ca9229en

Flegel TW (2009) Current status of viral diseases in Asian shrimp aquaculture. Israeli J Aquac Bamidgeh 61:229–239

Fontaine CT, Lightner DV (1975) Cellular response to injury in penaeid shrimp. Mar Fish Rev 37(5–6):4–10

Fowler AE, Leffler JW, Johnson SP, DeLancey LB, Sanger DM (2018) Relationships between meteorological and water quality variables and fisheries-independent white shrimp (*Litopenaeus setiferus*) catch in the ACE basin NERR, South Carolina. Estuaries and Coasts 41(1):79–88

Frischer ME, Lee RF, Price AR, Walters TL, Bassette MA, Verdiyev R, Torris MC, Bulski K, Geer PJ, Powell SA, Walker AN (2017) Causes, diagnostics, and distribution of an ongoing penaeid shrimp black gill epidemic in the US South Atlantic Bight. J Shellfish Res 36(2):487–500

Frischer ME, Fowler AE, Brunson JF, Walker AN, Powell SA, Price AR, Bulski K, Frede RL, Lee RF (2018) Pathology, effects, and transmission of black gill in commercial penaeid shrimp from the South Atlantic Bight. J Shellfish Res 37(1):149–158

Gambill JM, Doyle AE, Lee RF, Geer PJ, Walker AN, Parker LG, Frischer ME (2015) The mystery of Black Gill: shrimpers in the South Atlantic face off with acryptic parasite. Current: J Mar Educ 29(4):2–8

Goodwin TW (1960) Biochemistry of pigments. In: Waterman TH (ed) The physiology of Crustacea, Metabolism and growth, vol 1. Academic Press, New York, pp 101–140

Gopalakrishnan A, Parida A (2005) Incidence of loose shell syndrome disease of the shrimp *Penaeus monodon* and its impact in the grow-out culture. Curr Sci 88:1148–1154

Holmblad T, Söderhäll K (1999) Cell adhesion molecules and antioxidative enzymes in a crustacean, possible role in immunity. Aquaculture 172(1–2):111–123

Hou D, Huang Z, Zeng S, Liu J, Wei D, Deng X, Weng S, Yan Q, He J (2018) Intestinal bacterial signatures of white feces syndrome in shrimp. Appl Microbiol Biotechnol 102(8):3701–3709

Howell BK, Matthews AD (1991) The carotenoids of wild and blue disease affected farmed tiger shrimp (*Penaeus monodon*, Fabricius). Biochem and Mol Biol 98(2–3):375–379

Hunter B, Magarelli PC Jr, Lightner DV, Colvin LB (1979) Ascorbic acid-dependent collagen formation in penaeid shrimp. Comp Biochem Physiol B 64(4):381–385

Ishikawa Y (1967) On the filamentous bacteria which grow on the gills of cultured Kuruma prawn. Fish Pathol 2:68–72. In Japanese

Keleti G, Sykora JL, Lippy EC, Shapiro MA (1979) Composition and biological properties of lipopolysaccharides isolated from *Schizothrix calcicola* (Ag.) Gomont (Cyanobacteria). Appl Environ Microbiol 38(3):471–477

Khoa LV, Hatai K, Aoki T (2004) *Fusarium incarnatum* isolated from black tiger shrimp, *Penaeus monodon* Fabricius, with black gill disease cultured in Vietnam. J Fish Dis 27:507–515

Latscha T (1989) The role of astaxanthin in shrimp pigmentation. In: Advances in tropical aquaculture, pp 319–325

Lavilla-Pitogo CR, Lio-Po GD, Cruz-Lacierda ER, Alapide-Tendencia EV, de la Peña LD (2000a) Diseases of penaeid shrimps in The Philippines. Aquaculture Department, Southeast Asian Fisheries Development Center, Iloilo, Philippines, pp 47–48

Lavilla-Pitogo CR, Lio-Po GD, Cruz-Lacierda ER, Alapide-Tendencia EV, de la Peña LD (2000b) Diseases of penaeid shrimps in The Philippines, Aquaculture Department, Southeast Asian Fisheries Development Center, Iloilo, Philippines, p 56

Liao IC (1984) Status and problems of grass prawn culture in Taiwan. In: Liao IC, Hirano R (eds) TML Conference Proceedings 1. Tungkang Marine Laboratory, Tunkang, Taiwan, pp 81–98

Liao IC, Yang FR, Lou SW (1977) Preliminary report on some diseases of cultured prawn and their control method. JCRR Fish Ser 29:28–33

Lightner DV (1978) Possible toxic effects of the marine bluegreen alga, Spirulina subsalsa, on the blue shrimp, *Penaeus stylirostris*. J Invertebr Pathol 32:139–150

Lightner D (1985) A review of the diseases of cultured penaeid shrimps and prawns with emphasis on recent discoveries and developments. In: Proc. First. Int. Conf. Cult. 1984. Penaeid prawns/shrimp. SEAFDEC, Philippines, pp 79–103

Lightner DV (1993) Non-infectious diseases of Crustacea with an emphasis on cultured penaeid shrimp. In: Couch JA, Fournie JW (eds) Advances in fisheries science: pathobiology of marine and estuarine Ž. Organisms. CRC Press, Boca Raton, pp 343–358

Lightner DV (1996) A handbook of pathology and diagnostic procedures for diseases of Penaeid shrimp. World Aquaculture Society, Baton Rouge

Lightner DV, Redman RM (1985) Necrosis of hepatopancreas in *Penaeus monodon* and *Penaeus stylirostris* (arthropoda and decapoda) with red disease. J Fish Dis 8:181–188

Lightner DV, Salser BR, Wheeler RS (1974) Gas-bubble disease in the brown shrimp (*Penaeus aztecus*). Aquaculture 4:81–84

Lightner DV, Colvin LB, Brand C, Danald DA (1977) Black death, a disease syndrome of penaeid shrimp related to a dietary deficiency of ascorbic acid. Proc World Maric Soc 8:611–624

Lightner DV, Hunter B, Magarelli PC Jr, Colvin LB (1979) Ascorbic acid nutritional requirement and role in wound repair in penaeid shrimp. Proc World Maric Soc 10:513–528

Lightner DV, Redman RM, Price RL, Wiseman MO (1982) Histopathology of aflatoxicosis in the marine shrimp *Penaeus stylirostris* and *P. vannamei*. J Invertebr Pathol 40:279–291

Loka J, Janakiram P, Geetha GK, Sivaprasad B, Veerendrakumar M (2012) Loose shell syndrome (LSS) of cultured *Penaeus monodon*-microbiological and histopathological investigations. Indian J Fish 59(3):117–123

Magarelli PC, Hunter B, Lightner DV (1979) Black death: an ascorbic acid deficiency disease in penaeid shrimp. Comp Biochem Physiol 63A:103–108

Mckee C (1981) The toxic effect of five strains of blue-green algae on *Penaeus stylirostris* Stimpson. M.S. thesis, School of Renewable Natural Resources, Univ. of Arizona, Tucson, p 70

Melena J, Tomalá J, Panchana F, Betancourt I, Gonzabay C, Sonnenholzner S, Amano Y, Bonami J-R (2012) Infectious muscle necrosis etiology in the Pacific White Shrimp (*Penaeus vannamei*) cultured in Ecuador. Braz J Vet Pathol 5(1):31–36

Page J (2012) Georgia marine research, surveys, and development. In: Geer PJ, Belcher C (eds) Annual report to USFWS sportfish restoration project. GA DNR Coastal Resources Division, Brunswick, GA, p 17

Poulos BT, Tang KFJ, Pantoja CR, Bonami JB, Lightner DV (2006) Purification and characterization of infectious myonecrosisvirus of penaeid shrimp. J Gen Virol 87:987–996

Raja K, Gopalakrishnan A, Singh R, Vijayakumar R (2015) Loose Shell Syndrome (LSS) in *Litopenaeus vannamei* grow-out ponds and its effect on growth and production. Fish Aquac J 6(151):2

Rao KR, Doughtie DG (1984) Histopathological changes in grass shrimp exposed to chromium, pentachlorophenyl and dithiocarbamates. Mar Environ Res 14:371–395

Smiley JE, Drawbridge MA, Okihiro MS, Kaufmann RS (2011) Acute effects of gas supersaturation on juvenile cultured white seabass. Trans Am Fish Soc 140(5):1269–1276

Soderhall K, Cerenius L (1998) Role of the prophenoloxidase-activating system in invertebrate immunity. Curr Opin Immunol 10:23–28

Spaargaren DH (1996) Diseases in cultures of tiger prawns, *Penaeus monodon* Fabricius, 1798 (Decapoda, Natantia). Crustaceana 69(8):1018–1024

Sriurairatana S, Boonyawiwat V, Gangnonngiw W, Laosutthipong C, Hiranchan J, Flegel TW (2014) White feces syndrome of shrimp arises from transformation, sloughing and aggregation of hepatopancreatic microvilli into vermiform bodies superficially resembling gregarines. PLoS One 9:e99170

Stentiford GD, Neil DM, Peeler EJ, Shields JD, Small HJ, Flegel TW, Vlak JM, Jones B, Morado F, Moss S, Lotz J, Bartholomay L, Behringer DC, Hauton C, Lightner DV (2012) Disease will limit future food supply from the global crustacean fishery and aquaculture sectors. J Invertebr Pathol 110:141–157

Supplee VC, Lightner DV (1976) Gas-bubble disease due to oxygen supersaturation in raceway-reared California brown shrimp. Progr Fish-Culturist 38(3):158–159

Tang KFJ, Pantoja CR, Redman RM, Lightner DV (2007) Development of in situ hybridization and RT-PCR assay for the detection of a nodavirus (PvNV) that causes muscle necrosis in *Penaeus vannamei*. Dis Aquat Org 75:183–190

Tang KF, Han JE, Aranguren LF, White-Noble B, Schmidt MM, Piamsomboon P, Risdiana E, Hanggono B (2016) Dense populations of the microsporidian *Enterocytozoon hepatopenaei* (EHP) in feces of *Penaeus vannamei* exhibiting white feces syndrome and pathways of their transmission to healthy shrimp. J Invertebr Pathol 140:1–7

Tangprasittipap A, Srisala J, Chouwdee S, Somboon M, Chuchird N, Limsuwan C, Srisuvan T, Flegel TW, Sritunyalucksana K (2013) The microsporidian *Enterocytozoon hepatopenaei* is not the cause of white feces syndrome in whiteleg shrimp *Penaeus* (*Litopenaeus*) *vannamei*. BMC Vet Res 9:139

Diseases of Bivalve Molluscs

N. K. Sanil, G. Suja, and P. K. Asokan

Abstract

Bivalve molluscs, such as oysters, mussels, and clams, are economically vital organisms that serve as major protein sources and support fisheries, particularly in coastal regions. Driven by rising consumer demand, the bivalve aquaculture industry is rapidly expanding, significantly contributing to global food security and coastal economies. However, bivalves are susceptible to diseases caused by various pathogens including bacteria, fungi, viruses, and parasites. Intensified aquaculture practices and environmental changes have increased the risk of disease outbreaks. Disease diagnosis in bivalves is challenging owing to the limited visible symptoms and their enclosed shell structure, which conceals clinical signs. Bacterial diseases, particularly those caused by *Vibrio* species, are common and often lead to mass mortality in hatcheries and wild populations. Herpesvirus and iridovirus infections have repeatedly devastated bivalve aquaculture, causing substantial economic losses. Parasitic diseases significantly threaten wild and farmed bivalves, and protozoans play a major role in causing severe mortality. Five of the seven diseases listed by the World Organisation for Animal Health (WOAH) in bivalves have parasitic etiologies. Protozoan parasites, especially *Perkinsus*, *Bonamia*, *Marteilia*, and *Haplosporidium*, have caused major disease outbreaks and have contributed to the global decline in bivalve populations. *Perkinsus* infections have been linked to mass mortality in farmed green mussels in India. Several other protozoan and metazoan parasites also impact bivalve health with varying degrees of pathogenicity. Many of these pathogens spread via passive or horizontal transmission in high-density aquaculture environments, complicating disease management. This review focuses on the major pathogens affecting bivalves in Indian waters. Effective disease man-

N. K. Sanil (✉) · G. Suja · P. K. Asokan
ICAR-Central Marine Fisheries Research Institute, Kochi, India
e-mail: nksanil@gmail.com; suj1078@gmail.com; asokanpk@gmail.com

R. Kooloth Valappil et al. (eds.), *Aquatic Animal Health Management*,
https://doi.org/10.1007/978-981-96-7987-4_13

agement in bivalves requires a comprehensive understanding of the causative agents, their biology, transmission dynamics, and the host immune response. Early detection using specific and rapid diagnostic methods, implementation of suitable biosecurity measures, and maintaining good husbandry practices are crucial for the timely intervention and control of diseases in bivalves.

Keywords

Bivalves · Aquaculture · Diagnosis · Disease management · Virus · Bacteria · Parasites

1 Introduction

Bivalves are a diverse class of invertebrates belonging to the phylum Mollusca, and include oysters, clams, mussels, and scallops. They are filter feeders that inhabit a wide range of environments (marine to freshwater) and thrive in various habitats (sandy or muddy to rocky reefs and estuaries). Bivalves are an important source of protein and are a major fishery worldwide, contributing significantly to the coastal economies of many countries. A wide range of potentially exploitable biologically active products and metabolites, including various antimicrobial peptides and gene-encoded products, have been isolated from bivalves (Zannella et al. 2017).

Robust consumer demand and sustained high prices have accelerated the growth of bivalve aquaculture. Globally, molluscan aquaculture has undergone significant expansion, reaching 17.7 million tons by 2020 (FAO 2022). The unique feature of filter-feeding minimizes external feed inputs in bivalve farming, thereby reducing major input costs and environmental impacts. Moreover, by removing suspended organic matter from the water through continuous filtration, bivalve farming helps in nutrient cycling and reduces the eutrophication of estuaries, bays, and near-shore waters. In the Indian context, bivalves have always been considered a subsistence food for marginalised coastal communities. However, certain regions, such as Goa and northern Kerala, are exceptions, where bivalves are considered a delicacy and command high consumer demand (Mohamed 2013). Bivalve farming in India is limited to the farming of green mussels (*Perna viridis*) and edible oysters (*Crassostrea madrasensis*) along the west coast, and pearl oysters (*Pinctada fucata*) along the east coast. Consequently, little attention has been paid to diseases of bivalve molluscs in India.

The rapid expansion of bivalve aquaculture has greatly increased the risk of emerging diseases. Over the past few decades, it has been adversely affected by several endemic, exotic, emerging, and re-emerging diseases, impacting economic and socioeconomic development in many countries (Berthe 2008; Renault 2008; Aranguren and Figueras 2016). Intensification, unscientific farming practices, and increased transboundary movement of stocks have fueled the momentum of emerging diseases and their spread across continents (Berthe 2005). Additionally, shipping activities and ballast water also contribute to the horizontal spread of diseases (Minchin and Gollasch 2002; Wolff 2005; Berthe 2008).

Pathogens associated with infectious diseases in bivalves include viruses, bacteria, fungi, and protozoans (Renault 1996). Many of these are opportunists, taking advantage of the stress caused by farming practices and climatic variations. Summer mortality, commonly observed in farmed bivalves, is closely linked to rising temperatures and the animal's reproductive state, and involves complex interactions between pathogens and environmental factors (Cheyney et al. 1998; Berthelin et al. 2000; Malham et al. 2009). In culture operations, physical stress arising from crowding increases the risk of infection and facilitates the spread of diseases, leading to greater amplification of pathogenic agents compared to natural stocks. Non-infectious diseases and environmental factors, such as thermal and salinity stress, coupled with suboptimal transport and handling conditions, can also contribute to disease outbreaks and mortality (Gagnaire et al. 2007; Yu et al. 2009). Disease outbreaks in farming systems can also impact wild populations by altering disease dynamics through elevated pathogen levels (Raynard et al. 2007). Various researchers have extensively studied and reported on the diseases and pathologies of commercially important bivalves (Sindermann 1990; Bower et al. 1994; Shumway 1996; McGladdery et al. 2006; Meyers and Burton 2009).

Bivalve health is closely associated with immunological status, which can be adversely affected by stress and diseases. Linkages between environmental stress, compromised immunological status, and disease have been established in many invertebrates. The development of a disease depends on many factors, including host physiology, environmental conditions, and/or the cumulative effect of one or more adversities (Shumway 1996, Rahman et al. 2019). As filter feeders, bivalves encounter an extraordinarily large number of microbes. Unlike higher animals, they possess a form of nonadaptive innate immunity with physico-chemical barriers acting as the first line of defence. Pallial mucus acts as a primary deterrent against most invaders. The cross-linked glycoprotein meshwork formed by mucus effectively traps microbes at the entry level. Pallial mucus also contains haemocytes and bioactive molecules, including agglutinins, lectins, immune recognition proteins, and hydrolytic enzymes that interact with pathogens. Breaching these primary barriers initiates cellular and humoral defence mechanisms mediated mainly by circulating hemocytes. Bidirectional movement of hemocytes across epithelial barriers has been reported in bivalves. At the outset, recognition factors detect invaders and initiate cytokines to attract haemocytes through chemotaxis, followed by phagocytosis activation and the production and/or release of a wide range of antimicrobial compounds. The failure of these protective mechanisms eventually leads to pathogen establishment. Despite the absence of haematopoietic organs and tissues, recent research in bivalves suggests the presence of various immune recognition molecules, a certain level of specific immune priming, and even maternal transfer of immune protection (Allam and Raftos 2015; Wang et al. 2015; Canesi and Pruzzo 2016; Rahman et al. 2019).

Unlike fish and crustaceans, bivalves are predominantly bottom dwellers, either attached to the substratum or buried in sand or mud, and their ability to respond to diseases or exhibit external signs or symptoms attributable to specific disease states is limited. Restricted visual access within the shell cavity limits the direct observation of living animals. In the absence of any pathognomonic signs, the observer must depend on relatively subjective and inconsistent criteria to evaluate the health of individual animals underwater, making bivalve disease diagnosis a challenging task. Generally, adductor muscular tone, based on the strength and speed of shell closure and mantle retraction, are the two common criteria used to evaluate bivalve responses to external stimuli. The gaping of shell valves is frequently considered a cardinal sign of disease or morbidity in bivalves. A contracted mantle and/or inability to deposit new shell materials may indicate disease or compromised health. Bivalve mass mortality events often remain undetected in natural environments and are consequently underreported.

2 Diseases in Bivalve Molluscs

Most disease problems in bivalve molluscs revolve around opportunistic pathogens, taking advantage of the animals weakened by stress and suboptimal growth conditions. Among the pathogens associated with infectious diseases in bivalves, protozoans are the most frequently encountered, often leading to epizootics. The World Organisation for Animal Health (WOAH) has listed seven important diseases in bivalves.

From a diagnostic perspective, many standard techniques employed in higher animals are of limited value in bivalves. The inability of molluscs to produce antibodies limits the use of classical serological techniques. Histology is considered the standard diagnostic method for screening bivalves as it provides a wide spectrum of information, including the presence of pathogens and histopathological alterations in host tissues. However, histopathology is time-consuming, and the accurate interpretation of results requires high technical skill and expertise, and fails to provide species-level identification of the infectious agent(s). Presently, the best method for confirmatory diagnosis is molecular diagnosis, which is reasonably fast and offers high sensitivity and specificity. Polymerase chain reaction (PCR), followed by sequencing, has been widely used for the specific diagnosis of bivalve pathogens. Other methods, such as transmission electron microscopy and in situ hybridisation (ISH), can yield detailed information, but these techniques are time-consuming and therefore not considered suitable for routine diagnosis.

2.1 Bacterial Diseases

Through continuous filtration of the aquatic environment, bivalves can accumulate numerous bacterial pathogens, many of which have been linked to significant mortality in hatcheries and natural beds. The pathogenicity and virulence of bacterial infections often vary depending on the nature of host-pathogen interactions and infection dynamics (Travers et al. 2015). Most bacterial diseases in bivalves are generally associated with larval stages, and members of the genera *Vibrio*, *Pseudomonas*, and *Aeromonas* are frequently associated with mortality outbreaks in bivalve larvae (Sindermann 1990; Paillard et al. 2004). The genera *Vibrio, Nocardia,* and *Roseovarius* have a broad host range, infecting all life stages from larvae to adults, while others, such as *Rickettsia*, are more selective and target specific hosts (Renault 2008; Beaz-Hidalgo et al. 2010; Barbosa-Solomieu et al. 2015; Travers et al. 2015; Zannella et al. 2017). In many cases, bivalves may act as asymptomatic carriers of diseases. The diagnostic methods for bacterial diseases include macroscopic examination, microscopy, culture techniques, morphological and biochemical tests, and molecular methods. Extensive reviews of bacterial diseases in bivalves have been provided by various authors (Paillard et al. 2004; Travers et al. 2015).

Vibrios are natural inhabitants of estuarine and marine environments and have a cosmopolitan distribution. They are gram-negative, typically short, rod-shaped, facultatively anaerobic bacteria with a single polar flagellum (Percival and Williams 2014) and have a broad host range, including terrestrial and marine vertebrates and invertebrates (Paillard et al. 2004; Thompson et al. 2004). Of the 130 confirmed species, all except a few are halophilic. Although vibrios can tolerate a wide range of salinities, they thrive in warmer waters, and elevated temperatures are often linked to increased *Vibrio* infections in bivalves. *Vibrio* species are commonly isolated from filter-feeding shellfish, often at concentrations far exceeding those in the surrounding waters (Froelich et al. 2016).

Histopathological manifestations of *Vibrio* infections include hemocytic infiltration in connective tissues and adductor muscle lesions in adults, as well as velum and tissue necrosis in larvae. Mortalities due to vibriosis have been reported from both hatcheries and field conditions, and are primarily associated with vibrios belonging to the Splendidus (16 species), and Harveyi clades, as well as with *V. aestuarianus, V. tubiashii, V. coralliilyticus, V. tapetis* and other *Vibrio* spp. (Paillard et al. 2004; Travers et al. 2015; Zannella et al. 2017).

The 'Brown ring' disease of clams (*Ruditapes* spp.), leading to characteristic brown deposits as a ring, is caused by *V. tapetis* (Paillard et al. 1994). The infection starts at the periostracal lamina of the shell, leading to abnormal calcification and brown deposits as a ring between the pallial line and shell margin. Infection further proceeds to the soft tissues, where bacterial proliferation occurs in the pallial fluids and host circulatory system, leading to septicaemia and death of the host. Under field conditions, numerous factors, including seasons, habitats, virulence, composition of microbial populations, and their interactions can influence mortality (Travers et al. 2015; Zannella et al. 2017).

Nocardiosis reported in the Pacific oyster, *Crassostrea gigas,* is caused by the gram-positive bacterium *Nocardia crassostreae* and results in the phenomenon called 'summer mortality' as it occurs during the summer and fall months when water temperature rises over 20 °C. The condition is reported from the west coast of the US and Japan and can result in up to 35% mortality. Infections are characterized by yellow to green lesions in the mantle, gills, and adductor muscles (Paillard et al. 2004).

Another major disease causing significant seasonal mortality in hatchery-produced *C. virginica* from the Atlantic coast is Roseovarius Oyster disease (ROD) or Juvenile oyster disease (JOD). The disease is characterized by reduced growth, fragile and uneven shell margins, left valve cupping, mantle retraction, watery tissues and lesions, and proteinaceous deposits on the inner shell surfaces. A species of marine α-proteobacteria, the *Roseobacter* group, is suspected to be the etiological agent (Paillard et al. 1996; Boettcher et al. 2000). Bacterial orders such as Rickettsiales, Chlamydiales, and Mycoplasmatales are reported to cause intracellular pleomorphic inclusions in marine bivalves (Harshbarger et al. 1979; Bower and McGladdery 2010). Inclusions within the epithelial and connective cells of branchial and alimentary tissues elicit little host response or pathology. However, instances of infections leading to mortality have also been reported (McGladdery 1999).

2.2 Fungal Diseases

Information on fungal infections in bivalves is limited. However, there are reports on the presence of various species of filamentous fungi in bivalves, including farmed mussels. Nevertheless, the role of fungi as etiological agents of bivalve diseases remains to be established (Santos et al. 2020; Oliveira et al. 2024).

2.3 Viral Diseases

Several viral diseases have been reported in bivalves, the most notable of which are caused by the Herpesviridae and Iridoviridae families. The highly infectious nature of pathogens, altered climatic conditions, intensive culture systems, and unregulated stock movement across geographical regions have all contributed to the rapid spread of many viruses. The gill disease, "Maladie des branchies" was the first viral disease reported in bivalves affecting the Portuguese oyster *Crassostrea angulata* from France. Later, a herpes viral disease was reported in the eastern oyster *C. virginica*. Since then, several disease outbreaks of viral aetiology have been reported in various bivalves (Bower et al. 1994; Barbosa-Solomieu et al. 2015; Villalba 2016). The infected host cell degenerates and is subsequently destroyed, liberating progeny viruses that infect new cells (Gosling 2015).

The primary classification of viruses is generally based on viral particle morphology, which is often derived from electron microscopic observations. Definitive identification of these ultramicroscopic infectious agents is performed based on the structure and nature of the genome, mode of replication, chemical composition, and configuration of the nucleic acids. For the diagnosis of viral diseases, highly sensitive molecular techniques such as PCR, ISH, and immunochemistry have been adopted, along with histology and electron microscopy (Gosling 2015). The unavailability of molluscan cell lines constrains efforts to isolate and characterize bivalve viruses. Furthermore, molecular characterization has not been attempted for many viruses; hence, the exact taxonomic position remains uncertain in many cases.

2.3.1 Herpesviruses

Herpesviruses are responsible for the high mortality rates and subsequent losses in several bivalve species (Lauckner 1983; Hine et al. 1992). Herpes-like viral infections were first reported in *C. virginica* on the east coast of the USA (Farley et al. 1972). After two decades, occasional incidents of herpes-like viral infections leading to severe mortality of *C. gigas* larvae have been reported in France and other European countries (Renault et al. 2001). Subsequently, mortalities caused by herpes virus infections, especially during the summer months, have been reported worldwide in many bivalve species (Arzul and Renault 2002), and the etiological agent has been characterised as Ostreid Herpesvirus type 1 (OsHV-1) (Davison et al. 2009). The OsHV-1 particles have an enveloped icosahedral structure with electron-dense cores and a diameter of approximately 120 nm (OIE 2019). OsHV-1 is the only member of the genus *Ostreavirus* in the family Malacoherpesviridae (order Herpesvirales). Other genomic variants of OsHV-1 have been reported in various bivalve species from different geographical regions. OsHV-1 μVar is responsible for the massive mortality of *C. gigas* in Europe and elsewhere (Moss et al. 2007; Arzul et al. 2017). Spat and larvae are more susceptible than adults, and infected larvae exhibit reduced swimming and feeding and lesions in the velar and mantle tissues, leading to mortality (Renault and Novoa 2004; Zannella et al. 2017). Adults are less sensitive, and may act as disease carriers. Generally, disease outbreaks occur during spring and summer, suggesting the role of higher temperatures in triggering infections (Barbosa-Solomieu et al. 2015).

2.3.2 Iridoviruses

Among the iridoviruses that target bivalve molluscs, three are particularly important: Gill Necrosis Virus (GNV), Haemocyte Infection Virus (HIV), and Oyster Velar Virus Disease (OVVD). GNV is an icosahedral DNA virus responsible for gill necrosis and recurrent mortality of *C. angulata* and *C. gigas* in Europe. Initial symptoms manifest as yellow or green spots on the gills and labial palps, which progressively enlarge and darken to a brownish discolouration, leading to tissue perforations and destruction of gill filaments. Yellow or green pustules may also be observed in the mantle or adductor muscles (Arzul et al. 2017). HIV infects the haemocytes of *C. angulata* in France. The disease does not exhibit any external symptoms but leads to high mortality. The virus particles appear icosahedral, featuring an outer core measuring 350 nm and an inner core approximately 250 nm in diameter (Renault and Novoa 2004). OVVD has been associated with occasional mortality in oyster pediveliger larvae. The disease is characterized by the sloughing of velar epithelial cells, leading to blisters/blebs along the velar periphery, eventually interfering with the normal movement of the larvae (Renault and Novoa 2004; Arzul et al. 2017). Virions have an average length of 228 nm and possess a two-layered membranous structure (Comps 1988).

2.3.3 Picornavirus, Papovavirus, and Other Small Virus-Like Particles

Picorna-like viruses represent the most abundant group of RNA viruses found in coastal waters. Numerous studies have reported the presence of virus-like particles associated with disease and mortality in bivalves. Examples include the 'Akoya virus' responsible for mortality in Japanese pearl oysters (Miyazaki et al. 1999), and virus-like particles associated with granulomas and necrosis in *Pinctada margaritifera* (Comps et al. 1999). Virus-like particles associated with large foci of hemocytic infiltration have been observed in *Cerastoderma edule* in Spain (Carballal et al. 2003). Many of these species have been suggested to belong to Picornaviridae because of their size and other morphological features. Several authors have reported Papova-like viruses in various bivalve species (Farley 1976, 1985; Renault and Novoa 2004). Virus-like particles have been implicated in disseminated neoplasia in several bivalves (Elston et al. 1992).

2.4 Parasitic Diseases

2.4.1 Protozoan Infections

Protozoans are the most frequently described disease agents that cause severe mortality in wild and farmed bivalves (Andrews 1988; Soniat 1996). Five of the seven WOAH-listed diseases in bivalves have parasitic aetiologies (WOAH 2024). Major protozoan parasites infecting the wild and farmed populations of bivalves belong to the genera *Perkinsus, Bonamia, Marteilia*, and *Haplosporidium (=Minchinia),* while others, like gregarines, ciliates, etc., are considered to play only minor roles. 'Bonamiasis' caused by the protozoan *Bonamia ostreae* and 'marteliosis' caused by *Marteilia refringens* have destroyed the stocks of the European flat oyster, *Ostrea edulis* in France and other European countries (Culloty et al. 1999). *Haplosporidium nelsoni* and *Perkinsus marinus* have led to a decline in the fishery of the eastern oyster *C. virginica* in the USA (Gosling 2004; Villalba et al. 2004).

Infection with *Bonamia*

Bonamiasis, also known as 'microcell disease' or 'haemocyte disease' of flat oysters, is a fatal infection of haemocytes in the European flat oyster, *O. edulis*, and is caused by a microscopic haplosporidian parasite belonging to the genus *Bonamia* (Carnegie and Cochennec-Laureau 2004). Among the species of *Bonamia* infecting bivalves, *B. ostreae* and *B. exitiosa* have been listed as notifiable by WOAH. *B. exitiosa* was initially described from the oyster *Ostrea chilensis* from New Zealand and other parts of the world. Although they mainly infect oysters in the genus *Ostrea*, infections have also been reported in other atypical hosts (Dinamani et al. 1987; Hine et al. 2001; Carnegie 2017; Lane 2022; WOAH 2024).

The parasites appear as small (2–5 μm), spherical to ovoid organisms within the haemocytes, but have also been observed extracellularly in the gill, stomach, and other connective tissues. Clinically, gaping or dead animals are the first indications; however, these do not suggest any specific aetiology. In most cases, the infected animals do not exhibit any apparent clinical signs; nevertheless, some oysters may develop yellow discolouration and extensive lesions in the connective tissue of the gills, mantle, and digestive glands. As the disease becomes systemic, massive destruction of hemocytes ensues, leading to pathological and physiological changes followed by mortality.

Infection with *Haplosporidium*

Haplosporidians are a small group of pathogenic spore-forming protozoans that infect oysters in the Americas. Two species, *Haplosporidium nelsoni*, causing Multinucleate sphere unknown (MSX) disease and *H. costale*, causing Seaside organism (SSO) disease in many commercially important bivalve species, especially the eastern oyster, *C. virginica*, are considered important. Haplosporidians have also been reported in *C. gigas* in Europe and Asia. Multinucleated plasmodial stages measuring 5–50 μm in diameter are found throughout the body. Sporulation occurs in the digestive gland tubules, and the spores are released into the environment following the death of the host (Arzul and Carnegie 2015). The disease is characterised by a whitish discolouration of the digestive gland tubules, accompanied by other nonspecific signs, including emaciation, mantle recession, poor growth, and mortality in juvenile and adult oysters. The infection can be confirmed through histological examination. In heavy infections, hemolymph smears can detect the parasites. Molecular tools are also used for diagnosis. Higher temperatures and salinities are known to increase disease activity. The complete life cycle of *H. nelsoni* is unknown, and the role of an intermediate host in its life cycle cannot be ruled out (Burreson and Ford 2004; Arzul and Carnegie 2015).

Infection with *Marteilia*

Marteiliosis is caused by protists of the genus *Marteilia*, belonging to the phylum Paramyxea. Two important species of *Marteilia* that affect bivalves in Europe, South America, and Australia are *M. refringens* and *M. sydneyi*. *Marteilia refringens* is the causative agent of 'Aber disease' in *O. edulis*, while *M. sydneyi* causes 'QX disease' in *Crassostrea glomerata*. *Marteilia cochillia, M. lenghei, M. maurini, M. christenseni, M. tapetis, M. cocosarum,* and other *Marteilia* spp. are also known to cause disease in bivalves (Berthe et al. 2004; Carrasco et al. 2013, 2015; Adlard and Nolan 2015; Kang et al. 2019; Skujina et al. 2022).

Marteilia spp. generally infects the digestive system of bivalves. The parasite initially develops in the epithelium of the labial palps and stomach, and later undergoes sporulation in the digestive tubules. Infective stages are released into the digestive tract and discharged along with the feces. They exhibit a "cell-within-cell" developmental pattern due to endogenous budding (Berthe et al. 2004). The pathogenicity of *M. refringens* varies widely in different bivalve species and is known to cause serious mortality in oysters, whereas in mussels, they appear to be less pathogenic. Mortality and morbidity are usually associated with sporulation of the parasite in the digestive tissues and may reach 50–90% in oysters during the summer. Higher salinities are known to affect the development and proliferation of the parasite adversely. Similar to other protozoan infections, clinical signs of marteliosis include dead or gaping bivalves with pale digestive glands, thin and watery flesh, retracted mantles, and reduced growth rates; however, these signs are not specific for marteiliosis (Berthe et al. 2004). Wet mounts of squash/smears/imprints prepared from the digestive glands of gaping/freshly dead bivalves or faeces from infected live animals may reveal mature sporangia measuring 30–40 µm with refringent granules in advanced infections. Hematoxylin and eosin (H & E)-stained sections containing the gill, digestive gland, mantle, and gonad tissues are best suited for histological screening. Uninucleate young stages (4–40 µm) in the epithelium of labial palps and stomach, presporulating stages in the connective tissues and epithelia of digestive glands and stomach, and sporulating stages (exhibiting a "cell within the cell" structure) with refringent granules in the digestive gland tubules and lumen indicate infection with *Marteilia* (Berthe et al. 2004). PCR targeting the ITS1 region of the parasite can be successfully used for the specific diagnosis of *M. refringens*. Other molecular methods, including ISH assay, can also be used. No effective treatment is available to control this disease. The parasite is believed to have an indirect life cycle, and certain calanoid copepods may act as vectors, aiding its transmission (Boyer et al. 2013; Arzul et al. 2014). Further, *M. refringens* is listed by the WOAH as a notifiable molluscan pathogen (WOAH 2024).

Infection with *Perkinsus*

Perkinsosis, also known as 'dermo disease' or 'proliferative disease' is a serious disease infecting bivalve molluscs and is caused by the protozoan parasites of the genus *Perkinsus*. Although historically classified under Apicomplexa, molecular studies have shown that the genus *Perkinsus* is more closely related to dinoflagellates and is placed within the superphylum Alveolata. (Siddall et al. 1997; Saldarriaga et al. 2003; Villalba et al. 2004; Bachvaroff et al. 2011; Dungan and Reece 2020).

Members of the genus *Perkinsus* exhibit a common life cycle pattern. The trophozoite stage within the host's tissues is spherical, measuring 5–20 µm, with a large, slightly eccentric vacuole and a peripheral nucleus imparting a "signet ring" appearance. Mature trophozoites multiply either by binary fission or through a form of vegetative proliferation known as palintomy. In palintomy, a single trophozoite undergoes multiple rapid nuclear divisions followed by cytoplasmic division, resulting in the formation of a tomont. The tomont containing 32–64 daughter cells arranged in a rosette-like pattern, eventually ruptures and releases the daughter cells, which then mature into new trophozoites, perpetuating the cycle (Fig. 1a) (Perkins 1988, 1996; Sunila et al. 2001; Villalba et al. 2004; Ben-Horin et al. 2015). *Perkinsus* is transmitted directly between bivalve hosts, and all life cycle stages are infective. In dead and decaying hosts, the trophozoites in the tissues develop into prezoosporangia/hypnospores, which release numerous infective zoospores/dinospores into the water column (Fig. 1b & c). These free-living stages may survive for weeks in the water column and retain their infectivity. Trophozoites are also released directly from infected bivalves into the water column through pseudofeces and feces, which in turn infect other hosts (Fig. 1d) (Scanlon et al. 1997; Bushek et al. 2002; Casas et al. 2002; Villalba et al. 2004; Ben-Horin et al. 2015).

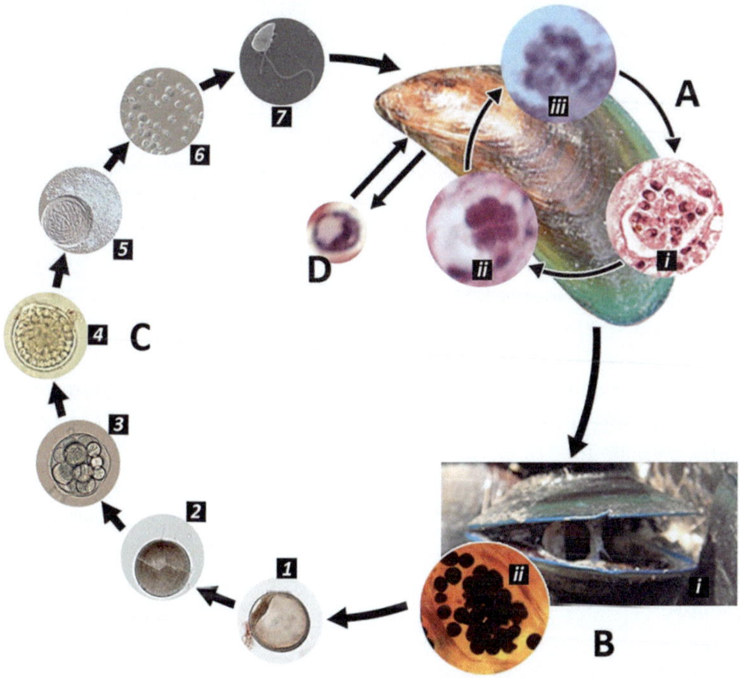

Fig. 1 Life cycle of *Perkinsus* spp. (**a**) *i*-mature trophozoites; *ii*-tomont; *iii*-immature trophozoites; (**b**) *i*-dead/decaying bivalve; *ii*-prezoosporangia; (**c**) sporulation of prezoosporangia; 1–6 development of zoospores; 7–infective zoospores; (**d**) infective trophozoites released through feces

Perkinsus spp. have been implicated as the causative organisms in numerous instances of mortality in bivalves worldwide. Of the seven reported species of *Perkinsus* (*P. marinus, P. olseni, P. quagwadi, P. chesapeaki, P. mediterraneus, P. honshuensis,* and *P. beihaiensis*), two species, *P. marinus* and *P. olseni* are WOAH notifiable (WOAH 2024). *Perkinsus marinus* (initially misidentified as the fungus, *Dermocystidium marinum*) mainly infects *C. virginica* along the coasts of America but has also been reported from other hosts. *Perkinsus olseni* has been known to infect more than 60 species of bivalves and enjoys a worldwide distribution. To date, three species of *Perkinsus* - *P. olseni, P. beihaiensis*, and *P. marinus* - have been reported from the Indian subcontinent (Sanil et al. 2010, 2012; Uma et al. 2021).

Perkinsus spp. infect the haemocytes and connective tissues of all organs and are often associated with hemocyte infiltration in tissues. Generally, infected bivalves do not exhibit any specific or conspicuous clinical signs. The primary indication is gaping, while mortality may also occur but is not specific to any particular aetiology. However, generalised clinical signs - including shrunken, thin, and watery-soft tissues; severe emaciation; pale digestive gland; and nodules or lesions in the gills, mantle, gonads, and digestive gland, are suggestive but not specific for perkinsosis.

Mortality varies among bivalve hosts, and is often associated with higher temperatures and salinity (Villalba et al. 2004; WOAH 2024).

The culture of host tissues in 'Ray's Fluid Thioglycollate Medium' (RFTM) is considered the gold standard for diagnosing *Perkinsus* infections. Connective tissues (gill, mantle, and rectal tissues) measuring 5–10 mm are excised, placed in RFTM, and incubated at 25 °C for 5 to 7 days in the dark. During incubation, *Perkinsus* trophozoites in bivalve tissues enlarge without proliferation (reaching 20–250 μm or more in size) and transform into round, enlarged, and refractile prezoosporangia/hypnospores (Ray 1966). The incubated tissues are macerated with a drop of Lugol's iodine (20–30% (v/v)) on a clean slide and allowed to sit for 10 min. When stained with Lugol's iodine, the prezoosporangia/hypnospores appear blue or bluish-black (Fig. 2a). Since all species of *Perkinsus* (except *P. qugwadi*) yield positive results in the RFTM assay, this technique cannot be used for species-level differentiation of *Perkinsus* spp. Whole-body assays using RFTM can be used to quantify parasite load in whole animals. (Villalba et al. 2004; Dungan and Reece 2020). *Perkinsus* infections can be diagnosed by histopathology; spherical, approximately 5–20 μm-sized cells with a large, slightly eccentric vacuole and a peripheral nucleus imparting a 'signet ring' appearance indicate the presence of *Perkinsus* (Fig. 2b). *Perkinsus* infections often elicit intense haemocyte infiltration, leading to inflammatory capsules in the surrounding tissues (Carella et al. 2023). However, histopathology is inadequate for detecting low-level infections and cannot identify parasites at the species level. Molecular diagnostic techniques are generally preferred for the confirmatory diagnosis of *Perkinsus*. PCR targeting the ITS region of the parasite followed by sequencing, species-specific PCR, ISH, and Fluorescent in situ hybridisation can provide a confirmatory diagnosis of the pathogen (Fig. 2c).

In India, *Perkinsus* spp. infections have been reported in 19 bivalve species, including the cultured green mussel (*Perna viridis*) and the oyster (*C. madrasensis*) (Sanil et al. 2010, 2012; Shamal et al. 2018), and have been linked to mass mortality in farmed bivalves during the summer months. The widespread movement of infected bivalves for trade, consumption, and culture contributes to the horizontal spread of the pathogen, establishing it as endemic in many regions. Bivalves such as the yellow clam (*Paphia malabarica*) and the invasive mussel (*Mytella strigata*) can tolerate high levels of *Perkinsus* infection without exhibiting mortality, thereby acting as lifelong carriers or reservoirs (Shamal et al. 2018). Coexistence of these

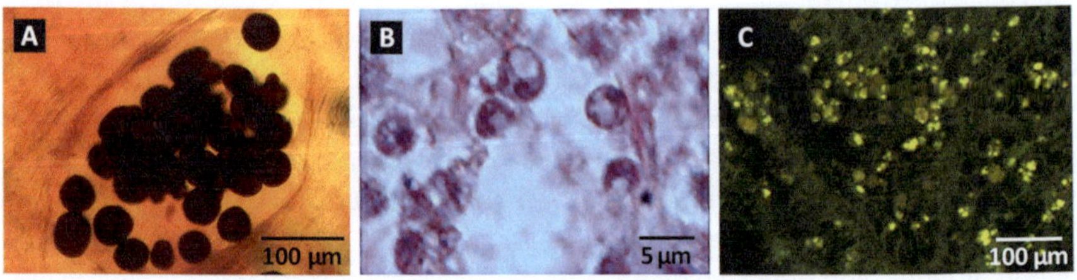

Fig. 2 *Perkinsus* spp. (**a**) prezoosporangia stained with Lugol's iodine; (**b**) trophozoites in tissues (**c**) fluorescent in situ hybridization

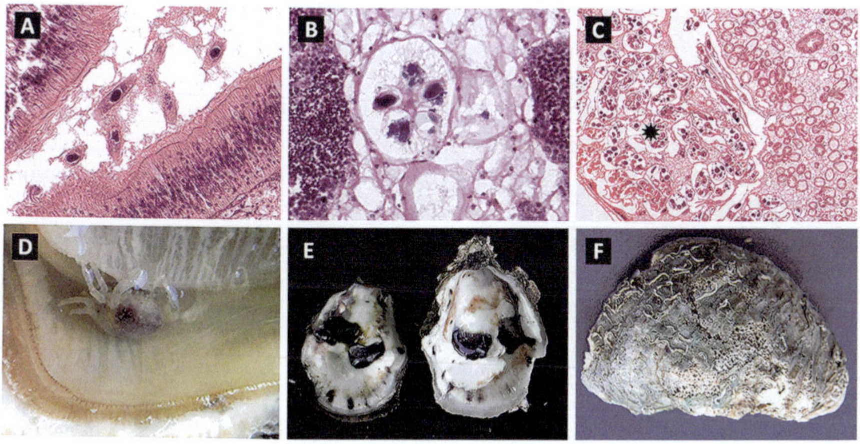

Fig. 3 (**a**) ciliates in the lumen of the digestive gland tubule; (**b**) oocysts of *Nematopsis* sp. (**c**) *larval stages of Bucephalid trematode in the digestive gland; (**d**) pea crab in the mantle cavity; (**e**) shell damage caused by *Polydora* spp. infestation (**f**) clinoid sponge infestation on the shell

carriers with cultured bivalves in the same environment enhances pathogen transmission. Currently, there is no effective treatment for perkinsosis, and control is largely limited to restricting the movement of stocks from infected areas.

Infection with Ciliates

Many ciliates are associated with bivalves; while most have a non-pathogenic, symbiotic relationship, some infect and cause disease, with ciliate-associated pathologies and mortality reported in various bivalve species. The ciliate *Ancistrocoma* sp. inhabits the lumen of the digestive tubules and feeds on the epithelial cells (Fig. 3a). *Sphenophrya* sp. is associated with the gill epithelium, palp, and mantle of many bivalves and often causes xenoma formation in bivalves. Trichodinid ciliates, frequently found attached to gills and occasionally to palps, are usually considered harmless in low numbers; however, heavy infestations can cause gill erosion and lead to mortality. Generally, clinical manifestations are not apparent, and infections can be detected through smear/squash observation and histology (Bower et al. 1994).

Infection with Gregarines

Gregarines of the genus *Nematopsis* (Apicomplexa: Porosporidae) are known to parasitise the gills, mantle, and sometimes the digestive glands of several commercially important marine bivalves. Decapod crustaceans are the definitive hosts, while marine molluscs act as intermediate hosts (Lauckner 1983; Canestri-Trotti et al. 2000; Padovan et al. 2003). Within the bivalve tissues, the parasite appears as resistant oocysts, enclosing a single uni-nucleated vermiform sporozoite (Fig. 3b). The prevalence, intensity, and pathology of *Nematopsis* spp. varies among bivalve hosts (Estevez et al. 1998; Padovan et al. 2003). Apparent clinical manifestations are generally absent. In histological sections, the parasite manifests as single or clustered intra-haemocytic oocysts enclosed by a thick hyaline wall. Each oocyst contains a basophilic, vermiform sporozoite within a parasitophorous vacuole, and the infected cells exhibit marked hypertrophy. Light infections with *Nematopsis* spp. are generally considered less pathogenic and are limited to the compression of adjacent cells/tissues, with minimal tissue damage in most cases (Suja et al. 2016). However, in cases of heavy infection, haemocyte infiltration, tissue damage, structural alterations, and obstruction of feeding have been reported in bivalves. Mortalities associated with heavy gill infections have been reported in cockles (Azevedo and Cachola 1992; Zainathan et al. 2022). Diagnosis is based on morphology and histopathology.

2.4.2 Metazoan Infections

Several metazoan parasites - including helminths, annelids, and crustaceans - exhibit a range of symbiotic relationships with bivalves, from mutualism to parasitism. Crustaceans, including copepods, amphipods, isopods, and decapods, have been reported infesting bivalves, typically causing mild pathological effects. Additionally, many metazoans exploit bivalves as intermediate hosts to complete their life cycles. Although not considered true parasites, numerous marine organisms, classified as foulers and borers also colonize the shells of bivalves, interfering with the animal's normal health (Alagarswami and Chellam 1976; Thomas 1979; Suja et al. 2017). The filter-feeding nature of bivalve molluscs facilitates direct, horizontal, and passive transmission of pathogens, especially in high-density culture systems. Metazoan parasites, including trematodes, cestodes, nematodes, and ectoparasitic crustaceans, can negatively impact bivalve health. While their disease potential and infection intensity are generally lower than those of protozoans, they can still significantly impact the growth, reproductive performance, and overall health of bivalves (Lauckner 1983; Gosling 2015).

Larval stages of trematodes are known to develop within the tissues of many bivalves. Sporocysts/redia of trematodes belonging to the families Gymnophalidae, Fellodistomidae, and Bucephalidae often develop in the digestive glands and gonads of bivalves (Fig. 3c). Parasitic castration due to larval trematode infection has been reported in several bivalve species. Gonadal damage may be physical, mechanical (ingestion, abrasion, and pressure), or chemical (secretion of lytic substances, toxins, or endocrinological antagonists) (Khamdan 1998; Winstead et al. 2004; Mladineo 2008; Darriba et al. 2010). Altered gonadal acinar architecture and drastic reduction in the number and area of gonadal acini, leading to castration, have been reported in *C. madrasensis* infected with the larval stages of *Bucephalus* sp. (Suja et al. 2014). Metacestode stages of tapeworms are known to infect bivalves, particularly in tropical and subtropical waters. Larval stages of *Tylocephalum* are frequently observed encapsulated within the digestive tissues of bivalves, though these infections typically do not cause significant harm to the host (Cake and Menzel 1980; Lauckner 1983; Gosling 2015). Although the larval stages of some nematodes have occasionally been reported in bivalves, they are not considered major pathogens affecting bivalve health (Gosling 2015).

Many crustaceans, including copepods, amphipods, isopods, and decapods, are closely associated with bivalves. Of particular significance are parasitic copepods of the genus *Mytilicola*, commonly known as 'red worms,' which infest the digestive tracts of various bivalve species. Unlike typical copepods, *Mytilicola* spp. have elongated, worm-like bodies with reduced appendages and simplified mouthparts. Some species may also infest gills. Heavy infections can compromise the health of bivalve hosts (Bower 2009; Gosling 2015). The pea crab, *Pinnotheres* spp., inhabits the mantle cavity of many bivalve species and feeds on the mucous strings of filtered food (Fig. 3d). This may cause gill erosion and interfere with bivalve growth (Bierbaum and Shumway 1988).

3 Foulers and Borers

The shells of bivalves serve as substrates for various marine organisms, including barnacles and polychaete worms. While these fouling organisms typically do not present a substantial threat, extensive colonisation may hinder bivalve shell functionality and nutrient acquisition (Bower et al. 1994). Certain shell-boring species, such as the mud worm *Polydora* spp. and the boring sponge *Cliona* spp., can damage shells and negatively impact the health of affected bivalves. *Polydora* spp. create burrows within bivalve shells, forming 'mud blisters' that are visually unappealing. Substantial infestations are often manifested as weakened, misshapen shells with darkened areas, compromised flesh quality, and diminished adductor muscle strength (Fig. 3e). Although such infestations do not directly cause mortality, the resultant unsightly shells and malodourous meat significantly reduce the market value of oysters. Clionid sponges are regarded as the most ubiquitous and detrimental macroborers of bivalve molluscs. Even minor infections can compromise the shell integrity and disrupt nacre formation in pearl oysters (Fig. 3f). In extreme instances, extensive shell erosion may result in mortality (Alagarswami and Chellam 1976; Thomas 1979; Lauckner 1983; Bower et al. 1994; Meyers and Burton 2009; Suja et al. 2017).

4 Conclusion

Given that bivalve molluscs are cultivated in open waters, conventional preventive and control measures, such as chemotherapeutics or vaccination, have proven ineffective. Effective management strategies encompass the utilisation of specific diagnostic tools, development of resistant organisms through genetic selection, restriction of infected stock movement, implementation of sound farming practices, adoption of appropriate biosecurity measures, and a comprehensive understanding of pathogens and their transmission mechanisms. Collectively, these approaches can contribute to the prevention and control of bivalve diseases.

References

Adlard RD, Nolan MJ (2015) Elucidating the life cycle of *Marteilia sydneyi*, the aetiological agent of QX disease in the Sydney rock oyster (*Saccostrea glomerata*). Int J Parasitol 45:419–426

Alagarswami K, Chellam A (1976) On fouling and boring organisms and mortality of pearl oysters in the farm at Veppalodai, Gulf of Mannar. Indian J Fish 23(1&2):10–22

Allam B, Raftos D (2015) Immune responses to infectious diseases in bivalves. J Invertebr Pathol 131:121–136

Andrews JD (1988) Epizootiology of the disease caused by the oyster pathogen *Perkinsus marinus* and its effects on the oyster industry. Am Fish Soc Special Publ 18:47–63

Aranguren R, Figueras A (2016) Moving from histopathology to molecular tools in the diagnosis of molluscs diseases of concern under EU legislation. Front Physiol 7(538):5–10

Arzul I, Carnegie RB (2015) New perspective on the haplosporidian parasites of molluscs. J Invertebr Pathol 131:32–42

Arzul I, Renault T (2002) Herpèsvirus et bivalves marins Virologie (Montrouge). 6(3):169–174

Arzul I, Chollet B, Boyer S, Bonnet D, Gaillard J, Baldi Y (2014) Contribution to the understanding of the cycle of the protozoan parasite *Marteilia refringens*. Parasitology 141:227–240

Arzul I, Corbeil S, Morga B, Renault T (2017) Viruses infecting marine molluscs. J Invertebr Pathol 147:118–135

Azevedo C, Cachola R (1992) Fine structure of the Apicomplexa oocyst of *Nematopsis* sp. of two marine bivalve mollusks. Dis Aquat Org 14:69–73

Bachvaroff TR, Handy SM, Place AR, Delwiche CF (2011) Alveolate phylogeny inferred using concatenated ribosomal proteins. J Eukaryot Microbiol 58:223–233

Barbosa-Solomieu V, Renault T, Travers MA (2015) Mass mortality in bivalves and the intricate case of the Pacific oyster, *Crassostrea gigas*. J Invertebr Pathol 131:2–10

Beaz-Hidalgo R, Balboa S, Romalde JL, Figueras MJ (2010) Diversity and pathogenicity of *Vibrio* species in cultured bivalve molluscs. Environ Microbiol Rep 2:34–43

Ben-Horin T, Bidegain G, Huey L, Narvaez DA, Bushek D (2015) Parasite transmission through suspension feeding. J Invertebr Pathol 131:155–176

Berthe F (2005) Diseases in mollusc hatcheries and their paradox in health management. In: Proceedings of the V Symposium on Diseases in Asian Aquaculture, Queensland, Australia, pp 239–248

Berthe FCJ (2008) New approaches to effective mollusc health management. In: Bondad-Reantaso MG, Mohan CV, Crumlish M, Subasinghe RP (eds) Diseases in Asian aquaculture VI, Fish Health Section. Asian Fisheries Society, Manila, Philippines, pp 343–352

Berthe F, Frederique L, Adlard RD, Figueras A (2004) Marteiliosis in Molluscs: a review. Aquat Living Resour 17:433–448

Berthelin C, Kellner K, Mathieu M (2000) Storage metabolism in the Pacific oyster (*Crassostrea gigas*) in relation to summer mortalities and reproductive cycle (West Coast of France). Comp Biochem Physiol B Biochem Mol Biol 125(3):359–369

Bierbaum R, Shumway SE (1988) Filtration and oxygen consumption in mussels, *Mytilus edulis*, with and without pea crabs, *Pinnotheres maculatus*. Estuaries 11(4):264–271

Boettcher KJ, Barber BJ, Singer JT (2000) Additional evidence that juvenile oyster disease is caused by a member of the Roseobacter group and colonization of nonaffected animals by *Stappia stellulata*-like strains. Appl Environ Microbiol 66(9):3924–3930

Bower SM (2009) Synopsis of infectious diseases and parasites of commercially exploited shellfish: *Mytilicola intestinalis* (Red Worm Disease) of Mussels. Available at: https://www.dfo-mpo.gc.ca/science/aah-saa/diseases-maladies/mirwdmu-eng.html (accessed 11 Nov 2024)

Bower SM, McGladdery SE (2010) Synopsis of infectious diseases and parasites of commercially exploited shellfish. Available at: https://www.dfo-mpo.gc.ca/science/aah-saa/diseasesmaladies/index-eng.html

Bower SM, McGladdery SE, Price IM (1994) Synopsis of infectious diseases and parasites of commercially exploited shellfish. Annu Rev Fish Dis 4:1–199

Boyer S, Chollet B, Bonnet D, Arzul I (2013) New evidence for the involvement of *Paracartia grani* (Copepoda, Calanoida) in the life cycle of *Marteilia refringens* (Paramyxea). Int J Parasitol 43(14):1089–1099

Burreson EM, Ford SE (2004) A review of recent information on the Haplosporidia, with special reference to *Haplosporidium nelsoni* (MSX disease). Aquat Living Resour 17(4):499–517

Bushek D, Ford SE, Chintala MM (2002) Comparison of *in vitro* cultured and wild-type *Perkinsus marinus*. III. Fecal elimination and its role in transmission. Dis Aquat Org 51:217–225

Cake EW, Menzel RW (1980) Infections of *Tylocephalum* metacestodes in commercial oysters and three predaceous gastropods of the Eastern Gulf of Mexico. Proc Natl Shellfishe Assoc 70:94–104

Canesi L, Pruzzo C (2016) Chapter 6 – Specificity of innate immunity in bivalves: A lesson from bacteria. In: Ballarin L, Cammarata M (Eds.) Lessons in immunity: From Single-Cell Organisms to Mammals. Academic Press, pp 79–91

Canestri-Trotti G, Baccarani EM, Paesanti F, Turolla E (2000) Monitoring of infections by protozoa of the genera *Nematopsis, Perkinsus* and *Porospora* in the smooth venus clam *Callista chione* from the North-Western Adriatic Sea (Italy). Dis Aquat Org 42:157–161

Carballal MJ, Villalba A, Iglesias D, Hine PM (2003) Virus-like particles associated with large foci of heavy hemocytic infiltration in cockles *Cerastoderma edule* from Galicia (NW Spain). J Invertebr Pathol 84(3):234–237

Carella F, Tejedor MF, Villari G, Andree KB, Vico GD (2023) The endoparasite *Perkinsus olseni* affecting the Mediterranean mussels (*Mytilus galloprovincialis*) in the Italian and Spanish waters: a new possible threat for mussel aquaculture and wild animal population. Front Mar Sci 10:1–9

Carnegie RB (2017) Bonamiosis of oysters caused by *Bonamia exitiosa*. ICES Identification Leaflets for Diseases and Parasites of Fish and Shellfish, No. 66. 6 pp

Carnegie RB, Cochennec-Laureau N (2004) Microcell parasites of oysters: recent insights and future trends. Aquat Living Resour 17:519–528

Carrasco N, Hine PM, Durfort M, Andree KB, Malchus N, Lacuesta B, González M, Roque A, Rodgers C, Furones MD (2013) *Marteilia cochillia* sp. nov., a new *Marteilia* species affecting the edible cockle *Cerastoderma edule* in European waters. Aquaculture 412-413:223–230

Carrasco N, Green T, Itoh N (2015) *Marteilia* spp. parasites in bivalves: a revision of recent studies. J Invertebr Pathol 131:43–57

Casas SM, Villalba A, Reece KS (2002) Study of perkinsosis in the carpet shell clam *Tapes decussatus* in Galicia (NW Spain). I. Identification of the aetiological agent and *in vitro* modulation of zoosporulation by temperature and salinity. Dis Aquat Org 50(1):51–65

Cheyney D, MacDonald B, Elston R (1998) Summer mortality of Pacific oysters, *Crassostrea gigas* (Thunberg): initial findings on multiple environmental stressors in Puget Sound, Washington

Comps M (1988) Epizootic diseases of oysters associated with viral infections. Am Fish Soc Spec Publ 18:23–37

Comps M, Herbaut C, Fougerouse A (1999) Virus-like particles in pearl oyster *Pinctada margaritifera*. Bull Eur Assoc Fish Pathol 19(2):85–88

Culloty SC, Novoa B, Pernas M, Longshaw M, Mulcahy MF, Feist SW, Figueras A (1999) Susceptibility of a number of bivalve species to the protozoan parasite *Bonamia ostreae* and their ability to act as vectors for this parasite. Dis Aquat Org 37(1):73–80

Darriba S, Iglesias D, Rui ZM, Rodriguez R, Lopez C (2010) Histopathological survey of symbionts and other conditions in razor clam *Ensis arcuatus* (Jeffreys, 1865) (Pharidae) of the coast of Galicia (NW Spain). J Invertebr Pathol 104:23–30

Davison AJ, Eberle R, Ehlers B, Hayward GS, McGeoch DJ, Minson AC, Pellett PE, Roizman B, Studdert MJ, Thiry E (2009) The order Herpesvirales. Arch Virol 154(1):171–177

Dinamani P, Hine PM, Jones JB (1987) Occurrence and characteristics of the haemocyte parasite *Bonamia* sp. in the New Zealand dredge oyster *Tiostrea lutaria*. Dis Aquat Org 3:37–44

Dungan CF, Reece KS (2020) *Perkinsus* sp. infections of marine molluscs. In: AFS-FHS (American Fisheries Society-Fish Health Section). FHS Blue Book: suggested procedures for the detection and identification of certain finfish and shellfish pathogens, Bethesda, Maryland

Elston RA, Moore JD, Brooks K (1992) Disseminated neoplasia of bivalve molluscs. Rev Aquatic Sci 6:405–466

Estevez J, Soto M, Rodríguez-Domínguez H, Arias CA (1998) Histopathology of bivalves in the Ria de Vigo (Galicia, NW Spain) infected by *Nematopsis* sp. (Protozoa:Apicomplexa). Res Rev Parasitol 58(1):63–65

FAO (2022) The state of world fisheries and aquaculture 2022. https://openknowledge.fao.org/server/api/core/bitstreams/9df19f53-b931-4d04-acd3-58a71c6b1a5b/content/sofia/2022/aquaculture-production.html

Farley CA (1976) Ultrastructural observations on epizootic neoplasia and lytic virus infection in bivalve mollusks. Prog Exp Tumor Res 20:283–294

Farley CA (1985) Viral gametocyte hypertrophy in oysters. ICES Identification Leaflets for Diseases and Parasites in Fish and Shellfish. No. 25

Farley CA, Banfield WG, Kasnic JRG, Foster WS (1972) Oyster herpes-type virus. Science 178(4062):75960

Froelich BA, Phippen B, Fowler P, Noble RT, Oliver JD (2016) Differences in abundances of total *Vibrio* spp., *V. vulnificus*, and *V. parahaemolyticus* in Clams and Oysters in North Carolina. Appl Environ Microbiol 83(2):e02265–16

Gagnaire B, Gay M, Huvet A, Daniel JY, Saulnier D, Renault T (2007) Combination of a pesticide exposure and a bacterial challenge: *in vivo* effects on immune response of Pacific oyster, *Crassostrea gigas* (Thunberg). Aquat Toxicol 84(1):92–102

Gosling E (2004) Diseases and parasites. In: Gosling E (ed) Bivalve molluscs biology, Ecology and Culture. Blackwell Science, Oxford, pp 370–411

Gosling E (2015) Disease and parasites. In: Gosling E (ed) Marine bivalve molluscs, 2nd edn. John Wiley & Sons, West Sussex, pp 429–477

Harshbarger JC, Otto SV, Chang SC (1979) Proliferative disorders in *Crassostrea virginica* and *Mya arenaria* from the Chesapeake Bay and intranuclear virus-like inclusions in *Mya arenaria* with germinomas from a Maine oil spill site. Haliotis 8:243–248

Hine PM, Wesney B, Hay BE (1992) Herpesviruses associated with mortalities among hatchery-reared larval pacific oysters *Crassostrea gigas*. Dis Aquat Org 12:135–142

Hine PM, Cochennec-Laureau N, Berthe FCJ (2001) *Bonamia exitiosus* n. sp. (Haplosporidia) infecting flat oysters *Ostrea chilensis* in New Zealand. Dis Aquat Org 47:63–72

Kang H, Itoh N, Limpanont Y, Lee H, Whang I, Choi K (2019) A novel paramyxean parasite, *Marteilia tapetis* sp. nov. (Cercozoa) infecting the digestive gland of Manila clam *Ruditapes philippinarum* from the southeast coast of Korea. J Invertebr Pathol 163:86–93

Khamdan SAA (1998) Occurrence of *Bucephalus* sp. trematode in the gonad of the pearl oyster *Pinctada radiata*. Environ Int 24(1–2):117–120

Lane HS (2022) Chapter 70 – Bonamiasis. In: Kibenge FSB, Baldisserotto B, Chong RSM (eds) Aquaculture pathophysiology, volume II. Academic Press, pp 533–546

Lauckner G (1983) Diseases of Mollusca: Bivalvia. In: Kinne O (ed) Diseases of marine animals: 2. Introduction, Bivalvia to Scaphopoda. Biologische Ansalt Helgoland, Hamburg, pp 477–961

Malham SK, Cotter E, O'Keeffe S, Lynch S, Culloty SC, King JW, Beaumont AR (2009) Summer mortality of the Pacific oyster, *Crassostrea gigas*, in the Irish Sea: the influence of temperature and nutrients on health and survival. Aquaculture 287(1–2):128–138

McGladdery SE (1999) Shellfish diseases (viral, bacterial and fungal). In: Woo PTK, Bruno DW (eds) Fish diseases and disorders, Volume 3: viral, bacterial and fungal infections, 2nd edn, pp 738–842

McGladdery SE, Sharon F, Bower SM, Getchell RD (2006) Chapter 11, Diseases and parasites of scallops. Dev Aquac Fish Sci 35:595–650

Meyers T, Burton T (2009) Diseases of wild and cultured shellfish in Alaska. Alaska Department of Fish and Game, Fish Pathology Laboratories, Anchorage

Minchin D, Gollasch S (2002) Vectors – how exotics get around. In: Leppäkoski E, Gollasch S, Olenin S (eds) Invasive aquatic species of Europe. Distribution, impacts and management. Kluwer Academic Publishers, Dordrecht, pp 183–192

Miyazaki T, Goto K, Kobayashi T, Kageyama T, Miyata M (1999) Mass mortalities associated with a virus disease in Japanese pearl oysters *Pinctada fucata martensii*. Dis Aquat Org 37:1–12

Mladineo I (2008) Risk assessment of parasitic/symbiotic organisms of the commercially important mytilid *Modiolus barbatus* (Linnaeus, 1758). Aquac Res 39:1705–1719

Mohamed KS (2013) Bivalve mariculture in India – progress in research and development. CMFRI, Kochi. eprints.cmfri.org.in

Moss JA, Burreson EM, Cordes JF, Dungan CF, Brown GD, Wang A, Wu X, Reece KS (2007) Pathogens in *Crassostrea ariakensis* and other Asian oyster species: implications for non-native oyster introduction to Chesapeake Bay. Dis Aquat Org 77(3):207–223

OIE (2019) Chapter 2.4.5. – Infection with ostreid herpesvirus 1 microvariants. 2019 © OIE – Manual of Diagnostic Tests for Aquatic Animals – 14/11/2019

Oliveira AGLD, Fernandes RB, Santos FDS, Silva IDSD, Gatti MJDA, Moraes AMLD, Mello-Silva CC, Santos CP (2024) Filamentous fungi associated with the brown mussel, *Perna perna* (Bivalvia: Mytilidae), off the coast of Rio de Janeiro, Brazil. Mar Biol Res 20(5–6):123–137

Padovan IP, Tavares LA, Corral L, Padovan PA, Azevedo C (2003) Fine structure of the oocyst of *Nematopsis mytella* (Apicomplexa, Porosporidae), a parasite of the mussel *Mytella falcata* and of the oyster *Crassostrea rizophorae* (Mollusca, Bivalvia) from the northeastern Atlantic coast of Brazil. Brazilian J Morphol Sci 20:141–145

Paillard C, Maes P, Oubella R (1994) Brown ring disease in clams. Annu Rev Fish Dis 4:219–240

Paillard C, Ashton-Alcox K, Ford SE (1996) Changes in bacterial densities and hemocyte parameters in oysters affected by Juvenile Oyster Disease. Aquat Living Resour 9:145–158

Paillard C, Roux FL, Borrego JJ (2004) Bacterial disease in marine bivalves, a review of recent studies: trends and evolution. Aquat Living Resour 17:477–498

Percival SL, Williams DW (2014) Chapter 12-*Vibrio*. In: Percival SL, Yates MV, Williams DW, Chalmers RM, Gray NF (eds) Microbiology of waterborne diseases: microbiological aspects and risks, 2nd edn. Academic Press, Amsterdam/Boston, pp 237–248

Perkins FO (1988) Structure of protistan parasites found in bivalve molluscs. Am Fish Soc Spec Publ 18:93–111

Perkins FO (1996) The structure of *Perkinsus marinus* (Mackin, Owen & Collier, 1950) Levine, 1978 with comments on taxonomy and phylogeny of *Perkinsus* spp. J Shellfish Res 15(1):67–87

Rahman MA, Henderson S, Miller-Ezzy P, Li XX, Qin JG (2019) Immune response to temperature stress in three bivalve species: Pacific oyster *Crassostrea gigas*, Mediterranean mussel *Mytilus galloprovincialis* and mud cockle *Katelysia rhytiphora*. Fish Shellfish Immunol 86:868–874

Ray SM (1966) A review of the culture method for detecting *Dermocystidium marinum*, with suggested modifications and precautions. Proc Natl Shellfish Assoc 54:55–69

Raynard R, Wahli T, Vatsos I, Mortensen S (2007) In: DIPNET (Ed.), Review of disease interactions and pathogen exchange between farmed and wild finfish and shellfish in Europe, p 459. http://www.dipnet.info/docs/doc.asp?id=48. Aberdeen

Renault T (1996) Appearance and spread of diseases among bivalve molluscs in the northern hemisphere in relation to international trade. Rev Sci Tech 15(2):551–561

Renault T (2008) Genomics and mollusc pathogens: trends and perspective. J Vet Clin Sci 1(2):4–14

Renault T, Novoa B (2004) Viruses infecting bivalve molluscs. Aquat Living Resour 17(4):397–409

Renault T, Lipart C, Arzul I (2001) A herpes-like virus infecting *Crassostrea gigas* and *Ruditapes philippinarum* larvae in France. J Fish Dis 24:369–376

Saldarriaga JF, McEwan ML, Fast NM, Taylor FJR, Keeling PJ (2003) Multiple protein phylogenies show that *Oxyrrhis marina* and *Perkinsus marinus* are early branches of the dinoflagellate lineage. Int J Syst Evol Microbiol 53:355–365

Sanil NK, Vijayan KK, Kripa V, Mohamed KS (2010) Occurrence of the protozoan parasite, *Perkinsus olseni* in the wild and farmed pearl oyster, *Pinctada fucata* (Gould) from the Southeast coast of India. Aquaculture 299:8–14

Sanil NK, Suja G, Lijo J, Vijayan KK (2012) First report of *Perkinsus beihaiensis* in *Crassostrea madrasensis* from the Indian subcontinent. Dis Aquat Org 98:209–220

Santos ALD, Medeiros JVFD, Grault CE, Santos MJS, Souza ALA, Carvalho RWD (2020) The fungus *Pestalotiopsis* sp., isolated from *Perna perna* (Bivalvia:Mytilidae) cultured on marine farms in South eastern Brazil and destined for human consumption. Mar Pollut Bull 153:110976

Scanlon CH, Ragone Calvo LM, Burreson EM (1997) The potential for transmission of *Perkinsus marinus* by fecal matter from the Eastern oyster, *Crassostrea virginica*. J Shellfish Res 16:332

Shamal P, Zacharia PU, Binesh CP, Pranav P, Suja G, Asokan PK, Pradeep MA, Rithesh R, Vijayan KK, Sanil NK (2018) *Perkinsus olseni* in the short neck yellow clam, *Paphia malabarica* (Chemnitz, 1782) from the southwest coast of India. J Invertebr Pathol 159:113–120

Shumway S (1996) Natural environmental factors. In: Kennedy VS, Newell RIE, Eble AF (eds) The Eastern Oyster *Crassostrea virginica*. Maryland Sea Grant College, College Park, pp 467–513

Siddall ME, Reece KS, Graves JE, Burreson EM (1997) Total evidence refutes the inclusion of *Perkinsus* species in the phylum Apicomplexa. Parasitology 115(2):165–176

Sindermann CJ (1990) Principal diseases of marine and shellfish, vol 1. Gulf Professional Publishing

Skujina I, Hooper C, Bass D, Feist SW, Bateman KS, Villalba A, Carballal MJ, Iglesias D, Cao A, Ward GM, Ryder DRG, Bignell JP, Kerr R, Ross S, Hazelgrove R, Macarie NA, Prentice M, King N, Thorpe J, Malham SK, McKeown NJ, Ironside JE (2022) Discovery of the parasite *Marteilia cocosarum* sp. nov. In common cockle (*Cerastoderma edule*) fisheries in Wales, UK and its comparison with *Marteilia cochillia*. J Invertebr Pathol 192(6):107786

Soniat TM (1996) Epizootiology of *Perkinsus marinus* disease of eastern oysters in the Gulf of Mexico. J Shellfish Res 15:35–44

Suja G, Sanil NK, Chinnadurai S, Vijayan KK (2014) Reproductive dysfunction in the edible oyster, *Crassostrea madrasensis* due to larval bucephalid infection – a case study. J Mar Biol Assoc India 55(2):5–10

Suja G, Kripa V, Mohamed KS, Shamal P, Sanil NK (2016) *Nematopsis* sp. (Apicomplexa: Porosporidae) infection in *Crassostrea madrasensis* and its associated histopathology. J Mar Biol Assoc India 58(1):29–33

Suja G, Kripa V, Mohamed KS, Lijo J, Mini KG, Sanil NK (2017) Parasites and pathological conditions in the edible oyster, *Crassostrea madrasensis* (Preston), from the east and west coasts of India. Parasitol Res 116(9):2569–2579

Sunila I, Hamilton RM, Dungan CF (2001) Ultrastructural characteristics of the *in vitro* cell cycle of the protozoan pathogen of oysters, *Perkinsus marinus*. J Eukaryot Microbiol 48(3):348–361

Thomas PA (1979) Boring sponges destructive to economically important molluscan beds and coral reefs in Indian seas. Indian J Fish 26:163–200

Thompson FL, Lida T, Swings J (2004) Biodiversity of Vibrios. Microbiol Mol Biol Rev 68(3):403–431

Travers MA, Miller KB, Roque A, Carolyn S, Friedman CS (2015) Bacterial diseases in marine bivalves. J Invertebr Pathol 131:11–31

Uma A, Karthik P, Raj JP, Rebecca G (2021) First report of *Perkinsus* sp. infection in the bivalve *Magallana bilineata* (Roding, 1798) from Pulicat Lake, Tamil Nadu, south-east coast region of India. Indian J Fish 68(2):137–142

Villalba A (2016) What's new in bivalve mollusc pathology – overview of articles published in the last two years. Bull Eur Assoc Fish Pathol 36(1):24–37

Villalba A, Reece KS, Ordás MC, Casas SM, Figueras A (2004) Perkinsosis in molluscs: a review. Aquat Living Resour 17:411–432

Wang L, Yue F, Song X, Song L (2015) Maternal immune transfer in mollusc. Dev Comp Immunol 48(2):354–359

Winstead JT, Volety AK, Tolley SG (2004) Parasitic and symbiotic fauna in oysters (*Crassostrea virginica*) collected from the Caloosahatchee River and estuary in Florida. J Shellfish Res 23:831–840

WOAH (2024) Manual of Diagnostic Tests for Aquatic Animals, eleventh edition 2024. https://www.woah.org/en/what-we-do/standards/codes-and-manuals/aquatic-manual-online-access/

Wolff WJ (2005) Development of human impact on suspension-feeding bivalves in coastal soft-bottom ecosystems. In: Dame RF, Olenin S (eds) The comparative roles of suspension-feeders in ecosystems, NATO Science Series IV: Earth and Environmental Series, vol 47. Springer, Dordrecht

Yu JH, Song JH, Choi MC, Park SW (2009) Effects of water temperature change on immune function in surf clams, *Mactra veneriformis* (Bivalvia: Mactridae). J Invertebr Pathol 102(1):30–35

Zainathan SC, Hong TK, Addis SNK, Musa N (2022) Chapter 75 – *Nematopsis* sp. infections. In: Kibenge FSB, Baldisserotto B, Chong RS (eds) Aquaculture pathophysiology, volume II. Crustacean and mollusks diseases. Academic Press, pp 569–575

Zannella C, Mosca F, Mariani F, Franci G, Folliero V, Galdiero M, Tiscar PG, Galdiero M (2017) Microbial diseases of bivalve mollusks: infections, immunology and antimicrobial defense. Mar Drugs 15(6):182

Aquatic Animal Health and Environment

Role of Environment in Aquatic Animal Health and Noninfectious Diseases in Finfish Aquaculture

S. Mohanty and P. K. Sahoo

Abstract

India is appreciably striving from underdeveloped to developing and yet a step closer to a developed nation. The country's economy is best supported by its increasing share of the world's fish production. Our sustained efforts to diversify and increase fish production have revolutionized the production trend. The biggest realisation from intensive research indicates that stress is the driver of every disease. Our understanding regarding fish immunity and environmental stress factors is that the invariable interaction of a multitude of factors surrounding fish is influenced differently, regulated by its physiological status, life stage, species, stocking density, nature and magnitude of stress. However, opportunistic pathogens overwhelm the immunologically compromised conditions, proliferate and establish infection when the H-P-E triad fails. Noninfectious diseases are often overlooked and masked due to their non-communicable/non-transmissible nature. The physicochemical parameters, as well as environmental pollutants, also play roles in dictating noninfectious diseases. In addition, the complex and dynamic interaction between these parameters is difficult to understand and probe. The nonspecific symptoms and clinical signs of pathogenic infection and noninfectious diseases are another challenge for pathologists. In a broad sense, lesions, imbalances, deformities, anorexia, anaemia, weight loss and respiratory failure may be the prominent manifestations of noninfectious diseased fishes. The omnipresent risk factors and threats from the environment are an ever-continuing process. Either deliberately or unknowingly, anthropogenic activities directly contribute to aquatic menace in the form of infectious and noninfectious diseases. Hence, constraints are many day by day in aquaculture, which warrants a one health programme and more understanding of epidemiological studies in

S. Mohanty · P. K. Sahoo (✉)
ICAR-Central Institute of Freshwater Aquaculture, Bhubaneswar, Odisha, India
e-mail: msnatashree@yahoo.com; pksahoo1@hotmail.com

R. Kooloth Valappil et al. (eds.), *Aquatic Animal Health Management*,
https://doi.org/10.1007/978-981-96-7987-4_14

order to prevent and diagnose diseases appropriately. The affirmation may be apparently impractical, but a reduction of such hitch factors from the culture system not only makes fish immunologically uncompromised but also limits the spread of pathogens and disease epidemiology.

Keywords

Noninfectious diseases · Stress · Environmental risk factors and morbidity

1 Introduction

The transformation of aquaculture practices from a traditional setup to a more diversified, technology and production-intensive farming system has tremendously contributed to the economic growth of the country, besides providing livelihood and nutritional security. However, this transformation has also coincided with increased incidences of diseases of both infectious and noninfectious aetiologies. In aquatic systems, the dynamic interaction between the environment, susceptible host and virulent pathogen is crucial, and any failure in this intricate balance leads to diseases.

Freshwater aquaculture, under the scenario of climate change in association with intensification, is encountering a spectrum of diseases and disease-causing agents, which often result in heavy mortality and significant production losses. Infectious diseases (caused by bacteria, viruses, parasites and fungi) spread from one host to the other and can even result in transboundary transmission. However, increased anthropogenic activities and environment-induced stress factors also downregulate the fish health and, thereby, increase its susceptibility towards infectious diseases. Environmental stress factors can influence the fish physiology and behavioural aspects. Basically, noninfectious diseases are non-communicable and non-contagious and are the consequences of adverse environmental conditions, nutritional disorders and genetic defects. (Erazo-Pagador 2001). Often, these diseases are of obscure aetiology, which categorizes them as miscellaneous noninfectious diseases (Roberts 2012). Both types of diseases cause fish mortality from a low to a high extent depending on the host immunity and virulence of opportunistic pathogens. Moreover, only very little and fragmented reports have been documented in the latter case. Establishing a health management programme includes not only infectious diseases but also noninfectious diseases, though the latter sometimes are overlooked due to the long list of infectious pathogens and having varied epidemiology, prevalence and pathology with time and environment. This chapter aims to connect readers to the available scattered information on the role of environment in the occurrence of diseases, especially of noninfectious aetiologies and to give new insights into the understanding of disease process in aquaculture.

2 Core Components of the H-E-P (Host–Environment–Pathogen) Triad

Disease in a production system indicates its deviation from normal and a reflection of suboptimal conditions. Though fish are lower vertebrates with less defensive weapons, unlike terrestrial animals, and are inevitably bathing in the suspension of microbes harboured in their environment, they do not always become sick. Illustrating this, a classic model of infectious disease was described by Snieszko (1974), which elaborates on the interplay of host, pathogen and the environment in the epidemiologic triad.

2.1 Host

Every individual varies with others residing in the same environment in its immunity. A host's susceptibility, as well as resistance, is being judged by its array of immune functions. The front-line defence, which provides immediate resistance to the entry of pathogens/insulting agents, includes skin, mucus and scale. The next line represents the nonspecific arm covering many enzymes, secretions, complement factors and cytokines, whereas the last line of immunity is the specific type that fights invaders to protect the host. It is not only the immune functions but also the age and life stages of the host also determine the susceptibility or resistance of a host.

2.2 Pathogen

Pathogens are infectious agents that, when they interact with the host, undergo cascades of reactions like colonization, infection, disease, recovery and death. Infectious agents adhere to the outermost layer, like skin, mucous membrane or gills, searching for an entry portal (Van Seventer and Hochberg 2017). Following colonization, the infectious agent tries to establish causing infection by disrupting host tissues. The disease is generally manifested as visible signs and symptoms. During these processes of progression, infectivity and virulence of the pathogen, along with the host's susceptibility and resistance, determine the disease outcome. In this way, a virulent pathogen can cause infectious disease in fish. Disease-causing agents can be of three types: physical, chemical and biological.

Extreme temperatures, as well as radiation with ultraviolet rays, come under physical agents that cause sunburn in fish. Chemical agents are also detrimental to fish. Oil and other contaminants, like aflatoxins, produced due to suboptimal storage of feed, also lead to hepatoma and subsequent fish mortality. Drugs and chemicals, with their irrational use, cause tremendous health implications in fish. Nutritional imbalances, which result in suboptimal supplementation of nutrients to fish, can be a prime factor leading to diseases. On the other hand, biological disease-causing agents include fungi, viruses, bacteria and parasites. Understanding the route of entry and the disease progression is important to draw remedial strategies for the elimination of these agents because pathogenicity and the extent of pathology vary from pathogen to pathogen and host to host.

2.3 Environment

The environment in the epidemiology triad plays an important role in the disease process. Aquatic organisms are sensitive to any sudden changes in the water environment, and fish being poikilothermic organisms, their physiology is synchronized with it. In addition, intensification of culture practices, which involve the application of a variety of inputs, high stocking and manipulation of environmental parameters, creates stress that will predispose the fish to diseases. Further, the everchanging climate also influences the immunity of aquatic life negatively and also leads to the evolution of virulent pathogens, thereby increasing the incidence of diseases. The add-on effects of toxic waste entry into aquatic ecosystems deteriorate the environment and affect the host's susceptibility to opportunistic pathogens.

2.4 Stress Factors and Aquatic Animal Health

As described, the health of aquatic animals is inexplicably linked to both biotic and abiotic factors in the environment. Opportunistic pathogens, such as *Pseudomonas fluorescens*, *Aeromonas hydrophila, Myxobacterium* sp., *Vibrio anguillarum,* are normally available in natural waters at all times but do not usually cause disease problems unless environmental stress arises (Raman et al. 2013). The strong affirmation regarding fish farming is that mere changes in the environment can induce stress in fish, which can compromise their immune capacity. Unfavourable conditions, such as overcrowding, temperature fluctuations, salinity variation, inadequate dissolved oxygen, high concentrations of obnoxious gases like hydrogen sulphide, carbon dioxide and ammonia, excessive or rough handling, sub-lethal levels of toxic materials, etc., encourage diseases. In the current trend of intensification of aquaculture practices, huge amounts of fertilizers, feeds and drugs are applied in the system, which creates fluctuations in the environment if ponds are not properly managed.

Stress is the driver of all diseases. Sometimes, stress is of less intensity, but for a longer duration, low stress levels can lead to immunologically compromised individuals. However, acute stress can directly result in disease conditions in aquatic organisms. Stress induces General Adaptation Syndrome (GAS), which comprises (a) alarm reaction, (b) the stage of resistance for adaptation and (c) the stage of exhaustion. Consequent upon exposure to stress factors, the central nervous system releases stress hormones like cortisol, adrenaline and noradrenaline, which can change behaviour, alter blood flow, increase heart rate, alter glucose and lipid metabolism (Green and Haukenes 2015), and eventually lead to an immunocompromised host.

3 Infectious Diseases

Fish diseases are broadly categorized into infectious and noninfectious diseases. Infectious diseases are caused by various biological agents, viz., bacteria, viruses, fungi and parasites, and they do spread among individuals. Viral diseases in food fish are very rare, except for the recent occurrence of tilapia lake virus in tilapia and betanodavirus in seabass. However, few major outbreaks in ornamental fish by cyprinid herpes virus-2 in goldfish, carp oedema virus in koi, red seabream iridovirus in seabream, infectious spleen and kidney necrosis virus in aquaria fishes are being reported (Bajpai et al. 2022). On the other hand, a large array of viruses infecting crustaceans are being reported.

Bacterial diseases are frequently observed in fish culture ponds, often elaborated as opportunistic pathogens that cause substantial loss. Mostly, the immunocompromised fish succumb to diverse bacterial pathogens. Major pathogenic groups elaborated to date are Aeromonads, Vibrios, *Flavobacterium* sp., *Edwardsiella* sp., Streptococci and *Mycobacterium* sp. Recently, the introduction of genome sequencing of a number of bacterial fish pathogens has significantly increased our understanding of the biology, host adaptation and virulence factors of important pathogens (Sudheesh et al. 2012), thus helping in developing diagnostics and vaccines. Although fungal spores and elements are common in aquatic environments, the incidences of fungal infections in aquaculture are very rare. All freshwater and a few marine water fish's eggs are susceptible to fungal infections. The most common and prevalent fungal diseases of fish are saprolegniasis, branchiomycosis, epizootic ulcerative syndrome (EUS) caused by *Aphanomyces invadans* and ichthyophoniasis (Choudhury et al. 2014).

When fish are infected with an external parasite, bacterial infection, or injured by handling, the fungi can colonize on the damaged tissue on the exterior of the fish. These areas appear to have a cottony growth or may appear as brown matted areas when the fish are removed from the water. On the other hand, parasitic infections are very high in incidence in aquaculture systems. Ciliates, like *Ichthyophthirius multifiliis,* causing ich diseases, and flagellates, myxozoans and monogeneans, including skin and gill flukes, trematodes, nematodes and crustaceans like *Argulus,* are life-threatening for fish (Idowu et al. 2017). Diagnosis by mere clinical signs is very difficult because similar clinical signs are seen in diverse pathogen types. Hence, molecular tools have given the upper hand in diagnosing fish disease.

Noninfectious diseases are caused by non-living or abiotic factors and do not spread like infectious diseases. The noninfectious disease leads to the breakdown of fish immunity, mostly and subsequently increases susceptibility towards infectious diseases. The perspective of noninfectious diseases is yet to be established and elucidated clearly. However, present literature says these diseases can arise from stressful environmental conditions, geography, malnutrition, genetic alteration, heavy metal and other pollutants toxicity. The worst consequence of these diseases may be fish mortality. However, due to its restrictedness at the individual level, these diseases seem to be less intense. The details of the environment-related noninfectious diseases are described below.

4 Noninfectious Fish Diseases

Noninfectious diseases are categorized into three groups: environmental, nutritional and genetic abnormalities.

4.1 Environmental Diseases

Major environmental diseases in fish are environmentally induced, which arise due to two factors: physicochemical parameters and physical factors (Erazo-Pagador 2001).

4.1.1 Fish Diseases Associated with Physicochemical Parameters of Water

(i) *Gas bubble disease (GBD):* Successful aquaculture depends on optimum physico-chemical water parameters, deviation of which results in compromised growth and diseases. GBD is an important disorder in cultured fish and is closely similar to hyperbaric problems in humans (Speare 1998). It is caused due to the supersaturation of dissolved nitrogen or oxygen, which may arise due to leaks in the pump, a damaged valve system, or a dense algal bloom. Fish undergoing this disorder exhibit loss of equilibrium and abnormal buoyancy, and affected fishes float at the water surface. In older fishes, the presence of bubbles in the abdominal cavity, chambers of eyes, skin, gills, swim bladder, digestive tract, fins and mouth is among other prominent features to diagnose the disease. However, in larval stages, bubbles are obvious in the yolk sac and sub-cuticle. A gas bubble in the eyes leads to blindness (Roberts 2012). The host may die due to blockage of the artery with a blood clot and expansion of trapped air in tissues, oedema, degeneration of the gill lamellae, etc. Further, water depth has an influential role in gas saturation in fish. Fishes residing in deeper water are more resistant to GBD (Weitkamp et al. 2003) than shallow water residents. Dissolved gas super-saturation (DGS) can produce that adversely affects fish and causes gas bubble trauma. DGS happens when partial pressures of atmospheric gases in solution exceed their respective partial pressures in the atmosphere. This phenomenon exists in both natural and man-made systems. Airlift aeration and gas injection systems are often used to increase the dissolved oxygen content of lakes as well as in hatcheries at depths where elevated hydrostatic pressures can quickly force gas in the form of bubbles, potentially a way to produce DGS (Edsall and Smith 1991). It can be prevented by monitoring dissolved oxygen (DO), avoiding algal blooms, maintaining pumps and sufficient water exchange.

(ii) *Swim Bladder Stress Syndrome:* Swim bladder stress syndrome (SBSS) is caused due to multiple factors. It is a disorder mainly due to dysfunction of the swim bladder, handling, high temperature and illumination and dense algal bloom. Affected fish larvae show large gas bubbles in the swim bladder. SBSS sometimes misleads to gas bubble disease. It differs in having bubbles only in the swim bladder. Hyperinflation of the swim bladder, high buoyancy and mass mortalities are also recorded in this disorder. SBSS and associated pathophysiology can be minimized by controlling algal bloom, sufficient water exchange and aeration of the system. Ferguson (2006) defined swim bladder stress or distress syndrome as characterized by an overinflated swim bladder showing buoyancy problems, abdominal enlargement and lack of histopathological lesions. Rekecki et al. (2016) presented interesting facts on pathology associated with swim bladder hyperinflation in burbot *Lota lota* L. larvae. Postmortem examination of fish showed a ventrally displaced, over-inflated swim bladder and association of *Vibrio* (*Listonella*) *anguillarum* with it. Histological examination revealed a distended swimbladder with a flattened epithelium and slight oedema in secondary gill filaments. SBSS and mortality could be avoided in sea bass larvae only when raised at ambient temperature (12.5 °C), natural light and reduced salinity (25‰) (Johnson and Katavic 1984). The researcher further indicated the multifactorial involvement of environmental factors in SBSS that needs more understanding for the effective treatment of SBSS.

(iii) *Asphyxiation/Hypoxia*: Dissolved oxygen (DO), when it drops to a critical level, causes asphyxiation/hypoxia. Affected fishes gather at the pond sides. Also, fish show gulping mouths and rapid opercular movement, and they swim on the water surface. Aeration and water exchange can increase the DO level and minimize the stress in fish. Low DO is a major impediment in aquaculture ponds, causing stress-induced immunosuppression, morbidity and mortality. The O_2 acquiring ability of an organism from its environment is considered an important determinant of hypoxia tolerance (Hughes 1973). Fish have aerobic metabolism that requires DO. Hence, it is a main limiting factor in fish farming, and even a concentration below 1–2 ppm for a few hours can adversely affect fish growth, resulting in fish death (Abdel-Tawwab et al. 2015). The mechanism of hypoxia tolerance has been well-studied by Mandic et al. (2009). Sculpins with higher hypoxia tolerance recorded low critical O_2 tensions (P_{crit}) and inhabited the O_2 variable intertidal zones. However, species with lower hypoxia tolerance were found in more O_2-stable subtidal zones or freshwater. Hypoxia tolerance is associated with an enhanced O_2 extraction capacity, 75 per cent of the variation in P_{crit}, routine O_2 consumption rate, mass-specific gill surface area and blood haemoglobin (Hb)–O_2-binding affinity (P_{50}). Organisms that possess a greater capacity for O_2 extraction can maintain a routine metabolic rate at lower O_2 tensions. Hence, manipulation of the respiratory cascade that enhances O_2 extraction capacity can be a possible approach for hypoxia tolerance, and the organism can be a potential candidate to be naturally selected under hypoxic conditions (Mandic et al. 2009).

(iv) *Sunburn:* Continuous exposure to solar radiation exerts discomfort in fish. Due to this reason, even in normal conditions, fishes swim deep towards the bottom of the pond, where comparative coolness provides them a better ambience (Roberts 2012). Sunburn disease is caused due to excessive levels of ultraviolet irradiation from sunlight. The UV spectrum and threats associated with three types of UV, A, B and C, are detailed in Sect. 14.4.1.4. under physical factors. The fish epidermis is devoid of melanin-containing pigments and also non-keratinized attributed to their enhanced susceptibility to UV B radiations. Prominent swollen malpighian cells with a large clear halo around the dark, shrunken nucleus, oedema, and grey colour ulcerative lesions on top of the head, pectoral, dorsal and upper tail fins are evident, which may sometimes encourage secondary opportunistic infections by bacteria as well as *Saprolegnia* sp. and *Achlya* sp. The infected fish show typical 'sunburn' cells (pyknotic or karyorrhectic nuclei) in the epithelial layer of the skin (Roberts 2012).

The other detrimental effect associated is photosensitization, where cells are not directly damaged by UV A but rather induced to absorb phototoxic compounds released from plants, animals and also from human activities. The freshwater ecosystem is also not resilient to climatic impact, anthropogenic activities and UV B radiations (Vincent and Belzile 2002). Phenothiazine, a compound that belongs to the photosensitizers group, causes black peel in salmonids and also Ulcerative Dermal Necrosis (UDN) associated with other photosensitizer compounds, e.g., natural porphyrins. The diagnostic signature for UDN lesions is epithelial necrosis with pemphigoid degeneration of the middle layer of Malpighian cells of the epidermis (Roberts 2012). Sunburn is usually an unattended component of fish pathology which also contributes to production loss. To prevent its lethal effects, sunshades over ponds, especially shallow ponds where the risk of radiation is very high, can minimize the radiation effect. It is mostly noticed that surface feeders like Catla often suffer from sunburn-related mortality during summer months due to massive exposure to radiation for its upper niche preference.

(v) *Alkalosis and Acidosis:* Alkalosis is a condition when the pH level is very high in ponds beyond the tolerance level of fish. Gross pathology includes corroded skin, gills and milky turbidity of the skin. To prevent alkalosis, pH level is to be maintained in the range of 6.5 to 8.5 at which fish homeostasis is not affected. Acidosis refers to a drop in pH of body fluid too low for the species to withstand. Blood acidosis can result from decreases in pH, whereas increases in pH can result in blood alkalosis (Roberts et al. 2006). Either of the cases is stressful to fish, which induces mucus secretion in the body. Affected fish show gasping over the water surface and rapid swimming. Under this condition, body growth and metabolism are undermined and often lead to death. Lime is applied to maintain the desired level of pH. Highly acidic (pH 3.5) or alkaline (pH 11) water leads to ionic imbalance and is lethal to fish (Parra and Baldisserotto 2007). The higher solubility of CO_2 in water compared to oxygen restricts aquatic animals from regulating internal acid-base through adjustments of blood P_{CO2} through ventilation (Randall and Cameron 1973).

4.1.2 Fish Diseases Associated with Physical Factors

Physical factors, including handling, transport, high stocking density and predation, may be a reason for injury in fish. Following any injury, the secondary bacterial, viral and parasitic infections may easily establish in the host and may lead to mortality.

4.1.3 Diseases Associated with Environment Hitch/Stress Factors

Physical Factors

(i) *Temperature*

Temperature is considered as one of the critical physical factors that influences not only metabolic activities but also the solubility of ammonia, chlorine, dissolved oxygen and their bioavailability in fish. A sudden change in temperature of 12 °C or more may lead to thermal shock and subsequent death in fish (Svobodova et al. 2017). Generally, at peak summer, organic matter decomposes at a faster rate, which involves a huge amount of dissolved oxygen in the water. Additionally, temperature has an indirect relation with oxygen solubility. Hence, both cases result in oxygen-depleted conditions, as is mostly evidenced in summer. Increased temperature increases oxidative stress, increases the prevalence of infection and heightens epidemics (Soares et al. 2014). However, the intensity of infection in higher temperatures varies between different parts of the world (Easterling et al. 2000). On the other hand, a rise in temperature can fasten the life cycle of some parasites like *Ichthiophthirius multifiliis,* causing ich disease in freshwater fishes and, hence, forms one of the control measures to get rid of this parasite. Malondialdehyde and catalase activities are good biomarkers of higher rearing temperature (Vinagre et al. 2012).

Likewise, lower temperature reduces the metabolic rate, and other chemical and biological processes. For example, nitrification slows down at temperatures lower than 5 °C (Svobodova et al. 2017). It also has a role in the spread of disease and the intensity of infection. For example, the winter syndrome is caused due to constant exposure of fishes to lower temperatures below 13 °C. Not only persistent lower temperature but also compromised immunity and undernutrition can aggravate the disease condition. Winter syndrome in severe outbreaks can cause mortality up to 30–50% (Coutteau et al. 2001). Fishes stressed with winter syndrome undergo skin pigmentation, become anaemic, show corneal cloudiness, distended abdomen, belly upside-down swimming and haemorrhages at fin bases. Distended intestines, gall bladder, pale liver and accumulation of ascitic fluids in internal organs are evidenced (Soares et al. 2014). To diminish this disorder, restriction in feeding, specifically beyond a temperature of 13 °C, is advised. Feeds incorporated with lower dietary lipid content in the fish feed before the winter months improve the fish immunological status. Optimal feeding rate and stocking density favour greater disease resistance and also prevent the stress exerted due to handling and transportation.

Environmental temperature plays a critical role for all aquatic organisms. The poikilothermic nature of fishes restricts them to control their body temperature, rather, it is environmentally influenced. However, fishes thrive well at specific temperatures. For example, cyprinids grow well at 18–28 °C and salmonids at 8–18 °C. Some fish, like channel catfish or koi, can tolerate a wide range of environmental temperatures (Ruth Francis-Floyd 2015). Sudden fluctuations in temperature can break down the host immunity and increase the pathogenicity and infectivity of pathogens. Increases in temperature adversely affect the solubility of dissolved oxygen content in water and increase the toxicity of contaminants that encourage infectious agents to proliferate. Especially virus and fungi at specific temperature ranges are able to cause disease and manifest clinical signs in fishes. Drop-in metabolic activities at lower temperatures reduces the ammonia excretion activity in gill, which tends to increase in ammonia concentration in the blood, leading to autointoxication and death (Svobodova et al. 1987). An obvious example of autointoxication is carps fed with nitrogen-rich feed if harvested in summer from the pond with a water temperature of 25 °C when stocked in tanks with a temperature of 15–18 °C (Svobodova et al. 2007a, b). Not only metabolic activities but also the digestion process is hindered at a colder temperature of 5–8 °C, leading to indigestion or partial digestion, thus causing gas production and accumulation. The worst consequence of this condition includes increased body pressure, imbalance, damage to internal organs and ultimate death (Svobodova et al. 2017). Lower rearing temperature decreases swimming activity (Hasler et al. 2009), which can be evidenced in gross observation.

Every organism tries to get rid of the adverse situation with some kind of built-in mechanism. The fishes are not the exception. Heat shock proteins (HSPs) are widely demonstrated as stress biomarkers for environment monitoring (Madeira et al. 2012). To withstand thermal stress and also from pollutants and pathogenic microorganisms (Iwama et al. 2005), special kinds of proteins called HSPs come into play, having some thermoregulatory functions. For example, the most documented HSP is HSP70, which can block cell death (Lewis et al. (2012). These also give a shield by protecting the membrane lipids and nucleic acids and reducing the production of hazardous unstable radicals (Svobodova et al. 2017).

Temperature-induced pathology includes haemolysis (Svobodova et al. 2008), accumulation of gases in the body cavity, hyperaemia and oedema in gills due to digestion disturbances and ammonia autointoxication. Larval stages are more delicate to resist sudden temperature changes. Hence, temperature adjustment while stocking seeds during release to the pond by matching the temperature of water in the polythene pack and outside pond water is important, where the seeds are to be released. Likewise, a similar temperature condition is to be maintained between the hatchery and pond water to minimize temperature fluctuation-induced thermal shock in brooders as well as seeds. Nitrogen-rich feed should be restricted to avoid autointoxication. Sudden weather changes influence fish health in almost all life stages with varied intensity. Hence, possible measures, especially stocking time, feeding schedule and harvesting, as well as hatchery operations, are to be monitored to match the temperature of the water where fish reside and where to be stocked.

(ii) *UV Radiation*

UV radiation is one of the prime environmental factors that impact fish health. Earth receives solar energy in three forms, viz., highly harmful ultraviolet C (200–280 nm), highly energetic and moderately harmful ultraviolet B (280–320 nm) and mildly energetic and less harmful ultraviolet A (320–400 nm) (McKenzie et al.

2007). Both UVA and UVB radiations can penetrate the water column, showing different attenuations contributed by dissolved organic matter and suspended organic particles in freshwater and saltwater ecosystems (Alves and Agusti 2020). Due to the high input of inorganic and organic material, freshwater ecosystems are susceptible to high UV absorption, which also depends on their level of eutrophication (Bracchini et al. 2004). Pavement cells (PVCs) constitute 90% of the gill surface epithelium that mainly functions for gaseous exchange in teleost gill. An exposure of 15 min to UV-B radiation causes significant damage in catla gills, specifically to PVCs, and subsequent exposure of mitochondrion-rich cells (MRCs) to UV affects respiratory functions and disturbs osmoregulation in catla (Sharma and Chakrabarty 2006). In addition, UV-B radiation enhances ROS production and antioxidant enzymes, affects the Hsp 70 level, and causes DNA damage (Singh et al. 2015). It also breaks down immunity and digestive enzymes, indicating its potent threat to aquatic organisms (Sharma et al. 2010). Solar UV radiation penetrates the aquatic systems and can affect the production of phytoplankton, zooplankton, microalgae and higher organisms like amphibians and fishes in the aquatic web of freshwater as well as marine water (Hader et al. 2007). The negative effects of UV radiation in fish include compromised growth, underscored development, behavioural changes, development of skin and eye lesions, breakdown of the immune system and disease resistance, DNA damage and other metabolic and physiological dysfunctions (Sandrini et al. 2009). The early life stages are most tempted to damage (Dahms and Lee 2010). The eggs and larvae (skin and gills) of many fish are prone to UV-B exposure. An increase in mortality, developmental abnormalities, behavioural and metabolic changes, lower hatching rate, embryonic malformation like enlarged pericardial sacs and spinal deformities have been previously studied (Dong et al. 2007). Zebrafish (*Danio rerio*) embryos of 4 h post-fertilization (hpf), when exposed for 2.4 h UVB, 295 nm, showed a reduction in their survival rate (more than 50%) at 6 days post-exposure with caudal notochord torsion and bending (Nunez et al. 2012), a prominent developmental malfunction. In the same species, embryos of 6–7 hpf, in the mid-gastrula stage, when exposed to UVB radiation of 31.1 kJ m^{-2}, led to mortality of more than 70% (Dong et al. 2007).

However, both UVA and UVB are detrimental to juveniles and adult fish (Sayed et al. 2016). In juveniles and adults, growth reduction, loss of body condition and feeding behaviour, loss of appetite, restless behaviour, changes in swimming activity and low predatory performance were evident. UVA and UVB exposure have wide negative health implications like sunburn on the skin, hyperpigmentation, a decrease in mucous development, inflammation, necrosis on the epidermis, damage in the corneal epithelium of the eyes, cataract, lipidosis, inflammatory lymphocytic infiltration, morphological malformations in red blood cells and also break down of innate and acquired immunity and cellular and molecular process impairment (Alves and Agusti 2020). The potential detrimental role of UVB on lymphocyte function, which is a prerequisite for IgM synthesis, is also evidenced. Plasma IgM concentration in common carp decreased significantly when exposed to a higher dose of UVB with 4 weeks of exposure. This decrease in the IgM levels reflects impairment in lymphocytes function due to UVB exposure (Jokinen et al. 2008,

2011). CPDs (cyclobutane pyrimidine dimers) are the major UV B-induced lesions followed by single- and double-strand breaks (Hader and Sinha 2005). In addition, damage to biological molecules such as DNA and proteins, the generation of reactive oxygen species and photosensitization consequences are potentially detrimental.

Multiple environmental factors influence aquatic life. Various research parallel to this has documented the enhanced deleterious effects of UVR by interactions with some environmental factors like increased temperature and hypoxia due to climate change or pollution (Hader et al. 2015; Hader and Barnes 2019). The interaction effect of UVR and the varied temperature has been studied in the zebrafish embryo. UVB radiation and temperature act synergistically at a low temperature of 24 °C. A high temperature of 30 °C disrupts embryonic metabolism and development, modulates immune system response and impairs embryonic development (Aksakal and Ciltas 2018). Like other organisms, fishes adopt multiple strategies like avoidance of UV radiation, production of UV-absorbing compounds and DNA damage repairing mechanisms to resist harmful effects. UV radiations are really alarming for fish productivity as well as fish health. Hence, research should be reoriented to establish the mechanism/mode of action of UV radiation to identify the effective molecules that can be targeted to block and minimize the radiation-induced risks in fish. It is also affirmative to select tolerant varieties of fish for multiple stress factors to maintain sustainability.

4.1.4 Chemical Stress Factors

(i) *pH*

pH is an important water quality parameter that affects homeostasis and survival in fish. The aquatic ecosystem and its inherent interactions with multiple factors are dynamic and ever-changing. pH values below 4.8 or above 9.2 are lethal to salmonids, and below 5.0 or above 10.8 are deadly in cyprinids (Svobodova et al. 2008). The carbonate cycle in the water body carries the important concept of pH, total alkalinity, total hardness and CO_2 due to their interdependent interactions. As understood, during daytime, owing to the photosynthesis in the presence of sunlight, dissolved oxygen level is high, and CO_2 fluctuates in an opposite trend. As CO_2 concentration changes, it shifts pH towards acidity or alkalinity. When CO_2 concentration decreases during daylight hours, pH rises due to less synthesis of carbonic acid. When CO_2 concentration increases during the night, pH drops towards acidic.

Higher acidity, when H^+ ion concentration is higher, increases gill epithelium permeability and leads to undergrowth and electrolyte loss. In contrast, alkaline pH values (9.5–10) slow down ammonia diffusion in the gills, and a higher quantum of ammonia is available in muscle (Golombieski et al. 2013). Reduced water pH values resulted in an increase in the levels of ammonia, glucose and glycogen in the liver of South American catfish (*Rhamdia quelen*) (Bolner et al. 2014). The impact of low pH on the survival of young guppies (*Poecilia reticulata*) has been studied by Araujo et al. (2008). They determined the median lethal time [the time after which

50% of the tested population died (LT50)] for varied pH values from 3 to 6.5. The lowest LT_{50} value was recorded at 1.5 h for pH 3, and gradually, the LT_{50} value increased along with increased pH, i.e. LT_{50} was 75 h at a pH of 5. This study indicates that increasing the pH value increases the survival time. The older fishes with bigger sizes can tolerate the sudden increase in pH as studied in channel catfish (*Ictalurus punctatus*) and their hybrids (Mischke and Chatakondi 2012). The pH sensitivity of fish depends on both the age of the fish and the oxygen availability in the water.

The pH value significantly affects the solubility and toxicity of substances like ammonia, hydrogen sulphide and metals (Svobodova et al. 2008). The most pronounced relation is between pH and ammonia. Increasing pH values along with rising water temperature sharply increase ammonia toxicity, whereas cyanide and hydrogen sulphide toxicity show a reverse trend, i.e., drops with rising water pH (Svobodova et al. 2007a). Under pH stress, fish produce more mucus on the skin, opercula and the gills with haemorrhagic gills. Extreme pH contributes to corneal opacity, damage to the skin and disintegration of fins (Svobodova et al. 2008). The highest pH values are mostly recorded due to heavy photosynthesis activity in eutrophicated ponds infested with bloom. Fishes have some strategies to adapt to varied, extreme pH conditions. Fish protect themselves against the effects of low or high pH, saving energy. They lower the frequency of respiratory movements and reduce the volume of water flowing over the gills (Svobodova et al. 2008). pH fluctuations in the water body should be strictly monitored with the ingress water and the runoff from the concrete house. A pH crash refers to an acute decrease in water pH, which is often preceded by a drop in alkalinity and may result in high fish mortalities (Roberts et al. 2006). To treat fluctuating pH, many commercial compounds are available, like sodium bicarbonate (baking soda) that can be added at 3 mg/L to temporarily improve buffering capacity (Noga 1996). Low alkalinity can be managed using crushed coral, crushed oyster shells, or limestone (Wildgoose 2001). Highly eutrophicated and algal bloom-flourished water bodies should not be selected for stocking fishes, which will create stress in fishes with very high pH conditions.

(ii) *Dissolved Oxygen*

Dissolved oxygen has always been the prime lifeline of an organism. The DO concentration in fish ponds depends on many factors: the photosynthesis of phytoplankton, respiration of aquatic organisms and diffusion of atmospheric O_2. The solubility of O_2 in water depends on various environmental conditions such as O_2 partial pressure, water temperature and salinity (Campbell 1990; Withers 1992). The normal level of oxygen is termed normoxia, and the lower value is hypoxia, whereas the higher side is called hyperoxia. Hypoxia is a prevailing condition in aquaculture ponds and a major stress factor (Karim et al. 2003), specifically in summer when the solubility of oxygen is very poor, during cloudy weather, early mornings due to oxygen depletion (Noga 2000), excessive feeding, eutrophication and algal blooms. Hypoxia leads to deformities (Sanchez et al. 2011), reduction of growth, feed consumption and feed conversion efficiency (Pichavant et al. 2001), breakdown of the immune system (Henrique et al. 2002) and ionic disturbances and accumulation of lactate (Vanlandeghem et al. 2010), making fish more susceptible to opportunistic pathogens.

Carps are susceptible to ammonia autointoxication and gill necrosis under hypoxic conditions. Fishes exposed to hypoxic environments look pale. The gills become hyperaemic, colour changes from blue to purple, gill filaments lose intactness and slight haemorrhage seen in the eyeball and gill cover (Svobodova et al. 2008). Cyprinids demand an optimum oxygen level of 6–8 mg/L, and below 2 mg/L, they are under stress due to suffocation (Dwyer et al. 2014). Oxygen deficiency may be a consequence of dropped photosynthesis in the culture system, which implies a lack of phytoplankton biomass that can be enhanced with supplementation of phosphorus fertilizer.

In contrast, hyperoxia refers to oxygen-supersaturated water, which most often arises in aquaculture ponds due to intense bloom and during seed transports. In such cases, the oxygen level is as high as 300%, which negatively affects fish during transport. Cyprinids have a critical oxygen saturation level, i.e., 250% and 300% in water (Svobodova et al. 2008). Oxygen supersaturation reduces ventilation rates (Shultz et al. 2011). The fish becomes unable to excrete wastes such as carbon dioxide (CO_2), which increases CO_2 concentration in circulation, leading to respiratory acidosis and nephrocalcinosis (Gómez 2000). Hyperoxia and high CO_2 levels generally happen in intensive fish-rearing conditions. The pathology associated with hyperoxia in fishes includes an accumulation of gases in the eyes, fins, skin and gills of fishes during transportation (Gultepe et al. 2011). Prominent necrosis with bubbles in the skin (Vatsos and Angelidis 2010), light red body colour, disintegrated edged gill filament (Svobodova et al. 2008) are obvious followed by secondary dermatomycosis, and death in hyperoxic exposed fish. Saroglia et al. (2002) reported significant changes in gill morphology in hyperoxic conditions. This condition leads to reduced Respiratory Surface Area (RSA) and increased Gill Diffusion Distance (GDD) in the gills of sea bass, reduced exchange of gases with water (O_2 and CO_2) and also reduction in the ion efflux rate. Brauner et al. (2000) documented a prominent change in acid-base equivalent exchanges across the gills during hyperoxia exposure in freshwater fishes. Hyperoxia notably impaired the hypo-osmoregulation process in Atlantic salmon smolts following seawater transfer.

To solve the hypoxia condition and increase the oxygen saturation level, aeration should be provided. Constant and meticulous monitoring for minimizing the organic load that depletes soluble oxygen from the water for its assimilation and aerobic decomposition is necessary.

(iii) *Nitrogenous Compounds*

Acute exposure to poor water quality can result in sudden and significant mortality in fish. Chronic exposure to suboptimal water quality conditions can cause immunosuppression and promote infectious diseases, leading to mortality (Roberts et al. 2008). Toxicities in water quality, especially the accumulation of nitrogenous compounds or its slow conversion to harmless compounds, increase the risk of fish mortality and economic loss. Ammonia is an important end product of nitrogen metabolism in fish. Freshwater bony fishes excrete about 90% of waste nitrogen substances in the form of ammonia, and 10% is excreted as urea (Svobodova et al. 2017), whereas freshwater cartilaginous fishes excrete nitrogenous waste substances solely in the form of urea (Wood 1993). Total ammonia in fish blood is a combination of ammonia from the external environment and the ammonia produced by protein metabolism. After the release of ammonia, it enters the nitrogen cycle, in which ammonia is converted to nitrite (NO_2) and then to nitrate (NO_3) owing to bacterial action. Among these, NO_2 is unstable, and NO_3 is removed during water exchange. The failure in the excretion of ammonia leads to high ammonia accumulation in fish blood and is termed autointoxication (Svobodova et al. 2007a, b). Ammonia nitrogen comes from protein-supplemented feed, industrial wastes and agricultural ingress. It has two forms: i. ionized ammonia NH_4^+ ion, ii. unionized ammonia/molecular NH_3, which coexist in the water body depending on the water pH and temperature. The molecular form of ammonia NH_3 is highly toxic and is produced at high pH and temperature, and often limits fish production. We typically measure the combination of NH_3 and NH_4^+ during laboratory water parameter analysis, referred to as total ammonia nitrogen (TAN), but not only NH_3. Overfeeding and defunct filtration systems encourage the accumulation of ammonia in the system. Ammonia toxicity affects neurologic and behavioural abnormalities, lethargy, anorexia, poor growth, secondary infections and gill hyperplasia/hypertrophy in fish.

The exchange of all ions happens through gills to the environment and vice versa. Gill is non-permeable to ammonium ions from the aquatic environment to the blood. But the vice versa trend of ammonia from blood to the external environment at the cost of Na + ions (Evans et al. 1999) has been evidenced. Reduced temperature and oxygen reduce metabolic rate and the efficiency of the Krebs citric acid cycle, which disrupts ammonia transport into the gills and leads to a rise in ammonia concentrations in the blood called autointoxication (Svobodova et al. 2017). Exposure of common carp (*C. carpio*) fry to ammonia at a concentration of 0.2 mg/L over 150 h produced alterations in the respiratory epithelium of the gills (Svobodova et al. 1987). The clinical symptoms of ammonia intoxication include muscle spasms, off-feeding, increased opercular movement frequency, restlessness and the movement of fish toward the water surface. Sleeping sickness in common carp, especially in the tank, is noticed in high ammonia concentrations, and the fish show no response to stimuli. This phase is followed by pale skin and, eventually, the death of fish (Svobodova et al. 2008). When NH_3 levels exceed 0.05 mg/L, it damages gills, and a level of 2 mg/L is lethal for many fish. Routine water changes and proper husbandry are very much essential in closed systems where fish are more susceptible and most often suffer from water quality-related toxicities. Ammonia is more toxic in warmer water at higher pH and lower salinity. At higher pH, more ammonia is accumulated in non-ionized form. For every 1-unit decrease in pH, there is a tenfold decrease in non-ionized ammonia (Noga 1996). Aqueous ammonia increases levels of blood and tissue ammonia, which further raises blood pH, causes osmoregulatory disturbance, increases tissue oxygen consumption and decreases blood oxygen transport (Noga 1996).

There are two undesirable conditions encountered in pet fishes due to high NH_3 concentrations, i.e., new tank syndrome and old tank syndrome.

(a) *New-Tank Syndrome:* It is a common problem in aquaria with poor maintenance. It occurs when NH_3 levels rise during the first 2–3 weeks after a new system is set up or the beginner hobbyist starts aquarium keeping without proper cleaning, when the biofilter may not be sufficiently developed to mobilize toxic molecular ammonia to a nontoxic, stable form. Overfeeding and overstocking, and lower aeration may increase the risk of NH_3 accumulation, which needs to be looked at as a priority. To prevent new-tank syndrome, aquarists adopt some strategies to accelerate and mature biofilters in the set-up. Nitrifying bacteria can be procured from the market, and a good biofiltration system can prevent new tank syndrome.

(b) *Old tank syndrome:* This condition arises at high NH_3 levels (TAN may be >20 mg/L), extremely low pH (usually <6, may be <5 in severe cases) and a complete absence of alkalinity. This condition prevails after a long time of aquarium keeping, but not immediately and when there is a complete failure of buffering capacity. Biofilter bacteria acidify the water through the nitrification process, but the buffering capacity (alkalinity) maintains the balanced system. When it is exhausted, the accumulation of organic acids causes a drop in pH, destroying healthy biofilter bacteria and raising NH_3 concentration. Hence, ammonia binders can be used to treat old tank syndrome treatments in aquarium tanks.

Nitrification is an important process of the nitrogen cycle where ammonia nitrogen is first oxidized to nitrites and subsequently to nitrates, which are less toxic to fish and stable in the system. If the nitrification process is inhibited, nitrites accumulate in the system, causing malfunction and even fish death (Svobodova et al. 2005). Freshwater fin fishes exposed to nitrite accumulate it not only in plasma (Jensen 2003) but also in tissues like gills, liver, brain and muscle (Margiocco et al. 1983). Nitrite enters the body through eosinophilic chloride cells of the gills, which participate in ion exchange between the organism and the surrounding aquatic environment (Harris and Coley 1991). Hence, sometimes the nitrite concentration of fish blood plasma may reach up to 60 times higher than the surrounding water (Fontenot et al. 1999).

Ammonia is oxidized to nitrite by Nitrosomonas bacteria and is a potential toxicant that has detrimental effects on multiple physiological functions. Clinical signs of nitrite toxicity in fish are respiratory burnt in nature and include increased opercular rate, piping, gasping at the surface and gathering in well-aerated areas. One of the worst and most serious consequences of nitrite accumulation is the oxidation of haemoglobin to methaemoglobin, which diminishes the oxygen transport system. The toxicity of nitrite in fish is also influenced by many inherent factors like pH, temperature, oxygen concentration, species, size and age, and individual fish susceptibility. Chloride concentration in water is the most important factor, and the uptake of chloride by cells is directly related to their susceptibility to nitrite toxicity. For example, fishes like channel catfish (*I. punctatus*), having high chloride uptake, are very sensitive to nitrites, and fishes like European eel (*Anguilla anguilla*) and common carp (*C. carpio*) are more nitrite resistant (Tomasso and Grosell 2005) due to less uptake of chloride cells. The most documented pathological condition in fish exposed to nitrite toxicity is 'brown blood disease or methemoglobinemia, which is detailed below. Various measures like the use of biofiltration, reducing stocking density and feeding, and water exchange can reduce the intensity of nitrite toxicity and can be prevented by adding sodium chloride to the water (Kajimura et al. 2023). Possibly, this is the reason, by virtue of chloride ions in seawater, marine fish are blessed to be less susceptible to nitrite toxicity.

(c) *Brown Blood Disease:* In fish, nitrites enter erythrocytes from blood plasma where these oxidize bivalent iron in haemoglobin to trivalent iron, producing methaemoglobin, a brown coloured substance, which imparts blood and gill brown colouration, hence termed as 'brown blood disease'. This substance can't transport oxygen (Bodansky 1951). Hence, with the formation of methaemoglobin, blood oxygen-carrying capacity is decreased. Fish blood commonly contains methaemoglobin up to 10%, even in the absence of nitrites (Soldatov 2002). Haemoglobin in fish is always tempted for auto-oxidation by oxygen than in mammals (Kiese 1974). Fish with this disease show hypoxia. Hence, rapid water exchange can substantially reduce this toxicity. Additionally, nitrites negatively affect the fish immune system when exposed acutely or to sublethal nitrite concentrations, thus making the fish more susceptible to infectious diseases. When nitrite-poisoned fishes are placed into clean water, methaemoglobin content usually drops to physiological levels within 24–72 h (Knudsen and Jensen 1997).

It has been studied that the presence of bromide, having similar properties as that of chloride, can reduce the toxicity and also increase the concentration of calcium. In contrast, a few parameters that aggravate nitrite toxicity are low oxygen availability, high temperature and life stages of fish. It has been proved that the older fishes are more sensitive to nitrite poisoning than the younger ones, as studied in common carp (Kroupova et al. 2010). This varied sensitivity level may be explained by the presence of methaemoglobin reductase enzyme converting methaemoglobin to haemoglobin activity, which is higher in small ones compared to older ones (Svobodova et al. 2017). In nitrite toxicity, methaemoglobin concentration values are ≥20–25% of total haemoglobin, and over 70% causes death (Svobodova et al. 2008). Nitrite toxicity is best diagnosed by measuring the increase in concentration of nitrite in the aquarium combined with elevated methemoglobin levels in the blood (Kajimura et al. 2023), plasma and tissues rather than the surrounding water. Hence, the transfer of intoxicated fish to clean water can help recover the fish to some extent. Some basic strategies and preventive measures like optimum feeding, maintenance of biofiltration system and vitamin C (Wise et al. 1988) supplemented feed are to be practised to prevent nitrite toxicity (Svobodova et al. 2008).

Nitrate is the oxidized product of nitrite and is more stable in aquatic ecosystems. Though adult fishes are comparatively resistant, eggs and fry stages are more susceptible. High levels of nitrate promote algal blooms and subsequent deterioration in other related parameters, that can result in other water quality abnormalities. Clinical signs of nitrate toxicity are poor growth, lethargy, anorexia and damage to fins (Wildgoose 2001). Frequent water exchange and correction of organic matter can reduce nitrate accumulation and toxicity.

(vi) *H_2S Toxicity*

Hydrogen sulphide (H_2S), a colourless, obnoxious gas and a potential toxicant to fish, usually increases in its level due to poor maintenance of ponds and tanks. It is produced from the reduction of sulphate ions under anaerobic conditions and is of more concern in marine and brackish water bodies due to the sufficient availability of sulphate ions (Noga 1996). The bottom of ponds, which are not regularly cleaned, encourages anoxic conditions in the bottom sediment. Heavy accumulation of H_2S on the bottom is sometimes reported as a black cloud. The source may be the effluents of wastewater treatment plants or agricultural runoff, accumulation of unconsumed feed and fertilizers, etc. Persistent anoxic conditions in the bottom layer are not only unfavourable for bottom dwellers but also spread the rotten egg smell in the water body, indicating H_2S release in water. Hydrogen sulphide is volatile through the process of diagnosis, and the sample needs to be instantly subjected to estimate this gas. The inorganic substance, when transformed into an organic substance having sulphur, releases H_2S gas. Hydrogen sulphide exists in two forms in water, i.e., HS (ionized sulphide ion) and H_2S (unionized hydrogen sulphide gas). More than 0.003 ppm of unionized hydrogen sulphide gas is highly dangerous to fish (Raman et al. 2013). Like ammonia, the percentage fractions of hydrogen sulphide depend directly on the pH of the water. Hence, pH is indirectly proportional to hydrogen sulphide and vice versa. Hence, the toxic effect of H_2S can be reduced by increasing pH units. The toxicity is also increased with reduced oxygen availability. An elevated level of hydrogen sulphide can be lethal to fish, and intoxicated fishes manifest compromised growth and malformation in hatchlings. Exogenous H_2S inhibits aerobic respiration and, when coupled with low oxygen (hypoxia), leads to fish kills (Lamberth et al. 2010). To mitigate the toxic effect of hydrogen sulphide, heavy aeration inducing the bottom disturbances, frequent netting, optimum feeding and stocking, water exchange, complete pond bottom drying following each cycle of culture and use of soil and water probiotics can be beneficial to minimize the toxicity and subsequent fish kill. Maintaining aerobic conditions in the bottom can reduce the toxicity of hydrogen sulfide and associated water quality deterioration, as well as pathology in water dwellers. Though $KMnO_4$ detoxifies H_2S toxicity, it cannot be used for marine water. Increasing pH and lowering temperature could be an option for the mitigation of this toxicity (Noga 1996).

(vii) *Heavy Metal Toxicity*

Heavy metals are metals and metalloids with an atomic density of more than 4 g/cm^3 or 5 times or more than water. The aquatic environment receives various metals from ever-increasing anthropogenic activities, industrial wastes, domestic waste, sewage, agricultural runoff and mining. Heavy metals such as Fe, Mn, Co, Ni, Cu, Zn and Cr are essential for different vital metabolic activities in a very low concentration, but exceeding concentrations create deleterious effects in fish, leading to the term 'toxicity'. Metal salts due to their higher solubility and bioavailability in solution often cause greater toxicity to aquatic organisms than the metals themselves. These metals stimulate the production of reactive oxygen species (ROS) in the living systems, a major cause of toxicity and oxidative stress (Javed and Usmani 2019). Heavy metals enter fish bodies in three ways: by gills, digestive tract and body surface. The toxic effects of heavy metals include lower individual growth rate, physiological functions, mortality and reproduction in fish. The morphological symptoms associated with heavy metal intoxication include restlessness, increased or decreased frequency of opercular beatings, secretion of slime, reduced sensitivity to stimulation, exhaustion, lateral disposition and death due to asphyxiation. The toxicity can be reduced by increasing water hardness, which allows many metals to precipitate, while soft water shows the reverse action (Raman et al. 2013). For aquarium keepers, it's always wise to include running water before placing it into the aquarium or special filtration, activated carbon to remove metals.

Heavy metals, due to their persistence in the aquatic environment for a longer period of time and unbiodegradable nature and risk of undergoing bio-magnification, have become a global concern and front-line study in fish ecotoxicology. Environmental toxicity is monitored using a few yardsticks like bioaccumulation, blood biochemistry, marker enzymes of tissue damage, oxidative stress, genotoxicity and histopathology. Because all these parameters are drastically affected, heavy metal intoxication is widely demonstrated in describing aquatic pollution (Javed and Usmani 2019). There are several indices that are considered to quantify the toxicity. The condition factor K is expressed as a weight (g)/length^3 (cm). Due to heavy metal intoxication, K decreases, which indicates poor growth and unsatisfactory well-being of fish. Since the liver is the major organ participating in the detoxification process, the lowered value of the hepatosomatic index (liver weight/the fish body weight X 100) indicates its potential impact on the liver. Looking at the biomagnification of heavy metals in the food web and its ultimate reach to the human food chain, it remains an ever-alarming issue. High accumulation of Ni in the integument of *Channa punctatus* (Javed et al. 2016), *Wallago attu* and *Labeo dyocheilus* (Yousafzai et al. 2010) and *Clarias gariepinus* (Javed and Usmani 2011) has been reported. Heavy metals not only retard growth and physiological functions but also act as carcinogens associated with the death of fish. Unfortunately, due to the continuous influx of metal wastes from the drainage, atmosphere and soil erosion into the aquatic environment, fish have become a guaranteed and permanent source of heavy metals. All-round monitoring for a clean water source for culture practice is thus highly desirable.

(viii) *Microplastics: A Major Aquatic Pollutant*

Diverse and intense human activities have unquestionably amplified the possibility and distribution of pollutants and contaminants in water bodies in various forms. Among the most deliberated and debatable issues is microplastics (MPs), which are now ubiquitously contaminating the environment. The uses of MPs are nowadays increasing in aquaculture, which has increased the risk of microplastic pollution. Uses of single-use plastics in all sectors, as well as domestic effluents, have multifold affected aquatic life. MPs can form biofilms and facilitate oncogenic viral transmission as a vector (Cole et al. 2011) and can accumulate various aquatic contaminants (Ziccardi et al. 2016). Ingestion of MPs has been confirmed in the laboratory studies. The availability of MPs in fish depends on the feeding types, and literature reveals that carnivorous fish consume fewer microplastics (Zhang et al. 2021). On the other hand, though the MP ingestion rate of omnivorous fish is lower than that of filter fish, its removal is difficult in omnivorous fish (Zhang et al. 2021). Continued exposure of fish to PVC or PE microplastics could impair fish immune parameters, which may be attributed to the oxidative stress produced in the fish leucocytes (Espinosa et al. 2018). Since fish species are being processed with the removal of the gastrointestinal tract (which forms a major source of MPs) before consumption by humans, the direct effect of MPs from fish to humans is minimal. However, small species like mola (*Amblypharyngodon mola*) and darkina (*Esomus danricus*) are being consumed as a whole in different parts of Asian countries, thus raising the concern of microplastic residues induced abnormalities. The presence of MPs has also been evidenced in other small freshwater species, such as the gudgeon *Gobio gobio* (Sanchez et al. 2014). Though MP pollution is alarming in the marine ecosystem, its spread to freshwater is never far if it goes irrationally applied. Encouraging alternate materials that are biodegradable or can be recycled can guarantee a sustainable practice of aquaculture.

(ix) *Biological (Algal) Toxicosis*

Algae, while forming a basic component in the aquatic food web, algal bloom depicts the uncontrolled accumulation of phytoplankton. The proliferation of algal bloom in culture ponds, lakes and reservoirs is a function of eutrophication, which is associated with various ecological, social and cultural consequences. Currently, the trend of intensification has encouraged the application of huge amounts of feeds, fertilizers and organic manures. Moreover, climate change has a significant effect on the flourishing phytoplankton population. Both act synergistically, and harmful algal bloom has become a global threat to aquaculture. Not only does it deteriorate the aesthetic value but it also acts as a stress factor for aquatic organisms by preventing sunlight and subsequent photosynthesis and depleting oxygen to even critical levels, leading to fish kill. Certain species of freshwater algae synthesize toxins secreted during the growing phase or consequent to lysis, which can be lethal to aquatic life. Unfortunately, the algal crash is inevitably associated with drastic oxygen depletion and hence goes unrecorded many times, and the actual cause remains masked. Algal toxicosis has been underexplored and overlooked in the freshwater ecosystem. No systematic study has been conducted to estimate loss incurred due to algal bloom-associated fish mortality and economic losses in India so far.

Microcystis aeruginosa is a cyanobacterium that forms harmful algal blooms in freshwater water bodies globally (Chorus 2005). It has created a concern worldwide because of its frequency, dominance and distribution in bloom episodes (Fristachi et al. 2008). Predominant cyanotoxin, microcystins (MCs) secreted primarily from *Microcystis aeruginosa* is a potent hepatotoxin that slows down protein synthesis by inhibiting protein phosphatase 1 and 2A and promotes tumour in fish (Van der Oost et al. 2003). The persistence and detoxification of MCs in aquatic animals is a major concern. A recent study in common carp, while analysing multiple environmental stress factors, reveals Cyanobacteria *Microcystis aeruginosa* contributing to the severity of spring viraemia of carp (SVCV) (Palikova et al. 2021). The early life stages of fish exposed to MCs cause perturbations to embryonic hatching, a decrease in survival and growth rate, enlarged and opaque yolk sacs, small heads, curved bodies and tails, hepatobiliary abnormalities and ultrastructural alterations in hepatocytes. In adults and juveniles, MCs exposure has been shown to display dose-dependent effects like affecting growth rate and osmoregulation and increasing enzyme alterations in the liver, intestine, kidneys, heart, spleen or gills (Malbrouck and Kestemont 2006). Optimum feeding, stocking and pond management strategies can minimise algal bloom incidences and their associated fish mortality.

4.2 Procedural Stress Factors

These stress factors are man-made and are generally overlooked but play a major role in impairing fish health. During the culture period, handling fish for various purposes during stocking, hauling, harvesting and treatment with drugs seems to cause stress in fish. Overcrowding is one of the conditions when fish are stocked beyond the optimum limit. This causes social stress, and the organisms compete for space, nutrients and oxygen. Hence, fish growth is undermined, and also the risk of possible infection increases due to a decrease in immunity level. Crowding due to high stocking acts as a potential stressor, causing alterations in physiological and biochemical functions in fish (Montero et al. 1999). High stocking density also induces haemoconcentration, affecting haematocrit, haemoglobin, red blood cell count and total plasma proteins as strategies to increase the oxygen-carrying capacity of the blood during high energy demand conditions (Montero et al. 1999).

Vaccination in fish is one of the important procedural stress factors in fish. Fish can be immunized by injection, immersion and oral administration. However, the vaccinated fish lose interest in feeding and may produce trauma at the site of injection. The intraperitoneal injection is a usual way of vaccine administration but can induce organ adhesions and sometimes deposition of melanin, spinal deformities and granulomatous peritonitis (Bjorge et al. 2011). Fish vaccination or antibiotic bath exposure is stressful for fish. Hence keeping fish submerged in clean water during vaccination along with anaesthetics is beneficial to minimize handling stress.

The other rigorous stress is transportation stress to fish seeds. If oxygen is supersaturated and the density of seeds is high, then the accumulation of gases, specifically ammonia, due to excretion, causes significant stress. In the wild, traumatic injury happens to be a normal incidence due to predator attacks like a stab, bite or scrap lesions. Such lesions may be extensive and extend up to the skeleton. Netting is a routine practice in a fish culture used for harvesting and sampling. However, incidental abrasion of fish due to net is sometimes detrimental, causing open wounds and subsequent attack by opportunistic pathogens.

4.3 Nutritional Imbalance or Deficiency Diseases

An inadequate diet in relation to protein, fat, carbohydrate, vitamins and minerals causes malnutrition. Nutritional deficiencies can be categorized into two types: deficiency of macronutrients (protein, carbohydrate, lipid and fibre) and deficiency of micronutrients (vitamins and minerals). Nutritional imbalance also indicates the excess of particular nutrients, like a condition of excessive carbohydrate or lipid levels, which may result in hepatocyte degeneration and increased mortality due to fatty livers (Roberts 2012). The use of vegetable products has several disadvantages, which include endogenous antinutritional factors. These anti-nutrients are harmful and manifest reduced palatability, decreasing nutrient utilization for growth, altering nutrient balances, inhibiting growth, increasing intestinal dysfunction, altering gut microflora, inducing pancreatic hypertrophy and immune modulation, causing hypoglycaemia and/or liver damage (Krogdahl et al. 2010). The most documented indicator of an amino acid deficiency problem is impairment of growth and reduced feed utilization. These include dorsal fin erosion, which has been associated with lysine deficiency, and cataracts with methionine and tryptophan deficiency. Diet or husbandry procedures affect spinal development like scoliosis, lordosis and sometimes opercular, mandibular or maxillary deformation (Roberts 2012). Likewise, vitamins and minerals, though needed in very small quantities in fish carry vital body functions. Vitamin-deficient diets can cause poor appetite, slower growth, increased mortalities and poor feed conversion efficiencies. Vitamin deficiencies can cause histological changes in the digestive gland cells (Rai and Reddy 2004). Nutritional diseases are sometimes difficult to diagnose. Some nutritional diseases are 'broken back disease' in catfish caused by vitamin C deficiency. The lack of dietary vitamin C leads to the malfunction of the spinal column. Another one is 'no blood disease' caused due to folic acid deficiency. Affected fish become anaemic and may die. The condition disappears when the deficient nutrients are supplemented in feed. Nutritional imbalance increases disease susceptibility and compromises immunity. Balanced nutrition not only translates to body mass but also contributes to increased productivity. Some of the following conditions associated with nutritional imbalance, viz., scoliosis and lordosis, fish scurvy, broken back syndrome, steatitis and white fat disease, and avitaminosis, have been described in fish (Shefat and Karim 2018).

4.4 Genetic Anomalies

Genetic abnormalities are not related to any pathogenic aetiology. Rather, it is a malfunction at the genomic level. Congenital malfunctions are evident at or before the birth of an organism and may involve jaw or skull malfunction, extra pectoral fins, spinal defects and shortening of the operculum (Roberts 2012). Skeletal deformities in larvae, as well as juveniles in freshwater and marine fishes, may be a consequence of genetic, xenobiotic physiological, nutritional and environmental factors (Lall and Lewis-McCrea 2007). Among these, genetic defects like hybridization or inbreeding and mutations play a vital role in deciding the quality of a generation (Sadler et al. 2001). In addition to spinal anomalies, some tumours are also caused due to hereditary predispositions. For example, malignant melanoma, a pigment-origin tumour in the platy-swordtail hybrid, is also a gene-mediated disorder (Roberts 2012). Interestingly, in hybrid catfish (*Clarias batrachus* female and *C. gariepinus* male), seminoma, a tumour associated with seminiferous tubules, is only recorded in male offspring, indicating the involvement of genes through hybridization (Sahoo et al. 2004). These findings of neoplastic susceptibility of clariid hybrids do not encourage its introduction and culture in aquaculture, though it has huge growth potential. Microphthalmia and anophthalmia of embryos, albinistic ophthalmia in Atlantic salmon are characterized by an orange colour due to the presence of xanthophore and iridophore observed histologically. Affected fishes show protruding eyes with traumatic haemorrhages approaching retinal displacement, orbital evolution and anterior senecia (Roberts 2012).

A major cause of genetic abnormalities is consanguinity which means the union of two individuals very closely related and manipulating genes through experiments can induce skeletal deformities (Piron 1978). In genomics, the spinal deformity is autosomal dominant and directed by various genes (Heary and Madhavan 2008). Previous literature says spinal deformity in rainbow trout is a consequence of the nonadditive gene effect of inbreeding (Kincaid 1976) and later supported by Evans and Neff (2009) for its contribution to larval spinal deformity in two populations of Chinook salmon (*Oncorhynchus tshawytscha*). In contrast, Gjerde et al. (2005) recorded the prevalence of spinal deformities due to additive genetic effects contributing to spinal deformities in 14–17 months old Atlantic salmon from full and half-sib families. However, Sullivan et al. (2007) claimed that spinal deformity is not a heritable disorder in Atlantic salmon and tilapia (*Oreochromis aureus*). A perusal of the literature indicates the complex aetiology associated with spinal deformity. Appropriate selection of broodstocks can minimize the incidences of spinal deformity cases in aquaculture.

4.5 Tumours

Tumours are synonymous with neoplasms refer to abnormal and uncontrollable growth of living tissue. These conditions are generally not associated with fish mortality (Soares et al. 2014) until multiple stress factors act on fish. Generally, fish suffer from benign tumours that do not invade the surrounding tissue and rarely form malignant that grow rapidly, metastasizing the neighbouring tissue. However, the economic loss of neoplastic fishes may be attributed to the under-preference of consumers and less marketability due to the degradation of aesthetic value as well as associated with loss due to poor growth. Viruses, toxins, parasites, genetic factors, and husbandry-related trauma may induce oncogenic alterations. All teleost fishes could potentially develop neoplasia. Clinical Signs of neoplasms include gross observation of an external or internal swelling, lump showing abnormal tissue growth. Neoplastic conditions are not transmissible except the case when caused by infectious viruses. The development of neoplasia due to environmental pollutants is today's most debated topic. These pollutants include all kinds of contaminants emitted to the ground, surface water, and atmosphere. Subsequently, fishes with constant exposure to these pollutants lead to the development of cancer. Hence these factors are synonymously called carcinogens by many authors. Some of the environmental pollutants causing cancer in fish are briefed below:

N-Methyl-N′-Nitro-N-Nitrosoguanidine (MNNG) causes neoplasia in gill, gas bladder, thyroid, gonad, pancreas, scale, skin, connective tissue, skeletal muscle of medaka, whereas channel catfish given sub-chronic aqueous exposure to MNNG at 6 months of age develop neoplasia in the skin, fin, gill, thymus and bone (Chen et al. 1996). Diethylnitrosamine (DEN) has been proven to be a potent inducer of hepatic neoplasms in medaka and rainbow trout. Likewise, benzopyrene, heavy metals such as arsenic, cadmium, chromium, nickel and lead are carcinogenic threats to humans as well as fish, among which arsenic and cadmium are the most harmful contaminants in fish. Accumulation of PAHs in the fish body and the residues of pesticides for a longer time leads to the development of different cancers. Diagnosis of tumours is done by histopathological and electron microscopic study besides using immunocytochemical markers. Horizontal transmission of neoplasia never happens. However, on the appearance of the neoplastic condition in fishes, the causes may be identified and avoided from the system. Aquaculture practices with absolutely clear water far away from industrial, domestic, hospital and wastewater treatment plants wastes are advisable to reduce the risk of inflow of deadly contaminants to the water body. Uncompromised immunity of fishes can protect fish from this stress and hitch factors up to a certain extent.

Though fish tumours have been studied extensively, their induction due to environmental toxicity is very difficult to establish (Roberts 2012). However, cavernous haemangioma (benign tumours of vascular origin) is common in humans and animals, including fish (Zhang et al. 2009; Sahoo et al. 2012). Fish is not an exception to this vascular malformation. In a case of haemangioma noticed in *Puntius sarana*, Sahoo et al. (2012) noticed the exophytic soft dermal masses with sero-sanguinous fluid at the base of the pelvic, pectoral and caudal fins. On histology, the tumours are characterised by vascular spaces filled with erythrocytes and surrounded by endothelial cells and pericytes.

Neoplasia is not only restricted to freshwater fishes but it has also been documented in brackish water and marine fishes. Poochirian et al. (2011) reported squamous cell carcinoma (SCC) in grey mullet comprising anaplastic epitheloid cells with fibroblast and necrotic cells. Although the tumours are uncommon in cultured fish, these are frequently found in polluted water bodies, mostly dams, reservoirs, seas or lakes.

5 Conclusion

Noninfectious diseases are often overlooked and masked due to their non-communicable/non-transmissible nature. Moreover, the diseases are difficult to diagnose, unlike those of pathogen-induced infections. The mortality associated with these diseases has not drawn much attention from economists, remarkably since it does not cause huge losses. Noninfectious diseases are mainly stress-induced. Stress in aquaculture covers a wide spectrum of activities, mainly associated with managerial aspects. The physicochemical parameters, as well as environmental pollutants, also take centre stage, dictating the noninfectious diseases. In a broad sense, these are attributed to poor cultural practices and ever-increasing anthropogenic activities, which generally give a qualitative impression of a lesser extent of severity. This may be the reason that sometimes noninfectious diseases are synonymously used by many authors as disorders. Clinically, lesions, imbalances, deformities, anorexia, anaemia, weight loss, ascites and suffocation are the prominent manifestations of noninfectious diseased fishes.

Inappropriate water quality parameters and poor husbandry are good drivers of stress, leading to significant changes in physiology and the breakdown of immunity. Risk factors and threats from the environment are an ever-continuing process with climate and lifestyle. Either deliberately or unknowingly, anthropogenic activities are directly contributing to aquatic menace in the form of infectious as well as non-infectious diseases. Hence, constraints are many in aquaculture, which warrants one health programme and more understanding of epidemiological studies in order to prevent the diseases appropriately.

Climate change may impose severe risks for aquatic animal health, not only by facilitating contagious aetiologies to proliferate but also the noninfectious diseases get widened scope. Though blood biochemical parameters, oxidative stress, hormonal profile and glucose are always gold standards for measuring stress response, the latest technologies like proteomics (e.g. HSPs), metabolomics and genomics can provide better insight to establish the mechanism in noninfectious diseases. Acute and chronic stressors in aquaculture are to be identified, which devalues the welfare index of aquaculture practice. Sustainable aquaculture can be well achieved with zero tolerance to stress inducers. The affirmation may be apparently impractical, but reduction of such hitch factors from the culture system not only makes an individual immunologically uncompromised but also limits the spread of pathogens and disease epidemiology.

References

Abdel-Tawwab M, Hagras AE, Elbaghdady HAM, Monier MN (2015) Effects of dissolved oxygen and fish size on Nile tilapia, Oreochromis niloticus (L.): growth performance, whole-body composition, and innate immunity. Aquac Int 23:1261–1274

Aksakal FI, Ciltas A (2018) The impact of ultraviolet B (UV-B) radiation in combination with different temperatures in the early life stage of zebrafish (*Danio rerio*). Photochem Photobiol Sci 17(1):35–41

Alves RN, Agusti S (2020) Effect of ultraviolet radiation (UVR) on the life stages of fish. Rev Fish Biol Fish 30(2):335–372

Araujo CV, Cohin-de-Pinho SJ, Chastinet CB, Santos JS, da Silva EM (2008) Discriminating the pH toxicity to *Poecilia reticulata* Peters, 1859 in the Dunas Lake (Camaçari, BA, Brazil). Chemosphere 73(3):365–370

Bajpai V, Divya P, Kirty S, Mohanty J, Sahoo PK (2022) Viral diseases in Indian freshwater and marine water pisciculture. Curr Sci 122(3):267–280

Bjorge MH, Nordgreen J, Janczak AM, Poppe T, Ranheim B, Horsberg TE (2011) Behavioural changes following intraperitoneal vaccination in Atlantic salmon (*Salmo salar*). Appl Anim Behav Sci 133(1–2):127–135

Bodansky FL (1951) Methemoglobinemia and methemoglobin-producing compounds. Pharmacol Rev 3:144–196

Bolner KC, Copatti CE, Rosso FL, Loro VL, Baldisserotto B (2014) Water pHand metabolic parameters in silver catfish (*Rhamdia quelen*). Biochem Syst Ecol 56:202–208

Bracchini L, Loiselle S, Dattilo AM, Mazzuoli S, Cózar A, Rossi C (2004) The spatial distribution of optical properties in the ultraviolet and visible in an aquatic ecosystem. Photochem Photobiol 80(1):139–149

Brauner CJ, Seidelin M, Madse SS, Jensen FB (2000) Effect of freshwater hyperoxia and hypercapnia and their influences on subsequent seawater transfer in Atlantic salmon (Salmo salar) smolts. Can J Fish Aquat Sci 57:2054–2064

Campbell NA (1990) Biology. Circulation and gas exchange. Chapter 38. Benjamin/Cummings Publishing Company, Redwood City, pp 683–705

Chen HC, Pan IJ, Tu WJ, Lin WH, Hong CC, Brittelli MR (1996) Neoplastic response in Japanese medaka and channel catfish exposed to N-methyl-N'-nitro-N-nitrosoguanidine. Toxicol Pathol 24(6):696–706

Chorus I (2005) Current approaches to cyanotoxin risk assessment, risk management and regulations in different countries. Fed. Environmental Agency

Choudhury TG, Singh SK, Parhi J, Barman D, Das BS (2014) Common fungal diseases of fish: a review. Environ Ecol 32(2):450–456

Cole M, Lindeque P, Halsband C, Galloway TS (2011) Microplastics as contaminants in the marine environment: a review. Mar Pollut Bull 62(12):2588–2597

Coutteau P, Robles R, De Nigris G, Cirillo A, Vestraete P, Tort L (2001) Nutritional solutions to winter syndrome in gilthead seabream, *Sparus aurata* verification at a land based farm. Aquafeed Int 2:30–33

Dahms HU, Lee JS (2010) UV radiation in marine ectotherms: molecular effects and responses. Aquat Toxicol 97(1):3–14

Dong Q, Svoboda K, Tiersch TR, Monroe WT (2007) Photobiological effects of UVA and UVB light in zebrafish embryos: evidence for a competent photo repair system. J Photochem Photobiol B Biol 88(2–3):137–146

Dwyer GK, Stoffels RJ, Pridmore PA (2014) Morphology, metabolism and behaviour: responses of three fishes with different lifestyles to acute hypoxia. Freshw Biol 59(4):819–831

Easterling DR, Meehl GA, Parmesan C, Changnon SA, Karl TR, Mearns LO (2000) Climate extremes: observations, modeling, and impacts. Science 289(5487):2068–2074

Edsall DA, Smith CE (1991) Oxygen-induced gas bubble disease in rainbow trout, *Oncorhynchus mykiss* (Walbaum). Aquac Res 22(2):135–140

Erazo-Pagador GE (2001) Southeast Asian Fisheries Development Centre Aquaculture Department SEAFDEC/AQD Institutional Repository. http://repository.seafdec.org.ph
Espinosa C, Beltrán JMG, Esteban MA, Cuesta A (2018) In vitro effects of virginmicroplastics on fish head-kidney leucocyte activities. Environ Pollut 235:30–38
Evans ML, Neff BD (2009) Non-additive genetic effects contribute to larval spinal deformity in two populations of Chinook salmon (*Oncorhynchus tshawytscha*). Aquaculture 296(1–2):169–173
Evans DH, Piermarini PM, Potts WTW (1999) Ionic transport in the fish gill epithelium. J Exp Zool 283(7):641–652
Ferguson HW (2006) Systemic Pathology of Fish (2nd edn), pp. 187–193. Scotian Press, London, UK
Fontenot QC, Isely JJ, Tomasso JR (1999) Characterization and inhibition of nitrite uptake in short nose sturgeon fingerlings. J Aquat Anim Health 11(1):76–80
Fristachi AJL, Sinclair JA, Hambrook-Berkman G, Boyer J, Burkholder J, Burns W, Carmichael A, du Four W, Frazier SL, Morton E, O'Brien S, Walker (2008) Occurrence of cyanobacterial harmful algal blooms working group report. In: Hudnell HK (ed) Proceedings of the interagency international symposium on cyanobacterial harmful algal blooms, Advances experimental medical biology, vol 619. Springer, New York, pp 37–97
Gjerde B, Pante MJR, Baeverfjord G (2005) Genetic variation for a vertebral deformity in Atlantic salmon (*Salmo salar*). Aquaculture 244(1–4):77–87
Golombieski JI, Gessi K, Alessandra JB, Ana PGA, Cândida T, Isabela AF, Maria AP, Tielle MA, Bernardo B (2013) Nitrogenous and phosphorus excretions in juvenile silver catfish (*Rhamdia quelen*) exposed to different water hardness, humic acid, and pH levels. Fish Physiol Biochem 39:837–849
Gómez S (2000) Nephrocalcinosis in Mediterreanean cultured white bream (*Diplodus sargus* L.). Bull Eur Assoc Fish Pathol 20(1):40–42
Green C, Haukenes A (2015) The role of stress in fish disease. South Region. Aquaculture Cent. Pub
Gultepe N, Ates O, Hisar O, Beydemir S (2011) Carbonic anhydrase activities from the rainbow trout correspond to the development of acute gas bubble disease. J Aquat Anim Health 23:134–139
Hader DP, Barnes PW (2019) Comparing the impacts of climate change on the responses and linkages between terrestrial and aquatic ecosystems. Sci Total Environ 682:239–246
Hader DP, Sinha RP (2005) Solar ultraviolet radiation-induced DNA damage in aquatic organisms: potential environmental impact. Mutat Res 571(1–2):221–233
Hader DP, Kumar HD, Smith RC, Worrest RC (2007) Effects of solar UV radiation on aquatic ecosystems and interactions with climate change. Photochem Photobiol Sci 6:267–285
Hader DP, Williamson CE, Wängberg SÅ, Rautio M, Rose KC, Gao K, Helbling EW, Sinha RP, Worrest R (2015) Effects of UV radiation on aquatic ecosystems and interactions with other environmental factors. Photochem Photobiol Sci 14(1):108–126
Harris RR, Coley S (1991) The effects of nitrite on chloride regulation in the crayfish *Pacifastacus leniusculus* Dana (Crustacea: Decapoda). J Comp Physiol 161:199–206
Hasler CT, Suski CD, Hanson KC, Cooke SJ, Philipp DP, Tufts BL (2009) Effect of water temperature on laboratory swimming performance and natural activity levels of adult largemouth bass. Can J Zool 87(7):589–596
Heary RF, Madhavan K (2008) Genetics of scoliosis. Neurosurgery 63(suppl_3):A222–A227
Henrique MMF, Gouillou-Coustans MF, Gomes E (2002) Effect of dietary ascorbic acid supplementation and chronic hypoxia on sea bream growth and vitamin C status. J Fish Biol 60(2):442–452
Hughes GM (1973) Respiratory responses to hypoxia in fish. Am Zool 13(2):475-489
Idowu TA, Adedeji HA, Sogbesan OA (2017) Fish disease and health management in aquaculture production. Int J Environ Agric Res 1(2):2–6
Iwama G, Vijayan M, Morgan J (2005) Stress and ion regulation in fish. Comp Biochem Physiol 141:181–182
Javed M, Usmani N (2011) Accumulation of heavy metals in fishes: a human health concern. Int J Environ Sci 2(2):659–670

Javed M, Usmani N (2019) An overview of the adverse effects of heavy metal ccontamination on fish health. Proc Natl Acad Sci India Sect B Biol Sci 89(2):389–403

Javed M, Ahmad I, Usmani N, Ahmad M (2016) Bioaccumulation, oxidative stress and genotoxicity in fish (*Channa punctatus*) exposed to a thermal power plant effluent. Ecotoxicol Environ Saf 127:163–169

Jensen FB (2003) Nitrite disrupts multiple physiological functions in aquatic animals. Comp Biochem Physiol 135:9–24

Johnson W, Katavic I (1984) Mortality, growth and swim bladder stress syndrome of sea bass (*Dicentrarchus labrax*) larvae under varied environmental conditions. Aquaculture 38:67–78

Jokinen IE, Markkula ES, Salo HM, Kuhn P, Nikoskelainen S, Arts MT, Browman HI (2008) Exposure to increased ambient ultraviolet B radiation has negative effects on growth, condition and immune function of juvenile Atlantic salmon (*Salmo salar*). Photochem Photobiol 84(5):1265–1271

Jokinen IE, Salo HM, Markkula E, Rikalainen K, Arts MT, Browman HI (2011) Additive effects of enhanced ambient ultraviolet B radiation and increased temperature on immune function, growth and physiological condition of juvenile (parr) Atlantic Salmon, *Salmo salar*. Fish Shellfish Immunol 30(1):102–108

Kajimura M, Takimoto K, Takimoto A (2023) Acute toxicity of ammonia and nitrite to Siamese fighting fish (Betta splendens). BMC Zool 8(1):1–7

Karim MR, Sekine M, Ukita M (2003) A model of fish preference and mortality under hypoxic water in the coastal environment. Mar Pollut Bull 47(1–6):25–29

Kiese M (1974) Methemoglobinemia: a comprehensive treatise. CRC Press, Cleveland

Kincaid HL (1976) Inbreeding in rainbow trout (*Salmo gairdneri*). J Fish Board Can 33(11):2420–2426

Knudsen PK, Jensen FB (1997) Recovery from nitrite-induced methaemoglobinaemia and potassium balance disturbances in carp. Fish Physiol Biochem 16(1):1–10

Krogdahl A, Penn M, Thorsen J, Refstie S, Bakke AM (2010) Important antinutrients in plant feedstuffs for aquaculture: an update on recent findings regarding responses in salmonids. Aquac Res 41(3):333–344

Kroupova H, Prokes M, Macova S, Penaz M, Barus V, Novotny L, Machova J (2010) Effect of nitrite on early-life stages of common carp (*Cyprinus carpio* L.). Environ Toxicol Chem 29:535–540

Lall SP, Lewis-McCrea LM (2007) Role of nutrients in skeletal metabolism and pathology in fish—an overview. Aquaculture 267(1–4):3–19

Lamberth SJ, Branch GM, Clark BM (2010) Estuarine refugia and fish responses to a large anoxic, hydrogen sulphide, ‘black tide’ event in the adjacent marine environment. Estuar Coast Shelf Sci 86:203–215

Lewis JM, Klein G, Walsh PJ, Currie S (2012) Rainbow trout (*Oncorhynchus mykiss*) shift the age composition of circulating red blood cells towards a younger cohort when exposed to thermal stress. J Comp Physiol 182:663–671

Madeira D, Narciso L, Cobral HN, Vinagre C, Diniz MS (2012) HSP70 production patterns in costal and estuarine organisms facing increasing temperatures. J Sea Res 73:137–147

Malbrouck C, Kestemont P (2006) Effects of microcystins on fish. Environ Toxicol Chem Int J 25(1):72–86

Mandic M, Todgham AE, Richard JG (2009) Mechanisms and evolution of hypoxia tolerance in fish. Proc R Soc B 276:735–744

Margiocco C, Arillo A, Mensi P, Shenone G (1983) Nitrite bioaccumulation in *Salmo gairdneri* and haematological consequences. Aquat Toxicol 3:261–270

McKenzie RL, Aucamp PJ, Bais AF, Björn LO, Ilyas M (2007) Changes in biologically-active ultraviolet radiation reaching the earth's surface. Photochem Photobiol Sci 6(3):218–231

Mischke CHC, Chatakondi N (2012) Effects of abrupt pH increases on survival of different ages of young channel catfish and hybrid catfish. N Am J Aquac 74:160–163

Montero D, Izquierdo MS, Tort L, Robaina L, Vergara JM (1999) High stocking density produces crowding stress altering some physiological and biochemical parameters in gilthead seabream, *Sparus aurata*, juveniles. Fish Physiol Biochem 20(1):53–60

Noga EJ (1996) Fish disease: diagnosis and treatment. Mosby, St. Louis, MO

Noga EJ (2000) Fish disease: diagnostic and treatment. Iowa State University Press, Ames, Iowa, p 367

Nunez ET, Sobrino C, Neale PJ, Ceinos RM, Du SJ, Rotllant J (2012) Molecular response to ultraviolet radiation exposure in fish embryos: implications for survival and morphological development. Photochem Photobiol 88:701–707

Palikova M, Kopp R, Kohoutek J, Blaha L, Mares J, Ondrackova P, Papezikova I, Minarova H, Pojezdal L, Adamovsky O (2021) Cyanobacteria *Microcystis aeruginosa* contributes to the severity of fish diseases: a study on spring viraemia of carp. Toxins 13(9):601(1-19)

Parra JEG, Baldisserotto B (2007) Effect of water pH and hardness on survival and growth of freshwater teleosts. In: Baldisserotto B, Mancera JM, Kapoor BC (eds) Fish osmoregulation. Science Publishers, Enfiled

Pichavant K, Person-Le-Ruyet J, Bayon NL, Severe A, Roux AL, Boeuf G (2001) Comparative effects of long-term hypoxia on growth, feeding and oxygen consumption in juvenile turbot and European Sea bass. J Fish Biol 59(4):875–883

Piron RD (1978) Spontaneous skeletal deformities in the zebra danio (*Brachydanio rerio*) bred for fish toxicity tests. J Fish Biol 13(1):79–83

Poochirian JK, Maruthamuthu N, Sahoo PK, Govindarajan T (2011) Squamous cell carcinoma in grey mullet, *Mugil cephalus*. L. (Perciformes: Mugilidae). Comp Clin Pathol 20(6):673–675

Rai A, Reddy HRV (2004) Evaluation of dietary essentiality of vitamins for *Penaeus monodon* (Fabricius). Indian J Exp Biol 42(01):58–62

Raman RP, Prakash C, Makesh M, Pawar NA (2013) Environmental stress mediated diseases of fish: an overview. Adv Fish Res 5:141–158

Randall DJ, Cameron JN (1973) Respiratory control of arterial pH as temperature changes in rainbow trout (Salmo gairdneri). Am J Physiol 225(4): 997–1002

Rekecki A, Meeus W, Chiers K, Adriaen J, Boyen F, Declercq A, Van den Broeck W, Decostere A (2016) Swimbladder hyperinflation in burbot Lota lota L. larvae. Aquac Res 47: 673–676

Roberts RJ (2012) Fish pathology, 4th edn. Wiley, Oxford, UK

Roberts RJ, Agius C, Saliba C, Bossier P, Sung Y (2006) Heat shock proteins (chaperones) in fish and shellfish and their potential role in health and welfare. J Fish Biol 69(4):1147–117

Ruth Francis-Floyd (2015) Enviwronmental diseases in aquatic systems – exotic and laboratory animals veterinary manual. https://www.msdvetmanual.com/exotic-andlaboratory-animals/aquatic-systems/environmental-diseases-in-aquatic systems#v23353151 1/6 DVM, MS, DACZM, Department of Large Animal Clinical Sciences, College of Veterinary Medicine, University of Florida

Sadler J, Pankhurst PM, King HR (2001) High prevalence of skeletal deformity and reduced gill surface area in triploid Atlantic salmon (*Salmo salar* L.). Aquaculture 198(3–4):369–386

Sahoo PK, Sahoo SK, Giri SS, Swain T, Sahu AK (2004) Seminoma in hybrid catfish [*Clarias batrachus* (Linnaeus) female x *Clarias gariepinus* (Burchell) male]. Indian J Exp Biol 42(6):626–627

Sahoo PK, Mohanty BR, Das A, Mahapatra KD, Saha JN (2012) Cavernous haemangioma in cultured olive barb, *Puntius sarana* (Hamilton). J Fish Dis 35(9):695–697

Sanchez RC, Obregon EB, Rauco MR (2011) Hypoxia is like an etiological factor in vertebral column deformity of salmon (*Salmo salar*). Aquaculture 316(1–4):13–19

Sanchez W, Bender C, Porcher JM (2014) Wild gudgeons (*Gobio gobio*) from French rivers are contaminated by microplastics: preliminary study and first evidence. Environ Res 128:98–100

Sandrini JZ, Trindade GS, Nery LEM, Marins LF (2009) Time-course expression of DNA repair-related genes in hepatocytes of zebrafish (*Danio rerio*) after UV-B exposure. Photochem Photobiol 85(1):220–226

Saroglia, M, Terova G, De Stradis A, Caputo A (2002) Morphometric adaptations of sea bass gills to different dissolved oxygen partial pressures. J of Fish Biol 60(6):1423–1430

Sayed AEDH, Watanabe-Asaka T, Oda S, Mitani H (2016) Apoptosis and morphological alterations after UVA irradiation in red blood cells of p53 deficient Japanese medaka (*Oryzias latipes*). J Photochem Photobiol B Biol 161:1–8
Sharma P, Chakrabarty D (2006) Effects of UV-B radiation on the gill morphology and function of Catla (Catla catla). J Env Biol 27(2):349–353
Sharma JG, Rao YV, Kumar S, Chakrabarti R (2010) Impact of UV-B radiation on the digestive enzymes and immune system of larvae of Indian major carp Catla catla. Int J Radiat Biol 3:181–186
Shefat SHT, Karim MA (2018) Nutritional diseases of fish in aquaculture and their management: A review. Acta Sci Pharm Sci 2(12):50–58
Shultz AD, Murchie KJ, Griffith C, Cooke SJ, Danylchuk AJ, Goldberg TL, Suski CD (2011) Impacts of dissolved oxygen on the behaviour and physiology of bonefish: implications for live-release angling tournaments. J Exp Mar Biol Ecol 402(1–2):19–26
Singh MK, Sharma JG. Chakrabarti R (2015) Simulation study of natural UV-B radiation on Catla catla and its impact on physiology, oxidative stress, Hsp 70 and DNA fragmentation. J Photochem Photobiology B: Biol 149:156–163
Snieszko SF (1974) The effects of environmental stress on outbreaks of infectious diseases of fishes. J Fish Biol 6(2):197–208
Soares F, Fernández I, Costas B, Gavaia P (2014) Non-infectious disorders of warmwater fish. In: Woo P, Bruno D (eds) Diseases and disorders of finfish in cage culture, 2nd edn. CAB eBooks. pp. 254
Soldatov AA (2002) Peculiarities of structure, polymorphism, and resistance to oxidation of fish hemoglobins. J Evol Biochem Physiol 38(4):392–400
Speare DJ (1998) Disorders associated with exposure to excess dissolved gases. In: Leatherland JF, Woo PTK (eds) Fish disease and disorders: non-infectious disorders. CABI Publishing, pp 207–224
Sudheesh PS, Al-Ghabshi A, Al-Mazrooei N, Al-Habsi S (2012) Comparative pathogenomics of bacteria causing infectious diseases in fish. Int J Evol Biol 2012:1–16
Sullivan M, Guy DR, Roberts RJ, Manchester NJ (2007) The aetiology of spinal deformity in Atlantic salmon, *Salmo salar* L.: influence of genetic factors on the frequency and severity in freshwater stages. J Fish Dis 30:753–758
Svobodova Z, Gelnarova J, Justyn J, Krupauer V, Machova J, Simanov L, Valentova V, Vykusova B, Wolgemuth E (1987) Toxikologie vodních živočichů. SZN, Praha
Svobodova Z, Machova J, Hamackova J, Huda J, Kroupova H (2005) Nitrite poisoning of fish in aquaculture facilities with water-recirculating systems: three case studies. Acta Vet Brno 74:129–137
Svobodova Z, Kolarova J, Navratil S, Vesely T, Chloupek P, Tesarcik J, Citek J (2007a) Nemoci sladkovodních a akvarijních ryb. Informatorium, Praha
Svobodova Z, Machova J, Kroupova H, Smutna M, Groch L (2007b) Ammonia autointoxication of common carp: case studies. Aquac Int 15:277–286
Svobodova Z, Machova J, Kroupova H (2008) Otravy ryb. In: Svobodová Z (ed) Veterinární toxikologie v klinické praxi. Profi Press, Praha
Svobodova Z, Machova J, Kroupova HK, Velisek J (2017) Water quality–disease relationship on commercial fish farms. In: Fish diseases. Academic Press, pp 167–185
Tomasso JR, Grosell M (2005) Physiological basis for large differences in resistance to nitrite among freshwater and freshwater-acclimated euryhaline fishes. Environ Sci Technol 39.98–102
Van der Oost R, Beyer J, Vermeulen NP (2003) Fish bioaccumulation and biomarkers in environmental risk assessment: a review. Environ Toxicol Pharmacol 13(2):57–149
Van Seventer JM, Hochberg NS (2017) Principles of infectious diseases: transmission, diagnosis, prevention, and control. In: International encyclopedia of public health, p 22
Vanlandeghem MM, Wahl DH, Suski CD (2010) Physiological responses of largemouth bass to acute temperature and oxygen stressors. Fish Manag Ecol 17(5):414–425
Vatsos IN, Angelidis P (2010) Water quality and fish diseases. J Hellenic Vet Med Soc 61:40–48

Vinagre C, Madeira D, Narciso L, Cabral HN, Diniz MS (2012) Effect of handling, confinement and crowding in HSP70 production in *Pachygrapsus marmoratus*, a model species for climate change experiments. J Sea Res 72(64–68):108

Vincent WF, Belzile C (2002) Ultraviolet radiation and arctic ecosystem. In: Hessen D (ed) Ecological studies. Springer, Berlin, Heidelberg. 153, 137.109

Wildgoose WH (2001) BSAVA manual of ornamental fish, 2nd edn. British Small Animal Veterinary Association, Quedgeley, Gloucester

Wise DJ, Tomasso JR, Brandt TM (1988) Ascorbic acid inhibition of nitrite-induced methemoglobinemia in channel catfish. Prog Fish Cult 50(2):77–80

Withers PC (1992) Comparative animal physiology. Saunders College Publishing, Sydney

Wood CM (1993) Ammonia and urea metabolism and excretion. In: Ewans DH (ed) Physiology of fishes. CRC Press, Boca Raton

Yousafzai AM, Chivers DP, Khan AR, Ahmad I, Siraj M (2010) Comparison of heavy metals burden in two freshwater fishes *Wallago attu* and *Labeo dyocheilus* with regard to their feeding habits in natural ecosystem. Pak J Zool 42(5):537–544

Zhang WJ, Wu LQ, Liu H, Ye LY, Xin Y, Grau GE, Lou JN (2009) Abnormal blood vessels formed by human liver cavernous hemangioma endothelial cells in nude mice aresuitable for drug evaluation. Microvasc Res 78:379–385

Zhang C, Wang J, Zhou A, Ye Q, Feng Y, Wang Z, Wang S, Xu G, Zou J (2021) Species-specific effect of microplastics on fish embryos and observation of toxicity kinetics in larvae. J Hazard Mater 403:123948

Ziccardi LM, Edgington A, Hentz K, Kulacki KJ, Kane Driscoll S (2016) Microplastics as vectors for bioaccumulation of hydrophobic organic chemicals in the marine environment: a state of the science review. Environ Toxicol Chem 35(7):1667–1676

Impact of Climate Change on Animal Health in Aquaculture Systems

Muralidhar M and Satheesha Avunje

Abstract

The increased incidence of diseases in aquaculture will lead to reduced profits, affecting economic viability, and there is a concern that climate change may further increase this risk. The specific impacts of climate change on animal diseases are not clearly understood, and it is important to comprehend the disease risks of farmed aquatic animals in the face of future climate change for the success and sustainability of aquaculture. The climatic scenarios revealed that India could experience increasing trends in annual mean temperature, an increase in the frequency of hot days and multiple-day heatwaves, variation in rainfall above and below the averages, and an increase in intensity and frequency of extreme weather events such as cyclones, torrential rains, and floods. These seasonal variations and extreme weather events have enormous negative consequences on potentially stress-farmed fish and shellfish species through deterioration in the water quality, suppression of their immune system, and increasing the incidence of disease outbreaks through alterations in the distribution, prevalence, and virulence of pathogens, resulting in crop losses and yield reduction. The present chapter throws light on the climate change scenarios and their impact on the physiology and immunity of cultured aquatic animals and their health, virulence of pathogens, host–pathogen interaction, and mitigation and adaptation measures to improve environmental and animal health to prevent the outbreak of diseases.

Muralidhar M (✉) · S. Avunje
ICAR-Central Institute of Brackishwater Aquaculture, Chennai, India
e-mail: muralichintu@rediffmail.com; s.avunje@gmail.com

R. Kooloth Valappil et al. (eds.), *Aquatic Animal Health Management*,
https://doi.org/10.1007/978-981-96-7987-4_15

Keywords

Disease · Flood · Heatwave · Mitigation · Pathogen · Precipitation · Temperature · Climate change

1 Introduction

Climate change is a risk to the quality and quantity of global food production (Beach and Viator 2008; Hamdan et al. 2015; Myers et al. 2017). Aquaculture increases the supply of nutritious foods in the diets of poor consumers without creating negative impacts on environmental services and ecological systems. The contribution of aquaculture to global fish production was 130.9 million tonnes (59%) out of the estimated 223.2 million tonnes of global fisheries and aquaculture production in 2022 (FAO 2024). Inland and marine/coastal aquaculture contributions to total farmed aquatic animals were 62.6 and 37.4 per cent, respectively. The 2023 United Nations Framework Convention on Climate Change (UNFCCC) Ocean Dialogue recognised the significant potential of aquatic foods for providing critical climate solutions and the importance of integrating them into both national and multilateral climate action-related processes (FAO 2024). However, the most urgent concern is the sustainability of aquaculture in the context of a rapidly growing human population and a changing climate, owing to the sector's significant contribution to global food security, nutrition, and livelihoods (Blanchard et al. 2017; Dabbadie et al. 2019; FAO 2024).

2 Climate Change Scenarios and Drivers

Assessment Report (AR) 5 of the Intergovernmental Panel on Climate Change (IPCC) reported a likely increase in the global surface temperature by 1.5 °C by 2100 relative to the period 1850–1900 (IPCC 2013). More floods, hurricanes and irregular monsoons have been experienced over the last 50 years, and these extreme weather events are expected to increase in frequency and intensity. Some of the developing world's largest rivers are drying up because of climate change. A change in the water flow was reported in one-third of 925 rivers assessed between 1948 and 2004 by the US-based National Center for Atmospheric Research (NCAR). Most of these rivers, including the Niger in West Africa, the Ganges in South Asia and the Yellow River in China, were dryer (Nightingale 2009). Climate change is likely to increase the frequency and intensity of climate processes, such as El Niño-Southern Oscillation (ENSO), and there are indications that all surface waters of oceans with some geographical variations are warming and increasing in salinity. Global mean sea level has risen by about 0.1–0.2 mm yr^{-1} over the past 3000 years and by 1–2 mm yr^{-1} since 1900, 1.3 to 2.3 mm per year over 1961 to 2003 and 2.4 to 3.8 mm per year from 1993 to 2003, and by 2080s sea level rise could cause the loss of as much as 22% of the world's coastal wetlands (Nicholls et al. 1999). In coastal areas, a rise in sea levels will exacerbate water resource constraints due to increased salinisation of groundwater supplies.

India is one of the countries more vulnerable to climate change (Cruz et al. 2007). The predictions of CC over India are increasing trends in annual mean temperature, more pronounced warming during post-monsoon and winter, an increase in the frequency of hot days and multiple-day heatwaves (Krishnan et al. 2020), decadal departures in rainfall above and below the long-term average (Kothyari and Singh 1996). The average climate is likely to be warmer by 1.7 to 2 °C for the 2030s and by 3.3 to 4.8 °C for the 2080s compared to the pre-industrial times, as for the Coupled Model Intercomparison Project (CMIP) 5 projections for India (Chaturvedi et al. 2012a, b). Water and air temperatures are expected to rise during the summer months and this will be more pronounced in southern states. The precipitation is likely to increase by 5 to 6 per cent and 6 to 14 per cent in the 2030s and 2080s, respectively. Recent decades have exhibited an increase in extreme rainfall events over northwest India during the summer monsoon and a decline in the number of rainy days along the east coast (Singh and Sontakke 2002; Krishnan et al. 2020).

3 Climate Change and Aquaculture

Climate change threatens the production and sustainability of aquaculture (Handisyde et al. 2006; Brander 2007; Ficke et al. 2007; Barange et al. 2018; Muralidhar et al. 2021). However, the effect will not be equal, as this sector is defined in time, space, and size (intensity of culture) and the environmental context in which it is practised (De Silva and Soto 2009; Merino et al. 2012; Bell et al. 2013; Sae-Lim et al. 2017; Barange et al. 2018; IPCC 2018; Zolnikov 2019). Several projections show that the entire aquaculture value chain is vulnerable to the effects of climate change (Cochrane et al. 2009; Fleming et al. 2014; Bueno and Soto 2017; Barange et al. 2018; Dabbadie et al. 2019). Climate variability, especially changes in temperature regimes, rainfall patterns, seasonal variations and increases in the intensity and/or frequency of extreme weather events (EWEs) like heavy rain, flooding, cyclones, heat waves, etc., have enormous negative consequences on aquaculture production systems (Muralidhar et al. 2012, 2013, 2021; Islam et al. 2022; Thuy et al. 2017). These changes are now evident in water resources, the basic input for aquaculture, and this primary resource is directly affected by climate change (Fleming et al. 2014; Blanchard et al. 2017; Zolnikov 2019).

Fish, being poikilothermic, are sensitive to temperature variations resulting from climate change (Sae-Lim et al. 2017; Ngoan 2018). With the predicted 1.5 °C rise in average global temperature in this century, increased mortalities are likely to occur for most fish, especially cold-water species, due to thermal stress (Hamdan et al. 2012; Gubbins et al. 2013). An increase in rearing water temperature by 1 °C for every 8 h to reach up to 41 °C for Indian white shrimp *Penaeus indicus* indicated increased mortality at temperatures beyond 38 °C and no survival at 41 °C (Hoang et al. 2020). Rainfall decreases the salinity of brackishwater ponds, which can lead to mortalities and affect farm production significantly (Pankhurst and Munday

2011; Jahan et al. 2019). High rainfall resulted in a rapid drop in salinity to levels that were lethal for *P. japonicus,* causing mass mortality of the farm crop (Preston et al. 2001). Aquaculture needs to consider the possible impact of future climate change on aquaculture, including aquatic animal diseases, and strategies to mitigate climate change impacts on aquaculture.

4 Climate Change and Aquatic Animal Health in Aquaculture Systems

Disease is a greater concern for aquaculture than it is for capture fisheries because diseases can proliferate when aquatic animals are reared nearby at high densities (Walker and Mohan 2009). The increased occurrence of diseases in aquaculture production systems will lead to reduced profits, affecting the social and economic sustainability of aquaculture production. Compared to temperate regions, aquaculture in the topics has the potential for greater economic loss due to climate change-mediated disease mortality (De Silva and Soto 2009; Mennerat et al. 2010; Leung and Bates 2013). There is concern that climate change may further increase the risk to aquaculture posed by diseases through alterations in the distribution, prevalence and virulence of pathogens (bacteria, viruses, fungi and parasites), affecting the health of farmed fish and shellfish (Halvell et al. 1999; Burge and Hershberger 2020). The precise impacts of global climate change on infectious aquatic animal diseases are hypothetical and uncertain. It is important to understand the disease risks of cultured animals in the future in the face of expected climate change events for the success and sustainability of future aquaculture.

4.1 Stress to Farmed Aquatic Animals Due to Pond Environmental Parameters

The variations in environmental parameters, in particular, temperature fluctuations, salinity changes, low pH and low dissolved oxygen, potentially stress aquaculture species and suppress their immune system, increasing the incidence of disease outbreaks and rates of pathogen transmission (Goggin and Lester 1995; Harvell et al. 2002; Vilchis et al. 2005). In aquaculture systems, increased water temperature influences other water quality parameters such as reduced oxygen dissolution rate, acidification, salinity changes, mineralisation and accumulation of toxic metabolites, higher pathogenic load, etc. Oxygen stress due to higher water temperatures can lead to fish suffocation and negatively impact the health of the animal (Chapman and McKenzie 2009). Thermal stratification can occur in deeper ponds due to unequal distribution of temperature, with a higher temperature near the surface layer and decreasing temperature with depth, resulting in a reduced heat budget and degradation of water quality with the formation of methane, hydrogen sulphide and ammonia. Increased temperatures will affect the physical, chemical and biological properties of freshwater lakes and rivers, with predominantly adverse impacts on water quality and many freshwater species. These problems are severe when climate change-linked environmental extremes not only stress the host but also favour the pathogen. Many anthropogenic stressors, such as nutrient and contaminant pollution, can also contribute to disease processes, which may intensify with seasonal variations and extreme weather events. A better understanding of the environmental influences on pathogen and host biology and the expression of all of these diseases is needed.

4.2 Harmful Algal Blooms

Temperature changes would change the plankton community structure. Dinoflagellates advanced their seasonal peak in response to warming, while diatoms showed no consistent pattern of change (Edwards and Richardson 2004). Climate change can promote the growth of harmful algae in aquaculture systems, leading to toxin production that can harm both aquatic animals and human consumers (Wasmund et al. 1998; Edwards and Richardson 2004; Edwards et al. 2006; Moore et al. 2008; De Silva and Soto 2009; Haberkorn et al. 2010; Basti et al. 2011; Anderson et al. 2012; Trainer et al. 2019). Farrell et al. (2015) reported that temperature variation can affect the metabolism of the most widespread harmful algae. Flagellates and dinoflagellates, taxonomic groups and other harmful species have been reported to contain potentially toxic or nuisance species that can be responsible for stress or kills in finfish and shellfish (Hinder et al. 2012; Gubbins et al. 2013; Basti et al. 2019). Elevated nutrient levels resulting from altered rainfall patterns can stimulate algal blooms containing toxins that accumulate in the oysters, posing a threat to public health (Nell 1993).

4.3 Effect of Temperature Fluctuations and Heatwave Stress on Physiology and Immunity of Farmed Aquatic Animals

Diseases in aquaculture, such as bacterial, parasitic, viral and fungal diseases, are likely to be affected by a changing temperature regime (Collins et al. 2020). Any increase and/or decrease in temperature beyond their tolerance limits (temperature stress) will have a significant influence on aquaculture animal species body metabolism, increased stress-related behaviour, leading to higher energy demands (Rodnick et al. 2004), causing physiological and health problems (Beitinger et al. 2000).

4.3.1 Temperature Fluctuations

Rising temperature is likely to accelerate the replication rate, virulence, life cycle longevity and transmission of pathogens among several finfish and shellfish species (Marcogliese 2008; Chiaramonte et al. 2016). Most tropical aquatic species have an optimal temperature range between 20 and 30 °C (Thuy et al. 2017). In aquaculture farms, sub-lethal temperature enhances metabolic activity, oxygen demand, ammonia excretion, etc., and increases the chances of crop collapse (Hewitt and Duncan

2001). Warm water disease outbreaks are predicted to occur more frequently when cultured species are exposed to thermal stress conditions. Conversely, cold water diseases, such as vibriosis and winter ulcer that affect Atlantic salmon, may gradually become extinct due to the emerging unfavourable conditions (Sae-Lim et al. 2017). Out of the 25 fish disease cases reported from Rahara and Kalyani farm, East Kolkata Wetlands and farmer's field of West Bengal from June 2017 to February 2019, 17 types of diseases occurred above 30 °C. Ulcer, haemorrhages, myxoboliasis, lymphocystis and deformities were found more at a temperature above 30 °C, and columnaris and protrusion of the eye were recorded in many cases below 30 °C (Anonymous 2020).

Rising water temperatures due to climate change can lead to heat stress in aquaculture species. This can weaken the immune system, making fish more vulnerable to pathogens, reduce growth rates, decrease reproductive success, increase susceptibility to diseases and even lead to mortality (Iwama et al. 1998; Barton 2002). At higher temperatures, reduced dissolved oxygen availability causes stress to the fish due to inefficient aerobic metabolism, thus aggravating oxidative stress and cellular damage (Hofmann and Somero 1995; Portner and Knust 2007; Madeira et al. 2018). This induces the aquatic animal to produce steroidal hormones (catecholamine and corticosteroids) as the first response, followed by a hormonal response, and these changes influence growth, immune response and reproductive performance (Islam et al. 2022). Thermal stress caused an increase in oxidative stress parameters, viz., oxygen generation capacity, lipid peroxidation and malondialdehyde and unstable antioxidant activities, viz., superoxide dismutase, catalase and glutathione peroxidase in *Penaeus vannamei* (Duan et al. 2018). These changes damage the intestinal mucosal membrane, making the animal vulnerable to infections with compromised immune responses. The yellowleg shrimp *Penaeus californiensis* registered a significant reduction in immune parameters in response to elevated temperature (Vargas-Albores et al. 1998), making the shrimp susceptible to disease. A similar observation was made in freshwater crayfish, *Pacifastacus leniusculus* and *Astacus astacus,* where the host experienced diminishing immune response with increased temperature from 4 to 22 °C that made them more vulnerable to white spot syndrome virus (WSSV) (Jiravanichpaisal et al. 2004). Water temperature fluctuation during shrimp farming has been a major concern for the WSSV outbreak (Esparza-Leal et al. 2010; Dayalan et al. 2024).

The blood parameters such as haemoglobin (Hb), red blood corpuscles (RBC) and packed cell volume (PCV) are considered important indicators to assess the reliable physiological status of fish under different temperatures. Fish reared at lower temperatures had a lower Hb concentration than the fish reared at elevated higher water temperatures. Chronic temperature stress may affect the neuroendocrine and osmoregulatory systems, altering cardiorespiratory performance and aerobic scope, as well as immune responses of several economically important species (Gazeau et al. 2014; Paukert et al. 2016; Stévant et al. 2017; Stewart et al. 2019). The metabolism and physiology, as well as feeding behaviour and growth performance of most finfish and shellfish species, are likely to be affected by chronic temperature stress (Marcogliese 2008; Akegbejo-Samsons 2009; Lemasson 2018).

As water temperatures change, some aquaculture species may be forced to migrate to different regions, potentially exposing them to new diseases and environmental conditions (Portner and Peck 2010). The increasing temperature pressures may promote the emergence of epizootic diseases in aquaculture that limit the success of aquaculture production systems in many countries of the world (Maulu et al. 2019). Alteration of rearing temperature to either lower or higher to optimum range causes stress for the fishes, and their adaptation to the new temperature regimen causes immune suppression. Shifting of Mozambique tilapia *Oreochromis mossambicus* from an optimum temperature of 27–35 °C significantly reduced innate immune response (Ndong et al. 2007). Teleosts reared within the range of optimal temperature showed a better immune response at higher temperature than at suboptimal temperature. Exposure of golden mahseer (*Tor putitora*) broodstock to an elevated temperature of about 2 °C above ambient temperature (21 °C) for a prolonged period caused reduced immunoglobulin level, lysozyme and myeloperoxidase activities (Akhtar et al. 2020).

4.3.2 Heatwave Stress

A heatwave condition causes excessive energy demand for maintaining cellular integrity and the need for higher oxygen levels. The heatwave stress response is more species-specific. Shin et al. (2018) reported that induced acute thermal stress (10 °C incremental temperature at 2 °C/h) reduced expression of superoxide dismutase (SOD), heat shock protein 70 (HSP 70), and HSP 90 making spotted sea bass (*Lateolabrax maculatus*) more vulnerable to diseases. Brook trout (*Salvelinus fontinalis*) recorded an enhanced stress response with increased temperature from 16 to 24 °C where HSP 70 expression enhanced 56 fold, plasma cortisol level increased by 18 times while daily temperature fluctuation of 8 °C caused enhancement of HSP 70 response by 400 fold suggesting a heat-stress at hyper-optimal temperature (Chadwick and McCormick 2017). Lin et al. (2011) recorded the role of heat shock protein 70 and NAD-dependent aldehyde dehydrogenase (ALDH), having a major role in controlling WSSV infection at higher temperatures, whereas the gene-knockdown shrimps succumbed to the disease. Contrarily, no significant physiological response was observed by camel shrimp (*Rhynchocinetes durbanensis*) and hermit crab (*Calcinus laevimanus*) exposed to heatwave conditions (Madeira et al. 2018).

Heatwave conditions not only influence the physiological condition of the fish but also the inhabiting microbes in the system. The heatwave causes significant variations in microbial growth rate, diversity and metabolite production, and disease outbreaks could be a serious concern for sustainable crop production in the event of climate change. Microbial multiplication rate is highly correlated with temperature, and it increases the growth rate of the microbes (Ratkowsky et al. 1982). Mass mortality of *Crassostrea gigas* was noticed when the rearing temperature was increased from 20 to 25 °C, whereas the mortality was limited under antibiotic administration, confirming bacteria as the causative agent for the mortality. A drastic increase in the load of the causative agents *Vibrio harveyi* and *Vibrio fortis* was noticed with altered temperatures (Green et al. 2019). Elevation of temperature on attenuation of

Aeromonas salmonicida and loss of its pathogenicity was observed by Ishiguro et al. (1981). The study observed the adaptability of the bacteria to higher temperatures, 30 °C, by losing an additional cell-wall layer. Marine heatwaves have been causing disease outbreaks in several cultured animals, such as red gorgonian *Paramuricea clavata* due to *Vibrio coralliilyticus* (Bally and Garrabou 2007), marine gastropods by *Vibrio harveyi* (Fukui et al. 2010). A significant shift in microbial diversity was observed as a result of the marine heatwave, followed by bacterial infection in the aquatic animals (Rubio-Portillo et al. 2021). A better understanding of the adaptability of cultured aquatic animals to heat wave stress is essential for better farming success.

4.4 Effect of Changing Precipitation Patterns and Flood on Cultured Aquatic Animal Health

Changes in precipitation patterns can lead to variations in salinity in coastal aquaculture systems, affecting the osmoregulation, respiratory physiology and health of aquatic organisms (Nilsson et al. 2012; Duarte et al. 2013), and animals need to adjust their osmoregulatory processes to maintain internal balance (Evans and Claiborne 2006). High salinity levels (hyperosmotic conditions) can cause water loss and dehydration in aquatic animals, leading to reduced growth, impaired organ function and stress-related health problems (Wright and Wood 2009). Randall and Brauner (1991) reported excessive water influx, potentially causing cell swelling, disrupted ion balance and damage to internal organs due to low salinity levels (hypoosmotic conditions). Animals experiencing salinity stress are often more susceptible to diseases and infections, as their weakened physiological condition compromises their immune responses (Bao et al. 2015; stripped catfish - Jahan et al. 2019; juvenile clams - Baker et al. 2005, 2007, 2013; oysters - Rodrick 2008; abalone - Morash and Alter 2016). In Virginia, the intensification of dermo disease in molluscs began in the 1980s with the onset of serious drought, which reduced stream flows and elevated bay salinities that caused the increase in *P. marinus* activity (Vogelbein et al. 2009). However, climate change may also have ameliorative effects on molluscan disease. Some climate models predict an increased intensity of summer storms, including tropical storms and hurricanes. The low salinity levels caused by these storms may lessen impacts from dermo disease and from MSX disease caused by *Haplo-sporidium nelson*, a mortal oyster pathogen intolerant of even short pulses of salinity less than 10. Further, the hard clam parasite, QPX, is already at its southern distributional limit in Virginia and may disappear with a warming climate.

Disease outbreak due to WSSV in shrimp farming is a common observation with the heavy downpours due to cyclonic storms and rain after a heatwave. Extremely heavy rainfall and floods significantly influence the physiological response of aquatic animals and increase their susceptibility to pathogenic organisms. The introduction of diseases or predators into aquaculture facilities, along with the flooded water, resulted in crop losses. The globalisation and increasing volume of international trade in live aquatic animals and their products created new pathways for the spread of transboundary aquatic animal pathogens (Rodgers et al. 2011). EUS since 2007 in Botswana, Namibia and Zambia, and WSD since 2011 in Mozambique are clear examples of how diseases can jump national and international borders and bring about devastating impacts on communities. EUS occurs in natural freshwater systems, and flooding can distribute the disease over thousands of kilometres.

5 Effect of Climate Change on Host–Pathogen Interaction

The influence of enhanced temperature as a result of climate change may have both detrimental and beneficial impacts depending on the geographical region and host–pathogen interaction. The alterations in the development, transmission, survival and virulence of pathogens can also be influenced by climate change. It is well known that environmental temperature plays a crucial role in the infectivity of pathogens that may be a virus, bacteria or parasites (Marcogliese 2008; Hata et al. 2010; Karvonen et al. 2010; Vezzulli et al. 2013) as well as the immune response of the host to the altered water temperature (Collet 2014). Increased temperature due to climate change could not only influence the immune response of the host but also affect the pathogenicity of infective micro- or macro-organisms. It is common to observe seasonal influence on disease outbreaks due to viral/bacterial/parasitic infections and the role of environmental parameters, predominantly temperature. The reason for the variability of the disease outbreak to environmental parameters is well corroborated with the infectivity of the pathogen and the immune response of the host (Bowden et al. 2007; Altizer et al. 2013). As an impact of climate change, tropical regions may record reduced parasitic load in the aquaculture farms in comparison to an increase in the temperate farms. Different pathogens have their range of temperature for pathogenicity or infectivity (Table 1). The benefit of increased temperature is not only resting the pathogens but also supporting hosts, depending on the host-pathogen association.

6 Effect of Climate Change on Vectors and Intermediate Hosts

The impact of climate change on the vectors and intermediary hosts, and the invasive nature of the vectors as a result of changed environmental parameters, could have a larger impact on the disease outbreaks in the aquaculture. Vectors play a major role in the retention and spread of pathogens in the wild as well as in the culture farms. Climate change-associated variations in abiotic parameters have a significant role in altered biodiversity in aquatic systems as well as the diversity of pathogen-carrying vectors. This could have a significant impact on disease outbreaks due to the invasion and spread of vector-borne diseases. Parasitic infections are much exposed to the external environment and the ones having at least one

Table 1 Pathogenicity of a few pathogens at different temperatures

S. no.	Pathogen	Temperature range	Pathogenicity/infectivity
1	Betanodavirus (genotype specific)	15–30 °C	Pathogenic fish virus infecting more than 100 fish species in marine and freshwater regions (Bandín and Souto 2020).
	(i) Barfin flounder nervous necrosis virus (BFNNV)	15–20 °C	–
	(ii) Tiger puffer nervous necrosis virus (TPNNV)	20 °C	–
	(iii) Striped jack nervous necrosis virus (SJNNV)	20–25 °C	European sea bass mortality due to SJNNV did not increase beyond 20 °C (Vendramin et al. 2014)
	(iv) Red-spotted grouper nervous necrosis virus (RGNNV)	25–30 °C	Found in Indian waters infecting several marine and brackishwater fish species (Shetty 2012; Banu et al. 2019) and reported to cause mass mortalities at a temperature below 30 °C (Sharma et al. 2019). Mortality of European sea bass was low at 20 °C (3.7%) but increased to 30% at higher temperatures of 25–30 °C (Toffan et al. 2016). RNA1 of RGNNV was responsible for its high temperature tolerance and variation in in mortality of the host at varying water temperature (Hata et al. 2010; Panzarin et al. 2014). Asian seabass mortality below 30 °C, coinciding with the winter season (Iwamoto et al. 1999) .
2	Haemorrhagic septicaemia virus (VHSV)	14 °C	High mortality at a lower temperature of 14 °C compared to >20 °C despite the virus being able to propagate at the higher temperature (Isshiki et al. 2002; Sano et al. 2009).
3	Red sea bream iridovirus (RSIV)	26–28 °C	Fish virus infecting over 30 fish species worldwide, (OIE 2021). Rock bream infected with RSIV and reared at 11 and 16 °C showed no mortality whereas the ones reared post-infection at 26 or 28 °C experienced high mortality. An increase in rearing water temperature from 18 to 26 °C caused 70% mortality of the fish whereas retention of the temperature triggered no mortality (Choi et al. 2006a, b; Oh et al. 2016; Jung et al. 2017).

(continued)

Table 1 (continued)

S. no.	Pathogen	Temperature range	Pathogenicity/infectivity
4	White spot syndrome virus (WSSV)	<33 °C	WSSV retained infectivity and caused mortality when the shrimp challenged at 27 °C for 24 h and shifted to 33 °C whereas shrimps did not experience mortality challenged at 27 °C for less than 24 h before shifting to 33 °C (Rahman et al. 2006, 2007). WSSV outbreak in *P.indicus* farms in Kingdom of Saudi Arabia due to rapid reduction in water temperature as a result of post-winter rains Kitto (2013). White spot disease in *P. vannamei* due to the virus prominence when the temperature is shifted from 33 to 27 °C after brief period of exposure (Rahman et al. 2007).
5	*Argulus* spp or *Lernaea* spp	<30 °C	Parasitic infestation, caused by *Argulus* spp or *Lernaea* spp, reaches peak in the winter season when temperature drops below 30 °C as the reproduction capacity of these parasites significantly diminishes at the higher temperature in the Indian subcontinent (Sahoo et al. 2013; Hossain et al. 2018). Infest on the host in the winter in the tropical climate while in the summer in temperate climate (Hossain et al. 2018; Sánchez-Hernández 2017).
6	*Ichthyophthirius multifiliis*, and *Flavobacterium columnare*	–	Increased incidences of diseases over the period to the increased rearing water temperature (Karvonen et al. 2010).
7	*Ichthyobodo necator* or *Chilodonella* spp	–	Reduced disease incidences at increased rearing water temperature (Karvonen et al. 2010).
8	Rhabdoviruses	–	Tend to recover from infection at higher temperature than at a lower temperature (Purcell et al. 2012).
9	Viral haemorrhagic septicaemia virus (VHSV)	<15 °C	Olive flounder (*Paralichthys olivaceus*) reported to be highly susceptible to at a lower temperature of 15 °C than at 20 °C within the optimum range of the fish owing to the quicker immune response by the host (Sano et al. 2009; Avunje et al. 2013).
10	*Vibrio vulnificus*	–	In oysters, it becomes more abundant as water temperature increases (Lipp and Rose 1997).
11	Sea lice	–	A warm water pathogen remains as a challenge in salmon culture and further warming is likely to worsen the infections in cold temperate conditions (Collins et al. 2020).

intermediate host. For example, trematodes have multiple hosts in their life cycle comprising first invertebrate mollusc, an invertebrate or vertebrate second intermediate host and final vertebrate host where it matures and reproduces (Herrmann and Poulin 2011; Lohmus and Bjorklund 2015). Increased water temperature has caused progenesis of the trematode worms at the intermediate host, truncating its life cycle and increasing the level of infestation (Herrmann and Poulin 2011). Multiple non-susceptible hosts in the aquatic system act as a natural reservoir for the spread of diseases (Small and Pagenkopp 2011).

7 Mitigation and Adaptation Measures

The 2015 Paris Climate Agreement recognises the need for effective and progressive responses to the urgent threat of climate change through mitigation and adaptation measures while taking into account the particular vulnerabilities of food production systems. Intensified adaptation efforts are needed despite the Paris Agreement, arrived at CoP 21 to limit warming to 1.5 °C (Runhaar et al. 2018). However, considering climate change variability across the regions, it is difficult to develop strategies for mitigation and adaptation (Ramarao et al. 2019). The strategies to improve environmental and animal health to prevent the outbreak of the disease in the scenario of climate change are:

- Selection of suitable culture species that is more compatible with varied environmental conditions and less amenable to infectious diseases, capable of thriving at lower dissolved oxygen or load of higher metabolites (ammonia, nitrite, etc.).
- *Ex: Diversification of species to euryhaline species such as Asian seabass, mullets, tilapia, milkfish,* etc., *that can tolerate a wider range of salinities can perform better than fishes such as carps or freshwater catfishes.*
- Identification of climate-resilient candidate species for aquaculture may reduce disease incidences
- *Ex: The use of brackishwater species instead of carps in seawater intruded areas of freshwater farm sites could reduce mortalities due to osmotic stress and disease outbreaks* (Dubey et al. 2017)*; Salinity tolerant striped catfish in the lower reaches of the Mekong to address the issues of seawater intrusion leading to salinity increases.*
- Establishment of a breeding programme for salinity and thermo-tolerance strains
- Moving the fish cages to cooler waters or lowering them to deeper waters when warm weather events are forecast.
- Identify climate-sensitive aquatic pathogens and outline the main climate change-related factors that are likely to cause changes in their distribution, prevalence and virulence of aquatic pathogens in climate-sensitive geographies.
- Develop and implement better management practices (BMPs) to address aquatic animal health risks specific to farming systems that are impacted by climate change (De Silva and Soto 2009; Galappaththi et al. 2020).

- *Ex: Monitoring water temperatures, providing proper management, and considering strategies such as water cooling systems, shade and selecting temperature-tolerant species to mitigate the negative impacts of temperature stress on animal health in aquaculture systems.*
- Science-based import risk analysis and development of biosecurity governance mechanisms in exporting and importing countries to manage transboundary diseases
- Enabling policy and investment support to implement aquatic animal disease management and biosecurity, to prevent future disease losses associated with climate change.
- Application of cutting-edge technologies, such as sensors, artificial intelligence for modelling and climate forecasting, and the adoption of models for predicting disease forecasting (Gomez-Zavaglia et al. 2020; Mustapha et al. 2021; Hu et al. 2022; Rather et al. 2024).
- A suitable legal approach to support better and effective implementation of the strategies to prevent production loss due to drastic changes in the environmental parameters and disease outbreaks.
- The preparedness for the impact of climate change through early warning systems regarding extreme weather events, financial support like insurance or credit facilities, conducting awareness programmes, etc.

8 Conclusion

It is important to examine the distribution of major aquatic animal pathogens in different climatic regions and to provide a few insights into possible long-term implications of diseases in aquaculture. There is a paucity of information as to how aquatic animal disease outbreak dynamics are mediated by climate-driven changes and what impact this will have on the future of aquaculture growth in the world. It is essential to examine how aquatic animal pathogens will respond to climate change and, in the changed scenario, how aquatic animal pathogens will interact with their farmed hosts. Based on the principles of the disease development process and the role of stress in disease development, it is possible to make some predictions on the likely impact of climate change on diseases in aquaculture. The effects of global climate change on infectious diseases of aquatic animals are hypothetical until more is known about the degree of change in temperature and other environmental factors. There is always a tendency to oversimplify the underlying mechanisms by which climate change influences disease emergence, transmission, and spread. Predictive climate change modelling (Muralidhar et al. 2023) will help to identify the most vulnerable aquaculture sectors based on geography and species adaptive range and disease risks to these sectors. The potential for increased fish escapes to pose a disease risk to wild fish and shellfish populations needs to be recognised, studied and managed.

References

Akegbejo SY (2009) Climate change impacts on fisheries production in land-water interface. IOP Conf Ser Earth Environ Sci 6(29):292022. IOP Publishing

Akhtar MS, Rajesh M, Kamalam BS, Ciji A (2020) Effect of photoperiod and temperature on indicators of immunity and wellbeing of endangered golden mahseer (*Tor putitora*) broodstock. J Therm Biol 93:102694

Altizer S, Ostfeld RS, Johnson PT, Kutz S, Harvell CD (2013) Climate change and infectious diseases: from evidence to a predictive framework. Science 341(6145):514–519

Anderson DM, Cembella AD, Hallegraeff GM (2012) Progress in understanding harmful algal blooms: paradigm shifts and new technologies for research, monitoring, and management. Annu Rev Mar Sci 4(1):143–176

Anonymous (2020). Annual Report 2020. ICAR-Central Institute of Freshwater Aquaculture, Bhubaneswar, India, pp 204

Avunje S, Oh MJ, Jung SJ (2013) Impaired TLR2 and TLR7 response in olive flounder infected with viral haemorrhagic septicaemia virus at host susceptible 15°C but high at non-susceptible 20°C. Fish Shellfish Immunol 34(5):1236–1243

Baker SM, Baker P, Heuberger D, Sturmer L (2005) Short-term effects of rapid salinity reduction on seed clams (*Mercenaria mercenaria*). J Shellfish Res 24(1):29–33

Baker S, Hoover E, Sturmer L (2007) The role of salinity in hard clam aquaculture: CIR1500/FA128, 1/2007. Edis, 2007(3)

Baker S, Hoover E, Sturmer L (2013) The role of salinity in hard clam aquaculture. University of Florida, Gainesville

Bally M, Garrabou J (2007) Thermodependent bacterial pathogens and mass mortalities in temperature benthic communities: a new case of emerging disease linked to climate change. Glob Change Biol 13:2078–2088

Bandín I, Souto S (2020) Betanodavirus and VER disease: a 30-year research review. Pathogens 9(2):106

Banu H, Pattanayak S, Sundaray JK, Sahoo PK (2019) Genetic diversity and latency status of betanodavirus in wild seeds of Asian seabass Lates calcarifer (Bloch) sampled along Indian coasts. Indian J Geo-Marine Sci 48(3):288–293

Bao Y, Wang R, Yang S, Li Y (2015) Effects of salinity on the immune response of the Pacific white shrimp *Litopenaeus vannamei*. Fish Shellfish Immunol 44(2):180–186

Barange M, Bahri T, Beveridge MC, Cochrane KL, Funge-Smith S, Poulain F (2018) Impacts of climate change on fisheries and aquaculture. United Nations' Food and Agriculture Organization 12(4):628–635

Barton BA (2002) Stress in fishes: a diversity of responses with particular reference to changes in circulating corticosteroids. Integr Comp Biol 42(3):517–525

Basti L, Endo M, Segawa S (2011) Physiological, pathological, and defense alterations in Manila clams (short-neck clams), *Ruditapes philippinarum*, induced by *Heterocapsa circularisquama*. J Shellfish Res 30(3):829–844

Basti L, Nagai K, Segawa S, Tanaka Y, Suzuki T, Nagai S (2019) Harmful algal blooms and shellfish aquaculture in changing environment. Bull Jpn Fish Res Educ Agen 49:73–79

Beach RH, Viator CL (2008) The economics of aquaculture insurance: an overview of the US pilot insurance program for cultivated clams. Aquac Econ Manag 12(1):25–38

Beitinger TL, Bennett WA (2000) Quantification of the role of acclimation temperature in temperature tolerance of fishes. Environ Biol Fish 58:277–288

Bell JD, Ganachaud A, Gehrke PC, Griffiths SP, Hobday AJ, Hoegh-Guldberg O, Johnson JE, Le Borgne R, Lehodey P, Lough JM, Matear RJ (2013) Mixed responses of tropical Pacific fisheries and aquaculture to climate change. Nat Clim Chang 3(6):591–599

Blanchard JL, Watson RA, Fulton EA, Cottrell RS, Nash KL, Bryndum BA, Büchner M, Carozza DA, Cheung WW, Elliott J, Davidson LN (2017) Linked sustainability challenges and trade-offs among fisheries, aquaculture and agriculture. Nat Ecol Evol 1(9):1240–1249

Bowden TJ, Thompson KD, Morgan AL, Gratacap RML, Nikoskelainen S (2007) Seasonal variation and the immune response: a fish perspective. Fish Shellfish Immunol 22:695–706

Brander KM (2007) Global fish production and climate change. Proc Natl Acad Sci 104(50):19709–19714

Bueno PB, Soto D (2017) Adaptation strategies of the aquaculture sector to the impacts of climate change 12, 99

Burge CA, Hershberger PK (2020) Climate change can drive marine diseases 2020: 83–94. Oxford University Press, Oxford

Chadwick JG, McCormick SD (2017) Upper thermal limits of growth in brook trout and their relationship to stress physiology. J Exp Biol 220(21):3976–3987

Chapman LJ, Mckenzie DJ (2009) Behavioral responses and ecological consequences. Fish Physiol 27:25–77

Chaturvedi RK, Jaideep J, Mathangi J, Bala G, Ravindranath NH (2012a) Multi-model climate change projections for India under representative concentration pathways. Curr Sci 103(7):1–12

Chaturvedi RK, Joshi J, Jayaraman M, Bala G, Ravindranath NH (2012b) Multi-model climate change projections for India under representative concentration pathways. Curr Sci 103:791–802

Chiaramonte L, Munson D, Trushenski J (2016) Climate change and considerations for fish health and fish health professionals. Fisheries 41(7):396–399

Choi HG, Kim YS, Kim JH, Lee SJ, Park EJ, Ryu J, Nam KW (2006a) Effects of temperature and salinity on the growth of *Gracilaria verrucosa* and *G. chorda*, with the potential for mariculture in Korea. J Appl Phycol 18:269–277

Choi HJ, Kim JW, Kim JH (2006b) Effects of elevated water temperature on the mortality and physiological responses of fish. Aquac Res 37(12):1030–1038

Cochrane K, De Young C, Soto D, Bahri T (2009) Climate change implications for fisheries and aquaculture. FAO Fisheries and aquaculture technical paper 530:212

Collet L, Ruelland D, Borrell-Estupina V, Servat E (2014) Assessing the long-term impact of climatic variability and human activities on the water resources of a meso-scale Mediterranean catchment. Hydrol Sci J 59(8):1457–1469

Collins C, Bresnan E, Brown L, Falconer L., Guilder J, Jones L, Kennerley A, Malham S, Murray A, Stanley M. (2020) Impacts of climate change on aquaculture. MCCIP Science Review 2020, 482–520.

Cruz RV, Harasawa H, Lal M, Wu S, Anokhin Y, Punsalmaa B, Honda Y, Jafari M, Li C, Huu N (2007) Asia climate change, 2007, impacts, adaptation and vulnerability. In: Parry ML, Canziani OF, Palutikof JP et al (eds) Contribution of working group II to the fourth assessment report of the intergovernmental panel on climate change. Cambridge University Press, Cambridge, pp 469–506

Dabbadie L, Aguilar-Manjarrez J, Beveridge MC, Bueno PB, Ross LG, Soto D (2019) Effects of climate change on aquaculture: drivers, impacts and policies. In: Impacts of climate change on fisheries and aquaculture 8449

Dayalan V, Kasivelu G, Raguraman V, Sharma AS (2024) Studies on temperature impact (sudden and gradual) of the white-leg shrimp Litopenaeus vannamei. Environ Sci Pollut Res 31, 38743–38750.

De Silva SS, Soto D (2009) Climate change and aquaculture: potential impacts, adaptation and mitigation. Climate change implications for fisheries and aquaculture: overview of current scientific knowledge. FAO Fisheries and Aquaculture Technical Paper 530:151–212

Duan Y, Wang Y, Zhang J, Xiong D (2018) Elevated temperature disrupts the mucosal structure and induces an immune response in the intestine of whiteleg shrimp *Litopenaeus vannamei* (Boone, 1931) (Decapoda: *Dendrobranchiata: Penaeidae*). J Crustac Biol 38(5):635–640

Duarte CM, Losada IJ, Hendriks IE, Mazarrasa I, Marbà N (2013) The role of coastal plant communities for climate change mitigation and adaptation. Nat Clim Chang 3(11):961–968

Dubey SK, Trivedi RK, Chand BK, Mandal B, Rout SK (2017) Farmers' perceptions of climate change, impacts on freshwater aquaculture and adaptation strategies in climatic change hotspots: a case of the Indian Sundarban delta. Environ Dev 21:38–51

Edwards M, Richardson AJ (2004) Impact of climate change on marine pelagic phenology and trophic mismatch. Nature 430(7002):881–884

Edwards M, Johns DG, Leterme SC, Svendsen E, Richardson AJ (2006) Regional climate change and harmful algal blooms in the northeast Atlantic. Limnol Oceanogr 51(2):820–829

Esparza LHM, Magallón B, Portillo CFJ, Perez GR, Alvarez RP, Escobedo BCM, Casillas HR (2010) Infection of WSSV-negative shrimp, *Litopenaeus vannamei*, cultivated under fluctuating temperature conditions. J World Aquacult Soc 41(6):912–922

Evans DH, Claiborne JB (2006) Osmotic and ionic regulation: cells and animals. CRC Press

FAO (2024) In brief to the state of world fisheries and aquaculture 2024. Blue Transformation in action, Rome

Farrell H, Seebacher F, O'Connor W, Zammit A, Harwood DT, Murray S (2015) Warm temperature acclimation impacts metabolism of paralytic shellfish toxins from *Alexandrium minutum* in commercial oysters. Glob Chang Biol 821(9):3402–3413

Ficke AD, Myrick CA, Hansen LJ (2007) Potential impacts of global climate change on freshwater fisheries. Rev Fish Biol Fish 17:581–613

Fleming A, Hobday AJ, Farmery A, van Putten EI, Pecl GT, Green BS, Lim-Camacho L (2014) Climate change risks and adaptation options across Australian seafood supply chains–a preliminary assessment. Clim Risk Manag 1:39–50

Fukui Y, Saitoh SI, Sawabe T (2010) Environmental determinants correlated to *Vibrio harveyi*-mediated death of marine gastropods. Environ Microbiol 12(1):124–133

Galappaththi EK, Ichien ST, Hyman AA, Aubrac CJ, Ford JD (2020) Climate change adaptation in aquaculture. Rev Aquac 12(4):2160–2176

Gazeau F, Alliouane S, Bock C, Bramanti L, López Correa M, Gentile M, Hirse T, Pörtner HO, Ziveri P (2014) Impact of ocean acidification and warming on the Mediterranean mussel (*Mytilus galloprovincialis*). Front Mar Sci 1:862

Goggin CL, Lester RJG (1995) Perkinsus, a protistan parasite of abalone in Australia: a review. Mar Freshw Res 46(3):639–646

Gomez-Zavaglia A, Mejuto JC, Simal-Gandara J (2020) Mitigation of emerging implications of climate change on food production systems. Food Res Int 134:109256

Green TJ, Siboni N, King WL, Labbate M, Seymour JR, Raftos D (2019) Simulated marine heat wave alters abundance and structure of Vibrio populations associated with the Pacific Oyster resulting in a mass mortality event. Microb Ecol 77:736–747

Gubbins M, Bricknell I, Service, M (2013) Impacts of climate change on aquaculture. MCCIP Sci Rev:318–327

Haberkorn H, Lambert C, Le Goïc N, Guéguen M, Moal J, Palacios E, Lassus P, Soudant P (2010) Effects of *Alexandrium minutum* exposure upon physiological and hematological variables of diploid and triploid oysters, Crassostrea gigas. Aquat Toxicol 97(2):96–108

Hamdan R, Kari F, Othman A, Samsi SM (2012) Climate change, socio-economic and production linkages in East Malaysia aquaculture sector. In: 2012 international conference on future environment and energy IPCBEE, vol 28

Hamdan R, Othman A, Kari F (2015) Climate change effects on aquaculture production performance in Malaysia: an environmental performance analysis. Int J Busin Soc 16(3):364

Handisyde NT, Ross LG, Badjeck MC, Allison EH (2006) The effects of climate change on world aquaculture: a global perspective. Aquaculture and Fish Genetics Research Programme, Stirling Institute of Aquaculture. Final Technical Report, DFID, Stirling. 151

Harvell CD, Kim K, Burkholder JM, Colwell RR, Epstein PR, Grimes DJ, Hofmann EE, Lipp EK, Osterhaus ADME, Overstreet RM, Porter JW, Smith GW, Vastra GR (1999) Emerging marine diseases--climate links and anthropogenic factors. Science 285(5433):1505–1510

Harvell CD, Mitchell CE, Ward JR, Altizer S, Dobson AP, Ostfeld RS, Samuel MD (2002) Climate warming and disease risks for terrestrial and marine biota. Science 296(5576):2158–2162

Hata N, Okinaka Y, Iwamoto T, Kawato Y, Mori KI, Nakai T (2010) Identification of RNA regions that determine temperature sensitivities in betanoda viruses. Arch Virol 155:1597–1606

Herrmann KK, Poulin R (2011) Life cycle truncation in a trematode: does higher temperature indicate shorter host longevity? Int J Parasitol 41(6):697–704

Hewitt DR, Duncan PF (2001) Effect of high-water temperature on the survival, moulting and food consumption of *Penaeus* (*Marsupenaeus*) *japonicus* (Bate, 1888). Aquac Res 32(4):305–313

Hinder SL, Hays GC, Edwards M, Roberts EC, Walne AW, Gravenor MB (2012) Changes in marine dinoflagellate and diatom abundance under climate change. Nat Clim Chang 2(4):271–275

Hoang T, Ho HC, Le NPT, Bui THH (2020) Effects of high temperature on survival and feed consumption of banana shrimp *Penaeus merguiensis*. Aquaculture 522:735152

Hofmann GE, Somero GN (1995) Evidence for protein damage at environmental temperatures: seasonal changes in levels of ubiquitin conjugates and hsp70 in the intertidal mussel *Mytilus trossulus*. J Exp Biol 198(7):1509–1518

Hossain MMM, Ferdoushi J, Rupom AH (2018) Biology of anchor worms *(Lernaea cyprinacea)*. J Entomol Zool Stud 6(1):910–917

Hu WC, Chen LB, Huang BK, Lin HM (2022) A computer vision-based intelligent fish feeding system using deep learning techniques for aquaculture. IEEE Sensors J 22(7):7185–7194

IPCC (2013) The Physical Science Basis: Contribution of Working Group I to the Fifth Assessment Report of the Intergovernmental Panel on Climate Change, T.F, D Qin, G.K. Plattner, M. Tignor, S.K. Allen, J. Boschung, A. Nauels, Y. Xia, V. Bex and P.M. Midgley (Eds.), Cambridge University Press, Cambridge, United Kingdom and New York, NY, USA.

IPCC (2018) Global Warming of 1.5°C. An IPCC Special Report on the impacts of global warming of 1.5° C above pre-industrial levels and related global greenhouse gas emission pathways, in the context of strengthening the global response to the threat of climate change, sustainable development, and efforts to eradicate poverty

Ishiguro S, Nakamura K, Kondo M (1981) Effects of temperature elevation on the attenuation of *Aeromonas salmonicida* and the loss of its pathogenicity. J Fish Dis 4(3):223–230

Islam MJ, Kunzmann A, Slater MJ (2022) Responses of aquaculture fish to climate change-induced extreme temperatures: a review. J World Aquacult Soc 53(2):314–366

Isshiki T, Nagano T, Miyazaki T (2002) Effect of water temperature on pathological states of Japanese flounder experimentally infected with viral hemorrhagic septicemia virus, an flounder isolate KRRV-9601. Fish Pathol 37(2):95–97

Iwama GK, Thomas PT, Forsyth RB, Vijayan MM (1998) Heat shock protein expression in fish. Rev Fish Biol Fish 8(1):35–56

Iwamoto T, Mori KI, Arimoto M, Nakai T (1999) High permissively of the fish cell line SSN-1 for piscine nodaviruses. Dis Aquat Org 39(1):37–47

Jahan A, Nipa TT, Islam SM, Uddin MH, Islam MS, Shahjahan M (2019) Striped catfish (*Pangasianodon hypophthalmus*) could be suitable for coastal aquaculture. J Appl Ichthyol 35(4):994–1003

Jiravanichpaisal P, Soderhäll K, Soderhäll I (2004) Effect of water temperature on the immune response and infectivity pattern of white spot syndrome virus (WSSV) in freshwater crayfish. Fish Shellfish Immunol 17(3):265–275

Jung MH, Nikapitiya C, Vinay TN, Lee J, Jung SJ (2017) Rock bream iridovirus (RBIV) replication in rock bream (*Oplegnathus fasciatus)* exposed for different time periods to susceptible water temperatures. Fish Shellfish Immunol 70:731–735

Karvonen A, Rintamaki P, Jokela J, Valtonen ET (2010) Increasing water temperature and disease risks in aquatic systems: climate change increases the risk of some, but not all, diseases. Int J Parasitol 40:1483–1488

Kitto MR (2013) Can the Blue Revolution with Shrimp in Saudi Arabia cross the disease Chasm? World Aquacult:57

Kothyari UC, Singh VP (1996) Rainfall and temperature trends in India. Hydrol Process 10(3):357–372

Krishnan R, Sanjay J, Gnanaseelan C, Mujumdar M, Kulkarni A, Chakraborty S (2020) Assessment of climate change over the Indian region: a report of the ministry of earth sciences (MOES), government of India, 226

Lemasson AJ (2018) Ocean acidification and warming impacts on native and non-native shellfish: a multidisciplinary assessment (Doctoral dissertation, University of Plymouth)

Leung TL, Bates AE (2013) More rapid and severe disease outbreaks for aquaculture at the tropics: implications for food security. J Appl Ecol 50:215–222

Lin YR, Hung HC, Leu JH, Wang HC, Kou GH, Lo CF (2011) The role of aldehyde dehydrogenase and hsp-70 in suppression of white spot syndrome virus replication at high temperature. J Virol 85(7):3517–3525

Lipp EK, Rose JB (1997) The role of seafood in foodborne diseases in The United States of America. Rev Sci Tech Office International des Epizooties 16(2):620–640

Lõhmus M, Björklund M (2015) Climate change: what will it do to fish–parasite interactions? Biol J Linn Soc 116:397–411

Madeira C, Leal MC, Diniz MS, Cabral HN, Vinagre C (2018) Thermal stress and energy metabolism in two circumtropical decapod crustaceans: responses to acute temperature events. Mar Environ Res 141:148–158

Marcogliese DJ (2008) The impact of climate change on the parasites and infectious diseases of aquatic animals. Rev Sci Tech Off Int Epiz 27:467–484

Maulu S, Munganga BP, Hasimuna OJ, Haambiya LH, Seemani B (2019) A review of the science and technology developments in Zambia's aquaculture industry. J Aquac Res Dev 10(567):2

Mennerat A, Nilsen F, Ebert D, Skorping A (2010) Intensive farming: evolutionary implications for parasites and pathogens. Evol Biol 37:59–67

Merino G, Barange M, Blanchard JL, Harle J, Holmes R, Allen I, Allison EH, Badjeck MC, Dulvy NK, Holt J, Jennings S (2012) Can marine fisheries and aquaculture meet fish demand from a growing human population in a changing climate? Glob Environ Chang 22(4):795–806

Moore SK, Trainer VL, Mantua NJ, Parker MS, Laws EA, Backer LC, Fleming LE (2008) Impacts of climate variability and future climate change on harmful algal blooms and human health. Environ Health 7:1–12

Morash AJ, Alter K (2016) Effects of environmental and farm stress on abalone physiology: perspectives for abalone aquaculture in the face of global climate change. Rev Aquac 8(4):342–368

Muralidhar M, Kumaran M, Jayanthi M, Muniyandi B, Ponniah AG, Nagothu US, White P, Eknath A (2012) Case study on the impacts of climate change on shrimp farming and developing adaptation measures for small-scale shrimp farmers in Krishna District, Andhra Pradesh, India. Network of Aquaculture Centers in Asia-Pacific, 126

Muralidhar M, Kumaran M, Kumar JA, Dayal JS, Jayanthi M, Saraswathy R, Murugan P., (2013) Climate change and coastal aquaculture in West Godavari District, Andhra Pradesh: Impacts, vulnerability, adaptations and mitigations for resilience. Journal of Agrometeorology. 15:116–22.

Muralidhar M, Kumaran M, Jayanthi M, Syama Dayal J, Ashok Kumar J, Saraswathy R, Nagavel A (2021) Impacts of climate change and adaptations in shrimp aquaculture: a study in coastal Andhra Pradesh, India. Aquat Ecosyst Health Manage 24(3):28–38

Muralidhar M, Ashok Kumar J, Suvana S, Jayanthi M, Vishwajeet P, Dayal S (2023) Dynamic modelling of coastal aquaculture systems: a review. Aquat Ecosyst Health Manage 26(3):40–52

Mustapha UF, Alhassan AW, Jiang DN, Li GL (2021) Sustainable aquaculture development: a review on the roles of cloud computing, internet of things and artificial intelligence (CIA). Rev Aquac 13(4):2076–2091

Myers SS, Smith MR, Guth S, Golden CD, Vaitla B, Mueller ND, Dangour AD, Huybers P (2017) Climate change and global food systems: potential impacts on food security and undernutrition. Annu Rev Public Health 38(1):259–277

Ndong D, Chen YY, Lin YH, Vaseeharan B, Chen JC (2007) The immune response of tilapia *Oreochromis mossambicus* and its susceptibility to *Streptococcus iniae* under stress in low and high temperatures. Fish Shellfish Immunol 22(6):686–694

Nell JA (1993) Farming the Sydney rock oyster (*Saccostrea commercialis*) in Australia. Rev Fish Sci 1(2):97–120

Ngoan LD (2018) Effects of climate change in aquaculture: case study in Thua Thien Hue Province, Vietnam. Biomed J Sci Tech Res 10:2018

Nicholls RJ, Hoozemans FM, Marchand M (1999) Increasing flood risk and wetland losses due to global sea-level rise: regional and global analyses. Glob Environ Chang 9:S69–S87

Nightingale A (2009) Warming up the climate change debate: a challenge to policy based on adaptation. J Forest Livelihood 8(1):84–89

Nilsson GE, Dixson DL, Domenici P, McCormick MI, Sorensen C, Watson SA, Munday PL (2012) Near-future carbon dioxide levels alter fish behaviour by interfering with neurotransmitter function. Nat Clim Chang 2:201–204

Oh SY, Kim WS, Oh MJ, Nishizawa T (2016) Multiplication rate of red seabream iridovirus (RSIV) in rock bream *Oplegnathus fasciatus* at different fish rearing temperatures. Fish Pathol 51(4):194–198

OIE (2021) Manual of diagnostic tests for aquatic animals. Red Sea Bream Iridoviral disease. Chapter, 2.3.7

Pankhurst NW, Munday PL (2011) Effects of climate change on fish reproduction and early life history stages. Mar Freshw Res 62(9):1015–1026

Panzarin V, Cappellozza E, Mancin M, Milani A, Toffan A, Terregino C, Cattoli G (2014) In vitro study of the replication capacity of the RGNNV and the SJNNV betanodavirus genotypes and their natural reassortants in response to temperature. Vet Res 45:1–11

Paukert CP, Lynch AJ, Whitney JE (2016) Effects of climate change on North American inland fishes: introduction to the special issue. Fisheries 41(7):329–330

Portner HO, Knust R (2007) Climate change affects marine fishes through the oxygen limitation of thermal tolerance. Science 315(5808):95–97

Pörtner HO, Peck M (2010) Climate change impacts on fish and fisheries: towards a cause and effect understanding. J. Fish Biol. 77, 1745–1779.

Preston NP, Jackson C, Thompson P, Austin M, Burford M, Rothlisberg P (2001) Prawn farm effluent: composition, origin and treatment. Fishing Research and Development Corporation Final Report 95, 162

Purcell MK, Laing KJ, Winton JR (2012) Immunity to fish rhabdoviruses. Viruses 4(1):140–166

Rahman MM, Escobedo-Bonilla CM, Corteel M, Dantas-Lima JJ, Wille M, Sanz VA, Pensaert MB, Sorgeloos P, Nauwynck HJ (2006) Effect of high-water temperature (33°C) on the clinical and virological outcome of experimental infections with white spot syndrome virus (WSSV) in specific pathogen-free (SPF) *Litopenaeus vannamei*. Aquaculture 261(3):842–849

Rahman MM, Corteel M, Wille M, Alday-Sanz V, Pensaert MB, Sorgeloos P, Nauwynck HJ (2007) The effect of raising water temperature to 33°C in *Penaeus vannamei* juveniles at different stages of infection with white spot syndrome virus (WSSV). Aquaculture 272(1–4):240–245

Rama Rao CA, Raju BMK, Islam A, Subba Rao AVM, Rao KV, Ravindra Chary G, Nagarjuna Kumar R, Prabhakar M, Sammi Reddy K, Bhaskar S, Chaudhari SK (2019) Risk and vulnerability assessment of Indian agriculture to climate change. ICAR-Central Research Institute for Dryland Agriculture, Hyderabad, p 124

Randall D, Brauner C (1991) Effects of environmental factors on exercise in fish. J Exp Biol 160(1):113–126

Rather MA, Ahmad I, Shah A, Hajam YA, Amin A, Khursheed S, Ahmad I, Rasool S (2024) Exploring opportunities of Artificial Intelligence in aquaculture to meet increasing food demand. Food Chem X 22:101309

Ratkowsky DA, Olley J, McMeekin TA, Ball A (1982) Relationship between temperature and growth rate of bacterial cultures. J Bacteriol 149(1):1–5

Rodgers CJ, Mohan CV, Peeler EJ (2011) The spread of pathogens through trade in aquatic animals and their products. Rev Sci Tech Off Int Epiz 30(1):241–256

Rodnick KJ, Gamperl AK, Lizars KR, Bennett MT, Rausch RN, Keeley ER (2004) Thermal tolerance and metabolic physiology among red band trout populations in South-Eastern Oregon. J Fish Biol 64(2):310–335

Rodrick GB (2008) Effect of temperature, salinity, and pesticides on oyster hemocyte activity. Florida Water Resour J 86:4–14

Rubio-Portillo E, Ramos-Espla AA, Antón J (2021) Shifts in marine invertebrate bacterial assemblages associated with tissue necrosis during a heat wave. Coral Reefs 40(2):395–404

Runhaar H, Wilk B, Persson A, Uttenbroek C, Wamsler C (2018) Mainstreaming climate adaptation: taking stock about what works from empirical research worldwide. Reg Environ Chang 18:1201–1210

Sae-Lim P, Kause A, Mulder HA, Olesen I (2017) Breeding and genetics symposium: climate change and selec0tive breeding in aquaculture. J Anim Sci 95(4):1801–1812

Sahoo PK, Jyotirmaya M, Sushil KG, Bikash RM, Banya K, Joykrushna J, Hema P (2013) Genetic diversity and species identification of Argulus parasites collected from major aquaculture regions of India using RAPD-PCR. Aquac Res 44:220–230

Sánchez-Hernández J (2017) *Lernaea cyprinacea* (Crustacea: Copepoda) in the Iberian Peninsula: climate implications on host–parasite interactions. Knowl Manag Aquat Ecosyst 418:11

Sano M, Ito T, Matsuyama T, Nakayasu C, Kurita J (2009) Effect of water temperature shifting on mortality of Japanese flounder *Paralichthys olivaceus* experimentally infected with viral hemorrhagic septicemia virus. Aquaculture 286(3–4):254–258

Sharma SK, Pradeep MA, Dube PN, Kumar TA, Kumar R, Swaminathan TR (2019) Betanodavirus-associated mortality in Asian seabass (*Lates calcarifer*, Bloch) cultured in indoor tanks and sea cages. Aquac Int 27:279–286

Shetty M, Maiti B, Shivakumar Santhosh K, Venugopal MN, Karunasagar I (2012) Betanodavirus of marine and freshwater fish: distribution, genomic organization, diagnosis and control measures. Indian J Virol 23:114–123

Shin MK, Park HR, Yeo WJ, Han KN (2018) Effects of thermal stress on the mRNA expression of SOD, HSP90, and HSP70 in the spotted sea bass (*Lateolabrax maculatus*). Ocean Sci J 53:43–52

Singh N, Sontakke NA (2002) On climatic fluctuations and environmental changes of the Indo-Gangetic plains, India. Clim Chang 52(3):287–313

Small HJ, Pagenkopp KM (2011) Reservoirs and alternate hosts for pathogens of commercially important crustaceans: a review. J Invertebr Pathol 106(1):153–164

Stévant P, Rebours C, Chapman A (2017) Seaweed aquaculture in Norway: recent industrial developments and future perspectives. Aquac Int 25(4):1373–1390

Stewart HA, Aboagye DL, Ramee SW, Allen PJ (2019) Effects of acute thermal stress on acid–base regulation, haematology, ion-osmoregulation and aerobic metabolism in Channel Catfish (*Ictalurus punctatus*). Aquac Res 50(8):2133–2141

Thuy NTT, Giang NTH, Hoai HTT, Van Dan T (2017) Impact of climate change on aquaculture in Phu Vang district, Thua Thien Hue province, Vietnam. Southeast Asian Regional Center for Graduate Study and Research in Agriculture (SEARCA), pp 1–51

Toffan A, Panzarin V, Toson M, Cecchettin K, Pascoli F (2016) Water temperature affects pathogenicity of different betanodavirus genotypes in experimentally challenged *Dicentrarchus labrax*. Dis Aquat Org 119(3):231–238

Trainer VL, Moore SK, Hallegraeff G, Kudela RM, Clement A, Mardones JI (2019) Pelagic harmful algal blooms and climate change: lessons from nature's experiments with extremes. Harmful Algae 91:101591

Vargas-Albores F, Hinojosa-Baltazar P, Portillo-Clark G, Magallon-Barajas F (1998) Influence of temperature and salinity on the yellowleg shrimp, *Penaeus californieinsis* Holmes, prophenoloxidase system. Aquac Res 29(8):549–553

Vendramin N, Toffan A, Mancin M, Cappellozza E, Panzarin V, Bovo G, Terregino C (2014) Comparative pathogenicity study of ten different betanodavirus strains in experimentally infected European sea bass, *Dicentrarchus labrax* (L.). J Fish Dis 37(4):371–383

Vezzulli L, Colwell RR, Pruzzo C (2013) Ocean warming and spread of pathogenic vibrios in the aquatic environment. Microb Ecol 65:817–825

Vilchis LI, Tegner MJ, Moore JD, Friedman CS, Riser KL, Robbins TT, Dayton PK (2005) Ocean warming effects on growth, reproduction, and survivorship of southern California abalone. Ecol Appl 15(2):469–480

Vogelbein WK, Burreson EM, Carnegie R, Shields JD (2009) Virginia Institute of Marine Science. Initiative for Coastal Climate Change Research. Climate change and aquatic animal health

in Virginia: effects and responses. VIMS climate change white papers. Virginia Institute of Marine Science
Walker PJ, Mohan CV (2009) Viral disease emergence in shrimp aquaculture: origins, impact and the effectiveness of health management strategies. Rev Aquac 1(2):125–154
Wasmund N, Nausch G, Matthäus W (1998) Phytoplankton spring blooms in the southern Baltic Sea—spatio-temporal development and long-term trends. J Plankton Res 20(6):1099–1117
Wright PA, Wood CM (2009) A new paradigm for ammonia excretion in aquatic animals: role of Rhesus (Rh) glycoproteins. J Exp Biol 212(15):2303–2312
Zolnikov TR (2019) Global adaptation and resilience to climate change. Springer International Publishing, p 135

Health Management in Aquaculture: Concepts, Tools, Technologies and Their Application

Health Management in Aquaculture: An Overview and Future Perspectives

K. Jeena, Anutosh Paria, Anirban Paul, P. K. Sahoo, and K. V. Rajendran

Abstract

Aquatic animal diseases form the most significant limiting factor in aquaculture production. As diseases in aquaculture result from a complex interaction between host, pathogen, environment and anthropogenic factors, and complete elimination of disease risk is almost impossible, a science-based health management strategy will be the cornerstone of sustainable aquaculture development. Addressing current challenges such as disease outbreaks, antimicrobial resistance and climate change requires a multidisciplinary approach combining strict biosecurity, stocking quality seed, using appropriate feed for complete nutrition, health monitoring and disease surveillance, and therapeutic and prophylactic strategies. The understanding of the concept of pathobiome and disease development processes in recent years has further strengthened health management strategies in aquaculture. Precise diagnosis plays a crucial role in health management, specifically in the stock maintenance and administration of appropriate therapeutic measures. Strict biosecurity and regulatory mechanisms can control the transboundary movements of infectious agents and curtail the indiscriminate use of drugs and chemicals. Future advancements in digital technologies, AI-based diagnosis and disease prediction, effective vaccines and genomic tools offer better ways to improve aquaculture health management.

K. Jeena · K. V. Rajendran (✉)
ICAR-Central Institute of Fisheries Education, Mumbai, India
e-mail: jeena@cife.edu.in; rajendrankv@hotmail.com

A. Paria
ICAR-National Bureau of Fish Genetic Resources, Lucknow, India
e-mail: paria.anutosh@gmail.com

A. Paul · P. K. Sahoo
ICAR-Central Institute of Freshwater Aquaculture, Bhubaneswar, India
e-mail: anirbancof@gmail.com; pksahoo1@hotmail.com

R. Kooloth Valappil et al. (eds.), *Aquatic Animal Health Management*,
https://doi.org/10.1007/978-981-96-7987-4_16

Keywords

Aquaculture · Health management · Diseases · Biosecurity

1 Introduction

In aquaculture, diseases have become the principal threat to the sustainable cultivation of various economically important cultured species. Among the diseases, the predominant share is attributed to diseases caused by infectious agents such as viruses, bacteria, fungi and parasites (Bondad-Reantaso et al. 2005). Transboundary movement of live fish and fish products, the introduction of wild populations to the culture environments, stocking beyond the carrying capacity of the culture environment, the introduction of aquatic animals into culture environments without screening of their health status, and inadequate biosecurity measures are the key factors that contribute to the disease outbreaks.

To minimize infectious disease outbreaks in aquaculture, it is important to negate the possible contributory factors through scientifically proven and feasible interventions. Although there are many studies on the development of effective therapeutic measures, identifying feasible and cost-effective therapeutics is one of the great challenges to aquaculturists. Therefore, preventing disease outbreaks is far more essential than relying on curative measures. Further, strategies related to the 'one pathogen and one disease' concept in aquatic animals are virtually impractical, as the water in which fish live can be considered a biological broth for many microbes. Hence, very often, the disease outcome in aquatic animals results from co-infection with multiple pathogens, including the common microflora of water, which are otherwise innocuous to healthy animals. So, designing health management strategies for aquatic animals is far more complex than for terrestrial animals and should include control and preventive measures with special emphasis on epidemiological approaches to identify possible risk factors in complex disease outcome situations.

2 Concept of Disease Development in Fish

Disease involves a complex interaction between the host, pathogen and environment. It has been observed frequently that even the less virulent pathogens can play a crucial role in the disease outcome. Therefore, diseases in aquaculture are an expression of a complex interaction between the host, pathogen and environment. This is illustrated in the Snieszko circle, otherwise known as the epidemiological triad (Snieszko 1972). However, as anthropogenic intervention has become a major factor that influences the onset and outcome of infectious diseases in modern aquaculture, these factors also need to be incorporated into the environmental compartment. Accordingly, to reflect the significance of anthropogenic factors in the disease outcome, the triad was updated by Shield (2013) to an epidemiological tetrad (Fig. 1). As elimination of pathogens from the aquatic environment is almost

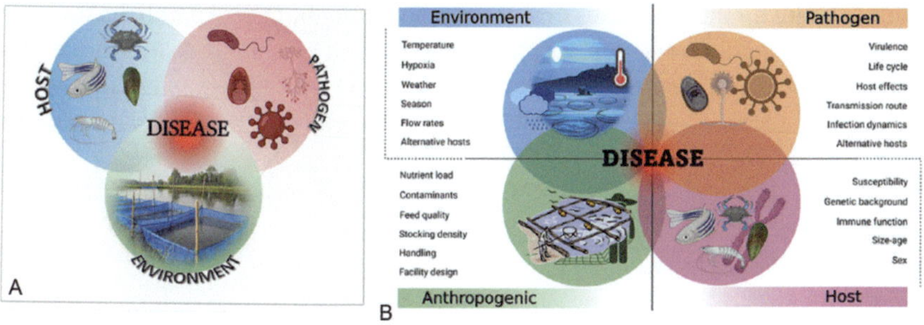

Fig. 1 (**a**) The epidemiological triad and (**b**) the epidemiological tetrad. (Adapted and modified from Dong et al. 2023)

impossible, the best possible measure to deal with diseases is through the management of the health of farmed animals. Therefore, in general, the concept of health management in aquaculture involves the application of science-based logical measures to minimize the risk of disease occurrence, considering all the determinants of disease development.

3 Better Management Practices for Aquatic Animal Health Management

Better Management Practices (BMPs), also termed 'best management practices', are structured management practices followed to increase the sustainability, efficiency and profitability of aquaculture operations while minimizing environmental impacts. In other words, it is aimed at increasing both the quality and quantity of products and ensuring food safety, animal health and environmental and socio-economic sustainability (Arthur et al. 2008). Health management approaches must be followed in each stage of aquaculture, from hatchery to harvest, to ensure that the operations are economically viable and ecologically responsible. Better management practices focus on preventive measures, early detection of diseases and sustainable farming techniques. Key BMPs for fish health management include:

3.1 Water Quality Parameters

Alteration or deviation from the optimal physico-chemical parameters of water leads to stress in fish and makes them vulnerable towards diseases of infectious and non-infectious origin (Devi et al. 2017). During adverse environmental conditions, the disease-causing agents that are otherwise innocuous may pose a threat to fish; this condition of increased susceptibility can be linked to the failure to mount effective immunological responses. Further, it is important to mention that depending on the culture system and species being cultured, values of these parameters (especially temperature and salinity) might vary. Hence, it is important to understand the optimal water quality conditions for the species being farmed. Apart from the major water quality parameters, certain other factors, such as organic load and the presence of toxic gases, including ammonia, nitrite and hydrogen sulphide, in the culture environment, should be monitored for efficient health management of fish (Nayan et al. 2020).

3.2 Seed Quality

Fish seed quality is one of the key components that determines the success of aquaculture. Three major attributes of fish seed are genetic, sanitary and physiological qualities. Genetic quality indicates the purity of the strain and level of improvement; sanitary quality pertains to the absence of diseases and physiological quality indicates their survival rates. Compromisation in any of these attributes can have a major impact on the overall yield from the culture system (Almekinders et al. 2019; Shikuku et al. 2021). The use of seeds from reputable, biosecure hatcheries ensures consistency in quality and reduces the risk of disease introduction into the culture systems.

3.3 Feed and Nutrition

To ensure the high health status against diseases in a particular group of fish, the level of nutrients should be calculated optimally because the nutrient requirement profile varies between the species (Oliva-Teles 2012). If the fish is fed continually with a diet deficient in proper nutrition, it can lead to growth retardation, increased susceptibility to diseases and nutritional deficiency-associated pathological manifestations. For efficient health management of aquatic animals, a rightly formulated live or artificial diet is important for not only preventing nutritional diseases but also curtailing the occurrence of infectious diseases (Shefat and Karim 2018).

3.4 Disease Diagnosis

Diagnosis of diseases in aquatic animals mostly follows a similar set of protocols that are followed in the case of terrestrial animals. Broadly, it can be categorized into two types, viz. presumptive diagnosis and confirmatory diagnosis. Presumptive diagnoses based on gross clinical observations and circumstantial evidence are not always conclusive and are difficult as many of the diseases produce similar or overlapping clinical symptoms or pathological lesions. However, it is important as it facilitates the immediate rectification of probable causes. Further, the presumptive diagnosis provides supplementary information to frame the hypothesis for initiating the confirmatory diagnosis. In contrast, confirmatory diagnosis is always necessary for identifying the definite disease-causing agent but involves sophisticated laboratory-based investigations (Raja and Jithendran 2015). The routine laboratory practices for ascertaining the disease-causing agents include parasitological examinations, bacteriology, mycology and histopathology. Furthermore, electron microscopy and virological, immunological and molecular diagnostic methods are employed to identify the specific cause of infection. An efficient and early diagnosis is essential in the best management approach to curtail the spread of diseases and for employing appropriate therapeutic measures during disease occurrence.

3.5 Biosecurity and Quarantine

'In general terms, biosecurity is a strategic and integrated approach to analyzing and managing relevant risks to human, animal (including aquatic), plant life and health and associated risks to the environment' (FAO 2007; Arthur et al. 2008). Biosecurity in aquaculture is a set of essential approaches for creating a disease-free environment in aquaculture systems (i.e. hatcheries, nursery rearing units and grow-out farms). The approaches specifically target to minimize the exposure of an aquaculture facility to biological or disease-causing agents by preventing or expelling disease-causing agents from the system and also include steps to prevent the spread of pathogens from one system to another (Aly and Fathi 2024). Aquaculture biosecurity is also comprised of routine monitoring of diseases, such as regular health inspections and employing prophylactic treatments before the introduction of animals into a new system. Three important components of biosecurity are quarantine, sanitation and disinfection.

Quarantine is an integral risk management measure in the overall health management programme. This can be applied, either alone or in combination, to reduce the risk posed by aquatic animal pathogens (Arthur et al. 2008). Quarantine in aquaculture is the process in which a group of animals are maintained in isolation with no direct or indirect contact with other animals and perform observations for a predetermined time. Proper quarantine protects established animal populations from exposure to potential pathogens, provides sufficient time for the newly introduced animals to acclimate to a new environment (water, feed and management) and minimizes handling stress. An appropriate quarantine period can allow the clinical manifestation of incubating diseases or latent infections.

Routine disinfection of the aquaculture facility and associated equipment with an appropriate disinfectant for a recommended period of time is critical and helps in reducing the risk of the spread of an infectious agent. A range of disinfectants that are commonly used in aquaculture systems includes chlorine-based, i.e. sodium or calcium hypochlorite; quaternary ammonium compounds such as benzalkonium chloride (BKC), iodophores, formaldehyde, etc.

3.6 Prophylactic Measures in Aquaculture

Prophylaxis is the best possible intervention for the health management and overall well-being of aquatic animals. Prophylactic strategies focus on preventing the disease occurrence either through the enhancement of the health status of farmed animals or minimizing the pathogen load in the culture environment (Vijayan et al. 2017). Some of the most widely sought-after prophylactic strategies in aquaculture include:

3.6.1 Probiotic and Prebiotic

In aquaculture, a probiotic organism can be regarded as a live or dead microorganism (bacteria, bacteriophages, microalgae and yeast), when administered via feed or the rearing water, it benefits the host by improving disease resistance, health status, growth performance, feed utilization and minimizing the effect of stressors on the host (Merrifield et al. 2010). Some of the commonly used probiotics in aquaculture are *Bacillus* sp., *Lactobacillus* sp., *Bifidobacterium* sp., *Streptomyces* sp., *Enterococcus* sp., *Lactococcus* sp., *Micrococcus* sp., *Saccharomyces cerevisiae, Tetrasehnis suecica, Spirulina platensis*, etc. (Llewellyn et al. 2014). Apart from the live organisms, non-viable microbial cells or crude cell extracts of a proven probiotic organism can confer similar health benefits to the host and are collectively termed as 'paraprobiotic' (Nataraj et al. 2020).

Prebiotics are generally of plant origin and cannot be readily utilized by animals; however, these can selectively stimulate the growth or proliferation of beneficial bacteria in the gut, thus improving the health of the animal. Some of the most readily available prebiotics are inulin, fructooligosaccharide and mannan oligosaccharide. Moreover, the term 'synbiotics', which implies the use of both probiotics and prebiotics in combination to obtain a faster outcome of the prophylactic strategy, is also being adopted in aquaculture.

3.6.2 Immunostimulants

Immunostimulants refer to a range of products of either natural or synthetic origin, which can enhance the potentiality of the immune system of a host to resist microbial infection non-specifically (Sakai 1999). Generally, immunostimulants are proven to potentiate the non-specific immune system of fish by stimulating immune mechanisms such as phagocytic activity, activation of the complement pathways, lysozyme activity, natural killer cell activity, facilitating the production of antimicrobial peptides and pro-inflammatory cytokines, etc.

3.6.3 Vaccination

Vaccination is increasingly being recognized as a vital tool in the prevention of infectious diseases in aquaculture, providing a safe and effective method to enhance fish immunity, reduce the need for antibiotics and improve survival rates (Gudding 2014). Understanding the immune response of aquatic animals is fundamental for vaccine development. Fish vaccines typically focus on stimulating the production of antibodies, which can neutralize pathogens. In shellfish, which rely predominantly on innate immunity, conventional vaccination is less effective. Aquaculture vaccines can be classified into several types, such as inactivated (killed), live attenuated, subunit, DNA, recombinant, etc., each having specific properties, mechanisms and applications. The routes of delivery, dosage of vaccines and the life stage of the host play important roles in the efficacy of vaccination. Over 50 commercial vaccines are available for major fish pathogens globally (Du et al. 2022).

3.6.4 Prophylactic Use of Chemicals or Disinfectants for Reducing the Pathogen Load

Aquaculture environments host a range of microbes naturally that can rapidly respond to environmental changes, often turning pathogenic and resulting in infectious disease outbreaks (Austin and Austin 2016). Prophylactic chemical treatments aim to reduce pathogen load in the culture system pre-emptively, aiming to decrease the incidence of disease outbreaks. Generally, the chemical-based disinfectants used in aquaculture are biocides that are antimicrobial in nature (Meade and Garvey 2018). Water quality, stocking density and environmental conditions heavily influence pathogen load in aquaculture systems, making regular prophylactic applications of disinfectants an essential part of disease management. Disinfectants and sanitizers are generally applied to surface waters and water systems to create environments that limit pathogen survival and replication (Boyd 2003). Some of the commonly used disinfectants employed as broad-spectrum prophylaxis are formalin, chlorine, hydrogen peroxide and potassium permanganate. Effective application strategies of these disinfectants involve the periodic treatment of tanks, ponds and equipment, as well as prophylactic baths for fish prior to their introduction to new systems (Tucker and Hargreaves 2008).

The prophylactic use of chemicals in aquaculture presents serious environmental concerns, including toxicity to non-target species and potential bioaccumulation within the ecosystem. Additionally, the frequent and excessive use of disinfectants and other chemicals can drive the development of resistance in aquatic pathogens, complicating future disease management strategies (Cabello 2006). To mitigate resistance risks, aquaculture facilities are encouraged to rotate disinfectants, avoid excessive use, and combine the use of chemical prophylaxis with probiotics or vaccines.

3.7 Therapeutics

Infectious diseases affecting aquatic animals during aquaculture production are caused by bacteria (54.9%), then viruses (22.6%), parasites (19.4%) and fungi (3.1%) (Kibenge et al. 2012). In aquaculture, as in other animal production sectors, veterinary medicines are used mainly to prevent and treat diseases (FAO 2019). Similar to the early diagnosis of infections for targeted treatment and effective disease control, timely treatment in the early stages of diseases is crucial for managing the health of aquatic animals. Therapeutic agents employed in aquaculture include antibiotic, antiparasitic, antifungal agents of biological and chemical origins. Plant-based therapeutics, or phytotherapeutics, has gained attention as a natural and eco-friendly approach to managing diseases in aquaculture. Extracts from plants like garlic, neem and aloe vera exhibit antimicrobial and antiparasitic properties (Elumalai et al. 2020).

3.7.1 Antibiotics

Antibiotics are commonly used in aquaculture as therapeutic, prophylactic or metaphylactic agents (Quesada et al. 2013) to treat bacterial infections and prevent secondary infections after initial disease outbreaks. The most commonly used antibiotics in aquaculture worldwide are oxytetracycline (tetracyclines), oxolinic acid, flumequine, sarafloxacin, enrofloxacin (quinolones), amoxicillin (β-lactams), erythromycin (macrolides), sulfadimethoxine (sulfonamides), ormetoprim (diaminopyrimidines) and florfenicol (amphenicols) (FAO 2019; Bondad-Reantaso et al. 2023). However, antibiotic usage, safe levels and residues, approval of antibiotic use, etc., vary across the countries. The misuse and overuse of antibiotics in aquaculture have increased antibiotic resistance, raising concerns about public health and environmental safety. Antibiotic residues in aquaculture products and the surrounding environment have also been detected, leading to strict regulations on antibiotic use in many countries (Aich et al. 2018). Therefore, alternative approaches, such as vaccination and the use of probiotics, are being explored to reduce reliance on antibiotics.

3.7.2 Antiparasitic and Antifungal Agents

Chemotherapeutants, mechanical, biological, immunological and genetic methods are available for control and treatment of parasites in aquaculture (Buchmann 2022). Commonly used disinfectants such as formalin, copper sulfate, sodium chloride, hydrogen peroxide, etc., can effectively reduce parasite loads (Hoai 2020). Moreover, plant extracts containing the volatile molecules in allium, thymus, origanum and coriander; bacterial surfactants; medicines such as nitroimidazoles (metronidazole, dimetridazole), anticoccidials, organophosphates (metrifonate, dichlorvos and azamethiphos), pyrethroids (deltamethrin), avermectin (emamectin benzoate), benzimidazoles (albendazole, flubendazole), praziquantel, the antibiotic fumagillin and chitin synthesis inhibitors (diflubenzuron, hexaflumuron, lufenuron and tefubenzuron) are also being used as antiparasitic treatments in aquaculture. Mechanical control attempts block the life cycle using mechanical filters, egg traps, delousing through heat treatment, brushing or flushing with freshwater, and the use of barriers to prevent the infective parasitic stages into a culture section. Further, biological control measures, including the introduction of cleaner fish and invertebrates into the culture system, are practised in order to control the parasites.

According to Wroński et al. (2024), antifungal agents represent a diverse array of pharmaceutical agents encompassing a broad spectrum of drugs designed to combat fungal infections in various organisms. These are formulated to target and inhibit the growth, reproduction or survival of fungal pathogens, aiming to alleviate or eradicate fungal-related ailments (Wroński et al. 2024). Fungal or oomycete infections can be devastating in aquaculture, particularly in egg and larval stages. Formalin, iodine, and malachite green have traditionally been used to control fungal pathogens (Van West 2006). However, due to environmental and safety concerns, some of the above-mentioned compounds, such as malachite green, are banned for use in aquaculture in various countries. However, there is a growing interest in developing environment-friendly antiparasitic and antifungal treatment measures (Paria et al. 2020).

The environmental impacts of therapeutic agents have led to increased regulatory scrutiny. Antibiotic and chemical residues can disrupt aquatic ecosystems and pose risks to other aquatic organisms (Costello 2009). Regulatory frameworks, such as those established by the Food and Agriculture Organization (FAO) and World Health Organization (WHO), provide guidelines for the responsible use of therapeutics in aquaculture, promoting practices that have minimal adverse effects on the environment and protect public health. Focusing on sustainable and innovative disease management strategies such as immunotherapy, nanotechnology and genetic selection is expected to offer new solutions for disease prevention and control. Additionally, integrated approaches that combine therapeutics, biosecurity measures and environmental monitoring can enhance the effectiveness of disease management and support sustainable aquaculture practices (Aly and Fathi 2024).

4 Pathobiome Approach

Fish and their surrounding microbiome interact dynamically, creating a complex environment in aquaculture systems (Bass et al. 2019). The 'pathobiome' concept represents a community-wide approach and encompasses all pathogenic agents integrated within their biotic environment (Vayssier-Taussat et al. 2014; Stentiford et al. 2017). It can also be described as the collection of potentially pathogenic microorganisms in a given community. The interplay between pathogen communities, aquatic organisms and water quality offers a more comprehensive understanding of disease development beyond the one pathogen–one disease model, particularly in complex aquatic environments (Rath et al. 2018). Factors that influence pathobiome structure are host species, environmental conditions and aquaculture practices (Bernardo-Cravo et al. 2020). Therefore, understanding pathobiomes can improve management strategies, including targeted probiotics, customized antimicrobials and enhanced environmental controls (Kamel et al. 2024). The pathobiome approach in disease management offers a comprehensive framework for managing complex diseases, promoting holistic aquaculture practices that target entire microbial communities.

5 Disease Surveillance and Reporting

The term 'disease surveillance' is used to describe a more active system and implies that some form of directed action will be taken if the data indicate a disease level above a certain threshold. The occurrence of diseases in aquaculture and the patterns of occurrence are difficult to predict and are uncertain. Hence, the application of 'disease surveillance' in the health management of aquatic animals plays a significant role. Accurate surveillance procedures followed by precise reporting are important for tackling aquatic animal diseases responsibly and efficiently. It is an important step towards building the essential information for formulating control and eradication measures and enabling early warning, risk assessment, contingency plans and emergency preparedness for aquatic animal diseases and epizootics (Sood et al. 2021). As stated in the World Organisation for Animal Health Aquatic Animal Health Code (WOAH 2024), surveillance is aimed at demonstrating the absence of infection or infestation, identifying the presence or distribution of infection or detecting exotic diseases or emerging diseases as early as possible.

5.1 Types of Disease Surveillance

Aquatic animal disease surveillance is broadly categorized as active surveillance and passive or farmer-based surveillance; however, over the years, to enhance the effect of surveillance, several targeted approaches, such as syndromic surveillance, sentinel surveillance, indirect surveillance, risk-based surveillance, etc., have been developed (Sood et al. 2024). Passive surveillance is the type of routine surveillance conducted to understand the extent of diseases and detect the change in disease status over a specified area. It is inexpensive and sustainable. Under this type of surveillance, generally, when fish farmers notice any abnormalities or disease, they report it to the concerned authorities. Therefore, passive surveillance is aimed at collecting information about all diseases. On the other hand, active surveillance aims to generate accurate information about a particular disease in a specific geographical region or farming condition through frequent intensive activities by the authorities. These involve the collection of data by trained personnel and take action in order to receive reports of disease cases, based on which the presence or absence of a particular disease can be established.

5.2 Zoning and Compartmentalization

Zonation is the process of delineating infected and uninfected populations within a specific geographical location or country or group of countries. However, compartmentalization is a tool for disease management largely depending on the system of production and the epidemiology of the disease(s) for which the compartment is being defined within a country (Zepeda et al. 2008). Zoning and compartmentalization are the concepts employed to assist in disease control or eradication within and across boundaries through the projection of disease-free status. These play major roles in the context of international trade due to the significance of disease freedom certifications.

6 Genetics and Genomic Approach

Apart from farm-level management measures to mitigate the disease impact, the application of genetics and genomic tools to improve the disease resistance of farmed animals is an innovative approach. These include selection for domestication, family-based selection, marker-assisted selection and genomic selection (Sciuto et al. 2022). The advances in sequencing technologies and novel bioinformatic tools have facilitated the selection approach for disease resistance. Further, advanced molecular tools helped in the rapid characterization of pathogens and their virulence, better understanding of host–pathogen interactions, elucidation of antimicrobial resistance genes and pathogen evolution and host immunity. Overall, these approaches will tremendously improve the efficiency of health management measures employed in aquaculture.

7 Artificial Intelligence (AI) and Fish Health

The integration of AI in monitoring fish health status has the potential to revolutionize the early diagnosis of diseases and their management. Various AI techniques, including computer vision, image processing, machine learning and data analytics, have been employed in aquaculture research and fish health monitoring (Mandal and Ghosh 2024). Some of the important AI-based approaches are the identification of diseases through image analysis using external clinical signs-based models (Zhang et al. 2020), identifying fish health abnormalities by AI-based algorithms using different fish health-related data, disease diagnosis by AI algorithms using empirical data of a specific disease and development of aquatic animal disease prediction models using time-series data.

8 Conclusion

Effective health management in aquaculture is pivotal for ensuring sustainable production, economic viability, disease prevention and environmental protection. The integration of biosecurity measures, disease surveillance and advancements in diagnostic tools has significantly enhanced the industry's ability to mitigate and manage emerging fish health challenges. However, the future of aquaculture health management lies in adopting innovative approaches such as precision aquaculture, genomic technologies for disease management and eco-friendly practices, including probiotics, herbal-based measures, AI-based tools and new-generation vaccines. Strengthening interdisciplinary research, fostering collaborations between stakeholders and implementing robust regulatory frameworks will be essential to address emerging diseases and global challenges like climate change. By prioritizing fish health and welfare, the aquaculture sector can achieve long-term resilience, productivity and sustainability.

References

Aich N, Ahmed N, Paul A (2018) Issues of antibiotic resistance in aquaculture industry and its way forward. Int J Curr Microbiol App Sci 7(8):26–41

Almekinders CJ, Walsh S, Jacobsen KS, Andrade-Piedra JL, McEwan MA, de Haan S, Kumar L, Staver C (2019) Why interventions in the seed systems of roots, tubers and bananas crops do not reach their full potential. Food Secur 11:23–42

Aly SM, Fathi M (2024) Advancing aquaculture biosecurity: a scientometric analysis and future outlook for disease prevention and environmental sustainability. Aquac Int 32:1–27

Arthur JR, Bondad-Reantaso MG, Subasinghe RP (2008) Procedures for the quarantine of live aquatic animals: a manual. FAO Fisheries Technical Paper. No. 502. FAO, Rome, 74

Austin B, Austin DA (2016) Bacterial fish pathogens: disease of farmed and wild fish. Springer. https://doi.org/10.1007/978-3-319-32674-0

Bass D, Stentiford GD, Wang HC, Koskella B, Tyler CR (2019) The pathobiome in animal and plant diseases. Trends Ecol Evol 34(11):996–1008

Bernardo-Cravo AP, Schmeller DS, Chatzinotas A, Vredenburg VT, Loyau A (2020) Environmental factors and host microbiomes shape host–pathogen dynamics. Trends Parasitol 36(7):616–633

Bondad-Reantaso MG, Subasinghe RP, Arthur JR, Ogawa K, Chinabut S, Adlard R, Tan Z, Shariff M (2005) Disease and health management in Asian aquaculture. Vet Parasitol 132(3–4):249–272

Bondad-Reantaso MG, MacKinnon B, Karunasagar I, Fridman S, Alday-Sanz V, Brun E, Le Groumellec M, Li A, Surachetpong W, Karunasagar I, Hao B, Dall'Occo A, Urbani R, Caputo A (2023) Review of alternatives to antibiotic use in aquaculture. Rev Aquac 15(4):1421–1451

Boyd CE (2003) Chemical use in aquaculture and its ecological consequences. Environ Pollut 129(3):377–387

Buchmann K (2022) Control of parasitic diseases in aquaculture. Parasitology 149:1985–1997

Cabello FC (2006) Heavy use of prophylactic antibiotics in aquaculture: a growing problem for human and animal health and for the environment. Environ Microbiol 8(7):1137–1144

Costello MJ (2009) The global economic cost of sea lice to the salmonid farming industry. J Fish Dis 32:115–118

Devi PA, Padmavathy P, Aanand S, Aruljothi K (2017) Review on water quality parameters in freshwater cage fish culture. Int J Appl Res 3(5):114–120

Dong HT, Chaijarasphong T, Barnes AC, Delamare-Deboutteville J, Lee PA, Senapin S, Mohan CV, Tang KFJ, McGladdery SE, Bondad-Reantaso MG (2023) From the basics to emerging diagnostic technologies: what is on the horizon for tilapia disease diagnostics? Rev Aquac 15(Suppl. 1):186–212

Du Y, Hu X, Miao L, Chen J (2022) Current status and development prospects of aquatic vaccines. Front Microbiol 13:1040336

Elumalai P, Kurian A, Lakshmi S, Faggio C, Esteban MA, Ringø E (2020) Herbal immunomodulators in aquaculture. Rev Fish Sci Aquacult 29(1):33–57

FAO (2007) Aquaculture development. 2. Health management for responsible movement of live aquatic animals. FAO Technical Guidelines for Responsible Fisheries. No. 5, Suppl. 2. Rome, FAO. 31 pp

FAO (2019) Aquaculture development. Recommendations for prudent and responsible use of veterinary medicines in aquaculture. FAO Technical Guidelines for Responsible Fisheries. No. 5. Suppl. 8. Rome

Gudding R (2014) Vaccination as a Preventive Measure. In: Gudding R et al (eds) Fish Vaccination 1st edition, John Wiley & Sons, Inc, oxford, 12–21. https://doi.org/10.1002/9781118806913.ch2

Hoai TD (2020) Reproductive strategies of parasitic flatworms (Platyhelminthes, Monogenea): the impact on parasite management in aquaculture. Aquac Int 28(1):421–447

Kamel M, Aleya S, Alsubih M, Aleya L (2024) Microbiome dynamics: a paradigm shift in combatting infectious diseases. J Person Med 14(2):217

Kibenge FS, Godoy MG, Fast M, Workenhe S, Kibenge MJ (2012) Counter measures against viral diseases of farmed fish. Antivir Res 95(3):257–281

Llewellyn MS, Boutin S, Hoseinifar SH, Derome N (2014) Teleost microbiomes: the state of the art in their characterization, manipulation and importance in aquaculture and fisheries. Front Microbiol 5:207

Mandal A, Ghosh AR (2024) Role of artificial intelligence (AI) in fish growth and health status monitoring: a review on sustainable aquaculture. Aquac Int 32(3):2791–2820

Meade E, Garvey M (2018) Efficacy testing of novel chemical disinfectants on clinically relevant microbial pathogens. Am J Infect Control 46(1):44–49

Merrifield DL, Dimitroglou A, Foey A, Davies SJ, Baker RT, Bøgwald J, Castex M, Ringø E (2010) The current status and future focus of probiotic and prebiotic applications for salmonids. Aquaculture 302:1–18

Nataraj BH, Ali SA, Behare PV, Yadav H (2020) Postbiotics-parabiotics: the new horizons in microbial biotherapy and functional foods. Microb Cell Factories 19:168

Nayan AA, Mozumder AN, Saha J, Mahmud KR, Azad AKA (2020) Early detection of fish diseases by analyzing water quality using machine learning algorithm. Int J Adv Sci Technol 29(5):14346–14358

Oliva-Teles A (2012) Nutrition and health of aquaculture fish. J Fish Dis 35:83–108

Paria A, Dev AK, Pradhan PK, Kumar R, Rathore G, Sood N (2020) Evaluation of therapeutic potential of selected antifungal chemicals and drugs against *Aphanomyces invadans*. Aquaculture 529:735643

Quesada SP, Paschoal JAR, Reyes FGR (2013) Considerations on the aquaculture development and on the use of veterinary drugs: special issue for fluoroquinolones-a review. J Food Sci 78(9):R1321–R1333

Raja RA, Jithendran KP (2015) Aquaculture disease diagnosis and health management. In: Perumal S et al (eds) Advances in marine and Brackishwater aquaculture. Springer India

Rath S, Rud T, Karch A, Pieper DH, Vital M (2018) Pathogenic functions of host microbiota. Microbiome 6:1–13

Sakai M (1999) Current research status of fish immunostimulants. Aquaculture 172(1–2):63–92

Sciuto S, Colli L, Fabris A, Pastorino P, Stoppani N, Esposito G, Prearo M, Esposito G, Ajmone-Marsan P, Acutis PL, Colussi S (2022) What can genetics do for the control of infectious diseases in aquaculture? Animals 12:2176. https://doi.org/10.3390/ani12172176

Shefat SHT, Karim MA (2018) Nutritional diseases of fish in aquaculture and their management: a review. Acta Sci Pharm Sci 2(12):50–58

Shields JD (2013) Complex etiologies of emerging diseases in lobsters (*Homarus americanus*) from Long Island sound. Canad J Fish Aquat Sci 70(11):1576–1587

Shikuku KM, Tran N, Joffre OM, Islam AHMS, Barman BK, Ali S, Rossignoli CM (2021) Lock-ins to the dissemination of genetically improved fish seeds. Agric Syst 188:103042

Snieszko SF (1972) The effects of environmental stress on outbreaks of infectious diseases of fishes. J Fish Biol 6(2):197–208

Sood N, Pradhan PK, Swaminathan TR, Rathore G, Jena JK, Lal KK (2021) National Surveillance Programme for aquatic animal diseases – a stepping stone for establishing disease governance system in India. Curr Sci 120(2):273–277

Sood N, Pradhan PK, Paria A, Kumar CB, Rathore G, Sarkar UK, Jena JK (2024) Aquatic animal disease surveillance and important disease affecting finfish and shellfish in India. In: Sarkar UK, Kumar TTA, Sood N, Singh RK, Kumar R, Tyagi LK (eds) Sustainable Management of Fish Genetic Resources. Springer, Singapore. https://doi.org/10.1007/978-981-97-5250-8_11

Stentiford GD, Sritunyalucksana K, Flegel TW, Williams BA, Withyachumnarnkul B, Itsathitphaisarn O, Bass D (2017) New paradigms to help solve the global aquaculture disease crisis. PLoS Pathog 13(2):1006160

Tucker CS, Hargreaves JA (2008) Environmental best management practices for aquaculture. Wiley. https://doi.org/10.1002/9780813818672

Van West P (2006) *Saprolegnia parasitica*, an oomycete pathogen with a fishy appetite: new challenges for an old problem. Mycologist 20(3):99–104

Vayssier-Taussat M, Albina E, Citti C, Cosson JF, Jacques MA, Lebrun MH, Le Loir Y, Ogliastro M, Petit MA, Roumagnac P, Candresse T (2014) Shifting the paradigm from pathogens to pathobiome: new concepts in the light of meta-omics. Front Cell Infect Microbiol 4:29

Vijayan KK, Makesh M, Otta SK, Patil PK, Poornima M, Alavandi SV (2017) Prophylaxis in aquaculture. ICAR-CIBA publication

WOAH (2024) WOAH - Aquatic Animal Health Code. https://www.woah.org/fileadmin/Home/eng/Health_standards/aahc/current/chapitre_aqua_ani_surveillance.pdf

Wroński M, Trawiński J, Skibiński R (2024) Antifungal drugs in the aquatic environment: a review on sources, occurrence, toxicity, health effects, removal strategies and future challenges. J Hazard Mater 465:133167. https://doi.org/10.1016/j.jhazmat.2023.133167

Zepeda C, Jones JB, Zagmutt FJ (2008) Compartmentalisation in aquaculture production systems. Rev Sci Tech-Office International des Epizooties 27(1):229–241

Zhang S, Yang X, Wang Y, Zhao Z, Liu J, Liu Y, Sun C, Zhou C (2020) Automatic fish population counting by machine vision and a hybrid deep neural network model. Animals 10(2):364

Health Management in Penaeid Shrimp Seed Production

P. Ravichandran

Abstract

Shrimp farming success hinges on two primary inputs: seed and feed. In traditional shrimp farming, seeds were naturally obtained from tidal waters, limiting scalability and reliability. During the early stages of scientific shrimp farming, particularly in Southeast Asia, tiger shrimp (*Penaeus monodon*) was the dominant species being cultivated. Farmers primarily relied on wild-caught seeds, and the scarcity of these seeds proved to be a major limiting factor in the development of the industry during the early 1980s. Although experimental hatcheries emerged in the late 1970s, it was not until the late 1980s that commercial hatcheries became established. Initially, egg production in hatcheries depended on fully mature wild spawners, with subsequent advancements enabling the maturation of immature wild broodstock. By the late 1990s, the shrimp farming sector faced a significant challenge due to the outbreak of bacterial and viral diseases. By 2000, most major shrimp-producing countries in Southeast Asia transitioned to farming *L. vannamei* by importing SPF broodstock, which allowed them to produce disease-free seed locally. In India, the government permitted the full-scale importation of *L. vannamei* SPF broodstock in 2009 under strict quarantine measures. SPF broodstock helped control vertical transmission of pathogens (from broodstock to larvae), though horizontal transmission (within hatcheries and farms) remained a challenge, necessitating high levels of biosecurity in shrimp hatcheries. Biosecurity measures focus on preventing pathogens from entering via seawater, wild broodstock, wet feeds and larval diets such as algae and *Artemia nauplii*, as well as controlling human and vehicle movement. A strict disinfection protocol is implemented for all materials entering the hatchery. During the 20 to 30-day larval rearing period, continuous health monitoring of

P. Ravichandran (✉)
ICAR-Central Institute of Brackishwater Aquaculture, Chennai, India
e-mail: ravichandranciba@gmail.com

R. Kooloth Valappil et al. (eds.), *Aquatic Animal Health Management*,
https://doi.org/10.1007/978-981-96-7987-4_17

the larvae at physical, behavioural, microscopic and molecular levels is essential to ensure a pathogen-free status. As seed quality is vital for sustaining shrimp farm production, maintaining high standards at every critical control point of larval production is crucial. Strict adherence to established biosecurity and health protocols ensures the production of healthy, high-quality shrimp larvae for successful shrimp farming.

Keywords

Penaeid shrimp · Seed production · Specific pathogen-free stock · Nucleus breeding centre · Selective breeding · Biosecurity

1 Introduction

Marine shrimp farming is an age-old practice in Asian countries. In India, the traditional systems were in vogue in the '*Bheries*' of West Bengal and '*Pokkali*' fields of Kerala. The source of seed for these extensive culture systems was the wild post-larvae that migrated from the marine shrimp breeding grounds to the brackishwater creek/canal. The post-larvae entered the culture area through the tidal exchange.

In the early stages of the development of scientific shrimp farming, manual collection of wild post-larvae was resorted to for standardizing the stocking density in ponds. Seasonal variations in wild seed availability were the major constraint, and the need for shrimp seed production under controlled conditions was very seriously felt by the sector.

2 Global Shrimp Farming Scenario

Globally, commercial shrimp farming started in the early 1970s. In 1970, about 24 countries reported shrimp aquaculture with a total production of about 9022 mt. By the year 2000, the production level had increased to 1,161,800 mt, and the species that contributed largely to the production was the tiger shrimp, *Penaeus monodon*. Since 2001, SPF *Litopenaeus vannamei* farming has been introduced in Southeast Asian countries, which resulted in a significant increase in global shrimp production by 2019. The total shrimp production reported by about 72 countries had reached 4.5 million MT, with *L. vannamei* contributing 75% of the total production.

2.1 Status of Shrimp Farming in India

Shrimp farming in India, till 2009, was primarily the monoculture of tiger shrimp, *Penaeus monodon*. During this time, about 1,90,000 ha of brackishwater area has been developed for shrimp culture. However, since 1995, the farming of *P. monodon* has been severely impacted by the widespread outbreak of the viral disease caused by white spot syndrome virus (WSSV). Due to this, there had been a stagnation in the development of shrimp farming, with a production level of 100,000 mt. Similar to India, most of the Southeast Asian countries like Thailand, Vietnam and Indonesia were also cultivating *P. monodon* during this period. However, as *P. monodon* was susceptible to WSSV and because of the availability of Specific pathogen-free (SPF) broodstock of non-native whiteleg shrimp, *Litopenaeus vannamei,* most of the countries shifted to this species since 2001–02 onwards. Following this, in India, *L. vannamei* was introduced in 2003 on a pilot scale. This was followed by a risk analysis study, and the commercial-scale production of *L. vannamei* since 2009 through the state-of-the-art aquatic quarantine located in Chennai. The introduction of SPF *L. vannamei* resulted in a significant increase in the production levels from 100,000 MT in 2008 to nearly 700,000 mt in 2018 in 172,000 ha ponds with a 90% contribution from *L. vannamei* (MPEDA 2019)

3 History of Penaeid Shrimp Hatchery Development

Although studies on the larval stages of shrimp, *Penaeus japonicus*, were initiated in 1933 in Japan, the research progress was hampered by the non-availability of larval stages from the wild. However, the laboratory-level experiments on larval rearing were made possible based on the observation that the mature females lay the eggs when held in aquarium conditions. Nevertheless, further success was hindered till 1940 because of the non-availability of suitable live feed for the larvae. Subsequently, the successful rearing of *P. japonicus* was realized after the technique of pure culture of live feed, *Skeletonema costatum,* was standardized (Hudinaga 1942).

It took nearly 20 years for the development of a successful culture of *P. japonicus* post-larvae in large outdoor tanks, a significant milestone in the history of shrimp hatchery development. This method is popularly known as the 'community culture' method or 'Big tank hatchery' (Hudinaga and Kittaka 1975).

Subsequently, the early model of hatchery system was modified in the Galveston Laboratory for the larval rearing of *P. aztecus, P. duorarum and P. setiferus* (Cook and Murphy 1966; Mock and Murphy 1970). This system was referred to as the 'Galveston method' or 'Small tank hatchery'. In this system, the rearing tanks were of 2-ton capacity (Mock and Murphy 1971).

As the hatchery development progressed in various shrimp farming countries, different modifications and combinations in these two systems evolved in countries like Taiwan (Liao 1985, 1986; Chiang and Liao 1985) and the Philippines (Villaluz et al. 1969; Parado-Estepa and Primavera 1988).

3.1 Community Culture or Big Tank Hatchery

The basic principle of this system was to maintain an appropriate concentration of the shrimp larvae and their food organisms in the same tank with the required light intensity, aeration and concentration of nutrients. The technology involved introducing mature females to lay their eggs in a tank filled with water up to 80 cm. The water was maintained at the same level until the larvae attained the mysis stage, followed by the addition of a small amount of fresh seawater daily. The daily addition of water in this way fills the tank to almost its full capacity (~2 m), by the time the shrimp attained the postlarval stage. Following this, depending on the water quality, 20–40% of the water was changed daily. The water exchange is essentially to remove metabolites and food residues accumulated during the rearing process and to maintain a relatively stable concentration of diatoms. The rearing system is simple to operate, not involving complex technical knowledge, and a postlarval survival of 10–20% could be achieved.

Though large-scale shrimp seed production is possible with this system, infection with bacteria, fungi and parasites often leads to production failures. And the emergence of these pathogens to an infection level is largely due to water quality deterioration. In order to overcome these problems, the system was modified in 1972 by introducing chemical sterilisation of water using sodium hypochlorite solution. Further, the potential water quality deterioration was avoided by producing live feed (*Chaetoceros)* meant for zoea and mysis stages independently, besides introducing formulated feed in suspension for the later stages to replace clam meat suspension (Shigueno 1975).

3.2 Galveston Method or Small Tank Hatchery

In this system, the culture of algae is carried out separately in algal culture rooms provided with 300-L to 1-ton tanks. The cultivated algae are then fed to the larvae at a pre-determined quantity (Mock 1974). Wild spawners are kept individually in spawning tanks, and after the spawning, the spawning debris is removed, and the eggs are kept for hatching. Hatched-out nauplii at the N4 stage were distributed in 2-ton larval rearing tanks. The seawater used for larval rearing was thoroughly cleaned of all suspended particles. Algae were fed during the zoea stage, and freshly hatched artemia nauplii were fed to the mysis and postlarval stages. At PL5, they were transferred to outdoor Nursery tanks and reared till the PL20 stage. Generally, in this system, 100–500 larvae per litre are stocked, and the survival rate of postlarvae achieved could range from 60 to 80 per cent (Mock and Murphy 1971).

3.2.1 Induction of Maturity in Captivity

Till the early 1970s, the hatchery production of penaeid shrimp larvae was exclusively dependent on fully mature spawners collected from the wild. However, one of the limitations was that females of most of the penaeid shrimp species do not mature in estuaries or captivity, though males, in the estuarine phase of their life cycle, have been observed to mature and mate. To overcome this, Adiyodi and Adiyodi (1970) demonstrated that eyestalk ablation could result in the removal of gonad-inhibiting hormone that can eventually lead to maturity in most decapod crustaceans. Based on this, during 1971–1976, various researchers employed the eyestalk ablation technique, which resulted in the successful maturation of different species of penaeid shrimps (Idyll 1971; Caillouet 1973; Arnstein and Beard 1980; Alikunhi et al. 1975; Wear and Santiag 1976). This procedure has become an established technique for the captive maturation of penaeid shrimps. However, although the closing of the life cycle of various penaeid species has been achieved under captivity (CNEXO Aquaculture team 1974; Alikunhi et al. 1975; Santiago 1977; Primavera 1978), one of the limitations is the poor quality of the eggs produced from the pond-reared broodstock.

3.2.2 Development of Specific Pathogen-Free (SPF) Broodstock

Successful closure of the life cycle in different penaeid shrimp species has been reported in the late 70s and 80s. (Forester and Beard 1974; Wickins and Beard 1978; Salser et al. 1978; Aquacop 1983; Moore and Brand 1993). In spite of such successes, the hatcheries generally avoid the use of captive pond-reared broodstock as a source of mother shrimps because of their poor reproductive performance compared to the wild spawners/broodstock. Wild shrimp adults and larvae were reported to be the carriers of viral pathogens, and the use of wild spawners and broodstock resulted in the production of infected seeds through vertical transmission (Chakraborty et al. 2002).

This necessitated ensuring the pathogen-free status of wild mother shrimps. Wild spawners or adults collected were transported individually, retained in quarantine, and tested for the major disease-causing viral pathogens; and the shrimps that are free of the pathogens will only be used for spawning or induced maturation. Till the year 2000, in many Asian and South Asian countries, the major shrimp species under culture was *Penaeus monodon,* and a number of viral diseases were reported during the period 1993 to 1996 (Lightner 2005; Lightner et al. 2009).

During this period in the Americas, the Pacific white shrimp, *L. vannamei,* was a major shrimp species under culture. The farms were mainly dependent on wild shrimp seed as the requirements were seasonal, and the use of hatchery-produced seed was uneconomical. Further, as the wild broodstock had a high prevalence of infectious hypodermal and hematopoietic necrosis virus (IHHNV), their performance declined within 2–3 months (Motte et al. 2003; Lightner et al. 2009). This necessitated the development of captive broodstock in strictly controlled biosecure systems, excluding the major pathogens.

With an objective to increase the local production of marine shrimp while decreasing the reliance on importation, the United States Marine Shrimp Farming Program (USMSFP) of the United States Department of Agriculture (USDA) was formed in 1984, involving seven Consortium Research. The objective of USMSFP, after a disease outbreak hit the breeding programme, was shifted towards the domestication and genetic improvement of Pacific whiteleg shrimp, *Li. vannamei,* in 1989. The process of development of SPF was followed as per the Guidelines developed for the introduction of non-indigenous marine organisms (ICES 1984).

The pathogens selected for exclusion in the Nucleus Breeding Centres (NBC) are based on the following criteria, which are also among those required for disease listing by the OIE:

(a) The pathogen(s) must be excludable.
(b) Adequate diagnostic and pathogen detection methods are available.
(c) The pathogen(s) poses a significant threat of causing disease and production losses.

The original selected pathogen list in the USMSFP SPF facility during 2008–2009 included seven OIE-listed shrimp viruses, two OIE-listed viruses under study and two unlisted viruses. It also included one OIE-listed bacterium (Lightner et al. 2009, 2011).

In present-day SPF facilities, all seven OIE-listed pathogens and other new emerging pathogens, such as EMS and EHP, are also included as per the demand of the SPF shrimp farmers/ importing countries.

The process used for the development of the SPF shrimp species included the following:

(a) *Identification of the founder stock:*
Stock from wild or farmed stock was collected and tested for the listed pathogens, and the pathogen-free stock was prepared as founder stock (F0) and reared in the primary quarantine.
(b) *Primary quarantine:*
F0 stocks were reared in primary quarantine and monitored continuously for disease signs, and tested periodically for specific pathogens. In case of any pathogen detection, the stock will be destroyed. The shrimps were reared in primary quarantine till they attained the maturing stage, and pathogen-free shrimps were transferred to secondary quarantine.
(c) *Secondary quarantine:*
At the secondary quarantine, the stocks were matured under captive conditions and selected for mating and production of the F1 generation. The F1 generation stocks were maintained in the quarantine till they attained adult status. During the whole process, continuous testing for listed pathogens will be done, and the shrimps free from all the viruses are transferred to the Nucleus Breeding Centre (NBC).
(d) *Nucleus Breeding Centre:*
In NBC, the stocks are maintained as different families and selective breeding is followed to avoid inbreeding. During the selective breeding process, the families are chosen based on specific traits like growth, survival or reproductive performance, which will lead to their improvement generation after generation. The post-larvae of the best families are passed on to the Broodstock Multiplication Centres (BMC) as Parent post-larvae (PPL).
(c) *Broodstock Multiplication Centre:*
BMC is the site where the PPLs are reared to SPF broodstock in a strict biosecure facility. Continuous monitoring of listed pathogens will be undertaken during the 3 to 4-month rearing period, and the disease-free SPF broodstock will be sold to commercial biosecure hatcheries for the production and sale of SPF/ High health post-larvae to the farmers.

3.2.3 Impact of Introduction of SPF Broodstock

USMSFP introduced SPF and high-health stocks of *L. vannamei,* and these have been cultivated successfully in U.S. shrimp farms in the mid-1990s. This resulted in a remarkable increase in production (almost double the production per crop, compared to the production achieved with the non-selected lines of *L. vannamei* previously) (Wyban et al. 1992; Pruder et al. 1995).

Since 1999, substantial quantities of SPF *L. vannamei* have been introduced into East Asia, resulting in notable performance in production. From 2000 to 2001, *L. vannamei* was introduced in almost all Southeast Asian countries. India introduced pilot-scale production of *L. vannamei* in two of the farms in 2003 and, after a detailed risk analysis, introduced it on a commercial scale in 2009 through a centralised quarantine. Global production of shrimps increased significantly from 2000 and has reached 4.5 million MT, with 75% of it being contributed by SPF *L. vannamei.* The success of the introduction of SPF *L. vannamei* in 2000 has led to the production of SPF broodstock of other species as well. With the closure of USMSFP, a number of SPF *L. vannamei* facilities have come up in the US and other countries.

4 Disease Outbreaks and Prevention of Transmission in Hatcheries

Globally, disease outbreaks in shrimp farms through bacterial and viral pathogens started as early as the late 1980s both in Asia and in the Americas, and it has been estimated that a loss of nearly USD 3 million per annum from 1994 (Lightner 2003). A review of disease emergence in farmed shrimps has been done by many researchers during the last two decades (Walker and Mohan 2009; Flegel 2006)

The WSSV pandemic started in Asian countries during the early 1990s, and it spread to almost all shrimp culture areas. Wild mother shrimps were identified as carriers, and vertical transmission was reported from broodstock to larvae. In India, till 2008, *P. monodon* was farmed in nearly 90% of the area under culture, and the prevention of the disease assumed greater importance both in the hatchery and farms. The improvements brought about in the hatcheries during that period for the production of disease-free, quality post-larvae included the following measures:

4.1 Prevention of Entry of Pathogens from External Source

4.1.1 Wild Broodstock and Spawners

In India till 2009, *Penaeus monodon* was the major penaeid shrimp species farmed, and the source of broodstock or spawners was from the wild. The wild-caught shrimps have been reported to be carriers of the known pathogens. Generally, wild-caught broodstock or spawners are brought collectively in small containers. However, this procedure has the disadvantage in that the process of collection, transportation and acclimatisation of broodstock leads to stress and cross-contamination. A high level of stress on broodstock during trawling or protracted periods of fishing will ultimately affect their performance.

In view of this, the following method was followed by the hatcheries to produce disease-free seeds:

- The seawater used for transporting the broodstock, as well as at the port collection centre, is to be filtered to eliminate pollutants and pathogens. Broodstock from different trawlers can be collected by collection boats and transported to the hatchery. After capture, the broodstock should be kept individually in separate containers in filtered and disinfected water. To avoid stress to the broodstock, they can be transported under oxygen packing at low temperatures (25 °C) with the stocking of 1 to 2 broodstock per 8 litres of water.
- Upon arrival at the hatchery, shrimp showing abnormal signs such as black melanised lesions in the body or white muscles or bright red colouration should be discarded. Apparently healthy broodstock should be disinfected with 100 ppm KMnO4 or liquid povidone for 1 minute before stocking in the quarantine tanks.
- The broodstock should be tested for listed pathogens, and only when found free of these pathogens, the animal can be taken into the maturation system. A small portion of the last pleopod can be used as a tissue source for DNA extraction that can be employed in PCR-based detection of DNA viruses. Samples of faeces can be used to detect bacterial or microsporidian pathogens.
- In the case of *L. vannamei* hatcheries, only SPF broodstock imported from overseas or broodstock from the Indian Broodstock Multiplication Centres is permitted to be used. Non-SPF broodstock is not permitted to be used in hatchery operations.

4.1.2 Seawater

- Each functional unit must have appropriate water treatment systems and preferably separate or isolated water supply systems and, if possible, aided with the re-circulatory system in specific sections to reduce water usage (Fig. 1).
- Initially, water having a salinity range of 29–32 ppt should be taken in the settlement tank and kept for 1–2 days. The water is then treated with chlorine (10–20 ppm for 12–24 h) and, subsequently, dechlorinated with sodium-thiosulphate. To the settlement tank, potassium permanganate (up to 2 ppm) can be added.
- To avoid chlorine or dissolved organic residues in the hatchery facility, an activated carbon filter is recommended. Further, for larval rearing, maturation, *Artemia* and algal units, suitable filter systems such as a cartridge filter (1–5 micron), UV or ozone filter can be employed along with EDTA (10–30 ppm) or Treflan (0.05–0.1 ppm).

4.1.3 Wet Feed

Polychaetes, squids and mussels are commonly used live feeds in hatcheries. However, as these are generally collected from different locations without proper quality checks, they are known to be a major source of contamination. Therefore, in order to reduce the level of contamination, the live polychaetes and molluscs should be depurated before feeding. To ensure the quality, these live feeds should be further

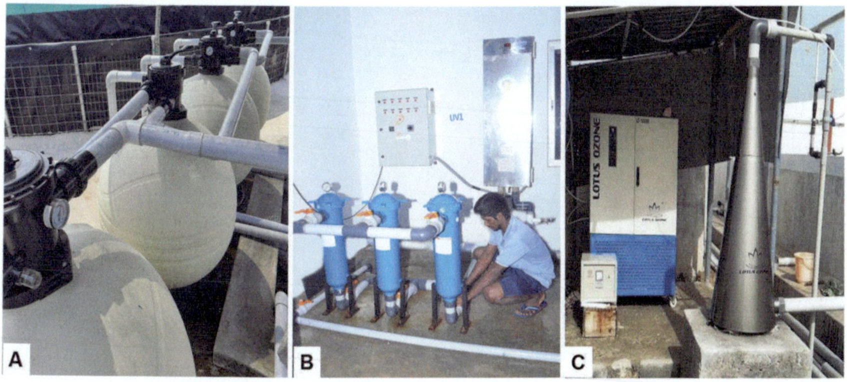

Fig. 1 Different filter systems used in hatcheries. (**a**) Rapid sand filters; (**b**) Cartridge filter and UV filter; (**c**) Ozonator

screened after depuration. If required, they should be disinfected using chemical disinfectants or pasteurisation. It is also recommended that separate isolated sections should be marked for handling the wet feed. Furthermore, to exclude pathogen contamination and transmission from live wet feeds to the broodstock, the use of SPF polychaetes is recommended.

4.1.4 Human Interference

All persons entering the production area should be disinfected by shower at the entrance of the production area, and hand-wash and leg-wash should be mandatory at every entry point. People who have visited other aquaculture units within the last 72 hours should not be allowed inside the production areas.

Vehicles entering the hatchery premises should be disinfected (Fig. 2).

4.1.5 Prevention of Pathogen Contamination Within the Production Area

Contamination of the pathogens within the different units of the hatchery needs to be avoided. The various units in the facility are: Quarantine, Maturation unit, Spawning and Hatching unit, Larval rearing unit, Nursery unit, Algal culture unit, Artemia hatching unit and Packing unit. Transmission of pathogens, if any, from these various units needs to be avoided by following these protocols:

(i) *Maturation*

- The optimum stocking density in the maturation tanks will vary according to the species. For *P. monodon,* the ideal stocking density is estimated to be 2–3 numbers/sq m, while for *L. vannamei,* it is 6–8 numbers/sq m. To enhance the reproductive performance and pigmentation, vitamins (vitamin C –2 g/kg and E –0.2 g/kg) or astaxanthin (0.1 g/kg) can be added to live feed like squid and molluscs before feeding to the broodstock.

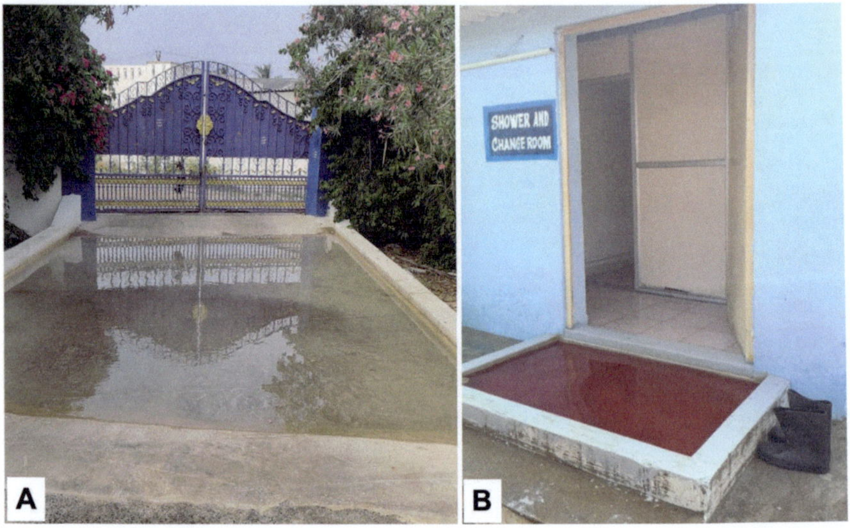

Fig. 2 (**a**) Vehicle tyre bath; (**b**) Foot bath and shower room

- After holding the broodstock for a week, females at the inter-moult stage (hard shell) can be subjected to eyestalk ablation. For this, one eye of the females is ablated using hot pincers or with a string tied to the eyestalk. However, prior to ablation, the animals should be kept in chilled seawater (20–25 °C). In the case of shrimp at pre-moult or post-moult, to withstand the stress due to ablation, animals can be left aside for a week or so.

(ii) *Spawning and Hatching Area*

- In the spawning/hatching tanks, water filtered using cartridge and UV filtration systems should be used. It is advisable to have individual spawning tanks to avoid contamination.
- To maintain the production of high-quality nauplii, the number of spawning achieved from each female broodstock before they are discarded varies according to the species. In the case of *P. monodon,* the animal should be spawned a maximum of 3 times, whereas it can be 10–12 times in the case of *L. vannamei*.
- As *P. monodon* has a closed thelycum, mating can occur only after moulting. A maximum of three spawnings within the period of moulting is expected.
- *L. vannamei* is an open thelycum species; hence, they can mate at any time during maturation.
- Immediately after spawning (usually between 8 pm and 2 am), the female should be transferred to a clean maturation tank. The eggs are collected, cleaned of spawning debris, disinfected by formalin dip and transferred to

individual tanks meant for egg-hatching containing clean seawater and 20–30 ppm EDTA and 0.05–0.01 ppm treflan.

- To increase the egg hatching rate, the hatching tank should be provided with sufficient illumination. This will help in identifying and collecting healthy nauplii that are attracted to light. The naupli should be subsequently washed and disinfected with formalin or other suitable disinfectant. The disinfected larvae can be transferred to tanks containing fresh, filtered and disinfected seawater.
- A hatching rate of <40% in a single spawn is considered unhealthy, and it is advisable to discard the animals.
- Healthy nauplii on reaching the N4-N5 stage are collected, disinfected and stocked in Larval Rearing Tanks (LRT).

(iii) *Larval Rearing Unit*

- Larval rearing of shrimps should be done without environmental stress and with adequate quality feed in the required quantity. This is essentially done by managing the water quality and a proper feeding schedule. In the case of *P. monodon,* a separate outdoor nursery-rearing system is used from the PL5 to the PL20 stage. For *L. vannamei,* generally, one tank system is used till the marketable size, PL9 to PL10.
- Generally, *P. monodon* nauplii are stocked at a low stocking density (50 to 100 numbers per litre (50–100,000/m^3), whereas *L. vannamei* nauplii are stocked up to 200 numbers per litre.
- Duration of stocking in the larval-rearing tank should not exceed 2–3 days. This will help in maximizing biosecurity and prevent contamination of the different stocks.
- Removal of unconsumed feed and faecal matter should be done every day.
- Water exchange, as per requirement, should be done daily after the start of feeding.
- The use of probiotics should be encouraged, and antibiotics should be avoided as prophylactic treatment.
- The best feed recommended for zoea is live algae (*Chaetoceros* sp, *Thalassiosira* sp or *Skeletonema* sp.). The algae should be fed at high concentrations (80–130,000 cells/mL) initially; however, this should be reduced thereafter during PL stages. Alternatively, in case live algae are not available due to the non-functioning of the outdoor algal unit, *Spirulina* in dried form and liquid or powdered/ microencapsulated dry diets (10–80 microns in size) can be fed every 2 to 4 hours duration (Fig. 3). The algal concentrate can also be used in larval rearing. This can be prepared by concentrating the algae when they are in full bloom by filtration using a fine mesh bag. The final cell densities (for *Chaetoceros* sp.) should be at 80–130,000 cells/ mL. For zoea and mysis (peaking at Zoea3) larvae, the final cell density of the algae should be 80–130,000 cells/mL. This should be reduced to 50–60,000 cells/mL during early PL stages.

Fig. 3 Algal rearing tanks

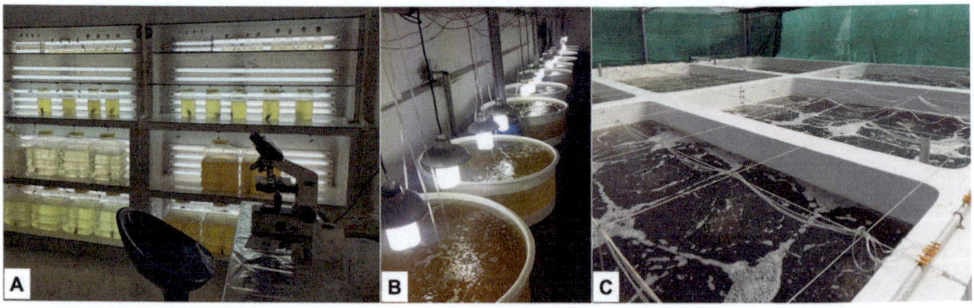

Fig. 4 (**a**) Algae pure culture; (**b**) Indoor culture; (**c**) Outdoor culture

(iv) *Algal Culture System*

- Algal culture is the most critical area where bacterial load is bound to increase because of the addition of nutrients and open-area mass culture. In terms of bacterial proliferation, outdoor mass culture is the most vulnerable area (Fig. 4).
- Pure culture algae alone are to be used in the indoor algal culture, and new cultures are to be brought in from time to time in a stipulated period. The culture vials and containers in the indoor algal culture system have to be sterilized for every culture. The working personnel in the indoor algae should wear lab coats, hand gloves and separate footwear to avoid outside contamination.
- Contamination in the outdoor tank should be avoided by maintaining good water quality and preventing airborne dust. Further, the use of suitable nutrients and inoculum is required.

(v) *Artemia Hatching Unit*

- *Artemia* nauplii are used as feed for shrimp, mysis and post-larvae, along with reduced levels of algae. The use of high-quality *Artemia* cysts with high hatching rates is always advantageous. To reduce contamination, *Artemia* cysts should be disinfected using hypochlorite solution. Otherwise, the hatched nauplii should be disinfected before releasing in LRT (Fig. 5).
- To improve the rate of hatching, instead of full-strength seawater, the use of 18–25 ppt seawater at 28–30 °C is practised. To reduce bacterial load in the hatching tanks containing *Artemia* cysts, either 60 ppm of chloramines-T or 5–10 ppm of suitable probiotics can be added. Active and positively phototactic *Artemia* nauplii alone should be collected and used as feed.

(vi) *Personnel Movement and Equipment Usage*

- Movement of technical personnel and staff from one unit to the other should be strictly restricted, and in case of urgency, proper disinfection protocols should be followed to prevent unit-to-unit contamination.
- Equipment and tools used in one unit should not be used in another unit. Tools like buckets, mugs, beakers, sample bottles, siphon tubes and filters should be maintained separately for each tank. At the end of every day's work, all the equipment and tools should be cleaned and disinfected.

Fig. 5 Artemia cyst hatching tanks

(vii) *Facility Maintenance*

- Total disinfection of the larval rearing tanks is required after every cycle. In the case of the maturation system, total disinfection should be done at least once in 3 months. Tanks used for spawning, egg hatching and holding of nauplii should be thoroughly cleaned after each use.
- The cleaning and disinfection procedures involve scrubbing with clean water and detergent to loosen all dirt and debris. Further, disinfection with hypochlorite solution (20–30 ppm active ingredient) and or a 10% solution of muriatic acid (pH 2–3) is recommended. Finally, to remove all traces of chlorine and or acid, thorough rinsing with clean water followed by drying should be carried out. Sun drying can be employed to sterilise outdoor tanks and small tanks.
- All materials used in the facility, such as filters, hoses, beakers, etc., and the equipment should be cleaned with a 10% muriatic acid solution by placing them in a tank and subsequently treated with hypochlorite solution.
- Periodic disinfection of the walls and floors of the hatchery facility should be carried out. Further, before commencing the new cycle, the stocking tanks should be thoroughly disinfected. For this, the tanks can be given an initial washing with a detergent, followed by rinsing with clean water and wiping down with 10% muriatic acid. Finally, before filling with treated water, the tanks should be rinsed one more time. However, according to the requirements of the facility, the procedure of disinfection may be modified suitably. People who are carrying out the chemical disinfection procedure should take appropriate care to avoid risk.
- Complete cleaning of all the pipelines in the facility should be done once a year during the annual closure.

(viii) *Wastewater Treatment System*

One of the important mandatory requirements in shrimp hatcheries is the wastewater treatment system. Effluent discharge from the hatchery into the open sea/backwaters without any treatment increases the risks of disease and pollution. Proper treatment and disposal of hatchery effluent will not only reduce disease problems within the hatcheries but also help avoid conflicts with other industries and users over water use.

- Each culture tank should have a separate harvest basin/area before its connection to the main drainage canal, instead of one harvest basin for 3 to 4 larval rearing tanks to avoid cross-contamination. Though this will result in increased floor area for the hatchery and the resultant cost, this will help in reducing the disease spread significantly.
- Proper care must be taken while using disinfectants, as their excess will result in residual effects. Therefore, while using chemicals such as chlorine, formalin, iodine, antibiotics, etc., before discharging the water, the chemicals should be allowed to dissipate completely.

Fig. 6 Wastewater treatment system

- Strict care should be taken not to discharge toilet waste into the water body without proper treatment. Further, the treatment of toilet wastes and hatchery wastes should not be allowed to come in contact with each other.
- Infected and dead shrimp, both broodstock and larvae, need to be disposed of after proper disinfection (chlorine at >50 ppm for 1 h). To completely eliminate the risk of disease spread, it is recommended to burn the infected shrimp after proper disinfection.

The process of wastewater treatment has three levels—Sedimentation and filtration, chlorination and dechlorination for disinfection, followed by aeration (Fig. 6).

4.2 Health Assessment Process in Shrimp Hatchery

Production of healthy shrimp seed in hatcheries requires continuous monitoring of the health of the larvae at various levels of production. As detailed monitoring of the whole population is impossible, three different types of health assessment are followed in shrimp hatcheries (FAO 2003, 2007).

4.2.1 Level 1 Health Assessment

The preliminary health assessment can be done using careful visual observations of the various life stages of the shrimp. This is mainly through visual observation with the naked eye. These observations are sufficient enough to decide on the health status of the larvae and also provide sufficient information to evaluate the need for more detailed examinations.

4.2.2 Swimming Activity

There will be dramatic but characteristic changes in the swimming activity of larvae through the larval cycle. Rapid and consistent forward swimming is the characteristic feature of zoeal stages, which feed on phytoplankton. In comparison, the mysis stage is characterized by backward swimming with intermittent flicks of their tails. They invariably remain in the water column and feed on phyto- and zooplankton using visual cues. On the contrary, post-larvae swim forward rapidly and consistently in search of food. They will remain in the water column due to the strong aeration present in the larval-rearing tank. Generally, a healthy stock of seed should have more than 95% of active post-larvae.

4.2.3 Phototaxis

Healthy shrimp nauplii show a strong positive phototactic response. Poor phototaxis and weak swimming behaviour are indicators of the unhealthy nature of the nauplii, and the entire batch can be discarded without any further examination.

Zoeal-stage larvae also have a strong positive phototactic response, and this can be tested by placing a sample of larvae in a translucent container and exposing them to a light source. Larvae are considered healthy when 70% to 95% of them move strongly towards the light, and if the response of less than 70 per cent, the larvae can be considered weak, and the batch can be rejected.

4.2.4 Faecal String

The health of the zoea 1 stages, when the zoea starts feeding almost exclusively on algae, can be assessed by the nature of the faecal strings. Long faecal strings seen projecting from the anus and found loose in the water column can be considered signs of good health. A well-fed stock can be identified when the majority of the larvae (90–100 per cent) exhibit long, continuous faecal strings all along the digestive tract and continue outside. However, when <70 per cent of the larvae have these strings, then it is evident that either the quality/quantity of the algal feed is not up to the required level or the digestive system of the larvae is poorly formed. Further checking at Level 2 is required for both the larvae and algal culture.

4.2.5 Luminescence

The presence of luminescent bacteria such as *Vibrio harveyi* generally results in larval luminescence. The larval tanks are observed under darkness. The tanks with quality larvae will not show any luminescence. The presence of any luminescence indicates a high load of potentially pathogenic Vibrios. Application of probiotics and/or changing water until the luminescence is gone are measures against bacterial luminescence. However, if luminescence is severe and if the treatment is not successful, it is recommended to destroy the batch of larvae.

4.2.6 White-Body Disease

The colour of the larval body will give an idea about the health status of PL. White body colouration with necrosis and whitening of hepatopancreas and midgut or whitening of the tail clearly indicate that the larvae are affected by the disease, and further testing at Level 2 and Level 3 should be taken up. If mortality is detected, the batch of larvae should be destroyed.

4.2.7 Size Uniformity/Size Variation

With reference to the uniformity of larval stages in a tank, most of the larvae should be within one moult stage of each other. The presence of many stages in a single tank indicates unsuitable water quality or disease conditions; however, confirmation of this may require further testing.

4.2.8 Intestinal Contents

The level of gut contents is visible in older larval stages as a dark line extending posteriorly from the hepatopancreas to the beginning of the uropod. This provides a clear indication of larval feeding and feed availability. An empty or partly filled gut indicates the larvae are underfed or unable to feed because of poor health, and remedial action is required.

4.3 Level 2 Health Assessment

Level 2 health assessment is based on microscopic observations and squash mounts. To make the assessment reliable, at least 20 larvae should be randomly examined per tank. Level 2 health assessment can be a good decision-making process of shrimp hatchery management.

4.3.1 Condition of the Hepatopancreas and Gut

The larval feeding and digestion process can be evaluated by microscopic observation of the condition of the hepatopancreas and gut. For this, wet mount preparation of samples of whole larvae should be observed under a microscope at 40× magnification. Hepatopancreas and midgut filled with discernible small lipid vacuoles, and the intestine showing strong peristalsis are indications of active feeding and digestion. A swollen hindgut (SHG) or a coiled posterior digestive tract in PL is not desirable. Underfed and/or diseased PL exhibit empty and pale hepatopancreas without lipid vacuoles, and this condition requires attention. If the majority of the PL (90 per cent or more) show abundant lipid vacuoles and/or a full gut without SHG, the batch can be considered very good; however, if the gut is less than 70 per cent full and/or the intestine is empty, or there is >10 per cent SHG, the batch should be rejected.

4.3.2 Necrosis

Necrotic changes in the limbs and body of the larvae, which can be identified by microscopic observation under low power, indicate cannibalism or possible bacterial infection in the larvae. The underlying reasons for the necrosis could be high bacterial load due to underfeeding or poor water quality. Therefore, water quality should be monitored constantly and improved if necessary.

4.3.3 Deformities

Deformities in nauplii indicate poor-quality spawning and hatching. In the later stages of larvae, deformities could be due to bacterial infections and stress. Characteristic deformities include bent, broken or missing setae on the limbs and/or their rostra. Further, an unusually bent tail and incomplete gut, which terminates before the anus, are observed. As no effective remedial measures are available, if the deformities are severe, it is recommended to discard the stock so that the infection is contained within the tank. However, if regular monitoring reveals frequent occurrences of any deformities, water quality and disease status need to be tested, and remedial measures need to be taken.

4.3.4 Epibiont Fouling

The common fouling organisms found in shrimp larvae include bacteria, fungi, protozoans, etc. The typical sites of attachment of these epibionts are the exoskeleton, especially around the gills. Poor water quality will be a predisposing factor for epibiont fouling. In the case of a low level of infection, fouling organisms may get removed during moulting; however, in severe cases, the fouling will persist or reoccur in the next stage. The remedial measures against fouling include the application of formalin (20–30 ppm) for 1 h with vigorous aeration, followed by a sufficient water exchange.

4.3.5 Bolitas

'Bolitas' is a syndrome characterised by the presence of small spheres within the digestive tract due to the detachment of epithelial cells from the intestine and hepatopancreas. Bacterial infection has been attributed as the cause of this condition, which can sometimes be fatal. Rapid stocking in the larval rearing tanks, the application of probiotics, along with good health and feeding management practices are reported to prevent 'bolitas'.

4.4 Level 3 Health Assessment

Level 3 techniques, which involve disease diagnosis employing molecular and immunological techniques, have become a common practice in shrimp hatcheries and also in laboratories associated with such hatcheries. The methods include polymerase chain reaction (PCR), dot-blot assay and immunodiagnostic tests for the screening of animals for viral diseases.

PCR tests for the OIE-listed pathogens are done in imported broodstock of *L. vannamei* in the aquatic quarantine facility, and only disease-free SPF stock is allowed in the country. For broodstock produced in Broodstock Multiplication Facilities, the necessary testing is done in approved laboratories.

In the case of wild spawners of *P. monodon,* they should be placed in quarantine inside the hatchery premises, and PCR testing should be done for the important pathogens of concern. Spawners that are free from all pathogens should only be taken to the Maturation section.

PLs, which are ready for sale, should be screened for pathogens of concern following the standard protocols by a competent health laboratory with experienced technicians following all the protocols for sampling, preservation and transport of the samples.

The testing should reveal the pathogen-free status of the stock, as evidenced by a negative PCR result. To make the PCR assay acceptable and confirmatory, the laboratory should include both negative and positive controls along with the samples in the PCR. If the samples are tested positive for a targeted pathogen, the stock should be destroyed following standard procedure.

4.5 Shrimp Seed Quality Criteria

The importance of shrimp seed quality evaluation has been well recognised since the early phase of the development of the shrimp farming industry (Bauman and Jamandre 1990). The quality of the post-larvae is extremely varied and decided by internal and external factors. The quality of the eggs and sperm of the parent broodstock is governed by the hormones. Apart from that, the environmental parameters of the water and the maintenance and management of the larval rearing process further impact the larval quality. A number of criteria have been recommended for the evaluation of the quality of post-larvae (Samocha et al. 1998).

The criteria and methodological tools that are used for larval quality assessment are well established (Bray and Lawrence 1992; Racotta et al. 2004). These include (1) morphology, (2) behaviour, (3) survival to stress test, (4) time needed to complete larval metamorphoses, (5) survival during larval development, (6) PL size, (7) screening of PL for major pathogens (Bray and Lawrence 1992; Racotta et al. 2004).

4.5.1 Morphology

Morphological criteria for the assessment of seed quality include size, weight, the occurrence of deformities, colour, muscle/gut ratio, morphology of the gill and digestive system, etc. (Racotta et al. 2003). However, visual observation of post-larvae has been used as the simplest and widely used criterion. Dark-coloured, active post-larvae are considered best for stocking. A clean carapace is an indication of fast growth and frequent moulting of the PL, and these larvae can be selected. However, the presence of injuries and necrosis is the indication of slow growth.

Good seed will have the following qualities:

- Transparent body and few pigmentation spots in the tail muscle
- Clean shell (indication of fast growth and frequent moulting)
- The presence of a clear, thick and smooth muscle
- A high index of gut fullness

- Dark-coloured body; relatively large hepatopancreas, having a large number of lipid vacuoles
- Branched nature of gill lamellae in post-larvae

4.5.2 Muscle Gut Ratio

Observations of the healthy wild seed have shown that the tail muscle diameter normally exceeds their hindgut diameter by a ratio of at least 4:1. Based on this, many hatcheries use the muscle-gut ratio as a criterion to assess the larval quality. For this, the measurement is taken halfway between the telson and the last abdominal segment. The shell from the gut down to the ventral side should be completely filled with muscle. A muscle-gut ratio of less than 4:1 may indicate poor larval quality.

4.5.3 Size Variation

Other larval quality criteria include increased growth and reduced variability in size during the postlarval stage (Castle et al. 1993). Further, uniform PL length is considered an early indicator of larval quality in hatchery. In *Litopenaeus vannamei*, the coefficient of variation (CV) in postlarval length is regarded as an important quality criterion (Clifford 1999). A larval population with a CV of less than 10% is considered excellent for stocking, while a stock with a CV greater than 15% should be rejected.

4.5.4 Behaviour

Larval behaviour is of predictive value in determining larval quality. For example, the positive phototactic behaviour of nauplii, an indicator of larval health, is widely used as a tool in the harvesting of nauplii. However, this criterion is found to have limitations at later stages of larval development, as the phototactic nature of the larvae would last only up to the protozoeal stage. In the later stage of development, no significant difference would be observed between larvae that had higher phototactic response and remnant larvae (Ibarra et al. 1998).

Swimming activity also indicates larval quality (Clifford 1992), as healthy shrimp post-larvae will swim against the water current in the tank. On the contrary, erratic swimming activity is an indicator of lower physiological activity and, thus, poor larval quality.

4.5.5 Survival to Stress Test

Stress tests have been recommended as a useful tool for checking the larval quality in both shrimp and fish (Wedemyer and McLeay 1981; Clifford 1992; Fegan 1992). Stress test basically determines the ability of the animals to maintain their physiological state and chances of survival after they are exposed to short but extreme external conditions (Fegan 1992). Larvae with poor health cannot withstand the stress conditions, so only animals that survive the test will be selected.

Salinity Stress Test

One of the significant stressors widely used to evaluate the quality or hardness of post-larvae is osmotic stress (Samocha et al. 1998). A direct correlation has been observed between the high survival observed in salinity stress and the high performance in grow-out systems. Therefore, the stress test can be applied to PL10 and PL20 to evaluate the expected grow-out performance.

Formalin Stress Test

Another widely used stressor in shrimp hatchery is acute formalin stress, and this has been a recommended procedure in *L. vannamei* hatcheries Samocha et al. (1998). The process involved exposing the post-larvae (PL7) to 100 to 200 ppm of formalin for 1–2 hours. Good quality seeds are those that exhibited a survival rate of more than 50%.

Pathogen Screening

Ensuring the disease-free status of the post-larvae is the most important criterion for seed quality. Testing for virus and bacterial pathogens during the course of larval rearing using molecular diagnostics such as Polymerase Chain Reaction (PCR) has become a mandatory requirement in shrimp hatcheries. Post-larvae should be tested for all the pathogens that are of concern in the country before sale, and only seeds that are found free of these pathogens should be sold to the farmers.

5 Conclusion

The two major inputs for successful shrimp farming are seed and feed. The traditional farming systems were dependent on the seeds that were collected from the tidal water. During the initial stages of scientific farming, the tiger shrimp, *Penaeus monodon,* was the major species cultivated in Southeast Asian countries, wherein seeds collected from the wild were stocked in ponds. Non-availability of seeds was a major constraint to the development of scientific farming during the early 80s. Though experimental hatcheries have been in operation since the late 70s, commercial hatcheries were set up only in the late 80s. At the beginning of hatchery technology, wild, fully mature spawners were the main source of egg production. As a next step, the induction of maturity of immature wild broodstock was achieved.

In the late 90s, the shrimp farming sector was seriously affected by bacterial and viral diseases, and the need for disease-free, high-quality broodstock was enormous. During the same time, Specific Pathogen-Free stocks of *Litopenaeus vannamei* were successful under the US Marine Shrimp Farming Project. By 2000, most of the major producers of shrimps in Southeast Asian countries shifted to *L. vannamei* culture by importing SPF broodstock and producing disease-free seed in local hatcheries. Vertical transmission of pathogens from the parent stock to the larvae was controlled because of the SPF broodstock. Horizontal transmission of the pathogens needs to be controlled both in the hatchery and on farms to ensure successful production. This required high biosecurity in the hatchery.

In India, the introduction of *L. vannamei* was initiated in 2003 through a pilot-scale project by two companies, and in 2009, after a risk analysis study, full-scale importation through government-controlled quarantine was permitted.

Biosecurity requirements include (a) prevention of entry of pathogens through seawater; (b) prevention of entry of pathogens through wild spawners, broodstock, wet feed, and larval diets such as algae, *Artemia* nauplii and other compounded feeds; (c) prevention of pathogen entry through personnel/vehicles. A strict disinfection protocol for all the material that enters the hatchery is carried out to prevent the entry of pathogens.

Further, during the course of larval rearing, which normally takes 20 to 30 days, continuous monitoring of the health status of the larvae should be done at three different levels—(1) Physical and behavioural, (2) Microscopic and (3) Molecular disease diagnostic methods - to ensure the pathogen-free status of the larvae. As seed quality is a major requirement for sustaining shrimp farm production, extreme care is required at all the critical control points of larval production, and strict adherence to the established protocols will help ensure the desired quality of shrimp larvae.

References

Adiyodi KG, Adiyodi RG (1970) Endocrine control of reproduction in decapod crustacea. Biol Rev 25:121–165

Alikunhi KH, Poernomo A, Adisukresno S, Budiono M, Busman S (1975) Preliminary observations on induction of maturation and spawning in *P. monodon* Fabricius and *Penaeus merguinesis* de Man by eyestalk extirpation. Bullet Shrimp Cult Res 1:1–11

Aquacop (1983) Constitution of broodstock maturation, spawning, and hatchery system for penaeid shrimps in the Centre Oceanologique du Pacifique. In: McVey JP, Moore JR (eds) CRC handbook of mariculture, Crustacean Aquaculture, vol 1. CRC Press, Boca Raton, pp 105–121

Arnstein DR, Beard TW (1980) Induced maturation of the prawn Penaeus orientalis Kishinouye in the laboratory by means of eyestalk removal. Aquaculture 5(4): 411–412. https://doi.org/10.1016/0044-8486(75)90060-5

Bauman RH, Jamandre DR (1990) A practical method for determining quality of *Penaeus monodon* (Fabricius) fry for stocking in grow-out ponds. In: New MB, de Saram H, Singh T (eds) Technical and economic aspects of skimp farming. Proceedings of the Aquatech '90 conference, 11 June-14 June 1990, Kuala Lumpur, Malaysia. INFOFISH, pp 124–137

Beard TW, Wickins JF (1980) Breeding of *Penaeus monodon* Fabricius in laboratory recirculation systems. Aquaculture 20(2):79–89

Bray WA, Lawrence AL (1992) Reproduction in *Penaeus* species in captivity. In: Fast AW, Lester LJ (eds) Marine shrimp culture: principles and practices. Elsevier, Amsterdam, pp 93–170

Caillouet CWJ (1973) Ovarian maturation of eyestalk ablation in pink shrimp *Penaeus duorarum* Burkenroad. Proc. 3rd Annual Workshop World Mariculture Society 3, pp 205–225

Castille FL, Samocha TM, Lawrence AL, He H, Frelier P, Jaenike F (1993) Variability in growth and survival of early postlarval shrimp, *Penaeus vannamei* (Boone). Aquaculture 113:65–81

Chakraborty A, Otta SK, Joseph B, Kumar S, Hossain MS, Karunasagar I, Venugopal MN, Karunasagar I (2002) Prevalence of white spot syndrome virus in wild crustaceans along the coast of India. Curr Sci 82(11):1392–1397

Chiang P, Liao TC (1985) The practice of grass prawn(*Penaeus monodon*) culture in Taiwan from 1968 to 1984. J World Maricult Soc 16:297–315

Clifford HC (1992) Marine shrimp farming: a review. In: Wybm J (ed) Proceedings of the special session on shrimp farming. World Aquaculture Society, Baton Roug, p 110

Clifford HC (1994) Semi-intensive sensation: a case study in marine shrimp pond management. World Aquacult 25(3):6–12; 98–104

Clifford HC (1999) Prevention of shrimp white spot virus disease: water management strategies. The Advocate:58–63

CNEXO Aquaculture Team (1974) Maturation and spawning in captivity of Penaeid prawns: *P. merguiensis* de Man, *P. japonicus* Bate, *P. aztecus* Ives and *Metapenaeus ensis* (de Haan). Rapport CNEXO, Paris, p 17

Cook HL, Murphy MA (1966) Rearing penaeid shrimp from eggs to postlarvae. Proc. Annu Conf Southeast Assoc Game Fish Comm 19:283–288

FAO (2003) Health management and biosecurity maintenance in white shrimp (*Penaeus vannamei*) hatcheries in Latin America. FAO Fisheries Technical Paper No. 450. Rome, FAO. 62 p

FAO (2007) Improving *Penaeus monodon* hatchery practices. Manual based on experience in India. FAO Fisheries Technical Paper. No. 446. Rome, FAO. 101 p

Fegan DF (1992) Recent developments and issues in the Penaeid shrimp hatchery industry. In: Wyban J (ed) Proceedings of the special session on shrimp farming. World Aquaculture Society, Baton Roug, pp 55–70

Flegel TW (2006) Detection of major penaeid shrimp viruses in Asia, a historical perspective with emphasis on Thailand. Aquaculture 258:1–33

Forester JRM, Beard TW (1974) Experiments to assess the suitability on nine species of prawns for intensive cultivation. Aquaculture 3:355–368

Hudinaga M (1942) Reproduction, development and rearing of *Penaeus japonicus* Bate. Japan J Zool 10(2):309–395

Hudinaga M, Kittaka J (1975) Local and seasonal influences on the large scale production method for penaeid shrimp larvae. Bull Japan Soc Sci Fish 41(8):843–854

Ibarra AM, Ramirez JL, Perez-Rostro CL (1998) Effect of positive phototropism selection in nauplii of white shrimp, *Penaeus vannmei*, on growth, survival and family variance of later developmental stages. Aquaculture '98. World Aquaculture Society, Las Vegas, p 261

ICES (1984) Guidelines for implementing the ICES code of practice concerning introductions and transfers of marine species. Cooperative Research Report No. 130. 20 pp.

Idyll CP (1971) Induced maturation of ovaries and ova in pink shrimp. Commer Fish Rev 33:20

Liao IC (1985) A brief review of the larval rearing techniques of penaeid prawns. In: Taki Y, Primavera JH, Llobrera JA (eds) Proceedings of the first international conference on the culture of penaeid prawns/shrimps; 1984 December 4-7; Iloilo City, Philippines. SEAFDEC Aquaculture Department, Iloilo, pp 65–78

Liao IC (1986) Recent trends in prawn hatchery design and management. Proceedings of the National Conference on Prawn and Fish Farming Technology; 1986 October; Makati, Metro Manila, Philippines

Lightner DV (2003) The penaeid shrimp viral pandemics due to IHHNV, WSSV, TSV and YHV: history in the Americas and current status. In: Sakai Y, McVey JP, Jang D, McVey E, Caesar M (eds) Proceedings of the thirty-second US Japan symposium on aquaculture. US-Japan cooperative program in natural resources (UJNR). U.S. Department of Commerce, N.O.A.A., Silver Spring, MD, USA, pp 6–24. and at http://www.lib.noaa.gov/japan/aquaculture/aquaculture_panel.htm

Lightner DV (2005) Biosecurity in Shrimp farming: pathogen exclusion through use of SPF stock and routine surveillance. J World Aquacult Soc 36:229–248

Lightner DV (2011) Status of shrimp diseases and advances in shrimp health management, pp. 121–134. In: Bondad-Reantaso MG, Jones JB, Corsin F, Aoki T (eds) Diseases in Asian aquaculture VII. Fish health section. Asian Fisheries Society, Selangor, p 385

Lightner DV, Redman RM, Arce SM, Moss SM (2009) Specific pathogen-free shrimp stocks in shrimp farming facilities as a novel method for disease control in crustaceans, chapter 16. In: Shumway S, Rodrick G (eds) Shellfish safety and quality. Woodhead Publishing Limited, Cambridge, UK, pp 384–424

Mock CR (1974) Larval culture of penaeid shrimp at the Galveston Laboratory. In: Shaw WN (ed) NOAA technical report NMFS CIRC-388, pp 33–40

Mock, CR, Murphy MA (1970) Techniques for raising penaeid shrimp from the egg to postlarvae. J World Aquacult Soc 1 (1-4): 141–156. https://doi.org/10.1111/j.1749-7345.1970.tb00022
Mock CR, Murphy MA (1971) Techniques for raising penaeid shrimp from the egg to postlarvae. J World Maricult Soc 2:143–156
Mock CR, Neal RA (1974) Penaeid shrimp hatchery systems. Proceedings of the FAO CARPAS symposium on aquaculture in Latin America; 1974. November 26-December 2; Montevideo, Uruguay.(CARPAS/6174/SE29)
Moore DW, Brand CW (1993) The culture of marine shrimp in controlled environment super intensive systems. In: McVey JP (ed) CRC handbook of Mariculture. 2nd edition volume 1 Crustacean aquaculture. CRC Press, Boca Raton, pp 315–348
Motte E, Yugcha E, Luzardo J, Castro F, Leclercq G, Rodrıguez J, Miranda P, Borja O, Serrano J, Terreros M, Montalvo K, Narváez A, Tenorio N, Cedeño V, Mialhe E, Boulo V (2003) Prevention of IHHNV vertical transmission in the white shrimp *Litopenaeus vannamei*. Aquaculture 219(1–4):57–70
MPEDA (2019) Production as in https://mpeda.gov.in/
Parado-Estepa F, Primavera JH (1988) Broodstock management and seed production of *Penaeus monodon* (Fabricius). In: Juario JV, Benitez LV (eds) Perspectives in aquaculture development in Southeast Asia and Japan; contributions of the SEAFDEC Aquaculture Department; proceedings; 1987 September 8-12; Iloilo City, Philippines. SEAFDEC Aquaculture Department, Iloilo, pp 149–168
Primvera JH (1978) Induced maturation and spawning in five-month-old *Penaeus monodon* Fabricius by eyestalk ablation. Aquaculture 13:355–359
Pruder GD, Brown CL, Sweeney JN, Carr WH (1995) High health shrimp systems: seed supply - theory and practice. In: Browdy CL, Hopkins JS (eds) Swimming through troubled water, proceedings of the special session on Shrimp farming, aquaculture '95. San Diego, California, 1-4 February 1995. World Aquaculture Society, Baton Rouge, LA, USA, pp 40–52
Racotta IS, Palacios E, Ibma AM (2003) Shrimp larval quality in relation to broodstock condition. Aquaculture 227:107–130
Racotta IS, Palacios E, Hernandez-Henera R, Bonilla A, Perez-Rostro CI, Ramirez JL (2004) Criteria for assessing larval and postlarval quality of Pacific white shrimp *Litopeneaus vannamei*. (Boone). Aquaculture 233:181–195
Salser B, Mahler L, Lightner DV, Ure J, Danald D, Brand C, Stamp N, Moore D, Colvin B (1978) Controlled-environment aquaculture of penaeids. In: Kaul PN, Sindermann CJ (eds) Drugs and food from the sea, myth or reality? The University of Oklahoma Press, Norman, Oklahoma, pp 345–355
Samocha TM, Guajardo H, Lawrence AL, Castilie FL, Speed M, McKee DA, Page KI (1998) A simple stress test for *Penaeus vannamei* postlarvae. Aquaculture 165:233–242
Santiago AC Jr (1977) Successful spawning of cultured *Penaeus monodon* Fabricius after eyestalk ablation. Aquaculture 11:185–196
Shigueno K (1975) Shrimp culture in Japan. Assoc. Intl. Tech. Promotion, Tokyo, p 153
Villaluz DK, Villaluz A, Ladrera B, Sheik M, Gonzaga A (1969) Reproduction, larval development and cultivation of sugpo (*Penaeus monodon* Fabricius). Philippine J Sci 98(3-4):205–233
Walker PJ, Mohan CV (2009) Viral disease emergence in shrimp aquaculture: origins, impact and effectiveness of health management strategies. Rev Aquac 1:125–154
Wear RG, Santiago A Jr (1976) Induction of maturity and spawning in *Penaeus monodon* Fabricius by unilateral eyestalk ablation (Decapoda: Natantia). Crustaceana 31:218–220
Wedemyer A, McLeay DJ (1981) Methods for determining the tolerance of fishes to environmental stress. In: Pickering AD (ed) Stress and fish. Academic Press, New York, pp 247–275
Wickins JF, Beard TW (1978) Prawn Culture Research. Ministry of Agriculture Fisheries and Food, Directorate of Fisheries Research. Laboratory Leaflet No. 42, Lowestoft. p 41
Wyban JA, Swingle J, Sweeney JN, Pruder JD (1992) Development and commercial performance of high health shrimp from SPF broodstock *Penaeus vannamei*. In: Wyban J (ed) Proceedings of the special session on Shrimp farming, Orlando, Florida, 22-25May 1992. World Aquaculture Society, Baton Rouge, pp 254–260

Disease Diagnosis in Aquaculture

K. Jeena, M. Makesh, Megha K. Bedekar, K. U. Shyam, Arun K. Dhar, and K. V. Rajendran

Abstract

Intensification of aquaculture practices and species diversification have resulted in disease outbreaks in farmed aquatic organisms. Disease outbreaks are the major bottleneck to the expansion and development of sustainable aquaculture. In the overall aquatic animal health management strategy to minimize disease risks, early detection of the disease-causing agents is essential. The earlier the disease is diagnosed and treated, the better it is for prognosis and prevention of secondary infections. According to the World Health Organization (WHO), an ideal diagnostic assay for pathogen detection should be rapid, specific, sensitive, instrument-free and cost-effective. The diagnostic procedures can be classified as presumptive and confirmatory based on the extent of precision of the results obtained. Presumptive diagnosis is the preliminary diagnosis of a disease based on clinical signs and symptoms and is made by gross examination and post-mortem findings of the diseased animal. Confirmatory diagnosis involves the definitive identification of the causative agent or its sequelae using techniques such as histopathology, electron microscopy, parasitological, bacteriological, virological, immunological and molecular techniques. FAO/NACA classifies

K. Jeena · M. K. Bedekar · K. V. Rajendran (✉)
ICAR-Central Institute of Fisheries Education, Mumbai, India
e-mail: jeena@cife.edu.in; megha.bedekar@cife.edu.in; rajendrankv@hotmail.com

M. Makesh
ICAR-Central Institute of Brackishwater Aquaculture, Chennai, India

K. U. Shyam
EU Reference Laboratory for Fish and Crustacean Diseases, DTU, Lyngby, Denmark

A. K. Dhar
Aquaculture Pathology Laboratory, The University of Arizona, Tucson, AZ, USA
e-mail: adhar@arizona.edu

R. Kooloth Valappil et al. (eds.), *Aquatic Animal Health Management*,
https://doi.org/10.1007/978-981-96-7987-4_18

diagnostic levels into three categories, levels I, II and III, based on the technical expertise, facilities and other infrastructures required for disease diagnosis. The sampling procedure, target samples and sample size depend on whether the sampling is for targeted surveillance or routine health monitoring and on the sensitivity and specificity of the diagnostic methods used. This chapter outlines various presumptive and confirmatory diagnostic tests used in aquaculture for assessing and managing aquatic animal health, including numerous modifications of existing methods and their respective advantages and disadvantages. Some of the latest developments in disease diagnosis, such as proteomics, protein–protein interactions, digital droplet PCR, the use of machine learning techniques for disease diagnosis and CRISPR-Cas systems, are also described. Additionally, point-of-care tests that can be used on-farm with minimal equipment are discussed. Finally, various diagnostic protocols reported for the major bacterial, fungal and viral diseases of finfish and shellfish are summarized.

Keywords

Aquaculture · Fish and shellfish diseases · Disease diagnosis · Aquatic animal pathogen

1 Introduction

Global fisheries and aquaculture production registered an impressive figure of 223.2 million tonnes in 2022, with an increase of 4.4% from 2020 (FAO 2024). As global fish production continues to grow, the intensification of culture practices and species diversification resulted in increased incidences of disease outbreaks in aquaculture. As a sequel of these, aquatic animal health management gained the utmost significance and became the focal point of aquaculture research. Aquatic animal diseases are the most important bottleneck for the growth and expansion of sustainable aquaculture (FAO 2020). Periodic disease outbreaks in aquaculture operations have been causing enormous losses to this emerging sector of agriculture and global trade (Meyer 1991). Disease-causing agents in aquaculture operations are of viral, bacterial, fungal and parasitic origins and are often modulated by environmental factors. Early detection of the disease-causing agents is critical in controlling the outbreaks and preventing further spread. To diagnose a disease, it is important to examine the clinical manifestations and identify the aetiological agent accurately. In aquaculture, failure to diagnose a disease will result in large-scale mortality as animals are often grown at a high density and for the ease of disease transmission in water. Indiscriminate use of drugs and chemicals without even knowing the aetiologic agent can contribute to environmental contamination and the emergence of drug-resistant pathogens. The sooner the disease is diagnosed and treated, the better it is for the prognosis of the case and for the prevention of other secondary infections (Mardones 2020). Therefore, accuracy/specificity, rapidity of detection and affordability are the cardinal themes in disease diagnostics.

2 Overview of Aquatic Animal Disease Diagnosis

Diagnostic procedures can be classified as presumptive and confirmatory based on the extent of precision of the results obtained. Presumptive diagnosis is the preliminary identification of an abnormality based on the clinical signs leading to an assumption of the causative agent. On the contrary, a confirmatory diagnosis involves the definitive identification of the causative agent or its sequelae with a high degree of diagnostic confidence. The latter requires highly sensitive laboratory equipment and processes.

A review of the literature reveals aquatic animal disease diagnosis has come a long way from the simple description of the clinical manifestations and form of the infectious agent with a microscope to molecular characterization and identification of the genotype of a pathogen. Disease history and clinical signs, examination of wet mount preparations and tissue impression smears constitute traditional methods of disease diagnosis. Classical methods of disease diagnosis are performed through the use of light microscopy, histopathology and transmission electron microscopy for identifying cellular changes upon microbial infection and morphological description of microorganisms. Until recently, the diagnosis of infections in aquatic animals was mainly carried out by histopathological examination of tissues derived from a diseased animal. Histopathology further facilitated the identification and localization of disease-causing agents with the use of gene probes and labelled antibody-based assays. Serology and tissue culture-based diagnoses, two promising diagnostic tools, are underused in the disease diagnosis of aquatic animals compared to their use in terrestrial animals (Jaramillo et al. 2017). Antibody-based diagnostic techniques, using polyclonal and monoclonal antibodies on different assay formats, such as enzyme-linked immunosorbent assay (ELISA) and lateral flow chromatography, are widely used for a wide array of pathogens.

With the development of modern biotechnological tools, disease diagnostics started relying more on molecular methods than conventional diagnostic tools. In the molecular diagnostic realm, disease diagnosis and characterization of the pathogens using non-radioactive probes, dot blot and in situ hybridization techniques were the first to be introduced. Subsequently, with the discovery of polymerase chain reaction (PCR) by Karry B. Mullis in the 1980s, disease diagnosis transitioned to PCR-based assays due to their specificity, sensitivity and speed of disease/pathogen detection. Additionally, PCR-based assays could reliably detect even a few copies of the pathogen genome, including pathogens that could not be cultured in vitro. Hence, the technique was rapidly adopted for disease diagnosis in humans, plants and animals, including aquatic animals. Today, PCR-based assays are widely used in disease surveillance, epidemiological studies and screening of sub-clinical/covert infections. The high sensitivity of PCR-based assays also has disadvantages, such as false positives due to contamination. In the early 2000s, real-time PCR and modifications of conventional PCR came into play, where the assay does not need gel electrophoresis of the amplicons for the interpretation of results. The elimination of post-PCR analysis (i.e. gel electrophoresis) coupled with higher sensitivity, ability to quantify pathogen load and real-time detection of the target pathogen even with less time than conventional PCR made the real-time PCR method a new gold standard in disease diagnosis. Today, real-time PCR assays and many modifications of this method using either dsDNA-binding dye such as SYBR-Green or TaqMan probe-based detection methods are available to detect and quantify a wide array of microbial pathogens infecting aquatic animals.

3 Levels of Diagnosis

FAO/NACA (2000) recommends the use of three levels of diagnostics (levels I, II and III) based on the resources, facilities and infrastructure required for disease diagnosis. Examination of the clinical manifestations in farmed animals, changes in the environmental factors during the culture period, clinical records and history of aquaculture practices form the level I diagnosis. Most importantly, level I diagnosis is performed at a farm level. Level II activities include parasitology, bacteriology, mycology and histopathology. Laboratories capable of performing level II diagnosis require basic equipment and personnel trained/experienced in aquatic animal pathology. The laboratory should maintain accurate diagnostic records and preserve and store specimens for level III diagnosis. Level III diagnosis includes laboratory observations using immunology, electron microscopy, virology and molecular biology tools. Level II and III diagnoses require aseptic conditions, sophisticated equipment and skilled personnel to perform laboratory practices and observations.

4 Sampling Procedures in Aquatic Animal Disease Diagnosis

On-farm sampling of aquatic animals is performed either targeting diseased animals (targeted surveillance) or for routine health monitoring. Sampling procedures aim to demonstrate disease freedom, precise diagnosis in case of mortalities or analyse the mortalities caused by diseases of unknown aetiology. Providing a detailed account of different approaches for aquatic animal disease surveillance and sampling procedures is not within the scope of this chapter. The topic is extensively described by Bondad-Reantaso et al. (2001), Tavornapanich et al. (2020) and other researchers in the field.

5 Gross Examination and Necropsy

5.1 Gross Examination

Diagnosis of diseases in aquaculture starts with the analysis of the history of activities carried out on the farm, including the source of seed, source and type of feed used, record of environmental conditions and gross and clinical signs exhibited by the farmed animals. Gross observation of the behaviour of aquatic animals, mortality pattern, if any, and surface examination of the body, skin, fins, eyes and gills will help in identifying any abnormality. Recording behavioural changes and clinical signs will help to identify the cause of a disease. The major behavioural signs of any diseased fish include lethargy, reduced intake or complete cessation of feeding, loss of balance, swimming near the water surface or sinking to the bottom, abnormal swimming movement such as corkscrewing, air gulping, flashing movement, rubbing against hard objects, congregation near the source of oxygen, etc. Major clinical signs include the appearance of white or black spots, changes in the size and colour of different organs, presence of ulcers, the presence of haemorrhage or abrasion on the skin, bloated and/or distended abdomen, abnormal gills and opercular movement, bulged and cloudy/opaque eyes, clamped /eroded fins, scale loss and large swelling or lump under the skin/ external body surface. Similarly, in the case of shrimp, a range of clinical manifestations can be observed, which include lethargy, reduced preening ability, broken antennae, muscular opacity, reddish discolouration, black gills, enlarged branchiostegites, stunted growth, etc. The appearance of black or white spots on the carapace and/or exoskeleton is also observed, depending on the nature of the infection. In the acute phase of infection, large-scale mortality can be observed. However, the clinical signs exhibited are primarily non-specific.

5.2 Necropsy

Post-mortem analysis or necropsy of the specimen is one of the essential steps in fish health examination. Initiation of the necropsy procedure involves anaesthetizing the fish using anaesthetics such as MS-222, benzocaine, lidocaine, 2-phenoxyethanol, clove oil, etc., as detailed by Jessup (2009); Ross and Ross (2009); Martins et al. 2019. Following anaesthesia, blood samples are collected based on the size and type of fish. The preferred blood collection route in finfish is the caudal vein, using appropriate needle sizes involving the use of anticoagulants based on the intended use of the blood samples (Midtlyng et al. 2000). In the case of crustaceans, haemolymph can be collected directly from the heart or from ventral sinuses. Post-blood collection, the external appearances are to be examined carefully and recorded. Necropsy is carried out aseptically to examine the internal organs, their appearance and integrity, the presence of any fluid in the body cavity and for collection of the internal organs and their preservation storage for further analysis. An impression smear or wet mounts from the mucus collected from skin, gills or intestine, or the squash preparations of the gill and other internal organs can provide information on the presence of any parasites when the sample is observed under the microscope. The adequate number of fishes required to be subjected to necropsy is based on the purpose, whether it is for disease investigations or routine health monitoring, i.e., active surveillance or passive surveillance, respectively (Bondad-Reantaso et al. 2001; Tavornapanich et al. 2020).

The information gathered following gross examination and necropsy will aid only in presumptive diagnosis. Confirmation of initial/ presumptive diagnosis is achieved by subsequent analysis of samples by histopathology, electron microscopy and a wide array of bacteriological, virological, immunological and molecular techniques. The following are the different types of diagnostic techniques used in aquatic animal health assessment.

6 Histology

Histology, the study of the microscopic structure of tissues, plays a major role in understanding normal tissue architecture and in identifying lesions caused by a wide variety of known and unknown pathogens (AFS-FHS 2010). The technique aims at the identification of the alterations in tissues in the context of any abnormalities of pathogenic as well as non-pathogenic origin and provides a powerful tool for identifying the cause of morbidity at the tissue and organ level (Kent et al. 2013). Histology findings form the basis of performing more advanced antibody and nucleic acid probe-based techniques such as immunohistochemistry and in situ hybridization.

Histology is a commonly used tool to aid in the diagnosis of disease (Kent et al. 2013), detection of any emerging or unsuspected pathogens (Saksida et al. 2012; Rodnik et al. 2008), or unknown pathogens (Rodnick et al. 2008). The method can enhance the sensitivity of screening and surveillance programmes due to its general attribute, i.e., non-specific in its detection pattern when compared to other microbiological and molecular tools (Kent et al. 2013). Histology can also be used to validate findings from molecular diagnosis, as molecular techniques cannot determine the viability of a pathogen in tissue compared to the presence of nucleic acids of the target pathogen in the tissue due to physical contamination. Hence, histology provides valuable information on the general health of an animal as well as the detection of a wide range of pathogens associated with disease conditions (Aranguren and Figueras 2016). Basic knowledge of the normal tissue structures and variations among different species and any departure from normal tissue architecture due to infection with a pathogen or physical or chemical stimuli are critical for disease diagnosis (Bruno et al. 2013).

7 Antibody-Based Tests for Disease Diagnosis

Unlike traditional/conventional diagnostic techniques that typically depend on physical examination and microscopy, antibody-based disease diagnosis depends on the detection of the antigen of a pathogen or the specific antibody produced following infection. Immunodiagnostic techniques make use of the specific interactions of antibodies with antigens and are more rapid, specific and sensitive than the traditional methods of disease diagnosis. These techniques are commonly used in surveillance and assessing vaccines' efficacy (Jaramillo et al. 2017). Antibody-based diagnostic techniques, or immunodiagnostics, complement the traditional techniques in disease diagnosis.

A variety of immunodiagnostic methods can successfully accomplish the detection of pathogens in both clinically and sub-clinically infected fish (Rose et al. 1989) without the need to isolate the pathogen (Adams and Thompson 2011). These tests detect the pathogen using immune sera (polyclonal (PAbs) or monoclonal (MAbs)) developed against specific pathogens or antigenic epitopes in laboratory animals such as mice or rabbits. Tissue sections, tissue samples and body fluids from infected/suspected animals can be used to detect antigens/antigenic epitopes using pathogen-specific antibodies.

Immunodiagnostic techniques include complement-fixation test (CFT), neutralization test, haemagglutination, enzyme-linked immunosorbent assay (ELISA), immunohistochemistry (IHC), precipitation tests, western blotting (WB) and dot blotting. In addition, Lateral flow immunochromatographic assays (LFIA) promise a highly sensitive and specific detection of one or more target antigens in a sample. The following section briefly describes different immunodiagnostic techniques currently used for disease detection in aquatic animals.

7.1 Enzyme-Linked Immunosorbent Assay (ELISA)

ELISA is the most powerful of all the immunochemical techniques. ELISA can be used to detect or measure pathogens and host antibody levels and is considered a gold standard of immunoassays. Compared to other serological test formats, ELISA offers advancements in simplicity, low cost and rapidity (Jaramillo et al. 2016). The technique makes use of enzyme-conjugated antibodies where the enzymes are covalently bound to the antibody in such a way that the enzyme's catalytic properties and the antibody's specificity remain unaltered (de la Peña 2001) and aids in the detection and quantification of the target by means of measuring the intensity of colour developed at the termination of the reaction. Enzymes used for the assay include peroxidase, alkaline phosphatase, β-galactosidase, etc., with the potential to yield coloured products on catalysing specific reactions, which are measurable even at lower concentrations.

A range of variations exists in the ELISA technique for the detection or quantitation of pathogens. A basic alteration is made in the methodology of the assay in order to enable the detection of an antigen or antibody while maintaining the basic principle and steps unaltered. Based on the variations, ELISA tests can be classified into four categories:

1. Direct ELISA
2. Indirect ELISA
3. Sandwich ELISA
4. Competitive ELISA

7.1.1 Direct ELISA

Direct ELISA aims to detect target antigens in a test sample. For this, initially, the antigen (sample in which the presence of the antigen has to be tested) is coated on

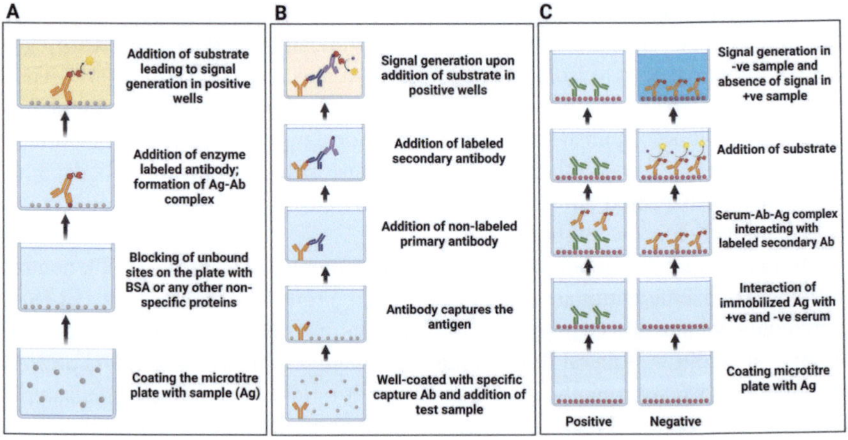

Fig. 1 A diagrammatic representation of Enzyme-linked immunosorbent assays (**a**) Direct ELISA; (**b**) Antigen Capture Indirect ELISA; (**c**) Competitive ELISA

the microtitre plate, followed by washing the plate to remove any unbound antigen. This step is followed by the addition of blocking agents to facilitate the blocking of free sites on the microtitre plate. Commonly used blocking agents in ELISA are bovine serum albumin (BSA), ovalbumin, aprotinin, skimmed milk powder or any other animal proteins. Enzyme-labelled antibody specific to the target antigen is then added, incubated and washed in order to remove any unbound antibody to the antigen. Following incubation, a substrate and/or chromophore is added, which results in the development of a coloured product in the case of a positive reaction. The intensity of the colour is measured as optical density (OD) using an ELISA reader or spectrophotometer at a particular wavelength, usually ranging from 450 to 540 nm. The intensity of the colour is directly proportional to the concentration of the target analyte. The quantification of the target is performed using a standard curve generated by known concentrations of a protein. The direct ELISA method is rapid compared to other variations described below because of its fewer steps. In addition, as no secondary antibody is used in the direct ELISA, the cross-reactivity of antibodies is eliminated. However, the sensitivity of direct ELISA is very low when compared to other ELISA methods (Fig. 1a).

7.1.2 Indirect ELISA

Indirect ELISA differs from direct ELISA due to the use of an enzyme-labelled secondary antibody specific to the primary antibody. The method involves the use of two antibodies: (i) a primary antibody that detects the antigen and binds to it and (ii) a labelled secondary antibody specific to the primary antibody. Hence the workflow in an indirect ELISA method involves the following steps: antigen (sample) is coated on a microtitre plate; free antigen binding sites on the plate are blocked with a blocking agent followed by the addition of primary antibody; unbound antibody is removed by washing, and a labelled secondary antibody which is specific to the primary antibody is added. Finally, the target antigen/analyte is detected by adding a substrate/chromogen to generate a coloured product in the case of a positive reaction. The intensity of colour is measured as OD using a spectrophotometer. The OD of the coloured product is directly proportional to the antigen concentration present in a test sample. Indirect ELISA is highly sensitive when compared to direct ELISA. However, a major drawback of indirect ELISA is antibody cross-reactivity, resulting in false positives (Fig. 1b).

7.1.3 Sandwich ELISA

Sandwich ELISA, also called antigen capture ELISA, aims at detecting unknown antigens in test samples by making use of two antibodies specific to two different epitopes of the test antigen. Here, the primary antibody is immobilized on the solid matrix first, and the free antibody binding sites on the matrix are blocked by adding blocking agents. A test sample containing antigen is then added and allowed to react with the antibody. This step is followed by a washing step to remove the unbound antigen. To the antigen-antibody complex, a secondary antibody (against the target antigen) that is labelled with an enzyme or any detection agent is added and allowed to interact. Following a washing step to remove unbound secondary antibody, a substrate for the enzyme is added and allowed to react. If the reaction is positive, a coloured product will be developed upon the addition of the substrate. The intensity of colour can be measured using a spectrophotometer or ELISA reader, which infers the concentration of the test target. Sandwich ELISA is the most sensitive and specific among different ELISA methods. An advantage of this method is that the antigen need not be purified prior to the assay. A major disadvantage is the cost of the assay, in addition to the requirement of antibodies targeting non-overlapping epitopes on the antigen and the time required to complete the assay (Engvall 2010).

7.1.4 Competitive ELISA

Competitive ELISA can be used to detect and quantify target antigens as well as antibodies in a test sample. The method involves a competition between the labelled and non-labelled analyte, test antigen or antibody. The reagents used and steps involved in the assay vary with respect to the target analyte. Briefly, to detect and quantify antibodies in a given sample, the antigen needs to be immobilized on a solid matrix. To this, a mixture of labelled antibodies of known concentration and a test sample containing non-labelled antibodies is added and allowed to react with the immobilized antigen on the solid matrix. During the reaction, the antibody in the test sample (serum in this case) will compete with the labelled antibody in binding to the antigen. Following this, a substrate for the conjugated enzyme is added to determine the reaction, and the intensity of colour developed is measured. If the test sample contains a high concentration of antibodies, they compete with the labelled antibodies to bind the antigen matrix, yielding lesser colour development and vice versa. In this assay, colour development is inversely proportional to the amount of analyte present in the test sample (Fig. 1c).

In ELISA, in order to differentiate background reactions from true positives, the cut-off value for OD is to be determined by including appropriate positive and negative control reactions. According to published literature, the sensitivity of ELISA techniques depends on the number of antigens/antibodies bound to the matrix, the specificity and avidity of the antibody, and its integrity even after labelling with detection agents or enzymes (de La Peña 2001). Washing steps enable the removal of unbound antibodies, resulting in the measurement of only labelled antibodies bound to the immobilized antigen.

7.2 Immunohistochemistry (IHC)

Immunohistochemistry is an important diagnostic and experimental tool used for the identification and localization of pathogens (Duncan and Seigel 2016) using specific antibodies conjugated with an enzyme for the detection of antigen–antibody complexes. Monoclonal antibody probes are preferred to polyclonal antibody probes because of improved specificity in binding to the antigens.

Histology is an essential prerequisite for performing IHC. IHC involves the incubation of thin paraffin-embedded tissue sections with the suspected antigen-specific antibody labelled (direct assay) or non-labelled (indirect) with an enzyme or any detection agent such as a fluorophore. Enzymes commonly used as labels are alkaline phosphatase (AP) and horseradish peroxidase (HRP). A range of chromogenic and chemiluminescent substrates can be used with these enzymes. Diaminobenzidine is the most widely used enzymatic substrate (Matos et al. 2010), and it gives a brown-coloured stable reaction product. The insoluble brown reaction product developed by enzyme-substrate interaction can be analysed using a light microscope. The reaction product developed on enzyme–substrate interaction is more long-lasting than fluorescent signals and thus is the gold standard in IHC (Duncan and Seigel 2016). Thus, chromogenic substrates are preferred over fluorescent and chemiluminescent substrates due to the durability of the coloured product (Fig. 2).

7.3 Flourescent Antibody Technique (FAT)

The fluorescent antibody technique utilizes fluorescent dyes to visualize the target antigen/protein. The technique is a rapid, simple and easy-to-perform assay. The basic requirement is formalin-fixed tissue sections, the target antigen-specific antibody tagged with a fluorophore, fluorescent/confocal microscope and skilled personnel to perform the assay. The technique enables the detection of microbes that are difficult to culture (Miles et al. 2003), pathogens in samples cultured from infected fish, detection, visualization and localization of viruses and bacteria directly on formalin-fixed tissue sections or imprints prepared from infected tissue (Adams and Thompson 2011). Fluorescent antibody technique involves the amplification of signal, resulting in a higher sensitivity compared to immunohistochemistry (IHC) (Im et al. 2019).

Fixation of tissue samples in an appropriate fixative in order to prevent post-mortem changes and to preserve the morphology is an essential and initial step in FAT. Neutral buffered formalin (NBF) and Davidson's fixative are used to preserve

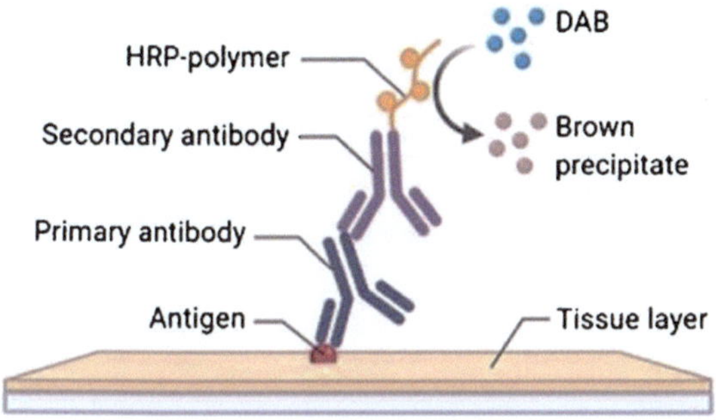

Fig. 2 An illustration of the basic format of immunohistochemistry (IHC) method. Tissue localization of antigen using a primary antibody and a species-specific secondary antibody conjugated with the enzyme horseradish peroxidase (HRP) is shown. Upon addition of a substrate (Diaminobenzidine), a coloured product is formed when the target pathogen is present and the reaction is considered positive

fish and shrimp tissue samples, respectively. Following fixation, the tissues are subjected to paraffin embedding. This step enables thin sectioning of tissues. These tissue sections are mounted onto glass slides, deparaffinized and rehydrated. Xylene is the most common medium for deparaffinization. Rehydration is done using decreasing grades of ethanol and distilled water washes (Harlow and Lane 1999). Antigen retrieval is an essential step to be performed in FAT in order to expose the epitopes on paraffin-fixed and treated tissue for its successful binding with the specific antibody. Heating sections in buffer (Myers 1989) and the use of proteases are needed for antigen retrieval without changing its conformation and tissue morphology (Im et al. 2019). Similar to other immunodiagnostic techniques, FAT can be performed in either direct or indirect (IFAT) format, which varies in the use of two antibodies.

In direct FAT, the primary antibody labelled with a fluorophore is used to react with the target antigen in the tissue sections/ samples. In contrast, in indirect FAT (IFAT), a secondary antibody labelled with a fluorophore and specific to the primary antibody is added to the target antigen in the initial step. This extra step of incubation with labelled secondary antibody is known to enhance the sensitivity, signal amplification and ability to detect several samples in the same assay (Vancells 2016). Most importantly, the primary antibody should be derived from a different animal than that of the sample, and the secondary antibody from a different animal than that of the primary antibody so that cross-reactivity of the antibodies (directly or indirectly) can be minimized. Signal amplification can be attained by incorporating fluorophore-protein complexes such as fluorophore-avidin/streptavidin-labelled secondary antibodies for the detection of biotinylated primary antibodies. Blocking plays a major role in preventing the antibody from binding the non-target epitopes on the tissue section. Commonly used blocking buffers are mentioned elsewhere in this chapter. Intermediate washing steps in order to wash off the unbound/free antibodies and prevent cross-reactivity are recommended. Further improvement of

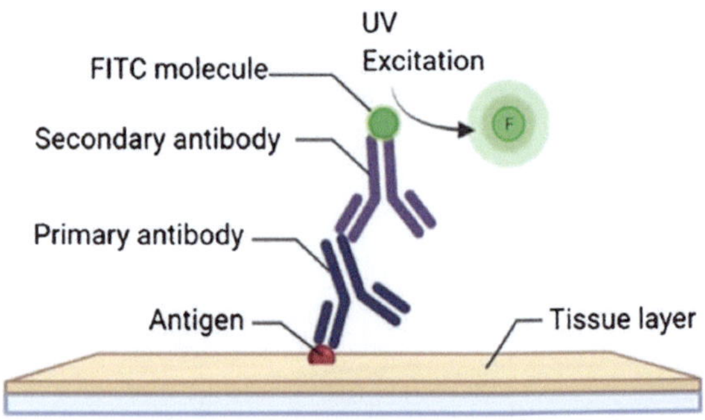

Fig. 3 An illustration of the basic format of fluorescent antibody test/ immunofluorescence technique. Antigen on the tissue section is localized using a primary antibody, and a secondary antibody conjugated with a fluorophore is used for detection

signal amplification in FAT can be achieved using the polyclonal antibodies in conjugation with enzyme labels such as fluorophore-conjugated streptavidin–biotin complex. FAT also enables the possibility of multiplexing to detect multiple pathogens simultaneously (Philips 2007) (Fig. 3).

7.4 Lateral Flow Immunoassay (LFIA)

Lateral flow immunoassay is a powerful and portable platform for the diagnosis of infectious diseases. It is the most extensively applied point-of-care testing (POCT) platform (Shi and Wang 2021). It is a low-cost, convenient, rapid method that requires a small sample volume, minimal specialized equipment and limited expertise with a quick turnaround time (Kozel and Burnham-Marusich 2017; Nardo et al. 2021). Laboratory-based diagnostic techniques require a long duration and costly equipment to perform disease testing, but disease diagnosis in an aquaculture farm setting often demands a fast and on-site detection of the disease (Hansen and Wahed 2020). The growing need for on-site disease detection led to an extensive research effort to reduce the turnaround time for disease detection without compromising the sensitivity (i.e. limits of detection) and specificity of detection.

Lateral flow immunoassay is based on the principle of chromatography and the immune recognition reaction. The assay is performed on a lateral flow device (LFD), a type of biosensor. Test and control lines are printed on a porous membrane, usually a Polyvinylidene Fluoride (PVDF) membrane or nylon membrane. A specific recognition reagent for the target analyte (i.e. specific antibody/antigen against the target antigen/antibody) is immobilized on the membrane in defined zones called reactive zones or detection sites. The sample and reagents flow on the membrane by capillary action, bind to the recognition agent and localize at the detection sites (Fig. 4). The LFIA technique has been developed for the detection of many viral and bacterial pathogens significant in aquaculture. Over the years, the technique has undergone significant improvements to enhance sensitivity by modifying the basic technique and incorporating different ways of detecting the analytes. Antibody/oligonucleotide conjugated nanoparticles, such as colloidal gold, silica nanoparticles, quantum dots and other nanomaterials, have been used for the visual

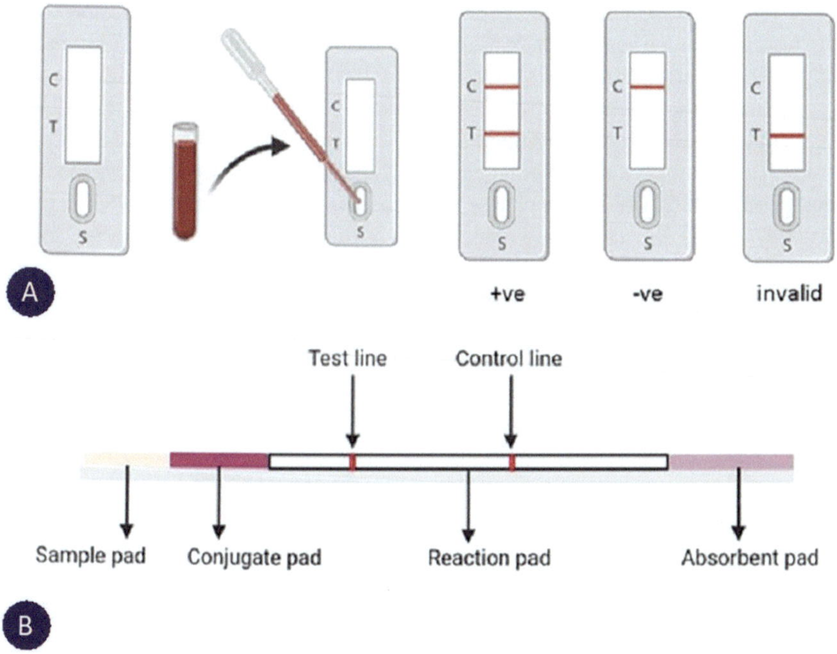

Fig. 4 A schematic diagram of the lateral flow immunoassay. (**a**) Parts of LFIA strip; (**b**) Description of the test strip, procedure for sample addition and the pattern of result interpretation. The test strip possesses a sample input window and an output window to visualize the results. Upon addition of 4–5 drops of the tissue homogenate/ cell supernatant on the sample pad through the window, the sample moves towards the reaction pad, which possesses the test and control zones. If the sample contains the antigen, it interacts with the immobilized antibody on the test zone, and colour develops. The control zone is kept for quality control purposes, and this zone possesses the antibody specific to the immobilized antibody at the test zone

detection of the analyte. Nanoparticle-based lateral flow biosensors for the detection of PCR products (Toubanaki et al. 2015, 2018), and nanostructured microfluidic assay (Yu et al. 2017) are also in use. Prosthuma-Trumpie et al. (2009) and Bahadir and Sezgintürk 2016) have comprehensively reviewed the basics, principles, designs and detection agents, strengths, weaknesses and opportunities of the LFIA technique.

8 Nucleic Acid-Based Tests for Disease Diagnosis

Traditional methods of diagnosis depend largely on the isolation of the pathogen in media or cells and identification using the phenotypic or serological properties of the pathogen and the pathological effects in the host tissues, among many other manifestations. The unique sequences of nucleic acids exist in every organism, and this characteristic is used in nucleic acid-based detection techniques for their identification. The invention of nucleic acid-based detection techniques has enabled the use of more precise and rapid tools in disease diagnosis, resulting in the direct detection and quantification of pathogens targeting their nucleic acid. The nucleic acid-based detection techniques have very low limits of detection and high specificity and enable the detection of non-culturable microorganisms whose in vitro culture media are unavailable (Cunningham 2004; Adams 2009; Adams and Thompson 2011). The assay uses short sequences of DNA/RNA specially designed complementary to the target sequence. These short sequences can be either probes or primers. They detect the target sequence by hybridizing it. Probes hybridize to the single-stranded complementary sequence on the target and can be detected using the fluorophore or enzyme labels on them. Short DNA/RNA sequences, commonly referred to as primers, are also used in a range of molecular techniques that require nucleic acid synthesis. Primers hybridize with the single-stranded complementary sequence of the target and provide a double-stranded 3' OH terminus, which acts as a starting point for the synthesis of a new strand complementary to the parent strand.

8.1 Probe-Based Nucleic Acid Detection

Nucleic acid probes are single-stranded fragments of DNA or RNA labelled with a marker that can hybridize (bind) with high specificity to DNA or RNA that has complementary sequences. Researchers began to use cloned DNA probes to detect viral nucleic acids in the 1970s. DNA probes for detection of heat-labile enterotoxin were first used by Moseley et al. (1982). Since then, a large number of DNA probes have been developed to detect various pathogens. Gene probes can be used in the detection of pathogens in tissue samples, environmental samples, pathogens that are difficult to culture, and those that exist in a dormant state, and to identify and differentiate closely related strains of any given pathogen (Adams and Thompson 2011).

Radioisotopes such as ^{32}P, ^{3}H, ^{35}S, ^{125}I and ^{14}C were initially used as labels to detect the hybridization. However, the short half-life, the requirement of specific radioactive detection systems, and associated health hazards limited their use and led to the invention of alternatives to radioisotope labelling and detection. Enzymes such as alkaline phosphatase and, peroxidase, and fluorophores are now used in the direct detection of the probes. However, direct detection was reported to have lower sensitivity due to limitations in the number of enzymes coupled per probe (Walker and Dougan 1989). Nucleic acid probes have become increasingly popular with the use of highly specific probes labelled with ligands like biotin in place of radioisotopes, which facilitated indirect detection of the hybridization and thus enhanced sensitivity. Detection of biotin is carried out either using avidin–enzyme conjugates or anti-avidin antibodies. Further, digoxigenin was reported as an alternative and superior to other non-radioactive ligands used in ISH (Komminoth 1992). Non-radiolabelled probes have advantages over radio-labelled probes in that they are safer and have longer shelf lives with higher sensitivity of detection.

The nucleic acid hybridization-based assay requires a probe labelled with the reporter molecule, the target nucleic acid (sample) and an optimized protocol to carry out the assay. Prerequisites for the detection of a target in a complex mixture

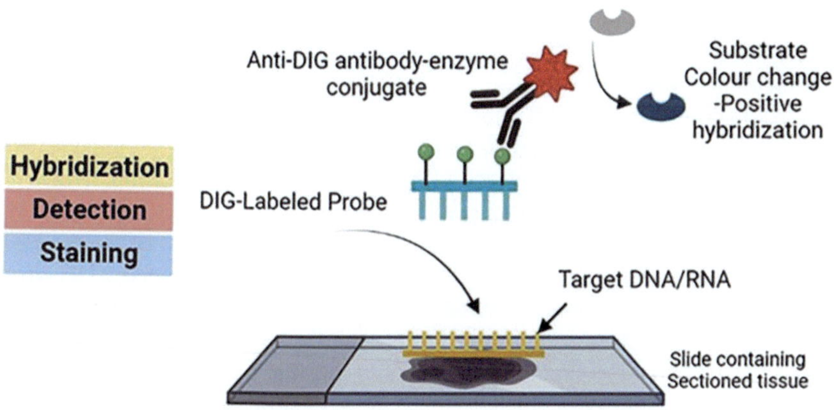

Fig. 5 A schematic representation of the principal steps in performing the in-situ hybridization technique. The tissue section immobilized on a glass slide is allowed to react with the labelled probe (DIG-labelled probe). If the target DNA/RNA is complementary to the added probe sequence, the probe hybridizes with the target sequence. Detection of probe-target hybridization is carried out by adding an anti-DIG antibody, which is conjugated to an enzyme. After the addition of the substrate, if a colour change occurs, it indicates a positive hybridization, whereas the absence of colour development indicates a negative reaction. Stringent washing steps are mandatory after every step to avoid background reactions or false positives

of nucleic acids are the denaturation of double strands, identifying the optimum temperature for probe hybridization in order to prevent mismatches and identifying the effective concentration of the probe for increased rates of hybridization. Hybridization can be performed in solutions or on solid support (dot-blot) or even on sections of tissues fixed on a glass slide (in-situ hybridization), the latter being a potential tool for pathogen tissue localization. In-situ hybridization allows locating the target analyte in a tissue section and avoids non-specific reactions, which results in false positive diagnosis in dot-blot assay (Lightner 1996) (Fig. 5).

8.2 Primer-Based Nucleic Acid Detection

Primers are short nucleic acid sequences that hybridize with complementary sequences and synthesize new strands, which are prerequisites in techniques that involve the detection of complementary sequences in the sample. Polymerase chain reaction (PCR) is the most widely used technique that involves the amplification of target DNA, starting with the binding of primers to complementary gene sequences. A few variants of PCR make use of both the primers and probes in the detection and quantitation of the amplicons. Besides PCR, primers have applications in reverse transcription (synthesis of cDNA from RNA), probe labelling and gene sequencing reactions (Van Pelt-Verkuil and te Witt 2019).

8.2.1 Polymerase Chain Reaction (PCR)

PCR is a method of in vitro enzymatic synthesis of specific fragments of DNA and amplification through repetitive cycles of oligonucleotide primer-directed DNA replication. Amplification of the target genetic element is achieved in three steps, viz., denaturation, where the target DNA is heated to 90–95 °C for strand separation, annealing at temperature 45–65 °C depending on the melting temperature (T_m) of the primers and extending the primers, complementarily bound to the target gene sequence by DNA polymerase at 72 °C resulting in the replication of target sequence. Upon annealing to the target strand, primers provide a double-stranded 3-OH terminus that acts as a starting site for the synthesis of complementary strands. With repeated heating and cooling cycles (thermal cycles), the denaturation of the double strands occurs and annealing of the primers to the single strands takes place. The annealing temperature is calculated based on the melting temperature (T_m) of the forward and reverse primers. During extension, the dNTPs, the building blocks of nucleic acids, are incorporated in the complementary strand, resulting in the replication of the target sequence. At the end of each cycle, the target DNA copy number doubles, and the synthesis of DNA continues to occur exponentially depending on the availability of primers, polymerase enzyme and its cofactors and the dNTPs. Denaturation, annealing and extension steps are carried out for 30–35 cycles to obtain the 2^n times replication of the template present in the sample, where 'n' is the number of cycles in the reaction. The final PCR product is visualized using agarose gel electrophoresis with ethidium bromide, a double-stranded DNA intercalating dye or SYBR Green, a double-stranded DNA minor groove binding dye. Reference markers used in the gel electrophoresis, along with the samples, aid in identifying the amplified DNA based on their size (number of base pairs) (Fig. 6).

The limit of sensitivity of PCR is theoretically a single molecule, making PCR a highly sensitive tool for the specific detection of the target DNA sequences in a sample (Wages Jr 2005). With the sensitivity it offers, PCR is considered the gold standard in disease diagnosis. With the development of this simple method of amplification of the target DNA sample by Saiki et al. (1985, 1988) and Mullis and Faloona (1987), PCR has been widely adopted in humans, as well as veterinary and fisheries biomedicine and diagnosis.

Variants of PCR

Since its invention, the technique of PCR has been revolutionizing the field of biomedicine, diagnostics and health management. The technique has been modified into different variants to suit a variety of applications.

Reverse Transcription (RT) PCR

Reverse transcription PCR involves the synthesis of complementary DNA (cDNA) from RNA, i.e. the process of reverse transcription. Synthesis of cDNA involves the synthesis of the first strand, which then serves as a template for the synthesis of the second strand. The first strand reaction is catalysed by reverse transcriptase, an RNA-dependent DNA polymerase with total RNA or purified mRNA as a template. Primers used in RT-PCR can be either of the following three: (i) oligo dT, (ii)

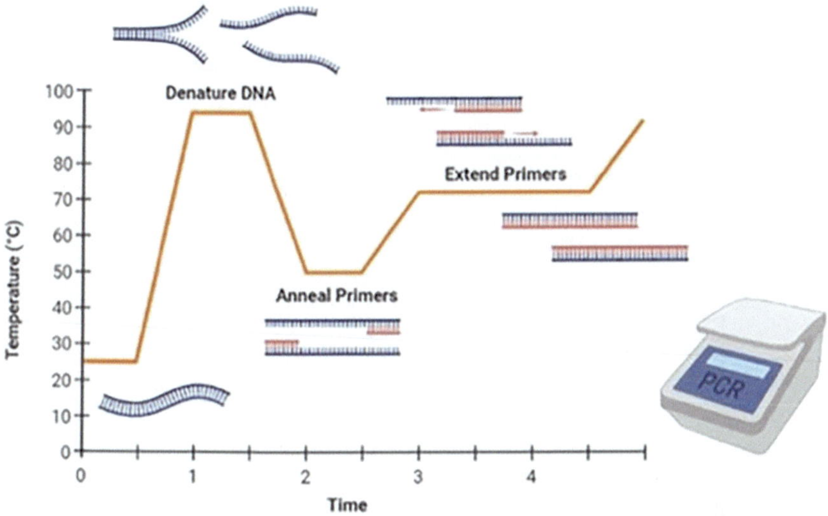

Fig. 6 Diagrammatic representation of a polymerase chain reaction. PCR is performed in a thermal cycler. The initial step of PCR is the denaturation of double-stranded DNA in the sample at an elevated temperature of 90–95 °C, resulting in the formation of single-stranded DNA. Following denaturation, the sample goes through a cooling to bring down the temperature to 45–65 °C, which facilitates the binding of primers to the single-stranded template. Then the temperature is gradually increased to 72 °C, which facilitates the synthesis of strands complementary to the template by primer extension. The extension step is the action of DNA polymerase present in the reaction mixture.

random hexamer and (iii) gene-specific primers. Oligo dT primers are designed to anneal to the poly(A) tail of mRNA. Oligo dT primers comprise 12–18 polythymidine repeats that bind to the poly(A) tail of eukaryotic and viral mRNAs. The primer is not recommended for reverse transcription of prokaryotic mRNA as it lacks the poly(A tail). Random hexamer primers comprise short oligonucleotide fragments of six nucleotides, lack specificity in binding and can randomly anneal to different regions in an mRNA. Gene-specific primers are target-oriented primers that are designed specifically to anneal to the target sequence on the RNA to amplify a particular gene sequence in the sample.

Multiplex PCR

Multiplex PCR reaction facilitates the simultaneous amplification of multiple target sequences of DNA using different sets of primers in a single reaction. Optimization of annealing temperatures and the amplicon sizes are the major criteria to be considered while developing a multiplex PCR. The primers used in a multiplex PCR should have similar melting temperatures (Tm) in order to facilitate the annealing of primers to their respective complementary target sequences at a particular temperature and allow the reaction to proceed. In order to visualize the results in an agarose gel, the amplicon sizes of the targeted products should be different but within the

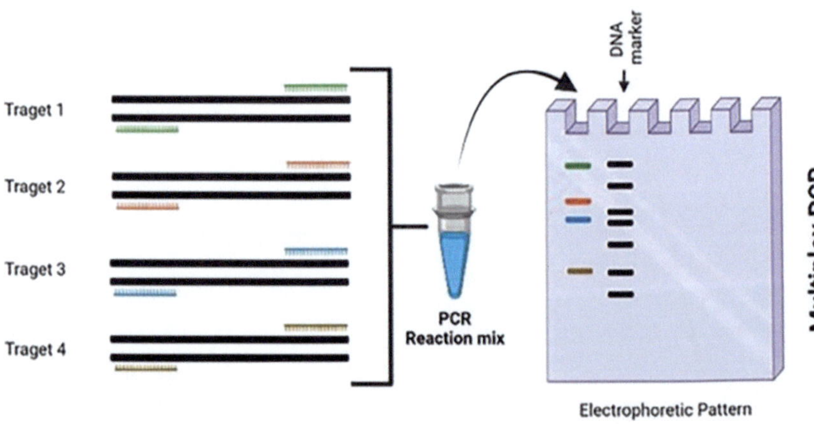

Fig. 7 Diagrammatic representation of the basic process of performing a multiplex PCR. Multiple primers targeting different regions of the same target or different targets are added to a single reaction mixture. After performing the reaction, multiple target amplicons with different sizes are visualized using gel electrophoresis

same size range, which facilitates the synthesis of each amplicon with the same efficiency and at equal rates (Mahony and Chernesky 1995) (Fig. 7). Advantages of multiplex PCR assays include the use of internal controls to monitor the false negative reactions; reduction in cost and time as a single reaction targets multiple sequences.

Nested PCR

Nested PCR involves two sequential steps of amplification using different sets of primers in the first step and the second step of the amplification process. The product of the first amplification serves as the template for the second step of PCR, and the primers used in the second round of PCR are designed to target an internal region of the first PCR product (Fig. 8). The use of two sets of primers increases specificity and sensitivity of detection of the target. Higher sensitivity is the result of a higher number of cycles performed due to the use of two rounds of amplification, and the higher specificity derives from the binding of two different sets of primers to two different regions of the same target template (Green and Sambrook 2019). Conventionally, the first set of primers is used to perform the first round of PCR, and an aliquot of this reaction is taken as the template for the second round of PCR using the second set of primers.

Due to the increased risk of contamination on the transfer of first-round amplification products to a second tube, a few modifications were developed for nested PCR. One of them is the single tube nested PCR assay (STNPCR) developed by Abath et al. (2002), where the nested primers are immobilized onto the cap of the microtube, thereby avoiding the need to open the reaction tube to add the second set of primers thus reducing the risk of contamination. Physical separation of amplification mixtures with wax or oil and designing the second primer set with higher

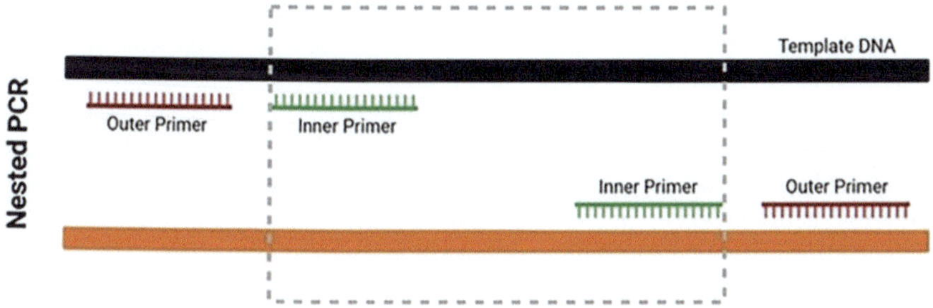

Fig. 8 A diagrammatic representation of the position of the different sets of primers used in a nested PCR

annealing temperatures are proven effective in reducing the risk of contamination (Tang and Stratton 2006).

Touchdown PCR

Touchdown PCR helps increase the specificity of the reaction by using higher annealing temperatures during the initial cycles and gradually reducing the temperatures towards the end of the cycles. The principle behind the touchdown PCR is that a higher annealing temperature allows only the specific base pairing of the primer and the template. Therefore, the DNA amplified during the initial cycles are highly specific and represents the intended target. The amplicons from the initial cycles are further amplified during subsequent cycles at lower annealing temperatures but with more efficiency.

While the different PCR methods discussed above are commonly used in disease diagnosis, many other modifications have also been developed for other applications. These include asymmetric PCR (sequencing and hybridization probing), assembly PCR (oligo synthesis), digital PCR (absolute quantification, analysis of copy number alterations, detecting rare mutations and next generation sequencing) (Pohl and Shih 2004); suicide PCR (paleogenetics; avoids false positive amplifications); and Variable Number Tandem Repeat PCR (genetics, biology research and forensics).

8.2.2 Loop-Mediated Isothermal Amplification (LAMP)

Loop-mediated isothermal amplification is an isothermal DNA amplification technique developed by Notomi et al. (2000). The technique is highly specific, efficient and rapid and can amplify a few copies of DNA to 10^9 copies in less than an hour (Notomi et al. 2000). This is achieved using *Bst* DNA polymerase, which has a strand displacement activity, and four sets of primers targeting a minimum of six distinct regions on the target sequence. The reaction is carried out in isothermal conditions, i.e. at one particular temperature without the need to change temperatures as in the case of PCR. The four sets of primers include inner (forward inner primer [FIP], backward inner primer [BIP]) and outer primers (forward outer primer F3, backward outer primer B3). A reaction is initiated with the binding of the inner primer with the target sequence and starts the synthesis of a new strand complementary to the target. FIP and BIP target two distinct sequences on the sense and antisense strands of the target DNA. The outer primer binds to its complementary region on the target and initiates the strand displacement reaction, displacing the complementary strand synthesized by the inner primer, and this forms a loop-like structure on one end where two regions appear to be complementary. This strand acts as a new template for the other inner primer, and the reaction proceeds, ultimately resulting in the formation of a dump-bell-shaped DNA. This dump-bell DNA is converted to stem-loop DNA by self-primed DNA synthesis. The stem-loop DNA acts as the starting material for the cycling reaction. Cycling reaction initiates the binding of FIP to the stem-loop at one end and BIP at the other, the synthesis of a complementary strand, and the displacement of the newly synthesized strand by the *Bst* DNA polymerase, finally resulting in the formation of elongated strands than the target and with stem-loops. Following this step is the recycling and further elongation of the elongated stem-loop DNA structures (steps elaborated in Notomi et al. 2000). The final product of the LAMP reaction is a mixture of stem-loop DNA of various lengths formed by annealing between alternately inverted repeats of the target sequence in the same strand (Fig. 9). The technique can also be used in the detection of RNA using reverse transcriptase, which is strand-displacing in nature. The

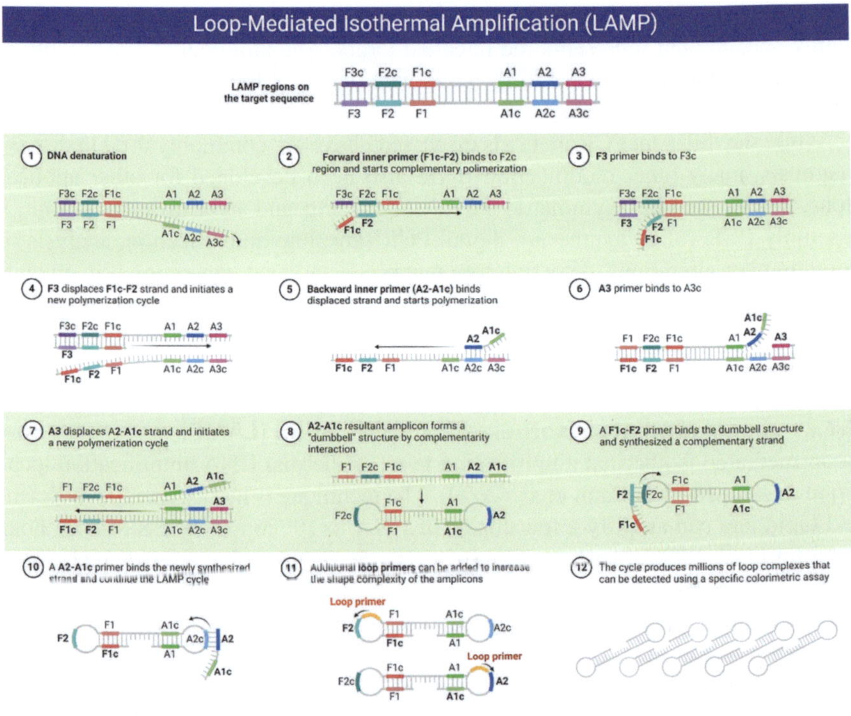

Fig. 9 A schematic representation of LAMP demonstrating major steps in the reaction

amplification process is completed within 1 hour, and the sensitivity of the reaction is similar to that of nested PCR (Notomi et al. 2015). Several modifications were developed to reduce the reaction time further. For example, Nagamine et al. (2002) reported the use of loop primers for a reduction in amplification time to half or one-third of the original LAMP method.

LAMP is a simple and easy-to-perform method, as it requires only a standard laboratory water bath or heating block to carry out the reaction. The technique is specific, as it uses four pairs of primers to target the template. In addition, multiplexing of LAMP (multi-LAMP) is also possible in differentiating microorganisms and in the detection of two or more targets in a single isothermal reaction. However, like other techniques, LAMP is associated with certain drawbacks. As the amplification is so rapid in the LAMP reaction, chances of contamination towards the final steps of the reaction are high. Also, as the amplicon detection method of measuring turbidity or fluorescence is not direct and measures even the non-target amplicons, there remains a possibility of false positive results.

Visualization of Lamplicons

The amplification products of the LAMP reaction (lamplicons) can be visualized using several methods. Agarose gel electrophoresis is the most commonly used visualization technique. DNA intercalating dyes like ethidium bromide or DNA minor groove-binding dyes like SYBR Green are used to stain lamplicons to visualize in an agarose gel. The LAMP products provide a ladder-like appearance in the agarose gel as the LAMP products consist of stem-loop DNA structures of varying sizes. During the LAMP reaction, a large amount of DNA is synthesized in a short time, resulting in the production of magnesium pyrophosphate. The magnesium pyrophosphate produces a white precipitate, which can be visualized by the naked eye. The accumulation of pyrophosphate as the reaction proceeds can also be quantified in real time by a calorimetric estimation of the turbidity. The amount of turbidity is proportional to the amount of DNA synthesized, which in turn depends on the quantity of the template present in the sample. The amplified products can also be detected and quantified using dsDNA binding dyes such as SYBR Green I and exposing the product to UV light in a transilluminator (Notomi et al. 2000). Researchers have been working towards simplification of the technique since its development. Techniques such as the use of microfluidic chips to perform the reaction, detection by electrophoresis and interpretation of the results on a single chip have been described (Hataoka et al. 2004). Detection of the single-copy gene using a bioluminescent assay in real-time (Hardinge et al. 2018) and the use of quenched fluorescent primers for reducing false positives (Hardinge and Murray 2019) are some other modifications of the original LAMP method.

8.2.3 Real-Time PCR

Research advancements in nucleic acid amplification methods have resulted in the improvement from end-point detection of PCR products to the real-time detection of the products as the reaction proceeds. Real-time PCR has the added advantage of being very sensitive, such that it can detect even one copy of the target gene and is amenable to high throughput. Real-time PCR assays have been developed for many aquatic animal pathogens.

There are three phases in a PCR: exponential, linear and plateau. In the exponential phase, amplification happens at an exponential order due to the abundance of reagents in the reaction mixture. As the reaction progresses, the reagents are used up, and the reaction enters a linear phase. Once the reagents become limited in availability, the reaction reaches a plateau phase where the rate of amplification diminishes and eventually stops. In a conventional PCR, the amplification is measured at the end phase, i.e., the plateau phase. In real-time PCR, the amplification of the target DNA is measured in real-time during the exponential phase as the reaction proceeds. This is a major advantage of real-time PCR over conventional PCR. The foundation of real-time PCR-based pathogen quantification is based on the idea that the quantity of PCR product formed at any given cycle number and the input target DNA have a quantitative relationship. Every cycle of real-time PCR detects the accumulation of amplicon, and this data is used to quantify the number of copies of templates in the sample. The data generated are then evaluated during the exponential phase of the PCR. In fact, real-time PCR employs a fluorescent molecule that reports an increase in the amount of DNA with a proportional increase in fluorescent signal. This is based on the fact that there is a quantitative relationship between the amount of starting target copies and the amount of PCR product at any given cycle number. The fluorescent chemistry employed for this purpose includes DNA-binding dyes and fluorescently labelled sequence-specific probes. The method is quicker in providing results and carries a lower risk of contamination, as it eliminates the need for post-PCR analysis of the amplicon, such as gel electrophoresis. Further, real-time PCR offers significant benefits in pathogen identification and quantification due to its improved sensitivity, repeatability and quantitative accuracy.

Baseline, Threshold and Cycle Threshold in Real-Time PCR

The fluorescence is captured in real time as the reaction proceeds. Initial cycles prior to target amplification with little change in the fluorescent signal are referred to as the baseline. The baseline is also denoted as background. When the fluorescence crosses the baseline/background, it is called the threshold. The threshold is always set at the exponential phase of the reaction. The increase in fluorescence above the threshold line is captured and displayed as an amplification plot on the screen of a computer attached to the real-time thermocycler. The amplification plot is a graphical representation of the reaction, and it initially contains the background and is followed by the increased fluorescence representing the exponential phase of the reaction, at which the line will reach above the baseline or threshold. This is followed by the linear phase and finally, the plateau. The cycle number at which the reaction outstretches the background is known as the 'cycle threshold or 'C_T value'.

Quantification of Target Copies

The starting copy number of the target and the amount of PCR product at any given cycle number are correlated in quantifying the target amplicons in real-time PCR. The C_T value and target copies are inversely related. The lower the C_T value of a given sample reaction, the higher the starting copy number is in a given sample and vice versa. The fluorescence of the reaction reaches above the background only after several rounds of amplification (higher C_T value). If the PCR efficiency is 100%, the amount of PCR product doubles with each cycle.

Real-time PCR assay can be used for absolute and relative quantification. Absolute quantification is used to determine the actual copy number of the target gene under investigation. Absolute quantification is carried out using a synthetic control (in-vitro synthesized RNA or plasmid DNA). The selection of an appropriate control is essential for the type of nucleic acid to be quantified. The control is made in such a way that the fragment will span the real-time PCR amplicon region. The number of copies in a synthetic control can be estimated using its concentration and Avogadro's number. To perform absolute quantification, a target template of known concentration is serially diluted (10-fold) over several orders of magnitude. The C_T value for each dilution is plotted against the copy number of the target to obtain a standard curve. The slope of the standard curve determines the PCR efficiency, and a slope of −3.32 indicates an efficiency of 100%. The C_T value of the unknown sample is plotted on the graph to determine the copy number of the target in a sample. The inclusion of a no-template control (NTC) in PCR enables the detection of contamination due to contaminated reagents or templates. An NTC contains all components of the amplification reaction except the templates. If there are primer dimers in the NTC, the primers should be redesigned. Melting curve analysis of NTCs can differentiate between primer dimer and amplification due to contaminating nucleic acids in the PCR reagents.

Two different chemistries of real-time assay are commonly employed in disease diagnosis (Fig. 10): SYBR Green, a double-stranded DNA minor groove-binding dye which specifically binds to all double-stranded DNA and TaqMan probe-based assays using a fluorogenic 5′ exonuclease assay. SYBR Green and TaqMan probe methods are discussed in the following sections.

SYBR Green Chemistry

SYBR Green is a DNA-binding dye that binds to the minor groove of the double-stranded DNA. As the amplification progresses during PCR, the quantity of double-stranded DNA synthesis also increases, and thus, the increase in binding of SYBR green occurs, resulting in increased fluorescence.

SYBR Green assays are simple, easy to perform and of low cost. Even though SYBR Green is a dsDNA binding dye, the chances of it binding primer dimers or non-specific amplicons cannot be avoided. Nevertheless, the non-specific fluorescence signal generated in the SYBR Green assay can be verified by performing a melting curve analysis. Since a target amplicon has a longer length than a primer dimer, it will have a higher melting temperature. By analysing the melting curve, it can be verified whether the increased signal intensity is due to the signals originating due to primer dimerization or due to the amplification of the targeted product. SYBR Green has limitations in the multiplex detection of pathogens.

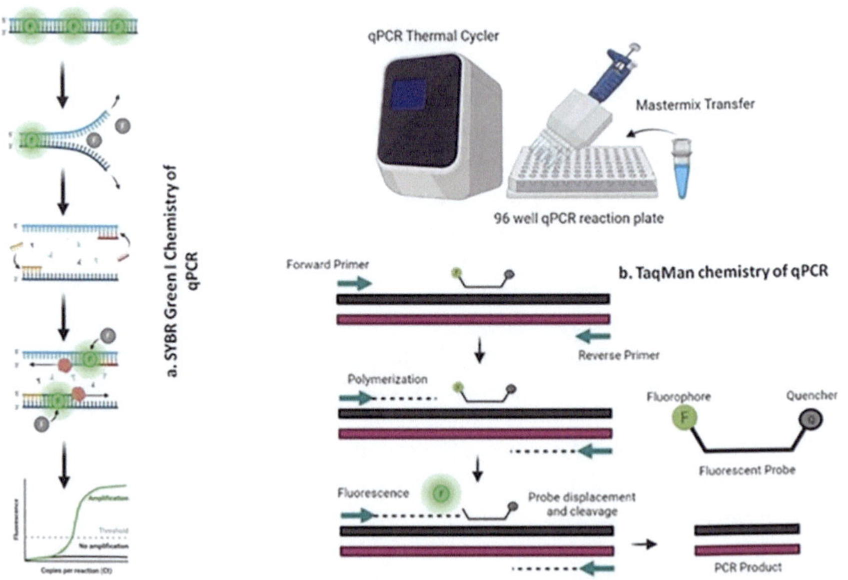

Fig. 10 Schematic representations of real-time PCR using SYBR Green I and TaqMan probe methods. The reaction mixture contains a template, forward and reverse primers, a probe (in the case of TaqMan chemistry), dNTPs, Taq DNA polymerase and a buffer. In the case of SYBR Green real-time PCR, the reaction mixture also contains SYBR Green I dye. Generally, the reactions are set up in a 96-well plate, and the plate is loaded onto a real-time PCR thermal cycler. As the reaction proceeds, fluorescence emission is captured, analysed and displayed on a screen

TaqMan Probe Chemistry

TaqMan assay uses a sequence-specific probe, usually 25–30 nucleotides long, possessing a reporter dye at the 5′ end and a quencher molecule at the 3′ end. The assay depends mainly on the 5′ exonuclease activity of the polymerase used in the reaction and the principle of Fluorescent Resonance Energy Transfer (FRET). The 5′ exonuclease cleaves the probe that remains bound to the DNA strand being amplified and interferes with the synthesis of a new DNA strand. FRET technology is based on the principle that when a high-energy dye is in close proximity to a low-energy dye, a transfer of energy will take place from high to low. The TaqMan probe is designed with a high-energy dye (Reporter) at the 5′ end and a low-energy molecule (Quencher) at the 3′ end. When excited by a light source, the fluorescence emission does not occur because of the close proximity of the reporter and the quencher. Examples of common quencher dyes include TAMRA, DABCYL and BHQ, whereas FAM, VIC and NED are commonly used reporter dyes.

During PCR, the probe anneals to a specific site downstream of the forward primer. When a new strand synthesis occurs, the 5′ exonuclease cleaves the probe, thereby setting the reporter and quencher apart, causing the cessation of energy transfer. This results in increased reporter emission. The instrumentation and the software used to carry out the fluorescence data capture and analysis during assay have the ability to capture and display the increase in the emission of fluorescence. Further, for a given sample, the increase in the reporter signal is directly proportional to the amount of product being synthesized. Therefore, the real-time detection of PCR amplification of a given target gene is made possible by combining 5′ exonuclease activity, FRET technology and specific instrumentation.

8.3 Digital Droplet PCR (ddPCR)

Digital droplet PCR is a unique third-generation PCR refinement that is known for its high specificity and sensitivity in absolute quantification of the target nucleic acid (Hindson et al. 2011; Li et al. 2018). Sykes put forward the concept in 1992, and the working principle at the single molecule level is based on the Poisson distribution. The ddPCR divides a PCR reaction mixture into a number of smaller reaction units as droplets. Each droplet is considered a separate reaction, and at the end of the reaction, each droplet is assessed as positive or negative. Using Poisson's distribution, the number of templates is positively correlated to the positive wells, so the exact copy number of templates can be calculated. The most prominent feature of this innovation is that it can detect low copies of DNA/RNA or even 'not-detectable' pathogens using conventional PCR. The method has a high accuracy and low variability and is less susceptible to reaction inhibition from the sample and a limit of detection of seven gene copies/mL (Ruelle et al. 2014; Te et al. 2015). In fish disease diagnosis, the method was found to take less time than other methods and cost half the price for real-time PCR. For example, Lewin et al. (2020) reported a ddPCR assay for the single and simultaneous detection of *Flavobacterium psychrophilum* and *Yersinia ruckeri* from farmed Atlantic salmon. The sensitivity was reported to be 0.0011 ng DNA for *F. psychrophilum* and 1.24 ng for *Y. ruckeri.*

8.4 Microarray Technology

Microarray technology allows the simultaneous detection of multiple nucleic acid sequences using probes spotted on a microchip (Fig. 11). The property of DNA to bind to its complementary sequence is employed in developing this technology. The probes are specific known sequences labelled and immobilized on the array, which can detect and hybridize a range of PCR-amplified complementary target sequences. Generally, fluorescent labels are used in the detection of hybridization, aided by software. Several advancements have been brought into the assay to facilitate high specificity in detection, viz., competitive hybridization, use of different substrate binders and labels, use of markers, and use of monoclonal antibodies for target protein detection, among others. The technique has been employed in gene expression studies, transcriptome analysis, cell biology, pathway analysis, pharmacology, etc. and is also being used in diagnosing a range of aquatic animal pathogens simultaneously and in the detection of emerging diseases and the pathogenic agents responsible for undiagnosed syndromes. Microarray technology has been developed for a

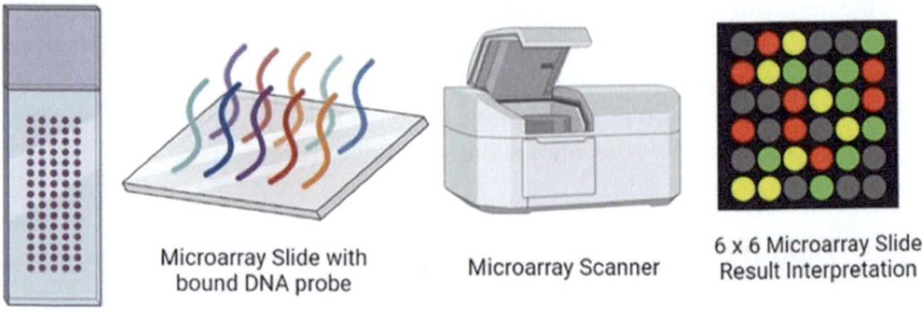

Fig. 11 An illustration depicting the microarray process. A microarray slide with bound DNA is fixed on distinct spots. The slide is then scanned using a microarray scanner, which reads the fluorescent signals emitted from the hybridized DNA probe. The result is displayed as grids with the intensity and colour of each spot that depicts the binding and expression of the target sequence in each sample

variety of bacterial pathogens and viruses infecting fish and shellfish and includes DNA microarray, protein array and antibody microarray. In addition, a naked-eye readable microarray that does not need laser scanning or imaging has been developed for the detection of multiple fish pathogens in aquaculture coupling 16s rDNA PCR and DNA hybridization technology by Chang et al. (2012). The technique is accurate, sensitive and time-saving in diagnosing pathogens that infect aquaculture species.

9 Recent Advancements in Fish Disease Diagnostics

9.1 Machine Learning Techniques

The machine learning technique is one of the latest diagnostic technologies employed in the field of aquaculture. Ahmed et al. (2022) developed fish disease detection using image-based machine learning. The most popular supervised machine learning technique is a support vector machine (SVM) (Suthaharan and Suthaharan 2016), as used by researchers to predict mortality in fish. Ahmed et al. reported the classification of salmon diseases, besides identifying whether the fish has an infection or not, with a machine vision-based technique. These researchers used image processing techniques to extort the features from images, following which an SVM was employed for the successful classification of infectious diseases.

9.2 Proteomics

Proteomics helps to analyse and accurately identify the highest number of proteins in a complex mixture. The method employs both a gel-based approach and a gel-free approach. In gel-based approaches, proteins are separated by one or two-dimensional gel electrophoresis (2-DE), and the separated proteins are identified by mass spectrometry. In gel-free approaches, the protein mixtures remain in solution prior to identification. Both approaches employ the digestion of proteins to peptides using a sequence-specific enzyme, typically trypsin. Peptide samples are further separated and analysed by liquid chromatography coupled to tandem mass spectrometry (LC-MS/MS). Peptide samples can also be analysed by matrix-assisted laser desorption/ionization (MALDI) time-of-flight (TOF) mass spectrometry. LC-MS/MS is the best-preferred approach in host–pathogen interaction studies; it allows multiple sample analysis and quantification of the target proteins. MALDI TOF MS is reported as best suited for the identification and diagnosis of microbes. The detailed explanation of the proteomics-based disease diagnosis approach is beyond the scope of this chapter.

9.3 Omics-Based Strategies and Protein–Protein Interaction (PPI) Networks

High-throughput technologies such as RNA sequencing, MS-based proteomics and metabolomics have aided in unravelling the genes and/or proteins involved in host-pathogen interactions during microbial infections. The combination of omics-based approaches with in vivo studies has enabled an understanding of microbial interactions at a single-cell level to the whole organismal level.

The protein–protein interaction (PPI) network helps in identifying the regulatory complexity of proteins in pathogens and hosts during infection, especially those with major roles in disease mechanisms. The PPI is useful in unravelling the pathogen–host interactions and the pathogenesis. Identification of health or disease biomarkers is one of the greatest outcomes of using PPI networks in studying infectious diseases. Such biomarkers could be used in developing preventive and therapeutic measures, as well as in developing vaccines. A comprehensive review detailing the use of PPI and the identification of biomarkers has been published by Nissa et al. (2021).

9.4 CRISPR-Cas System

The CRISPR-Cas system (Clustered Regularly Interspaced Short Palindromic Repeats and CRISPR-associated proteins) comprises an array of short, repetitive DNA sequences interspersed with unique spacers. These sequences are predominantly found in bacterial and archaeal genomes. Initially identified as a bacterial defence mechanism, the CRISPR-Cas system has rapidly evolved into a versatile tool, particularly in disease diagnostics. The pathogen detection method based on CRISPRs and CRISPR/Cas system has a very high specificity and sensitivity. For its ease of application, the CRISPR/Cas system has been used to diagnose diseases (Yuan et al. 2022). By understanding the natural ability of CRISPR-Cas systems to recognize and cleave specific nucleic acid sequences, innovative methods for the rapid and sensitive detection of pathogens have been developed.

The CRISPR-Cas system consists of two main components: CRISPR-associated (Cas) proteins with endonuclease activity and CRISPR RNA (crRNA) that guides the Cas proteins to their target sequences, enabling precise identification. The mechanism of the CRISPR-Cas system is broadly divided into three stages: adaptation or spacer acquisition, crRNA biogenesis and maturation and target interference. During adaptation, short fragments of the invader genome (pre-spacers) are incorporated into the CRISPR array. These fragments are then transcribed into long pre-crRNA and processed into short crRNA molecules (maturation). Subsequently, crRNA forms a complex with Cas proteins, creating a functional guided endonuclease. This endonuclease, guided by the crRNA or guide RNA (gRNA), recognizes the protospacer adjacent motif (PAM), hybridizes with the target DNA or RNA via its spacer sequence, and cleaves the target DNA or RNA sequences.

The CRISPR-Cas-based diagnostic process typically involves three steps: (A) target gene isolation with or without a pre-amplification step, (B) target cleavage by the CRISPR-Cas system and (C) endpoint detection using colourimetry, voltammetry or fluorescence.

9.4.1 Different Cas Proteins

The CRISPR-Cas system is classified into two main classes: Class 1 and Class 2. Class 1 is further divided into three types: I, III and IV. Class 2 consists of types II, V and VI. Signature proteins for types II, V and VI are Cas9, Cas12 and Cas13, respectively. Cas9, Cas12 (Cpf1), Cas13 and Cas14 from the Class 2 CRISPR-Cas system are widely used in diagnostics and research due to their unique target specificities. Cas9, the most commonly used for gene editing, targets double-stranded DNA (dsDNA). Cas13 targets single-stranded RNA (ssRNA) and exhibits collateral cleavage. Cas12 targets both dsDNA and ssRNA, also showing collateral cleavage activity. Cas14 targets single-stranded DNA (ssDNA). The characteristic properties of different Cas proteins are summarized in Table 1.

Table 1 Summary of the characteristic properties and attributes of Class II, V and VI Cas proteins

Type	II	V	VI
Effector protein	Cas9	Cas12a	Cas13a
Pre-RNA processing	No	Yes	No
tracrRNA	Yes	No	No
Spacer length	18–24 nt	18–24 nt	22–28 nt
Substrate	dsDNA	dsDNA, ssDNA	ssRNA
PAM/PFS	3′, G-rich (NGG)	5′, T-rich (TTTN)	3′, non-G-PFS
Cleavage pattern	Blunt	Sticky end or Staggered	Near U or A
Collateral cleavage	No	DNA (ss)	RNA (ss)

9.4.2 CRISPR-Cas based Gene Detection

The CRISPR-Cas system employs three main functional activities for detecting target genes: (1) sequence-specific binding to target DNA or RNA, (2) sequence-specific cleavage of target DNA or RNA, and (3) trans-cleavage activity. Catalytically inactive Cas proteins bind to the target gene complementary to the gRNA in sequence-specific binding. These methods often use deactivated Cas proteins (dCas9) with eliminated DNA cleavage activity through mutations in the cleavage-responsible domains (RuvC and HNH), displaying strong binding affinity to dsDNA.

The sequence-specific cleavage activity of the CRISPR-Cas system makes it a versatile tool for gene editing, where the target gene can be easily cleaved or modified. Most Cas proteins require a PAM site near the target sequence for cleavage activity. The CRISPR-Cas system can carry out the trans-cleavage activity, in which certain Cas proteins (such as Cas12a or Cas13a) non-specifically cleave neighbouring ssDNA or ssRNA after recognizing and cleaving the target sequence. This non-specific cleavage can be evaluated using colorimetry or fluorescence detection, such as a fluorophore-quencher reporter system.

9.4.3 Pre-Amplification Strategy

The initial concentration of nucleic acids determines the sensitivity of CRISPR-Cas assays. To increase sensitivity, isothermal amplification methods, such as recombinase polymerase amplification (RPA), nucleic acid sequence-based amplification (NASBA), loop-mediated isothermal amplification (LAMP), strand displacement amplification (SDA), helicase-dependent amplification (HDA) and exponential amplification reaction (EXPAR), are often incorporated. These methods operate at various temperatures (e.g. RPA at 37–42 °C, NASBA at 40–55 °C, LAMP at 60–65 °C). Pre-amplification allows for the detection of low copy numbers, even in the femtomolar range, enhancing sensitivity, specificity and quantification.

While pre-amplification is commonly used to enhance sensitivity in CRISPR-based diagnostics, amplification-free approaches are being explored to simplify assay workflows and reduce potential errors, enabling rapid detection suitable for point-of-care (POC) applications.

9.4.4 Quantification of Target

Quantitative determination of copy numbers in PCR can be performed either absolutely or relatively. In absolute quantification, exact copy numbers are evaluated using a standard curve obtained by serial dilutions of known concentrations. In relative quantification, the target concentration is expressed in relation to a reference. In CRISPR-Cas-based diagnostics, absolute quantification can be achieved by comparing the CRISPR-based readout to a standard curve generated using known target concentrations, as the CRISPR-Cas trans-cleavage activity correlates with target concentration. The CRISPR-Cas assay can be performed in a single tube where target amplification and detection are carried out simultaneously (one-pot method) or in two tubes where target amplification and detection are carried out separately (two-pot method). Since pre-amplification and CRISPR-based detection are carried out in a single tube, a real-time readout suitable for quantification is possible in the one-pot method. This is achieved by allowing collateral cleavage to occur simultaneously with target amplification when continuous signal measurement is incorporated. In contrast, in the two-pot method, saturation during the pre-amplification step (RPA) hinders quantitative readout.

To prevent saturation during a pre-amplification step in two-pot reactions, lower RPA primer concentrations can be used. Additional options for quantification include serial dilutions of the unknown sample compared to a known standard. Partitioning samples into many individual reactions, such as in microfluidic droplets or microchambers, enables digital readouts where counting the number of positive partitions allows the original target concentration to be quantified. Immobilizing CRISPR components on magnetic beads allows for quantitative detection using a portable magnetic particle quantification (MPQ) device, where the magnetic signal correlates with the target concentration. Engineered nanomaterials with enzyme-like activities (nanozymes) can be coupled with CRISPR to achieve quantitative detection without the need for pre-amplification, as the nanozyme catalytic activity is modulated upon target binding, enabling quantitative readout. Clinically useful assays typically require a lower limit of detection (LOD), necessitating pre-amplification. However, this approach demands the use of sophisticated laboratory equipment, which limits the use of this technique as a point-of-care (POC) assay.

In short, remarkable progress has been made in the area of disease diagnosis in aquaculture over the past three decades, and to summarize these advancements, a list of major aquatic animal diseases and their available diagnostic tools is provided in Table 2.

10 Conclusion

The major impediment in fast-expanding aquaculture is disease outbreaks, which result in major economic losses. Diagnosis of diseases with high specificity and sensitivity in a timely manner is essential for an effective disease management strategy. Over the years, remarkable progress has been made in disease diagnosis in aquaculture. The progress has been achieved in terms of improved specificity, sensitivity, rapidity, amenability to high throughput and feasibility in helping farm-level (point-of-care) diagnosis. Though disease diagnosis in aquaculture evolved from clinical observation and microscopic description of pathogens to molecular characterization and probe-based diagnosis, each procedure has its own merits and demerits, and none of them is totally exclusive. Therefore, disease diagnosis in aquatic animals should employ all available tools and information, including farm history, clinical observations, histopathology and molecular diagnosis to accurately diagnose a disease and prevent its spread.

Table 2 Summary of major aquatic animal diseases and available diagnostic techniques used in identifying those diseases

S.no	Name of the pathogen	Diagnostic methods available	References
Bacterial diseases of finfish			
1.	*Flavobacterium branchiophilum*	Wet mount, Gram staining, biochemical tests, bacterial culture IFAT	Bullock (1990), Bullock et al. (1994), Huh and Wakabayashi (1987), Ostland et al. (1994)
		ELISA	Lumsden et al. (1993), MacPhee et al. (1995)
		PCR	Toyama et al. (1996)
2.	*Flavobacterium columnare*	Wet mount, Gram staining, biochemical tests, bacterial culture TEM Cytology Haematology PCR	Davis (1922), Declerq et al. (2013), Griffin (1992), Decostere et al. (1999), Decostere (2002), La Frentz et al. (2012), Panangala et al. (2007)
3.	*Flavobacterium psychrophilum*	Gram staining	Holt (1988), Rangdale (1995), Madetoja et al. (2001)
		Biochemical tests	Otis (1984), Holt et al. (1993), Ostland et al. (2000)
		Bacterial culture ELISA RFLP ISH PCR	Anacker and Ordal (1959) Rangdale and Way (1995) Nilson and Strom (2002) Liu et al. (2001) Toyama et al. (1994), Izumi and Wakabayashi (1997), Urdaci et al. (1998), Wiklund et al. (2000)
4.	*Tenacibaculum maritimum*	Gram staining, biochemical tests Bacterial culture	Bernardet et al. (1990)
		Histopathology RAPD	Avendaño Herrera et al. (2004)
		PCR	Bader and Shotts (1998)
5.	*Edwardsiella ictaluri*	Gram staining, biochemical tests	Nagai et al. (2008)
		IFA PCR, Real-time PCR SEM Microarray	Ainsworth et al. (1986) Griffin et al. (2011) Nagai et al. (2008) Peatman et al. (2008)

(continued)

Table 2 (continued)

S.no	Name of the pathogen	Diagnostic methods available	References
6.	*Edwardsiella tarda*	Gram staining, biochemical tests Bacterial culture Histopathology	Reichley et al. (2017) Sakai et al. (2007) Acharya et al. (2007) Savan et al. (2004) Reichley et al. (2015)
		PCR Real-time PCR RAPD LAMP Serology Immunochromatography	Swain and Nayak (2003) Liu et al. (2015)
7.	*Vibrio anguillarum*	Gram staining	Sørensen and Larsen (1986), Egidius (1987), Austin and Austin (1999)
		Biochemical tests Histopathology PCR Fluorescently labelled monoclonal antibody/DAPI (4′,6-diamidino-2-phenylondole) double staining technique	Buller (2004) Yiagnisis et al. (2007) Izumi et al. (2016) Miyamoto and Eguchi (1997)
		Latex Agglutination test	Romalde et al. (1995)
8.	*Vibrio salmonicida*	Gram staining, biochemical Tests	Holm (1985), Holm et al. (1985), Egidius et al. (1986)
		Serology Real-time PCR	Espeiid et al. (1988) Chapela et al. (2018)
9.	*Moritella viscosa*	Gram staining, biochemical Tests, 16s rDNA sequencing Histopathology, Pulsed-field gel electrophoresis (PFGE), Western blotting. Real-time PCR	Colquhoun et al. (2004) Grove et al. (2010) Grove et al. (2008)
10.	*Aeromonas salmonicida*	Gram staining, biochemical tests Latex agglutination test qPCR RAPD pulsed-field gel electrophoresis	Bernoth (1997), Byers et al. (2002) McCarthy et al. (1975) Chapela et al. (2018) Miyata et al. (1995) Umelo and Trust (1998)
11.	*Aeromonas hydrophila*	Gram staining, biochemical tests Histopathology PCR qPCR dRAA-CRISPR/Cas12a	Shotts Jr and Rimler (1973) AlYahya et al. (2018) Yogananth et al. (2009) Griffin et al. (2013), Sebastiao et al. (2018) Lin et al. (2022)

(continued)

Table 2 (continued)

S.no	Name of the pathogen	Diagnostic methods available	References
12.	*Yersinia ruckeri*	Gram staining, biochemical tests, Bacterial culture, Serology	Tobback et al. (2007)
		ELISA, agglutination and immunofluorescence antibodies	Smith (1982)
		RFLP, LAMP, PCR CRISPR/Cas13a system (SHERLOCK)	Garcia et al. (1998), Gibello et al. (1999) Calderón et al. (2024)
13.	*Photobacterium damsela subsp. Piscicida*	Gram staining, biochemical Tests, PCR technique and plating method ELISA	Rajan et al. (2003)
		PCR-RFLP method	Romalde (2002) Zappulli et al. (2005)
15.	*Pseudomonas anguilliseptica*	Gram staining, biochemical tests, Bacterial culture	MacFaddin (1983), Wiklund and Bylund (1990), Thoesen (1994), Berthe et al. (1995)
		RAPD	López-Romalde et al. (2003)
16.	*Epitheliocystis organism*	Histopathology TEM PCR & Sequencing	Draghi et al. (2007) Schmidt-Posthaus et al. (2001), Stride et al. (2013) Stenium et al. (2010)
		qPCR ISH	Steigen et al. (2015), Taylor-Brown et al. (2017)
		Immunohistochemical staining (IHC)	Groff et al. (1996)
17.	*Francisella sp*	Gram staining, biochemical tests, Bacterial culture, Histopathology, PCR, 16s rRNA Sequencing	Soto et al. (2009)
18	*Rickettsia salmonis*	H&E, Gram, Giemsa, acridine orange, Gimenez, Machiavello, and PAS staining Cell lines Immunofluorescent antibody technique and immuno-histochemistry with peroxidase-anti peroxidase (PAP) Nested PCR	Fryer et al. (1990), Branson and Nieto (1991), Lannan and Fryer (1991) Lannan and Fryer (1991) Lannan et al. (1991), Alday-Sanz et al. (1994)
			Mauel et al. (1996)

(continued)

Table 2 (continued)

S.no	Name of the pathogen	Diagnostic methods available	References
19.	*Renibacterium salmoninarum*	Gram staining,	Goodfellow et al. (1985), Bruno and Munro (1986), Holt et al. (2000) Bethke et al. (2017)
		Biochemical tests	Ordal and Earp (1956), Sakai and Kobayashi (1992)
		Bacterial culture	Banner et al. (1991)
		16S rRNA sequencing FAT	Bullock and Stuckey (1975)
		qPCR	Nance et al. (2010), Powell et al. (2005)
20.	*Carnobacterium piscicola*	Gram staining, Biochemical test, 16s rRNA sequencing	Wallbanks et al. (1990)
		Western blotting	Bandin et al. (1993)
21.	*Lactococcus garviae (Enterococcus seriolicida)*	Bacterial culture Biochemical test PCR Slide agglutination test FAT	Austin and Austin (1999) Ravelo et al. (2001) Zlotkin et al. (1998), Aoki et al. (2000) Eldar et al. (1995) Kawahara and Kusuda (1987)
22.	*Streptococcus iniae*	Plate cultivation, biochemical techniques, enzyme reactions IFAT	Facklam (1991), Shoemaker and Klesius (1997), Shoemaker et al. (2000)
		16S rRNA sequencing	Klesius et al. (2006) Mishra et al. (2017)
23.	*Clostridium botulinum*	Gram staining, biochemical tests, PCR-enzyme-linked immunosorbent assay	Fach et al. (2002)
		qPCR	Getchell et al. (2006)
24.	*Mycobacterium marinum* *Mycobacterium fortuitum* *Mycobacterium chelonae*	Gram staining, biochemical tests 16S rRNA sequencing	Gangadharam and Jenkins (1997) Goodfellow et al. (1998), Roberts (2012)
		16S rRNA gene (PCR-RFLP) FRET probe assay	Talaat et al. (1997) Salati et al. (2010)
25.	*Nocardia asteroids* *Nocardia seriolae*	Gram staining, biochemical tests Histopathology 16S rRNA gene PCR LAMP	Chen et al. (2000) Laurent et al. (1999) Itano et al. (2006), Xia et al. (2015)

(continued)

Table 2 (continued)

S.no	Name of the pathogen	Diagnostic methods available	References
Bacterial diseases of shellfish			
1	AHPND	Gram staining, biochemical tests PCR Real-time PCR LAMP RPA CRISPR/ Cas12a assay	Lee et al. (2015), Soto-Rodriguez et al. (2015) Sirikharin et al. (2015) Han et al. (2015) Koiwai et al. (2016) Li et al. (2021)
2	NHP	Wet mount IHC ISH and PCR Real-time PCR	Bradley-Dunlop et al. (2004), Loy and Frelier (1996), Loy et al. (1996) Aranguren et al. (2010)
3	Luminous bacterial disease	Gram staining, biochemical tests, SEM	Lavilla-Pitogo et al. (1990)
		PCR, RAPD	Hernandez and Olmos (2004)
		Real-time PCR	Pang et al. (2019)
4	White muscle disease	Gram staining, biochemical tests, Histopathology, 16S rDNA sequencing, PCR	Wang et al. (2008)
Fungal diseases of finfish			
1	Saprolegniasis	Culture, PCR Squash preparation Imprint smear & staining Serological methods	Touhali and Al-uguali (2018) Pickering and Christie (1980) Krishna et al. (1990) Bullis et al. (1990)
2	Epizootic ulcerative syndrome	Culture & isolation Histopathology Antibody-based antigen detection methods	Lilley et al. (1998) Chinabut and Roberts (1999) Miles et al. (2003)
		PCR FISH	Phadee et al. (2004), Vandersea et al. (2006)
4	Branchiomycosis	Culture & isolation Histopathology	Khoo et al. (1998)
		Spore & hyphae staining PCR	Ibrahim (2011) El-Bouhy et al. (2014)
5	Ichthyophoniaisis	Light & scanning electron microscopy	Spanggaard et al. (1995)
		Histopathology PCR	Kocan et al. (2011) LaPatra et al. (2013)
		Culture	White et al. (2013)

(continued)

Table 2 (continued)

S.no	Name of the pathogen	Diagnostic methods available	References
Fungal diseases of shellfish			
1	Crayfish plague	Wet mount, culture & isolation PCR qPCR	Benisch (1940) Oidtmann et al. (1999) Tuffs and Oidtmann (2011).
2	Larval Mycosis		
Viral diseases of finfish			
1.	CyHV2	Histopathology PCR, cell culture, qPCR	Jung and Miyazaki (1995) Jeffery et al. (2007) Goodwin et al. (2006)
2.	KHV	Imprint staining ELISA Immunofluorescence (IF) and in situ hybridization (ISH) Taqman real—time PCR	Pikarsky et al. (2004), Shapira et al. (2005) Bergmann et al. (2009), Bergmann et al. (2010) Gilad et al. (2004)
3.	CCV	PCR Nested PCR Nucleic acid probe Cell culture SEM	Boyle and Blackwell (1991) Baek and Boyle (1996) Wise and Boyle (1985) Bowser et al. (1985) Ueno et al. (1992)
4.	OMV	Histopathology Cell culture TEM PCR	Furihata et al. (2005) Wolf (1988) Aso et al. (2001)
5.	Lymphocystis	Histopathology PCR Real-time PCR LAMP Immunoblot Cell culture SEM	Sheng and Zhan (2006) Cano et al. (2007) Valverde et al. (2016) Valverde et al. (2017) Cano et al. (2006) Alonso et al. (2005)
6.	RSBIV	Electron microscopy Histopathology Multiplex PCR Cell culture, Immunofluroscence test	Qin et al. (2001) Jung et al. (1997) Jeong et al. (2004) Nakajima et al. (1995)
7.	IPNV	Histopathology neutralization, indirect fluorescent antibody test or enzyme-linked immunosorbent assay), or molecular analysis (polymerase chain reaction, DNA probe or sequencing) methods	Bootland and Leong (1999) Arakawa et al. (1990), Arnzen et al. (1991), Deering et al. (1991), Dixon and Hill (1984), Jorgensen et al. (1991), LaPatra et al. (1989), Purcell et al. (2006), Winton and Einer-Jensen (2002)

(continued)

Table 2 (continued)

S.no	Name of the pathogen	Diagnostic methods available	References
8.	EHNV	ELISA, cell culture Immunogold labelling techniques, Electron microscopy Molecular characterization	Whittington and Steiner (1993) Hyatt (1991) Mao et al. (1997)
9.	IHNV	Histopathology Electron microscopy neutralization, indirect fluorescent antibody test or enzyme-linked immunosorbent assay), or molecular analysis (polymerase chain reaction, DNA probe or sequencing) CRISPR Cas12a	Bootland and Leong (1999) Wolf (1988) Arakawa et al. (1990), Arnzen et al. (1991), Deering et al. (1991), Dixon and Hill (1984), Jorgensen et al. (1991), Winton (1991) LaPatra et al. (1989), Purcell et al. (2006), Winton and Einer-Jensen (2002) Rong et al. (2024)
10.	VHSV	Immunohistochemistry Electron microscopy RT-PCR Real-time PCR	Evensen et al. (1994) Baroni et al. (1982) Snow et al. (2004) Garver et al. (2011), Jonstrup et al. (2013)
		Cell culture	Lorenzen and Lapatra (1999), Olesen and Vestergård Jørgensen (1992)
		ELISA	Olesen and Jørgensen (1991)
11.	SVC	RT-PCR & nested PCR ELISA Immunoperoxidase, fluorescent antibody and cell culture Semi nested PCR	Koutná et al. (2003) Dixon and Hill (1984) Faisal and Ahne (1984) Liu et al. (1998)
12.	ISA	Cell culture RT-PCR, taqman® real-time RT-PCR	Devold et al. (2000) Snow et al. (2006) OIE (2011)
13.	SAV	Histopathology IHC	Christie et al. (2007), Kerbart Boscher et al. (2006), McLoughlin and Graham (2007), Taksdal et al. (2007)
		Cell culture	Graham et al. (2008), Herath et al. (2009)
		RT-PCR, Real-time PCR	Hodneland and Endresen (2006), Fringuelli et al. (2008).

(continued)

Table 2 (continued)

S.no	Name of the pathogen	Diagnostic methods available	References
14.	TiLV	Cell culture, PCR RT-PCR Real-time SYBR Green assay RT-RPA CRISPR Cas12a	Eyngor et al. (2014), Liamnimitr et al. (2018) Dong et al. (2017) Tattiyapong et al. (2017) Sukonta et al. (2022)
1.1.15.	NNV	IFAT, IHC	Johansen et al. (2003), Nguyen et al. (1997)
		Cell culture	Frerichs et al. (1996), Iwamoto et al. (2000)
		PCR Real-time PCR ELISA, Nanoparticle-based lateral flow strip biosensor CRIPR Cas13	Nishizawa et al. (1994) Panzarin et al. (2010) Watanabe et al. (2000), Toubanaki et al. (2020) Huang et al. (2022)
Viral diseases of shellfish			
1	WSSV	Smear preparation western blot analysis, immunodot assay, indirect fluorescent antibody test (IFAT), immunohistochemistry (IHC) or enzyme-linked immunosorbent assay (ELISA) Taqman real-time PCR In situ hybridization (ISH)	Momoyama et al. (1995) Huang et al. (1995), Poulos et al. (2001), Sithigorngul et al. (2006), Yoganandhan et al. (2004) Durand and Lightner (2002) Nunan and Lightner (1997)
		Loop-mediated isothermal amplification (LAMP)	Kono et al. (2004)
		Pre-amplification PCR	Pan et al. (2017)
		Field usable lateral flow immunoassay	Kulabusan et al. (2017)
		RPA CRISPR/ Cas12a	Chaijarasphong et al. (2019)
		CRISPER-based SHERLOCK	Sullivan et al. (2019)
		CRISPR Cas12a	Wang et al. (2023)
2	MBV	Wet mount, Histopathology, Autofluorescence method with phloxine stain, Dot-blot hybridization	Bondad-Reantaso et al. (2001), Lightner (1988, 1996), Lightner et al. (1983) Couch (1991), Fegan et al. (1991), Johnson and Lightner (1988), Lu et al. (1996a, b), Mari et al. (1993b)
		Electron microscopy PCR	Chang et al. (1993), Lu et al. (1993), Umesha et al. (2003), Vickers et al. (1992), Vickers et al. (1994)

(continued)

Table 2 (continued)

S.no	Name of the pathogen	Diagnostic methods available	References
3	BP	Wet mount, Histopathology, Autofluorescence method with phloxine stain Electron microscopy	Bondad-Reantaso et al. (2001), Brock and Main (1994), Lightner (1996)
		Antibody-based antigen detection methods, ISH	Couch (1974), Couch (1991), Johnson and Lightner (1988) Lightner (1996)
		PCR	Wang et al. (1996)
5	IHHNV	Histopathology	Bell and Lightner (1988), Lightner (1996)
		DNA probes for dot-blot and ISH applications In situ hybridization (ISH) procedure	Lightner (1996), Lightner and Redman (1998a, b)
		PCR	Bell et al. (1990)
		Real-time PCR (qPCR)	Mari et al. (1993a), Lightner (1996) Tang et al. (2007), Nunan et al. (2000)
		CRPAC-CRISPR	Dhar et al. (2001), Tang and Lightner (2001) Aiamsa at et al. (2025)
6	HPV	Histopathology ISH, Dot-blot and PCR methods	Bell and Lightner (1988) Phromjai et al. (2001). Phromjai et al. (2002), Manjanaik et al. (2005) La Fauce et al. (2007)
7	YHV	Wet mounts Electron microscopy/ cytopathology	Lightner (1996)
		Antibody-based antigen detection Reverse-transcription polymerase chain reaction (RT-PCR)	Loh et al. (1998), Lu et al. (1995) Mohr et al. (2015)
		In situ hybridization Bioassay RR-Cas	Tang and Lightner (1999) Lu et al. (1995) Aiamsa at et al. (2023)
9	TSV	Antibody-based antigen detection	Erickson et al. (2002), Erickson et al. (2005)
		Bioassay method	Hasson et al. (1999), Lightner (1996), White et al. (2002)
		RT-PCR	Nunan et al. (1998)
		Real-time PCR CRISPR Cas13a	Tang et al. (2004)
		LAMP CRISPR Cas12b (SHERLOCKv2)	Liu et al. (2023) Major et al. (2023)

(continued)

Table 2 (continued)

S.no	Name of the pathogen	Diagnostic methods available	References
10	DIV-1	In situ hybridization PCR nested-PCR two TaqMan probe-based real-time	Qiu et al. (2017) Xu et al. (2016) Qiu et al. (2017) Qiu et al. (2018, 2020) Chen et al. (2019)
11	MoV	Histopathology In situ hybridization	Cowley et al. (2005)
		PCR	Cowley et al. (2002)
		TaqMan real-time RT-PCR	Rajendran et al. (2006)
12	LSNV	Histopathology, TEM, PCR, ISH	Pratoomthai et al. (2008)
13	IMNV	Antibody-based antigen detection	Kunanopparat et al. (2011)
		Nested RT-PCR	Andrade et al. (2007), Poulos et al. (2006)
		Real-time qRT-PCR	Poulos and Lightner (2006) Andrade et al. (2007)
14	MrNV	Direct detection methods	Romestand and Bonami (2003), Sri Widada et al. (2003), Yoganandhan et al. (2005)
		ELISA TAS-ELISA RT-PCR Sequencing Agent purification	Romestand and Bonami (2003) Qian et al. (2006) Sudhakaran et al. (2007) Sambrook and Russell (2001) Bonami et al. (2005)
		LAMP RT LAMP	Pillai et al. (2006) Puthawibool et al. (2010), Haridas et al. (2010)

References

Abath FGC, Melo FL, Werkhauser RP, Montenegro L, Montenegro R, Schindler HC (2002) Single-tube nested PCR using immobilized internal primers. BioTechniques 33(6):1210–1214

Acharya M, Maiti NK, Mohanty S, Mishra P, Samanta M (2007) Genotyping of *Edwardsiella tarda* isolated from freshwater fish culture system. Comp Immunol Microbiol Infect Dis 30(1):33–40

Adams A, Thompson KD (2011) Development of diagnostics for aquaculture: challenges and opportunities. Aquac Res 42:93–102

Ahmed MS, Aurpa TT, Azad MAK (2022) Fish disease detection using image based machine learning technique in aquaculture. J King Saud Univ Comput Inf Sci 34(8):5170–5182

Aiamsa-At P, Nonthakaew N, Phiwsaiya K, Senapin S, Chaijarasphong T (2023) CRISPR-based, genotype-specific detection of yellow head virus genotype 1 with fluorescent, lateral flow and DNAzyme-assisted colorimetric readouts. Aquaculture 574:739696

Aiamsa-at P, Sunantawanit S, Chumroenvidhayakul R, Nakarin F, Sanguanrat P, Sritunyalucksana K, Chaijarasphong T (2025) A novel dual CRISPR-Cas assay for detection of infectious hypodermal and hematopoietic necrosis virus (IHHNV) in penaeid shrimp without false positives from its endogenous viral elements (EVEs). Aquaculture 595:741452

Ainsworth AJ, Capley G, Waterstreet P, Munson D (1986) Use of monoclonal antibodies in the indirect fluorescent antibody technique (IFA) for the diagnosis of Edwardsiella ictaluri. J Fish Dis 9(5):439–444

Alday-Sanz V, Rodger H, Turnbull T, Adams A, Richards RH (1994) An immunohistochemical diagnostic test for rickettsial disease. J Fish Dis 17:189–191

Alonso MC, Cano I, Garcia-Rosadoe, Castro D, Lamas J, Barja JL, Borrego JJ (2005) Isolation of lymphocystis disease virus from sole, Solea senegalensis Kaup, and blackspot sea bream, Pagellus bogaraveo (Brunnich). J Fish Dis 28(4):221–228

AlYahya SA, Ameen F, Al-Niaeem KS, Al-Sa'adi BA, Hadi S, Mostafa AA (2018) Histopathological studies of experimental Aeromonas hydrophila infection in blue tilapia, Oreochromis aureus. Saudi J Biol Sci 25(1):182–185

Anacker RL, Ordal EJ (1959) Studies on the myxobacterium *Chondrococcus columnaris*: I. Serological typing. J Bacteriol 78:25–32

Andrade TPD, Srisuvan T, Tang KFJ, Lightner DV (2007) Real-time reverse transcription polymerase chain reaction assay using TaqMan probe for detection and quantification of infectious myonecrosis virus (IMNV). Aquaculture 264:9–15

Aoki T, Park CI, Yamashita H, Hirono I (2000) Species-specific polymerase chain reaction primers for *Lactococcus garvieae*. J Fish Dis 23:1–6

Arakawa CK, Deering RE, Higman KH, Oshima KH, O'hara PJ, Winton JR (1990) Polymerase chain reaction (PCR) amplification of a nucleoprotein gene sequence of infectious hematopoietic necrosis virus. Dis Aquat Org 8:165–170

Aranguren R, Figueras A (2016) Moving from histopathology to molecular tools in the diagnosis of molluscs diseases of concern under EU legislation. Front Physiol 7:538

Aranguren LF, Tang KFJ, Lightner DV (2010) Quantification of the bacterial agent of necrotizing hepatopancreatitis (NHP-B) by real-time PCR and comparison of survival and NHP load of two shrimp populations. Aquaculture 307:187–192

Arnzen JM, Ristow SS, Hesson CP, Lientz J (1991) Rapid fluorescent antibody tests for infectious hematopoietic necrosis virus (IHNV) utilizing monoclonal antibodies to the nucleoprotein and glycoprotein. J Aquat Anim Health 3:109–113

Aso Y, Wani J, Klenner DAS, Yoshimizu M (2001) Detection and identification of *Oncorhynchus masou* virus (OMV) by polymerase chain reaction (PCR). Bullet Facul Fish Hokkaido Univ 52(2):111–116

Austin B, Austin DA (1999) Bacterial fish pathogens, diseases of farmed and wild fish, 3rd edn. Springer-Praxis, Chichester

Avendaño-Herrera R, Magariños B, Toranzo AE, Beaz R, Romalde JL (2004) Species-specific polymerase chain reaction primer sets for the diagnosis of *Tenacibaculum maritimum* infection. Dis Aquat Org 62(1–2):75–83

Bader JA, Shotts EB (1998) Identification of Flavobacterium and Flexibacter species by species-specific polymerase chain reaction primers to the 16S ribosomal RNA gene. J Aquat Anim Health 10:311–319

Baek YS, Boyle JA (1996) Detection of channel catfish virus in adult channel catfish by use of nested polymerase chain reaction. J Aquat Anim Health 8:97–103

Bahadır EB, Sezgintürk MK (2016) Lateral flow assays: principles, designs and labels. TrAC Trends Anal Chem 82:286–306

Bandín I, Santos Y, Barja JL, Toranzo AE (1993) Detection of a common antigen among *Renibacterium salmoninarum*, *Corynebacterium aquaticum*, and *Carnobacterium piscicola* by the Western blot technique. J Aquat Anim Health 5(3):172–176

Banner CR, Rohovec JS, Fryer JL (1991) A new value for mol percent guanine+ cytosine of DNA for the salmonid fish pathogen *Renibacterium salmoninarum*. FEMS Microbiol Lett 79(1):57–59

Baroni A, Galesso R, Bovo G (1982) Ultrastructural observations on the virus of viral haemorrhagic septicaemia in culture. J Fish Dis 5:439–444

Bell TA, Lightner DV (1988) A handbook of normal Penaeid Shrimp histology. World Aquaculture Society, Baton Rouge, p 114
Bell TA, Lightner DV, Brock JA (1990) A biopsy procedure for the nondestructive determination of infectious hypodermal and hematopoietic necrosis virus (IHHNV) infection in *Penaeus vannamei*. J Aquat Anim Health 2(2):151–153
Benisch J (1940) Kuenstlich hervorgerufener Aphanomyces Befall bei Wollhandkrabben. Zeitschrift fuer Fischerei 38:71–80
Bergmann SM, Schutze H, Fischer U, Fichtner D, Riechardt M, Meyer K, Schrudde D, Kempter J (2009) Detection of koi herpes-virus (KHV) genome in apparently healthy fish. Bull Eur Assoc Fish Pathol 29:145–152
Bergmann SM, Lutze P, Schutze H, Fischer U, Dauber M, Fichtner D, Kempter J (2010) Goldfish (*Carassius auratus*) is a susceptible species for koi herpesvirus (KHV) but not for KHV disease. Bull Eur Assoc Fish Pathol 30:74–84
Bernardet JF, Campbell AC, Buswell JA (1990) *Flexibacter maritimus* is the agent of 'black patch necrosis' in Dover sole in Scotland. Dis Aquat Orgn 8:233–237
Bernoth EM (1997) Diagnosis of furunculosis: the tools. In: Bernoth E-M, Ellis AE, Midtlyng PJ, Olivier G, Smith P (eds) Furunculosis: multidisciplinary fish disease research. Academic Press, London, pp 98–158
Berthe FCJ, Michel C, Bernardet JF (1995) Identification of *Pseudomonas anguilliseptica* isolated from several fish species in France. Dis Aquat Org 21:151–155
Bethke J, Quezada J, Poblete-Morales M, Irgang R, Yáñez A, Oliver C, Avendaño-Herrera R (2017) Biochemical, serological, and genetic characterisation of *Renibacterium salmoninarum* isolates recovered from salmonids in Chile. Bull Eur Assoc Fish Pathol 37:169–180
Bonami JR, Shi Z, Qian D, Sri Widada J (2005) White tail disease of the giant freshwater prawn, Macrobrachium rosenbergii: separation of the associated virions and characterization of MrNV as a new type of nodavirus. J Fish Dis 28(1):23–31
Bondad-Reantaso MG, McGladdery SE, East I, Subasinghe RP (2001) Asia diagnostic guide to aquatic animal diseases. FAO fisheries technical paper 402, supplement 2. FAO, Rome, Italy, p 240
Bootland LM, Leong JC (1999) Infectious hematopoietic necrosis virus. In: Fish diseases and disorders, Volume 3: viral, bacterial and fungal infections, pp 57–121
Bowser PR, Munson AD, Jarboe HH, Francis-Floyd R, Waterstrat PR (1985) Isolation of channel catfish virus from channel catfish *Ictalurus punctatus* (Rafinesque), broodstock. J Fish Dis 8:557–561
Boyle J, Blackwell J (1991) Use of polymerase chain reaction to detect latent channel catfish virus. Am J Vet Res 52:1965–1968
Bradley-Dunlop DJ, Pantoja C, Lightner DV (2004) Development of monoclonal antibodies for detection of necrotizing hepatopancreatitis in penaeid shrimp. Dis Aquat Org 60:233–240
Branson EJ, Nieto D (1991) Description of a new disease condition occurring in farmed coho salmon, *Oncorhynchus kisutch* (Walbaum), in South America. J Fish Dis 14:147–156
Brock JA, Main K (1994) A guide to the common problems and diseases of cultured *Penaeus vannamei*. Oceanic Institute, Makapuu Point, Honolulu, p 242
Bruno DW, Munro ALS (1986) Uniformity in the biochemical properties of *Renibacterium salmoninarum* isolates obtained from several sources. FEMS Microbiol Lett 33:247–250
Bruno DW, Noguera PA, Poppe TT (2013) A colour atlas of salmonid diseases, 2nd edn. Springer
Buller NB (2004) Bacteria from fish and other aquatic animals: a practical identification manual. CABI Publishing, Wallingford, Oxfordshire
Bullis RA, Noga EJ, Levy MG (1990) Immunological relationship of the fish pathogenic oomycytes and unrelated fungi. J Aquat Anim Health 2:223–227
Bullock GL (1990) Bacterial gill disease of freshwater fishes. Fish disease Leaflet 84. US Fish & Wildlife Publications. 130. https://digitalcommons.unl.edu/usfwspubs/130
Bullock GL, Stuckey HM (1975) Fluorescent antibody identification and detection of *Corynebacterium* causing kidney disease in salmonids. J Fish Res Board Can 32:2224–2227

Bullock G, Herman R, Heinen J, Noble A, Weber A, Hankins J (1994) Observations on the occurrence of bacterial gill disease and amoeba gill infestation in rainbow trout cultured in a water recirculation system. J Aquat Anim Health 6:310–317

Byers HK, Gudkovs N, St. Crane MJ (2002) PCR-based assays for the fish pathogen *Aeromonas salmonicida*. I. Evaluation of three PCR primer sets for detection and identification. Dis Aquat Org 49:129–138

Calderón IL, Barros MJ, Fernández-Navarro N, Acuña LG (2024) Detection of nucleic acids of the fish pathogen *Yersinia ruckeri* from planktonic and biofilm samples with a CRISPR/Cas13a-based assay. Microorganisms 12(2):283

Cano I, Alonso MC, Garcia-Rosado E, Saint-Jean SR, Castro D, Borrego JJ (2006) Detection of lymphocystis disease virus (LCDV) in asymptomatic cultured gilt-head seabream (*Sparus aurata*, L.) using an immunoblot technique. Vet Microbiol 113(1–2):137–141

Cano I, Ferro P, Alonso MC, Bergmann SM, Römer-Oberdörfer A, Garcia-Rosado E, Castro D, Borrego JJ (2007) Development of molecular techniques for detection of lymphocystis disease virus in different marine fish species. J Appl Microbiol 102(1):32–40

Chaijarasphong T, Thammachai T, Itsathitphaisarn O, Sritunyalucksana K, Suebsing R (2019) Potential application of CRISPR-Cas12a fluorescence assay coupled with rapid nucleic acid amplification for detection of white spot syndrome virus in shrimp. Aquaculture 512:734340

Chang PS, Lo CF, Kou GH, Shen SN (1993) Purification and amplification of DNA from *Penaeus monodon*-type baculovirus (MBV). J Invertebr Pathol 62:116–120

Chang CI, Hung PH, Wu CC, Cheng TC, Tsai JM, Lin KJ, Lin CY (2012) Simultaneous detection of multiple fish pathogens using a naked-eye readable DNA microarray. Sensors 12(3):2710–2728

Chapela MJ, Ferreira M, Ruiz-Cruz A, Martin-Varela I, Fernández-Casal J, Garrido-Maestu A (2018) Application of real-time PCR for early diagnosis of diseases caused by *Aeromonas salmonicida*, *Vibrio anguillarum*, and *Tenacibaculum maritimum* in turbot: a field study. J Appl Aquac 30(1):76–89

Chen SC, Lee JL, Lai CC, Gu YW, Wang CT, Chang HY, Tsai KH (2000) Nocardiosis in sea bass, *Lateolabrax japonicus*, in Taiwan. J Fish Dis 23:299–307

Chen X, Qiu L, Wang HL, Zou PZ, Dong X, Li FH, Huang J (2019) Susceptibility of *Exopalaemon carinicauda* to the infection with shrimp hemocyte iridescent virus (SHIV 20141215), a strain of Decapod iridescent virus 1 (DIV1). Viruses 11(4):387

Chinabut S, Roberts RJ (1999) Pathology and histopathology of epizootic ulcerative syndrome (EUS). Aquatic Animal Health Research Institute, Department of Fisheries. Royal Thai Government, Bangkok, Thailand, p 33 (ISBN 974-7604-55-8)

Christie KE, Graham DA, McLoughlin MF, Villoing S, Todd D, Knappskog D (2007) Experimental infection of Atlantic salmon *Salmo salar* pre-smolts by i.p. injection of new Irish and Norwegian salmonid alphavirus (SAV) isolates: a comparative study. Dis Aquat Org 75:13–22

Colquhoun DJ, Hovland H, Hellberg H, Haug T, Nilsen H (2004) *Moritella viscosa* isolated from farmed Atlantic cod (*Gadus morhua*). Bull Eur Assoc Fish Pathol 24(2):109–114

Couch JA (1974) An enzootic nuclear polyhedrosis virus of pink shrimp: ultrastructure, prevalence, and enhancement. J Invertebr Pathol 24:311–331

Couch JA (1991) Baculoviridae. Nuclear polyhedrosis viruses. Part 2. Nuclear polyhedrosis viruses of invertebrates other than insects. In: Adams JR, Bonami JR (eds) Atlas of invertebrate viruses. CRC Press, Boca Raton, pp 205–226

Cowley JA, McCulloch R, Spann K, Cadogan L, Walker P (2002) Preliminary molecular and biological characterisation of Mourilyan virus (MoV): a new bunya-related virus of penaeid prawns. In: 5th symposium on diseases in Asian aquaculture; handbook and abstracts, 103

Cowley JA, McCulloch RJ, Spann KM, Cadogan LC, Walker PJ (2005) Preliminary molecular and biological characterisation of Mourilyan virus (MoV): a new bunya-related virus of penaeid prawns. In: Diseases in Asian aquaculture V. Proceedings of the 5th symposium on diseases in Asian aquaculture. Fish Health Section, Asian Fisheries Society, Manila, The Philippines, pp 113–124

Cunningham CO (2004) Use of molecular diagnostic tests in disease control: making the leap from laboratory to field application. In: Leung K-Y (ed) Current trends in the study of bacterial and

viral fish and Shrimp diseases (molecular aspects of fish and marine biology), vol 3. World Scientific Publishing, Singapore, pp 292–312
Davis HS (1922) A new bacterial disease of fresh-water fishes. US Government Printing Office, Washington, DC
De la Peña LD (2001) Immunological and molecular biology techniques in disease diagnosis. In: Lio-Po GD, Lavilla CR, Cruz-Lacierda ER (eds) Health Management in Aquaculture. Aquaculture Department, Southeast Asian Fisheries Development Center, Tigbauan, Iloilo, Philippines, pp 137–158
Declercq AM, Haesebrouck F, Van den Broeck W, Bossier P, Decostere A (2013) Columnaris disease in fish: a review with emphasis on bacterium-host interactions. Vet Res 44(1):1–17
Decostere A (2002) Flavobacterium columnare infections in fish: the agent and its adhesion to the gill tissue. Verhandelingen-Koninklijke Academie voor Geneeskunde van België 64(6):421–430
Decostere A, Haesebrouck F, Van Driessche E, Charlier G, Ducatelle R (1999) Characterization of the adhesion of Flavobacterium columnare (*Flexibacter columnaris*) to gill tissue. J Fish Dis 22(6):465–474
Deering RE, Arakawa CK, Oshima KH, O'hara PJ, Landolt ML, Winton JR (1991) Development of a biotinylated DNA probe for detection and identification of infectious hematopoietic necrosis virus. Dis Aquat Org 11:57–65
Devold M, Krossoy B, Aspehaug V, Nylund A (2000) Use of RT-PCR for diagnosis of infectious salmon anemia virus (ISAV) in carrier sea trout, *Salmo trutta*, after experimental infection. Dis Aquat Org 40:9–18
Dhar AK, Roux MM, Klimpel KR (2001) Detection and quantification of infectious hypodermal and hematopoietic necrosis virus and white spot virus in shrimp using real-time quantitative PCR and SYBR green chemistry. J Clin Microbiol 39:2835–2845
Dixon PF, Hill BJ (1984) Rapid detection of fish rhabdoviruses by the enzyme-linked immunosorbent assay (ELISA). Aquaculture 42:1–12
Dong HT, Siriroob S, Meemetta W, Santimanawong W, Gangnonngiw W, Pirarat N, Khunrae P, Rattanarojpong T, Vanichviriyakit R, Senapin S (2017) Emergence of tilapia lake virus in Thailand and an alternative semi-nested RT-PCR for detection. Aquaculture 476:111–118
Draghi A II, Bebak J, Popov VL, Noble AC, Geary SJ, West AB, Byrne P, Frasca S Jr (2007) Characterization of a Neochlamydia-like bacterium associated with epitheliocystis in cultured Arctic charr Salvelinus alpinus. Dis Aquat Org 76(1):27–38
Duncan SM, Seigel GM (2016) High-contrast enzymatic immunohistochemistry of pigmented tissues. J Biol Methods 3(3):e47. https://doi.org/10.14440/jbm.2016.122
Durand SV, Lightner DV (2002) Quantitative real-time PCR for the measurement of white spot syndrome virus in shrimp. J Fish Dis 25:381–389
Egidius E (1987) Vibriosis: pathogenicity and pathology. A review. Aquaculture 67(1–2):15–28
Egidius E, Wiik R, Andersen K, Hoff KA, Hjeltnes B (1986) *Vibrio salmonicida* sp. nov., a new fish pathogen. Int J Syst Bacteriol 36:518–520
El-Bouhy ZM, Shaheen AA, Hassanin ME, Mahboub HH (2014) Branchiomyosis in Nile tilapia (*Oreochromis niloticus*) in Behiera governorate with trials for treatment. Zagazig Veterin J 42(3):29–42
Eldar A, Shapiro O, Bejerano Y, Bercovier H (1995) Vaccination with whole-cell vaccine and bacterial protein extract protects tilapia against *Streptococcus difficile* meningo-encephalitis. Vaccine 13:867–870
Engvall E (2010) The ELISA, enzyme-linked immunosorbent assay. Clin Chem 56(2):319–320
Erickson HS, Zarain-Herzberg M, Lightner DV (2002) Detection of Taura syndrome virus (TSV) strain differences using selected diagnostic methods: diagnostic implications in penaeid shrimp. Dis Aquat Org 52:1–10
Erickson HS, Poulos BT, Tang KFJ, Bradley-Dunlop D, Lightner DV (2005) Taura syndrome virus from Belize represents a unique variant. Dis Aquat Org 64:91–98
Espeiid S, Holm KO, Hjeimeland K, Jorgensen T (1988) Monoclonal antibodies against *Vibrio salmonicida*: the causative agent of coldwater vibriosis ('Hitra disease') in Atlantic salmon, *Salmo salar*. J Fish Dis 11:207–214

Evensen Ø, Meier W, Wahli T, Olesen NJ, Jørgensen PEV, Håstein T (1994) Comparison of immunohistochemistry and virus cultivation for detection of viral hemorrhagic septicaemia virus in experimentally infected rainbow trout, *Oncorhynchus mykiss*. Dis Aquat Org 20:101–109

Eyngor M, Zamostiano R, Tsofack JEK, Berkowitz A, Bercovier H, Tinman S, Lev M, Huryitz A, Galeotti M, Eldar A (2014) Identification of a novel RNA virus lethal to tilapia. J Clin Microbiol 52(12):4137–4146

Fach P, Perelle S, Dilasser F, Grout J, Dargaignaratz C, Botella L, Gourreau JM, Carlin F, Popoff MR, Broussolle V (2002) Detection by PCR-enzyme-linked immunosorbent assay of *Clostridium botulinum* in fish and environmental samples from a coastal area in northern France. Appl Environ Microbiol 68(12):5870–5876

Facklam R (1991) Streptococcus and related catalase negative gram positive cocci. Manual of clinical microbiology, 5th edn. American Society of Microbiology, Washington, DC, pp 238–257

Faisal M, Ahne W (1984) Spring viremia of carp virus (SVCV): comparison of immunoperoxidase, fluorescent antibody and cell culture isolation techniques for detection of antigen. J Fish Dis 7:57–64

FAO (2020) The state of world fisheries and aquaculture 2020. Sustainability in action, Rome

FAO (2024) The state of world fisheries and aquaculture 2024 – Blue transformation in action, Rome

FAO/NACA (2000) Asia Regional technical guidelines on health management for the responsible movement of live aquatic animals and the Beijing Consensus and Implementation Strategy. FAO Fishery Technical Paper No. 402. Rome, FAO, 53

Fegan DF, Flegel TW, Sriurairatana S, Waiyakruttha M (1991) The occurrence, development and histopathology of monodon baculovirus in Southern Thailand. Aquaculture 96:205–217

Frerichs GN, Rodger HD, Peric Z (1996) Cell culture isolation of piscine neuropathy nodavirus from juvenile sea bass, *Dicentrarchus labrax*. J Gen Virol 77:2067–2071

Fringuelli E, Rowley HM, Wilson JC, Hunter R, Rodger H, Graham DA (2008) Phylogenetic analyses and molecular epidemiology of European salmonid alphaviruses (SAV) based on partial E2 and nsP3 gene nucleotide sequences. J Fish Dis 31:811–823

Fryer JL, Lannan CN, Garces LH, Larenas JJ, Smith PA (1990) Isolation of a rickettsiales-like organism from diseased coho salmon (*Oncorhynchus kisutch*) in Chile. Fish Pathol 25:107–114

Furihata M, Suzuki K, Hosoe A, Miyazaki T (2005) Histopathological study on Oncorhynchus masou virus disease (OMVD) of cultured rainbow trout in natural outbreaks and artificial infection. Fish Pathol 40(4):161–167

Gangadharam PR, Jenkins PA (1997) Mycobacteria: I basic aspects, vol 1. Springer Science and Business Media

Garcia JA, Dominguez L, Larsen JL, Pedersen K (1998) Ribotyping and plasmid profiling of *Yersinia ruckeri*. J Appl Microbiol 85(6):949–955

Garver KA, Hawley LM, McClure CA, Schroeder T, Aldous S, Doig F, Snow M, Edes S, Baynes C, Richard J (2011) Development and validation of a reverse transcription quantitative PCR for universal detection of viral hemorrhagic septicemia virus. Dis Aquat Org 95:97–112

Getchell RG, Culligan WJ, Kirchgessner M, Sutton CA, Casey RN, Bowser PR (2006) Quantitative polymerase chain reaction assay used to measure the prevalence of *Clostridium botulinum* type E in fish in the lower Great Lakes. J Aquat Anim Health 18(1):39–50

Gibello A, Blanco MM, Moreno MA, Cutuli MT, Domenech A, Domínguez L, Fernández Garayzabal JF (1999) Development of a PCR assay for detection of Yersinia ruckeri in tissues of inoculated and naturally infected trout. Appl Environ Microbiol 65:346–350

Gilad O, Yun S, Zagmutt-Vergara FJ, Leutenegger CM, Bercovier H, Hedrick RP (2004) Concentrations of a Koi herpesvirus (KHV) in tissues of experimentally infected *Cyprinus carpio* koi as assessed by real-time TaqMan PCR. Dis Aquat Org 60:179–187

Goodfellow M, Embley TM, Austin B (1985) Numerical taxonomy and emended description of *Renibacterium salmoninarum*. J Gen Microbiol 131:2739–2752

Goodfellow M, Magee JG, Gangadharam PRJ, Jenkins PA (1998) Mycobacteria: I basic aspects. Taxonomy of mycobacteria. Chapman and Hall, New York, pp 1–71

Goodwin AE, Merry GE, Sadler J (2006) Detection of the herpesviral hematopoietic necrosis disease agent (Cyprinid herpesvirus 2) in moribund and healthy goldfish: validation of a quantitative PCR diagnostic method. Dis Aquat Org 69(2–3):137–143

Graham DA, Wilson C, Jewhurst H, Rowley H (2008) Cultural characteristics of salmonid alphaviruses – influences of cell line and temperature. J Fish Dis 31:859–868

Green MR, Sambrook J (2019) Nested polymerase chain reaction (PCR). Cold Spring Harb Protoc. https://doi.org/10.1101/pdb.prot095182

Griffin BR (1992) A simple procedure for identification of *Cytophaga columnaris*. J Aquat Anim Health 4:63–66

Griffin MJ, Mauel MJ, Greenway TE, Khoo LH, Wise DJ (2011) A real-time polymerase chain reaction assay for quantification of *Edwardsiella ictaluri* in catfish pond water and genetic homogeneity of diagnostic case isolates from Mississippi. J Aquat Anim Health 23(4):178–188

Griffin MJ, Goodwin AE, Merry GE, Liles MR, Williams MA, Ware C, Waldbieser GC (2013) Rapid quantitative detection of *Aeromonas hydrophila* strains associated with disease outbreaks in catfish aquaculture. J Vet Diagn Invest 25(4):473–481

Groff JM, LaPatra SE, Munn RJ, Anderson ML, Osburn BI (1996) Epitheliocystis infection in cultured white sturgeon (*Acipenser transmontanus*): antigenic and ultrastructural similarities of the causative agent to the chlamydiae. J Vet Diagn Invest 8:172–180

Grove S, Reitan LJ, Lunder T, Colquhoun D (2008) Real-time PCR detection of Moritella viscosa, the likely causal agent of winter-ulcer in Atlantic *salmon Salmo salar* and rainbow trout *Oncorhynchus mykiss*. Dis Aquat Org 82(2):105–109

Grove S, Wiik-Nielsen CR, Lunder T, Tunsjø HS, Tandstad NM, Reitan LJ, Marthinussen A, Sørgaard M, Olsen AB, Colquhoun DJ (2010) Previously unrecognised division within Moritella viscosa isolated from fish farmed in the North Atlantic. Dis Aquat Org 93(1):51–61

Han JE, Tang KFJ, Pantoja CR, White BL, Lightner DV (2015) qPCR assay for detecting and quantifying a virulence plasmid in acute hepatopancreatic necrosis disease (AHPND) due to pathogenic *Vibrio parahaemolyticus*. Aquaculture 442:12–15

Hansen S, Abd El Wahed A (2020) Point-of-care or point-of-need diagnostic tests: time to change outbreak investigation and pathogen detection. Tropi Med Infect Dis 5:151

Hardinge P, Murray JAH (2019) Reduced false positives and improved reporting of loop-mediated isothermal amplification using quenched fluorescent primers. Sci Rep 9:7400

Hardinge P, Kiddle G, Tisi L, Murray JA (2018) Optimised LAMP allows single copy detection of 35Sp and NOSt in transgenic maize using Bioluminescent Assay in Real Time (BART). Sci Rep 8(1):1–17

Haridas DV, Pillai D, Manojkumar B, Nair CM, Sherief PM (2010) Optimisation of reverse transcriptase loop-mediated isothermal amplification assay for rapid detection of Macrobrachium rosenbergii noda virus and extra small virus in Macrobrachium rosenbergii. J Virol Methods 167(1):61–67

Harlow E, Lane D (1999) Using antibodies: a laboratory manual. Cold Spring Harbor, New York

Hasson KW, Lightner DV, Mohney LL, Redman RM, Poulos BT, White BM (1999) Taura syndrome virus (TSV) lesion development and the disease cycle in the Pacific white shrimp *Penaeus vannamei*. Dis Aquat Org 36:81–93

Hataoka Y, Zhang L, Mori Y, Tomita N, Notomi T, Baba Y (2004) Analysis of specific gene by integration of isothermal amplification and electrophoresis on poly(methyl methacrylate) microchips. Anal Chem 76:3689–3693

Herath T, Costa J, Thompson K, Adams A, Richards R (2009) Alternative cell line for the isolation of salmonid alphavirus 1. Icel Agric Sci 22:19–27

Hernandez G, Olmos J (2004) Molecular identification of pathogenic and nonpathogenic strains of *Vibrio harveyi* using PCR and RAPD. Appl Microbiol Biotechnol 63(6):722–727

Hindson BJ, Ness KD, Masquelier DA, Belgrader P, Heredia NJ, Makarewicz AJ, Bright IJ, Lucero MY, Hiddessen AL, Legler TC, Kitano TK (2011) High-throughput droplet digital PCR system for absolute quantitation of DNA copy number. Anal Chem 83(22):8604–8610

Hodneland K, Endresen C (2006) Sensitive and specific detection of salmonid alphavirus using real-time PCR (TaqMan). J Virol Methods 131:184–192

Holm KO (1985) Isolering og karakterisering av Vibriobakterier fra laks (*Salmo salar* L.) med HUrasyke. Thesis, University of Tromsø

Holm KO, Strom E, Stensvag K, Raa J, Jorgensen T (1985) Characteristics of a *Vibrio* sp. associated with the 'Hitra disease' of Atlantic salmon in Norwegian fish farms. Fish Pathol 20:125–129

Holt RA (1988) *Cytophaga psychrophila*, the causative agent of bacterial cold-water disease in salmonid fish. PhD thesis, Oregon State University, Corvallis

Holt RA, Rohovec JS, Fryer JL (1993) Bacterial cold-water disease. In: Inglis V, Roberts RJ, Bromage NR (eds) Bacterial diseases of fish. Blackwell Scientific Publications, Oxford, pp 3–23

Holt JG, Krieg NR, Sneath PHA, Staley JT, Williams ST (2000) Bergey's manual of determinative bacteriology, 9th edn. Lippincott Williams & Wilkins, Philadelphia

Huang J, Yu J, Wang X-H, Song X-L, Ma C-S, Zhao F-Z, Yang C-H (1995) Survey on the pathogen and route of transmission of baculoviral hypodermal and hematopoietic necrosis in shrimp by ELISA of monoclonal antibody. Marine Fish Res (Chinese J) 16:40–50

Huang F, Shan J, Liang K, Yang M, Zhou X, Duan X, Jia X, Zhao H, Qin Q, Wang Q (2022) A new method to detect red spotted grouper neuro necrosis virus (RGNNV) based on CRISPR/Cas13a. Aquaculture 555:738217

Hyatt AD (1991) Immunogold labelling techniques. In: Harris R (ed) Electron microscopy in biology: a practical approach. IRL Press, Oxford, UK, pp 59–81

Ibrahim KS (2011) Isolation and pathological study of Branchiomycosis from the commercial pond of common carp (*Cyprinus carpio*) fish, in Governorate of Duhok/Iraq. Iraqi J Veterin Med 35(1):1–9

Im K, Mareninov S, Diaz M, Yong WH (2019) An introduction to performing immunofluorescence staining. Methods Mol Biol 1897:299–311

Itano T, Kawakami H, Kono T, Sakai M (2006) Detection of fish nocardiosis by loop-mediated isothermal amplification. J Appl Microbiol 100:1381–1387

Iwamoto T, Nakai T, Mori K, Arimoto M, Furusawa I (2000) Cloning of the fish cell line SSN-1 for piscine nodaviruses. Dis Aquat Org 43:81–89

Izumi S, Suzuki K (2016) A novel PCR detection method for major fish pathogenic bacteria of *Vibrio anguillarum*. Turk J Fish Aquat Sci 16(3):571–578

Izumi S, Wakabayashi H (1997) Use of PCR to detect *Cytophaga psychrophila* from apparently healthy juvenile ayu and coho salmon eggs. Fish Pathol 32(3):169–173

Jaramillo D, Hick P, Deece K, Tweedie A, Kirkland P, Arzey E, Whittington RJ (2016) Comparison of ELISA formats for detection of antibodies specific for nervous necrosis virus (*Betanodavirus*) in the serum of immunized barramundi (*Lates calcarifer*) and Australian bass (*Macquaria novemaculeata*). Aquaculture 451:33–38

Jaramillo D, Peeler EJ, Laurin E, Gardner IA, Whittington RJ (2017) Serology in finfish for diagnosis, surveillance, and research: a systematic review. J Aquat Anim Health 29(1):1–14

Jeffery KR, Bateman K, Bayley A, Feist SW, Hulland J, Longshaw C, Stone D, Woolford G, Way K (2007) Isolation of a cyprinid herpesvirus 2 from goldfish, *Carassius auratus* (L.), in the UK. J Fish Dis 30(11):649–656

Jeong JB, Park KH, Kim HY, Hong SH, Chung J-K, Komisar JL, Jeong HD (2004) Multiplex PCR for the diagnosis of red sea bream iridoviruses isolated in Korea. Aquaculture 235:139–152

Jessup DA (2009) Anaesthetic and sedative techniques for aquatic animals. J Wildl Dis 45(2):552–554

Johansen R, Amundsen M, Dannevig BH, Sommer AI (2003) Acute and persistent experimental nodavirus infection in spotted wolffish (*Anarhichas minor*). Dis Aquat Org 57:35–41

Johnson PT, Lightner DV (1988) Rod-shaped nuclear viruses of crustaceans: gut-infecting species. Dis Aquat Org 5:123–141

Jonstrup SP, Kahns S, Skall HF, Boutrup TS, Olesen NJ (2013) Development and validation of a novel Taqman-based real-time RT-PCR assay suitable for demonstrating freedom from viral haemorrhagic septicaemia virus. J Fish Dis 36(1):9–23

Jorgensen PEV, Olesen NJ, Lorenzen N, Winton JR, Ristow SS (1991) Infectious hematopoietic necrosis (IHN) and viral hemorrhagic septicemia (VHS): detection of trout antibodies to the causative viruses by means of plaque neutralization, immunofluorescence, and enzyme-linked immunosorbent assay. J Aquat Anim Health 3:100–108

Jung SJ, Miyazaki T (1995) Herpesviral haematopoietic necrosis of goldfish, *Carassius auratus* (L.). J Fish Dis 18(3):211–220

Jung S, Miyazaki T, Miyata M, Danayadol Y, Tanaka S (1997) Pathogenicity of iridovirus from Japan and Thailand for the red sea bream *Pagrus major* in Japan, and histopathology of experimentally infected fish. Fish Sci 63:735–740

Kawahara E, Kusuda R (1987) Direct fluorescent antibody technique for diagnosis of bacterial diseases in eel. Nippon Suisan Gakkaishi 53:395–399

Kent ML, Benda S, St-Hilaire S, Schreck CB (2013) Sensitivity and specificity of histology for diagnoses of four common pathogens and detection of nontarget pathogens in adult Chinook salmon (*Oncorhynchus tshawytscha*) in fresh water. J Vet Diagn Invest 25(3):341–351

Kerbart Boscher S, McLoughlin M, Le Ven A, Cabon J, Baud M, Castric J (2006) Experimental transmission of sleeping disease in one-year-old rainbow trout (*Oncorhynchus mykiss* (Walbaum)), induced by sleeping disease virus. J Fish Dis 29:263–273

Khoo L, Leard AT, Waterstrat PR, Jack SW, Camp KL (1998) Branchiomyces infection in farm-reared channel catfish, *Ictalurus punctatus* (Rafinesque). J Fish Dis 21(6):423–432

Klesius P, Evans J, Shoemaker C, Yeh H, Goodwin AE, Adams A, Thompson K (2006) Rapid detection and identification of *Streptococcus iniae* using a monoclonal antibody-based indirect fluorescent antibody technique. Aquaculture 258(1–4):180–186

Kocan R, Dolan H, Hershberger P (2011) Diagnostic methodology is critical for accurately determining the prevalence of *Ichthyophonus* infections in wild fish populations. J Parasitol 97(2):344–348

Koiwai K, Tinwongger S, Nozaki R, Kondo H, Hirono I (2016) Detection of acute hepatopancreatic necrosis disease strain of *Vibrio parahaemolyticus* using loop-mediated isothermal amplification. J Fish Dis 39:603–606

Komminoth P (1992) Digoxigenin as an alternative probe labeling for in situ hybridization. Diagn Mol Pathol 1(2):142–150

Kono T, Savan R, Sakai M, Itami T (2004) Detection of white spot syndrome virus in shrimp by loop-mediated isothermal amplification. J Virol Methods 115:59–65

Koutná M, Vesely T, Psikal I, Hulová J (2003) Identification of spring viraemia of carp virus (SVCV) by combined RT-PCR and nested PCR. Dis Aquat Org 55:229–235

Kozel TR, Burnham-Marusich AR (2017) Point-of-care testing for infectious diseases: past, present, and future. J Clin Microbiol 55:2313–2320

Krishna L, Gupta VK, Katoch RC, Singh D (1990) Saprolegniasis in Indian major carps-an investigation. Indian Veterin J 67:554–555

Kulabhusan PK, Rajwade JM, Sugumar V, Taju G, Sahul Hameed AS, Paknikar KM (2017) Field-usable lateral flow immunoassay for the rapid detection of white spot syndrome virus (WSSV). PLoS One 12(1):0169012

Kunanopparat A, Chaivisuthangkura P, Senapin S, Longyant S, Rukpratanporn S, Flegel TW, Sithigorngul P (2011) Detection of infectious myonecrosis virus using monoclonal antibody specific to N and C fragments of the capsid protein expressed heterologously. J Virol Methods 171:141–148

La Fauce KA, Layton R, Owens L (2007) TaqMan real-time PCR for detection of hepatopancreatic parvovirus from Australia. J Virol Methods 140(1–2):10–16

LaFrentz BR, LaPatra SE, Shoemaker CA, Klesius PH (2012) Reproducible challenge model to investigate the virulence of *Flavobacterium columnare* genomovars in rainbow trout *Oncorhynchus mykiss*. Dis Aquat Org 101:115–122

Lannan CN, Fryer JL (1991) Recommended methods for inspection of fish for the salmonid rickettsia. Bull Eur Assoc Fish Pathol 11:135–136

Lannan CN, Ewing SA, Fryer JL (1991) A fluorescent antibody test for detection of the rickettsia causing disease in Chilean salmon. J Aquat Anim Health 3:29–234

LaPatra SE, Kocan RM (2013) PCR testing for diagnosis of *Ichthyophonus hoferi*. Dis Aquat Org 106(3):273–274

Lapatra SE, Roberti KA, Rohovec JS, Fryer JL (1989) Fluorescent antibody test for the rapid diagnosis of infectious hematopoietic necrosis virus. J Aquat Anim Health 1:29–36
Laurent FJ, Provost F, Boiron P (1999) Rapid identification of clinically relevant *Nocardia* species to genus level by 16S rRNA gene PCR. J Clin Microbiol 37(1):99–102
Lavilla-Pitogo CR, Baticados MCL, Cruz-Lacierda ER, Leobert D (1990) Occurrence of luminous bacterial disease of *Penaeus monodon* larvae in The Philippines. Aquaculture 91(1–2):1–13
Lee CT, Chen IT, Yang YT, Ko TP, Huang YT, Huang JY, Huang MF, Lin SJ, Chen CY, Lin SS, Lightner DV, Wang AH, Wang HC (2015) The opportunistic marine pathogen *Vibrio parahaemolyticus* becomes virulent by acquiring a plasmid that expresses a deadly toxin. Proc Natl Acad Sci USA 112:10798–10803
Lewin AS, Haugen T, Netzer R, Tøndervik A, Dahle SW, Hageskal G (2020) Multiplex droplet digital PCR assay for detection of Flavobacterium psychrophilum and Yersinia ruckeri in Norwegian aquaculture. J Microbiol Methods 177:106044
Li H, Bai R, Zhao Z, Tao L, Ma M, Ji Z, Jian M, Ding Z, Dai X, Bao F, Liu A (2018) Application of droplet digital PCR to detect the pathogens of infectious diseases. Biosci Rep 38(6):BSR20181170
Li C, Lin N, Feng Z, Lin M, Guan B, Chen K, Liang W, Wang Q, Li M, You Y, Chen Q (2021) CRISPR/Cas12a based rapid molecular detection of acute Hepatopancreatic necrosis disease in Shrimp. Front Veterin Sci 8:819681. https://doi.org/10.3389/fvets.2021.819681
Liamnimitr P, Thammatorn W, U-thoomporm S, Tattiyapong P, Surachetpong W (2018) Non-lethal sampling for Tilapia Lake Virus detection by RT-qPCR and cell culture. Aquaculture 486:75–80
Lightner DV (1988) Diseases of cultured penaeid shrimp and prawns. In: Sindermann CJ, Lightner DV (eds) Disease diagnosis and control in North American marine aquaculture, pp 8–127
Lightner DV (1996) A handbook of Shrimp pathology and diagnostic procedures for diseases of cultured Penaeid Shrimp. World Aquaculture Society, Baton Rouge, p 304
Lightner DV, Redman RM (1998a) Shrimp diseases and current diagnostic methods. Aquaculture 164:201–220
Lightner DV, Redman RM (1998b) Strategies for the control of viral diseases of shrimp in the Americas. Fish Pathol 33:165–180
Lilley JH, Callinan RB, Chinabut S, Kanchanakhan S, Macrae IH, Phillips MJ (1998) Epizootic ulcerative syndrome (EUS) technical handbook. Aquatic Animal Health Research Institute, Bangkok, p 88
Lin Z, Lu J, Wu S, Lin X, Zheng L, Lou Y, Xiao X (2022) A novel detection method for the pathogenic *Aeromonas hydrophila* expressing aerA gene and/or hlyA gene based on dualplex RAA and CRISPR/Cas12a. Front Microbiol 13:973996
Liu CT, Schmidt NT, Stone DM (1998) Detection of SVCV in fish tissues by semi-nested PCR: 4th Symposium on viruses of lower vertebrates, Weymouth, UK, 12–15 May 1998
Liu H, Izumi S, Wakabayashi H (2001) Detection of *Flavobacterium psychrophilum* in various organs of ayu *Plecoglossus altivelis* by in situ hybridization. Fish Pathol 36:7–11
Liu H, Wang Y, Xiao J, Wang Q, Liu Q, Zhang Y (2015) An immunochromatographic test strip for rapid detection of fish pathogen *Edwardsiella tarda*. Bioresour Bioprocess 2:1–8
Liu J, Li Y, Cao H, Yao S, Hu K, Zhao Q, He R, Zhu N, Yu X, Fang S, Huang J (2023) Amplification-free CRISPR-Cas13a assay for detection of Taura syndrome virus. Aquacult Rep 30:101552. https://doi.org/10.1016/j.aqrep.2023.101552
Loh PC, Cesar E, Nadala BJ, Tapay LM, Lu Y (1998) Recent developments in immunologically-based and cell culture protocols for the specific detection of shrimp viral pathogens. In: Flegel TW (ed) Advances in Shrimp biotechnology, pp 255–259
López-Romalde S, Magariños B, Núñez S, Toranzo AE, Romalde JL (2003) Phenotypic and genetic characterization of *Pseudomonas anguilliseptica* strains isolated from fish. J Aquat Anim Health 15(1):39–47
Lorenzen N, Lapatra SE (1999) Immunity to rhabdoviruses in rainbow trout: the antibody response. Fish Shellfish Immunol 9:345–360
Loy JK, Frelier PF (1996) Specific, non-radioactive detection of the NHP bacterium in *Penaeus vannamei* by in situ hybridization. J Vet Diagn Invest 8:324–331

Loy JK, Frelier PF, Varner P, Templeton JW (1996) Detection of the etiologic agent of necrotizing hepatopancreatitis in cultured *Penaeus vannamei* from Texas and Peru by polymerase chain reaction. Dis Aquat Org 25:117–122

Lu CC, Tang KFJ, Kou GH, Chen SN (1993) Development of a *Penaeus monodon*-type baculovirus (MBV) DNA probe by polymerase chain reaction and sequence analysis. J Fish Dis 16:551–559

Lu Y, Tapay LM, Loh PC, Brock JA, Gose RB (1995) Distribution of yellow-head virus in selected tissues and organs of penaeid shrimp *Penaeus vannamei*. Dis Aquat Org 23(1):67–70

Lu CC, Tang KFJ, Chen SN (1996a) Morphogenesis of the membranous labyrinth in penaeid shrimp cells infected with *Penaeus monodon* baculovirus (MBV). J Fish Dis 19:57–364

Lu Y, Tapay LM, Loh PC (1996b) Development of a nitrocellulose-enzyme immunoassay for the detection of yellow-head virus from penaeid shrimp. J Fish Dis 19(1):9–13

Lumsden JS, Ostland VE, Byrne PJ, Ferguson HW (1993) Detection of a distinct gill-surface antibody response following horizontal infection and bath challenge of brook trout *Salvelinus fontinalis* with *Flavobacterium branchiophilum*, the causative agent of bacterial gill disease. Dis Aquat Org 16:21–27

MacFaddin JF (1983) Biochemical tests for identification of medical bacteria, 2nd edn. Williams and Wilkins, Baltimore

MacPhee DD, Ostland VE, Lumsden JS, Ferguson HW (1995) Development of an enzyme-linked immunosorbent assay (ELISA) to estimate the quantity of *Flavobacterium branchiophilum* on the gills of rainbow trout *Oncorhynchus mykiss*. Dis Aquat Org 21:13–23

Madetoja J, Hanninen ML, Hirvela-Koski V, Dalsgaard I, Wiklund T (2001) Phenotypic and genotypic characterization of *Flavobacterium psychrophilum* from Finnish fish farms. J Fish Dis 24:469–479

Mahoney JB, Chernesky MA (1995) Multiplex polymerase chain reaction. In: Molecular methods for viral detection. Academic Press, Inc, New York, pp 219–237

Major SR, Harke MJ, Cruz-Flores R, Dhar AK, Bodnar AG, Wanamaker SA (2023) Rapid detection of DNA and RNA shrimp viruses using CRISPR-based diagnostics. Appl Environ Microbiol. https://doi.org/10.1128/aem.02151-22

Manjanaik B, Umesha KR, Karunasagar I, Karunasagar I (2005) Detection of hepatopancreatic parvovirus (HPV) in wild shrimp from India by nested polymerase chain reaction (PCR). Dis Aquat Org 63(2–3):255–259

Mao J, Hedrick RP, Chinchar VG (1997) Molecular characterization, sequence analysis and taxonomic position of newly isolated fish iridoviruses. Virology 229:212–220

Mardones FO (2020) Preventive medicine of aquatic animals. In: Aquaculture health management, design and operation approaches. Acadmeic Press, pp 163–186

Mari J, Bonami JR, Lightner DV (1993a) Partial cloning of the genome of infectious hypodermal and hematopoietic necrosis virus, an unusual parvovirus pathogenic for penaeid shrimps; diagnosis of the disease using a specific probe. J Gen Virol 74:2637–2643

Mari J, Bonami JR, Poulos B, Lightner DV (1993b) Preliminary characterization and partial cloning of the genome of a baculovirus from *Penaeus monodon* (PmSNPV = MBV). Dis Aquat Org 16:207–215

Martins T, Valentim A, Pereira N, Antunes LM (2019) Anaesthetics and analgesics used in adult fish for research: a review. Lab Anim 53(4):325–341

Matos LD, Trufelli DC, de Matos MGL, da Silva Pinhal A (2010) Immunohistochemistry as an important tool in biomarkers detection and clinical practice. Biomark Insights 5:9–20

Mauel MJ, Giovannoni SJ, Fryer JL (1996) Development of polymerase chain reaction assays for detection, identification, and differentiation of *Piscirickettsia salmonis*. Dis Aquat Org 26:189–195

McCarthy DH, Stevenson JP, Roberts MS (1975) Some blood parameters of the rainbow trout (*Salmo gairdneri* Richardson) II. The Shasta variety. J Fish Biol 7(2):215–219

McLoughlin MF, Graham DA (2007) Alphavirus infections in salmonids – a review. J Fish Dis 30:511–531

Meyer FP (1991) Aquaculture disease and health management. J Anim Sci 69(10):4201–4208

Midtlyng PJ, Bleie H, Helgason S, Jansson E, Larsen JL, Olesen NJ, Olsen AB, Vennerstrøm P (2000) Nordic manual for the surveillance and diagnosis of infectious diseases in farmed Salmonids. Nordic Council of Ministers

Miles DJC, Thompson KD, Lilley JH, Adams A (2003) Immunofluorescence of the epizootic ulcerative syndrome pathogen, *Aphanomyces invadans*, using a monoclonal antibody. Dis Aquat Org 55:77–84

Mishra A, Nam GH, Gim JA, Seong M, Choe Y, Lee HE, Jo A, Kim S, Kim DH, Cha HJ, Kang HY (2017) Comparative evaluation of 16S rRNA gene in world-wide strains of *Streptococcus iniae* and *Streptococcus parauberis* for early diagnostic marker. Gen Genom 39(7):779–791

Miyamoto N, Eguchi M (1997) Direct detection of a fish pathogen, *Vibrio anguillarum* serotype JO-1, in freshwater by fluorescent antibody technique. Fish Sci 63(2):253–257

Miyata M, Aoki T, Inglis V, Yoshida T, Endo M (1995) RAPD analysis of *Aeromonas salmonicida* and *Aeromonas hydrophila*. J Appl Bacteriol 79(2):181–185

Mohr PG, Moody NJG, Hoad J, Williams LM, Bowater RO, Cummins DM, Cowley JA, Crane MSTJ (2015) New yellow head virus genotype (YHV7) in giant tiger shrimp *Penaeus monodon* indigenous to northern Australia. Dis Aquat Org 115:263–268

Momoyama K, Hiraoka M, Inouye K, Kimura T, Nakano H (1995) Diagnostic techniques of the rod-shaped nuclear virus infection in the kuruma shrimp, *Penaeus japonicus*. Fish Pathol 30:263–269

Moseley SL, Echeverria P, Seriwatana J, Tirapat C, Chaicumpa W, Sakuldaipeara T, Falkow S (1982) Identification of enterotoxigenic *Escherichia coli* by colony hybridization using three enterotoxin gene probes. J Infect Dis 145(6):863–869

Mullis KB, Faloona FA (1987) Specific synthesis of DNA in vitro via a polymerase-catalyzed chain reaction. In: Ray W (ed) Methods in enzymology (155) 335–350. Elsivier

Myers JD (1989) Development and application of immunocytochemical staining techniques: a review. Diagn Cytopathol 5(3):318–330

Nagai T, Iwamoto E, Sakai T, Arima T, Tensha K, Iida Y, Iida T, Nakai T (2008) Characterization of *Edwardsiella ictaluri* isolated from wild ayu *Plecoglossus altlvelis* in Japan. Fish Pathol 43(4):158–163

Nagamine K, Hase T, Notomi TJMCP (2002) Accelerated reaction by loop-mediated isothermal amplification using loop primers. Mol Cell Probes 16(3):223–229

Nakajima K, Maeno Y, Fukudome M, Fukuda Y, Tanaka S, Matsuoka S, Sorimachi M (1995) Immunofluorescence test for the rapid diagnosis of red sea bream iridovirus infection using monoclonal antibody. Fish Pathol 30:115–119

Nance SL, Riederer M, Zubkowski T, Trudel M, Rhodes LD (2010) Interpreting dual ELISA and qPCR data for bacterial kidney disease of salmonids. Dis Aquat Org 91(2):113–119

Nguyen HD, Mushiake K, Nakai T, Muroga K (1997) Tissue distribution of striped jack nervous necrosis virus (SJNNV) in adult striped jack. Dis Aquat Org 28:87–91

Nilson WB, Strom MS (2002) Detection and identification of bacterial pathogens of fish in kidney tissue using terminal fragment length polymorphism (T-RFLP) analysis of 16S rRNA genes. Dis Aquat Org 53:149–156

Nishizawa T, Mori K, Nakai T, Furusawa I, Muroga K (1994) Polymerase chain reaction (PCR) amplification of RNA of striped jack nervous necrosis virus (SJNNV). Dis Aquat Org 18:103–107

Nissa MU, Pinto N, Parkar H, Goswami M, Srivastava S (2021) Proteomics in fisheries and aquaculture: an approach for food security. Food Control 127:108125

Notomi T, Okayama H, Masubuchi H, Yonekawa T, Watanabe K, Amino N, Hase T (2000) Loop-mediated isothermal amplification of DNA. Nucleic Acids Res 28(12):E63

Notomi T, Mori Y, Tomita N, Kanda H (2015) Loop-mediated isothermal amplification (LAMP): principle, features, and future prospects. J Microbiol 53(1):1–5

Nunan LM, Lightner DV (1997) Development of a non-radioactive gene probe by PCR for detection of white spot syndrome virus (WSSV). J Virol Methods 63:193–201

Nunan LM, Poulos BT, Lightner DV (1998) Reverse transcription polymerase chain reaction (RT-PCR) used for the detection of Taura Syndrome Virus (TSV) in experimentally infected shrimp. Dis Aquat Org 34:87–91

Nunan LM, Poulos BT, Lightner DV (2000) Use of polymerase chain reaction for the detection of infectious hypodermal and hematopoietic necrosis virus in penaeid shrimp. Mar Biotechnol 2:319–328

Oidtmann B, Schmid I, Rogers D, Hoffmann R (1999) An improved isolation method for the cultivation of the crayfish plague fungus, *Aphanomyces astaci*. Freshwater Crayfish 12:303–312

Olesen NJ, Jørgensen PEV (1991) Rapid detection of viral haemorrhagic septicaemia virus in fish by ELISA. J Appl Ichthyol 7:183–186

Olesen NJ, Vestergård Jørgensen PE (1992) Comparative susceptibility of three fish cell lines to Egtved virus, the virus of viral haemorrhagic septicaemia (VHS). Dis Aquat Org 12:235–237

Ordal EJ, Earp BJ (1956) Cultivation and transmission of the etiological agent of bacterial kidney disease. Proc Soc Exp Biol Med 92:85–88

Ostland VE, Lumsden JS, MacPhee DD, Ferguson HW (1994) Characteristics of *Flavobacterium branchiophilum*, the cause of salmonid bacterial gill disease in Ontario. J Aquat Anim Health 6:13–26

Ostland VE, Byrne PJ, Hoover G, Ferguson HW (2000) Necrotic myositis of rainbow trout, *Oncorhynchus mykiss* (Walbaum): proteolytic characteristics of a crude extracellular preparation from *Flavobacterium psychrophilum*. J Fish Dis 23:329–336

Otis EJ (1984) Lesions of Cold Water Disease in Steelhead Trout (*Salmo gairdneri*): The Role of *Cytophaga psychrophila* Extracellular Products. MSc thesis, University of Rhode Island, Kingston

Pan X, Zhang Y, Sha X, Wang J, Li J, Dong P, Liang X (2017) Highly sensitive detection of low-abundance white spot syndrome virus by a pre-amplification PCR method. J Microbiol Biotechnol 27(3):471–479

Panangala VS, Shoemaker CA, Klesius PH (2007) TaqMan real-time polymerase chain reaction assay for rapid detection of *Flavobacterium columnare*. Aquac Res 38:508–517

Pang J, Wang Q, Fei Y, Zhu P, Qiao L, Huang H, Dang C, Gao W (2019) A real-time recombinase polymerase amplification assay for the rapid detection of *Vibrio harveyi*. Mol Cell Probes 44:8–13

Panzarin V, Patarnello P, Mori A, Rampazzo E, Cappellozza E, Bovo G, Cattoli G (2010) Development and validation of a real-time TaqMan PCR assay for the detection of betanodavirus in clinical specimens. Arch Virol 155:1193–1203

Peatman E, Terhune J, Baoprasertkul P, Xu P, Nandi S, Wang S, Somridhivej B, Kucuktas H, Li P, Dunham R, Liu Z (2008) Microarray analysis of gene expression in the blue catfish liver reveals early activation of the MHC class I pathway after infection with *Edwardsiella ictaluri*. Mol Immunol 45(2):553–566

Phadee P, Kurata O, Hatai K, Hirono I, Aoki T (2004) Detection and identification of fish-pathogenic *Aphanomyces piscicida* using polymerase chain reaction (PCR) with species-specific primers. J Aquat Anim Health 16:220–230

Phillips ML (2007) How to maximize immunofluorescence multiplexing. The Scientist. Available at: http://www.thescientist.com/?articles.view/articleNo/25735/title/How-to-Maximize-Immunofluorescence-Multiplexing/

Phromjai J, Sukhumsirichart W, Pantoja C, Lightner DV, Flegel TW (2001) Different reactions obtained using the same DNA detection reagents for Thai and Korean hepatopancreatic parvo virus of penaeid shrimp. Dis Aquat Org 46:153–158

Pickering AD, Christie P (1980) Sexual differences in the incidence and severity of ectoparasitic infestation of the brown trout, *Salmo trutta* L. J Fish Biol 16:669–683

Pikarsky E, Ronen A, Abramowitz J, Levavi-Sivan B, Hutoran M, Shapira Y, Steinitz M, Perelberg A, Soffer D, Kotler M (2004) Pathogenesis of acute viral disease induced in fish by carp interstitial nephritis and gill necrosis virus. J Virol 78:9544–9551

Pillai D, Bonami JR, Sri Widada J (2006) Rapid detection of Macrobrachium rosenbergii nodavirus (MrNV) and extra small virus (XSV), the pathogenic agents of white tail disease of

Macrobrachium rosenbergii (De Man), by loop-mediated isothermal amplification. J Fish Dis 29(5):275–283

Pohl G, Shih LM (2004) Principle and applications of digital PCR. Expert Rev Mol Diagn 4(1):41–47

Posthuma-Trumpie GA, Korf J, van Amerongen A (2009) Lateral flow (immuno) assay: its strengths, weaknesses, opportunities and threats. A literature survey. Anal Bioanal Chem 393:569–582

Poulos BT, Lightner DV (2006) Detection of infectious myonecrosis virus (IMNV) of penaeid shrimp by reverse-transcriptase polymerase chain reaction (RT-PCR). Dis Aquat Org 73:69–72

Poulos BT, Pantoja CR, Bradley-Dunlop D, Aguilar J, Lightner DV (2001) Development and application of monoclonal antibodies for the detection of white spot syndrome virus of penaeid shrimp. Dis Aquat Org 47:13–23

Poulos BT, Tang KFJ, Pantoja CR, Bonami JR, Lightner DV (2006) Purification and characterization of infectious myonecrosis virus of penaeid shrimp. J Gen Virol 87:987–996

Powell M, Overturf K, Hogge C, Johnson K (2005) Detection of *Renibacterium salmoninarum* in Chinook salmon, *Oncorhynchus tshawytscha* (Walbaum), using quantitative PCR. J Fish Dis 28(10):615–622

Pratoomthai B, Sakaew W, Sriurairatana S, Wongprasert K, Withyachumnarnkul B (2008) Retinopathy in stunted black tiger shrimp *Penaeus monodon* and possible association with Laem-Singh virus (LSNV). Aquaculture 284(1–4):53–58

Purcell MK, Hart SA, Kurath G, Winton JR (2006) Strand-specific, real-time RT-PCR assays for quantification of genomic and positive-sense RNAs of the fish rhabdovirus, infectious hematopoietic necrosis virus. J Virol Methods 132:18–24

Puthawibool T, Senapin S, Flegel TW, Kiatpathomchai W (2010) Rapid and sensitive detection of *Macrobrachium rosenbergii* nodavirus in giant freshwater prawns by reverse transcription loop-mediated isothermal amplification combined with a lateral flow dipstick. Mol Cell Probes 24(5):244–249

Qian D, Liu W, Jianxiang W, Yu L (2006) Preparation of monoclonal antibody against *Macrobrachium rosenbergii* Nodavirus and application of TAS-ELISA for virus diagnosis in post-larvae hatcheries in east China during 2000–2004. Aquaculture 261:1144–1150

Qin QW, Lam TJ, Shen H, Chang SF, Ngoh GH, Chen CL (2001) Electron microscopic observations of a marine fish iridovirus isolated from brown-spotted grouper, *Epinephelus tauvina*. J Virol Methods 98:17–24

Qiu L, Chen MM, Wan XY, Li C, Zhang QL, Wang RY, Cheng DY, Dong X, Yang B, Wang XH, Xiang JH, Huang J (2017) Characterization of a new member of *Iridoviridae*, Shrimp hemocyte iridescent virus (SHIV), found in white leg shrimp (*Litopenaeus vannamei*). Sci Rep 7(1):11834

Qiu L, Chen MM, Wan XY, Zhang QL, Li C, Dong X, Yang B, Huang J (2018) Detection and quantification of Shrimp hemocyte iridescent virus by TaqMan probe based real-time PCR. J Invertebr Pathol 154:95–101

Qiu L, Chen X, Guo XM, Gao W, Zhao RH, Zhang QL, Yang B, Huang J (2020) A TaqMan probe based real-time PCR for the detection of Decapod iridescent virus 1. J Invertebr Pathol 173:107367

Rajan PR, Lin JY, Ho MS, Yang HL (2003) Simple and rapid detection of *Photobacterium damselae* ssp. *piscicida* by a PCR technique and plating method. J Appl Microbiol 95(6):1375–1380

Rajendran KV, Cowley JA, McCulloch RJ, Walker PJ (2006) A TaqMan real-time RT-PCR for quantifying Mourilyan virus infection levels in penaeid shrimp tissues. J Virol Methods 137(2):265–271

Rangdale RE, Way K (1995) Rapid identification of *Cytophaga psychrophila* from infected spleen tissue using an enzyme-linked immunosorbent assay (ELISA). Bull Eur Assoc Fish Pathol 15:213–216

Ravelo C, Magariños B, Romalde JL, Toranzo AE (2001) Conventional versus miniaturized systems for the phenotypic characterization of *Lactococcus garvieae* strains. Bull Eur Assoc Fish Pathol 21:136–144

Reichley SR, Ware C, Greenway TE, Wise DJ, Griffin MJ (2015) Real-time polymerase chain reaction assays for the detection and quantification of *Edwardsiella tarda, Edwardsiella piscicida,* and *Edwardsiella piscicida*–like species in catfish tissues and pond water. J Vet Diagn Invest 27(2):130–139

Reichley SR, Ware C, Steadman J, Gaunt PS, García JC, LaFrentz BR, Thachil A, Waldbieser GC, Stine CB, Buján N, Arias CR (2017) Comparative phenotypic and genotypic analysis of Edwardsiella isolates from different hosts and geographic origins, with emphasis on isolates formerly classified as *E. tarda*, and evaluation of diagnostic methods. J Clin Microbiol 55(12):3466–3491

Roberts RJ (2012) Fish pathology. Wiley

Rodnick KJ, St-Hilaire S, Battiprolu PK (2008) Evidence for sex and/or habitat differences in swimming performance, parasite infestation, tissue biochemistry, and morphology of juvenile coho salmon in Oregon's West Fork Smith River. Trans Am Fish Soc 137:1571–1590

Romalde JL (2002) *Photobacterium damselae* subsp. *piscicida*: an integrated view of a bacterial fish pathogen. Int Microbiol 5(1):3–9

Romalde JL, Magariños B, Fouz B, Bandin I, Núñez S, Toranzo AE (1995) Evaluation of BIONOR Mono-kits for rapid detection of bacterial fish pathogens. Dis Aquat Org 21(1):25–34

Romestand B, Bonami JR (2003) A sandwich enzyme-linked immunosorbent assay (S-ELISA) for detection of MrNV in the giant freshwater prawn, *Macrobrachium rosenbergii* (de Man). J Fish Dis 26:71–75

Rong F, Wang H, Tang X, Xing J, Sheng X, Chi H, Zhan W (2024) The development of RT-RPA and CRISPR-Cas12a based assay for sensitive detection of infectious hematopoietic necrosis virus (IHNV). J Virol Methods 326:114892

Rose AS, Ellis AE, Adams A (1989) An assessment of routine Aeromonas salmonicida carrier detection by ELISA. Bull Eur Assoc Fish Pathol 9:65–67

Ross LG, Ross B (2009) Anaesthetic and sedative techniques for aquatic animals. Wiley

Ruelle J, Yfantis V, Duquenne A, Goubau P (2014) Validation of an ultrasensitive digital droplet PCR assay for HIV-2 plasma RNA quantification. J Int AIDS Soc 17:19675

Saiki RK, Scharf S, Faloona F, Mullis KB, Horn GT, Erlich HA, Arnheim N (1985) Enzymatic amplification of β-globin genomic sequences and restriction site analysis for diagnosis of sickle cell anemia. Science 230(4732):1350–1354

Saiki RK, Gelfand DH, Stoffel S, Scharf SJ, Higuchi R, Horn GT, Mullis KB, Erlich HA (1988) Primer-directed enzymatic amplification of DNA with a thermostable DNA polymerase. Science 239(4839):487–491

Sakai M, Kobayashi M (1992) Detection of *Renibacterium salmoninarum*, the causative agent of bacterial kidney disease in salmonid fish, from pen-cultured coho salmon. Appl Environ Microbiol 58(3):1061–1063

Sakai T, Iida T, Osatomi K, Kanai K (2007) Detection of type 1 fimbrial genes in fish pathogenic and non-pathogenic *Edwardsiella tarda* strains by PCR. Fish Pathol 42(2):115–117

Saksida S, Marty GD, Jones SR (2012) Parasites and hepatic lesions among pink salmon, *Oncorhynchus gorbuscha* (Walbaum), during early seawater residence. J Fish Dis 35:137–151

Salati F, Meloni M, Fenza A, Angelucci G, Colorni A, Orru G (2010) A sensitive FRET probe assay for the selective detection of *Mycobacterium marinum* in fish. J Fish Dis 33(1):47–56

Sambrook J, Russell DW (2001) Chapter 12 DNA sequencing. In: Molecular cloning: a laboratory manual, 3rd edn. Cold Spring Harbor Laboratory Press, Cold Spring Harbor, pp 1–120

Savan R, Igarashi A, Matsuoka S, Sakai M (2004) Sensitive and rapid detection of edwardsiellosis in fish by a loop-mediated isothermal amplification method. Appl Environ Microbiol 70(1):621–624

Schmidt-Posthaus H, Bernet D, Wahli T, Burkhardt-Holm P (2001) Morphological organ alterations and infectious diseases in brown trout *Salmo trutta* and rainbow trout *Oncorhynchus mykiss* exposed to polluted river water. Dis Aquat Org 44(3):161–170

Sebastiao F, Lemos EM, Pilarski F (2018) Development of an absolute quantitative real-time PCR (qPCR) for the diagnosis of *Aeromonas hydrophila* infections in fish. Acta Sci Microbiol 1:23–29

Shapira Y, Magen Y, Zak T, Kotler M, Hulata G, Evavi-Sivan B (2005) Differential resistance to koi herpesvirus (KHV)/carp interstitial nephritis and gill necrosis virus (CNGV) among common carp (Cyprinus carpio *L.*) strains and crossbreds. Aquaculture 245:1–11

Sheng XZ, Zhan WB (2006) Histopathological studies on the target organs of lymphocystis disease virus of fish. Periodical Ocean Univ China 36:749–753

Shi X, Wang Z (2021) The current advances and future perspectives of lateral flow immunoassay for infectious diseases. Rev Med Microbiol 32(3):183–189

Shoemaker CA, Klesius PH (1997) Streptococcal disease problems and control—a review. In: Fitzsimmons K (ed) Tilapia Aquaculture, vol 2. Northeast Regional Agricultural Engineering Service, Ithaca, pp 671–682

Shoemaker CA, Evans JJ, Klesius PH (2000) Density and dose: factors affecting mortality of *Streptococcus iniae* infected tilapia (*Oreochromis niloticus*). Aquaculture 188(3–4):229–235

Shotts EB Jr, Rimler R (1973) Medium for the isolation of *Aeromonas hydrophila*. Appl Microbiol 26(4):550–553

Sirikharin R, Taengchaiyaphum S, Sanguanrut P, Chi TD, Mavichak R, Proespraiwong P, Nuangsaeng B, Thitamadee S, Flegel TW, Sritunyalucksana K (2015) Characterization and PCR detection of binary, Pir-like toxins from *Vibrio parahaemolyticus* isolates that cause acute hepatopancreatic necrosis disease (AHPND) in shrimp. PLoS One 10:0126987

Sithigorngul W, Rukpratanporn S, Pecharaburanin N, Longyant S, Chaivisuthangkura P, Sithigorngul P (2006) A simple and rapid immunochromatographic test strip for detection of white spot syndrome virus (WSSV) of shrimp. Dis Aquat Org 72:101–106

Smith PD (1982) Analysis of the hyperosmotic and bath methods for fish vaccination—comparison of uptake of particulate and non-particulate antigens. Dev Comp Immunol 2:181–186

Snow M, Bain N, Black J, Taupin V, Cunningham CO, King JA, Skall HF, Raynard RS (2004) Genetic population structure of marine viral haemorrhagic septicaemia virus (VHSV). Dis Aquat Org 61:11–21

Snow M, McKay P, McBeath AJA, Black J, Doig F, Kerr R, Cunningham CO, Nylund A, Devold M (2006) Development, application and validation of a TaqMan® real-time RT-PCR assay for the detection of infectious salmon anaemia virus (ISAV) in Atlantic salmon (*Salmo salar* L.). In: Vannier P, Espeseth D (eds) New diagnostic technology: applications in animal health and biologics controls. Developments in biologicals, Basel, Karger 126, pp 133–145

Sørensen UBS, Larsen JL (1986) Serotyping of *Vibrio anguillarum*. Appl Environ Microbiol 51:593–597

Soto E, Hawke JP, Fernández D, Morales JA (2009) *Francisella* sp., an emerging pathogen of tilapia, *Oreochromis niloticus* (L.), in Costa Rica. J Fish Dis 32(8):713–722

Soto-Rodríguez SA, Gomez-Gil B, Lozano-Olvera R, Betancourt-Lozano M, Morales-Covarrubias MS (2015) Field and experimental evidence of *Vibrio parahaemolyticus* as the causative agent of acute hepatopancreatic necrosis disease of cultured shrimp (*Litopenaeus vannamei*) in northwestern Mexico. Appl Environ Microbiol 81:1689–1699

Spanggaard B, Huss HH, Bresciani J (1995) Morphology of *Ichthyophonus hoferi* assessed by light and scanning electron microscopy. J Fish Dis 18(6):567–577

Sri Widada J, Durand S, Cambournac I, Qian D, Shi Z, Dejonghe E, Richard V, Bonami JR (2003) Genome-based detection methods of *Macrobrachium rosenbergii* nodavirus, a pathogen of the giant freshwater prawn, *Macrobrachium rosenbergii*: dot-blot, in situ hybridization and RT-PCR. J Fish Dis 26:583–590

Steigen A, Karlsbakk E, Plarre H, Watanabe K, Overgard AC, Brevik O, Nylund A (2015) A new intracellular bacterium, *Candidatus Similichlamydia labri* sp. nov. (Chlamydiaceae) producing epitheliocysts in ballan wrasse, *Labrus bergylta* (Pisces, Labridae). Arch Microbiol 197:311–318

Steinum T, Kvellestad A, Colquhoun DJ, Heum M, Mohammad S, Grontvedt RN, Falk K (2010) Microbial and pathological findings in farmed Atlantic salmon, *Salmo salar*, with proliferative gill inflammation. Dis Aquat Org 91:201–211. https://doi.org/10.3354/dao02266

Stride MC, Polkinghorne A, Miller TL, Groff JM, Lapatra SE, Nowak BF (2013) Molecular characterization of *Candidatus Parilichlamydia carangidicola*, a novel Chlamydia-like epitheliocystis agent in yellowtail kingfish, *Seriola lalandi* (Valenciennes), and the proposal of a new family, *Candidatus Parilichlamydiaceae* fam. nov. (order Chlamydiales). Appl Environ Microbiol 79:1590–1597

Sudhakaran R, Ishaq Ahmed VP, Haribabu P, Mukherjee SC, Sri Widada J, Bonami JR, Sahul Hameed AS (2007) Experimental vertical transmission of *Macrobrachium rosenbergii* nodavirus (MrNV) and extra small virus (XSV) from brooders to progeny in *Macrobrachium rosenbergii* and Artemia. J Fish Dis 30:27–35

Sukonta T, Senapin S, Taengphu S, Hannanta-anan P, Kitthamarat M, Aiamsa-at P, Chaijarasphong T (2022) An RT-RPA-Cas12a platform for rapid and sensitive detection of tilapia lake virus. Aquaculture 560:738538

Sullivan TJ, Dhar AK, Cruz-Flores R, Bodnar AG (2019) Rapid, CRISPR-based, field-deployable detection of white spot syndrome virus in shrimp. Sci Rep 9(1):1–7

Suthaharan S (2016) Support vector machine. In: Machine learning models and algorithms for Big Data classification. Integrated series in information systems, vol 36. Springer, Boston

Swain P, Nayak SK (2003) Comparative sensitivity of different serological tests for seromonitoring and surveillance of *Edwardsiella tarda* infection of Indian major carps. Fish Shellfish Immunol 15(4):333–340

Taksdal T, Olsen AB, Bjerkaas I, Hjortaas MJD, Dannevig BH, Graham DA, McLoughlin MF (2007) Pancreas disease in farmed Atlantic salmon, *Salmo salar* L., and rainbow trout, *Oncorhynchus mykiss* (Walbaum), in Norway. J Fish Dis 30:545–558

Talaat AM, Reimschuessel R, Trucksis M (1997) Identification of mycobacteria infecting fish to the species level using polymerase chain reaction and restriction enzyme analysis. Vet Microbiol 58(2–4):229–237

Tang KFJ, Lightner DV (1999) A yellow head virus gene probe: nucleotide sequence and application for in situ hybridization. Dis Aquat Org 35:165–173

Tang KF, Lightner DV (2001) Detection and quantification of infectious hypodermal and hematopoietic necrosis virus in penaeid shrimp by real-time PCR. Dis Aquat Org 44(2):79–85

Tang KFJ, Wang J, Lightner DV (2004) Quantitation of Taura syndrome virus by real-time RT-PCR with a TaqMan assay. J Virol Methods 115:109–114

Tang KFJ, Navarro SA, Lightner DV (2007) A PCR assay for discriminating between infectious hypodermal and hematopoietic necrosis virus (IHHNV) and virus-related sequences in the genome of *Penaeus monodon*. Dis Aquat Org 74:165–170

Tattiyapong P, Sirikanchana K, Surachetpong W (2017) Development and validation of a reverse transcription quantitative polymerase chain reaction for tilapia lake virus detection in clinical samples and experimentally challenged fish. J Fish Dis 41:255

Tavornapanich S, Brun E, Dverdal Jansen M (2020) Guidelines for on-farm sampling for targeted surveillance to certify disease freedom, diagnosis in case of mortalities, and for analysis of mortalities caused by unknown aetiology. In: Zrncic S (ed) Diagnostic manual for the main pathogens in European seabass and gilthead seabream. Aquaculture 15–20

Taylor-Brown A, Pillonel T, Bridle A, Qi W, Bachmann NL, Miller TL, Polkinghorne A (2017) Culture-independent genomics of a novel chlamydial pathogen of fish provides new insight into host-specific adaptations utilized by these intracellular bacteria. Environ Microbiol 19:1899–1913

Te SH, Chen EY, Gin KYH (2015) Comparison of quantitative PCR and droplet digital PCR multiplex assays for two genera of bloom-forming cyanobacteria, Cylindrospermopsis and Microcystis. Appl Environ Microbiol 81(15):5203–5211

Thoesen JC (1994) Suggested procedures for the detection and identification of certain fish pathogens, 4th edn. American Fisheries Society, Fish Health Section, Bethesda

Tobback E, Decostere A, Hermans K, Haesebrouck F, Chiers K (2007) *Yersinia ruckeri* infections in salmonid fish. J Fish Dis 30(5):257–268

Toubanaki DK, Karagouni E (2018) Towards a dual lateral flow nanobiosensor for simultaneous detection of virus genotype-specific PCR products. J Anal Methods Chem 7691014. https://doi.org/10.1155/2018/7691014

Toubanaki DK, Margaroni M, Karagouni E (2015) Nanoparticle-based lateral flow biosensor for visual detection of fish nervous necrosis virus amplification products. Mol Cell Probes 29:158–166

Toubanaki DK, Margaroni M, Prapas A, Karagouni E (2020) Development of a nanoparticle-based lateral flow strip biosensor for visual detection of whole nervous necrosis virus particles. Sci Rep 10(1):1–12

Touhali IS, Al-uguali IS (2018) Isolation and identification of *Saprolegnia parasitica* and other fungi from farmed fishes in the province of Wasit/Iraq. J Glob Pharm Technol 10:135–142

Toyama T, Kita-Tsukamoto K, Wakabayashi H (1996) Identification of *Flexibacter maritimus*, *Flavobacterium branchiophilum* and *Cytophaga columnaris* by PCR targeted 16S ribosomal DNA. Fish Pathol 31(1):25–31

Tuffs S, Oidtmann B (2011) A comparative study of molecular diagnostic methods designed to detect the crayfish plague pathogen, *Aphanomyces astaci*. Vet Microbiol 153(3–4):343–353

Ueno Y, Kitao T, Chen SN, Aoki T, Kou GH (1992) Characterization of a herpes-like virus isolated from cultured Japanese eels in Taiwan. Fish Pathol 27(1):7–17

Umelo E (1998) Physical map of the chromosome of *Aeromonas salmonicida* and genomic comparisons between Aeromonas strains. Microbiology 144(8):2141–2149

Umesha RK, Uma A, Otta SK, Karunasagar I, Karunasagar I (2003) Detection by PCR of hepatopancreatic parvovirus (HPV) and other viruses in hatchery-reared *Penaeus monodon* postlarvae. Dis Aquat Org 57:141–146

Urdaci MC, Chakroun C, Faure D, Bernardet JF (1998) Development of a polymerase chain reaction assay for identification and detection of the fish pathogen *Flavobacterium psychrophilum*. Res Microbiol 149(7):519–530

US Fish and Wildlife Service (2010) American Fisheries Society–Fish Health Section (AFS-FHS): 2010, Standard procedures for aquatic animal health inspections. AFS-FHS blue book: suggested procedures for the detection and identification of certain finfish and shellfish pathogens

Valverde EJ, Cano I, Labella A, Borrego JJ, Castro D (2016) Application of a new real-time polymerase chain reaction assay for surveillance studies of lymphocystis disease virus in farmed gilthead seabream. BMC Vet Res 12(1):1–8

Valverde EJ, Cano I, Castro D, Paley RK, Borrego JJ (2017) Rapid and sensitive detection of lymphocystis disease virus genotype VII by loop-mediated isothermal amplification. Food Environ Virol 9(1):114–122

van Pelt-Verkuil E, te Witt R (2019) Primers and probes. Molecular diagnostics: part 1: technical backgrounds and quality aspects, 51–95

Vancells JC (2016) Direct vs indirect immunofluorescence. Available at: http://www.abcam.com/secondary-antibodies/direct-vs-indirect-immunofluorescence

Vandersea MW, Litaker RW, Yonnish B, Sosa E, Landsberg JH, Pullinger C, Moon-Butzin P, Green J, Morris JA, Kator H, Noga EJ, Tester PA (2006) Molecular assays for detecting *Aphanomyces invadans* in ulcerative mycotic fish lesions. Appl Environ Microbiol 72:1551–1557

Vickers JE, Lester RJG, Spradbrow PB, Pemberton JM (1992) Detection of Penaeus monodon-type baculovirus (MBV) in digestive glands of postlarval prawns using polymerase chain reaction. Dis Asian Aquacult 1:127–133

Vickers JE, Lester RJG, Spradbrow PB, Pemberton JM (1994) Evidence for homology between polyhedrin genes of baculoviruses of a prawn and an insect. J Invertebr Pathol 63:207–208

Wages JM Jr (2005) Polymerase chain reaction. In: Encyclopedia of analytical science, p 243

Walker J, Dougan G (1989) DNA probes: a new role in diagnostic microbiology. J Appl Bacteriol 67(3):229–238

Wallbanks S, Martinez-Murcia AJ, Fryer JL, Phillips BA, Collins MD (1990) 16S rRNA sequence determination for members of the genus *Carnobacterium* and related lactic acid bacteria and description of *Vagococcus salmoninarum* sp. nov. Int J Syst Evol Microbiol 40(3):224–230

Wang SY, Hong C, Lotz JM (1996) Development of a PCR procedure for the detection of *Baculovirus penaei* in shrimp. Dis Aquat Org 25:123–131

Wang PC, Lin YD, Liaw LL, Chern RS, Chen SC (2008) *Lactococcus lactis* subspecies *lactis* also causes white muscle disease in farmed giant freshwater prawns *Macrobrachium rosenbergii*. Dis Aquat Org 79(1):9–17

Wang P, Yang L, Guo B, Shang Z, Gao S, Zhang X (2023) Highly sensitive detection of white spot syndrome virus with an RPA-CRISPR combined one-pot method. Aquaculture 567:739296

Watanabe K, Nishizawa T, Yoshimizu M (2000) Selection of broodstock candidates of barfin flounder using an ELISA system with recombinant protein of barfin flounder nervous necrosis virus. Dis Aquat Org 41:219–223

White BL, Schofield PJ, Poulos BT, Lightner DV (2002) A laboratory challenge method for estimating Taura syndrome virus resistance in selected lines of Pacific white shrimp *Penaeus vannamei*. J World Aquacult Soc 33:341–348

White VC, Morado JF, Crosson LM, Vadopalas B, Friedman CS (2013) Development and validation of a quantitative PCR assay for *Ichthyophonus* spp. Dis Aquat Org 104(1):69–81

Whittington RJ, Steiner KA (1993) Epizootic haematopoietic necrosis virus (EHNV): improved ELISA for detection in fish tissues and cell cultures and an efficient method for release of antigen from tissues. J Virol Methods 43:205–220

Wiklund T, Bylund G (1990) *Pseudomonas anguilliseptica* as a pathogen of salmonid fish in Finland. Dis Aquat Org 8:13–19

Wiklund T, Madsen L, Bruun MS, Dalsgaard I (2000) Detection of Flavobacterium psychrophilum from fish tissue and water samples by PCR amplification. J Appl Microbiol 88(2):299–307

Winton JR (1991) Recent advances in the detection and control of infectious hematopoietic necrosis virus (IHNV) in aquaculture. Annu Rev Fish Dis 1:83–93

Winton JR, Einer-Jensen K (2002) Molecular diagnosis of infectious hematopoietic necrosis and viral hemorrhagic septicemia. In: Cunningham CO (ed) Molecular diagnosis of salmonid diseases, pp 49–79

Wise JA, Boyle JA (1985) Detection of channel catfish virus in channel catfish, *Ictalurus punctatus* (Rafinesque): use of nucleic acid probe. J Fish Dis 8:417–424

Wolf K (1988) Fish viruses and fish viral diseases. Cornell University Press, Ithaca, New York

Xia L, Zang H, Lu Y, Cai J, Wang B, Jian J (2015) Development of a loop-mediated isothermal amplification assay for rapid detection of *Nocardia salmonicida*, the causative agent of nocardiosis in fish. J Microbiol Biotechnol 25:321–327

Xu L, Wang T, Li F, Yang F (2016) Isolation and preliminary characterization of a new pathogenic iridovirus from redclaw crayfish *Cherax quadricarinatus*. Dis Aquat Org 120(1):17–26

Yiagnisis M, Vatsos IN, Kyriakou C, Alexis M (2007) First report of *Vibrio anguillarum* isolation from diseased big scale sand smelt, Atherina boyeri *Risso* 1810, in Limnos, Greece

Yoganandhan K, Syed Musthaq S, Narayanan RB, Sahul Hameed AS (2004) Production of polyclonal antiserum against recombinant VP28 protein and its application for the detection of white spot syndrome virus in crustaceans. J Fish Dis 27:517–522

Yoganandhan K, Sri Widada J, Bonami JR, Sahul Hameed AS (2005) Simultaneous detection of *Macrobrachium rosenbergii* nodavirus and extra small virus by a single tube, one-step multiplex RT-PCR assay. J Fish Dis 28:65–69

Yogananth N, Bhakyaraj R, Chanthuru A, Anbalagan T, Nila KM (2009) Detection of virulence gene in *Aeromonas hydrophila* isolated from fish samples using PCR technique. Glob J Biotechnol Biochem 4:51–53

Yu X et al (2017) A nanostructured microfluidic immunoassay platform for highly sensitive infectious pathogen detection. Small 13:7700425

Yuan B, Yuan C, Li L, Long M, Chen Z (2022) Application of the CRISPR/Cas system in pathogen detection: a review. Molecules 27(20):6999

Zappulli V, Patarnello T, Patarnello P, Frassineti F, Franch R, Manfrin A, Castagnaro M, Bargelloni L (2005) Direct identification of *Photobacterium damselae* subspecies *piscicida* by PCR-RFLP analysis. Dis Aquat Org 65(1):53–61

Zlotkin A, Eldar A, Ghittino C, Bercovier H (1998) Identification of *Lactococcus garvieae* by PCR. J Clin Microbiol 36:983–985

Prospects of Vaccine in Shrimp Aquaculture

A. S. Sahul Hameed and S. Vimal

Abstract

One of the major constraints in the shrimp culture system is diseases and parasites affecting the stock and responsible for huge economic loss. Of the infectious diseases of shrimp, viral and bacterial infections have been found to be a serious problem and responsible for severe economic loss in the shrimp culture industry worldwide. Various strategies are being followed to prevent or control the diseases in shrimp culture systems, and these strategies include good nutrition in improving aquatic animal health and the use of efficient vaccines, immunostimulants, non-specific immune-enhancers and probiotics. Most of the research has been concentrated on the development of vaccines against *Vibrio* and white spot syndrome virus (WSSV). The research carried out so far has resulted in the development of different types of vaccines, such as inactivated vaccines, recombinant vaccines and DNA vaccines against *Vibrio* and WSSV. This paper summarizes the status of different vaccines developed against viral and bacterial pathogens of shrimp.

Keywords

Shrimp · Vaccines · Immunostimulants · Virus · Bacteria

1 Introduction

Shrimp is the most favoured crustacean species for aquaculture because of its high value both in domestic and world markets. It represents the second leading group of exported species in value terms. The global shrimp production reached almost 4

A. S. Sahul Hameed (✉) · S. Vimal
OIE Reference Laboratory, C. Abdul Hakeem College, Melvisharam, Tamil Nadu, India
e-mail: cah_sahul@hotmail.com; vimalbiotechs@gmail.com

R. Kooloth Valappil et al. (eds.), *Aquatic Animal Health Management*,
https://doi.org/10.1007/978-981-96-7987-4_19

million tonnes in 2018. *Penaeus vannamei* makes up to 77% of total shrimp production from the aquaculture sector, followed by *Penaeus monodon* and *Macrobrachium rosenbergii*, in terms of volumes of total crustacean production. Nearly 75 to 80% of the shrimp production originated in Asia-Pacific regions. However, disease outbreaks are being increasingly recognized as a major constraint to aquaculture production and trade and are affecting the economic development of the sector in many countries, including India. Disease is now considered to be the most important limiting factor in aquaculture production worldwide. The economic loss due to diseases has been estimated at about US$ 1 billion in developing countries of Asia in 1990 alone. Since then, losses have been increasing. A 1995 estimate suggests that aquatic animal disease and environment-related problems may cause annual losses to aquaculture production in Asian countries to the tune of more than US$ 2 billion per year. According to recent reports, global losses due to shrimp disease are more than US$ 2.8 billion. Besides, the World Bank has invested Rs. 1240 crores for conducting research on shrimp diseases. In India, the loss due to WSSV has been estimated to be about USD 150 million per year, and the loss continues to the same extent (Kalaimani et al. 2013). The above figures provide an indication of the overall economic significance of aquatic animal diseases. Viral and bacterial pathogens are very important as they are responsible for huge economic losses in aquaculture systems worldwide.

Any practical method to control the diseases of viral and bacterial origin in culture systems would have enormous practical benefits for shrimp farmers and hatchery operators. In the field condition, various control strategies, such as improvement of environmental conditions, stocking of specific pathogen-free post-larvae and increased disease resistance by oral immunostimulants, are being applied to control diseases. Vaccines have been tried at the laboratory level to protect the shrimp from viral and bacterial infections by many workers.

2 Vaccines for Shrimp

Shrimps have an innate immune system, characterized by a lack of immunoglobulin and memory, but are efficient enough to protect and preserve themselves from all intruding antigens. Specialized haemocytes are responsible for immune defence by carrying out activities such as haemocytic infiltration, encapsulation, phagocytosis and production of antimicrobial substances to remove or neutralize the intruding pathogens. In shrimp, vaccines have been developed by many workers to target mainly *Vibrio* and white spot syndrome virus at an experimental stage. Different types of vaccines, such as inactivated viruses or bacteria, recombinant vaccines, and DNA vaccines, have been developed to prevent viral and bacterial infections in shrimp for laboratory experiments. As shrimp lack a true adaptive immune system and rely solely on innate immune responses, the possibility of vaccinating shrimp is not feasible. However, some studies have reported protection of shrimp using vaccines against bacterial (Itami et al. 1989; Alabi et al. 1999) or viral pathogens

(Witteveldt et al. 2004a, b; Jha et al. 2006a, b), suggesting that protective responses can be inducible (Loker et al. 2004; Kurtz and Franz 2003).

2.1 Bacterial Vaccines

Vibrio harveyi is a serious bacterial pathogen to all life stages of penaeid shrimp and is responsible for severe mortality in hatcheries and grow-out ponds. Mass mortalities of penaeid shrimp resulting from *V. harveyi* infection have been frequently reported in shrimp hatcheries and grow-out ponds worldwide. Larval mortalities associated with the presence of *V. harveyi* have been reported in *P. monodon* and *P. vannamei* in Indonesia (Sunaryanto and Mariam 1986), Thailand (Jiravanichpaisal and Miyazaki 1994), India (Karunasagar et al. 1994; Abraham et al. 1997; Stalin and Srinivasan 2017), Philippines (Baticados et al. 1990; Lavilla-Pitogo et al. 1990), Australia (Pizzutto and Hirst 1995), Taiwan (Song and Lee 1993; Liu et al. 1996) and Ecuador (Robertson et al. 1998). Some chemicals such as malachite green, formalin and antibiotics are being used to control *V. harveyi* in the hatcheries and grow-out ponds (Raffi and Suresh 2011). Medicinal plants were screened for their antibacterial activity against *V. harveyi* in different countries by various workers and were found to be alternatives to antibiotics to control *V. harveyi* in aquaculture systems (Immanuel et al. 2004; Sivaram et al. 2004; Citarasu et al. 2006; Rameshthangam and Ramasamy 2007; Bai et al. 2009). Bacteriophages were found to be an effective alternative to antibiotics to control infectious bacterium such as *V. harveyi* in hatcheries (Ruangpan et al. 1999; Oakey and Owens 2000; Karunasagar et al. 2007).

Besides these control measures, an effective vaccine in the form of heat- or formalin-killed *Vibrio* was found to be effective in protecting the shrimp against vibriosis (Kou et al. 1989; Itami et al. 1989, 1992). This method of vaccine preparation is being used widely to produce vaccine against *Vibrio* species by many workers (Lin et al. 2013; Patil et al. 2013; Ray et al. 2017) because it is simple and cost-effective. Teunissen et al. (1998) developed a polyvalent vaccine prototype against *V. alginolyticus* and the results of vaccination trials showed significant enhancement in the resistance in challenged shrimp against vibriosis. Fresh or freeze-dried *V. harveyi* vaccines protect vaccinated larvae of *P. indicus* against different strains of *V. harveyi* by immersion method (Alabi et al. 1999). Biofilm developed from *V. alginolyticus* has been used as a vaccine to protect the shrimp from vibriosis (Sharma et al. 2010). Studies revealed that the combination of vaccine and carboxymethyl β-1,3-glucans (CMBG) increases the efficacy of the vaccine to protect the shrimp from vibriosis (Devaraja et al. 1998; Klannukarn et al. 2004). Different routes of administration have been tested to vaccinate larvae, post-larvae and adult shrimp, which included intramuscular injection (Itami et al. 1989; Goarant and Boglio 2000; Klannukarn et al. 2004; Powell et al. 2011), oral route through feeding (Powell et al. 2011; Patil et al. 2013; Ray et al. 2017) and immersion method (Azad et al. 2005; Hettiarachchi et al. 2005). The only commercial vaccine available for vibriosis is the AquaVac™Vibromax™ manufactured by Schering-Plough Animal Health Corporation, UK. This vaccine contains inactivated *Vibrio* pathogens and protects shrimp against different species of *Vibrio*. The post-larvae of *P. monodon* and *L. vannamei* were fed with Vibromax™-enriched *Artemia* for ten days continuously and challenged with *V. parahaemolyticus* and the results showed that Vibromax™ enhanced the growth and survival of shrimp post-larvae (Wongtavatchai et al. 2010).

2.2 Viral Vaccines

In shrimp, the white spot syndrome virus (WSSV) is the only pathogen targeted for vaccine development to control because it is a major viral pathogen and responsible for huge economic loss in the shrimp culture sector worldwide (Flegel 1997; Lightner 1996). White spot syndrome was first recognized in 1992–1993 in North East Asia (Takahashi et al. 1994; Chou et al. 1995) and has spread to all shrimp culture areas of the Indo-Pacific. Cumulative mortalities in infected populations may reach 100% within 2 to 10 days of the onset of clinical signs. It affects penaeid shrimp and other crustaceans such as marine, brackish and freshwater crayfishes, crabs and lobsters (Sahul Hameed et al. 2001, 2003; Syed Musthaq et al. 2006a, b). Various preventive and control measures have been employed to control WSSV infection (Lightner 2003). Vaccination could be one of the potential strategies to overcome WSSV infection in the culture system. Vaccines in the form of inactivated WSSV, recombinant viral proteins and plasmid encoding viral genes have been employed to prevent WSSV infection in shrimp by different workers and results showed protection against the virus to some extent, even though shrimp lack a true adaptive immune system (Witteveldt et al. 2004a, b, 2006; Bright Singh et al. 2005; Jha et al. 2006a, b; Vaseeharan et al. 2006; Fu et al. 2010; Caipang et al. 2008; Rajesh Kumar et al. 2008, 2009; Mavichak et al. 2009). These reports suggest that contrary to previous belief that invertebrates rely entirely on the innate immune system, some aspects of specific immunity, like inducibility, appear to be present in some cases (Loker et al. 2004; Kurtz and Franz 2003).

2.2.1 Whole Virus Vaccine

Many workers have used inactivated whole virus vaccine to vaccinate shrimp against WSSV (Namikoshi et al. 2004; Bright Singh et al. 2005; Zhu et al. 2009; Amar and Faisan 2011; Yogeeswaran et al. 2012; Heidarieh et al. 2014; Sudheer et al. 2015). Namikoshi et al. (2004) have used formalin/heat-inactivated WSSV or recombinant viral protein rVP28 for the first time to vaccinate by intramuscular injection in *Marsupenaeus japonicus* against WSSV. Bright Singh et al. (2005) reported the resistance induced by oral administration of formalin-inactivated WSSV in *Fenneropenaeus indicus* against WSSV. The formalin-killed virus vaccine was found to induce resistance against WSSV in shrimp better than the heat-killed virus (Namikoshi et al. 2004). Zhu et al. (2009) have used binary ethylenimine to inactivate WSSV and tested its efficacy in protecting crayfish, *Procambarus clarkia,* from WSSV infection. Vaccination using a combination of inactivated WSSV and an immunostimulant increased the protection rate in shrimp against WSSV (Yogeeswaran et al. 2012). Sudheer et al. (2015) reported the up-regulation of humoral immune genes (prophenoloxidase, alpha 2-macroglobulin (α2M), crustin and *Pm*RACK, and the cell-mediated immune genes (caspase and Rab7) in gill tissue upon vaccination with inactivated WSSV and subsequent virus challenge. Amar et al. (2021) conducted a field trial to test the efficacy of formalin-inactivated WSSV to prevent virus infection in grow-out ponds, and a successful harvest was made without mass mortality and WSSV infection.

2.2.2 Recombinant Vaccines

There are several recombinant proteins effectively used to vaccinate shrimp against WSSV. Different expression systems such as viruses, bacteria, algae, yeast and transgenic animals have been used to produce recombinant viral proteins for the purpose of vaccination (Witteveldt et al. 2004b; Fu et al. 2010; Syed Musthaq et al. 2009; Jha et al. 2007; Mu et al. 2012; Feng et al. 2014). Injections with r-VP26 or r-VP28 protein induced a higher resistance against WSSV (Namikoshi et al. 2004; Lo et al. 1996; Jha et al. 2006a, b). Shrimp were fed with pellet feed coated with inactivated bacteria overexpressed with VP19 and VP28 of WSSV, and the results showed significant protection in shrimp vaccinated with VP28 (Witteveldt et al. 2004a, b). Oral administration of recombinant VP28 protein increased the survival of *P. monodon* and *L. vananmei* after the WSSV challenge (Du et al. 2006; Witteveldt et al. 2006).

Two proteins, VP19 and VP28, expressed in yeast (*Pichia pastoris*), were used as oral vaccines in different forms to induce resistance against WSSV in crayfish, and the results showed more than 60% relative percent survival in vaccinated animals (Jha et al. 2007). Shrimp vaccinated with recombinant VP292 protein of WSSV showed resistance with 52% relative percentage survival after the WSSV challenge (Vaseeharan et al. 2006). Two recombinant proteins, VP26 and VP28, were used to vaccinate *Marsupeneaeus japonicus* orally, and vaccinated shrimp were challenged with WSSV by oral, immersion, and intramuscular (IM) routes to compare protection efficacy against WSSV (Satoh et al. 2001). Fu et al. (2010) reported that the rVP28 protein expressed in *Bacillus subtilis* was administered orally to crayfish, *Procambarus clarkii* to enhance the resistance against WSSV. The rVP28 protein expressed in *B. subtilis* offered better protection when compared to the rVP28 protein produced in *E. coli* (Fu et al. 2010). Syed and Syed and Kwang (2011) demonstrated the recombinant VP28 protein expressed in baculovirus to vaccinate the shrimp by oral and immersion routes and observed high survival of vaccinated shrimp after challenge by both routes of vaccination. Ning et al. (2011) used the VP28 protein expressed on the surfaces of *B. subtilis* spores to vaccinate crayfish by oral route and observed a high survival rate in the vaccinated group after the challenge with WSSV. Yang et al. (2012) reported that two WSSV envelope proteins, VP28 and VP36B, in *E. coli* were used to vaccinate the shrimp, and the results showed high survival in the rVP28 group in contrast to the rVP36B associated with high activity level of phenoloxidase and superoxide dismutase. Lanh et al. (2021) investigated the possibility of the VP28 protein of WSSV expressed in *Chlamydomonas reinhardtii* to control WSSV infection in shrimp, and results indicated 70% survival in vaccinated shrimp. The N-terminal and C-terminal fragments of VP28 expressed in *E. coli were used to vaccinate L. vannamei by oral and injection routes, and the results revealed significant protection in shrimp vaccinated with the* C-terminal VP28 compared to N-terminal VP28 (Zhi-guang et al. 2012). Sun et al. (2013) have used live bacteria overexpressed with recombinant VP28 to enhance the resistance in shrimp against WSSV. The pellet feed coated with *Bacillus subtilis* spores containing VP26 fusion protein was fed to shrimp, *L. vannamei,* to induce resistance against WSSV infection, and the results showed 100% protection by oral vaccination with the constructed recombinant spores (Valdez et al. 2014). Solís-Lucero et al. (2016) used recombinant phage expressed with recombinant VP28 protein and used to vaccinate shrimp against WSSV, and its efficacy was assessed by immunoassay and challenge studies. Taengchaiyaphum et al. (2017) have produced a multimeric recombinant VP28 vaccine (4XrVP28) and tested its efficacy, and the results showed high relative percent survival (81%) in shrimp vaccinated with multimeric recombinant vaccine compared to shrimp vaccinated with monomeric vaccines (67%). The multimeric vaccines provide high protection when compared to monomeric vaccines as observed in vertebrates (Jardine et al. 2013; Hadifar et al. 2014). Taju et al. (2018) tested the efficiency of chitosan tripolyphosphate nanoparticles to deliver the r-VP28 protein in shrimp by oral route and confirmed the tissue distribution of VP28 protein and immune response at the gene level in treated shrimp. Crayfish injected with recombinant V28 protein of WSSV enhanced the survival and innate immunity after the WSSV challenge when compared to the non-vaccinated group (Wang et al. 2023).

2.2.3 DNA Vaccines

The protection conferred by these recombinant protein vaccines has been found to be short-lived. Hence, long-term antigen expression through genetic immunization could be a useful strategy to adopt against viral diseases. Vaccine research is currently oriented towards replacing recombinant protein vaccines with more effective and safer approaches, such as DNA vaccines. Immunization with antigen-encoding plasmid DNA can elicit strong and long-lasting humoral and cellular immune responses. This approach also offers economic, environmental and safety advantages, which are particularly attractive for the aquaculture industry (Heppell and Davis 2000; Kwang 2000). DNA vaccination using plasmid encoding genes of WSSV has been tried and found to be useful to protect the shrimp from WSSV infection in the long term, and encouraging immunological responses to DNA vaccine was also reported (Rout et al. 2007; Rajesh Kumar et al. 2008; Ning et al. 2009; Li et al. 2010). DNA vaccine using the VP 28 gene of WSSV has already been shown to give resistance in black tiger shrimp against WSSV for about 7 weeks by experimental challenge with the virus (Namita Rout et al. 2007). Rajesh Kumar et al. (2008) used a DNA construct with VP28 gene to vaccinate the shrimp and the results showed a significantly higher survival rate in vaccinated shrimp compared to control groups. Ning et al. (2009) have used the attenuated *Salmonella enterica* serovar Typhimurium (*Salmonella typhimurium*) as a bactofection vehicle for the oral delivery of a DNA vaccine plasmid with VP28 gene of WSSV to crayfish for immunization in crayfish and the results showed significant protection in the vaccinated animal when compared to control groups.

3 Immunotherapy Against WSSV

Neutralization efficiency of antibodies raised against different proteins of WSSV against the virus has been studied by many workers (Van Hulten et al. 2001; Robalino et al. 2006; Syed Musthaq et al. 2006a, b; Lu et al. 2008, 2009). van Hulten et al. (2001) reported that a specific polyclonal antibody raised against VP28 protein was able to neutralize WSSV infection in *P. monodon* in a concentration-dependent manner by intramuscular injection of neutralized WSSV. Polyclonal antibody raised against recombinant VP28 neutralized WSSV and 100% survival was observed in shrimp injected with neutralized WSSV (Syed Musthaq et al. 2006a, b). Monoclonal and polyclonal antibodies raised against VP28 were tested for their efficiency in neutralizing the WSSV, and results showed strong inactivation of WSSV by some rabbit sera in a manner independent of anti-vp28 antibodies (Robalino et al. 2006). Lu et al. (2008) reported that egg yolk immunoglobulin (IgY) from hens immunized with an inactivated vaccine or DNA vaccine was able to neutralize WSSV and observed high survival in WSSV-challenged shrimp, and the same results were obtained in crayfish also (Lu et al. 2009). These results indicate that passive immunotherapy with antibodies would be a valuable method to control WSSV infection in shrimp.

4 Immune Response to Vaccines

Immunological response in shrimp to the vaccine has been investigated by many workers (Rajesh Kumar et al. 2008, 2009; Kulkarni et al. 2014; Sudheer et al. 2015; Chen et al. 2016; Ancy et al. 2016). Rajesh Kumar et al. (2008) reported high levels of immunological parameters such as prophenol oxidase, superoxide dismutase and superoxide anion in the heamolymph of DNA-vaccinated shrimp when compared to control shrimp. Kulkarni et al. (2014) reported changes in the expression levels of crustacyanin, serine proteases, myosin light chain and ER protein 57 in orally vaccinated shrimp. Up-regulation in the expression of caspase, penaeidin, crustin, astakine, syntenin, PmRACK, Rab7, STAT and C-type lectin in shrimp orally fed with rVP28 protein was observed, and this up-regulation due to VP28 administration resulted in reduction in the expression of VP28 gene and delay in mortality after WSSV challenge (Ancy et al. 2016). Chen et al. (2016) reported the expression of calreticulin and heat shock protein 70 in rVP28-vaccinated shrimp. The high levels of immunological parameters found in the vaccinated shrimp are correlated to the high survival of WSSV-challenged shrimp compared to the unvaccinated control groups. Prophenol oxidase activity, antioxidative enzymes, as well as antibiotic peptides are all important components necessary for the full functioning of the invertebrate immune system and essential components of an organism's self-maintenance (Anderson et al. 1996; Torbjorn and Kenneth 1999). The phenol oxidase system is the prime immune defence of invertebrates (Sritunyalucksana and Soderhall 2000), and it is involved in phagocytosis, which in turn is the common process in the cellular defence of crustaceans. Shelby and Popham (2006) have reported phenoloxidase having virucidal activity by inactivating the insect virus *Helicoverpa zea* single nucleopolyhedro virus (*Hz*SNPV) in vitro in *Heliothis virescens.* The high level of phenoloxidase in DNA-vaccinated shrimp observed by Rajeshkumar et al. (2008, 2009) suggested that it might have virucidal activity against WSSV as observed by Shelby and Popham (2006) in *H. virescens larvae.* Superoxide dismutase (SOD) is one of the main antioxidant defence enzymes generated in response to oxidative stress.

5 Conclusion

In conclusion, even though shrimp have innate immunity, vaccines or immunostimulants provide significant protection to shrimp against viral and bacterial pathogens and also enhance the immunity of shrimp. Further studies need to be carried out on the practical application of these vaccines in field conditions to obtain more data.

References

Abraham TJ, Palaniappan R, Dhevendaran K (1997) Epibiotic infestation of luminous bacteria in penaeid shrimp *Penaeus indicus* (H. Milne Edwards) larvae. Indian J Geo-Mar Sci 26:209–211

Alabi AO, Jones DA, Latchford JW (1999) The efficacy of immersion as opposed to oral vaccination of *P. indicus* larvae against *V. harveyi*. Aquaculture 178:1–11

Amar EC, Faisan JP (2011) Efficacy of an inactivated vaccine and nutritional additives against white spot syndrome virus (WSSV) in shrimp (*Penaeus monodon*). Isr J Aquac-Bamidgeh 63:1–9

Amar EC, Joseph P, Faisan J, Rolando Gapasin SJ (2021) Field efficacy evaluation of a formalin-inactivated white spot syndrome virus (WSSV) vaccine for the preventive management of WSSV infection in shrimp grow-out ponds. Aquaculture 531:735907

Ancy TN, Sudheer S, Kiron V, Bright Singh IS, Narayanan RB (2016) Expression profile of key immune-related genes in *Penaeus monodon* juveniles after oral administration of recombinant envelope protein VP28 of white spot syndrome virus. Microb Pathog 96:72–79

Anderson ED, Mourich DV, Leong JA (1996) Gene expression in rainbow trout (*Oncorhynchus mykiss*) following intramuscular injection of DNA. Mol Mar Biol Biotechnol 5:105–113

Azad IS, Panigrahia A, Gopala C, Paulpandia S, Mahimaa C, Ravichandrana P (2005) Routes of immunostimulation vis-a-vis survival and growth of *P. monodon* postlarvae. Aquaculture 248:227–234

Bai DQ, Li R, Xing KZ, Guo YL, Chen CX, Qiao XT, Mao HT, Zhu GX (2009) In vitro antibacterial activity of herbal medicines and combinations of herbal medicines and antibiotics against *Edwardsiella tarda*. Isr J Aquac-Bamidgeh 61:27–34

Baticados MCL, Lavilla-Pitogo CR, Cruz-Lacierda ER, de la Pena LD, Sunaz NA (1990) Studies on the chemical control of luminous bacteria *Vibrio harveyi* and *V splendidus* isolated from diseased *Penaeus monodon* larvae and rearing water. Dis Aquat Org 9:133–139

Bright Singh IS, Manjusha M, Somnath PS, Philip R (2005) *Fenneropenaeus indicus* is protected from white spot disease by oral administration of inactivated white spot syndrome virus. Dis Aquat Org 66:265–270

Caipang CMA, Verjan N, Ooi EL, Kondo H, Hirono I, Aoki T, Kiyono H, Yuki Y (2008) Enhanced survival of shrimp, *Penaeus (marsupenaeus) japonicus* from white spot syndrome disease after oral administration of recombinant VP28 expressed in *Brevibacillus brevis*. Fish Shellfish Immunol 25:315–320

Chen LH, Lin SW, Liu KF, Chang CI, Hseu JR, Tsai JM (2016) Comparative proteomic analysis of *Litopenaeus vannamei* gills after vaccination with two WSSV structural proteins. Fish Shellfish Immunol 49:306–314

Chou HY, Huang CY, Wang CH, Chiang HC, Lo CF (1995) Pathogenicity of a baculovirus infection causing white spot syndrome in cultured penaeid shrimp in Taiwan. Dis Aquat Org 23:165–173

Citarasu T, Sivaram V, Immanuel G, Rout N, Murugan V (2006) Influence of selected Indian immunostimulant herbs against white spot syndrome virus (WSSV) infection in black tiger shrimp, *Penaeus monodon* with reference to haematological, biochemical and immunological changes. Fish Shellfish Immunol 21:372–384

Devaraja TN, Oha SK, Shubha G, Karunasagar I, Tauro P (1998) Immunostimulation of shrimp through oral administration of Vibrio bacterin and yeast glucans. In: Flegel TW (ed) Advances in shrimp biotechnology. National Centre for Engineering and Biotechnology, Bangkok, pp 167–170

Du HZ, Xu X, Wu W, Dai W (2006) Increased resistance to white spot syndrome virus in Procambarus clarkii by injection of envelope protein VP28 expressed using recombinant baculovirus. Aquaculture 260(1–4):39–43

Feng SY, Feng WP, Zhao L, Gu HH, Li QH, Shi K, Guo SX, Zhang NN (2014) Preparation of transgenic *Dunaliella salina* for immunization against white spot syndrome virus in crayfish. Arch Virol 159:519–525

Flegel TW (1997) A special topic review: major viral diseases of the black tiger prawn (*Penaeus monodon*) in Thailand. World J Microbiol Biotechnol 13:433–442

Fu LL, Shuai JB, Xu ZR, Li JR, Li WF (2010) Immune responses of *Fenneropenaeus chinensis* against white spot syndrome virus after oral delivery of vp28 using Bacillus Subtilis as vehicles. Fish Shellfish Immunol 28:49–55

Goarant C, Boglio E (2000) Changes in hemocyte counts in *L. stylirostris* subjected to sublethal infection and to vaccination. J Vet Anim Sci 31:123–129

Hadifar F, Ignjatovic J, Tarigan S, Indriani R, Ebrahimie E, Hasan NH, McWhorter A, Putland S, Ownagh A, Hemmatzadeh F (2014) Multimeric recombinant M2e protein-based ELISA: a significant improvement in differentiating avian influenza infected chickens from vaccinated ones. PLoS One 9:e108420

Heidarieh M, Sedeh FM, Soltani M, Rajabifar S, Afsharnasab M, Dashtiannasab A (2014) White spot syndrome virus inactivation study by using gamma irradiation. Chin J Oceanol Limnol 32:1024–1028

Heppell J, Davis HL (2000) Application of DNA vaccine technology to aquaculture. Adv Drug Deliv Rev 43:29–43

Hettiarachchi M, Pathiage S, Hettiarachchi G (2005) Isolation of the bacterium, *V. harveyi* from cultured shrimp, *P. monodon* and production of vaccines against the bacterium. J Natl Sci Found 33:257–263

Immanuel G, Vincy VC, Bai A, Palavesam M, Marian P (2004) Effect of butanolic extracts from terrestrial herbs and seaweeds on the survival: growth and pathogen (*Vibrio parahaemolyticus*) load on shrimp *Penaeus indicus* juveniles. Aquaculture 236:53–65

Itami T, Takahashi Y, Nakamura Y (1989) Efficacy vaccination against vibriosis in cultured kuruma prawns *Penaeus* japonicus. J Aquat Anim Health 1:238–242

Itami T, Yan Y, Takahashi Y (1992) Studies on vaccination against vibrios in cultured kuruma prawn, *Penaeus japonicus*. Effect of different vaccine preparations and oral vaccination efficacy. J Shimonoseki Univ Fish 40:139–144

Jardine J, Julien JP, Menis S, Ota T, Kalyuzhniy O, McGuire A, Sok D, Huang PS, MacPherson S, Jones M, Nieusma T, Mathison J, Baker D, Ward AB, Burton DR, Stamatatos L, Nemazee D, Wilson IA, Schief WR (2013) Rational HIV immunogen design to target specific germline B cell receptors. Science 340:711–716

Jha RK, Roung XZ, Shen J, Juan BS, Yu SJ, Fen LW (2006a) The efficacy of recombinant vaccines against white spot syndrome virus in *Procambarus clarkii*. Immunol Lett 105:68–76

Jha RK, Zi Roung X, Bai SJ, Sun JY, Li WF, Shen J (2006b) Protection of *Procambarus clarkii* against white spot syndrome virus using vaccine expressed in *Pichia pastoris*. Fish Shellfish Immunol 22:295–307

Jha RK, Roung XZ, Shen J, Bai SJ, Sun Y, Li WF (2007) Protection of Procambarus clarkii against white spot syndrome virus using recombinant oral vaccine expressed in *Pichia pastoris*. Fish and Shellfish Immunology 22:295–307

Jiravanichpaisal P, Miyazaki T (1994) Histopathology, biochemistry and pathogenicity of *V. harveyi* infecting black tiger prawn *P. monodon*. J Aquat Anim Health 6:27–35

Kalaimani N, Avisankar T, Chakravarthy N, Raja S, Santiago TC, Ponniah AG (2013) Economic losses due to disease incidences in shrimp farms of India. Fish Technol 50:80–86

Karunasagar I, Pai R, Malathi GR, Karunasagar I (1994) Mass mortality of *Penaeus monodon* larvae due to antibiotic resistant *Vibrio harveyi* infection. Aquaculture 128:203–209

Karunasagar I, Shivu MM, Girisha SK, Krohne G, Karunasagar I (2007) Biocontrol of pathogens in shrimp hatcheries using bacteriophages. Aquaculture 268:288–292

Klannukarn SS, Wongprasert K, Khanobdee K, Meeratana P, Taweepreda P, Withyachumnarnkul B (2004) Vibrio bacterin and carboxymethyl β-1,3-glucans protect P. monodon from *V. harveyi* infection. J Aquat Anim Health 16:238–245

Kou GH, Chen SN, Huang SL (1989) Studies on bacterial infection and vaccination trials for culture *Penaeus monodon* in Taiwan. In: Disease of fish and shellfish, abstract of the 4th EAFP international conference, 24–28 September 1989, Santiago de Compostela, Spain, p 94

Kulkarni AD, Caipang CMA, Kiron V, Rombout JHWM, Fernandes JMO, Brinchmann MF (2014) Evaluation of immune and apoptosis related gene responses using an RNAi approach in vaccinated Penaeus monodon during oral WSSV infection. Mar Genomics 18:55–65

Kurtz J, Franz K (2003) Innate defence: evidence for memory in invertebrate immunity. Nature 425:37–38

Kwang J (2000) Fishing for vaccines. Nat Biotechnol 18:1145–1146

Lanh PT, Nguyen HM, Duong BTT, Hoa NT, Thom LT, Tam LT, Thu HT, Nha VV, Hong DD, Mouradov A, Koyande AK, Show PL, Quyen DV (2021) Generation of microalga *Chlamydomonas reinhardtii* expressing VP28 protein as oral vaccine candidate for shrimps against White Spot Syndrome Virus (WSSV) infection. Aquaculture 540:736737. https://doi.org/10.1016/j.aquaculture.2021.73673

Lavilla-Pitogo CR, Baticados MCL, Cruz-Lacierda ER, de la Pena LD (1990) Occurrence of luminous bacterial disease of *Penaeus monodon* larvae in The Philippines. Aquaculture 91:1–13

Li X, Liu QH, Hou L, Huang J (2010) Effect of VP28 DNA vaccine on white spot syndrome virus in *Litopenaeus vannamei*. Aquac Int 18:1035–1044

Lightner DV (ed) (1996) A handbook of shrimp pathology and diagnostic procedures for diseases of cultured penaeid shrimp. World Aquaculture Society, Baton Rouge

Lightner DV (2003) Exclusion of specific pathogens for disease prevention in a penaeid shrimp biosecurity program. In: Lee C-S, O'Bryen PJ (eds) Biosecurity in aquaculture production systems: exclusion of pathogens, other, undesirables. The World Aquaculture Society, Baton Rouge, pp 81–116

Lin YC, Chen JC, Wan Zabidii WM, Dedi Fazriansyah P, Huang CL, Li CC, Hsieh JF (2013) Vaccination enhances early immune responses in white shrimp *L. vannamei* after secondary exposure to *V. alginolyticus*. PLoS One 8:e69722

Liu PC, Lee KK, Chen SN (1996) Pathogenicity of differentisolates of *Vibrio harveyi* in tiger prawn, *Penaeus monodon*. Lett Appl Microbiol 22:413–416

Lo CF, Ho CH, Peng SE, Chen CH, Hsu HE, Chiu YL, Chang CF, Liu KF, Su MS, Wang CH, Kou GH (1996) White spot syndrome associated virus (WSBV) detected in cultured and captured shrimp, crabs and other arthropods. Dis Aquat Org 27:215–225

Loker ES, Adema CM, Zhang SM, Kepler TB (2004) Invertebrate immune systems—not homogeneous, not simple, not well understood. Immunol Rev 198:10–24

Lu Y, Liu J, Jin L, Li X, Zhen Y, Xue H (2008) Passive protection of shrimp against white spot syndrome virus (WSSV) using specific antibody from egg yolk of chickens immunized with inactivated virus or a WSSV-DNA vaccine. Fish Shellfish Immunol 25:604–610

Lu Y, Liu J, Jin L (2009) Passive immunization of crayfish (*Procambius clarkiaii*) with chicken egg yolk immunoglobulin (IgY) against White Spot Syndrome Virus (WSSV). Appl Biochem Biotechnol 159:750

Mavichak R, Kondo H, Hirono I, Aoki T, Kiyono H, Yuki Y (2009) Protection of Pacific white shrimp, *Litopenaeus vannamei* against white spot virus following administration of N-terminus truncated recombinant VP28 protein expressed in gram-positive bacteria, *Brevibacillus choshinensis*. Aquac Sci 57:83–90

Mu Y, Lan JF, Zhang XW, Wang XW, Zhao XF, Wang JX (2012) A vector that expresses VP28 of WSSV can protect red swamp crayfish from white spot disease. Dev Comp Immunol 36:442–449

Namikoshi A, Wu JL, Yamashita T, Nishizawa T, Nishioka T, Armoto M, Muroga K (2004) Vaccination trials with *Penaeus japonicus* to induce resistance to white spot syndrome virus. Aquaculture 229:25–35

Ning JF, Zhu W, Xu JP, Zheng CY, Meng XL (2009) Oral delivery of DNA vaccine encoding VP28 against white spot syndrome virus in crayfish by attenuated Salmonella typhimurium. Vaccine 27:1127–1135

Ning D, Leng X, Li Q, Xu W (2011) Surface-displayed VP28 on *Bacillus subtilis* spores induce protection against white spot syndrome virus in crayfish by oral administration. J Appl Microbiol 111:1327–1336

Oakey HJ, Owens L (2000) A new bacteriophage, VHML, isolated from a toxin producing strain of *Vibrio harveyi* in tropical Australia. J Appl Microbiol 89:702–709

Patil PK, Gopal C, Panigrahi A, Rajababu D, Pillai SM (2013) Oral administration of formalin killed *V. anguillarum* cells improves growth and protection against challenge with *V. harveyi* in banana shrimp. Lett Appl Microbiol 58:213–218

Pizzutto M, Hirst RG (1995) Classification of isolates of *Vibrio harveyi* virulent to *Penaeus monodon* larvae by protein profile analysis and M13 DNA fingerprinting. Dis Aquat Org 21:61–68

Powell A, Pope EC, Eddy FE, Roberts EC, Shields RJ, Francis MJ, Smith P, Topps S (2011) Enhanced immune defences in Pacific white shrimp (*L. vannamei*) post-exposure to a vibrio vaccine. J Invertebr Pathol 107:95–99

Raffi SM, Suresh TV (2011) Screening of chloramphenicol in wild and cultured shrimp *Penaeus monodon* by competitive enzyme linked immunosorbent assay. In: International conference on chemical, biological and environment sciences (ICCEBS'2011), Bangkok

Rajesh Kumar S, Ishaq Ahamed VP, Sarathi M, Nazeer Basha A, Sahul Hameed AS (2008) Immunological responses of *Penaeus monodon* to DNA vaccine and its efficacy to protect shrimp against white spot syndrome virus (WSSV). Fish Shellfish Immunol 24:467–478

Rajeshkumar S, Venkatesan C, Sarathi M, Sarathbabu V, Thomas J, Basha KA (2009) Oral delivery of DNA construct using chitosan nanoparticles to protect the shrimp from white spot syndrome virus (WSSV). Fish Shellfish Immunol 26:429–437

Rameshthangam P, Ramasamy P (2007) Antiviral activity of bis(2- methylheptyl)phthalate isolated from *Pongamia pinnata* leaves against white spot syndrome virus of *Penaeus monodon* Fabricius. Virus Res 126:38–44

Ray AK, Gopal C, Solanki HG, Ravisankar T, Patil PK (2017) Effect of orally administered vibrio bacterin on immunity, survival and growth in tiger shrimp (*P. monodon*) grow-out culture ponds. Lett Appl Microbiol 65:475–481

Robalino J, Payne C, Parnell P, Shepard E, Grimes AC, Metz A (2006) Inactivation of white spot syndrome virus (WSSV) by normal rabbit serum: implications for the role of the envelope protein VP28 in WSSV infection of shrimp. Virus Res 118:55–61

Robertson PAW, Calderon J, Carrera L, Stark JR, Zherdmant M, Austin B (1998) Experimental *Vibrio harveyi* infections in *Penaeus vannamei* larvae. Dis Aquat Org 32:151–155

Rout N, Kumar S, Jaganmohan S, Murugan V (2007) DNA vaccines encoding viral envelope proteins confer protective immunity against WSSV in black tiger shrimp. Vaccine 25:2778–2786

Ruangpan L, Danayadol Y, Direkbusarakom S, Siurauratana S, Flegel TW (1999) Lethal toxicity of *Vibrio harveyi* to cultivated *Penaeus monodon* induced by a bacteriophage. Dis Aquat Org 35:195–201

Sahul Hameed AS, Yoganandhan K, Sathish S, Murugan V, Rasheed M, Jayaraman K (2001) Experimental pathogenicity of white spot syndrome virus (WSSV) in two freshwater crabs (*Partelphusa hydrodomous* and *P. pulvinata*). Aquaculture 201:179–186

Sahul Hameed AS, Balasubramanian G, Syed Musthaq S, Yoganandhan K (2003) Experimental infection of twenty species of Indian marine crabs with white spot syndrome virus (WSSV). Dis Aquat Org 57:157–161

Satoh J, Mushiake K, Mori K, Arimoto M, Imaizumi K (2001) Control of penaeid acute viremia (PAV) in seed production of *Penaeus japonicus*. Bull Natl Res Inst Aquac (Jpn) 5:95–99

Sharma SRK, Shankar SM, Sathyanarayana ML, Sahoo AK, Patil R, Narayanaswamy HD, Rao S (2010) Evaluation of immune response and resistance to diseases in tiger shrimp, *P. monodon* fed with biofilm of *V. alginolyticus*. Fish Shellfish Immunol 29:724–732

Shelby K, Popham HJ (2006) The role of plasma phenoloxidase in resistance of *Heliothis virescens* larvae to baculovirus infection. J Insect Sci 6:13

Sivaram V, Babu MM, Immanuel G, Murugadass S, Citarasu T, Marian MP (2004) Growth and immune response of juvenile greasy groupers (*Epinephelus tauvina*) fed with herbal antibacterial active principle supplemented diets against *Vibrio harveyi* infections. Aquaculture 237:9–20

Solís-Lucero G, Manoutcharian K, Hernandez-L J, Opez Ascencio F (2016) Injected phage-displayed-VP28 vaccine reduces shrimp *Litopenaeus vannamei* mortality by white spot syndrome virus infection. Fish Shellfish Immunol 55:401–406

Song YL, Lee SP (1993) Characterization and ecological implication of luminous Vibrio harĺeyi isolated from tiger shrimp *Penaeus monodon*. Bull Inst Zool Acad Sin 32:217–220

Sritunyalucksana K, Soderhall K (2000) The prophenoloxidase and clotting system in crustaceans. Aquaculture 191:53–69

Stalin N, Srinivasan P (2017) Efficacy of potential phage cocktails against *Vibrio harveyi* and closely related *Vibrio* species isolated from shrimp aquaculture environment in the south east coast of India. Vet Microbiol 207:83–96

Sudheer NS, Poulose G, Thomas A, Viswanath K, Kulkarni A, Narayanan RB (2015) Expression profile of bio-defense genes in *Penaeus monodon* gills in response to formalin inactivated white spot syndrome virus vaccine. Antivir Res 117:60–68

Sun Y, Li F, Chi Y (2013) Enhanced resistance of marine shrimp *Exopalamon carincauda* Holthuis to WSSV by injecting live VP28-recombinant bacteria. Acta Oceanol Sin 32:52–58

Sunaryanto A, Mariam A (1986) Occurrence of a pathogenic bacteria causing luminescence in penaeid larvae in Indonesian hatcheries. Bull Brackish Water Aquac Dev Centre 8:64–70

Syed MS, Kwang J (2011) Oral vaccination of baculovirus-expressed VP28 displays enhanced protection against white spot syndrome virus in *Penaeus monodon*. PLoS One 6(11):e26428

Syed Musthaq S, Sudhakaran R, Balasubramanian G, Sahul Hameed AS (2006a) Experimental transmission and tissue tropism of white spot syndrome virus (WSSV) in two species of lobsters, *Panulirus homarus* and *Panulirus ornatus*. J Invertebr Pathol 93:75–80

Syed Musthaq S, Sudhakaran R, Ishaq Ahmed VP, Balasubramanian G, Sahul Hameed AS (2006b) Tandem repetitive DNA sequences in Indian white spot syndrome virus (WSSV) isolates and suitability of VP28 gene to detects different repeat groups of WSSV. Aquaculture 256:34–41

Syed Musthaq S, Selvaraj Madhan, Sahul Hameed AS, Jimmy K (2009) Localization of VP28 on the baculovirus envelope and its immunogenicity against white spot syndrome virus in Penaeus monodon. Virology 391:315–324

Taengchaiyaphum S, Nakayama H, Srisala J, Khiev R, Aldama-Cano DJ, Thitamadee S, Sritunyalucksana K (2017) Vaccination with multimeric recombinant VP28 induces high protection against white spot syndrome virus in shrimp. Dev Comp Immunol 76:56–64

Taju G, Kumar DV, Majeed SA, Vimal S, Tamizhvanan S, Kumar SS, Sivakumar S, Basha AN, Haribabu P, Kannabiran K, Hameed ASS (2018) Delivery of viral recombinant VP28 protein using chitosan tripolyphosphate nanoparticles to protect the whiteleg shrimp, *Litopenaeus vannamei* from white spot syndrome virus infection. Int J Biol Macromol 107:1131–1141

Takahashi Y, Itami T, Maeda M, Fujii R, Tomonaga S, Supamattaya K, Boonyaratpalin S (1994) Electron microscopic evidence of bacilliform virus infection in kuruma shrimp. Fish Pathol 29:121–125

Teunissen OSP, Faber R, Booms GHR, Latscha T, Boon JH (1998) Influence of vaccination on vibriosis resistance of the giant black tiger shrimp *P. monodon Fabricius*. Aquaculture 164:359–366

Torbjorn H, Kenneth S (1999) Cell adhesion molecules and antioxidative enzymes in a crustacean, possible role in immunity. Aquaculture 172:111–123

Valdez A, Yepiz-Plascencia G, Ricca E, Olmos J (2014) First *Litopenaeus vannamei*WSSV 100% oral vaccination protection using CotC::Vp26 fusion protein displayed on *Bacillus subtilis* spores surface. J Appl Microbiol 117:347–357

Zhi-guang Q, Qing-hui L, Jie H (2012) Efficiency of two fragments of VP28 against white spot syndrome virus in *Litopenaeus vannamei*. Aquaculture 338–341:2–12

Van Hulten MC, Witteveldt J, Snippe M, Vlak JM (2001) White spot syndrome virus envelope protein VP28 is involved in the systemic infection of shrimp. Virology 285:228–233

Vaseeharan B, Prem Anand T, Murugan T, Chen JC (2006) Shrimp vaccination trials with the VP292 protein of white spot syndrome virus. Lett Appl Microbiol 43:137–142

Wang Q, Xu Y, Xiao C, Zhu F (2023) The effect of white spot syndrome virus (WSSV) envelope protein VP28 on innate immunity and resistance to white spot syndrome virus in *Cherax quadricarinatus*. Fish Shellfish Immunol 137:*108795*. https://doi.org/10.1016/j.fsi.2023.108795

Witteveldt J, Cifuentes C, Vlak JM, Van Hulten MCW (2004a) Protection of *Penaeus monodon* against white spot syndrome virus by oral vaccination. J Virol 78:2057–2061

Witteveldt J, Vlak JM, Van Hulten MCW (2004b) Protection of *Penaeus monodon* against white spot syndrome virus using a WSSV subunit vaccine. Fish Shellfish Immunol 16:571–579

Witteveldt J, Vlak JM, van Hulten MCW (2006) Increased tolerance of *Litopenaeus vannamei* to white spot syndrome virus (WSSV) infection after oral application of the viral envelope protein VP28. Dis Aquat Org 70:167–170

Wongtavatchai J, Lopez-Doriga MV, Francis MJ (2010) Effect of AquaVacTM VibromaxTM on size and health of post larva stage of pacific white shrimp *L. vannamei* and black tiger shrimp *P. monodon*. Aquaculture 308:75–81

Yang JY, Chang C-I, Liu K-F, Hseu J-R, Chen L-H, Tsai J-M (2012) Viral resistance and immune responses of the shrimp *Litopenaeus vannamei* vaccinated by two WSSV structural proteins. Immunol Lett 148:41–48

Yogeeswaran A, Velmurugan S, Punitha SMJ, Babu MM, Selvaraj T, Kumaran T, Citarasu T (2012) Protection of *Penaeus monodon* against white spot syndrome virus by inactivated vaccine with herbal immunostimulants. Fish Shellfish Immunol 32:1058–1067

Zhu F, Du HH, Miao ZG, Quan HZ, Xu ZR (2009) Protection of *Procambarus clarkii* against white spot syndrome virus using inactivated WSSV. Fish Shellfish Immunol 26:685–690

Bacterial Vaccines for Fish

Megha K. Bedekar and Arya Singh

Abstract

Globally, aquaculture has faced severe economic challenges with various infectious diseases, particularly those of bacterial origins. Even though multiple preventative and therapeutic approaches are being employed simultaneously to identify and treat these diseases, vaccinations with broad spectrum activity remain an essential preventive tool in aquaculture. However, treatments such as probiotics and antibiotics appear to be less successful when disease-causing microorganisms become resistant to routinely used medications and foods containing antibiotics and when new mutant strains emerge. Therefore, vaccine development is considered an efficient method of addressing the microorganisms that cause diseases in aquatic organisms. In addition to offering some perspectives on how the area is anticipated to develop in the future, this chapter summarizes the commercially available bacterial fish vaccines.

Keywords

Vaccine · Antibiotic · Probiotic · Therapeutic · Prophylactic

M. K. Bedekar (✉)
ICAR-Central Institute of Fisheries Education, Mumbai, Maharashtra, India
e-mail: megha.bedekar@cife.edu.in

A. Singh
Acharya Narendra Deva University of Agriculture & Technology,
Faizabad, Uttar Pradesh, India
e-mail: singharya4950@gmail.com

R. Kooloth Valappil et al. (eds.), *Aquatic Animal Health Management*,
https://doi.org/10.1007/978-981-96-7987-4_20

1 Introduction

Aquaculture is the food industry with the fastest growth rate worldwide, with a global production of 94.4 million tons in 2022. This sector has grown to be a significant supplier of seafood as wild stocks have decreased and the capture fishing sector has declined. It sustains the livelihood of millions of people globally and makes a major contribution to food security and nutrition, particularly in areas with some of the highest rates of food insecurity. High-value species like salmon and trout make up 16% of the overall value and 7% of the total volume of fish farmed worldwide. Fish farming is now dominated by cyprinids, including barbels and carps. Norway, Chile, the United Kingdom, and Canada are the world's top producers of Atlantic salmon, one of the species farmed most intensively worldwide. High-value marine species, including European sea bass and sea bream, halibut, flounder, cod, tuna, eel, and amberjack/yellowtail, will likely rise in tandem with the development of suitable intensive aquaculture systems.

Aquaculture development has surged during the last 50 years as a result of the use of contemporary technology and scientific advancements. It is possible to meet demand and greatly enhance the quantity and quality of fish in aquaculture systems by employing modern biotechnology techniques to boost fish output (Sudheesh et al. 2012). Compared to other agricultural enterprises, aquaculture is more varied in terms of species, food, culture techniques, products, disease conditions, and environments (de Bruijn et al. 2018). According to Rohani et al. (2022), ecosystem-based management, production system design, and the advancement of sustainable production are the main areas of focus for industry. Advances in science and technology have helped aquaculture in nearly every way. Aquaculture production has benefited greatly from numerous technological advancements. Better reproductive technologies, for instance, have made it possible to complete the life cycles of aquaculture species, allowing for the diversification of species in the industry (Naylor et al. 2021). Over 60 aquaculture species have seen significant improvements in features of commercial value through selective breeding using quantitative genetics (Miccoli et al. 2021). Genomic selection has the enormous potential to change genetic improvement initiatives and aquaculture industry production methods (Muktar et al. 2016). Feed conversion rate (FCR) and feed cost have increased due to improved feed formulations that are tailored to the nutritional needs of individual fish species (Mohd-Aris et al. 2019). Disease control technologies have somewhat reduced the number of diseases in aquaculture.

Thousands of fish are being raised in cramped quarters and near to one another as aquaculture continues to expand essentially, tilapias from Africa (Nile tilapia, *Oreochromis niloticus*), chinese carps (*Hypophthalmichthys molitrix, Hypophthalmichthys nobilis* and *Ctenopharyngodon idella*), atlantic salmon (*Salmo salar),* pangasius catfishes (*Pangasius* spp.), largemouth black bass (*Micropterus salmonids*), turbot (*Scophthalmus maximus*), piarapatinga (*Piaractus brachypomus*), pacu (*Piaractus mesopotamicus*), and rainbow trout (*Oncorhynchus mykiss*). However, the intensive growth of food fish has resulted in outbreaks of several bacterial infections, driving up the rate at which diseases spread and causing annual losses for the aquaculture sector.

2 Diseases in Aquaculture

Aquaculture facilities are highly vulnerable to disease outbreaks due to the presence of several opportunistic bacterial, viral, and parasitic pathogens and parasites that can be found in the environment or on certain fish as asymptomatic carriers. The aquaculture industry is very concerned about infections in fish that cause disease outbreaks since the morbidity and mortality rates can be very high and generate substantial financial losses. Cultivated aquatic animals are extremely stressed by intensive farming practices, which impairs their natural immune systems' capacity to fend off disease-causing bacterial and viral infections. Regardless of whether one or more fish species are raised in dense populations, proper husbandry and overall management—which includes biosecurity, nutrition genetics, system management, and water quality—are essential for aquaculture production in all intensive culture farming methods. These factors also impede the development of an efficient, stable, and cost-effective aquaculture process. The interaction between the pathogen, host, and environment determines the emergence and progression of fish disease, as shown in Fig. 1. Stressful environments, such as dense populations, temperature swings, and hypoxia, can speed up the spread of harmful germs and cause significant disease outbreaks. Salmon farming in Norway suffered enormous losses in the 1980s as a result of bacterial infections, primarily caused by *Vibrio* spp., and the industry's complete collapse was only averted by the extensive use of antibiotics.

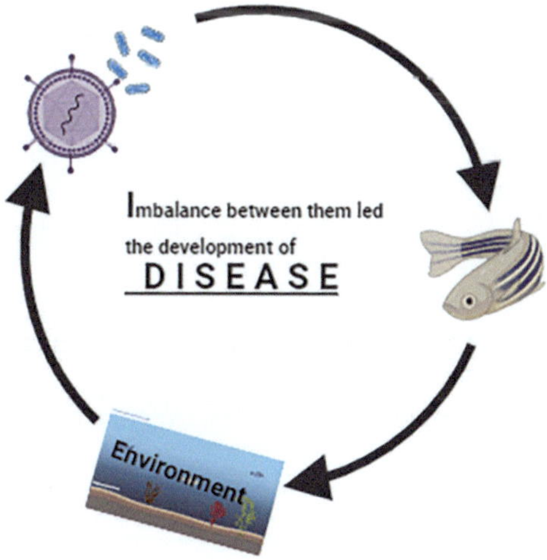

Fig. 1 Interaction between the pathogen, host, and environment leads to disease condition in the host

3 Bacterial Pathogens in Fish and Shellfish

Numerous infections with bacteria in fish species, including enteric red mouth, *Aeromonas septicemia*, edwardsiellosis, bacterial gill disease, columnaris disease, streptococcosis, *Pseudomonas septicemia*, bacterial kidney disease, and vibriosis, have yet to be reported in the aquaculture sector. However, the majority of the worldwide economic losses in aquaculture production are determined to be attributed to a few numbers of these pathogens. *Aeromonas* spp. is prevalent in bacterial infections found in freshwater and tropical habitats that can induce bacterial haemorrhages in farmed fish (Ramesh and Souissi 2018). Among the earliest recognized infections in fish, *Aeromonas salmonicida*, is present worldwide in aquaculture regions with fresh and marine waters. It is linked to recurrent clinical symptoms of infection, such as skin ulcers and haemorrhages (Menanteau-Ledouble et al. 2016). *Vibrio* sp. is responsible for the majority of bacterial diseases among shellfish, including early mortality syndrome, acute hepatopancreatic necrosis disease (AHPND), and vibriosis in shrimp hatcheries and grow-out ponds.

4 Disease Management

Drug-supplemented diets are used in both preventive and curative measures to prevent diseases in farmed fish (Cabello 2006). Bacteria pathogens spread swiftly throughout the population of farmed fish because of the high density of fish utilized in commercial large-scale farms. Therefore, the usage of enormous quantities of antibiotics was the only thing that kept the business from collapsing completely. In fact, the overuse of antibiotics has resulted in the emergence of resistant strains and the requirement to transition to alternative antibiotics (Park et al. 2012). Antibiotic-resistant bacteria-related diseases are hard to cure, and there are not many novel antimicrobial substances being developed as drugs. Thus, substitutes for antimicrobial compounds that provide similar or enhanced protection to aquatic animals are urgently needed (Pridgeon and Klesius 2012).

Antibiotics may hinder fish larvae's defence mechanisms and slow down their growth (Defoirdt et al. 2011). Research has shown that between 60% and 73% of antibiotics and associated chemical medications administered to farmed fish were expelled with the faeces, in addition to the sensitization reaction and other unfavourable side effects on fish (Rigos et al. 2004). The best strategy for enhancing aquatic animal health over the long run will be an integrated one that takes into account the pathogen, the host, and the environment. From an ethical, environmental, and economic perspective, vaccination has been the best way to control disease-causing bacteria since its discovery. A functioning immune system is necessary for vaccinations.

5 Vaccines

Fish diseases continue to pose a significant challenge to the global commercial aquaculture industry despite various breakthrough therapeutic techniques. There are some obvious disadvantages of antibiotics or chemotherapy to treat diseases, including medication resistance problems and safety concerns (Sneeringer et al. 2019). Vaccination helps to ensure the environmental, social, and economic sustainability of aquaculture worldwide and is an efficient means of preventing bacterial and viral infections.

A biological product designed to increase immunity against a particular disease or combination of diseases is called a vaccine. Many vaccinations have been developed that have considerably decreased the impact of bacterial and some viral diseases in fish since the earliest reports of fish vaccination for disease prevention in the 1940s (Gudding and Goodrich 2014). Every year, millions of fish are vaccinated, and in certain parts of the world, immunization has replaced antibiotics. Since vaccination became available, for instance, the use of antibiotics in Norwegian salmon farming has drastically decreased (Rodger 2016). Vaccination is now the most sustainable and economical way to prevent infectious fish diseases. It has been crucial to the success of salmon farming and plays a significant role in large-scale commercial fish farming. Commercial vaccines are available for many of the farmed fishes, namely, channel catfish, European sea bass and sea bream, Japanese amberjack and yellowtail, tilapia, and Atlantic cod, in addition to salmon and trout. Fish vaccinations that were developed empirically and based on inactivated bacterial infections have generally shown to be quite effective.

5.1 Historical Perspectives

Biological substances known as vaccines work by stimulating the immune system to react to a particular antigen that is obtained from an infectious pathogen that causes diseases (Czochor and Turchick 2014). Vaccination is a crucial component in aquaculture. Immunization has long been recognized as an effective means of treating and preventing a large range of viral and bacterial illnesses. Numerous monovalent and multivalent bacterial and viral vaccines have been effectively produced and brought to market (Mohamed and Soliman 2013). Ma et al. (2019) and Horzinek et al. (1997) reported that immunization has emerged as one of the most sustainable and economical ways to control a variety of infectious fish diseases.

Protective immunity was initially reported by Snieszko et al. in 1938, who immunized carp with the bacteria *Aeromonas punctata* in order to avoid disease. In 1942, Duff published the first English-language report on the use of oral and parenteral inoculation to protect rainbow trout (*Oncorhynchus mykiss*) against *Aeromonas salmonicida*. Since the first fish vaccine was used to prevent infections in the 1940s, numerous immunizations have been developed that dramatically reduce the prevalence of hazardous bacterial and viral diseases in fish (Snieszko and Friddle 1949). In order to prevent enteric-red mouth and treat yersiniosis in fish, the US Department of Agriculture licensed the first fish vaccination in 1976. This was followed by a vibriosis vaccine (Kumar et al. 2015). These immunizations were given by immersion and were based on inactivated whole-cell preparations. These vaccines have shown to be effective in preventing a number of major bacterial illnesses, and their inexpensive cost of production makes them ideal for application in the aquaculture sector. Even now, the most highly effective commercial vaccinations on the market include those for bacterial diseases, yersiniosis, and vibriosis. In 1970, vaccination against infectious bacterial illnesses was introduced to the farming community. The foundation of vaccination, particularly DNA vaccination, is the delivery of a plasmid that encodes the vaccine against the antigen rather than the antigen itself. Recombinant viral proteins are the primary basis for attenuated bacterial capsule polysaccharides, bacteria, viruses, and toxins, as well as monoantigen vaccines. According to Perrie et al. (2008), the host's humoral and cellular immune responses are triggered when the plasmid is expressed in the somatic cells.

5.2 Mechanisms

Fish vaccinations only make up a small portion of the total fish health management program; proper animal husbandry is still essential for aquaculture development and profitability. By introducing pathogens into the fish, the vaccination works as a prophylactic measure to boost its immune system. Since treating illnesses in aquacultures is typically ineffective, the preventive strategy becomes more important. Fish immune systems are exposed to either a portion of a pathogen or the pathogen in its entirety (antigen), and their immune systems are then given time to mount a defence. This is how fish vaccines operate, as given below in Fig. 2. Fish that receive vaccinations also develop a memory that helps them respond more quickly to subsequent infections by the disease-causing bacteria.

5.3 Fish Vaccine Development

Aquaculture now relies heavily on immunization, with relatively little antibiotic usage. Fish vaccine research involves constant collaboration between academic institutions, the pharmaceutical sector, and regulatory bodies, much like the development of vaccines for humans and animals. Small local enterprises were responsible for developing and commercializing the majority of fish vaccines until the early 1990s. In the 1990s and up to this point, five large animal health corporations have bought out or partnered with smaller businesses that specialize in aquaculture vaccinations. The leading manufacturers of fish vaccines are Bayer Animal Health (Bayotek)/Microtek, Inc. (Germany/Canada), Pharmaq (Norway; was a part of Alpharma Animal Health until 2004), Schering-Plough Animal Health (United States), Novartis Animal Health (Switzerland), and Intervet International (The Netherlands). These companies presently focus on the salmon and trout industries in Northern Europe, Chile, Canada, and the USA, where the cost of a vaccination is justified by the benefit of a healthy population. Most Japanese firms develop and distribute the vaccines used in Japan. Additionally, commercial vaccinations are offered for the catfish business in the USA and, to a lesser extent, for tilapia, sea bass, and European sea bream. Certain locally produced, limited-use vaccinations are also accessible in China, Russia, Spain, and Germany (Fig. 3).

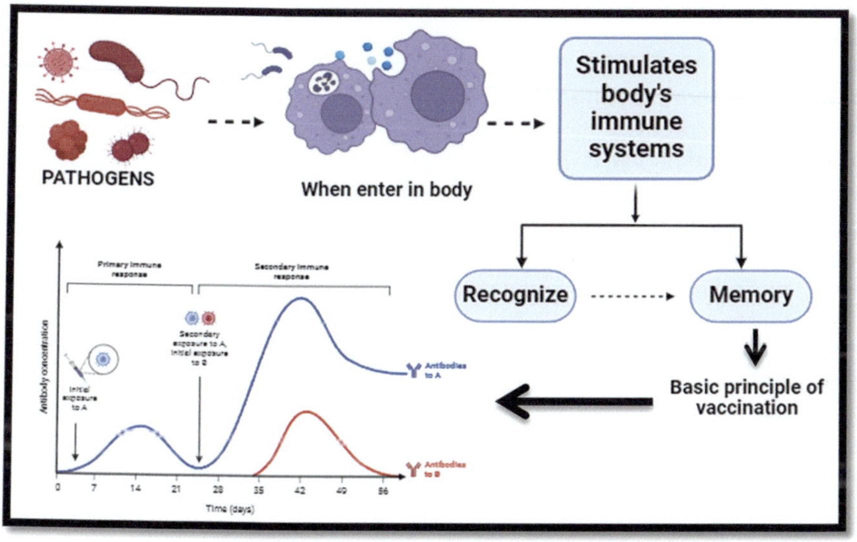

Fig. 2 Pictorial presentations of the basic principles of a vaccine (Biorender.com)

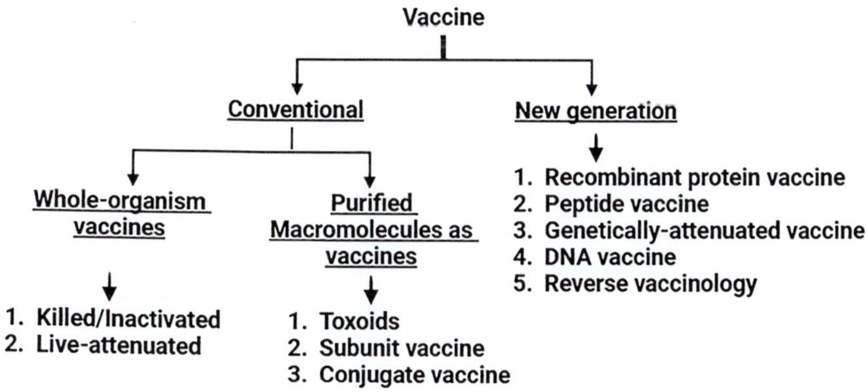

Fig. 3 Types of fish vaccines

5.4 Conventional Fish Vaccines

Most traditional fish vaccines are made from inactivated whole organisms, but there are now a number of commercially available live attenuated or subunit protein vaccines that contain adjuvants. Most licensed vaccinations used in aquaculture today are made with traditional techniques and ideas that are akin to those created centuries ago by Jenner and Pasteur (Tizard 1999). The usage of chemical therapies and feed-delivered antibiotics has decreased as a result of the successful adoption of these vaccinations, which have also increased productivity for commercial aquaculture.

5.4.1 Inactivated/Killed Vaccine

Inactivated or destroyed vaccines are typically derived from pathogenic microorganisms that are no longer able to proliferate or infect new hosts. The microbial agent's resistance to infection is unaffected by these changes, which can be caused by physical, chemical, or radiation treatments. Compared to other vaccination kinds, inactivated vaccines may cause poorer or shorter-lived immunity since they do not persist in the environment or in the vaccinated fish. Nevertheless, they are generally considered to be safe. The majority of early immunization experiments in aquaculture employed dead vaccines. When Duff studied the oral vaccination of cutthroat trout (*Oncorhynchus clarkii*), he used a killed *Aeromonas salmonicida* vaccine, which was the first fish vaccine ever known to be used.

In the USA, there are eight approved fish vaccines, four of which are lethal vaccinations. These include an infectious salmon anaemia (ISA) viral vaccine, a *Y. ruckeri* bacterin for use in salmonids, and an *A. salmonicida–Vibrio anguillarum–V. ordalii–V. salmonicida* bacterin for use in salmonids, and an *A. salmonicida* bacterin for use in salmonids and Koi carp. In Japan and Europe, there are also killed vaccines to prevent Streptococcus or Lactococcus spp. infections in rainbow trout, commonly known as amberjack (*Seriola dumerili*) and yellowtail (*Seriola quinqueradiata*). In European sea bass and sea bream cultivation, a lethal vibriosis vaccine is also available and utilized in conjunction with *Photobacterium damselae* (subsp. *piscicida*). There is now only one bacterin (ERM vaccine) that is offered for sale as an oral vaccine.

5.4.2 Live Vaccines

The development of modified live vaccines uses a virus or bacterium that has naturally low pathogenicity or attenuated virulence towards the target fish species. Pathogens can be rendered less potent through chemical or physical methods, serial passage in cell culture, culture under non-optimal circumstances, or genetic modification. Live vaccines have been shown to elicit stronger cellular responses associated with both innate and adaptive immunity than killed preparations, owing to their capacity to proliferate or penetrate the host (Levine and Sztein 2004). These cell-mediated immune responses are thought to produce a potent antibody response by nearly exactly mimicking an infection by a natural pathogen. The addition of an adjuvant is usually not necessary to improve the efficacy of live vaccinations because they are generally quite effective and maintain characteristics of a natural infection. Live vaccinations are more likely to be given orally or through immersion in terms of aquaculture's economic uses. Consequently, the administration method is more flexible than it is for inactivated vaccines, which need the use of adjuvants (Tlaxca et al. 2015).

In the United States, three modified live aquaculture vaccines are currently licensed. A *Flavobacterium columnare* vaccine against columnaris in catfish, an *E. ictalurii* vaccine against enteric septicemia of catfish (ESC), and an Arthrobacter vaccine against bacterial kidney disease (BKD) for use in salmonids are among them (Klesius and Pridgeon 2014). Renogen against BKD is a live Arthrobacter vaccine that has been licensed in Chile and Canada (Shoemaker et al. 2009). It is made up of non-pathogenic soil bacteria that offer cross-protection to the virulent *Renibacterium salmoninarum*.

5.5 Subunit Vaccines

Subunit vaccines have the benefit of just needing antigenic components to be effective; additionally, since they cannot replicate in the host, there is no risk of pathogenicity to the host or non-target species (Hansson et al. 2000). According to Holten-Andersen et al. (2004), subunit vaccines have the ability to be manufactured in a well-defined state, target immune responses towards certain microbiological determinants, permit the insertion of non natural components, and can be freeze-dried, so enabling non-refrigerated transit and storage Although subunit vaccines have many positive attributes, they frequently lack the potency of whole cell preparations—either dead or living—in their capacity to elicit a strong immune response. This is because there aren't many components that can activate the immune system and are represented, and there isn't a whole cell vaccination that involves multiple antigen exposure or replication. A subunit vaccination can be made in several ways. Direct isolation and purification of immunogenic components from the target pathogen is possible, or different recombinant expression vectors can be used to create specific immunogenic proteins.

5.6 Recombinant Vaccines

In these vaccines, only the immunogenic areas of the pathogen can be produced in the heterologous host and utilized (Adams et al. 2008). According to Ire et al. (2005), fish can be experimentally infected with a live pathogen after being vaccinated with the relevant antigen. This allows researchers to investigate the enhanced protection caused by the associated antigen and may provide insight into the survival rate of the vaccinated fish. These days, several researchers are utilizing recombinant DNA technology to enhance vaccines that are essential and have significant potential impacts on the aquaculture sector. In order to create protective immune responses against infections such as *Aeromonas hydrophila*, a number of researchers have employed recombinant technology (Poobalane et al. 2010).

5.7 Reverse Vaccinology

Reverse vaccinology is a new technology that has gained attention in light of biotechnology's recent developments. Developing a vaccination against a variety of dangerous organisms can be aided by the most recent technologies. According to Rappuoli (2000), the process of creating a successful and effective vaccine takes several years. The production of these vaccines has sped up from 5 to 10 years because of the concept of vaccination. It uses a bioinformatics approach to identify immunogenic sequences, and it also predicts sections that this program expresses as recombinant proteins (Dadar et al. 2016).

Photobacterium damselae subsp. piscicida is one marine species to which this technology has recently been applied. There are serious side effects, according to numerous studies on reverse vaccination (Andreoni et al. 2016). Two important intracellular infectious fish pathogens, *Flavobacterium columnare* and *Edwardsiella tarda*, which cause columnaris and edwardsiellosis, respectively, have been successfully vaccinated using software (Mahendran et al. 2016).

5.8 DNA Vaccines

A plasmid with an expression gene that codes for a specific antigenic protein makes up DNA vaccines. A strong immunological response is expected to be triggered when this protein is expressed in the host. The creation of plasmids is enhanced in bacterial cells, and the target gene is surrounded by components that stimulate and terminate its expression in eukaryotic cells (Kurath 2008). It is possible to boost humoral and cellular immunity with DNA vaccines greatly. DNA vaccine development can move forward swiftly and be rather easy if a protective antigen is found. When it comes to preventing viral infections, DNA vaccines are frequently more effective. They have been particularly successful in preventing fish rhabdovirus infections.

DNA vaccines have been effective in preventing fish exposure to intracellular microorganisms such as *Mycobacterium marinum* (Pasnik and Smith 2006). Rainbow trout were used in experiments to test the first known DNA vaccine against IHN for use in aquaculture (Kurath 2008). Other experimental DNA vaccines targeting a wide range of fish species and aquatic infections have been produced throughout the last 10 years. Only a small quantity of these have, however, been produced and released onto the market (Table 1).

Table 1 Different DNA vaccines developed for fish (Ma et al. 2019)

Bacteria	Fish species	Gene antigen inserted	Administration route	Status	References
Piscirickettsia salmonis	Coho salmon	Full expression library	Intramuscular	Experimental	Miquel et al. (2003)
Aeromonas veronii	Spotted S. bass	Omp38 and Omp48	Intramuscular	Experimental	Vazquez-Juarez et al. (2005)
Mycobacterium marinum	Hybrid striped bass	Ag85A	Intramuscular	Commercial	Pasnik and Smith (2005)
Flavobacterium psychrophilum	Rainbow trout	Hsp60 and 70	Intramuscular	Experimental	Plant et al. (2009)
Streptococcus iniae	Japanese flounder	Sia10	Intramuscular	Experimental	Sun et al. (2012)
Vibrio parahaemolyticus	Black sea bream	OmpK	Oral	Experimental	Li et al. (2013)

6 Methods of Administration

The best way to vaccinate fish would be by oral injection using an antigen found in their food, and considerable work has gone into creating these vaccines. However, because of antigen oxidation in the stomach, typical oral vaccinations have been associated with weak and inconsistent responses (Hart et al. 1988). Various strategies to prevent the antigen from breaking down, like entrapping it in liposomes or alginate beads (Ire et al. 2005), neutralizing stomach secretions, or using biofilm vaccines (Azad et al. 2000), have shown some encouraging outcomes.

The two major ways to administer vaccines to fish are by injecting the vaccine into their body cavity (usually intraperitoneally) or submerging them in a diluted suspension of the vaccine. Immersion vaccinations are inexpensive, simple to deliver to small fish, and effective against a variety of bacterial diseases (Table 2). On the other hand, vaccination of fry is challenging since injecting fish with a vaccine needs a lot of work and requires the fish to be a specific size. However, because at least one of the components needs to be injected and an oil adjuvant is needed for it to be effective, a lot of fish vaccines available today are multicomponent injection vaccines. The benefits of injectable immunizations are that comparatively only a small amount of vaccine is required, and accurate dosage and vaccination of each fish can be administered.

Table 2 Advantages and disadvantages of different vaccine deliveries

S. No.	Vaccine	Advantage	Disadvantage
1.	Oral	(i) The easiest method because feeding is a normal part of production (ii) Stress on fish is minimal	(i) A coating agent is often needed to avoid breakdown in the fish digestive system (ii) Conveys relatively short immunity, may require additional vaccination
2.	Immersion	(i) Relatively easy to perform with minimal interruption to the production schedule (ii) Stress on fish is minimal	(i) Smaller, younger fish may have immature immune systems that require a second vaccination (ii) May not convey protection as effective as injection for some pathogens
3.	Injection	(i) Effective for many disease pathogens (ii) Much longer protection duration	(i) Requires more time and skilled personnel (ii) Causes the most stress on fish

Source: Plant and LaPatra (2011)

7 Limitations in Vaccine Development

Fish are the first class of animals to possess all of the essential components of the immune system found in higher vertebrates. Despite having less developed immune systems than those of mammals, fish and fish seem to have more in common than not (Secombes 2003). The predominant antibody type in the few fish species that have been examined is an immunoglobulin (Ig)M-like isotype, which normally appears in its secretory form as a tetramer (Pilstrom and Bengten 1996). Fish have no isotypes that match mammalian IgG, IgA, or IgE, and their secondary humoral immune response, if it exists at all, is weaker than that of mammals. As a result, it is likely that the ways in which various animals respond to vaccinations and fight against infections vary significantly. The ability to investigate pathogen and vaccine-induced immunity is limited by an incomplete understanding of the immune systems of various fish species. A crucial barrier to the development of commercial vaccines is cost-effectiveness in the field, as it is for all veterinary vaccines. Fish appear to lack maternal immunity, which further reduces the likelihood that parental vaccination can protect offspring (Tanaka et al. 1999).

8 Current Status of Fish Bacterial Vaccines and Their Applications

8.1 Vibriosis

It has been reported that vibriosis is a serious bacterial disease that affects marine fish worldwide. The following vibrios are responsible for some of the most serious diseases in marine animals: *Vibrio anguillarum, Aliivibrio salmonicida, V. ordalii,*

V. harveyi, V. parahaemolyticus, and *V. vulnificus* biotype 2 (Woo et al. 2002). A novel whole-cell oral polyvalent vaccine based on feed was recently created to protect the Asian sea bass, *Lates calcarifer*, against vibriosis. This vaccine shows promise for mass fish vaccination programs in aquaculture (Mohamad et al. 2021). The outcomes showed that the vaccination was effective in provoking both innate and adaptive immune responses in Asian sea bass that are resistant to vibriosis.

8.2 Bacterial Cold-Water Disease (BCWD)

Flavobacterium psychrophilum (syn., *Flexibacter psychrophilus* and *Cytophaga psychrophila*) have been causing BCWD or peduncle disease since 1948 in salmonids. A few nations have begun employing autogenous vaccines derived from single isolates since there are no commercial vaccinations available to prevent this illness (Toranzo et al. 2009). In rainbow trout, oil-adjuvanted IP vaccinations significantly reduced the risk of contracting this disease (La Frentz et al. 2002). According to a recent study, there are currently no effective vaccinations against this bacterial cold-water illness (Takeuchi et al. 2021).

8.3 Enteric Septicemia of Catfish

Edwardsiella ictaluri is the etiological agent that causes intestinal septicemia in catfish. Oral and immersion forms of the first commercial bacterins were approved for use (Muktar et al. 2016). A defective O-antigen strain of attenuated *E. ictaluri* was developed, and it was able to produce acquired immunity in channel catfish for at least 4 months. This immunity was then acquired through bath immersion without the need for a booster shot (Klesius and Shoemaker 1998). Comparably, a different investigation on the mutant WzM-L3 live attenuated vaccination was suggested to have the ability to shield the *Pangasius hypophthalmus* from ESC (Triet et al. 2019).

8.4 Yersiniosis

The bacteria *Yersinia ruckeri* (EC 2003) is the cause of this freshwater enteric red mouth (ERM) disease in rainbow trouts. In freshwater and infrequently in marine environments, it can also have an impact on other fish, such as Atlantic salmon. Currently, *Y. ruckeri*, biotype 1 with serovar I (Hagerman strain) can be protected using a newly created formalin-inactivated whole-cell vaccine (Toranzo et al. 2009).

ERM immersion vaccines have been shown to protect rainbow trout, *Oncorhynchus mykiss*, based on both biotypes 1 and 2 with serotype O1 against the pathogen *Y. ruckeri* (Yang et al. 2021). The fish can be protected from illness by both vaccination biotypes 1 and 2, they demonstrated.

8.5 Bacterial Kidney Disease (BKD)

Renibacterium salmoninarum, a member of the diplobacillus group, is the bacterium that causes BKD (Muktar et al. 2016). Salmonids have been reported to be susceptible to this chronic systemic disease, which kills fish in both freshwater and marine habitats.

There is evidence that this bacterium can elicit an immunological response, despite the fact that numerous immunization experiments utilizing live attenuated, recombinant, or classical bacterins have been shown (Wood and Kaattari 1996). However, the vaccine's ability to prevent is questionable since viruses are spread vertically, as well as because of the intracellular nature and immunosuppressive function of the p57 protein (Toranzo et al. 2009; Muktar et al. 2016). There have been reports of Novartis in South Africa (S.A.) licensing a commercial live vaccination for the prevention of BKD under the trade name 'Renogen' (Toranzo et al. 2009).

9 Commercial Fish Vaccines

Many of the severe bacterial infections that are currently plaguing the fish farming sectors can be prevented with vaccines. Numerous live-attenuated, inactivated, and DNA vaccines have been created and are presently used in extensive fish farming ventures (Table 3). The US Department of Agriculture licensed the first effective commercial bacterial vaccination, which was developed to treat yersiniosis against enteric red mouth (Kumar et al. 2015), followed by a vaccine for vibriosis and was introduced in the USA in the late 1970s. It was developed using whole-cell inactivation as a basis and administered using immersion techniques (Sommerset et al. 2005). These vaccinations have been shown to be effective in preventing a number of the most common bacterial illnesses, and their affordability makes them perfect for use in the aquaculture sector. Vibriosis has a global distribution in a wide range of fish species.

More than 10 serotypes of *Listonella anguillarum*—formerly known as *Vibrio anguillarum*—the bacterial species that causes classical vibriosis, have been identified so far. Vaccines made for salmonids cannot be utilized for cold-water marine fish species like cod and halibut because differing serotypes may be connected to disease outbreaks in these fish species.

Fish vaccine development has been carried out globally since 1990 in a manner similar to that of human and veterinary vaccines, requiring close coordination between R&D, the pharmaceutical sector, and regulatory bodies in pertinent regions. Schering-Plough Animal Health (USA), Pharmaq (Norway), Intervet International (The Netherlands), Novartis Animal Health (Switzerland), and Bayer Animal Health (Bayotek)/Microtek, Inc. (Germany/Canada) are the top producers of fish vaccinations. Aquaculture production of salmon and trout dominates the worldwide commercial market for these businesses (Sommerset et al. 2005).

Worldwide, there are more than 26 approved fish vaccines that can be used on a variety of fish species. The USDA reports that 77 distinct species of fish are

Table 3 Bacterial fish vaccines, commercially available

Disease	Pathogen	Host	Vaccine type	Delivery methods	Target antigens	Country/ Regions	Status	Make	References
Furunculosis	*Aeromonas salmonicida*	Salmonids	Inactivated	Immersion, IP	Inactivated *Aeromonas salmonicida* spp.	USA, Canada, Chile, Europe, Australia	–	MSD Animal Health	Thornton et al. (1991)
Yersiniosis	*Yersinia ruckeri*	Atlantic salmon, rainbow trout	Inactivated lipopolysaccharide	Immersion	Inactivated yersiniosis	USA	Experimental	Elanco (Aqua Health)	Siwicki et al. (1998), Skov et al. (2012), Jaafar et al. (2018)
Vibriosis	*Vibrio anguillarum, Vibrio salmonicida, Vibrio ordalii*	Salmonids, grouper, sea bass, sea bream, cod, yellowtail, ayu, halibut	Inactivated/killed	Immersion, IP	Inactivated vibriosis	USA, Canada, Japan, Europe, Australia	Commercial (AquaVac Vibrio Oral), ALPHA MARINE™ Vibrio	Merck Animal Health	Angelidis et al. (2006) and Galindo-Villegas et al. (2013)
Bacterial kidney disease (BKD)	*Renibacterium salmoninarum*	Salmonids	Avirulent live culture	IP	*Arthrobacter davidanieli*	Canada, Chile, USA	–	Renogen	Salonius et al. (2005)
Enteric septicemia of catfish (ESC)	*Edwardsiella ictaluri*	Catfish	Inactivated outer membrane proteins	IP	Inactivated *Edwardsiella ictaluri*	Vietnam	AQUAVAC-ESCO	Pharmaq	Thinh et al. (2009)

(continued)

Table 3 (continued)

Disease	Pathogen	Host	Vaccine type	Delivery methods	Target antigens	Country/ Regions	Status	Make	References
Enteric redmouth disease (ERM)	*Yersinia ruckeri*	Salmonids rainbow trout	Attenuated	Immersion	Inactivated *Yersinia ruckeri*	USA	Experimental	Elanco (Aqua Health)	Costa et al. (2011) and Villumsen et al. (2014)
Columnaris disease	*Flavobacterium columnaris*	Fresh-water finfish species, Nile tilapia, bass, bream, salmon, turbot	Attenuated	Immersion	Attenuated *Flavobacterium columnare*	USA	Experimental	Merck Animal Health	Shoemaker et al. (2011) and Mohammed et al. (2013)
Mycobacteriosis	*Mycobacterium marinum*	Pejerrey, turbot, Pacific and Atlantic salmon, snake-head fish, European sea bass and red drum, tilapia	Avirulent heat-killed	IP			Experimental		Ravid-Peretz et al. (2019) and Ziklo et al. (2018)

Bacterial cold-water disease (BCWD)	*Flavobacterium psychrophilum*	Salmonids	Live attenuated	Immersion			B.17-ILM		Bøgwald and Dalmo (2019)
Motile Aero-monas septicaemia (MAS)	*Aeromonas hydrophila, A. caviae, A. sobria*		Inactivated	IP		Asia, Europe, USA		Pharmaq	

Ma et al. (2019), Bøgwald and Dalmo (2019), Miccoli et al. (2021)

presently immunized against over 22 different bacterial species and six different virus species (Tizard 2021). Several countries, including Japan and Korea, have granted licenses and made fish vaccinations available for purchase (Hwang et al. 2020).

10 Recent Studies on Fish Vaccine Development

There are currently no effective vaccines for many bacterial diseases, despite several bacterial vaccines being commercially available for use in aquaculture operations. As a result of developments in reverse vaccinology, biotechnology, and molecular biology, subunit vaccines, plasmid DNA vaccines, recombinant live vector vaccines, and recombinant protein vaccines have been developed. Following testing in fish trials, these vaccines have been approved for commercialization (Mzula et al. 2019).

The focus of the vaccine experiments was on killed vaccines, the earliest in tradition. According to Gudding et al. (2014), the first fish vaccination to be documented was an oral vaccine containing dead *Aeromonas salmonicida* administered to cutthroat trout (*Oncorhynchus clarkii*). Following the vaccine's effectiveness, formalin-killed immersion vaccinations were developed to prevent salmon and trout vibriosis. A similar method was used to generate earlier salmonid vaccines, which were administered by immersion and prevented bacterial infections in Atlantic salmon (*Salmo salar*). *A. hydrophila* infection can be effectively controlled by biofilm vaccination; this method incorporates both protective and non-protective proteins and may induce a heterologous adaptive immune response in fish that get the immunization (Kaur et al. 2021).

Reverse vaccinology is a new technology that has gained attention in light of biotechnology's recent developments. Developing a vaccination against a variety of dangerous organisms can be aided by the most recent technologies. According to Rappuoli (2000), the process of creating a successful and effective vaccine takes several years. The production of vaccines has sped up from 5 to 10 years because to the concept of vaccination. Using a reverse vaccinology study of the strong antigenic target contents of particular pathogens, subunit vaccines were created to prevent fish nocardiosis in largemouth bass (*Micropterus salmoides*). The findings indicated that although the vaccines had varying effects, they were very promising for preventing nocardial prophylaxis (Ho et al. 2018).

Under varied experimental conditions, it was shown that a multicomponent vaccine protected trout against three relevant bacterial diseases (furunculosis, vibriosis, and yersiniosis). This suggests that the vaccine induces specific antibody responses to different bacterial antigens and regulates effective expressions of various genes involved in the immune response (Marana et al. 2019). Combination vaccines with several inactivated diseases have been reported on numerous times. Following challenges with *V. alginolyticus*, *V. parahaemolyticus*, and *Photobacterium damselae* subsp., a combination of these three inactivated bacterins demonstrated an RPS > 80% in cobia fish (Lin et al. 2006).

The red hybrid tilapia (*Oreochromis* sp.) was immunized with an inactivated recombinant vaccine that encoded the *S. agalactiae* cell wall surface anchoring family protein. Orally inoculated fish showed a robust and significantly elevated IgM antibody immune response with an effectiveness of 70% RPS in blood, mucus, and gut lavage fluid samples (Bøgwald and Dalmo 2021).

11 Conclusions and Recommendations

The widespread decrease in the use of antibiotics can be attributed to the efficaciousness of fish vaccinations. However, the true problem in developing a fish vaccine still lies in integrating all the factors that affect the plan from conception to delivery. The creation of new vaccinations against bacterial infectious pathogens has been greatly aided by technological breakthroughs in recent times, as they have decreased the likelihood of disease outbreaks and associated losses in the aquaculture sector. Future fish vaccinations developed to combat various infectious pathogens, such as bacteria, viruses, and other agents, ought to be both economically and environmentally sustainable. They ought to be easily accessible, appropriate for large-scale production, and reasonably priced, even for small fish farmers. Long into the future, new vaccinations will still be required, and it will be crucial to use all of the biotechnology at our disposal to address the problems posed by developing diseases. For this, further investigation must be done.

Key Issues

- At present, the majority of vaccinations are formulated using inactivated pathogens that were developed empirically. There are a few vaccines made using recombinant subunits as well.
- The commercial fish vaccine market is dominated by injectable multivalent oil-emulsion vaccinations for salmonids.
- In comparison to vaccines used in larger vertebrates, most fish vaccines require a comparatively high antigenic load.
- A lack of complete understanding of fish immune systems and species diversity prevents the development of new vaccines or vaccine delivery systems based on non-empirical methods.

References

Adams A, Aoki T, Berthe C, Grisez L, Karunasagar I (2008) Recent technological advancements on aquatic animal health and their contributions toward reducing disease risks- a review. In: Diseases in Asian aquaculture, 6th edn. Fish Health Section, Asian Fisheries Society, Colombo, pp 71–88

Andreoni F, Amagliani G, Magnani M (2016) Selection of vaccine candidates for fish pasteurellosis using reverse vaccinology and an in vitro screening approach. Methods Mol Biol 1404:181–192

Angelidis P, Karagiannis D, Crump EM (2006) Efficacy of a Listonella anguillarum (syn. *Vibrio anguillarum*) vaccine for juvenile sea bass *Dicentrarchus labrax*. Dis Aquat Org 71(1):19–24. https://doi.org/10.3354/dao071019

Azad IS, Shankar KM, Mohan CV, Kalita B (2000) Uptake and processing of biofilm and free-cell vaccines of *Aeromonas hydrophila* in indian major carps and common carp following oral vaccination-antigen localization by a monoclonal antibody. Dis Aquat Org 43:103–108

Bøgwald J, Dalmo RA (2019) Review on immersion vaccines for fish: an update 2019. Microorganisms 7(12):627

Bøgwald J, Dalmo RA (2021) Protection of teleost fish against infectious diseases through oral administration of vaccines: Update. Int J Mol Sci 22:10932

Cabello FC (2006) Heavy use of prophylactic antibiotics in aquaculture: a growing problem for human and animal health and for the environment. Environ Microbiol 8(7):1137–1144

Costa AA, Leef MJ, Bridle AR, Carson J, Nowak BF (2011) Effect of vaccination against yersiniosis on the relative percent survival, bactericidal and lysozyme response of Atlantic salmon, *Salmo salar*. Aquaculture 315(3–4):201–206

Czochor J, Turchick A (2014) Introduction. Yale J Biol Med 87(4):401–402

Dadar M, Dhama K, Vakharia VN, Hoseinifar SH, Karthik K, Tiwari R, Khandia R, Munjal A, Salgado- Miranda C, Joshi SK (2016) Advances in aquaculture vaccines against fish pathogens: global status and current trends. Rev Fish Sci Aquacult 25(3):184–217

de Bruijn I, Liu Y, Wiegertjes GF, Raaijmakers JM (2018) Exploring fish microbial communities to mitigate emerging diseases in aquaculture. FEMS Microb Ecol 94:161

Defoirdt T, Sorgeloos P, Bossier P (2011) Alternatives to antibiotics for the control of bacterial disease in aquaculture. Curr Opin Microbiol 14(3):251–258

Duff DCB (1942) The oral immunization of trout against *Bacterium salmonicida*. J Immunol 44:87–94

E.C (European Commission) (2003) The use of fish by-products in aquaculture, Report of the scientific committee on animal health and animal welfare, scientific co-operations and networks, Ireland, pp 25–29

Galindo-Villegas J, Mulero I, García-Alcazar A, Muñoz I, Peñalver-Mellado M, Streitenberger S et al (2013) Recombinant TNFα as oral vaccine adjuvant protects European sea bass against vibriosis: insights into the role of the CCL25/CCR9 axis. Fish Shellfish Immunol 35(4):1260–1271

Gudding R, Goodrich T (2014) The history of fish vaccination. In: Gudding R, Lillehaug A, Evensen Ø (eds) Fish Vaccination, 1st edn. Wiley, New York, pp 1–11

Gudding R, Lillehaug A, Evensen Ø (2014) Fish vaccination. In: Fish vaccination. Wiley, Hoboken, pp 1–383

Hansson M, Nygren PAK, Stahl S (2000) Design and production of recombinant subunit vaccines. Biotechnol Appl Biochem 32:95–107

Hart S, Wrathmell AB, Harris JE, Grayson TH (1988) Gut immunology in fish: a review. Dev Comp Immunol 12:453–480

Ho PY, Chen YC, Maekawa S, Hu HH, Tsai AW, Chang YF, Wang PC, Chen SC (2018) Efficacy of recombinant protein vaccines for protection against *Nocardia seriolae* infection in the large-mouth bass *Micropterus salmoides*. Fish Shellfish Immunol 78:35–41

Holten-Andersen L, Doherty TM, Korsholm KS, Andersen P (2004) Combination of the cationic surfactant dimethyl dioctadecyl ammonium bromide and synthetic mycobacterial cord factor as an efficient adjuvant for tuberculosis subunit vaccines. Infect Immun 72:1617

Horzinek MC, Schijns VECJ, Denis M, Desmettre P, Babiuk LA (1997) General description of vaccines. In: Pastoret PP, Blancou J, Vannier P, Verschueren C (eds) Veterinary vaccinology. Elsevier Press, Amsterdam, pp 132–152

Hwang JY, Kwon MG, Seo JS, Hwang SD, Jeong JM, Lee JH, Jeong AR, Jee BY (2020) Current Use and Management of Commercial Fish Vaccines in Korea. Fish Shellfish Immunol 102:20–27

Ire T, Watarai S, Iwasaki T, Kodama H (2005) Protection against experimental *Aeromonas salmonicida* infection in carp by oral immunisation with bacterial antigen entrapped in liposomes. Fish Shellfish Immunol 18:235–242

Jaafar RM, Al-Jubury A, Chettri JK, Dalsgaard I, Kania PW, Buchmann K (2018) Secondary immune response of rainbow trout following repeated immersion vaccination. J Fish Dis 41(1):117–123. https://doi.org/10.1111/jfd.12682

Kaur B, Kumar BTN, Tyagi A, Holeyappa SA, Singh NK (2021) Identification of Novel Vaccine Candidates in the WholeCell *Aeromonas hydrophila* Biofilm Vaccine through Reverse Vaccinology Approach. Fish Shellfish Immunol 114:132–141

Klesius PH, Pridgeon JW (2014) Vaccination against enteric septicemia of catfish. In: Fish vaccination. Wiley, Chichester, pp 211–225

Klesius PH, Shoemaker CA (1998) Development and use of modified live *Edwardsiella ictaluri* vaccine against enteric septicemia of catfish. J Adv Vet Res 41:523–537

Kumar G, Ledouble SM, Saleh M, El-Matbouli M (2015) Yersinia ruckeri, the causative agent of enteric redmouth disease in fish. Vet Res 46:103. https://doi.org/10.1186/s13567-015-0238-4

Kurath G (2008) Biotechnology and DNA vaccines for aquatic animals. Rev Sci Tech/Office international des épizooties 27:175–196

La Frentz BR, La Patra SE, Jones GR, Congleton JL, Sun B (2002) Characterization of serum and mucosal antibody responses and relative percent survival in rainbow trout (Oncorhynchus mykiss) (Walbaum), following immunization and challenges with *Flavobacterium psychrophilum*. J Fish Dis 25(12):703–713

Levine MM, Sztein MB (2004) Vaccine development strategies for improving immunization: the role of modern immunology. Nat Immunol 5:460

Li L, Lin SL, Deng L, Liu ZG (2013) Potential use of chitosan nanoparticles for oral delivery of DNA vaccine in black seabream *Acanthopagrus schlegelii* to protect from *Vibrio parahaemolyticus*. J Fish Dis 36:987–995

Lin JHY, Chen TY, Chen MS, Chen HE, Chou RL, Chen TI, Su MS, Yang HL (2006) Vaccination with Three Inactivated Pathogens of Cobia (*Rachycentron canadum*) Stimulates Protective Immunity. Aquaculture 255:125–132

Ma J, Bruce TJ, Jones EM, Cain KD (2019) A review of fish vaccine development strategies: conventional methods and modern biotechnological approaches. Microorganisms 7(11):569. https://doi.org/10.3390/microorganisms7110569

Mahendran R, Jeyabaskar S, Sitharaman G, Michael RD, Paul AV (2016) Computer-aided vaccine designing approach against fish pathogens *Edwardsiella tarda* and *Flavobacterium columnare* using bioinformatics softwares. Drug Des Devel Ther 10:1703–1714

Marana MH, Sepúlveda D, Chen D, Al-Jubury A, Jaafar RM, Kania PW, Henriksen NH, Krossøy B, Dalsgaard I, Lorenzen N et al (2019) A pentavalent vaccine for rainbow trout in Danish aquaculture. Fish Shellfish Immunol 88:344–351

Menanteau-Ledouble S, Kumar G, Saleh M, El-Matbouli M (2016) Aeromonas salmonicida: updates on an old acquaintance. Dis Aquat Org 120:49–68

Miccoli A, Manni M, Picchietti S, Scapigliati G (2021) State-of-the-art vaccine research for aquaculture use: the case of three economically relevant fish species. Vaccine 9:140

Miquel A, Muller I, Ferrer P, Valenzuela PDT, Burzio LO (2003) Immuno response of Coho salmon immunized with a gene expression library from *Piscirickettsia salmonis*. Biol Res 36:313–323

Mohamad A, Zamri-Saad M, Amal MNA, Al-saari N, Monir MS, Chin YK, Yasin I-S (2021) Vaccine efficacy of a newly developed feed-based whole-cell polyvalent vaccine against vibriosis, streptococcosis and motile aeromonad septicemia in Asian seabass, *Lates calcarifer*. Vaccines 9(4):368

Mohamed LA, Soliman WS (2013) Development and efficacy of fish vaccine used against some bacterial diseases in farmed Tilapia. J Nat Sci 11(6):120–128

Mohammed H, Olivares-Fuster O, La Frentz S, Arias CR (2013) New attenuated vaccine against columnaris disease in fish: choosing the right parental strain is critical for vaccine efficacy. Vaccine 31(45):5276–5280

Mohd-Aris A, Muhamad-Sofie MHN, Zamri-Saad M, Daud HM, Yasin Ina-Salwany M (2019) Live vaccines against bacterial fish diseases: a review. Vet World 12:1806–1815

Muktar Y, Tesfaye S, Tesfaye B (2016) Present status and future prospects of fish vaccination: a review. Res Rev J Vet Sci Technol 7(2):1000299

Mzula A, Wambura PN, Mdegela RH, Shirima GM (2019) Current state of modern biotechnological-based *Aeromonas hydrophila* vaccines for aquaculture: a systematic review. Biomed Res Int 2019:3768948

Naylor RL, Hardy RW, Buschmann AH, Bush SR, Cao L, Klinger DH, Little DC, Lubchenco J, Shumway SE, Troell M (2021) A 20-year retrospective review of global aquaculture. Nature 591:551–563

Park YH, Hwang SY, Hong MK, Kwon KH (2012) Use of antimicrobial agents in aquaculture. Rev Sci Tech l'OIE 31(1):189–197

Pasnik DJ, Smith SA (2005) Immunogenic and protective effects of a DNA vaccine for *Mycobacterium marinum* in fish. Vet Immunol Immunopathol 103:195–206

Pasnik DJ, Smith SA (2006) Immune and histopathologic responses of. DNA-vaccinated hybrid striped bass. *Morone saxatilis M. chrysops* after acute. *Mycobacterium marinum* infection. Dis Aquat Org 73:33–41

Perrie Y, Mohammed AR, Kirby DJ, McNeil SE, Bramwell VW (2008) Vaccine adjuvant systems: enhancing the efficacy of sub-unit protein antigens. Int J Pharm Pharm Sci 364(2):272–280. https://doi.org/10.1016/j.ijpharm.2008.04.036

Pilstrom L, Bengten E (1996) Immunoglobulin in fish: genes, expression and structure. Fish Shellfish Immunol 6:243–262

Plant KP, LaPatra SE (2011) Advances in fish vaccine delivery. Dev Comp Immunol 35(12):1256–1262

Plant KP, LaPatra SE, Cain KD (2009) Vaccination of rainbow trout, Oncorhynchus mykiss (Walbaum), with recombinant and DNA vaccines produced to *Flavobacterium psychrophilum* heat shock proteins 60 and 70. J Fish Dis 32:521–534

Poobalane S, Thompson KD, Ardo L, Verjan N, Han H-J, Jeney G, Hirono I, Aoki T, Adams A (2010) Production and efficacy of an *Aeromonas hydrophila* recombinant S-layer protein vaccine for fish. Vaccine 28(20):3540–3547

Pridgeon JW, Klesius PH (2012) Major bacterial diseases in aquaculture and their vaccine development. CAB Rev 48:1–16

Ramesh D, Souissi S (2018) Antibiotic resistance and virulence traits of bacterial pathogens from infected freshwater fish, *Labeo rohita*. Microb Pathog 116:113–119

Rappuoli R (2000) Reverse vaccinology. Curr Opin Microbiol 3(5):445–450

Ravid-Peretz S, Colorni A, Sharon G, Ucko M (2019) Vaccination of European sea bass *Dicentrarchus labrax* with avirulent *Mycobacterium marinum* (iipA::kan mutant). Fish Shellfsh Immunol 90:317–327

Rigos G, Tyrpenou AE, Nengas I, Alexis M, Athanassopoulou F, Troisi GM (2004) Poor bioavailability of oxytetracycline in sharpsnout sea bream *Diplodus puntazzo*. Aquaculture 235(1–4):489–497

Rodger HD (2016) Fish disease causing economic impact in global aquaculture. In: Adams A (ed) Fish vaccines, 1st edn. Springer, Basel, pp 1–34

Rohani MF, Islam SM, Hossain MK, Ferdous Z, Siddik MA, Nuruzzaman M, Padeniya U, Brown C, Shahjahan M (2022) Probiotics, prebiotics and synbiotics improved the functionality of aquafeed: upgrading growth, reproduction, immunity and disease resistance in fish. Fish Shellfish Immunol 120:569–589

Salonius K, Siderakis C, MacKinnon AM, Grifths SG (2005) Use of Arthrobacter davidanieli as a live vaccine against *Renibacterium salmoninarum* and Piscirickettsia salmonis in salmonids. Dev Biol 121:189–197

Secombes CJ (2003) Adaptive immunity in teleost II (cellular immunity). In: 3rd international symposium on fish vaccinology. Abstract book, 45. Bergen, Norway, 9–11 April

Shoemaker CA, Klesius PH, Evans JJ, Arias CR (2009) Use of modified live vaccines in aquaculture. J World Aquacult Soc 40:573–585

Shoemaker CA, Klesius PH, Drennan JD, Evans JJ (2011) Efficacy of a modified live Flavobacterium columnare vaccine in fish. Fish Shellfish Immunol 30(1):304–308

Siwicki AK, Morand M, Terech-Majewska E, Niemczuk W, Kazun K, Glabski E (1998) Influence of immunostimulants on the effectiveness of vaccines in fsh: in vitro and in vivo study. J Appl Ichthyol 14(5):225–227

Skov J, Kania PW, Holten-Andersen L, Fouz B, Buchmann K (2012) Immunomodulatory effects of dietary beta1,3-glucan from Euglena gracilis in rainbow trout (Oncorhynchus mykiss) immersion vaccinated against *Yersinia ruckeri*. Fish Shellfsh Immunol 33(1):111–120

Sneeringer S, Bowman M, Clancy M (2019) The US and EU animal pharmaceutical industries in the age of antibiotic resistance; USDA Economic Research Service Report Number 264; USDA, Washington, DC

Snieszko SF, Friddle SB (1949) Prophylaxis of furunculosis in brook trout (*Salvelinus fontinalis*) by oral immunization and sulfamerazine. Prog Fish-Cult 11:161–168

Snieszko S, Piotrowska W, Kocylowski B Marek K (1938) Badania bakteriologiczne i serogicz-zne nad bakteriami posocznicy karpi. Memoires de l'Institut d'Ichtyobiologie et Pisciculture de la Station de Pisciculture Experimentale a Mydlniki de l'Universite Jagiellonienne a Cracovie Nr, 38

Sommerset I, Krossøy B, Biering E, Frost P (2005) Vaccines for Fish in Aquaculture. Expert Rev Vaccines 4:89–101

Sudheesh PS, Al-Ghabshi A, Al-Mazrooei N, Al-Habsi S (2012) Comparative pathogenomics of bacteria causing infectious diseases in fish. Int J Evol Biol 2012:45726

Sun Y, Zhang M, Liu CS, Qiu R, Sun L (2012) A divalent DNA vaccine based on Sia10 and OmpU induces cross protection against *Streptococcus iniae* and *Vibrio anguillarum* in Japanese flounder. Fish Shellfish Immunol 32:1216–1222

Takeuchi M, Fujiwara-Nagata E, Katayama T, Suetake H (2021) Skin bacteria of rainbow trout antagonistic to the fish pathogen *Flavobacterium psychrophilum*. Sci Rep 11:7518

Tanaka T, Furukawa K, Suzuki Y, Aida K (1999) Transfer of maternal antibody from mother to egg may have no protective meaning for larvae of Red sea bream *Pagrus major*, a marine teleost. Fish Sci 65:240–243

Thinh NH, Kuo TY, Hung LT, Loc TH, Chen SC, Evensen O, Schuurman HJ (2009) Combined immersion and oral vaccination of Vietnamese catfish (*Pangasianodon hypophthalmus*) confers protection against mortality caused by *Edwardsiella ictaluri*. Fish Shellfsh Immunol 27(6):773–776

Thornton JC, Garduno RA, Newman SG, Kay WW (1991) Surface-disorganized, attenuated mutants of aeromonas-salmonicida as furunculosis live vaccines. Microb Pathog 11(2):85–99. https://doi.org/10.1016/0882-4010(91)90002-r

Tizard I (1999) Grease, anthraxgate, and kennel cough: a revisionist history of early veterinary vaccines. Adv Vet Med 41:7–24

Tizard IR (2021) Fish vaccines. In: Vaccines for veterinarians. Elsevier, Amsterdam, pp 281–292

Tlaxca JL, Ellis S, Remmele RL Jr (2015) Live attenuated and inactivated viral vaccine formulation and nasal delivery: potential and challenges. Adv Drug Deliv Rev 93:56–78

Toranzo A, Romalde J, Magarinos B, Barja J (2009) Present and future of aquaculture vaccines against fish bacterial diseases. In: Rogers C, Basurco B (eds) The use of veterinary drugs and vaccines in Mediterranean aquaculture, CIHEAM Opt Méditerranéennes 86, pp 155–176

Triet TH, Tinh BTT, Hau LV, Huong TV, Binh NQ (2019) Development and potential use of an *Edwardsiella ictaluri* wzz mutant as a live attenuated vaccine against enteric septicemia in *Pangasius hypophthalmus* (Tra catfish). Fish Shellfish Immunol 87:87–95

Vazquez-Juarez RC, Gomez-Chiarri M, Barrera-Saldaña H, Hernandez-Saavedra N, Dumas S, Ascencio F (2005) Evaluation of DNA vaccination of spotted sand bass (*Paralabrax maculato fasciatus*) with two major outer-membrane protein encoding genes from *Aeromonas veronii*. Fish Shellfish Immunol 19:153–163

Villumsen KR, Neumann L, Ohtani M, Strøm HK, Raida MK (2014) Oral and anal vaccination confers full protection against enteric redmouth disease (ERM) in rainbow trout. PLoS One 9(4):93845

Woo P, Burno DW, Lim L (2002) Diseases and disorders of finfish in cage culture. CABI Publishing, Walling Ford

Wood PA, Kaattari SL (1996) Enhanced immunogenicity of *Renibacterium salmoninarum* in Chinook salmon after removal of the bacterial cell surface associated 57 KDa proteins. Dis Aquat Org 25(1–2):71–79

Yang H, Zhujin D, Marana MH, Dalsgaard I, Rzgar J, Heidli M et al (2021) Immersion vaccines against *Yersinia ruckeri* infection in rainbow trout: comparative effects of strain differences. J Fish Dis 44(12):1937–1950

Ziklo N, Colorni A, Gao LY, Du SJ, Ucko M (2018) Humoral and cellular immune response of European seabass *Dicentrarchus labrax* vaccinated with heat-killed *Mycobacterium marinum* (iipA::kan Mutant). J Aquat Anim Health 30(4):312–324

Antimicrobial Usage in Aquaculture: Potential Implications

Gaurav Rathore and Badireddy Madhusudana Rao

Abstract

Farmed aquatic animals significantly contribute to global fish production, with aquaculture surpassing capture fisheries production in 2022. However, the farming of aquatic finfish and shellfish species is besieged with problems associated with diseases, leading to both food and economic losses. The inappropriate use of antimicrobials by aquaculturists to control diseases in farmed aquatic animals is an important driver for the emergence and dissemination of antimicrobial resistance (AMR). Additionally, the presence of antimicrobial residues in farmed aquatic animals raises serious food safety concerns. This chapter focuses on the use of antimicrobials in aquaculture, the potential implications of their usage and how AMR impacts aquatic animal health and public health.

Keywords

Antimicrobial use · Antimicrobial resistance · Aquaculture

G. Rathore (✉)
ICAR- National Bureau of Fish Genetic Resources, Lucknow, India
e-mail: grathore69@gmail.com

B. Madhusudana Rao
ICAR-Central Institute of Fisheries Technology, Visakhapatnam Research Centre, Visakhapatnam, India
e-mail: bmrcift@gmail.com

R. Kooloth Valappil et al. (eds.), *Aquatic Animal Health Management*,
https://doi.org/10.1007/978-981-96-7987-4_21

1 Introduction

Aquaculture is a rapidly growing industry, and aquaculture production has increased at an average annual percent rate of 6.24% per year since 1994 across the world. In 2022, the total global aquaculture production was 130.9 million tonnes, valued at USD 312.8 billion, of which farmed aquatic animals contributed 94.4 million tonnes with a value of USD 295.7 billion (FAO 2024). In 2022, for the first time in history, the global aquaculture production of aquatic animals surpassed capture fisheries production, and the share of farmed aquatic animals is further set to increase to 54% in 2032 (FAO 2024). However, the sustainability and growth of aquaculture are affected by bacterial diseases, which results in huge economic losses to the industry and farming community. Bacterial disease outbreaks in aquaculture are usually treated with antimicrobials, mainly antibiotics. Rational antimicrobial usage (AMU) under the supervision of a trained fish health professional is effective in controlling bacterial diseases in aquaculture and minimizing disease losses. However, the use of antibiotics, especially in low- and middle-income countries, is often indiscriminate. The main drivers of irrational use of antimicrobials are lack of proper disease diagnosis, absence of approved antibiotics and treatment guidelines for prevalent diseases, prescription by untrained persons, easy access to antibiotics, sub-standard drugs and poor awareness of prudent use of antimicrobials among stakeholders. Over-dosing or under-dosing and extra-label use of antibiotics is also quite widespread. In addition, treatment failures can further worsen the outcome of the disease, leading to the use of newer or preserved classes of antimicrobials in aquaculture. The aim of this chapter is to review the current information about the effects of AMU in aquaculture.

2 Use of Antimicrobials in Aquaculture

Demand for fish and seafood is growing day by day, and to meet the requirement, one has to make a paradigm shift from extensive to semi-intensive and intensive systems. However, the intensification increases water quality issues, stress in culturing animals as well as disease outbreaks. Disease outbreaks in aquaculture are caused by bacterial (54.9%), viral (22.6%), mycotic (3.1%) agents and parasitic (19.4%) agents (Dhar et al. 2014). As bacterial problems are generally encountered more frequently, the administration of antimicrobials can be the most effective treatment option. The term 'antimicrobial' refers to any compound, which includes antibiotics, sanitizers, disinfectants or any other substances that act against microorganisms. Antibacterials or antibiotics are types of antimicrobials effective against bacterial pathogens. Antibiotics are commonly used for the treatment of bacterial infections in aquaculture. There are no antibiotics specifically developed for aquaculture (FAO 2017). The first antibiotics used in fish farming were sulfonamides potentiated with trimethoprim which prevented bacterial growth by inhibiting the folate pathway. Global consumption of antimicrobials in aquaculture was 10,259 tonnes in 2017, of which more than 50% of consumption was in China (57.9%). Fish species-wise consumption of antibiotics was estimated as 157 mg/kg for catfish, 103 mg/kg for trout, 59 mg/kg for tilapia, 46 mg/kg for shrimp and 27 mg/kg for salmon. The projected antimicrobial consumption in 2030 was 236,757 tonnes, of which the share of aquatic animal use was the lowest (5.7%), followed by human use (20.5%) and terrestrial animal use (73.7%) (Schar et al. 2020). The quantity and the class of antibiotics used in aquaculture vary with the countries. In salmon fish farming, the quantity of antibiotics used per tonne of fish production varies from 0.02–1.3 g in Norway to 660–700 g in Chile (Watts et al. 2017). India accounted for 11.3% (~1150 tonnes) of antimicrobial consumption in 2017, and the share is expected to remain the same in 2030 (Schar et al. 2020).

3 About Antibiotics

Based on the mode of action, antibiotics can be of two types: Bactericidal, which can kill the bacteria or Bacteriostatic, which will inhibit the growth of bacteria. Bactericidal antibiotics include penicillins and cephalosporins that target the bacterial cell wall or polymyxins, which break the cell membrane, whereas quinolones and sulfonamides class of antibiotics interfere with essential bacterial enzymes. Aminoglycosides class of antibiotics such as streptomycin, kanamycin, tobramycin, gentamicin and neomycin are also bactericidal in nature and inhibit bacterial protein synthesis. Bacteriostatic antibiotics belonging to macrolides, lincosamides and tetracyclines work by inhibiting bacterial protein synthesis. Antibiotics can also be classified on the basis of bacterial target specificity. Broad-spectrum antibiotics are effective on both Gram-negative and Gram-positive bacteria, whereas narrow-spectrum antibiotics affect either Gram-negative or Gram-positive bacteria.

3.1 Choice of Antibiotic for Treatment

Antibiotic treatment should be carefully chosen depending on the following factors to ensure the best levels of efficacy, safety and responsibility:

(i) Type of bacterial pathogen causing disease in aquaculture.
(ii) Antibiotic susceptibility testing of the bacterial pathogen or minimum inhibitory concentrations (MIC) evaluation.
(iii) Pharmacokinetic (PK) and pharmacodynamics (PD) properties of antibacterial compounds. PK and PD parameters of antimicrobials are used to optimize the dosing of antimicrobials to maximize their effects and minimize toxicity to hosts. PK/PD is also used to determine breakpoints for categorizing the MIC of an isolate as susceptible, intermediate or resistant. Thus, breakpoints are always host-specific, bacteria-specific and specific for a particular antibiotic (Fig. 1).
(iv) Cost of treatment.

In animals, antimicrobials are administered for four types of applications:

- Therapeutic treatment of disease: Administration of antimicrobials only to diseased animals after disease diagnosis.

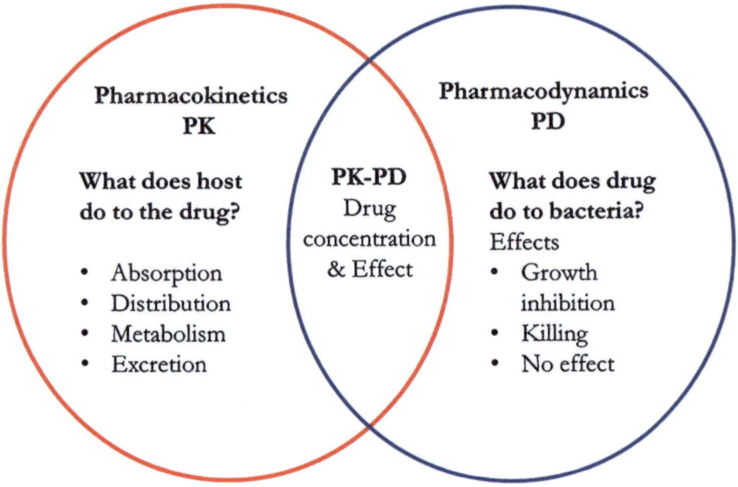

Fig. 1 Interaction between pharmacokinetics (PK) and pharmacodynamics (PD) for antibiotics for determining breakpoints. (Source: CLSI 2018)

- Metaphylaxis: The administration of antimicrobials to animals when they are perceived to be in contact with animals diagnosed with disease.
- Prophylaxis: Mass administration of antimicrobials to animals to prevent disease when risk is established.
- Growth Promotion: Administration of antimicrobials to animals to boost feed efficiency and increase weight gain.

However, in aquaculture, most antimicrobial treatments are metaphylactic as they are administered to particularly large populations including both infected and uninfected individuals. Metaphylaxis means 'the administration of a medicinal product to a group of animals after a diagnosis of clinical disease, with the aim of treating the clinically sick animals and controlling the spread of the disease to animals in close contact and at risk and which may already be sub-clinically infected' (Regulation of the European Parliament and of the Council, 2019/4) (Rigos et al. 2021).

3.2 Common Antimicrobials Used in Aquaculture

A list of commonly used antimicrobials and banned antimicrobials in different countries is given in Tables 1 and 2. Some of the commonly used antimicrobials in aquaculture are as follows:

3.2.1 β-Lactams

Some of the commonly used antibiotics of this class are ampicillin or amoxicillin. They are relatively heat-labile and unstable and can degrade in the presence of heavy metals, oxidizing and reducing agents. Medicated feed prepared with β-lactams must be top coated and should not be stored for long to avoid photodecomposition. These antibiotics are rapidly absorbed and metabolized. However, they are not recommended for use in aquaculture in most EU countries due to a lack of PK-PD data on fish. Dose rates are typically 50–75 mg/kg body weight for 7–10 days.

3.2.2 Tetracyclines

Tetracyclines are broad-spectrum low-cost bacteriostatic agents that inhibit bacterial protein synthesis. Oxytetracycline is the most common tetracycline used in aquaculture for the treatment of bacterial infections. However, the bioavailability of oxytetracycline is only 9–20% (Rigos et al. 2003, 2004). Oxytetracycline is light-sensitive and will turn dark brown after degradation. Degraded drug is nephrotoxic

Table 1 List of antimicrobials being used in aquaculture in different countries

Class	Indonesia	Malaysia	Singapore	Vietnam	China	EU	USA
Tetracyclines	CTC, OTC	TC, OTC	TC, OTC, CTC	OTC, DOX	DOX	OTC, CTC, DOX	OTC
Macrolides	ERY	ERY	–	ERY	–	ERY	–
Quinolones	ENR	–	ENR	–	ENR OA, FLU	OA, FLU, SAR, ENR, DAN,	–
β-Lactam	AMO	–	–	AMO	–	AMO	–
Sulfonamides	–	SMR	SMZ, SDM	–	SDZ, SML, SDM	SDZ, SMX	–
Potentiated sulfonamides	–	–	–	SDZ-TRI; SMX-OMP; SMX-TRI; SDM-TRI; SML-TRI	SDZ, SML, SDM, with TRI	SDZ-TRI, SMX-TRI	SMX-TRI
Phenicols	–	–	FLO	FLO	THI FLO	FLO	FLO
Aminoglycosides	–	–	–	NEO	–	–	–

TC tetracycline, *OTC* oxytetracycline, *CTC* chlortetracycline, *DOX* doxycycline, *OA* oxolinic acid, *FLU* flumequine, *ENR* enrofloxacin, *THI* thiamphenicol, *DAN* danofloxacin, *NEO* neomycin, *ERY* erythromycin, *SAR* sarafloxacin, *FLO* florfenicol, *AMO* amoxicillin, *SDZ* sulfadiazine, *SMX* sulfadimethoxine, *OMP* ormetoprim, *SDM* sulfadimidine, *SML* sulfamethoxazole, *SMZ* sulfamethazine, *SMR* sulfamerazine, *TRI* trimethoprim

Table 2 List of some of the banned/prohibited antimicrobials in aquaculture in different countries

Antibiotic	Indonesia	Malaysia	Philippines	Vietnam	China	India[a]
Amphenicols	Chloramphenicol thiamphenicol, florfenicol	Chloramphenicol	Chloramphenicol	–	Chloramphenicol	Chloramphenicol
Nitroimidazoles	Dimetridazole, metronidazole, fluconazole, tinidazole	Nitroimidazole	Nitroimidazole,	Dimetridazole, metronidazole, other nitroimidazoles		Nitroimidazoles
Nitrofurans	Nitrofurantoin, nifurpirinol, furazolidone, nifurtoinol	Nitrofurans	Nitrofurans	Nitrofuran (including furazolidone)	Nitrofuran (including furazolidone)	Nitrofuran (including furazolidone)
Macrolides	Virginiamicyn, tylosin, spiramycin	–	–	–	Erythromycin and tylosin	–
Quinolones	–	–	–	Enrofloxacin and Ciprofloxacin	Norfloxacin and Ciprofloxacin	–
Glycopeptides	–	–	–	Glycopeptides	–	Glycopeptides

[a] Some of the other banned antibiotics are neomycin, nalidixic acid, sulfamethoxazole, sulfonamide drugs (except approved sulfadimethoxine, sulfabromomethazine and sulfaethoxypyridazine)

for human beings, and therefore, they must be handled with care. Oxytetracycline is suitable for bath immersion as it is stable in water. Hard water containing higher divalent cations (Ca^{++}, Mg^{++}) can cause loss of activity of tetracyclines. The bioavailability of tetracyclines through the oral route is poor in cyprinids and is shed by the fish through faeces. The unabsorbed antibiotic has the tendency to bind to pond organic matter and can persist in sediment for a long time. Compared to oxytetracycline, doxycycline is more effective in euryhaline fish. The treatment dose for oral administration of oxytetracycline is 75 mg/kg fish for 7–10 days.

3.2.3 Fluoroquinolones

Fluoroquinolones such as enrofloxacin and oxolinic acid are effective against a broad spectrum of systemic infections by Gram-negative bacteria. They are bactericidal and inhibit bacteria by interfering with bacterial DNA gyrase in a dose-dependent manner. They can be used as bath or medicated feed for the treatment of bacterial infections caused by Gram-negative and Gram-positive bacteria. The oral dose of enrofloxacin is 5–10 mg/kg fish for 7–10 days.

3.2.4 Erythromycin

Erythromycin is a macrolide antibiotic, mainly effective against Gram-positive bacteria, such as infections caused by streptococci. The oral dose is 25–50 mg/kg fish for 7 days.

3.2.5 Phenicols

Phenicol class of antimicrobials are bacteriostatic broad-spectrum that inhibit bacterial protein synthesis. Chloramphenicol has been banned for use in fish and animal production due to serious side effects. Instead, florfenicol is approved for use in aquaculture in the US and EU countries. This is very effective against *Aeromonas* and Flavobacterial infections. The oral dose is 10 mg/kg fish for 7–10 days.

3.2.6 Sulfonamides

Sulfonamides are potentiated in combination with diaminopyrimidines such as trimethoprim and ormetoprim to increase their antibacterial activity. Sulfamethoxazole, sulfadiazine or sulfadimethoxine are some of the commonly used sulfonamides. Potentiated sulfonamides are bactericidal in nature with broad spectrum activity. They act as competitive inhibitors of folic acid metabolism for the killing of bacteria. Diaminopyrimidines inhibit tetrahydrofolic acid formation in bacteria. An oral dose of potentiated sulfadiazine along with trimethoprim is 25/5 mg/kg fish for 5–10 days.

3.3 Methods for Application of Antimicrobials

There are four different methods of application of antimicrobials in aquaculture, which are as follows:

1. Bath or dip method: Antimicrobials are dissolved in water at a particular concentration, and sick animals are placed in the solution for a specific duration. This method is used for the treatment of external infection and may not be effective for systemic infections due to poor absorption of the antimicrobials.
2. Topical application is mostly recommended for the treatment of external infection in brood fish or ornamental fish to treat ulcers or injuries.
3. Injection by either intraperitoneal or intramuscular is the least preferred method as it is stressful and time-consuming, although very effective.
4. Oral application of the antimicrobial agent through medicated feed. In this method, a known dose of antimicrobial agent is mixed in commercial feed and given to affected fish for treatment. Many times, if the fish is in an advanced stage of the disease, this method may not work as the sick fish may not consume the medicated feed. The dosage of the antimicrobial in the feed depends on the ratio of the active ingredient, daily feeding rate and biomass of fish being treated. Medicated feeds are normally prepared in small quantities as long-term storage can lead to a loss in the efficacy of the antimicrobial agent.

3.4 Failure of Antimicrobial Treatment in Aquaculture

The reason for the failure of antimicrobial treatments in aquaculture could be due to wrong diagnosis and the fish are not infected with a bacterial pathogen. Secondly, if the fish is suffering from co-infection of a bacteria and a virus, then antibiotic treatment will cause elimination of the bacteria, but the virus may continue to cause mortality. A third reason could be the wrong choice of antibiotic, as many bacterial pathogens can have inherent resistance to antibiotics used for treatment. Fourthly, a sub-optimal dose of an antimicrobial agent or a shorter duration of application may not be sufficient to eliminate the bacteria. This can also be another reason why fish may not respond to antibiotics. Another reason for treatment failure is an emerging issue of acquired antimicrobial resistance (AMR) in bacterial pathogens, which is the ability of microorganisms to resist the effects of antimicrobials.

3.5 Potential Implications of Antimicrobial Usage in Aquaculture

There are two major implications of antimicrobial usage in aquaculture, which are as follows:

A. The development of AMR is an emerging concern for human and animal health associated with antimicrobial usage.
B. The presence of antibiotic residues in the tissues of fish/shellfish is another concern related to food safety and aquatic animal trade.

3.5.1 Development of AMR in the Aquatic Environment: A Potential Impact of AMU

In aquaculture, antimicrobials are mostly administered through medicated feed to both healthy and diseased fish together, which is called as metaphylaxis. Unconsumed medicated food, unabsorbed antimicrobials or their active metabolites enter the aquatic environment through faeces and other secretions and can accumulate in the sediments and water. In addition, agricultural run-off, animal manure and human pollution can also introduce antimicrobials to aquaculture. These antimicrobials can affect entire aquatic fauna, including microbes, till their degradation. Studies have shown that biologically active concentrations of antimicrobials such as oxytetracycline, sulfonamides and quinolones persist in sediments and can also spread downstream (Liu et al. 2019). These antimicrobials present at sub-therapeutic levels for prolonged periods in the water, and the sediments exert selection pressure on bacteria, promoting mutagenesis and the emergence of antimicrobial resistance in bacterial populations of aquatic environments (Santos and Ramos 2018). The susceptible bacterial population is replaced by resistant organisms resulting in the alteration of microbial diversity. The spread of AMR in aquatic environments is further aided by horizontal gene transfer (HGT) of AMR genes (ARGs) within aquatic bacteria through mobile genetic elements. Several types of mobile genetic elements have been documented in bacterial species for the spread of ARGs. Some of these prominently include plasmids, integrons, transposons, insertion sequences and genomic islands. In addition, bacteriophages also facilitate the spread of ARGs in bacterial communities. Wastewater discharges from aquaculture could contribute to the spread of AMR in aquatic environments. The microorganism community in the aquatic environment, either pathogenic or non-pathogenic, can be influenced by antimicrobial abuse, as evidenced by the presence of large numbers of antimicrobial resistance genes in them. Aquatic environments, especially aquaculture systems, are termed genetic hotspots as their environmental factors favour HGT (Watts et al. 2017). These days, the focus of research is on studying the resistome (a collection of all AMR genes in a microbial community) of different habitats to know the pathways of AMR. Considering these adverse impacts of antimicrobial usage in aquaculture, worldwide attention is being paid towards the prudent use of antimicrobials in aquaculture, along with increased emphasis on monitoring of AMR in aquaculture.

How AMR Impacts Aquatic Animal Health and Public Health?

Aquatic Animal Health

Disease outbreaks by multiple drug-resistant bacterial pathogens in aquaculture make treatment of disease by common antimicrobials difficult or impossible. This causes economic losses to the aqua-farmers. AMR in aquatic animal pathogens has been documented worldwide (Miller and Harbottle 2018). Some of the frequently isolated bacterial pathogens of aquatic animals showing AMR are motile aeromonas (Preena et al. 2021), plasmid-mediated tetracycline resistance in *Edwardsiella* spp. (Loch et al. 2017), Streptococci carrying tetracycline and macrolide resistance genes (Nguyen et al. 2017), penicillin and cefotaxime-resistant Streptococci (Li et al. 2020), Chilean isolates of *Flavobacterium psychrophilum* showing high levels of resistance to OTC, FLO and OXA (Henríquez-Núñez et al. 2012), quinolone-resistant *F. columnare* (Chokmangmeepisarn et al. 2021) and vibrios isolated from shrimps (Vaiyapuri et al. 2021). These recent studies clearly prove that AMR in aquaculture is an emerging threat that can affect the sustainability of the sector.

Public Health

It is quite likely that indiscriminate antimicrobial use in aquaculture may lead to the development of AMR in aquatic animal pathogens and public health pathogens, which often reach aquatic environments through human-animal interactions. HGT may also contribute to the direct spread of AMR in public health pathogens or through commensal organisms. These pathogens can be transmitted to human beings through contaminated water, food or open wounds. Disease outbreaks caused by these multi-drug-resistant pathogens in humans cause high morbidity and mortality. These infections are often difficult to treat and involve higher costs and prolonged treatments (Bello-López et al. 2019). In aquatic environments, bacteria colonise various substrate surfaces such as rocks, microplastics, etc., forming multispecies microbial communities referred to as biofilms. These biofilms provide a conducive environment for HGT of ARGs between bacteria in the aquatic environments through three major mechanisms: conjugation, natural transformation and bacteriophage infection. In addition, two new mechanisms of HGT have been also reported. These are membrane vesicles (MVs) (Biller et al. 2014) and nanotubes (Dubey and Ben-Yehuda 2011; Dubey et al. 2016). MVs are typically 20- to 400-nm-diameter lipid-bilayer-enclosed particles released from bacteria (Toyofuku et al. 2015, 2019), which are reported to mediate the exchange of ARGs in a wide range of bacteria under laboratory conditions (Domingues and Nielsen 2017). Nanotubes are membranous tube-like structures distinguishable from conjugation pili that are known to transfer non-conjugative plasmid-carrying ARGs between bacteria (Dubey and Ben-Yehuda 2011; Dubey et al. 2016). The exchange of ARGs between aquatic bacteria and human pathogens, especially zoonotic pathogens, is a cause of concern (Abe et al. 2020). Tetracycline resistance gene *Tet(M)* was reported to be transferred from fish pathogen, *Lactococcus garvieae* to human *Enterococcus faecalis* (Neela et al. 2009). AMR has been reported in several zoonotic pathogens such as *Mycobacterium, Streptococci, Aeromonas, Vibrio*, diarrhoeagenic *E. coli*, *Salmonella, Plesiomonas, Listeria, Clostridium*, etc. Bi-directional transmission of such pathogens to human beings through contact with aquatic animals or via consumption of contaminated food products poses a risk to human health (Cabello et al. 2013). In this context, quantitative microbial risk assessment (QMRA) has been proposed to evaluate the potential public health risks posed by a given pathogen. QMRA requires information on the occurrence, exposure and dose-response of a particular pathogen (Amarasiri et al. 2020). This approach can be researched for pathogens of aquaculture concern and zoonotic importance.

3.6 Surveillance of AMR in Aquaculture

Currently, there are limited data available on AMR in the aquaculture sector in India. Most of them are individual studies with limited geographical coverage, samples and questionable quality. Thus, it is important to quantify the burden of AMR in aquaculture through structured surveillance with pan-India coverage. Surveillance AMR programme in aquaculture can serve the following purposes:

- Document baseline resistance levels in different geographical regions, target fish species and target pathogens.
- Describe the spread of resistant bacterial strains and resistance genes.
- Identify temporal and spatial trends in resistance.
- Generate hypotheses about sources and reservoirs of resistant bacteria.
- Understand the association between antimicrobial usage and resistance.
- Identify risk factors and clinical outcomes of infections caused by AMR bacteria.
- Guide evidence-based policies and guidelines to control antimicrobial use in aquaculture.

3.7 Objectives of AMR Surveillance Programmes

The objectives of the AMR surveillance programme in aquaculture could be:

1. To protect public health through active surveillance. In this, healthy fish at a farm or at the market are analysed for AMR in public health pathogens and indicator organisms.
2. To protect animal health through passive surveillance. In this, diseased fish will be used for the isolation of aquatic animal pathogens and subsequently analysed for AMR.

Based on the objectives of the surveillance programme, the target fish species, target pathogens and panel of antimicrobials to be used for antimicrobial susceptibility testing are selected.

3.8 Indian Scenario on AMR Surveillance in Aquaculture: INFAAR

Limited data are available on AMR in the aquaculture sector in India, which is based on individual studies with limited geographical coverage and samples. Thus, it is important to quantify the burden of AMR in food-producing animals and aquaculture through structured surveillance with pan-India coverage. Implementation of the AMR surveillance programme can document resistance levels in different production systems, describe the spread of resistant bacterial strains and resistance genes, identify trends in resistance and generate hypotheses about sources and reservoirs of resistant bacteria. Considering the above, ICAR, in technical support with FAO, initiated INFAAR (Indian Network for Fisheries and Animal Antimicrobial Resistance) in August 2018 for AMR surveillance. Together, 20 centres (17 ICAR and 3 State Agricultural Universities) were partners in the programme till 2023, of these, 9 centres were from fisheries and 11 from the animal sector. From 2024, the network has expanded to 31 centres. The priority microorganisms under the fisheries component of INFAAR are *E. coli*, *Staphylococcus aureus*, Coagulase-negative *Staphylococcus* species (CONS) and *Aeromonas* species in farmed freshwater fish; and *E. coli, S. aureus*, CONS and *Vibrio* species including *Vibrio parahaemolyticus* in farmed shrimps and cage cultured marine fish. Target freshwater fish for AMR analysis are carps and *Pangasionodon hypophthalmus*, while farmed shrimp, mainly *Penaeus vannamei* and cage-cultured seabass (*Lates calcarifer*), are the target species for brackishwater aquaculture and mariculture, respectively. Standard operating procedures for isolation, identification and antibiotic sensitivity testing of microorganisms are being used by all partners. The results are analysed through WHONET software. The information generated by INFAAR will form the baseline data on AMR in aquaculture, covering major aquaculture states and important aquaculture commodities (Rathore et al. 2020). This will lead to the identification of strategies to prevent and reduce the development and spread of AMR in farmed animals and fish.

3.8.1 Antibiotic Residues in Aquatic Animals and Their Products: Potential Impact of AMU

The use of antibiotics in aquaculture leads to the accumulation of antibiotic or their metabolites in the edible tissue of fish. Non-adherence to the specified withdrawal period after administering the permitted antibiotic results in the presence of antibiotic residues above the maximum residue limit (MRL) in the fish tissue, making it unsafe for human consumption. The withdrawal period refers to the minimum period of time (in days) from administering the last dose of antibiotic and the harvest of fish/shrimp to ensure that antibiotic residue in animal tissue is below the MRL. In fish, the withdrawal period is expressed as 'degree days' as environmental temperature influences the metabolic rate of fish. The Gazette of India notification (GOI 2012) states that if the withdrawal time has not been validated for any antibiotic, then the withdrawal period should not be less than 500 degree days for fish. The daily water temperature in degrees Celsius (°C) needs to be added till the total reaches 500. The fish should be harvested after this period so that the antibiotic level in the edible tissue will be lower than the prescribed MRL. On the other hand, the mere presence of 'prohibited antibiotics' in fish tissue is considered a food safety hazard. The prohibited antibiotics do not have MRLs but have the minimum required performance limit (MRPLs). The European Union has established MRPLs in aquaculture products as 1 μg/kg for nitrofuran metabolites and 0.3 μg/kg for chloramphenicol, and the same criteria are being followed in the Indian domestic market (FSSAI, 2011). Though MRPLs are fixed for prohibited antibiotics, countries adopt a 'zero tolerance' towards the presence of these antibiotics or their metabolites. India has prohibited the use of antibiotics such as chloramphenicol, nitrofurans,

Table 3 Harmful effect of antibiotic residues in food (Badireddy et al. 2023)

Antibiotic/metabolite	Effect on human health
Chloramphenicol	Aplastic anaemia, bone marrow toxicity
Nitrofurans and their metabolites	Genotoxic and carcinogenic
Sulfonamides	Skin allergies
Gentamicin	Mutagenic, nephropathic, hepatotoxic
β-Lactams (penicillins, cephalosporins)	Allergy, dermatitis, cutaneous eruptions, anaphylaxis and gastrointestinal symptoms
Tetracycline	Staining of teeth in young children, poor development of foetuses, gastrointestinal disorders, pro-inflammatory, cytotoxic and immune-pathological effects.

neomycin, nalidixic acid, sulfonamide drugs (except approved sulfadimethoxine, sulfabromomethazine and sulfaethoxypyridazine) and glycopeptides in aquaculture. The consumption of fish meat containing antibiotic residues, either the parent compound or their metabolites, is harmful to consumers (Table 3). Consumption of meat containing β-lactam antibiotics causes allergic reactions; chloramphenicol causes aplastic anaemia, and nitrofurans and metabolites are carcinogenic.

4 Regulations Regarding Use of Antibiotics in Aquaculture in India

The demand from consumers for antibiotic-free farm-raised meat products from both terrestrial and aquatic animals is increasing worldwide. Many countries have laid food safety requirements regarding the presence of antibiotic residues in fish meat sold in domestic markets. The Gazette of India Extraordinary, vide order No.729 (E) of the Ministry of Commerce and Industry, has prohibited the use of 20 antibiotics/other pharmacologically active substances (Table 4) in hatcheries, aquaculture ponds, feed manufacturing units, pre-processing units or processing units of shrimps, prawns or any other variety of fish and fishery products.

Recently, the Indian government, based on section 26A of the Drugs and Cosmetics Act, 1940, prohibited the use of colistin in aquaculture and directed the manufacturer to label the container with the words 'not to be used in food-producing animals, poultry, aqua farming and animal feed supplements' (GOI 2019).

4.1 Maximum Residue Limit (MRL)

The maximum level of antibiotic that may be present in the edible animal tissue without causing any harm to the consumer is known as MRL. In the case of finfish, the muscle with adherent skin in natural proportion is considered edible tissue, and in the case of shrimp, abdominal meat without shell is taken as edible tissue. The Government of India has established MRLs for four antibiotics, viz., tetracycline,

Table 4 Regulations regarding use of antibiotics in aquaculture in India

Antibiotics/pharmacologically active substance	
Prohibited list	Antibiotics with maximum residue limits (MRL)
Chloramphenicol Nitrofurans (furaltadone, furazolidone, Furylfuramide, nifuratel, Nifuroxime, nifurprazine, Nitrofurantoin, nitrofurazone) Neomycin Nalidixic acid Sulfamethoxazole Glycopeptides Sulfonamide drugs (except approved sulfadimethoxine, sulfabromomethazine and sulfaethoxypyridazine) Chloroform Chloropromazine Colchicine Dapsone Dimetridazole Metronidazole Ronidazole Ipronidazole and other nitroimidazoles Clenbuterol Diethylstibestrol (DES) *Aristolochia* spp. Ref: EIC (2002)	Tetracycline (0.1 ppm)* Oxytetracycline (0.1 ppm) Oxolinic acid (0.3 ppm) Trimethoprim (0.05 ppm) **ppm = parts per million,* *1 ppm = 1 mg/kg* *1 ppm = 1000 ppb* Ref: EIC (2002), FSSAI (2011)

oxytetracycline, oxolinic acid and trimethoprim (EIC 2002; FSSAI, 2011), and the MRLs ranged between 50 and 100 ppb (Table 1). The European Union (EU 2010) has fixed MRLs for several antibiotics namely amoxicillin, ampicillin, benzyl penicillin, chlortetracycline, cloxacillin, colistin, danofloxacin, dicloxacillin, difloxacin, enrofloxacin, erythromycin, florfenicol, flumequine, lincomycin, neomycin, oxacillin, oxolinic acid, oxytetracycline, paromomycin, sarafloxacin, spectinomycin, sulfonamides, tetracycline, thiamphenicol, tilmicosin, trimethoprim and tylosin (Table 5). The MRLs range between 50 ppb and 1000 ppb, depending on the antibiotic. In the case of terrestrial animals, apart from meat, MRLs are fixed for fat, liver and kidney. However, in the case of fish, there are no MRLs for these tissues.

4.2 Antibiotic Residues and International Fish Trade

Antibiotic residues were reported in farmed shrimp (e.g., *Penaeus vannamei*) exported from India to the European Union, the United States and Japan, which has led to the rejection of frozen shrimp consignments. The Rapid Alert System for Food and Feed (RASFF) of the European Union (RASFF portal; https://ec.europa.

Table 5 European Union's maximum residue limits (MRLs) of veterinary medicinal products in foodstuffs of animal origin, including fish

Antibiotic class	Antibiotic residue	MRL in fish
β-Lactam antibiotics	Amoxicillin	50 μg/kg
	Ampicillin	50 μg/kg
	Benzylpenicillin	50 μg/kg
	Cloxacillin	300 μg/kg
	Dicloxacillin	300 μg/kg
	Oxacillin	300 μg/kg
Tetracyclines	Chlortetracycline (sum of parent drug and its 4-epimer)	100 μg/kg
	Oxytetracycline (sum of parent drug and its 4-epimer)	100 μg/kg
	Tetracycline (sum of parent drug and its 4-epimer)	100 μg/kg
Quinolones and fluoroquinolones	Danofloxacin	100 μg/kg
	Difloxacin	300 μg/kg
	Sum of enrofloxacin and ciprofloxacin	100 μg/kg
	Flumequine	600 μg/kg
	Oxolinic acid	100 μg/kg
	Sarafloxacin	30 μg/kg
Macrolides	Erythromycin A	200 μg/kg
	Tilmicosin	50 μg/kg
	Tylosin A	100 μg/kg
Phenicols	Sum of florfenicol and its metabolites	1000 μg/kg
	Thiamphenicol	50 μg/kg
Aminoglycoside	Neomycin B	500 μg/kg
	Paromomycin	500 μg/kg
Aminocyclitol	Spectinomycin	300 μg/kg
Lincosamide	Lincomycin	100 μg/kg
Polymixins	Colistin	150 μg/kg
Folate pathway inhibitors	Sulfonamides (all substances belonging to the sulfonamide group)	100 μg/kg
	Trimethoprim	50 μg/kg

eu/food/safety/rasff/portal_en) is a key tool that ensures the flow of information regarding public health risks detected in the food chain. During the period from January 2010 to December 2020, a total of 89 consignments of shrimp exported from India to the European Union were rejected due to the presence of antibiotics or their metabolites in shrimp meat. The rejections were reported every year from 2010 to 2020 (Fig. 2), ranging from 4 to 15 rejections per year.

Of the total rejections, 82% of the rejections were due to the presence of nitrofurans, followed by chloramphenicol (13.5%) and tetracycline (4.5%), indicating indirectly that the use of these antibiotics in the shrimp culture system. Nitrofurans

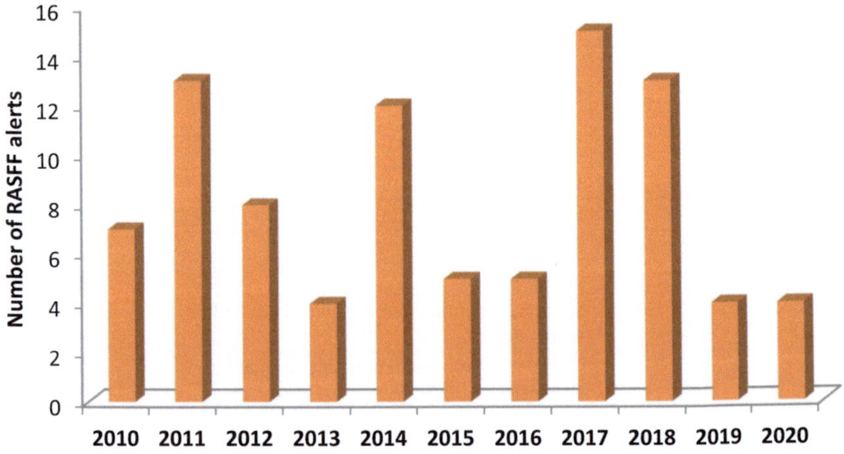

Fig. 2 Rejections of farmed shrimp exported from India to the European Union due to the presence of antibiotics and their metabolites

are a group of synthetic, broad-spectrum antibacterial agents that include compounds such as furazolidone, furaltadone, nitrofurazone and nitrofurantoin. These compounds are called parent compounds. Nitrofurans parent compounds metabolise rapidly after ingestion by the shrimp to form corresponding tissue-bound metabolites (Table 6). These metabolites, viz., AOZ, AMOZ, AHD and SEM, persist for several weeks in the edible tissue and serve as markers for the use of the corresponding parent compound during the aquaculture phase.

The annual trend in the total RASFF alerts due to the presence of antibiotic residues in farmed shrimp exported from India was mainly influenced by Furazolidone-AOZ (Fig. 3). The RASFF notifications indirectly indicate the use of the furazolidone in the shrimp culture in India during the last fifteen years (Rao and Prasad 2015).

4.3 National Monitoring Programme on Antibiotic Residues in Shrimp

The National Residue Control Plan (NRCP) of India is a system for monitoring residues of antibiotics/veterinary medicinal products and environmental contaminants in farmed shrimp, shrimp feed and shrimp seed. Samples for NRCP are collected from hatcheries, aquaculture farms, processing plants and feed mills of different states in India. Two categories of substances, namely Group A (substances having anabolic effect and unauthorized substances) and Group B (Veterinary drugs and environmental contaminants), are tested under the NRCP. The Group A substances are diethyl stilbesterol, estradiol, progesterone, chloramphenicol and nitrofurans. The Group B substances include tetracyclines, sulfonamides, quinolones/

Table 6 Nitrofuran parent compounds and their corresponding metabolites

Nitrofuran parent compounds	Nitrofuran metabolite
Furazolidone	AOZ (3-amino-2-oxazolidinone)
Furaltadone	AMOZ (3-amino-5-morpholinomethyl-1,3-oxazolidinone)
Nitrofurantoin	AHD (1-aminohydantoin)
Nitrofurazone	SEM (semicarbazide)

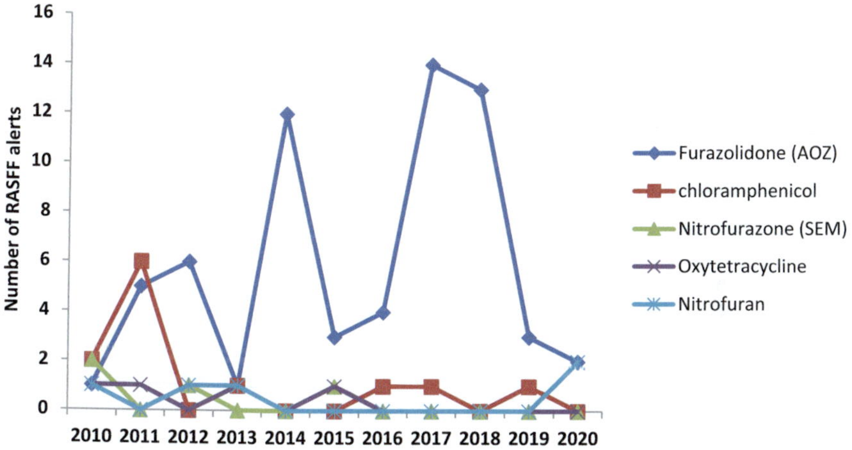

Fig. 3 Trend in the antibiotic-wise RASFF alerts in farmed shrimp exported from India to the European Union

fluoroquinolones, ivermectin, organochlorine compounds, polychlorinated biphenyls, mycotoxins and dyes.

The NRCP-2018 results indicate that of the total farmed shrimp samples (n = 1720) tested, only 0.9% of the shrimp samples were positive for the presence of antibiotics that included chloramphenicol (0.47%), Furazolidone-AOZ (0.3%), nitrofurazone-SEM (0.06%) and nitrofurantoin-AHD (0.06%). The results also indicated that 32% of shrimp hatchery seed samples were positive for the presence of antibiotic residues. However, all the shrimp feed samples tested were negative for the presence of antibiotics.

5 Antibiotic Residue Testing

Antibiotic residues are tested by employing screening and confirmatory methods. The commonly employed screening method is the enzyme-linked immunosorbent assay (ELISA). This method is employed for 'pre-harvest testing' (PHT) to screen for the presence of chloramphenicol and nitrofuran metabolites, viz., AOZ, AMOZ, AHD and SEM, and certify that this fish/shrimp harvested from that pond is fit for processing. Liquid chromatography Tandem Mass Spectrometry (LC-MS-MS) is a confirmatory and quantitative method employed for 'pre-export testing' to determine whether the processed fish/shrimp are fit for human consumption. Pre-export testing is performed to detect residues of chloramphenicol, nitrofuran metabolites (AOZ, AAMOZ, AHD, SEM), enrofloxacin (sum of enrofloxacin and ciprofloxacin), oxolinic acid, difloxacin, sparfloxacin, flumequine, sulfonamides (sulfadiazine, sulfamethoxazole, sulfamethoxypyridine, sulfamethazole, sulfamethazine, sulfamerazine, sulfapyridine, sulfadimethoxine, sulfachloropyradizine, sulfathiazole) and tetracyclines (parent drug and 4-epimer) in fish and shellfish meat. In the event of detection of antibiotics or their metabolites (above MRL/MRPL) in fish products, the faulty products are labelled as unfit for human consumption and are rejected or destroyed.

6 Conclusions

Antibiotics are commonly used for the treatment of bacterial infections in aquaculture. Judicious use of antimicrobials in aquaculture can help in the control of infectious diseases and also minimize economic losses caused by bacterial diseases. However, the use of prohibited antimicrobials or the improper use of antimicrobials in aquaculture could lead to the development of AMR in the bacterial pathogens of aquatic animals and pathogens of public health significance. The emergence of AMR in aquaculture can be minimized by following good aquaculture practices and optimal disease prevention and control measures. It is also essential to monitor AMR in aquaculture through structured surveillance programmes so as to understand the pathways of entry of AMR and geographical hotspots in aquaculture for initiating corrective measures. Another possible implication of the use of antimicrobials in aquaculture is the accumulation of antibiotic residues in the edible tissues of farmed fish and shellfish. Antibiotic residues make the cultured products unfit for consumption or export, not only causing monetary loss but also affecting the reputation of the country. Declining trends in RASSF alerts and NRCP monitoring indicate the lower presence of nitrofurans, chloramphenicol and tetracycline residues in the shrimp aquaculture system of India, and hence, the mitigation strategies need to be focused on these antibiotics. Continuous monitoring of the antibiotic residues in farmed fish and shellfish provides vital information on the quantity of antibiotic consumption and also the class of antibiotics. This data is pertinent to take corrective actions so as to ensure the food safety of fish and shellfish produced through aquaculture. In quality-driven aquaculture, it is pertinent to produce safe fish for human consumption and in an economically sustainable manner for the farmer as well as the environment. Research on alternatives to antimicrobials, such as antimicrobial peptides, bacteriophages or vaccines, needs to be promoted as they could help reduce the use of antimicrobials in aquaculture and their associated adverse effects.

References

Abe K, Nomura N, Suzuki S (2020) Biofilms: hot spots of horizontal gene transfer (HGT) in aquatic environments, with a focus on a new HGT mechanism. FEMS Microbiol Ecol 96(5):fiaa031

Amarasiri M, Sano D, Suzuki S (2020) Understanding human health risks caused by antibiotic resistant bacteria (ARB) and antibiotic resistance genes (ARG) in water environments: current knowledge and questions to be answered. Crit Rev Environ Sci Technol 50(19):2016–2059

Badireddy MR, Vaiyapuri M, Mothadaka MP, Chandragiri RN (2023) Status of AMR in food sector: implications for food safety and food security with special reference to fisheries. In: Mothadaka MP, Vaiyapuri M, Rao Badireddy M, Nagarajrao Ravishankar C, Bhatia R, Jena J (eds) Handbook on antimicrobial resistance. Springer, Singapore, pp 95–130. https://doi.org/10.1007/978-981-16-9723-4_7-1

Bello-López JM, Cabrero-Martínez OA, Ibáñez-Cervantes G, Hernández-Cortez C, Pelcastre-Rodríguez LI, Gonzalez-Avila LU, Castro-Escarpulli G (2019) Horizontal gene transfer and its association with antibiotic resistance in the genus *Aeromonas* spp. Microorganisms 7(9):363

Biller SJ, Schubotz F, Roggensack SE, Thompson AW, Summons RE, Chisholm SW (2014) Bacterial vesicles in marine ecosystems. Science 343(6167):183–186

Cabello FC, Godfrey HP, Tomova A, Ivanova L, Dölz H, Millanao A, Buschmann AH (2013) Antimicrobial use in aquaculture re-examined: its relevance to antimicrobial resistance and to animal and human health. Environ Microbiol 15(7):1917–1942

Chokmangmeepisarn P, Thangsunan P, Kayansamruaj P, Rodkhum C (2021) Resistome characterization of *Flavobacterium columnare* isolated from freshwater cultured Asian sea bass (*Lates calcarifer*) revealed diversity of quinolone resistance associated genes. Aquaculture 544:737149

Dhar AK, Manna SK, Thomas Allnutt FC (2014) Viral vaccines for farmed finfish. Virusdisease 25(1):1–17

Domingues S, Nielsen KM (2017) Membrane vesicles and horizontal gene transfer in prokaryotes. Cur Opin Microbiol 38:16–21

Dubey GP, Mohan GBM, Dubrovsky A, Amen T, Tsipshtein S, Rouvinski A, Rosenberg A, Kaganovich D, Sherman E, Medalia O, Ben-Yehuda, S (2016) Architecture and characteristics of bacterial nanotubes. Dev Cell 36(4):453–461

Dubey GP, Ben-Yehuda S (2011) Intercellular nanotubes mediate bacterial communication. Cell 144(4):590-600

EIC (2002) Compendium of Fish and Fishery Products Export Inspection Council, (Ministry of Commerce and Industry, Govt. of India) S.O. 729 (E) dated 21st August 1995; subsequently amended vide No. Order S.O.722 (E) dated 10th July, 2002

EU (2010) European Union Commission regulation No 37/2010 of 22 December 2009 on pharmacologically active substances and their classification regarding maximum residue limits in foodstuffs of animal origin. Official Journal of the European Union L15:1–72

FAO (2017) Antimicrobial resistance (AMR) in aquaculture, Committee on Fisheries, Subcommittee on aquaculture, Ninth Session, Food and Agriculture Organization of the United Nations, Rome, 24–27 October 2017, COFI:AQ/IX/2017/SBD.11

FAO (2024) The State of World Fisheries and Aquaculture 2024—Blue Transformation in action. Rome. https://doi.org/10.4060/cd0683enFSSAI 2011. Food safety and standards (contaminants, toxins and residues) regulations, 2011 (version 3, 31/08/2018), Ministry of Health and Family Welfare, Govt. of India

FSSAI (2011) Food Safety and Standards (Contaminants, toxins and Residues) Regulations, 2011 (version 3, 31/08/2018), Ministry of Health and Family Welfare, Government of India

GOI (2012) Drugs and cosmetics act, 1940 and the drugs and cosmetic rules, 1945 (rule 97) The Gazette of India Extraordinary, Part II, Section 3, dated 17 Jan, 2012

GOI (2019) The Gazette of India: extraordinary, Part II, Sec 3 (ii) No 2377, Ministry of Health and Family Welfare, Government of India, Notification dated 9 July 2019

Henríquez-Núñez H, Evrard O, Kronvall G, Avendaño-Herrera R (2012) Antimicrobial susceptibility and plasmid profiles of Flavobacterium psychrophilum strains isolated in Chile. Aquaculture 354:38–44

Li C, Sapugahawatte DN, Yang Y, Wong KT, Lo NWS, Ip M (2020) Multidrug-resistant Streptococcus agalactiae strains found in human and fish with high penicillin and Cefotaxime non-susceptibilities. Microorganisms 8(7):1055

Liu X, Lv K, Deng C, Zhimin Y, Shi J, Johnson AC (2019) Persistence and migration of tetracycline, sulfonamide, fluoroquinolone, and macrolide antibiotics in streams using a simulated hydrodynamic system. Environ Pollut 252:1532–1538

Loch TP, Hawke JP, Reichley SR, Faisal M, Del Piero F, Griffin MJ (2017) Outbreaks of edwardsiellosis caused by *Edwardsiella piscicida* and *Edwardsiella tarda* in farmed barramundi (*Lates calcarifer*). Aquaculture 481:202–210

Miller RA, Harbottle H (2018) Antimicrobial drug resistance in fish pathogens. Microbiol Spectr 6(1):6–1. https://doi.org/10.1128/microbiolspec.arba-0017-2017

Neela FA, Nonaka L, Rahman MH, Suzuki S (2009) Transfer of the chromosomally encoded tetracycline resistance gene tet (M) from marine bacteria to *Escherichia coli* and *Enterococcus faecalis*. World J Microbiol Biotechnol 25(6):1095–1101

Nguyen TT, Thu HT, Nguyen M-AT, Byadgi O, Wang P-C, Yoshida T, Chen S-C (2017) Genetic diversity, virulence genes, and antimicrobial resistance of *Streptococcus dysgalactiae* isolates from different aquatic animal sources. Aquaculture 479:256–264

Preena PG, Dharmaratnam A, Kumar VJR, Swaminathan TR (2021) Plasmid-mediated antimicrobial resistance in motile aeromonads from diseased Nile tilapia (*Oreochromis niloticus*). Aquac Res 52(1):237–248

Rao BM, Prasad MM (2015) Residues of veterinary medicinal products (antibiotics) in shrimp exported from India to the European Union (EU). Fishing Chimes 34:6–11

Rathore G, Lal KK, Bhatia R, Jena JK (2020) INFAAR–A research platform for accelerating laboratory-based surveillance of antimicrobial resistance in fisheries and aquaculture in India. Curr Sci 119(12):1884–1885

Rigos G, Nengas I, Tyrpenou AE, Alexis M, Troisi GM (2003) Pharmacokinetics and bioavailability of oxytetracycline in gilthead sea bream (Sparus aurata) after a single dose. Aquaculture 221(1-4):75–83

Rigos G, Nengas I, Alexis AEM, Athanassopoulou F, Troisi GM (2004) Poor bioavailability of oxytetracycline in sharpsnout sea bream Diplodus puntazzo. Aquaculture 235(1-4):489–497

Rigos G, Kogiannou D, Padrós F, Cristofol C, Florio D, Fioravanti M, Zarza C (2021) Best therapeutic practices for the use of antibacterial agents in finfish aquaculture: a particular view on European seabass (*Dicentrarchus labrax*) and gilthead seabream (*Sparus aurata*) in Mediterranean aquaculture. Rev Aquac 13(3):1285–1323

Santos L, Ramos F (2018) Antimicrobial resistance in aquaculture: current knowledge and alternatives to tackle the problem. Int J Antimicrob Agents 52(2):135–143

Simner PJ, and Miller L (2018) Understanding Pharmacokinetics (PK) and Pharmacodynamics (PD). CLSI AST News Update, Volume 3, Issue 1

Schar D, Klien EY, Laxminarayan R, Gilbert M, van Boeckel TP (2020) Global trends in antimicrobial use in aquaculture. Sci Rep 10:21878. https://doi.org/10.1038/s41598-020-78849-3

Toyofuku M, Tashiro Y, Hasegawa Y, Kurosawa M, Nomura N (2015) Bacterial membrane vesicles, an overlooked environmental colloid: biology, environmental perspectives and applications. Adv Colloid Interface Sci 226:65–77

Toyofuku M, Nomura N, Eberl L (2019) Types and origins of bacterial membrane vesicles. Nat Rev Microbiol 17(1):13–24

Vaiyapuri M, Pailla S, Badireddy MR, Pillai D, Nagarajarao RC, Mothadaka MP (2021) Antimicrobial resistance in Vibrios of shrimp aquaculture: incidence, identification schemes, drivers and mitigation measures. Aquac Res 52:2923. https://doi.org/10.1111/are.15142

Watts JEM, Schreier HJ, Lanska L, Hale MS (2017) The rising tide of antimicrobial resistance in aquaculture: sources, sinks and solutions. Mar Drugs 15:158

Application of Probiotics, Immunostimulants and Bacteriophages in Aquaculture

Subhendu Kumar Otta and Indrani Karunasagar

Abstract

Aquaculture practice is progressing at a rapid pace and has already overtaken capture fisheries to meet the demand of the growing population. Disease has been the major constraint on the way to progress in this rapidly growing industry. Increasing stocking density has been the main focus for the overall increase of production and making aquaculture practice more profitable. Thereby, this industry often faces disease problems. However, the use of antibiotics and other chemicals for preventing or controlling the disease is not encouraged as it deteriorates the environment and develops drug-resistant strains. Biological agents such as probiotics, immunostimulants and bacteriophages have been used successfully to control different microbial diseases in aquaculture systems. All these products have been found to be safe in many aspects including the environment and human consumption. While controlling diseases in aquatic animals, these products also prevent the spread of pathogens to human beings and livestock. Research in many of these areas is still at the infant stage, and more effort is required to make these applications safe as well as economical. There are no strict distinctions between all these biologicals as far as their function is considered. All three biologicals mentioned can act as immunostimulants as they possess immunomodulatory properties. The limitations in the use of these products should be considered carefully, and effort should be made along those lines to make these products more suitable for the successful aquaculture practice.

S. K. Otta (✉)
ICAR-Central Institute of Brackishwater Aquaculture, Chennai, India
e-mail: ottasubhendu@gmail.com

I. Karunasagar
FAO Ref Centre for Antimicrobial Resistance, Nitte University, Mangalore, India
e-mail: indrani.karunasagar@nitte.edu.in

R. Kooloth Valappil et al. (eds.), *Aquatic Animal Health Management*,
https://doi.org/10.1007/978-981-96-7987-4_22

Keywords

Probiotics · Immunostimulant · Bacteriophage · Aquaculture · Growth · Immunity

1 Introduction

The role of fisheries in food security and livelihood, as well as a part of income generation and societal development, is well recognised (FAO 2016). As a result of this, aquaculture has expanded to several folds with intensification and several scientific interventions. Particularly, the practice of shrimp farming has been a normal activity in many parts of the world. This industry has been progressing at a rapid pace due to high profit value within a short period of time. Annual increases in production of almost three million tonnes of shrimp have been recorded during the last two decades (FAO 2019). The demand for fish as a cheap source of animal protein is continuously increasing. Production from aquaculture is almost completely utilised for human consumption (FAO 2016) and, thus, meets the protein requirement of the increasing population to a major extent. Aquaculture is thus considered the fastest-growing food production technology, accounting for more biomass than some of the animal sectors (Edwards et al. 2019).

An increase in stocking density per unit area to get more production, the use of high-quality artificial feed to have better growth and the application of several chemicals and biologicals to tackle disease problems are some of the major components of present-day aquaculture intensification. As far as present-day aquaculture is concerned, the concepts of environmentally responsible and sustainable aquaculture are no longer a topic of debate (Engle and D'Abramo 2018). Those concepts have been suitably modified with an eye to achieving better production. From a comparatively high protein diet (crude protein 30–40%) used for aquaculture, about 75% are just directly released into the aquatic environment as organic nitrogen and ammonia (Boyd and Tucker 1998; Ebeling et al. 2006), and additionally, faeces and other dead materials such as planktons also decompose to add extra toxic substances which all are harmful to aquaculture practice (Valencia-Castañeda et al. 2018; Zhou and Boyd 2015). Additionally, there is a long list of chemicals and biologicals that are being added to the aquatic environments either to improve water quality or manage disease problems (Rico et al. 2012), and all these possibly may further deteriorate the water quality if not used properly.

Disease is one of the major constraints that can severely damage the aquaculture industry. As per the investigation, almost 50% of loss in aquaculture owes to disease outbreaks, and most of these are in high-intensity culture practices (Assefa and Abunna 2018). All kinds of microorganisms have been reported to be responsible for some form of disease in aquatic organisms. Viruses, bacteria and microsporidia continue to threaten the shrimp aquaculture industry worldwide (Thitamadee et al. 2016). Recently, similar loss estimation has been reported to be huge in the Indian shrimp farming industry (Patil et al. 2020). Emerging pathogens are a major concern as they can cause huge losses before these are characterised for any action to be taken.

Whether it is the improvement of the environment or the health of aquatic animals, the use of chemicals and antibiotics is not advocated. Following the concept of 'prevention is better than cure', wherever possible, useful biologicals should be considered instead of any chemicals. Though vaccines are preferred for the prevention of diseases, it is more specific for the control of specific diseases and for some lower animals like invertebrates, it is difficult to develop any vaccine as they lack memory. Similarly, though antibiotics are useful in controlling a number of bacterial diseases, their use has been restricted. Several restrictions and regulations have been put down for the use of antimicrobials because of the development of drug-resistant strains (Thornber et al. 2020). Any single approach will definitely not be helpful in managing the disease problems, and therefore, multiple approaches that do not affect either the environment or the consumers should be thought about. For a long time, different biologicals such as probiotics, immunostimulants and phage have been found to be useful either for general health improvement, disease prevention or both. However, improper use of these can either be harmful to both the environment and fish growth or incur unnecessary expenditure. This chapter aims to cover some of their application aspects and summarises the latest developments in the field.

2 Probiotics in Aquaculture

Originating from the Greek words 'Pro' and 'Bios', meaning 'for life', probiotics are generally known as microbial derivatives, which are normally supplied through feed to improve the intestinal microbiota and confer resistance to the host through immunomodulation. The first attempt to provide 'probiotics' a proper definition was by Parker (1974), where it was mentioned as organisms and substances that affect the microbial population of the intestine. Subsequently, this definition has been modified in a number of ways by various authors (Sharifuzzaman and Austin 2017). Considering the special status of the aquatic environment, probiotics were again defined as 'live, dead or even component of a microbial cell, which when administered either to feed or to the environment, benefit the host by improving health status, stress response or general vigour, growth performance, feed utilisation and these are achieved at least in part via improving the microbial balance of either the host or the ambient environment' (Merrifield et al. 2010).

The use of probiotics has been widespread in aquaculture practices throughout the world. Different bacterial species, microalgae and yeast either as single species or in combination, have been tried and many such products are also commercially available (Hoseinifar et al. 2018).

2.1 Source of Probiotics for Aquaculture Use

It is better to isolate a probiotic strain from the host itself or at least from the environment in which it is cultured. This will help these microorganisms to adapt, survive, grow well and physiologically function at optimal levels. Thus, probiotic strains collected from the host, water or sediment of the culture environment can be a good source (Lauzon et al. 2010; Sharifuzzaman et al. 2014; Lazado et al. 2015). However, many of the strains from terrestrial environments, particularly lactic acid bacteria, which have already proved to be safe and useful for humans and livestock, are being used successfully for aquaculture practice also. Similarly, strains collected from the gut of healthy terrestrial animals can also be used for aquaculture practice, provided they prove to be effective. However, their efficacy in a different environment, like a marine ecosystem, may not be as expected and, therefore, should initially be verified (Vazquez et al. 2003; Ren et al. 2013).

2.2 Selection of Probiotics

There are several basic criteria for the selection of probiotics. As has been mentioned earlier, it is preferred to be from a similar environment as the host. The prime aim of the use of probiotics is to have a beneficial effect. In no circumstances, the organism should be pathogenic either to the host or other living organisms of the environment. It should also not possess any virulence or antibiotic-resistance genes. Similarly, the use of probiotics should not have any toxic effect on the host or ecosystem. The probiotic organism should have the desired property of getting attached to the appropriate target location and function as expected (Veld et al. 1994).

The development of a final product involves a number of steps and a multidisciplinary approach; however, initial basic research is required to characterise the strains. The next step is to conduct small-scale lab experiments to prove the real effect. This is followed by large-scale application to repeat the effects. Once this is done, the economy is calculated to find out the possibility of developing it into a suitable product (Nikapitiya 2013). Outlines of the overall process are depicted in Fig. 1.

2.3 Probiotic Organisms Used in Aquaculture Practice

A wide range of organisms have offered promise to be used as probiotics in aquaculture practices. The diversity of these organisms is found to be wider than that of human or livestock use. Similarly, the source of these organisms is also very diverse as both aquatic and terrestrial environments have been exploited for selection. Organisms such as Gram-positive bacteria, Gram-negative bacteria, yeast and algae have been effectively used in aquaculture as probiotics. A list of such organisms is provided in Table 1.

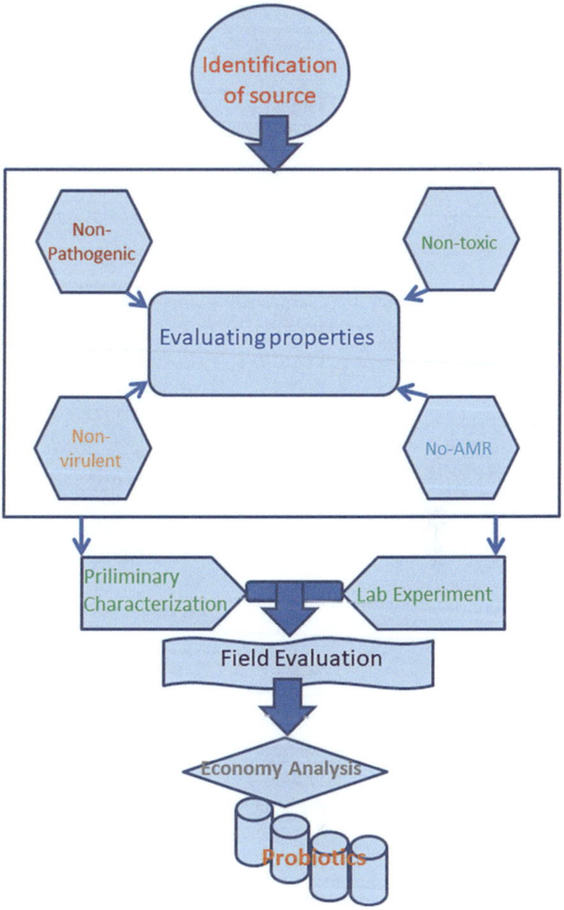

Fig. 1 Outline of the development of probiotics

2.4 Modes of Application of Probiotics in an Aquaculture System

An aquatic ecosystem is a complex environment that deals with several variabilities. Some of these variables are very critical and must be taken into consideration while standardising the application of any inputs in this system. While temperature, salinity and pH are some of the critical environmental factors, the type and stage of the host, stocking density, etc., are host factors that need consideration.

The use of accurate doses is very important for the probiotics to be effective. The dose of probiotics is generally assessed by number. A minimum specific number and above that only becomes effective in providing beneficial effects. Like the lower dose, which becomes ineffective, overdosing may also have adverse effects. Therefore, estimating and determining the accurate dose is an important aspect of the successful application of probiotics (Vine et al. 2006).

Table 1 Probiotic species used in aquaculture

Probiotic	Probiotic type	Used against pathogen	Aquatic organism
Aeromonas hydrophila	Gram –ve bacteria	*Aeromonas salmonicida*	Rainbow trout, goldfish
Aeromonas sobria	Gram –ve bacteria	*Aeromonas bestiarum, Lactobacillus gravieae, Streptococcus iniae*	Rainbow trout
Aeromonas veronii	Gram –ve bacteria	*A. hydrophila*	Common carp
Arthrobacter sp.	Gram +ve bacteria	*Vibrio parahaemolyticus*	Shrimp
Alteromonas macleodii, Neptunomonas sp.	Gram –ve bacteria	*Vibrio splendidus, V. coralliiyticus, V. pectenicida*	Mussel, scallop, oyster
Bacillus amyloliquefaciens	Gram +ve bacteria	*Yersinia ruckeri, Clostridium perfringens type D*	Tilapia, Nile tilapia
Bacillus circulans	Gram +ve bacteria	*A. hydrophila*	*Catla catla*
Bacillus subtilis	Gram +ve bacteria	*A. hydrophila, Aeromonas* sp., *Edwardsiella ictaluri, Streptococcus agalactiae, Vibrio harveyi*	Indian major carp, rainbow trout, channel catfish, striped catfish, shrimp
Bacillus pumilus	Gram +ve bacteria	*A. hydrophila*	Tilapia
Bacillus licheniformis	Gram +ve bacteria	*Streptococcus iniae*	Tilapia
Bacillus subtilis, B. licheniformis	Gram +ve bacteria	*Y. ruckeri, Streptococcus iniae*	Trout, olive flounder
Bacillus sp.	Gram +ve bacteria	*V. harveyi*	Shrimp
Brochothrix thermosphacta	Gram +ve bacteria	*A. bestiarum, V. anguillarum, V. ordalli*	Rainbow trout
Carnobacteria inhibens	Gram +ve bacteria	*A. salmonicida, V. ordalii, Y. ruckeri*	Salmon, trout
C. divergens	Gram +ve bacteria	*V. anguillarum*	Cod
Clostridium butyricum	Gram +ve bacteria	*Vibriosis*	Rainbow trout, Chinese drum
Debaryomyces hansenii	Yeast	*A. hydrophila*	Grouper
Enterobacter amnigenus, Enterobacter sp.	Gram –ve bacteria	*Flavobacterium psychrophilum*	Rainbow trout
E. cloacae	Gram –ve bacteria	*Y. ruckeri*	Rainbow trout
Enterococcus faecium	Gram +ve bacteria	*E. tarda*	Eel

(continued)

Table 1 (continued)

Probiotic	Probiotic type	Used against pathogen	Aquatic organism
E. casseiflavus	Gram +ve bacteria	*S. iniae*	Rainbow trout
E. gallinarum	Gram +ve bacteria	*V. anguillarum*	Sea bass
E. faecalis	Gram +ve bacteria	*A. salmonicida*	Rainbow trout
E. faecium, L. gravieae	Gram +ve bacteria	*V. harveyi, V. parahaemolyticus*	Shrimp
E. faecium, L. pentosus	Gram +ve bacteria	*V. parahaemolyticus*	Shrimp
Flavobacterium sasangense	Gram –ve bacteria	*A. hydrophila*	Common carp
Kocuria sp.	Gram +ve bacteria	*V. anguillarum* and *V. ordalii*	Rainbow trout
Lactobacillus acidophilus	Gram +ve bacteria	*Pseudomonas fluorescens, S. iniae, Staphylococcus xylosus, A. hydrophila, S. agalactiae, V. Alginilyticus*	Nile tilapia, cat fish, shrimp
L. acidophilus, L. bulgaricus, S. cremoris	Gram +ve bacteria	*V. alginolyticus*	Shrimp
L. brevis	Gram +ve bacteria	*A. hydrophila*	Tilapia
L. lactis	Gram +ve bacteria	*S. iniae*	Olive flounder
L. pentosus	Gram +ve bacteria	*E. tards*	Eel
L. plantarum	Gram +ve bacteria	*Lactococcus garvieae, V. alginolyticus, V. harveyi*	Rainbow trout, shrimp
L. rhamnosus	Gram +ve bacteria	*A. salmonicida, E. tarda*	Rainbow trout, tilapia
L. sakei	Gram +ve bacteria	*E. tarda*	Rock bream
Leuconostoc mesenteroides	Gram +ve bacteria	*Furunculosis, A. salmonicida*	Rainbow trout, brown trout
Micrococcus luteus	Gram +ve bacteria	*A. hydrophila, A. salmonicida*	Nile tilapia, rainbow trout
Pediococcus acidilactici	Gram +ve bacteria	*Vertebral column compression syndrome, V. nigripulchritudo*	Rainbow trout, shrimp
P. pentosaceus	Gram +ve bacteria	*V. anguillarum*	Grouper
Phaeobacter gallaeciensis	Gram –ve bacteria	*V. anguillarum*	Cod
Phaeobacter gallaeciensis, Pseudoalteromonas	Gram –ve bacteria	*V. coralliiyticus* and *V. splendidus*	Scallop
Phaffa rhodozyma	Yeast	*Vibriosis*	Shrimp

(continued)

Table 1 (continued)

Probiotic	Probiotic type	Used against pathogen	Aquatic organism
Pseudomonas aeruginosa	Gram –ve bacteria	*A. hydrophila, V. parahaemolyticus, V. harveyi*	Rohu, zebra fish, shrimp
P. chlororaphis	Gram –ve bacteria	*A. sobria*	Perch
P. fluorescens	Gram –ve bacteria	*V. anguillarum*	Rainbow trout
Pseudomonas sp.	Gram –ve bacteria	*F. psychrophilum, V. harveyi*	Rainbow trout, shrimp
Rhodococcus sp.	Gram +ve bacteria	*V. anguillarum*	Rainbow trout
Roseobacter sp.	Gram –ve bacteria	*V. anguillarum*	Turbot
Saccharomyces cerevisiae	Yeast	*A. hydrophila, Y. ruckeri, Vibriosis*	Tilapia, rainbow trout, shrimp
Shewanella putrefaciens	Gram –ve bacteria	*V. anguillarum, Photobacterium damsel*	Seabream, sole
S. xiamenensis	Gram –ve bacteria	*A. hydrophila*	Grass carp
Streptococcus phocae	Gram +ve bacteria	*V. harveyi*	Shrimp
Streptomyces sp.	Gram –ve bacteria	*V. harveyi*	Shrimp
Vibrio alginolyticus	Gram –ve bacteria	*A. salmonicida*	Salmon
Zooshikella sp.	Gram +ve bacteria	*S. inane*	Olive flounder

For detailed information, refer to Sharifuzzaman and Austin (2017) and Hoseinifar et al. (2018)

There can be several modes of application for probiotics depending on aquatic animal stock and specific requirements. Immersion application can also have a positive effect (Feliatra et al. 2018), but it is tedious to apply. Therefore, it needs to be restricted only to specific requirements. Though this will be tedious, saving valuable stock or maintaining the economy becomes more important and inevitable in adopting this method. Otherwise, it is a general practice to add the probiotics directly to the rearing water in case of large aquaculture practices, though a large amount of probiotics will be required for this. In other cases, the probiotics are either made as an ingredient in feed preparation or coated to already prepared feed and then supplied to aquatic animals. For a more valuable and effective application, probiotics can be bioencapsulated with live feeds such as artemia or rotifer and then supplied to larvae (Sharifuzzaman and Austin 2017).

2.5 Beneficial Effect of Probiotics in Aquaculture Practice

Many probiotic bacteria produce lots of natural metabolites that are either toxic to kill competitive organisms or inhibit their action through some mechanism. These antagonistic products, either in the gut or surface of the host or in water, will help in preventing the proliferation of pathogenic microorganisms (Lara-Flores 2011). In addition to antibacterial, probiotics have also been shown to be antiviral and anti-parasitic. Though the exact mechanism of such action is not known, it is speculated to be due to a similar antagonistic effect as against bacteria (Nikapitiya 2013). Probiotic organisms contain different nutrients and vitamins and also secrete different enzymes such as lipase, protease, chitinase and amylase. All these together help to improve feed efficiency and digestion of food thereby ultimately improving the growth (Lara-Flores et al. 2010; Sharifuzzaman and Austin 2017). The cell wall of probiotic organisms also contain immunomodulators such as lipopolysaccharide, peptidoglycan or beta-glucan and thereby enhance the immune system, particularly the non-specific immunity (Mohapatra et al. 2013; Austin and Austin 2012) and thereby provide resistance against many pathogens. Competitive exclusion is another aspect through which the probiotic organisms compete with other pathogens or harmful microorganisms to settle in the gut of aquatic animals and thereby reduce disease occurrence (Knipe et al. 2020).

3 Use of Immunostimulants in Aquaculture Practice

Immunostimulants are defined as substances that can stimulate the immune system, particularly the innate or non-specific one, through direct interaction with certain cells associated with it. This has been one of the most important areas in health care, livestock and aquaculture. Lots of applied medical research is being continuously carried out in this area for further improvement. The use of immunostimulants alone or along with others can increase the immune system of aquatic animals and thus prevent infectious diseases (Devaraja et al. 1998; Sahoo 2007). The use of immunostimulants can strengthen the immune system in general and thereby increase the resistance of aquatic animals to tolerate a wide range of stresses such as sampling, handling, transport, crowding, vaccination and reproduction-related procedures. There is no clear-cut distinction between different biologicals, and even probiotics or prebiotics can also be used as immunostimulants. A wide range of biologicals and their derivatives and many synthetic substances have been proven to have immunostimulant properties (Song et al. 2014; Mohapatra et al. 2013) that have been used in aquaculture practice. Even nucleic acids such as double-stranded RNA can stimulate the non-specific immune system and thus be considered immunostimulants (Taju et al. 2015). Earlier, it was also suggested that the immunostimulants be classified based on the source of their origin (Sakai 1999). Sometimes the nutrition provided to the organisms is also considered immunostimulant. Particularly many of the vitamins and micronutrients present in the diet stimulate the immune system. However, these substances just act as a substrate, and whether to include these under immunostimulant is under controversy (Suchner et al. 2000). A large number of commercial products (Table 2) have been developed in several names, claiming to be immunostimulants, and are also being widely used in aquaculture practices (Barman et al. 2013).

The earlier attempt to define immunostimulants was solely based on their activity in the mononuclear phagocytic system (Seljelid 1990). However, based on the later development of research, it was understood that the pattern recognition receptors (PPRs) in different leucocytes are different, and the immune response mainly depends on the type of ligands that bind to different PPRs. Based on this, the definition of immunostimulants was further modified as 'An immunostimulant is a naturally occurring compound that modulates the immune system by increasing the host's resistance against diseases that in most circumstances are caused by pathogens' (Bricknell and Dalmo 2005).

Table 2 List of commercially available immunostimulants which have been used for aquatic animals

Commercial name of the immunostimulant	Company	Active ingredient in the product
Agrimos	Agrimerica, USA Santel, France	Oligosaccharide
Aqua-Mune	Park Tonks	Glucan
AQUAVAC® Ergosan™	Merck Animal Health	Algins and polysaccharide
Bio-Mos®	Alltech	Outer cell wall of yeast
BZT® PRE GE	UnitedTech	Glucan and oligosaccharide
Calcium Spirulan	Kelly Moorhead, Hawaii	Spirulina polysaccharide
DS 1999	International Aquaculture Biotechnologies Ltd.	Bacterin
Elorisan	BUGICO, Switzerland	Organic silicon
EcoActiva™	Ecostadt	Glucan
Immustim	Immundyne, USA	Glucan
Laminam	Pronova, Norway	Glucan
Lactoferrin	DMV International, The Netherland	Lactoferrin
Levamisole	Janssen Pharmaceutical, Belgium	Tetrahydro-6-Phenylimidazolthiazole hydrochloride
Lysozyme hydrochloride	Belove, Belgium	Lysozyme
Macroguard	Biotec-Mackzymal, Norway	Glucan
Penstim	AURUM Aquaculture Ltd., USA	Glucan
Sanoguard®	INVE Aquaculture Health	Immunostimulants, vitamins, nucleotides
SP 604	Alltech Inc., USA	Oligosaccharide, yeast, probiotics
Vannagen®	Chemoforma Ltd.	Nucleotides
Vitastim	Taito Co ltd, Japan	Glucan

For details, refer Barman et al. (2013) and Apines-Amar and Amar (2015)

3.1 Mode of Application

A number of methods have been used for the administration of immunostimulants to aquatic animals. Most of the laboratory-based experiments involve injection and immersion methods. Though the injection method requires a very small quantity of immunostimulants and proves to be the most effective, it is one of the most expensive methods and has serious limitations for mass application in aquaculture practice. Additionally, it requires large size of animals to handle and administer. Some of the most valuable stocks, such as broodstocks, can be immunised through this process. The immersion method of administration is also practically difficult to apply in aquaculture practice. This requires lots of standardisation based on the number of animals and the duration of possible treatment. Here, larvae or broodstocks in limited numbers can be used for the administration of immunostimulants. The supply of immunostimulants as part of feed through oral administration is one of the most practical methods for general aquaculture practice. Though it is possible in most of the aquaculture practices, this is difficult, when the animals do not accept artificial feed.

3.2 Types of Immunostimulants

As mentioned earlier, large variants of immunostimulants are being used for aquaculture practice both for finfish and shellfish. All these belong to wider groups of complex carbohydrates, herbal products, nutrients, microorganisms and antimicrobial peptides.

Complex carbohydrates include a large number of polysaccharides and oligosaccharides. Glucan, in the form of ß-glucan, has been the most widely used immunostimulant in aquaculture, both for finfish and shellfish. Apart from providing disease resistance through immunostimulation, it also increases growth for the animals (Dalmo and Bogwald 2008). Peptidoglycans, which are subunits within the cell wall of bacteria, are also considered suitable immunostimulants and like ß-glucans, these also provide better growth in addition to protection from diseases (Zhou et al. 2006). Chitosan, which is found in the shell of different shellfishes, is an alkaline polysaccharide, and has been found to be useful as a growth promoter, immunity enhancer and water purifier (Wang et al. 2017). A variety of prebiotics such as fructooligosaccharide, mannan oligosaccharide, inulin, etc., which are generally considered immunosaccharides, have been used for aquaculture, and different modes of action have been described (Song et al. 2014).

Many terrestrial plants and aquatic plants, particularly seaweeds, contain various bioactive compounds and extracts of such plants, which have proved to be immunostimulants, growth promoters, anti-stress agents and antimicrobials (Immanuel et al. 2004). Similarly, different Chinese and many other herbs have been used as immunostimulants and found to be useful (Apines-Amar and Amar 2015; Vallejos-Vidal et al. 2017).

Different nutrient components such as vitamins (such as A, C and E), carotenoids and trace elements (Zinc, Copper, Manganese and Selenium) have been shown to stimulate the immune system and provide disease resistance. Application of vitamins along with carotenoids could enhance several immune parameters (Amar et al. 2001). Deficiency of trace elements was found to reduce the activity of several cells associated with the immune system, indicating their role in immune enhancement (Apines-Amar et al. 2004).

Different hormones or hormone components have also been found to have an effect on the immune system of aquatic animals. Growth hormone, either directly or in recombinant form, has been proven to have immunostimulant activity apart from various other beneficial effects and protect aquatic animals against diseases (Kang et al. 2005; Yada 2007). Similarly, incorporation of lactoferrin in feed has been reported to enhance immune indices and provide protection against pathogens (Chand et al. 2006).

Evidence of different synthetic products, such as levamisole and CpG-oligodeoxynucleotide, having immunomodulatory effects has been reported (Li and Gatlin III 2006; Huxley et al. 2011).

Vibrio bacterin in inactivated form has been shown to have an immunomodulatory effect and provides protection against pathogen challenge (Devaraja et al. 1998). As has been discussed earlier, different live bacteria, such as *Bacillus* sp. and *Lactobacillus* sp., which are generally used as probiotics, can also have immunomodulatory effects on aquatic animals.

3.3 Beneficial Effects of Immunostimulant

In general, immunostimulants have been used to enhance fish health status and thereby increase production. Enhanced activity of phagocytic cells, natural killer cells, complement activity, lysozyme activity and antibody response increases the overall disease resistance capacity and thereby provides protection from infection. In the case of higher vertebrates like fish, it can enhance both specific and non-specific defence systems to provide better protection. Similarly, in the case of lower vertebrates like shrimp, it stimulates the non-specific immune system and increases the disease resistance capacity. Immunostimulants are also known to increase the efficacy of vaccines. Immunostimulants can improve growth and enhance survivability under stressful conditions.

4 Use of Bacteriophage in Aquaculture Practice

The use of bacteriophages, or in simple terms, phages, has a long history in treating bacterial diseases. Subsequently, it has made its way into aquaculture practice and provides great hope as a biological control of a number of bacterial diseases. This is particularly important at a point where the use of antibiotics is strictly discouraged because of the development of drug-resistant strains. Therefore, this therapy has been popular in all types of medical uses, such as in humans, livestock and aquaculture. These viruses are very specific to the host and, therefore, safe for use in the aquatic environment as they do not disturb the natural biodiversity, and only the pathogens can be targeted. Among the two types of phage, temperate and lytic, the former is usually avoided as there is the possibility of antimicrobial and virulence gene transfer through the horizontal gene transfer process, and therefore, lytic phages are the preferred ones for safe use. Phages belonging to different families, such as myoviridae, podoviridae and siphoviridae, have been used for aquaculture practice (Kowalska et al. 2020).

4.1 Phage Therapy Opportunities in Aquaculture

Because of self-replicating and self-limiting entities, bacteriophages are considered safe for the environment and, hence, suitable for use in an aquatic environment (Kowalska et al. 2020). This bacteria-eating virus acts directly only on pathogens and, thereby, protects the crop and, at the same time, prevents the spread of pathogens to human beings (Sharma et al. 2016). As bacteriophages are very specific to their host, the use of a phage cocktail is preferable to target multiple pathogens, and with the discovery of more numbers of phages, this has now been possible (Cooper et al. 2016). Since most of the reported aquaculture researches (Table 3) indicate a positive effect of phage therapy as expected, it appears to have an encouraging future (Culot et al. 2019; Kowalska et al. 2020). However, phage therapy is at the very beginning phase in aquaculture, and much more research is still required regarding its suitability (Jassim and Limoges 2014).

4.2 Important Steps in Developing a Phage Therapy

The first and foremost step is to isolate an appropriate phage based on the target pathogen. Though phages can be isolated from different environments, natural aquaculture environments such as farms, hatcheries, rivers, reservoirs, estuaries, etc. should be preferred. Once the phage is isolated through the standard protocol, it needs to be propagated and characterised by various techniques. The therapeutic efficiency of the phage should be verified initially in laboratory-based experiments and subsequently directly in the field, such as a farm or hatchery. Care should be taken to see that the isolated phage intended to be used in aquatic environments should not contain any virulence genes, antimicrobial resistance genes or any other toxic genes. As has already been mentioned, lytic phage should be selected for therapeutic use. Phage can be directly added to the control environment, treated by immersion, administered through injection or orally through feed and anal intubation. Exact dose determination is important and low concentration, or low multiplicity of infection (MOI) is preferred to make it economical (Gon Choudhury et al. 2017; Culot et al. 2019).

Table 3 Examples of different phage therapy in aquaculture practice

Aquatic organisms	Pathogen	Phage, source	Mode of administration
Finfish			
Atlantic salmon	*Vibrio anguillarum* (VA)	CHOED, from bivalve	Directly added to water
Atlantic salmon	*Aeromonas salmonicida* (AS)	O, R and B, from fish farms and reservoirs	Injection, oral and immersion
Atlantic cod and turbot	VA	KVP40, from seawater	Directly added to egg
Zebrafish	VA	NA, from sewage	Directly added to water
Brook trout	AS	HER 110, from river	Directly added to water
African sharptooth catfish	*Pseudomonas aeruginosa*	NA, waste water	Direct swab to skin lesions
Ayu fish	*Pseudomonas plecoglossicida*	Myoviridae and Podoviridae, disease fish and pond water	Through feed
Rainbow trout	*Flavobacterium columnare*	FCL-2, fish farm	Directly added to water
Rainbow trout	*Aeromonas hydrophila* and *Pseudomonas fluorescens*	BAFADOR®, commercial	Immersion and feeding
European eel	*Aeromonas hydrophila* and *Pseudomonas fluorescens*	BAFADOR®, commercial	Immersion and feeding
Senegalese sole	AS	AS-A, sewage water	Directly added to water
Shellfish			
Abalone	*Vibrio harveyi* (VH)	Siphoviridae family, hatchery water and oyster tissue	Immersion for 2 hours
Artemia	*V. parahaemolyticus* (VP)	Vpms1, from clam	Directly added to water
Artemia	Different *Vibrio* sp.	φSt2 and φGrn1, coastal water	Directly added to water
Oyster meat	VP	Phage OMN, from seawater	Applied directly to oyster
Oyster	VP	Phage VPp1, sewage water	Directly added to water
Sea cucumber	*V. alginolyticus*	NA, from drain pipe of aquatic market	Directly added to water

(continued)

Table 3 (continued)

Aquatic organisms	Pathogen	Phage, source	Mode of administration
Sea cucumber	*V. splendidus*	Phages VS-1, PVS-2, PVS-3, from hatchery raw sewage	Through feeding
Sea cucumber	*V. cyclitrophicus*	Phage vB_VcyS_Vc1, from hatchery raw sewage	Feeding, immersion or injection
Shrimp larvae	VH	NA, from shrimp farm and seawater	Added directly to water
Shrimp larvae	VH	Viha10 and Viha8, from oyster tissue and shrimp hatchery	Added directly to water
Shrimp larvae	VH	VHM1 and Viha8, from sediments of shrimp and aquaculture environment	Directly added to water
Shrimp	VP	Phage pVp-1, from coastal water	Through pellet feed or immersion
Shrimp	VP	Phage A3S and Vpms1, from shrimp and clam culture	Directly added to water

For detailed information, readers are requested to follow Gon Choudhury et al. (2017), Culot et al. (2019), and Kowalska et al. (2020)

4.3 Benefits of Phage Therapy

Bacteriophages are abundant in the environment and, therefore, easy to find and inexpensive to isolate. Considering the aquatic environment, the safety and maintenance of biodiversity are the measure issues. As phage is isolated from a similar environment and very specific to the host, these are considered safe from both points of view as they do not interact with other natural flora. Phage is preferred over antibiotics as they can target the most difficult organisms, such as AMR and biofilm bacteria. Phage can also act as an immunostimulant and provide resistance against bacterial infections. Phage can multiply itself once the host is available, and therefore, excess or multiple applications are not necessary. Bacteriophages are also non-toxic to human beings, livestock and the environment. There is a possibility that phage can overcome the occurrence of AMR due to the indiscriminate use of antibiotics.

4.4 Limitations of Phage Therapy

As phage has a very narrow host range, infections due to multiple pathogens or systemic infections are difficult to treat. If the proper life cycle of phage is not accurately determined and used, a chance of virulence gene or toxic gene is possible. It is also possible that in case the phage changes its life cycle at a certain point, it may create a difficult situation. Clearance of phage by the immune system is a challenge, and it should be seriously thought about. If proper formulation is not done, phage may not reach the site of infection to do the desired effect. As phages are live organisms, these may be sensitive to various physical and chemical agents and, therefore, may not be stable. Proper regulatory guidelines should be developed regarding the use of phage therapy.

5 Conclusion

Through scientific innovations, aquaculture practice is expected to be modified in future to accommodate higher stocking density. Therefore, it is expected that many of the existing and emerging diseases will surely affect the progress and the overall production. As bacterial diseases are becoming major challenges, as can be seen in the case of acute hepatopancreatic necrosis disease (AHPND), safe control of such infections is necessary. Probiotics, immunostimulants and bacteriophages are considered biological agents as these are prepared from naturally existing organisms and thereby considered as safe for aquaculture practice. All three products have been found to be effective in controlling disease in aquatic animals, and therefore, attempts have been made to develop commercial products. However, still, there are many limitations in all these three areas to make it more effective for aquaculture practice. Advanced research in these areas is necessary for the bright future of aquaculture practice.

References

Amar EC, Kiron V, Satoh S, Watanabe T (2001) Influence of various dietary synthetic carotenoids on bio-defense mechanisms in rainbow trout, *Oncorhynchus mykiss* (Walbaum). Aquac Res 32:162–173

Apines-Amar MJS, Amar EC (2015) Use of immunostimulants in shrimp culture. In: Caipang CMA, Bacano-Maninggas MBI, Fagutao FF (eds) Biotechnological advances in shrimp health management in the Philippines, research signpost. Research Signpost, Trivandrum, Kerala

Apines-Amar MJS, Satoh S, Kiron V, Watanabe T (2004) Effects of supplemental amino acid–chelated trace elements on the immune response of rainbow trout subjected to bacterial challenge. J Aquat Anim Health 16:53–57

Assefa A, Abunna F (2018) Maintenance of fish health in aquaculture: review of epidemiological approaches for prevention and control of infectious disease of fish. Vet Med Int:5432497

Austin B, Austin DA (2012) Bacterial fish pathogens: disease of farmed and wild animals, 5th edn. Springer, Dordrecht

Barman D, Nen P, Mandal SC, Kumar V (2013) Immunostimulants for aquaculture health management. J Mar Sci Res Dev 3:134

Boyd CE, Tucker CS (1998) Pond aquaculture water quality management. Springer, New York. https://doi.org/10.1007/978-1-4615-5407-3

Bricknell I, Dalmo RA (2005) The use of immunostimulants in fish larval aquaculture. Fish Shellfish Immunol 19:457–472

Chand RK, Sahoo PK, Kumari J, Pillai BR, Mishra BK (2006) Dietary administration of bovine lactoferrin influences the immune ability of the giant freshwater prawn *Macrobrachium rosen-*

bergii (de Man) and its resistance against *Aeromonas hydrophila* infection and nitrite stress. Fish Shellfish Immunol 21:119–121

Cooper CJ, Mirzaei MK, Nilsson AS (2016) Adapting drug approval pathways for bacteriophage-based therapeutics. Front Microbiol 7:1–15

Culot A, Grosset N, Gautier M (2019) Overcoming the challenges of phage therapy for industrial aquaculture: a review. Aquaculture 513:734423

Dalmo RA, Bogwald J (2008) ß-glucans as conductors of immune symphonies. Fish and shellfish immunology 25:384–396

Devaraja TN, Otta SK, Shubha G, Karunasagar I, Tauro P, Karunasagar I (1998) Immunostimulation of shrimp through oral administration of *Vibrio* bacterin and yeast glucan. In: Flegel TW (ed) Advances in shrimp biotechnology. National Center for Genetic Engineering and Biotechnology, Bangkok

Ebeling JM, Timmons MB, Bisogni JJ (2006) Engineering analysis of the stoichiometry of photoautotrophic, autotrophic, and heterotrophic removal of ammonia-nitrogen in aquaculture systems. Aquaculture 257:346–358

Edwards P, Zhang W, Belton B, Little DC (2019) Misunderstandings, myths, and mantras in aquaculture: its contribution to world food supplies has been systematically over reported. Mar Policy 106:103547

Engle C, D'Abramo L (2018) Showcasing research focusing on sustainability of aquaculture enterprises and global food security. J World Aquacult Soc 47(3):311–313

FAO (2016) The state of world fisheries and aquaculture, contributing to food security and nutrition for all. Food and Agriculture Organization of the United Nations, Rome, p 2016

FAO (2019) GLOBEFISH Highlights January 2019 issue, with January–September 2018. Statistics – a quarterly update on world seafood markets. Globefish Highlights:4–68

Feliatra F, Elizal E, Lukistyowati I, Melina D, Ramadhani M (2018) Effectiveness of immersion with probiotics in improving the health of Nile tilapia (*oriochromis niloticus*). Asian J Anim Vet Adv 13:43–51

Gon Choudhury T, Tharabenahalli NV, Gita S, Paria A, Parhi J (2017) Advances in bacteriophage research for bacterial disease control in aquaculture. Rev Fish Sci Aquac 25(2):113–125

Hoseinifar SH, Sun Y-Z, Wang A, Zhou Z (2018) Probiotics as means of disease control in aquaculture, a review of current knowledge and future perspectives. Front Microbiol 9:2429

Huxley VAJ, Suthan P, Joseph MV, Lipton AP (2011) Immunomodulatory potential of Chitosan and Levamisole against bacterial diseases of *Penaeus monodon*. Asian J Anim Sci 6(1):33–40

Immanuel G, Vincybai VC, Sivaram V, Palavesam A, Marian MP (2004) Effect of butanolic extracts from terrestrial herbs and seaweeds on the survival, growth and pathogen (*Vibrio parahaemolyticus*) load on shrimp *Penaeus indicus* juveniles. Aquaculture 236(1–4):53–65

Jassim SAA, Limoges RG (2014) Natural solution to antibiotic resistance: bacteriophages "The Living Drugs". World J Microbiol Biotechnol 30:2153–2170

Kang W-K, Lee B-U, Chun H-J, Hwang S-D, Kim B-C, Kim H-S (2005) Method of enhancing growth and immunity of shrimp larvae using recombinant bovine growth hormone. International Publication Number WO 2005/115166 A1. World Intellectual Property Organization, 16pp

Knipe H, Temperton B, Lange A, Bass D, Tyler CR (2020) Probiotics and competitive exclusion of pathogens in shrimp aquaculture. Aquac Rev 13:324–352. https://doi.org/10.1111/raq.12477

Kowalska JD, Kazimierczak J, Sowinska PM, Wojcik EA, Siwicki AK, Dastych J (2020) Growing trend of fighting infections in aquaculture environment – opportunities and challenges of phage therapy. Antibiotics 9:301

Lara-Flores M (2011) The use of probiotic in aquaculture: an overview. Int Res J Microbiol 2(12):471–478

Lara-Flores M, Olivera-Castillo L, Olvera-Novoa MA (2010) Effect of the inclusion of a bacterial mix (*Streptococcus faecium* and *Lactobacillus acidophilus*), and the yeast (*Saccharomyces cerevisiae*) on growth, feed utilization and intestinal enzymatic activity of Nile tilapia (*Oreochromis niloticus*). Int J Fish Aquac 2:93–101

Lauzon HL, Gudmundsdottir S, Steinarsson A et al (2010) Effects of bacterial treatment at early stages of Atlantic cod (*Gadus morhua* L.) on larval survival and development. J Appl Microbiol 108:624–632

Lazado CC, Caipang CMA, Estante EG (2015) Prospects of host-associated microorganisms in fish and penaeids as probiotics with immunomodulatory functions. Fish Shellfish Immunol 45:2–12

Li P, Gatlin DM III (2006) Nucleotide nutrition in fish: current knowledge and future applications. Aquaculture 251:141–152

Merrifield DL, Dimitroglou A, Foey A, Davies SJ, Baker RT, Bøgwald J et al (2010) The current status and future focus of probiotic and prebiotic applications for salmonids. Aquaculture 302:1–18

Mohapatra S, Chakraborty T, Kumar V, De Boeck G, Mohanta KN (2013) Aquaculture and stress management: a review of probiotic intervention. J Anim Physiol Anim Nutr 14:1–26

Nikapitiya C (2013) Marine bacteria as probiotics and their applications in aquaculture. In: Kim S-K (ed) Marine microbiology: bioactive compounds and biotechnological applications, 1st edn. Wiley-VCH Verlag GmbH & Co. KGaA

Parker R (1974) Probiotics, the other half of the antibiotic story. Anim Nutr Health 29:4–8

Patil PK, Geetha R, Ravisankar T, Avunje S, Solanki HG, Abraham TJ, Vinoth SP, Jithendran KP, Alavandi SV, Vijayan KK (2020) Economic loss due to diseases in Indian shrimp farming with special reference to *Enterocytozoon hepatopenaei* (EHP) and white spot syndrome virus (WSSV). Aquaculture 533:736231

Ren P, Xu L, Yang Y et al (2013) *Lactobacillus planarum* subsp. *plantarum* JCM 1149 *vs. Aeromonas hydrophila* NJ-1 in the anterior intestine and posterior intestine of hybrid tilapia *Oreochromis niloticus* ♀ × *Oreochromis aureus* ♂: an *ex vivo* study. Fish Shellfish Immunol 35:146–153

Rico A, Satapornvanit K, Haque MM, Min J, Nguyen PT, Telfer TC, van den Brink PJ (2012) Use of chemicals and biological products in Asian aquaculture and their potential environmental risks: a critical review. Rev Aquac 4(2):75–93

Sahoo PK (2007) Role of Imunostimulants in disease resistance of fish. CAB Rev Perspect Aquac Vet Sci Nat Natl Resour 2(045):1–3

Sakai M (1999) Current research status of fish immunostimulants. Aquaculture 172:63–92

Seljelid R (1990) Immunomodulators – medicine for the 90-ies? In: Wadström T et al (eds) Workshop on pathogenesis of wound and biomaterial-associated infections, August 30–September 01, 1989. Springer, New York, pp 107–113

Sharifuzzaman SM, Austin B (2017) Probiotics for disease control in aquaculture. In: Austin B, Newaj-Fyzul A (eds) Diagnosis and control of diseases of fish and shellfish, 1st edn. Wiley

Sharifuzzaman SM, Al-Harbi AH, Austin B (2014) Characteristics of growth, digestive system functionality, and stress factors of rainbow trout fed probiotics *Kocuria* SM1 and *Rhodococcus* SM2. Aquaculture 418/419:55–61

Sharma S, Chatterjee S, Datta S, Prasad R, Dubey D, Prasad RKR, Vairale MG (2016) Bacteriophages and its applications: an overview. Folia Microbiol 62:17–55

Song SK, Beck BR, Kim D, Park J, Kim J, Kim HD, Ringo E (2014) Prebiotics as immunostimulants in aquaculture: a review. Fish Shellfish Immunol 40:40–48

Suchner U, Kuhn KS, Fürst P (2000) The scientific basis of immune-nutrition. Proc Nutr Soc 59:553–563

Taju G, Madan N, Majeed SA, Kumar TR, Thamizvanan S, Otta SK, Hameed ASH (2015) Immune responses of white leg shrimp, *Litopenaeus vannamei* (Boone, 1931) to bacterially expressed dsRNA specific to VP28 gene of white spot syndrome virus. J Fish Dis 38(5):451–465

Thitamadee S, Prachumwat A, Srisala J, Jaroenlak P, Salachan PV, Sritunyalucksana K, Flegel TW, Itsathitphaisarn O (2016) Review of current disease threats for cultivated penaeid shrimp in Asia. Aquaculture 452:69–87

Thornber K, Verner-Jeffreys D, Hinchliffe S, Rahman MM, Bass D, Tyler CR (2020) Evaluating antimicrobial resistance in the global shrimp industry. Reviews in Aquaculture 12(2):966–986. https://doi.org/10.1111/raq.12367

Valencia-Castañeda G, Frías-Espericueta MG, Vanegas-Pérez RC, Pérez-Ramírez JA, Chávez-Sánchez MC, Páez-Osuna F (2018) Acute toxicity of ammonia, nitrite and nitrate to shrimp *Litopenaeus vannamei* postlarvae in low-salinity water. Bull Environ Contam Toxicol 101:229–234

Vallejos-Vidal E, Reyes-Lopez F, Mackenzie S (2017) Immunostimulant diets and oral vaccination in fish. In: Austin B, Newaj-Fyzul A (eds) Diagnosis and control of diseases of fish and shellfish, 1st edn. Wiley

Vazquez JA, Cabo ML, Gonzalez MP, Murado MA (2003) Survival of lactic acid bacteria in seawater: a factorial study. Curr Microbiol 47:508–513

Veld H, Havenaar JHJ, Marteau PH (1994) Establishing a scientific basis for probiotic R&D. Trends Biotechnol 12:6–8

Vine NG, Leukes WD, Kaiser H (2006) Probiotics in marine larviculture. FEMS Microbiol Rev 30:404–427

Wang W, Sun J, Liu C, Xue Z (2017) Application of immunostimulants in aquaculture: current knowledge and future perspectives. Aquac Res 48:1–23

Yada T (2007) Growth hormone and fish immune system. Gen Comp Endocrinol 152:353–358

Zhou L, Boyd CE (2015) An assessment of total ammonia nitrogen concentration in Alabama (USA) ictalurid catfish ponds and the possible risk of ammonia toxicity. Aquaculture 437:263–269

Zhou J, Song X, Huang J, Wang X (2006) Effects of dietary supplementation of A3a-peptidoglycan on innate immune responses and defense activity of Japanese flounder (*Paralichthys olivaceus*). Aquaculture 251:172–181

Bioremediators in Shrimp Aquaculture

K. K. Krishnani, R. Abisha, and M. Muralidhar

Abstract

Shrimp aquaculture is a significant food production enterprise for farmers and entrepreneurs as shrimp fetch a huge demand and price in the domestic as well as international markets. The farming of shrimp in brackish and coastal waters has gained a remarkable position. However, various constraints exist during the culture period, and the major one is the deterioration of water quality. Shrimp aquaculture is presently focused on intensive culture practices, and this necessitates huge application of inputs like feed, fertilizers, and therapeutics. Along with these, uneaten feed, excreta, metabolic by-products, etc., result in poor water quality comprising an excess concentration of nitrogenous and sulphurous compounds. This makes the farmed animal more susceptible to diseases caused by water quality deterioration and pathogens. To overcome these constraints, a suitable method is bioremediation, an advanced eco-friendly, sustainable technique of utilizing the biological material, particularly microbes and their enzymes, as bioremediators to accomplish environmental and disease management in shrimp aquaculture. Bioremediators are in different forms: bacteria, plant-based products, microalgae, macroalgae, biofilm/periphyton, mat-forming, biofloc, probiotics, cell-free enzyme, and green water. This chapter focuses on

K. K. Krishnani (✉)
ICAR-Indian Institute of Agricultural Biotechnology, Garhkhatanga, Ranchi, India
e-mail: krishnanik@hotmail.com

R. Abisha
ICAR-Central Institute of Fisheries Education, Mumbai, India
e-mail: abisha.aqcpa907@cife.edu.in

M. Muralidhar
ICAR-Central Institute of Brackishwater Aquaculture, Chennai, India
e-mail: muralichintu@rediffmail.com

R. Kooloth Valappil et al. (eds.), *Aquatic Animal Health Management*,
https://doi.org/10.1007/978-981-96-7987-4_23

the different bioremediators that are mainly used in the remediation of chemical and microbial contaminants in shrimp aquaculture.

Keywords

Shrimp aquaculture · Contaminants · Plant-based bioremediators · Microbial products · Bioremediation

1 Introduction

Aquaculture, the fastest-growing food production sector in the world, has the potential to meet the nutritional demands of the growing world population (Moriarty 1999). There exist various aquaculture strategies such as monoculture, polyculture, and integrated farming in different habitats like freshwater, brackish water, and coastal waters. Shrimp farming is one of the fastest-growing aquaculture activities in India and ranks second in global aquaculture production. Different shrimp species, such as *Penaeus vannamei*, *P. monodon*, and *P. indicus* are cultured worldwide. To achieve more productivity, shrimp aquaculture has been focussed on more production from lesser areas which can be achieved by adopting intensive farming practices where there is an increased usage of inputs like feed, fertilizers, and therapeutics. Despite getting more production from intensified culture practice, it leads to negative impacts by generating excessive amounts of waste in the system, which arises from uneaten feed particles, faecal matter, fertilizers, and other additives such as prophylactic and therapeutic inputs used in the system (Tacon 2002). In shrimp aquaculture, the generation of nitrogenous, phosphorous compounds and organic matter is of prime concern. The water quality deterioration in the culture system can cause major stress to farmed animals and ultimately lead to disease outbreaks. It is noted that the shrimp could assimilate about 23–31% nitrogen and 10–13% phosphate from the total inputs (Granada et al. 2018). Thus, the remaining results in the emergence of major constraints as wastewater with excess nitrogenous, and phosphate forms. In aquaculture practices, the types of wastes generated in the system are generally alike, but there exists variation in the quantity and quality of components that are based on the species cultured and the type of culture practice adopted. The wastewater generated from shrimp culture constitutes an increased concentration of organic matter due to waste excreted and the leftover feed in the system (Krasaesueb et al. 2019). Thus, the effluent/wastewater from the shrimp culture unit has to be managed efficiently and effectively in terms of cost with minimal impact on the environment to achieve more sustainable shrimp production. This is crucial for the growth and success of the shrimp industry (Kang et al. 2011). The current approach to improve water quality in aquaculture is the application of microbes/enzymes to the ponds, known as 'bioremediation' (Sharma et al. 2018). When macro and microorganisms and/or their products are used as additives to improve water quality, they are referred to as bioremediators or bioremediating agents (Moriarty 1998). This results in a lower accumulation of organic matter in the pond bottom, better penetration of oxygen into the sediment and a generally better environment for the farmed stock (Rao and Karunasagar 2000). A very effective technique applied in shrimp aquaculture for environmental and health management is bioremediation, which is an eco-friendly method where living organisms like microbes and their enzymes, plants, fungi, and algae are utilized to convert the organic and inorganic pollutants into compounds that cause less or zero impact in the system and the environment (Antony and Philip 2006). In this approach, the microbes were stimulated by the presence of nutrients (such as NH_3, NO_3, etc.) and other chemical compounds that enable them to remove the contaminants or to break down the toxic, hazardous substance into less or non-toxic compounds.

In the present scenario of climate change, the temperature of water bodies increases, which leads to increases in the toxicity of metallic and ammonical pollutants (Abisha et al. 2022). In aquaculture, ammonia nitrogen is the final product of protein catabolism and metabolism, which is the most common environmental limiting factor (Arunkumar et al. 2023).

Bioremediators are nutrient extractors that remediate effluents. There are different bioremediators, i.e. micro and macro-organisms like beneficial microbes, microalgae, macroalgae, biofloc, and biofilm/periphyton. This approach is purely biological, and there is no usage of chemicals; therefore, this could be a suitable alternative for the remediation of shrimp culture wastewater instead of using costly and toxic chemicals. Shrimp aquaculture faces challenges like high nutrient loading in the system i.e. water quality deterioration, the incidence of toxic gases such as ammonia (NH_3), nitrite (NO_2), H_2S etc., excess of organic matter, and disease occurrence as a result of intensive stocking and feeding practices (Ravichandran 2012) that the application of various bioremediators can control. Probiotics and Green water technology come under the bioremediation approach and are effective against the usage of antibiotics in shrimp farming to treat diseases. Microbes such as denitrifying, nitrifying, photosynthetic, and organic matter-degrading bacteria have evolved as potential bioremediators in shrimp aquaculture (Azim et al. 2003; Krishnani 2010). The application of the bioremediation technique helps in reducing the accumulation of organic matter in the bottom of shrimp ponds, provides better oxygen penetration into the sediment and, overall, facilitates a suitable environment for the culture of shrimp. Currently, more attention has been given to bioremediation and the usage of microbes in pollutant remediation as it has the potential to degrade both natural and artificial substances. This chapter describes the different bioremediators used in environmental and health management in shrimp aquaculture. The schematic representation of bioremediation in shrimp aquaculture is depicted in Fig. 1.

2 Bioremediators for Environmental Management of Shrimp Aquaculture

In shrimp aquaculture, the prime constraint to the environment is the deterioration of water quality. Both the culture water and the wastewater of shrimp farms contain ammonia, the major nitrogenous compound rendering undesirable effects on the

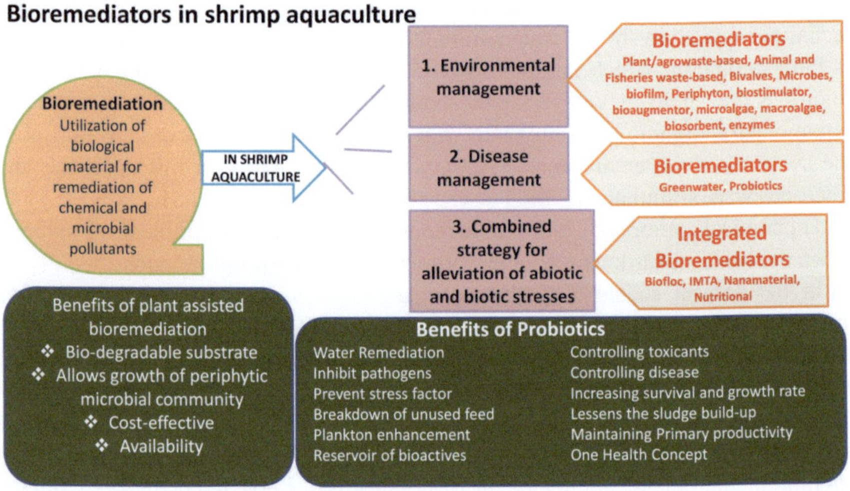

Fig. 1 Bioremediators in shrimp aquaculture

cultured animal and the system. This ammonia exists in unionized and ionized forms and is influenced by pH and temperature. However, unionized ammonia remains toxic compared to the ionized form. Undesirable levels of unionized ammonia affect oxygen transport and osmoregulation in aquatic species. The excess concentration of nitrogenous compounds (ammonia, nitrite, nitrate), phosphorus, and organic matter in the system is generated from the uneaten feed, excreta, and usage of therapeutics, which are ultimately the results of intensive culture practice, which employs large inputs of feed, and fertilizers. Ammonia is the resultant of protein metabolism, whereas nitrite is formed as an intermediate product when ammonia is transformed into nitrate as the end product of nitrification in the system. Similarly, phosphorus has also resulted from the metabolic process, uneaten feed, and faeces containing undigested phosphorus. These pollutants, when reaching higher concentrations than their safe level, will affect the growth and survival of aquatic animals as well as cause an impact on the environment by degrading the water and soil quality. Therefore, it is recommended to maintain the following levels in the shrimp culture: ammonia <0.1 ppm, nitrite <0.3 ppm, nitrate 0.2–10 ppm and phosphate 0.005–0.2 ppm (Boyd 1998). The shrimp culture water also gets contaminated by heavy metals when coastal water is polluted with agricultural waste, sewage, etc. Concentrations of these pollutants in the culture water must be reduced to create a better living environment for the cultivable species so that better shrimp production can be achieved successfully, causing minimal or no environmental impact. Bioremediators performing bioremediation functions have been mostly preferred in recent days as they are a sustainable way of improving environmental quality. Bioremediators have performed well in the remediation of the nitrogenous and sulphurous compounds and suspended and dissolved organic matter in the aquaculture system. There is a diverse microbial community, such as nitrifying bacteria, denitrifying bacteria, and purple and green sulphur bacteria, which breaks down the pollutants and makes them harmless to the system. The main actions performed by bioremediators include: (i) optimization of nitrification and denitrification rate to reduce the level of nitrogenous compounds, (ii) maximizing the oxidation of sulphide, carbon mineralization, and primary productivity and (iii) maintenance of desirable microbial community and avoiding the undesirable one. These pollutants also serve as nutrients that are highly available to the algae and microbes present in the environment.

2.1 Plant-Assisted Bioremediation

Plant-assisted bioremediation is commonly defined as the use of green or higher terrestrial plants to degrade, remove, transform, phyto-stabilize, phyto-extract, bioadsorb/accumulate, phyto-volatilize, or immobilize contaminants, including petroleum hydrocarbons, chlorinated solvents, pesticides, explosives, heavy metals, and radionuclides in soil and water in polluted soils, groundwater, and wastewater in treatment wetlands. The use of plants in conjunction with plant-associated bacteria (rhizosphere or endophytic) offers greater potential for bioremediation of organic compounds and, in some cases, inorganic pollutants than using plants alone in bioremediation. Candidate plants for phytoremediation should have characteristics such as high biomass production, extensive root system, and the ability to tolerate high concentrations of pollutants and to withstand environmental stress.

The use of medicinal and aromatic plants (MAPs) in aquaculture has gained popularity due to their demonstrated antimicrobial activity and ability to promote the growth and development of farmed species. Kate et al. (2023) also demonstrated the effectiveness of *Minata sepicata* in the bioremediation of ammonia from aquaculture water. Plants facilitate the biodegradation of contaminants by releasing root exudates and other compounds to the surrounding soil as well as providing a surface for the colonization of microbes and, thereby, contributing to the increased number and metabolic activity of microorganisms (rhizosphere effect) and enhanced bioavailability of the contaminant. Plants provide nutrients for endophytic bacteria and induce catabolic gene expression. Endophytic bacteria degrade organic pollutants, reducing phytotoxicity in this way producing plant growth-promoting hormones. This process is often referred to as plant-assisted bioremediation or periphyton-based bioremediation, which can be classified based on the use of living and nonliving materials.

2.1.1 Plant/Agro-Waste-Based Bioremediator (Biostimulator)

Agriculture, being a major contributor to the world's food demand, also generates a considerable quantity of waste by-products. These wastes can be utilized as a resource for another process, thus reducing the waste and increasing the usage of the product. The different by-products and wastes from the agriculture sector include rice husk and straw, sugarcane bagasse, groundnut husk, etc. These by-products are easily available and economical to utilize it for other purposes. Bagasse, a lignocellulosic fibrous waste, is the by-product generated after extracting the juice from the sugarcane and is widely used as a bioremediator in shrimp aquaculture, taking advantage of its low cost and availability (Krishnani et al. 2006).

Most bioremediation technologies involve naturally occurring microorganisms and the improvement of nutrient conditions and environmental factors for the process of biodegradation. Biostimulation involves the addition of electron acceptors, nutrients, or electron donors to stimulate the activity of indigenous microorganisms. Krishnani et al. (2006) adopted a bagasse-assisted bioremediation technique to remediate ammonia from the shrimp culture wastewater. In this, five different bagasse products (raw bagasse fibre, dried bagasse fibre, oven-dried bagasse powder, charred bagasse powder, and bagasse charred with acid) were developed. The oven-dried bagasse powder at the dose of 1–6 mg/L within 24 hours achieved the most efficient ammonia removal. Biostimulation technology is simple and cost-effective without much technical sophistication, integration of which in zero-exchange and water reuse systems could be advantageous for water and area saving, reduced risk of contamination and better environmental control for the circular economy. This bioremediator is a highly preferable alternative for ammonia removal in shrimp aquaculture as in this type of remediation, the indigenous microbes present in the system are harnessed in the bioremediation of pollutants/contaminants. Similarly, bagasse products are administered in the remediation process of chromium to detoxify Cr (VI) from high saline coastal water, where bagasse charred with sulphuric acid serves as the best chromium detoxifying product (Krishnani et al. 2004). The coconut husk, a fibrous cellulose material, has been used for the removal of heavy metal chromium from seawater, which shows the potential of agro-waste products in remediating the pollutants to achieve a better cost-effective, harmless environment management practice (Parimala et al. 2004). Several other plants/agro-waste-based products are also involved in the environmental management of shrimp aquaculture. The neem plant-based product has found application in removing high concentrations of ammonia from brackish water. The neem seed powder, neem seed oil, and commercial neem products have been used to determine its efficiency in removing ammonia. The study revealed that neem seed oil and commercial neem products are effective in the remediation of ammonia from brackish water (Krishnani et al. 2002). Similarly, the extract (oil) of custard apple seed has been used to effectively remediate the nitrogenous toxicants from the brackish water (Krishnani et al. 2003). The extracts of some plants also possess an antimicrobial property that can be utilized in the remediation process. Cassava waste positively impacted the removal of nitrogenous metabolites in shrimp aquaculture, along with enhancement in shrimp growth (Lalitha et al. 2017). Plant/agro-waste-based products play a vital role in water quality improvement because of their cost-effective nature and availability, and they also enhance shrimp production. The diversity of nitrifying, denitrifying, nitrogen-fixing, methane-oxidizing, and sulphur-oxidizing bacteria, and bacteria implicated in the bioremediation of nitrogenous and sulphurous fluxes in coastal aquaculture soils has been examined using functional metagenomics (Krishnani et al. 2009b, 2010a, b, 2015) which help in the predictability and reliability of biostimulator for the application in coastal shrimp aquaculture.

2.1.2 Biofilm-Based Bioremediator (Periphyton)

Aquatic microbial communities can grow on immersed substrates as periphyton or in suspension as floc. Organisms such as algae, zoogleal, and filamentous bacteria growing on aquatic macrophytes and other submerged substrates/surfaces are termed periphyton (Keshavanath et al. 2001). Periphyton grown on these substrates are excellent natural food for certain fish species and support enhanced fish production. Periphytic bacteria in the biofilm play a key role in the production of enzymes, degradation of organic matter and environmental toxicants. Hence, periphyton improves production and water quality as well (Azim et al. 2003). Periphyton represents the entire biota that colonizes immersed mats. Fixed microbial communities present many advantages in comparison to planktonic communities in terms of pond productivity, carbon and nutrient absorption from water, consumption by shrimp etc. Zhang (2010) studied the effects of artificial substrates on the spatial distribution of shrimp in intensive culture conditions and suggested that the difference in shrimp growth and survival was mainly due to living space added with the addition of artificial substrates. Therefore, a better understanding of the effect of artificial substrate on shrimp performance is necessary. Based on this concept, substrate-based aquaculture is a simple and efficient technology to increase productivity by putting some easily degradable biological substances like water hyacinth, paddy straw or bagasse, which can provide good surface area for periphyton. Sruthisree et al. (2015) evaluated periphyton and biofilm on different substrates such as bamboo, PVC pipe, paddy straw and coconut fronts in the *Penaeus monodon* culture ponds and the use of paddy straw as substrates found to increase biomass in the culture system.

Bagasse is the best option as it serves as a biodegradable substrate which harbours higher periphytic biomass than non-degradable ones. This could be because biodegradable substrates provide a better surface structure for periphytic species to attach and aid in nutrient-leaching beneficial for the growth of periphyton, predominantly consisting of bacteria, thus providing no residues in the environment and improving the water quality. Krishnani et al. (2013) have successfully demonstrated the removal of ammonia up to 52% in the biostimulated culture ponds due to a biofilm mode of growth of nitrifying consortia onto bagasse with the result of achieving higher shrimp production under field conditions. They quantified the autotrophic ammonia-oxidizing bacterial biofilm population using a real-time PCR assay targeting the *ammonia monooxygenase* (*amoA*) gene, which was found to be in the range of 10^4–10^5 and 10^5–10^6 *amoA* gene copies/g of the substrate based on the amount of aeration and duration of the treatment with bagasse-biostimulator applied in the ponds. A real-time PCR method has been developed by Krishnani et al. (2010a, b) for quantification of AOB in brackish water and biofilm. Bagasse-biostimulation technology is advantageous due to its cost-effectiveness, and it is a simple technique that can easily be adopted by aqua farmers using locally available sugarcane bagasse, thereby ensuring water remediation and good yield. Substrate-based aquaculture, the concept of biofilm, and biocapsules are increasingly being accepted in coastal shrimp aquaculture.

2.1.3 Algal Bioremediators

Algae are divided into two main types: microalgae and macroalgae. Macroalgae, also known as seaweeds, are multicellular plants that grow in fresh or salt water. Based on the pigmentation, they are classified into three groups: brown, red and green algae (Anastopoulos and Kyzas 2015). On the other hand, microalgae are microscopic and photosynthetic unicellular organisms/plants that also grow in fresh or salt water. The four groups of microalgae, such as diatoms, green algae, golden algae and blue-green algae (cyanobacteria), are classified based on their pigmentation, arrangement(s) of their photosynthetic membranes or other morphological features.

Microalgae have a great capacity to remediate nitrogenous compounds from shrimp culture wastewater. The microalgae mode of bioremediation varies based on the pollutant and the algal species. Either the algal cell wall or extracellular polysaccharides excreted by the algal cells adsorb with the pollutants. The method is ideal for the bioremediation of heavy metals and other toxic metallic compounds such as arsenic, boron, copper, manganese, and zinc (Saavedra et al. 2018; Ibuot et al. 2019). The microalgae *Tetraselmis chui* had been utilized for the remediation of nitrogen and phosphorus from the super-intensive shrimp pond wastewater (Tahir et al. 2020). Macroalgae are widely used in the remediation of shrimp effluents, as they possess the capacity to absorb nutrients like nitrogen and phosphorous from the water serving as an instrument in bioremediation (Neori et al. 2004). Macroalgae like *Ulva* spp. and *Gracillaria* sp. are widely utilized in treating pollutants like nitrogenous and phosphorous compounds in shrimp aquaculture. The *Ulva* spp. are most commonly preferred over macroalgae because of their high biomass production potential and biofiltration capacity. The seaweed *Ulva reticulata* has been used in the bioremediation of shrimp hatchery effluent, particularly from shrimp broodstock, which resulted in a 100% reduction of ammonia and nitrite, 33% nitrate, and 89% orthophosphate. This study has proved the potential of *U. reticulata* in the remediation of shrimp effluents when used directly in the culture system as a biofilter (Rabiei et al. 2014). Likewise, a study has been performed utilizing the species *Ulva lactuca* as a bioremediator in the *P. vannamei* culture tank where the water from shrimp culture is exchanged for the growth *U. lactuca* in a recirculation system (i.e. seaweed *U. lactuca* water exchange) which absorbs the inorganic nutrients from the shrimp culture water. The type of seaweed-based bioremediation is advantageous over shrimp culture with clean water exchange because of its better biofiltration capacity as well as higher biomass production, which can be used as a feed additive for the shrimps (Elizondo-González et al. 2018). Similarly, *Gracilaria* sp. is also used for water quality management in shrimp culture. The macroalgae *Gracilaria caudata* has been used in the shrimp pond to mitigate the excess concentration of nitrogenous and phosphorus compounds and were examined to determine the growth and storage capacity of nutrients in the macroalgae. Marinho-Soriano et al. (2009) observed a reduction of 59.5% NH_4–N, 49.6% NO_3–N, and 12.3% and PO_4–P in 4 h with a modest growth rate and they suggested the use of the macroalgae *G. caudata* as a bioremediator in the *P. vannamei* pond. Similarly, *G. tenuistipitata*-based *P. vannamei* culture has been carried out by Sarkar et al. (2019), where co-culture of seaweed with shrimp showed improved water quality. Likewise, another two species of *Gracilaria*, i.e. *G. edulis* and *G. changii*, are cultured in shrimp wastewater in an outdoor recirculating system where the macroalgae serve as a biofilter, removing the nutrients from the system as an energy source for its biomass production. These two species achieved nutrient removal of around 71% ammonia and 57% nitrate from the system, thus suggesting that these species are biofilters in shrimp farming (Mawi et al. 2020). It was found that two species of brown algae (*Padina australis* and *Sargassum hemiphyllum*) were tolerant of low salinity, which makes them suitable for use in brackish water aquaculture (Tanaka et al. 2020). These findings revealed the potential biofiltration capacity of different macroalgae in environmental management following a sustainable approach by converting the waste generated in the system into a resource by biomass production. It is suggested that shrimp culture be integrated with seaweeds, which would effectively aid in maintaining the water quality to cope with the intensive culture impacts. However, the biomass of macroalgae should also be managed appropriately to avoid excess biomass in the system which affects the survival of the aquatic animal.

Marinho-Soriano et al. (2011) successfully demonstrated that the macroalgae *Gracilaria caudata* and the microcrustacean *Artemia franciscana*, in combination, were effective in removing nutrients in nitrogenated forms from aquaculture effluents. They concluded that the use of these organisms as biofilters has the potential to be an ecologically correct practice and may contribute to improving the water quality of coastal regions. The fast growth rates and broad geographical distribution of *Ulva ohnoi* along the coastline make this an ideal species to target for bioremediation activities at land-based intensive aquaculture, which generates nutrient-rich wastewaters, if untreated, can cause eutrophication of coastal waters and adversely impact downstream biological communities (Wang et al. 2018; Lawton et al. 2013). Algal bioremediation is widely accepted as an efficient and cost-effective wastewater treatment method for the removal of excessive dissolved nutrients, particularly for land-based aquaculture operations, where production is often limited by strict environmental regulations around the water quality of point-source discharges.

2.1.4 Biosorption (Bioadsorbent)

Biosorption is the ability of biological materials to bind heavy metals from contami nated soil and water through metabolically mediated or via spontaneous physico-chemical pathways of uptake or as a property of certain types of inactive, non-living microbial biomass (Volesky and Holan 1995; Shamim 2018; Krishnani and Ayyappan 2006). It is a complex process that depends on different factors like cell physiology, physicochemical factors such as pH, temperature, contact time, ionic strength, metal concentration, metal speciation and chemistry, membrane composition of microorganisms, etc. (Krishnani et al. 2008a, b). Biosorption of different heavy metals using microorganisms like fungi, algae or bacteria have also been studied by different groups. Krishnani et al. (2008a, b, 2009a, 2010a, b, 2015, 2021) investigated the biosorption mechanism of nine different heavy metals onto rice husk, paddy straw, bagasse and plant bark-derived biosorbents and they ascertained the role of calcium and magnesium present in the biosorbents in ion-exchange mechanism and the maximum adsorption capacities. For the first time, Krishnani et al. (2015, 2021) identified Whewellite ($C_2CaO_4{\cdot}H_2O$) in the neem, mango and guava barks, which indicated that calcium ions are electrovalently bonded with carboxylate ions facilitating the ion exchange mechanism with metal ions. Krishnani et al. (2004) have demonstrated that bagasse charred with acid is most effective for detoxification of Cr (VI) into Cr (III) in an acidic medium due to their electron-donating capacity. In contrast, the removal of Cr (VI) in the treatment with other products in an alkaline medium was due to the biostimulation of native microbial communities in coastal waters. The biosorption has potential application in the decontamination of heavy metals.

2.2 Microbial Bioremediation

2.2.1 Bacterial Bioremediator (Bioaugmentor)

Microorganisms are nature's original recycler and inevitable organism in any ecosystem that plays a prime role in bioremediation. Utilizing microorganisms to detoxify soils, culture water, and effluent is getting such wide acceptance that 'bioremediation' has now become a common buzzword in waste management. This type of bioremediator utilizes pollutants like nitrogen and phosphorous as the energy source and has great potential for immobilizing these contaminants. Microbial remediation has been regarded as the most preferred bioremediation method because of its cost-effective nature (Gadd 2000). It is also known as bioaugmentation, where exogenous microbes are applied over the site of action for microbial degradation of toxic metabolic compounds. It is also recommended to use the non-native microbes of the ecosystem, i.e. the microbes that are exogenous and genetically engineered, which can be introduced into the system to perform the bioremediation process. The microbial bioremediator involves in the environmental management of shrimp aquaculture. Initially, analysis of the system in search of potential microbes for remediating pollutants is needed. This isolation of potential indigenous microbes will aid in carrying out the remediation process successfully. Microbial bioremediation, being the most efficient, safe, and cost-effective method, is well-accepted for mitigating toxic metabolic wastes in aquaculture.

Nitrification is a process where ammonia is oxidized with oxygen to form nitrite and further oxidation to convert nitrite to nitrate. Autotrophic bacteria grouped as ammonia oxidizing bacteria (AOB) and nitrite oxidizing bacteria (NOB) are involved in the aerobic nitrification process, where ammonia is oxidized to nitrite and nitrite to nitrate, respectively (Chávez-Crooker and Obreque-Contreras 2010). Ghosh et al. (2010) evaluated the potential of indigenous microflora present in shrimp culture systems using microcosm assay for the bioremediation of total ammoniacal nitrogen (TAN) under laboratory conditions; however, the supply of TAN in the experimental system was maintained similar to the concentration generated through excretion and uneaten feed in the shrimp aquaculture. The findings of the study predicted that almost 85–99% of TAN was removed effectively by the indigenous microbes. This paved the way for further studies to bring out the role of microbes in bioremediation in shrimp aquaculture. Another work has been done by Devaraja et al. (2013) where indigenous *Bacillus* spp. such as *Bacillus subtilis*, *B. licheniformis*, and *B. pumilus* were selected and isolated via a holistic approach as a means of bioremediation in *P. monodon* postlarval rearing system. The *Bacillus* sp. is advantageous as it is easy to mass culture, and has the capacity to tolerate ammonia concentration and the ability to facilitate inhibition of pathogenic microbes like *Vibrio* sp. in the system (Far et al. 2009). *Bacillus* acts on the pollutants by utilizing it as an energy source, and the production of enzymes by them helps to break down the contaminants into a non-toxic form. The above-mentioned three species are effective in reducing TAN from shrimp culture water and improving the growth and survival of *P. monodon* postlarvae without performing water exchange in the culture practice. Thus, *Bacillus* species have been identified as potential bioremediator in water quality management in shrimp aquaculture (Liu et al. 2010; Banerjee et al. 2010). Baskaran et al. (2020) reported the composition of microbial enrichment developed for the aerobic oxidation of ammonia and nitrite from aquaculture farming systems. The study reported domination of α-proteobacteria and γ proteobacteria in AOB and NOB consortia developed, respectively.

Hydrogen sulphide is considered one of the most toxic metabolic wastes in the aquatic system. Naturally, a closed or sequential association between photolithotrophic, chemolithotrophic bacteria, and archaea results in the biological formation of the sulphur cycle (Ghosh and Dam 2009). The sulphide oxidation can be an aerobic or anaerobic process. In the anaerobic condition, chemolithotrophic sulphur oxidizing bacteria (SOB) use nitrate or nitrite to accept electrons released from the process of sulphide oxidation. In the aerobic oxidation process, photoautotrophic SOB (aerobic sulphide oxidation) derives energy from the light energy while oxygen is employed as an electron acceptor released as a result of sulphide oxidation (Pokorna and Zabranska 2015). For biological removal of toxic hydrogen sulphide, colourless chemolithotrophic SOB are preferred the most due to their high affinity for sulphide and oxygen, and they can be compared at par with chemical sulphide oxidation in bioreactors or natural environment (Janssen et al. 1998). SOB cover obligate or facultative chemolithotrophs, chemolithoheterotrophs, or sulphur-oxidizing chemoorganoheterotrophs that have a diverse mechanism of sulphide oxidation (Robertson and Kuenen 2006). These colourless SOBs are Gram-negative bacteria that occur in a significantly diverse ecosystem with temperatures ranging between 4 and 90 °C and a pH range of 1–9 (Tang et al. 2009). The filamentous chemolithotrophic SOBs like *Beggiatoa* spp. and *Thiothrix* spp. oxidize H_2S to S^- inside the cell and release in the form of sulphate when present in excess (Pokorna and Zabranska 2015). Obligate chemolithotrophs like *Thiobacillus* and *Thiomicrospira* utilize reduced sulphur compounds as an energy source and CO_2 as a carbon source, while the facultative chemolithotrophic SOB is capable of extracting energy from sulphide compounds and use CO_2 as a carbon source.

The microbes like actinomycetes also serve as bioremediators in the environmental management of shrimp farming. Actinomycetes have been isolated from the marine environment and from pond sediments of the shrimp rearing system to determine their potential in the remediation process. These microbes were found to be a potential bioremediator by rendering a significant reduction in BOD in the *P. monodon* larval rearing system, thus improving the water quality (Babu et al. 2018). Research studies have identified the role of different microbes in the bioremediation process. Hence, identifying a suitable potential microbe and understanding its bioremediation action in coastal/brackish water would help to carry out shrimp farming in a sustainable way. Bacterial species belonging to genera *Bacillus*, *Pseudomonas*, *Acinetobacter*, *Cellulomonas*, *Rhodoseudomonas*, *Nitrosomonas*, and *nitritrobacter* are known to help in the mineralization of organic wastes.

2.2.2 Microbial Mat-Forming Biofilm-Based Bioremediator

Another effective approach in environmental management in shrimp aquaculture is using of microbial mats. The microbial mats develop a biofilm which helps in reducing/degrading the pollutants. Biofilm is epiphytic microbial communities that get attached to submerged substrates and perform the action of bioremediation. The microbial mats, i.e. the microbial biofilms growing on the substrate, play a vital role in effluent treatment in shrimp aquaculture, significantly reducing the organic matter as well as the nitrogen compounds. A constructed microbial mat developed using glass wool has been used for treating shrimp culture effluent as an ex-situ bioremediation approach. The natural marine microbial communities were immobilized on the glass wool and administered to reduce the level of pollutants from shrimp culture effluent. The findings from the study showed that the constructed microbial mat treating the *P. vannamei* culture water reduces 97% of ammonia nitrogen and 95% of nitrate nitrogen. It also reduces the organic matter from the effluent (Paniagua-Michel and Garcia 2003). To understand the usage of microbial mats in bioremediation, it is necessary to know about the microbial communities for better knowledge of the process. Therefore, metagenomic analysis of biofilms on *P. vannamei* effluents was performed by Mendoza et al. (2020) for the first time using BLAST. In this study, non-woven geotextile carpets were used as a substrate to immobilize microbial communities. The results of metagenomics conveyed a great diversity of microbes, including bacteria, fungi, protozoa, archaea, and viridiplantae, and revealed the role of 12 species in the nitrification process in shrimp effluent. The microbial communities that adhere to the substrate also serve as a food source for the culture of animal shrimp. The study elucidated the potential of the application of microbial mats as one of the sound technologies that can be used for bioremediation in shrimp aquaculture. Microbial mats are laminated, cohesive microbial communities composed of a consortium of bacteria dominated by photoautotrophic cyanobacteria/blue-green algae (Nisbet and Fowler 1999).

2.3 Bivalve/Molluscan Bioremediator (Biofilters)

The environmental management in shrimp aquaculture can be performed not only by plant-based products but also by animal-based products. The critical nutrients in water such as ammonia and nitrite must be maintained within their optimal concentration so that a healthy habitat will appear. Animal-based products such as clam lipids and squid lipids were used to determine the removal efficiency of nitrogenous compounds from the brackish water. The results showed that squid lipids are more effective in nitrite removal than clam lipids. Both animal extracts are effective in removing nitrogenous pollutants like ammonia and nitrite from the brackish water under laboratory conditions (Krishnani et al. 2003). This can be used directly in the culture environment after undergoing further toxicological studies. Thus, this animal-based product also emerges as one of the alternative bioremediators used in shrimp aquaculture because of its eco-friendly nature.

Bivalves/molluscs serve as one of the best alternative bioremediators for treating aquaculture effluents. Molluscs are filter-feeding organisms that extract nutrients like suspended organic and inorganic particles and algae from the aquatic system. These organisms have the potential to filter out the substances from the water, and when nutrients are absorbed, they expel out as pseudo faeces. To maintain recycle of energy inside an ecologically balanced aquaculture farm, lower trophic organisms like molluscs are used to consume the wastewater generated from the aquaculture of carnivorous and omnivorous fishes (Neori 2007). Accordingly, several findings have been made predicting the suitability of molluscs in the remediation process in shrimp aquaculture. The adult black clam (*Chione fluctifraga*) has been used in the remediation of aquaculture effluent from *Penaeus vannamei* culture unit (Martínez-Córdova et al. 2011a). It filters the total suspended solids and organic suspended solids from shrimp culture effluent, thus reducing its concentration in the effluent (Martínez-Córdova et al. 2011b). Hence, the farming of black clams possessing bioremediation capacity in the shrimp effluent will reduce the suspended solids as well as generate income by its production besides good shrimp production. Likewise, the ark shell bivalve (*Scapharca subcrenata*) has influenced the growth of potential microbes in the bioremediation system of mariculture effluents, thus reducing around 84% TAN, 81% nitrite, and 72% nitrate from the system (Lukwambe et al. 2020). Similarly, oyster species and cockle are also involved in the remediation of shrimp effluents. The bivalves Cortez oyster (*Crassostrea corteziensis*) and the mud cockle (*Anadara tuberculosa*) are administered in the shrimp effluent treatment process where the efficiency of the bivalve was determined by analysing different filtration variables at three different flux velocities. The results suggested the potential of the two species as bioremediators as it reduces a huge amount of suspended solids and also assimilates a higher concentration of organic suspended solids (Peña-Messina et al. 2009). This preliminary evaluation might pave the way towards utilizing native species by considering their ecological and economic characteristics in the bioremediation of shrimp aquaculture effluent. This concept/technique is gaining interest nowadays as it is cost-effective, environment friendly, and socially acceptable, focusing on sustainability in the effluent management in aquaculture systems.

2.4 Enzymatic Bioremediator (Recombinant Protein)

Emerging pollutants in aquatic environments are linked to abiotic stresses and subsequently deteriorate the ecosystem with serious hazards. Conventional methods for removing such xenobiotic, persistent organic pollutants are ineffective and result in more toxic metabolites. Expression cloning is one of the most basic techniques of molecular biology to study protein function, wherein DNA coding for a protein of interest is cloned into an expression vector, which may have special promoter elements to drive the production of the protein of interest. Enzymatic bioremediation is a cost-effective technique involving the use of cell-free enzymes for the rapid biodegradation of contaminants (Wackett and Robinson 2020). The most representative enzymes involved in bioremediation include esterases, laccases, hydrolases, dehalogenases, dehydrogenases, proteases, and lipases, which have shown promising potential for degradation of polymers, aromatic hydrocarbons, halogenated compounds, dyes, detergents, and agrochemical-pesticides such as organochlorines, organophosphates, synthetic pyrethroids, and carbamates, favoured by various mechanisms such as oxidation, reduction, elimination, and ring-opening (Pandey et al. 2010). Biodegradation of pollutants can be enhanced by utilizing genetically engineered microorganisms that produce several recombinant enzymes. Recombinant microorganisms with expanded bioremediation capacity are increasingly accepted, though application in the aquaculture sector is yet to take shape.

Transgenic plants can be developed by transferring genes from organisms that have the potential for degradation/mineralization of xenobiotic pollutants to candidate plants to improve the ability of plants to degrade/metabolize xenobiotic pollutants. Genes involved in the degradation of xenobiotic pollutants can be isolated from bacteria/fungi/ animals/plants and introduced into candidate plants using Agrobacterium-mediated or direct DNA methods of gene transfer (Eapen et al. 2005). Specific catabolic genes essential for the degradation of a contaminant are overexpressed in a plant, resulting in enhanced phytoremediation. For example, transgenic tobacco, rockcress, mustard, poplar, rice, and potato have been reported to be able to improve phytoextraction, phytovolatilization and phytodegradation of heavy metals and organic contaminants like explosives, chlorinated solvents, etc. The most recent and very promising approach to improve phytoremediation ability is the construction of plants with enhanced secretion of enzymes capable of degrading xenobiotics into the rhizosphere. The advantage of this method is that the plants do not need to take up the pollutants to detoxify them; instead, the secreted enzymes can degrade the pollutants in the rhizospheric zone.

A recombinant Elastin-like biopolymer (ELP) composed of a polyhistidine domain with Ag^+ has been developed, and its bactericidal activity against pathogenic bacteria has successfully been demonstrated by Krishnani et al. (2014). This study has an application in formulating biopolymer-based antibacterial in diverse fields of healthcare and disease management in aquaculture systems.

3 Bioremediators for Health Management of Shrimp Aquaculture

One major constraint prevailing in shrimp aquaculture is the outbreak of diseases. In order to effectively manage diseases, three factors must be considered: host, environment, and pathogen. Microbial attack on the farmed animal is due to deteriorated water quality and soil conditions. Water quality should be maintained in the culture system to avoid other challenges in the system. The presence of an excess concentration of pollutants will provide favourable conditions for the microbes to invade the system as the culture animals were under stress. Diseases are mainly caused by microbes such as bacteria, viruses, fungi, and protozoans. The disease attack will lead to production loss from the culture unit. Hence, management practices utilizing antibiotics were practised earlier. The use of antibiotics was avoided due to disadvantages such as the development of antibiotic-resistant bacteria and the residues of antibiotics in sediments and farmed animals. Therefore, later on, sustainable approaches to using beneficial microbes were adopted by the farmers, and such an approach is generally called bioremediation. Several bioremediation strategies are used for the biocontrol of diseases in the aquaculture system, and, among them, green water technology and probiotics serve as excellent bioremediators for health management in shrimp aquaculture.

3.1 Green Water-Based Bioremediator

In penaeid shrimp culture systems and hatcheries, vibriosis is responsible for the mass mortalities in larvae, postlarvae and grow-out shrimp. Vibriosis is caused by bacteria such as *Vibrio harveyi*, *V. splendidus*, *V. parahaemolyticus*, *V. anguillarum*, *V. vulnificus*, *V. fischeri*, etc., which are difficult to control in aquaculture and related aquatic environment when the stocking density is high (Karunasagar et al. 1994). Green water technology is an innovative approach towards controlling pathogens in shrimp aquaculture. In India, the Green water system for improving the health of shrimp was carried out by Krishnani et al. (2012a), where herbivorous fishes serve as bioaugmentors yielding green water that can be used in shrimp farming. The green water contains a diverse group of beneficial indigenous microorganisms, which help improve the water quality and also prevent the proliferation of luminous bacteria, and improve the shrimp production from the system. In this study, herbivorous fishes like milkfish were stocked in cages inside the shrimp ponds, and thus, indigenous microbes were utilized. Similarly, tilapia-produced green water was also used as a biocontrol agent for luminous bacteria in *P. monodon* culture ponds (Corre Jr et al. 2005; Tendencia et al. 2015). It is also suggested that the use of green water produced from tilapia protects WSSV in *P. monodon* culture (Tendencia et al. 2012). The green water (tilapia-produced) also possesses the potential to inhibit *Vibrio harveyi* in *P. vannamei* culture (Wibowo et al. 2015). Even in the tank culture of shrimp, tilapia-produced green water is effective in reducing the *Vibrios* (Cadiz et al. 2016). This green water-based bioremediation is a valuable technology as it has great potential in disease management as well as water quality management. The bacterial diversity present in green water has been examined and revealed its role in water quality improvement. It is suggested to use herbivorous fishes like tilapia, milkfish, mullet, etc., for the production of green water, which could be utilized in shrimp farming. The mucus produced by freshwater fishes has an antimicrobial property that aids in the health management of shrimp farming. This is a very effective and efficient alternative approach to bioremediation in shrimp aquaculture, as it helps in both disease management and environmental management, produces dual crops from the system, and increases the income generated from the system by practising a sustainable, environmentally safe culture practice. A hypothesis has been proposed to utilize green water technology as a bioaugmentation technology based on the integration of fish green slime beneficial bacteria and their bioactives for controlling priority pathogens in freshwater and brackish water shrimp aquaculture systems (Krishnani et al. 2021). This method has the following advantages over the previous practices: (i) Economically and environmentally viable technology; (ii) Avoidance of probiotics; (iii) Area-saving and less sophisticated; (iv) Higher production of shrimp; (v) Environmental and health management benefits; (vi) Compatibility of culturing herbivorous fish along with shrimp.

Unculturable bacterial diversity associated with green slime has been examined for the first time using metagenomics (Kathiravan and Krishnani 2012, Krishnani and Kathiravan 2015; 2014; Kathiravan et al. 2012; Krishnani et al. 2010a, b). Nucleic acid methods based on sequencing of clone libraries have provided sequence and the phylogenetic information of individual clones, which indicated that there is an abundance of functional genes implicated in denitrification, nitrogen fixation, methane oxidation and sulphide oxidation in greenwater systems of coastal aquaculture. The adoption of molecular techniques led to the realization that microbial populations relevant to coastal aquaculture are much more diverse than previously thought using traditional culture methods. Krishnani et al. (2021) has recommended that besides tiger shrimp, *P. monodon*, this technology can also have future potential application in zero-water exchange systems of white shrimp *P. indicus* and whiteleg shrimp *P. vannamei* with integration of grey mullet (*Mugil cephalus*) or milkfish (*Chanos chanos*) in pens for controlling pathogenic vibrios and achieving higher shrimp production in brackish water/coastal aquaculture and inland saline aquaculture. This can also be applied in the pen culture of Indian major carps or any other ideal candidate herbivorous fish species in the freshwater aquaculture system cultured with the giant freshwater prawn *Macrobrachium rosenbergii* for controlling biotic stresses in the prawn.

3.2 Probiotic Bioremediator (Probionts)

Water quality and disease control are directly related and controlled by microbial activity in the system. Microbial intervention in aquaculture system works because of the competitive exclusion of the indigenous microflora and the antagonistic characteristics against potential pathogens. Probiotic is an antonym of antibiotics, and it involves multiplying and competing with live beneficial microorganisms with harmful ones by suppressing their growth and contributing to intestinal microbial balance in the host (Nayak 2010). Probiotics are the best alternative to antibiotics in shrimp aquaculture and are a bioremediation strategy that offers effective control over disease-causing pathogens in the system. There exist two types of probiotics: soil probiotics and water probiotics. The action of beneficial bacteria over disease-causing pathogens makes it a promising tool in health management in aquaculture practice. The probiotics not only serve as biocontrol over pathogens but also aid in improving the growth and survival of the cultured animal, apart from water quality improvement. Jha (2011) has reported using probiotics in shrimp aquaculture, which was effective against pathogenic *Vibrios* and toxic gases like NH_3, NO_2, etc. The most widely used probiont in aquaculture includes *Bacillus* sp., *Lactobacillus* sp., etc. *Bacillus* sp. can tolerate adverse environmental conditions and aids in the proliferation of beneficial bacteria and suppression of pathogenic forms. Different strains of *Bacillus* find application as probiotics in aquaculture. The strain *Bacillus subtilis* BT23 was used to treat pathogens in the *P. monodon* culture unit against *Vibrio* sp. The beneficial bacterial strain was isolated from the shrimp pond, and examined its efficiency of bioremediation under in vivo and in vitro conditions. The results of the study elucidated the potential of utilizing beneficial microbes over antibiotics in shrimp farming (Vaseeharan and Ramasamy 2003). Similarly, two *Bacillus subtilis* strains, L10 and G1, have been administered in the *P. vannamei* juvenile rearing water to enhance the resistance against *V. harveyi* infection, water quality, and the growth performance of juveniles. The results revealed the potential of these two strains to render beneficial effects in disease management and environmental management in shrimp aquaculture (Zokaeifar et al. 2014). A finding by Karthik et al. (2014) showed that *Lactobacillus* sp. (AMET1506) could be effectively used in the bioremediation approach as a probiont to control *Vibrios* in *P. monodon* culture. Likewise, microbes such as *Alteromonas* showed protection against the pathogen *Vibrio harveyi* and increased the survival rate of *P. indicus* larvae. Non-pathogenic isolates of *Vibrio alginolyticus* were found to have the potential to suppress pathogenic forms of *Vibrio* like *V. harveyi* and *V. parahaemolyticus*. The use of probiotic bioremediators in shrimp aquaculture is a promising sustainable technique to manage the system from adverse water quality problems and disease outbreaks. This approach paves the way towards adopting an ecologically safe farming practice.

Apart from probiotics, a recent concept of one health is prebiotics, which are basically resistant to attack by the enzymes of the host body (endogenous) so that they can reach the site of action to promote the proliferation of gut microflora by acting as the foods for probiotics. Prebiotics have been reported to have several beneficial effects such as increased disease resistance and improved nutrient availability, which in turn have much potential to increase the efficiency and sustainability of aquaculture production. Decades of research on the application of prebiotics and probiotics in aquaculture have been primarily driven by the need to reduce or eliminate antibiotics from production systems and improve sustainability. Compounds produced by beneficial bacteria that restrict pathogens include natural antibiotics, bacteriocins, siderophores, lysozymes, proteases, hydrogen peroxide, organic acids, and other bactericidal bioactives.

4 Combined Strategy for Aquatic Bioremediation for Environmental and Health Management of Shrimp Aquaculture

The usage of a single type of bioremediator in shrimp aquaculture failed to reduce all types of chemical and microbial contaminants in the system. Hence, it is recommended to use a combination of different bioremediators, which would bring solutions to multiple emerging constraints simultaneously in the aquaculture environment. This approach, which combines different bioremediators should be emphasized as it is economically efficient to enhance the environmental quality and health of farmed animals, and it renders more production and more income by recycling the wastes generated in the system and, thereby, making the system environmentally sustainable. There are several research findings on the integrated bioremediation approach that give insight into the widely practised bioremediation methods.

4.1 Biofloc-Based Bioremediator (Immunomodulator)

Biofloc technology is an emerging practice of culturing aquatic animals at high densities. In this system, flocs are developed consisting of bacteria (nitrifying, heterotrophic), fungi, ciliates, microalgae, flagellates, zooplankton, faeces, and exoskeleton of dead animals (Gao et al. 2012). It is a promising technique as it circulates the waste generated in the system into microbial floc consisting primarily of heterotrophic bacteria. The detritus and microbial biomass called flocs serve as natural food for the shrimp, thus reducing the feed conversion ratio and also improving the water quality by degrading the wastes generated in the system by the action of microbes (Brito et al. 2016). Biofloc is a sustainable zero water exchange practice where the water quality is maintained by balancing the carbon:nitrogen (C:N) ratio in the system. The carbon can be added to the system, whereas the nitrogenous compounds are obtained from the feed supplied as well as from the waste excreted by the animal. By balancing this C:N ratio at the rate of 10–15:1, the nitrogenous compounds will get converted into microbial biomass, which renders several advantages, including improvement in water quality in the culture unit. Thus, this biofloc serves as a bio-remediator in the shrimp culture system. However, the identification of organisms in the biofloc by examining its composition is of prime importance as it helps to understand the action of microbes in the system. A study has been conducted by Manan et al. (2017) to determine the composition of biofloc to identify its role and interaction among floc particles in the bioremediation in shrimp aquaculture, particularly in *P. vannamei* farming system. The aggregated flocs developed in the system comprising different microorganisms like heterotrophic bacteria effectively degrade the organic matter present in the shrimp culture system, and nematodes, algae, etc., aggregate and act as natural food as well as help in the remediation of pollutants in the system. Thus, biofloc is regarded as a natural bioremediator in zero water exchange shrimp aquaculture. It has been proven that bacteria that are present naturally in the biofloc system culturing shrimp are adequate to maintain the water quality in the system, providing excellent bioremediation as a means of environmental management, and there is no additional requirement for probiotics in the biofloc system (Ferreira et al. 2017).

In a biofloc-based culture system, ideal levels of protein sources and suitable and low-cost carbon sources have been evaluated and standardized by Panigrahi et al. (2018, 2019a, b). They observed that shrimp have an innate immune system that responds against the pathogen in biofloc-based culture system, which can be attributed to the beneficial microbiota and their cell components and metabolites acting as immunostimulants. Panigrahi et al. (2020) have evaluated and compared various biofloc generation protocols for nursery rearing with higher stocking density for growth performances, physicochemical parameters, microbial dynamics, feed utilization, nutritional advantages and immunomodulatory effects on gene expression level in *Penaeus vannamei* early postlarvae.

In the biofloc system of *P. vannamei*, the mullet has been integrated, which produced protection for *P. vannamei* against *Vibrio* sp. and also causes a reduction in sludge production in biofloc as the sludge is consumed by mullet (Borges et al. 2020). There are different combined strategies of bioremediation; however, identifying a suitable approach with regard to one's farming practice is of prime importance. It is better to choose bioremediators based on availability, cost-effectiveness, and remediation potential.

4.2 Integrated Multi-Trophic Aquaculture (IMTA)

The integrated strategy has recently gained focus on treating aquaculture wastewater owing to its sustainable nature. IMTA is regarded as a suitable and very efficient bioremediation approach to limit aquaculture nutrients, organic matter outputs and pathogens through biomitigation. The use of the biofilter alone was not optimal for the remediation of organic matter in shrimp effluents. Besides, the capacity of macroalgae to extract inorganic nitrogen was insufficient. The co-cultured species are used as biofilters, and each level has its own independent commercial value, providing both economic and environmental sustainability. IMTA combines inorganic extractors like macroalgae and organic extractive species like bivalves along with shrimp culture to cope with the abiotic and biotic stressors in the aquatic environment. Macroalgae possess a good assimilation capacity for inorganic pollutants like nitrogenous compounds and phosphorous. An integrated approach combining different bioremediators like clam, biofilm, and macrophytes was adopted in shrimp farms as a strategy to reduce the excessive nutrient concentration from the system. Lukwambe et al. (2019) observed a greater reduction in nitrogen, phosphorus and particulate organic matter in shrimp farms predicting the significant action of different bioremediators in environmental management. The experimental result obtained by Song et al. (2016) has shown that *Argopecten irradians*, *Gracilaria lemaneiformis* and the biofilter efficiently reduced the contents of SS, dissolved inorganic carbon (DIC) and dissolved organic carbon (DOC) in the breeding wastewater. Similarly, another combination of biofilters like scallops and seaweed in shrimp culture also resulted in considerable enhancement of water quality in a closed circulation system when compared to culturing a single species (Song et al. 2016). The first innovative approach to bioremediation has been successfully performed by Martínez-Córdova et al. (2011a, b), where the shrimp effluents were subjected to bioremediation using black clam (*Chione fuctifraga*) and benthic microalgae (*Navicula* sp.) and this treated effluent was reused to farm *P. vannamei*. This shows the great potential of bioremediators in finding a new solution for aquaculture issues sustainably. The combination of Artemia and macroalgae, such as *Artemia franciscana* and *Gracilaria caudate*, respectively, were used as bioremediators to determine their nutrient removal efficiency from aquaculture effluents. The results predicted this combination as an alternative bioremediator in coastal waters (Marinho-Soriano et al. 2011). Likewise, another integrated approach combining copepod (*Oithona rigida*) and macroalgae (*Amphora* sp. and *Picochlorum maculatam*) as bioremediators in treating *P. vannamei* effluents has been performed, and its efficiency in environmental management illustrated by Dinesh Kumar et al. (2016).

The biofloc system itself is a self-sufficient bioremediation approach in shrimp aquaculture. However, the effluents from the biofloc system were treated using the bioremediator clam (*Anomalocardia brasiliana*), which revealed the efficiency of clam as an excellent bioremediator in removing the solid wastes from the system (Brito et al. 2018). The integration of filter-feeding organisms capable of uptake of potential pathogenic microorganisms and particles from the water column may present an opportunity to decrease disease outbreaks and also control pathogens (Granada et al. 2016).

4.3 Nanobioremediator

The use of nanotechnology in coastal shrimp aquaculture is still in its infancy. Silver nanoparticles (AgNPs) have emerged as a powerful weapon against antibiotic-resistant microorganisms. Krishnani et al. (2012b) investigated the ammonia and antimicrobial activity of silver ion-exchanged zeolite against *Escherichia coli*, *Vibrio harveyi*, *Vibrio cholerae* and *Vibrio parahaemolyticus* and the minimum inhibitory concentration was found to be 40 and 50–60μg/mL, respectively. With continued exploration and refinement in nanobiotechnological interventions, nanostructured materials could play a significant role in coastal shrimp aquaculture. Biogenic synthesis of nanoparticles is advantageous in comparison to the physical and chemical synthesis methods as it is cost-effective, environmentally viable and involves biomolecule-based capping agents that provide specific functional groups on the surface of nanoparticles (Das et al. 2023; Chakraborty et al. 2023). Chakraborty et al. (2024) have successfully demonstrated a novel approach to the valorization of nanosilver for the detoxification of hexavalent chromium under neutral to alkaline environmental conditions, offering a comprehensive framework and expanding applications in the field of nanobioremediation and circular bioresource utilization.

4.4 Nutri-Bioremediator

Recently, extensive research has been done on nutritional remediation of environmental and microbial contaminants. Nutritional systems have been developed by Kumar et al. (2014, 2016a, b, 2017), wherein dietary lecithin and pyridoxine, have successfully been demonstrated as important nutritional components in developing immunity and providing protection in brackish water fish against cellular stress caused by an organochlorine pesticide and temperature. The nutritional components, viz. methyl donors, pyridoxine, yeast RNA, omega 3 fatty acid, and microbial levan, can be very important nutritional components that may have future applications in the mitigation of multiple stresses in coastal shrimp aquaculture.

5 Conclusion

Shrimp farming, the burgeoning industry, is facing challenges owing to adverse environmental conditions like poor water and soil quality and the emergence of diseases. Farmers follow different chemical and biological methods to overcome these challenges. The best choice of treatment is the adoption of bioremediation in shrimp farming which renders several advantages over chemical methods. The bioremediation approach plays a vital role in the nitrification and denitrification processes in the aquaculture system and promotes the environment and health status of the cultured animal. The approach of integrating different biological components in the growing system of shrimps, as well as the adoption of single bioremediators serve as excellent alternatives in shrimp aquaculture towards environmental and health management. The use of probiotics and green water is effective against the pathogen attack, which also facilitates improvement in water quality. Likewise, the utilization of plant-based products, biofloc, microalgae, and macroalgae are well-suitable bioremediators in shrimp aquaculture. However, it is necessary to understand the indigenous microbial community, which is the prime player in any type of bioremediation. Thus, understanding the microbial community using a multi-omics approach will further add an advantage over the bioremediation process of environmental and disease management in shrimp aquaculture. The adoption of the bioremediation process in the system would enhance the carrying capacity of the system and bring restoration to the environment.

References

Abisha R, Krishnani KK, Sukhdhane K, Verma AK, Brahmane M, Chadha NK (2022) Sustainable development of climate-resilient aquaculture and culture-based fisheries through adaptation of abiotic stresses: a review. J Water Clim Change 13(7):2671–2689

Anastopoulos I, Kyzas GZ (2015) Progress in batch biosorption of heavy metals onto algae. J Mol Liq 209:77–86

Antony SP, Philip R (2006) Bioremediation in shrimp culture systems. NAGA, WorldFish Center Q 29:62–66

Arunkumar D, Krishnani KK, Kumar N, Sarkar B, Upadhyay AK, Sawant PB, Chadha NK, Abisha R (2023) Mitigating abiotic stresses using natural and modified stilbites synergizing with changes in oxidative stress markers in aquaculture. Environ Geochem Health 45(7):4565–4581

Azim ME, Milstein A, Wahab MA, Verdegam MCJ (2003) Periphyton–water quality relationships in fertilized fishponds with artificial substrates. Aquaculture 228:169–187

Babu DT, Archana K, Kachiprath B, Solomon S, Jayanath G, Singh IB, Philip R (2018) Marine actinomycetes as bioremediators in *Penaeus monodon* rearing system. Fish Shellfish Immunol 75:231–242

Banerjee S, Khatoon H, Shariff M, Yusoff FM (2010) Enhancement of *Penaeus monodon* shrimp postlarvae growth and survival without water exchange using marine *Bacillus pumilus* and periphytic microalgae. Fish Sci 76:481–487

Baskaran V, Patil PK, Antony ML, Avunje S, Nagaraju VT, Ghate SD, Nathamuni S, Dineshkumar N, Alavandi SV, Vijayan KK (2020) Microbial community profiling of ammonia and nitrite oxidizing bacterial enrichments from brackishwater ecosystems for mitigating nitrogen species. Sci Rep 10:1–1

Borges BA, Rocha JL, Pinto PHO, Zacheu T, Chede AC, Magnotti CCF, Cerqueira VR, Arana LAV (2020) Integrated culture of white shrimp *Litopenaeus vannamei* and mullet *Mugil liza* on biofloc technology: Zootechnical performance, sludge generation, and *Vibrio spp.* reduction. Aquaculture:735234

Boyd CE (1998) Water quality for pond aquaculture. Res Dev Ser 43:1–37

Brito LO, dos Santos IGS, de Abreu JL, de Araújo MT, Severi W, Gàlvez AO (2016) Effect of the addition of diatoms (*Navicula spp.*) and rotifers (*Brachionus plicatilis*) on water quality and growth of the *Litopenaeus vannamei* postlarvae reared in a biofloc system. Aquac Res 47:3990–3997

Brito LO, Cardoso Junior LDO, Lavander HD, Abreu JLD, Severi W, Gálvez AO (2018) Bioremediation of shrimp biofloc wastewater using clam, seaweed and fish. Chem Ecol 34:901–913

Cadiz RE, Traifalgar RFM, Sanares RC, Andrino-Felarca KGS, Corre VL Jr (2016) Comparative efficacies of tilapia green water and biofloc technology (BFT) in suppressing population growth of green *Vibrios* and *Vibrio parahaemolyticus* in the intensive tank culture of *Penaeus vannamei*. Aquac Aquar Conserv Legis 9:195–203

Chakraborty P, Krishnani KK, Mulchandani A, Sarkar DJ, Das BK, Pani PK, Sawant PB, Kumar N, Sarkar B, Poojary N, Mallik A, Pal P (2023) Toxicity assessment of poultry-waste biosynthesized nanosilver in *Anabas testudineus* (Bloch, 1792) for responsible and sustainable aquaculture development-a multi-biomarker approach. Environ Res 235:116648

Chakraborty P, Krishnani KK, Mulchandani A, Pani PK, Sarkar DJ, Sawant PB, Kumar N, Sarkar P, Mallik A, Pal P, Nagendrasai K, Das BK (2024) Speciation specific chromium bioaccumulation and detoxification in fish using hydrogel microencapsulated biogenic nanosilver and zeolite synergizing with biomarkers. Environ Geochem Health 46:298

Chávez-Crooker P, Obreque-Contreras J (2010) Bioremediation of aquaculture wastes. Curr Opin Biotechnol 21:313–317

Corre VL Jr, Janeo R, Ronquillo JO, Kurokura H (2005) Use of green water technology as biocontrol of luminous bacteria in intensive shrimp (*Penaeus monodon*) grow-out culture. UPV J Nat Sci 10:51–60

Das K, Krishnani KK, Upadhyay AK, Shukla SP, Prasad KP, Chakraborty P, Sarkar B (2023) Fish waste capped and colloidal nanosilver and its valorization as natural zeolite conjugates for application in aquaculture. J Dispers Sci Technol 45(7):1281–1295

Devaraja T, Banerjee S, Yusoff F, Shariff M, Khatoon H (2013) A holistic approach for selection of *Bacillus spp*. as a bioremediator for shrimp postlarvae culture. Turk J Biol 37:92–100

Eapen S, D'Souza SF (2005) Prospects of genetic engineering of plants for phytoremediation of toxic metals, Biotechnol Adv 23(2): 97–114

Elizondo-González R, Quiroz-Guzmán E, Escobedo-Fregoso C, Magallón-Servín P, Peña-Rodríguez A (2018) Use of seaweed *Ulva lactuca* for water bioremediation and as feed additive for white shrimp *Litopenaeus vannamei*. PeerJ 6:4459

Far HZ, Saad CRB, Daud HM, Harmin SA, Shakibazadeh S (2009) Effect of *Bacillus subtilis* on the growth and survival rate of shrimp (*Litopenaeus vannamei*). Afr J Biotechnol 8(14)

Ferreira MGP, Melo FP, Lima JPV, Andrade HA, Severi W, Correia ES (2017) Bioremediation and biocontrol of commercial probiotic in marine shrimp culture with biofloc. Lat Am J Aquat Res 45:167–176

Gadd GM (2000) Bioremedial potential of microbial mechanisms of metal mobilization and immobilization. Curr Opin Biotechnol 11:271–279

Gao L, Shan HW, Zhang TW, Bao WY, Ma S (2012) Effects of carbohydrate addition on *Litopenaeus vannamei* intensive culture in a zero-water exchange system. Aquaculture 342:89–96

Ghosh W, Dam B (2009) Biochemistry and molecular biology of lithotrophic sulfur oxidation by taxonomically and ecologically diverse bacteria and archaea. FEMS Microbiol Rev 33:999–1043

Ghosh S, Sasmal D, Jawahar Abraham T (2010) Microcosm evaluation of indigenous microflora of traditional shrimp farming system as bioremediators. Indian J Fish 57:97–101

Granada L, Sousa N, Lopes S, Lemos MFL (2016) Is integrated multitrophic aquaculture the solution to the sectors' major challenges? – a review. Rev Aquac 8(3):283–300

Granada L, Lopes S, Novais SC, Lemos MF (2018) Modelling integrated multi-trophic aquaculture: optimising a three trophic level system. Aquaculture 495:90–97

Ibuot AA, Gupta SK, Ansolia P, Bajhaiya AK (2019) Heavy metal bioremediation by microalgae. In: Chang Y-C (ed) Microbial biodegradation of xenobiotic compounds. CRC Press, Florida, p 57

Janssen AJ, Meijer S, Bontsema J, Lettinga G (1998) Application of the redox potential for controling a sulfide oxidizing bioreactor. Biotechnol Bioeng 60:147–155

Jha AK (2011) Probiotic technology: an effective means for bioremediation in shrimp farming ponds. J Bangladesh Acad Sci 35:237–240

Kang YH, Park SR, Chung IK (2011) Biofiltration efficiency and biochemical composition of three seaweed species cultivated in a fish-seaweed integrated culture. Algae 26:97–108

Karthik R, Hussain AJ, Muthezhilan R (2014) Effectiveness of *Lactobacillus sp* (AMET1506) as probiotic against vibriosis in *Penaeus monodon* and *Litopenaeus vannamei* shrimp aquaculture. Biosci Biotechnol Res Asia 11:297–305

Karunasagar I, Pai R, Malathi GR, Karunasagar I (1994) Mass mortality of *Penaeus monodon* larvae due to antibiotic-resistant *Vibrio haeveyi* infection. Aquaculture 128:203–209

Kate GU, Krishnani KK, Kumar N, Sukhdhane K, Verma AK, Brahmane MP, Kumar J (2023) Abiotic and biotic stress alleviating effects of the medicinal and aromatic plant-derived product on striped catfish *Pangasianodon hypophthalmus*. Fish Shellfish Immunol 135:108625

Kathiravan V, Krishnani KK (2012) Diversity of diazotrophic bacteria in greenwater system of coastal aquaculture. J Environ Appl Biores 1(1):5–11. ISSN 2319-8745

Kathiravan V, Krishnani KK (2014) Diversity of denitrifying bacteria in greenwater system of coastal aquaculture. Int Aquat Res 6(3):135–145

Kathiravan V, Krishnani KK, Azad IS, Kailasam M, Ponniah AG (2012) Diversity of nitrite oxidizers in greenwater system of coastal aquaculture by metagenomic approach targeting *nxrb* genes. Biotechnol Indian J 6(11):359–366. ISSN 0974-7435

Keshavanath P, Gangadhar B, Ramesh TJ, Van Rooji JM, Beveridge MCM, Baird DJ, Verdegem MCJ, van Dam AA (2001) The potential of artificial substrates to enhance production of herbivorous fish in Indian freshwater ponds—preliminary trials. Aquac Res 32:189

Krasaesueb N, Incharoensakdi A, Khetkorn W (2019) Utilization of shrimp wastewater for poly-β-hydroxybutyrate production by *Synechocystis sp.* PCC 6803 strain ΔSphU cultivated in photobioreactor. Biotechnol Rep 23:e00345

Krishnani KK (2010) Detection and diversity of nitrifying and denitrifying bacteria in coastal aquaculture. Aquaculture 302:57–70

Krishnani KK, Ayyappan S (2006) Heavy metals remediation of water using plants and lignocellulosic agrowastes. Rev Environ Contam Toxicol 188:59–84

Krishnani KK, Kathiravan V (2015) Functional gene based multiplex PCR to detect bacteria implicated in biotransformation of nitrogen, sulfur and methane in aquatic environment. Indian J Biotechnol 14:482–488

Krishnani KK, Gupta BP, Joseph KO, Muralidhar M, Nagavel A (2002) Studies on the use of neem products for removal of ammonia from brackishwater. J Environ Sci Health 37:893–904

Krishnani KK, Gupta BP, Joseph KO, Muralidhar M, Sarada C, Nagavel A, Parimala V (2003) Decontamination of nitrogenous toxicants from brackish water using plant and animal extracts. Bull Environ Contam Toxicol 71:0196–0203

Krishnani KK, Parimala V, Meng X (2004) Detoxification of chromium (VI) in coastal water using lignocellulosic agricultural waste. Water SA 30:541–545

Krishnani KK, Parimala V, Gupta BP, Azad IS, Meng X, Abraham M (2006) Bagasse-assisted bioremediation of ammonia from shrimp farm wastewater. Water Environ Res 78:938–950

Krishnani KK, Meng X, Boddu VM (2008a) Fixation of heavy metals onto lignocellulosic sorbent prepared from paddy straw. Water Environ Res 80(11):2165–2174

Krishnani KK, Meng X, Christodoulatos C, Boddu VM (2008b) Biosorption mechanism of nine different heavy metals onto biomatrix from rice husk. J Hazard Mater 153:1222–1234

Krishnani KK, Meng X, Dupont L (2009a) Metal ions binding onto lignocellulosic biosorbent. J Environ Sci Health A Tox Hazard Subst Environ Eng 44(7):688–699

Krishnani KK, Shekhar MS, Gupta BP, Gopikrishna G (2009b) Sequence similarity based identification of nitrifying bacteria in coastal aquaculture for bioremediation predictability. Asian Fish Sci 22:41–49

Krishnani KK, Gopikrishna G, Pillai SM, Gupta BP (2010a) Abundance of sulfur oxidizing bacteria in coastal aquaculture using *soxB* gene analyses. Aquac Res 41(9):1290–1301

Krishnani KK, Kathiravan V, Natarajan M, Kailasam M, Pillai SM (2010b) Diversity of sulfur oxidizing bacteria in greenwater system of coastal aquaculture. Appl Biochem Biotechnol 162:1225–1237

Krishnani KK, Kailasam M, Kathiravan V, Nagavel A (2012a) Aquatic bioaugmentation in coastal aquaculture – development and demonstration of green water technology. Fish Chimes 31

Krishnani KK, Zhang Y, Yan Y, Xiong L, Boopathy R, Mulchandani A (2012b) Bactericidal and ammonia removal activity of silver ion exchanged zeolite. Bioresour Technol 117C:86–91

Krishnani KK, Nagavel A, Kathiravan V, Ponnusamy K, Solanki, Pillai SM, Ponniah AG (2013) Evaluation of bagasse-assisted biostimulation in coastal aquaculture through field and molecular approaches. Res Rev Biosci 7(7): 277–293

Krishnani KK, Hao J, Meng X, Mulchandani A (2014) Bactericidal activity of elastin-like polypeptide biopolymer with polyhistidine domain and silver. Colloids Surf B: Biointerfaces 119:66–70

Krishnani KK, Boddu VM, Moon DH, Ghadge SV, Sarkar B, Brahmane MP, Choudhary K, Kathiravan V, Meng X (2015) Metals bioaccumulation mechanism in neem bark. Bull Environ Contam Toxicol 95(3):414–419

Krishnani KK, Choudhary K, Veera BM, Moon DH, Meng X (2021) Heavy metals biosorption mechanism of partially delignified products derived from mango (*Mangifera indica*) and guava (*Psidium guiag*) barks. Environ Sci Pollut Res 28:32891–32904

Kumar N, Minhas PS, Ambasankar K, Krishnani KK, Rana RS (2014) Dietary lecithin potentiates thermal tolerance and cellular stress protection of milk fish (*Chanos chanos*) reared under low dose endosulfan-induced stress. J Therm Biol 46:40–46

Kumar N, Ambasankar K, Krishnani KK, Gupta SK, Bhushan S, Minhas PS (2016a) Acute toxicity, biochemical and histopathological responses of endosulfan in *Chanos chanos*. Ecotoxicol Environ Saf 131:79–88

Dinesh Kumar S, Santhanam P, Kumar N, Ananth S, Nihtya P, Dhanalakshmi B, Kim M-K (2016) Bioremediation of shrimp (Litopenaeus vannamei) cultured effluent using copepod (Oithona rigida) and microalgae (Picochlorum maculatam and Amphora sp.)—an integrated approach. Desal Water Treatment 57:26257–26266. 10.1080/19443994.2016.1163509

Kumar N, Ambasankar K, Krishnani KK, Kumar P, Akhtar MS, Bhushan S (2016b) Dietary pyridoxine potentiates thermal tolerance, heat shock protein and protect against cellular stress of Milk fish (*Chanos chanos*) under endosulfan-induced stress. Fish Shellfish Immunol 55:407–414

Kumar N, Ambasankar K, Krishnani KK, Gupta SK, Minhas PS (2017) Dietary pyridoxine promotes growth and cellular metabolic plasticity of *Chanos chanos* fingerlings exposed to endosulfan induced stress. Aquac Res 48(5):2074–2087

Lalitha N, Muralidhar M, Saraswathy R, Kumararaja P, Nagavel A (2017) Effect of cassava waste on bioremediation of nitrogenous metabolites and shrimp growth. J Environ Biol 38:611

Lawton RJ, Mata L, de Nys R, Paul NA (2013) Algal bioremediation of waste waters from land-based aquaculture using ulva: selecting target species and strains. PLoS One 8(10):e77344. https://doi.org/10.1371/journal.pone.0077344

Liu KF, Chiu CH, Shiu YL, Cheng W, Liu CH (2010) Effects of the probiotic, *Bacillus subtilis* E20, on the survival, development, stress tolerance, and immune status of white shrimp, *Litopenaeus vannamei* larvae. Fish Shellfish Immunol 28(5):837–844

Lukwambe B, Zhao L, Nicholaus R, Yang W, Zhu J, Zheng Z (2019) Bacterioplankton community in response to biological filters (clam, biofilm, and macrophytes) in an integrated aquaculture wastewater bioremediation system. Environ Pollut 254:113035

Lukwambe B, Nicholaus R, Zhao L, Yang W, Zhu J, Zheng Z (2020) Microbial community and interspecies interaction during grazing of ark shell bivalve (*Scapharca subcrenata*) in a full-scale bioremediation system of mariculture effluents. Mar Environ Res 158:104956

Manan H, Moh JHZ, Kasan NA, Suratman S, Ikhwanuddin M (2017) Identification of biofloc microscopic composition as the natural bioremediation in zero water exchange of Pacific white shrimp, *Penaeus vannamei*, culture in closed hatchery system. Appl Water Sci 7:2437–2446

Marinho-Soriano E, Panucci RA, Carneiro MAA, Pereira DC (2009) Evaluation of *Gracilaria caudata J. Agardh* for bioremediation of nutrients from shrimp farming wastewater. Bioresour Technol 100:6192–6198

Marinho-Soriano E, Azevedo CAA, Trigueiro TG, Pereira DC, Carneiro MAA, Camara MR (2011) Bioremediation of aquaculture wastewater using macroalgae and Artemia. Int Biodeterior Biodegradation 65:253–257

Martínez-Córdova LR, López-Elías JA, Leyva-Miranda G, Armenta-Ayón L, Martinez-Porchas M (2011a) Bioremediation and reuse of shrimp aquaculture effluents to farm whiteleg shrimp, *Litopenaeus vannamei*: a first approach. Aquac Res 42:1415–1423

Martínez-Córdova LR, López-Elías JA, Martínez-Porchas M, Bernal-Jaspeado T, Miranda-Baeza A (2011b) Studies on the bioremediation capacity of the adult black clam, *Chione fluctifraga*, of shrimp culture effluents. Rev Biol Mar Oceanogr 46:105–113

Mawi S, Krishnan S, Din MFM, Arumugam N, Chelliapan S (2020) Bioremediation potential of macroalgae *Gracilaria edulis* and *Gracilaria changii* co-cultured with shrimp wastewater in an outdoor water recirculation system. Environ Technol Innov 17:100571

Mendoza O, Pretell K, Ramírez B, Sandoval J, Caballero JL, Dorado G (2020) Metagenomic analyses of biofilms on whiteleg shrimp (*Litopenaeus vannamei*) effluents: implications for worldwide aquaculture bioremediation and environmental sustainability in the current trend of climate change and global warming–state of the art and experimental proof of concept. Rev Fish Sci Aquac 29:1–14

Moriarty DJW (1998) Control of luminous Vibrio sp. in penaeid aquaculture ponds. Aquaculture 164:351–358

Moriarty DJ (1999) Disease control in shrimp aquaculture with probiotic bacteria. In: Proceedings of the 8th international symposium on microbial ecology. Atlantic Canada Society for Microbial Ecology, Halifax, pp 237–243

Nayak SK (2010) Probiotics and immunity: a fish perspective. Fish Shellfish Immunol 29(1):2–14

Neori A (2007) Essential role of seaweed cultivation in integrated multi-trophic aquaculture farms for global expansion of mariculture: an analysis. J Appl Phycol 20:567–570

Neori A, Chopin T, Troell M, Buschmann AH, Kraemer GP, Halling G, Shpigel M, Yarish C (2004) Integrated aquaculture: rationale, evolution and state of the art emphasizing seaweed biofiltration in modern aquaculture. Aquaculture 231:362–391

Nisbet E, Fowler C (1999) Archaea metabolic evolution of microbial mats. Proc Royal Soc Biol Sci Ser B 266: 2375–2382

Pandey G, Dorrian SJ, Russell RJ, Brearley C, Kotsonis S, Oakeshott JG (2010) Cloning and biochemical characterization of a novel carbendazim (methyl-1H-benzimidazol-2-ylcarbamate)-hydrolyzing esterase from the newly isolated Nocardioides sp. strain SG-4G and its potential for use in enzymatic bioremediation. Appl Environ Microbiol 76:2940–2945

Paniagua-Michel J, Garcia O (2003) Ex-situ bioremediation of shrimp culture effluent using constructed microbial mats. Aquac Eng 28:131–139

Panigrahi A, Saranya C, Sundaram M, Vinoth Kannan SR, Das RR, Satish Kumar R, Rajesh P, Otta SK (2018) Carbon: Nitrogen (C:N) ratio level variation influences microbial community of the system and growth as well as immunity of shrimp (*Litopenaeus vannamei*) in biofloc based culture system. Fish Shellfish Immunol 81:329–337

Panigrahi A, Sundaram M, Saranya C, Kumar RS, Dayal JS, Saraswathy R, Otta SK, Anand PS, Rekha PN, Gopal C (2019a) Influence of differential protein levels of feed on production performance and immune response of pacific white leg shrimp in a biofloc–based system. Aquaculture 503:118–127

Panigrahi A, Sundaram M, Saranya C, Swain S, Dash RR, Dayal JS (2019b) Carbohydrate sources deferentially influence growth performances, microbial dynamics and immunomodulation in Pacific white shrimp (*Litopenaeus vannamei*) under biofloc system. Fish Shellfish Immunol 86:1207–1216

Panigrahi AC, Saranya K, Ambiganandam M, Sundaram MR, Sivakumar KP, Vasagam K (2020) Evaluation of biofloc generation protocols to adopt high density nursery rearing of *Penaeus vannamei* for better growth performances, protective responses and immuno modulation in biofloc based technology. Aquaculture 522:735095

Parimala V, Krishnani KK, Gupta BP, Jayanthi M, Abraham M (2004) Phytoremediation of chromium from seawater using five different products from coconut husk. Bull Environ Contam Toxicol 73:31–37

Peña-Messina E, Martínez-Córdova LR, Bückle-Ramírez LF, Segovia-Quintero MA, Zertuche-González JA (2009) A preliminary evaluation of physiological filtration variables for *Crassostrea corteziensis* (Hertlein, 1951) and *Anadara tuberculosa* (Sowerby, 1833) in shrimp aquaculture effluents. Aquac Res 40:1750–1758

Pokorna D, Zabranska J (2015) Sulfur-oxidizing bacteria in environmental technology. Biotechnol Adv 33:1246–1259

Rabiei R, Phang SM, Yeong HY, Lim E, Ajdari D, Zarshenas G, Sohrabipour J (2014) Bioremediation efficiency and biochemical composition of *Ulva reticulata* Forsskal (Chlorophyta) cultivated in shrimp (*Penaeus monodon*) hatchery effluent. Iran J Fish Sci 13:621–639

Rao SPS, Karunasagar I (2000) Incidence of bacteria involved in nitrogen and sulphur cycles in tropical shrimp culture ponds. Aquac Int 8:463–472

Ravichandran P (2012) Overview of *Litopenaeus vannamei* culture practice and importance of govt. regulation measures. In: Jithendran KP (ed) Training manual on management of emerging diseases of shrimp with special reference to Pacific White Shrimp, *Litopenaeus vannamei* 64

Robertson LA, Kuenen JG (2006) The colorless sulfur bacteria. Prokaryotes 2:985–1011

Saavedra R, Muñoz R, Taboada ME, Vega M, Bolado S (2018) Comparative uptake study of arsenic, boron, copper, manganese and zinc from water by different green microalgae. Bioresour Technol 263:49–57

Sarkar S, Rekha PN, Balasubramanian CP, Ambasankar K (2019) Bioremediation potential of the brackishwater macroalga *Gracilaria tenuistipitata* (Rhodophyta) co-cultured with Pacific white shrimp *Penaeus vannamei* (boone). J Coast Res 86:248–254

Shamim S (2018) Biosorption of heavy metals. Biosorption 2:21–49

Sharma B, Dangi AK, Shukla P (2018) Contemporary enzyme based technologies for bioremediation: a review. J Environ Manag 210:10–22

Sruthisree C, Nayak H, Gowda G, Naveen BT (2015) Evaluation of periphyton and biofilm growth on different substrates in shrimp culture pond. J Exp Zool India 18: 625–630

Song X, Yang Q, Ren JS, Sun Y, Wang X, Sun F (2016) Integrated bioremediation techniques in a shrimp farming environment under controlled conditions. Acta Oceanol Sin 35:88–94. https://doi.org/10.1007/s13131-016-0812-2

Tacon AG (2002) Thematic review of feeds and feed management practices in shrimp aquaculture. Report prepared under the World Bank, NACA, WWF and FAO Consortium Program on Shrimp Farming and the Environment 69

Tahir A, Rukminasari N, Yaqin K, Lukman M (2020) Increasing CO_2 concentration impact upon nutrient absorption and removal efficiency of supra intensive shrimp pond wastewater by marine microalgae *Tetraselmis chui*. Int J Phytoremediation 23:1–8

Tanaka Y, Ashaari A, Mohamad FS, Lamit N (2020) Bioremediation potential of tropical seaweeds in aquaculture: low-salinity tolerance, phosphorus content, and production of UV-absorbing compounds. Aquaculture 518:734853

Tang K, Baskaran V, Nemati M (2009) Bacteria of the sulphur cycle: an overview of microbiology, biokinetics and their role in petroleum and mining industries. Biochem Eng J 44:73–94

Tendencia EA, Bosma RH, Sorio LR (2012) Effect of three innovative culture systems on water quality and white spot syndrome virus (WSSV) viral load in WSSV-fed *Penaeus monodon* cultured in indoor tanks. Aquaculture 350–353:169–174

Tendencia EA, Bosma RH, Verdegem MC, Verreth JA (2015) The potential effect of greenwater technology on water quality in the pond culture of *Penaeus monodon Fabricius*. Aquac Res 46:1–13

Vaseeharan BA, Ramasamy P (2003) Control of pathogenic *Vibrio spp*. by *Bacillus subtilis* BT23, a possible probiotic treatment for black tiger shrimp *Penaeus monodon*. Lett Appl Microbiol 36:83–87

Volesky B, Holan ZR (1995) Biosorption of heavy metals. Biotechnol Prog 11(3):235–250

Wackett LP, Robinson SL (2020) The ever-expanding limits of enzyme catalysis and biodegradation: polyaromatic, polychlorinated, polyfluorinated, and polymeric compounds. Biochem J 477:2875–2891

Wang L, Sun J, Zheng W, Huang T, Zhang Y, Wu Z, He F (2018) Effects of a combined biological restoration technology on nitrogen and phosphorus removal from eutrophic water. Pol J Environ Stud 27:2293–2301

Wibowo AM, Fadjar M (2015) Utilization of Tilapia Mucus to inhibit *Vibrio harveyi* on Vannamei (*Litopenaeus vannamei*). J Life Sci Biomed 5:141–148

Zokaeifar H, Babaei N, Saad CR, Kamarudin MS, Sijam K, Balcazar JL (2014) Administration of *Bacillus subtilis* strains in the rearing water enhances the water quality, growth performance, immune response, and resistance against *Vibrio harveyi* infection in juvenile white shrimp, *Litopenaeus vannamei*. Fish Shellfish Immunol 36:68–74

Zhang Bo, Lin Wenhui, Huang Jianrong, Wang Yajun (2010) Effects of artificial substrates on the growth, survival and spatial distribution of Litopenaeus vannamei in the intensive culture condition. Iranian J Fish Sci 9: 293–304

Finfish Cell Lines and Their Application in Viral Disease Management

T. R. Swaminathan, R. Srinath, S. Visnuvinayagam, and A. Sudhagar

Abstract

Fish cell lines are essential tools in aquaculture research, offering platforms for studying viral pathogens, vaccine development, and toxicology. They aid in understanding immune responses, host–pathogen interactions, and cellular processes, contributing to disease management and sustainable aquaculture. Fish cell cultures have emerged as an effective biological alternative to using animals in research. An extensive number of 960 fish cell lines have been documented by Cellosaurus, an encyclopaedia of cell lines as of January 2025. These in vitro models facilitate vaccine development, drug screening, and pollutant assessment, reducing reliance on live animals while ensuring reproducible results. Beyond experimental trials, fish cell cultures are cost-effective alternatives to mammalian systems for recombinant protein production, and vaccines. Techniques like CRISPR/Cas9 and 3D cultures enhance their therapeutic potential. The use of unauthenticated cell lines would result in the generation of spurious data, and the International Cell Line Authentication Committee (ICLAC) - Register of Misidentified Cell Lines currently lists 593 cell lines. However, challenges such as limited species-specific cell lines, authentication issues, and cross-contamination persist, emphasizing the need for standardized protocols and expanded repositories to maximize their research and industry applications. Advancing fish cell line research and addressing existing challenges will unlock their full potential, driving innovations in aquaculture and biotechnology.

T. R. Swaminathan (✉) · S. Visnuvinayagam
ICAR Central Institute of Fisheries Technology, Kochi, India
e-mail: rajathanga@yahoo.co.in; visnuvinayagam@yahoo.co.in

R. Srinath · A. Sudhagar
Centre for Peninsular Aquatic Genetic Resources, ICAR National Bureau of Fish Genetic Resources, Kochi, India
e-mail: srinath15151@gmail.com; arunsaqua@gmail.com

R. Kooloth Valappil et al. (eds.), *Aquatic Animal Health Management*,
https://doi.org/10.1007/978-981-96-7987-4_24

Keywords

Cell lines · Aquaculture · Virus · Vaccine · Fish

1 Introduction

Aquaculture is a cornerstone of global food security, providing a sustainable protein source and significantly contributing to economies worldwide. However, the sector faces challenges, particularly from infectious diseases caused by viral pathogens, which result in high mortality rates and economic losses. Fish cell lines, derived from various fish species, are versatile models for studying interactions between fish and viral pathogens. The first fish cell line, the Rainbow trout gonadal cell line (RTG-2), was developed in 1962. These cells replicate natural biological functions, enabling researchers to investigate viral entry, replication, and manipulation of host cells. Studies using susceptible fish cell lines have shed light on how viruses like infectious pancreatic necrosis virus (IPNV) and viral haemorrhagic septicaemia virus (VHSV) invade host cells and exploit cellular resources, crucial for developing interventions to prevent infections.

Fish cell lines play a vital role in vaccine development. They facilitate the propagation of target viral pathogens needed for vaccine production. Moreover, cell lines provide a controlled environment to evaluate vaccine candidates, screen immunogenic components, and assess efficacy and safety before live fish trials. Additionally, cell lines allow researchers to explore adjuvants and delivery mechanisms to improve vaccine efficacy. Beyond vaccines, fish cell lines are instrumental in antiviral drug discovery. They facilitate screening of compounds that inhibit viral replication or enhance immune responses. Understanding virus-host interactions through cell lines helps to design targeted measures to manage viral diseases in aquaculture.

Fish cell lines also have broader applications in fish health management. They aid in disease diagnosis, toxicological studies, and assessing environmental stressors. For instance, cell lines help determine the effects of toxins like pesticides or heavy metals, providing data for environmental regulations. As global aquaculture grows and the frequent emergence of viral diseases causes considerable economic losses in farmed fishes, fish cell line research remains integral to combating viral diseases and ensuring the industry's sustainability. Advancing viral disease understanding and developing innovative management strategies will equip the aquaculture sector to meet the rising global demand for seafood sustainably.

In this chapter, we discuss the status of finfish cell lines and their application in the virus isolation and their management. Further, we highlight some of the constraints and legal issues pertaining to finfish cell line research.

2 Overview of Fish Cell Lines

Zebrafish was introduced as a model organism for in vivo studies of vertebrate development (Grunwald and Eisen 2002). This development was driven by the need to address viral infections affecting farmed juvenile trout (RTG2) (Wolf and Quimby 1962). To identify the pathogen and its biological niche, researchers cultured cells from the organs of infected fish, leading to the discovery of infectious pancreatic necrosis virus (IPNV) (Nicholson 1989).

Fish cell lines have predominantly been derived from normal tissues, including the skin, gills, heart, liver, kidney, spleen, swim bladder, and brain. Among these, embryos and fins are the most widely used tissues for primary culture due to their high regenerative capacity. Other cell lines have been developed from the ovary, skin, fin, vertebrae, and scales. Primary cell cultures are often considered more representative of in vivo tissue conditions (Freshney 2006). However, their use is hindered by challenges such as inconsistent initiation and lack of uniformity in the cultures, which limit their reliability and broader applications (Bols et al. 1994). In contrast, established cell lines are typically derived through processes like spontaneous transformation, oncogenic immortalization, or from malignant tumours. These transformations result in continuously proliferating, or immortal, cell lines that offer greater reproducibility and long-term usability (Freshney 2006).

Embryonic and larval cells are among the easiest to cultivate due to their mitotic activity. Historically, challenges in obtaining eggs or fry of certain fish species, particularly pelagic spawners, limited their availability. However, advancements in aquaculture have enabled the routine cultivation of many species previously unavailable in embryonic forms. As a result, numerous embryonic stem-like cell lines have been established in recent years. Extensive studies on small model fish species like zebrafish (*Danio rerio*) and medaka (*Oryzias latipes*) have contributed significantly to the development of embryonic stem (ES) cell lines and gene-targeting techniques. These models facilitate the integration of embryological, genetic, and molecular analyses in vertebrate development. ES-like cell lines have been successfully established in zebrafish and medaka (Lakra et al. 2011). Additionally, a pluripotent cell line, LJES1, was developed from blastula-stage embryos of *Lateolabrax japonicus*, capable of differentiating into various cell types upon treatment with retinoic acid (Chen et al. 2003). Continuous embryonic cell lines, such as SISE from blastula-stage embryos of sea bass (*Lates calcarifer*) (Parameswaran et al. 2006) and the pluripotent SBES cell line from sea bass and catfish (*Heteropneustes fossilis*) (Lakra et al. 2010; Parameswaran et al. 2006), have also been successfully developed, highlighting the progress in establishing robust cell lines for aquaculture research.

Fish cell cultures can be maintained as primary cultures or cell lines, initiated from various tissues using explant or dissociation methods. The simpler explant method involves placing tissue pieces in a culture medium for cells to outgrow, while the dissociation method uses enzymes like trypsin or collagenase to create a single-cell suspension (Collodi and Barnes 1990). Cells require a solid substrate to grow, with initial cultures termed primary cultures, which can be passaged indefinitely or for a limited period. Commonly used media include EMEM and L-15, supplemented with 10–20% foetal bovine serum (FBS), although fish sera were used historically but discontinued due to cytotoxicity and limited availability (Taju et al. 2013). Attachment factors like laminin and gelatin are sometimes added, and antibiotics are used initially to prevent contamination but removed later for pathogen studies. For subculturing, cells are detached with 0.25% trypsin or a trypsin–EDTA combination, while papain is preferred for isolating brain cells due to better viability (Souto et al. 2020).

Once a cell line is developed, it undergoes authentication using methods like growth rate analysis, chromosomal analysis, and sequence analysis (*16S rRNA* and *COI*), with specific functional markers like *Oct4* and *Nanog* for stem cell lines (Swaminathan et al. 2010). Transfection studies assess transfection efficiency, and cell lines must remain contamination-free and undergo periodic characterization to avoid genetic changes that could alter function. Proper maintenance and characterization ensure the reliability and utility of fish cell lines in research and applications in pathogen characterization and management.

Cellosaurus, a knowledge base on cell lines, is an encyclopaedia of information on cell lines, continuously documenting all cell lines used in in vitro research and cataloguing information of 960 finfish cell lines as of 28 January 2025 (Bairoch 2018). So, this large number of fish cell lines raises the importance of their long-term storage in specialized repositories in different locations in the world. The repositories can act as a 'protection' to secure the loss of cell lines in case of any disastrous event. There are many leading cell line repositories in the world including European Collections of Cell Cultures (ECACC), American Type Culture Collection (ATCC), German Collection of Microorganisms and Cell Cultures (DSMZ), Italian Biobank of Veterinary Resources (BVR), Health Science Research Resources Bank (HSRRB), Japanese Collection of Research Bioresources (JCRB), RINKEN Gene Bank, UK Stem Cell Bank, WiCell, China Centre for Type Collection (CCTCC), CellBank Australia, etc., which have been providing authenticated cell line to researchers across the world. The particulars of fish cell lines maintained in some of the reputable repositories are given in Table 1.

Table 1 Status of cell line repository in different parts of the world with reference to fish cell line

Repository name	Country	Fish cell line
Cellosaurus—a knowledge resource on cell lines		**960**
European Collection of Authenticated Cell Cultures (ECACC)	UK	25
American Type Culture Collection (ATCC)	USA	14
CellBank	Australia	17
Leibniz Institute German Collection of Microorganisms and Cell Cultures (DSMZ)	Germany	4
China Center for Type Culture Collection (CCTCC)	China	5
Italian Biobank of Veterinary Resources (BVR)	Italy	11
National Repository of Fish Cell Lines (NRFC)	India	71
National Centre for Veterinary Type Cultures	India	–
National Centre for Cell Science (NCCS)	India	2

A key advantage of fish cell lines is their ability to reveal species-specific responses to viral infections. Certain viruses exhibit organ- or tissue-specificity, emphasizing the importance of establishing cell lines from various tissues and species. Between 2010 and 2020, 280 new cell lines were developed, increasing the total to 783. In India, the National Repository of Fish Cell Lines (NRFC) was established in 2010 at ICAR-National Bureau of Fish Genetic Resources (ICAR-BFGR), Lucknow, with support from the Department of Biotechnology, Government of India and currently maintaining ~81 fish lines. It aims to develop, deposit, maintain, and distribute authenticated fish cell lines. Designated under the Biological Diversity Act, 2002, it is Asia's largest national repository for fish genetic resources. NRFC facilitates research and supports biodiversity conservation by providing cell lines on request (Thangaraj et al. 2021a).

Cell line authentication must be conducted before freezing a master cell bank (MCB) or seed stock using DNA-based methods to confirm the origin and check for misidentification (Capes-Davis et al. 2010). Short tandem repeat (STR) profiling is recommended to ensure uniqueness and allow tracking through transfers. For inbred mouse cell lines, allele-specific tests for introduced genetic alterations should be used when STR profiling is insufficient. It has to be confirmed that cell lines are suitable for their specific purpose, as prolonged passaging can lead to the loss of key characteristics. Routine tests, such as karyotyping, can reveal chromosomal changes, while molecular assays like copy number variation (CNV) or RNA profiling, though costly, provide additional insights. Basic characterization, including imaging at different densities and calculating population doubling time, should be performed upon arrival. For new cell lines, confirming their origin using intermediate filament proteins (e.g. cytokeratins for epithelial cells) or specific markers (e.g. A2B5 for glial cells) is essential, requiring multiple markers for reliable characterization (Aarattuthodi and Dharan 2021).

Misidentification of cell lines occurs due to cross-contamination, poor handling practices, or errors such as mislabelling or thawing incorrect samples, violating Good Cell Culture Practices (GCCP). Sources of contamination include inadequately treated feeder cells, improperly filtered conditioned media, or the coexistence of rapidly growing cell lines with slower ones. This issue, highlighted since the 1960s, often involves commonly misidentified lines like HeLa, MCF-7, or T24 being used under different names (Gartler 1968; Nelson-Rees and Flandermeyer 1976, 1977). Constant vigilance, adherence to GCCP, and regular authentication are essential, with a list of 593 misidentified lines available from the International Cell Line Authentication Committee (Capes-Davis et al. 2010).

3 Applications of Fish Cell Lines in Aquaculture

Fish cell cultures offer a viable alternative to animal use in research, addressing social and ethical concerns while providing advantages such as easy drug dosing, reduced waste, cost-efficiency, and reproducible results (Kelly et al. 1978; Nicholson 1989; Sommerset et al. 2005). These cultures thrive under varied temperatures, osmolarity conditions, and bicarbonate-buffered media with minimal equipment requirements (Lannan 1994; Wolf and Quimby 1976). They have applications in virology, toxicology, radiation, and thermal biology, similar to mammalian cell lines, and can enhance knowledge in endocrinology, physiology, and immunology, benefiting fish health, growth, and reproduction (Bols and Lee 1991; Hightower and Renfro 1988). With over 30,000 fish species, there is significant untapped potential for developing species and tissue-specific cell models for advancing research (Babich and Borenfreund 1991; Verma et al. 2020). A comprehensive list of available fish cell lines can be accessed through repositories like ATCC and ECACC.

3.1 Virology

Aquaculture is a highly active food-producing industry, and over 350 different species of aquatic animals are farmed, including 34 finfish, 8 crustaceans and 12 molluscan species. However, the rise of intensive aquaculture and more global trade of aquatic animals and fisheries products led to the emergence of many new diseases in fish, resulting in huge economic losses in aquaculture and threats to wild aquatic animals. In addition, improvements in aquatic animal disease surveillance have led to the discovery of several emerging and re-emerging viruses. Because of the risk of spread through commercial trade in finfish, many of the diseases are listed by the World Organization for Animal Health (WOAH) to safeguard the global aquatic animal trade to minimize the spread of diseases. The important details on the isolation of finfish viruses affecting farmed fishes are mentioned in Table 2.

Virus isolation is the most sensitive technique for the detection of pathogens and their management, including the production of vaccines and antigens for serological diagnosis to control disease spread. A range of fish cell lines that can sustain the growth of these viruses has been progressively developed. Many of these cell lines are used worldwide by laboratories involved in fish disease diagnosis and research. Many viruses replicate in the cell, and infected cells gradually develop evidence of cell damage, and new virions infect nearby cells, causing the destruction of the monolayer. These changes, including enlargement and rounding of cells, formation of syncytia, multinucleated giant cells, vacuolations in the cells, inclusion bodies, etc., are known as cytopathic effects (CPE). These CPE could be examined either in the unstained cells or in stained cells in some cases. The typical CPE produced by major finfish viruses on susceptible cell lines is given in Fig. 1. The confirmatory diagnosis of viral pathogens will be done by a two-step procedure of isolation and growth of the virus in cell culture followed by virus identification by real-time RT-PCR, enzyme-linked immunosorbent assays (ELISAs), and indirect fluorescent antibody test (IFAT) with specific monoclonal or polyclonal antibodies or by conventional RT-PCR and sequencing of the amplicon.

Table 2 Viral pathogens of finfish and their isolation using fish cell lines

Disease	Causative agent	Recommended cell line	Cytopathic effect (CPE)	Reference
Infection with epizootic haematopoietic necrosis virus	Epizootic haematopoietic necrosis virus is a species of the genus Ranavirus in the family Iridoviridae	Bluegill fry (BF-2), Fathead minnow (FHM), Epithelioma papulosum cyprini (EPC), Chinook salmon embryo (CHSE-214)	Focal lysis surrounded by rounded granular cells monolayer detachment and disintegration	WOAH (2024)
Infection with infectious salmon anaemia virus (ISAV)	Infectious salmon anaemia virus is a species of the genus Isavirus and family Orthomyxoviridae	Atlantic salmon kidney (ASK) cells Atlantic salmon head kidney (SHK-1) CHSE-214	Extensive vacuolation, rounding and detachment of cells	
Infection with infectious haematopoietic necrosis virus	Salmonid novirhabdovirus (commonly known as infectious haematopoietic necrosis virus [IHNV]) is a species of the genus Novirhabdovirus and family Rhabdoviridae	EPC FHM CHSE-214	Grape-like clusters of rounded cells, plaque formation, round and shrinkage of the plaque	
Infection with koi herpesvirus (KHV)	Cyprinid herpesvirus-3 (CyHV-3), is a species of the genus Cyprinivirus in the family Alloherpesviridae	Common Carp Brain (CCB) Koi Fin-1 (KF-1) Koi caudal fin (KCF-1) EPC	Enlargement of cells, syncytia and vacuolated cytoplasm detachment	
Red sea bream iridoviral disease (RSIVD)	Red sea bream iridovirus is a species of the genus Megalocytivirus and family Iridoviridae	Grunt fin (GF) Tail fin of red sea bream (CRF-1) Spotted knifejaw fin (SKF-9)	Distinct cytopathic effect (CPE) was not identified. Incubate at 25 °C for 10 days	

(continued)

Table 2 (continued)

Disease	Causative agent	Recommended cell line	Cytopathic effect (CPE)	Reference
Infection with salmonid alphavirus (SAV)	Salmonid alphavirus is a species of the genus Alphavirus and family Togaviridae	CHSE-214	Typical CPE due to SAV appears as plaques of pyknotic, vacuolated cells	
Infection with spring viraemia of carp virus	Carp sprivivirus (commonly known as spring viraemia of carp virus [SVCV]), of the genus Sprivivirus and the family Rhabdoviridae	EPC FHM	Rounding, detachment and lysis of cells	
Infection with viral haemorrhagic septicaemia virus (VHSV)	Viral haemorrhagic septicaemia virus is a species of the genus Novirhabdovirus and family Rhabdoviridae	BF-2 CHSE-214 EPC FHM Rainbow trout gonad (RTG-2) Rainbow trout gill (RTgill-W1) Rainbow trout spleen macrophages (RTS11)	Rounding of cells, granular cell grape-like clusters, destruction of monolayer	
Koi ranavirus (KIRV)	Koi ranavirus is a species of the genus *Ranavirus* of the family *Iridoviridae*	SNKD2a	Small foci of rounded cells, aggregation of the round cells, destruction of monolayer	Riji John et al. (2023)
Similar damselfish virus (SRDV)	Similar damselfish virus is a species of the genus *Ranavirus* of the family *Iridoviridae*	Snakehead kidney cell line (SNKD2a) Snakehead fish caudal peduncle cell lines (SBCP2 and SBCP2a)	Formation of granulated cells, rounding of cells at the edge of foci, localized cell death and complete damage to the monolayer	
Lymphocystis disease virus (LCDV)	Lymphocystis disease virus is the species of genus *Lymphocystiviruses* of the family Iridoviridae	SAF-1 Flounder (*Paralichthys olivaceus*) gill (FG) Grass carp ovary (GCO) Grass carp kidney (GCK)	Syncretic cell patches, cell aggregation and fusion	Sheng et al. (2020)

(continued)

Table 2 (continued)

Disease	Causative agent	Recommended cell line	Cytopathic effect (CPE)	Reference
Cyprinid herpesvirus 2 (CyHV-2)	*Cyvirus cyprinidallo2*, also known as *Cyprinid herpesvirus 2* (CyHV-2) is a species of virus in the genus *Cyprinivirus*, family *Alloherpesviridae*	FHM, EPC TO-2 KF-1, BF-2, Goldfish fin (GF-1), CCB, Standard Ryukin Takafumi (SRTF), Ryukin fin (RKF), Cyprinus carpio koi fin (CCKF), Fantail Goldfish fin (FtGF) Fan-tail Goldfish Gill (FtGG), Fantail Goldfish liver (FtGL) Fantail Goldfish brain (FtGB Gibel Carp Brain (GiCB)	Pyknosis, granulation, cyto-plasmic vacuolization, syncytium formation in focal areas, appearance of rounded bright cells lysis, and destruction of monolayer	Thangaraj et al. (2021b)
Infectious spleen and kidney necrosis virus (ISKNV)	Infectious spleen and kidney necrosis virus is a species of the genus Megalocytivirus and family Iridoviridae	SKF-9 GF Dwarf gourami fin (DGF) *Astronotus ocellatus* fish (AOF)	Cell rounding and enlargement, vacuolation, shrinkage and rounding	Jeong and Kim (2023), Jitrakorn et al. (2020), and Kawato et al. (2017)
Carp oedema virus (CEV)	Carp oedema virus (CEV) is classified as a member of the family Poxviridae belonging to the subfamily Chordopoxvirinae	CCB EPC	Cell shrinkage, cell rounding, cell–cell fusion and formation of giant multinucleated cells, cytoplasmic inclusion bodies and vacuolation	Abdelsalam et al. (2024)

(continued)

Table 2 (continued)

Disease	Causative agent	Recommended cell line	Cytopathic effect (CPE)	Reference
Viral nervous necrosis (VNN)	Fish nodaviruses, members of the genus *Betanodavirus* under the family *Nodaviridae*	Grouper fin cells (GF-1) Grouper spleen (GS) cells striped snakehead whole fry cells (SSN-1) Grouper fin cells (GF-1) Sahul India Sea bass kidney cell line (SISK)	Rounding, grape-like clusters, detachment, and meshwork, swelling with large bubbles from the plasma membrane, obvious vacuolation in the cytoplasm	Qin et al. (2020)
Tilapia lake virus (TiLV)	*Tilapia tilapinevirus*, under genus *Tilapinevirus* and family *Amnoonviridae*	E-11 cells TiB cells, *Oreochromis niloticus* liver (OnlL) *Oreochromis niloticus* brain (OnlB) OnH E-11 Clone of SSN-1	Cytoplasmic vacuolation, plaque formation, cell aggregation and detachment	Thangaraj et al. (2018)
Tilapia parvovirus (TiPV)	TiPV is a novel parvovirus under the proposed genus *Chapparvovirus* in the family *Parvoviridae*	onlL Snakehead kidney (CSK), Snub nose pompano fin (SPF) Tilapia heart (TH)	Granularity, elongation, rounding and detachment of cells	Rajendran et al. (2023)

3.2 Vaccine Development

Infectious pathogens, particularly viruses, cause significant economic losses in aquaculture due to the lack of effective therapeutics and vaccines. Vaccination has proven to be an effective preventive strategy, inducing neutralizing antibodies in fish such as tilapia, catfish, Atlantic salmon, and sea bream (Hanif et al. 2005). Cell cultures play a vital role in vaccine development by enabling virus propagation for attenuated or inactivated vaccines (Ma et al. 2019). Attenuated vaccines, such as the koi herpesvirus (CyHV3) vaccine and channel catfish virus (CCV) vaccine, are developed through serial passaging in fish cell cultures, which weakens the virus while ensuring safety and efficacy (Awad et al. 1989; Ronen et al. 2003; Perelberg et al. 2005). These vaccines have shown strong protective immunity and no reversion to pathogenicity, as seen in immunized fish with CyHV3 and CCV (Wise et al.

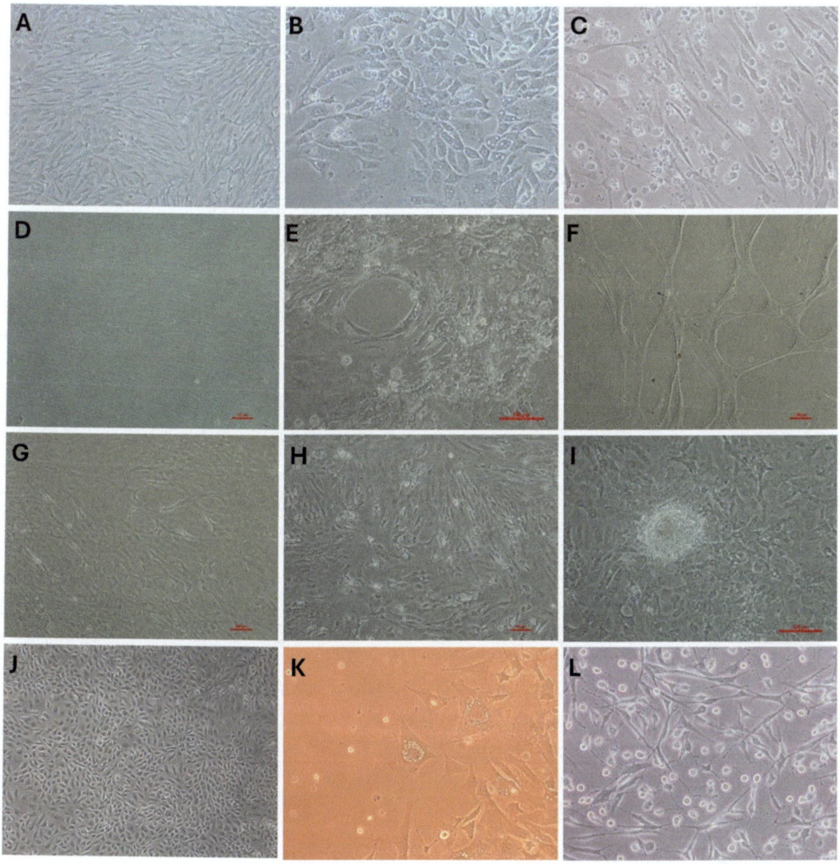

Fig. 1 Various cytopathic effects (CPE) caused by different finfish viruses in the susceptible fish cell lines. (**a**) The non-infected healthy FtGF cells; (**b**) CPE in FtGF cells infected with CyHV-2 showing more vacuolation of cells at 3 dpi; (**c**) CPE in FtGF cells following infection with CyHV-2 showing rounding and elongation of cells and cell death at 5 dpi; (**d**) The non-infected healthy onlB cells; (**e**) CPE in onlB cells infected with TiLV showing syncytial formation on cells 3 dpi; (**f**) CPE in onlB cells infected with TiLV showing plaque formation on cells 5 dpi; (**g**) The non-infected healthy AOF cells; (**h**) CPE in AOF cells following infection with ISKNV showing more vacuolation and rounding of cells at 3 dpi; (**i**) CPE in AOF cells following infection with ISKNV showing vacuolation and aggregation and fusion of infected cells at 5 dpi; (**j**) The non-infected healthy CCKF cells; (**k**) CPE in CCKF cells following infection with CyHV-2 showing more multiple vacuolations in the infected cells at 3 dpi; (**l**) CPE in CCKF cells following infection with CyHV-2 showing more elongation and rounding of cells at 5 dpi

2015; Aarattuthodiyil et al. 2020). The use of cell cultures for developing heat and formalin-inactivated vaccines against the viral pathogen (CyHV2) offers a promising approach to enhancing aquaculture health and production (Dharmaratnam et al. 2022, 2023).

3.3 Discovery of Antiviral Agents

Several antiviral agents and strategies have shown potential against fish viruses (Marecki et al. 2019). Acyclovir, a guanosine derivative, has demonstrated efficacy against channel catfish virus (CCV) and cyprinid herpesvirus-3 (CyHV-3) by suppressing viral DNA synthesis and replication (Troszok et al. 2018; Hao et al. 2021). Exopolysaccharides from *Arthrospira platensis* inhibited KHV replication in cell cultures (Reichert et al. 2017). Broad-spectrum antivirals like LJ001 and its analogue JL122 blocked viral entry by preventing virus-host membrane fusion (Balmer et al. 2017, 2018). Polyinosinic-polycytidylic acid (poly I:C) induced antiviral states in fish cell lines by producing interferons that inhibit viral replication (Jensen et al. 2002; Peng et al. 2010). Antiviral compounds were screened using grouper fin (GF-1) cell lines, where dasyscyphin C from *Eclipta prostrata* inhibited betanodavirus propagation. Immune gene regulation by antimicrobial peptides was studied in trout macrophages (Dewitteorr et al. 2007; Verma et al. 2013). These findings highlight the promise of cell culture systems in screening and evaluating antiviral agents for fish viruses.

3.4 Cell Toxicology

Fish cell cultures are extensively utilized to evaluate drug effects, toxicity, and environmental pollutants, offering an alternative to animal testing by reducing costs and variability. Liver and kidney-derived cell cultures are particularly valuable for studying cytotoxicity, genotoxicity, and metabolic pathways (Babich and Borenfreund 1991; Segner 1998). These cultures enable rapid screening of compounds and mechanisms of toxicity, as well as identifying metabolic targets and pathways (Castaño et al. 2003; Rehberger et al. 2018). Key advancements include the use of *Poeciliopsis lucida* hepatoma cell lines (PLHC-1) to study xenobiotic efflux and toxicity of drugs and chemicals like alkylbenzenes and phthalates (Babich and Borenfreund 1991; Caminada et al. 2008). Chromium cytotoxicity was analysed using fish cell lines from *Etroplus suratensis*, *Lates calcarifer*, and *Catla catla* (Taju et al. 2017). Cell lines like RTG-2 and PLHC-1 were employed for nanomaterial toxicity studies, including cerium, silicon, zinc, and carbon compounds (Bermejo-Nogales et al. 2017; Klingelfus et al. 2019). DNA damage caused by environmental genotoxicants was assessed using comet assays in cell lines such as RTG-2 and RTL-W1 (Žegura and Filipič 2019).

In endocrine disruption research, vitellogenin (VTG) induction in hepatocytes serves as a biomarker for estrogenic activity, helping identify endocrine-disrupting compounds. EDCs like bisphenol A (BPA) and pesticides with suspected hormonal activity are detected using in vitro hepatocyte assays (Pelissero et al. 1993). Fish cell cultures also evaluate bacterial and fungal toxins. For example, extracellular products of *Aeromonas* sp. and *Vibrio anguillarum* showed toxicity to SICH cells derived from *Catla catla* heart tissue (Ishaq Ahmed et al. 2009). EPC cell lines were used to study *Renibacterium salmoninarum* virulence (Mcintosh et al. 1997), and rainbow trout cell lines like RGB and RGG demonstrated susceptibility to *Photobacterium damselae* toxins (Ku et al. 2009). Fish cell lines provide a more relevant, cost-effective, and ethical tool for environmental monitoring compared to human cell lines (Rodrigues et al. 2020). The MTT assay (3-(4,5-dimethylthiazol-2-yl)-2,5-diphenyltetrazolium bromide assay) is a widely used method to assess cell viability and metabolic activity by measuring mitochondrial activity, which is proportional to the number of viable cells (Kumar et al. 2024). However, the colour development was not observed properly in the MTT assay in fish cells, i.e. weaker colour formation was noticed in fish cell lines compared to the mammalian cell lines. The lack of a properly optimized protocol for the fish cell line and variations in each cell line for the optimization is required for a better comparison of the existing result.

3.5 Fish Stem Cell Research

Embryonic stem (ES) cells, derived from developing embryos, are pluripotent and have significant applications in biodiversity conservation and marine biotechnology (Till and McCulloch 1961; Evans and Kaufman 1981; Thomson et al. 1998). Fish ES cell lines are utilized for efficient foreign DNA transfer into germ cells, making them valuable for biotechnological applications. Hong et al. (2004) developed a spermatogonia cell line from medaka fish, which could undergo spermiogenesis, suggesting potential applications in spermatogenesis and hybrid catfish production. Additionally, gene targeting and germ cell transplantation have shown progress in these areas (Hong et al. 2011). Embryonic germ cell transplantation was successfully used for surrogate production in salmonids (Yoshizaki et al. 2010). ES cell lines have been established from species like medaka (*Oryzias latipes*) and zebrafish (*Danio rerio*) (Ciarlo and Zon 2016; Yi et al. 2009). Tumours also serve as sources for ES cells, such as the Epithelioma Papulosum Cyprini (EPC) and rainbow trout liver (RTH-149) cell lines (Fijan et al. 1983; Lee et al. 1993). Furthermore, zebrafish blastula cells can express genes introduced via transfection, supporting genotype and phenotype alterations in cultured cells (Collodi et al. 1992). Marine fish cell lines are less readily available compared to other types, and the collection of these lines is limited. Similarly, embryonic stem cells (ESCs) and mesenchymal stem cells (MSCs) from marine fish are scarce. A thorough understanding of these ESCs, induced pluripotent stem cells (iPSCs), and MSCs would support advancements in regenerative medicine. However, research in this area faces a significant barrier due to the limited accessibility for marine fish. Investigating rare marine species could offer new opportunities to advance science.

4 Future Potential

The lack of species- and tissue-specific fish cell cultures hinders pathogen isolation, and many established cell lines are unavailable in repositories, leaving fish as a largely unexplored source for cell line development. Establishing cell lines as standard research tools requires standardized protocols, including media, reagents, equipment, quality control, proper characterization, and documentation. Utilizing 3D cell culture technology, as demonstrated for long-term maintenance of organoids from snake venom glands for venom production (Puschhof et al. 2021), can be applied to fish venom research. Fish venom from species like sharks, catfish, and stonefish contains bioactive compounds with enzymatic, antimicrobial, cytotoxic, and anticancer properties, making it valuable for pharmacological, therapeutic, and pesticidal applications (Ortiz et al. 2015). Additionally, organoids offer a physiologically relevant platform to model complex diseases and evaluate pathogen behaviour in specific tissues, advancing disease management in fish health (Masters et al. 2001). Fish cell lines have higher buffering capacity compared to the mammalian cell culture; in addition, they can grow in hypoxic conditions. Fish muscle has a unique oxygen-carrying capacity for metabolic activity (Verde et al. 2008), especially in some artic fish species *Chaenocephalus aceratus,* which does not have haemoglobin and it has higher mitochondria compared to all fish species (Johnston 1987). Owing to two of these features, the fish cell line can meet the requirement of cellular meat as per consumer preference (Rubio et al. 2019).

5 Legal Issues

5.1 Ethical Approval

All research involving fish cell lines should adhere to ethical standards to ensure responsible scientific practices and animal welfare. Although fish cell lines eliminate the need for live animal experimentation, researchers must still comply with institutional and regulatory guidelines. For projects involving fish cell lines, researchers should follow their institution's specific policies, often overseen by Research and Development (R&D) Departments, to secure necessary approvals. Ethical considerations include sourcing cell lines responsibly, maintaining transparency about experimental objectives, and ensuring that derived data contribute to advancing science and reducing live animal use. While the Human Tissue Act 2004 of the UK governs the handling of human tissue, similar ethical and regulatory principles can be applied to research involving fish cell lines. Fish cell lines derived from primary tissues of fish are not subject to the same licensing and consent requirements as human tissues under the Act. However, the responsible sourcing, use, and storage of fish cell lines require adherence to institutional and ethical standards to ensure scientific integrity and animal welfare.

The act of deriving fish cell lines from primary fish tissues aligns with the principle of minimizing harm and maximizing the utility of biological materials. While primary fish tissue requires ethical oversight for collection, fish cell lines, once established, offer a sustainable resource that reduces the reliance on live fish for research purposes. Unlike primary tissues, the storage of fish cell lines for research does not typically require additional licensing, provided that their use complies with national regulations and institutional policies. Researchers working with fish cell lines should familiarize themselves with relevant Codes of Practice or institutional guidelines, particularly those that address consent for tissue collection, disposal standards, and proper documentation. The adoption of fish cell lines supports the principles of the 3Rs (Replacement, Reduction, and Refinement) by offering an ethical alternative to live fish studies while advancing research in aquaculture, disease modelling, and environmental sciences.

5.2 Ownership Rights

Ownership and patent rights related to fish cell lines and their derivatives can involve multiple stakeholders, including the original tissue source providers (e.g. fish farms or aquaculture facilities), researchers, research institutions, funding agencies, and collaborating commercial entities. Determining control over ownership, intellectual property, and rights requires clear agreements and negotiations among these parties to ensure transparency and fairness. Universities, research institutes, and funding agencies often have dedicated technology transfer or research offices to facilitate these negotiations. These offices help define the terms of ownership and manage intellectual property arising from fish cell line research, particularly in cases where the work may lead to commercial applications, such as disease-resistant strains, vaccine development, or aquaculture innovations. Establishing clear agreements before research begins is crucial to avoid disputes and ensure that all parties involved benefit equitably from the advancements and commercialization of fish cell line technologies.

5.3 Material Transfer Agreement (MTA)

A Material Transfer Agreement (MTA) is a legally binding contract that governs the transfer of research materials, such as fish cell lines, between organizations. The MTA defines key aspects such as ownership, intellectual property, and patent rights, ensuring clarity and compliance for both parties. It must be signed by the legal representatives of both parties before the materials are transferred. For fish cell lines, the MTA should specify the purpose of use, prohibit unauthorized third-party transfer, and outline restrictions on commercial exploitation unless explicitly agreed upon. A new agreement may be required to address ownership and intellectual property rights if a sub-line or derivative is created from the transferred cell line. While ethical approval requirements differ from those involving human tissue, the MTA should confirm that the fish tissues used to derive the cell lines were ethically sourced in accordance with institutional and national guidelines. Universities and research institutes typically manage MTAs through their research or technology transfer offices to ensure proper documentation and compliance with all applicable regulations.

5.4 Legal Guidelines in India

The '*Rules for the Manufacture, Use, Import, Export, and Storage of Hazardous Micro-organisms, Genetically Engineered Organisms or Cells, 1989*' were established to ensure biosafety in India, aiming to protect the health, the environment, and nature regarding gene technology and microorganisms (Ministry of Environment, Forest & Climate Change 1989). These rules cover the research, large-scale applications, and use of genetically modified organisms (GMOs) and hazardous microorganisms, including their importation, exportation, and sale. Implemented through key committees such as the Recombinant DNA Advisory Committee (RDAC), Institutional Biosafety Committee (IBSC), and Genetic Engineering Appraisal Committee (GEAC), the framework ensures regulatory oversight and monitoring. Specific activities, including the import, export, production, and sale of GMOs, require approval from the GEAC. Approval is also necessary for the release of GMOs into the environment, food production, and related applications. Institutions involved with GMOs must establish an Institutional Biosafety Committee, while district and state-level committees monitor safety and compliance. The GEAC is the central authority for granting approvals, setting the period of approval to a maximum of 4 years, with a two-year renewal. Violations of these rules can result in penalties, and committees are authorized to intervene in emergencies or accidents. Individuals dissatisfied with the decisions of the GEAC or state committees may appeal to the Ministry of Environment, Forest, and Climate Change. In some cases, exemptions may be granted for specific GMOs or activities.

6 Conclusion

Over the past four decades, cell culture systems have evolved from being a research tool to a fundamental component in several life science fields, including stem cell and targeted therapies. The growing demand for cell culture-based products has driven improvements in process development, with ongoing efforts to enhance cell lines for better growth, product synthesis, and metabolic characteristics supported by genomic and proteomic techniques. Fish cell cultures have proven to be an effective alternative to animal research, addressing ethical concerns and offering new avenues for disease diagnosis and therapeutic development. Despite the increasing importance of fish cell lines, there remains a significant gap in the number of cell lines available globally. More contributions to repositories would benefit the scientific community by providing valuable resources for species- and tissue-specific studies. The scarcity of knowledge about fish cell lines contributes to the risk of misidentification, which can result in irreproducible data, wasted resources, and scientific errors. Therefore, improving monitoring and maintaining strong documentation is essential for ensuring the reliability and reproducibility of data in fish cell line research (Horbach and Halffman 2017). The development of fish cell lines for recombinant protein expression and therapeutic applications is still largely empirical, with future research focused on optimizing cell selection procedures and culture conditions to improve product quality. There is a growing emphasis on the use of primary cells, especially in combination with advanced technologies like 3D cell culture, which have shown promise in biomedical research. However, concerns about misidentification and cross-contamination of cell lines highlight the need for rigorous authentication and standardized protocols to ensure the reliability of research outcomes. Despite some challenges, fish cell cultures are increasingly recognized for their potential in advancing aquaculture production, particularly in disease diagnosis, vaccine development, and antiviral agent identification.

References

Aarattuthodi S, Dharan V (2021) Applications of fish cell cultures. In: Biotechnological advances in aquaculture health management. Springer Nature Singapore, Singapore, pp 123–163. https://doi.org/10.1007/978-981-16-5195-3_7

Aarattuthodiyil S, Griffin MJ, Greenway TE, Khoo LH, Byars TS, Lewis M, Steadman J, Wise DJ (2020) An orally delivered, live-attenuated *Edwardsiella ictaluri* vaccine efficiently protects channel catfish fingerlings against multiple *Edwardsiella ictaluri* field isolates. J World Aquac Soc 51:1354–1372. https://doi.org/10.1111/jwas.12693

Abdelsalam EEE, Bláhová Z, Baloch AA, Piačková V (2024) Potential to grow carp oedema virus (genogroup I) in monolayers of carp-derived primary cells with further implication in cell analysis. J Fish Dis 47:e13934. https://doi.org/10.1111/jfd.13934

Awad MA, Nusbaum KE, Brady YJ (1989) Preliminary studies of a newly developed subunit vaccine for channel catfish virus disease. J Aquat Anim Health 1:233–237. https://doi.org/10.1577/1548-8667(1989)001<0233:PSOAND>2.3.CO;2

Babich H, Borenfreund E (1991) Cytotoxicity and genotoxicity assays with cultured fish cells: a review. Toxicol In Vitro 5:91–100. https://doi.org/10.1016/0887-2333(91)90052-F

Bairoch A (2018) The cellosaurus, a cell-line knowledge resource. J Biomol Tech 29:25–38. https://doi.org/10.7171/jbt.18-2902-002

Balmer BF, Powers RL, Zhang T-H, Lee J, Vigant F, Lee B, Jung ME, Purcell MK, Snekvik K, Aguilar HC (2017) Inhibition of an aquatic rhabdovirus demonstrates promise of a broad-spectrum antiviral for use in aquaculture. J Virol 91. https://doi.org/10.1128/JVI.02181-16

Balmer BF, Getchell RG, Powers RL, Lee J, Zhang T, Jung ME, Purcell MK, Snekvik K, Aguilar HC (2018) Broad-spectrum antiviral JL122 blocks infection and inhibits transmission of aquatic rhabdoviruses. Virology 525:143–149. https://doi.org/10.1016/j.virol.2018.09.009

Bermejo-Nogales A, Fernández-Cruz ML, Navas JM (2017) Fish cell lines as a tool for the ecotoxicity assessment and ranking of engineered nanomaterials. Regul Toxicol Pharmacol 90:297–307. https://doi.org/10.1016/j.yrtph.2017.09.029

Bols NC, Lee LEJ (1991) Technology and uses of cell cultures from the tissues and organs of bony fish. Cytotechnology 6:163–187. https://doi.org/10.1007/BF00624756

Bols N, Barlian A, Chirino-Trejo M, Caldwell S, Goegan P, Lee L (1994) Development of a cell line from primary cultures of rainbow trout, *Oncorhynchus mykiss* (Walbaum), gills. J Fish Dis 17:601–611. https://doi.org/10.1111/j.1365-2761.1994.tb00258.x

Caminada D, Zaja R, Smital T, Fent K (2008) Human pharmaceuticals modulate P-gp1 (ABCB1) transport activity in the fish cell line PLHC-1. Aquat Toxicol 90:214–222. https://doi.org/10.1016/j.aquatox.2008.08.013

Capes-Davis A, Theodosopoulos G, Atkin I, Drexler HG, Kohara A, MacLeod RAF, Masters JR, Nakamura Y, Reid YA, Reddel RR, Freshney RI (2010) Check your cultures! A list of cross-contaminated or misidentified cell lines. Int J Cancer 127:1–8. https://doi.org/10.1002/ijc.25242

Castaño A, Bols N, Braunbeck T, Dierickx P, Halder M, Isomaa B, Kawahara K, Lee LEJ, Mothersill C, Pärt P, Repetto G, Sintes JR, Rufli H, Smith R, Wood C, Segner H (2003) The use of fish cells in ecotoxicology: the report and recommendations of ECVAM workshop 47. Altern Lab Anim 31:317–351. https://doi.org/10.1177/026119290303100314

Chen S-L, Sha Z-X, Ye H-Q (2003) Establishment of a pluripotent embryonic cell line from sea perch (*Lateolabrax japonicus*) embryos. Aquaculture 218:141–151. https://doi.org/10.1016/S0044-8486(02)00570-7

Ciarlo CA, Zon LI (2016) Embryonic cell culture in zebrafish. Methods Cell Biol 133:1–10. https://doi.org/10.1016/bs.mcb.2016.02.010

Collodi P, Barnes DW (1990) Mitogenic activity from trout embryos. Proc Natl Acad Sci USA 87:3498–3502. https://doi.org/10.1073/pnas.87.9.3498

Collodi P, Kame Y, Ernst T, Miranda C, Buhler DR, Barnes DW (1992) Culture of cells from zebrafish (*Brachydanio rerio*) embryo and adult tissues. Cell Biol Toxicol 8:43–61. https://doi.org/10.1007/BF00119294

Dewitteorr S, Hsu H, Bols N (2007) Induction of homotypic aggregation in the rainbow trout macrophage-like cell line, RTS11. Fish Shellfish Immunol 22:487–497. https://doi.org/10.1016/j.fsi.2006.06.008

Dharmaratnam A, Sudhagar A, Das S, Nair RR, Nithianantham SR, Preena PG, Lekshmi N, Swaminathan TR (2022) Immune gene expression and protective effects in goldfish (*Carassius auratus L.*) immunized with formalin-inactivated cyprinid herpesvirus-2 (CyHV-2) vaccine. Microb Pathog 164:105452. https://doi.org/10.1016/j.micpath.2022.105452

Dharmaratnam A, Sudhagar A, Swaminathan TR (2023) Evaluation of protective effects of heat-inactivated cyprinid herpesvirus-2 (CyHV-2) vaccine against herpesviral hematopoietic necrosis disease (HVHND) in goldfish (*Carassius auratus*). Fish Shellfish Immunol 132:108460. https://doi.org/10.1016/j.fsi.2022.108460

Evans MJ, Kaufman MH (1981) Establishment in culture of pluripotential cells from mouse embryos. Nature 292:154–156. https://doi.org/10.1038/292154a0

Fijan N, Sulimanović D, Bearzotti M, Muzinić D, Zwillenberg LO, Chilmonczyk S, Vautherot JF, de Kinkelin P (1983) Some properties of the Epithelioma papulosum cyprini (EPC) cell line from carp *cyprinus carpio*. Ann Inst Pasteur Virol 134:207–220. https://doi.org/10.1016/S0769-2617(83)80060-4

Freshney RI (2006) Culture of cells for tissue engineering. Wiley, Hoboken, p 512. https://doi.org/10.1002/0471741817

Gartler SM (1968) Apparent HeLa cell contamination of human heteroploid cell lines. Nature 217:750–751. https://doi.org/10.1038/217750a0

Grunwald DJ, Eisen JS (2002) Headwaters of the zebrafish — emergence of a new model vertebrate. Nat Rev Genet 3:717–724. https://doi.org/10.1038/nrg892

Hanif A, Bakopoulos V, Leonardos I, Dimitriadis G (2005) The effect of sea bream (*Sparus aurata*) broodstock and larval vaccination on the susceptibility by *Photobacterium damsela subsp. piscicida* and on the humoral immune parameters. Fish Shellfish Immunol 19:345–361. https://doi.org/10.1016/j.fsi.2004.12.009

Hao K, Yuan S, Yu F, Chen XH, Bian WJ, Feng YH, Zhao Z (2021) Acyclovir inhibits channel catfish virus replication and protects channel catfish ovary cells from apoptosis. Virus Res 292:198249. https://doi.org/10.1016/j.virusres.2020.198249

Hightower LE, Renfro JL (1988) Recent applications of fish cell culture to biomedical research. J Exp Zool 248:290–302. https://doi.org/10.1002/jez.1402480307

Hong Y, Liu T, Zhao H, Xu H, Wang W, Liu R, Chen T, Deng J, Gui J (2004) Establishment of a normal medaka fish spermatogonial cell line capable of sperm production in vitro. Proc Natl Acad Sci 101:8011–8016. https://doi.org/10.1073/pnas.0308668101

Hong N, Li Z, Hong Y (2011) Fish stem cell cultures. Int J Biol Sci 7:392–402. https://doi.org/10.7150/ijbs.7.392

Horbach SPJM, Halffman W (2017) The ghosts of HeLa: how cell line misidentification contaminates the scientific literature. PLoS One 12:e0186281. https://doi.org/10.1371/journal.pone.0186281

Ishaq Ahmed VP, Sarath Babu V, Chandra V, Nambi KS, Thomas J, Bhonde R, Sahul Hameed AS (2009) A new fibroblastic-like cell line from heart muscle of the Indian major carp (*Catla catla*): development and characterization. Aquaculture 293:180–186. https://doi.org/10.1016/j.aquaculture.2009.05.012

Jensen I, Larsen R, Robertsen B (2002) An antiviral state induced in Chinook salmon embryo cells (CHSE-214) by transfection with the double-stranded RNA poly I:C. Fish Shellfish Immunol 13:367–378. https://doi.org/10.1006/fsim.2002.0412

Jeong Y-J, Kim K-I (2023) A new cell line derived from the caudal fin of the dwarf gourami (*Trichogaster lalius*) and its susceptibility to fish viruses. Biology (Basel) 12. https://doi.org/10.3390/biology12060829

Jitrakorn S, Gangnonngiw W, Bunnontae M, Manajit O, Rattanarojpong T, Chaivisuthangkura P, Dong HT, Saksmerprome V (2020) Infectious cell culture system for concurrent propagation and purification of Megalocytivirus ISKNV and nervous necrosis virus from Asian Sea bass (*Lates calcarifer*). Aquaculture 520:734931. https://doi.org/10.1016/j.aquaculture.2020.734931

Johnston IA (1987) Respiratory characteristics of muscle fibres in a fish (*Chaenocephalus Aceratus*) that lacks haem pigments. J Exp Biol 133:415–428. https://doi.org/10.1242/jeb.133.1.415

Kawato Y, Yamashita H, Yuasa K, Miwa S, Nakajima K (2017) Development of a highly permissive cell line from spotted knifejaw (*Oplegnathus punctatus*) for red sea bream iridovirus. Aquaculture 473:291–298. https://doi.org/10.1016/j.aquaculture.2017.02.027

Kelly RK, Souter BW, Miller HR (1978) Fish cell lines: comparisons of CHSE-214, FHM, and RTG-2 in assaying IHN and IPN viruses. J Fish Res Board Can 35:1009–1011. https://doi.org/10.1139/f78-164

Klingelfus T, Disner GR, Voigt CL, Alle LF, Cestari MM, Leme DM (2019) Nanomaterials induce DNA-protein crosslink and DNA oxidation: a mechanistic study with RTG-2 fish cell line and Comet assay modifications. Chemosphere 215:703–709. https://doi.org/10.1016/j.chemosphere.2018.10.118

Ku C, Teng Y-C, Wang C-S, Lu C (2009) Establishment and characterization of three cell lines derived from the rockfish grouper *Epinephelus quoyanus*: use for transgenic studies and cytotoxicity testing. Aquaculture 294:147–151. https://doi.org/10.1016/j.aquaculture.2009.05.010

Kumar A, Rai Y, Bhatt AN (2024) Anti-cancer drug-mediated increase in mitochondrial mass limits the application of metabolic viability-based MTT assay in cytotoxicity screening. Cytotechnology 76:301–311. https://doi.org/10.1007/s10616-024-00618-1

Lakra WS, Swaminathan TR, Rathore G, Goswami M, Yadav K, Kapoor S (2010) Development and characterization of three new diploid cell lines from *Labeo rohita* (Ham.). Biotechnol Prog 26:1008–1013. https://doi.org/10.1002/btpr.418

Lakra WS, Swaminathan TR, Joy KP (2011) Development, characterization, conservation and storage of fish cell lines: a review. Fish Physiol Biochem 37:1–20. https://doi.org/10.1007/s10695-010-9411-x

Lannan CN (1994) Fish cell culture: a protocol for quality control. J Tissue Cult Methods 16:95–98. https://doi.org/10.1007/BF01404817

Lee LEJ, Clemons JH, Bechtel DG, Caldwell SJ, Han K-B, Pasitschniak-Arts M, Mosser DD, Bols NC (1993) Development and characterization of a rainbow trout liver cell line expressing cytochrome P450-dependent monooxygenase activity. Cell Biol Toxicol 9:279–294. https://doi.org/10.1007/BF00755606

Ma J, Bruce TJ, Jones EM, Cain KD (2019) A review of fish vaccine development strategies: conventional methods and modern biotechnological approaches. Microorganisms 7:569. https://doi.org/10.3390/microorganisms7110569

Marecki JC, Aarattuthodiyil S, Byrd AK, Penthala NR, Crooks PA, Raney KD (2019) N-Naphthoyl-substituted indole thio-barbituric acid analogs inhibit the helicase activity of the hepatitis C virus NS3. Bioorg Med Chem Lett 29:430–434. https://doi.org/10.1016/j.bmcl.2018.12.026

Masters JR, Thomson JA, Daly-Burns B, Reid YA, Dirks WG, Packer P, Toji LH, Ohno T, Tanabe H, Arlett CF, Kelland LR, Harrison M, Virmani A, Ward TH, Ayres KL, Debenham PG (2001) Short tandem repeat profiling provides an international reference standard for human cell lines. Proc Natl Acad Sci 98:8012–8017. https://doi.org/10.1073/pnas.121616198

Mcintosh D, Flaño E, Grayson TH, Gilpin ML, Austin B, Villena AJ (1997) Production of putative virulence factors by *Renibacterium salmoninarum* grown in cell culture. Microbiology (N Y) 143:3349–3356. https://doi.org/10.1099/00221287-143-10-3349

Ministry of Environment, Forests and Climate Change (1989) Rules for the Manufacture, Use/Import/Export and Storage of Hazardous Microorganisms/Genetically Engineered Organisms or Cells. New Delhi, India. Accessed on June 18, 2025. http://geacindia.gov.in/resource-documents/biosafety-regulations/acts-and-rules/Rules-for-the-manufacture-use-import-export-and-storage-1989.pdf

Nelson-Rees WA, Flandermeyer RR (1976) HeLa cultures defined. Science 191:96–98. https://doi.org/10.1126/science.1246601

Nelson-Rees WA, Flandermeyer RR (1977) Inter- and intraspecies contamination of human breast tumor cell lines HBC and BrCa5 and other cell cultures. Science (1979) 195:1343–1344. https://doi.org/10.1126/science.557237

Nicholson BL (1989) Fish cell culture: an update. Adv Cell Cult 7:1–18. https://doi.org/10.1016/B978-0-12-007907-0.50007-4

Ortiz E, Gurrola GB, Schwartz EF, Possani LD (2015) Scorpion venom components as potential candidates for drug development. Toxicon 93:125–135. https://doi.org/10.1016/j.toxicon.2014.11.233

Parameswaran V, Shukla R, Bhonde RR, Sahul Hameed AS (2006) Splenic cell line from sea bass, Lates calcarifer: establishment and characterization. Aquaculture 261:43–53. https://doi.org/10.1016/j.aquaculture.2006.07.034

Pelissero C, Flouriot G, Foucher JL, Bennetau B, Dunoguès J, Le Gac F, Sumpter JP (1993) Vitellogenin synthesis in cultured hepatocytes; an in vitro test for the estrogenic potency of chemicals. J Steroid Biochem Mol Biol 44:263–272. https://doi.org/10.1016/0960-0760(93)90086-C

Peng C-A, Wang C-H, Wang W-L (2010) Rapid antiviral assay using QD-tagged fish virus as imaging nanoprobe. J Virol Methods 169:412–415. https://doi.org/10.1016/j.jviromet.2010.07.027

Perelberg A, Ronen A, Hutoran M, Smith Y, Kotler M (2005) Protection of cultured *Cyprinus carpio* against a lethal viral disease by an attenuated virus vaccine. Vaccine 23:3396–3403. https://doi.org/10.1016/j.vaccine.2005.01.096

Puschhof J, Post Y, Beumer J, Kerkkamp HM, Bittenbinder M, Vonk FJ, Casewell NR, Richardson MK, Clevers H (2021) Derivation of snake venom gland organoids for in vitro venom production. Nat Protoc 16:1494–1510. https://doi.org/10.1038/s41596-020-00463-4

Qin Y, Liu J, Liu W, Shi H, Jia A, Lu Y, Liu X (2020) First isolation and identification of red-grouper nervous necrosis virus (RGNNV) from adult hybrid Hulong grouper (*Epinephelus fuscoguttatus* × *Epinephelus lanceolatus*) in China. Aquaculture 529:735662. https://doi.org/10.1016/j.aquaculture.2020.735662

Rajendran KV, Sood N, Rao BM, Valsalam A, Bedekar MK, Jeena K, Pradhan PK, Paria A, Swaminathan TR, Verma DK, Sood NK (2023) Widespread occurrence of Tilapia parvovirus in farmed Nile tilapia *Oreochromis niloticus* from India. J Fish Dis. https://doi.org/10.1111/jfd.13871

Rehberger K, Kropf C, Segner H (2018) In vitro or not in vitro: a short journey through a long history. Environ Sci Eur 30:23. https://doi.org/10.1186/s12302-018-0151-3

Reichert M, Bergmann SM, Hwang J, Buchholz R, Lindenberger C (2017) Antiviral activity of exopolysaccharides from *Arthrospira platensis* against koi herpesvirus. J Fish Dis 40:1441–1450. https://doi.org/10.1111/jfd.12618

Riji John K, Sivasankar P, George MR (2023) Status of emerging and reemerging fish viral infections in India. In: Viral outbreaks – global impact and newer horizons. IntechOpen. https://doi.org/10.5772/intechopen.109012

Rodrigues ET, Varela AT, Pardal MA, Sardão VA (2020) Cell-based assays as an alternative for the study of aquatic toxicity of pharmaceuticals. Environ Sci Pollut Res 27:7145–7155. https://doi.org/10.1007/s11356-019-07384-0

Ronen A, Perelberg A, Abramowitz J, Hutoran M, Tinman S, Bejerano I, Steinitz M, Kotler M (2003) Efficient vaccine against the virus causing a lethal disease in cultured *Cyprinus carpio*. Vaccine 21:4677–4684. https://doi.org/10.1016/S0264-410X(03)00523-1

Rubio N, Datar I, Stachura D, Kaplan D, Krueger K (2019) Cell-based fish: a novel approach to seafood production and an opportunity for cellular agriculture. Front Sustain Food Syst 3. https://doi.org/10.3389/fsufs.2019.00043

Segner H (1998) Fish cell lines as a tool in aquatic toxicology. In: Fish ecotoxicology. Birkhäuser Basel, Basel, pp 1–38. https://doi.org/10.1007/978-3-0348-8853-0_1

Sheng X, Zhong Y, Zeng J, Tang X, Xing J, Chi H, Zhan W (2020) Lymphocystis disease virus (Iridoviridae) enters flounder (*Paralichthys olivaceus*) gill cells via a caveolae-mediated endocytosis mechanism facilitated by viral receptors. Int J Mol Sci 21:4722. https://doi.org/10.3390/ijms21134722

Sommerset I, Krossøy B, Biering E, Frost P (2005) Vaccines for fish in aquaculture. Expert Rev Vaccines 4:89–101. https://doi.org/10.1586/14760584.4.1.89

Souto EB, Campos JR, Da Ana R, Martins-Gomes C, Silva AM, Souto SB, Lucarini M, Durazzo A, Santini A (2020) Ocular cell lines and genotoxicity assessment. Int J Environ Res Public Health 17:2046. https://doi.org/10.3390/ijerph17062046

Swaminathan TR, Lakra WS, Gopalakrishnan A, Basheer VS, Khushwaha B, Sajeela KA (2010) Development and characterization of a new epithelial cell line PSF from caudal fin of Green chromide, *Etroplus suratensis* (Bloch, 1790). In Vitro Cell Dev Biol Anim 46:647–656. https://doi.org/10.1007/s11626-010-9326-y

Taju G, Abdul Majeed S, Nambi KSN, Sahul Hameed AS (2013) Development and characterization of cell line from the gill tissue of *Catla catla* (Hamilton, 1822) for toxicological studies. Chemosphere 90:2172–2180. https://doi.org/10.1016/j.chemosphere.2012.11.027

Taju G, Abdul Majeed S, Nambi KSN, Sahul Hameed AS (2017) Application of fish cell lines for evaluating the chromium induced cytotoxicity, genotoxicity and oxidative stress. Chemosphere 184:1–12. https://doi.org/10.1016/j.chemosphere.2017.05.151

Thangaraj RS, Ravi C, Kumar R, Dharmaratnam A, Valaparambil Saidmuhammed B, Pradhan PK, Sood N (2018) Derivation of two tilapia (*Oreochromis niloticus*) cell lines for efficient propagation of Tilapia Lake Virus (TiLV). Aquaculture 492:206–214. https://doi.org/10.1016/j.aquaculture.2018.04.012

Thangaraj RS, Narendrakumar L, Prasannan Geetha P, Shanmuganathan AR, Dharmaratnam A, Nithianantham SR (2021a) Comprehensive update on inventory of finfish cell lines developed during the last decade (2010–2020). Rev Aquac 13:2248–2288. https://doi.org/10.1111/raq.12566

Thangaraj RS, Nithianantham SR, Dharmaratnam A, Kumar R, Pradhan PK, Thangalazhy Gopakumar S, Sood N (2021b) Cyprinid herpesvirus-2 (CyHV-2): a comprehensive review. Rev Aquac 13:796–821. https://doi.org/10.1111/raq.12499

Thomson JA, Itskovitz-Eldor J, Shapiro SS, Waknitz MA, Swiergiel JJ, Marshall VS, Jones JM (1998) Embryonic stem cell lines derived from human blastocysts. Science 1979(282):1145–1147. https://doi.org/10.1126/science.282.5391.1145

Till JE, McCulloch EA (1961) A direct measurement of the radiation sensitivity of normal mouse bone marrow cells. J Radiat Res 14:213. https://doi.org/10.2307/3570892

Troszok A, Kolek L, Szczygieł J, Wawrzeczko J, Borzym E, Reichert M, Kamińska T, Ostrowski T, Jurecka P, Adamek M, Rakus K, Irnazarow I (2018) Acyclovir inhibits Cyprinid herpesvirus 3 multiplication in vitro. J Fish Dis 41:1709–1718. https://doi.org/10.1111/jfd.12880

Verde C, Giordano D, di Prisco G (2008) The adaptation of polar fishes to climatic changes: structure, function and phylogeny of haemoglobin. Int Union Biochem Mol Biol Life 60:29–40. https://doi.org/10.1002/iub.1

Verma D, Kim EA, Swaminathan S (2013) Cell-based screening assay for antiviral compounds targeting the ability of herpesvirus posttranscriptional regulatory proteins to stabilize viral mRNAs. J Virol 87:10742–10751. https://doi.org/10.1128/JVI.01644-13

Verma A, Verma M, Singh A (2020) Animal tissue culture principles and applications. In: Animal biotechnology. Elsevier, pp 269–293. https://doi.org/10.1016/B978-0-12-811710-1.00012-4

Wise DJ, Greenway TE, Byars TS, Griffin MJ, Khoo LH (2015) Oral vaccination of channel catfish against enteric septicemia of catfish using a live attenuated *Edwardsiella ictaluri* isolate. J Aquat Anim Health 27:135–143. https://doi.org/10.1080/08997659.2015.1032440

WOAH (2024) World Organisation for Animal Health manual of diagnostic tests for aquatic animals. https://www.woah.org/en/what-we-do/standards/codes-and-manuals/aquatic-manual-online-access/. Accessed 30 Jan 2025

Wolf K, Quimby MC (1962) Established eurythermic line of fish cells in vitro. Science 135:1065–1066. https://doi.org/10.1126/science.135.3508.1065

Wolf K, Quimby MC (1976) Primary monolayer culture of fish cells initiated from minced tissues. Tissue Cult Assoc Manual 2:445–448. https://doi.org/10.1007/BF00918338

Yi M, Hong N, Hong Y (2009) Generation of medaka fish haploid embryonic stem cells. Science 1979(326):430–433. https://doi.org/10.1126/science.1175151

Yoshizaki G, Ichikawa M, Hayashi M, Iwasaki Y, Miwa M, Shikina S, Okutsu T (2010) Sexual plasticity of ovarian germ cells in rainbow trout. Development 137:1227–1230. https://doi.org/10.1242/dev.044982

Žegura B, Filipič M (2019) The application of the Comet assay in fish cell lines. Mutat Res Genet Toxicol Environ Mutagen 842:72–84. https://doi.org/10.1016/j.mrgentox.2019.01.011

Crustacean Cell Line Development: Status, Constraints, and the Way Forward

C. P. Sreevidya, S. Manomi, I. S. Bright Singh, and Jayesh Puthumana

Abstract

Crustaceans are among the most diverse and ancient arthropods, thriving across widespread habitats and playing a pivotal role in aquatic food production. Their significant contribution to global food security and economic development has positioned them as key resources in meeting the demands of a growing global population. However, challenges such as enhancing the production of economically valuable species and managing infections arising from high-density aquaculture remain critical. A major limitation is the inadequate understanding of host–pathogen interactions at cellular and molecular levels, highlighting the need for advanced tools like crustacean cell lines for disease diagnosis and management. Despite efforts spanning over a century, the successful establishment of crustacean cell lines remains elusive. This highlights the necessity for a comprehensive retrospective analysis of previous research to extract actionable insights. This chapter aims to consolidate the current status, advancements, and applications of crustacean cell culture, focusing on studies conducted between 2010 and 2025. By addressing key impediments and proposing a way forward, this review seeks to guide future research efforts, advancing the development of crustacean cell lines as indispensable tools for aquatic animal health management and aquaculture sustainability.

Keywords

Crustacean cell culture · Culture medium · Primary cell culture · Cell line · Immortalization

C. P. Sreevidya · S. Manomi · I. S. Bright Singh · J. Puthumana (✉)
National Centre for Aquatic Animal Health, Cochin University of Science and Technology, Cochin, India
e-mail: sreevidya@cusat.ac.in; manomisarasan@gmail.com; isbsingh@gmail.com; jayesh@cusat.ac.in

R. Kooloth Valappil et al. (eds.), *Aquatic Animal Health Management*,
https://doi.org/10.1007/978-981-96-7987-4_25

1 Introduction

Crustaceans, the most diverse group of arthropods, play critical roles in ecological and economic contexts. Large decapods such as shrimp, prawns, lobsters, and crabs are major contributors to global food production, while smaller planktonic crustaceans, including copepods, krill, and water fleas, form vital links in aquatic food webs by connecting photosynthetic phytoplankton to higher trophic levels. Beyond their ecological importance, crustaceans hold tremendous potential in the medical and biotechnological sectors. Their unique biomolecules, such as Limulus amoebocyte lysate (LAL) and chitin, have applications ranging from bacterial endotoxin detection to therapeutic innovations (Kumar et al. 2016). Additionally, crustaceans contribute significantly to nutrient cycling by breaking down organic material from dead organisms (Szaniawska 2018). Most importantly, with an ever-growing demand for aquatic food products, the conservation and sustainable management of crustaceans have become increasingly critical. However, disease outbreaks, particularly viral infections, threaten aquaculture production, causing up to 40% annual losses in shrimp farming alone (Bondad-Reantaso et al. 2012). Current approaches to disease management rely on culling infected stocks and decontaminating culture systems, which are reactive and unsustainable. Unlike vertebrates, crustaceans lack adaptive immunity, relying on haemocytes for innate immune defence against pathogens (Johansson et al. 2000). This highlights the need for in-depth studies on host immune responses and pathogen mechanisms at the cellular level.

A deeper understanding of crustacean biology, particularly at the cellular and molecular levels, is necessary to develop effective disease prevention strategies, besides elucidating toxicity response, developing drugs, and studying regenerative biology. Developing permissive crustacean cell cultures would enable advanced research on viral pathogenesis, host–pathogen interactions, and immune responses, paving the way for more effective disease management strategies in aquaculture. Unfortunately, the lack of stable crustacean cell lines has hindered progress in this area. Despite over a century of efforts, challenges such as replicative senescence, telomere shortening, and the absence of spontaneous in vitro transformation have prevented the establishment of immortalized crustacean cell lines (Anoop et al. 2021a). Nevertheless, significant progress has been made in developing primary cell cultures from various tissues, including shrimp lymphoid tissue, crab testicular tissue, and crayfish haematopoietic tissue (Shashikumar and Desai 2011; Xu et al. 2018). These cultures have shown promising results, with some sustaining viability for months and undergoing multiple passages, yet they remain finite and unsuitable for large-scale applications.

In recent years, primary cell cultures have also been explored for novel applications, such as cultivated meat production, leveraging the regenerative capacity of crustaceans (Musgrove et al. 2024). These advancements underscore the importance of standardizing methodologies for crustacean cell culture development, optimizing growth conditions, and exploring molecular techniques to overcome current limitations.

2 Overview of Crustacean Cell Line Research

2.1 Selection of Species and Tissue Sources in Crustacean Cell Line Research

The selection of species for developing crustacean cell culture systems is primarily based on the purpose of the study, focusing on economically important species such as shrimps, prawns, crabs, and crayfish. Over the past 15 years, significant research efforts have concentrated on these species, as they are commercially valuable and contribute to the aquaculture industry (Claydon et al. 2010; George and Dhar 2010; Goswami et al. 2010; Hu et al. 2010; Jose et al. 2010; George et al. 2011; Jiang et al. 2011; Jose et al. 2012; Huang et al. 2013; Jayesh et al. 2013; Han et al. 2015; Jayesh et al. 2015; Li et al. 2015a, b; Maikaeo et al. 2015; Puthumana et al. 2015a, b, 2016; Thansa et al. 2016; Wajsenzon et al. 2016; Pu et al. 2017; Vieira-Girão et al. 2017; Thansa et al. 2018; Zhang et al. 2018; Sivakumar et al. 2019b).

In terms of tissue source, various body parts have been explored for cell culture development. Haemolymph has been the most commonly used tissue, accounting for 39% of the studies, followed by tissues like hematopoietic cells, lymphoid tissues, ovary, hepatopancreatic cells, and others, such as testicular and neuronal cells. Even though the success rates vary depending on the body part and medium used, efforts over the decades have provided crucial information regarding optimal culture conditions and media for crustacean cell cultures. For instance, the first primary culture of *Penaeus monodon* was reported in 1986, followed by significant advancements in the optimization of culture systems and medium for penaeid species (Ellender et al. 1992; Frerichs 1996; Mulford and Austin 1998). In 1931, the first long-term crustacean cell culture was established using saline solutions, and by 1994, the first stable transformation event using oncogenic mammalian virus genes was achieved in crustacean cells (Fischer-Piette 1931; Tapay 1994). The continuous development in genetic modification and transfection techniques has expanded the potential of crustacean cell lines as models for pathogen studies and other in vitro experiments, with enhanced cell viability and long-term culture capabilities (Claydon and Owens 2009; Li et al. 2011).

2.2 Progress and Attempts in Crustacean Cell Culture Development

The order Decapoda encompasses the infraorders Astacidea (crayfish), Caridea and Dendrobranchiata (shrimps and prawns, collectively referred to as Natantia in older classifications), and Brachyura (crabs), including the most economically significant crustaceans (Xu et al. 2018). This review examines cell culture research focused on these commercially important groups. While no permanent cell lines have been established to date (Musgrove et al. 2024) due to the lack of spontaneous in vitro transformations in crustacean cells (Anoop et al. 2021b), significant efforts from different countries (Fig. 1) have been made using various crustacean groups especially the decapods (Fig. 2a) during the past 15 years (Fig. 2b). Among the 67 studies reviewed (Table 1), 37% concentrated on primary cell culture development and standardization of culture conditions, 25% focused on viral studies, 6% addressed toxicity assessments, 18% explored transfection-based studies, and the remaining 13% covered other related topics (Fig. 2c). Additionally, 57% of the studies targeted shrimp and prawn cultures, 21% focused on crabs, 10% on crayfish, and 5% on species from other phyla.

The body parts of crustaceans have been examined in various media and supplemented under different culture conditions. Among the selected species from the Decapoda order, 34 publications specifically focused on haemolymph culture, with 17 of these studies centred on shrimps and prawns. Lymphoid cell culture is exclusive to shrimp and prawns, whereas testicular tissue culture has been explored in crabs and crayfish. Additionally, cultures of testicular cells, muscle cells, epidermal cells, heart cells, neural cells, and hepatopancreatic cells have been documented in lesser numbers. In contrast, more publications have focused on haemolymph, haematopoietic cells, lymphoid cells, and ovarian cells (Fig. 3a).

Modifications in L-15 medium have been used more frequently, accounting for approximately 61% of the studies over the last 15 years, followed by 6% for M199 medium, 3% for DMEM medium, and 3% for modified Grace insect medium. Additionally, 24% of studies employed other media. Interestingly, shrimp and prawns have been tested across all these media (Fig. 3b). The concentration of serum used in these media ranges from 5% to 20%. Among the selected publications, 51% support the use of 10% FBS, 27% support 20%, and 19% and 2% support 15% and 5%, respectively (Fig. 3c).

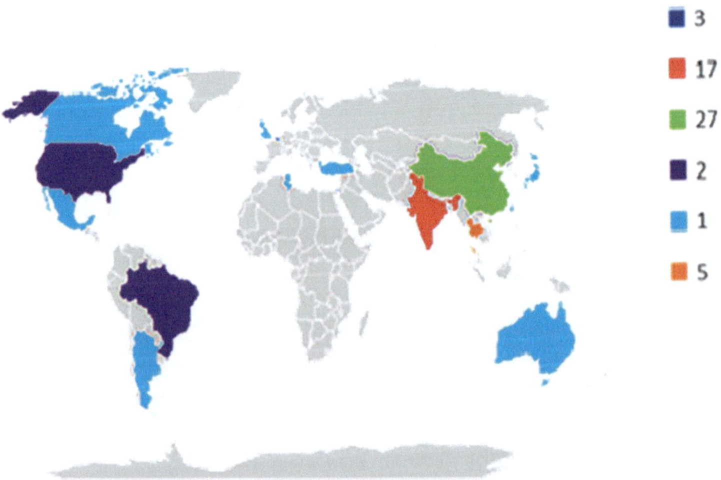

Fig. 1 Number of publications in selected decapods reported from different countries over the period of 2010–2025

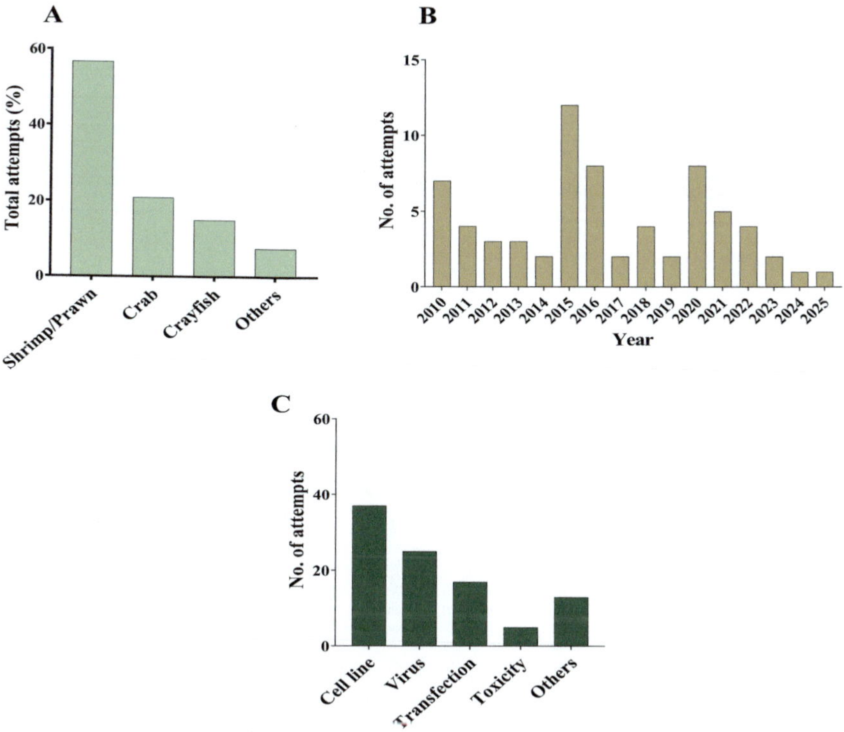

Fig. 2 Crustacean cell culture advancements from 2010 to 2025, with representative figures showing: (**a**) attempts at cell culture development from decapods, (**b**) number of publications, (**c**) applications of decapod cell cultures

2.3 Progress and Challenges in Crustacean Primary Cell Culture Development and Subculture

One of the significant challenges in crustacean cell culture is the absence of a species-specific culture medium, leading researchers to adapt or modify general-purpose media. Leibovitz's L-15 medium has been the most widely used, accounting for 61% of studies between 2010 and 2025 (Claydon et al. 2010; Jose et al. 2012; Sivakumar et al. 2019a). Short-term neuronal cultures in crabs and lobsters were initially maintained using unmodified L-15 medium (Cooke et al. 1989), but subsequent studies introduced modifications to prolong cell survival (Jiang et al. 2011; Sivakumar et al. 2019a). Other commonly used media include Medium 199 (M199) (Xu et al. 2010; Chaulet et al. 2012), Grace's insect medium (George and Dhar 2010), and Dulbecco's Modified Eagle Medium (DMEM) (Han et al. 2015; Wang et al. 2015b). Specialized media, such as Shrimp Cell Culture Medium (SCCM), Medium for *Scylla paramamosain* (MSP), and Medium for Penaeid Shrimp (MPS), have been explored to address species-specific needs (Hu et al. 2010; Zeng et al. 2010; Jayesh et al. 2013).

Table 1 Summary of reviewed publications on crustacean cell culture from 2010 to 2025

S. no.	Organism	Body part	Medium	Life span	Reference
1.	*Macrobrachium rosenbergii*	Egg/embryo	DMEM: F12 supplemented with 10% FBS	NR	Sudarshan et al. (2023)
2.	*Penaeus monodon*	Lymphoid organ	SCCM	Cell line	Anoop et al. (2021a)
3.	*M. rosenbergii*	Haemocyte	Libovitz-15 (L-15) culture medium supplemented with 1 μM phenylthiourea, 100–500 μg/ml penicillin, 100–500 μg/ml streptomycin, 250 μg/ml amphotericin B, 50 μg/ml gentamicin, 2 mM l-glutamine and 10% foetal bovine serum (FBS)	10 days	Thansa et al. (2021)
4.	*Metapenaeus ensis*	Haemolymph	Gelatin-containing optimal basic medium (SBM-12)	275 days	Zhao et al. (2023)
5.	*M. rosenbergii*	Heart	L-15 supplemented with 20% FBS along with 1% prawn haemolymph serum, 0.1% glucose, 0.5% NaCl, and antibiotics	More than 30 days	Goswami et al. (2010)
6.	*Penaeus vannamei*	Haemocyte	2 × grace's supplemented insect medium supplemented with 10% FBS, 10% shrimp, muscle extract, and antibiotics	Few days	George et al. (2011)
7.	*Cherax quadricarinatus*	Haemocyte	L-15 medium containing 5 μM 2-mercaptoethanol, 1 μM phenylthiourea, 2 mM L-glutamine and a crude astakine	Few days	Jeswin et al. (2016)

8.	*P. vannamei*	Lymphoid organ	2 × L-15 supplemented with 20% FBS, 10% Chen's salt, and 10% of eye extract or 3% of ovary extract	10 days	Li et al. (2015a, b)
9.	*Artemia sinica*	Egg/embryo	L-15 medium supplemented with 20% FBS and 100 μg carboxymethyl-chitooligosaccharide, 20 ng basic fibroblast growth factor 80 ng insulin-like growth factor-II, and antibiotics	200 days	Jiang et al. (2011)
10.	*Eriocheir sinensis*	Haemocyte	3 × L-15 medium	5 days	Hong et al. (2013)
11.	*Scylla serrata*	Haemocyte	2 × L-15 medium supplemented with amino acid, sugar and 15% FBS, and antibiotics	21 days	Sivakumar et al. (2019a, b)
12.	*P. monodon*	Haemolymph	SCCM	8 days	Jayesh et al. (2013)
13.	*P. vannamei*	Lymphoid organ	2 × L-15 supplemented with 20% FBS, 10% Chen's salt, with 10% of eye extract or 3% of ovary extract and antibiotics	NR	Li et al. (2015a, b)
14.	*M. ensis*	Lymphoid organ	1.5 × L-15-based supplemented with NaCl, glucose, $NaHCO_3$, taurine, proline, 15% FBS, 10% shrimp muscle extract, bFGF, EGF, and antibiotics	20 days	Han et al. (2013)
15.	*U. cordatus*	Neural cells of the eyestalk	2× L-15 medium with L-glutamine, 10% foetal bovine serum, and antibiotics	7 days	Wajsenzon et al. (2016)

(continued)

Table 1 (continued)

S. no.	Organism	Body part	Medium	Life span	Reference
16.	*P. monodon*	Haemocyte	L-15 medium with 10% FBS supplemented with antibiotics	NR	Claydon et al. (2010)
17.	*L. vannamei*	Eye stalk, hepatopancreas, muscle, ovary, and haemocytes	Grace's insect medium supplemented with 10% foetal bovine serum and 10% shrimp muscle extract and antibiotics	48–66 days	George and Dhar (2010)
18.	*Skogsbergia lerneri*	Epidermal tissue	M199 with supplements, 10% FBS, 10% non-essential amino acids GlutaMAX and antibiotics	28 days	Morgan et al. (2020)
19.	*Scylla paramamosain*	Cerebral ganglion	Modified Medium-199 with 20% FBS and antibiotics	9 days	Xu et al. (2010)
20.	*P. monodon*	Lymphoid organ	SCCM with 10% FBS and antibiotics	NR	Puthumana et al. (2015a)
21.	*P. monodon*	Lymphoid organ	SCCM with 10% FBS and antibiotics	NR	Puthumana et al. (2015b)
22.	*C. quadricarinatus*	Abdominal muscle	M199 medium with 10% bovine serum antiproteases, EDTA, and antibiotics	NR	Chaulet et al. (2012)
23.	*S. paramamosain*	Hepatopancreas	Medium for *S. paramamosain* (MSP) supplemented with FBS, crab muscle extracts, and antibiotics	13 days	Zeng et al. (2010)
24.	*S. serrata*	Testicular tissues	L-15-crab saline medium supplemented with epidermal growth factor, glucose, and antibiotics	150 days	Shashikumar and Desai (2011)

25.	*P. monodon*	Haemocyte	L-15 medium supplemented with 5% lactalbumin solution and 10% FBS, shrimp meat extract, and antibiotics	NR	Thansa et al. (2016)
26.	*P. monodon*	Lymphoid organ	SCCM	Hybrid cell line	Anoop et al. (2021b)
27.	*M. rosenbergii*	Egg/embryo	Opti-MEM medium	NR	Molcho et al. (2022)
28.	*M. rosenbergii*	Haemocyte	L-15 medium	NR	Rotem-Dai et al. (2021)
29.	*P. monodon*	Haemolymph	L-15 medium	45 days	Thansa et al. (2018)
30.	*C. quadricarinatus*	Haemocyte	L-15 medium containing 5% crayfish plasma and antibiotics	NR	Xu et al. (2018)
31.	*P. clarkii*	Haemocyte	L-15 medium with 15% FBS and antibiotics	NR	Li et al. (2011)
32.	*P. monodon*	Lymphoid organ	SCCM with 10% FBS and antibiotics	14 days	Puthumana et al. (2015a, b)
33.	*C. quadricarinatus*	Haemocyte	L-15 medium supplemented with meat extract, phenylthiourea and L-glutamine, and antibiotics	NR	Shi et al. (2018)
34.	*M. ensis*	Lymphoid organ	DMEM with FBS and antibiotics	NR	Han et al. (2015)
35.	*Penaeus chinensis*	Ovary	Medium for penaeid shrimp (MPS) with 20% FBS and antibiotics	21 days	Hu et al. (2010)
36.	*P. monodon*	Lymphoid organ	SCCM with 10% FBS and antibiotics	NR	Puthumana et al. (2016)
37.	*M. ensis* and *Marsupenaeus japonicus*	Lymphoid organ	DMEM with 10% foetal calf serum and antibiotics	NR	Pu et al. (2017)
38.	*Artemia franciscana*	Egg/embryo	1.2 × L-15 medium supplemented with 15% FBS	10 days	Duan et al. (2022)

(continued)

Table 1 (continued)

S. no.	Organism	Body part	Medium	Life span	Reference
39.	*C. quadricarinatus*	Haemocyte	L-15m81 medium	NR	Zhao et al. (2016)
40.	*P. vannamei*	Haemocyte	2× L-15 medium composed of 10.5% (v/v) Chen's salts (CS), 10% (v/v) foetal calf serum (FCS), and penicillin/streptomycin (P/S: 100 units mL^{-1}/100 μg mL^{-1}) (pH 7.5)	NR	Parrilla-Taylor et al. (2022)
41.	*P. vannamei*	Lymphoid organ	2× L-15 supplemented with 20% FBS, Chen's salt, with eye extract/ovary extract and antibiotics	10 days	Li et al. (2015a)
42.	*P. vannamei*	Haemocyte, heart, lymphoid organ, hepatopancreas, gill, eye stalk, and muscle	2× L-15, with L-proline, glucose, 15% FBS, and antibiotics	20 days	Sivakumar et al. (2019b)
43.	*Eriocheir sinensis*	Haemocyte	L-15 supplemented with 15% foetal bovine serum along with glucose, NaCl, and antibiotics	NR	Gu et al. (2014)
44.	*E. sinensis*	Haemocyte	L-15 medium and antibiotics	NR	Xu et al. (2018)
45.	*P. monodon*	Lymphoid organ	2× L-15 supplemented with glucose, MEM vitamins, tryptose phosphate broth, 20% FBS, IGF-1, and antibiotics	NR	Jose et al. (2012)
46.	*P. vannamei*	Haemocyte	2× L-15 supplemented with 20% FBS, glucose, MEM vitamins, tryptose phosphate broth and phenylthiourea and antibiotics	8 days	Vieira-Girão et al. (2017)
47.	*P. clarkii* and *P. vannamei*	Haemocyte	L-15 medium supplemented with 15% FBS, glucose and NaCl and antibiotics	NR	Huang et al. (2013)

48.	*P. vannamei*	Haemocyte	L-15 supplemented with glucose,MEM vitamins, tryptose phosphate broth, 20% FBS and antibiotics	NR	Jose et al. (2010)
49.	*S. serrata*	Haemocyte and testis	2× L-15 prepared in crab saline, supplemented with 5% foetal bovine serum and testis in 3× L-15 supplementedwith non-essential amino acid mixture, lipid concentrateand antibiotic–antimycotic solution	21 days	Deepika et al. (2014)
50.	*Fenneropenaeus merguiensis*	Lymphoid organ	2× L-15 medium and 10% foetal bovine serum and antibiotics	NR	Maikaeo et al. (2015)
51.	*E. sinensis*	Haemocyte	L-15 medium 20% foetal bovine serum and antibiotics	35 days	Liang et al. (2012)
52.	*P. monodon*	Haemocyte	L-15 supplemented with 20% FBS, glucose, and antibiotics	14 days	Thedcharoen et al. (2020)
53.	*C. quadricarinatus*	Haemocyte	L-15 medium supplemented with meat extract, phenylthiourea, L-glutamine, and antibiotics	NR	Meng et al. (2020)
54.	*P. clarkii*	Haemocyte	L-15 supplemented with 15% FBS along with glucose, NaCl, and antibiotics	NR	Wu et al. (2015)
55.	*Apostichopus japonicus*	Intestine	L-15 medium	NR	Wang et al. (2020)

(continued)

Table 1 (continued)

S. no.	Organism	Body part	Medium	Life span	Reference
56.	*M. japonicus*	Lymphoid organ	Medium 199 (M199, ×1), NaCl (14 mg/mL), $MgCl_2·6H_2O$ (3.3 mg/mL), $MgSO_4·7H_2O$ (3.0 mg/mL), $CaCl_2·2H_2O$ (0.9 mg/mL), KCl (0.4 mg/mL), $NaHPO_4·12H_2O$ (0.1 mg/mL), l-glutamine (0.15 mg/mL), d-glucose (1.0 mg/mL), and 4-(2-hydroxyethyl)-1-piperazineethanesulfonic acid (HEPES) (5.9 mg/mL). The medium was supplemented with 10% FBS and an antibiotic mixture (× 1) containing penicillin, amphotericin B, and streptomycin	NR	Tsuchiya et al. (2023)
57.	*P. vannamei*	Haemocyte	2× L-15 supplemented with 20% FBS 20% shrimp muscle extract and antibiotics	NR	Ding et al. (2020)
58.	*P. monodon*	Haemocyte	2× L-15, 5% lactalbumin solution, 10% of FCS and shrimp meat extract together with antibiotics	NR	Thansa et al. (2016)
59.	*E. sinensis*	Testicular tissues	DMEM media with high glucose and antibiotics	30 days	Wang et al. (2015a, b)
60.	*S. paramamosain*	Haemolymph	L-15 medium with 10% foetal calf serum and antibiotics	NR	Wang et al. (2015a, b)
61.	*P. vannamei*	Haemocyte	2× L-15 with 10% FBS, 10% shrimp muscle extract and glucose and antibiotics	NR	Tong et al. (2020)
62.	*P. vannamei*	Haemocyte	Insect-XPRESSTM	NR	Zhang et al. (2018)

63.	*Chionoecetes opilio*	Haemolymph	Nephrops medium with 10% foetal calf serum in a balanced nephrops saline and antibiotics	112 days	Gaudet et al. (2015)
64.	*Ruditapes decussatus*	Haemocyte	L-15 medium	NR	Ladhar-Chaabouni et al. (2021)
65.	*E. sinensis*	Haemocyte	L-15 and glucose and antibiotics	NR	Huang et al. (2020)
66.	*Astacus leptodactylus*	Hepatopancreas, muscle, gill, intestine, and gonads	L-15 medium	NR	Yurdakok-Dikmen et al. (2020)
67.	*Daphnia magna*	Egg/embryo	Schneider's insect medium with 15% FBS	60 days	Sreevidya et al. (2025)

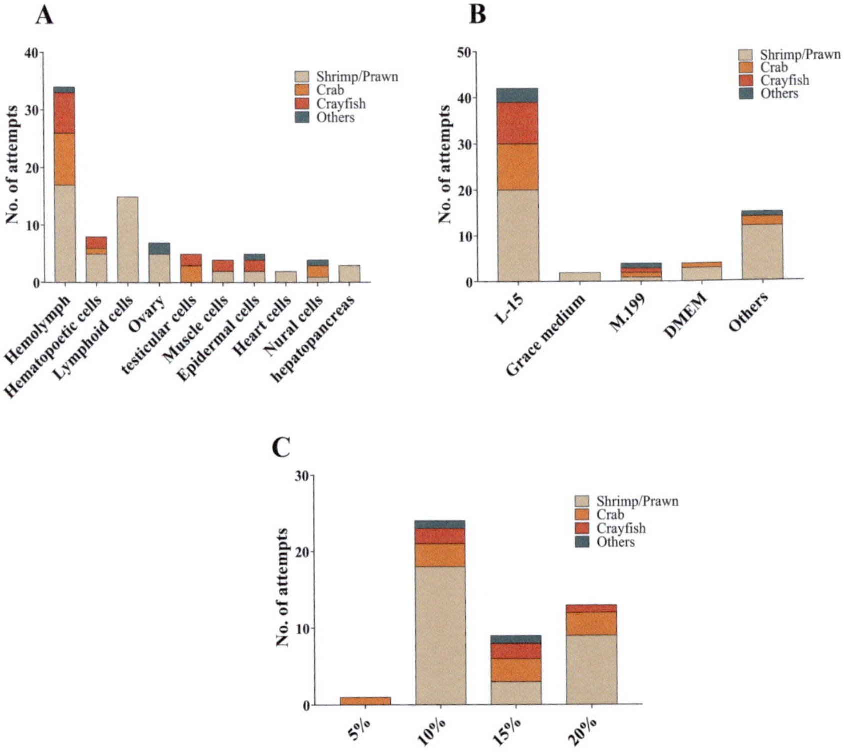

Fig. 3 Crustacean cell culture advancements from 2010 to 2025, with representative figures showing: (**a**) Tissue sources used in selected decapods, (**b**) Tissue culture media employed in selected decapods, and (**c**) Percentage of serum used

Adding supplements like body extracts has significantly improved primary culture longevity and proliferation. Prawn haemolymph serum (Goswami et al. 2010; Jose et al. 2012), muscle extract (George and Dhar 2010; Zeng et al. 2010), and shrimp meat extract (Thansa et al. 2016, 2018) have shown positive effects. Other beneficial additives include crab saline (Deepika et al. 2014), eye and ovary extracts (Li et al. 2015a, b), and foetal bovine serum (FBS) (Freshney 2006), alternatives like horse serum and crayfish plasma have also been tested (Shashikumar and Desai 2011; Xu et al. 2018).

Growth factors, such as Insulin-like Growth Factor I (IGF-I), have stimulated protein synthesis in *C. quadricarinatus* muscle cells (Chaulet et al. 2012) and *P. monodon* lymphoid cells (Jose et al. 2012; Jayesh et al. 2015). Similarly, Epidermal Growth Factor (EGF) has been shown to promote cell growth and proliferation (Shashikumar and Desai 2011). Incorporating extracellular matrix (ECM) components like poly-L-ornithine for neural cells (Wajsenzon et al. 2016) and collagen I and IV for epithelial cells (Morgan et al. 2020) further enhanced adhesion and differentiation.

The development of optimal cell culture media for crustaceans has been significantly advanced by studying the micro-components of their body fluids, particularly haemolymph. Insights into the biochemical composition of crustacean haemolymph have facilitated the formulation of specialized media that support in vitro studies. Over the past decade, several specialized media formulations have been developed and refined for crustacean cell cultures.

One notable example is the SCCM (Shrimp Cell Culture Medium), formulated based on the haemolymph composition of *P. monodon* (Jayesh et al. 2013). This medium exemplifies the importance of incorporating species-specific haemolymph components to support the growth and functionality of crustacean cells. Similarly, the modified L-15 medium, tailored for crustacean cell culture, was developed to address the specific nutritional and physiological needs of these organisms (Sivakumar et al. 2019b). Other significant contributions include the MSP medium, designed for the hepatopancreatic cells of *S. paramamosain* (Zeng et al. 2010), and the MPS medium, optimized for penaeid shrimp (Hu et al. 2010). These media formulations highlight the crucial role of species-specific and tissue-specific requirements in achieving successful in vitro cell culture systems for crustaceans. Such advancements underscore the necessity of a detailed understanding of crustacean physiology and biochemistry, paving the way for more refined and efficient culture systems in the future.

Subculturing crustacean cells has historically been challenging due to poor cell reattachment following passaging. Early attempts using enzymatic and mechanical dissociation often ended in failure or limited success. For example, enzymatic dissociation with trypsin or collagenase caused significant damage to cells, while mechanical dissociation disrupted cell structures (Han et al. 2013). Despite these challenges, some advancements have been made. For example, effective dissociation of *P. vannamei* hepatopancreatic tissue was achieved with collagenase I and IV (George and Dhar 2010). Cerebral ganglion cells were successfully dissociated using collagenase type I and trypsin (Xu et al. 2010). However, enzymatic damage limited cell survival, as noted in multiple studies (Han et al. 2013; Xu et al. 2018). Similarly, mechanical dissociation, although less commonly employed, has been moderately effective in certain cases, enabling limited passaging (Zhao et al. 2023, 2024; Sreevidya et al. 2025). It has shown promise for loosely adherent lymphoid tissue, which exhibited higher subculture success rates than other tissues (Jose et al. 2012; Han et al. 2013). Innovative methods like enzyme-free dissociation solutions, such as ECDS and HyQTase, have demonstrated improved subculture efficiency and cell viability compared to traditional enzymatic or mechanical approaches (Han et al. 2013). These methods may offer a more viable alternative to routine subculturing. While changing half of the medium during subculture, as recommended by Hink (1993), is a widely accepted practice for maintaining cell viability (Jiang et al. 2011; Sivakumar et al. 2019a), further research is essential to overcome the challenges and optimize subculturing techniques.

3 Exploring the Role of Crustacean Cell Culture in Aquatic Biotechnology and Virus Research

Over the past 15 years, significant advancements have been made in crustacean cell culture techniques, primarily focusing on improving cell survival, proliferation, and application in various research fields such as virology, toxicology, immunology, and developmental biology. The primary challenge has been the maintenance of cell viability for extended periods. While earlier studies struggled to maintain viable cultures beyond a few days (George and Dhar 2010; Goswami et al. 2010), recent advancements have demonstrated successful maintenance of crustacean cell lines for extended durations, often surpassing 20 days (Zeng et al. 2010; Han et al. 2015; Sreevidya et al. 2025). For example, testicular tissues from *S. serrata* (Shashikumar and Desai 2011), heart tissue from *M. rosenbergii* (Goswami et al. 2010), and embryonic cells from *Artemia sinica* (Jiang et al. 2011) have been cultured for over a month with significant improvements in viability, marking considerable progress in cell culture technology.

In vitro cultures of primary cells from species like *P. monodon* (Thansa et al. 2016), *L. vannamei*, and other crustaceans have been instrumental in investigating cellular processes such as proliferation, mitotic division, and cell death patterns (Jose et al. 2011; Jayesh et al. 2013). These studies have expanded our understanding of cellular behaviour under various environmental and experimental conditions, and they also offer insights into cellular responses to viral infections, toxic compounds, and ecological stressors (Wang et al. 2015a; Sivakumar et al. 2019a). Furthermore, crustacean cell cultures have proven valuable for studying the pathogenesis of diseases caused by pathogens like *Hematodinium* sp., *Spiroplasma* sp., and *Aeromonas hydrophila* (Liang et al. 2012; Gaudet et al. 2015).

Cell cultures have also played a key role in the study of viral infections in aquaculture species, which are a major cause of economic loss in the industry. In the last decade, several attempts were made to develop cell cultures from different crustacean species to investigate the mechanisms of viral infection and replication. For instance, white spot syndrome virus (WSSV), infectious myonecrosis virus (IMNV), and Taura syndrome virus (TSV) have been studied in crustacean cell lines, with replication peaks observed at various time points post-infection (George et al. 2011; Li et al. 2015a; Vieira-Girão et al. 2017). These studies have shed light on the interaction between viral pathogens and host cells, including the mechanisms of internalization and replication (Marsh and Pelchen-Matthews 2000; Pelkmans and Helenius 2003). For example, haematopoietic cells in crayfish have been shown to be highly permissive to viral replication (Wu et al. 2015), and cytopathogenic effects such as refractile granule accumulation have been observed in infected cells (Jose et al. 2012).

In response to these virus challenges, advanced techniques such as RNA interference (RNAi) have been explored to mitigate virus replication in crustaceans. RNAi technology, which targets viral genes by introducing complementary small interfering RNA (siRNA), has been successfully applied to silence key genes in viruses like WSSV, TSV, and yellow head virus (YHV) (Xu et al. 2019; Itsathitphaisarn et al.

2017). This has led to the development of DNA vaccines for YHV in *P. monodon*, where RNAi was used to silence viral genes in primary haemocyte cultures (Thedcharoen et al. 2020). Additionally, plant extracts, such as those from *Emilia sonchifolia*, have demonstrated antiviral properties by inhibiting WSSV and YHV infections in primary cell cultures, highlighting the potential of natural compounds as immune modulators (Maikaeo et al. 2015; Ding et al. 2020).

In the pursuit of immortalizing crustacean cell lines, various transfection techniques have been employed, but challenges remain in overcoming the rapid mitotic arrest observed in these cells. Retrovirus- and baculovirus-based systems have been tested using viral promoters like WSSV-Ie1, IHHNV-P2, and shrimp beta-actin promoters to improve transfection efficiency (Puthumana et al. 2016; Pu et al. 2017). Despite these efforts, transfection efficiency has remained low, largely due to the rapid cell cycle arrest, which hinders successful immortalization (Han et al. 2015; Shi et al. 2016). Nevertheless, continuous optimization of these methods, such as improving electroporation techniques (Li et al. 2011; Shi et al. 2018), offers hope for future success.

Overall, the development of crustacean cell cultures has paved the way for a deeper understanding of cellular responses to environmental stressors, pathogens, and biotechnological applications. While there are still challenges in immortalization and high-efficiency transfection, these cultures remain invaluable tools for advancing research in aquaculture, virus management, and ecological sustainability. The integration of modern genetic engineering tools, such as CRISPR/Cas systems, and refined culture conditions will likely overcome many of the current limitations, opening new avenues for crustacean cell culture applications in biotechnology and aquaculture.

4 Advancements and Challenges in Developing Immortalized Crustacean Cell Lines

Establishing immortalized crustacean cell lines is a long-standing goal in aquatic biotechnology, with significant implications for fields like aquaculture, biotechnology, and virology. Despite extensive efforts over the past 15 years, including various approaches such as viral transduction, electroporation, lipofection, and cell fusion, achieving complete success has remained elusive. While advancements have been made, several biological and technical challenges continue to hinder the development of stable, immortalized crustacean cell lines.

For species like *P. monodon*, several strategies have been attempted. Thansa et al. (2018) tested lipofection and electroporation with specific markers to immortalize hematopoietic tissue cells, achieving successful transfection, but the cells failed to survive long-term. Similarly, attempts to introduce the SV40LT oncogene into primary lymphoid cells using electroporation led to rapid cell death (Puthumana et al. 2015b; Xu et al. 2018). These setbacks underline the difficulty of overcoming cellular mechanisms like apoptosis and cell cycle arrest, which continue to hinder the immortalization process. Baculovirus-mediated transduction has shown partial success by inducing telomerase activity (Puthumana et al. 2015b, 2016), and retrovirus-mediated transduction using a triple-pseudotyped system has enhanced transgene expression (Han et al. 2015; Pu et al. 2017). However, stable immortalization has not been achieved, likely due to inadequate expression of the necessary cellular factors for long-term viability.

In *M. rosenbergii*, attempts to transduce embryonic cells using a lentiviral platform carrying a WSSV promoter and the MrRas oncogene demonstrated enhanced proliferation, but the cells eventually succumbed to apoptosis and cell cycle arrest (Rotem-Dai et al. 2021). These studies reinforce the challenge of balancing gene expression while maintaining cellular integrity in crustacean cells.

Gene delivery techniques in other crustaceans, such as *P. clarkii* and *C. quadricarinatus*, have also been explored using electroporation, lipofection, and jetPEI-based methods (Li et al. 2011; Xu et al. 2018). Despite advances in delivery efficiency, the inability to achieve long-term survival or immortalization has been a consistent outcome, highlighting that gene delivery efficiency alone is insufficient to overcome the broader challenges of immortalization.

A critical barrier to the development of immortalized crustacean cell lines is the lack of an appropriate cell culture medium. While Grace's medium, developed for insect cell culture in 1962, has supported over 500 insect cell lines, no comparable medium exists for crustacean cells. Common media, such as L-15, designed for fish cells, can support crustacean cells to some extent but fails to sustain them through multiple passages. Despite various modifications, these media do not meet the long-term requirements for crustacean cell cultures, making the development of an optimized medium essential.

Contamination remains another significant challenge in crustacean cell culture. Although several antibiotic treatments and disinfection protocols have been developed, contamination persists, often undermining the success of experiments. Antibiotics and antimicrobials can reduce contamination, but they do not fully prevent it (Fogh et al. 1950). Therefore, reliable disinfection and culture protocols are crucial for establishing stable, uncontaminated cell cultures.

Subculturing techniques also present difficulties in maintaining cell health and longevity. Despite the widespread use of various methods, primary cell cultures often fail to transition into continuous cell lines due to issues with cell proliferation and survival. The gradual decline in cell division observed in many studies is attributed to suboptimal media and handling techniques, emphasizing the need for improved cultivation methods.

Immortalization strategies have been explored as a potential solution for developing continuous cell lines, but most have not succeeded. Key reasons include the inability to form monolayers, inefficiency in proliferative capacity, non-specificity of promoters, and the toxic effects of transfection reagents. These challenges prevent the stable immortalization of crustacean cells, highlighting the need for refined techniques and a deeper understanding of crustacean cell biology.

Furthermore, misidentification of crustacean cell lines remains a serious concern. The neural ganglia of *O. limosus* were once believed to represent the only crustacean cell line, but subsequent studies failed to authenticate this (Lee et al.

2011). Misidentifications can occur due to incorrect identification of the tissue of origin or contamination, emphasizing the importance of rigorous cell line authentication to avoid confusion (Capes-Davis et al. 2010).

In addition to technical challenges, gaps in our understanding of the molecular mechanisms involved in virus-host interactions in crustaceans also hinder progress. While some viral pathogenic processes have been studied, the molecular pathways behind innate immunity evasion and intracellular transport of viruses remain poorly understood. Addressing these questions will not only enhance our understanding of crustacean virology but also provide a foundation for designing antiviral vaccines and therapeutic strategies.

In conclusion, while significant progress has been made in the development of immortalized crustacean cell lines, the field faces several persistent challenges. These include issues with cell culture media, contamination, subculture protocols, and the complex biological mechanisms underlying cell survival and immortalization. Despite these challenges, advancements in gene-editing technologies and a deeper understanding of crustacean biology hold promise for overcoming these obstacles. The successful development of immortalized crustacean cell lines would open up a range of exciting possibilities in research and biotechnology, with far-reaching implications for aquaculture, virology, and beyond.

4.1 Crustacean Cell Line Development in India

Active research on cell line development has been undertaken by many laboratories in India during the past several years, and significant contributions have been made by the National Centre for Aquatic Animal Health (NCAAH) in developing shrimp cell lines for effective investigation of shrimp viruses, and much progress has been made to date. In this, the first successful attempt was a primary haemocyte culture from *P. monodon* developed using a modified 2X Leibovitz L-15 medium, which served as a suitable in vitro model for WSSV titration and viral and immune-related gene expression besides assessing cytotoxicity of aquaculture drugs and chemicals and also the cytotoxicity and genotoxicity of aquatic pollutants such as heavy metals and pesticides (Jose et al. 2010, 2011). This was followed by the development of primary culture from a lymphoid organ of *P. monodon* as a platform for research in virus-cell interaction, virus morphogenesis, up and downregulation of shrimp immune-related genes, and also for the discovery of novel drugs to combat WSSV in shrimp culture (Jose et al. 2012).

In another significant study, the haemolymph composition of *P. monodon* was evaluated for the first time, leading to the formulation of a novel cell culture medium (seawater-based), named Shrimp Cell Culture Medium (SCCM), which showed better proliferation when compared to the conventionally used commercial media. Shrimp cell dissociation 'cocktail' was also designed for passaging the primary cell culture, which showed better survival (Jayesh et al. 2013). As an initial step to immortalize shrimp cells using oncogenes, two recombinant baculoviral vectors (BacIe1-GFP and BacP2-GFP) with shrimp viral promoters WSSV-Ie1 and

IHHNV-P2 were constructed with green fluorescent protein (GFP) as the transgene and successfully expressed in shrimp cells in vivo and in vitro (Puthumana et al. 2016). For in vitro transformation, *P. monodon* lymphoid cells were transfected with pSV3-neo vector encoding SV40-T oncogene and transduced with BacP2-12S E1A-GFP encoding Adenovirus type 12 early region 1A (12S E1A) oncogene. The SV40T transformed cells showed an abnormal nucleus and activated mitotic division. Transduction efficiency in primary lymphoid cells was more than 80%, and high levels of oncogenic 12S E1A expression were observed after 14 days post-transduction, and the proliferating cells survived for more than 90 days with GFP expression, however, without in vitro transformation and immortalization (Puthumana et al. 2015a, b). In another attempt, human oncogene H-*ras* was cloned into a baculoviral vector downstream to shrimp viral promoter IHHNV-P2 and upstream to GFP and tried in vitro transduction at an early dividing stage of the cells by simply incubating recombinant baculovirus with minced tissue explants, resulting in fibroblast-like primary cell culture that showed H-ras and GFP expression. The cell aggregates formed existed for many days but without further division (Anoop et al. 2021a, b).

The breakthrough progress made was the immortalization of shrimp lymphoid cells by Poly ethylene glycol (PEG)-mediated hybridization with the metabolically blocked continuous cell line Sf9 (*Spodoptera frugiperda* ovarian cells) leading to the development of 'PmLyO-Sf9' (Fig. 4). This is the first successful shrimp hybrid cell line developed, with shrimp components in it, and has been found to be capable of undergoing more than 300 subcultures so far. The cell line is deposited in the National Cell Line Repository of ICAR–National Bureau of Fish Genetic Resources, Lucknow, India (Anoop et al. 2021a, b). Most importantly, this hybrid cell line is susceptible to major shrimp viral pathogen WSSV (Sathyabhama et al. 2021) and supported the propagation of another important viral pathogen, IHHNV (Dominic et al. 2023) and MBV [data yet to be published]. PmLyO-Sf9 has been successfully used for the screening and cytotoxicity analysis of drug candidates against WSSV (Namitha et al. 2023; Dharani et al. 2024) and IHHNV infections (Ameer et al. 2024). Subsequently, a novel *Daphnia magna* embryonic cell culture system with a remarkable longevity of two months was developed (Fig. 5) and used for the toxicity study of bisphenol A (Sreevidya et al. 2025).

5 Perspectives and the Way Forward in Crustacean Cell Line Development

The research on crustacean cell line development has made significant strides, but several challenges remain to be addressed to realize their full potential in aquaculture, biotechnology, and other fields. One key factor influencing the success of cell culture initiation is the tissue origin. Previous studies have highlighted that embryonic tissues are the most promising for cell culture, largely due to their totipotency, which provides a higher likelihood of success in culture establishment (Fan and

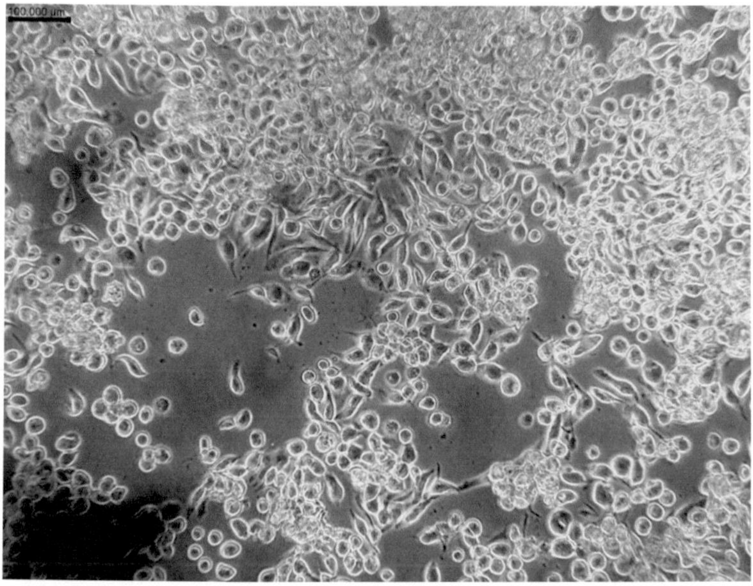

Fig. 4 Crustacean hybrid cell line PmLyO-Sf9, developed by the fusion of Sf9 cells and lymphoid cells from *Penaeus monodon*

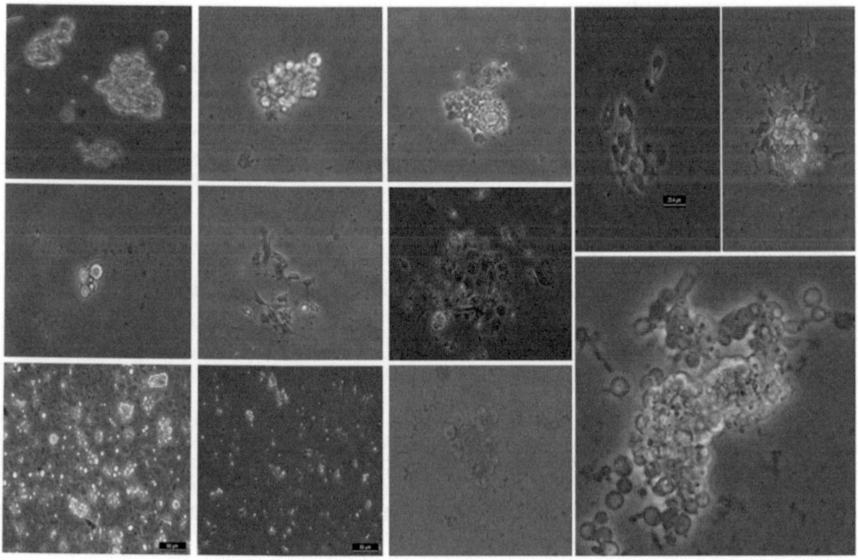

Fig. 5 Primary cell cultures derived from the freshwater crustacean *Daphnia magna*

Wang 2002; Odintsova 2009; Jiang et al. 2011; Meng et al. 2020). Researchers in this field can take advantage of this by utilizing advanced techniques such as induced pluripotent stem cell (iPSC) generation, which could offer new opportunities for establishing sustainable and versatile cell lines.

In recent years, the focus of tissue culture research has shifted towards more sophisticated technologies that improve the fidelity and efficiency of cell culture systems. One notable advancement is the development of three-dimensional (3D) cell culture techniques. Kenny et al. (2007), Li et al. (2017), Zhang et al. (2019), Ao et al. (2021), Zhou et al. (2021), and Marchini and Gelain (2022) demonstrated that 3D culture provides a more accurate representation of the in vivo environment, offering better microenvironmental characteristics and enhanced cellular responses compared to traditional two-dimensional cultures which is further evidenced by the increased longevity of haemocytes from *P. trituberculatus* (Guo et al. 2023). These methods help address the limitations of conventional cell culture systems and are increasingly being applied in drug discovery, toxicity testing, and disease modelling (Park et al. 2021).

The integration of advanced technologies such as CRISPR gene-editing, optimized transfection protocols, and sophisticated instruments further accelerates cell line development. By maximizing transfection efficiency while minimizing toxicity, these technologies offer solutions to key barriers in gene delivery and genome modification in tissue cultures (Ishibashi et al. 2020; Amiri et al. 2023; Geoffroy et al. 2023). Additionally, advancements in co-culture systems and multi-organ chips hold promise for more accurate and predictive studies, enhancing the precision of research and improving the relevance of findings.

Despite the progress in human cell culture systems, such as 3D culturing, co-culturing, and organoids, there remains a significant gap in applying these advanced methods to crustacean cell culture. The use of 3D cultures and organoids, which are already well-established in human cell research, could provide substantial benefits in the crustacean domain. Such techniques would allow for the replication of more complex organ structures and functions, leading to better insights into cellular behaviour and interactions. Moreover, as crustaceans contribute significantly to the global economy, particularly in aquaculture, advancing these technologies will have a direct impact on both economic growth and societal development.

In summary, the future of crustacean cell line development hinges on adopting advanced tissue culture technologies such as 3D cell culture, CRISPR-based gene editing, and co-culture systems. These innovations are not only critical for overcoming existing barriers in crustacean cell culture but also for ensuring the accuracy and relevance of research outcomes. Bridging the gap between human cell culture advances and crustacean cell culture will not only advance scientific understanding but will also have profound implications for the aquaculture industry and broader biotechnological applications.

6 Conclusion

Establishing a stable crustacean cell line holds significant potential to advance research in crustacean biology and contribute to global food security. Over the years, efforts have focused on overcoming key challenges such as the development or modification of culture media, which remains a major bottleneck, improving subculturing techniques, achieving immortalization through genetic manipulation, and enhancing cell longevity. Technologies like CRISPR-Cas systems and RNA interference (RNAi) offer promising solutions to address these challenges, while innovations in 3D cell culture and induced pluripotent stem cells (iPSCs) present new opportunities for in vitro studies in this field.

References

Ameer A, Diya DD, Gopakumar L, Murray F, Madiyan S, Abdul Khader M, Santhini P, Singh B (2024) Detection of antiviral moieties in the leaves of *Rhizophora mucronata* functional on the shrimp pathogen *Penaeus stylirostris* penstyldensovirus 1 (PstDVI)/IHHNV as the prospective drugs of choice in aquaculture. Aquacult Int 32:8791–8808. https://doi.org/10.1007/s10499-024-01590-5

Amiri S, Adibzadeh S, Ghanbari S, Rahmani B, Kheirandish MH, Farokhi-Fard A et al (2023) CRISPR-interceded CHO cell line development approaches. Biotechnol Bioeng 120(4):865–902

Anoop BS, Puthumana J, Vazhappilly CG, Kombiyil S, Philip R, Abdulaziz A, Singh ISB (2021a) Immortalization of shrimp lymphoid cells by hybridizing with the continuous cell line Sf9 leading to the development of 'PmLyO-Sf9'. Fish Shellfish Immunol 113:196–207. https://doi.org/10.1016/j.fsi.2021.03.023

Anoop BS, Puthumana J, Sukumaran V, Vazhappilly CG, Kombiyil S, Philip R, Singh ISB (2021b) A novel approach of transducing recombinant baculovirus into primary lymphoid cells of *Penaeus monodon* for developing continuous cell line. Mar Biotechnol 23:517–528

Ao DS, Xu YE, Sun X et al (2021) Use of a collagen hydrogel to create a three-dimensional culture model of HepG2.2.15 cells. J Pathog Biol 16(5):531–535, 539

Bondad-Reantaso MG, Subasinghe RP, Josupeit H et al (2012) The role of crustacean fisheries and aquaculture in global food security: past, present and future. J Invertebr Pathol 110:158–165. https://doi.org/10.1016/j.jip.2012.03.010

Capes-Davis A, Theodosopoulos G, Atkin I et al (2010) Check your cultures! A list of cross-contaminated or misidentified cell lines. Int J Cancer 127:1–8. https://doi.org/10.1002/ijc.25242

Chaulet A, Medesani DA, Freitas J, Cervino A, Cervino N, Rodríguez EM (2012) Induction of somatic growth in juvenile crayfish *Cherax quadricarinatus* (Decapoda, Parastacidae) by ecdysone and insulin growth factor. Aquaculture 370–371:1–6

Claydon K, Owens L (2009) Attempts at immortalization of crustacean primary cell cultures using human cancer genes. In Vitro Cell Dev Biol Anim 44:451–457. https://doi.org/10.1007/s11626-008-9141-x

Claydon K, Roper KG, Owens L (2010) Attempts at producing a hybridised *Penaeus monodon* cell line by cellular fusion. Fish Shellfish Immunol 29:539–543. https://doi.org/10.1016/j.fsi.2010.05.012

Cooke I, Graf R, Grau S et al (1989) Crustacean peptidergic neurons in culture show immediate outgrowth in simple medium. Proc Natl Acad Sci USA 86:402–406. https://doi.org/10.1073/pnas.86.1.402

Deepika A, Makesh M, Rajendran KV (2014) Development of primary cell cultures from mud crab *Scylla serrata*, and their potential as an in vitro model for the replication of white spot syndrome virus. In Vitro Cell Dev Biol Anim 50:406–416. https://doi.org/10.1007/s11626-013-9718-x

Dharani G, Singh B, Ameer A, Dominic DV, Diya D, Anoop BS, Namitha R, Panchalingam S, Govindaraju K (2024) Seaweed assisted preparation of Fe_2O_3 nanoparticles and their antiviral studies against White Spot Syndrome Virus (WSSV): in vitro and in silico approach. Aquaculture 593:741365. https://doi.org/10.1016/j.aquaculture.2024.741365

Ding Y, Pan L, Gao G, Huang H (2020) In vitro and in vivo immunologic potentiation of herb extracts on shrimp (*Litopenaeus vannamei*). Fish Shellfish Immunol 107:556–566

Dominic DVD, Sajeevan TP, Ameer A, Muneer A, Eldho L, Anoop BS, Jayesh P, Joseph V, Philip R, Singh ISB (2023) In vitro propagation of infectious hypodermal hematopoietic necrosis virus (*Penaeus stylirostris* penstyldensovirus) in PmLyO-Sf9 cells. Arch Virol 168(8):211. https://doi.org/10.1007/s00705-023-05827-1

Duan H, Jin Y, Shao X et al (2022) Stable primary embryonic cells of *Artemia* are suitable for tracing the process of *Vibrio anguillarum* and *V. parahaemolyticus* infection. Aquaculture 560:738598

Ellender RD, Najafabadi AK, Middlebrooks BL (1992) Observations on the primary culture of hemocytes of *Penaeus*. J Crustac Biol 12:178–185. https://doi.org/10.2307/1549072

Fan TJ, Wang XF (2002) In vitro culture of embryonic cells from the shrimp, *Penaeus chinensis*. J Exp Mar Biol Ecol 267:175–184. https://doi.org/10.1016/S0022-0981(01)00364-1

Fischer-Piette E (1931) Culture de tissus de Crustacés, la glande lymphatique du Homard. Archives Zoologiques Expérimentales et Générales 74:33–52

Fogh J, Holmgren N, Ovici P (1950) A review of cell culture contaminations: *Aeruginosa*, *Pseudomonas fluorescens*, and *Aleali*. In Vitro 7:26–41

Frerichs GN (1996) In vitro culture of embryonic cells from the freshwater prawn *Macrobrachium rosenbergii*. Aquaculture 143:227–232. https://doi.org/10.1016/0044-8486(96)01281-1

Freshney RI (2006) Basic principles of cell culture. Cult Cells Tissue Eng:3–22

Gaudet PH, Cawthorn RJ, Buote MA et al (2015) In vitro cultivation of *Hematodinium* sp. isolated from Atlantic snow crab, *Chionoecetes opilio*: partial characterization of late developmental stages. Parasitology 142:598–611. https://doi.org/10.1017/S0031182014001656

Geoffroy M, Pili L, Buffa V, Caroff M, Bigot A, Gicquel E et al (2023) Crispr-cas9 ko cell line generation and development of a cell-based potency assay for raav-fkrp gene therapy. Cells 12(20):2444

George SK, Dhar AK (2010) An improved method of cell culture system from eyestalk, hepatopancreas, muscle, ovary, and hemocytes of *Penaeus vannamei*. In Vitro Cell Dev Biol Anim 46:801–810. https://doi.org/10.1007/s11626-010-9343-x

George SK, Kaizer KN, Betz YM, Dhar AK (2011) Multiplication of Taura syndrome virus in primary hemocyte culture of shrimp (*Penaeus vannamei*). J Virol Methods 172:54–59. https://doi.org/10.1016/j.jviromet.2010.12.020

Goswami M, Lakra WS, Rajaswaminathan T, Rathore G (2010) Development of cell culture system from the giant freshwater prawn *Macrobrachium rosenbergii* (de Man). Mol Biol Rep 37:2043–2048. https://doi.org/10.1007/s11033-009-9659-3

Gu W, Yao W, Zhao Y et al (2014) Establishment of spiroplasma-infected hemocytes as an in vitro laboratory culture model of Chinese mitten crab *Eriocheir sinensis*. Vet Microbiol 171:215–220. https://doi.org/10.1016/j.vetmic.2014.03.016

Guo L, Zhao Y, Guo H (2023) Long-term in-vitro culture and subculture of the hemocytes of swimming crab *Portunus trituberculatus*. J Oceanol Limnol 41:1918–1939

Han Q, Li P, Lu X et al (2013) Improved primary cell culture and subculture of lymphoid organs of the greasyback shrimp *Metapenaeus ensis*. Aquaculture 410–411:101–113. https://doi.org/10.1016/j.aquaculture.2013.06.024

Han Q, Dong D, Zhang X et al (2015) Problems with the use of liposome- and retrovirus-mediated gene transfer methods in the primary lymphoid cells of the oka organs of the greasyback shrimp, *Metapenaeus ensis* (De Haan, 1844). Crustaceana 88:1351–1365. https://doi.org/10.1163/15685403-00003498

Hong Y, Yang X, Cheng Y et al (2013) Effects of pH, temperature, and osmolarity on the morphology and survival rate of primary hemocyte cultures from the Mitten Crab, *Eriocheir sinensis*. In Vitro Cell Dev Biol Anim 49:716–727. https://doi.org/10.1007/s11626-013-9658-5

Hu GB, Wang D, Chen JX (2010) Retroviral delivery of simian virus 40 large T antigen gene into primary cultured ovary cells of the penaeid shrimp, *Penaeus chinensis*: indirect evidence of retroviral integration. J World Aquacult Soc 41(1):137–143

Huang ZJ, Kang ST, Leu JH, Chen LL (2013) Endocytic pathway is indicated for white spot syndrome virus (WSSV) entry in shrimp. Fish Shellfish Immunol 35:707–715. https://doi.org/10.1016/j.fsi.2013.05.028

Huang Y, Hong Y, Huang Z, He H (2020) Cytotoxicity induced by abamectin exposure in haemocytes of Chinese mitten crab, *Eriocheir sinensis*. Environ Toxicol Pharmacol 77:103384. https://doi.org/10.1016/j.etap.2020.103384

Ishibashi A, Saga K, Hisatomi Y, Li Y, Kaneda Y, Nimura K (2020) A simple method using CRISPR-Cas9 to knock-out genes in murine cancerous cell lines. Sci Rep 10(1):22345

Itsathitphaisarn O, Thitamadee S, Weerachatyanukul W, Sritunyalucksana K (2017) Potential of RNAi applications to control viral diseases of farmed shrimp. J Invertebr Pathol 147:76–85. https://doi.org/10.1016/j.jip.2016.11.006

Jayesh P, Jose S, Philip R, Bright Singh IS (2013) A novel medium for the development of in vitro cell culture system from *Penaeus monodon*. Cytotechnology 65:307–322. https://doi.org/10.1007/s10616-012-9491-9

Jayesh P, Philip R, Singh ISB (2015) Multifactorial interaction of growth factors on *Penaeus monodon* lymphoid cells and the impact of IGFs in DNA synthesis and metabolic activity in vitro. Cytotechnology 67:559–571. https://doi.org/10.1007/s10616-014-9697-0

Jeswin J, Xie XL, Ji QL et al (2016) Proteomic analysis by iTRAQ in red claw crayfish, *Cherax quadricarinatus*, hematopoietic tissue cells post white spot syndrome virus infection. Fish Shellfish Immunol 50:288–296. https://doi.org/10.1016/j.fsi.2016.01.035

Jiang G, Xu X, Jing Y et al (2011) Comparative studies on sorting cells from *Artemia sinica* at different developmental stages for in vitro cell culture. In Vitro Cell Dev Biol Anim 47:341–345. https://doi.org/10.1007/s11626-011-9389-4

Johansson MW, Keyser P, Sritunyalucksana K, Söderhäll K (2000) Crustacean haemocytes and haematopoiesis. Aquaculture 191:45–52. https://doi.org/10.1016/S0044-8486(00)00418-X

Jose S, Mohandas A, Philip R, Bright Singh IS (2010) Primary hemocyte culture of *Penaeus monodon* as an in vitro model for white spot syndrome virus titration, viral and immune related gene expression and cytotoxicity assays. J Invertebr Pathol 105(3):312–321. https://doi.org/10.1016/j.jip.2010.08.006

Jose S, Jayesh P, Mohandas A et al (2011) Application of primary haemocyte culture of *Penaeus monodon* in the assessment of cytotoxicity and genotoxicity of heavy metals and pesticides. Mar Environ Res 71:169–177. https://doi.org/10.1016/j.marenvres.2010.12.008

Jose S, Jayesh P, Sudheer NS et al (2012) Lymphoid organ cell culture system from *Penaeus monodon* (Fabricius) as a platform for white spot syndrome virus and shrimp immune related gene expression. J Fish Dis 35:321–334. https://doi.org/10.1111/j.1365-2761.2012.01348.x

Kenny PA, Lee GY, Myers CA et al (2007) The morphologies of breast cancer cell lines in three-dimensional assays correlate with their profiles of gene expression. Mol Oncol 1(1):84–96

Kumar V, Roy S, Sahoo AK, Kumar V (2016) Horseshoe crabs: biomedical importance and its potential use in developing health-care products. Indian J Geo-Marine Sci 45:1234–1244

Ladhar-Chaabouni R, Ayadi W, Sahli E, Mokdad-Gargouri R (2021) Establishment of primary cell culture of *Ruditapes decussatus* haemocytes for metal toxicity assessment. In Vitro Cell Dev Biol Anim 57:477–484

Lee LEJ, Bufalino MR, Christie AE et al (2011) Misidentification of OLGA-PH-J/92, believed to be the only crustacean cell line. Vitr Cell Dev Biol Anim 47:665–674. https://doi.org/10.1007/s11626-011-9447-y

Li F, Ke W, Yan X, Xu L (2011) Gene transfection and expression in the primary culture of crayfish hemocytes. Fish Shellfish Immunol 31:161–163. https://doi.org/10.1016/j.fsi.2011.04.002

Li W, Desmarets LMB, De Gryse GMA et al (2015a) Virus replication cycle of white spot syndrome virus in secondary cell cultures from the lymphoid organ of *Litopenaeus vannamei*. J Gen Virol 96:2844–2854. https://doi.org/10.1099/vir.0.000217

Li W, Van Tuan V, Van Thuong K et al (2015b) Eye extract improves cell migration out of lymphoid organ explants of *L. vannamei* and viability of the primary cell cultures. Vitr Cell Dev Biol Anim 51:651–654. https://doi.org/10.1007/s11626-015-9882-2

Li T, Zhang XX, Li J et al (2017) Research advance in scaffolds of three-dimensional cell culture. Beijing Biomed Eng 36(4):433–437

Liang T, Ji H, Du J et al (2012) Primary culture of hemocytes from *Eriocheir sinensis* and their immune effects to the novel crustacean pathogen *Spiroplasma eriocheiris*. Mol Biol Rep 39:9747–9754

Maikaeo L, Chotigeat W, Mahabusarakam W (2015) *Emilia sonchifolia* extract activity against white spot syndrome virus and yellow head virus in shrimp cell cultures. Dis Aquat Org 115:157–164. https://doi.org/10.3354/dao02891

Marchini A, Gelain F (2022) Synthetic scaffolds for 3D cell cultures and organoids: applications in regenerative medicine. Crit Rev Biotechnol 42(3):468–486. https://doi.org/10.1080/07388551.2021.1932716

Marsh M, Pelchen-Matthews A (2000) Endocytosis in viral replication. Traffic 1:525–532. https://doi.org/10.1034/j.1600-0854.2000.010701.x

Meng C, Liu LK, Li DL et al (2020) White spot syndrome virus benefits from endosomal trafficking, substantially facilitated by a valosin-containing protein, to escape autophagic elimination and propagate in the crustacean *Cherax quadricarinatus*. J Virol 94. https://doi.org/10.1128/jvi.01570-20

Molcho J, Manor R, Shamsian M, Sudarshan G, Ofir R, Parker D, Sagi A (2022) On genome editing in embryos and cells of the freshwater prawn *Macrobrachium rosenbergii*. Aquaculture 558:738391

Morgan SR, Paletto L, Rumney B et al (2020) Establishment of long-term ostracod epidermal culture. In Vitro Cell Dev Biol Anim 56:760–770. https://doi.org/10.1007/s11626-020-00508-8

Mulford AL, Austin B (1998) Development of primary cell cultures from *Nephrops norvegicus*. Methods Cell Sci 19:269–275. https://doi.org/10.1023/A:1009787223797

Musgrove L, Russell FD, Ventura T (2024) Considerations for cultivated crustacean meat: potential cell sources, potential differentiation and immortalization strategies, and lessons from crustacean and other animal models. Crit Rev Food Sci Nutr 65:1–25

Namitha R, Baskaran A, Anoop BS, Diya DV, Ameer A, Kaari M, Radhakrishnan M, Panchalingam S, Bhaskar PV, Govindaraju K, Singh B, Dharani G (2023) Synthesis and characterization of gold nanoparticles using *Brevibacterium casei* (SOSIST-06) isolated from Southern Ocean water samples and their in vitro and in silico anti-WSSV activity. Aquaculture 579:740205. https://doi.org/10.1016/j.aquaculture.2023.740205

Odintsova NA (2009) Stem cells of marine invertebrates: regulation of proliferation and induction of differentiation in vitro. Cell Tissue Biol 3:403–408. https://doi.org/10.1134/S1990519X09050010

Park Y, Huh KM, Kang SW (2021) Applications of biomaterials in 3D cell culture and contributions of 3D cell culture to drug development and basic biomedical research. Int J Mol Sci 22(5):2491

Parrilla-Taylor DP, Elizondo-González R, Gutierrez-Rivera JN et al (2022) Gene expression in primary hemocyte culture of the Pacific white shrimp *Penaeus vannamei* infected with different white spot syndrome virus (WSSV) strains. Latin Am J Aquat Res 50(4):562–572

Pelkmans L, Helenius A (2003) Insider information: what viruses tell us about endocytosis. Curr Opin Cell Biol 15:414–422. https://doi.org/10.1016/S0955-0674(03)00081-4

Pu L, Wang J, Zhang X, Guo H (2017) Development of pseudotyped retroviral system for effective gene transfer and expression in penaeid shrimp cells. Aquaculture 467:198–210. https://doi.org/10.1016/j.aquaculture.2016.04.014

Puthumana J, Jose S, Philip R, Singh ISB (2015a) Cellular and molecular markers in monitoring the fate of lymphoid cell culture from *Penaeus monodon* Fabricius (1798). Fish Shellfish Immunol 47:893–901. https://doi.org/10.1016/j.fsi.2015.05.033

Puthumana J, Prabhakaran P, Philip R, Singh ISB (2015b) Attempts on producing lymphoid cell line from *Penaeus monodon* by induction with SV40-T and 12S EIA oncogenes. Fish Shellfish Immunol 47:655–663. https://doi.org/10.1016/j.fsi.2015.08.010

Puthumana J, Philip R, Bright Singh IS (2016) Transgene expression in *Penaeus monodon* cells: evaluation of recombinant baculoviral vectors with shrimp-specific hybrid promoters. Cytotechnology 68:1147–1159. https://doi.org/10.1007/s10616-015-9872-y

Rotem-Dai N, Weil S, Greenshpan Y, Abehsera S, Manor R, Ofir R, Sagi A (2021) Lentiviral-transduced ectopic expression of androgenic hormone in a crustacean hematopoietic primary cell culture. Front Mar Sci 8:677679

Sathyabhama AB, Puthumana J, Kombiyil S, Philip R, Singh ISB (2021) 'PmLyO-Sf9 – WSSV complex' could be a platform for elucidating the mechanism of viral entry, cellular apoptosis and replication impediments. Virology 553:102–110. https://doi.org/10.1016/j.virol.2020.10.014

Shashikumar A, Desai PV (2011) Development of cell line from the testicular tissues of crab *Scylla serrata*. Cytotechnology 63:473–480. https://doi.org/10.1007/s10616-011-9365-6

Shi Y, Xiang J, Zhou G et al (2016) The Pacific white shrimp *β-actin* promoter: functional properties and the potential application for transduction system using recombinant baculovirus. Mar Biotechnol 18:349–358. https://doi.org/10.1007/s10126-016-9700-1

Shi H, Ruan L, Söderhäll I et al (2018) Transfection of crayfish hematopoietic tissue cells. Dev Comp Immunol 88:70–76. https://doi.org/10.1016/j.dci.2018.07.005

Sivakumar S, Raja Swaminathan T, Kumar R, Kalaimani N (2019a) The development and characterization of a cell culture system from Indian mud crabs *Scylla serrata*. J Aquat Anim Health 31:244–258. https://doi.org/10.1002/aah.10073

Sivakumar S, Swaminathan TR, Anandan R, Kalaimani N (2019b) Medium optimization and characterization of cell culture system from *Penaeus vannamei* for adaptation of white spot syndrome virus (WSSV). J Virol Methods 270:38–45. https://doi.org/10.1016/j.jviromet.2019.04.016

Sreevidya CP, Kumar M, Balakrishnan S, Kunjiraman S, Sarasan M, Magnuson JT, Puthumana J (2025) Establishment of a cell culture from *Daphnia magna* as an in vitro model for (eco) toxicology assays: case study using Bisphenol A as a representative cytotoxic and endocrine disrupting chemical. Aquat Toxicol 278:107173. https://doi.org/10.1016/j.aquatox.2024.107173

Sudarshan G, Weil S, Rotem-Dai N, Manor R, Greenshpan Y, Goldstein O, Sagi A (2023) Enhanced proliferation in a prawn embryonic primary cell culture ectopically expressing mutated Ras. Front Mar Sci 9:1100971

Szaniawska A (2018) Function and importance of crustaceans. Baltic Crustaceans:185–188

Tapay LM (1994) Transformation of primary cultures of shrimp. 40:73–78

Thansa K, Yocawibun P, Suksodsai H (2016) The cellular death pattern of primary haemocytes isolated from the black tiger shrimp (*Penaeus monodon*). Fish Shellfish Immunol 57:243–251. https://doi.org/10.1016/j.fsi.2016.08.045

Thansa K, Rungsiwiwut R, Kitiyanant N, Taengchaiyaphum S (2018) Optimisation of electroporation and lipofection protocols to derive the black tiger shrimp cell line (*Penaeus monodon*). Fish Shellfish Immunol 81:204–213. https://doi.org/10.1016/j.fsi.2018.07.030

Thansa K, Kruangkum T, Pudgerd A, Chaichandee L, Amparyup P, Suebsing R, Sritunyalucksana K (2021) Establishment of hematopoietic tissue primary cell cultures from the giant freshwater prawn *Macrobrachium rosenbergii*. Cytotechnology 73:141–157

Thedcharoen P, Pewkliang Y, Kiem HKT et al (2020) Effective suppression of yellow head virus replication in *Penaeus monodon* hemocytes using constitutive expression vector for long-hairpin RNA (lhRNA). J Invertebr Pathol 175, Article 104346

Tong R, Pan S, Pan L, Zhang L (2020) Effects of biogenic amines on the immune response and immunoregulation mechanism in hemocytes of *Litopenaeus vannamei* in vitro. Mol Immunol 128:1–9

Tsuchiya A, Okamura Y, Kono T, Hikima JI, Sakai M (2023) Transcriptomic analysis of cultured kuruma shrimp primary lymphoid organ cells for identifying key cascades hindering sustainable culture. Fish Sci 89:449–462

Vieira-Girão PRN, Falcão CB, Rocha IRCB et al (2017) Antiviral activity of Ctn[15-34], a cathelicidin-derived eicosapeptide, against infectious myonecrosis virus in *Litopenaeus van-*

namei primary hemocyte cultures. Food Environ Virol 9:277–286. https://doi.org/10.1007/s12560-017-9285-5
Wajsenzon IJR, de Carvalho LA, Biancalana A et al (2016) Culture of neural cells of the eyestalk of a mangrove crab is optimized on poly-l-ornithine substrate. Cytotechnology 68:2193–2206. https://doi.org/10.1007/s10616-015-9942-1
Wang SP, Chen FY, Dong LX et al (2015a) A novel innexin2 forming membrane hemichannel exhibits immune responses and cell apoptosis in *Scylla paramamosain*. Fish Shellfish Immunol 47:485–499. https://doi.org/10.1016/j.fsi.2015.09.028
Wang YL, Sun WJ, He L et al (2015b) Morphological alterations of all stages of spermatogenesis and acrosome reaction in Chinese mitten crab *Eriocheir sinensis*. Cell Tissue Res 360:401–412. https://doi.org/10.1007/s00441-014-2092-5
Wang T, Chen X, Xu K, Zhang B, Huang D, Yang J (2020) Apoptosis induction and detection in a primary culture of sea cucumber intestinal cells. JoVE (J Visualized Exp) 155:e60557
Wu J, Li F, Huang J et al (2015) Crayfish hematopoietic tissue cells but not hemocytes are permissive for white spot syndrome virus replication. Fish Shellfish Immunol 43:67–74. https://doi.org/10.1016/j.fsi.2014.12.013
Xu Y, Ye H, Ma J et al (2010) Primary culture and characteristic morphologies of neurons from the cerebral ganglion of the mud crab, *Scylla paramamosain*. In Vitro Cell Dev Biol Anim 46:708–717. https://doi.org/10.1007/s11626-010-9327-x
Xu X, Duan H, Shi Y et al (2018) Development of a primary culture system for haematopoietic tissue cells from *Cherax quadricarinatus* and an exploration of transfection methods. Dev Compar Immunol 88:45–54. https://doi.org/10.1016/j.dci.2018.07.006
Xu W, Jiang X, Huang L et al (2019) RNA interference technology. In Comprehensive Biotechnology (pp. 560–575). Elsevier. https://doi.org/10.1016/B978-0-444-64046-8.00282-2
Yurdakok-Dikmen BEG, Turgut Y, Gunal AÇ, Uyar REC, Kuzukıran O, Filazi AYH, Erkoc F (2020) In vitro effects of selected endocrine disruptors (DEHP, PCB118, BPA) on narrow-clawed crayfish (*Astacus leptodactylus*) primary cells. In Vitro Cell Dev Biol Anim 56:783–791
Zeng H, Ye H, Li S et al (2010) Hepatopancreas cell cultures from mud crab, *Scylla paramamosain*. In Vitro Cell Dev Biol Anim 46:431–437. https://doi.org/10.1007/s11626-009-9259-5
Zhang Y, Gao G, Lin R et al (2018) Transcriptome analyses reveal *Litopenaeus vannamei* hemocytes response to lipopolysaccharide. Fish Shellfish Immunol 76:187–195. https://doi.org/10.1016/j.fsi.2018.03.002
Zhang LN, Liu XF, Li Y et al (2019) 3D culture technology combined with MSC in improving the ability of hematopoietic stem cells to produce platelets in vitro. J Chin Clin Med 11(21):35–37
Zhao MR, Meng C, Xie XL et al (2016) Characterization of microRNAs by deep sequencing in red claw crayfish *Cherax quadricarinatus* haematopoietic tissue cells after white spot syndrome virus infection. Fish Shellfish Immunol 59:469–483
Zhao Y, Guo L, Guo H (2023) Routine development of long-term primary cell culture and finite cell line from the hemolymph of greasyback shrimp (*Metapenaeus ensis*) and virus susceptibility. Aquaculture 563:739007
Zhao Y, Xue T, Wang J, Guo H (2024) Development of reliable in vitro long-term culture systems for hematopoietic cells of crayfish *Cherax quadricarinatus* and virus susceptibility assay. Aquaculture 589:741016
Zhou Y, Zhou JH, Xu X et al (2021) Matrigel/umbilical cord-derived mesenchymal stem cells promote granulosa cell proliferation and ovarian vascularization in a mouse model of premature ovarian failure. Stem Cells Dev 30(15):782–796. https://doi.org/10.1089/SCD.2021.0005

Applications of Nanotechnology in Aquatic Animal Health Management

Priyabrat Swain, Satya Narayan Sahoo, and Pushpa Choudhary

Abstract

Aquaculture is one of the fastest-growing food production sectors in the world, playing a crucial role in fulfilling the increased demand for animal protein requirements and ensuring future food security. To meet the challenge of global food security, production and productivity have increased multifold due to intensification in the aquaculture systems. Disease prevalence, environmental degradation and chemical contamination are major factors hindering the outcomes. In this regard, nanotechnology has opened new avenues for its application to cope with these challenges in aquaculture and increase production, efficiency and sustainability. Nanotechnology has a significant role to play in improving the efficiency of the aquaculture system. It has an array of applications in fish health management, viz., as an antimicrobial agent for disinfection, treatment of water pollutants, delivery of vaccines, drugs, nutraceuticals and dietary supplements, as nanosensors for pathogen detection and control of biofouling agents, etc. This chapter discusses the various aspects of nanotechnology in aquatic animal health management.

Keywords

Nanotechnology · Aquatic animal health · Antimicrobial agent · Drug delivery · Nanosensor

---

P. Swain (✉) · S. N. Sahoo · P. Choudhary
ICAR-Central Institute of Freshwater Aquaculture, Bhubaneswar, India
e-mail: pswainy2k@yahoo.co.in; nsahoo.icar@gmail.com; pushpafpm@gmail.com

R. Kooloth Valappil et al. (eds.), *Aquatic Animal Health Management*,
https://doi.org/10.1007/978-981-96-7987-4_26

1 Introduction

Nanotechnology is the branch of science and engineering dealing with materials at the nanoscale of 1–100 nm range. The use of nanotechnology is gaining importance day by day as it possesses defined mechanical, chemical and biological properties. Nanotechnology has emerged as an effective tool to address various aspects of aquaculture, particularly in areas of genetics, biotechnology, nutrition, breeding, disease management and environmentally sustainable productivity. The nanonutrient-enriched feed has significantly enhanced fish growth, disease resistance and mitigation of pollutants, thereby increasing profits. Linking biotechnological techniques with nanotechnology has made advancements in genetics and molecular biology research and applications in fish. Nanotechnology devices, such as DNA nanovaccines, nanosensors and drug delivery vehicles, have shown much impact on fish health, breeding and disease prevention (Narsale et al. 2024). Nanotechnology has enormous applications in aquaculture health, such as the detection of pathogens, antimicrobial activity, delivery of nutraceuticals and vaccines, water remediation and preservation of products (Fajardo et al. 2022). This is becoming a well-accepted and fast-growing technology with potential commercial impact. Thus, nanotechnology offers many opportunities for aquaculture, although it has some limitations in response to environmental safety and ethical and regulatory issues.

2 What Are Nanoparticles?

The nanotechnology term was coined in 1974 by Professor Norio Taniguchi of Tokyo Science University. Nanotechnology involves the use of nanoparticles (NPs) in the size range of 1–100 nm and consists of clusters of atoms. The word 'Nano' is a Greek word synonymous with dwarf and refers to one billionth of a meter or 10^{-9}. So the Nano stands for extremely small particle. Nano-biotechnology, an integration between nanotechnology and biotechnology, has emerged as an effective tool for the synthesis of biosynthetic and eco-friendly nanomaterials. The metallic nanoparticles are most promising in antimicrobial applications, due to their large surface area to volume ratio, thus tackling the growth of antimicrobial resistance and the development of resistant strains. Among different nanomaterials with antimicrobial properties, such as silver, gold, zinc, copper, titanium, magnesium and alginate, silver is the most effective antimicrobial against bacteria, viruses, fungi and parasites in animals and humans (Khezerlou et al. 2018).

3 Production of Nanoparticles

Mainly, nanoparticles are synthesized based on two methods, such as 'Top-down' and 'bottom-up' approaches. The top-down method is reducing the size of larger particles of matter to nanometre size using physical or chemical means. The use of laser-based techniques and vaporization followed by cooling is also the top-down

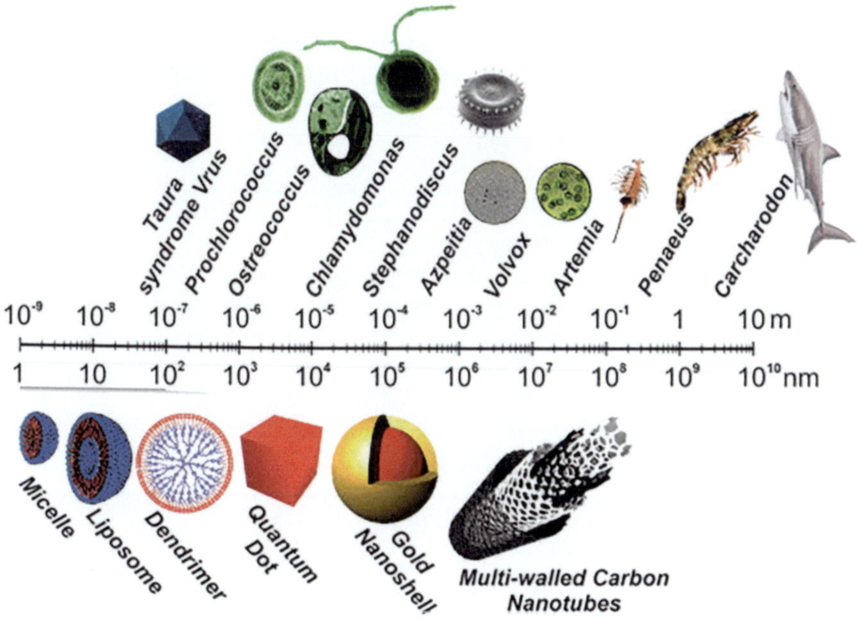

Fig. 1 Relative size of nanoparticles compared to the size of different aquatic organisms. (Source: Fajardo et al. 2022)

process of nanomaterial manufacture. The bottom-up manufacturing includes layer-by-layer deposition, crystallization, self-assembly, solvent extraction/evaporation, biomass reactions and microbial synthesis. The bottom-up approach produces more complex molecular structures of compounds with effective biological activities than the top-down approach. Many nanoparticles with better activity have also been synthesized by organic methods (Swain et al. 2016a). The functional activities of nanoparticles are primarily influenced by the particle size, i.e. the smaller the size, the better the effect (Fig. 1). Thus, the nanoparticles have gained popularity for their unique physical, chemical and biological properties, which can easily be changed by reducing their size. Considering these unique properties, organic and inorganic nanoparticles have been produced for use in aquaculture (Vijayaram et al. 2023; Izquierdo et al. 2017).

4 Nanotechnology Application in Aquatic Animal Health

Maintaining the optimum health of aquatic animals is required for higher productivity. Nanotechnology plays a crucial role in the diagnosis and control of pathogens, pond sanitization, water treatment and efficient delivery of nutrients and drugs in aquaculture (Fig. 2).

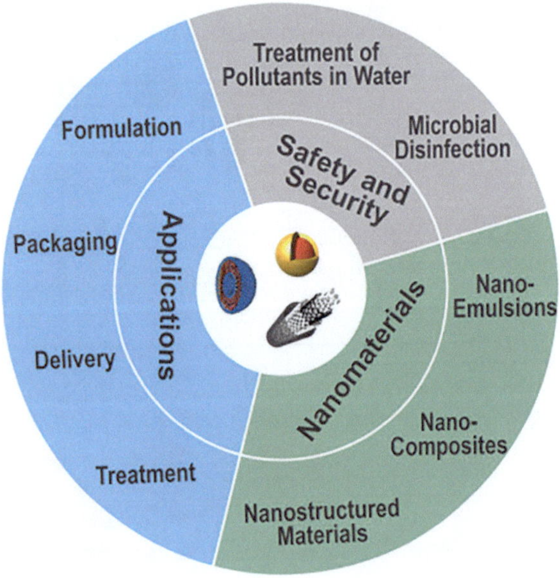

Fig. 2 Current key applications of nanotechnology in the aquaculture. (Source: Fajardo et al. 2022)

DNA-nanovaccines have been applied to improve the immunity of fish against pathogens (Huo et al. 2023). Iron nanoparticles have been employed to improve fish growth (Mohammadi and Tukmechi 2015; Behera et al. 2014; Thangapandiyan et al. 2020). Other applications of nanoparticles include the reduction of the microbial load of aquaculture gadgets and increasing the shelf life of fish feed (Azari et al. 2020).

5 Drug Delivery for Health Management

Nanoparticles, due to their smaller size and increased bioavailability, have multiple advantages in drug administration. The poriferous nanomaterials are used as a delivery matrix for drugs and chemicals. Mesoporous silica particles are used for the controlled, sustained release of drugs and genes (Strømme et al. 2009; Zhou et al. 2018). Oral delivery of nanoparticles, as well as direct target tissue release, are being practised for better bioavailability of drugs with low absorption rates, stabilization of drugs in the gut and improved absorption capability (Fajardo et al. 2022). The common delivery agents include alginate, chitosan, poly (lactic-co-glycolic acid) (PLGA), calcium phosphate nanoparticles, etc. Thus, nanotechnology plays an important role in the delivery of nutrients, hormones and other biological materials to fish (Handy et al. 2008; Shaw and Handy 2011). Several nanoparticles such as Zn, Se and Fe are less toxic to the different cells of the fish system and play a crucial role in the control of bacterial, fungal and viral infections in fish (Vimal et al. 2012; Munawar et al. 2021; Nasr-Eldahan et al. 2021).

6 Delivery of Vaccines

Vaccines are essential in aquaculture to protect from pathogens and safeguard the health and productivity of host animals. Oral, immersion and injectable methods are the most acceptable methods of fish immunization. However, traditional injection practices with oil/water formulations often lead to adverse effects and occasional fish mortality (Shah and Mraz 2020). Nanodelivery systems are safer and more effective alternatives for vaccine delivery in fish. Various encapsulation techniques, including alginate particles, have been developed and tested for oral vaccine delivery to *Oncorhynchus mykiss* and *Paralichthys olivaceus* (Radhakrishnan et al. 2023). Its mechanical stability and mucoadhesive properties make it ideal for oral administration, enabling direct contact with epithelial cell walls (Sosnik 2014). Alginate, a copolymer of β-D-mannuronic acid (M) and α-L-guluronic acid (G) and derived from brown algae or bacterial polysaccharides, has much application in fish immunization. Alginate nanoparticles are generally produced by emulsification and are one of the easiest methods for NP production (Reis et al. 2017). Chitosan (CS), derived from crustaceans and insects, is the second most abundant biopolymer for its unique biological properties, such as biodegradable, bioadhesive, biocompatible and non-toxic properties. Its formulations are widely used in drug delivery, bio-nanosensors and edible coatings. Moreover, the oral administration of nanochitosan-encapsulated DNA vaccine has been shown to positively enhance the immune response against specific fish pathogens. Calcium phosphate nanoparticles adsorbed with antigen have also been found to stimulate both innate and adaptive immunity in *Labeo rohita* (Behera and Swain 2011). Nanoparticles are considered an effective tool for vaccine delivery to farmed fishes against pathogenic diseases.

7 Nanosensors in the Detection of Pathogens

Nanobiosensor detects extremely low concentrations of bacteria, viruses, parasites, pollutants and heavy metals in water (Kapil et al. 2022). Nanosensors used for fish pond cleaning and stock scrutiny are highly sensitive carbon nanotubes. This is very important in detecting and predicting disease outbreaks and early detection and removal of pathogens at large aquaculture systems. Nanosensors detect a wide range of pathogenic agents such as *Aeromonas salmonicida, Vibrio parahaemolyticus, Streptococcus iniae, Aphanomyces invadans,* Nervous necrosis virus *and Penaeus vannamei* nodavirus, and efficient electrical nanosensor can detect even a single virus particle (Patolsky et al. 2004). Antibody-coated gold nanoparticles are also able to detect a particular biomolecule of interest.

8 Anti-microbial Disinfection

Metal NPs such as silver, gold, copper oxide, zinc oxide, titanium dioxide, magnesium oxide, etc., have been reported to have antibacterial properties against important fish pathogens, namely, *Aeromonas hydrophila, A. salmonicida, A. veronii, Vibrio parahaemolyticus* and *Streptococcus iniae* (Radhakrishnan et al. 2023). The mode of action of metal NPs against different bacteria is breaking the cell membrane by electrostatic interaction and disrupting the ion transport. These NPs also break the double strand of the DNA via electrostatic interactions, interfere with enzymatic activity and ribosome assembly, triggering higher oxidative stress with more production of reactive oxygen species (ROS) leading to damage of DNA, lipids, proteins and other micro molecules (Fajardo et al. 2022). Silver, due to its wide range of chemical, biochemical and antimicrobial properties, is considered one of the most effective among different metallic nanoparticles (Nangmenyi and Economy 2014). Colloidal silver nanoparticles are widely used against a broad spectrum of pathogens, including viruses, protozoa, parasites, fungi and bacteria. Silver nanoparticles destroy different strains of bacteria, namely, *Staphylococcus aureus*, *Klebsiella pneumoniae* and *Escherichia coli,* and are much used in tackling antimicrobial resistance (Jeong et al. 2005). Several other nanoparticles like zinc oxide (ZnO), copper oxide (CuO), silver (Ag) and silver-doped titanium dioxide ($Ag–TiO_2$) have shown broad-spectrum antibacterial, antifungal and anti-algal activities in aquaculture use (Swain et al. 2014).

9 Treatment of Water Pollutants

Nanotechnology has been used to treat polluted water, which is one of the main problems in sustainable aquaculture. Nanomaterials, like metal oxide nanoparticles, are commonly used for efficient heavy metal removal from water due to their high reactivity and large surface area. New synthesis methods and practical applications, such as composite materials or granular oxides, to enhance heavy metal removal effectiveness and understand underlying mechanisms such as XAS and NMM have evolved (Hua et al. 2012; Kumar and Chawla 2014). The impact of humic acid and fulvic acid on heavy metal removal using a range of nanomaterials from aqueous solutions has been studied involving iron-based, carbon-based and photocatalytic nanomaterials (Tang et al. 2014). Chitosan nanoparticles are utilized as adsorbents for heavy metal removal. Recent research focuses on removing heavy metals from clays like kaolinite, bentonite and montmorillonite using nanochitosan-clay composites for metal ion removal (Nasr Eldahan et al. 2021). Chitosan-magnetite nanocomposites were also suggested for removing heavy metals from aqueous solutions (Namdeo and Bajpai 2008). Magnetic konjac glucomannan aerogels have been developed to remove arsenite contamination from groundwater. Graphene oxide and graphene nanosheets are also used for the removal of pollutants from water. Moreover, graphene oxide-titanium oxide nanocomposites have been effective for the adsorption and removal of organic compounds from residual water.

10 Delivery of Nutraceuticals and Dietary Supplements

Nanotechnology has played a crucial role in nutrition and dietary intake of trace minerals. It has improved absorption, assimilation and bioavailability in the body of fish. Nanoparticles of micronutrients, incorporated in fish feeds, can enter cells more efficiently and, therefore, raise the absorption rate (Zhou et al. 2009; Ogunkalu 2019). Nanoselenium is more effective in fish feeds to enhance antioxidant status, muscle selenium concentration and weight gain in crucian carp (*Carassius auratus gibelio*) (Zhou et al. 2009). Chitosan-vitamin C nanocomposite has also enhanced the immunity of *Oreochromis niloticus* (Ibrahim et al. 2021). The fish fed with nanonutrients have shown increased muscle contents of Zn, Se, Fe, Cu and Mn compared to the commercial diets. Iron supplements have been reported to improve fish growth and weight (Khan et al. 2020). Nano-Fe as a feed additive has improved the haematology and immunity parameters of fish, *Labeo rohita* (Behera et al. 2014). Feeding of feed incorporated with zinc oxide and selenium nanoparticles also increases growth performance, enzyme activities and immunity in *Labeo rohita* (Swain et al. 2019). Feeding of dietary ZnO-NP for 45 days has also improved growth, serum biochemical parameters, immune function and enzyme activities in rohu (Mondal et al. 2020).

11 Control of Biofouling Agents

Aquaculture productivity can be enhanced through nanotechnology applications by improving biofouling management for disease control. By incorporating metal oxide nanoparticles like CuO, ZnO and SiO_2 into coatings or paints, biofouling from unwanted bacteria, invertebrates (such as mussels and barnacles) and algae (like seaweeds and diatoms) can be monitored and controlled effectively (Munawar et al. 2021). The antifouling technology can be applied in fishing and aquaculture nets, tanks and new marine product packaging materials. Coatings of aquaculture cement tanks with CuO nanoparticles also improve the health, hatching and survival of *L. rohita* fry, reduce the microbial load of water and do not affect the water quality parameters (Swain et al. 2016b).

12 Nanofilms, Nanoemulsions, Nanomicelles and Nanoliposomes

Nanocomposite films or nanofilms are natural biopolymers of polysaccharides, lipids and proteins. Edible coating techniques are used to incorporate them into foods along with antimicrobial films. These packaging materials are being used now as replacements for petrochemical plastics. They are anti-carcinogenic in nature, environmental-friendly and edible. Others, like carbon nanotubes, graphene, graphene oxide and fullerene, also have very good antimicrobial properties that can be used in food and feed packaging (Azizi-Lalabadi et al. 2020). Products using silver nanomaterials are becoming very popular in packaging due to their high antimicrobial properties (Vasile 2018). Chitosan nanoparticle also possesses antimicrobial properties, and nanocomposite films with chitosan nanoparticles, gelatin and oregano essential oil show very good antimicrobial activities (Hosseini et al. 2016).

Nanoemulsions are also new tools for the production of edible films. Nanoemulsion of essential oil has the antibacterial property found to inhibit *E. coli* (Otoni et al. 2014). Micelle is a colloidal particle formed by loosely bound aggregation of several tens or hundreds of atoms, ions, or molecules. It is essential for the absorption of fat-soluble vitamins and lipids within the body. Bile salts secreted from the liver produce micelles of fatty acids. Micelle form of phytochemicals increases its solubility and absorptivity in the intestine. Dietary intake of phytochemical, silymarin, is a more effective supplement in olive flounder (Kim et al. 2023).

Nanodendrimers are unimolecular micelles with a hydrophobic interior and hydrophilic exterior that have been developed and applied as drug-delivery vehicles through intravenous, subcutaneous, intraperitoneal, oral and ocular routes for natural products like polyglycerol, polylysine, polyamidoamine, polypropylene, etc., in humans (An et al. 2023).

A liposome is a tiny bubbled vesicle made up of phospholipids and similar material to a cell membrane. Nanoemulsions are the new areas of production of edible films. Nanoliposomes are prepared from gelatin, alginate and poly (lactic-co-glycolic acid) (PLGA) and have several applications in fish health and nutrition (Jiménez et al. 2014; Acevedo-Fani et al. 2015; Wu et al. 2015)

13 Nanotoxicity and Environmental Impact

The production and use of nanoparticles have gained momentum in the last decades and may constitute a new source of pollutants to the environment (Shah and Mraz 2020), having issues of toxicity to human, animal and environmental health. The majority of the nanomaterials come into the water, thus raising the concern about their deposition and toxicity to aquatic ecosystems. The most studied nanoparticles are silver and titanium, whereas *Danio rerio* is the most suitable fish species used in biomedical and aquaculture research (Bai and Tang 2020). Metal and metal oxide nanoparticles (Cu, Ag, Zn, Ti, Au, ZnO, CuO and TiO_2) that attract a lot of attention from the researchers have been found to be toxic if administered in higher doses. Some of these metal nanoparticles, like Cu, may also release free ions from the surface of the particle and cause further toxicity. Many pharmacological studies like body distribution, metabolism and excretion are lacking, and further, there is lacking of methods for measuring NPs in tissues for many nanoparticles. Nanomaterials have the potential to produce DNA mutation and major structural damage to mitochondria, resulting in cell death (Majumder and Dash 2017; Meghani et al. 2020). Different research findings have suggested acute toxicity of nanometals lethality to *Danio rerio, Labeo rohita* and *Cyprinus carpio* (Bai and Tang 2020; Khan et al. 2015) even in traces. Studies suggest that nanometals may also cause a range of sublethal effects in fish, like respiratory toxicity, disturbances to trace elements in tissues, inhibition of Na^+K^+-ATPase and enzymatic processes and oxidative stress. The dissolved forms of some nanometals (e.g. Cu) are even more acutely toxic to some fish and gills, liver, intestine and brain, and some other organs are the important site of nanotoxicity. Nanoparticles like Ag-NPs are toxic to early life stages of fish, and the nanoforms of some metals (Cu and ZnO) may be more toxic to embryos or juveniles. Along with fishes, other aquatic animals are frequently exposed to nanoparticles (e.g. Ag and ZnO), and exposure has been demonstrated to cause toxicity in aquatic animals like amphibians, *Polypedates maculatus* tadpoles, and the species *P. maculatus* has been used as a bio-indicator for ecotoxicological studies of some metal nanoparticles (Murthy et al. 2022). Besides these, the disposal of nanoparticles also adds concerns to the environment and health (Sajid et al. 2015; Donia and Carbone 2019) since there is the possibility that it may result in genetic mutations (Suppan 2011). Hence, to make nanotechnology more sustainable, further research is to be undertaken to assess the health and environmental risks associated with the exposure, uptake, accumulation, release and deposition of nanomaterial.

14 Conclusion

Current approaches to controlling various fish diseases using chemicals and drugs have numerous drawbacks and necessitate effective alternative, eco-friendly and economically feasible methods. In this aspect, nanotechnology and nanotools can be regarded as novel candidates for addressing fish welfare challenges and ensuring better fish health management. However, evidence also shows that nanometals can affect physiology in fish, causing a range of sublethal effects and also affecting organs like the gill, liver, intestine, brain, etc. in fish. Therefore, future studies need to be aimed at balancing the efficacy of the nanoparticles and reducing their toxicity to humans, animals and the environment.

References

Acevedo-Fani A, Salvia-Trujillo L, Rojas-Graü MA, Martín-Belloso O (2015) Edible films from essential-oil-loaded nanoemulsions: physicochemical characterization and antimicrobial properties. Food Hydrocoll 47:168–177

An H, Deng X, Wang F, Xu P, Wang N (2023) Dendrimers as nanocarriers for the delivery of drugs obtained from natural products. Polymers 15(10):2292

Azari A, Anvar SAA, Ahari H, Sharifan A, Motallebi Moghanjoghi AA (2020) Study on Nanosilver-TiO2 photocatalytic nanocomposite coating with extrusion technique for increasing shelf life of Nile Tilapia (Oreochromis niloticus). Iran J Fish Sci 19(5):2618–2633

Azizi-Lalabadi M, Hashemi H, Feng J, Jafari SM (2020) Carbon nanomaterials against pathogens; the antimicrobial activity of carbon nanotubes, graphene/graphene oxide, fullerenes, and their nanocomposites. Adv Colloid Interf Sci 284:102250

Bai C, Tang M (2020) Toxicological study of metal and metal oxide nanoparticles in zebrafish. J Appl Toxicol 40(1):37–63

Behera T, Swain P (2011) Antigen adsorbed calcium phosphate nanoparticles stimulate both innate and adaptive immune response in fish, *Labeo rohita* H. Cell Immunol 271(2):350–359

Behera T, Swain P, Rangacharulu PV, Samanta M (2014) Nano-Fe as feed additive improves the hematological and immunological parameters of fish, *Labeo rohita* H. Appl Nanosci 4:687–694

Donia DT, Carbone M (2019) Fate of the nanoparticles in environmental cycles. Int J Environ Sci Technol 16:583–600

Fajardo C, Martinez-Rodriguez G, Blasco J, Mancera JM, Thomas B, De Donato M (2022) Nanotechnology in aquaculture: applications, perspectives and regulatory challenges. Aquacult Fish 7(2):185–200

Handy RD, Henry TB, Scown TM, Johnston BD, Tyler CR (2008) Manufactured nanoparticles: their uptake and effects on fish – a mechanistic analysis. Ecotoxicology 17:396–409

Hosseini SF, Rezaei M, Zandi M, Farahmandghavi F (2016) Development of bioactive fish gelatin/chitosan nanoparticles composite films with antimicrobial properties. Food Chem 194:1266–1274

Hua M, Zhang S, Pan B, Zhang W, Lv L, Zhang Q (2012) Heavy metal removal from water/wastewater by nanosized metal oxides: a review. J Hazard Mater 211:317–331

Huo X, Tang L, Liu Q, Zhu W, Zhang J, Hu M, Zhao F, Wang P, Yuan G, Yang C, Su J (2023) Oral pcDNA3. 1-VP4/VP56-FlaC DNA vaccine encapsulated by chitosan/sodium alginate nanoparticles confers remarkable protection against GCRV infection in grass carp. Aquaculture 577:739996

Ibrahim RE, Amer SA, Farroh KY, Al-Gabri NA, Ahmed AI, El-Araby DA, Ahmed SA (2021) The effects of chitosan-vitamin C nanocomposite supplementation on the growth performance, antioxidant status, immune response, and disease resistance of Nile tilapia (*Oreochromis niloticus*) fingerlings. Aquaculture 534:736269

Izquierdo MS, Ghrab W, Roo J, Hamre K, Hernández-Cruz CM, Bernardini G, Terova G, Saleh R (2017) Organic, inorganic and nanoparticles of Se, Zn and Mn in early weaning diets for gilthead seabream (*Sparus aurata*; Linnaeus, 1758). Aquac Res 48(6):2852–2867

Jeong SH, Hwang YH, Yi SC (2005) Antibacterial properties of padded PP/PE nonwovens incorporating nano-sized silver colloids. J Mater Sci 40:5413–5418

Jiménez A, Sánchez-González L, Desobry S, Chiralt A, Tehrany EA (2014) Influence of nanoliposomes incorporation on properties of film forming dispersions and films based on corn starch and sodium caseinate. Food Hydrocoll 35:159–169

Kapil S, Bhattu M, Vinayak A, Pal N, Sharma V (2022) Nanobiosensors' potentialities for environmental monitoring. In: Nanobiosensors for environmental monitoring: fundamentals and application. Springer, Cham, pp 41–74

Khan MS, Jabeen F, Qureshi NA, Asghar MS, Shakeel M, Noureen A (2015) Toxicity of silver nanoparticles in fish: a critical review. J Biodivers Environ Sci 6(5):211–227

Khan MZH, Hossain MMM, Khan M, Ali MS, Aktar S, Moniruzzaman M, Khan M (2020) Influence of nanoparticle-based nano-nutrients on the growth performance and physiological parameters in tilapia (*Oreochromis niloticus*). RSC Adv 10(50):29918–29922

Khezerlou A, Alizadeh-Sani M, Azizi-Lalabadi M, Ehsani A (2018) Nanoparticles and their antimicrobial properties against pathogens including bacteria, fungi, parasites and viruses. Microb Pathog 123:505–526

Kim HS, Jo S, Yun KS, Lee KJ (2023) Effects of dietary micelle silymarin on the growth performance, feed utilization and health of olive flounder (*Paralichthys olivaceus*). Aquac Int 31(6):3419–3436

Kumar R, Chawla J (2014) Removal of cadmium ion from water/wastewater by nano-metal oxides: a review. Water Qual Expo Health 5:215–226

Majumder D, Dash G (2017) Application of nanotechnology in fisheries and aquaculture. Everyman Sci 51(6):358–364

Meghani N, Dave S, Kumar A (2020) Introduction to nanofood. In: Hebbar U, Ranjan S, Dasgupta N, Kumar Mishra R (eds) Nano-food engineering. Food engineering series. Springer, Cham, pp 1–23

Mohammadi N, Tukmechi A (2015) The effects of iron nanoparticles in combination with *lactobacillus casei* on growth parameters and probiotic counts in rainbow trout (*Oncorhynchus mykiss*) intestine. J Vet Res 70(1):47–53

Mondal AH, Behera T, Swain P, Das R, Sahoo SN, Mishra SS, Das J, Ghosh K (2020) Nano zinc vis-à-vis inorganic Zinc as feed additives: effects on growth, activity of hepatic enzymes and non-specific immunity in rohu, *Labeo rohita* (Hamilton) fingerlings. Aquac Nutr 26(4):1211–1222

Munawar N, Aleem M, Abru S, Laraib T, Zainab SN, Manzoor S, Ali A (2021) Interaction and applications of nanoparticles in fishes and aquaculture. Sch Bull 7(6):150–155

Murthy MK, Mohanty CS, Swain P, Pattanayak R (2022) Assessment of toxicity in the freshwater tadpole *Polypedates maculatus* exposed to silver and zinc oxide nanoparticles: a multi-biomarker approach. Chemosphere 293:133511

Namdeo M, Bajpai SK (2008) Chitosan–magnetite nanocomposites (CMNs) as magnetic carrier particles for removal of Fe (III) from aqueous solutions. Colloids Surf A Physicochem Eng Asp 320(1–3):161–168

Nangmenyi G, Economy J (2014) Nanometallic particles for oligodynamic microbial disinfection. In: Street A, Sustich R, Duncan J, Savage N (eds) Nanotechnology applications for clean water: solutions for improving water quality. William Andrew, William Andrew, Norwich, pp 283–295

Narsale SA, Debbarma S, Prakash P, Sheikh S, Shinde S, Baidya S (2024) Exploring the potential of nanotechnology in fisheries and aquaculture: opportunities and implications. J Adv Biol Biotechnol 27(3):1–17

Nasr-Eldahan S, Nabil-Adam A, Shreadah MA, Maher AM, El-Sayed Ali T (2021) A review article on nanotechnology in aquaculture sustainability as a novel tool in fish disease control. Aquac Int 29:1459–1480

Ogunkalu OA (2019) Utilization of nanotechnology in aquaculture and seafood sectors. Eurasian J Food Sci Technol 3(1):26–33

Otoni CG, de Moura MR, Aouada FA, Camilloto GP, Cruz RS, Lorevice MV, de Soares FFN, Mattoso LH (2014) Antimicrobial and physical-mechanical properties of pectin/papaya puree/cinnamaldehyde nanoemulsion edible composite films. Food Hydrocoll 41:188–194

Patolsky F, Zheng G, Hayden O, Lakadamyali M, Zhuang X, Lieber CM (2004) Electrical detection of single viruses. Proc Natl Acad Sci 101(39):14017–14022

Radhakrishnan A, Vaseeharan B, Ramasamy P, Jeyachandran S (2023) Oral vaccination for sustainable disease prevention in aquaculture – an encapsulation approach. Aquac Int 31(2):867–891

Reis CP, Neufeld RJ, Veiga F, Ribeiro AJ (2017) Preparation of drug-loaded polymeric nanoparticles. In: Nanomedicine in cancer. Jenny Stanford Publishing, pp 171–214

Sajid M, Ilyas M, Basheer C, Tariq M, Daud M, Baig N, Shehzad F (2015) Impact of nanoparticles on human and environment: review of toxicity factors, exposures, control strategies, and future prospects. Environ Sci Pollut Res 22:4122–4143

Shah BR, Mraz J (2020) Advances in nanotechnology for sustainable aquaculture and fisheries. Rev Aquac 12(2):925–942

Shaw BJ, Handy RD (2011) Physiological effects of nanoparticles on fish: a comparison of nanometals versus metal ions. Environ Int 37(6):1083–1097

Sosnik A (2014) Alginate particles as platform for drug delivery by the oral route: state-of-the-art. Int Sch Res Notice 2014(1):926157

Strømme M, Brohede U, Atluri R, Garcia-Bennett AE (2009) Mesoporous silica-based nanomaterials for drug delivery: evaluation of structural properties associated with release rate. Wiley Interdiscip Rev Nanomed Nanobiotechnol 1(1):140–148

Suppan S (2011) Racing ahead: US agri-nanotechnology in the absence of regulation. Institute for Agriculture and Trade Policy, Minneapolis, pp 3–20

Swain P, Nayak SK, Sasmal A, Behera T, Barik SK, Swain SK, Mishra SS, Sen AK, Das JK, Jayasankar P (2014) Antimicrobial activity of metal based nanoparticles against microbes associated with diseases in aquaculture. World J Microbiol Biotechnol 30:2491–2502

Swain S, Barik SK, Behera T, Nayak SK, Sahoo SK, Mishra SS, Swain P (2016a) Green synthesis of gold nanoparticles using root and leaf extracts of *Vetiveria zizanioides* and *Cannabis sativa* and its antifungal activities. BioNanoScience 6:205–213

Swain P, Sasmal A, Nayak SK, Barik SK, Mishra SS, Mohapatra KD, Swain SK, Saha JN, Sen AK, Jayasankar P (2016b) Evaluation of selected metal nanoparticles on hatching and survival of larvae and fry of Indian major carp, rohu (Labeo rohita). Aquac Res 47(2):498–511

Swain P, Das R, Das A, Padhi SK, Das KC, Mishra SS (2019) Effects of dietary zinc oxide and selenium nanoparticles on growth performance, immune responses and enzyme activity in rohu, *Labeo rohita* (Hamilton). Aquac Nutr 25(2):486–494

Tang WW, Zeng GM, Gong JL, Liang J, Xu P, Zhang C, Huang BB (2014) Impact of humic/fulvic acid on the removal of heavy metals from aqueous solutions using nanomaterials: a review. Sci Total Environ 468:1014–1027

Thangapandiyan S, Alisha AA, Anidha K (2020) Growth performance, hematological and biochemical effects of iron oxide nanoparticles in *Labeo rohita*. Biocatal Agric Biotechnol 25:101582

Vasile C (2018) Polymeric nanocomposites and nanocoatings for food packaging: a review. Materials 11(10):1834

Vijayaram S, Tsigkou K, Zuorro A, Sun YZ, Rabetafika H, Razafindralambo H (2023) Inorganic nanoparticles for use in aquaculture. Rev Aquac 15(4):1600–1617

Vimal S, Taju G, Nambi KN, Majeed SA, Babu VS, Ravi M, Hameed AS (2012) Synthesis and characterization of CS/TPP nanoparticles for oral delivery of gene in fish. Aquaculture 358:14–22

Wu J, Liu H, Ge S, Wang S, Qin Z, Chen L, Zheng Q, Liu Q, Zhang Q (2015) The preparation, characterization, antimicrobial stability and in vitro release evaluation of fish gelatin films incorporated with cinnamon essential oil nanoliposomes. Food Hydrocoll 43:427–435

Zhou X, Wang Y, Gu Q, Li W (2009) Effects of different dietary selenium sources (selenium nanoparticle and selenomethionine) on growth performance, muscle composition and glutathione peroxidase enzyme activity of crucian carp (*Carassius auratus gibelio*). Aquaculture 291(1–2):78–81

Zhou Y, Quan G, Wu Q, Zhang X, Niu B, Wu B, Huang Y, Pan X, Wu C (2018) Mesoporous silica nanoparticles for drug and gene delivery. Acta Pharm Sin B 8(2):165–177

Application of Genetics and Genomics in Aquatic Animal Health Management

G. Gopikrishna

Abstract

The importance of using genetically improved strains of fish that are resistant to disease is elaborated. The reasons why aquaculture in general is still lagging behind in using genetically improved strains of fishes are discussed. Examples of genetically improved aquatic species that are cultured globally are presented. The prerequisites for initiating a genetic improvement program are highlighted, citing examples from aquatic species. The genetic basis for disease resistance is discussed with examples both from India and globally. The progress made in India, especially with regard to freshwater species, such as carps and scampi, is discussed. The shrimp industry of India today depends on an imported selectively bred strain of Pacific white shrimp for commercial culture and export, a species that dominates even the global shrimp industry in terms of production. Genomic research carried out in tiger shrimp and carp in India is detailed. While the selection of individuals for breeding utilizes computed breeding values, there appears to be a higher advantage when genomic information is incorporated into the selection process. Instances of genetics and genomics aiding selection programs in Atlantic salmon and Genetically Improved Farmed Tilapia (GIFT) are also explained. The advantages of using strains of fish genetically improved for disease resistance and advanced genomic tools, such as CRISPR for gene editing, as future methods to improve disease resistance in fish are elucidated.

Keywords

Genetic improvement · Genomics · Disease resistance · Gene editing

G. Gopikrishna (✉)
ICAR-Central Institute of Brackishwater Aquaculture, Chennai, India
e-mail: gknani@yahoo.com

R. Kooloth Valappil et al. (eds.), *Aquatic Animal Health Management*,
https://doi.org/10.1007/978-981-96-7987-4_27

1 Introduction

Aquaculture and fisheries constitute an important food-producing sector contributing substantially to food security, nutrition, and socioeconomic development and also providing livelihood options to fishers. Aquaculture is an important activity that substantially contributes to India's agriculture sector. In 2022–2023, the contribution of the fisheries sector to the Indian economy was 1.1% and to agriculture, it was 6.72% (Handbook on Fisheries Statistics 2023). Another important aspect is the contribution of this sun-rise sector to the Indian economy by garnering precious foreign exchange obtained in terms of the export of captured and cultured fish. India in 2022–2023 exported 1.73 million MT of seafood worth INR 8090 million or US$ 8.09 billion (Handbook on Fisheries Statistics 2023). Aquaculture has in fact been given the much-needed prominence in national policies, and the government has taken several steps to boost the fisheries sector to give a fillip to the blue revolution. The Indian fisheries sector is diverse with subsectors: marine, freshwater, cold water, and brackish water. Each of them is distinct, and the methods for culturing fish are unique. The marine sector constitutes primarily capture fisheries, and recently, sea cage farming of certain selected species, such as cobia and silver pompano, has been initiated. In the freshwater sector, the most significant contributor is the Indian major carps followed by several other species, such as Indian catfish, scampi, and other riverine fishes. Today, brackish water aquaculture is dominated by an imported shrimp species, *Penaeus vannamei*. In addition, tiger shrimp *Penaeus monodon* and the Indian white shrimp *Penaeus indicus* are being farmed, although their contribution is much less relative to *P. vannamei*. Under finfishes, farming of Asian seabass is prominent, and recently, milkfish has also been farmed in a limited way. In cold water fisheries that dominate the hilly regions of the country, among the many species, the most promising is the Himalayan Golden Mahseer. The vastly rich fisheries fauna of the eight states of the Northeast region need special mention as this hugely untapped region has immense potential to enhance fish farming and production in that region. Fishes, when cultured in ponds, sea cages, and raceways, are exposed to the prevailing environmental conditions. In addition to proper nutrition, the management of aquatic animal health plays a significant role. For a culture operation to succeed, the initial focus must be on quality seed. Once this aspect is addressed, nutrition plays a vital role. Ongoing culture brings its own problems in terms of health, and there should be proper means and ways to identify and treat diseases. Any culture operation will be successful only if a synergy exists between genetics, nutrition, and health. For this precise reason, inputs from a team of personnel from each of these disciplines need to be in place to ensure a successful crop. Notably, compared to terrestrial animals, control and prevention of diseases in aquatic animals is a challenging proposition. However, the use of genetically improved strains of fish in terms of disease resistance is one of the ways the aquatic animal health problem can be addressed effectively. Nevertheless, globally, less than 10% of fish produced through aquaculture is based on genetically improved fish; in India, this figure is far too low, except for carps and scampi. Therefore, genetically improved strains of fish need to be produced to support the long-term sustainability of the sector. In this background, an attempt has been made in the following sections to elucidate the role of genetics and breeding in aquatic animal health management specific to India, although the role of genetics in disease resistance in other aquatic species farmed globally will be briefly discussed.

1.1 Role of Genetics in Aquaculture

Plants and terrestrial farm animals have had the advantage of undergoing planned genetic improvement programs for a very long time. The word "breeding" means a planned mating of plants/animals with known information on their parents and grandparents or pedigree. In effect, it is the scientific execution of a planned breeding strategy. Initially in livestock, in the early seventeenth century, several breeds of animals were mated to retain the species characteristics. It was only after Mendel's theory of inheritance was enunciated in the 1900s that the concepts of genetics and quantitative genetics came into existence. The credit for enunciating animal breeding plans goes to Sewall Wright and Jay Laurence Lush, who, in the early nineteenth century, elucidated the principles of animal breeding plans in animals (Gjedrem and Baranski 2009). The increased global production of food in terms of rice, wheat, maize, potatoes, milk, fruits, etc., is a result of the conventional breeding programs that have been in vogue for quite some time in this century. Unfortunately, aquaculture has been lagging in this aspect, and it is only in the early seventies that selective breeding principles were first applied in Norway to rainbow trout in 1973 and Atlantic Salmon in 1975 (Gjedrem and Baranski 2009). These programs have in fact come a long way and have added disease resistance as well as other traits too. The selective breeding programs in Norway stand out as a fine example of how a fish can be bred for increased growth and improved disease resistance. In the late nineties, efforts have been focused on genetically improving the Pacific white shrimp *P. vannamei,* and today's specific pathogen-free and genetically improved *P. vannamei* is a direct result of this endeavor. The improved strains from the USA are finding their way to several tropical countries where aquaculture is being practiced. India also imports this improved strain since 2009 with adequate checks and balances through regulatory bodies. It would be pertinent to mention here that the improvement in Pacific white shrimp could be attributed to the ease of reproduction in this species, the main factor being the open thelycum in females that facilitate the production of a large number of families with relative ease. In 1988, a tilapia breeding program was initiated by CGIAR researchers at the WorldFish Center and their partners in Norway and the Philippines, which resulted in the improved strain known as Genetically Improved Farmed Tilapia (GIFT) (https://www.cgiar.org/innovations/genetically-improved-farmed-tilapia-gift/). This strain significantly contributed to boosting fish productivity both at small-scale and commercial levels, benefitting millions worldwide. Today GIFT is farmed in 14 countries across five continents (https://worldfishcenter.org/project/genetically-improved-farmed-tilapia-gift).

1.2 Why Is Genetic Improvement Necessary?

Genetic improvement relies heavily on information obtained from individuals in a population and their pedigree. Improvement is possible for any quantitative trait that exhibits polygenic inheritance, which means that they are primarily controlled by several genes. In aquaculture, body weight is the most important measurable trait because it translates into heavier weight fetching higher economic returns in a commercial operation. When fishes are selected for heavier body weight, there would be traits that get improved simultaneously. Survival is one trait that gets a boost when fishes are selected for heavy body weight. In essence, the genetic correlation between traits comes into play when selecting for heavier body weight. When the genetic correlation between traits is high, in the positive direction, selecting for one trait automatically improves the other, a phenomenon known as correlated response. Another important aspect is that the improvement obtained in one generation is carried over to the next generation, thus the increment in a selected trait is cumulative. Furthermore, any fish species that is to be genetically improved should have its life cycle closed, meaning that it is mandatory that there is ease in reproduction and that any number of families be produced as per requirement. The breeding efficiency should be more than 90%, which means that if we allow 10 pairs of fish to mate, at least 9 pairs need to produce families successfully. This is a major barrier that has to be overcome, as closing the life cycle of any fish species is a challenging proposition and takes time and effort to execute. The closing of the life cycle is preceded by collecting the specific species of fish from the wild, domesticating them, and allowing them to breed in captivity such that fertilization is successful and families are produced without any hindrance. This is the first step while planning for a breeding program and only under such a scenario can we think of initiating genetic improvement. The main advantage of having a genetic improvement program is that we have a source of improved strains of fish that exhibit faster growth, have resistance to a particular disease, and can be used for propagation/culture. Selected strains have a uniform growth, and the economic returns are higher. In aquatic animals, we have the advantage of producing families of either half or full siblings due to the external fertilization and plenty of offspring to choose from. However, there are some disadvantages too. Once a genetic improvement program starts, it has to be maintained, and the cost of such an operation is very high. The gains obtained over several generations need to be kept intact by continuing with the program and by exercising selection each generation, an activity that becomes long-term in nature. Another aspect is that progeny from different families need to be kept separately till they are physically tagged, which means that there is an additional burden of maintaining the families separately up to tagging, which imposes an added financial cost. However, once the families are tagged, they could be communally reared, which is a requirement for obtaining scientifically sound data. In terrestrial farm animals, there are examples of how such programs are carried out in collaboration with the farmers/private sector so that the difficulties they face are addressed scientifically each generation. A similar arrangement needs to be explored in India to facilitate the genetic improvement of food fish. Houston (2017), in a review, opined that the route to improved disease resistance in the vast majority of the farmed fish in the world is to work towards the implementation of selective breeding as ~90% of the world aquaculture relies on unimproved stock.

1.3 Genetic Basis of Disease Resistance

Before we delve deep into the genetic basis of disease resistance, it is important that we comprehend the concept of heritability. Polygenic traits or characters are governed by several genes, and each gene contributes to the expression of the trait, which, when added up, gives the additive genetic variance that can be estimated statistically using appropriately designed experiments. Heritability can, therefore, be defined as the portion of the phenotypic variance that is attributable to the additive genetic variance of the trait and is a property of a population. When heritability nears zero, it means that the trait is not inherited, and selection would be ineffective in improving the trait. However, when heritability ranges from 0.1 to 0.9, it indicates that selection would be effective in improving the trait under question. This ratio gives the breeder an idea as to what extent the trait under consideration could be transmitted genetically by the parents to the progeny generation. Heritability estimates in some select aquatic species are as follows: 0.3 for body weight in Tilapia (Ponzoni et al. 2003), 0.23 in rohu for growth (Gjerde et al. 2004), 0.32 in female kuruma shrimp for harvest weight (Gopikrishna et al. 2012), and 0.27 for harvest weight in tiger shrimp (Krishna et al. 2011). Growth is generally a quantitative trait that is controlled by several genes and is amenable to selection. However, it has generally been observed that reproductive traits and a few disease-resistance traits have very low estimates of heritability, some nearing zero. Gitterle et al. (2005, 2006) reported extremely low or zero heritability estimates for resistance to white spot syndrome virus (WSSV) in Pacific white shrimp, whereas Hayes et al. (2010) reported a heritability estimate of 0.000010 for the same trait in tiger shrimp from India. This indicates that we cannot genetically select shrimp in a breeding program for resistance to WSSV due to the heritability estimate nearing zero. Since this trait is not amenable to selection, the only other alternative is to unravel quantitative trait-linked (QTL) loci linked to this trait. In this regard, Robinson and co-workers (2014a, b) and Baranski et al. (2014) carried out an experiment to unravel QTLs for resistance to WSSV in tiger shrimp in India in a collaborative project with the Norwegian Institute of Food, Fisheries and Aquaculture Research (NOFIMA), Norway, and ICAR-Central Institute of Brackishwater Aquaculture, Chennai, India. For the first time globally, single nucleotide polymorphisms (SNPs) were utilized to construct a linkage map consisting of 44 linkage groups. An Illumina iSelect genotyping array containing 6000 SNPs was developed and was utilized to genotype 1024 offspring belonging to seven full sibling families. A total of 3959 transcribed SNPs were mapped to 44 linkage groups. The female map was 28% longer than the male map with a higher recombination rate observed in females compared to males during meiosis. These same seven full sibling families of tiger shrimp were challenge tested with WSSV, and 3959 SNPs were used to scan the tiger shrimp genome for loci associated with resistance to WSSV. Data was recorded in terms of *hours of survival* postinfection. Nine QTLs were unraveled that were significantly associated with *hours of survival*. It was further observed that most of these SNPs were found to occur in transcripts showing homology to genes with putative immune functions of interest. The markers, however, need to be verified in a population of tiger shrimp and, if they in fact affect the trait, could be used for marker-assisted selection for improving resistance to WSSV in penaeids. The genomic information from this approach has substantially increased the DNA marker resources in tiger shrimp and would be a useful resource for QTL mapping and genome-wide association studies for commercially important traits in this species. In contrast, resistance to Taura syndrome virus (TSV) has a low to moderate (0.30 heritability, Argure et al 2002). This virus has been rampant in *P. vannamei* in the USA (Hawaii & Florida) and in Ecuador (Latin America), and the shrimp breeders have adopted a novel strategy for improving resistance to TSV by challenge testing families using cohorts/sibs and thereafter selecting only those families that exhibit resistance to TSV, for further breeding. This strategy has been successful, and it is observed that shrimp selected for high resistance to TSV also have higher growth rates and survival (Moss et al. 2011). The cohort/sibs utilized for challenge tests do not find a place in the breeding program. However, there is a flipside if we are using sib selection because it utilizes only the variance between families and tends to ignore one-half of the additive genetic variance, which means that within-family variance is not taken into consideration (Hill 2013), resulting in a suboptimal genetic gain for the trait in question. Gjedrem and Robinson (2014) have reported from selection experiments and breeding programs in aquaculture that substantial genetic variation does exist for resistance to infectious pancreatic necrosis (IPN) in Atlantic salmon and resistance to *Vibrio salmonicida* in rainbow trout, and this could be realized through execution of challenge testing studies. Robinson et al. (2014a, b), while working on the carp *Labeo rohita* in a collaborative project at ICAR-Central Institute of Freshwater Aquaculture, Bhubaneswar, challenged the fish with *Aeromonas hydrophila*, an endemic motile pathogenic bacteria that causes hemorrhaging and ulceration when fish are under stress. Fish were challenged with this bacterium, and the mortality was recorded in terms of *hours of survival*. A genome-wide association study (GWAS) revealed that 21 single nucleotide polymorphisms (SNPs) were mapped to 10 linkage groups and covered 12 distinct regions, showing significant associations with the trait *hours of survival*. Most of these SNPs showed homology to genes of known immune function. It was also observed that the enzyme ceruloplasmin was 4.58 times more highly expressed in resistant than in susceptible lines of rohu. The level of ceruloplasmin could, therefore, be used as an indirect marker for the selection of *A. hydrophila*-resistant rohu (Sahoo et al. 2013).

1.4 Disease Resistance in Aquaculture

Disease resistance could be described as the ability of the host to reduce infection or its consequences by containing the replication of the pathogen. The ability of the host, once infected, to restrict the detrimental effect of the pathogen by maintaining the production performance is termed tolerance. Råberg et al. (2007), while working on mice as a model system, elucidated the terms resistance and tolerance in relation to parasites. The authors reported that resistance and tolerance were negatively genetically correlated, and organisms, such as plants, can evolve these two mechanisms for defense. Doeschl-Wilson et al. (2012) also described the resistance and tolerance of the host to external infections both at the individual as well as at population levels using two mathematical models. The authors concluded that host genetic resistance and tolerance to infectious pathogens are highly desirable targets for genetic improvement. Robinson et al. (2017), while assessing disease resistance in fish, emphasized that it is not practical to disentangle resistance and tolerance. Bishop and Woolliams (2014) studied the cases of resistance of IPN in Atlantic salmon and bovine tuberculosis (bTB) in dairy cattle and concluded that if a disease is of less importance to the breeder and if there are other satisfactory ways of handling it, then there may not be a necessity to go in for improving resistance. The authors opined that costly endemic diseases that cannot be controlled by other means should be subject to genetic improvement for resistance. When challenge studies are carried out, there would be two categories of animals—those that overcome the infection (resistant) and those that die (susceptible). There could be a wide array of diseases—bacterial, fungal, viral, etc. that come into play during the culture of aquatic animals. Gjedrem (1998) opined that diseases in aquaculture result in mortality, poor quality of a product at slaughter and pollution through the use of antibiotics. This is why aquaculture is perceived as an industry with very high risk. Fish/shellfish during culture are maintained either in a pond or hapas or in cages at sea. The animals in such a situation are exposed to an environment over which man has no control (farming in sea cages). Even animals cultured in ponds are exposed to the vagaries of nature. Pathogens could enter the pond water or seawater without any hindrance. In shrimp culture, it is observed that birds drop infected shrimp/fish into culture ponds, which can quickly endanger the health of the existing shrimp and result in serious disease outbreaks. Disease and survival are related, and usually, when disease strikes, mortality increases and survival dips. Hence, improvement in survival usually takes place with enhanced weight gain. There does exist a relationship between growth and survival. Several authors have reported positive genetic correlations between growth and survival. Some of them are 0.37 in Atlantic Salmon fingerlings (Rye et al. 1990), 0.20 in Nile tilapia fingerlings (Eknath et al. 1998), 0.40 to 0.42 in pond/tank survival of *P. vannamei* (Gitterle et al. 2005); 0.05 in *P. monodon* (Krishna et al. 2011), and 0.19 in *L. rohita* (Gjerde et al. 2019). Another aspect to be borne in mind is that it may be difficult to bring out a strain of genetically improved fish/shellfish, which exhibit faster growth as well as resistance to all diseases. This is because resistance to a particular disease "A" needs to be addressed first, and once this is achieved, one can move on to the next disease "B." It would be extremely difficult to select for resistance to any number of diseases simultaneously and, hence, is not a practical option. The appropriate and pragmatic way would be to prioritize the diseases in terms of economic loss and choose and tackle those diseases that impact production to a greater extent.

1.5 Tackling Disease Resistance Through the Application of Genomics

The conventional selective breeding/genetic improvement programs utilize the information on the individuals and also those from their pedigree. Usually, the trait selected for is measured, and breeding values are computed with the aid of the estimated heritability. The animals to be chosen as parents for the next generation are selected on the basis of these estimated breeding values. The animals possessing higher breeding values would be chosen as parents for mating and the generation of the progeny.

Of late, there has been considerable advancement in DNA sequencing, which has opened up an arena where a lot of information at the molecular level has been made available. This information, after proper analysis, could aid in sharpening the selection procedure, which can ultimately lead to higher genetic gains. The selection of an individual in a breeding program by using genomic information can be termed genomic selection. This approach was first proposed by Meuwissen et al. (2001), after which plant and animal breeders have used it to enhance genetic gains. Instead of solely relying on the individual as well as pedigree data to estimate the breeding value, information on DNA sequence is used to estimate the genomic relationship between individuals at several thousands of positions throughout the genome of the organism. This technology gives a highly refined and more accurate breeding value of the broodstock that is to be selected as parents for producing the next generation and is termed as the "genomic breeding value" (Lillehammer et al. 2020). The authors opined that genomic selection for resistance to WSSV could be carried out in this way to select shrimp that exhibit greater resistance, and a major advantage is that the necessity of conducting a challenge test on the selected shrimp could be circumvented. The main focus here is that when the shrimp are selected based on the genomic breeding values, the predictability is increased, and it was found that survival was also enhanced. Lillehammer et al. (2020) carried out challenge tests in *P. vannamei* by exposing the shrimp to WSSV and thereafter used genomic breeding values to select broodstock for further breeding. This was made possible by genotyping all the challenged shrimps for 18,643 SNPs. A very interesting observation in their study was that the progeny obtained from mating of parent shrimp having high genomic breeding values for resistance to WSSV exhibited a higher survival than those progeny from mating of parent shrimp with low genomic breeding values. They concluded that this could possibly be a very useful procedure for developing shrimp with increased resistance to WSSV. Such a procedure could have far-reaching implications in the global shrimp industry as decades of research have already gone into comprehending this highly virulent virus, and a solution still evades the industry. Zenger et al. (2019) opined that in aquaculture species that have been improved viz. Atlantic salmon (*Salmo salar*), Nile tilapia (*Oreochromis niloticus*), and the Pacific white shrimp (*P. vannamei*), selection for growth has dramatically increased efficiencies, helping establish these species as global commodities. It would be pertinent to mention here that fast-growing and SPF Pacific white shrimp has found its way to several countries all around the world, and the present global shrimp production is dominated by this species alone.

In the Atlantic salmon industry, amoebic gill disease (AGD) causes serious economic losses. (Robledo et al. 2018). This disease is caused by *Neoparamoeba perurans* and affects other fish species, such as rainbow trout, Chinook salmon, and in some cases, turbot (Young et al. 2008). Robledo et al. (2018) challenged Atlantic salmon with the amoeba, using gill damage and amoebic load as indicator traits for host resistance. Both these traits exhibited medium heritability and also showed a high positive correlation, indicating that these traits may be good measurements of host resistance to AGD. The authors further found two regions in the 18th chromosome of Atlantic salmon associated with these two traits and suggested that incorporation of this information into a breeding program for genomic selection would yield AGD-resistant stocks.

Another good example of how genetics offers solutions to either increasing resistance or even eradicating is given the manner in which infectious pancreatic necrosis (IPN) in Atlantic salmon was addressed. This disease led to more than 90% mortality, especially when the fingerlings were in hatcheries or had just been transferred to sea cages for farming (Houston et al. 2020). Atlantic salmon has two phases in its life cycle: the initial phase is spent in freshwater, and the second phase is spent in seawater. It was reported that IPN had moderate to high heritability (Storset et al. 2007). Companies breeding salmon initiated family selection to alleviate this problem. Houston et al. (2008, 2010), Moen et al. (2009), and Gheyas et al. (2010), through their studies, unraveled a major QTL on chromosome 26 of Atlantic salmon that had a very significant effect on IPN by explaining 80 to 100% of the genetic variation for this trait in experimental trials both in freshwater and seawater. The requirement to carry out challenge tests was circumvented by high-throughput sequencing and unravelling SNPs that affect IPN (Houston et al. 2012; Moen et al. 2015). Hence, a marker-assisted selection procedure was put in place wherein fish were selected for a favorable allele, in this case, resistance to IPN (Houston et al. 2010; Moen et al. 2015). This resulted in a significant reduction in the occurrence of IPN and, in some cases, was nearing zero (Hjeltnes 2014; Norris 2017).

Another example is the Tilapia lake virus (TiLV), a pathogen affecting the Nile tilapia (*Oreochromis niloticus*) and is considered an important one as tilapia is a species farmed globally (Barría et al. 2021). The authors used data on GIFT infected with this virus in ponds. The fish were categorized as either survived or dead and genotyped using a 65 K SNP array. A significant QTL could be unraveled for this trait on chromosome *Oni22*. The authors could identify the individual homozygous for the susceptible and resistant alleles. They also found candidate genes for host response to viral infections within this QTL. These markers could thus be utilized to select tilapia resistant to TiLV and use them to propagate the next generation, obviating the need for carrying out any challenge tests in the broodstock fish.

An important outcome of developing strains of fishes genetically resistant to certain diseases is that one can circumvent the application of antibiotics during culture, which creates a lot of problems both at the farm level and also during export.

In aquaculture farms, the antibiotics administered to fishes can accumulate in the aquatic environment, creating problems. The antibiotic residues in the body of the fish can result in the rejection of entire consignments when exported to other countries. Presently, the standards regarding antibiotic residues in fishes and fish products imported are getting more stringent globally, especially those set by the European countries. We do hear instances of fish consignments getting rejected because of the presence of even traces of antibiotics. This can be detrimental to the farmers who depend on aquaculture for their livelihood. Using genetically improved strains of fish that are resistant to certain diseases for culture would, therefore, be more pragmatic, and there would be no problems related to antibiotic residues as these do not enter the food chain.

1.6 Improving Disease Resistance Through Gene-Editing

Gratacap et al. (2019) suggested an advanced technique known as gene editing to fix alleles specific to disease resistance in fishes. The technology is named CRISPR/Cas9, which is "*clustered regularly interspaced short palindromic repeats*." and Cas9 stands for CRISPR-associated protein 9. The authors suggest combining in vivo and in vitro approaches to identify, test, and apply disease-resistant alleles in aquaculture species. By doing so, a fish line or strain with allele(s) fixed for resistance to a particular disease could be produced. The possibility of using this technique in future aquatic animal breeding programs does exist. However, there is a concern that such animals might be classified as GMOs and that there could be an obstacle to their propagation. For GMOs to be consumed by humans, clearance by regulatory authorities is mandatory and such a task would in fact be an uphill one.

1.7 Present Scenario in India

The future of aquaculture production would be dependent on the way genetics, and genomics are utilized for various programs. In India, *Labeo rohita, Labeo catla,* and *Macrobrachium rosenbergii* are the species that are presently undergoing genetic improvement at ICAR-Central Institute of Freshwater Aquaculture, Bhubaneswar. In addition, there is a selective breeding program in *Clarias magur* on a smaller scale at the ICAR-Central Institute of Fisheries Education at Balabhadrapuram near Kakinada in Andhra Pradesh. As already indicated, such programs are of a long-term nature and have to be continued in a systematic manner as the genetic gains obtained in each generation have to be maintained. There is a huge cost in terms of finance, manpower, and maintenance. It is not desirable to unnecessarily add more species, and if any new ones have to be added, they need to be restricted to the ones that have their life cycle closed and have substantial consumer preference. However, in the case of penaeid shrimp, as of date, India does not have a breeding program, although preliminary studies on genetic parameters in tiger shrimp *P. monodon* have been carried out in a collaborative ICAR-CIBA-CIFE-NORAD project from

2005–2009. The native Indian white shrimp holds great promise in this regard, and it would be desirable to initiate a selective breeding program to cater to the needs of the Indian shrimp industry, as today it is wholly dependent on the imported *P. vannamei*. Dependence on a single species does not augur well for the Indian shrimp industry, and it is desirable to have an alternative in place. *P. indicus* fits into this perfectly, and it would be possible to develop strains for high growth quite easily and also exhibit resistance to WSSV using the information already available for tiger shrimp in Indian conditions. The advantage of having a genetic improvement program in place is that as selection progresses over generations, resistance to any existing disease or any new disease that might crop up in future can be added as a selection trait in due course along with the increment in body weight, which will always remain the main objective for a fish breeding program.

2 The Way Forward

Disease resistance is a trait that needs to be handled with caution as it is specific to each disease. The lessons from long-term genetic improvement programs in Atlantic salmon, Rainbow trout, Rohu, Tilapia, etc., do indicate that disease resistance is dynamic in nature. There exists a certitude that new diseases might emerge in the future course of the selective breeding programs, and methods need to be developed to tackle such diseases as and when they come into play. We would be on a strong footing if the correlation between growth and resistance is positive, but if found to be negative, it might be difficult to find a solution, and the primary trait, i.e. growth, will have to be toned down. The best way to tackle diseases would be to strengthen the selective breeding program by incorporating biotechnological tools for unravelling SNPs and QTLs and applying the information to maximize genetic gains. Today, the cost of sequencing and manufacturing SNP arrays has come down drastically compared to the early nineties, and this is an additional advantage we possess. Unravelling SNPs, followed by searching for QTLs, would be a wise choice to take care of resistance traits. The genomic information could also be channeled effectively into the breeding programs by estimating genomic breeding values instead of breeding values alone to select the parental population and obtain significant genetic gains. The advantages we have in aquatic species, such as external fertilization and high coefficient of variation for growth, put the aquaculturist in a better position to harness the benefits.

3 Conclusions

Disease resistance is a complex trait as it behaves differently for various diseases. It may not be possible to find a panacea for this trait. Any disease has to be studied individually and dealt with accordingly. It is quite clear that resistance to certain diseases such as TSV has sufficient additive genetic variation controlling them, whereas some other diseases (WSSV) may be bereft of this. As most of the selective breeding programs in aquatic species deal primarily with growth, the resistance to disease as a trait for any disease comes into play during the course of selection. In such a case, it would be desirable to find out whether the trait is heritable or not. If heritable, selection could be carried out provided the genetic correlation with growth is not negative. Otherwise, molecular biological tools could be deployed to aid in detecting SNPs, leading to unravelling QTLs that have a significant effect on the trait and then using these to select fish that become resistant over the generations. In genetic improvement programs, using genomic breeding values that incorporate genomic information instead of the breeding values alone could certainly pave the way for realizing substantial genetic gains for disease-resistant traits.

References

Argue BJ, Arce SM, Lotz JM, Moss SM (2002) Selective breeding of Pacific white shrimp (*Litopenaeus vannamei*) for growth and resistance to Taura syndrome virus. Aquaculture 2002(204):447–460

Baranski M, Gopikrishna G, Robinson NA, Katneni VK, Shekhar MS, Shanmugakarthik J, Jothivel S, Gopal C, Ravichandran P, Kent M, Arnyasi M, Ponniah AG (2014) The development of a high density linkage map for black tiger shrimp (*Penaeus monodon*) based on cSNPs. PLoS One 9(1):e85413. https://doi.org/10.1371/journal.pone.0085413

Barría A, Trịnh TQ, Mahmuddin M, Peñaloza C, Papadopoulou A, Gervais O, Mohan CV, Benzie JA, Houston RD (2021) A major quantitative trait locus affecting resistance to Tilapia lake virus in farmed Nile tilapia (*Oreochromis niloticus*). Heredity 127:334–343. https://doi.org/10.1038/s41437-021-00447-4

Bishop SC, Woolliams JA (2014) Genomics and disease resistance studies in livestock. Livest Sci 166:190–198

Doeschl-Wilson AB, Bishop SC, Kyriazakis I, Villanueva B (2012) Novel methods for quantifying individual host response to infectious pathogens for genetic analyses. Front Genet 3:266

Eknath AE, Dey MM, Rye M, Gjerde B, Abella TA, Sevilleja R, Tayamen MM, Reyes RA, Bentsen HB (1998) Selective breeding of Nile tilapia for Asia. In: Proceedings of the 6th world congress on genetics applied to livestock production, vol 27, pp 89–96

Gheyas AA, Houston RD, Mota-Velasco JC, Guy DR, Tinch AE, Haley CS, Woolliams JA (2010) Segregation of infectious pancreatic necrosis resistance QTL in the early life cycle of Atlantic Salmon (*Salmo salar*). Anim Genet 41:531–536. https://doi.org/10.1111/j.1365-2052.2010.02032.x

Gitterle T, Rye M, Salte R, Cock J, Johansen H, Lozano C, Suárez JA, Gjerde B (2005) Genetic (co)variation in harvest body weight and survival in *Penaeus vannamei* under standard farming conditions. Aquaculture 243:83–92

Gitterle T, Johansen H, Erazo C, Lozano C, Cock J, Salazar M, Rye M (2006) Response to multi-trait selection for harvest weight, overall survival, and resistance to white spot syndrome virus (WSSV) in *Penaeus (Litopenaeus) vannamei*. IAGA, 26–30 June 2006, Montpelier, Abstract: 35

Gjedrem T (1998) Diseases and parasites in fish. Genetic improvement of resistance. GIFT. Final Report (March 1988 to December 1997). ICLARM Publication: Attachment 11, 27pp

Gjedrem T, Baranski M (2009) Selective breeding in aquaculture: an introduction. In: 1 reviews: methods and technologies in fish biology and fisheries 10. Springer. https://doi.org/10.1007/978-90-481-2773-3_1

Gjedrem T, Robinson N (2014) Advances by selective breeding for aquatic species: a review. Agric Sci 5:1152–1158. https://doi.org/10.4236/as.2014.512125

Gjerde B, Terjesen BF, Barr Y, Lein I, Thorland I (2004) Genetic variation for juvenile growth and survival in Atlantic cod (*Gadus morhua*). Aquaculture 236:167–177

Gjerde B, Mahapatra KD, Reddy PVGK, Sahu JN, Jana RK, Meher PK, Sahoo M, Khaw HL, Gjedrem T, Rye M (2019) Genetic parameters for growth and survival in rohu carp *Labeo rohita*. Aquaculture 503:381–388. https://doi.org/10.1016/j.aquaculture.2019.01.029

Gopikrishna G, Gopal C, Shekhar MS, Vinaya Kumar K, Kannappan S, Ponniah AG (2012) Growth performance and White Spot Syndrome Virus resistance in families of kuruma shrimp (*Marsupenaeus japonicus*). Indian J GeoMarine Sci 41(1):920–924

Gratacap RL, Wargelius A, Edvardsen RB, Houston RD (2019) Potential of genome editing to improve aquaculture breeding and production. Trends Genet 35(9):672. https://doi.org/10.1016/j.tig.2019.06.006

Handbook on Fisheries Statistics (2023) Department of Fisheries, Ministry of Fisheries, Animal Husbandry & Dairying, Govt of India

Hayes BJ, Gitterle T, Gopikrishna G, Gopal C, Krishna G, Jahageerdar S, Lozano C, Alavandi S, Paulpandi S, Ravichandran P, Rye M (2010) Limited evidence for genetic variation for resistance to White Spot Syndrome virus (WSSV) in Indian populations of *Penaeus monodon*. Aquac Res 41:872877

Hill WG (2013) On estimation of genetic variance within families using genome-wide identity-by-descent sharing. Genet Sel Evol 45:32. https://doi.org/10.1186/1297-9686-45-32

Hjeltnes B (ed) (2014) The health situation in Norwegian aquaculture 2013. Norwegian Veterinary Institute, Fish Health Report of 2013

Houston RD (2017) Future directions in breeding for disease resistance in aquaculture species. Rev Bras Zootec 46(6):545–551

Houston RD, Haley CS, Hamilton A, Guy DR, Tinch AE, Taggart JB, McAndrew BJ, Bishop SC (2008) Major quantitative trait loci affect resistance to infectious pancreatic necrosis in Atlantic salmon (*Salmo salar*). Genetics 178:1109–1111

Houston RD, Haley CS, Hamilton A, Guy DR, Mota-Velasco JC, Gheyas AA, Tinch AE, Taggart JB, Bron JE, Starkey WG, McAndrew BJ, Verner-Jeffreys DW, Paley RK, Rimmer GSE, Tew IJ, Bishop SC (2010) The susceptibility of Atlantic salmon fry to freshwater infectious pancreatic necrosis is largely explained by a major QTL. Heredity 105:318–327

Houston RD, Davey JW, Bishop SC, Lowe NR, Mota-Velasco JC, Hamilton A, Guy DR, Tinch AE, Thomson ML, Blaxter ML, Gharbi K, Bron JE, Taggart JB (2012) Characterisation of QTL-linked and genome-wide restriction site-associated DNA (RAD) markers in farmed Atlantic salmon. BMC Genomics 13:244

Houston RD, Bean TP, Macqueen DJ, Gundappa MK, Jin HY, Jenkins TL, Selly SLC, Martin SAM, Stevens JR, Santos EM, Davie A, Robledo D (2020) Harnessing genomics to fast-track genetic improvement in aquaculture. Nat Rev Genet 21:389–409. https://doi.org/10.1038/s41576-020-0227-y

Krishna G, Gopikrishna G, Gopal C, Jahageerdar S, Ravichandran P, Kannappan S, Pillai SM, Paulpandi S, Kiran RP, Saraswati R, Venugopal G, Kumar D, Gitterle T, Lozano C, Rye M, Hayes B (2011) Genetic parameters for growth and survival in *Penaeus monodon* cultured in India. Aquaculture 318:74–78

Lillehammer M, Bangera R, Salazar M, Vela S, Erazo EC, Suarez A, Cock J, Rye M, Robinson NA (2020) Genomic selection for white spot syndrome virus resistance in whiteleg shrimp boosts survival under an experimental challenge test. Sci Rep 10:20571. (2020). https://doi.org/10.1038/s41598-020-77580-3

Meuwissen T, Hayes B, Goddard M (2001) Prediction of total genetic value using genome-wide dense marker maps. Genetics 157:1819

Moen T, Baranski M, Sonesson AK, Kjøglum S (2009) Confirmation and fine-mapping of a major QTL for resistance to infectious pancreatic necrosis in Atlantic salmon (*Salmo salar*): population-level associations between markers and trait. BMC Genomics 10:368

Moen T, Torgersen J, Santi N, Davidson WS, Baranski M, Ødegård J, Kjøglum S, Velle B, Kent M, Lubieniecki KP, Isdal E, Lien S (2015) Epithelial Cadherin Determines Resistance to

Infectious Pancreatic Necrosis Virus in Atlantic Salmon. Genetics 200(4):1313–1326. https://doi.org/10.1534/genetics.115.175406

Moss DR, Arce SM, Otoshi CA, Moss SM (2011) Shrimp breeding for resistance to Taura Syndrome Virus. Global Aquaculture Advocate, January/February 2011

Norris A (2017) Application of genomics in salmon aquaculture breeding programs by Ashie Norris: who knows where the genomic revolution will lead us? Mar Genomics 36:13–15

Ponzoni RW, Hamzah AB, Saadiah ST, Kamaruzzaman N (2003) Phenotypic and genetic parameters for live weight in two environments in a selected line of Nile tilapia in Malaysia. IAGA, Genetics in Aquaculture VIII. 9–15 November 2003, Puerto Varas, Chile, 83

Råberg L, Sim D, Read AF (2007) Disentangling genetic variation for resistance and tolerance to infectious diseases in animals. Science 318:812–814

Robinson NA, Gopikrishna G, Baranski M, Katneni VK, Shekhar MS, Shanmugakarthik J, Jothivel S, Gopal C, Ravichandran P, Gitterle T, Ponniah AG (2014a) QTL for white spot syndrome virus resistance and the sex determining locus in the Indian black tiger shrimp (*Penaeus monodon*). BMC Genomics 15:731

Robinson N, Baranski M, Mahapatra KD, Saha JN, Das S, Mishra J (2014b) A linkage map of transcribed single nucleotide polymorphisms in rohu (*Labeo rohita*) and QTL associated with resistance to *Aeromonas hydrophila*. BMC Genomics 15:541. https://doi.org/10.1186/1471-2164-15-54

Robinson NA, Gjedrem T, Quillet E (2017) Chapter 2: Improvement of disease resistance by genetic methods. Fish Dis 2:21–50. https://doi.org/10.1016/B978-0-12-804564-0.00002-8

Robledo D, Matika O, Hamilton A, Houston RD (2018) Genome wide association and genomic selection for resistance to amoebic gill disease in Atlantic Salmon. G3: Genes Genomes Genet 8:1195–1203. https://doi.org/10.1534/g3.118.200075

Rye M, Lillevik KM, Gjerde B (1990) Survival in early life of Atlantic salmon and rainbow trout: estimates of heritabilities and genetic correlations. Aquaculture 89:209–216

Sahoo PK, Das S, Mahapatra KD, Saha JN, Baranski M, Odegard J, Robinson N (2013) Characterization of the ceruloplasmin gene and its potential role as an indirect marker for selection to *Aeromonas hydrophila* resistance in rohu, Labeo rohita. Fish Shellfish Immunol 34(5):1325–1334

Storset A, Strand C, Wetten M, Kjøglum S, Ramstad A (2007) Response to selection for resistance against infectious pancreatic necrosis in Atlantic salmon (*Salmo salar* L.). Aquaculture 272(Suppl. 1):S62–S68

Young ND, Dyková I, Snekvik K, Nowak BF, Morrison RN (2008) *Neoparamoeba perurans* is a cosmopolitan aetiological agent of amoebic gill disease. Dis Aquat Org 78:217–223. https://doi.org/10.3354/dao01869

Zenger KR, Khatkar MS, Jones DB, Khalilisamani N, Jerry DR, Raadsma HW (2019) Genomic selection in aquaculture: application, limitations and opportunities with special reference to marine shrimp and pearl oysters. Front Genet 9:693. https://doi.org/10.3389/fgene.2018.00693

Live Feeds in Shrimp Aquaculture: An Overview

K. P. Sandeep, T. Sivaramakrishnan, K. Ambasankar, and K. K. Vijayan

Abstract

Live feeds are an inevitable source of nutrition to aquaculture, especially for successful larval rearing. During the critical phase of larval transition in finfish and shellfish, live feeds, rich in essential nutrients and easy to digest, play a vital role in survival and growth. Common live feeds include phytoplankton and zooplankton, such as microalgae, rotifers, Artemia, copepods, cladocerans, and polychaete worms. Microalgae, particularly autotrophic species, such as diatoms and flagellates, are fundamental in larval nutrition, providing proteins, essential fatty acids, and other key nutrients. Heterotrophic microalgae, such as Thraustochytrids, are also important for their rich polyunsaturated fatty acids (PUFA) content. Rotifers are another critical live feed, especially for marine fish larvae, due to their suitable size and reproductive rates. Artemia, known for its optimal size and nutritional value, is widely used in aquaculture, with techniques developed for enriching Artemia to enhance its nutrient profile. Polychaete worms are used in shrimp broodstock diets for their high protein and fatty acid content, aiding reproductive performance. Copepods, despite challenges in mass production, are valued for their superior nutritional composition and suitability for various larval stages. Cladocerans, though less common, are also used for their balanced nutrient profile and ability to improve water quality. Advancements in live feed culture have significantly improved aquaculture practices. The development of artificial media and enrichment techniques has facilitated the mass culture of microalgae and other live feeds, supporting the growth and survival of various aquaculture species. Ensuring the availability of high-quality live feeds remains essential for the success of aquaculture systems.

K. P. Sandeep (✉) · T. Sivaramakrishnan · K. Ambasankar · K. K. Vijayan
ICAR-Central Institute of Brackishwater Aquaculture, Chennai, India
e-mail: sandkp@gmail.com; sivaraman.fish@gmail.com; ambasankarciba@gmail.com; vijayankk@gmail.com

R. Kooloth Valappil et al. (eds.), *Aquatic Animal Health Management*,
https://doi.org/10.1007/978-981-96-7987-4_28

Keywords

Live feeds · Larval nutrition · Shrimp production · Essential nutrients

1 Introduction

Hatchery-produced seeds are the major driver in finfish and shellfish aquaculture, while the use of proper live feed determines the successful larval rearing in the hatchery, which includes the transition from endogenous to exogenous feeding by the larvae. Generally, there are two types of fish larvae: precocial and altricial. Precocial larvae, when the yolk sac is exhausted, look like mini adults with fully developed fins and a mature digestive system. They can ingest and digest formulated feeds as a first feed, e.g., salmon and trout. In the case of altricial larvae, when the yolk sac is exhausted, they remain in a relatively undeveloped state. The rudimentary digestive system will be present in them, and much of the digestion takes place in the hindgut epithelial cells. So, the larvae will be incapable of processing formulated diets. Fishes, shrimps, and other cultivable aquatic animals at the time of their first feeding are quite fragile and delicate creatures. This stage of their life is known as the "critical phase," where they require the right type of food for their survival and growth. If this requirement is not met, larvae may not be able to survive. So, it is necessary to give the larvae a specialized feed, the "live feed," at this stage.

Mainly, live feeds consist of phytoplankton and zooplankton grazed upon by economically important fish or shellfish. Live feed is essential due to several factors such as rich essential nutrients, broad spectrum composition of food, better intake due to the movement, auto-digestion characteristics, better nutrient assimilation in larvae, stimulate feeding due to soft texture, and attractability. This is why live feed organisms are called *living capsules of nutrition*. The commonly used live feeds in brackish water aquaculture and mariculture are microalgae, rotifers, artemia, copepods, cladocerans, polychaete worms, etc.

1.1 Brief History of Live Feed Culture

Microalgae culture was started by bacteriologists during the late 1800s (Bold 1950). Subsequently, significant advancement in microalgae culture was reported by Gross (1937), as the successful culture of diatoms and dinoflagellates was accomplished. In the middle of the twentieth century, the formulation of artificial media (Provasoli et al. 1957) and the subsequent development of the "*f*" *medium* for the enrichment of seawater (Guillard and Ryther 1962) improved the microalgae culture technologies. After these advancements, various culture methods developed by researchers for monospecific algal cultures supported the quick expansion of finfish and shellfish, especially shrimp hatcheries, across the globe. During the 1960s, Japanese researchers discovered that rotifer (*Brachionus plicatilis*) could be used as the first food for larvae of both freshwater and marine fish species (Hirata 1979), which was previously considered a pest in culture ponds. This advancement clearly allowed the culture of many more species whose hatched-out larvae with very small mouth gapes were insufficient for the ingestion of the larger *Artemia* nauplii. Subsequent research by various researchers around the globe to find a superior live feed to increase the survival of fish and shellfish larvae has contributed to the advancement of successful larval-rearing technologies of aquaculture species.

2 Live Feeds Used in Shrimp Aquaculture

In nature, shrimp larvae fed on phytoplankton, zooplankton, and bioflocs of various sizes and with different biochemical compositions. This makes them suitable for meeting the overall nutritional requirements. At the same time, the feeding pattern used in commercial shrimp hatcheries depends on a limited number of live feed organisms as well as on a combination of live and artificial food. Most of the commercial shrimp hatcheries use different types of live food organisms, such as different species of microalgae (Gallardo-Rodríguez et al. 2012; Sandeep et al. 2021) and newly hatched *Artemia* nauplii as shrimp larval diet (Sorgeloos 1980; Kumlu and Jones 1995; Lavens and Sorgeloos 1996). Apart from these live food organisms, different researchers have documented various live feeds that can be given to Zoea 2, which include rotifers (Emmerson 1984; Samocha et al. 1989; Naessens et al. 1995), nematodes (Wilkenfeld et al. 1984; Biedenbach et al. 1989; Focken et al. 2006), and copepods (Louis et al. 2006). Some of the common live-feed organisms are given in Fig. 1.

2.1 Microalgae

Microalgae are one of the live feeds and are considered to be the starting point of all the live feeds. Microalgae play a pivotal role in aquaculture as a means of enriching zooplankton for feeding finfish and shellfish larvae (Borowitzka 1997; Muller-Feuga et al. 2003; Zmora and Richmond 2004). In addition to protein and energy supply, these live feeds provide other key nutrients, such as vitamins, essential polyunsaturated fatty acids (PUFA), pigments, and sterols, which are transferred through the food chain (Brown et al. 1997; Brown 2002; Becker 2004). Therefore, microalgae are an inevitable source of nutrition for finfish and shellfish larvae.

2.1.1 Autotrophic Microalgae

Marine microalgae are floating microscopic unicellular organisms that are generally free-living, pelagic in the size range of 2 to 20 μm, and they use sunlight for photosynthesis. Microalgae are essential food sources in the rearing of all stages of marine bivalve molluscs (clams, oysters, scallops, etc.), the larval stages of marine gastropods (abalone), and shrimps (*Penaeus vannamei, P. indicus, P. monodon,* etc.). Larval stages of penaeid shrimp go through a succession of moults and metamorphoses. During the first naupliar stages (Nauplius 1 to 6), there is only endogenous

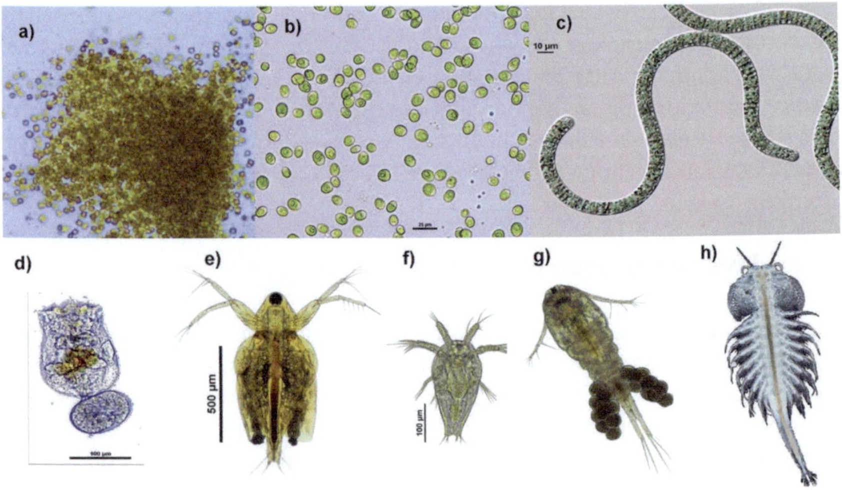

Fig. 1 Various live feed organisms used in aquaculture

feeding. After the first metamorphosis, larvae go through 3 zoea stages. At this stage, as the larvae show predominant filter-feeding behavior, the growth and survival of larvae are totally dependent on the supply of live feed such as microalgae. Subsequent larval stages, such as mysis and postlarvae (PL) of shrimp, can ingest microalgae. Furthermore, the presence of microalgae in the culture system improves survival and growth during the PL stages. Since the beginning, Indian shrimp hatcheries have been using *Skeletonema costatum*, *Chaetoceros gracilis,* and *Tetraselmis* sp.; however, recently, many hatcheries shifted to *Thalassiosira* sp. Mixed-algae diets have resulted in higher survival and faster development of larvae than the component species alone (Jamali et al. 2015; Crisp et al. 2018; Sandeep et al. 2021).

The most commonly used species of microalgae in shrimp larval nutrition are diatoms (*Skeletonema costatum*, *Chaetoceros gracilis*, *C. calcitrans*, *Thalassiosira pseudonana,* and *T. weissflogii*), flagellates (*Isochrysis galbana*, *Tetraselmis*, and *Chlorella marina*), etc. (Muller-Feuga 2000). Mass-cultured microalgae are the primary food source for the larval stages of shrimps. These also serve as the primary diet of zooplankton reared as food for advanced larval stages and postlarval stages. Only a small number of microalgal strains are routinely cultured in shrimp hatcheries depending on certain conditions, such as availability of strain, method of culture, cell physical characteristics, nutritional composition, digestibility, and absence of toxins (Guedes and Malcata 2012; Tredici et al. 2009; Muller-Feuga 2004; Muller-Feuga et al. 2003).

Microalgae are an inevitable source of nutrition for most shrimp larvae in captive rearing (Pulz and Gross 2004; Hemaiswarya et al. 2011). In penaeid shrimps, different microalgae have been used by the shrimp industry for the last few decades in search of better feeding, survival, and ease of production (D'Souza and Loneragan 1998; D'Souza and Kelly 2000; Jamali et al. 2015; Sandhya et al. 2020; Sandeep et al. 2021). Researchers have reported that diatoms, such as *Skeletonema costatum*, *Chaetoceros gracilis* and *Thalassiosira* spp., were found to be ingested and digested readily by shrimp larvae (Kent et al. 2011). The variations in survival and growth of shrimp larvae fed with different microalgae were reported by various researchers (Chu and Lui 1990; Naranjo et al. 1995). These differences are attributed to the size, shape, digestibility, antimicrobial properties, and nutritional composition of the algal cells (Pulz and Gross 2004; Ferreira et al. 2009).

2.1.2 Heterotrophic Microalgae

Apart from autotrophic microalgae, some heterotrophic microalgae are also used in aquaculture. They are promising candidates for the commercial production of long-chain ω-3 fatty acids, unlike their autotrophic counterparts (Lewis et al. 1999; Gupta et al. 2012; Ugalde et al. 2018). Long-chain polyunsaturated fatty acids (LC-PUFAs) are an important component in aquaculture and livestock feeds. Thraustochytrids are one of the potent organisms and rich sources of PUFA (Lewis et al. 1999; Martins et al. 2013). Thraustochytrids are osmo-heterotrophic, oleaginous, eukaryotic, unicellular monocentric fungi-like protists, classified under the kingdom Heterokonta, and class Labyrinthulomycetes (Leander et al. 2004; Damare 2009). As one of the crucial bottlenecks in marine larval nutrition is the dietary supply of LC-PUFAs, Thraustochytrids can be utilized for the same. *Aurantiochytrium* sp. is a Thraustochytrid widely studied in aquaculture. The superior nutritive quality (total protein: 50–55% and total fat: 26%, good fatty acid profile) makes *Aurantiochytrium* sp. a promising candidate for shrimp aquaculture nutrition, especially in *Penaeus monodon* (Jaseera et al. 2021).

2.1.3 Critical Points in Microalgae Culture

Indoor microalgae culture systems are prone to contamination from different sources, such as intake water, aeration tubes or air lines, glassware, handling, cotton plugs, etc. Sterilization is a process for establishing an aseptic condition by removing or killing of all microorganisms. Sterilization is very important in phycological research, especially when maintaining living organisms as isolated strains in culture. The intake water should be adequately sterilized. The use of the sterile technique, in combination with sterile equipment and supplies, minimizes contamination, resulting in more precise experiments free of potential variables caused by unwanted organisms. The indoor culture should be handled only by a qualified culturist to reduce the chance of contamination.

2.2 *Artemia*

Artemia is a primitive arthropod with a segmented body with broad leaf-like appendages. They are commonly known as brine shrimp. In adverse conditions, they produce dormant cysts, and the cyst will hatch only when the conditions are favorable. *Artemia* is widely used as live feed in aquatic larviculture because of optimal size, easy culturing, high nutritional value, scope for enrichment, convenience in production and storage, acceptability, and suitability for a wide variety of larval forms. Due to these advantages, *Artemia* nauplii have been adopted as a standard diet in the commercial hatcheries of finfish and shellfish. In aquaculture, *Artemia* is being used for larval nutrition of more than 85% of marine species. In shrimp hatcheries, freshly hatched *Artemia* nauplii are used for advanced larval stages, such as mysis and postlarvae (PL). Some studies suggested the usage of umbrella-stage *Artemia* to feed Zoea 2 of *P. vannamei* and demonstrated that this could replace live, frozen, heat-killed, or blended *Artemia* nauplii used in many commercial shrimp hatcheries (Cobo et al. 2015). Apart from using freshly hatched *Artemia* nauplii, the benefits of using enriched nauplii (24 h), dry decapsulated cysts, juveniles, and adult biomass were shown in earlier studies (Bengtson et al. 1991). Naessens et al. (1997) have shown successful replacement of polychaetes, which have been identified as a critical live feed in the maturation diet of *P. vannamei* (Bray and Lawrence 1992), by frozen *Artemia* biomass. Nutritionally, *Artemia* sp. contains up to 52.7% protein but is low in essential fatty acids, such as eicosapentaenoic acid (EPA), docosahexaenoic acid (DHA), and vitamins. As *Artemia* sp. is a non-selective filter feeder, the nutritional quality depends on the quality of its culture media. Hence, enrichment of *Artemia* nauplii is considered a viable option to increase the nutrient quality and meet the nutritional demands of the larvae.

2.3 Rotifers

Rotifers are tiny zooplanktons belonging to the genus *Brachionus* and have been used for larviculture since the 1970s (Lubzens et al. 2001). It has become the most important live feed for marine fish larvae as they possess attributes, such as suitable size, the possibility to culture at high densities and high reproductive rate (due to parthenogenetic reproduction dependent on the environment), high salinity and temperature, slow swimming, and filter feeding behavior (Lavens and Sorgeloos 1996). Two major species used in aquaculture are *Brachionus plicatilis* and *B. rotundiformis*. In the case of rotifers, their nutritional quality is greatly influenced by the feed they consume. Hence, in commercial hatcheries, the rotifers are enriched with specific microalgae or enrichment media to ensure the nutritional quality required for the larvae. Rotifers are the first feed for many marine finfish. They have been used in shrimp (*Penaeus indicus* and *P. semisulcatus*) larval rearing along with *Artemia* nauplii (Emmerson 1984; Samocha et al. 1989), and Lovett and Felder (1988) effectively used the rotifer *Brachionus plicatilis* as a substitute for *Artemia* in *Macrobrachium rosenbergii* larval nutrition. It is proved that the inclusion of rotifers in the larval feeding regime of *P. vannamei* from Z2 to M3/PL1 at densities of 20–40/mL/day improved the performance of larvae (Naessens et al. 1995). Moreover, the usage of enriched rotifers in the feeding of mysis to PL5 in *P. vannamei* significantly increased the growth and survival (Najmi et al. 2018). Sambhu et al. (2014) suggested that the enrichment of rotifers in cod liver oil and yeast for 12 h at 30 °C is ideal for enhancing the growth and survival of *P. indicus* larvae. Moreover, this study demonstrated that rotifers are essential for larval survival and for the development of early larval stages of *P. indicus*, but enriched rotifers can also be successfully used for larval rearing of *P. indicus.* Lately, commercial hatcheries have used rotifers for selective enrichment with probiotics, which can reduce the cost compared to the addition of huge volumes of probiotics to culture water.

2.4 Polychaete Worms

Polychaetes belong to Phylum Annelida; they are aquatic multi-segmented worms. Polychaetes are widely distributed from abyssal depths to shallow estuaries, rocky shores, and free-swimming in open water bodies. Polychaetes are the most abundant organisms in mud and sand habitats along with sediment-dwelling molluscs and crustaceans. Polychaetes are widely given to shrimp broodstock as maturation diet as they contain high levels of proteins, lipids, n-3 polyunsaturated fatty acids (PUFA) especially arachidonic acid, eicosapentaenoic acid, docosahexaenoic acids, etc., and hormonally active compounds that are responsible for the ovarian maturation, enhance vitellogenesis and reproductive performance of penaeid shrimps, and at the same time, improve the quality and viability of young ones (Naessens et al. 1997; Coman et al. 2007; Meunpol et al. 2007; Leelatanawit et al. 2014).

Different species of polychaete worms have proven their potential to be used as shrimp maturation diet, and they are *Perinereis helleri* (Palmer et al. 2014), *Palola* sp. (Pamungkas and Glasby 2015), *Nereis virens*, *Perinereis nuntia* (Techaprempreecha et al. 2011), and *Marphysa* sp. (Meunpol et al. 2005). Development of grow-out procedure and disease-free polychaete production in captive conditions is the need of the hour as the natural collection of polychaetes from the wild depletes the natural stock, apart from harboring pathogens that can be transmitted to hatcheries (Kannappan et al. 2021).

2.5 Copepods

Copepods are a diverse group of animals belonging to the class Crustacea, with more than 11,500 species residing in different environmental conditions (Boxshall and Halsey 2004; Humes 1994). Generally, copepods are cylindrical in shape with a trunk comprising 10 segments, consisting of head, thorax, and abdomen, and an adult copepod measures between 0.5 and 5 mm. Thus, the availability of different size groups makes copepods a potential live diet for finfish and shellfish larvae. There are three major groups of free-living copepods: cyclopoids, calanoids, and harpacticoids.

The marine copepods are known as "nutritionally superior live feeds" for many of the commercially important cultivable species in aquaculture, as they are rich in protein, lipid (especially EPA and DHA), carbohydrates, and enzymes (amylase, protease, exonuclease, and esterase). These are essential for the survival, growth, digestion, and metamorphosis of finfish and shellfish larvae (Støttrup 2003). Copepod nauplii are used as the first feed for fish larvae since the rotifers are inadequate in providing the required nutrition to larvae. Apart from the advantage of superior nutritional composition, the size variation and peculiar swimming pattern make them important live feed in larviculture compared to rotifers and *Artemia* nauplii (Ajiboye et al. 2011; Drillet et al. 2011). However, the greatest constraint is the availability of a sufficient quantity of copepods of the desired stage or size at a specific time period. Studies proved that marine copepods (*Oithona rigida, Pseudodiaptomus* sp., and *Macrosetella gracilis*) can be successfully used as live feed in larval rearing of tiger shrimp *P. monodon* in aqua hatcheries (Ananthi et al. 2011). The greatest advantage of using copepods as live feeds in shrimp larviculture is that, as the copepods are rich in many polyunsaturated fatty acids (Fig. 2), enrichment is not required like other live feeds (*Artemia* and Rotifers). Nauplii of *T. biminiensis* copepod were proved as a potential live food in larviculture of *P. vannamei* (de Lima and Souza-Santos 2007) as mysis and postlarval stages of *P. vannamei* showed increased ingestion rate of copepods.

The major challenge in copepod production is the lack of feasible technologies for intensive mass-scale production. Obtaining adequate copepods of desired stages at a specific time has been one of the major bottlenecks for many years. Most of the larviculture practices use copepods collected from wild or blooms induced in confined areas, such as lagoons and large ponds. The collection of copepods from the sea is not a viable option as there can be variability in abundance that may impede the staple supply; moreover, wild copepods may carry parasites that can be a threat

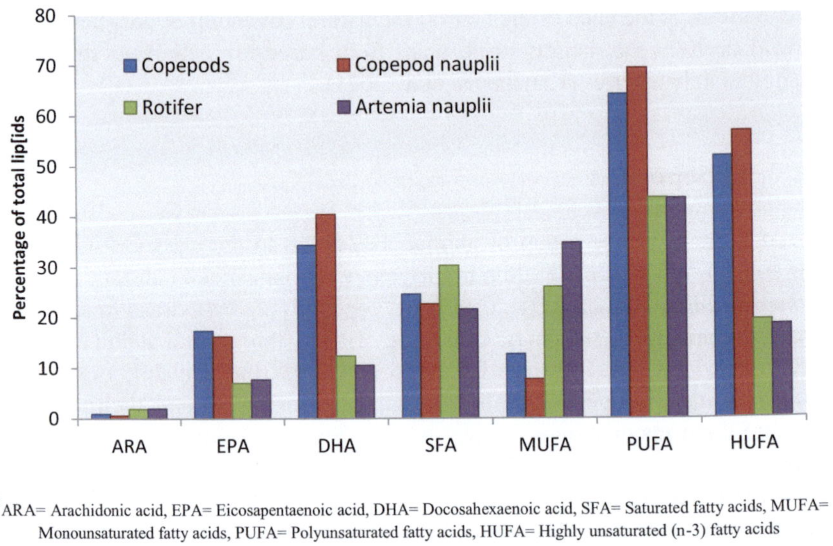

Fig. 2 Fatty acid composition of various live feeds. (Source: van der Meeren et al. 2008). *ARA = arachidonic acid, EPA = eicosapentaenoic acid, DHA = docosahexaenoic acid, SFA = saturated fatty acids, MUFA = monounsaturated fatty acids, PUFA = polyunsaturated fatty acids, and HUFA = highly unsaturated (n-3) fatty acids

to fish larvae. Therefore, specific culture strategies should be developed and followed for each copepod group.

2.6 Cladocera

Cladocera are a diverse group of tiny crustaceans commonly found in a variety of aquatic habitats, ranging from ponds to oceans. Most cladocerans consume algae and microbes (Havel et al. 2009). Cladocerans are nutritionally balanced with many essential nutrients for larvae, which get easily ingested and digested. They are used as live feeds in larviculture and to improve the water quality by minimizing the need of artificial feeding (Lavens and Sorgeloos 1996; He et al. 2001). The population density of cladoceran was enhanced by dietary manipulations and made possible by increasing the fecundity and decreasing the reproductive period (Hakima et al. 2013). Aravind et al. (2021) showed successful partial replacement of *Artemia* nauplii with a cladoceran, *Eurycercus beringi,* during the early postlarval stages of *Penaeus indicus*. This study revealed the potential of cladoceran to be used as live feed in shrimp larval rearing.

2.7 Other Potential Live Feeds

Some unconventional live feeds are also used in shrimp larval rearing, apart from the above-discussed live food organisms. As the food ingested by the shrimp larvae in nature is diverse and different from what is being used in hatcheries, the use of unconventional live feeds in shrimp larval rearing is a potential area. Protozoans and nematodes were used in shrimp larval nutrition, and protozoans, such as flagellates and ciliates, are rich in several nutrients (Loureiro et al. 2012). The high abundance of protozoan and nematodes in intensive shrimp culture has a positive correlation with shrimp growth (Martinez-Cordova, 2002; Loureiro et al. 2012). *Panagrellus redivivus,* a free-living soil nematode, is used as a live food source for shrimp larvae as a replacement for *Artemia*. Samocha and Lewinsohn (1977) reported the successful use of the *Panagrellus redivivus* in rearing postlarvae of *Penaeus semisulcatus* and *Metapenaeus stebbingi*. Further studies showed that *P. redivivus* can be a potential replacement for *Artemia* in the larval rearing of *P. vannamei* (Focken et al. 2006).

3 Shrimp Health and Live Feeds

Even though the incorporation of live feed in shrimp larval rearing and maturation is almost inevitable, there is growing evidence that many of these live food organisms are carriers of some pathogenic microorganisms (Tacon 2017). Various studies have shown that live feed can be potential carriers of pathogens in aquaculture (Table 1) when there are compromises on biosecurity.

Table 1 Diseases transmitted through live feeds in shrimp aquaculture

Sl no	Disease	Carrier of pathogen	Reference
1	White spot syndrome virus (WSSV)	Polychaete worms	Desrina (2016) Laoaroon et al. (2005) Vijayan et al. (2005)
		Artemia biomass	Chang et al. (2002); Sahul Hameed et al. (2002); Li et al. (2003); Parenrengi (2004); Alday-Sanz (2016)
		Phytoplankton and rotifer	Jiang et al. (2019)
		Copepods and amphipods	Song et al. (2001)
2	Pathogenic bacteria	*Artemia*	Høj et al. (2009); Abraham and Palaniappan (2004)
		Zooplankton	Karunasagar et al. (2016)
		Polychaetes and bivalves	Songsangjinda et al. (2016)
3	*Enterocytozoon hepatopenaei*	Polychaetes	Krishnan et al. (2021)
		Artemia biomass	Han et al. (2013)

As rotifers are filter feeders, the microbial community of the rotifers is often the microbial community of the culture water. The dose of enrichment has to be standardized based on the requirement of larvae, as the excess doses will create a negative effect (Haché and Plante 2011). Many hatcheries practice the collection of polychaete worms from the wild to feed the shrimp as a maturation diet. However, the worms are potential carriers of pathogens as they accumulate the viral pathogen in their digestive tract by consuming virus particles in the sediment. Subsequently, the pathogens will be transmitted to the broodstock and larvae (Vijayan et al. 2005). Furthermore, recent studies revealed that the polychaetes are apparently capable of carrying infectious spores for some period after ingestion of *Enterocytozoon hepatopenaei* (EHP)-contaminated feed or soil sediments (Krishnan et al. 2021).

4 Biosecurity in Live Feed Production in Shrimp Aquaculture: The Way Forward

Biosecurity, or "hazard reduction through environmental manipulation" (Plumb 1992), is mostly referred to as a practice that can reduce the number of pathogenic organisms that enter a system or facility. At large, biosecurity is the protection of cultured animals from diseases and preventing the spread of diseases across boundaries. This concept is becoming crucial with the intensification of aquaculture production systems.

As discussed in the earlier sections, the risk of diseases can be reduced by following proper biosecurity protocol in live feed culture and utilization. For example, in the case of rotifer culture, the following practices should be followed:

- Frequent cleaning/ disinfection of all used materials.
- After each harvest, the tanks, aeration tubing, and harvesting devices must be disinfected using sodium hypochlorite solution or any other potential disinfectants.
- Maintain the stock cultures in a separate room from the mass cultures to avoid cross-contamination.
- In case contamination persists or stock cultures are contaminated, discard the culture and disinfect the containers.

As similar practices can be followed in all other live feed cultures, the hatcheries should have trained technical personnel to ensure necessary biosecurity protocols in live feed culture. In the case of polychaetes, there is an urgent need to produce disease-free polychaetes in captive conditions to protect the natural stocks and prevent the pathogen transfer from the wild to the culture system and vice versa.

5 Importance of Live Feeds in Shrimp Nursery and Grow-Out Culture

The dependency on live feeds is more during the larval and postlarval stages of shrimp. Various workers reported that postlarvae of penaeid shrimp fed on a variety of zooplankton, including rotifers, copepods, protozoan ciliates, small worms, and larval stages of various aquatic invertebrates. In contrast, juveniles and adults mainly fed on organic detritus and small animals in pond systems (De et al. 2020). Sivakami (1988) reported that the zooplankton abundance and sudden blooms of phytoplankton in pond conditions accelerate the detritus formation. The dominance of protozoan and rotifers in shrimp ponds positively correlates with enhanced growth of shrimps (Neto et al. 2009). Ciliated protozoans play a crucial role in the energy flow of the ecosystems, as they consume bacteria, algae, and fungi and act as a food source for fish and shrimp larvae and juveniles (Decamp et al. 1999; Nagano and Decamp 2004). Therefore, the abundance of zooplankton and phytoplankton will support the growth of penaeid shrimps in aquaculture ponds throughout the culture period.

6 Conclusion

Research conducted over many decades has proven without a doubt that live feeds are the inevitable source of food in aquaculture, especially in larval nutrition. As the shrimp farming industry is one of the most important aquaculture production systems in terms of value and ensuring nutritional security to a large portion of the population, the sustainability and success of the industry largely depend on the high-quality seed produced in the hatchery. The mass culture of live feeds, as per the requirement, is the backbone of any hatchery. In recent years, researchers have focused on various potential live feeds that were not used earlier, and these types of studies have accelerated shrimp maturation and seed production considerably. Certain live feed organisms provide nutritional support to the early stages of larvae and impart certain health benefits to them. Lack of biosecurity in live feed production systems had invited many pathogenic microorganisms to the larval rearing and maturation systems, resulting in disease conditions inflicting heavy economic losses. The only option to eliminate these pathogens transmitted through live feeds is to strictly follow the biosecurity protocols developed and suggested by various regulatory authorities. Implementing strict biosecurity protocols in live feed production systems in finfish and shellfish hatchery rearing systems will not only ensure disease-free fish and shrimp seed but also increase aquaculture production.

References

Abraham TJ, Palaniappan R (2004) Distribution of luminous bacteria in semi-intensive penaeid shrimp hatcheries of Tamil Nadu, India. Aquaculture 232(1–4):81–90

Ajiboye OO, Yakubu AF, Adams TE, Olaji ED, Nwogu NA (2011) A review of the use of copepods in marine fish larviculture. Rev Fish Biol Fish 21(2):225–246

Alday-Sanz V (2016) Biosecurity risks and challenges in the shrimp industry. Paper presented at the first international technical seminar/workshop on early mortality syndrome (EMS) and acute Hepatopancreatic necrosis syndrome (AHPNS): government, scientist and farmer responses (under the FAO technical cooperation Programme [TCP/INT/3502]). Panama City, Florida, USA, June 22–24, 2015. FAO Fisheries and Aquaculture Report

Ananthi P, Santhanam P, Nandakumar R, Ananth S, Jothiraj K, Kumar SD, Prasath BB, Jayalakshmi T (2011) Production and utilization of marine copepods as live feed for larval rearing of tiger shrimp *Penaeus monodon* with special emphasis on astaxanthin enhancement. Indian J Natl Sci Int Bimonthly 976:0997

Aravind R, Anand PS, Vinay TN, Biju IF, Sandeep KP, Raymond JAJ, Rajamanickam S, Balasubramanian CP, Vijayan KK (2021) Population growth and mass production of brackish water cladoceran *Eurycercus beringi* sp. nov. under different diet and salinity regime, and its role in *P. indicus* larval rearing. Reg Stud Marine Sci 44:101777

Coman GJ, SJ Arnold, TR Callaghan, NP Preston 2007 Effect of two maturation diet combinations on reproductive performance of domesticated *Penaeus monodon*. Aquaculture 263: 75–83

Becker W (2004) Microalgae in human and animal nutrition. In: Richmond A (ed) Handbook of microalgal culture. Blackwell, Oxford, pp 312–351

Bengtson DA, Léger P, Sorgeloos P (1991) Use of Artemia as a food source for aquaculture. In: Browne RA, Sorgeloos P, Trotman CAN (eds) Artemia biology. CRC Press, Boca Raton, pp 255–285

Biedenbach JM, Smith LL, Thomson TK, Lawrence AL (1989) Use of nematode *Pangrellus redivivus* as an Artemia replacement in a larval penaeid diet. J World Aquacult Soc 20(2):61–71

Bold HC (1950) Problems in the cultivation of algae. In: Brunel J, Prescott GW, Tiffany LH (eds) The culturing of algae. Charles F. Kettering Foundation, New York, pp 11–17

Borowitzka MA (1997) Algae for aquaculture: opportunities and constraints. J Appl Phycol 9:393–401

Boxshall GA, Halsey SH (2004) An introduction to copepod diversity. Ray Society

Bray WA, Lawrence AL (1992) Reproduction of *Penaeus* species in captivity. In: Fast A, Lester LJ (eds) Marine shrimp culture: principles and practices. Elsevier, Amsterdam, pp 93–170

Brown MR (2002) Nutritional value of microalgae for aquaculture. In: Cruz-Suárez LE, Ricque-Marie D, Tapia-Salazar M, Gaxiola-Cortés MG, Simoes N (eds) Avancesen Nutrición Acuícola VI. Memorias del VI Simposium Internacional de Nutrición Acuícola. 3 al 6 de Septiembre del 2002. Cancún, Quintana Roo, México

Brown MR, Jeffrey SW, Volkman JK, Dunstan GA (1997) Nutritional properties of microalgae for mariculture. Aquaculture 151:315–331

Chang YS, Lo CF, Peng SE, Liu KF, Wang CH, Ko GH (2002) WSSV PCR–positive Artemia cysts yield PCR–negative nauplii that fail to transmit WSSV when fed to shrimp postlarvae. Dis AquatOrg 49:1–10

Chu KH, Lui LK (1990) Survival and development of *Metapenaeus ensis* larvae reared on algal and artificial diets. In: Hirano R, Hanyu I (eds) The second Asian fisheries forum. Asian Fish Soc, Manila

Cobo MDL, Wouters R, Wille M, Sonnenholzner S, Sorgeloos P (2015) Evaluation of frozen Umbrella-stage Artemia as first animal live food for *Litopenaeus vannamei* (Boone) larvae. Aquac Res 46(9):2166–2173

Crisp JA, D'Souza FM, Tweedley JR, Partridge GJ, Moheimani NR (2018) Performance of mixed species and mono-specific algal diets for culture of larval Western school prawns, *Metapenaeus dalli*. J World Aquacult Soc 49(5):845–856

D'Souza FM, Kelly GJ (2000) Effects of a diet of a nitrogen-limited alga (*Tetraselmis suecica*) on growth, survival and biochemical composition of tiger prawn (*Penaeus semisulcatus*) larvae. Aquaculture 181(3–4):311–329

D'Souza FML, Loneragan NR (1998) Effects of monospecific and mixed-algae diets on survival, development and fatty acid composition of penaeid prawn (*Penaeus* spp.) larvae. Mar Biol 133:621–633

Damare V (2009) Ecological studies and molecular characterization of thraustochytrids and aplanochytrids from oceanic water column. Ph.D Thesis. Goa University, India, p 223

de Lima LC, Souza-Santos LP (2007) The ingestion rate of *Litopenaeus vannamei* larvae as a function of *Tisbe biminiensis* copepod concentration. Aquaculture 271(1–4):411–419

De D, Sandeep KP, Kumar S, Raja RA, Mahalakshmi P, Sivaramakrishnan T, Ambasankar K, Vijayan KK (2020) Effect of fish waste hydrolysate on growth, survival, health of *Penaeus vannamei* and plankton diversity in culture systems. Aquaculture 524:735240

Decamp O, Warren A, Sanchez R (1999) The role of ciliated protozoa in subsurface flow wetlands and their potential as bioindicators. Water Sci Technol 40(3):91–98

Desrina H (2016) Polychaetes as potential risk for the shrimp pathogen transmission. Paper presented at the second international technical seminar/workshop on acute Hepatopancreatic necrosis syndrome (AHPND): there is a way forward (under the FAO technical cooperation Programme [TCP/INT/3501 and TCP/INT/3502]). Bangkok, Thailand, June 23–25, 2016. FAO Fisheries and Aquaculture Report

Drillet G, Hansen BW, Kiørboe T (2011) Resting egg production induced by food limitation in the calanoid copepod *Acartia tonsa*. Limnol Oceanogr 56(6):2064–2070

Emmerson WD (1984) Predation and energetics of *Penaeus indicus* (Decapoda: Penaeidae) larvae feeding on *Brachionus plicatilis* and Artemia Nauplii. Aquaculture 38:201–209

Ferreira M, Coutinho P, Seixas P, Fábregas J, Otero A (2009) Enriching rotifers with "premium" microalgae *Nannochloropsis gaditana*. Mar Biotechnol 11(5):585–595

Focken U, Schlechtriem C, Von Wuthenau M, García-Ortega A, Puello-Cruz A, Becker K (2006) *Panagrellus redivivus* mass produced on solid media as live food for *Litopenaeus vannamei* larvae. Aquac Res 37(14):1429–1436

Gallardo-Rodríguez J, Sánchez-Mirón A, García-Camacho F, López-Rosales L, Chisti Y, Molina-Grima E (2012) Bioactives from microalgal dinoflagellates. Biotechnol Adv 30(6):1673–1684

Gross F (1937) Notes on the culture of some marine plankton organisms. J Mar Biol Assoc UK 21:753–768

Guedes AC, Malcata FX (2012) Nutritional value and uses of microalgae in aquaculture. In: Muchlisin ZA (ed) Aquaculture, pp 59–78. https://doi.org/10.5772/1516

Guillard RRL, Ryther JH (1962) Studies of marine planktonic diatoms. I. *Cyclotella nana* Hustedt and *Detonula confervacea* Cleve. Can J Microbiol 8:229–239

Gupta A, Barrow CJ, Puri M (2012) Omega-3 biotechnology: thraustochytrids as a novel source of omega-3 oils. Biotechnol Adv 30:1733–1745

Haché R, Plante S (2011) The relationship between enrichment, fatty acid profiles and bacterial load in cultured rotifers (*Brachionus plicatilis* L-strain) and Artemia (*Artemia salina* strain franciscana). Aquaculture 311:201–208

Hakima B, Khémissa C, Boudjéma S (2013) Effects of food limitation on the life history of *Simocephalus expinosus* Koch (Cladocera: Daphniidae). Zoology 5(2):25–31

Han QX, Wang YQ, Zhang Y, Keesing J, Liu DY (2013) Effects of intensive scallop mariculture on macrobenthic assemblages in Sishili Bay, the northern Yellow Sea of China. Hydrobiologia 718:1–15

Havel JE, Medley KA, Dickerson KD, Angradi TR, Bolgrien DW, Bukaveckas PA, Jicha TM (2009) Effect of main-stem dams on zooplankton communities of the Missouri River (USA). Hydrobiologia 628(1):121–135

He ZH, Qin JG, Wang Y, Jiang H, Wen Z (2001) Biology of *Moina mongolica* (Moinidae, Cladocera) and perspective as live food for marine fish larvae. Hydrobiologia 457(1):25–37

Hemaiswarya S, Raja R, Kumar RR, Ganesan V, Anbazhagan C (2011) Microalgae: a sustainable feed source for aquaculture. World J Microbiol Biotechnol 27:1737–1746

Hirata H (1979) Rotifer culture in Japan. In: Jaspers E et al (eds) Cultivation of fish fry and its live food, pp 361–375

Høj L, Bourne DG, Hall MR (2009) Localization, abundance and community structure of bacteria associated with Artemia: effects of nauplii enrichment and antimicrobial treatment. Aquaculture 293(3–4):278–285

Humes AG (1994) How many copepods? In: Ecology and morphology of copepods. Springer, Dordrecht, pp 1–7

Jamali H, Ahmadifard N, Abdollahi D (2015) Evaluation of growth, survival and body composition of larval white shrimp (*Litopenaeus vannamei*) fed the combination of three types of algae. Int Aquatic Res 7:115–122

Jaseera KV, Ebeneezar S, Sayooj P, Nair AV, Kaladharan P (2021) Dietary supplementation of microalgae, *Aurantiochytrium* sp. and co-feeding with Artemia enhances the growth, stress tolerance and survival in *Penaeus monodon* (Fabricius, 1798) post larvae. Aquaculture 533:736176

Jiang N, Pan X, Gu Z, Liu W, Si K, Zhou Y, Zhou Y, Zhai L, Fan Y, Zeng L (2019) Proliferation dynamics of WSSV in crayfish, *Procambarus clarkii*, and the host responses at different temperatures. J Fish Dis 42(4):497–510

Kannappan S, Sivagnanam S, Poochirian Jithendran K, Praveena PE, Balasubramanian CP, Kizhakedath Vijayan K (2021) Development of indoor grow-out practices for polychaete, *Marphysa gravelyi* with a note on biochemical composition. Aquacult Res. https://doi.org/10.1111/are.15266

Karunasagar I, Karunasagar I, Raghunath P (2016) Ecology, virulence, and detection of pathogenic and pandemic *Vibrio parahaemolyticus*. Front Microbiol 7:156

Kent M, Browdy CL, Leffler JW (2011) Consumption and digestion of suspended microbes by juvenile Pacific white shrimp *Litopenaeus vannamei*. Aquaculture 319(2–3):363–368

Krishnan AN, Kannappan S, Aneesh PT, Praveena PE, Jithendran KP (2021) Polychaete worm-a passive carrier for *Enterocytozoon hepatopenaei* in shrimp. Aquaculture 545:737187

Kumlu M, Jones DA (1995) The effect of live and artificial diets on growth, survival, and trypsin activity in larvae of *Penaeus indicus*. J World Aquacult Soc 26(4):406–415

Laoaroon S, Boonnat A, Poltana P, Kanchanaphum P, Gangnonngiw W, Nash G, Withyachumnarnkul B (2005) Infectivity of white spot syndrome virus (WSSV) to the polychaete *Pereneis nuntia* and a possibility of WSSV transmission from the polychaete to the black tiger shrimp Penaeus monodon. In: Walker P, Lester R, Bondad-Reantaso MG (eds) Diseases in Asian aquaculture V. Fish Health Section, Asian Fisheries Society, Manila, pp 353–361

Lavens P, Sorgeloos P (1996) Manual on the production and use of live food for aquaculture (No. 361). Food and Agriculture Organization (FAO)

Leander CA, Porter D, Leander BS (2004) Comparative morphology and molecular phylogeny of aplanochytrids (Labyrinthulomycota). Eur J Protistol 40:317–328

Leelatanawit R, Uawisetwathana U, Khudet J, Klanchui A, Phomklad S, Wongtripop S, Angthoung P, Jiravanichpaisal P, Karoonuthaisiri N (2014) Effects of polychaetes (*Perinereis nuntia*) on sperm performance of the domesticated black tiger shrimp (*Penaeus monodon*). Aquaculture 433:266–275

Lewis TE, Nichols PD, McMeekin TA (1999) The biotechnological potential of thraustochytrids. Mar Biotechnol 1(6):580–587

Li Q, Zhang J, Chen Y, Yang F (2003) White spot syndrome virus (WSSV) infectivity for Artemia at different developmental stages. Dis Aquat Org 57(3):261–264

Louis RD, Perez EI, Sangha R, Puello-Cruz A (2006) Successful culture of larvae of *Litopenaeus vannamei* fed a microbound formulated diet exclusively from either stage PZ2 or M1 to PL1. Aquaculture 261(4):1356–1362

Loureiro CK, Junior WFBW, Abreu PCOVD (2012) The use of protozoan, rotifers and nematodes as live food for shrimp raised in BFT system. http://repositorio.furg.br/handle/1/3831

Lovett DL, Felder DL (1988) Evaluation of the rotifer *Brachionus plicatilis* as a substitute for Artemia in feeding larvae of *Macrobrachium rosenbergii*. Aquaculture 71(4):331–338

Lubzens E, Zmora O, Barr Y (2001) Biotechnology and aquaculture of rotifers. Hydrobiologia 446(447):337–353

Martins D, Custódio L, Barreira L, Pereira H, Ben-Hamadou R, Varela J, Abu-Salah K (2013) Alternative sources of n-3 long-chain polyunsaturated fatty acids in marine microalgae. Mar Drugs 11:2259–2281

Meunpol O, Meejing P, Piyatiratitivorakul S (2005) Maturation diet based on fatty acid content for male *Penaeus monodon* (Fabricius) broodstock. Aquac Res 36(12):1216–1225

Meunpol O, Iam-Pai S, Suthikrai W, Piyatiratitivorakul S (2007) Identification of progesterone and 17α-hydroxyprogesterone in polychaetes (*Perinereis* sp.) and the effects of hormone extracts on penaeid oocyte development in vitro. Aquaculture 270(1–4):485–492

Muller-Feuga A (2000) The role of microalgae in aquaculture: situation and trends. J Appl Phycol 12:527–534

Muller-Feuga A (2004) Microalgae for aquaculture: the current global situation future trends. In: Richmond A (ed) Handbook of microalgal culture: biotechnology and applied phycology, Oxford, pp 352–364

Muller-Feuga A, Moal J, Kaas R (2003) The microalgae of aquaculture. In: Støttrup JG, McEvoy LA (eds) Live feeds in marine aquaculture. Blackwell Science, pp 206–243

Naessens E, Wouters R, Cobo M, Vargas V, Pedrazzoli A, Van Hauwaert A, Lavens P (1995) Effect of n-3 HUFA and DHA/EPA ratio in enriched live feed on fatty acid composition and culture performance of *Penaeus vannamei* larvae. In Fish and shellfish Larviculture symposium. Special Publication European Aquaculture Society 24, pp 213–216

Naessens E, Lavens P, Gomez L, Browdy CL, McGovern-Hopkins K, Spencer AW, Kawashigashi D, Sorgeloos P (1997) Maturation performance of *Penaeus vannamei* co-fed Artemia biomass preparations. Aquaculture 155:87–101

Nagano N, Decamp O (2004) Ingestion of a ciliated protozoa by first-feeding larval stage of Pacific white shrimp, *Litopenaeus vannamei* (Boone). Aquac Res 35(5):516–518

Najmi N, Yahyavi M, Haghshenas A (2018) Effect of enriched rotifer (*Brachionus plicstilis*) with probiotic lactobacilli on growth, survival and resistance indicators of western white shrimp (*Litopenaeus vannamei*) larvae. Iran J Fish Sci 17(1):11–20

Naranjo J, Porchas MA, Robles M, Magallon FJ, Valdez J, Salinas C, Villareal H (1995) Survival, metamorphosis and growth of brown shrimp *Penaeus californiensis* larvae, fed with different microalgae. In: Aquaculture ¢95 conference. February 1–4, San Diego, California. World Aquaculture Society. Baton Rouge. Louisiana, 235 (Abstract. No. A503)

Neto FP, Neumann-Leitão S, Casé M, Sant'Anna EE, Cavalcanti EH, Schwamborn R, Gusmão LMO, Melo PAMC (2009) Zooplankton from shrimp culture ponds in Northeastern Brazil. Trans Ecol Environ 122:251–260

Palmer PJ, Wang S, Houlihan A, Brock I (2014) Nutritional status of a nereidid polychaete cultured in sand filters of mariculture wastewater. Aquac Nutr 20(6):675–691

Pamungkas J, Glasby CJ (2015) Taxonomy of reproductive Nereididae (Annelida) in multispecies swarms at Ambon Island, Indonesia. ZooKeys 520:1

Parenrengi A (2004) Prevalence, intensity, and transmission of white spot syndrome virus (WSSV) in tiger shrimp, *Penaeus monodon* culture. J Penelitian Perikanan Indonesia 10(5):103–110

Plumb JA (1992) Disease control in aquaculture. In: Shariff M, Subasinghe RP, Arthur JR (eds) Diseases in Asian aquaculture I. Fish Health Section, Asian Fisheries Society, Manila, pp 3–17

Provasoli L, McLaughlin JJA, Droop MR (1957) The development of artificial media for marine algae. Arch Microbiol 25:392–428

Pulz O, Gross W (2004) Valuable products from biotechnology of microalgae. Appl Microbiol Biotechnol 65(6):635–648

Sahul Hameed A, Murthi B, Rasheed M, Sathish S, Yoganandhan K, Murugan V, Jayaraman K (2002) An investigation of Artemia as a possible vector for white spot syndrome virus (WSSV) transmission to *Penaeus indicus*. Aquaculture 204(1–2):1–10

Sambhu C, Samy AD, Kitto MR (2014) Early larval rearing of white shrimp, *Fenneropenaeus indicus* by feeding enriched rotifer diets. Int J Marine Sci 30:9–12

Samocha T, Lewinsohn CH (1977) A preliminary report on rearing penaeid shrimps in Israel. Aquaculture 19:291–292

Samocha TM, Uziel N, Browdy CL (1989) The effect of feeding two prey organisms, nauplii of Artemia and rotifers, *Brachionus plicatilis* (Muller), upon survival and growth of larval marine shrimp, *Penaeus semisulcatus* (de Haan). Aquaculture 77(1):11–19

Sandeep KP, Avunje S, Dayal JS, Balasubramanian CP, Sawant PB, Chadha NK, Ambasankar K, Vijayan KK (2021) Efficiency of different microalgae as monospecific and bispecific diets in larval rearing of *Penaeus indicus* with special reference to growth, nutrient composition and antimicrobial activity of microalgae. Aquac Res 52:5146. https://doi.org/10.1111/are.15382

Sandhya SV, Sandeep KP, Vijayan KK (2020) In vivo evaluation of microbial cocktail of microalgae-associated bacteria in larval rearing from zoea I to mysis I of the Indian white shrimp, *Penaeus indicus*. J Appl Phycol 32(6):3949–3954

Sivakami S (1988) Observations on the effect of fertilizer and feed applications on the growth of *Penaeus indicus* H. Milne Edwards. Indian J Fish 35(1):18–25

Song X, Shi C, Huang J, Zhang L (2001) Investigation and study on carrying status of white spot syndrome virus (WSSV) in shrimps and other aquatic animals by dot-blot hybridization. J Fish Sci China 8(4):36–40

Songsangjinda P, Chaweepack T, Khomvilai C, Thawonsuwan J (2016) Government actions on EMS/AHPND in Thailand. Paper presented at the first international technical seminar/workshop on early mortality syndrome (EMS) and acute Hepatopancreatic necrosis syndrome (AHPNS): government, scientist and farmer responses (under the FAO technical cooperation Programme [TCP/INT/3502]). Panama City, Florida, USA, June 22–24, 2015. FAO Fisheries and Aquaculture Report

Sorgeloos P (1980) Life history of the brine shrimp Artemia. In: Persoone G, Sorgeloos P, Roels O, Jaspers E (eds) The brine shrimp Artemia, Vols 1–3. Universa Press, Wetteren, pp XIX–XXIII

Støttrup JG (2003) Production and nutritional value of copepods. In: Stottrup JG, McEvoy LA (eds) Live feeds in marine aquaculture. Blackwell Science, Oxford, UK, pp 145–205

Tacon AG (2017) Biosecure shrimp feeds and feeding practices: guidelines for future development. J World Aquacult Soc 48(3):381–392

Techaprempreecha S, Khongchareonporn N, Chaicharoenpong C, Aranyakananda P, Chunhabundit S, Petsom A (2011) Nutritional composition of farmed and wild sandworms, *Perinereis nuntia*. Anim Feed Sci Technol 169(3–4):265–269

Tredici MR, Biondi N, Ponis E (2009) Advances in microalgal culture for aquaculture feed and other uses. In: Burnell G, Allan G (eds) New technologies in aquaculture: improving production efficiency, quality and environmental management, Cambridge, pp 611–676

Ugalde V, Armenta RE, Kermanshahi-pour A, Sun Z, Berryman KT, Brooks MS (2018) Improvement of culture conditions for cell biomass and fatty acid production by marine thraustochytrid F24-2. J Appl Phycol 30:329–339

van der Meeren T, Olsen RE, Hamre K, Fyhn HJ (2008) Biochemical composition of copepods for evaluation of feed quality in production of juvenile marine fish. Aquaculture 274:375–397
Vijayan KK, Raj VS, Balasubramanian CP, Alavandi SV, Sekhar VT, Santiago TC (2005) Polychaete worms- a vector for white spot syndrome virus (WSSV). Dis Aquat Org 63(2–3):107–111
Wilkenfeld JS, Lawrence AL, Kuban FD (1984) Survival, metamorphosis and growth of penaeid shrimp larvae reared on a variety of algal and animal foods. J World Maricult Soc 15:31–49
Zmora O, Richmond A (2004) Microalgae for aquaculture: microalgae production for aquaculture. In: Richmond A (ed) Handbook of microalgal culture: biotechnology and applied phycology. Blackwell Science Ltd, pp 365–379

Sustainable Intensification in Aquaculture

A. Panigrahi and P. S. Shyne Anand

Abstract

Aquaculture is widely recognized as a vital resource for satisfying nutritional security and livelihood options for the growing world population when the capture fishery resources are stagnating. Although aquaculture started with traditional farming models, the spiraling demand for quality protein sources to meet the ever-increasing population resulted in a paradigm shift from conventional to semi-intensive or intensive farming. Intensification in aquaculture nevertheless doubled the income with better nutritional security; the sector started to face myriad challenges, such as eutrophication, single species-oriented culture, the escalating price of fish meal and feed ingredients, disease outbreaks, antibiotic residue, traceability, etc. This social and environmental insecurity urgently calls for farming practices that are economically profitable to meet the growing demand for food and eco-friendly to ensure future sustainable aquaculture production. Sustainable integrated aquaculture considerably differs from extensive or intensive stand-alone farming models as it integrates ecosystems with diversified species having different trophic levels. Thus, this farming system can reduce the risk associated with intensive monoculture, disease outbreaks, price volatility of aquatic products, etc. As these farming models have better synergism and are environmentally sound, they can be considered a panacea for small-scale farmers in developing countries. The majority of sustainable aquaculture embraces the age-old integrated agriculture-aquaculture farming models, such as paddy-cum-fish culture, livestock-cum-fish culture, etc. Due to technological advancement, many farming models, such as recirculatory aquaculture (RAS), integrated multitrophic aquaculture (IMTA), biofloc-based farming system (BFT), polyculture,

A. Panigrahi (✉) · P. S. Shyne Anand
ICAR-Central Institute of Brackishwater Aquaculture, Chennai, India
e-mail: apanigrahi2k@gmail.com; shyne.anand@gmail.com

R. Kooloth Valappil et al. (eds.), *Aquatic Animal Health Management*,
https://doi.org/10.1007/978-981-96-7987-4_29

etc., integrate sustainability in intensification and are gaining popularity among the farming community and stakeholders due to their multiple advantages.

Keywords

Aquaculture · Sustainable intensification · Biofloc · Recirculatory aquaculture · IMTA

1 Introduction

Aquaculture is a sunrise food-producing sector contributing to the lion's share (130.9 million tons) of total global fish production, 223.2 million tons (FAO 2024). Aquaculture is widely recognized as a crucial resource for satisfying nutritional security and livelihood options for the growing world population, where capture fishery resources are stagnating. India secured the second position in global fish production, with shrimp being the major export commodity. Indian shrimp industry produced 9.3lakh tons of shrimp in 2023 and earned Rs 63,969.14 crore (USD 8.09 billion) during 2022–23 from seafood export due to technological intervention and intensified farming practices (MPEDA 2023).

The traditional farming system that prevailed during the 20th century was characterized by a low input-based farming system with low productivity and was sustainable due to low intensification and inadequate input-based management practices. However, the growing population, coupled with spiraling demand for seafood and other protein sources, and technological interventions results in shifting farming practices from traditional to semi-intensive or intensive farming methods. Although intensification doubled income, employment, and nutritional security, farming's impact on the surrounding environment became a significant concern. For example, shrimp farming, although considered the economic engine of brackish water aquaculture due to undoubted growth and employment generation, farming sophistication also raises concern about its sustainability prospects (Naylor et al., 2000). Along with intensification, disease outbreaks, environmental degradation, eutrophication, traceability, antibiotic issues, biodiversity threats, the escalating price of feed ingredients, other accessory input cost, etc., became daunting challenges in aquaculture. For example, intensive cage farming results in higher nutrient load discharges to the surrounding ecosystem, leading to eutrophication, altering the delicate balance in the ecosystem, and eventually hampering biodiversity. Similarly, mass mortality of native fauna is reported due to eutrophication near intensive cage farming (Wei et al. 2020; Sievers et al. 2022.

Apart from the environmental issues, quality and food safety, antibiotic residues, traceability, biodiversity impact on native stock, etc., also came up as of utmost public concern. This social and environmental insecurity urgently calls for farming practices that are economically profitable to meet the growing demand for food and eco-friendliness to ensure sustainability (Campanati et al. 2022; Dong et al. 2022). Thus, sustainable integrated aquaculture aims to establish socioeconomic sustainability in intensive farming models using an ecosystem-based approach (Chary et al. 2024).

2 Concepts and Principles

The generation of nutrients or waste is an inherent part of any farming system (Farrant et al. 2021; Villar-Navarro et al. 2021). The concept behind sustainable integrated aquaculture arises when effluent loads generated from intensive farming systems eutrophicate natural water bodies and result in environmental concerns. This concept was first adopted in 1972 by Ryther et al. (1972) to mitigate the aquaculture system's eutrophication issues. Sustainable integrated aquaculture considerably differs from extensive or intensive stand-alone farming models as it integrates ecosystems with diversified species. Thus, this farming system can reduce the risk associated with intensive monoculture, price volatility of aquatic products, etc. These farming models have better synergism and are environmentally sound; therefore, they can be considered a panacea for small-scale farmers in developing countries. The majority of sustainable aquaculture practices embrace the age-old integrated agriculture-aquaculture farming models or semi-intensive farming systems with commercially important fish or shrimp as feed species and dedicated feed management practices. Integration of these farming systems with secondary aquaculture or non-fed species such as molluscs or seaweeds can give better economic value and environmental safety with nutrient recycling (Granada et al. 2016; Cubillo et al. 2016). Being a self-regulatory species, these extractive species, such as molluscs or seaweeds, contribute a major share in global aquaculture production (FAO 2024). Sustainable integrated farming systems consist of sustainable and eco-friendly farming models based on open water and land-based SI A farming practices (Barrington et al. 2009). The low input based on traditional integrated agriculture-aquaculture systems, such as paddy cum fish culture, polyculture, etc., are sustainable traditional farming models. Similarly, intensive farming systems, such as integrated multitrophic aquaculture (IMTA), recirculatory aquaculture (RAS), integrated aquaponics, biofloc-based farming models, etc., are also gaining popularity due to better economic returns from these farming practices. Thus, the strategies toward sustainable intensification should encompass a system of farming practices that incorporates environmental sustainability standards, efficient resource utilization with a sound aquaculture value chain system for a circular blue economy, and minimum greenhouse gas emission effects supported with effective government policies and planning (Figs. 1 and 2).

3 Background: Historical Perspectives

The concept of sustainable intensification (SI) was raised among African smallholder agriculture in 1997 (Pretty 1997). Later, the first World Development Report published in the year 2008 on agricultural productivity gave crucial

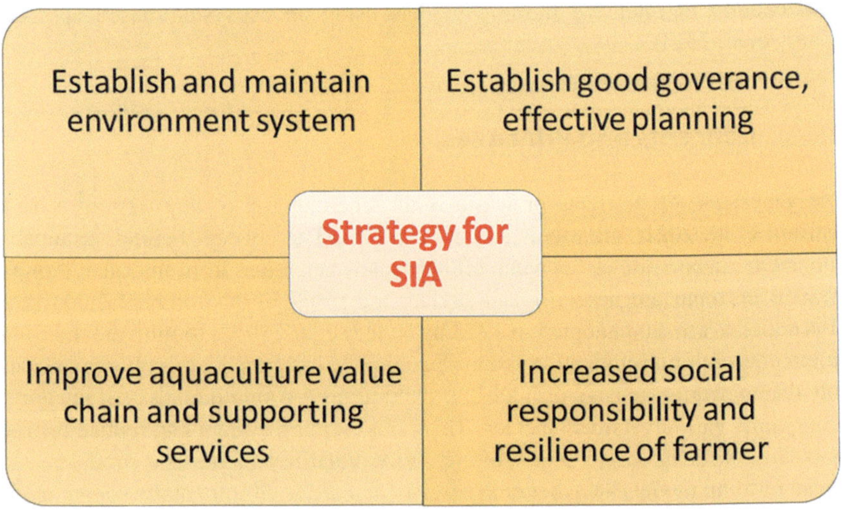

Fig. 1 Diagrammatic illustration of strategies toward sustainable intensification

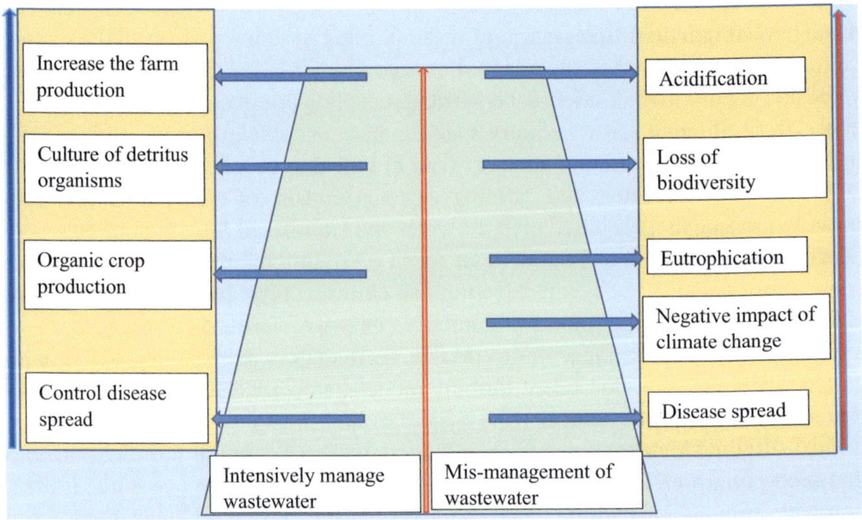

Fig. 2 Management strategies toward sustainable intensification goal

importance to sustainable intensification when climate change and the population explosion became a daunting challenge (Kesavan and Swaminathan 2009; Barrington et al. 2009). The main focus of the theme was to produce more with less environmental impact and maximum societal benefit. An increase in farming productivity through technological advancement and innovation was targeted in an eco-friendly manner to achieve this goal. It is suggested that more holistic attributes of food systems that may be demand-driven and other societal drivers (Cook et al. 2016) need to be considered. According to Garnett et al. (2015), sustainable intensification can be achieved through (a) increasing production to meet greater demand for food, (b) intensification to utilize the area judiciously with minimum environmental impact, (c) food security is as vital as environmental sustainability, and (d) treating SI as a goal and not as an approach. Furthermore, SI should complement other development goals, such as human welfare, animal welfare, nutrition, conserving biodiversity, and broader sustainable development. As ecology is gaining importance, technological innovations, before being widely promoted, should pass through stringent screening.

Area-based management used to have a significant effect in the UK and Norwegian salmon sectors. However, openness or porosity challenges the sustainable intensification of modern commercial aquaculture systems. For example, many aquaculture farms, with their proximity to natural open water bodies, make both farms and the ecosystem vulnerable to the potential source of hazards, such as contaminants, pathogens, genetic impurity, eutrophication, biodiversity threat, or exotic species introduction, suggesting that SI should be enriched by the value chain (Pelletier et al. 2009; Newton and Little 2017).

4 Sustainable Intensification in Aquaculture Assessment Criteria

The SIA criteria include improved input use, improved production efficiency, increased socioeconomic benefits, improved governance, and environmental sustainability. Types of SIA assessment include territorial diagnoses of aquaculture system sustainability, meta-diagnosis of the aquaculture system studied, and environmental diagnoses of the aquaculture system based on the LCA method (Dongyu 2024; Lazard et al. 2014; FAO 2018).

These recent aquaculture development assessment criteria include:

1. Closed/semi-closed intensive farming system.
2. Development and dissemination of specific pathogen-free (SPF) *Penaeus monodon, P. indicus, P. japonicus, and Macrobrachium rosenbergii fir diversified shrimp farming*
3. Biofloc and periphyton-based shrimp and fish farming.
4. High-density Genetically Improved Farmed Tilapia (GIFT) farming.
5. Group approach in implementing better management practice (BMP)/ Good AquaculturePractices in shrimp farming.
6. Multitrophic marine aquaculture system (shrimp-mollusc-seaweed).
7. Integrated multitrophic aquaculture (IMTA): Shrimp-seaweed-mollusc system.
8. Improved participation and contribution of women in aquaculture intensification.

4.1 SIA Practices: Technology Development and Dissemination

Sustainable intensification strategies, which primarily result from the scientific and technological progress, are intensive culture of SPF *Pvannamei* farming, SPF *P. monodon* farming, genetically improved Common carp and Jayanthi Rohu varieties, BFT-based shrimp and fish, RAS, IMTA, integrated trout in high altitudes, polyculture, high-density GIFT culture, marine cage farming, etc. Establishment of Successful intensive farming system model is highly needed for its implementation and technology dissemination in the farming community. Many success stories based on the principles of sustainability have been reported earlier (Chary et al. 2024).

4.1.1 Semi-Closed Intensive *P. vannamei* (SPF *Pv*) Shrimp Farming

Biosecured semi-intensive farming methods in India and Southeast Asian countries are based on the zero water exchange principles with biosecurity. Better management practices (BMPs) followed with semi-closed intensive shrimp farming have an average productivity of 8–10 tons per ha. The success of the farming model can be credited to zero tolerance for the antibiotic, customized indigenous-based probiotics (Sundaram et al. 2017; Panigrahi et al. 2020a, b), farm-level biosecurity, optimum use of inland low and high saline farming for shrimp culture, better comprehensive national surveillance program, low protein-based feeds, and automatic feeder sand eco-based technologies (Panigrahi et al. 2019a b; Anand et al. 2019). Given the scope and potential for future expansion of zero water exchange shrimp farming, a sustainable intensification effort with the adoption of BMP and regulation can lead to maximizing export potential. Indian shrimp farming's long-term success will continue to rely on the farmers' adoption of better management practices (BMPs). Adopting critical quality inputs and integration with secondary aquaculture species, such as seaweeds and bivalves, for nutrient recycling in ETP ponds can enhance sustainability.

4.1.2 Intensive Farming of Specific Pathogen-Free *Penaeus monodon* (SPF *Pm*)

Black tiger shrimp, *P. monodon,* was the most widely farmed shrimp before the onset of the White Spot Syndrom Virus (WSSV) outbreak during the first half of the 1990s. The white spot syndrome virus and other diseases dominated black tiger shrimp farming in India (Karunasagar et al. 1998; Dash et al. 2017; Otta et al. 2018). Farmers started looking for alternative species or SPF varieties, and the introduction of SPF *P. vannamei* gave a much-needed boost to disease-struck Indian shrimp farming. However, a single species-oriented culture resulted in multiple challenges, such as the emergence of new diseases, the escalating price of imported broodstock, etc. Recently, with the aim to diversify shrimp farming, a selective breeding program for producing an SPF (specific pathogen-free) fast-growing and disease-resistant variety of *P. monodon* has been initiated. It is reported that the growth rate and WSSV resistance have been improved after each generation. The practice developed elsewhere for selectively bred *P. monodon* and SPF broodstock production can be disseminated to other countries to improve their own native *P. monodon* through a public-private participation model.

4.1.3 Biofloc and Periphyton-Based Shrimp and Fish Farming

Biofloc-based farming is becoming popular and can achieve higher productivity through sustainability, which is befitting to the SIA concept. This technology is based on management through heterotrophic microflora promoted and maintained through C.N. ratio manipulation (Samocha et al. 2019; Anand et al. 2014; Panigrahi et al. 2018; 2022). It is based on retaining waste and its conversion to microbial floc as a natural food within the culture system. Science-based BFT farming is promoted by optimizing inputs, protein content in the feed, C: N ratio, and floc microbial constitutions. Maximizing productivity from minimum natural resources is recognized as one of the most vibrant futuristic technologies. It encourages scientific farming among farming communities, with farmers exposed to many parameters, situations, and how to produce in controlled conditions. BFT encourages farmers to start shrimp or fish culture diversified farming units such as inland areas or homestead farming units throughout the year across all geographic locations. It is practiced in both freshwater and brackish water systems with a wide range of species, such as shrimp (*P. vannamei, P indicus,* and *P. monodon)*, Tilapia, Pangasius, Amur carps, and other carps, and native air-breathing fishes such as Heteropneustes fossilis (singhi), Anabas testudineus (koi), and Clarius sp are also reported to have better growth performance in the tank-based farming system. Total production of 400–600 kg of Tilapia/Pungas and 300–400 kg of Koi, Singi, or 50–60 kg of shrimp can be achieved from a 10-ton tank system. ICAR-Central Institute of Brackishwater Aquaculture (ICAR-CIBA),Chennai has standardized the technology protocols for biofloc-based nursery and grow-out culture technology for penaeid shrimps (Panigrahi et al. 2019a, b, 2020b; Anand et al. 2013, 2014) and selected finishes. About two million nursery seeds or 1200 kg of shrimp per year is achievable from a 100-ton tank. Substrate integrated biofloc based system reported to yield an average grow-out yield of 6–8 kg/m^3 shrimp with around 130–144% rate of return.

The use of submerged substrate in the natural environment to aggregate fish is an age-old practice in different parts of Africa and Asian countries. Acdja-based fisheries prevalent in West Africa, where dense tree branches are placed in lagoons to attract fish, is a pioneering example of the periphyton-based farming system (Welcomme 1972). "Katha fisheries" in Bangladesh and "Samarahs" or brush parks in Cambodia, Athkatu fishery in Sri Lanka, prevalent for centuries, employ the same principle (Wahab and Kibria 1994). Floating islands formed through the dense growth of aquatic weeds and grasses in the Lok Tak of Manipur are examples of this farming system. In most of the substrate-based aquaculture systems, submerged substrates in culture ponds provide sites for the development of periphyton, a group of autotrophic and heterotrophic microorganisms. Periphyton serves as an excellent natural food for different species of fish or shrimps. Apart from providing natural food as autotrophic and heterotrophic periphytic colonizing organisms, it improves environmental conditions through ecological and biological processes. It is reported that aquatic animals are mechanically more efficient to graze a two-dimensional layer of periphyton than to filter algae from a three-dimensional planktonic environment (van Dam et al. 2002). Thus, periphyton in the aquatic ecosystem enhances the system's nutrient-transfer efficiency due to the shift from less stable phytoplankton to a more stable periphyton community. Integration of the periphyton in biofloc-based farming systems can sustainably enhance fish or shrimp production (Anand et al. 2013, 2019).

4.1.4 High-Density Genetically Improved Farmed Tilapia (GIFT) Farming

Nile tilapia is the third most widely cultivated fish after carps and salmon globally and is being farmed in more than 120 countries, contributing a major share of global production (Cai et al. 2018; FAO 2024). After witnessing the spectacular success in breeding programs for salmon and trout established in Norway in the 1970s, WorldFish and partners created an improved tilapia strain, called the Genetically Improved Farmed Tilapia (GIFT), from 8 different Nile tilapia strains by pioneering a systematic breeding method that began in 1988. The GIFT is an important candidate species for tropical aquaculture due to its fast growth rate, short duration, tolerance to water quality variation, disease resistance, and easy culture techniques with better adaptability to wide water quality parameters. Intensive tilapia farming with nursery rearing at 75/m^2 and grow-out farming at 3–5/m^2 reported a yield of 10–12 tons per ha in a six-month duration (Eknath and Hulata 2009). This allows farmers to take two crops per year. Presently, the monosex all-male culture of Genetically Modified Farmed Tilapia) is farmed globally. In India, an increase in demand for GIFT tilapia is going on with the availability of seed from RGCA, and a production of 8–10 tons/ha can be obtained in one crop of 6 months. Improving technical knowledge and economic performance on tilapia ponds is possible with sustainable intensification using better seed and feed quality, better business and operational planning with value addition, and an ideal domestic market chain. Tailor-made bio-economic models need to be constructed based on specific farming conditions to guide farmers' business and operational planning.

4.1.5 Genetically Improved "Jayanti" Rohu (*Labeo rohita*) and Common Carp

Jayanti Rohu™ was developed through a selective breeding program of *Labeo rohita* from different founder populations of North Indian Rivers. The improved strain is reported to give a 17% growth gain per generation for growth traits. Disease resistance against *Aeromonas hydrophila* was also added as the second trait of selection to the ongoing rohu breeding program. The superiority with Jayanti rohu (increased by 23 per cent) was proved over normal rohu in the farmers' fields, reducing production cost by 26.6 per cent with the reduced rearing period (53 days in a crop). Similarly, the genetically improved strain of common carp (Jian carp x Huanghe carp) reported 28% genetic growth gain after the fourth-generation selection process. This genetic gain through the selection process will help to improve productivity from aquaculture farms with the same intensification level as nonselected seed stocks. As naturalized aquaculture species in the Asia-Pacific countries, they are essential for nutritional security as almost the total production is consumed within the same countries.

4.1.6 In-Pond Raceway System (IPRS)

The In-Pond Raceway System (IPRS) is a strategy to increase fish production with reduced environmental impact. The IPRS's concept is to concentrate fed fish in cells or "raceways" within a pond and provide them with constant water circulation to maintain optimal water quality and to improve feed management. This system can reduce solid waste loading in the pond by concentrating and removing it from the downstream end of the raceway units. The approach uses continual aeration and mixing of the full pond volume to accelerate the assimilation of organic wastes and nutrients produced by fish culture and feeding. As a continual, aerated water flow is maintained, a more stable and predictable water quality and culture environment is achieved. A stocking rate of 113 per m^3 with 86 per cent survival and 57 kg/m^3 harvest biomass could reach 69 tons/ha/year of trout employing this model. Though initiated in the USA, this technology is now getting promoted across the Asia-Pacific countries. However, the techno-economic feasibility needs to be evaluated before farm-level adoption.

4.1.7 Recirculatory Aquaculture System (RAS)

As sustainability has a close link with the environment, the RAS is considered an environment-friendly aquaculture farming system (Connor, 2015) The RAS uses much less water and land facilities compared to flow through the farming system. Moreover, with well-developed filtration technology, greater control over disease outbreaks, water quality parameters, and effluent water quality is possible in this system. Thus, with better biosecurity standards, the RAS is considered an environmentally sustainable farming system. Moreover, as the RAS can be established based on market accessibility, the carbon footprint from the farm produce can be controlled. According to statistics, when a traditional farming system uses 20 liter of water per kg fish, the RAS system uses only 1/eighth of the land area and 1/sixth of water usage and ten fold more production (50–150 kg/ton) using advanced water treatment and with better feed and disease management. However, the wide adoption of this system needs to be promoted as the system has its limitations in large-scale adoption due to high upfront investment in infrastructure, intensive technical monitoring, the need for trained and skilled technicians, etc. The economically viable fishes, such as Asian seabass/Bhetki (*Lates calcarifer*), Cobia (*Rachycentron canadum*), silver/Indian Pompano (*Trachinotus Blochii/ Trachinotus mookalee*), GIFT tilapia (*Oreochromis niloticus*), Pearl spot (*Etroplus suratensis*), Pangasius (*Pangasianodon hypophthalmus*), and Rainbow trout (*Oncorhynchus mykiss*) in hilly/cold water region, and Pacific white shrimp (*P. vannamei*) are ideal candidates for RAS-based systems. RAS produces fresh products, allowing better traceability (egg to fish) to make this sustainable farming system one of the best economic business models.

4.1.8 Low Input, Low-Cost Feed-Based Polyculture Farming

Polyculture is the farming of two or more compatible aquatic species to achieve higher productivity and is the dominant farming practice and a sustainable livelihood among Asian aquafarmers. Many of the traditional brackish water farming methods practiced in India and Southeast Asian countries, such as *bheries, ghers,* and Pokkali-fish farming, are based on polyculture principles. The ICAR-CIBA has taken the initiative to improve and intensify the polyculture farming methods using technological advancement, better managerial measures, and formulated farm-made feed. Standardized farming methods, such as mullet–shrimp farming (Biswas et al. 2012), mud crab–fish-based polyculture farming (Anand et al. 2018), low-cost polyculture feed (De et al. 2018), etc., have given encouraging results. The polyculture experiments conducted by ICAR-CIBA in indoor systems, on-farm and at farmers' ponds using indigenous brackish water fish and shrimp species, and low-cost, farm-made feed from locally available ingredients (crude protein—29.77 per cent) yielded productivity up to 4 tons per ha. The polyculture farm-made feed containing different unconventional ingredients (sunflower cake and mung husk) with 10 per cent fishmeal performed at par with feed containing 22 per cent fishmeal used in farmers' ponds (De et al. 2018). The economic analysis revealed a net profit of about USD 4693.5/ha with a cost-benefit ratio of 1.85. A production improvement of two to four-fold compared to conventional systems was observed, and this improved polyculture practice was disseminated to 57 farmers in West Bengal, India, covering 30 ha. To cater to the increased demand for farm-made feed, the farmers formed a cooperative society that established the necessary infrastructure with financial help. The low-input, low-cost farm-made feed-based polyculture can undergo science-based sustainable adaptations when shared with other Southeast Asian countries and even Pacific Island countries that share similar aquaculture practices using the local ingredients. Due to diversity, the polyculture system enhances productivity by minimizing risk. However, the effluent mitigating role is limited in this system compared to RAS or IMTA models.

4.1.9 Integrated Multitrophic Aquaculture (IMTA): Shrimp-Seaweed-Mollusc System

Integrated multitrophic aquaculture is a farming practice where byproducts from one species are recycled to form input for other species. In this farming model, nutrient loads generated from fed species, such as finfish or shrimp, act as fertilizers or inputs for the nonfed secondary species, such as molluscs or seaweeds, cocultured in the system (Chopin et al. 2012). Extractive organisms, such as molluscs, can filter organic particles and, thus, provide both an economic and a mitigation value. Extractive species used for farming are mussels, oysters, clams, sea urchins, or polychaetes, which feed on organic waste, whereas macroalgae, such as *Ulva*, *Gracilaria*, *Saccharina*, and *Laminaria,* extract the inorganic matter. The IMTA model is different from polyculture, where species combination with other trophic levels is used, and nutrient load is mitigated (Fig. 3).

Secondary species, such as seaweeds, are selected based on desirable characteristics, such as their high surface area to volume ratio, nutrient sequestration ability,

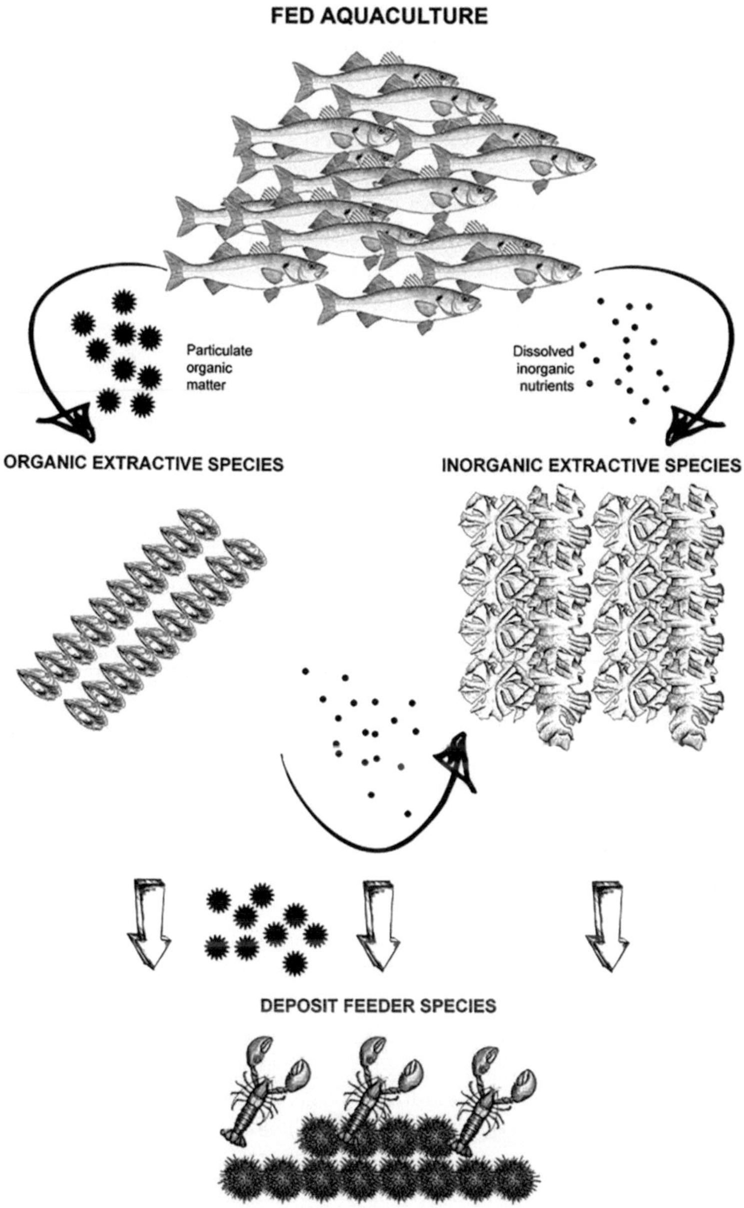

Fig. 3 Schematic diagram of IMTA model

ability to withstand environmental parameters, disease, epiphyte resistance, etc. As seaweed and its byproducts have broad applications for human consumption, therapeutic or pharmaceutical applications (Edwards 2015) and can be used as fertilizer, fodder, or even biofuel (FAO 2018), it increases the overall economic farm return. Additionally, seaweeds due to their antibacterial activity against pathogenic fish bacteria. Similarly, bivalve species must be selected to suit the local ecology, potential markets, etc. Commonly used bivalves for IMTA farming along the Indian coast are mussel, *Perna viridis,* and oyster *Crassostrea madrasensis.* It is reported that 95% of particles released from aquaculture systems range from 5 to 200 μm and generally settle at the bottom. Thus, the benefits of IMTA can be summarized as: (1) bio-mitigation of effluent from farming and (2) diversification, which reduce the overdependence of single species, resulting in creating risk and financial burden to the farmers. IMTA farming can enhance the domestic market chain and improve the livelihood options of small-scale farmers. Thus, the diversification of farming with complementary species can reduce intensive agriculture pressure and minimize price volatility, economic risk during market fluctuation, and crop losses due to disease outbreaks or inclement weather.

ICAR-Central Institute of Brackishwater Aquaculture (ICAR-CIBA) took the initiative to implement the IMTA system in the Sindhudurg District of Maharashtra, India one of the first initiatives in this direction on the West coast. Both land-based and open-water IMTA systems with shrimp and finfish as fed species and backwater oysters as non-fed species were attempted with impressive growth and productivity (Balasubramanian et al. 2018). Similarly, an integrated rearing system approach in farming mud crabs, shrimp, fish, oysters, and periphyton in a brackish water pond in Sundarbans gave promising results (Christina et al. 2019). It is reported that seaweeds can remove 35–100% of dissolved nitrogen produced by fed species when incorporated in a land-based system. The farming of bivalves (*C. madrasensis*) and finfish (*E. suratensis)* resulted in better productivity and low nutrient load when integrated cages were installed in backwaters. (Viji et al. 2014; Viji 2015). Similarly, integration of mangrove-based IMTA models in shrimp farming is reported to yield better productivity and nitrogen removal efficiency compared to conventional farming (Kumara et al. 2024). Intensive cage culture can be made sustainable through the model of open-sea IMTAs. It is noticed that the cage culture of *E. suratensis,* within the bivalve racks culture, resulted in high survival rates and growth of the finfish in the cage. Similarly, the culture of *Gracilaria* sp. with *P. indicus* in backwaters results in nutrient removal by seaweed. The estuarine oyster, *Crassostrea cuttackensis,* and green seaweed, *Enteromorpha intestinalis,* as the extractive species at varied densities enhanced the productivity and culture environment in brackish water integrated multitrophic aquaculture (Naskar et al. 2022, 2023). Similarly, de Morais et al. 2023 reported integration of seaweed (*Ulva ohnoi*) in biofloc-based shrimp and tilapia culture improved nitrogen dynamics in the system. It is also noticed that sea cucumbers (holothurians), detritus feeders, sea urchin *Strongylocentrotus droebachiensis*, and deposit feeders were found to be the ideal candidates for the IMTA system in Japan and many parts of Asia (Granada et al. 2016). Similarly, annelid polychaetes also form a complementary species in IMTA due to their biofiltration and sediment aeration ability (Brown et al. 2011). However, it is important to keep ecological carrying capacity standards to safeguard environmental issues. For example, bivalve culture needs to be optimized by considering ecological carrying capacity to prevent plankton depletion from water bodies (Chary et al. 2024)

4.1.10 Sustainable Marine Cage Farming

Integrated farming in the open sea needs advanced technological and sophisticated engineering investment. To avoid conflict regarding the usage of available water bodies, close collaboration and coordination among the various stakeholders are required (Kolda et al. 2020; Sievers et al. 2022). As intensive cage farming generates enormous nutrient loads that affect the native fauna, complementary species must be integrated with these farming systems to mitigate the eutrophication issue. Cage culture of salmon with raft culture of mussels or oysters was reported to increase productivity with control over nutrient discharge (Rosaa et al., 2020; Granada et al. 2016). In India, highly valued marine species, such as Asian sea bass and Pompano, are reported to be highly profitable in cage culture. Along the east coast of India, the introduction of IMTA in open sea *R. canadum* cage farming yielded 50% higher production of seaweed, *Kappaphycus alvarezii*; similarly, raft culture of green mussels, *P. viridis* resulted in a positive impact when integrated with open sea cage culture (Jeeva et al., 2022; Divu et L., 2023).

4.1.11 Intensive Bivalve Farming for Empowering Women

Many farmers successfully adopted intensive bivalve farming after the technology was standardized with lots of science-driven innovations, such as breeding, environmental manipulation, spatial calendar, etc. Furthermore, the value-addition in ready-to-cook or ready-to-eat products made marketing more manageable and opened new vistas in the domestic market. Environmental sustainability can be established through synergy development among researchers, financiers, promoters, and farmers. This technology has the potential to motivate groups for new technology adoption and sustainable intensification of the untapped coastal resources of other countries with similar ecology and habitats.

4.1.12 Integrated Trout Farming in High Altitude

Small farmers with seed production technology and the locally available ingredients for feed are successfully producing rainbow trout in Nepal's Himalayan region (168 tons/ha from 90 farmers spread across a few districts). The effluent of this intensive trout farming gives revenue of USD 6/sqm, as it is used for vegetable farming, thus making it an integrated farming unit. Furthermore, value-addition in terms of ecotourism and the development of sport fisheries in the region are added advantages. The system also provides livelihood to farmers with small land-holding, making it practicable as a family business or small- or large-scale business and can be replicated elsewhere in other countries. However, the environmental impact of such an economic activity in such an eco-sensitive mountainous area needs to be assessed.

5 Factors to Boost SIA Practices

Evolving eco-based farming techniques, such as organic aquaculture, biofloc/periphyton-based aquaculture, and biosecure zero water-based farming systems, can enhance shrimp productivity and double the farmer's income.

Aquatic production: Diversified crops, such as algal culture, cleaner fish integration, and fishes with better resilience to climate change, natural disaster tolerance, and salinity fluctuation adaptability, can be considered complementary species for SIA.

Approaches to measuring the impacts: The direct measures, such as fish catch, catch stability, growth performance, yield, and indirect measures, such as fish and feed diversity, conversion efficiency, pest prevalence, etc.

Strategy and action plan for SIA: SIA is identified as one of the most needed approaches to have sustained growth in aquaculture.

Establishment of good governance, effective planning and management, qualifying environment, improving management along the aquaculture value chain, strengthening supporting services such as R&D, capacity building, social responsibility, and equitable benefits and focus on increasing the resilience of aquaculture farmers are pertinent to enhancing farm productivity with better economic output.

- Well-developed biosecurity management or farm-level HACCP plan and overall health management system through regular health monitoring.
- Explore the potential and under-exploited resources for sustainable integration and diversification of aquaculture.
- *Quality seed development* is one of the critical components *determining* farm success.
- Exploration of novel ingredients and recycling of pond sludge as fertilizers and byproducts to support the circular economy.
- Assessment of carrying capacity and stocking based on the system capacity and ecological carrying capacity to safeguard environmental sustainability.
- Conserve biodiversity, habitat preservation, and soil regenerative practices with minimum nutrient emission.
- Sustainability in the shrimp sector can be achieved by adopting better management practices in broodstock management, establishing aquatic quarantine facilities, broodstock multiplication centers, etc.
- A diversified farming system can focus on the domestic market, and a well-developed domestic market chain will help farmers overcome market fluctuations.
- Technology dissemination among inland areas and rural areas where vast resources remain untapped can be achieved through the sustainable and scientific integration of intensive farming techniques.
- Encourage the farming community to practice climate-resilient practices, crop calendars, multi-cropping patterns, polyculture, BMP adoption, cluster-based farming, etc.

- Educate and train farming communities about government schemes, insurance credit, subsidy, geo-tagging, ecolabelling, or organic farming potential to promote aquaculture among the rural community.

6 The Way Forward

Sustainable integrated aquaculture has tremendous opportunity. The cocultivation of multiple species and the use of open water bodies need to be tackled with better registration, awareness about the farming system among the local fishermen community with better policies to attract entrepreneurs and various stakeholders, open water leasing policies, etc. The development of multitrophic aquaculture systems where appropriate species from different trophic levels are grown in combination (including fish, bivalves, algae, etc.) has received only limited attention; further research is required to create more synergistic relationships in resource use and recycling. In many cases, open water integrated farming licensing and regulation are often ad hoc and depend on local priorities and political demand. Above all, many industrial open water bodies are under the threat of industrial pollution. As bivalves are secondary products of this system, they can accumulate contaminants, such as metals and pharmaceuticals, apart from nutrients due to their filter-feeding ability. This can pose a risk; hence, human health and food safety issues must be addressed.

7 Conclusion

Diversification is the central focal point of sustainability. Diversified farming systems that are robust can help farmers tide over market price fluctuations and disease outbreaks. In the era where the growing human population across the globe with an increase in demand for quality protein sources and concern over public health and environmental issues suggest the need to intensify aquaculture with sustainability as a central point. The intensive farming system's sustainability needs to be achieved shoulder to shoulder with government bodies, various stakeholders, farming communities, legislators, and consumer demand. In this context, diversified and sustainable intensive farming methods, such as RAS, biofloc, polyculture, IMTA, mixotrophic model, location-specific BMPs, nursery integration, and two-phase shrimp farming models, can have a great way to achieve the sustainability goal without compromising the productivity from the farming system.

References

Anand PSS, Kumar S, Panigrahi A, Ghoshal TK, Syama Dayal J, Biswas G, Sundaray JK, De D, Ananda Raja R, Deo AD, Pillai SM, Ravichandran P (2013) Effects of C: N ratio and substrate integration on periphyton biomass, microbial dynamics and growth of *Penaeus monodon* juveniles. Aquac Int 21(2):511–524

Anand PSS, Kohli MPS, Kumar S, Sundaray JK, Roy SD, Venkateshwarlu G, Sinha A, Pailan GH (2014) Effect of dietary supplementation of biofloc on growth performance and digestive enzyme activities in *Penaeus monodon*. Aquaculture 418:108–115

Anand PSS, Balasubramanian CP, Christina L, Panigrahi A, Gopal C, Ghoshal TK, Vijayan KK (2018) Comparison of mud crab based brackishwater polyculture systems with different finfish species combinations in Sundarban, India. Aquacult Res 49(9):2965–2976

Anand PSS, Balasubramanian CP, Christina L, Kumar S, Biswas G, De D, Ghoshal TK, Vijayan KK (2019) Substrate based black tiger shrimp, Penaeus monodon culture: stocking density, aeration and their effect on growth performance, water quality and periphyton development. Aquaculture 507:411–418

Balasubramanian CP, Mhaska SS, Sukumaran K, Panigrahi A, Vasagam K, Kumararaja P, Thigale D, Sawant R, Vijayan KK, Vasudevan N (2018) Development of integrated multitrophic aquaculture (IMTA) for tropical brackishwater species in Sindhudurg District, Maharashtra, west coast of Indian. J Fish 65(1):59–64

Barrington K, Chopin T, Robinson S (2009) Integrated multi-trophic aquaculture (IMTA) in marine temperate waters. In: Soto D (ed) Integrated mariculture: a global review. FAO fisheries and aquaculture technical paper. No. 529. FAO, Rome, pp 7–46

Brown N, Eddy S, Plaud S (2011) Utilization of waste from a marine recirculating fish culture system as a feed source for the polychaete worm, *Nereis virens*. Aquaculture 322–323(C):177–183

Biswas G, Ananda Raja R, De D, Sundaray JK, Ghoshal TK, Anand S, Kumar S, Panigrahi A, Thirunavukkarasu AR, Ponniah AG (2012) Evaluation of productions and economic returns from two brackishwater polyculture systems in tide-fed ponds. J Appl Ichthyol, 28(1), pp.116–122

Cai J, Leung P, Luo Y, Yuan X, Yuan Y (2018) Improving the performance of tilapia farming under climate variation: perspective from bioeconomic modelling. FAO Fish Aquacult Tech Paper 608:I–64

Campanati C, Willer D, Schubert J, Aldridge DC (2022) Sustainable intensification of aquaculture through nutrient recycling and circular economies: more fish, less waste, blue growth. Rev Fish Sci Aquacult 30(2):143–169

Chary K, van Riel AJ, Muscat A, Wilfart A, Harchaoui S, Verdegem M, Filgueira R, Troell M, Henriksson PJ, de Boer IJ, Wiegertjes GF (2024) Transforming sustainable aquaculture by applying circularity principles. Rev Aquac 16(2):656–673

Chopin T, Cooper JA, Reid G, Cross S, Moore C (2012) Open-water integrated multitrophic aquaculture: environmental biomitigation and economic diversification of fed aquaculture by extractive aquaculture. Rev Aquac 4(4):209–220

Christina L, Balasubramanian CP, Shyne Anand PS, Ghoshal TK, Kumar P, Vijayan KK (2019) Integrated rearing system approach in the farming of mud crab, shrimp, fish, oyster and periphyton in brackishwater pond. Aquac Res 51:4165. https://doi.org/10.1111/are.14758

Cubillo AM, Ferreira JG, Robinson SMC, Pearce CM, Corner RA, Johansen J (2016) Role of deposit feeders in integrated multitrophic aquaculture — a model analysis. Aquaculture 453:54–66

Cook SJ, Allison EH, Beard TD, Arlinghaus R, Arthington AH, Bartley DM, Cowx IG, Fuentevilla C, Leonard NJ, Lorenzen K, Lynch AJ (2016) On the sustainability of inland fisheries: finding a future for the forgotten. Ambio 45: pp.753–764

Connor R (2015) The United Nations world water development report 2015: water for a sustainable world vol 1, UNESCO publishing

Dash P, Avunje S, Tandel RS, Sandeep KP, Panigrahi A (2017) Biocontrol of luminous Vibriosis in shrimp aquaculture: a review of current approaches and future perspectives. Rev Fish Sci Aquacult 25(3):245–255

de Morais APM, Santos IL, Carneiro RFS, Routledge EAB, Hayashi L, de Lorenzo MA, do Nascimento Vieira F (2023) Integrated multitrophic aquaculture system applied to shrimp, tilapia, and seaweed (Ulva ohnoi) using biofloc technology. Aquaculture 572:739492

De D, Ghoshal TK, Biswas G, Mukherjee S, Kumar S, Shyne Anand PS, Raja RA, Vijayan KK (2018) Evaluation of growth performance in *Mugil cephalus* juveniles fed diets incorporated with fermented plant feedstuffs replacing fishmeal or diets supplemented with fish gut bacteria. J Food Proc Technol 9(4):1–9

Divu D, Mojjada SK, Muktha M, Azeez PA, Tade MS, Subramanian A, Shree J, Chellappan A, Suresh Babu PP, Anuraj A, Rohit P (2023) Mapping of potential sea-cage farming sites through spatial modelling: preliminary operative suggestions to aid sustainable mariculture expansion in India. Indian J Fish 70(4):1–17

Garnett T, Mathewson S, Angelides P, Borthwick F (2015) Policies and actions to shift eating patterns: what works. Foresight, 515(7528), pp.518–522.

Dong SL, Dong YW, Cao L, Verreth J, Olsen Y, Liu WJ, Fang QZ, Zhou YG, Li L, Li JY, Mu YT (2022) Optimization of aquaculture sustainability through ecological intensification in China. Rev Aquac 14(3):1249–1259

Dongyu Q (2024) 2024 the state of world fisheries and aquaculture-blue transformation in action. The state of world fisheries and Aquaculture R1–232

Edwards P (2015) Aquaculture environment interactions: past, present, and likely future trends. Aquaculture 447(C):2–14

Eknath AE, Hulata G (2009) Use and exchange of genetic resources of Nile tilapia (*Oreochromis niloticus*). Rev Aquac 1(3–4):197–213

FAO (2018) The state of world fisheries and aquaculture 2018 – meeting the sustainable development goals. Rome. Licence: CC BY-NC-SA 3.0 IGO

FAO (2024) The state of world fisheries and aquaculture 2024. Sustainability in action. Food and Agriculture Organization of the United Nations

Ferreira JG, Sequeira A, Hawkins AJS, Newton A, Nickell TD, Pastres R et al (2009) Analysis of coastal and offshore aquaculture: application of the FARM model to multiple systems and shellfish species. Aquaculture 289(1–2):32–41

Granada L, Sousa N, Lopes S, Lemos MFL (2016) Is integrated multitrophic aquaculture the solution to the sectors' major challenges? – a review. Rev Aquac 8(3):283–300

Jeeva JC, Ghosh S, Raju SS, Megarajan S, Vipinkumar VP, Edward L, Narayanakumar R (2022) Success of cage farming of marine finfishes in doubling farmers' income: a techno-social impact analysis. Curr Sci 123(8):1031–1037

Jeney Z, Jian Z (2009) Use and exchange of aquatic resources relevant for food and aquaculture: common carp (Cyprinus carpio L.). Reviews in Aquaculture, 1(3-4), pp.163–173.

Karunasagar I, Otta SK, Karunasagar I (1998) Disease problems affecting cultured penaeid shrimp in India. Fish Pathol 33:413–419

Kesavan P, Swaminathan M (2009) The Royal Society. Science And Sustainable Food Security: Selected Papers Of MS Swaminathan, 3, p.346.

Kolda A, Gavrilović A, Jug-Dujaković J, Ljubešić Z, El-Matbouli M, Lillehaug A, Lončarević S, Perić L, Knežević D, Lušić DV, Kapetanović D (2020) Profiling of bacterial assemblages in the marine cage farm environment, with implications on fish, human and ecosystem health. Ecol Indic 118:106785

Kumara R, Syamala K, Shyne Anand PS, Chadha NK, Sawant PB, Chithira P, Muralidhar AP (2024) Halophyte and bivalve-based integrated multi-trophic aquaculture (IMTA): effect on growth, water quality, digestive and antioxidant enzymes of Penaeus monodon and Chanos chanos reared in brackishwater ponds. Aquac Int 32(2):1927–1953

Lazard J, Rey-Valette H, Aubin J, Mathé S, Chia E, Caruso D, Mikolasek O, Blancheton JP, Legendre M, René F, Levang P (2014) Assessing aquaculture sustainability: a comparative methodology. Int J Sustain Dev World Ecol 21(6):503–511

MPEDA (2023) Marine product export developemnt authority of India, Annual Report,392pp

Naskar S, Biswas G, Kumar P, De D, Sawant PB, Das S, Roy U (2022) Effects of estuarine oyster, Crassostrea cuttackensis as the extractive species at varied densities on productivity and culture environment in brackishwater integrated multi-trophic aquaculture (BIMTA) system. Aquaculture 554:738128

Naylor RL, Goldburg RJ, Primavera JH, Kautsky N, Beveridge MC, Clay J, Folke C, Lubchenco J, Mooney H, Troell M, (2000) Effect of aquaculture on world fish supplies. Nature, 405(6790), pp.1017-1024.

Naskar S, Biswas G, Kumar P, De D, Das S, Sawant PB, Chadha NK, Behera P (2023) The green seaweed, Enteromorpha intestinalis: An efficient inorganic extractive species for environmental remediation and improved performances of fed species in brackishwater integrated multitrophic aquaculture (BIMTA) system. Aquaculture 569:739359

Newton RW, Little DC. Mapping the impacts of farmed Scottish salmon from a life cycle perspective. Int J Life Cycle Assess. 2018 May;23:1018–29.

Otta SK, Praveena PE, Raj RA, Saravanan P, Priya MS, Amarnath CB, Bhuvaneswari T, Panigrahi A, Ravichandran P (2018) Pythium insidiosum as a new opportunistic fungal pathogen for Pacific white shrimp, Litopenaeus vannamei. Indian J GeoMarine Sci 47(5):1036–1041

Panigrahi A, Saranya C, Sundaram M, Vinoth Kannan SR, Das RR, Satish Kumar R, Rajesh P, Otta SK (2018) Carbon: nitrogen (C:N) ratio level variation influences microbial community of the system and growth as well as immunity of shrimp (*Litopenaeus vannamei*) in biofloc based culture system. Fish Shellfish Immunol 81:329–337

Panigrahi A, Sundaram M, Saranya C, Kumar RS, Dayal JS, Saraswathy R, Otta SK, Anand PS, Rekha PN, Gopal C (2019a) Influence of differential protein levels of feed on production performance and immune response of pacific white leg shrimp in a biofloc–based system. Aquaculture 503:118–127

Panigrahi A, Sundaram M, Saranya C, Swain S, Das RR, Dayal JS (2019b) Carbohydrate sources deferentially influence growth performances, microbial dynamics and immunomodulation in Pacific white shrimp (*Litopenaeus vannamei*) under biofloc system. Fish Shellfish Immunol 86:1207–1216

Panigrahi A, Das RR, Sivakumar MR, Saravanan A, Saranya C, Sudheer NS, Vasagam KK, Mahalakshmi P, Kannappan S, Gopikrishna G (2020a) Bio-augmentation of heterotrophic bacteria in biofloc system improves growth, survival, and immunity of Indian white shrimp *Penaeus indicus*. Fish Shellfish Immunol 98:477–487

Panigrahi A, Naveenkumar R, Das RR (2022) Immunoprophylactic measures in aquaculture. In Adv Fish Biotechnol (pp. 263-288).

Panigrahi A, Saranya C, Ambiganandam K, Sundaram M, Sivakumar MR, Vasagam KK (2020b) Evaluation of biofloc generation protocols to adopt high density nursery rearing of *Penaeus vannamei* for better growth performances, protective responses. Aquaculture 522:735095

Pelletier N, Tyedmers P, Sonesson U, Scholz A, Ziegler F, Flysjo A, Kruse S, Cancino B, Silverman H (2009) Not all salmon are created equal: life cycle assessment (LCA) of global salmon farming systems.

Pretty JN (1997) November. The sustainable intensification of agriculture. In Natural resources forum Vol 21, No. 4, pp. 247-256. Oxford, UK: Blackwell Publishing Ltd.

Rosaa J, Lemosc MFL, Crespo D, Nunesa M, Freitas A, Ramos F, Pardala MÂ, Leston S (2020) Integrated multitrophic aquaculture systems—potential risks for food safety. Trends Food Sci Technol 96:79–90

Ryther JH, Dunstan WM, Tenore KR, Huguenin JE (1972) Controlled eutrophication: increasing food production from the sea by recycling human wastes. Bioscience 22(3):144–152

Sievers M, Korsøen Ø, Warren-Myers F, Oppedal F, Macaulay G, Folkedal O, Dempster T (2022) Submerged cage aquaculture of marine fish: a review of the biological challenges and opportunities. Rev Aquac 14(1):106–119

Samocha TM (2019) Sustainable biofloc systems for marine shrimp. Academic press.

Sundaram M, Panigrahi A, Ganesh J, Rekha PN, Sivagnanam S, Rajamanickam S, Gopal C (2017) Evaluation of different probiotic strains for growth performance and immunomodulation in Pacific white shrimp Boone, 1931. Indian J Fish (Special Issue) 64:89–96

van Dam AA, Beveridge MC, Azim ME, Verdegem MC (2002) The potential of fish production based on periphyton. Reviews in fish biology and fisheries 12: pp.1–31.

Viji CS (2015) Studies on integrated multitrophic aquaculture in a tropical estuarine system in Kerala, India. Ph.D. Thesis. Central Institute of Fisheries Education, Mumbai, p 128

Viji CS, Chadha NK, Kripa V, Prema D, Prakash C, Sharma R, Jenni B, Mohamed KS (2014) Can oysters control eutrophication in an integrated fish-oyster aquaculture system? J Marine Biol Assoc India 56(2):67–73

Villar-Navarro E, Garrido-Perez C, Perales JA (2021) Recycling "waste" nutrients back into RAS and FTS marine aquaculture facilities from the perspective of the circular economy. Sci Total Environ 762:143057

Wei Y, Wei Q, An D (2020) Intelligent monitoring and control technologies of open sea cage culture: a review. Comput Electron Agric 169:105119

Wahab MA, Kibria MG (1994) Katha and kua fisheries-unusual fishing methods in Bangladesh. Aquaculture News 18(24): pp.175–188.

Welcomme RL (1972) An evaluation of the acadja method of fishing as practised in the coastal lagoons of Dahomey (West Africa). J Fish Biol, 4(1), pp.39–55.

Wikinson S, Ellis W, Weimin M (2014) Proceedings of the Regional Consultation on Sustainable Intensification of Aquaculture in Asia and the Pacific.

Aquatic Animal Health Policy and Regulations

Aquatic Animal Health Regulations and Policy in India

J. K. Jena, N. Sood, and P. K. Pradhan

Abstract

Aquatic animal health regulations and policy are important for ensuring the sustainable development of aquaculture and meeting food security. Regulatory bodies, namely, the Department of Fisheries, Ministry of Fisheries, Animal Husbandry and Dairying, and Marine Products Export Development Authority under the Ministry of Commerce and Industry and State Fisheries Departments, have a crucial role in developing and implementing policies related to surveillance of aquatic animal populations, quarantine measures, usage of aquaculture inputs, aquaculture biosecurity, health certification for imports and exports, and compliance with international trade regulations. In addition, the Indian Council of Agricultural Research under the Department of Agricultural Research and Education, the Government of India, and the Colleges of Fisheries in different states have been providing research support for issues concerning aquatic animal health. The existing regulations, including the Prevention and Control of Infectious and Contagious Diseases of Animal Act 2009, Livestock Importation Act 1898, Coastal Aquaculture Authority Act, 2005, etc., provide a framework for controlling the spread of diseases and regulating the importation and management of live animals and coastal aquaculture activities in a manner that protects aquatic animals. The ongoing National Surveillance Programme for Aquatic Animal Diseases has been informing the competent authority of the disease situation in the country, providing early warning of disease emergencies and

J. K. Jena (✉)
Indian Council of Agricultural Research, New Delhi, India
e-mail: jkjena2@rediffmail.com

N. Sood · P. K. Pradhan
ICAR-National Bureau of Fish Genetic Resources, Lucknow, India
e-mail: nsoodnbfgr@gmail.com; pradhanpk1@gmail.com

R. Kooloth Valappil et al. (eds.), *Aquatic Animal Health Management*,
https://doi.org/10.1007/978-981-96-7987-4_30

strengthening international confidence in the aquatic animal health status of the country. Coordinated efforts of all the regulatory bodies and research organisations are important for strengthening the implementation of existing regulations and policies to prevent the introduction and control the spread of diseases affecting aquatic animals, thereby facilitating the aquatic animal trade.

Keywords

Aquatic animal health · Control · Diseases · Importation · Policy · Prevention · Regulations · Reporting · Surveillance

1 Introduction

India takes pride in occupying the second position in aquaculture production globally, with the present production reaching almost two-thirds of the total fish production of 17.545 million metric tonnes in 2022–2023 (DoF, GoI 2023). Importantly, freshwater aquaculture contributes about 90% of aquaculture production. The Indian major carps (IMC), namely, catla (*Labeo catla*), rohu (*L. rohita*) and mrigal (*Cirrhinus mrigala*), contribute over two-thirds of freshwater aquaculture production. The other significant contributors to the total aquaculture production have been common carp (*Cyprinus carpio*), grass carp (*Ctenopharyngodon idella*), silver carp (*Hypophthalmichthys molitrix*) and striped catfish (*Pangasianodon hypophthalmus*) in freshwater and Pacific white shrimp (*Penaeus vannamei*) in brackishwater aquaculture. With the increased emphasis on species diversification, several indigenous species of regional importance are becoming popular among farmers.

The fisheries sector has been contributing to the overall economy of the country over the years, sharing 1.1% of the National Gross Value Added (GVA) and 6.72% of the Agricultural GVA (DoF, GoI 2023). It contributes about 15.23% of the country's agricultural exports, with a value of INR 639.69 billion in 2022–2023 (US$ 8.09 billion). The sector continues to be important for providing food, nutrition and livelihood security for millions. Although the annual growth rate of the fisheries sector in the country has been quite impressive (~8%) during the last three decades, the per capita consumption of fish remained less than 10 kg, in comparison to per capita consumption of 20.5 kg at the global level (FAO 2020). To meet the demand, India has set a target of 24 million tonnes fish production by 2030. With production from capture fisheries stagnating, the bulk of the increase has to be contributed by aquaculture, and therefore, intensification and diversification are being emphasised to meet the target. While the country envisages achieving the targeted production through increased focus on intensification, area expansion, diversification and adoption of new culture systems and practices, the increasing threat of new and existing diseases is a major concern.

2 Fish Diseases

Diseases are recognised as a significant threat to the success and sustainable growth of the aquaculture sector all around the globe, including India. In the country, finfish diseases are reported to be caused predominantly by parasites, bacteria and oomycetes. In the last few years, several viral diseases have been reported in the country. On the contrary, viral and microsporidian diseases have been responsible for huge production losses in shrimp culture. Besides, in molluscs, heavy mortalities have been attributed to a few protistan parasites. Some of the major diseases encountered in the aquaculture sector are discussed.

2.1 Finfish Diseases

The reports of diseases in aquaculture have been increasing over the years. Some of the major diseases affecting finfish in the country are discussed subsequently. The viral diseases affecting finfish include viral encephalopathy and retinopathy (VER), Goldfish haematopoietic necrosis disease (GHND), Koi sleepy disease (KSD), Tilapia lake virus disease (TiLVD), infection with ranavirus, Red sea bream iridoviral disease (RSIVD), infectious spleen and kidney necrosis disease (ISKND), infection with rhabdovirus, and Tilapia parvovirus disease (TiPVD). The causative agent of VER is a *Betanodavirus*, and the disease has been reported from marine as well as brackishwater fish species, including Asian seabass (Azad et al. 2005). The disease affects early life stages, especially larvae and juveniles, and the mortality rate can reach up to 100%. Cyprinid herpesvirus-2, the aetiological agent of GHND, results in large-scale mortalities in goldfish (Sahoo et al. 2016). The carp edema virus responsible for causing KSD can affect adult koi *Cyprinus carpio koi* (Swaminathan et al. 2016). There are reports of mass mortalities in *Cyprinus carpio koi* due to infection with ranavirus (George et al. 2015). TiLVD, caused by *Tilapinevirus tilapiae*, has been responsible for huge mortalities in tilapia (Behera et al. 2018) and is considered a threat to the tilapia industry. RSIVD has been associated with mortalities in seabass cultured in cages (Girisha et al. 2019). ISKND has been reported from different ornamental fish species and can result in mortalities (Girisha et al. 2021). Infection with rhabdovirus has been detected in snakehead suffering from mortality with surface ulcers and haemorrhages (John et al. 2021). TiPV has been detected in tilapia displaying characteristic clinical signs in association with TiLV and/or *Aeromonas* spp. and also apparently healthy tilapia (Rajendran et al. 2023). Importantly, till now, viral diseases have not been reported from IMC. It is to be noted that under the National Surveillance Programme on Aquatic Animal Diseases (NSPAAD), Infectious pancreatic necrosis virus (IPNV), Viral haemorrhagic septicemia virus (VHSV) and Spring viraemia of carp virus (SVCV) have not been detected in IMC samples collected from different parts of the country.

Bacterial diseases are known to result in huge mortalities in finfish. The motile aeromonas septicemia is one of the most prevalent diseases of fish. The disease is caused by *Aeromonas hydrophila*, *A. sobria*, *A. caviae* and other motile *Aeromonas* sp. Most of the freshwater fishes, including IMC, are affected by this disease, which is observed in all the age groups (Jena 2006; Rathore et al. 2006). Edwardsiellosis, caused by *Edwardsiella tarda*, is also reported in several freshwater fishes including IMC, magur, koi carps and murrels (Mohanty and Sahoo 2007). *Flavobacterium columnare*, the causative agent of columnaris disease, affects farmed fishes including IMC and goldfish (Verma and Rathore 2013; Verma et al. 2015). Epitheliocystis was diagnosed in farmed *P. hypophthalmus,* suffering from mortalities, and the causative agent was identified to be *Candidatus* Actinochlamydia pangasiae sp. nov. (Sood et al. 2018). In addition, infection with *Lactococcus garvieae* has been associated with haemorrhagic septicemia in rainbow trout (Shahi et al. 2018).

Among oomycete diseases, infection with *Aphanomyces invadans* is a serious disease of farmed and wild freshwater and brackishwater fishes. The disease is a major problem in North and Northeast India during the winter months (Pradhan et al. 2014). Saprolegniasis is another major disease caused by a water mould *Saprolegnia parasitica*. Most of the freshwater fishes, including carps, barbs, murrels, catfishes, tilapia, trouts, etc., are affected. Over the last few years, heavy mortalities due to saprolegniosis have been observed in pangas culture during the winter months in North India (Ravindra et al. 2022).

Parasitic diseases are a serious concern to the aquaculture sector and account for significant production losses. Argulosis is the most important parasitic disease in freshwater aquaculture, particularly cultured carp species (Sahoo et al. 2013). Infestation with *Lernaea* sp. is also a common problem in carps, including IMC, common carp, goldfish, etc. In addition, monogenetic trematodes, such as *Dactylogyrus* sp. and *Gyrodactylus* sp., can result in mortality, particularly in the early life stages of fish. Furthermore, some of the protozoan ciliates (*Ichthyophthirius* sp. and *Trichodina* sp.) and flagellates (*Ichthyobodo* sp. and *Cryptobia* sp.) are also common ectoparasites of fish.

2.2 Shellfish Diseases

Shellfish diseases are reported to cause huge losses in aquaculture. Infection with white spot syndrome virus (WSSV), also known as White spot disease, is reported to affect all the life cycle stages of *P. monodon* and *P. vannamei* and is the most devastating shrimp disease for the last three decades. The disease can result in 100% mortality (Rajendran et al. 1999). Besides, slow growth in shrimps has been attributed to infection with several viruses including infectious hypodermal and hematopoietic necrosis virus (IHHNV), Monodon baculovirus (MBV), Laem-Singh virus (LSNV) and hepatopancreatic parvovirus (HPV) (Moger et al. 2011; Rai et al. 2009; Kumar et al. 2011; Prakasha et al. 2007). Infectious myonecrosis is reported to cause muscle opaqueness and mortality in *P. vannamei* (Sahul Hameed et al. 2017; CIBA 2019). In addition, *Macrobrachium rosenbergii* nodavirus (MrNV), the causative agent of White tail disease, has been reported to result in heavy mortalities in the early life stages of freshwater prawns (Sahul Hameed et al. 2004). Recently, infection with Wenzhou shrimp virus 8 and *Macrobrachium rosenbergii* golda virus has been reported in whiteleg shrimp and freshwater prawns, respectively. Furthermore, several *Vibrio* species including *V. harveyi, V. parahaemolyticus, V. alginolyticus, V. vulnificus, V. fluvialis, V. anguillarum* and *V cholerae* have been associated with huge production losses and mortalities, particularly in shrimp hatcheries (Karunasagar et al. 2004; Kannapiran et al. 2009). In recent years, there have been several reports of retarded growth and huge production losses due to Hepatopancreatic microsporidiosis, which is caused by the microsporidian *Enterocytozoon hepatopenaei* (EHP) (Rajendran et al. 2016; Biju et al. 2016; Patil et al. 2021).

In molluscs, protistan parasites, particularly *Perkinsus olseni,* have been reported to infect several molluscan species, such as oysters, scallops, pearl oysters, clams and mussels, and result in huge losses (Sanil et al. 2010). Sanil et al. (2012) have also reported infection with *P. beihaiensis* in *Crassostrea madrasensis.*

3 Economic Losses Due to Diseases

There are a few studies regarding the estimation of economic losses in Indian aquaculture due to aquatic animal diseases. Previously, annual losses due to diseases in Andhra Pradesh, a leading aquaculture-producing state, were estimated to be about 10% of the production cost, amounting to INR 40 million (Gopal Rao et al. 1992). Importantly, in carp culture, the loss due to argulosis was reported to be INR 3.0 billion per annum (Sahoo et al. 2013). As per an estimate, the annual loss due to diseases in shrimp aquaculture in the country was estimated to be approximately INR 71.0 billion in the year 2018–19 (Patil et al. 2021), and most of this loss was due to infection with EHP and WSSV. EHP infection was found to account for a revenue loss of INR 39.77 billion, whereas the loss due to WSSV infection was estimated to be INR 16.70 billion.

4 Organisations Responsible for Aquatic Animal Health Management and Policy

The authorities/organisations associated with the planning and implementation of programmes concerning the management of aquatic animal health are the Department of Fisheries (DoF), Ministry of Fisheries, Animal Husbandry and Dairying, Government of India; Indian Council of Agricultural Research (ICAR); State Fisheries Departments (SFD); and Marine Products Export Development Authority (MPEDA). Organisations, such as the Coastal Aquaculture Authority (CAA) and National Fisheries Development Board (NFDB) under the DoF, are also involved in the operation of programmes on aquatic animal health.

4.1 Department of Fisheries

The DoF at the Central Government is headed by the Union Minister of Fisheries, Animal Husbandry and Dairying, who in turn is assisted by two Ministers of State at present. The administrative head of DoF is the Secretary, who is supported by the Joint Secretary (Marine Fisheries) and Joint Secretary (Inland Fisheries). The Joint Secretary (Inland Fisheries) is also the National Focal Point of the World Organisation for Animal Health (WOAH) for matters related to aquatic animal diseases. Besides, the Joint Secretary also looks after the administrative matters of the Department and is the Chairman of the National Committee to Oversee and Regulate the Introduction of Exotic Aquatic Species in Indian Waters. Both the Joint Secretaries are assisted by one Joint Commissioner and three Deputy Commissioners of Fisheries. The Joint Commissioner deals with matters related to states/union territories, deep-sea fisheries, the Fisheries and Aquaculture Infrastructure and Development Fund, the Kisan Credit Card, the Inland Fisheries Policy, etc. The Deputy Commissioners of Fisheries are responsible for matters related to fisheries trade, Coastal Aquaculture Authority, technical matters of State Fisheries Departments (SFDs) and institutes directly under the Department of Fisheries (DoF 2021).

4.2 State Fisheries Departments

There is a Fisheries Department in each state, with a Director/Commissioner as its head. The Department is responsible for the implementation of state and central schemes. Aquatic animal health is a state subject; therefore, the responsibility of investigating disease cases reported by the farmers and suggesting management measures lies with the SFD. Additionally, SFDs have the responsibility to submit reports about diseases encountered in their respective state on quarterly basis to the Department of Fisheries, Government of India, for compilation at the national level.

4.3 Indian Council of Agricultural Research

ICAR is an autonomous organisation under the Department of Agricultural Research and Education (DARE), Ministry of Agriculture and Farmers Welfare, Government of India. The Council is responsible for coordination, guidance and management of research and education in agriculture, including fisheries and animal sciences, in the country. The ICAR has eight subject matter divisions, including Fisheries Science. The research programmes in fisheries and aquaculture are implemented through 8 fisheries research institutes of the Division. Besides, Fisheries Universities and several Colleges of Fisheries under State Veterinary as well as Agricultural Universities offer Bachelor of Fisheries Science (B.F.Sc.), Master of Fisheries Science (M.F.Sc.) and doctoral (Ph.D.) programmes in different disciplines of fisheries science, including aquatic animal health (ICAR 2021).

4.4 Coastal Aquaculture Authority

The Coastal Aquaculture Authority (CAA) was established as per the Coastal Aquaculture Authority Act, 2005, to regulate the activities related to aquaculture in coastal areas and other aspects of coastal aquaculture. The CAA Act is intended to ensure that coastal aquaculture does not cause any adverse impacts on the coastal environment by formulating guidelines. The CAA has the following powers: (a) to regulate the construction and operation of aquaculture farms in the coastal areas; (b) to check coastal aquaculture farms for knowing their environmental impacts; (c) to register aquaculture farms in coastal areas; (d) to remove or demolish coastal aquaculture farms that cause pollution; and (e) to perform functions prescribed from time to time.

With regard to *P. vannamei* culture, CAA has prepared guidelines for regulating hatcheries for the production of seed and grow-out farms for its culture. As per the guidelines, all the hatcheries should be registered with CAA. Besides, the hatcheries must get approval for importing Specific Pathogen-Free (SPF) *P. vannamei* broodstock from suppliers shortlisted by CAA. Such hatcheries must adhere to the CAA guidelines regarding basic biosecurity measures, including intake and treatment of water, wastewater discharge, disinfection of fomites, seed production and sale, reporting disease incidence and maintenance of records. Furthermore, all the farmers who intend to culture *P. vannamei* must register with CAA, and they need to adhere to the guidelines for farming *P. vannamei*, which include basic biosecurity measures, such as bird fencing, water reservoirs, separate implements for each pond, protocols for discharge of wastewater and maintenance of farm records.

CAA has prescribed maximum permissible residue levels of antibiotics and pharmacologically active compounds, substances having anabolic effects, veterinary drugs, environmental contaminants, pesticides and heavy metals in fish and fishery products. In addition, it has developed standards for treating wastewater from hatcheries, farms, feed mills and processing plants. It has also prepared a list of banned antibiotics and pharmacologically active compounds in coastal aquaculture. The ban would also apply to substances notified by the Government of India from time to time (CAA 2021).

4.5 National Fisheries Development Board

NFDB, established in 2006, is an autonomous organisation under the Department of Fisheries, Government of India. The major objective is to increase fish production as well as productivity and comprehensively coordinate fishery development. Some of the important programmes of NFDB related to aquatic animal health include funding support for the Aquatic Quarantine Facility (AQF), Chennai, which is operated by MPEDA-RGCA to check the Specific Pathogen Free (SPF) status of *P. vannamei* imported broodstock. NFDB had also funded the first phase of the NSPAAD, which was implemented in 19 states and 3 Union Territories. NFDB has also been supporting the establishment of Aquatic Animal Health Laboratories in different states. Furthermore, it has also developed an Aquatic Animal Health and Quality Testing Laboratory at its headquarters for providing diagnostic services to fish farmers and for monitoring the residues of aquaculture drugs and environmental contaminants in fish and fishery products as well as in feed samples (NFDB 2021).

4.6 Marine Products Export Development Authority

The Marine Products Export Development Authority (MPEDA), established in 1972, operates under the Ministry of Commerce and Industry, Government of India. It serves to regulate marine product exports and implement strategies for ensuring sustainable exports of quality seafood from the country. Under the MPEDA, two societies, namely the Rajiv Gandhi Centre for Aquaculture (RGCA) and the National Centre for Sustainable Aquaculture (NaCSA), have been established to develop shrimp aquaculture sustainably and holistically. RGCA acts as the research and development arm of MPEDA for promoting exports of Indian seafood through the development of sustainable aquaculture technologies. The AQF for *P. vannamei* is operated by RGCA to quarantine the broodstock of *P. vannamei* imported to the country. This facility, located in Chennai, is a state-of-the-art facility for quarantine of the whiteleg shrimp *P. vannamei* imported to India. In addition, the Central Aquaculture Pathology Laboratory of RGCA, located at Sirkali, Tamil Nadu, provides disease diagnosis services to the farmers. It is also one of the collaborating centres of the NSPAAD.

NaCSA serves as to an outreach organisation to improve the livelihood generation for small-scale shrimp farmers. NaCSA helps to shrimp farmers form societies, educates them regarding better management practices (BMP) and also promotes a cluster approach in shrimp farming. Cluster refers to interdependent ponds that culture the same candidate species, have the same production system and share the same water sources and effluent discharge systems. The cluster-based approach to shrimp farming helps the farmers to coordinate various farm activities including farm operations, procurement of quality seed as well as stocking, water exchange and harvesting at the same time. Notably, in case of a disease outbreak, the farmers immediately bring it to the notice of the aqua society, which convenes an emergency meeting to decide the future actions.

In addition, MPEDA is establishing a national traceability practice for fish and fishery products whereby the enrolled farms (with GPS coordinates) can trace the products from the farm to the consumer to detect any unwanted/unauthorised compounds. The enrolled farmers need to maintain details of all the inputs, including the source of seed and feed, management measures and harvesting records. The traceability system would help to track the source of infectious agents and antibiotic residues in affected consignments (MPEDA 2021).

5 Existing Aquatic Animal Health Regulations in India

The country has regulations concerning aquatic animal health for preventing and controlling diseases encountered in aquaculture which are discussed below;

5.1 The Prevention and Control of Infectious and Contagious Diseases in Animals Act 2009

This Act serves to prevent, control and eradicate infectious and contagious diseases affecting animals. While it seeks to prevent disease outbreaks from spreading from one state to another state, it also helps to fulfil the country's international obligations regarding the trade in animals and animal products. Realising the importance of a national approach to prevent and control animal diseases, the Act stresses active participation of all the State Governments. As per the Act, it is mandatory to comply with the general obligations and recommendations of the World Organization for Animal Health (WOAH). Besides, the Act classifies all aquatic animal diseases listed in the WOAH Aquatic Animal Health Code as Scheduled Diseases.

The Act was amended by the Department of Animal Husbandry, Dairying, and Fisheries (DADF) through notification S.O. 995(E) on April 1, 2014, to expand the definition of 'animal' to include aquatic animals, such as fish, molluscs, crustaceans and amphibians. Furthermore, through notification S.O. 996(E) on April 1, 2014, the list of Scheduled Diseases was expanded to include diseases of national concern in addition to the WOAH-listed diseases.

This is to mention that the Act was prepared to consider terrestrial animals, as evidenced by some of the clauses like the appointment of veterinary officers, segregation of infected animals and detection of animals suffering from scheduled diseases at check posts and quarantine posts. Therefore, efforts are being made to amend the existing bill or introduce a new bill specifically for the prevention and control of diseases of aquatic animals.

5.2 Live-Stock Importation Act 1898

This Act aims to regulate the import of livestock that are susceptible to infectious or contagious diseases. As per the Act, the Central Government can issue notifications in the Official Gazette for regulating, restricting or prohibiting the import of specific livestock or related materials that can carry infectious agents. The 'livestock' includes horses, cattle, camels, sheep and any other species according to Section 2(b) of the Act. The definition of livestock has subsequently been expanded to include aquatic animals, such as live fish, crustaceans and molluscs vide notification S.O. 1495(E) dated June 10, 2014. In addition, the definition of 'livestock products' was expanded to include eggs and seeds of all aquatic animals through notification S.O. 2666(E) dated October 16, 2014.

As per the Act, excluding live shrimp, the import of livestock requires an import license issued by the Directorate General of Foreign Trade (DGFT). Besides, imports must comply with sanitary conditions outlined in the country's Veterinary Health Certificate, and the same should be issued by the official veterinarian of the exporting country. If required, the imported livestock has to undergo quarantine in accordance with the prescribed government regulations (India Code 2021).

5.3 Policy Documents on Aquatic Exotics and Quarantine

For regulating the import of aquatic animals, two key policy documents were prepared, which have been approved by the Ministry of Agriculture, Government of India for implementation. The National Strategic Plan for Aquatic Exotics and Quarantine provides a framework for regulating the introduction of exotic aquatic animals and translocation of aquatic species within the country, thereby mitigating the ecological and disease risks (Ponniah et al. 2002). Furthermore, the 'Aquatic Exotics and Quarantine Guidelines', a complementary document to the 'National Strategic Plan', provides the procedures for the evaluation of introduction proposals, management of exotics already present in India, implementation of quarantine and surveillance and reporting of aquatic animal diseases (Ponniah and Sood 2002).

5.4 Procedures and Requirements for Import of Aquatic Animals

The National Committee to Oversee and Regulate the Introduction of Exotic Aquatic Species in Indian Waters is responsible for evaluating proposals for the importation of exotic aquatic species. The Joint Secretary (Inland Fisheries), Department of Fisheries (DoF) chairs the committee, whereas Directors from ICAR Fisheries Research Institutes and experts are the members. The importer who wants to bring exotic aquatic species has to submit an application in the prescribed format to DoF, Government of India. After receipt of the proposal, the committee conducts ecological and disease risk assessments, and if the proposal is deemed to have no serious ecological or disease risks, then it is approved and subject to compliance with relevant guidelines.

For import, the consignment must be accompanied by a health certificate from the competent authority of the exporting country stating that the consignment is free from WOAH-listed diseases or other diseases of concern. On arrival, all the certificates are verified, and the consignment is cleared at the port of entry and transferred to an accredited quarantine facility. The shrimp consignments are quarantined at the Aquatic Quarantine Facility (AQF), RGCA, in Chennai, whereas the other aquatic animals are quarantined at the importer's farm. It is to be noted that currently, only hatcheries approved by the Coastal Aquaculture Authority (CAA) are permitted to import SPF broodstock of *P. vannamei* from the approved suppliers. The supplier has to submit a certificate confirming that the imported broodstock has undergone

12 days of pre-quarantine and tested negative for WSSV, YHV, IHHNV, IMNV, TSV, BP, NHPB, EHP and DIV1 within 10 days before shipment. After arrival at the quarantine facility, the consignment undergoes 5 days of quarantine, and only shrimps testing negative for the above pathogens are released to the importer. Concerning the quarantine of finfish, a committee constituted by the competent authority inspects the consignments at the importer's farm, which is designated as the quarantine facility and collects samples for screening the desired pathogens. Furthermore, the consignment is held at the importer's farm for a stipulated observation period, and after successful completion of the quarantine, the consignment is released. The importer must take care to prevent the accidental escape of the imported aquatic animals into natural waters and maintain detailed records of the imported consignment.

5.5 National Surveillance Programme for Aquatic Animal Diseases (NSPAAD)

The main driver for the NSPAAD is the significant threat that the endemic and emerging diseases pose to the growth of aquaculture due to intensification, diversification and increased trade in aquatic animals. It is important to note that early detection of emerging/exotic pathogens is critical for their control, and knowledge about endemic pathogens is essential for developing management strategies. Recognising the importance, the Department of Animal Husbandry, Dairying and Fisheries (DADF) under the Ministry of Agriculture funded a National Surveillance Programme for Aquatic Animal Diseases with funding support through NFDB, Hyderabad, in 2013. The programme was initiated in 14 states of aquaculture importance; thereafter, it was extended to cover 19 states and 3 Union Territories through a network of 31 collaborating centres; and the programme was coordinated by the ICAR-National Bureau of Fish Genetic Resources (NBFGR), Lucknow. Phase II of the programme is funded under Pradhan Mantri Matsya Sampada Yojana (PMMSY) of the Department of Fisheries, Government of India, and it is implemented with additional involvement of SFD and MPEDA. NSPAAD has been successful in strengthening passive surveillance through the organisation of awareness programmes for the stakeholders and training programmes for state fisheries officers and research scholars. Besides, it has enabled a robust country-wide network of aquatic animal disease diagnostic laboratories and the development of the diagnostic capability to detect WOAH/Network of Aquaculture Centres in Asia-Pacific (NACA)-listed and emerging aquatic animal pathogens. The success of the programme is evident from the first-time report of 13 pathogens in the country till 2024. Importantly, the programme has helped to dispel rumours about the emergence of Acute Hepatopancreatic Necrosis Disease (AHPND) in India, and the country now has the evidence to be considered free of AHPND.

Furthermore, under the programme, a mechanism has been developed for confirmation of exotic/emerging pathogens, as well as providing alerts/advisories following first-time reports of the pathogen(s). It is important to note that investigation of disease cases has been helping in suggesting management measures to the farmers, thereby reducing the production losses. The transparent reporting of diseases to international organisations has helped enhance the country's credibility. Importantly, a database similar to the National Animal Disease Referral Expert System has been developed under the programme. NSPAAD has been helping in improved understanding of the disease status of aquatic animals in the country, and it has reached a stage where it can accurately identify the presence/absence of diseases in the country. In addition, the programme has been helping to provide scientific input to the government regarding aquatic animal health issues and trade (Sood et al. 2021).

5.6 Disease Reporting Mechanism

Quarterly Aquatic Animal Disease (QAAD) reporting falls under India's reporting obligations to WOAH and NACA. The reporting is based on disease information data received from State Fishery Departments on the presence of WOAH/NACA-listed diseases in aquaculture species of the concerned state. DoF shares disease information for all the states with ICAR-NBFGR. This information, along with disease information from NSPAAD collaborating centres, is compiled by NSPAAD Coordinating Institute, i.e. ICAR-NBFGR and submitted to the DoF, and subsequently shared with NACA/WOAH.

6 Way Forward

Aquaculture is a potential sector for ensuring food and nutritional security and achieving the United Nation's sustainable development goals of poverty alleviation, zero hunger and good health. In India, while the demand for fish is steadily increasing, it is envisaged that the demand-supply gap can be met only through aquaculture production. The major hurdle to sustainable aquaculture development is the disease outbreaks. Therefore, more significant efforts are necessary to manage aquatic animal health to keep the pace of aquaculture growth. Considering the same, many countries, such as Australia and the United States, have developed and implemented National Aquatic Animal Health Plan to manage the risk of aquatic animal diseases. Hence, to protect aquatic animal health and to realise the full potential of aquatic animal production, the country must develop and implement necessary strategies, and for the same, some of the suggested action points are as follows:

- There is a need to develop a National Plan on Aquatic Animal Health. The primary goal of the plan should be to have strategies for controlling and managing aquatic animal diseases and preventing the incursion of exotic diseases in the country. The important activities should include import risk analysis, disease surveillance, reporting and diagnostic services; improving availability of approved aquaculture drugs; improving biosecurity, training and awareness;

developing emergency disease preparedness and contingency planning; and defining roles and responsibilities of the stakeholders.

- A National Advisory Committee comprising aquatic animal health experts, policymakers, aquaculture entrepreneurs and personnel from the aquaculture industry should be constituted to provide scientific advice and decide priorities on issues related to aquatic animal health.
- A National Centre on Aquatic Animal Health should be established with referral diagnostics, disease informatics and a repository for pathogens/infected tissues. The Centre should develop a list of aquatic animal diseases of national concern based on internationally accepted criteria in consultation with aquatic animal health experts in the country, and this list should be periodically revised. The Centre should keep a watch on emerging aquatic animal diseases globally and should be equipped to deal with such scenarios. Besides, referral laboratories need to be established in different regions. In addition, there should be a state-level referral laboratory in each state at State Agricultural Universities or SFD. There should be a strong network among all the disease diagnostic laboratories in the public and private sectors, and there is a need for accreditation of all these diagnostic laboratories.
- The National Surveillance Programme for Aquatic Animal Diseases should be institutionalised, and surveillance should be an integral activity of SFD similar to that of the Animal Husbandry Departments in different states. Their participation would strengthen the farmer-based reporting system, as this reporting is continuous and gives comprehensive coverage of the population. This would also widen the coverage of the ongoing surveillance programme and enable rapid detection of exotic and emerging diseases. Furthermore, for active disease surveillance, information regarding all the aquaculture farms and hatcheries, along with GPS coordinates, is essential. The latter would also help in establishing a traceability system. Moreover, the surveillance data should include the estimate of economic losses due to different diseases, which would be helpful for the prioritisation of diseases and undertaking control measures. Reporting of aquatic animal diseases of national concern should be mandatory.
- Generic contingency planning for exotic diseases should be prepared with needed actions, requirements and resources. The roles and responsibilities of all the involved organisations, including stakeholders, must be clearly defined, and in case of suspicion/confirmation of exotic/emerging diseases, the contingency plan should be implemented.
- Import risk analysis of live aquatic animals should be mandatory to prevent the incursion of exotic pathogens associated with the introduction of a particular aquatic animal.
- The existing 'Prevention and Control of Infectious and Contagious Diseases of Animal Act 2009' needs to be amended to address the issues concerning aquatic animal diseases. In addition, under the Live-Stock Importation Act, rules and guidelines should be incorporated with specific reference to aquatic animals, which would help in better implementation of the Act.

- A Fisheries and Aquaculture Council of India with State Councils should be established for empanelling aquatic animal health professionals to prescribe medicines to treat aquatic animal diseases. Furthermore, a list of approved drugs along with recommended dose, duration, route of application and withdrawal period should be prepared.

7 Conclusion

Aquatic animal health management is crucial due to the substantial economic losses caused by fish diseases. To reduce the same, the organisations responsible for aquatic animal health management and policy making must work very closely. Additionally, the regulatory framework for implementing the existing acts needs to be strengthened. There is also a pressing need to develop a National Plan for Aquatic Animal Health, establish a National Advisory Committee and create a National Centre on Aquatic Animal Health to manage and control aquatic animal diseases in the country. These measures will help to achieve sustainable growth of aquaculture, thereby contributing significantly to food and nutritional security.

References

Azad IS, Shekar MS, Thirunavukkarasu AR, Poornima M, Kailasan M, Rajan JJS, Ali SA, Abraham M, Ravichandran P (2005) Nodavirus infection causes mortalities in hatchery produced larvae of *Lates calcarifer*: first report from India. Dis Aquat Org 63:113–118

Behera BK, Pradhan PK, Swaminathan TR, Sood N, Paria P, Das A, Verma DK, Yadav MK, Dev AK, Parida PK, Das BK, Jena JK (2018) Emergence of Tilapia Lake Virus associated with mortalities of farmed Nile Tilapia *Oreochromis niloticus* (Linnaeus 1758) in India. Aquaculture 484:168–174

Biju N, Sathiyaraj G, Raj M, Shanmugam V, Baskaran B, Govindan U, Kumaresan G, Kasthuriraju KK, Chellamma TS (2016) High prevalence of *Enterocytozoon hepatopenaei* in shrimps *Penaeus monodon* and *Litopenaeus vannamei* sampled from slow growth ponds in India. Dis Aquat Org 120(3):225–230

CAA (2021) Coastal aquaculture authority. Available at http://caa.gov.in/. Accessed 16 June 2021

CIBA (2019) Annual report. ICAR-Central Institute of Brackishwater Aquaculture, Chennai. 240 pp

Department of Animal Husbandry and Dairying (2021) The Prevention and Control of Infectious and Contagious Diseases in Animals Act 2009. Available at: http://www.dahd.nic.in/sites/default/filess/Gazette_20-03-09.pdf. Accessed 24 June 2021

DoF (2021) Department of Fisheries. Available at: http://dof.gov.in/sites/default/files/2021-01/Organizational_Structure.pdf. Accessed 6 June 2021

DoF, GoI (2023) Handbook on fisheries statistics 2023. Fisheries Statistics Division, Department of Fisheries, Ministry of Fisheries, Animal Husbandry & Dairying, Govt. of India, New Delhi. 291 pp

FAO (2020) FishStatJ – software for fishery and aquaculture statistical time series v4.01.4 Available at: http://www.fao.org/fishery/statistics/software/fishstatj/en. Accessed 5 June 2021

George MR, John KR, Mansoor MM, Saravanakumar R, Sundar P, Pradeep V (2015) Isolation and characterization of a ranavirus from koi, *Cyprinus carpio* L., experiencing mass mortalities in India. J Fish Dis 38(4):389–403

Girisha SK, Puneeth TG, Nithin MS, Kumar BTN, Ajay SK, Vinay TN, Suresh T, Venugopal MN, Ramesh KS (2019) Red sea bream iridovirus disease (RSIVD) outbreak in Asian seabass (*Lates calcarifer*) cultured in open estuarine cages along the west coast of India: first report. Aquaculture 520:734712

Girisha SK, Kushala KB, Nithin MS, Puneeth TG, Naveen Kumar BT, Vinay TN, Suresh T, Ajay SK, Venugopal MN, Ramesh KS (2021) First report of the infectious spleen and kidney necrosis virus (ISKNV) infection in ornamental fishes in India. Transbound Emerg Dis 68(2):964–972

Gopal Rao K, Mohan CV, Seenappa D (1992) The use of chemotherapeutic agents in fish culture in India. In: Shariff M, Subasinghe RP, Arthur JR (eds) Diseases in Asian aquaculture I. Fish Health Section, Asian Fisheries Society, Manila, pp 505–517

ICAR (2021) Indian Council of Agricultural Research. Available at: https://icar.org.in/content/fisheries_division. Accessed 16 June 2021

India Code (2021) The Livestock Importation Act, 1898. https://www.indiacode.nic.in/handle/123456789/1362//simple-search?page-token=6229d5d31f6b&page-token-value=4e3d4ffc7c8fed0cecd78dacec07c835&nccharset=31AAAE8B&query=Livestock+Importation+Act. Accessed 16 June 2021

Jena JK (2006) Cultured aquatic species information programme. *Labeo rohita*. Cultured aquatic species information programme. FAO Fisheries and Aquaculture Department [online], Rome. Updated 21 February 2006

John KR, George MR, Mageshkumar P, Mansoor MM, Kaviarasu D, Sivasankar P, Selvamagheswaran M, Petchimuthu M (2021) Isolation and characterisation of a novel rhabdovirus from snakehead fish showing surface ulcerations and mortality from India: first report. Aquaculture 544:737120

Kannapiran E, Ravindran J, Chandrasekar R, Kalaiarasi A (2009) Studies on luminous, *Vibrio harveyi* associated with shrimp culture system rearing *Penaeus monodon*. J Environ Biol 30:791–795

Karunasagar I, Karunasagar I, Umesha RK (2004) Microbial diseases in shrimp aquaculture. In: Ramaiah N (ed) Marine microbiology: facets & opportunities. National Institute of Oceanography, Goa, pp 121–134

Kumar TS, Krishnan P, Makesh M, Chaudhari A, Purushothaman CS, Rajendran KV (2011) Natural host-range and experimental transmission of Laem-Singh virus (LSNV). Dis Aquat Org 96:21–27

Moger R, Mohan SA, Venugopal MN, Karunasagar I (2011) Presence of multiple infections in newly introduced White leg shrimp, *Penaeus vannamei* and assessment of their risk to aquaculture in India. In: Book of abstracts, 8th symposium on diseases in Asian Aquaculture, 21–25 November 2011. College of Fisheries, Mangalore, p 166

Mohanty BR, Sahoo PK (2007) Edwardsiellosis in fish: a brief review. J Biosci 32:1331–1344

MPEDA (2021) State-wise details of shrimp and scampi production. Marine products export development authority. Ministry of Commerce and Industry, Government of India. Available at https://mpeda.gov.in/?page_id=651. Accessed 24 June 2021

NFDB (2021) National Fisheries Development Board. Available at https://nfdb.gov.in/. Accessed 16 June 2021

Patil PK, Geetha R, Ravisankar T, Avunje S, Solanki HG, Abraham TJ, Vinoth SP, Jithendran KP, Alavandi SV, Vijayan KK (2021) Economic loss due to diseases in Indian shrimp farming with special reference to *Enterocytozoon hepatopenaei* (EHP) and white spot syndrome virus (WSSV). Aquaculture 533:736231

Ponniah AG, Sood N (2002) Aquatic exotics and quarantine guidelines. NBFGR Special Publication No. 4

Ponniah AG, Unnithan VK, Sood N (2002) National strategic plan for aquatic exotics and quarantine. NBFGR Special Publication No. 3

Pradhan PK, Rathore G, Sood N, Verma DK, Chaudhary DK, Abidi R, Punia P, Jena JK (2014) Emergence of epizootic ulcerative syndrome: large-scale mortalities of cultured and wild fish species in Uttar Pradesh, India. Curr Sci 106:1711–1718

Prakasha BK, Ramakrishna RP, Karunasagar I, Karunasagar I (2007) Detection of Laem-Singh virus (LSNV) in cultured *Penaeus monodon* from India. Dis Aquat Org 77:83–86

Rai P, Pradeep B, Karunasagar I, Karunasagar I (2009) Detection of viruses in *Penaeus monodon* from India showing signs of slow growth syndrome. Aquaculture 289:231–235

Rajendran KV, Vijayan KK, Santiago TC, Krol RM (1999) Experimental host range and histopathology of white spot syndrome virus (WSSV) in shrimps, prawns, crabs and lobsters from India with a note on the carrier/reservoir hosts. J Fish Dis 22:183–191

Rajendran KV, Shivam S, Praveena PE, Rajan JJS, Kumar TS, Avunje S, Alavandi SV (2016) Emergence of *Enterocytozoon hepatopenaei* (EHP) in farmed *Penaeus (Litopenaeus) vannamei* in India. Aquaculture 454:272–280

Rajendran KV, Sood N, Rao BM, Valsalam A, Bedekar MK, Jeena K, Pradhan PK, Paria A, Swaminathan TR, Verma DK, Sood NK (2023) Widespread occurrence of Tilapia parvovirus in farmed Nile tilapia *Oreochromis niloticus* from India. J Fish Dis. https://doi.org/10.1111/jfd.13871

Rathore G, Singh V, Kumar G, Swaminathan TR, Mahanta PC (2006) Antibiotic sensitivity and characterization of *Aeromonas hydrophila* isolated from fish and water samples. J Ecophysiol Occup Health 6:41–43

Ravindra, Ahmad N, Verma DK, Kantharajan G, Kumar CB, Paria A, Swaminathan TR, Rathore G, Sood N, Pradhan PK, Lal KK, Jena JK (2022) Mortalities in cultured *Pangasianodon hypophthalmus* due to oomycete *Saprolegnia parasitica* infection in Uttar Pradesh, India. Aquacult Rep 23:101047

Sahoo PK, Mohanty J, Garnayak SK, Mohanty BR, Kar B, Prasanth H, Jena JK (2013) Estimation of loss due to argulosis in carp culture ponds in India. Indian J Fish 60(2):99–102

Sahoo PK, Swaminathan TR, Abraham TJ, Kumar R, Pattanayak S, Mohapatra A, Rath SS, Patra A, Adikesavalu H, Sood N, Pradhan PK, Das BK, Jayasankar P, Jena JK (2016) Detection of goldfish haematopoietic necrosis herpes virus (Cyprinid herpesvirus-2) with multi-drug resistant *Aeromonas hydrophila* infection in goldfish: first evidence of any viral disease outbreak in ornamental freshwater aquaculture farms in India. Acta Trop 161:8–17

Sahul Hameed AS, Yoganandhana K, Sri Widada J, Bonami JR (2004) Studies on the occurrence of *Macrobrachium rosenbergii* nodavirus and extra small virus-like particles associated with white tail disease of *M. rosenbergii* in India by RT-PCR detection. Aquaculture 238:127–133

Sahul Hameed AS, Abdul Majeed S, Vimal S, Madan N, Rajkumar T, Santhosh Kumar S, Sivakumar S (2017) Studies on the occurrence of infectious myonecrosis virus in pond-reared *Litopenaeus vannamei* (Boone, 1931) in India. J Fish Dis 40:1823–1830

Sanil NK, Vijayan KK, Kripa V, Mohamed KS (2010) Occurrence of the protozoan parasite, *Perkinsus olseni* in the wild and farmed Pearl Oyster, *Pinctada fucata* (Gould) from the Southeast coast of India. Aquaculture 299:8–14

Sanil NK, Suja G, Lijo J, Vijayan KK (2012) First report of *Perkinsus beihaiensis* in *Crassostrea madrasensis* from the Indian subcontinent. Dis Aquat Org 98:209–220

Shahi N, Mallik SK, Sahoo M, Chandra S, Singh AK (2018) First report on characterization and pathogenicity study of emerging *Lactococcus garvieae* infection in farmed rainbow trout, *Oncorhynchus mykiss* (Walbaum), from India. Transbound Emerg Dis 65:1039–1048

Sood N, Pradhan PK, Verma DK, Yadav MK, Ravindra D, Swaminathan TR, Sood NK (2018) *Candidatus* Actinochlamydia pangasiae sp. nov. (Chlamydiales, Actinochlamydiaceae), a bacterium associated with epitheliocystis in *Pangasianodon hypophthalmus*. J Fish Dis 41(2):281–290

Sood N, Pradhan PK, Swaminathan TR, Rathore G, Jena JK, Lal KK (2021) National surveillance programme for aquatic animal diseases – a stepping stone for establishing disease governance system in India. Curr Sci 120(2):273–277

Swaminathan TR, Kumar R, Dharmaratnam A, Basheer VS, Sood N, Pradhan PK, Sanil NK, Vijayagopal P, Jena JK (2016) Emergence of carp edema virus in cultured ornamental koi carp, *Cyprinus carpio koi*, in India. J Gen Virol 97:3392–3399

Verma DK, Rathore G (2013) Molecular characterization of *Flavobacterium columnare* isolated from a natural outbreak of columnaris disease in farmed fish, *Catla catla* from India. J Gen Appl Microbiol 59:417–424

Verma DK, Rathore G, Pradhan PK, Sood N, Punia P (2015) Isolation and characterization of *Flavobacterium columnare* from freshwater ornamental goldfish *Carassius auratus*. J Environ Biol 36:433–439

National Surveillance Programme for Aquatic Animal Diseases (NSPAAD): Enhancing Aquatic Animal Health Preparedness in India

Kuldeep K. Lal, Neeraj Sood, T. Raja Swaminathan, Dev K. Verma, P. K. Pradhan, and J. K. Jena

Abstract

Fish production in India has shown an impressive average annual growth of about 8% during the last four decades. The country is the third-largest fish-producing country and one of the largest exporters of shrimps. However, a recent study indicates huge economic losses due to shrimp diseases. This highlights the significant threat of diseases to the growth of aquaculture. Early detection is considered crucial for controlling the spread of diseases, and this is outlined in 'The Prevention and Control of Infectious and Contagious Diseases in Animals Act, 2009'. To fulfil the spirit of the Act, the Government of India initiated a National Surveillance Programme for Aquatic Animal Diseases (NSPAAD) in 2013 with the objectives of improving the collection of information on aquatic animal diseases, rapidly detecting new and exotic diseases, and meeting interna-

K. K. Lal (✉)
ICAR-Central Institute of Brackishwater Aquaculture, Chennai, India

ICAR-National Bureau of Fish Genetic Resources, Lucknow, India
e-mail: kuldeepklal@gmail.com

N. Sood · D. K. Verma · P. K. Pradhan
ICAR-National Bureau of Fish Genetic Resources, Lucknow, India
e-mail: nsoodnbfgr@gmail.com; pradhanpk1@gmail.com

T. Raja Swaminathan
ICAR-National Bureau of Fish Genetic Resources, Lucknow, India

Peninsular Aquatic Genetic Resources Centre, ICAR-National Bureau of Fish Genetic Resources, Kochi, India
e-mail: rajathanga@yahoo.co.in

J. K. Jena
Indian Council of Agricultural Research, New Delhi, India
e-mail: jkjena2@rediffmail.com

R. Kooloth Valappil et al. (eds.), *Aquatic Animal Health Management*,
https://doi.org/10.1007/978-981-96-7987-4_31

tional disease reporting obligations. Currently, the programme is being implemented through 32 Collaborating Centres, State Fisheries Departments and the Marine Products Export Development Authority (MPEDA), and is coordinated by ICAR-National Bureau of Fish Genetic Resources, Lucknow. The major emphasis of NSPAAD has been to strengthen the passive surveillance system and undertake need-based active surveillance. The strengthening of passive disease surveillance is evident in the first time report of thirteen new diseases from the country. The active surveillance carried out under programme allows us to state that three major pathogens of finfish, i.e. spring viremia of carp virus, infectious pancreatic necrosis virus and viral haemorrhagic septicaemia virus, and three shrimp pathogens, namely, yellow head virus genotype 1, Taura syndrome virus and *Vibrio parahaemolyticus* AHPND strain in shrimp, have not been detected in India and provides transparent scientific evidence for this. Notably, NSPAAD has made a significant impact in terms of developing a network of aquatic animal health laboratories, enhancing diagnostic capability for aquatic animal pathogens, issuing alerts to stakeholders following the detection of new diseases and providing scientific advice to farmers. Currently, the Government of India is strengthening the disease surveillance programme, which would have a major impact on reducing disease losses and assist in the sustainable growth of the aquaculture sector in the country.

Keywords

Aquatic animal diseases · Surveillance · Reporting · ReportFishDisease App

1 Introduction

Fish is considered to be one of the cheapest sources of quality protein, essential fatty acids and minerals. India ranks third in fish and aquaculture production globally, with an estimated production of about 17.54 million tonnes in 2022–23. The total exports from this sector were approximately INR 63969.14 crores (DoF, GoI 2023). With the production from capture fisheries stagnating, an increase in aquaculture production is considered to play an important role in ensuring food and nutritional security in the coming years. The growth of the aquaculture sector mainly relies on expansion, intensification and diversification of farming practices. Intensification generally leads to the emergence of pathogens, whereas global trade in live aquatic animals results in the spread of pathogens to newer places. Notably, diseases are the most important constraint to the growth of aquaculture, resulting in huge economic losses. There are a few studies regarding the estimation of economic losses due to diseases in Indian aquaculture. The annual losses due to shrimp diseases in the country in 2018–19 were estimated to be about INR 7100 crores (Patil et al. 2021). Infection with *Enterocytozoon hepatopenaei* (EHP) and white spot syndrome virus (WSSV) were found to cause losses to the tune of INR 3977 crores and INR 1670 crores, respectively. Previously, economic losses due to argulosis, an ectoparasitic disease, were reported to be about INR 30,000/– per ha per year in carp culture

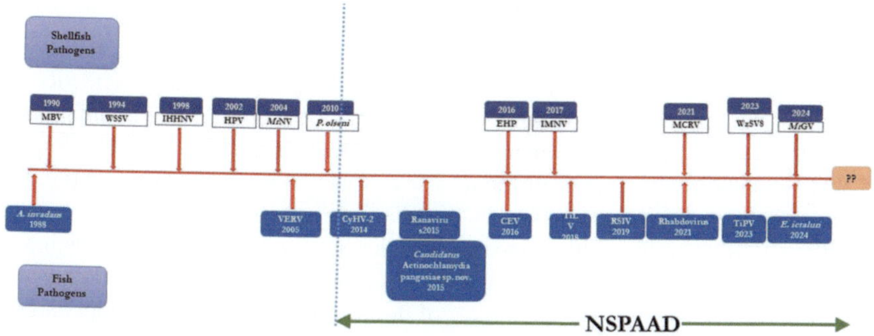

Fig. 1 The timeline of reporting of aquatic animal pathogens for the first time from India. *A. invadans*—*Aphanomyces invadans*, MBV—Monodon baculovirus, WSSV—White spot syndrome virus, IHHNV—infectious hypodermal and hematopoietic necrosis virus, HPV—hepatopancreatic parvovirus, *Mr*NV—*Macrobrachium rosenbergii* nodavirus, VERV—viral encephalopathy and retinopathy virus, *P. olseni*—*Perkinsus olseni*, CyHV-2—Cyprinid herpesvirus-2, EHP—*Enterocytozoon hepatopenaei*, CEV—Carp edema virus, IMNV—infectious myonecrosis virus, TiLV—Tilapia lake virus, RSIV—Red sea bream iridovirus, Rhabdovirus, MCRV—Mud crab reovirus, TiPV—Tilapia parvovirus, WzSV8—Wenzhou shrimp virus 8, *Mr*GV- *Macrobrachium rosenbergii* Golda Virus, BNP- Bacillary necrosis of pangasius

ponds (Sahoo et al. 2013). Besides, a number of diseases also have been responsible for economic losses to the sector during the last few decades, and the timeline of reporting some of the important pathogens is depicted in Fig. 1.

Early detection of disease outbreaks through active vigilance and rapid identification of the aetiological agent is important for early warning of diseases, planning and monitoring of disease control programmes, provisioning sound aquatic animal health advice to farmers and meeting disease reporting obligations to international organisations. This can be achieved through structured aquatic animal disease surveillance. Some of the basic concepts of disease surveillance are discussed below.

Surveillance is the systematic, ongoing collection, collation and analysis of information related to animal health and the timely dissemination of information to farmers, disease control agencies or other authorities. It is necessary to understand the health status of the animals in the country, so that problems can be identified and actions can be taken. Surveillance is a cost-effective approach and it reduces economic losses by allowing rapid response to disease outbreaks.

2 Purposes of Disease Surveillance

The major purposes for undertaking surveillance include measuring the level of disease, finding cases of disease, demonstrating freedom from disease and early detection of new and emerging diseases (Cameron 2012). The first two purposes are for surveillance of diseases that are present in the country, whereas the latter two purposes are for surveillance of diseases that are absent in the country. The details about the purposes of disease surveillance are given below.

2.1 Measuring Level of Disease

The most common measures of level of disease are prevalence and incidence. A single measure provides an indication of the level of the disease in a population at one point in time. When surveillance is combined with data on the economic impact of disease, single measures of different diseases can be used to determine the diseases with the greatest impact, thereby prioritizing diseases for undertaking control activities. Furthermore, multiple measures of disease for comparison provide information regarding the spatial and temporal distribution of disease. Spatial distribution of diseases, i.e. measuring the level of disease in different geographical areas, helps in knowing the risk factors that influence the disease. The temporal distribution of diseases provides information about the efficacy of control programs and early detection of changes in the level of disease.

2.2 Finding Cases of Disease

In some situations, surveillance is undertaken to detect individual cases of disease or individual outbreaks of disease. This is the most common with surveillance as part of the disease control programme. For example, in the early stages of disease control programs, stocking with specific pathogen-free seed in the farms is used to reduce the level of disease, and in later stages, efforts may be made to find infected farms and take appropriate management measures.

2.3 Early Detection of Disease Incursions

The ability to detect incursions of disease is a basic surveillance requirement of each country. The ability to detect new disease incursions as quickly as possible is important for effective control. Early detection is crucial as the cost and effectiveness of eradication or control of a new disease outbreak are directly related to the delay in detection.

2.4 Demonstrating Freedom from Disease

When disease is not present in a country, there may be several benefits, including the ability to export animals. However, in order to avail these benefits, the national authorities need to be confident that the disease is truly absent. Besides, freedom from disease can be used to stop disease control measures. Furthermore, if the importing country has demonstrated that it is free from infection, it can prevent animal import from an infected country. Importantly, finding a single infected animal is sufficient to prove that the disease is present in the country. However, demonstrating freedom from disease is a probabilistic approach, i.e. how likely the country is free from infection.

It is to be noted that the priority of surveillance should always be given to national-level information needs rather than international considerations. The disease information from surveillance should help the national authorities to make appropriate, well-informed decisions about disease control. On the contrary, international organizations, such as the World Organisation for Animal Health (WOAH), share disease information between member countries in order to help them make appropriate decisions to limit the risk of disease spread. The minimum surveillance capabilities of each country should be to describe the important diseases that are present, and detect the occurrence of new, emerging or exotic diseases. Importantly, the trading partners are likely to be very cautious about importing from a country with an unknown disease status.

3 Types of Surveillance System

Surveillance systems are often divided into two primary categories: passive and active surveillance.

3.1 Passive Surveillance

Where farmers make the primary observation, it is passive surveillance. They may ask aquatic animal health experts to visit to examine the animals on the farm and provide advice for treatment, but the first person to collect information about the disease is the farmer. The primary reason for generating information by the farmer is to get assistance to solve a disease problem, and surveillance represents the secondary use of the information. Farmer-based reporting systems are the most common and probably the most important form of surveillance in any country. These systems are often the means by which new diseases are first discovered because there is high population coverage, and general surveillance is capable of detecting any clinical disease. The passive disease reporting system is:

- Continuous as farmers are in contact with their animals on a daily basis. In case of any disease occurrence, they will recognize it quickly.
- Comprehensive as all the farmed animals in the country are covered by this surveillance system.
- Relatively inexpensive.

However, passive surveillance is not considered good for diseases that show no clinical signs or reliably measuring the level of disease, or confidently prioritizing diseases for undertaking control measures. A number of other approaches to surveillance have been developed, based on farmer observations, to overcome some of the weaknesses of farmer passive reporting. These include:

- *Syndromic surveillance*: In this surveillance, farmers' observations are recorded and analysed to detect abnormal patterns, which can trigger an investigation. The primary observer is again the farmer, but this approach can help to optimize the use of resources by launching a detailed investigation only when there is evidence that something unusual is happening.
- *mSurveillance*: The Internet is a valuable tool for rapid and early reporting of disease events. Over the last few years, mobile phone coverage has expanded rapidly, and a very high proportion of the population has access to mobile internet. Therefore, fish disease surveillance through the use of mobile phones can help in rapid communication, which is essential for effective surveillance.
- *Indirect surveillance*: This approach is based on the use of existing data that can act as an indirect indicator of disease, such as records of the sales of antibiotics. Similarly, the use of feed additives, such as probiotics, may increase when farmers notice a disease problem.

3.2 Active Surveillance

When an aquatic animal health expert makes the first observation, it means that the expert is examining the animal without first being informed about the disease problem by the farmer. The purpose of the examination is surveillance, and this represents active surveillance. This normally takes the form of a structured survey and is designed to collect specific information to answer a specific question. It involves setting up a sampling plan, calculating the appropriate sample size and analysing the data to ensure that there is no bias. This surveillance overcomes many of the weaknesses of passive surveillance systems as it can detect subclinical disease, collect high-quality, focused information in response to a specific question and be designed to meet the specific needs of the survey.

Active surveillance, also known as representative survey, is designed to measure the level of disease accurately and also for planning and monitoring control programs. These surveys are suitable for measuring the level of disease without bias and comparing the level of disease between two geographical areas and different time points. However, representative surveys are not good for early detection of new disease outbreaks and case finding as part of a disease control program. These surveys are expensive and difficult to organize, cover a relatively small sample of the population, are carried out over a defined period and, therefore, are not continuous. Over the last few years, the use of risk assessment methods in active surveillance has led to the emergence of risk-based surveillance to improve the efficiency of surveillance (Oidtmann et al. 2013).

Risk-Based Surveillance Risk-based surveillance is generally considered to be a more efficient way to demonstrate freedom from disease. These surveys can achieve a higher sensitivity with a lower sample size by focusing on populations that have a higher risk of being infected. This surveillance involves the use of knowledge of risk factors to improve the probability of finding a disease or infection. The risk factors can be spatial factors (climate and population density), host factors (age and species), management factors (bio security and antimicrobial usage) and other factors (e.g. history of cases or risky practices) (Wells et al. 2009). The most important advantage of risk-based surveillance is the low costs related to obtaining a defined level of confidence.

Risk-based surveys are considered good for demonstrating freedom from subclinical diseases and diseases with unclear clinical signs. However, these surveys are not good for measuring the level of disease, early detection of new or exotic diseases and also case detection. The other disadvantage of risk-based surveillance is that the observed prevalence cannot be easily extrapolated to the general population (Williams et al. 2009).

4 Aquatic Animal Disease Surveillance in India

India is a geographically vast country with both federal and state administrative structures. The states have divergent levels of capabilities and priorities in the fisheries sector. Therefore, establishing a coordinated disease surveillance programme was a challenge. In addition, preventing the spread of diseases between the states is crucial, as outlined in 'The Prevention and Control of Infectious and Contagious Diseases in Animals Act, 2009' and its subsequent amendment in 2014. To fulfil the spirit of the legislative act and meet international obligations of disease reporting, the country needed an institutionalized network and procedures that could capture the emergence of diseases and ensure their quick reporting. Therefore, a pilot-level network was necessary to establish procedures and disease diagnostic capabilities with the aim that the lessons learnt during the course of this would lead to the development of a framework for a surveillance mechanism. Keeping the same in consideration, the Department of Animal Husbandry, Dairying and Fisheries, Ministry of Agriculture, Government of India, took a visionary step in the year 2013 by approving a National Surveillance Programme for Aquatic Animal Diseases (NSPAAD) with funding through the National Fisheries Development Board, Hyderabad. This programme was initiated in 14 states and later expanded to cover 19 states and 3 Union Territories. The programme involved 31 NSPAAD Collaborating Centres and was coordinated by ICAR-National Bureau of Fish Genetic Resources (Fig. 2a). Since 2022, Phase 2 of NSPAAD has been implemented pan-India with the involvement of 32 collaborating centres, including the ICAR Fisheries Research Institutes and the Colleges of Fisheries. Besides, State Fisheries Departments and Marine Products Export Development Authority have been included to strengthen the programme, and it is funded under Pradhan Mantri Matsya Sampada Yojana by the Department of Fisheries, Government of India (Fig. 2b).

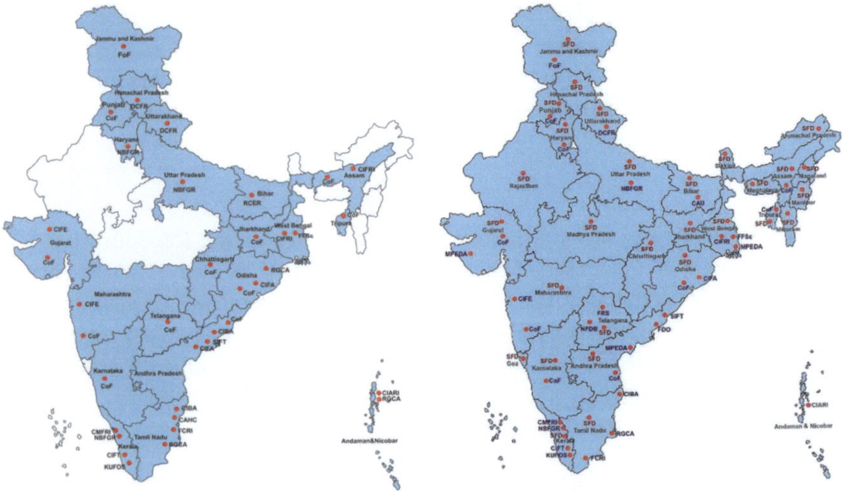

Fig. 2 States and union territories being covered under NSPAAD and collaborating centres responsible for implementing surveillance in each state. (**a**) Phase 1 and (**b**) Phase 2

4.1 Implementation of NSPAAD

In the initial stages of the NSPAAD, a 'Diagnostic Manual' containing internationally agreed protocols for diagnosing aquatic animal diseases was prepared. Simultaneously, diagnostic capability for WOAH-listed diseases was strengthened by procuring positive controls from WOAH-Referral Laboratories, and the capacity for the detection of emerging aquatic animal pathogens is continuously being upgraded. Notably, the investigators involved in the project were sensitized to surveillance and its underpinning science, epidemiology, through 'Orientation Training Workshop' involving international resource persons. Moreover, the knowledge of project investigators and research scholars is regularly updated by organizing international events and training programmes. The major emphasis of the NSPAAD has been to strengthen passive disease surveillance in the country. In addition, NSPAAD Collaborating Centres also undertake need-based active surveillance for some of the important pathogens. The mechanism for implementation of the NSPAAD project is given in Fig. 3.

4.2 Salient Achievements of NSPAAD

The programme has been successful in strengthening the passive surveillance system in the country. Till 2024, about 850 awareness programmes involving over forty thousand fish farmers have been organized. Besides, to strengthen the diagnostic capability of officers of State Fisheries Departments, about 125 training programmes involving three thousand officers have been organized whereby the officers are

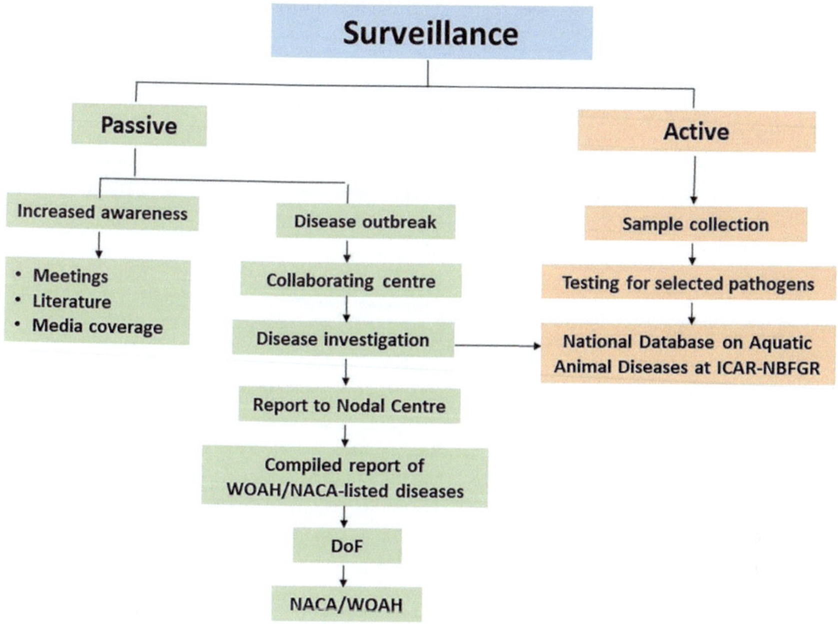

Fig. 3 Flow chart showing the mechanism of implementation of the National Surveillance Programme for Aquatic Animal Diseases. DoF-Department of Fisheries, NACA—Network of Aquaculture Centres in Asia-Pacific and WOAH-World Organisation for Animal Health

sensitized about major diseases and their diagnosis as well as management measures to minimize the losses. Under the passive disease surveillance, over 4000 disease cases have been investigated, and scientific advice is being provided to the farmers to manage the diseases in order to minimize the losses due to aquatic animal diseases. Over the years, there has been an increase in the disease reporting by the farmers following the implementation of the NSPAAD project.

Under the NSPAAD, thirteen aquatic animal diseases have been reported for the first time in India till 2024 (Table 1). These diseases include Goldfish haematopoietic necrosis disease (Sahoo et al. 2016); Koi sleepy disease (Swaminathan et al. 2016); hepatopancreatic microsporidiosis (Rajendran et al. 2016); infectious myonecrosis (Sahul Hameed et al. 2017); Tilapia lake virus disease (Behera et al. 2018); epitheliocystis (Sood et al. 2018); infection with red sea bream iridovirus (Girisha et al. 2020); infection with rhabdovirus (John et al. 2021); infection with mud crab reovirus (Ganesan et al. 2023); infection with tilapia parvovirus (Rajendran et al. 2023), infection with Wenzhou shrimp virus 8, infection with *Macrobrachium rosenbergii* golda virus and Bacillary necrosis of pangasius. Besides, infection with *Perkinsus olseni* (OIE-listed pathogen) has also been reported in a new host, i.e. short neck yellow clam *Paphia malabarica* (Shamal et al. 2018). It is important to mention that the first reports of new diseases have been published in peer-reviewed

Table 1 New aquatic animal diseases reported for the first time in India under NSPAAD till 2024

Sl. no.	Disease	Pathogen	Host
1.	Goldfish haematopoietic necrosis disease	Cyprinid herpesvirus 2 (CyHV-2)	Goldfish
2.	Koi sleepy disease	Carp edema virus (CEV)	Koi carp
3.	Tilapia lake virus disease	Tilapia lake virus (TiLV)	Tilapia
4.	Infection with Red Sea bream iridovirus	Red Sea bream iridovirus (RSIV)	Sea bass
5.	Epitheliocystis	*Candidatus* Actinochlamydia pangasiae sp. nov.	Striped catfish
6.	Infection with infectious myonecrosis virus	Infectious myonecrosis virus (IMNV)	Pacific white shrimp
7.	Hepatopancreatic microsporidiosis	*Enterocytozoon hepatopenaei* (EHP)	Pacific white shrimp
8.	Infection with rhabdovirus	Rhabdovirus	Snakehead
9.	Infection with mud crab reovirus	Mud crab reovirus (MCRV)	Mud crab
10.	Infection with tilapia parvovirus	Tilapia parvovirus (TiPV)	Tilapia
11.	Infection with Wenzhou shrimp virus 8	Wenzhou shrimp virus 8 (WzSV8)	Pacific white shrimp
12.	Infection with *Macrobrachium rosenbergii* Golda Virus	*Macrobrachium rosenbergii* golda virus (MrGV)	Freshwater prawn
13.	Bacillary necrosis of pangasius	*Edwardsiella ictaluri*	Pangas

international journals. From these reports, it is evident that the NSPAAD has strengthened India's aquatic animal disease surveillance and reporting system.

Importantly, NSPAAD has helped to dispel rumours of acute hepatopancreatic necrosis disease (AHPND), a WOAH-listed disease, from the country. The disease was first reported in China in 2009. Subsequently, the disease had spread to Thailand by 2012, at a time when the country was at the top position as the major exporter of shrimp. The disease had a devastating effect on the shrimp industry in Thailand where it resulted in a 65% decrease in shrimp production within a year. There were speculations about the occurrence of AHPND in India in 2013. Hence, an emergency response team was constituted under NSPAAD, which carried out targeted surveillance to determine the status of the disease. AHPND was not detected in any of the samples. Subsequently in 2017, following reports of AHPND in three bordering districts of a neighbouring country, targeted active surveillance was undertaken in bordering districts of West Bengal, and all the screened samples were found to be negative for *Vibrio parahaemolyticus* AHPND strain. Now, India has the evidence to be considered free of AHPND. In addition, the active surveillance carried out under the project provided transparent scientific evidence for the disease-free status of three major pathogens of finfish and three shrimp pathogens in India (Table 2). The pathogens, namely spring viremia of carp virus, infectious pancreatic necrosis virus and viral haemorrhagic septicaemia virus, have not been detected in finfish. Similarly, in the case of shrimp, yellow head virus genotype 1, Taura syndrome virus, and *Vibrio parahaemolyticus* AHPND strain have not been detected in the country.

Table 2 Major aquatic animal diseases not detected in India

Sl. no.	Disease	Pathogen	Host	Status
1.	Acute hepatopancreatic necrosis disease	*Vibrio parahaemolyticus* AHPND strain (Vp_{AHPND})	Shrimps	WOAH-listed
2.	Infection with spring viremia of carp virus	Spring viremia of carp virus (SVCV)	Carps	WOAH-listed
3.	Infection with infectious pancreatic necrosis virus	Infectious pancreatic necrosis virus (IPNV)	Wide range of fish species	Previously, WOAH-listed
4.	Infection with viral haemorrhagic septicaemia virus	Viral haemorrhagic septicaemia virus (VHSV)	Wide range of fish species	WOAH-listed
5.	Infection with yellow head virus genotype 1	Yellow head virus genotype I (YHV1)	Shrimps	WOAH-listed
6.	Infection with Taura syndrome virus	Taura syndrome virus (TSV)	Shrimps	WOAH-listed

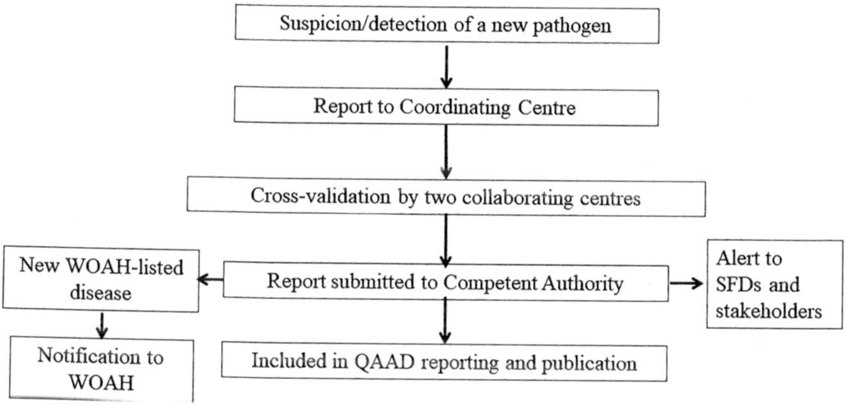

Fig. 4 Mechanism for first time reporting of a new disease under the National Surveillance Programme for Aquatic Animal Diseases. WOAH—World Organisation for Animal Health

4.3 Strengthening the Disease Reporting System Under NSPAAD

Under the programme, a strong country-wide network of aquatic animal health laboratories, a mechanism for rapid and accurate confirmation of exotic and emerging diseases and an early warning system through alerts following first-time reports of emerging and exotic diseases have been established (Fig. 4). Diagnosis of disease cases and scientific advice to the farmers have helped in reducing production losses. Importantly, a National Database on Aquatic Animal Diseases has been developed under the programme. The transparent reporting of aquatic animal diseases has enhanced the credibility of the country in terms of reporting aquatic animal diseases to international organizations. This has been recognized by NACA, and India's aquatic disease surveillance programme has been highlighted in the QAAD report for the Asia and Pacific Region (October to December 2016) published jointly by the Network of Aquaculture Centres of Asia-Pacific OIE and FAO (1494556312_quarterly-aquatic-animal-disease-report-october-december-2016.pdf). Under the programme, several articles in peer-reviewed journals with ISI impact factors have been published.

To strengthen farmer-based reporting of the diseases, recently, an android-based GPS-enabled mobile application named 'ReportFishDisease (RFD)' has been launched. The RFD app can be downloaded from Google Play Store and is available in Hindi, English, and 11 regional languages. Using the app, the farmers can share reports of diseases affecting aquatic animals on their farms with officers of State Fisheries Departments and fish health experts in real time and get quick scientific advice based on the presumptive diagnosis. Importantly, the information can be used for mapping the disease cases on a time and geographic scale.

5 Conclusion

The main driver for the NSPAAD is the potential threat posed by the existing and emerging aquatic animal diseases to the aquaculture industry due to ongoing intensification, diversification and increase in international trade in aquatic animals. Early detection of emerging diseases is considered a key to their control, and knowledge about the existing disease is essential in developing national control or containment strategies or declaring disease freedom. The programme provides a system for doing that by strengthening passive disease surveillance based on reporting by farmers. Phase 2 covers all the states of the country, involves the State Fisheries Departments (SFD). Moreover, MPEDA has been involved in strengthening shrimp disease surveillance in the country. The programme tends to improve disease reporting by increasing awareness among the farming community and popularization of the RFD app, training of officers of SFD and strengthening of transport mechanisms of disease samples to diagnostic laboratories. Furthermore, it extends the capacity of the NSPAAD to identify the presence, distribution and frequency of existing aquatic animal diseases and, importantly, helps to know the diseases that are absent in the country. NSPAAD has been assisting the competent authority in better understanding the disease situation in aquatic animals in the country and has reached the stage where it can identify diseases that are present and those that are absent in the country.

References

Behera BK, Pradhan PK, Swaminathan TR, Sood N, Paria P, Das A, Verma DK, Kumar R, Yadav MK, Dev AK, Parida PK, Das BK, Lal KK, Jena JK (2018) Emergence of Tilapia Lake Virus associated with mortalities of farmed Nile Tilapia *Oreochromis niloticus* (Linnaeus 1758) in India. Aquaculture 484:168–174

Cameron A (2012) Manual of basic animal disease surveillance. Interafrican Bureau for Animal Resources (IBAR). ISBN:978-9966-1659-1-6

DoF, GoI (2023) Handbook on fisheries statistics 2023. Fisheries Statistics Division, Department of Fisheries, Ministry of Fisheries, Animal Husbandry & Dairying, Government of India, New Delhi. 291 pp

Ganesan S, Baskaran B, Raj M, Marimuthu S, Krishnasamy V, Lamech R, Mandal A, Shanmuganathan K, Marimuthu PN (2023) Reovirus occurrence in mud crab farming systems and wild-caught brooders located in eastern coastal area of India. Aquac Int 31:739–758

Girisha SK, Puneeth TG, Nithin MS, Naveen KBT, Ajay SK, Vinay TN, Suresh T, Venugopal MN, Ramesh KS (2020) Red sea bream iridovirus disease (RSIVD) outbreak in Asian seabass (*Lates calcarifer*) cultured in open estuarine cages along the west coast of India: first report. Aquaculture 520:734712

John KR, George MR, Mageshkumar P, Mansoor MM, Kaviarasu D, Sivasankar P, Selvamagheswaran M, Petchimuthu M (2021) Isolation and characterisation of a novel rhabdovirus from snakehead fish showing surface ulcerations and mortality from India: first report. Aquaculture 544:737120

Oidtmann B, Peeler E, Lyngstad T, Brun E, Bang Jensen B, Stärk KD (2013) Risk-based methods for fish and terrestrial animal disease surveillance. Prev Vet Med 112(1–2):13–26

Patil PK, Geetha R, Ravisankar T, Avunje S, Solanki HG, Abraham TJ, Vinoth SP, Jithendran KP, Alavandi SV, Vijayan KK (2021) Economic loss due to diseases in Indian shrimp farming with special reference to *Enterocytozoon hepatopenaei* (EHP) and white spot syndrome virus (WSSV). Aquaculture 533:736231

Rajendran KV, Shivam S, Ezhil PP, Joseph SRJ, Sathish KT, Avunje S, Jagadeesan V, Prasad BSVANV, Pande A, Navaneeth KA, Alavandi SV, Vijayan KK (2016) Emergence of *Enterocytozoon hepatopenaei* (EHP) in farmed *Penaeus* (*Litopenaeus*) *vannamei* in India. Aquaculture 454:272–280

Rajendran KV, Sood N, Rao BM, Valsalam A, Bedekar MK, Jeena K, Pradhan PK, Paria A, Swaminathan TR, Verma DK, Sood NK (2023) Widespread occurrence of Tilapia parvovirus in farmed Nile tilapia *Oreochromis niloticus* from India. J Fish Dis. https://doi.org/10.1111/jfd.13871

Sahoo PK, Mohanty J, Garnayak SK, Mohanty BR, Kar B, Prasanth H, Jena JK (2013) Estimation of loss due to argulosis in carp culture ponds in India. Indian J Fish 60(2):99–102

Sahoo PK, Swaminathan TR, Abraham TJ, Kumar R, Pattanayak S, Mohapatra A, Rath SS, Patra A, Adikesavalu H, Sood N, Pradhan PK, Das BK, Jayasankar P, Jena JK (2016) Detection of goldfish haematopoietic necrosis herpes virus (Cyprinid herpesvirus-2) with multi-drug resistant *Aeromonas hydrophila* infection in goldfish: first evidence of any viral disease outbreak in ornamental freshwater aquaculture farms in India. Acta Trop 161:8–17

Sahul Hameed AS, Abdul Majeed S, Vimal S, Madan N, Rajkumar T, Santhoshkumar S, Sivakumar S (2017) Studies on the occurrence of infectious myonecrosis virus in pond-reared *Litopenaeus vannamei* (Boone, 1931) in India. J Fish Dis 40:1823–1830

Shamal P, Zacharia PU, Binesh CP, Pranav P, Suja G, Asokan PK, Pradeep MA, Rithesh R, Vijayan KK, Sanil NK (2018) *Perkinsus olseni* in the short neck yellow clam, *Paphia malabarica* (Chemnitz, 1782) from the southwest coast of India. J Invertebr Pathol 159:113–120

Sood N, Pradhan PK, Verma DK, Yadav MK, Ravindra D, Swaminathan TR, Sood NK (2018) *Candidatus* Actinochlamydia pangasiae sp. nov. (Chlamydiales, Actinochlamydiaceae), a bacterium associated with epitheliocystis in *Pangasianodon hypophthalmus*. J Fish Dis 41:281–290

Swaminathan TR, Kumar R, Dharmaratnam A, Basheer VS, Sood N, Pradhan PK, Sanil NK, Vijayagopal P, Jena JK (2016) Emergence of carp edema virus in cultured ornamental koi carp, *Cyprinus carpio koi*, in India. J Gen Virol 97:3392–3399
Wells SJ, Ebel ED, Williams MS, Scott AE, Wagner BA, Marshall KL (2009) Use of epidemiologic information in targeted surveillance for population inference. Prev Vet Med 89:43–50
Williams MS, Ebel ED, Wells SJ (2009) Population inferences from targeted sampling with uncertain epidemiologic information. Prev Vet Med 89:25–33

Issues in International Trade of Seafood

Iddya Karunasagar

Abstract

Seafood is one of the most traded food commodities globally. International trade in seafood has been growing at a steady pace, and seafood is contributing greatly to the foreign exchange earnings of low- and middle-income countries. India is among the top five exporters, and when it comes to shrimp, India is the second largest exporter after Ecuador. Seafood trade is well-regulated internationally, and importing countries have stringent requirements regarding quality and safety. World Trade Organisation (WTO) Sanitary and Phytosanitary (SPS) Agreement and Technical Barriers to Trade (TBT) Agreements apply to seafood trade. SPS agreement covers both food safety as well as animal health, and the agreement recommends that the countries harmonise their requirements with standards set by international organisations, such as the Codex Alimentarius Commission for Food Safety and the World Organisation for Animal Health (WOAH) for Animal Health. Most animal health regulations are applicable to international trade in live aquatic animals. Based on import risk analysis for prawns, Australia requires health certification confirming the absence of pathogens, such as white spot syndrome virus and yellow head virus, for the import of prawns into Australia. The seafood imported by major importers such as the EU, USA and Japan are subjected to risk-based inspection, including testing of a small proportion of samples. Non-compliance would lead to import alerts or import refusals. An analysis of the causes of import alerts and import refusals in the EU, USA and Japan indicates different risk perceptions and testing priorities in different markets. Residues of veterinary drugs are a major cause of such import alerts or import refusals. This chapter discusses issues related to international seafood trade,

I. Karunasagar (✉)
FAO Ref Centre for Antimicrobial Resistance, Nitte University, Mangalore, India
e-mail: iddya.karunasagar@nitte.edu.in

R. Kooloth Valappil et al. (eds.), *Aquatic Animal Health Management*,
https://doi.org/10.1007/978-981-96-7987-4_32

mainly in reference to animal health issues, and presents an analysis of the causes of import alerts in EU, USA and Japanese markets.

Keywords

SPS Agreement · Codex Alimentarius Commission · Aquatic Animal Health Code · Import risk analysis · Health certificate

1 Introduction

Aquatic products are an important commodity in international trade, and according to FAO State of the World Fisheries and Aquaculture 2024 (FAO 2024), the net seafood trade (exports minus imports) in low- and middle-income countries reached USD 45 billion greater than that of all agricultural commodities combined. According to the FAO outlook for fisheries 2022–2032, world production of aquatic animals is expected to reach 205 million tonnes in 2032, 111 million tonnes from aquaculture and 94 million tonnes from capture fisheries. This would require a 17% increase in aquaculture production. Most aquaculture production will be for food fish, and 60% of aquatic food for human consumption will come from aquaculture. In India, the total fish production by aquaculture is estimated to be 12 million tonnes in 2023, and India aims to increase this to 18 million tonnes by 2030.

Trade in aquatic products is very well regulated, and WTO international agreements, such as the Sanitary and Phytosanitary (SPS) Agreement and Technical Barriers to Trade (TBT) agreement, will apply to international seafood trade. As per the SPS agreement, countries are to harmonise their national standards with those set by international organisations, such as the Codex Alimentarius Commission (JOINT FAO/WHO body), the World Organization for Animal Health (WOAH, formerly OIE) and the International Plant Protection Convention (IPPC). If any country has to set standards other than those in line with these international standard-setting bodies, it should be based on a scientific risk assessment carried out following the guidelines of these international bodies. Based on SPS and TBT agreements, a number of seafood trading countries have implemented regulations that cover food safety and animal health in the seafood trade. This chapter presents the current situation concerning international trade, regulations related to aquatic animal health and food safety in major importing countries and challenges exporting countries face in meeting strict international standards.

2 Global Seafood Trade and Position of India

According to FAO State of the World Fisheries and Aquaculture (FAO 2024), the global seafood trade reached USD195 billion in 2022, a 19% increase over pre-pandemic levels in 2019. China has become the main exporter of aquatic animal products, accounting for 12% of the total value, followed by Norway (8%), Vietnam

(6%), Ecuador (5%) and Chile (4%). India slipped to sixth position among exporters in 2022, having been the fourth largest exporter in 2019. This has been attributed to a drop in shrimp prices and shrimp production values not increasing compared to countries, such as Ecuador. Nevertheless, shrimp accounted for 70% of USD 7.9 billion generated by seafood exports in India in 2022. As per the Ministry of Commerce, Government of India, the exports during the financial year 2023–24 were 17.82 MT valued at USD 7.39 billion.

As per FAO statistics (FAO 2024), globally, the United States is the single largest importing country, accounting for 17% of global imports, followed by China (12%), Japan (8%), Spain (5%) and France (4%). The seafood imports into the USA were valued at USD 32.1 billion, and major suppliers of seafood for the USA were Canada (13%), Chile (12%), India (10%), Indonesia (9%) and Vietnam (7%). Together, these suppliers accounted for over half of seafood imports into the USA. The major commodities imported by the USA were shrimps (22%), salmonids (22%) and tunas (8%). With an import value of USD 22.5 billion in 2022, China became the second-largest importing country. However, processing outsourcing by other countries, increasing domestic demand and raw material requirements by the aquaculture and animal feed industry contributed to increased seafood imports into China. The main suppliers were Ecuador, Russian Federation and Vietnam, and the major commodities were shrimps, cods, lobsters and crabs. Japan was the third-largest importing country, with imports worth USD 15.1 billion in 2022. As a region, the European Union (27 countries) is the single largest market, accounting for USD 62.7 billion, and this includes USD 33.2 billion in imports from non-member states, and the rest is accounted for by intra-EU trade flow. Among suppliers of seafood for the EU, Norway was the major country, accounting for 27%, followed by China (6%), Ecuador (6%), Morocco (5%) and the United Kingdom and Northern Ireland (5%). Most of the imports (about 80%) are salmonids, cods, shrimp, tuna and squids.

In terms of value, salmon remains the largest traded commodity (20% of value), followed by shrimp and prawns (17%), cods, hakes and haddocks (9%), as well as tuna, bonitos and billfishes (9%) and cephalopods (7%). Due to the high value of finfish, such as salmons, tuna and cods, global trade is dominated by finfish, which account for 65% of global trade value. Crustaceans account for 23% in value, with shrimp and prawns contributing to 17% and squids and cuttlefish account for 7% of global trade value. According to FAO data, global shrimp exports declined during 2018–2020 but showed a 22% increase in 2021 and an 8.7% increase in 2022, when the value reached USD 32.3 billion. FAO fish price index indicated that fish prices dropped during the pandemic period but rose again (https://www.fao.org/fishery/en/fishstat/fishpriceindex). Average fish prices in the international markets in 2022 were 19% higher than in 2021, which was 6.4% higher than in 2020. But in 2023, the prices dropped. The trend was similar to shrimp prices, which dropped 16% in 2023. Due to increasing production costs, this drop in price has negatively impacted the global shrimp industry.

3 Seafood Trade Performance of India

Indian seafood exports touched an all-time high of 17.82 million metric tonnes worth USD 7.38 billion in 2023–24. This is a 2.67% increase in volume (17.35 million metric tonnes in 2022–23), but in value, this was an 8.8% decrease compared to export value of USD 8.1 billion in 2022–23). The trend in export value followed the global seafood price index brought out by FAO, which indicates lower value following the pandemic and then a rise in 2022 (https://www.fao.org/fishery/en/fishstat/fishpriceindex) and a drop in 2023. The export trend during the period 2013–14 to 2023–24 is illustrated in Fig. 1.

As shown in Table 1, in terms of volume, products from capture fisheries accounted for 62% of exported products, but in value, aquaculture products contributed to 62–63%. The value per kg of aquaculture products (USD 6.82 to 7.72) was higher than for products from capture fisheries (2.52–2.79). This can be attributed to the fact that shrimp accounted for 40.19% in terms of volume of exported products and 66.12% in terms of value.

Frozen shrimp exports during 2023–24 were 716,004 MT valued at USD 4.88 billion. USA was the major buyer (297,571 MT), followed by China (148,483 MT), European Union (89,297), South East Asia (52,254), Japan (35,906), Middle East (28,571) and other countries (63,521). The export of shrimp *Penaeus vannamei* during 2023–24 increased marginally (0.33%) in volume (623,432 MT to 625,475 MT), but the value declined 11.56% (USD 4809.99 million to USD 4253.86 million). The major market (53.52%) was USA followed by China (15.78%), European Union (9.01%), South East Asia (5.81), Japan (3.41), Middle East (2.9%) and other countries (9.57%). The average export value from January 2023 to March 2024 is illustrated in Fig. 2.

Export of black tiger shrimp *P. monodon* increased by 24.91% in terms of volume and 8.28% in value in dollars (USD 321.23 million to USD 347.84 million). The main markets for *P. monodon* were China (28.43%), the USA (18.21%), the EU

Fig. 1 Seafood export value in India. (Source: Ministry of Commerce, Government of India)

Table 1 Contribution of aquaculture and capture fisheries to seafood exports in India

	Aquaculture		Capture fisheries	
Year	2022–23	2023–24	2022–23	2023–24
Quantity	38%	38%	62%	62%
Value (USD)	63%	62%	37%	38%
Value (USD/kg)	7.72	6.82	2.79	2.52

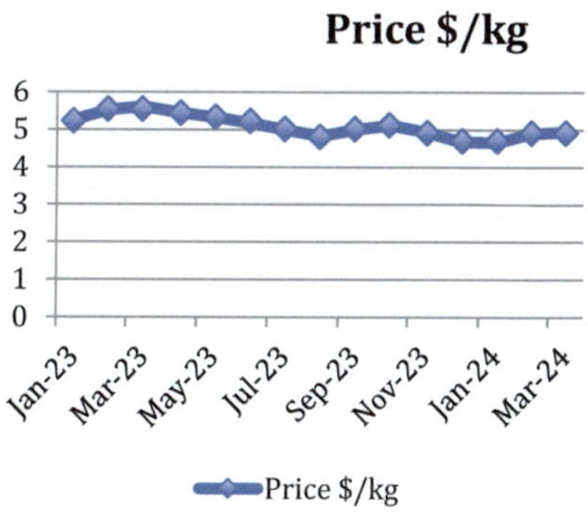

Fig. 2 Price ($/kg) of vannamei shrimp during January 2023 to March 2024. (Source: Ministry of Commerce, Government of India)

(18.06), Japan (13.12%), Middle East (8.27), South East Asia (6.45%), other countries (7.46%).

Most of the shrimp exports from India are in the frozen form. Value-added products formed only a small proportion of exported shrimp, but over the years, this segment has increased (Fig. 3).

The value-added shrimp component reached a peak in 2022 (74,029 MT) but declined to 62,914 MT in 2023. In 2020, during the pandemic period, there was an increase (53,426 MT) compared to 2019 (37,141 MT). It is interesting to note that though total shrimp exports declined in 2020, the export of value-added products increased.

4 International Regulations Related to Seafood Trade

World Trade Organisation (WTO) deals with international agreements and issues related to trade between Member countries. Regarding agricultural commodities, including aquaculture products in trade, the issues could be related to food safety and the transboundary movement of pathogens and pests with food commodities. In this connection, the WTO Member countries signed the Sanitary and Phytosanitary (SPS) Agreement in 1995, which covers food safety and plant and animal health. The WTO SPS Agreement recognises the sovereign rights of member countries to implement measures to protect human, animal and plant life and health within their territory (WTO, 1995). However, it specifies that the measures should not be arbitrary but should be based on risk assessments performed following international

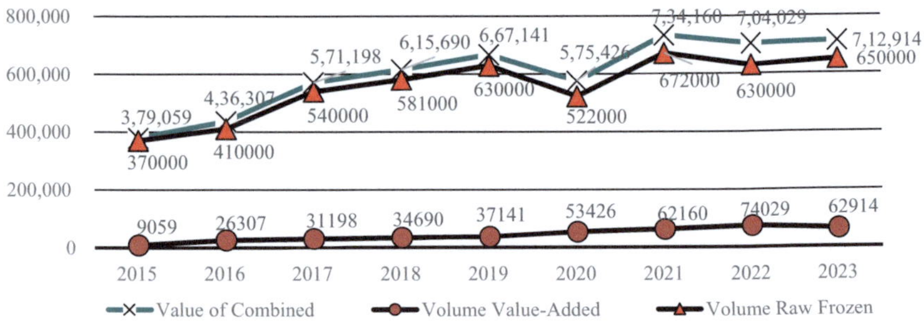

Fig. 3 Export volumes of total frozen shrimp and value-added component in this total. (Source: Ministry of Commerce, Government of India)

best practices. The SPS Agreement underlines the importance of harmonising the measures and national standards to those set by international standard-setting bodies: Codex Alimentarius Commission, a Joint FAO/WHO body for food safety, World Organisation for Animal Health (WOAH, formerly OIE) for animal health and International Plant Protection Convention (IPPC) for plant health. International trade will also depend on standards and regulations in areas other than health and food safety, and these are covered under the WTO Technical Barriers to Trade (TBT) Agreement, which came into effect in 1995 along with the SPS agreement. The aim of the TBT agreement is to ensure that technical regulations, standards and conformity assessment procedures are non-discriminatory and do not cause unnecessary obstacles to trade. Some examples of international regulations related to aquatic animal health and international trade from international bodies, such as the Codex Alimentarius Commission and the World Organisation of Animal Health (WOAH), are discussed below:

5 Standards and Guidelines from the World Organisation of Animal Health (WOAH)

The standards from WOAH cover mainly animal health and also touch upon the interface between animal health and food safety. There are mainly two standard documents from WOAH: the Aquatic Animal Health Code and the Diagnostic Manual for Aquatic Animal Diseases (WOAH, 2024). Aquatic Animal Health Professionals need to be aware of the provisions of both these standards. The Aquatic Animal Health Code covers aspects such as criteria for listing diseases, aquatic animal disease surveillance, import risk analysis, quality of aquatic animal health services, disease prevention and control, trade measures, import/export procedures, health certification, antimicrobial use in aquatic animals and welfare of farmed fish.

6 EU Regulations Related to Aquatic Animal Health Applicable to International Trade

Based on the WOAH Aquatic Animal Health Code, countries can have their national regulations. For example, the EU has an Animal Health Law (EU regulation 2016/429) covering both terrestrial and aquatic animals (EU, 2016). Part I of this law covers general rules, listing of diseases and categorisation of diseases. Part II covers disease notification and reporting, surveillance, eradication programmes and declaration of disease-free status as per WOAH guidelines. Part III covers disease awareness, preparedness and control programmes for diseases of different categories. Part IV covers farm registrations, approval, traceability and movements. Part V covers entry into the EU and export. Part I, covering listing and delisting, provides the actions needed for diseases of different categories. Diseases categorised as 'A' do not occur in the EU, and these warrant immediate eradication. Category B diseases include listed diseases that must be controlled in all Member states with the goal of eradication throughout the Union action; hence, the action needed is compulsory eradication. Category C would include listed diseases of relevance to some Member States and measures needed to prevent them from spreading to areas that are disease-free or that have eradication programmes. In such cases, the action is optional eradication; for diseases of the D category, movement should follow the regulation, and category E diseases require notification. The Animal Health Law fixes the responsibilities of the operators (e.g. implementing biosecurity, prudent and responsible use of veterinary medicines), veterinarians and aquatic animal health professionals (preventing disease, early diagnosis and raising awareness about antimicrobial resistance), laboratories and others handling disease agents (biosecurity, notification of suspicions and cases to competent authority).

The categorisation of diseases of aquatic animals as per Commission implementing regulation (EU) 2018/1882 (EU, 2018) are as follows:

Category A diseases: Epizootic hematopoietic necrosis, infection with *Mikrocytos mackini*, infection with *Perkinsus marinus*, Taura syndrome and yellow head disease.

Category C diseases: Viral haemorrhagic septicaemia (VIIS), infectious hematopoietic necrosis (IHN), infectious salmon anaemia HRP deleted, infection with *Bonamia exitiosa*/Ostreae, infection with *Marteilia refringens* and white spot syndrome.

Category E diseases: Koi herpes virus.

According to Animal Health Law Article 237 (1), aquatic animals and products of aquatic animal origin can enter the EU only if accompanied by an Animal Health certificate issued by the Competent Authority. According to Article 237 (2), the health certificate should have been verified and signed by the official veterinarian in the Third country in compliance with the certification requirement laid down in Article 149 (3) or 216 (3). According to Article 273 (4), the Commission is empowered to adopt a delegated act in accordance with Article 264 concerning derogation of Health certificate requirements for consignments where the risk to human health and animal health is considered insignificant due to their intended use (for human consumption). This is further clarified in Commission Delegated Regulation (EU) 2020/692 (EU, 2020). Article 170 of this regulation states:

- Consignments of aquatic animals and products of animal origin from aquatic animals other than live aquatic animals shall only be permitted to enter the Union if the aquatic animals and products of animal origin of the consignment come from a third country or territory or zone or compartment thereof which complies with the following requirements.
- It must be free from the following listed diseases.
- Category A and Category B listed diseases.
- Relevant category C diseases: when the aquatic animals or products of animal origin are destined for Member States, zones or compartments which have disease-free status or an approved eradication programme for the specific diseases.

Para 78 states that aquatic animals may enter the Union for different purposes. Given the disease risk associated with the movement of live animals, such animals entering the Union for human consumption should be treated in the same way as if they were entering the Union for other purposes, such as farming or release into the wild. Products of animal origin from aquatic animals other than live aquatic animals represent a lower risk than aquatic animals, and the measures to be taken in case of such products entering the Union for further processing are, therefore, less rigorous than those that apply to live animals.

Para 83 of EU Regulation 2020/692 states that it is not mandatory in all situations to apply the requirement that aquatic animals of listed species and products of animal origin from those animals originate from a third country or territory or zone or compartment thereof free from disease. Certain risk mitigation measures can be taken to facilitate entry into the Union for certain aquatic animals and their products that do not have such an origin. Certain risk mitigation measures are acceptable for aquatic animals of listed species, and given the lower level of risk associated with such movements, different, less stringent risk mitigation measures are acceptable for products of animal origin from aquatic animals other than live aquatic animals. Para 85 of the EU Regulation 2020/692 clarifies that the risk mitigation measures allowing such trade to occur may consist of assigning the products to a disease control aquatic food establishment (an establishment that has the facility to treat fish waste and wastewater) in the Union for further processing, or of ensuring that the products consist of fish that is slaughtered and eviscerated before despatch to the Union. In either case, the risk posed by the products of animal origin is considered negligible.

Article 172 specifies derogation of requirements elaborated in Article 170. By way of derogation from Article 170, the requirements laid down in that Article shall not apply to the following categories of aquatic animals:

- Aquatic animals that are destined for an aquatic food establishment where they are to be processed for human consumption.
- Aquatic animals intended for research purposes, which are destined for confined establishments and have been approved for that purpose by the competent authority of the Member State of destination.

Hence, frozen fish and shrimp entering the EU for human consumption do not have requirements similar to those for live aquatic animals in relation to aquatic animal health. The format of the health certificate required for products intended for human consumption is the FISH-CRUST-HC certificate provided in Commission Implementing Regulation (EU) 2020/2235 (https://eur-lex.europa.eu/eli/reg_impl/2020/2235/oj/eng).

7 Regulations in Australia Regarding the Import of Prawns for Human Consumption

Australia has special requirements related to aquatic animal health for prawn imports. This is because the black tiger shrimp *Penaeus monodon* is farmed in Australia, particularly Queensland. Aquaculture production of prawns (*P. monodon* and *Fenneropenaeus merguiensis*) in Queensland in 2022–23 was 9825.5 tonnes, valued at USD 212.9 million (Queensland Government 2024). Australia had outbreaks of white spot disease in natural waters due to the use of infected imported prawns intended for human consumption as bait for fishing. Following this, Biosecurity Australia revised the import risk analysis (Department of Agriculture, Fisheries and Forestry 2023). As per the revised requirement, uncooked prawns, marinated prawns from other countries and Australian prawns processed overseas (excluding those processed in an Australian government-approved supply chain) are grouped under the product class ‘uncooked prawns’. All consignments should be accompanied by a new health certificate issued by the government of the exporting country confirming that the prawns had no visible signs of disease. All batches in every consignment should test negative for WSSV and yellow head virus prior to export following test methods recognised by the World Organisation of Animal Health (WOAH, formerly OIE). Prawns are to be frozen with heads and shells removed (last shell segment and tail fans permitted) and deveined. The consignment should have secure seals for inspection, and upon arrival, every batch in a consignment should be tested at an approved laboratory for WSSV and YHV.

In the case of uncooked breaded, battered and crumbed (BBC) prawns, the competent authorities from the exporting country should issue a health certificate confirming that the prawns are free from visible signs of disease and have undergone par-cooking steps after the prawns were coated to solidify and adhere the coating to the prawn. Prawns are to be frozen with the head and shell removed. The Australian authorities have to confirm that during the import permit application process, the par-cooking step has been assessed. The entire consignment should have inspected seals to ensure that they were intact while arriving. Cooked prawn and prawn products should be accompanied by a health certificate issued by the competent authority of the exporting country confirming that the prawns have been cooked and, as a result of the cooking process, all protein in the prawn meat has coagulated, and no raw prawn meat remains. Randomly selected seals should be intact, and the consignment should undergo fully supervised inspection in Australia to confirm that all requirements for cooked prawns have been met.

8 Food Safety Standards Set by Codex Alimentarius Commission

As per the WTO SPS agreement, countries are advised to harmonise their food safety standards with those set by the Codex Alimentarius Commission, a Joint FAO/WHO standard-setting body. Codex has several types of documents—standards that the products should meet and Codes of Practice and Guidelines that the Food Business operators need to follow to achieve the standards, maximum residue limits for food additives, residues of veterinary drugs and pesticide residues. For fish and fishery products, the Codex Code of Practice for Fish and Fishery products (CXC 52–2003 updated 2019) provides guidelines on good aquaculture practices to be followed (Codex Alimentarius Commission 2019). Codex Guidelines for the design and implementation of National Regulatory Food Safety Assurance Programmes associated with the use of veterinary drugs in food-producing animals (Codex Alimentarius Commission 2014) provides guidance on the approval of veterinary drugs, responsibilities of competent authorities, pharma industry, veterinary professionals, farm operators, feed and other input providers in responsible use of veterinary medicine and implementation of national residue management programme. Codex general standards on food additives (https://www.fao.org/fao-who-codexalimentarius/codex-texts/dbs/gsfa/en/) provide limits for additives in food processing. Codex maximum residues of veterinary drugs in foods are provided in CXM 2–2023 (Codex Alimentarius Commission 2023), and this document also provides risk management recommendations for veterinary drugs for which there are no MRL, e.g. chloramphenicol nitrofurantoin. Codex online database for maximum residue limits for pesticide residues in foods (https://www.fao.org/fao-who-codexalimentarius/codex-texts/dbs/pestres/en/) has information important for products of aquaculture.

9 National Seafood Import Regulations and Their Impact on Seafood Trade

Major seafood importing countries and markets have their own requirement for import, e.g. EU requirements (https://food.ec.europa.eu/animals/animal-products-movements/products-animal-origin-human-consumption_en), requirements for USA (https://www.fao.org/in-action/globefish/markets%2D%2D-trade/food-safety-regulations-for-fisheries-and-aquaculture-products/united-states-of-america/

en) and requirements for import into Japan (https://www.fao.org/in-action/globefish/markets%2D%2D-trade/food-safety-regulations-for-fisheries-and-aquaculture-products/japan/en). When seafood consignments arrive at the port of entry in these countries, they are subject to risk-based inspection. Most consignments are cleared after a documentary check, but a small proportion undergoes physical inspection and testing. Compliance failures could lead to increased testing. The importing countries and the EU document cases of compliance failures and data on such failures would give an idea of the challenges faced by the exporting countries.

10 Causes of Rapid Alerts Associated with Seafood in the EU

EU has a system of Rapid Alerts on Foods and Feeds (RASFF). Since the product entering any one of the EU member countries can move freely within the Union, the EU has devised a system of alerting all members whenever any compliance failure is detected, so that other members are aware and can take care to see that the product does not gain entry in another port. The details are available in the RASFF portal (https://webgate.ec.europa.eu/rasff-window/screen/search).

The causes of rapid alerts associated with seafood during 2016–2023 are indicated in Table 2. The major group 'others' includes a number of parameters, such as package deficiency, HACCP failure, labelling deficiency and others. The next largest group was due to chemical contamination followed by microbiological issues, histamine, biotoxins and parasites. There seems to be a declining trend in the total alerts as well as alerts due to chemical and other causes.

The major microbiological cause was the non-compliance of *Listeria monocytogenes* criteria (Table 3). This was followed by *E. coli* and Norovirus criteria. There were alerts due to the presence of *Vibrio cholerae* in shrimp, particularly in 2023 (30 notifications). *Salmonella* was another important cause of import alert. Occasionally, Clostridium botulinum and Hepatitis A virus led to rapid alerts. The large number of non-compliance due to *Listeria monocytogenes* is worrying. In fact, EU microbiological criteria for seafood, as per the Commission Regulation (EU) 2023/2005, do not include criteria for *Vibrio* spp. The notifications related to these indicate

Table 2 Causes of RASFF notifications associated with seafood during 2016–2023

	Year								
Causes	2016	2017	2018	2019	2020	2021	2022	2023	Total
Others	96	115	139	116	77	92	91	53	779
Chemical	114	159	73	68	62	58	45	48	627
Microbiological	62	41	75	44	57	37	52	71	439
Histamine	20	30	13	15	7	6	15	11	117
Toxins	23	11	7	10	10	8	6	6	70
Parasites	3	3	1	8	14	20	11	9	69
Total	**307**	**359**	**308**	**261**	**227**	**221**	**220**	**198**	**2101**

Table 3 Causes of microbiological alerts in seafood in EU during 2016–2023

Causes	Year 2016	2017	2018	2019	2020	2021	2022	2023	Total
Listeria monocyte-genes	19	11	24	19	20	17	26	8	144
E. coli	22	12	21	12	12	10	4	8	101
Norovirus	4	7	21	3	22	4	7	13	81
Vibrio spp.	6	2	1	1	2	1	6	34	53
Salmonella	7	6	7	6	1	5	9	7	48
Clostridium botulinum	1	1	0	2	0	0	0	1	5
Hepatitis A virus	3	0	1	1	0	0	0	0	5
Not specified	0	2	0	0	0	0	0	0	2
Total	**62**	**41**	**75**	**44**	**57**	**37**	**52**	**71**	**439**

Table 4 Rapid alerts due to chemical parameters in EU during 2016–2023

Causes	Year 2016	2017	2018	2019	2020	2021	2022	2023	Total
Mercury	61	95	42	38	30	23	22	18	329
Residues of veterinary drugs	29	35	21	8	8	4	10	10	125
Cadmium	15	25	6	11	13	13	7	8	98
Additives	1	0	2	1	5	7	2	1	19
Benzo(a)pyrene	8	0	0	4	1	0	1	3	17
Sulphite	0	0	0	0	0	4	2	8	14
Chlorate	0	0	0	5	4	0	1	0	10
Irradiation	0	0	0	0	0	5	0	0	5
Contaminants	0	0	2	1	0	0	0	0	3
Lead	0	3	0	0	0	0	0	0	3
Carbon monoxide	0	1	0	0	1	0	0	0	2
Arsenic	0	0	0	0	0	1	0	0	1
Sodium carbonate	0	0	0	0	0	1	0	0	1
Total	**114**	**159**	**73**	**68**	**62**	**58**	**45**	**48**	**627**

testing beyond the official criteria in the EU. It is not clear whether the organism detected was choleragenic *V. cholerae* O1/O139 or otherwise. It is to be pointed out that non-O1/O139 may be widely present as natural flora in water coastal environments. The causes of microbiological alerts in seafood in the EU and the major chemical parameters that were involved in non-compliance during 2016–23 are indicated in Tables 3 and 4, respectively.

Mercury exceeding limits was the major one, followed by residues of veterinary drugs. This includes residues of permitted drugs exceeding limits as well as the detection of residues of banned drugs, such as chloramphenicol, nitrofurans and dyes like methylene blue and malachite green. Though there is a declining trend, the continued detection of residues of banned antibiotics and dyes indicates the need to better regulate their application in aquaculture in producing countries. India is one of the countries facing issues related to antibiotic residues in the EU market. In

2023, there were 14 alerts related to seafood from India, and 29% of these were due to non-compliance with regulations related to residues of veterinary drugs. Depending on the compliance rate, the EU increases the testing frequency, and for quite some time now, 50% of Indian seafood undergoes testing in the EU. India needs to make more efforts to come out of this situation.

11 Causes of Import Alerts in the United States of America

US Import alert notifications are available in the US FDA portal (https://www.fda.gov/industry/actions-enforcement/import-alerts) (Table 5).

The data in Table 5 indicates that the causes of import alerts are similar to that in the EU, but the numbers are much higher. This may be a reflection of the differences in risk perceptions and testing frequency in the EU and the USA. Even in the USA, the largest number was 'other causes' that included packaging, hygiene failures (unfit for human consumption), labelling, adulteration, allergens and others. In the EU, chemical causes led to a higher number of alerts compared to microbiological, but in the USA, this was the other way around. The number of alerts due to toxins in the USA was much lower compared to that in the EU.

The details of microbiological alerts in the USA are given in Table 6. Unlike in the EU, *Salmonella* far outnumbered *Listeria monocytogenes*. This could be due to increased testing for *Salmonella* in the USA compared to the EU. Alert notifications due to *E. coli* were very low compared to the EU, and there were no notifications for norovirus or *Vibrio* spp. in the USA, though the total number of microbiological alerts in the USA was more than four times the EU RASFF notifications related to microbiological issues. The reason could be due to differences in risk perception and testing frequency.

As given in Table 7, the causes of import alerts related to residues of veterinary drugs in the USA were very high compared to the EU. In Table 7, data for the banned drugs—chloramphenicol and nitrofurans—are shown separately from other veterinary drugs. The large number of import notifications under this category is a matter of great concern that needs to be addressed by the aquaculture-producing countries. India faced 44 import refusals in the USA in 2023, and 41% of these were related to non-compliance with regulations related to residues of veterinary drugs.

Table 5 Causes of import alerts associated with seafood in USA 2016–23

	Year								
Causes	2016	2017	2018	2019	2020	2021	2022	2023	Total
Others	1553	1114	917	864	419	476	428	471	6242
Microbiological	253	281	215	230	219	400	243	168	2069
Chemical	233	156	241	151	72	90	93	129	1165
Histamine	32	25	24	59	20	46	23	22	251
Toxins	1	1	0	0	0	0	0	0	2
Total	**2072**	**1577**	**1457**	**1304**	**730**	**1012**	**787**	**790**	**9729**

Table 6 Causes of microbiological alerts in USA during 2016–23

	Year								
Causes	2016	2017	2018	2019	2020	2021	2022	2023	Total
Salmonella	212	246	202	177	166	384	201	154	1742
Listeria monocytogenes	41	33	38	52	53	16	41	14	288
Hepatitis A	0	2	35	1	0	0	0	0	38
E. coli	0	0	0	0	0	0	1	0	1
Total	**253**	**281**	**275**	**230**	**219**	**400**	**243**	**168**	**2069**

Table 7 Details of alerts due to chemical causes in seafood in USA during 2016–2023

	Year								
Causes	2016	2017	2018	2019	2020	2021	2022	2023	Total
Residues of veterinary drugs	160	117	192	81	49	44	43	41	727
Nitrofurans	45	14	21	55	15	42	33	50	275
Additives	12	14	13	9	0	3	8	15	74
Chloramphenicol	10	10	3	5	5	0	5	21	59
Pesticide	0	1	12	1	3	0	4	2	23
Sulphites	0	0	0	0	0	1	0	0	1
Total	**227**	**156**	**241**	**151**	**72**	**90**	**93**	**129**	**1159**

If India has to continue its dominance in the USA market, the issue related to residues of veterinary drugs needs to be addressed urgently.

12 Causes of Import Refusals for Seafood in Japan

Japan is the third-largest seafood importing country after the USA and China. The Japanese Ministry of Health, Labour and Welfare of Japan provides information on imported food, concerned regulations, monitoring plans and violations of the Food Sanitation Act (https://www.mhlw.go.jp/stf/seisakunitsuite/bunya/kenkou_iryou/shokuhin/yunyu_kanshi/index_00017.html). The data in Table 8 shows that the total number of import refusals in Japan is much lower than in the USA and EU. Microbiological causes accounted for the largest portion of violations, followed by chemical causes.

The microbiological causes of violation of the Food Sanitation Act are detailed in Table 9. Most violations are due to the presence of coliform bacteria, followed by the presence of live bacteria in products where they should not be there. The presence of *E. coli* caused import violations, and the contribution of *Salmonella* and *Vibrio* was very small. This is in contrast to data from the USA, where *Salmonella* was the major cause, followed by *Listeria nomocytogenes*. The latter does not figure in the Japanese market, possibly due to the lower import of products such as cold smoked fish. Coliform is not an adequate indicator for faecal contamination, and in both the EU and the USA, *E. coli* are used as indicators, but possibly, due to the

Table 8 Causes of violation of Food Sanitation Act due to seafood in Japan

	Year								
Causes	2016	2017	2018	2019	2020	2021	2022	2023	Total
Microbiological	94	99	81	77	47	94	81	63	636
Chemical	42	36	24	27	32	25	32	31	249
Toxins	0	0	0	4	7	2	1	2	16
Others	0	8	1	1	0	0	0	0	10
Total	**136**	**143**	**106**	**109**	**86**	**121**	**114**	**96**	**911**

Table 9 Details of microbiological causes of violation of the Food Sanitation Act in Japan during 2016–23

	Year								
Causes	2016	2017	2018	2019	2020	2021	2022	2023	Total
Coliform	51	50	42	43	23	59	45	27	340
Live bacteria	30	29	19	23	16	24	21	25	187
E. coli	10	19	17	10	8	11	12	10	97
Vibrio spp.	2	1	0	1	0	0	2	1	7
Salmonella	1	0	3	0	0	0	1	0	5
Total	**94**	**99**	**81**	**77**	**47**	**94**	**81**	**63**	**636**

practice of raw fish consumption in Japan, the presence of coliforms would be considered a violation of Food Sanitation Act.

Details of import violations due to chemical causes in Japan are indicated in Table 10. Residues of veterinary drugs accounted for the largest proportion, but still much lower than the numbers seen in the USA. Heavy metals, such as mercury, do not appear on the list from Japan but are a major component in the EU market. India faced 10 import refusals from Japan in 2023, and 90% of these were due to non-compliance with regulations related to residues of veterinary drugs.

Overall, the analysis of import refusals and rapid alerts in the EU, the USA and Japan indicates that the seafood exporting countries need to pay more attention to international best practices regarding health management in aquaculture and prudent and responsible use of veterinary drugs. The detection of residues of banned veterinary drugs such as chloramphenicol, nitrofurans and dyes like leucomalachite green indicate failure to implement regulations related to antimicrobial use in aquaculture.

13 Conclusion

International trade in seafood has been steadily growing, but these need to meet the stringent regulations in importing countries. Importing countries are implementing animal health regulations as recommended by the Aquatic Animal Health Code of the World Organisation of Animal Health and food safety regulations harmonised with the Codex Alimentarius Commission. Animal health-related issues do not

Table 10 Details of import refusals due to chemical causes in Japan during 2016–23

Causes	Year								Total
	2016	2017	2018	2019	2020	2021	2022	2023	
Residues of veterinary drugs	35	31	15	15	14	11	20	22	163
Additives	4	4	8	9	15	8	11	8	67
Pesticides	2	1	0	3	3	6	1	0	16
Irradiation	1	0	0	0	0	0	0	1	2
Contaminants	0	0	1	0	0	0	0	0	1
Total	**42**	**36**	**24**	**27**	**32**	**25**	**32**	**31**	**249**

seem to be a constraint for international trade since most products are in frozen form for human consumption. The case study of Australian import regulations for prawns would be a good example of implementing a regulation based on national risk assessment. Analysis of the causes of import alerts and import refusals would give an idea of the main concerns of the importing countries. Though the major issues are the same in the three major markets—USA, EU and Japan, each importing country seems to have different testing frequencies and different emphases for the parameters. Microbiological issues, residues of veterinary medicines and contamination with heavy metals are impacting seafood trade in the three major markets analysed in this chapter.

References

Codex Alimentarius Commission (2014) Guidelines for the design and implementation of National Regulatory Food Safety Assurance Programmes associated with the use of veterinary drugs in food producing animals CXG 71–2009. https://www.fao.org/fao-who-codexalimentarius/codex-texts/guidelines/en/

Codex Alimentarius Commission, 2019. Code of practice for fish and fishery products CXC 52-2003 last amended 2019. https://www.fao.org/fao-who-codexalimentarius/codex-texts/codes-of-practice/en/

Codex Alimentarius Commission (2023) Maximum residue limits and risk management recommendations on residues of veterinary drugs in foods CXM 2-2023. https://www.fao.org/fao-who-codexalimentarius/codex-texts/maximum-residue-limits/tr/

Department of Agriculture Fisheries and Forestry (2023) Review of the biosecurity risks of prawns imported from all countries for human consumption – final report. https://www.agriculture.gov.au/sites/default/files/documents/prawn-final-report.pdf

EU (2016) Regulation (EU) 2016/429 of the European Parliament and the Council of 9 March 2016 on transmissible animal diseases and amending and repealing certain acts in the area of animal health (The Animal Health Law). https://eur-lex.europa.eu/eli/reg/2016/429/oj

EU (2018) Commission implementing regulation (EU) 2018/1882 of 3 December 2018 on the application of certain disease prevention and control rules to categories of listed diseases establishing a list of species and groups of species posing a considerable risk for the spread of those diseases. https://eur-lex.europa.eu/legal-content/HR/TXT/PDF/?uri=CELEX:32018R1882

EU (2020) Commission Delegated Regulation (EU) 2020/692 Of 30 January 2020 supplementing regulation (EU) 2016/429 of the European Parliament and of the Council as regards rules for the entry into the Union and the movement and handling after entry of consignments of certain animals, germinal products and products of animal origin. https://eur-lex.europa.eu/legal-content/EN/TXT/PDF/?uri=CELEX:32020R0692

FAO (2024) The state of world fisheries and aquaculture 2024. Blue Transformation in action. Rome. https://doi.org/10.4060/cd0683en

Queensland Government (2024) 2022-23 aquaculture production summary report. https://www.publications.qld.gov.au/dataset/3e2c107d-c49e-4b75-9994-63e513280824/resource/d5603d18-096c-418c-883f-8f97c2b818c5/download/2022-23-aquaculture-production-summary-report.pdf

WOAH (2024) The Aquatic Animal Health Code. https://www.woah.org/en/what-we-do/standards/codes-and-manuals/aquatic-code-online-access/

WTO (1995) The WTO Agreement on the application of Sanitary and Phytosanitary Measures (SPS Agreement). https://www.wto.org/english/tratop_e/sps_e/spsagr_e.htm

Building Financial Resilience in Shrimp Aquaculture: Crop Insurance to Cover Aquatic Animal Disease Losses

T. Ravisankar and Kuldeep K. Lal

Abstract

Shrimp aquaculture is vital to the global seafood trade but is susceptible to significant risks from weather, diseases, and market fluctuations. Insurance is a crucial tool for de-risking and supporting shrimp farmers, yet its adoption remains limited due to high premiums and operational challenges. This chapter explores global aquaculture insurance models, highlights disease and risk advancements in management, and examines recent initiatives in shrimp insurance in India. Recommendations include tailored policies, technological integration, and government assistance to enhance sector resilience.

Keywords

Crop insurance · Shrimp aquaculture · Aquatic animal disease

1 Introduction

Shrimp farming is a key component of global aquaculture, contributing significantly to food security and economic growth. Major producers, such as India, China, and Vietnam, lead the sector, providing employment and foreign exchange earnings. Despite its potential and profitability, shrimp farming faces risks from climatic events, disease outbreaks, and market volatility. Effective financial tools, such as insurance, are essential to ensure the stability and sustainability of the industry.

Insurance provides a financial safety net for shrimp farmers, helping them manage risks, recover losses, and maintain production capacity. This chapter reviews

T. Ravisankar (✉) · K. K. Lal
ICAR-Central Institute of Brackishwater Aquaculture, Chennai, India
e-mail: trsciba@gmail.com; kuldeepklal@gmail.com

R. Kooloth Valappil et al. (eds.), *Aquatic Animal Health Management*,
https://doi.org/10.1007/978-981-96-7987-4_33

global shrimp insurance models, examines India's efforts to revitalize shrimp crop insurance, and proposes strategies for enhancing insurance adoption and efficiency in the aquaculture sector.

2 Key Risks in Shrimp Aquaculture

- *Natural hazards*: Shrimp aquaculture is threatened by natural disasters, such as floods, cyclones, and extreme weather events, leading to infrastructure damage, crop loss, and operational disruptions.
- *Disease outbreaks*: Diseases, such as white spot syndrome virus (WSSV) and *Enterocytozoon hepatopenaei* (EHP), are significant risks, causing high mortality rates and economic losses. Effective disease management is crucial for minimizing these impacts.
- *Market risks*: As a globally traded commodity, shrimp prices are influenced by international market dynamics. Fluctuations in supply and demand can lead to price volatility, affecting farmers' profitability and sustainability.

3 Insurance Models and Global Trends

1. *Indemnity-based insurance*: This traditional model compensates for direct losses by covering specific or all perils, such as disease outbreaks, natural disasters, and operational risks. While effective, it often involves higher administrative costs and lengthy claim processes.
2. *Index-based insurance*: This innovative model uses predefined indices, such as weather parameters, to trigger payouts. It offers advantages, such as reduced transaction costs, quicker settlements, and minimized moral hazard. However, its effectiveness depends on accurately correlating indices with actual losses.

Globally, aquaculture insurance adoption varies (Munich Re 2009). Developed nations, such as the USA and Norway, have robust insurance systems supported by government subsidies (Brown et al. 2019; Kostyuchenko et al. 2019). In contrast, developing nations face challenges, such as high premiums, limited awareness, and insufficient government involvement (Nguyen et al. 2022a, b; Beukes and Venter 2018) Countries, such as China and Japan, have achieved significant insurance penetration due to strong institutional support, while nations, such as Bangladesh and Morocco (Rahman et al. 2020; Benjelloun et al. 2018; Oyinbo et al. 2019), struggle with low uptake rates.

4 India's Experience and Initiatives

India is one of the largest producers of shrimp, with production nearing one million metric tons annually. The sector's growth has been driven by advancements in farming practices including better management practices, bio-security and cluster farming, government policies, and the introduction of Specific Pathogen Free (SPF) exotic species, such as Pacific white shrimp (*Penaeus vannamei*). Recirculatory Aquaculture Systems (RAS), Super Intensive Precision Shrimp Farming (SIP) Technology. However, challenges, such as disease outbreaks and extreme weather events, underscore the need for effective risk management through insurance.

5 Past Challenges

India's early attempts at shrimp crop insurance faced setbacks due to high premiums, lack of comprehensive coverage, lack of availability of domain experts for survey and loss assessment and distrust in claim settlements. Many policies failed to cover diseases, a significant risk factor in shrimp farming, leading to low adoption rates.

6 Recent Developments

- *Pilot insurance schemes*: Under the *Pradhan Mantri Matsya Sampada Yojana* (PMMSY), pilot insurance schemes are being tried in states, such as Andhra Pradesh, Tamil Nadu, and Gujarat. These initiatives aim to provide financial protection to farmers and promote insurance adoption. The *Pradhan Mantri Matsya Kisan Samridhi Sah-Yojana* (PM-MKSSY) includes aquaculture insurance to support fish farmers. It provides a one-time incentive of up to Rs. 1,00,000 per farmer for insurance coverage, targeting at least 1 lakh hectares of aquaculture farms. This initiative aims to mitigate risks from disease and unforeseen events, enhance production, and ensure sustainable growth in the fisheries sector.
- *New insurance products*: Major insurance companies, such as Oriental Insurance and Agricultural Insurance Company, Agricultural Insurance Company, and Tata-AIG have launched shrimp crop insurance policies developed in consultation with the ICAR-Central Institute of Brackishwater Aquaculture (ICAR-CIBA), covering weather and disease risks. Another insurance company, Future Generali, has also launched a parametric crop insurance product in technical support with ICAR-CIBA. These products incorporate modern risk assessment tools and offer flexible premium rates (PIB 2024; OICL 2023).
- *Technological integration*: Digital tools, such as the CIBA ShrimpApp, ReportfishdiseaseApp, and another beta version of app dedicated to shrimp crop insurance, are being experimented with to streamline farm monitoring, risk assessment, and claim processing. Technologies, such as IoT and satellite imaging, enable precise loss estimation, reducing disputes and enhancing transparency.

7 Expanded Initiatives in Technology

Recent advancements in technology have revolutionized risk management in aquaculture. IoT sensors and real-time monitoring systems allow for precise tracking of pond conditions, including temperature, pH, and oxygen levels. These innovations reduce the likelihood of catastrophic losses and facilitate better disease management. Remote sensing and satellite imagery enable insurers to assess damage quickly and accurately, leading to expedited claim settlements. AI-based predictive models help forecast risks, giving farmers a chance to implement preventive measures. Such technologies are becoming central to India's efforts to modernize its aquaculture sector.

8 Challenges and Opportunities

8.1 Challenges

- High premiums: Many small-scale farmers find insurance unaffordable without subsidies.
- Complex claims processes: Lengthy and complicated procedures deter farmers from opting for insurance.
- Technical barriers: Lack of expertise in loss assessment and verification hampers efficient claims processing.
- Trust deficit: Past delays in claim settlements have eroded farmers' confidence in insurance providers.
- Limited awareness: Many farmers remain unaware of the benefits of insurance or how to access available schemes.

8.2 Opportunities

- Technological advancements: Leveraging IoT, AI, and satellite imaging can improve risk assessment accuracy and expedite claim settlements.
- Policy innovations: Index-based insurance models, group insurance schemes, and affinity insurance can reduce costs and increase coverage.
- Government support: Subsidies, tax exemptions, and disaster relief measures can make insurance more accessible to small-scale farmers.
- Awareness and training: Educating farmers about insurance benefits and risk management practices can drive adoption.
- Public-private partnerships: Collaboration between government bodies and private insurers can lead to the development of more effective and affordable insurance products.

9 Recommendations

9.1 For Governments

- Develop crop monitoring databases and aquaculture vulnerability maps.
- Provide premium subsidies to bridge the gap between farmers' willingness to pay and insurers' costs.
- Integrate aquaculture insurance with institutional credit to encourage uptake.
- Include aquaculture losses in disaster relief programs to ensure comprehensive coverage.
- Foster collaboration between research institutions and insurers to design tailored products.
- Reduce taxes and levies on insurance premiums to make policies more affordable.

9.2 For Farmers

- Adopt Better Management Practices (BMPs) to minimize risks.
- Maintain detailed farm records to facilitate faster claim processing.
- Leverage collective bargaining through farmer associations to negotiate better insurance terms.
- Understand policy terms and conditions to avoid disputes during claims.
- Utilize digital platforms and mobile applications for accessing insurance and advisory services.

9.3 For Insurers

- Develop tailored insurance products with input from aquaculture experts.
- Use parametric models for quicker payouts and reduced administrative costs.
- Promote group and affinity insurance to increase coverage and reduce premiums.
- Establish transparent grievance redressal mechanisms to rebuild trust.
- Collaborate with technology providers to integrate advanced risk assessment tools.
- Offer incentives for farmers who adopt BMPs and comply with regulatory standards.

10 Conclusion

Shrimp aquaculture insurance is indispensable for mitigating risks and fostering sector resilience. Collaborative efforts among governments, insurers, and farmers are essential to create affordable, effective insurance solutions. The aquaculture sector can achieve sustainable growth by embracing technological innovations, promoting awareness, and addressing existing challenges. The integration of advanced tools, such as IoT and AI, combined with policy support and industry collaboration, can transform the insurance landscape. As global demand for shrimp continues to rise, robust insurance mechanisms will play a pivotal role in ensuring the industry's long-term viability and stability. By adopting these measures, stakeholders can create a more resilient and prosperous shrimp farming ecosystem, benefiting millions of farmers worldwide.

References

Benjelloun M, Cherkaoui M, Benslimane A (2018) Aquaculture insurance and its role in risk management in Morocco. Mar Policy 95:120–126

Beukes C, Venter W (2018) Challenges and prospects of aquaculture insurance in Namibia. J Fish Manag Ecol 25(3):213–222

Brown D, Mitchell S, Green R (2019) Historical development and current practices in aquaculture insurance in Australia. Aust Aquacult J 41(2):112–125

Kostyuchenko N, Trofimova T, Volodin A (2019) Aquaculture insurance in Russia: current state and future directions. Russian J Fish Aquacult 5(2):85–92

Munich Re (2009) Agro Microinsurance: Innovative Insurance Solutions for the Agricultural Sector. Available at: https://www.munichre.com

Nguyen HT, Tran TT, Pham DP (2022a) The development of aquaculture insurance in Vietnam: challenges and opportunities. J Aquacult Develop 14(1):45–58

Nguyen P, Tran H, Pham T (2022b) Aquaculture insurance landscape in Vietnam. Aquac Int 30(5):789–805

OICL (2023) Oriental Insurance Company Limited Product Details. [Online] Available at: https://orientalinsurance.org.in/

Oyinbo O, Chamberlain D, Ajayi O (2019) Insurance adoption among fish farmers in Nigeria: current status and prospects. Agric Econ Rev 11(3):197–210

PIB (2024) Press information bureau. "Press Release: Fisheries and Aquaculture Insurance" Ministry of Fisheries, Animal Husbandry & Dairying January 16, 2024. Accessed October 1, 2024. https://pib.gov.in/PressReleasePage.aspx?PRID=1996575

Rahman MM, Alam S, Hossain MS (2020) Challenges of aquaculture insurance in Bangladesh: policy perspectives. Asian Fish Sci 33(4):250–262

MIX
Papier aus verantwortungsvollen Quellen
Paper from responsible sources
FSC® C105338

If you have any concerns about our products,
you can contact us on
ProductSafety@springernature.com

In case Publisher is established outside the EU,
the EU authorized representative is:
Springer Nature Customer Service Center GmbH
Europaplatz 3, 69115 Heidelberg, Germany

Printed by Libri Plureos GmbH
in Hamburg, Germany